MATHEMATICS:

ANALYSIS AND APPROACHES

HIGHER LEVEL

COURSE COMPANION

Natasha Awada
Paul Belcher
Jennifer Chang Wathall
Phil Duxbury
Jane Forrest
Tony Halsey
Josip Harcet
Rose Harrison
Lorraine Heinrichs
Ed Kemp
Paul La Rondie
Palmira Mariz Seiler
Jill Stevens
Ellen Thompson
Marlene Torres-Skoumal

OXFORD
UNIVERSITY PRESS

Great Clarendon Street, Oxford, OX2 6DP, United Kingdom

Oxford University Press is a department of the University of Oxford. It furthers the University's objective of excellence in research, scholarship, and education by publishing worldwide. Oxford is a registered trade mark of Oxford University Press in the UK and in certain other countries

First published in 2019

British Library Cataloguing in Publication Data
Data available

978-0-19-842717-9

10 9 8 7

Paper used in the production of this book is a natural, recyclable product made from wood grown in sustainable forests. The manufacturing process conforms to the environmental regulations of the country of origin.

Printed in India by Multivista Global Pvt. Ltd

Acknowledgements

The publisher would like to thank the following authors for contributions to digital resources:

Alexander Aits
Tom Edinburgh
Jim Fensom
Josip Harcet
Rose Harrison
Lorraine Heinrichs
Neil Hendry
Georgios Ioannadis
Alissa Kamilova
Ed Kemp
Martin Noon
Ellen Thompson
Marlene Torres-Skoumal
Daniel Wilson-Nunn

Cover: bjdlzx/iStockphoto. All other photos © Shutterstock, except: **p145**: mgkaya/iStockphoto; **p146(r)**: Everett Collection Inc/Alamy Stock Photo; **p218(tr)**: NASA/SCIENCE PHOTO LIBRARY; **p246(t)**: Blackfox Images/Alamy Stock Photo; **p680(l)**: dpa picture alliance/Alamy Stock Photo; **p680(r)**: Xinhua/Alamy Stock Photo; **p743(m)**: Oxford University Press ANZ/Brent Parker Jones Food Photographer/Sebastian Sedlak Food Stylist.

Every effort has been made to contact copyright holders of material reproduced in this book. Any omissions will be rectified in subsequent printings if notice is given to the publisher.

Course Companion definition

The IB Diploma Programme Course Companions are designed to support students throughout their two-year Diploma Programme. They will help students gain an understanding of what is expected from their subject studies while presenting content in a way that illustrates the purpose and aims of the IB. They reflect the philosophy and approach of the IB and encourage a deep understanding of each subject by making connections to wider issues and providing opportunities for critical thinking.

The books mirror the IB philosophy of viewing the curriculum in terms of a whole-course approach and include support for international mindedness, the IB learner profile and the IB Diploma Programme core requirements, theory of knowledge, the extended essay and creativity, activity, service (CAS).

IB mission statement

The International Baccalaureate aims to develop inquiring, knowledgable and caring young people who help to create a better and more peaceful world through intercultural understanding and respect.

To this end the IB works with schools, governments and international organisations to develop challenging programmes of international education and rigorous assessment.

These programmes encourage students across the world to become active, compassionate, and lifelong learners who understand that other people, with their differences, can also be right.

The IB learner profile

The aim of all IB programmes is to develop internationally minded people who, recognising their common humanity and shared guardianship of the planet, help to create a better and more peaceful world. IB learners strive to be:

Inquirers They develop their natural curiosity. They acquire the skills necessary to conduct inquiry and research and show independence in learning. They actively enjoy learning and this love of learning will be sustained throughout their lives.

Knowledgeable They explore concepts, ideas, and issues that have local and global significance. In so doing, they acquire in-depth knowledge and develop understanding across a broad and balanced range of disciplines.

Thinkers They exercise initiative in applying thinking skills critically and creatively to recognise and approach complex problems, and make reasoned, ethical decisions.

Communicators They understand and express ideas and information confidently and creatively in more than one language and in a variety of modes of communication. They work effectively and willingly in collaboration with others.

Principled They act with integrity and honesty, with a strong sense of fairness, justice, and respect for the dignity of the individual, groups, and communities. They take responsibility for their own actions and the consequences that accompany them.

Open-minded They understand and appreciate their own cultures and personal histories, and are open to the perspectives, values, and traditions of other individuals and communities. They are accustomed to seeking and evaluating a range of points of view, and are willing to grow from the experience.

Caring They show empathy, compassion, and respect towards the needs and feelings of others. They have a personal commitment to service, and act to make a positive difference to the lives of others and to the environment.

Risk-takers They approach unfamiliar situations and uncertainty with courage and forethought, and have the independence of spirit to explore new roles, ideas, and strategies. They are brave and articulate in defending their beliefs.

Balanced They understand the importance of intellectual, physical, and emotional balance to achieve personal well-being for themselves and others.

Reflective They give thoughtful consideration to their own learning and experience. They are able to assess and understand their strengths and limitations in order to support their learning and professional development.

Contents

Introduction vii

How to use your enhanced online course book ix

1 From patterns to generalizations: sequences, series and proof 2

1.1 Sequences, series and sigma notation 4

1.2 Arithmetic and geometric sequences and series 10

1.3 Proof 33

1.4 Counting principles and the binomial theorem 51

Chapter review 68

Modelling and investigation activity 70

2 Representing relationships: functions 72

2.1 Functional relationships 75

2.2 Special functions and their graphs 81

2.3 Classification of functions 102

2.4 Operations with functions 108

2.5 Function transformations 117

Chapter review 142

Modelling and investigation activity 144

3 Expanding the number system: complex numbers 146

3.1 Quadratic equations and Inequalities 149

3.2 Complex numbers 161

3.3 Polynomial equations and Inequalities 175

3.4 The fundamental theorem of algebra 184

3.5 Solving equations and inequalities 195

3.6 Solving systems of linear equations 200

Chapter review 214

Modelling and investigation activity 216

4 Measuring change: differentiation 218

4.1 Limits, continuity and convergence 220

4.2 The derivative of a function 236

4.3 Differentiation rules 249

4.4 Graphical interpretation of the derivatives 262

4.5 Applications of differential Calculus 278

4.6 Implicit differentiation and related rates 288

Chapter review 298

Modelling and investigation activity 302

Paper 3 question and comments 304

5 Analysing data and quantifying randomness: statistics and probability 306

5.1 Sampling 308

5.2 Descriptive statistics 317

5.3 The justification of statistical techniques 335

5.4 Correlation, causation and linear regression 347

Chapter review 360

Modelling and investigation activity 364

6 Relationships in space: geometry and trigonometry 366

6.1 The properties of three-dimensional space 369

6.2 Angles of measure 378

6.3 Ratios and identities 384

6.4 Trigonometric functions 410
6.5 Trigonometric equations 420
Chapter review 437
Modelling and investigation activity 440

7 Generalizing relationships: exponents, logarithms and integration 442

7.1 Integration as antidifferentiation and definite integrals 444
7.2 Exponents and logarithms 460
7.3 Derivatives of exponential and logarithmic functions; tangents and normals 483
7.4 Integration techniques 488
Chapter review 514
Modelling and investigation activity 516

8 Modelling change: more calculus 518

8.1 Areas and volumes 520
8.2 Kinematics 528
8.3 Ordinary differential equations (ODEs) 533
8.4 Limits revisited 550
Chapter review 565
Modelling and investigation activity 568

9 Modelling 3D space: Vectors 570

9.1 Geometrical representation of Vectors 573
9.2 Introduction to vector algebra 586
9.3 Scalar product and its properties 597
9.4 Vector equation of a line 605
9.5 Vector product and properties 614
9.6 Vector equation of a plane 621
9.7 Lines, planes and angles 628
9.8 Application of vectors 633
Chapter review 643
Modelling and investigation activity 646

10 Equivalent systems of representation: more complex numbers 648

10.1 Forms of a complex number 650
10.2 Operations with complex numbers in polar form 656
10.3 Powers and roots of complex numbers in polar form 664
Chapter review 676
Modelling and investigation activity 678

11 Valid comparisons and informed decisions: probability distributions 680

11.1 Axiomatic probability systems 683
11.2 Probability distributions 696
11.3 Continuous random variables 706
11.4 Binomial distribution 712
11.5 The normal distribution 718
Chapter review 727
Modelling and investigation activity 730

12 Exploration 732

Practice exam paper 1 746

Practice exam paper 2 749

Practice exam paper 3 752

Answers 755

Index 837

Digital contents

Digital content overview

Click on this icon here to see a list of all the digital resources in your enhanced online course book. To learn more about the different digital resource types included in each of the chapters and how to get the most out of your enhanced online course book, go to page ix.

Syllabus coverage

This book covers all the content of the Mathematics: analysis and approaches HL course. Click on this icon here for a document showing you the syllabus statements covered in each chapter.

Practice exam papers

Click on this icon here for an additional set of practice exam papers.

Worked solutions

Click on this icon here for worked solutions for all the questions in the book

Introduction

The new IB diploma mathematics courses have been designed to support the evolution in mathematics pedagogy and encourage teachers to develop students' conceptual understanding using the content and skills of mathematics, in order to promote deep learning. The new syllabus provides suggestions of conceptual understandings for teachers to use when designing unit plans and overall, the goal is to foster more depth, as opposed to breadth, of understanding of mathematics.

What is teaching for conceptual understanding in mathematics?

Traditional mathematics learning has often focused on rote memorization of facts and algorithms, with little attention paid to understanding the underlying concepts in mathematics. As a consequence, many learners have not been exposed to the beauty and creativity of mathematics which, inherently, is a network of interconnected conceptual relationships.

Teaching for conceptual understanding is a framework for learning mathematics that frames the factual content and skills; lower order thinking, with disciplinary and non-disciplinary concepts and statements of conceptual understanding promoting higher order thinking. Concepts represent powerful, organizing ideas that are not locked in a particular place, time or situation. In this model, the development of intellect is achieved by creating a synergy between the factual, lower levels of thinking and the conceptual higher levels of thinking. Facts and skills are used as a foundation to build deep conceptual understanding through inquiry.

The IB Approaches to Teaching and Learning (ATLs) include teaching focused on conceptual understanding and using inquiry-based approaches. These books provide a structured inquiry-based approach in which learners can develop an understanding of the purpose of what they are learning by asking the questions: why or how? Due to this sense of purpose, which is always situated within a context, research shows that learners are more motivated and supported to construct their own conceptual understandings and develop higher levels of thinking as they relate facts, skills and topics.

The DP mathematics courses identify twelve possible fundamental concepts which relate to the five mathematical topic areas, and that teachers can use to develop connections across the mathematics and wider curriculum:

Approximation	Modelling	Representation
Change	Patterns	Space
Equivalence	Quantity	Systems
Generalization	Relationships	Validity

Each chapter explores two of these concepts, which are reflected in the chapter titles and also listed at the start of the chapter.

The DP syllabus states the essential understandings for each topic, and suggests some content-specific conceptual understandings relevant to the topic content. For this series of books, we have identified important topical understandings that link to these and underpin the syllabus, and created investigations that enable students to develop this understanding. These investigations, which are a key element of every chapter, include factual and conceptual questions to prompt students to develop and articulate these topical conceptual understandings for themselves.

A tenet of teaching for conceptual understanding in mathematics is that the teacher does not **tell** the student what the topical understandings are at any stage of the learning process, but provides investigations that guide students to discover these for themselves. The teacher notes on the ebook provide additional support for teachers new to this approach.

A concept-based mathematics framework gives students opportunities to think more deeply and critically, and develop skills necessary for the 21st century and future success.

Jennifer Chang Wathall

Investigation 3

1 On your GDC, plot the graphs of each of these functions:

a $f(x)=\frac{1}{x}$ b $g(x)=\frac{2}{x}$ c $h(x)=\frac{3}{x}$

d $j(x)=\frac{-1}{x}$ e $k(x)=\frac{-2}{x}$ f $m(x)=\frac{-3}{x}$

2 **Factual** What features of each graph in question **1** are:

a the same for each graph? b different for each of the three graphs?

3 **Conceptual** What effect does changing the magnitude of the parameter k in the function $f(x)=\frac{k}{x}$ have on the graph of $y=f(x)$?

In every chapter, investigations provide inquiry activities and factual and conceptual questions that enable students to construct and communicate their own conceptual understanding in their own words. The key to concept-based teaching and learning, the investigations allow students to develop a deep conceptual understanding. Each investigation has full supporting teacher notes on the enhanced online course book.

Gives students the opportunity to reflect on what they have learned and deepen their understanding.

Reflect Why does zero not have a reciprocal?

Developing inquiry skills

Does mathematics always reflect reality? Are fractals such as the Koch snowflake invented or discovered?

Think about the questions in this opening problem and answer any you can. As you work through the chapter, you will gain mathematical knowledge and skills that will help you to answer them all.

Every chapter starts with a question that students can begin to think about from the start, and answer more fully as the chapter progresses. The developing inquiry skills boxes prompt them to think of their own inquiry topics and use the mathematics they are learning to investigate them further.

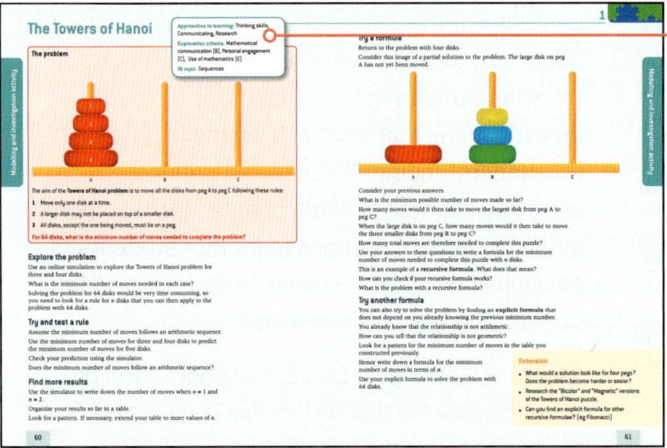

The modelling and investigation activities are open-ended activities that use mathematics in a range of engaging contexts and to develop students' mathematical toolkit and build the skills they need for the IA. They appear at the end of each chapter.

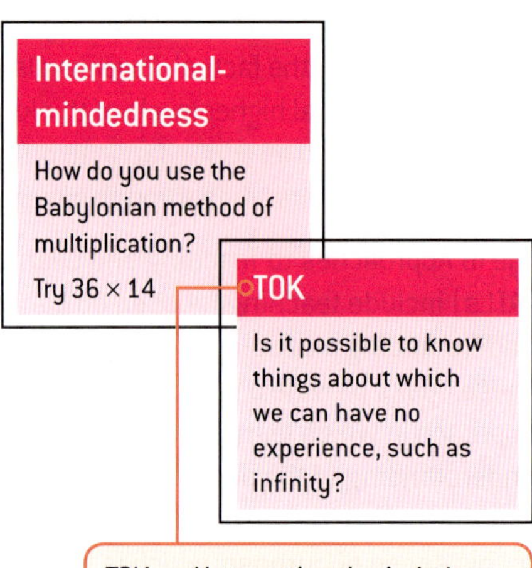

International-mindedness
How do you use the Babylonian method of multiplication?
Try 36 × 14
TOK
Is it possible to know things about which we can have no experience, such as infinity?

TOK and International-mindedness are integrated into all the chapters.

The chapters in this book have been written to provide logical progression through the content, but you may prefer to use them in a different order, to match your own scheme of work. The Mathematics: analysis and approaches Standard and Higher Level books follow a similar chapter order, to make teaching easier when you have SL and HL students in the same class. Moreover, where possible, SL and HL chapters start with the same inquiry questions, contain similar investigations and share some questions in the chapter reviews and mixed reviews – just as the HL exams will include some of the same questions as the SL paper.

How to use your enhanced online course book

Throughout the book you will find the following icons. By clicking on these in your enhanced online course book you can access the associated activity or document.

Prior learning

Clicking on the icon next to the "Before you start" section in each chapter takes you to one or more worksheets containing short explanations, examples and practice exercises on topics that you should know before starting, or links to other chapters in the book to revise the prior learning you need.

Additional exercises

The icon by the last exercise at the end of each section of a chapter takes you to additional exercises for more practice, with questions at the same difficulty levels as those in the book.

Animated worked examples

This icon leads you to an animated worked example, explaining how the solution is derived step-by-step, while also pointing out common errors and how to avoid them.

Click on the icon on the page to launch the animation. The animated worked example will appear in a second screen.

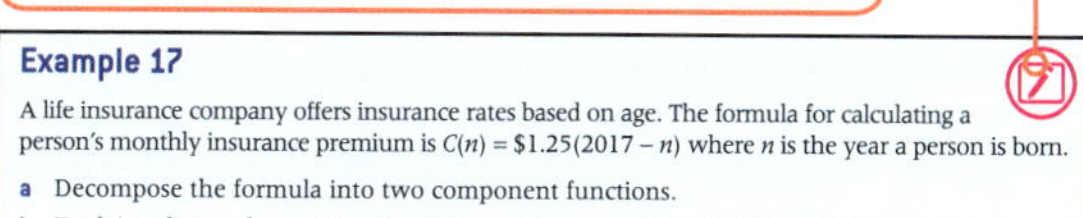

Example 17

A life insurance company offers insurance rates based on age. The formula for calculating a person's monthly insurance premium is $C(n) = \$1.25(2017 - n)$ where n is the year a person is born.

a Decompose the formula into two component functions.

b Explain what each component represents in terms of the insurance rates.

Click here for a transcript of the audio track.

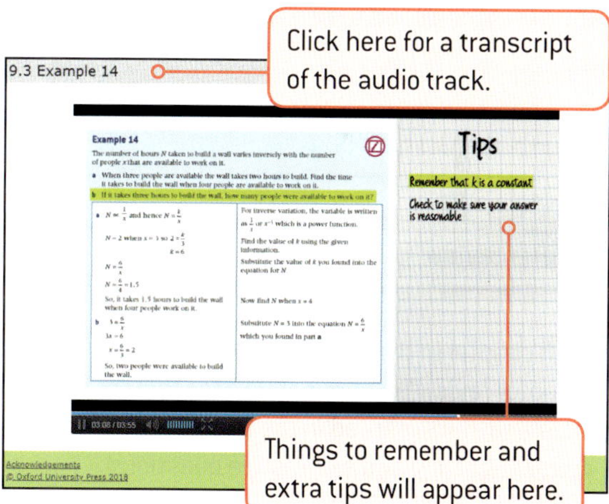

Things to remember and extra tips will appear here.

Graphical display calculator support

Supporting you to make the most of your TI-Nspire CX, TI-84+ C Silver Edition or Casio fx-CG50 graphical display calculator (GDC), this icon takes you to step-by-step instructions for using technology to solve specific examples in the book.

Click on the icon for the menu and then select your GDC model.

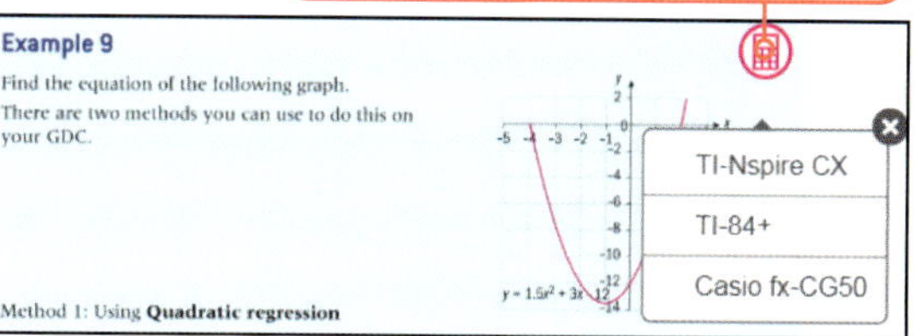

Teacher notes

This icon appears at the beginning of each chapter and opens a set of comprehensive teaching notes for the investigations, reflection questions, TOK items, and the modelling and investigation activities in the chapter.

Assessment opportunities

This Mathematics: analysis and approaches enhanced online course book is designed to prepare you for your assessments by giving you a wide range of practice. In addition to the activities you will find in this book, further practice and support are available on the enhanced online course book.

End of chapter tests and mixed review exercises

This icon appears twice in each chapter: first, next to the "Chapter summary" section and then next to the "Chapter review" heading.

Click here for an end-of-chapter summative assessment test, designed to be completed in one hour.

Chapter summary

Click here for the mixed review, a summative assessment consisting of exercises and exam-style questions, testing the topics you have covered so far.

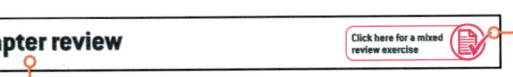

Each chapter in the printed book ends with a "Chapter review", a summative assessment of the facts and skills learned in the chapter, including problem-solving and exam-style questions.

Exam-style questions

Exam-style questions

15 **P1:** Find an expression for derivative of each of the following functions.

a $f(x) = x^4 - 2x^3 - x^2 + 3x - 4$

(1 mark)

Plenty of exam practice questions, in Paper 1 (P1) or Paper 2 (P2) style. Each question in this section has a mark scheme in the worked solutions document found on the enhanced online course book, which will help you see how marks are awarded.

The number of darker bars shows the difficulty of the question (one dark bar = easy; three dark bars = difficult).

Click here for further exam practice

Exam practice exercises provide exam style questions for Papers 1, 2, and 3 on topics from all the preceding chapters. Click on the icon for the exam practice found at the end of chapters 4, 6, 8, and 11 in this book.

Introduction to Paper 3

The new HL exams will have three papers, and Paper 3 will have just two extended response problem-solving questions. This introduction, on page 304, explains the new Paper 3 format, and gives an example of a Paper 3 question, with notes and guidance on how to interpret and answer it.

There are more Paper 3 questions in the exam practice exercises on the Enhanced Online Course Book, at the end of Chapters 4, 6, 8, 11.

Answers and worked solutions

Answers to the book questions

Concise answer to all the questions in this book can be found on page 755.

Worked solutions

Worked solutions for **all** questions in the book can be accessed by clicking the icon found on the Contents page or the first page of the Answers section.

Answers and worked solutions for the digital resources

Answers, worked solutions and mark schemes (where applicable) for the additional exercises, end-of-chapter tests and mixed reviews are included with the questions themselves.

1 From patterns to generalizations: sequences, series and proof

Concepts

- Patterns
- Generalization

Microconcepts

- Arithmetic and geometric sequences and series
- Introduction to limits
- Sum of series
- Permutations and combinations
- Proof
- Binomial theorem

You do not have to look far and wide to find visual patterns—they are everywhere!

Can these patterns be explained mathematically?

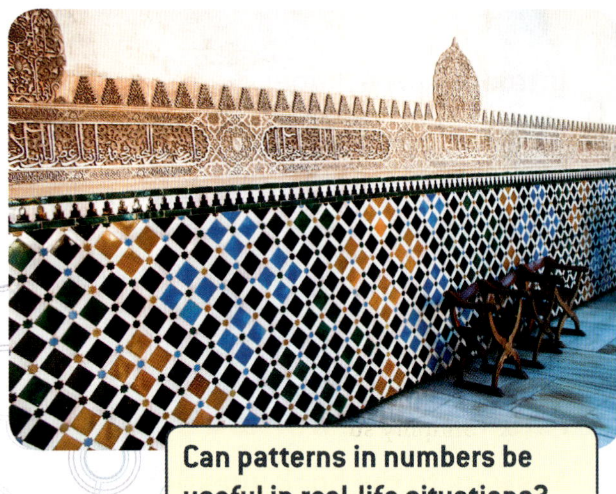

Can patterns in numbers be useful in real-life situations?

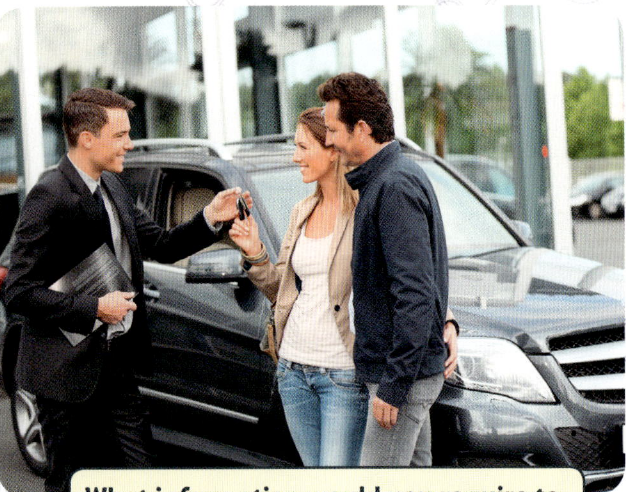

What information would you require to choose the best loan offer? What other scenarios could this be applied to?

If you take out a loan to buy a car, how can you determine the total amount it will cost?

The diagrams shown here are the first four iterations of a fractal called the Koch snowflake.

What do you notice about:

- How each pattern is created from the previous one?
- the perimeter as you move from the first iteration through the fourth iteration? How is it changing?

What changes would you expect in the fifth iteration?

How would you measure the perimeter at the fifth iteration if the original triangle had sides of 1m in length?

What happens if you start with a square instead of an equilateral triangle?

If this process continues forever, how can an infinite perimeter enclose a finite area?

Developing inquiry skills

Does mathematics always reflect reality? Are fractals such as the Koch snowflake invented or discovered?

Think about the questions in this opening problem and answer any you can. As you work through the chapter, you will gain mathematical knowledge and skills that will help you to answer them all.

Before you start

You should know how to:

1 Solve linear algebraic equations.

eg $x - 3(x+5) = 20 - 3x$

$\Rightarrow x - 3x - 15 = 20 - 3x$

$\Rightarrow -2x - 15 = 20 - 3x$

$\Rightarrow x = 35$

2 Simplify surds.

eg simplify $\frac{\sqrt{2}}{1-\sqrt{2}}$

$$\frac{\sqrt{2}}{1-\sqrt{2}} = \frac{\sqrt{2}\left(1+\sqrt{2}\right)}{\left(1-\sqrt{2}\right)\left(1+\sqrt{2}\right)} = \frac{\sqrt{2}+2}{1-2} = -2-\sqrt{2}$$

3 Manipulate algebraic fractions.

eg simplify $\frac{x+3}{x} = \frac{2}{x+1} - \frac{3x}{x-1}$

$$= \frac{(x+3)(x+1)(x-1)+2x(x-1)-3x^2(x+1)}{x(x+1)(x-1)}$$

$$= \frac{(x+3)(x^2-1)+2x^2-2x-3x^3-3x^2}{x(x^2-1)}$$

$$= \frac{x^3-x+3x^2-3+2x^2-2x-3x^3-3x^2}{x(x^2-1)}$$

$$= \frac{-2x^3+2x^2-3x-3}{x(x^2-1)}$$

Skills check

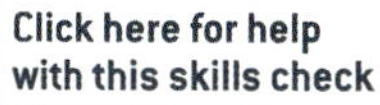

1 Solve the following equations:

a $3x + 5(x-4) = 20x + 4$

b $\frac{x+1}{2x-1} = \frac{x-3}{2x+1}$

2 Simplify the following:

a $\frac{1+\sqrt{2}}{1-\sqrt{2}}$

b $\frac{2\sqrt{2}}{1-\sqrt{3}}$

3 Simplify:

$$\frac{x}{x+1} - \frac{1}{2x-1} = \frac{2}{x-1}$$

1.1 Sequences, series and sigma notation

Opening investigations

You are going to start this chapter by doing some simple arithmetic with the aim of recognizing patterns. The challenge is for you to understand and explain the patterns that emerge. In Investigation 2, you will be asked to propose a conjecture, which is a rule generalizing findings based on observed patterns.

Investigation 1

Work out the following products:

$1 \times 1 \quad 11 \times 11 \quad 111 \times 111 \quad 1111 \times 1111$

1 What pattern do you see emerging?

2 Does this continue as you make the string of 1's longer?

3 Can you predict when this pattern stops and explain why this happens?

International-mindedness

Where did numbers come from?

Investigation 2

This diagram represents the floor of a room covered with square tiles. It has a total of nine tiles along the main diagonals (shaded), and five tiles on each side. 25 tiles are used to cover the floor completely.

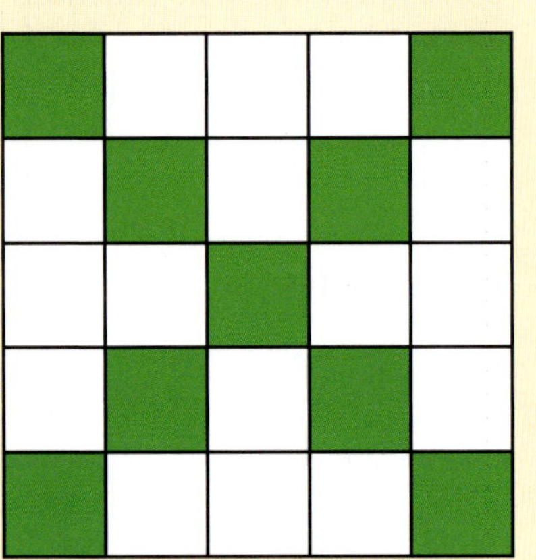

Another room has a total of 13 square tiles along the diagonals.

1 How many square tiles are there on each side in this other room?

2 How many tiles are needed to completely cover the floor?

3 What if the total number of tiles along the diagonals is 15?

4 What if there is a total of 135 tiles along the diagonals?

5 What if the total number of squares along the diagonals is an even number?

6 Continue to generate data to help you form a conjecture. Can you explain why this rule holds true?

7 How can you write the generalization concisely?

8 Why is an algebraic expression more useful than generating numerical values?

A **sequence** is a list of numbers that is written in a defined order, ascending or descending, following a specific rule. Each of the numbers making up a sequence is called a **term** of that sequence. Sometimes a sequence is also referred to as a **progression**.

Look at the following sequences of numbers and identify the rule which would help you obtain the next term.

i 7, 5, 3, 1, …

ii 2, 4, 8, 16, …

iii 1, 3, 9, 27, …

Sequences may be **finite** or **infinite**.

The sequence 7, 5, 3, 1, −1, −3 is a finite sequence with six terms, whereas the sequence 7, 5, 3, 1, −1, −3, … is an infinite sequence with an infinite number of terms. The distinction is indicated by the ellipsis (…) at the end of the sequence.

A sequence is sometimes written in terms of the general term as $\{u_r\}$, where r can take values 1, 2, 3, …
If the sequence is finite then r will terminate at some point.

The sequence $\{u_r\} = \{3r - 1\}$, where $r \in \mathbb{Z}^+$ represents the infinite sequence 2, 5, 8, 11, …, whereas the sequence $\{u_r\} = \left\{\frac{1}{r^2}\right\}$, where $r \in \mathbb{Z}^+$, $r \le 5$, represents the finite sequence $1, \frac{1}{4}, \frac{1}{9}, \frac{1}{16}, \frac{1}{25}$.

HINT

$\{u_r\}$ represents the sequence whereas u_r represents the rth term.

All the terms in a sequence added together are called a **series**. Like sequences, series can be finite or infinite.

The series obtained by adding the six terms of the sequence 7, 5, 3, 1, −1, −3 is $7 + 5 + 3 + 1 - 1 - 3 = 12$. This is a finite series. The sum $1 + 3 + 9 + 27 + 81 + \ldots$ continues indefinitely and is an infinite series.

The set of positive integers $\mathbb{Z}^+$ can be written as {1, 2, 3, 4, 5, …, r, …} where the letter r is used to represent the general term. If the positive integers which are multiples of 5 are considered, then the set {5, 10, 15, 20, …5r, …} is obtained. In this case the general term is $5r$ where r is any positive integer. The **harmonic series** is the infinite sum of the reciprocals of positive integers, ie $1 + \frac{1}{2} + \frac{1}{3} + \ldots + \frac{1}{r} + \ldots$

TOK

Do the names that we give things impact how we understand them?

Series can be represented in compact form using sigma (Σ) notation. This makes use of the general term written in terms of r, which often represents a positive integer.

The sum of the first 10 positive integers can be written as follows using sigma notation:

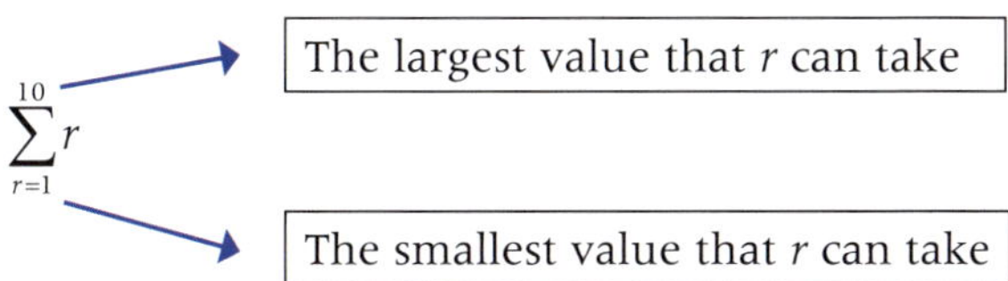

Read this as "The sum of r, from $r = 1$ to $r = 10$."

If you want to write the sum of the positive multiples of 5 less than 100, then you first need to think of the general term, which is $5r$, and then establish the range of values that r can take. The smallest positive multiple of 5 is 5 in which case $r = 1$, and since you want the largest multiple of 5 to be 100, the largest value that r can take is 20 because $100 = 5 \times 20$.

$$5+10+15+\ldots+100=\sum_{r=1}^{20}5r$$

Sometimes you will also have to interpret a sum given in sigma notation and expand it into individual terms. For example:

$$\sum_{r=0}^{4}(2r+1)=(2\times0+1)+(2\times1+1)+(2\times2+1)+(2\times3+1)+(2\times4+1)=1+3+5+7+9$$

HINT

In this case the series starts $r = 0$.

In Example 1 you will learn how to look for a pattern and write the general term.

Example 1

For each of the following sequences, write the next three terms and find the general term:

a 2, 7, 12, 17, … **b** 2, 6, 12, 20, … **c** $\frac{1}{2}, \frac{2}{3}, \frac{3}{4}, \frac{4}{5}, \ldots$ **d** 5, 10, 20, 40, …

a The next three terms of this sequence are 22, 27, 32.	Note that at each step you add 5 to get the next term.
The sequence can be written as: $2, 2+5, 2+10, 2+15$ $= 2, 2+(1\times5), 2+(2\times5), 2+(3\times5), \ldots, 2+(r-1)\times5$	Write the sequence using the pattern noticed.
The general term is $2+(r-1)\times5 = 5r-3$, where r can take the values 1, 2, 3, …	
b The next three terms are 30, 42, 56.	Note that the given terms can be written as: $1\times2, 2\times3, 3\times4, 4\times5, \ldots$
The sequence can be written as $1\times2, 2\times3, 3\times4, 4\times5, \ldots, r\times(r+1), \ldots$	
The general term is $r\times(r+1)$, where r can take the values 1, 2, 3, …	
c The next three terms are $\frac{5}{6}, \frac{6}{7}, \frac{7}{8}$.	The pattern here is easy to follow.
The sequence can be written as: $\frac{1}{2}, \frac{2}{3}, \frac{3}{4}, \frac{4}{5}, \ldots, \frac{r}{r+1}, \ldots$	
The general term is $\frac{r}{r+1}$, where r can take the values 1,2,3, …	
d The next three terms are 80, 160, 320.	Each term is obtained by multiplying the previous term by 2.
The general term is $5\times2^{r-1}$, where r can take the values 1, 2, 3, …	

HINT

You can check the answers by putting $r = 5,6,7$ in the general term obtained in each case.

Example 2 shows how to find the terms of a sequence represented by its general term.

Example 2

Write down the first three terms of each of the following sequences:

a $\{u_r\} = \{5r - 2\},\ r \in \mathbb{Z}^+$ **b** $\{u_r\} = \left\{\frac{(-1)^r}{r^2}\right\},\ r \in \mathbb{Z}^+$

a $u_1 = 5 \times 1 - 2 = 3$ $u_2 = 5 \times 2 - 2 = 8$ $u_3 = 5 \times 3 - 2 = 13$ 3, 8, 13	Substitute values 1, 2 and 3 for r.
b $u_1 = \frac{(-1)^1}{1^2} = -1$ $u_2 = \frac{(-1)^2}{2^2} = \frac{1}{4}$ $u_3 = \frac{(-1)^3}{3^2} = -\frac{1}{9}$ $-1,\ \frac{1}{4},\ -\frac{1}{9}$	Substitute values 1, 2 and 3 for r.

Example 3 shows how to represent a given sequence by its general term after recognizing a pattern.

Example 3

Write each of the following sequences using the general term:

a 3, 6, 9, 12, ... **b** 2, −10, 50, −250 **c** $\frac{1}{3}, \frac{2}{5}, \frac{3}{7}, \frac{4}{9}, \ldots$

a 3, 6, 9, 12, ... $\{u_r\} = \{3r\},\ r \in \mathbb{Z}^+$	This is an infinite sequence of the positive multiples of 3.
b 2, −10, 50, −250 $\{u_r\} = \{2(-5)^{r-1}\},\ r \in \mathbb{Z}^+,\ r \le 4$	This finite sequence can be written as: $2,\ 2\times(-5),\ 2\times 25,\ 2\times(-125)$ which can be rewritten in terms of powers of −5: $= 2\times(-5)^0,\ 2\times(-5)^1,\ 2\times(-5)^2,\ 2\times(-5)^3$
c $\frac{1}{3}, \frac{2}{5}, \frac{3}{7}, \frac{4}{9}, \ldots$ $\{u_r\} = \left\{\frac{r}{2r+1}\right\},\ r \in \mathbb{Z}^+$	In this infinite sequence, the numerators are the positive integers and the denominators are successive odd integers greater than 1.

Example 4 shows how to expand a series written in sigma notation.

Example 4

For each of the following series written in sigma notation, write the first five terms:

a $\sum_{r=1}^{10} r(r-1)$ **b** $\sum_{r=1}^{\infty} (-1)^r r^2$ **c** $\sum_{r=1}^{\infty} \frac{r+1}{2r-1}$

a $\sum_{r=1}^{10} r(r-1) = 1\times 0 + 2\times 1 + 3\times 2 + 4\times 3 + 5\times 4 + \ldots$	Substitute $r = 1$ to 5 for the first through to the fifth term.
$= 0 + 2 + 6 + 12 + 20 + \ldots$	Simplify.
b $\sum_{r=1}^{\infty} (-1)^r r^2$ $= (-1)^1 \times 1^2 + (-1)^2 \times 2^2 + (-1)^3 \times 3^2 + (-1)^4 \times 4^2 + (-1)^5 \times 5^2 + \ldots$ $= -1 + 4 - 9 + 16 - 25 + \ldots$	
c $\sum_{r=1}^{\infty} \frac{r+1}{2r-1} = \frac{1+1}{2-1} + \frac{2+1}{4-1} + \frac{3+1}{6-1} + \frac{4+1}{8-1} + \frac{5+1}{10-1}$ $= 2 + 1 + \frac{4}{5} + \frac{5}{7} + \frac{6}{9} + \ldots$	

In Example 5 you will see how a given series can be written in sigma notation.

TOK

Is mathematics a language?

Example 5

Write each of the following series in sigma notation:

a $3 + 11 + 19 + 27 + 35$ **b** $1 - 1 + 1 - 1 + 1 - 1 + \ldots$ **c** $-6 + 12 - 24 + 48 - 96 + 192$

a $3 + 11 + 19 + 27 + 35$ $= \sum_{r=1}^{5} 8r - 5$	This is a finite series which can be written as: $3 + (3 + 8) + (3 + 16) + (3 + 24) + (3 + 32)$ $= 3 + (3 + 1 \times 8) + (3 + 2 \times 8) + (3 + 3 \times 8) + (3 + 4 \times 8)$ The general term is $3 + (r - 1) \times 8 = 8r - 5$.
b $1 - 1 + 1 - 1 + 1 - 1 + \ldots$ $= \sum_{r=1}^{\infty} (-1)^{r-1}$	This is an infinite series. Each term oscillates between -1 and $+1$ and the general term is $(-1)^{r-1}$.
c $-6 + 12 - 24 + 48 - 96 + 192$ $\sum_{r=1}^{6} (-1)^r 6r$	This is a finite series with oscillating signs and each term is the next multiple of 6.

Exercise 1A

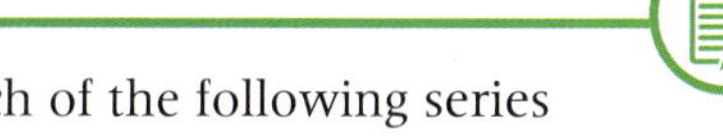

1 For each of the following sequences, write the next three terms and find the general term:

a $3, 4.5, 6, 7.5, \ldots$

b $17, 14, 11, 8, \ldots$

c $3, 9, 27, 81, \ldots$

d $\frac{1}{4}, \frac{4}{7}, \frac{7}{10}, \frac{10}{13}, \ldots$

e $\frac{1}{2}, \frac{1}{12}, \frac{1}{30}, \frac{1}{56}, \ldots$

2 Write down the first five terms of each of the following sequences:

a $u_r = 3 - 2r$ **b** $u_r = \frac{r}{2r+1}$

c $u_r = 2r + (-1)^r r$ **d** $u_r = (-1)^r \times 2$

e $u_r = \frac{3}{2^{r-1}}$

3 Write each of the following sequences using the general term:

a $5, 10, 15, 20, \ldots$

b $6, 14, 22, 30, \ldots$

c $\frac{1}{2}, \frac{1}{4}, \frac{1}{8}, \frac{1}{16}, \ldots$

d $1, -\frac{1}{3}, \frac{1}{9}, -\frac{1}{27}, \ldots$

e $0, 3, 8, 15, \ldots$

4 Write each of the following series in full:

a $\sum_{r=1}^{4} 2r(1-r)$ **b** $\sum_{r=0}^{5} (-1)^r r^2$

c $\sum_{r=1}^{5} \frac{r}{3r-1}$ **d** $\sum_{r=1}^{4} 5$

e $\sum_{r=0}^{3} (r^2 - 3)$

5 For each of the following series written in sigma notation, write the first five terms:

a $\sum_{r=1}^{\infty} \frac{r+1}{r^2}$ **b** $\sum_{r=1}^{\infty} \frac{(-1)^r}{2r^2-1}$

c $\sum_{r=1}^{20} r(5r-1)$ **d** $\sum_{r=0}^{5} (2^r - 3)$

e $\sum_{r=1}^{\infty} r^r$

6 Write each of the following series in sigma notation:

a $8 + 5 + 2 - 1 - 4$

b $3 + 10 + 21 + 36 + 55$

c $0 + \frac{1}{3} + \frac{1}{2} + \frac{3}{5} + \frac{2}{3} + \frac{5}{7} + \ldots$

d $1 + 9 + 25 + 49 + 81$

e $3k + 6k + 9k + 12k + 15k$

Developing inquiry skills

Now go back to the opening question. Suppose the length of each side of the first triangle is 81 cm. Can you work out the length of each side of the figure in each iteration? Tabulate your results and try to find a pattern and then make a conjecture.

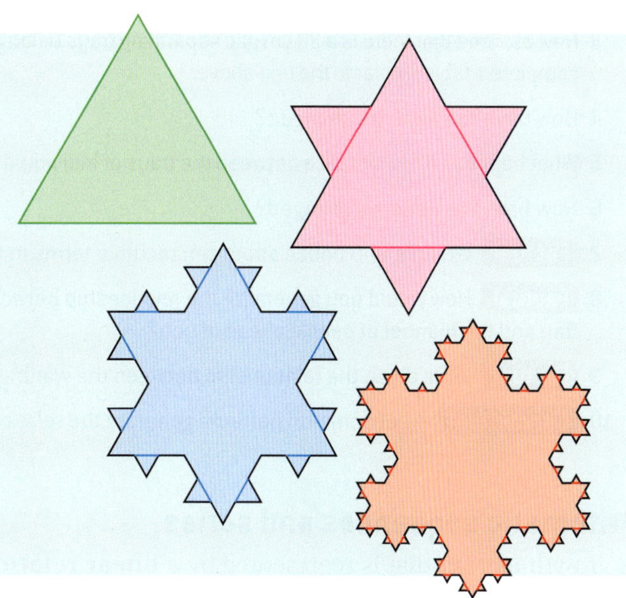

1.2 Arithmetic and geometric sequences and series

Investigation 3

Whenever you go through airport security you have to place your hand luggage, coat, phone, etc into a tray that goes on a conveyer belt which then takes it through an x-ray scanner.

When answering the following questions, you can assume the following:

- Trays are placed on the conveyer belt with no gaps between them.
- The length of each tray is 60 cm.
- The conveyer belt is moving at 10 cm per second.
- Each person uses three trays.

1 Copy and complete the following table:

Number of people ahead of you	Distance of your first tray to machine, d (m)	Waiting time, T (s)
0	0	0
1	1.8	
2		36
⋮	⋮	⋮
n		

2 What patterns do you see emerging?

3 Now assume that there is a 30 cm gap separating trays belonging to different passengers. Construct and complete a table similar to the one above.

4 How have the patterns changed?

5 What happens if the distance between the trays of individual passengers changes to 50 cm? 60 cm? 80 cm?

6 How have the patterns changed?

7 **Factual** What do you notice about consecutive terms in the second and third columns?

8 **Factual** How would you generalize the relationship between the distance from the machine to your first tray and the number of people ahead of you?

9 **Factual** Write down the relationship between the waiting time and the number of people ahead of you.

10 **Conceptual** What common patterns generate the relationships developed in this investigation?

Arithmetic sequences and series

A growth pattern that is represented by a **linear relationship** is also known as an arithmetic sequence, which is defined as follows:

If the difference between two consecutive numbers in a sequence is constant then it is an **arithmetic sequence** or an **arithmetic progression**. The constant difference is called the **common difference** and is denoted by d.

Consider how an arithmetic sequence with first term u_1 and common difference d grows:

First term u_1

Second term $u_2 = u_1 + d$

Third term $u_3 = u_2 + d = u_1 + 2d$

Fourth term $u_4 = u_3 + d = u_1 + 3d$

This leads to the general term $u_n = u_1 + (n-1)d$.

An arithmetic sequence with first term u_1 and common difference d has **general term** $u_n = u_1 + (n-1)d$.

HINT

A recursive equation is one in which the next term is defined as a function of earlier terms. In the case of an arithmetic sequence the recursive equation is $u_n = u_{(n-1)} + d$.

The next four examples show you how to use the general term formula to answer different types of questions.

Example 6

The fourth term of an arithmetic sequence is 18 and the common difference is –5. Determine the first term and the nth term.

$u_4 = u_1 + 3\times(-5) = 18$ $\Rightarrow u_1 = 18 + 15 = 33$ $u_n = 33 + (n-1)\times(-5)$ $\Rightarrow u_n = 38 - 5n$	Using $u_n = u_1 + (n-1)d$.

Example 7

Find the number of terms in the following arithmetic sequences:

a 20, 23, 26, …, 83 **b** 34, 30, 26, …, –30 **c** $6a, 4a, 2a, \ldots, -22a$

a $u_1 = 20,\ d = 3$ $u_n = 17 + 3n = 83$ $\Rightarrow n = 22$ **b** $u_1 = 34,\ d = -4$ $u_n = 38 - 4n = -30$ $\Rightarrow n = 17$ **c** $u_1 = 6a,\ d = -2a$ $u_n = 8a - 2an = -22a$ $\Rightarrow n = 15$	$d = 23 - 20 = 3$ Using $u_n = u_1 + (n-1)d$. Solve the linear equation to obtain n.

Example 8

Three numbers are consecutive terms of an arithmetic sequence. The sum of the three numbers is 45, and their product is 3240. Find the three numbers.

Let the three numbers be $u - d, u, u + d$	
$3u = 45$	Taking the sum of the numbers.
$\Rightarrow u = 15$	
$u(u^2 - d^2) = 3240$	Taking the product.
$\Rightarrow 15^2 - d^2 = \frac{3240}{15} = 216$	Substitute $u = 15$ and divide by 15.
$\Rightarrow d^2 = 225 - 216 = 9$	
$\Rightarrow d = \pm 3$	The two values of d produce two possible sequences: 12, 15, 18 or 18, 15, 12
The three numbers are 12, 15 and 18.	

Example 9

The second term of an arithmetic sequence is 20 and the seventh term is 55. Find the first term and the common difference of the sequence.

$u_2 = u_1 + d = 20$	$u_n = u_1 + (n - 1)d$
$u_7 = u_1 + 6d = 55$	
$\Rightarrow 5d = 35 \Rightarrow d = 7$	
$u_1 = 20 - 7 = 13$	Solving simultaneously.
Or	
$u_7 = u_2 + 5d \Rightarrow 5d = 55 - 20$	Write u_7 in terms of u_2.
$\Rightarrow 5d = 35 \Rightarrow d = 7$	Solve for d.
$u_1 = 20 - 7 = 13$	

The sum of an arithmetic sequence

Investigation 4

Miss Sandra, the Grade 5 teacher, pairs up her students and gives each pair 55 cards numbered from 1 to 55. She tells the students that she wants them to use these cards to find the sum of the numbers $1 + 2 + 3 + \ldots + 55$.

Michela and Grisha start by laying out the cards in ascending order. Michela takes away the first card and the last card and notes that their sum is 56. Grisha then takes the first and last card from the cards that remain and notes that their sum is also 56. They continue to do this until just one card is left.

1. Which card will this be?
2. Using the information above, how would you determine the sum of the first 55 positive integers?
3. What if you wanted to find the sum of the first 1000 positive integers?
4. **Factual** Explain the importance of the actual number of terms added.
5. Repeat the process for finding the sum of:
   a. the first 100 even numbers
   b. the positive multiples of 3 less than 1000.
6. **Conceptual** How was Michela's and Grisha's method more efficient?

International-mindedness

Karl Friedrich Gauss (1777–1855) was a renowned German mathematician. It is said that when he was in primary school his teacher challenged him to find the sum of the numbers from 1 to 100. To the teacher's amazement, Gauss gave the correct answer almost immediately. He came to the answer by using the method used in investigation 4.

Reflect on Investigation 4 and explain how the method used is equivalent to the direct derivation for the sum of an arithmetic series containing n terms, with first term u_1 and common difference d as shown below.

$$\begin{array}{rcccccccccc} S_n = u_1 & + & u_1+d & + & u_1+2d & + & \dots & + & u_1+(n-2)d & + & u_1+(n-1)d \\ S_n = u_1+(n-1)d & + & u_1+(n-2)d & + & u_1+(n-3)d & + & \dots & + & u_1+d & + & u_1 \\ \hline 2S_n = 2u_1+(n-1)d & + & 2u_1+(n-1)d & + & 2u_1+(n-1)d & + & \dots & + & 2u_1+(n-1)d & + & 2u_1+(n-1)d \end{array}$$

$$\Rightarrow 2S_n = n[2u_1+(n-1)d]$$

$$\Rightarrow S_n = \frac{n}{2}[2u_1+(n-1)d]$$

This can be rewritten as follows:

$$S_n = \frac{n}{2}[2u_1+(n-1)d]$$

$$= \frac{n}{2}[u_1+u_1+(n-1)d]$$

$$= \frac{n}{2}[u_1+u_n]$$

TOK

How is intuition used in mathematics?

The sum of a finite arithmetic series is given by

$S_n = \frac{n}{2}[2u_1+(n-1)d] = \frac{n}{2}[u_1+u_n]$ where n is the number of terms in the series, u_1 is the first term, d is the common difference and u_n is the last term.

Example 10

The first term of an arithmetic series is 5 and the last term is –51. The series has 15 terms. Find:

a the common difference

b the sum of the series.

a $-51 = 5 + 14d$ $d = \frac{-56}{14} = -4$	Using $u_n = u_1 + (n-1)d$.
b $S_{15} = \frac{15}{2}[5 + (-51)] = -345$	Using $S_n = \frac{n}{2}[u_1 + u_n]$.

Example 11

The first term of an arithmetic series is –7 and the fourth term is 23. The sum of the series is 689. Find the number of terms in the series.

$u_1 = -7$ $u_1 + 3d = 23 \Rightarrow d = \frac{23+7}{3} = 10$	Using $u_4 = u_1 + 3d$.
$S_n = 689 = \frac{n}{2}[-14 + (n-1)\times 10]$	
$\Rightarrow 10n^2 - 24n - 1378 = 0$ $\Rightarrow 5n^2 - 12n - 689 = 0$ $\Rightarrow (5n+53)(n-13) = 0$ $\Rightarrow n = 13$, since $n \in \mathbb{Z}^+$	Rearrange and solve for n.

Reflect Why can n not be a rational or a negative number?

Example 12

Find the value of $\sum_{r=1}^{28} 5r - 4$.

$u_1 = 1$ $u_{28} = 140 - 4 = 136$	Substitute $r = 1$ and $r = 28$ to find the first and last terms.
$S_{28} = \frac{28}{2}(1 + 136) = 1918$	Using the formula $S_n = \frac{n}{2}[u_1 + u_n]$.

Example 13

The sum of an arithmetic series is given by $S_n = n(2n - 3)$. Find the common difference and the first three terms of the series.

$S_1 = u_1 = -1$	Using $S_n = n(2n - 3)$.
$S_2 = u_1 + (u_1 + d) \Rightarrow -2 + d = 2$	Using $S_2 = u_1 + u_2$.
$d = 4$	
$u_1 = -1$	
$u_2 = 3$	
$u_3 = 7$	

Exercise 1B

1 Find the nth term of each of these sequences:

a 3, 8, 13, 18, …

b 101, 97, 93, 89, …

c $a - 3, a + 1, a + 5, a + 9, \ldots$

d −20, −5, 10, 25, …

2 Find the terms indicated in each of these arithmetic sequences:

a 5, 11, 17, 23, … 15th term

b 10, 3, −4, −11, … 11th term

c $a, a + 2, a + 4, a + 6, \ldots$ 17th term

d 16, 12, 8, 4, … $(n + 1)$th term

3 Find the number of terms in each of these arithmetic sequences:

a 16, 11, 6, …, −64

b −108, −101, −94, …, 60

c −15, −19, −23, …, −95

d $2a + 5, 2a + 3, 2a + 1, \ldots, 2a - 23$

4 Determine the first term and the common difference of the arithmetic sequences that are generated by each of the following nth terms:

a $u_n = 5n - 7$

b $u_n = 3n + 11$

c $u_n = 6 - 11n$

d $u_n = 2a + 2n + 1$

5 The sixth term of an arithmetic sequence is 37 and the common difference is 7. Find the first term and the nth term.

6 The fifth term of an arithmetic sequence is 0 and the 15th term is 180. Find the common difference and the first term.

7 The sum of three consecutive terms of an arithmetic sequence is 24 and their product is −640. Find the three numbers.

8 Jung Ho earned €38 000 when he started his first job in the year 2000. He received a raise of €500 each consecutive year. Determine how much he earned in 2017? Evaluate in which year he would earn 50% more than his original salary for the first time.

9 Find the value of each of the following series:

a $3 - 3 - 9 - 15 - 21 - \ldots - 93$

b $31 + 40 + 49 + \ldots + 517$

c $(a - 1) + (a + 2) + (a + 5) + \ldots + (a + 146)$

10 Find the value of each of the following sums:

a $\sum_{r=1}^{50}(3r - 8)$ **b** $\sum_{r=1}^{100}(7 - 8r)$

c $\sum_{r=1}^{20}(2ar - 1)$, where a is a constant

11 Find the sums of the following sequences up to the term indicated:

a 4, −1, −6, ... 15th term

b 3, 11, 19, ... 10th term

c 1,−4, −9, ... 20th term

12 Calculate the sum of an arithmetic series with 25 terms given that the fifth term is 19 and 10th term is 39.

13 The third term of an arithmetic sequence is −8, and the sum of the first 10 terms of the sequence is −230. Find:

a the first term of the sequence

b the sum of the first 13 terms.

14 The sum of an arithmetic series is given by $S_n = 6n - 3n^2$. Find the common difference and the first four terms of the series.

15 Calculate the sum of all the odd numbers less than 300.

Investigation 5

The diagram below shows the first two iterations when constructing Sierpinski's triangle, named after the Polish mathematician Waclaw Sierpinski who first described it in 1915.

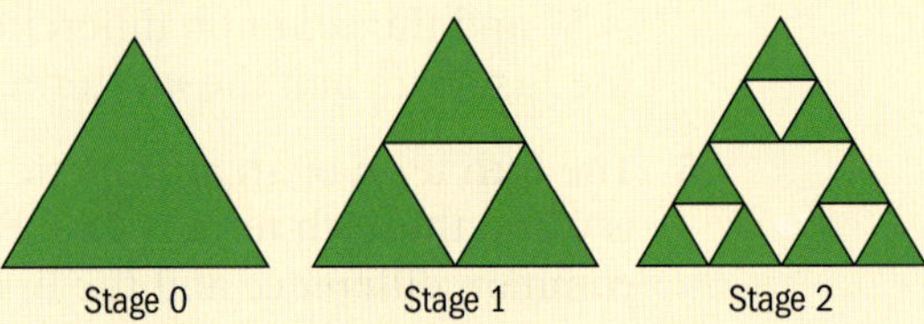

1 Construct the next iteration (Stage 3).

2 Copy and fill out the table below by following these instructions:

- Count the number of green triangles at each stage.
- If the sides of the triangle in stage 0 are each 1 unit long, what are the lengths of the sides of the green triangles at each of the following three stages? (Express your answers as rational numbers.)
- Now assume that the area of the triangle at Stage 0 is 1 unit². What is the area of each green triangle at each of the next three stages? (Leave answers in fractional form.)

Stage	0	1	2	3
Number of green triangles	1			
Length of one side of one green triangle	1			
Area of each green triangle	1			

3 **Factual** What patterns emerge from each of the three rows of the table?

4 **Factual** What do these three patterns have in common?

5 Based on your results, form a conjecture to obtain the numbers if you were to extend the table further to stages 4, 5, 6, etc.

6 **Conceptual** How would you compare the sets of numbers obtained?

TOK

Is all knowledge concerned with identification and use of patterns?

Geometric sequences and series

In Investigation 5 you should have noticed that when filling out the table you would need to multiply the numbers in each row by a particular constant to obtain the following column. In other words, the ratio of a particular term to the previous term is a constant. Such sequences are known as geometric sequences.

If the ratio of two consecutive terms in a sequence is constant then it is a **geometric sequence** or a **geometric progression**. The constant ratio is called the **common ratio** and denoted by r.

HINT

The recursive equation for a geometric sequence is $u_n = u_{n-1} \times r$.

Consider how a geometric sequence with first term u_1 and common ratio r grows:

First term	u_1
Second term	$u_2 = u_1 r$
Third term	$u_3 = u_2 r = u_1 r^2$
Fourth term	$u_4 = ur = u_1 r^3$

This leads to the general term $u_n = u_1 r^{n-1}$.

A geometric sequence with first term u_1 and common ratio r has **general term** $u_n = u_1 r^{n-1}$, $r \neq 1, 0, -1$, $u_1 \neq 0$.

Curiosities in geometric patterns

- What happens if you have a sequence with first term u_1 and common ratio 1?
- What if the common ratio is 0?
- And what happens if the common ratio is −1?

In the first case, the sequence is just made up of constant terms u_1. This is called a uniform sequence.

The next case is a sequence with first term u_1 and all the other terms are 0, which is a rather uninteresting sequence.

The third case leads to what is known as an oscillating sequence: $u_1, -u_1, u_1, -u_1, \ldots$

This oscillating sequence becomes particularly interesting if $u_1 = 1$, which then leads to the sequence 1, −1, 1, −1, 1, ...

If you try to take the sum of this series you run into some curious and interesting results.

You want to look at the sum $S = 1 - 1 + 1 - 1 + 1 - 1 + \dots$

There are various ways of looking at this sum. Possibly the most intuitive way of finding this sum is by grouping the terms into pairs as follows:

$$S = (1 - 1) + (1 - 1) + (1 - 1) + (1 - 1) + \dots = 0 + 0 + 0 + 0 + \dots = 0$$

But what happens if you pair the terms starting from the second term instead of the first?

$$S = 1 + (-1 + 1) + (-1 + 1) + (-1 + 1) + (-1 + 1) + \dots$$
$$= 1 + 0 + 0 + 0 + 0 \dots = 1$$

Yet another result is obtained if you look at the series from a different perspective:

$$S = 1 - (1 - 1 + 1 - 1 + 1 - 1 + \dots)$$
$$= 1 - S$$
$$\Rightarrow 2S = 1$$
$$\Rightarrow S = \frac{1}{2}$$

Why does this paradox arise and which is the correct answer? You have once more stumbled on the concept of infinity. If the number of terms were to be made finite, then the result would be 0 if there are an even number of terms, and 1 if the number of terms were odd, but an infinite sum never ends.

The next examples show how to use the general term formula for a geometric sequence to answer different types of questions.

International-mindedness

The series $S = 1 - 1 + 1 - 1 + 1 - 1 + \dots$ is known as Grandi's series, after the Italian mathematician Guido Grandi (1671–1742). You may want to look into the history and research on this sum by various mathematicians after its first appearance in Grandi's book published in 1703.

Example 14

Find the common ratio and write the next two terms of each sequence:

a 2.5, 5, 10, … **b** 9, 3, 1, … **c** $x, 2x^3, 4x^5, \dots$

a $r = \frac{5}{2.5} = 2$ The next two terms are 20, 40.	Find r by calculating $\frac{u_2}{u_1}$. Use the recursive equation to find the next two terms.
b $r = \frac{3}{9} = \frac{1}{3}$ The next two terms are $\frac{1}{3}, \frac{1}{9}$.	
c $r = \frac{2x^3}{x} = 2x^2$ The next two terms are $8x^7$, $16x^9$.	

Example 15

Find the number of terms in each of these geometric sequences:

a 0.15, 0.45, 1.35, ..., 12.15 **b** 440, 110, 27.5, ..., 0.4296875

a $u_1 = 0.15,\ r = \frac{0.45}{0.15} = 3$

Determine the value of r by computing $\frac{u_2}{u_1}$.

$u_n = 0.15 \times 3^{n-1} = 12.15$

Use $u_n = u_1 r^{n-1}$ to find n.

$\Rightarrow 3^{n-1} = 81 = 3^4$

$\Rightarrow n = 5$

This sequence has 5 terms.

b $u_1 = 440,\ r = \frac{110}{440} = 0.25$

$u_n = 440 \times 0.25^{n-1} = 0.4296875$

$\Rightarrow n - 1 = 5$

$\Rightarrow n = 6$

This sequence has 6 terms.

Use technology to find the value of n.

x	$440 \times (0.25)^x$
1	110
2	27.5
3	6.875
4	1.71875
5	0.4296875
6	0.1074219
7	0.0268555

Example 16

The first term of a geometric sequence is 4 and the common ratio is –2. Determine which term has the value of –2048?

$4 \times (-2)^{n-1} = -2048$

$\Rightarrow n = 10$

Use technology to find the value of n.

x	$4 \times (-2)^{x-1}$
1	4
2	–8
3	16
4	–32
5	64
6	–128
7	256
8	–512
9	1024
10	–2048

HINT

This time the formula uses $(x - 1)$ in the exponent so the answer is $n = 10$.

Example 17

The fourth term of a geometric sequence is 54 and the sixth term is 486. Determine the possible values of the common ratio.

$\left.\begin{array}{l} u_4 = u_1 \times r^3 = 54 \\ u_6 = u_1 \times r^5 = 486 \end{array}\right\} \Rightarrow r^2 = \frac{486}{54} = 9$ $r = \pm 3$ **Or** $u_6 = u_4 \times r^2$ $\Rightarrow r^2 = \frac{486}{54} = 9$ $\Rightarrow r = \pm 3$	Use $u_n = u_1 r^{n-1}$. Divide the two expressions to obtain r^2.

Example 18

The first term of a geometric sequence is 16 and the common ratio is $\frac{1}{2}$.
Find the biggest term that is smaller than $\frac{1}{1000}$.

$16 \times \left(\frac{1}{2}\right)^{n-1} < 0.001$ $\Rightarrow n = 15$ $u_{15} = 16 \times \left(\frac{1}{2}\right)^{14}$ Alternatively, you can use your GDC.	Use technology to find the value of n.

x	$16 \times (0.5)^{(x-1)}$
1	16
2	8
3	4
4	2
.	.
.	.
13	0.00390625
14	0.001953125
15	0.000976563
16	0.000488281

The sum of a geometric sequence

When trying to find the value of the series $S = 1 + 3 + 9 + 27 + 81 + 243$, Max notices that this is a geometric series with common ratio 3, and that if he were to multiply the series by 3, he could more easily calculate the sum as follows:

$$\left.\begin{array}{l} 3S = 3 + 9 + 27 + 81 + 243 + 729 \\ S = 1 + 3 + 9 + 27 + 81 + 243 \end{array}\right\} \Rightarrow 3S - S = 2S = 729 - 1 = 728$$

$$S = 364$$

Max then tried to generalize this result for a finite geometric series with common ratio r and having n terms as follows:

$$\left.\begin{aligned} S_n &= u_1 + u_1r + u_1r^2 + u_1r^3 + \ldots + u_1r^{n-1} \\ rS_n &= u_1r + u_1r^2 + u_1r^3 + \ldots + u_1r^{n-1} + u_1r^n \end{aligned}\right\} \Rightarrow (1-r)S_n = u_1 - u_1r^n$$

$$\Rightarrow S_n = \frac{u_1(1-r^n)}{1-r}, \quad r \neq 1$$

HINT

This formula can also be written as follows:

$$S_n = \frac{u_1(r^n - 1)}{r-1}, \quad r \neq 1.$$

This makes calculations easier when $r > 1$.

The sum of a finite geometric series is given by

$$S_n = \frac{u_1(1-r^n)}{1-r}, \quad r \neq 1$$

where n is the number of terms, u_1 is the first term and r is the common ratio.

Investigation 6

In the diagram, AB represents a piece of string which is 100 cm long.

The string is cut in half and one of the halves, CD, is placed underneath. The remaining half is now cut in half and one of the halves, DE, is placed next to CD. The process is continued as shown in the diagram.

A B
C D
C D E
C D E F
C D E F G

1 Copy and complete the table below.

Line segment	Length of string segment (cm)	Total length of segments (cm)
CD	50	50
DE	25	75
EF		
FG		

2 **Factual** As this process continues indefinitely, what do you notice about the length of the line segments? What about the total length of segments?

3 **Factual** What type of sequence is this?

Modelling this scenario mathematically:

$\text{CD} = 50 \text{ cm}$

$\text{DE} = 50 \text{ cm} \times \frac{1}{2} = 25 \text{ cm}$

$\text{EF} = \text{DE} \times \frac{1}{2} = 50 \text{ cm} \times \left(\frac{1}{2}\right)^2 = 12.5 \text{ cm}$

$\text{CD} + \text{DE} + \text{EF} + \text{FG} = 50 + 50 \times \left(\frac{1}{2}\right) + 50 \times \left(\frac{1}{2}\right)^2 + 50 \times \left(\frac{1}{2}\right)^3$

Continued on next page

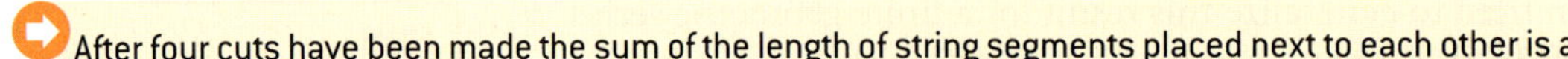

After four cuts have been made the sum of the length of string segments placed next to each other is a geometric sequence with four terms. Show that if n cuts are made this sum becomes $\dfrac{50\left(1-\left(\frac{1}{2}\right)^n\right)}{1-\left(\frac{1}{2}\right)}$.

Enter this into a table as shown below to see what happens as n gets bigger.

n	$\dfrac{50\times(1-(0.5)^n)}{(1-0.5)}$
1	50
2	75
3	87.5
4	
5	
6	
7	
8	

4 What would happen if you repeated this experiment, but this time you cut CD to be $\frac{2}{3}$ of AB and DE to be $\frac{2}{3}$ of the remaining piece of string?

5 Repeat the process using CD to be $\frac{3}{4}$ of AB and DE to be $\frac{3}{4}$ of the remaining piece of string. What if the fraction used was $\frac{4}{5}$?

6 Write a short reflection on your results which includes answers to the following questions:

- **Factual** Why were you asked to change the length of the string cut?
- **Conceptual** How has this process helped you analyse the situation?
- **Conceptual** How can the sum of an infinite series converge to a finite number?

Convergent and divergent series

An infinite geometric series is **convergent** when the sum tends to a finite value as the number of terms gets bigger. If a geometric series does not converge it is said to be **divergent**.

In Investigation 6, the series always converged to 100 cm, the length of the original piece of string.

Investigation 7

In Investigation 6, you would have noticed that you had a geometric series in each case. You will now investigate a general geometric series in order to understand which conditions will make the series converge.

For a geometric series, you know that $S_n = \dfrac{u_1(1-r^n)}{1-r}$, $r \neq 1$.

Use technology to copy and complete the following table:

n	3^n	$(-2)^n$	$(1.5)^n$	$(0.5)^n$	$(-0.2)^n$	$(-0.75)^n$
1	3	−2	1.5	0.5	−0.2	−0.75
2						
3						
4						
5						
6						
7						
8						
9						
10						

1 Extend the table for different values of r^n and larger values of n.

2 **Factual** What is the value of the common ratio?

3 **Conceptual** What role does the value of the common ratio play in a geometric series?

4 Use your results to justify the following statements:

a $r > 1 \Rightarrow r^n$ increases as n gets larger.

b $0 < r < 1 \Rightarrow r^n$ decreases as n gets larger.

c $r < -1 \Rightarrow r^n$ has a large absolute value, but its sign oscillates.

d $-1 < r < 0 \Rightarrow r^n$ has a small absolute value but its sign oscillates.

e When the value of r is close to (but still less than) 1, the value of r^n decreases more slowly but still gets close to zero when n gets larger.

f $-1 < r < 1 \Rightarrow S_n \to \frac{u_1}{1-r}$ as the value of n gets larger.

The sum of n terms of a geometric series is $S_n = \frac{u_1(1-r^n)}{1-r}$, $r \neq 1$.

When $-1 < r < 1$, r^n approaches zero for very large values of n. The series therefore converges to a finite sum given by $S = \frac{u_1}{1-r}$.

The next examples demonstrate how to use the formulae for sums of finite and infinite geometric series.

Example 19

A geometric series has first term 3 and common ratio 2. Find the sum of the first five terms.

$S_5 = \frac{3(1-2^5)}{1-2} = \frac{3(1-32)}{-1} = 93$	Use the formula $S_n = \frac{u_1(1-r^n)}{1-r}$.

Example 20

Calculate the geometric series given by $\sum_{i=1}^{7} 2\times\left(\frac{1}{2}\right)^i$.

$u_1 = 1, r = \frac{1}{2}$	Enter $i = 1$ and $i = 2$ to find u_1 and u_2 and hence r.
$S_7 = \frac{1\left(1-\left(\frac{1}{2}\right)^7\right)}{1-\frac{1}{2}} = \frac{1\left(1-\frac{1}{128}\right)}{\frac{1}{2}} \simeq 1.98$	Use the formula $S_n = \frac{u_1(1-r^n)}{1-r}$.
Or Using a GDC: $\sum_{i=1}^{7} 2\times\left(\frac{1}{2}\right)^i \simeq 1.98$	The sum can be found using technology.

Example 21

Find two possible geometric sequences where the sum of the first two terms is 20 and the sum of the first four terms is 1640, and write the general term of each sequence.

$\left.\begin{array}{l} S_2 = u_1 + u_2 = u_1(1+r) \\ S_4 = \frac{u_1(1-r^4)}{1-r} \end{array}\right\} \frac{S_4}{S_2} = \frac{(1-r^4)}{(1-r)(1+r)}$	$\frac{(1-r^4)}{(1-r)(1+r)} = \frac{(1-r^2)(1+r^2)}{(1-r^2)} = (1+r^2)$
$\Rightarrow \frac{S_4}{S_2} = \frac{1640}{20} = 1+r^2$ $\Rightarrow r^2 = 81$ $\Rightarrow r = \pm 9$	Find the ratio of S_4 to S_2. Simplify and solve for r.
$r = 9 \Rightarrow u_1 = \frac{20}{10} = 2$ or $r = -9 \Rightarrow u_1 = \frac{20}{-8} = -\frac{5}{2}$ The two possible geometric sequences are given by the general terms: $u_n = 2\times 9^{n-1}$ or $u_n = \left(-\frac{5}{2}\right)\times(-9)^{n-1}$	Calculate u_1 for each value of r.

Example 22

The sum of the first n terms of a geometric sequence is given by $S_n = 7^n - 1$. Find the first term and the common ratio of the sequence.

$S_1 = 7 - 1 = 6 \Rightarrow u_1 = 6$ $S_2 = 49 - 1 = 48 \Rightarrow 6 + u_2 = 48 \Rightarrow u_2 = 42$ $r = \frac{42}{6} = 7$	Use the formula given to find S_1 and S_2.

Example 23

Determine how many terms are required for the sum of the geometric series given by $\sum_{i=1}^{n} 3 \times 2^i$ to exceed 1000.

$u_1 = 6,\ r = \frac{u_2}{u_1} = 2$	
$\frac{6(1-2^n)}{1-2} > 1000$	$S_n = \frac{u_1(1-r^n)}{1-r}$
$\Rightarrow 2^n - 1 > \frac{1000}{6}$	Use technology to produce a table:
$\Rightarrow 2^n > 167.\dot{6}$ Using the table, $n = 8$. When 8 or more terms are added, the sum exceeds 1000.	

n	2^n
1	2
2	4
3	8
4	16
5	32
6	64
7	128
8	256

Example 24

For what values of x does the series $\sum_{i=1}^{\infty}\left(1+\frac{x}{2}\right)^i$, $x \neq -2$, converge? Find the sum when $x = -1.5$.

$S = \left(1+\frac{x}{2}\right) + \left(1+\frac{x}{2}\right)^2 + \left(1+\frac{x}{2}\right)^3 + \ldots$	Write the first three terms of the series.
This is a geometric series where $u_1 = r = \left(1+\frac{x}{2}\right)$	Identify r.

Continued on next page

For convergence: $-1 < \left(1+\frac{x}{2}\right) < 1$ $\Rightarrow -2 < \frac{x}{2} < 0$ $\Rightarrow -4 < x < 0$, but x cannot be -2	Use condition for convergence, ie $-1 < r < 1$.
When $x = -1.5$, the series converges to $S = \frac{-0.5}{1.5} = -\frac{1}{3}$.	Use the formula $S = \frac{u_1}{1-r}$.

Example 25

A geometric series converges to 8. The second term of the series is $-\frac{5}{2}$. Find the common ratio.

$\left.\begin{array}{l} S = \frac{u_1}{1-r} = 8 \\ u_2 = u_1 r = -\frac{5}{2} \end{array}\right\} \Rightarrow 8(1-r)\times r = -\frac{5}{2}$	Use the formula $S = \frac{u_1}{1-r}$.
$16r^2 - 16r - 5 = 0$ $\Rightarrow (4r+1)(4r-5) = 0$ $\Rightarrow r = -\frac{1}{4}$ or $r = \frac{5}{4}$ But, since the series converges, $-1 < r < 1$ Hence the answer is $r = -\frac{1}{4}$.	Simplify and factorize.

Exercise 1C

1 Write down the fifth term and the general term of each of the following sequences:

a 1, 3, 9, … **b** 8, 4, 2, …

c $\frac{x}{2}, \frac{x^3}{2}, \frac{x^5}{2}, \ldots$ **d** −3, 3, −3, 3, …

2 Determine the common ratio and write the terms indicated in each of the following sequences:

a 63, 21, 7, … 6th term

b $243, \frac{81}{2}, \frac{27}{4}, \ldots$ 7th term

c $\frac{a}{2}, -\frac{a}{6}, \frac{a}{18}, \ldots$ 5th term

3 Determine the number of terms in each of the following sequences:

a 0.02, 0.06, 0.18, …, 393.66

b $64, 32, 16, \ldots, \frac{1}{128}$

4 The fourth term of a geometric sequence is 6 and the seventh term is 48. Find the common ratio and the first term of the sequence.

5 The third term of a geometric sequence is 6 and the fifth term is 54. Find the two possible values of the common ratio and the sixth term of each sequence.

6 The first term of a geometric sequence is 9 and the fifth term is 16. Show that there are two possible sequences and find their common seventh term.

7 The numbers $3a+1$, $a+2$, and $a-4$ are three consecutive terms of a geometric sequence. Find the two possible values of the common ratio.

8 The numbers $a-1$, $a+1$, and $a-2$ are the fourth, fifth and sixth terms, respectively, of a geometric sequence. Find the common ratio and the first term of the sequence.

9 Find the following series for the number of terms stated:

a $3-1+\frac{1}{3}-\frac{1}{9}+\ldots$ 6 terms

b $8+4+2+1+\ldots$ 10 terms

c $0.1+0.03+0.009+0.0027+\ldots$ 15 terms

d $0.1-0.03+0.009-0.0027+\ldots$ 15 terms

10 Calculate:

a $\sum_{i=1}^{6} 7^{3-i}$ b $\sum_{i=0}^{n-1} 5\times 10^{i}$

11 Show that a geometric sequence with first term 3 and seventh term $\frac{1}{243}$ has two possible sums to infinity and find them.

12 The sum of n terms of a certain series is given by $S_n=\left(-\frac{1}{2}\right)^n-1$.

a Find the first three terms of the series.

b Show that the terms of the series are in geometric progression.

13 The second term of a geometric series is 28 and the third term is $28(1-a)$.

Find the common ratio, given that the series converges and the sum of the first three terms is 147.

14 A length of material measures 2 m. It is cut into three lengths which are in geometric progression. The longest piece is twice as long as the shortest piece. Find the common ratio of the sequence and the exact length of the shortest piece.

15 Write the first four terms of the series $\sum_{i=0}^{\infty}(-1)^i\left(\frac{x}{2}+1\right)^i$. Determine for what values of x this series converges. Find the value of the series when $x=-0.8$.

Developing inquiry skills

Go back to the original question about Koch's snowflake and try to address the following, assuming that the length of each side of the original triangle is 81 cm:

- Calculate the perimeter of the snowflake at each iteration.
- Calculate the area of the snowflake at each iteration.
- Tabulate the results and explain the number patterns that you observe.
- Create a model that helps you generalize the perimeter and area at any iteration.

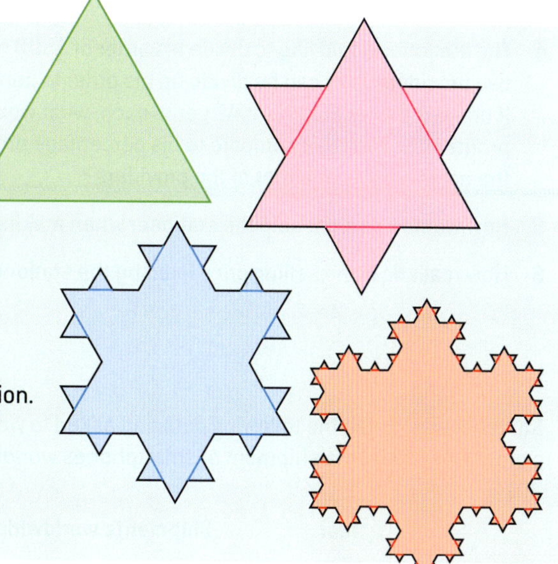

Although it might not be obvious, you have actually been exposed to arithmetic and geometric sequences and series in previous mathematics classes. Usually this was in the form of solving word problems.

TOK

How do mathematicians reconcile the fact that some conclusions conflict with intuition?

When you apply knowledge that was obtained by abstracting generalizations of mathematical concepts to real-life situations you are actually modelling the situation mathematically. The following investigations illustrate how arithmetic sequences can be hidden in everyday practices.

Investigation 8

Before the start of the school year, a stationer needs to stock up with notebooks. He has been given the following offers:

Provider A: Notebooks in packs of 20, at an offer of 6 for the price of 4, where each packet of 20 notebooks costs €10.

Provider B: Notebooks in packs of 100, at an offer of 3 for the price of 2, where each pack of 100 costs €48.

The stationer is considering stocking between 500 and 3000 notebooks, in multiples of 100.

1 The stationer first looks at the offer made by Provider A and realizes that he would get the cheaper rate when the number of notebooks ordered are in a particular arithmetic sequence. Show that the stationer is correct.

2 The stationer's wife tells him that the argument is true for the offer from Provider B. Is the wife also correct?

3 They then compare costs incurred when buying notebooks from Provider A and from Provider B. They notice that for certain numbers of notebooks ordered it would be cheaper or the same rate if they were to order from Provider A. Determine which numbers they are referring to. How is this list of numbers different from the previous two answers?

4 The stationer would like to divide his order of 1500 notebooks between the two providers. How can he divide up his order to minimize his cost price? If he sells the notebooks at 45 cents each, what would be his percentage profit? How does this compare to his percentage profit had he ordered all the notebooks from either of the providers?

5 How would this work help the stationer when making the order?

6 How realistic is the selling price fixed by the stationer?

Investigation 9

Students were given the following data and asked to work in groups to create a growth model for the shipment of smartphones worldwide from 2011 to 2016.

Year	Shipments worldwide in billions
2011	0.52
2012	0.74
2013	1.05

Students in group A decided that the data could be modelled using an arithmetic sequence by taking the average difference.

Students in group B decided that the data could be modelled using a geometric sequence by taking an average common ratio.

1 Create each growth model and determine the number of smartphones shipped in the years 2014 to 2016 as predicted by each model.

2 The actual shipments for the years 2014 to 2016 are given below. Which seems to be the better model for growth?

Year	Shipments worldwide in billions
2014	1.32
2015	1.46
2016	1.51

3 Use your models to predict shipment numbers up to and including the year 2025.

4 How can you determine whether an arithmetic series grows faster than a geometric series? Which model looks more realistic and why?

The following examples illustrate how arithmetic and geometric sequences and series can be used to solve problems.

Example 26

The number of Facebook users at the end of 2008 was 145 million and growing at a rate of 3% per week. At the end of 2010 the number of Facebook users was 608 million.

a If the rate of growth had remained constant at 3% per week, determine the number of users at the end of 2010.

b The growth rate of 3% per week remained steady for 6 months and then dropped to 1.1% per week. This growth rate was maintained for another six months but then it dropped to 0.75% per week. Assuming that this rate was sustained for a whole year, show that this model better describes the recorded numbers.

c If the rate of growth dropped to 0.6% at the beginning of 2011 and remained steady, determine how long it would take for the number of users to reach 1 billion.

a The number of users after 2 years is $145 \times (1.03)^{104} \approx 3136$ million.	A growth of 3% weekly can be modelled using a geometric sequence where the value after 2 years (104 weeks) is given by $145 \times (1.03)^{104}$.
b The number of users after 6 months is $145 \times (1.03)^{26} \approx 313$ million. The number of users at the end of 2009 is $312.706 \times (1.011)^{26} \approx 416$ million. The number of users at the end of 2010 is $415.59 \times (1.0075)^{52} \approx 613$ million.	All answers are given to 3 significant figures.

Continued on next page

c The number of users at the end of 2010 is 613 million and the growth model is given by $613 \times (1.006)^n$.

You want $613 \times (1.006)^n = 1000$.

Using technology and a table to solve for n gives:

The number of users will reach 1 billion during the year 2092, which is 82 years after the end of 2010.

Or

Use a GDC to solve the equation.

Year	Number of years	Number of users 613×1.006^n
2011	1	616.678
2012	2	620.378068
2013	3	624.1003364
2014	4	627.8449384
2015	5	631.6120081
2016	6	635.4016801
.	.	.
.	.	.
.	.	.
2090	80	989.2337664
2091	81	995.169169
2092	82	1001.140184

Example 27

When Jacob turned 18 he had access to the money his grandparents had invested in a savings account. He decided to reinvest \$10 000 at a compound interest rate of 3% each year. He decided to add \$200 to this investment on his next birthday and each following birthday until he turned 25. Evaluate how much money was in his account just after his 25th birthday. Evaluate the total interest gained over this time.

On his 18th birthday Jacob had \$10 000.

Just before his 19th birthday he had 10 000(1.03)

Amount in bank + interest for one year.

On his 19th birthday the amount was 10 000(1.03) + 200.

Value after 1 year + \$200.

Just before his 20th birthday: $(10\,000(1.03) + 200) \times 1.03$

Just after his 20th birthday:

$(10\,000(1.03) + 200) \times 1.03 + 200$

$= 10\,000(1.03)^2 + 200(1.03)^1 + 200$

After 7 years, just after his 25th birthday the amount would be:

Using the pattern from years 1 and 2.

$$10\,000(1.03)^7 + 200(1.03)^6 + 200(1.03)^5 + \ldots + 200$$
$$= 10\,000(1.03)^7 + 200(\underbrace{(1.03)^6 + (1.03)^5 + \ldots + (1.03) + 1}_{\text{Geometric series}})$$

Geometric series with $u_1 = 1$, $r = 1.03$ and $n = 7$.

$= 12\,298.73 + 200\left(\dfrac{1-(1.03)^7}{1-1.03}\right) = 12\,298.73 + 200(7.662..)$

$= \$13\,831$ (to the nearest dollar)

Money invested without interest $= 10000 + 200 \times 7 = 11400$

Total interest $= 13831 - 11400 = \$2431$

Exercise 1D

1 In 2010, a shop sold 220 televisions. Every six months the shop sold five more televisions, so that it sold 230 televisions in 2011, 240 in 2012, etc.

- **a** Evaluate how many televisions the shop sold in 2017.
- **b** Calculate the total number of televisions sold between 2010 and 2017 inclusive.
- **c** The selling price of a television was €600 in 2010 but the selling price fell by €20 each year. In a particular year the number of televisions sold by the shop was half of the selling price of each television. Determine in which year this occurred.

2 Jane started working in 2000. On each successive year she received a salary increase equivalent to 1.5% of her previous salary. In 2011 her salary was €49 650. Determine her starting salary to the nearest €.

3 Carla traced her family tree back four generations. Carla's parents are the first generation back and her first set of ancestors. Carla's four grandparents are the second generation back and her second set of ancestors.

- **a** How many ancestors are in Carla's family tree?
- **b** Determine how many generations back she would have to trace to find more than 1 million ancestors.

4 Prisana is testing a recipe for a cake and tries it out several times, adjusting the amount of flour and sugar used each time. In the first recipe she uses 200 g of flour, and she decides to increase the weight of flour by 20 g in each trial. She has time to try out the recipe just 10 times. How many kg of flour would she need?

She has 1.5 kg of sugar available and knows that the amount of sugar needed is usually half the amount of flour by weight. After the first two trials she decides to change the amount of sugar in each trial according to a geometric sequence. Evaluate the number of trials this would allow her to carry out. Explain how reliable her model is for using a geometric growth model for sugar content.

5 The first diagram shows a sequence of squares.

Starting with the largest square, the midpoints are joined to form the second square of the sequence. This process can be continued infinitely (in theory, but not in practice!)

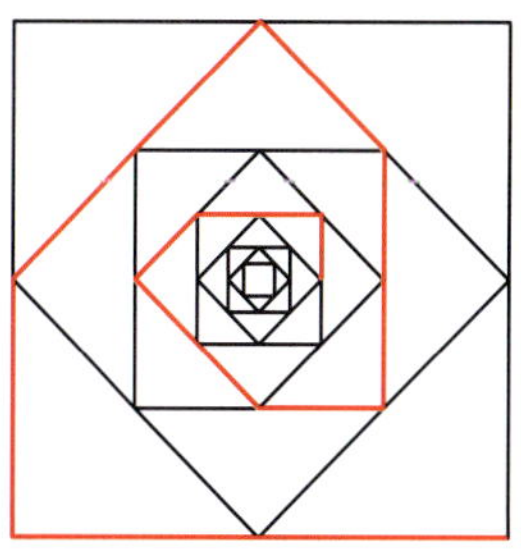

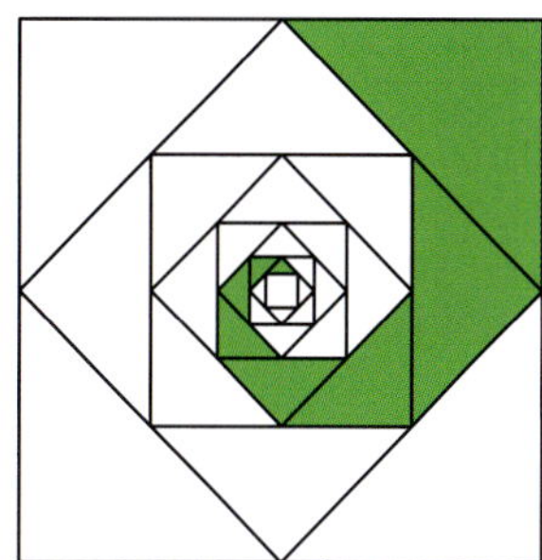

a If the sides of the largest square have a length of 2 units, calculate the side lengths of the second, third and fourth squares.

A spiral is formed by joining segments shown as red lines in the second diagram.

b Use your answers in part **a** to find the length of the spiral shown.

c Explain what happens to the length of the spiral if you continue the process infinitely.

A different spiral is formed by shading the triangles as shown in the third diagram.

d Find the total area of the shaded triangles.

e Determine the total area of the spiral formed if the process of forming squares and shading triangles is continued infinitely.

6 Gyuhun takes out a loan of \$1500 to furnish his new student apartment.

The terms of the loan are that Gyuhun will pay equal monthly instalments. Interest is calculated monthly and is charged at 12% p.a. The loan is to be repaid in two years.

a Calculate the amount that Gyuhun has to pay each month if the first repayment is made one month after the money is borrowed and after interest is calculated.

b Evaluate how much, to the nearest dollar, Gyuhun actually has to pay for furnishing the apartment?

HINT

Per annum (p.a.) is a term often used in financial contexts and means "for each year".

7 An architect is designing a cinema that should hold 570 seats. The front row of the cinema should hold 30 seats and each consecutive row is to hold six seats more than the previous row.

a Evaluate how many rows the cinema should hold.

For optimal viewing, the floor of the cinema needs to be stepped so that the step between each row is 15 cm high and 95 cm deep. Assuming that the front row is 3 m away from the screen and the ceiling is 2.4m above the last step, determine:

b the horizontal distance from the screen to the top row

c the maximum height of the ceiling, assuming that it is not slanted.

8 Ayla, Brynna and Cindy each receive €200 from their parents with the condition that they promise to invest it for at least 10 years. Ayla invests her money in an account with Rapid Bank that offers 5% simple interest annually. Brynna says it is better to invest in an account that offers compound interest because it grows quicker, so she uses an account with Quick Bank that offers 3.5% interest compounded annually. Cindy is not sure which offer is best so she invests €100 with Rapid Bank and €100 with Quick Bank. Answer the following questions giving your answers to the nearest euro.

TOK

Do all societies view investment and interest in the same way?

What is your stance?

a Evaluate how much each investment is worth after 10 years.

b How much is each investment worth after 25 years?

c Determine after how many years the three investments yield approximately the same amount.

9 At the start of 2010 Karim had \$5 000 to invest. He decided to invest part of the money in a savings account that offered 1.5% simple interest per year. He added \$1 000 to this amount and fixed it for 10 years in bonds that offered 2.5% compound interest per year. The rest of the money he invested in shares.

a After one year the money invested in shares made a loss of 1%. Given that the total amount of money invested increased by \$75, determine how much money Karim invested in each.

b At the end of the first year, Karim decided to sell the shares at their current value and reinvest the money in the savings account. Evaluate the total value of his investment at the end of 2020.

c Evaluate how much more money he would have made if he had divided up the $5 000 equally between the savings account and bonds at the very start? (Give all your answers to the nearest dollar.)

10 A pharmaceutical company has developed a drug that fights a bacterial infection. The drug is to be administered four times per day, every six hours. It was found that six hours after administering, 37.5% of the original amount was still in the bloodstream. The maximum safe level of the drug in the bloodstream is 8 mg ml^{-1}.

a Construct a model to represent the amount of drug in the bloodstream at the end of day 1, ie immediately after the fourth administration.

b The company advises that the drug should not be administered for more than 10 days. Evaluate the maximum amount that should be administered to ensure that the amount of drug in the bloodstream does not exceed the safety level.

c The drug starts being effective when the amount of drug in the bloodstream is 7 mg ml^{-1}. Determine how many times the drug must be administered for this level to be reached.

Developing your toolkit

Now do the Modelling and investigation activity on pages 70–71.

1.3 Proof

Investigation 10a

Each of the three diagrams below, not drawn to scale, consists of a square ABCD of different sizes. Line PR is perpendicular to AB and DC and line SQ is perpendicular to AD and BC. Copy and complete the table on the next page.

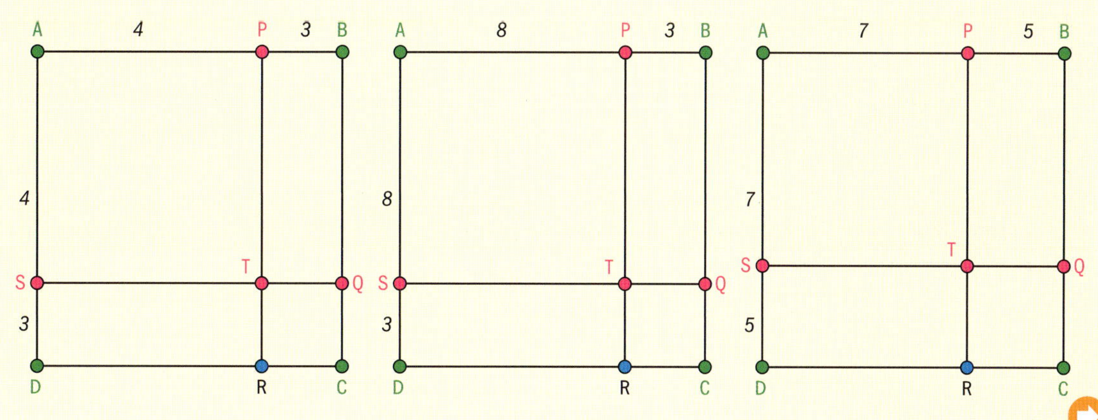

Continued on next page

Area ABCD	Area APTS	Area BPQT	Area STRD	Area TQCR
$(3+4)^2=49$	4^2	3×4		

1 Describe the relationship between the areas of square ABCD, rectangles PBCT and TRDS, and square TQCR.

2 Rewrite the relationship above replacing words with numbers.

3 Now rewrite the relationship for the diagram on the right.

4 **Conceptual** What do you call this relationship? Why?

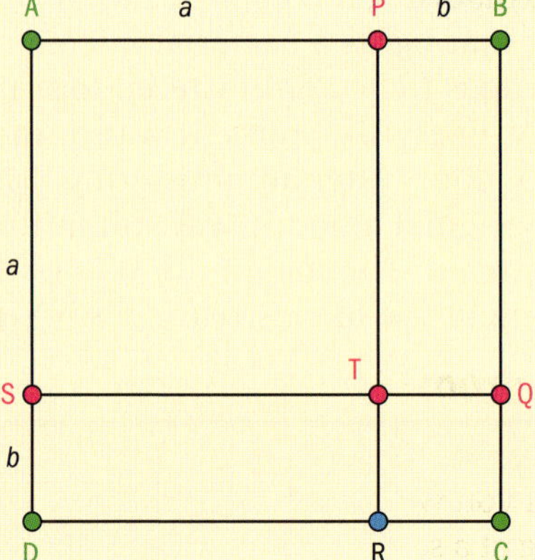

Investigation 10b

Each of the three diagrams below, not drawn to scale, consists of two squares ABCD and PQRS. Use this information to copy and complete the table for each diagram.

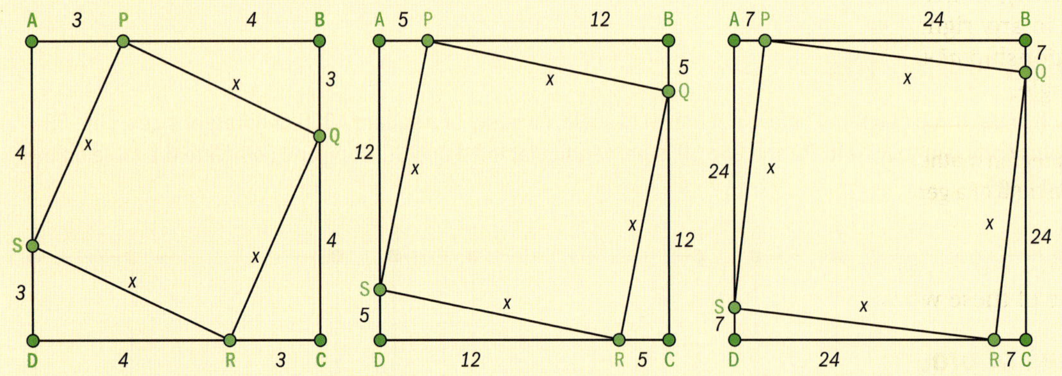

Area ABCD	Area PQRS	Area ΔPBQ	Area PQRS + 4 × area ΔPBQ
$(3+4)^2=49$	x^2	6	
	x^2		
	x^2		

1 Describe the relationship between the areas of the squares and the triangles.

2 Rewrite the relationship above replacing words with numbers and making the area of PQRS the subject.

3 **Factual** What do you call each of these relationships? Why?

4 Determine the value of x in each case.

5 Now rewrite the relationship for the diagram on the right.

6 **Conceptual** What do you call this relationship? Why?

7 **Conceptual** How would you describe the difference between an equation and an identity?

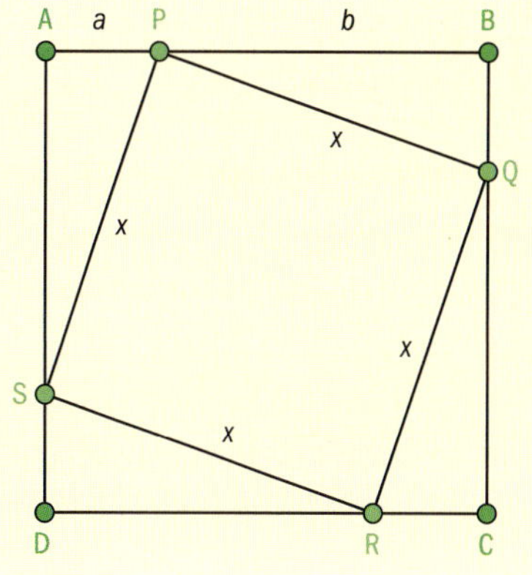

What is proof?

In Investigation 10a, you should have noticed that by making the question visual you were able to prove the identity $(a + b)^2 \equiv a^2 + 2ab + b^2$. By looking at a square which is divided into two smaller squares and two rectangles and comparing areas you came to a valid conclusion, and by then representing numbers with variables you could show that this identity is valid for all values of a and b. You can say that you have proved the statement $(a + b)^2 = a^2 + 2ab + b^2$ is true for all $a, b \in \mathbb{R}^+$ because the perpendiculars PR and SQ could be placed anywhere along the sides of square ABCD, which could also be as large or as small as you wanted.

Similarly, in Investigation 10b, you were able to validate the statement that in any right-angled triangle the lengths of the sides obey the relationship $a^2 + b^2 = c^2$, where c is the length of the hypotenuse of the triangle.

A **proof** in mathematics often consists of a logical set of steps that validates the truth of a general statement beyond any doubt.

TOK

What is the role of the mathematical community in determining the validity of a mathematical proof?

There are many ways of presenting a proof and you will be looking at some of these ways in this section.

Types of proof

Investigation 11

Copy and complete the table below and then suggest a conjecture.

1	
1 + 3	
1 + 3 + 5	
1 + 3 + 5 + 7	

Continued on next page

$1+3+5+7+9$	
$1+3+5+7+9+11$	

Now let's look at the same sequence visually.

Below you will find two visuals that represent a 4×4 square differently. In the first one, if you add the squares shaded white and green alternately you will obtain the sequence $1+3+5+7=16$.

The second visual is made up of the sequences $1+2+3+4$, and the sequence $1+2+3$, and when placed next to each other in the orientation shown they form a 4×4 square giving a sum of 16.

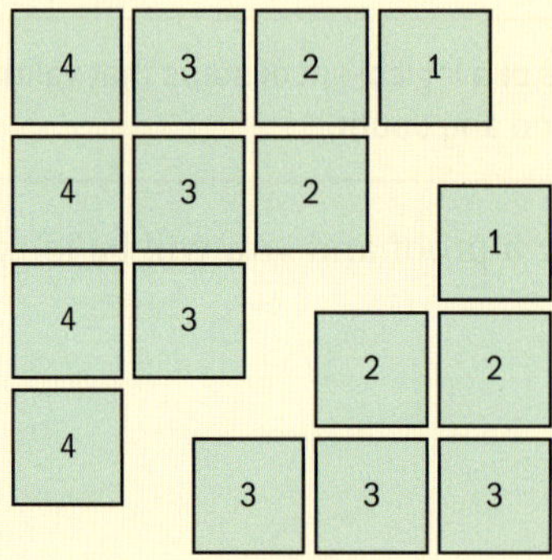

Combining the two visuals you can say that:

$$\begin{aligned}4^2 &= 1+3+5+7\\ &= (1+2+3+4)+(1+2+3)\\ &= \frac{4}{2}(1+4) + \frac{3}{2}(1+3)\\ &= 10 + 6\end{aligned}$$

Seeing the question from this perspective should make your conjecture more valid. However, this alone is not a proof, because you cannot really generate a general square. A more rigorous proof of the conjecture is required and this is found in Example 28.

Example 28

Show that $1 + 3 + 5 + 7 + \ldots + (2n - 1) = n^2$.

$S = 1 + 3 + 5 + \ldots + (2n-3) + (2n-1)$ $S = (2n-1) + (2n-3) + (2n-5) + \ldots + 3 + 1$	Write out the sum in reverse order.
$2S = \underbrace{2n + 2n + 2n + \ldots + 2n + 2n}_{n \text{ times}}$ $\Rightarrow 2S = 2n^2$ $\Rightarrow S = n^2$	Add the two sums and simplify.

A **direct proof** is a way of showing the truth of a given statement by constructing a series of reasoned connected established facts. In a direct proof the following steps are used:

- Identify the given statement.
- Use axioms, theorems, etc, to make deductions that prove the conclusion of your statement to be true.

The proof given in Example 28 consists of a set of reasoned steps that leads to the required result. Note that you did not just quote the sum of an arithmetic sequence with first term 1 and common difference 2, although by the definition of direct proof this would have been valid.

Example 29

Show that:

a the sum of an odd and even positive integer is always odd

b the sum of two even numbers is always even

c the sum of two odd numbers is always even.

a Let m and n be an odd and an even positive integer respectively. $\Rightarrow m = 2p - 1$ and $n = 2s$ where $p, s \in \mathbb{Z}^+$ $\Rightarrow m + n = 2p - 1 + 2s = 2(p + s) - 1$ This is an odd number since $p + s \in \mathbb{Z}^+$.	An odd positive integer can be written as $2k - 1$ where $k \in \mathbb{Z}^+$. An even positive integer can be written as $2k$ where $k \in \mathbb{Z}^+$.
b Let m and n be two even positive integers. $\Rightarrow m = 2p$ and $n = 2s$ where $p, s \in \mathbb{Z}^+$ $\Rightarrow m + n = 2(p + s)$ This is an even number since $p + s \in \mathbb{Z}^+$.	

Continued on next page

c Let m and n be two odd positive integers. $\Rightarrow m = 2p - 1$ and $n = 2s - 1$ where $p, s \in \mathbb{Z}^+$ $\Rightarrow m + n = 2p + 2s - 2$ $\Rightarrow m + n = 2(p + s - 1)$ This is an even number since $p + s - 1 \in \mathbb{Z}^+$.	

Example 30

Show that $\left(x+\left(\frac{a}{2}\right)\right)^2-\left(\frac{a}{2}\right)^2 \equiv x^2+ax$.

LHS $\equiv \left(x^2+2\left(\frac{a}{2}\right)x+\frac{a^2}{4}\right)-\frac{a^2}{4}$ $\equiv x^2+ax$	Expand and simplify.

In Example 30 you started from the left hand side and showed that this is equivalent to the right hand side. When writing down a proof it is very important to work on one side of the statement only. A proof is also valid if you work on each side consecutively to obtain the same result. This method is shown in the example below.

Example 31

Prove that $(n + 4)^2 - 3n - 4 = (n + 1)(n + 4) + 8$.

LHS $= n^2 + 8n + 16 - 3n - 4$ $= n^2 + 5n + 12$ RHS $= n^2 + n + 4n + 4 + 8$ $= n^2 + 5n + 12$ LHS = RHS Therefore $(n + 4)^2 - 3n - 4 = (n + 1)(n + 4) + 8$	Note that you work on each side separately and not in the same line.

Example 32

Prove that if the sum of the digits of a four-digit number is divisible by 3, then the four-digit number is also divisible by 3.

Let n be a four-digit number such that $n = a_3a_2a_1a_0$. You know that:	
$\left.\begin{array}{l} a_3 = p \times 10^3 \\ a_2 = q \times 10^2 \\ a_1 = r \times 10^1 \\ a_0 = s \times 10^0 \end{array}\right\} 0 \leq p, q, r, s \leq 9 \text{ and } p \neq 0$	Since you are given that n is a four-digit number, $p \neq 0$.
and $p + q + r + s = 3k$, $k \in \mathbb{Z}$. $\Rightarrow n = p \times 10^3 + q \times 10^2 + r \times 10^1 + s \times 10^0$ and $s = 3k - p - q - r$	You are given that the sum of the digits is a multiple of 3.
Therefore $n = p \times 10^3 + q \times 10^2 + r \times 10^1 + 3k - p - q - r$ $= (p \times 10^3 - p) + (q \times 10^2 - q) + (r \times 10^1 - r) + 3k$ $= p(10^3 - 1) + q(10^2 - 1) + r(10^1 - 1) + 3k$ $= 999p + 99q + 9r + 3k$ $= 3(333p + 33q + 3r + k)$ Since $(333p + 33q + 3r + k) \in \mathbb{Z}$ it follows that n is divisible by 3.	

Example 33

Show that $\frac{1}{2} - \frac{1}{4} + \frac{1}{8} - \frac{1}{16} + \frac{1}{32} - \frac{1}{64} + \ldots = \frac{1}{3}$.

LHS $= \left(\frac{1}{2} + \frac{1}{8} + \frac{1}{32} + \ldots\right) - \left(\frac{1}{4} + \frac{1}{16} + \frac{1}{64} + \ldots\right)$	Separate the given series into one series with positive terms and one with negative terms.
$= \underbrace{\left(\frac{1}{2} + \frac{1}{2}\left(\frac{1}{4}\right) + \frac{1}{2}\left(\frac{1}{4}\right)^2 + \ldots\right)}_{\text{Converging geometric series}} - \underbrace{\left(\frac{1}{4} + \frac{1}{4}\left(\frac{1}{4}\right) + \frac{1}{4}\left(\frac{1}{4}\right)^2 + \ldots\right)}_{\text{Converging geometric series}}$	Note that this is the difference of two converging geometric series.
$= \frac{\frac{1}{2}}{1 - \frac{1}{4}} - \frac{\frac{1}{4}}{1 - \frac{1}{4}} = \frac{2}{3} - \frac{1}{3} = \frac{1}{3}$	Use the formula for the sum of converging geometric series and simplify.

Exercise 1E

1 Prove that $(a + b)^2 + (a - b)^2 = 2(a^2 + b^2)$.

2 Show that the product of two odd numbers is always an odd number.

3 Prove that a four-digit number is divisible by 9 if the sum of its digits is divisible by 9. Hence identify which of the numbers 3978, 5453, 7898, 9864, 5670 are divisible by 9 without carrying out any division.

4 Show that $(a^2 + b^2)(c^2 + d^2) = (ad + bc)^2 + (bd - ac)^2$.

5 Prove that $\frac{1}{3} - \frac{2}{9} + \frac{1}{27} - \frac{2}{81} + \frac{1}{243} - \frac{2}{729} + \ldots = \frac{1}{8}$.

6 Prove that the difference between the squares of two consecutive numbers is always an odd number.

7 Show that $\frac{1}{n-1} - \frac{1}{n} + \frac{1}{n+1} = \frac{n^2+1}{n(n^2-1)}$. Hence determine the value of $\frac{1}{5} - \frac{1}{6} + \frac{1}{7}$.

8 The diagram here shows a trapezium ABCD that has been divided into three triangles. Use your knowledge of areas and the diagram to show that $a^2 + b^2 = c^2$.

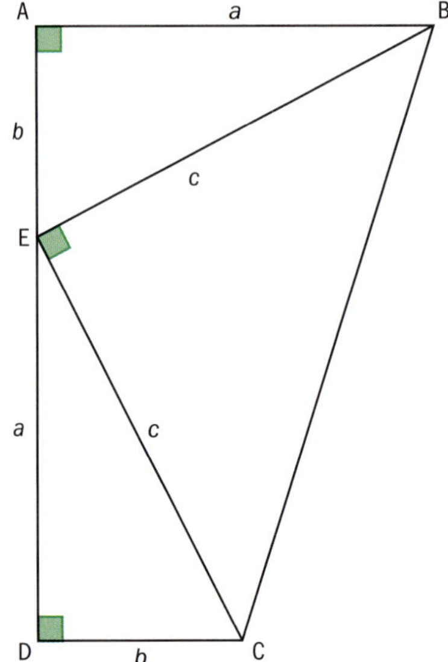

Proof by contradiction

Investigation 12

You will now look at the statement below and answer the questions:

If $n \in \mathbb{Z}$ and $5n + 2$ is even, then n is even.

1 Write a direct proof of this statement.

2 Now assume that n is an odd number and rewrite the expression $5n + 2$ to reflect this.

3 Simplify your expression and explain why this can never be an even number.

4 How is the second method different to a direct proof?

In Investigation 12 you managed to find a different argument to prove the statement by using the **contrapositive.** You started by assuming that the second part of the statement is false and showed that this led to a contradiction, ie the first part of the statement was also false. In this case, the statement could easily be proved directly but this is not always the case, and sometimes you will need to use the second method to prove a statement correct.

HINT

A proof by contradiction is built on the logical reasoning that If a proposition is true then its contrapositive is also true. That is, if in a statement a implies b then it is also true that if b is not true, this implies that a is not true.

Investigation 13

In geometry, the triangle inequality states that in any triangle ABC, the sum of the lengths of any two sides must be greater than or equal to the length of the remaining side.

Applying this to the triangle on the right you get the following:

$a+b \geq c$

$a+c \geq b$

$b+c \geq a$

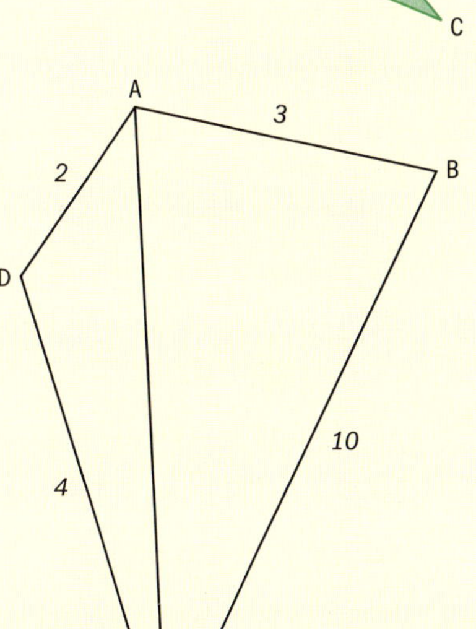

1 Explain the situation when:

$a+b=c$

$a+c=b$

$b+c=a$

2 Look at triangle ADC in the quadrilateral on the right and write the triangle inequality in terms of AD and DC.

3 Now apply the triangle inequality to triangle ABC in terms of AC and AB.

4 Look at the inequalities obtained in questions **2** and **3** and comment on your result.

5 Now draw diagonal BD and repeat steps **3** and **4** and **5** for triangles ABD and DBC.

6 What happens if you change the order of the sides?

7 What do you conclude from this investigation?

8 How else could you have come to the same conclusion?

When setting out a **proof by contradiction** you follow the following steps:

- Identify what is being implied by the statement.
- Assume that the implication is false.
- Use axioms, theorems, etc … to arrive at a contradiction.
- This proves that the original statement must be true.

TOK

What do mathematicians mean by mathematical proof, and how does it differ from good reasons in other areas of knowledge?

In other words, the assumption contradicts either a given statement or something you already know to be true, or in some cases both.

International-mindedness

Reductio ad absurdum (reduce to absurdity) is a term used to describe logical reasoning that attempts to disprove a statement by showing that it leads to an absurd result. In fact, this method can be traced back to the Greek philosopher Aristotle where he talks about "reduction to the impossible" in his book *Prior Analytics*.

The following examples will help you understand this method of proof so that you can then apply it in set tasks.

Example 34

Prove by contradiction:

a if the integer n is odd then n^2 is also odd

b if n^2 is even then n is also even

a Assume that n^2 is even. $\Rightarrow n^2 = 2k,\ k \in \mathbb{Z}$ $\Rightarrow n \times n = 2k$ But this cannot be true if n is odd as you know that the product of two odd numbers is also odd. Hence given that n is odd n^2 is also odd.	Start by making an assumption that the resulting statement is false.
b Assume that n is an odd integer. $\Rightarrow n = 2k \pm 1,\ k \in \mathbb{Z}$ $\Rightarrow n^2 = (2k \pm 1)^2 = 4k^2 \pm 4k + 1$ $\Rightarrow n^2 = 2(2k^2 \pm 2k) + 1 = 2p + 1$ Which is an odd number. Hence given that n^2 is an even integer n is also an even integer.	You want to prove that n^2 even $\Rightarrow n$ even. To use the contrapositive you start with n and deduce the contradiction ie that n^2 is not even.

Example 35

Show that $\sqrt{2}$ is irrational.

Assume that $\sqrt{2} = \frac{m}{n}$ where $m, n \in \mathbb{Z}$ and m, n have no common factors. $\Rightarrow 2 = \frac{m^2}{n^2}$ $\Rightarrow m^2 = 2n^2$	Make the assumption that the statement is false, ie $\sqrt{2}$ is rational.
This means that m is an even integer. $\Rightarrow m = 2k$ and $m^2 = 4k^2$ This leads to: $m^2 = 4k^2 = 2n^2$ $\Rightarrow n^2 = 2k^2$ This means that n is an even integer.	Here you are using the fundamental theorem of arithmetic which states that every integer bigger than 1 is either prime or a multiple of primes and since m is an even integer it must contain a prime factor of 2.

International-mindedness

How did the Pythagoreans find out that $\sqrt{2}$ is irrational?

But it was assumed that m, n have no common factors and now you have found that they have a common factor of 2. Hence, you cannot find $m, n \in \mathbb{Z}$ that have no common factors to make $\sqrt{2} = \frac{m}{n}$ a rational number.	

Example 36

Prove that there is no $x \in \mathbb{R}$ such that $\frac{1}{x-2} = 1 - x$.

Assume there exists a real number a such that $\frac{1}{a-2} = 1 - a$.	Assume that a real solution $x = a$ exists.
$\Rightarrow 1 = (a-2)(1-a)$ $\Rightarrow 1 = -a^2 + 3a - 2$ $\Rightarrow a^2 - 3a + 3 = 0$	Solve for a.
$\Rightarrow a = \frac{3 \pm \sqrt{9-12}}{2} \notin \mathbb{R}$	Apply the quadratic formula and conclude that a is not a real number since the number under the square root is negative. Show that a cannot be a real number.

Example 37

Prove that if $m, n \in \mathbb{Z}$, then $m^2 - 4n - 7 \neq 0$.

Assume that $m^2 - 4n - 7 = 0$. $\Rightarrow m^2 = 4n + 7$	Assume that the statement is false.
$\Rightarrow m^2 = 4n + 6 + 1$	Rearrange the equation.
$\Rightarrow m^2 = 2(2n + 3) + 1$	Since $n \in \mathbb{Z} \Rightarrow 2n + 3 \in \mathbb{Z}$.
This means that m^2 is an odd integer. But this means that m is an odd integer since an even integer squared is even. Let $m = 2p + 1$ where $p \in \mathbb{Z}$. Then $(2p+1)^2 - 4n - 7 = 0$ $\Rightarrow 4p^2 + 4p + 1 - 4n - 7 = 0$ $\Rightarrow 2p^2 + 2p - 2n - 3 = 0$ $\Rightarrow 2(p^2 + p - n) = 3$ Since $p, n \in \mathbb{Z}$ this cannot be true as 3 is not even. Hence $m^2 - 4n - 7 \neq 0$ given that $m, n \in \mathbb{Z}$.	

Sometimes you encounter statements that seem to be true, and for every example that you consider the statement seems to hold. Although such examples are good to verify a statement or conjecture, they are not sufficient to *prove* the statement. It takes only one example that contradicts the statement to justify that the statement is wrong.

A **counterexample**, or **counterclaim**, is an acceptable "proof" of the fact that a given statement is false.

Some examples follow to demonstrate how counterexamples can be used.

Example 38

Show by a counterexample that the following statements are not true.

a If $n \in \mathbb{Z}$ and n^2 is divisible by 4, then n is divisible by 4.

b If $n \in \mathbb{Z}$ then $n^2 + 1$ is a prime number.

c If an integer is a multiple of 10 and 12 then it is a multiple of 120.

a When $n = 6$, $n^2 = 36$ which is divisible by 4, but 6 is not divisible by 4. **b** $n = 3 \Rightarrow n^2 + 1 = 10$ which is not a prime number. **c** 60 is a multiple of both 10 and 12 but it is not a multiple of 120.	Again, the statement is true for many integers but you only need one counterexample.

Exercise 1F

In questions **1** to **9**, prove the statements by contradiction.

1 For all $n \in \mathbb{Z}$, if n^2 is odd then n is also odd.

2 $\sqrt{3}$ is irrational.

3 $\sqrt[5]{2}$ is irrational.

4 For all $p, q \in \mathbb{Z}$, $p^2 - 8q - 11 \neq 0$.

5 For all $a, b \in \mathbb{Z}$, $12a^2 - 6b^2 \neq 0$.

6 If $a, b, c \in \mathbb{Z}$, where c is an odd number and $a^2 + b^2 = c^2$, then either a or b is an even number.

7 If $n, k \in \mathbb{Z}$, then $n^2 + 2 \neq 4k$.

8 If p is an irrational number and q is a rational number, then $p + q$ is also irrational.

9 Given that m and n are positive integers with $m>n$, it follows that $m^2 - n^2 \neq 1$.

10 Show by a counterexample that the following statements are not true in general:

a $(m + n)^2 \neq m^2 + n^2$

b If a positive integer is divisible by a prime number, then the number is not prime.

c $2^n - 1$ is a prime number for all $n \in \mathbb{N}$.

d $2^n - 1$ is a prime number for all $n \in \mathbb{Z}^+$.

e The sum of three consecutive positive integers is always divisible by 4.

f The sum of four consecutive positive integers is always divisible by 4.

Proof by induction

Investigation 14

Look at the diagrams and answer the questions.

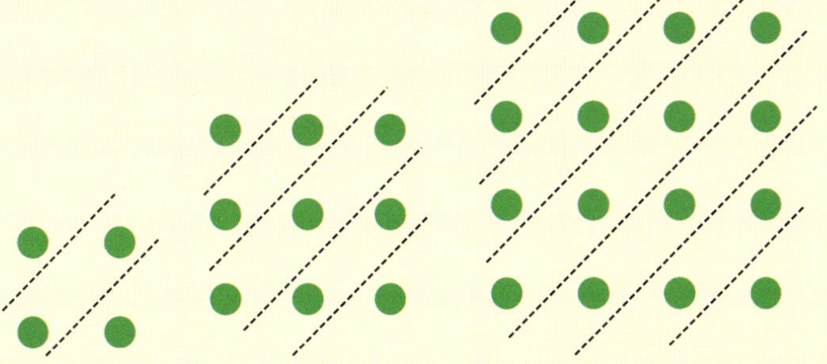

1 Each of the three diagrams represents a series. Write them down.

2 If the diagrams were to continue, what would the next three diagrams be?

3 Write a conjecture based on your findings.

4 Prove your conjecture using a direct proof.

In Investigation 14, you used a visual representation of a series to make a conjecture about a special series and then prove it. In this case, you were able to prove the conjecture directly, but there are times when such a direct proof is not possible. Sometimes you need to revert to a different proof which is called **proof by induction.**

To illustrate the principle of proof by induction, imagine two dominoes placed standing at a distance less than half their length, as illustrated in the diagram on the right. If the first domino is knocked over it will fall and cause the second domino to fall with it. This is the starting point of the process and is called the **basic step**.

Now assume that the domino in the kth place falls if the domino before it (in the $(k-1)$th position) falls. This assumption is the second step in the process.

If you were to add another domino at the end of the k dominoes, this last domino will also fall. This analogy represents the final step of the process which is called the **inductive step**.

You can then finalize your argument by stating that at the start of the process it was shown that the first domino caused the second domino to fall. You can now use the second and third step that a third domino placed behind the first step will also fall, and again using the two steps, a fourth domino will also topple over. You can continue repeating this process as many times as you want. In other words, you have shown that you can have as many dominoes as you like and they will all fall if the first domino knocks over the second domino.

You apply the dominoes analogy to mathematics to prove the statement in Investigation 14. It should be noted that the visual could have started from a previous step, ie with just one green circle.

In Example 39, you are going to use this as the basic step, so that you have the proof for all positive integer values of n.

Example 39

Prove by mathematical induction that $1 + 2 + 3 + \ldots + (n-1) + n + (n-1) + \ldots + 3 + 2 + 1 = n^2$.

$P(n)$: $1 + 2 + 3 + \ldots + (n-1) + n + (n-1) + \ldots + 3 + 2 + 1 = n^2$, $n \in \mathbb{Z}^+$	You start by making a statement $P(n)$ which you need to prove true for certain values of n (usually positive integers).
When $n = 1$ LHS $=1$ RHS $=1$ Since LHS = RHS $\Rightarrow P(1)$ is true	This is the basic step.
Assume that $P(n)$ is true for some value of k, $k \geq 1$, $k \in \mathbb{Z}^+$ ie $1 + 2 + 3 + \ldots + (k-1) + k + (k-1) + \ldots + 3 + 2 + 1 = k^2$	This is where you make the assumption. Note the wording just before you substitute for n in the statement.
When $n = k + 1$ LHS $= 1 + 2 + 3 + \ldots + (k-1) + k + (k+1) + k + (k-1) + \ldots + 3 + 2 + 1$ $= 1 + 2 + 3 + \ldots + (k-1) + k + (k-1) + \ldots + 3 + 2 + 1 + (k+1) + k$ $= k^2 + (k+1) + k$ $= k^2 + 2k + 1$ $= (k+1)^2$	In the inductive step you *must* use the assumption to show that the statement is also true for $k+1$.
Since $P(1)$ was shown to be true and it was also shown that if the statement is true for some $n = k$, $k \in \mathbb{Z}^+$, it is also true for $n = k + 1$, it follows by the principle of mathematical induction that the statement is true for all positive integers.	This final statement completes the proof and should always be included.

Why is the basic step important?

The principle of mathematical induction is very rigorous, provided that all the steps are used. If the basic step is left out, you can end up with erroneous results. Suppose you were asked whether 10^n is a multiple of 7 and you try using mathematical induction without the basic step. Here is what you would obtain:

Assume $10^k = 7a$ for some n, $a \in \mathbb{Z}^+$.

You then move to the inductive step $10^{k+1} = 10 \times 10^k$ and using the assumption would give $10^{k+1} = 10 \times 10^k = 10 \times 7a = 7(10a)$. Since a is a positive integer, $10a$ is a positive integer also, so 10^n is a multiple of 7 for all positive integers. Of course, you know that this is not true because $10 = 2 \times 5 \Rightarrow 10^n = 2^n \times 5^n$ and since 2 and 5 are prime numbers 7 will never divide 10^n exactly.

Incorrect use of the inductive step

Proof by induction is also not valid if the assumption is not used in the inductive step, as shown below to prove that $11^n - 6$ is a multiple of 5.

$P(n)$: $11^n - 6 = 5a$, where $n, a \in \mathbb{Z}^+$.

Basic step:

When $n = 1$, LHS $= 11 - 6 = 5$.

Therefore, the statement is true for $n = 1$.

Assumption:

Assume that the statement is true for some $k \in \mathbb{Z}^+$, $k \geq 1$.

ie $11^k - 6 = 5b$, where $n, b \in \mathbb{Z}^+$.

Inductive step:

When $n = k + 1$:

LHS

$= 11^{k+1} - 6$

$= 11 \times 11^k - 6$

$= (5 + 6) \times 11^k - 6$ Write 11 as 5 + 6

$= 5 \times 11^k + 6 \times 11^k - 6$ Distribute 11^k

$= 5 \times 11^k + 6(11^k - 1)$ 6 is a common factor

$= 5 \times 11^k + 6(11 - 1)(11^{k-1} + 11^{k-2} + 11^{k-3} + \dots + k + 1)$

$= 5 \times 11^k + 60(11^{k-1} + 11^{k-2} + 11^{k-3} + \dots + k + 1)$

$= 5(11^k + 12(11^{k-1} + 11^{k-2} + 11^{k-3} + \dots + k + 1))$

$= 5m$

HINT

In the chapter review at the end of this chapter you will be guided to prove that $a^n - b^n = (a - b)(a^{n-1} + a^{n-2}b + a^{n-3}b^2 + \dots + ab^{n-2} + b^{n-1})$

This is the result required and the mathematics above is correct. However, the assumption was not used to obtain the result. **Hence the proof by mathematical induction is incorrect.** The correct solution is shown in Example 40.

Example 40

Use mathematical induction to prove that $11^n - 6$ is a multiple of 5.

$P(n)$: $11^n - 6 = 5a$, where $n, a \in \mathbb{Z}^+$	Write the statement that you want to prove.
When $n = 1$	Basic step, show statement is true for $n = 1$.

Continued on next page

LHS = 11 − 6 = 5	
Therefore, the statement is true for $n = 1$.	
Assume that the statement is true for some $k \in \mathbb{Z}^+$, $k \geq 1$.	Make assumption and then make 11^k the subject of the formula.
ie $11^k - 6 = 5b$, where n, $b \in \mathbb{Z}^+$	
$\Rightarrow 11^k = 5b + 6$	
When $n = k + 1$:	Inductive step.
LHS	
$= 11^{k+1} - 6$	Let $n = k + 1$.
$= 11 \times 11^k - 6$	Rewrite the LHS in terms of 11^k.
$= 11(5b + 6) - 6$	Use the assumption.
$= 55b + 66 - 6$ $= 55b - 60$ $= 5(11b - 12)$	Simplify to obtain the required result, in this case write it as a product with a factor of 5.
Since $P(1)$ was shown to be true and it was also shown that if the statement is true for some $n = k$, $k \in \mathbb{Z}^+$, it is also true for $n = k + 1$, it follows by the principle of mathematical induction that the statement is true for all positive integers.	Write the final statement.

Example 41

Prove the following statements using mathematical induction.

a The sum of the first n terms of an arithmetic sequence with first term u_1 and common difference d is given by $S_n = \frac{n}{2}(2u_1 + (n-1)d)$.

b The sum of the first n terms of a geometric sequence with first term u_1 and common ratio r is given by $S_n = \frac{u_1(1-r^n)}{1-r}$.

a $P(n)$: $S_n = \frac{n}{2}(2u_1 + (n-1)d)$	
LHS: When $n = 1$, $S_1 = u_1$	Basic step.
RHS: When $n = 1$, $\frac{1}{2}(2u_1 + (1-1)d) = u_1$	
LHS = RHS	
Therefore $P(1)$ is true.	
Assume that $S_k = \frac{k}{2}(2u_1 + (k-1)d)$ for some $k \in \mathbb{Z}^+$.	Assumption. Substitute k in the statement.
When $n = k + 1$	Inductive step.

$S_{k+1} = S_k + u_{k+1}$ $= \frac{k}{2}\left(2u_1 + (k-1)d\right) + u_1 + kd$	Use assumption to obtain result $S_{k+1} = \frac{k+1}{2}\left(2u_1 + ((k+1)-1)d\right)$.
$= ku_1 + \frac{k}{2}(k-1)d + u_1 + kd$ $= u_1(k+1) + kd\left(\frac{k-1}{2} + 1\right)$ $= u_1(k+1) + kd\left(\frac{k+1}{2}\right)$ $= \left(\frac{k+1}{2}\right)(2u_1 + kd)$ $= \left(\frac{k+1}{2}\right)(2u_1 + ((k+1)-1)d)$	Use algebra to simplify.
Since $P(1)$ was shown to be true and it was also shown that if the statement is true for some $n = k$, $k \in \mathbb{Z}^+$, it is also true for $n = k + 1$, it follows by the principle of mathematical induction that the statement is true for all positive integers.	Write the final statement.
b $P(n)$: $S_n = \frac{u_1(1-r^n)}{1-r}$	
LHS: When $n = 1$, $S_1 = u_1$ RHS: When $n = 1$, $\frac{u_1(1-r)}{1-r} = u_1$ LHS = RHS Therefore $P(1)$ is true.	Basic step.
Assume that for some $k \in \mathbb{Z}^+$ $S_k = \frac{u_1(1-r^k)}{1-r}$	Assumption. Substitute k in the statement.
When $n = k + 1$ $S_{k+1} = S_k + u_{k+1}$ $= \frac{u_1(1-r^k)}{1-r} + u_1 r^k$ $= \frac{u_1(1-r^k) + u_1 r^k(1-r)}{1-r}$	Inductive step. Use assumption to obtain result. $S_{k+1} = \frac{u_1(1-r^{k+1})}{1-r}$.
$= \frac{u_1(1-r^k + r^k - r^{k+1})}{1-r}$ $= \frac{u_1(1-r^{k+1})}{1-r}$	Use algebra to simplify.
Since $P(1)$ was shown to be true and it was also shown that if the statement is true for some $n = k$, $k \in \mathbb{Z}^+$, it is also true for $n = k + 1$, it follows by the principle of mathematical induction that the statement is true for all positive integers.	Write the final statement.

Example 42

Use mathematical induction to prove that $3^{2n} + 7$ is divisible by 8 for all $n \in \mathbb{N}$.

$P(n)$: $3^{2n} + 7 = 8A$ When $n = 0$, LHS $= 3^0 + 7 = 8$ So $P(0)$ is true.	Note that since the statement holds for $n \in \mathbb{N}$ you have to start with 0.
Assume that the statement is true for some $k \in \mathbb{N}$, $k \geq 0$. ie $3^{2k} + 7 = 8A$ $\Rightarrow 3^{2k} = 8A - 7,\ A \in \mathbb{Z}^+$	Assumption.
When $n = k + 1$: LHS $= 3^{2(k+1)} + 7$ $= 9 \times 3^{2k} + 7$ $= 9(8A - 7) + 7$ $= 72A - 63 + 7$ $= 72A - 56$	Use the assumption.
$= 8(9A - 7)$ $= 8B$	$(9A - 7) \in \mathbb{Z}^+$, since $A \in \mathbb{Z}^+$.
Since $P(0)$ was shown to be true and it was also shown that if the statement is true for some $n = k$, $k \in \mathbb{Z}^+$, it is also true for $n = k + 1$, it follows by the principle of mathematical induction that the statement is true for all positive integers.	Write the final statement.

Exercise 1G

1 Use the diagrams to answer the questions below.

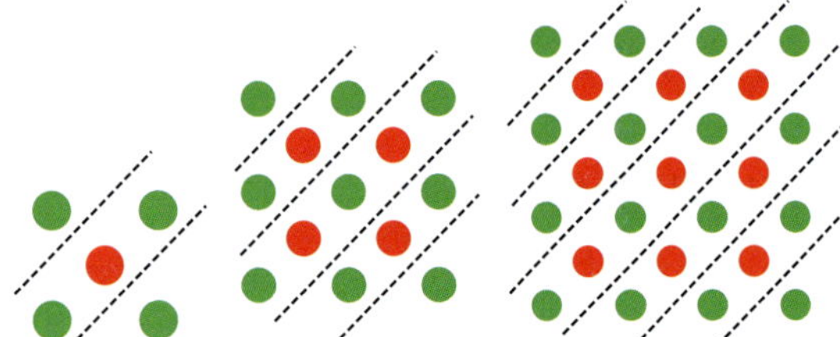

a Each of the three diagrams represents a sequence.

i Write down a sequence based on the line divisions.

ii Write down a sequence based on colour.

b If the diagrams were to continue what would the next two terms be?

c Write a conjecture based on your findings.

d Prove your conjecture using a direct proof.

e Prove your conjecture using the principle of mathematical induction.

2 Use mathematical induction to prove the following statements:

a $1^2 + 2^2 + 3^2 + \ldots + n^2 = \frac{1}{3}n(n+1)(n+\frac{1}{2})$

b $1 - 4 + 9 - 16 + \ldots + (-1)^{n+1}n^2 = (-1)^{n+1}\frac{n(n+1)}{2}$

c $\sum_{i=0}^{n} 2^i = 2^{n+1} - 1$

d $9^n - 1$ is divisible by 8 for all $n \in \mathbb{N}$.

e $1^3 + 2^3 + 3^3 + \ldots + n^3 = \frac{n^2(n+1)^2}{4}$

f $n^3 - n$ is divisible by 3 for all $n \in \mathbb{N}$.

g $\frac{1}{1\times 2} + \frac{1}{2\times 3} + \frac{1}{3\times 4} + \ldots + \frac{1}{n(n+1)} = \frac{n}{n+1}$

h $n^3 - n$ is a multiple of 6 or all $n \in \mathbb{Z}^+$.

i $2^{n+2} + 3^{2n+1}$, $n \in \mathbb{Z}^+$ is divisible by 7.

j $1^2 + 3^2 + 5^2 + \ldots + (2n-1)^2 = \frac{n(2n-1)(2n+1)}{3}$

k $\sum_{r=1}^{n} r(r+1) = \frac{n}{3}(n+1)(n+2)$

l $\sum_{r=1}^{n} \frac{1}{r(r+1)} = \frac{n}{n+1}$

3 Given the statements below, decide how best you can prove or disprove the statement. Some statements can be proved in more than one way; you should attempt both ways and then decide which is the more elegant proof, and why.

a Prove that $(4n + 3)^2 - (4n - 3)^2$ is divisible by 12 for all positive integers n.

b $n^2 + 37n + 37$ is a prime number.

c $1^3 + 3^3 + 5^3 + \ldots + (2n - 1)^3 = n^2(2n^2 - 1)$

d $1\times 2 + 2\times 3 + 3\times 4 + \ldots + (n-1)\times n = \frac{n(n^2-1)}{3}$

e $n^3 - n$ is divisible by 3 for all values of $n \in \mathbb{Z}^+$.

1.4 Counting principles and the binomial theorem

Investigation 15

Mary creates a fun game for practising some mathematics. She arranges 10 cups, numbers them as shown in the diagram on the right, and places one marble just outside cup number 1. She then writes the following instructions.

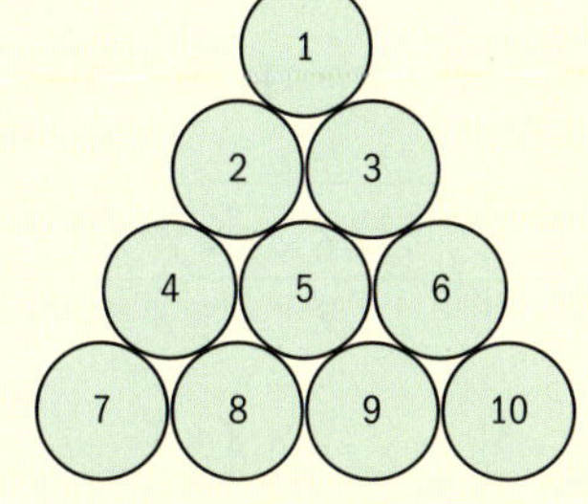

Instructions for play:

- The number of marbles you place in each cup is equal to the number of the cup multiplied by the number of marbles in the previous cup.
- The starting point is cup 1, where you will multiply the number on the cup by the number of marbles outside of cup 1.

1 If you were to follow these instructions, find how many marbles would be placed in:

a cup number 2 **b** cup number 3
c cup number 5 **d** cup number 8.

2 If Mary places another two rows underneath this arrangement, how many cups would there be in total?

3 How can you represent the number of marbles that would be placed in the last cup?

4 Comment on your results.

5 How would you represent the number of marbles in the last cup if there were n cups in total?

Factorial notation

The numbers in Investigation 15 got large very quickly. You can denote these numbers mathematically as follows:

The first marble outside the cup = $u_0 = 1$.

For the other cups, $u_n = n \times u_{n-1}$.

If you were to build up each term of this sequence, you would end up with the result $u_n = n \times (n-1) \times (n-2) \times \ldots \times 3 \times 2 \times 1$.

Some mathematical problems about arrangements and combinations involve counting techniques that use this sequence. A simpler way to denote the sequence is to use factorial notation where $u_n = n!$, $u_0 = 0! = 1$.

Here are the first five factorial numbers:

$0! = 1$

$1! = 1 \times 0! = 1$

$2! = 2 \times 1! = 2$

$3! = 3 \times 2! = 6$

$4! = 4 \times 3! = 24$

$$n! = n \times (n-1)! = n \times (n-1) \times (n-2) \times \ldots \times 3 \times 2 \times 1$$

This pattern lends itself to calculating expressions with very large numbers, as shown in the following examples.

Example 43

Find the value of these expressions:

a $\frac{7!}{5!}$ **b** $\frac{3!}{5!}$ **c** $\frac{8! \times 4!}{10!}$ **d** $\frac{7! \times 5!}{10! \times 6!}$

a $\frac{7!}{5!} = \frac{7 \times 6 \times \cancel{5} \times \cancel{4} \times \cancel{3} \times \cancel{2} \times \cancel{1}}{\cancel{5} \times \cancel{4} \times \cancel{3} \times \cancel{2} \times \cancel{1}} = 42$

b $\frac{3!}{5!} = \frac{\cancel{3} \times \cancel{2} \times \cancel{1}}{5 \times 4 \times \cancel{3} \times \cancel{2} \times \cancel{1}} = \frac{1}{20}$

c $\frac{8! \times 4!}{10!} = \frac{4!}{10 \times 9} = \frac{4 \times 3 \times 2 \times 1}{10 \times 9} = \frac{4}{15}$

d $\frac{7! \times 5!}{10! \times 6!} = \frac{1}{10 \times 9 \times 8 \times 6} = \frac{1}{4320}$

You can also find these using technology as follows:

```
7!÷5!
                    42
3!÷5!
                  0.05
```

```
(8!×4!)÷10!
                  4/15
(7!×5!)÷(10!×6!)
                1/4320
```

Example 44

Simplify the following.

a $\dfrac{n(n+1)!}{n!}$ **b** $\dfrac{n!-(n-1)!}{(n+1)!}$

a $\dfrac{n(n+1)!}{n!} = \dfrac{n(n+1)\times n!}{n!} = n(n+1)$	Rewrite $(n+1)!$ as $(n+1) \times n!$
b $\dfrac{n!-(n-1)!}{(n+1)!} = \dfrac{n(n-1)!-(n-1)!}{(n+1)\times n\times(n-1)!}$ $= \dfrac{\cancel{(n-1)!}\,(n-1)}{(n+1)\times n\times\cancel{(n-1)!}} = \dfrac{(n-1)}{n(n-1)}$	Rewrite $n!$ as $n(n-1)!$ and $(n+1)!$ as $(n+1) \times n \times (n-1)!$

Permutations and combinations

Investigation 16a

Angela is creating invitation cards for her birthday party. She has three images which she wants to use on her invitation. She puts the images in a row as shown below.

Pool Party

She would like to consider all the possible ways of arranging these images in one line on a card and then choose her favourite.

1 How many arrangements can she choose from?

2 How did you come up with your answer?

3 Angela realizes that she should also include the address, and she still wants to leave the four objects in a line. In how many ways can this be done?

4 She then decides that she does not need to include her address on invitations to her cousins, and realizes that if she makes all invitations individual she will have just enough different invitation cards for all her guests. How many people is she going to invite to her party?

5 What happens if she wants to invite another friend to the party? Explain your answer.

Most probably, when responding to Investigation 16a, all the different arrangements (called **permutations**) were listed. Here is another way of reasoning out the responses:

Suppose you want to find the total number of arrangements of the letters A, B and C. You have three letters to choose from for the first letter.

Having chosen this first letter, you have two choices for the second letter, and then you are left with only one letter to complete the whole set. In other words, Angela has $3 \times 2 \times 1$ ways of designing the invitation cards for her cousins. You can think of this method as filling boxes as shown here, starting from left to right.

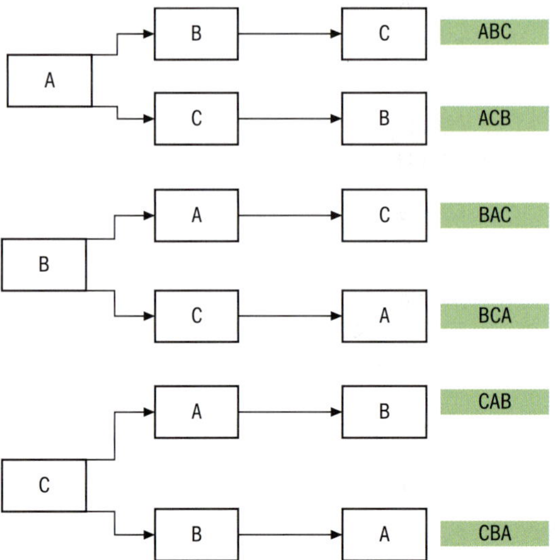

This reasoning can be extended to deduce that the number of ways in which n distinct objects can be arranged in a row is

$n \times (n - 1) \times (n - 2) \times \ldots \times 3 \times 2 \times 1 = n!$

> The number of ways of arranging n distinct objects in a row is $n!$

Now suppose that you have five different letters and you want to find the number of possible arrangements of just three of these letters. You can choose the first letter in five ways, the second letter in four ways and the third letter in three ways, giving:

$5 \times 4 \times 3 = \dfrac{5 \times 4 \times 3 \times 2 \times 1}{2 \times 1} = \dfrac{5!}{(5-3)!}$ ways.

Using the same reasoning, you can deduce that the number of ways of arranging three objects chosen out of n distinct objects would be $n \times (n-1) \times (n-2) = \dfrac{n!}{(n-3)!}$. And generalizing even further:

> The number of permutations of r objects out of n distinct objects is given by
> $^{n}P_{r} = \dfrac{n!}{(n-r)!}$.

TOK

How many different tickets are possible in a lottery?

What does this tell us about the ethics of selling lottery tickets to those who do not understand the implications of these large numbers?

Investigation 16b

Let's revisit Investigation 16a. Angela now wants to choose a photo of the pool for her third image. She has five pool photos and decides to choose two different images, one for the invitation to relatives and another one for the invitation to friends.

1 In how many ways can she choose the first photo?

2 In how many ways can she choose the second photo?

3 It is irrelevant which photo to use on the two sets of invitations. In how many ways can she choose two photos out of five?

4 Why is this answer different to arranging two photos chosen out of five?

Investigation 16c

Let's consider what happens if you want to choose three letters out of five and represent these on a chart similar to the one discussed for permutations. If the first letter chosen is A then you have the chart shown here.

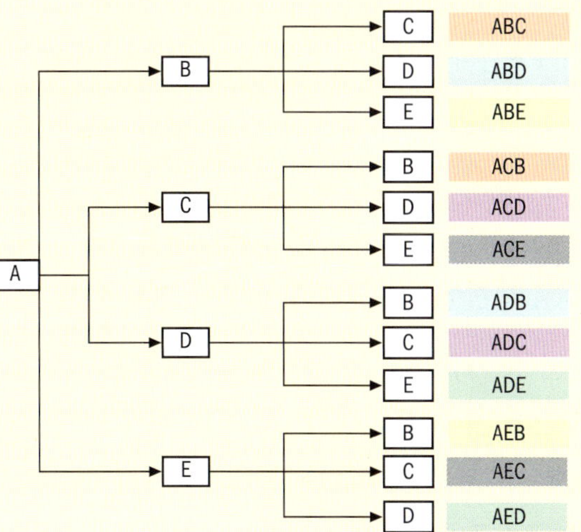

1 What do you notice about the colour coded arrangements on the right?
2 What would you expect to notice if the first letter chosen had been B?
3 What if you were to consider all the possible permutations?
4 If the order of choosing the letters is not important, how can you derive the number of combinations?

When the order of arrangements is not relevant you speak about combinations.

The number of ways of choosing (when order is not important) r objects from n distinct objects is ${}^{n}C_{r} = \dfrac{n!}{r!(n-r)!}$

EXAM HINT

In some books you may find the notation $\binom{n}{r}$ for ${}^{n}C_{r}$.

In this book you will use the latter which is the notation you will encounter in IB exams.

Example 45

a In how many ways can the letters of the word *candle* be arranged?

b In how many ways can a group of four boys and three girls be arranged?

c In how many ways can a group of four boys and three girls be arranged if no girls are to be next to each other?

a There are six letters in the word *candle*, which can be arranged in $6! = 720$ ways.

b Seven children can be arranged in $7! = 5040$ ways.

c $4!$ ways of arranging the boys.

$3!$ ways of arranging the girls.

If the girls are to be separated, then each pair of girls must be separated by at least one boy.

There are five positions each girl can fill, at each end or the slots between boys, as shown in the following diagram, giving the number of combinations ${}^{5}C_{3}$.

Using a calculator:

```
6!
                720
7!
               5040
```

Continued on next page

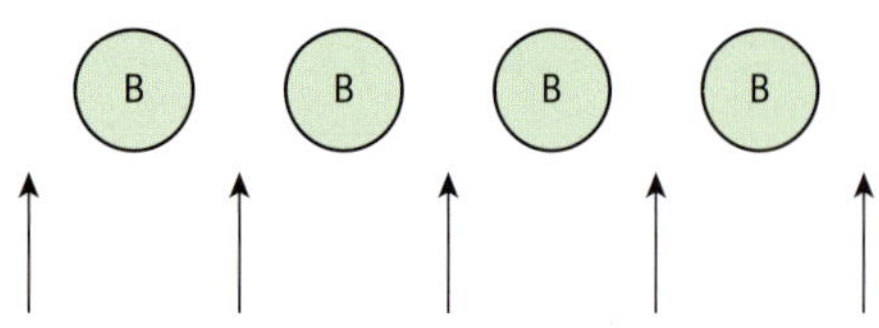 The total number of arrangements is therefore $4! \times 3! \times {}^5C_3 = 1440$. So total number of arrangements is $4! \times 3! \times 10 = 1440$	

Example 46

How many four-digit numbers can be made using each of the following digits only once?

a 5, 6, 7 and 8 **b** 5, 6, 7, 8 and 9 **c** 7, 8, 9 and 0

a ${}^4P_4 = 4! = 24$ **b** ${}^5P_4 = \dfrac{5!}{(5-4)!} = 120$	There are 4! ways of creating a four-digit number with these digits using each digit only once.
c $3 \times {}^3P_3 = 3 \times 3! = 18$	The number cannot start with 0. So, there are three ways of choosing the first digit, and the next three digits can be chosen from the three remaining digits which include 0.

Example 47

There are eight boys and five girls who attend the Senior Mathematics Club. Find how many ways the teacher can choose a team of six students to represent the school in a competition if:

a There are no gender restrictions.

b The team is to be made up of three girls and three boys.

c At least two of each gender are included in the team.

a ${}^{13}C_6 = \dfrac{13!}{6!(13-6)!} = 1716$	You have to use combinations as the order of choosing is not important.
b ${}^8C_3 \times {}^5C_3 = 560$	You now need to choose three boys out of eight and three girls out of five.
c You cannot have a team with only one girl or with no girls. You also cannot have a team with only one boy.	

From the total number of ways of choosing the team (answer **a**) you need to exclude these three combinations, ie $^{13}C_6 - {}^8C_5{}^5C_1 - {}^8C_6{}^5C_0 - {}^8C_1{}^5C_5 = 1400$ **Or** There are three ways to consider, two boys and four girls or three boys and three girls or four boys and two girls, giving $^8C_2 \times {}^5C_4 + {}^8C_3 \times {}^5C_3 + {}^8C_2 \times {}^5C_4 = 1400$	You cannot have a team with no boys as there are not enough girls to form a team. Use technology to calculate the answer.

Exercise 1H

1 Copy and complete the table below, simplifying expressions as shown in the first row.

$8! - 6!$	$6!(56-1) = 39\,600$
$9! + 8!$	
$7! - 6!$	
$6! + 5!$	
$(n+1)! - n!$	
$n! - (n-1)!$	
$n! + (n-1)!$	
$(n+1)! + n!$	

2 Find the value of:

a $\frac{8!}{4 \times 6!}$ **b** $\frac{4! \times 5!}{3! \times 6!}$ **c** $\frac{10! \times 8!}{11! \times 6!}$

3 Simplify the following:

a $\frac{(n+1)!}{n!-(n+1)!}$ **b** $\frac{n!+(n+1)!}{n!}$ **c** $\frac{(n!)^2-1}{n!-1}$

4 Show that $\frac{(2n+2)!(n!)^2}{\left[(n+1)!\right]^2(2n)!} \equiv \frac{2(2n+1)}{n+1}$.

5 Solve for $n \in \mathbb{Z}^+$, ${}^nC_2 = 66$.

6 Solve the equation $16(n-1)! = 5n! + (n+1)!$ where $n \in \mathbb{Z}^+$.

7 On a bookshelf there are four mathematics books, three science books, two geography books and four history books. The books are all different.

a In how many different ways can the books be arranged on the shelf?

b In how many ways can the books be arranged so that books of the same subject are grouped together?

8 A safe has two dials, one with 26 letters and one with the digits 0 to 9.

In order to open the safe, Rose has to choose a code consisting of three distinct letters followed by two distinct digits. Determine how many different safe codes are possible.

9 A delegation of five students is to be selected for a Model United Nations conference. There are 10 boys and 13 girls to choose from.

a In how many different ways can a delegation be chosen if there are no restrictions?

b If the team is to include at least one girl and one boy, in how many ways can the delegation be selected?

10 a How many four-digit numbers can be made using the digits 0, 1, 3, 4, 5, 8 and 9?

b How many of the four-digit numbers have no repeated digits?

c How many four-digit even numbers can be made using the digits?

d How many of these even numbers are divisible by 5?

11 Graeme is training for a 10 km run. He has six different routes to choose for his training and he trains four times a week. He calculates that he will just manage to run a different set of routes each week leading up to his next race.

How many weeks are there before Graeme's race?

12 A group of 12 people want to go to a concert. They can travel in a small car that takes one driver and one passenger and two cars each taking one driver and four passengers. If there are five drivers in the group, in how many different ways can they travel?

The binomial theorem

Investigation 17

Copy and complete this table by using repeated algebraic multiplication.

$(1+x)^i$	Constant	Coefficient of x	Coefficient of x^2	Coefficient of x^3	Coefficient of x^4	Coefficient of x^5
$(1+x)^0$	1	-	-	-	-	-
$(1+x)^1$	1	1	-	-	-	-
$(1+x)^2$						
$(1+x)^3$						
$(1+x)^4$						
$(1+x)^5$						

1 Comment on any patterns that you recognize.

2 Rearrange the numbers so that rather than forming a right-angled triangle, they form an isosceles triangle with 1 at the top vertex.

3 What new patterns do you notice now?

4 If you extended this pattern, what would you get in the next row?

5 Verify your response to question **4** using algebraic multiplication.

Investigation 18

Consider the expansion $(1+x)^3=(1+x)(1+x)(1+x)$.

1 How is the constant term obtained in this expansion?

2 How is the term in x obtained?

3 What about the term in x^2?

4 And the term in x^3?

5 Summarize your responses using mathematical notation.

6 Repeat the process for the expansion $(a+x)^3=(a+x)(a+x)(a+x)$

7 Write a conjecture for obtaining the expansion of $(a+x)^n=\underbrace{(a+x)(a+x)(a+x)\dots(a+x)}_{n \text{ factors}}$

8 **Conceptual** How does the binomial theorem use combinations to obtain a binomial expansion?

9 **Conceptual** How is binomial theorem related to Pascal's triangle?

The binomial theorem states that for all $n \in \mathbb{Z}^+, a, x \in \mathbb{R}$.

$$(a+x)^n = {}^nC_0a^n + {}^nC_1a^{n-1}x + {}^nC_2a^{n-2}x^2 + \ldots + {}^nC_ra^{n-r}x^r + \ldots + {}^nC_nx^n$$
$$= \sum_{r=0}^{n} {}^nC_ra^{n-r}x^r$$

Investigation 19

1 Write the expansion of $(1+x)^n$ using combinations.

2 Find how many terms this expansion has when:

a $n = 4, 6, 10$ b n is even

c $n = 3, 5, 7$ d n is odd.

3 The pattern found in Investigation 17 is known as Pascal's triangle. The following properties of this pattern were found:

- There is a line of symmetry going down the middle of the numbers.
- Each row starts and ends with 1 and each of the other numbers is the sum of the two numbers above it to either side.

The expansion of $(1+x)^5$ can be written as follows:

$(1+x)^5 = {}^5C_0x^0 + {}^5C_1x + {}^5C_2x^2 + {}^5C_3x^3 + {}^5C_4x^4 + {}^5C_5x^5$

a Use this expansion and the one for $(1+x)^6$ to verify the two properties above.

b Write the expansions for $(1+x)^n$ and $(1+x)^{n+1}$.

c Use the same method as in part **a** to show that the two properties above hold.

d **Conceptual** How can you explain the patterns in Pascal's triangle by considering the general expansion of the binomial expansion?

TOK

Why do we call this Pascal's triangle when it was in use before Pascal was born?

Are mathematical theories merely the collective opinions of different mathematicians, or do such theories give us genuine knowledge of the real world?

Example 48

Find the values of a, b and c in the following identities:

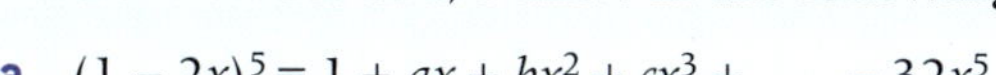

a $(1-2x)^5 \equiv 1 + ax + bx^2 + cx^3 + \ldots - 32x^5$

b $\left(1+\dfrac{x}{3}\right)^a \equiv 1 + bx + cx^2 + \ldots + \left(\dfrac{x}{3}\right)^8$

c $(2+ax)^7 \equiv b + 224x + cx^2 + 70x^3 + \ldots$

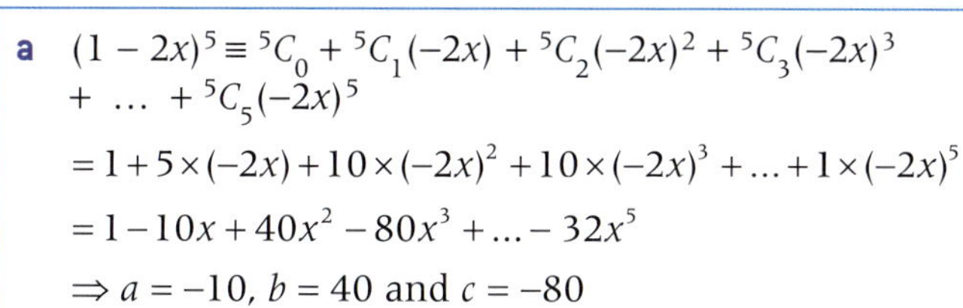

a $(1-2x)^5 \equiv {}^5C_0 + {}^5C_1(-2x) + {}^5C_2(-2x)^2 + {}^5C_3(-2x)^3 + \ldots + {}^5C_5(-2x)^5$	You should know the first five rows of Pascal's triangle.
$= 1 + 5\times(-2x) + 10\times(-2x)^2 + 10\times(-2x)^3 + \ldots + 1\times(-2x)^5$	$(1-2x)^5 = (1+(-2x))^5$
$= 1 - 10x + 40x^2 - 80x^3 + \ldots - 32x^5$	
$\Rightarrow a = -10,\ b = 40$ and $c = -80$	

Continued on next page

b $\left(1+\frac{x}{3}\right)^a \equiv 1+bx+cx^2+\ldots+\left(\frac{x}{3}\right)^8$

$\Rightarrow a = 8$

$$\left(1+\frac{x}{3}\right)^8 \equiv 1+{}^8C_1\left(\frac{x}{3}\right)+{}^8C_2\left(\frac{x}{3}\right)^2+\ldots+\left(\frac{x}{3}\right)^8$$

$$=1+\frac{8!}{1!7!}\times\frac{x}{3}+\frac{8!}{2!6!}\times\frac{x^2}{9}+\ldots+\left(\frac{x}{3}\right)^8$$

$$=1+\frac{8}{3}x+\frac{28}{9}x^2+\ldots+\left(\frac{x}{3}\right)^8$$

$\Rightarrow b=\frac{8}{3}$ and $c=\frac{28}{9}$

c $(2+ax)^7 \equiv 2^7+{}^7C_1\times 2^6\times(ax)+{}^7C_2\times 2^5\times(ax)^2 + {}^7C_3\times 2^4\times(ax)^3+\ldots$

$$=128+\frac{7!}{1!6!}\times 2^6ax+\frac{7!}{2!5!}\times 2^5a^2x^2+\frac{7!}{3!4!}\times 2^4a^3x^3+\ldots$$

$$=128+448ax+672a^2x^2+560a^3x^3+\ldots$$

$\Rightarrow 448a = 224$

$\Rightarrow a=\frac{1}{2}$, $b=128$ and $c=168$

Example 49

Find the coefficient of x^3y^3 in the expansion of $(x+3y)^6$.

The general term in the expansion of $(x+3y)^6$ is given by ${}^6C_r x^{6-r}(3y)^r$.

You want ${}^6C_r x^{6-r}(3y)^r = Ax^3y^3 \Rightarrow r=3$.

Then $A={}^6C_3\times 3^3=\frac{6!}{3!3!}\times 27=540$.

Example 50

Use the binomial theorem to expand $(2x+3y)^5$. Hence find the value of 2.03^5 correct to 5 decimal places.

$$(2x+3y)^5=(2x)^5+5(2x)^4(3y)+10(2x)^3(3y)^2+10(2x)^2(3y)^3+5(2x)(3y)^4+(3y)^5$$

$$=32x^5+240x^4y+720x^3y^2+1080x^2y^3+810xy^4+243y^5$$

When $x=1$, $y=0.01$ you obtain

$$(2.03)^5=32+2.40+0.0720+0.001\,080+0.000\,008\,10+0.000\,000\,0243$$

$$\simeq 32.47309 \text{ (to 5 decimal places)}$$

Example 51

Find the term independent of x in the expansion of $\left(x^2 - \frac{1}{2x}\right)^6$.

The general term in the expansion of $\left(x^2 - \frac{1}{2x}\right)^6$ is given by: $^6C_r\left(x^2\right)^{6-r}\left(-\frac{1}{2x}\right)^r = {^6C_r}\left(-\frac{1}{2}\right)^r x^{12-2r-r}$	Write the general term of the expansion.
For the term independent of x: $12 - 3r = 0 \Rightarrow r = 4$	For the term independent of x, the total power of x must be 0.
$^6C_4\left(x^2\right)^{6-4}\left(-\frac{1}{2x}\right)^4 = {^6C_4}\left(-\frac{1}{2}\right)^4 x^0 = \frac{6!}{4!2!} \times \frac{1}{16} = \frac{15}{16}$	Give your answer as an exact fraction.

Exercise 1I

1 Write the first four terms in the binomial expansion of:

a $\left(1-\frac{x}{3}\right)^{11}$ **b** $\left(1+\frac{x}{2}\right)^{7}$ **c** $\left(x+\frac{2}{x}\right)^{8}$

2 In each of the following binomial expressions, write down the required term.

a fifth term of $(a - 2b)^{10}$

b third term of $\left(a+\frac{4}{a^2}\right)^{11}$

c fourth term of $\left(x-\frac{2y}{x}\right)^{8}$

3 Find the term independent of x in the expansion of $\left(x-\frac{2}{x^2}\right)^{12}$.

4 Use the binomial theorem to expand $\left(2-\frac{x}{5}\right)^{4}$. Hence find the value of $(1.99)^4$ correct to 5 decimal places.

5 Find the term in x^6 in the expansion of $\left(x^2-\frac{1}{x}\right)^{6}$.

6 **a** Expand $\left(x+\frac{y}{x}\right)^{5}$.

b Find the coefficient of x^3y^2 in the expansion of $(2x+y)\left(x+\frac{y}{x}\right)^{5}$.

7 Write in factorial notation:

a the coefficient of x^4 in the expansion of $(1 + x)^{n+1}$

b the coefficient of x^3 in the expansion of $(1 + 2x)^n$.

c Find n, given that these two coefficients are equal.

8 **a** Express $\left(\sqrt{3}-\sqrt{2}\right)^5$ in the form of $a\sqrt{3}+b\sqrt{2}$ where $a, b \in \mathbb{Z}$.

b Express $\left(\sqrt{2}-\frac{1}{\sqrt{5}}\right)^4$ in the form $a+b\sqrt{10}$, $a, b \in \mathbb{Q}$.

c Express $\left(1+\sqrt{5}\right)^7-\left(1-\sqrt{5}\right)^7$ in the form $a\sqrt{5}$, $a \in \mathbb{Z}$.

9 Find the value of the following by choosing an appropriate value for x in the expansion of $(1 + x)^n$.

a $^nC_0 - 2 \times {^nC_1} + 4 \times {^nC_2} - 8 \times {^nC_3} + \ldots + (-1)^r 2^r \times {^nC_r} + \ldots + (-1)^n 2^n \times {^nC_n}$

b $^nC_0 + {^nC_1} + {^nC_2} + {^nC_3} + \ldots + {^nC_r} + \ldots + {^nC_n}$

Generalization of the binomial expansion

It was around 1665 that Isaac Newton generalized the binomial theorem to allow for negative and fractional exponents. Let's try to examine this using some facts which were established earlier in this chapter.

Consider the geometric series $1 + x + x^2 + x^3 + \dots$, where x is not equal to 0.

For which values of x does this series converge?

What is the sum to infinity for this series when it converges?

You can write the answers to these two questions as follows:

For $-1 < x < 1$, $S = \frac{1}{1-x}$

In other words:

$$\frac{1}{1-x} = (1-x)^{-1} = 1 + x + x^2 + x^3 + \dots$$

If you want to expand $(1 - x)^{-2}$ you could say that this is equivalent to $((1 - x)^{-1})^2 = (1 - x)^{-1}(1 - x)^{-1}$

$= (1 + x + x^2 + x^3 + \dots)(1 + x + x^2 + x^3 + \dots)$

$= 1 + x + x^2 + x^3 + x^4 + \dots$ multiplying the terms in the left bracket by 1

$+ x + x^2 + x^3 + x^4 + \dots$ multiplying the terms in the left bracket by x

$+ x^2 + x^3 + x^4 + \dots$ multiplying the terms in the left bracket by x^2

...

$= 1 + 2x + 3x^2 + 4x^3 + \dots$

$\Rightarrow (1 - x)^{-2} = 1 + 2x + 3x^2 + 4x^3 + \dots$

Similarly, you can repeat the process to obtain the expansion of $(1 - x)^{-3}$:

$(1-x)^{-3} = (1-x)^{-2}(1-x)$

$= (1 + 2x + 3x^2 + 4x^3 + \dots)(1 + x + x^2 + x^3 + \dots)$

$= 1 + 2x + 3x^2 + 4x^3 + \dots$

$+ x + 2x^2 + 3x^3 + \dots$

$+ x^2 + 2x^3 + \dots$

...

$= 1 + 3x + 6x^2 + 10x^3 + \dots$

Newton generalized this result for negative and rational exponents of the binomial theorem as follows:

The binomial expansion for $(1 - x)^{-n}$ for $n \in \mathbb{Z}^+$ and $-1 < x < 1$ is given by the infinite series:

$$(1-x)^{-n} = 1 + nx + \frac{n(n+1)}{2!}x^2 + \frac{n(n+1)(n+2)}{3!}x^3 + \dots + \frac{n(n+1)(n+2)\dots(n+r-1)}{r!}x^r + \dots$$

TOK

Is it possible to know things about which we can have no experience, such as infinity?

International-mindedness

We cannot simply take it for granted that we can multiply out two infinite series easily. Gustav Dirichlet, a German mathematician, proved how this can be done.

The binomial expansion for $(1+x)^{\alpha}$ for $\alpha = \frac{p}{q} \in \mathbb{Q}$ and $-1 < x < 1$ is given by the infinite series:

$$(1+x)^{\alpha} = 1 + \alpha x + \frac{\alpha(\alpha-1)}{2!}x^2 + \frac{\alpha(\alpha-1)(\alpha-2)}{3!}x^3 + \ldots + \frac{\alpha(\alpha-1)(\alpha-2)\ldots(\alpha-r+1)}{r!}x^r + \ldots$$

Investigation 20

1 Show that the generalized form for negative integer powers given above can be written as

$$(1-x)^{-n} = \sum_{r=0}^{\infty} {}^{n+r-1}C_r x^r$$

The table below includes the results obtained above.

$(1-x)^i$	Constant	Coefficient of x	Coefficient of x^2	Coefficient of x^3	Coefficient of x^4	Coefficient of x^5
$(1-x)^{-1}$	1	1	1	1	1	1 …
$(1-x)^{-2}$	1	1	3	4	5	6…
$(1-x)^{-3}$	1	3	6	10…		
$(1-x)^{-4}$						
$(1-x)^{-5}$						

2 Use Newton's generalization to verify that the coefficients shown in the table are correct.

3 Apply Newton's generalization to copy and complete the table.

4 Show that the generalized form of the binomial theorem for fractional powers can be written as

$$(1+x)^{\alpha} = \sum_{r=0}^{\infty} {}^{\alpha}C_r x^r$$

5 Use the generalization to find the expansion of $\sqrt{(1+x)}$ and $\sqrt{(1-x)}$.

Example 52

Expand the following up to the term in x^3.

a $\sqrt{(1+2x)}$, for $|x| < \frac{1}{2}$ **b** $\frac{2}{(1-3x)}$, for $|x| < \frac{1}{3}$

a $\sqrt{(1+2x)} = (1+2x)^{\frac{1}{2}}$

$= 1 + \frac{1}{2}(2x) + \left(\frac{1}{2}\right)\left(-\frac{1}{2}\right)\frac{(2x)^2}{2!} + \left(\frac{1}{2}\right)\left(-\frac{1}{2}\right)\left(-\frac{3}{2}\right)\frac{(2x)^3}{3!} + \ldots$

$= 1 + x - \frac{1}{2}x^2 + \frac{1}{2}x^3$

Using $(1+x)^{\alpha} = 1 + \alpha x + \frac{\alpha(\alpha-1)}{2!}x^2 + \frac{\alpha(\alpha-1)(\alpha-2)}{3!}x^3 + \ldots$

Simplify.

Continued on next page

b $\frac{2}{(1-3x)} = 2(1-3x)^{-1}$ $= 2\left(1+(1)(3x)+\frac{(1)(2)}{2!}(3x)^2+\frac{(1)(2)(3)}{3!}(3x)^3+\ldots\right)$ $= 2\left(1+3x+9x^2+27x^3+\ldots\right)$ $= 2+6x+18x^2+54x^3+\ldots$	Using $(1-x)^{-n} = 1+nx+\frac{n(n+1)}{2!}x^2 + \frac{n(n+1)(n+2)}{3!}x^3+\ldots$ $n \in \mathbb{Z}^+$ and $\lvert x\rvert < \frac{1}{3}$

Example 53

Use the binomial expansion to show that $\sqrt{\frac{1+x}{1-x}} \simeq 1+x+\frac{1}{2}x^2$, $\lvert x\rvert < 1$.

$\sqrt{\frac{1+x}{1-x}} = (1+x)^{\frac{1}{2}}(1-x)^{-\frac{1}{2}}$	
$(1+x)^{\frac{1}{2}} = 1+\left(\frac{1}{2}\right)x+\left(\frac{1}{2}\right)\left(-\frac{1}{2}\right)\frac{x^2}{2!}+\ldots$ $= 1+\frac{x}{2}-\frac{x^2}{8}+\ldots$	Using $(1+x)^{\alpha} = 1+\alpha x+\frac{\alpha(\alpha-1)}{2!}x^2 + \frac{\alpha(\alpha-1)(\alpha-2)}{3!}x^3+\ldots$ where $\alpha = \frac{1}{2}$ You only need the expansion until the term in x^2.
$(1-x)^{-\frac{1}{2}} = 1+\left(-\frac{1}{2}\right)(-x)+\left(-\frac{1}{2}\right)\left(-\frac{3}{2}\right)\frac{(-x)^2}{2!}+\ldots$ $= 1+\frac{x}{2}+\frac{3x^2}{8}+\ldots$	Using $(1+x)^{\alpha} = 1+\alpha x+\frac{\alpha(\alpha-1)}{2!}x^2 + \frac{\alpha(\alpha-1)(\alpha-2)}{3!}x^3+\ldots$ where $\alpha = -\frac{1}{2}$, and noting that you have $-x$ inside the brackets. You only need the expansion until the term in x^2.
$(1+x)^{\frac{1}{2}}(1-x)^{-\frac{1}{2}} \simeq \left(1+\frac{x}{2}-\frac{x^2}{8}+\ldots\right)\left(1+\frac{x}{2}+\frac{3x^2}{8}+\ldots\right)$	Multiply the two expansions up to and including terms in x^2.
$(1+x)^{\frac{1}{2}}(1-x)^{-\frac{1}{2}} \simeq \begin{cases} 1+\frac{x}{2}+\frac{3x^2}{8} \\ +\frac{x}{2}+\frac{x^2}{4} \\ -\frac{x^2}{8} \quad +\ldots \simeq 1+x+\frac{x^2}{2} \end{cases}$	

TOK

"Mathematics may be defined as the economy of counting. There is no problem in the whole of mathematics which cannot be solved by direct counting."

– E. Mach

To what extent do you agree with this quote?

Exercise 1J

1 Expand the following up to the term in x^3, given that $|x| < \frac{1}{2}$.

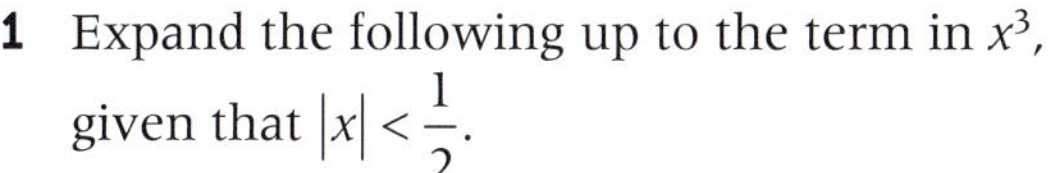

a $\dfrac{1}{1+x}$ **b** $\dfrac{1}{(1-2x)^2}$

c $\dfrac{2}{1+2x}$ **d** $\dfrac{2}{(1-x)^3}$

2 Find the first four terms of each of the following expansions where $|x| < \dfrac{1}{10}$:

a $\sqrt{1+2x}$ **b** $(1+x)^{\frac{3}{2}}$

c $(1-3x)^{-\frac{1}{2}}$ **d** $2(1+x)^{\frac{1}{3}}$

3 Show that $\sqrt{\dfrac{1-x}{1+x}} \simeq 1 - x + \dfrac{x^2}{2} - \dfrac{x^3}{2}$, where $|x| < 1$.

4 Show that $\dfrac{x}{(1+x)^2} \simeq x - 2x^2 + 3x^3 - 4x^4 + \ldots, |x| < 1.$

5 Find the first four terms of the binomial expansion of $(2-3x)^{-3}$, $|x| < \dfrac{2}{3}$.

6 **a** Find the first four terms of the binomial expansion of $\sqrt{1-4x}$, $|x| < \dfrac{1}{4}$.

b Show that the exact value of $\sqrt{1-4x}$ when $x = \dfrac{1}{100}$ is $\dfrac{2\sqrt{6}}{5}$.

c Hence, determine $\sqrt{6}$ to 5 decimal places.

7 **a** Find the first three terms of the binomial expansion of $\dfrac{1}{\sqrt{1-2x}}$ where $|x| < \dfrac{1}{2}$.

b Hence or otherwise, obtain the expansion of $\dfrac{(2+3x)^3}{\sqrt{1-2x}}$, $|x| < \dfrac{1}{2}$ up to and including the term in x^3.

Chapter summary

- Sequences may be **finite** or **infinite**.
- If the difference between two consecutive numbers in a sequence is constant then it is an **arithmetic sequence** or an **arithmetic progression**. The constant difference is called the **common difference** and is denoted by d.
- An arithmetic sequence with first term u_1 and common difference d has **general term** $u_n = u_1 + (n-1)d$.
- The sum of a finite arithmetic series is given by $S_n = \dfrac{n}{2}[2u_1 + (n-1)d] = \dfrac{n}{2}[u_1 + u_n]$ where n is the number of terms in the series, u_1 is the first term, d is the common difference and u_n is the last term.

Continued on next page

- If the ratio of two consecutive terms in a sequence is constant then it is a **geometric sequence** or a **geometric progression**. You call the constant ratio the **common ratio** and denote it by r.
- A geometric sequence with first term u_1 and common ratio r has **general term** $u_n = u_1 r^{n-1}, r \neq 1, 0, -1, u_1 \neq 0$.
- The sum of a finite geometric series is given by

  $$S_n = \frac{u_1(1-r^n)}{1-r}, \quad r \neq 1$$

  where n is the number of terms, u_1 is the first term and r is the common ratio.
- The sum of n terms of a geometric series is $S_n = \frac{u_1(1-r^n)}{1-r}, \quad r \neq 1$.

  When $-1 < r < 1$, r^n approaches zero for very large values of n. The series therefore converges to a finite sum given by $S = \frac{u_1}{1-r}$.
- A **proof** in mathematics often consists of a logical set of steps that validates the truth of a general statement beyond any doubt.
- A **direct proof** is a way of showing the truth of a given statement by constructing a series of reasoned connected established facts. In a direct proof the following steps are used:
  - Identify the given statement.
  - Use axioms, theorems, etc, to make deductions that prove the conclusion of your statement to be true.
- When setting out a **proof by contradiction** you follow the following steps:
  - Identify what is being implied by the statement.
  - Assume that the implication is false.
  - Use axioms, theorems, etc ... to arrive at a contradiction.
  - This proves that the original statement must be true.
- A **counterexample**, or **counterclaim**, is an acceptable "proof" of the fact that a given statement is false.
- $n! = n \times (n-1)! = n \times (n-1) \times (n-2) \times \ldots \times 3 \times 2 \times 1$
- The number of ways of arranging n distinct objects in a row is $n!$
- The number of permutations of r objects out of n distinct objects is given by ${}^nP_r = \frac{n!}{(n-r)!}$.
- The number of ways of choosing (when order is not important) r objects from n distinct objects is ${}^nC_r = \frac{n!}{r!(n-r)!}$

- The binomial expansion for $(1-x)^{-n}$ for $n \in \mathbb{Z}^+$ and $-1 < x < 1$ is given by the infinite series:

$$(1-x)^{-n} = 1 + nx + \frac{n(n+1)}{2!}x^2 + \frac{n(n+1)(n+2)}{3!}x^3 + \ldots + \frac{n(n+1)(n+2)\ldots(n+r-1)}{r!}x^r + \ldots$$

- The binomial expansion for $(1+x)^{\alpha}$ for $\alpha = \frac{p}{q} \in \mathbb{Q}$ and $-1 < x < 1$ is given by the infinite series:

$$(1+x)^{\alpha} = 1 + \alpha x + \frac{\alpha(\alpha-1)}{2!}x^2 + \frac{\alpha(\alpha-1)(\alpha-2)}{3!}x^3 + \ldots + \frac{\alpha(\alpha-1)(\alpha-2)\ldots(\alpha-r+1)}{r!}x^r + \ldots$$

Developing inquiry skills

Return to the chapter opening problem. The enclosed area of the Koch snowflake can be found using the sum of an infinite series.

In the second iteration, since the sides of the new triangles are $\frac{1}{3}$ the length of the sides of the original triangle, their areas must be $\left(\frac{1}{3}\right)^2 = \frac{1}{9}$ of its area.

If the area of the original triangle is 1 square unit, then the total area of the three new triangles is $3\left(\frac{1}{9}\right)$.

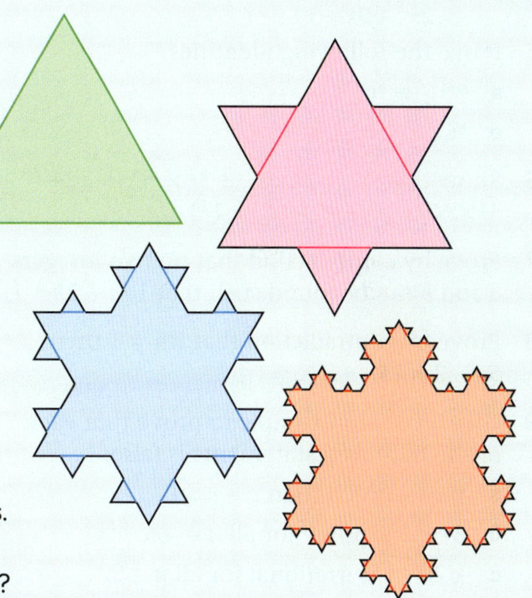

i Find the total area for the third and fourth iterations.

ii How can you use what you have learned in this section to find the total area of the Koch snowflake?

iii How does the area of a Koch snowflake relate to the area of the initial triangle?

Chapter review

1 Show that there are two geometric sequences such that the second term is 9 and the sum of the first three terms is 91. Write the fourth term of each sequence.

2 Find the sum of the series $1 + 2 + 3 + 4 + 5 + 7 + 8 + 9 + 11 + 13 + 15 + 16 + 17 + \ldots + 64$.

3 Three numbers a, b and c form an arithmetic sequence. The numbers b, c and a form a geometric sequence. Find the three numbers given that they add up to 36.

4 **a** Prove the identity

$$\frac{1}{1+x} - \frac{1}{3\left(1+\frac{2}{3}x\right)} \equiv \frac{x+2}{2x^2+5x+3}.$$

b Hence, use the binomial expansion to find the first four terms of the expansion of $\frac{x+2}{2x^2+5x+3}$.

5 Prove the following identities:

a ${}^{n+1}C_2 \equiv {}^nC_2 + n$

b ${}^nC_2 \times {}^{n-2}C_{k-2} \equiv {}^nC_k \times {}^kC_2$

6 Show that ${}^nC_0 + 3 \times {}^nC_1 + 3^2 \times {}^nC_2 + \ldots + 3^r \times {}^nC_r + \ldots + 3^n \times {}^nC_n = 2^{2n}$

7 Prove by contradiction that no two integers a and b can be found such that $14a + 7b = 1$.

8 Prove by contradiction that if $x = 3$ then $5x - 7 \neq 13$.

9 Give a counterexample to prove that each of the following statements is false:

a If $a^2 - b^2 < 0$ then $a - b > 0$.

b $3^n + 2$ is prime for all $n \in \mathbb{Z}^+$.

c $\sqrt{2n-1}$ is irrational for all $n \in \mathbb{Z}^+$.

d $2^n - 1$ is prime for all $n \in \mathbb{Z}^+$.

10 Prove by mathematical induction that $(1 \times 1!) \times (2^2 \times 2!) \times (3^3 \times 3!) \times (4^4 \times 4!) \times \ldots \times (n^n \times n!) = (n!)^{n+1}$

11 Use mathematical induction to prove that $n^3 + 2n$ is a multiple of 3.

12 Use mathematical induction to prove the following statements:

a $\sum_{r=1}^{n} r = \frac{n(n+1)}{2}$ **b** $\sum_{r=1}^{n} r^2 = \frac{n(n+1)(2n+1)}{6}$

c $\sum_{r=1}^{n} r^3 = \frac{n^2(n+1)^2}{4}$

Hence prove that

$$\sum_{r=1}^{n} r(r+1)(r+2) = \frac{n(n+1)(n+2)(n+3)}{4}$$

13 **a** In how many ways can the letters of the word *harmonics* be arranged?

b Determine how many numbers bigger than 30 000, less than 9 999 999 and divisible by 5 can be formed using the digits 0, 1, 2, 3, 5, 8 and 9.

c In how many ways can a committee of five people be selected from seven men and four women, so that there is at least one male and one female and there are more women than men on the committee?

14 Let $a = x + y$ and $b = x - y$.

a Write $a^2 - b^2$ in terms of x and y and hence show that $a^2 - b^2 = (a - b)(a + b)$.

b Use the binomial theorem to write a^3 and b^3 in terms of x and y.

c Use your results to part **b** to show that $a^3 - b^3 = (a - b)(a^2 + ab + b^2)$.

d Use the binomial theorem to write a^4 and b^4 in terms of x and y and use your result to factorize $a^4 - b^4$.

e Use your results to make a conjecture for the factors of $a^n - b^n$.

f Prove your conjecture using mathematical induction.

15 Given that the coefficients of x^{r-1}, x^r and x^{r+1} in the expansion of $(1 + x)^n$ are in arithmetic sequence, show that $n^2 + 4r^2 - 2 - n(4r + 1) = 0$.

Hence find three consecutive coefficients of the expansion of $(1 + x)^{14}$ which form an arithmetic sequence.

16 Given that

$$\frac{2+x-7x^2}{(1-2x)(1-x^2)} \equiv \frac{A}{(1-2x)} + \frac{B}{(1+x)} + \frac{C}{(1-x)},$$

determine the values of A, B and C.

Hence use the binomial theorem to find the expansion of $\frac{2+x-7x^2}{(1-2x)(1-x^2)}$ in ascending powers of x up to and including the term in x^3.

Exam-style questions

17 **P2:** Find the coefficient of the term in x^5 in the binomial expansion of $(3+x)(4-2x)^8$. (4 marks)

18 **P1:** The coefficient of x^2 in the binomial expansion of $(1+3x)^n$ where $n \in \mathbb{Q}$ is 495.

Determine the possible values of n. (6 marks)

19 **P2:** Find the value of $\sum_{n=0}^{n=15}\left(1.6^n - 12n + 1\right)$, giving your answer correct to 1 decimal place. (6 marks)

20 **P1:** Prove the binomial coefficient identity $\binom{n}{k} = \binom{n-1}{k} + \binom{n-1}{k-1}$. (6 marks)

21 **P2:** Find the sum of all integers between 500 and 1400 (inclusive) that are not divisible by 7. (7 marks)

22 **P1:** Prove by contradiction that for all $n \in \mathbb{Z}^+$, if $n^3 + 3$ is odd, then n is even. (7 marks)

23 **P1:** Prove, by mathematical induction, that $5^{2n-1} + 1$ is divisible by 6 for all $n \in \mathbb{Z} + n \in (\mathbb{Z})^+$. (8 marks)

24 **P2:** **a** Find the first four terms, in ascending powers of x, of the binomial expansion of $\sqrt[3]{1-x}$, $|x| < 1$. (4 marks)

b Use your answer to part **a** to find an approximation for $\sqrt[3]{63}$ to six decimal places. You must show all your working. (5 marks)

25 **P2:** Seven women and two men are chosen to sit in a row and have their photograph taken.

a How many different ways can they be arranged? (1 mark)

b How many ways can they be arranged if the men must sit together? (2 marks)

c How many ways can they be arranged if the men must sit apart? (2 marks)

d How many ways can they be arranged if there must be at least two women separating the men? (3 marks)

The Towers of Hanoi

Approaches to learning: Thinking skills, Communicating, Research

Exploration criteria: Mathematical communication (B), Personal engagement (C), Use of mathematics (E)

IB topic: Sequences

Modelling and investigation activity

The problem

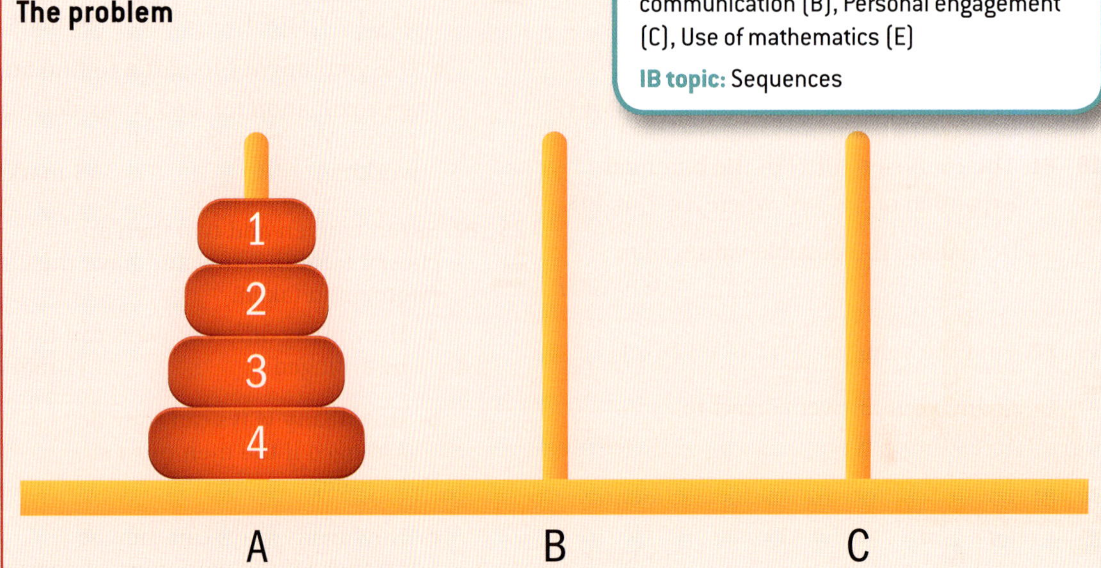

The aim of the **Towers of Hanoi problem** is to move all the disks from peg A to peg C following these rules:

1. Move only one disk at a time.
2. A larger disk may not be placed on top of a smaller disk.
3. All disks, except the one being moved, must be on a peg.

For 64 disks, what is the **minimum** number of moves needed to complete the problem?

Explore the problem

Use an online simulation to explore the Towers of Hanoi problem for three and four disks.

What is the minimum number of moves needed in each case?

Solving the problem for 64 disks would be very time consuming, so you need to look for a rule for n disks that you can then apply to the problem with 64 disks.

Try and test a rule

Assume the minimum number of moves follows an arithmetic sequence.

Use the minimum number of moves for three and four disks to predict the minimum number of moves for five disks.

Check your prediction using the simulator.

Does the minimum number of moves follow an arithmetic sequence?

Find more results

Use the simulator to write down the number of moves when $n = 1$ and $n = 2$.

Organize your results so far in a table.

Look for a pattern. If necessary, extend your table to more values of n.

Try a formula

Return to the problem with four disks.

Consider this image of a partial solution to the problem. The large disk on peg A has not yet been moved.

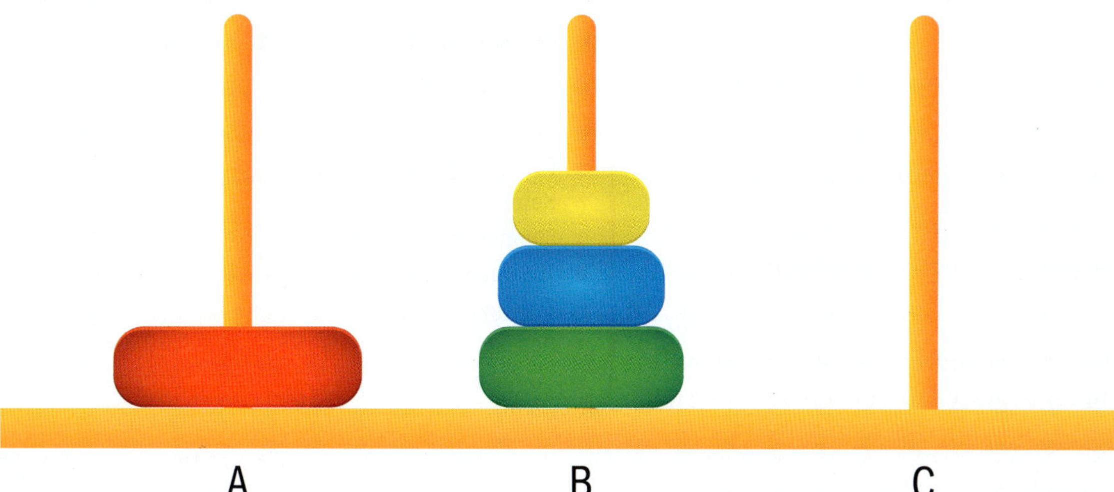

Consider your previous answers.

What is the minimum possible number of moves made so far?

How many moves would it then take to move the largest disk from peg A to peg C?

When the large disk is on peg C, how many moves would it then take to move the three smaller disks from peg B to peg C?

How many total moves are therefore needed to complete this puzzle?

Use your answers to these questions to write a formula for the minimum number of moves needed to complete this puzzle with n disks.

This is an example of a **recursive formula**. What does that mean?

How can you check if your recursive formula works?

What is the problem with a recursive formula?

Try another formula

You can also try to solve the problem by finding an **explicit formula** that does not depend on you already knowing the previous minimum number.

You already know that the relationship is not arithmetic.

How can you tell that the relationship is not geometric?

Look for a pattern for the minimum number of moves in the table you constructed previously.

Hence write down a formula for the minimum number of moves in terms of n.

How does an explicit formula differ from a recursive formula?

Use your explicit formula to solve the problem with 64 disks.

Extension

- What would a solution look like for four pegs? Does the problem become harder or easier?
- Research the "Bicolor" and "Magnetic" versions of the Towers of Hanoi puzzle.
- Can you find an explicit formula for other recursive formulae? (eg Fibonacci)

2 Representing relationships: functions

Concepts

- Representation
- Relationships

Microconcepts

- Function, domain, range
- Key features of graphs
- Composite functions
- The quadratic function
- Transformations
- Odd and even functions
- Partial fractions

Relations and functions are among the most important and abundant of all mathematical patterns. Understanding the behaviour of functions is essential to modelling real-life situations. When you drive a car, your speed is a function of time. The amount of energy you have is a function of how many calories you consume, or the amount of time you sleep, or the general state of your health. In Chapter 1 you learned that the amount of money you earn on your savings is a function of the interest rate you receive from the bank, or the number of times the interest rate is compounded, or the length of time you keep your money in the savings account. In this chapter you will model a variety of problems using different forms of functions.

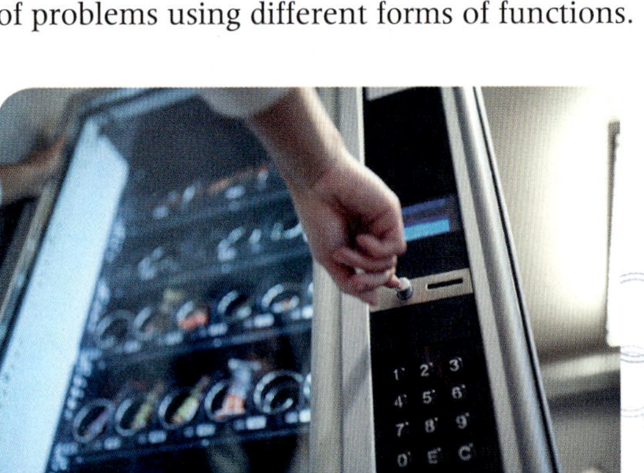

What kind of relationships exist between two quantities or variables?

How far will a car drive on a fuel tank of fuel?

Can shadows be modelled using functions?

Is the relationship between runner and water the same as car and fuel?

One of the most important concepts in economics is supply and demand. Supply refers to the amount of goods available for people to purchase, whereas demand is the actual amount of goods people will buy at a given price. An example of a supply and demand relationship is shown in the graph.

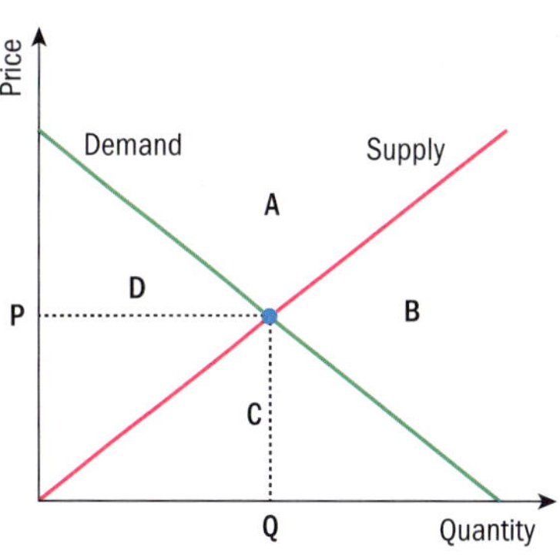

- How does the graph help you explain the relationship between the two variables "quantity" and "price"?
- Which variable would you label as "independent" and which as "dependent"?
- How do you interpret the special point where the two lines meet?
- How can you interpret the areas A, B, C and D in terms of the given variables?
- How can you express the relationship shown in the graph between quantity and price in other ways, for example, numerically or algebraically?

Developing inquiry skills

Imagine you work in a grocery store and you are asked to assign a price to a box of apples. What kind of inquiry questions would you ask? For example, how many apples are in each box? How can the value of one apple be determined? Might the value of the box change over time? What further information do you need?

Think about the questions in this opening problem and answer any you can. As you work through the chapter, you will gain mathematical knowledge and skills that will help you to answer them all.

Before you start

You should know how to:

1 Use technology to graph a function and identify key features of the graph.
eg Graph $y = x^2 + 3x - 4$, clearly labelling all intercepts, zeros and vertices.

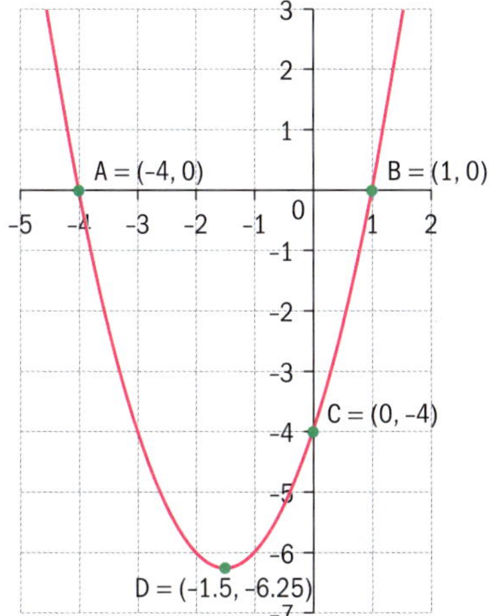

Skills check

Click here for help with this skills check

1 Graph the following, clearly labelling all intercepts, zeros and vertices.

a $y = 2\sqrt{x-1}$

b $y = -x^2 + 5x - 6$

Continued on next page

2 Use technology to find the points of intersection of two functions.

eg Find the point(s) of intersection of $y = x^2 - 4$ and $y = 2x - 1$.

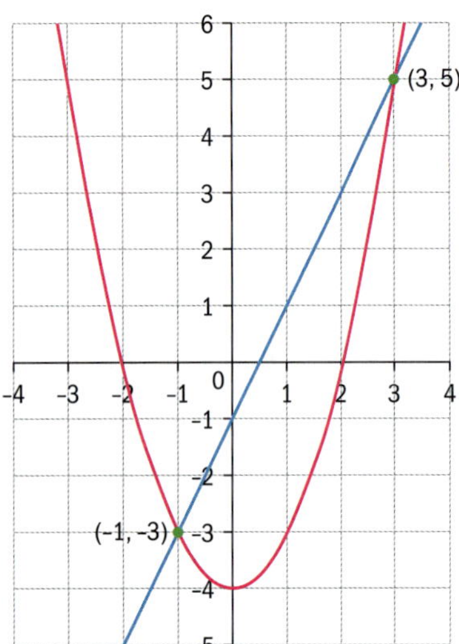

2 Find the points of intersection of:

a $y = x^2 - 3$; $y = 3 - x^2$

b $y = x^3 + 2x^2$; $-x + 2y - 3 = 0$

3 Change a quadratic from standard form to vertex form, and find the vertex.

eg Change the quadratic $y = 2x^2 - 4x + 1$ to the vertex form, $y = a(x - h)^2 + k$.

$y = 2x^2 - 4x + 1$

$= 2(x^2 - 2x + 1) + 1 - 2$

$y = 2(x - 1)^2 - 1$

vertex is $(1, -1)$

3 Change these quadratics from standard form to vertex form.

a $y = x^2 - 2x + 3$

b $y = 1 - 6x - x^2$

c $y = 3x^2 + 6x + 1$

2.1 Functional relationships

Investigation 1

Various relations are sorted into two columns below. Analyse the relations in each column and answer the questions that follow.

Column 1	Column 2
a i $\{(3,2), (3,5), (4,1), (4,2)\}$ ii $\{(1,2), (3,2), (1,5), (3,5)\}$	**a** i $\{(3,6), (2,5), (4,1), (7,2)\}$ ii $\{(1,2), (3,5), (2,2), (4,5)\}$
b Countries: Belgium, Switzerland, USA, Italy → Brands of chocolate: Neuhaus, Leonidas, Toblerone, Lindt, Hershey, Baci	**b** 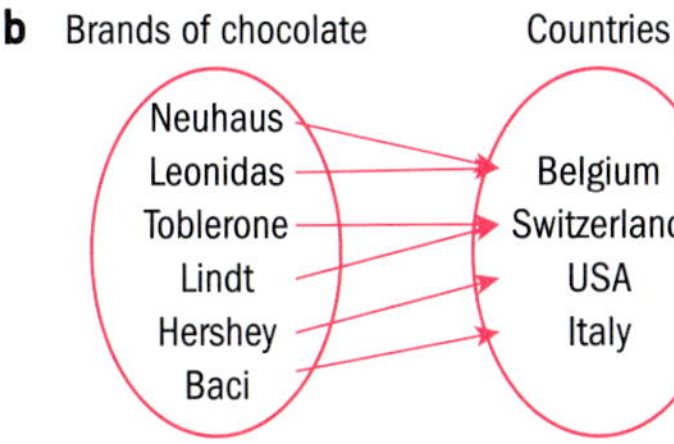
c 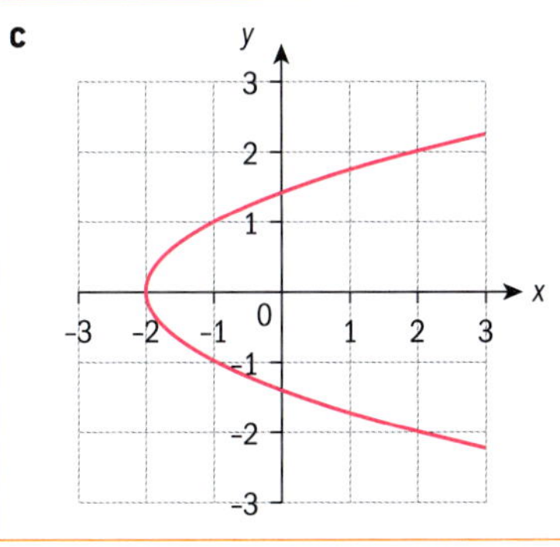	**c** 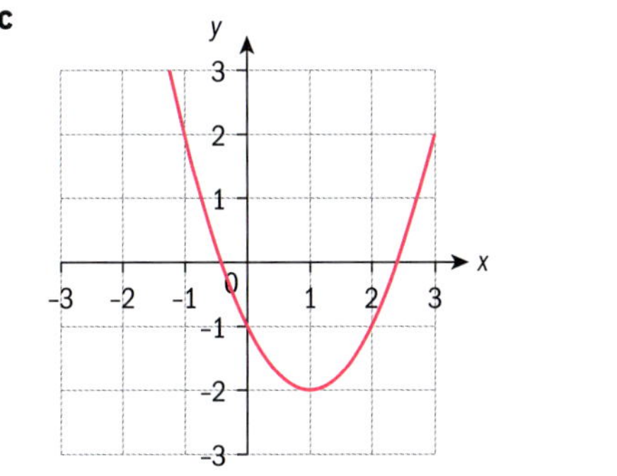
d 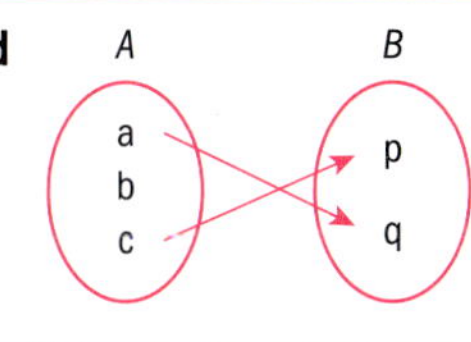	**d** 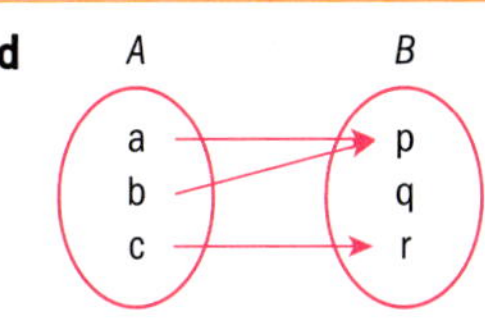
e Path of an ant walking on a piece of paper. 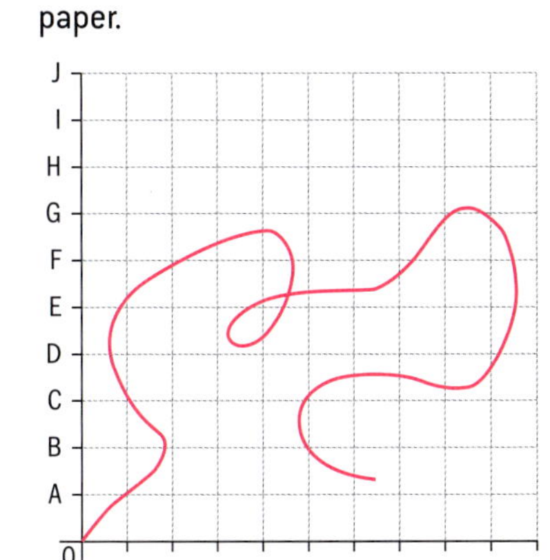	**e** 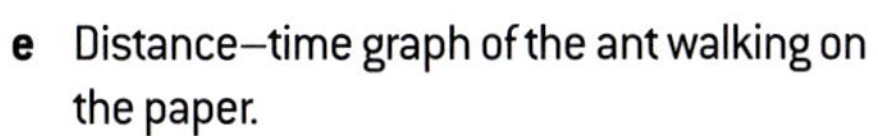 Distance–time graph of the ant walking on the paper. 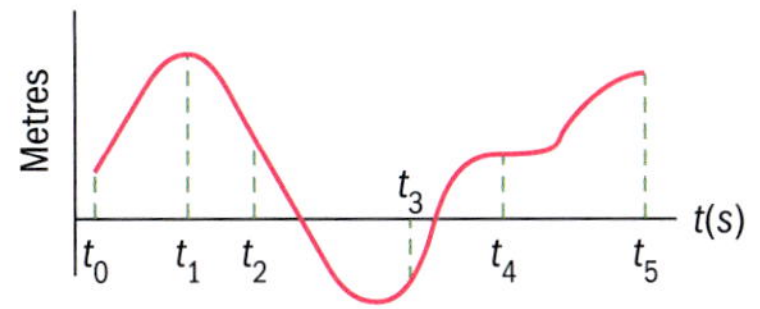

Continued on next page

1 **Factual** What is a relation?

The relations in the second column are called functions.

2 How does a function differ from a relation?

3 **Conceptual** What is a function?

4 Give three new examples (various forms) of functions.

5 In what different ways do you think relations and functions can be used?

6 Why do you think a function is useful to model real-life situations?

7 **Conceptual** Why is it useful to be able to express functions in different forms?

HINT

Think about the compound interest examples in Chapter 1. What would happen if two investments of the same amount, with the same rate and same time frame, gave two different returns?

A relation R is a set of ordered pairs (x, y) such that $x \in A, y \in B$ and sets A and B are non-empty.
A function f is a relation in which every x-value has a unique y-value.

HINT

The variables x and y are often used for ordered pairs, but any variables can be used in defining a function. For example, the area of a circle is a function of its radius, or $A = f(r) = \pi r^2$. The function would map $\mathbb{R}^+ \mapsto \mathbb{R}^+$, since the domain is restricted to the set of positive real numbers. If the domain is not restricted by its context, or otherwise, then the domain of a function is assumed to be the largest set of x-values for which the range will have real values. This set is called the natural, or implied, domain of the function f.

The set of x-values is called the **domain** of a function. The set of y-values that the domain is mapped to is called the **range** of the function. Since the y-values (output) depend on the x-values (input), y is called the dependent variable, and x is the independent variable. The independent variable x is also called the **argument** of the function.

A relation R is a function f if:

- f acts on **all** elements of the domain.
- f is well-defined, ie, it pairs each element of the domain with one and only one element of the range. Therefore, if f contains two ordered pairs (a, b_1) and (a, b_2), then $b_1 = b_2$.

In general, if y is a function of x, you can write $y = f(x)$, or $x \mapsto f(x)$.

Example 1

Determine, giving reasons, which of these relations are functions. For those that are functions, write down the domain and range.

a $\{(-2, 0), (-3, 3), (-4, 8), (0, 0)\}$

b $\{(5, -1), (2, 2), (5, 3), (10, 4)\}$

c $y = 4 - 3x$ **d** $y^2 = x$ **e** $f: \mathbb{R} \to \mathbb{R}; f \mapsto \frac{1}{x+1}$

a Relation is a function since no two y-values have the same x-value. $D_f = \{-4, -3, -2, 0\}$; $R_f = \{0, 3, 8\}$	D_f stands for the domain of f, and is the set of x-values; R_f stands for the range of f, and is the set of y-values.

b Relation is not a function since $(5, -1)$ and $(5, 3)$ have the same x-value.	The relation is not well-defined.
c Relation is a function since no two y-values have the same x-value. $D_f = \mathbb{R}$; $R_f = \mathbb{R}$	Both conditions are satisfied. All non-vertical lines are graphs of functions.
d Relation is not a function since both -1 and 1 get mapped to 1.	f is not well-defined. You need only find one counter-example.
e Relation is not a function since for $-1 \in \mathbb{R}$ f is undefined, as $f(-1) = \frac{1}{0}$.	The first condition is not met, as -1 is a real number and is not mapped to any real number.

In Example **1e,** the given relation is not a function since not all elements of the domain can be mapped to the set of real numbers. In order for the relation to be a function, you would have to restrict the domain to exclude the element -1, ie, $D_f = \{x \mid x \in \mathbb{R}, x \neq -1\}$. The range would then have to exclude the real number 0 since the numerator of the rational function is non-zero, $R_f = \{y \mid y \in \mathbb{R}, y \neq 0\}$.

Investigation 2

Use the graphs of the relations to answer the questions.

1

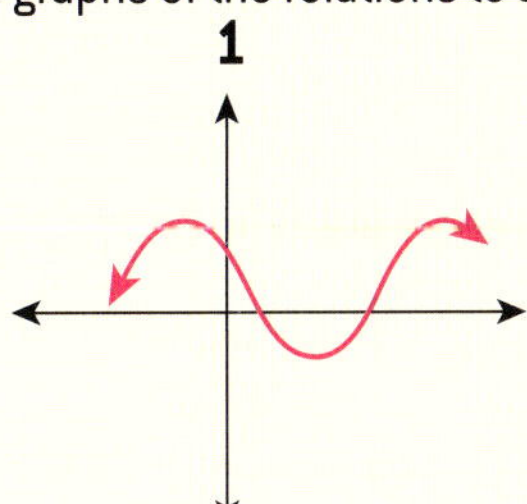

2

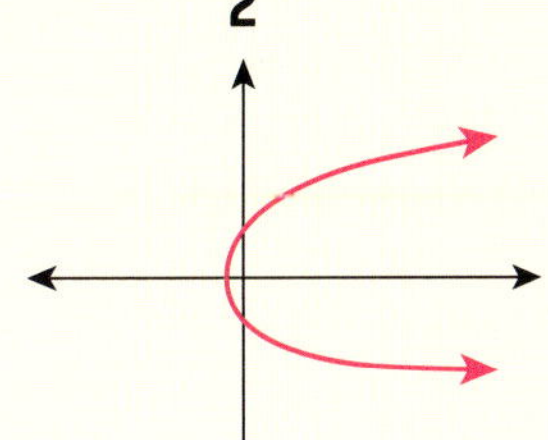

3

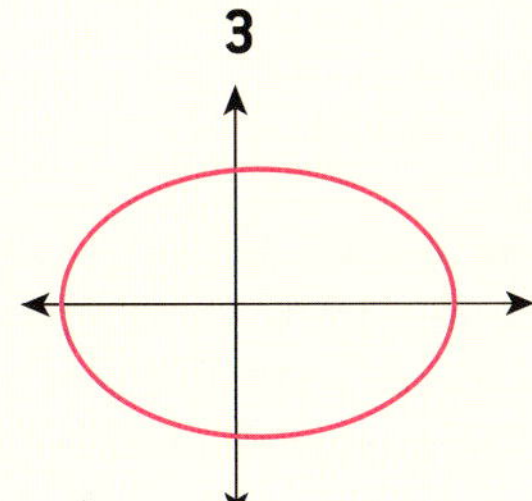

4

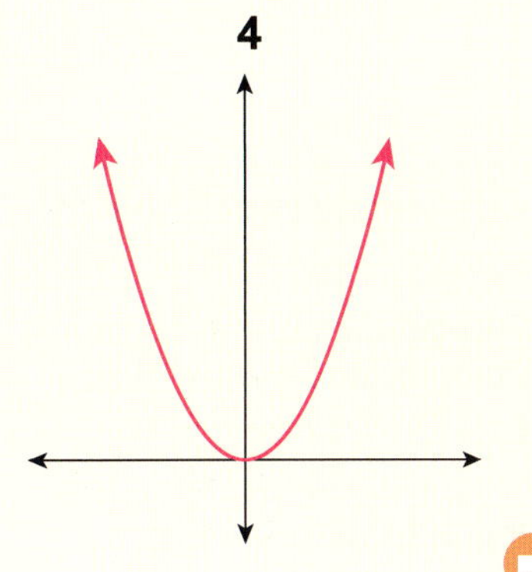

Continued on next page

1 Which graph(s):

i represent a function ii do not represent a function?

Imagine a vertical line is drawn on each graph.

2 What do the graphs of **i** have in common, and the graphs of **ii** have in common, in relation to the vertical lines?

3 Formulate a test using vertical lines that can be used to determine whether a graph represents a function.

4 Justify your test using the definition of a function.

5 **Conceptual** Why is using vertical lines to test whether the graph of a relation represents a function effective?

Example 2

Determine which of these graphs show relations that are functions. For those that are functions, state the domain and range of the function.

a

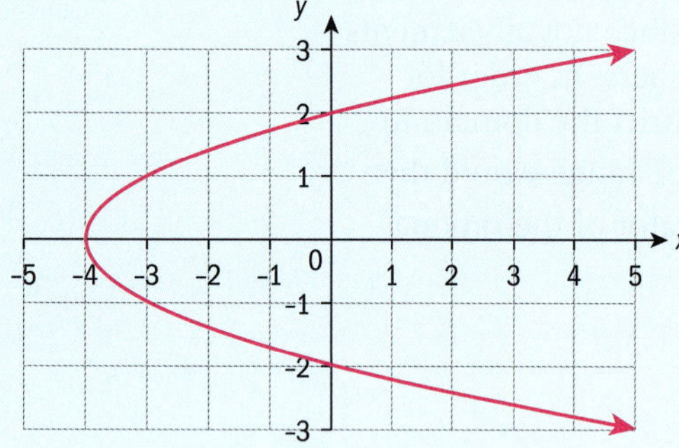

b

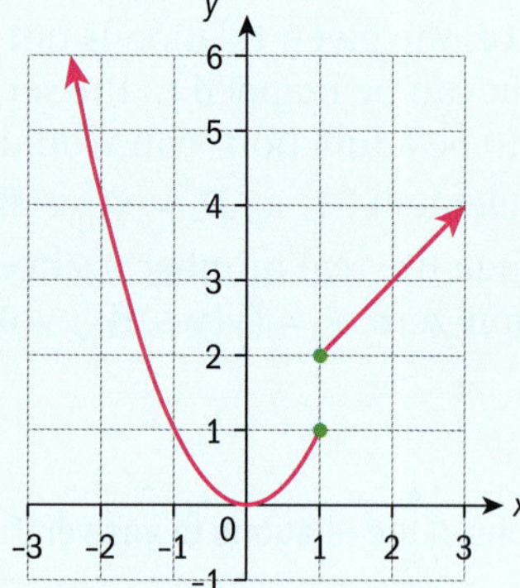

c

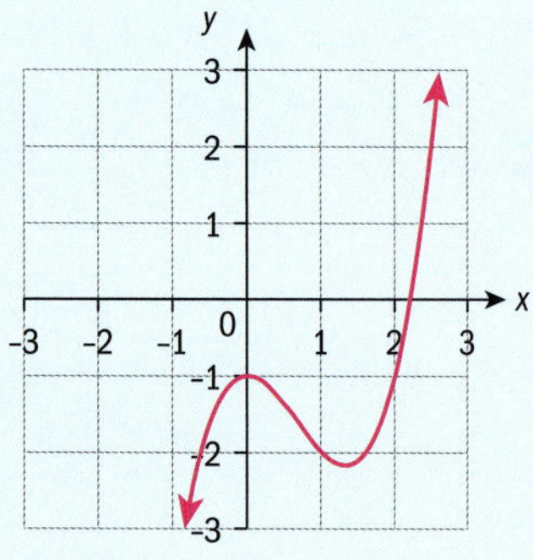

d

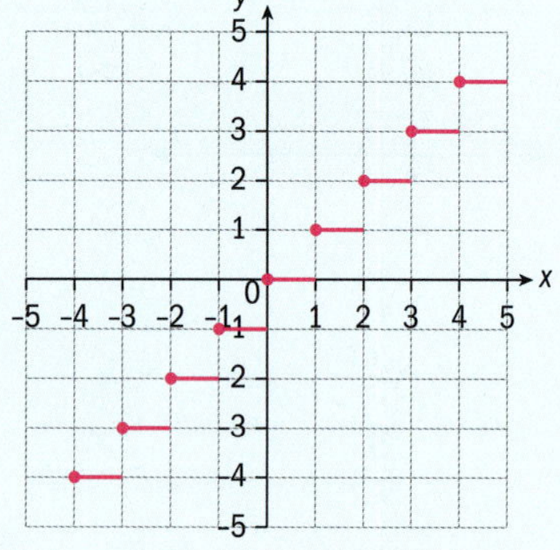

Graphs **c** and **d** pass the vertical line test so they represent functions. Both the domain and range of **c** are the set of real numbers. The domain of **d** is the set of real numbers, and its range is the set of integers. Graphs **a** and **b** do not pass the vertical line test, therefore they are not graphs of functions.	Imagine vertical lines drawn on the graphs of the relations. The graphs where the vertical lines intersect in only one point are graphs of functions.

Example 3

Use these graphs to identify the domain and range of the functions they represent.

1

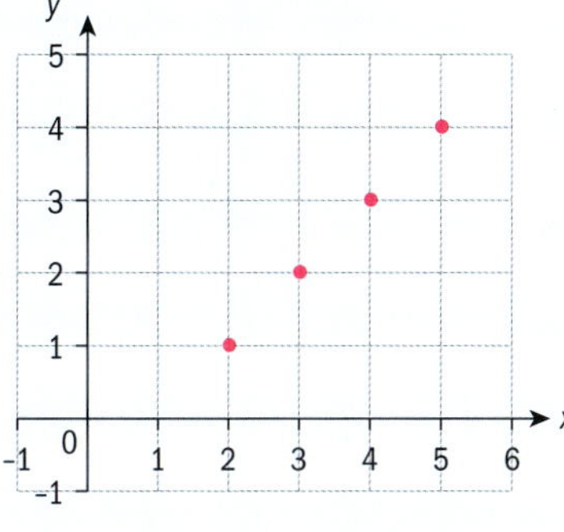

2

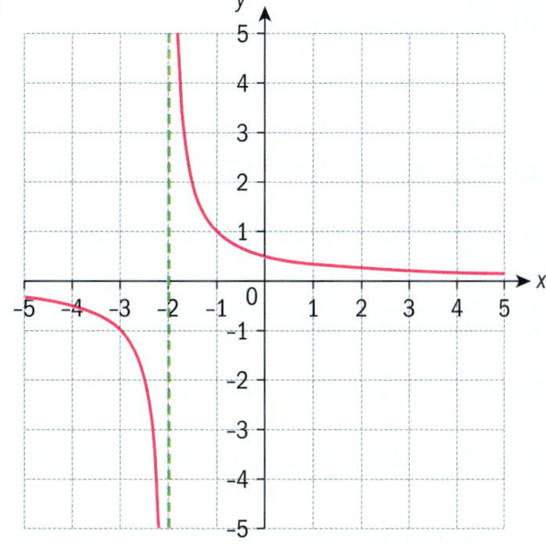

3

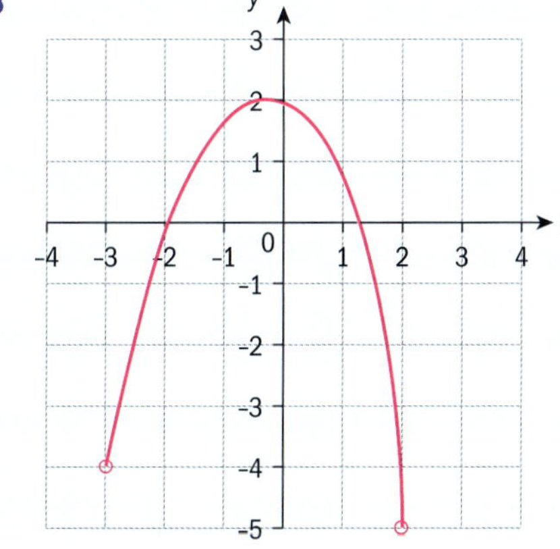

4

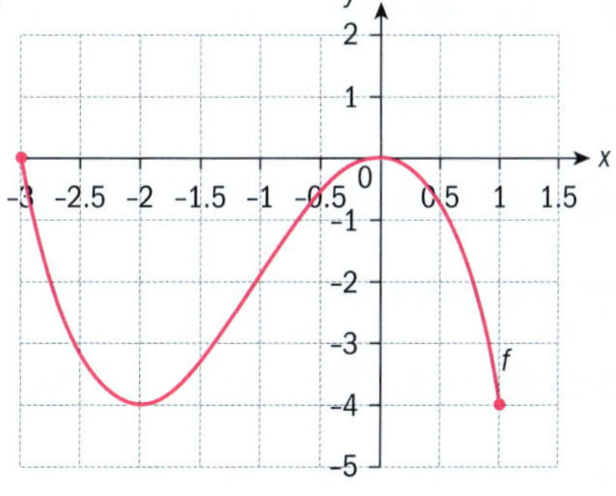

1 Domain = {2, 3, 4, 5}; Range = {1, 2, 3, 4}	You can also use set-builder notation, $D = \{x \mid x \in \mathbb{Z},\ 2 \le x \le 5\}$; $R = \{y \mid y \in \mathbb{Z},\ 1 \le y \le 4\}$
2 $D = \{x \mid x \ne -2\}$; $R = \{y \mid y \ne 0\}$	The vertical asymptote is $x = -2$ and the horizontal asymptote is $y = 0$.
3 $D = \{x \mid x \in\ ]-3, 2[\}$; $R = \{y \mid y \in\ ]-5, 2]\}$	The brackets indicate an interval; the inverted bracket indicates that the number is not a part of the interval.
4 $D = -3 \le x \le 1$; $R = -4 \le y \le 0$	Using brackets this would read $x \in [-3, 1]$; $y \in [-4, 0]$.

TOK

Around the world you will often encounter different words for the same object, like trapezium and trapezoid or root and surd.

Sometimes more than one type of symbol might have the same meaning such as interval and set notation.

To what extent does the language we use shape the way we think?

HINT

If a number set is not specifically stated for the domain, then it is assumed that it is the set of real numbers.

Exercise 2A

1 Determine which of the following relations are functions. For those that are functions, state the domain and range. For those that are not functions, state the reason why.

a $\{(1, 2), (2, 3), (3, 4), (4, 0)\}$

b $\{(-2, 1), (-1, 1), (0, 1), (1, 1)\}$

c $\{(2, -2), (2, -1), (2, 0), (2, 1)\}$

d $\{(\pi, \pi), (-\pi, \pi^3), (\pi, \pi^\pi)\}$

e

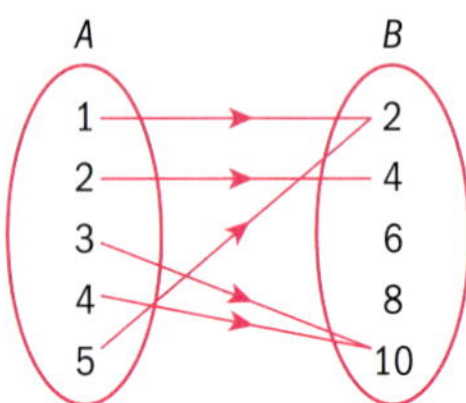

f

x	4	−5	11	−5	23
y	−3	1	1	0	6

g

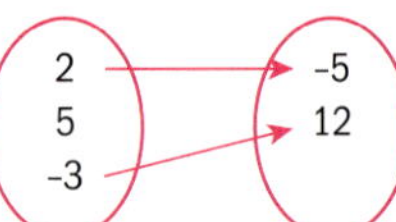

h

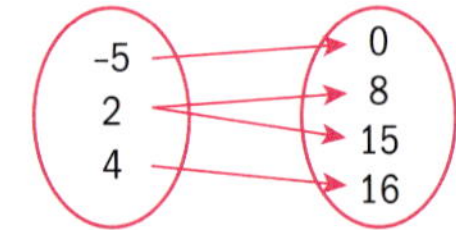

2 Determine which of the following graphs represent functions. For those that are functions, state the domain and range. For those that do not represent functions, give a reason why.

a

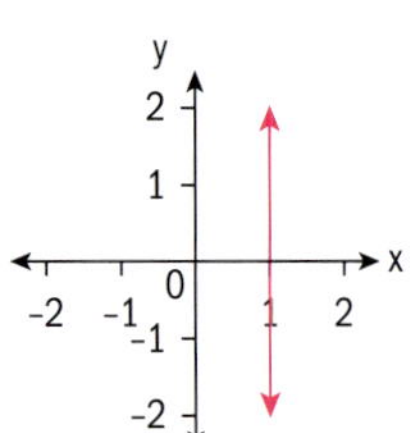

b

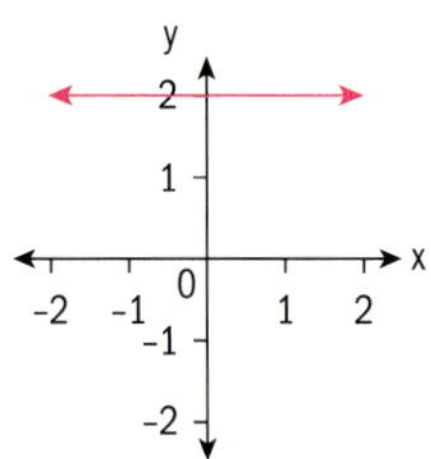

c

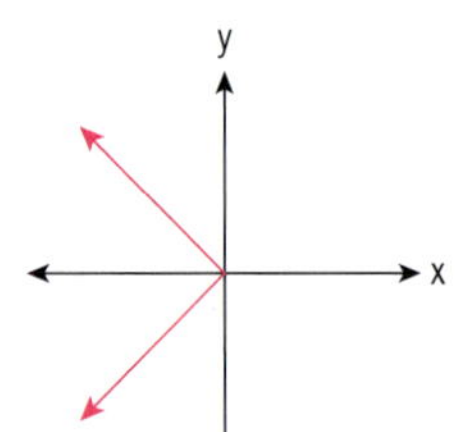

d

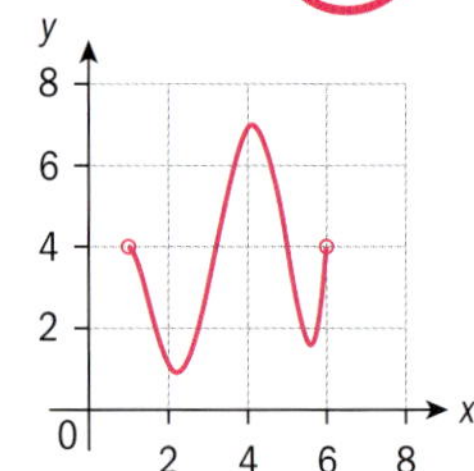

e

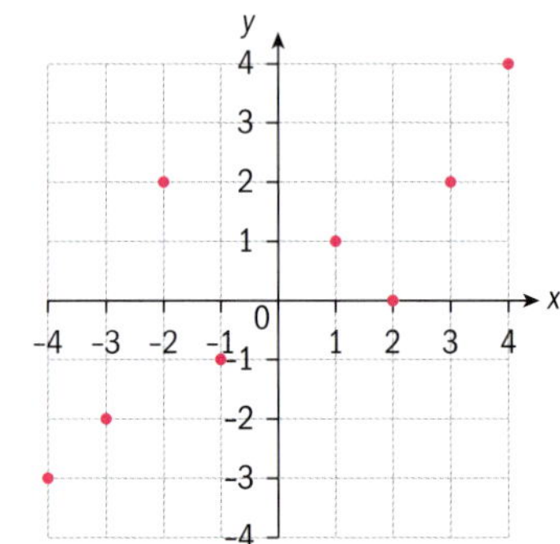

f

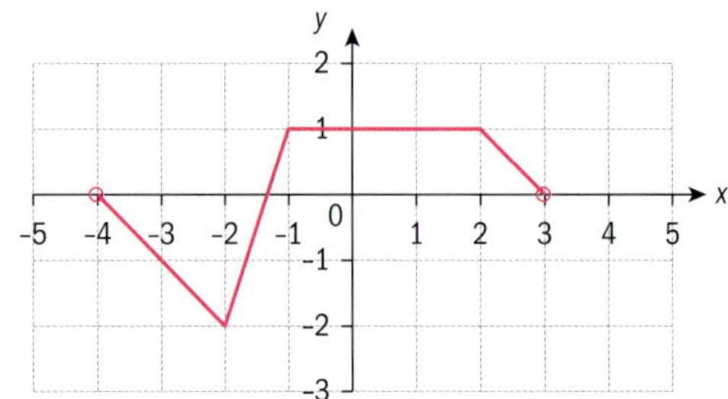

g

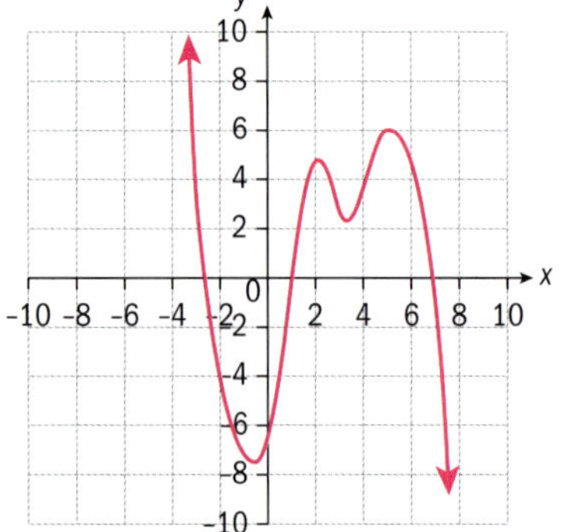

Developing inquiry skills

Refer back to the opening problem. How does this section help you to answer the first two questions of the opening problem?

2.2 Special functions and their graphs

Quadratic functions

Investigation 3

A projectile is hurled into the air from the ground and its height is recorded every second until it hits the ground again.

Time (s)	Height (m)
0	0
1	35
2	60
3	75
4	80
5	75
6	60
7	35
8	0

1 Calculate the first and second differences of the heights. What do you notice?

2 Graph the data points and connect all the points with a smooth curve.

3 Find the function that matches your graph in the form $y = a(x-p)(x-q)$, where p and q are the x-intercepts.

4 How long is the projectile in the air?

5 What is the maximum height that the projectile reaches, and how long does it need to reach this height?

6 How can you rewrite the function from **3** so that it shows the vertex of the function, ie the maximum height, and the time it takes to reach this height?

7 **Factual** Which forms of a quadratic function highlight different information in a given problem?

8 **Conceptual** Why is it useful to be able to express quadratic functions in different equivalent forms?

HINT

Use another point on the curve to find a.

The curve that models the problem in Investigation 3 is a parabola. It is the graph of a quadratic function. A quadratic function has the general form $y = ax^2 + bx + c$; $a \neq 0$. The variables are x and y, and the **parameters** are a, b and c. The parameters determine the shape of the particular quadratic function.

International-mindedness

Over 2000 years ago, Babylonians and Egyptians used quadratics to work with land area.

A quadratic function can be expressed in different forms. In all forms, all parameters are real numbers, and $a \neq 0$.

Standard form: $y = ax^2 + bx + c$.

Vertex form: $y = a(x - h)^2 + k$, where (h, k) is the vertex.

Intercept form: $y = a(x - p)(x - q)$, where p and q are the x-intercepts, or zeros of the function.

A function can either be concave up or concave down. You will learn analytical ways of determining the concavity of a function in Chapter 4.

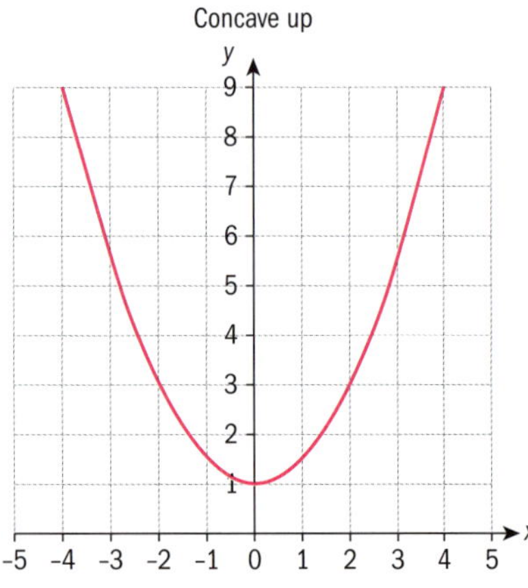

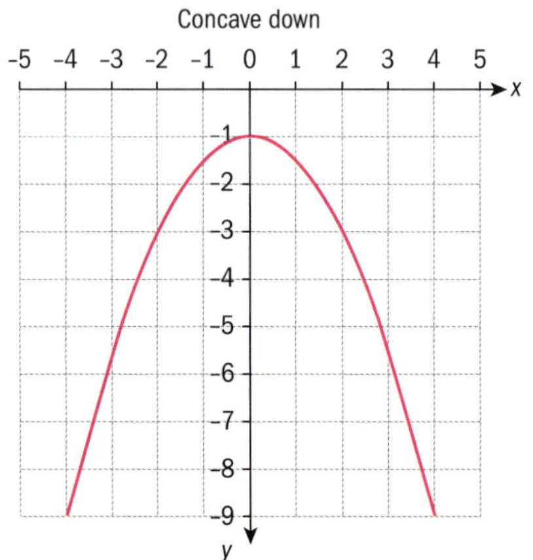

Investigation 4

1. Create quadratics of the form $y = a(x - h)^2 + k$ and graph them.
2. Describe the effect that the parameters h and k have on the graph of $y = x^2$.
3. Graph various quadratic functions in any of the three forms, changing the values and signs of the leading coefficient a.
4. Deduce the sign of a for the parabola to be: **i** concave up **ii** concave down.
5. What happens to the function when $a = 0$?
6. How do the vertex (h, k) and the leading coefficient a help you determine the range of the function?
7. **Conceptual** What features of a graph of a function do the parameters represent?

As you can see in the graphs, a parabola is symmetrical about the vertical line drawn through its vertex. Therefore, the x-coordinate of the vertex is the same as the equation of its axis of symmetry, $x = -\frac{b}{2a}$.

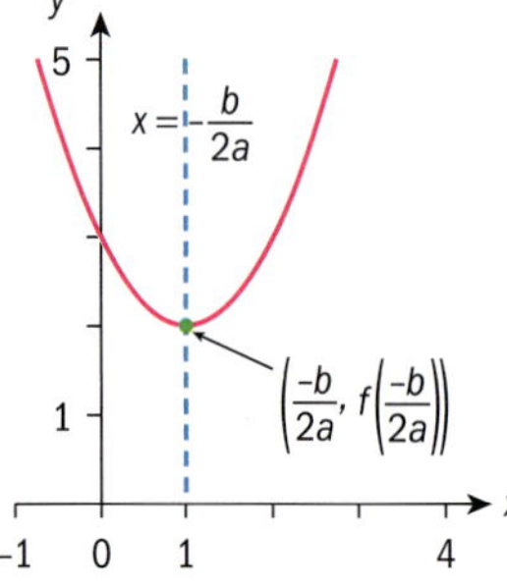

Since the axis of symmetry is also the x-coordinate of the vertex (h, k), the equation of the axis of symmetry is also $x = h$.

International-mindedness

Cartesian coordinates are named after Frenchman René Descartes.

Example 4

State the concavity of the function $y = -2x^2 + 4x + 1$ and find the:

a equation of the axis of symmetry **b** vertex

c domain and range **d** vertex form of the quadratic function.

Concave down.	Since $a < 0$, the parabola is concave down.
a $x = -\dfrac{4}{2(-2)} = 1$	Use the formula for the axis of symmetry, $x = -\dfrac{b}{2a}$.
b $f(1) = 3$ Vertex: $(1, 3)$	Substitute the value from part **a** into the quadratic and solve for y.
c Domain: Real numbers Range: $y \leq 3$	The vertex of a parabola that is concave down is a maximum. Hence, the set of y-values will be at most 3.
d $a = -2,\ h = 1,\ k = 3 \Rightarrow y = -2(x - 1)^2 + 3$	Substitute the parameters into the vertex form $y = a(x - h)^2 + k$.

Example 5

A calculator company's profits in euros are modelled by the function $P(x) = -20x^2 + 1400x - 12\,000$ where x is the selling price per calculator. Find the selling price that generates the maximum profits, and the maximum profits. Justify your answer.

Vertex: $x = -\dfrac{1400}{2(-20)} = 35$ The selling price is 35 euros.	Use the formula for the axis of symmetry, $x = -\dfrac{b}{2a}$.
$P(35) = -20(35)^2 + 1400(35) - 12\,000 = 12\,500$ The maximum profits are 12 500 euros.	Substitute this value into the quadratic.
Since $a < 0$, the quadratic is concave down, and the vertex is a maximum point.	The concavity will indicate whether the vertex is a maximum or minimum point.

Investigation 5

Using mathematical functions (and vectors, which you will learn about in a later chapter), mathematicians work with film producers to create the figures for animated films. A basic example is the face shown here, which is made up of only parabolas.

Continued on next page

1 Experiment with different forms of the quadratic functions to arrive at the functions of the various parabolas shown, and define their restricted domains.

2 **Conceptual** Why is it helpful to move between different forms of the quadratic to model the given problem?

3 Define the window used for the picture.

4 **Conceptual** When using technology, how does the choice of viewing window affect your ability to communicate patterns effectively?

5 Graph your functions and their domains to see whether they render the given picture.

HINT

Keep a record of your answers to this investigation as you will need to refer to them later in the chapter.

Did you know?

You are learning about functions in this chapter, some of which have curves that can be obtained by taking slices of a cone at different angles. In the year 1000 CE the Islamic mathematician al-Kuhi first described an instrument for drawing the different curves.

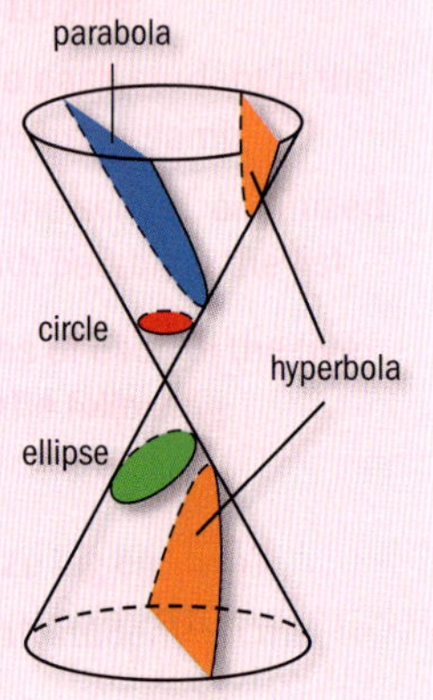

Example 6

Find the equation of the quadratic function whose graph is the parabola.

a

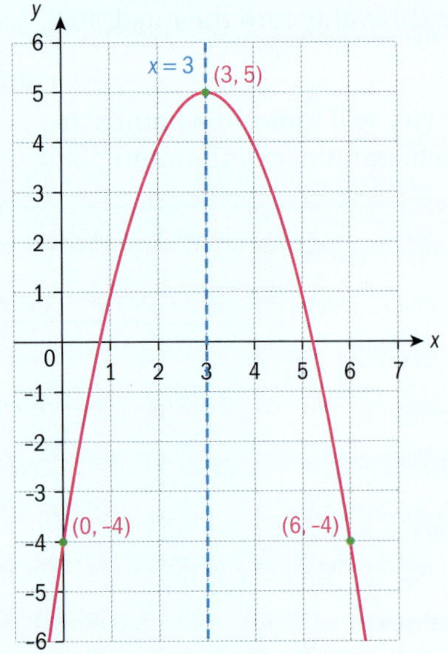

b

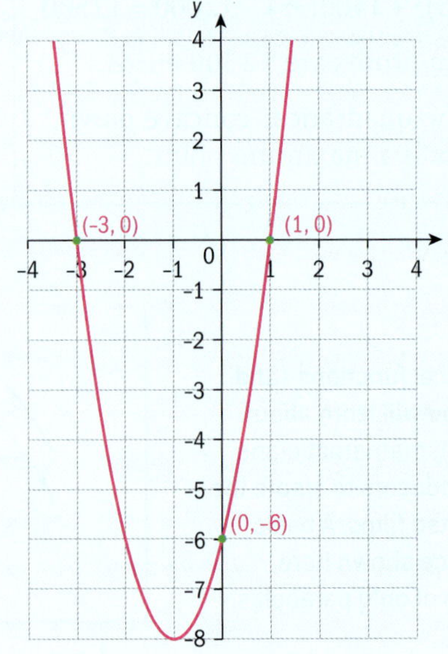

a Vertex is $(3, 5) \Rightarrow y = a(x-3)^2 + 5$	Since the vertex is given, use the vertex form $y = a(x-h)^2 + k$.
$-4 = a(0-3)^2 + 5 \Rightarrow a = -1$ $y = -(x-3)^2 + 5$	Select any other point on the parabola and substitute its values into the vertex form and solve for a.
b Zeros are $(-3, 0)$ and $(1, 0)$ $\Rightarrow y = a(x+3)(x-1)$	Since the zeros are given, use the intercept form $y = a(x-p)(x-q)$.
$-6 = a(0+3)(0-1) = -3a \Rightarrow a = 2$ $y = 2(x+3)(x-1)$	Substitute another point into the intercept form and solve for a.

Constant, linear and quadratic functions belong to the family of *polynomial* functions. The largest exponent on the variable of a polynomial function is called its degree. A polynomial of degree 1 is a linear function, and a polynomial of degree 2 is a quadratic function. The term without a variable is called the constant term. Polynomial functions are formally defined in Chapter 3.

Did you know?

Some mathematicians say that the degree of a constant function is 0, while others say it is undefined. You have learned that a real non-zero number to the zero power is 1, eg $5^0 = 1$. However, when asking the GDC to graph, for example, $f(x) = 1 \times x^0$ and evaluate it at $x = 0$, it states that at this point f is undefined!

This is because mathematicians are still in disagreement about the value of 0^0.

Do some research on the value of 0^0. What do you think?

Exercise 2B

1 For each quadratic function, determine:

i the equation of the axis of symmetry

ii the vertex.

iii State its concavity, domain and range.

a $y = x^2 + 6x + 8$ **b** $y = 10 + 3x - x^2$

c $y = 3x^2 - 12x - 5$ **d** $y = 7 - 4x - 2x^2$

2 Use the information shown in each graph to write the equation of the quadratic function it represents. Select the most appropriate form for finding the function.

a

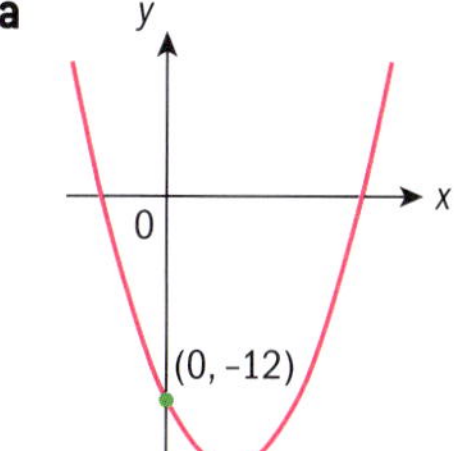

b

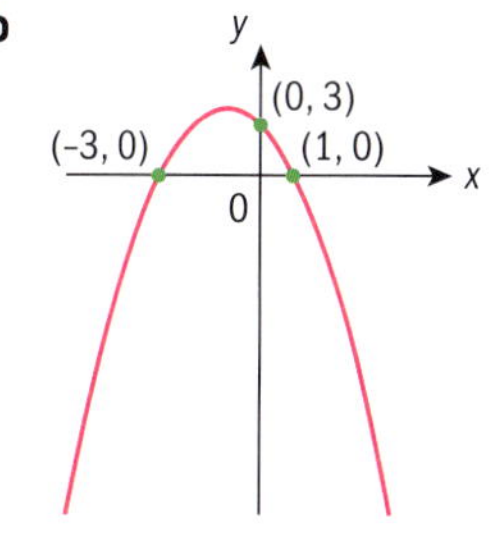

c

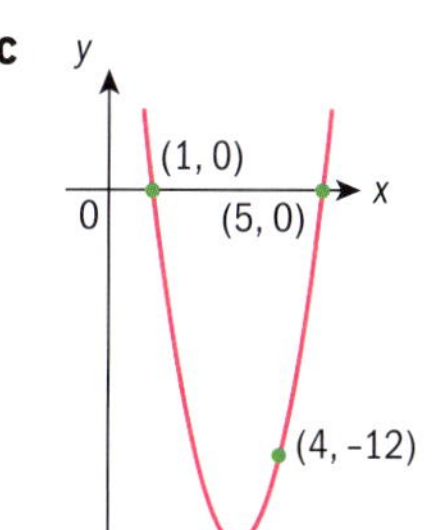

d

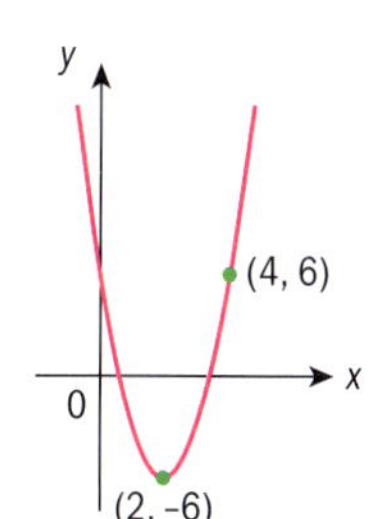

e

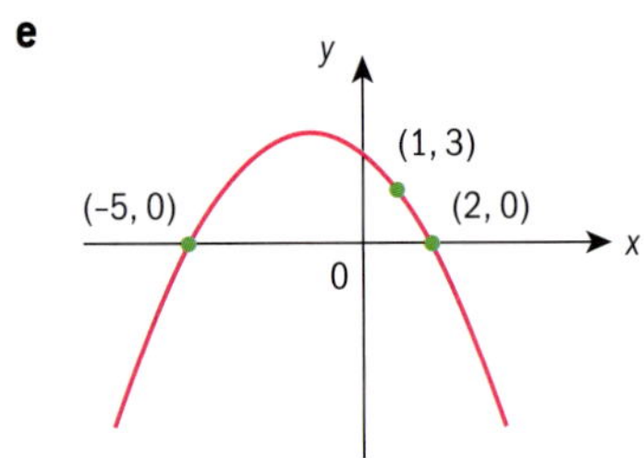

f

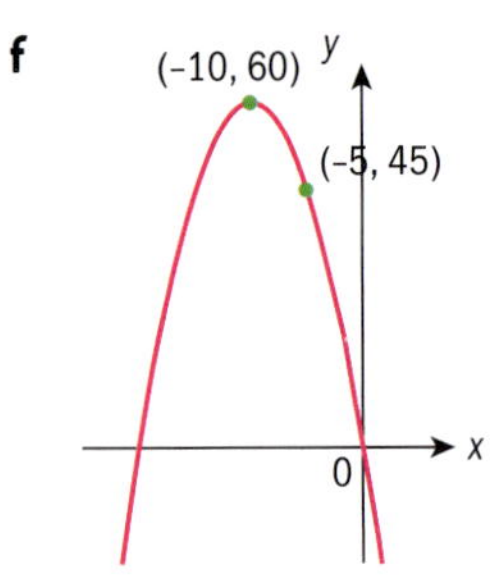

Rational functions

You have worked with rational algebraic expressions by performing the four arithmetic operations on them. You will now work with rational functions whose numerator and denominator are polynomials.

A **rational function** is any function which can be expressed as a quotient $y = \frac{f(x)}{g(x)}$, $g(x) \neq 0$, where $f(x)$ and $g(x)$ are polynomial functions.

The reciprocal function, $y = \frac{1}{x}$, $x \neq 0$, is a rational function, since its numerator 1 is a constant polynomial, and its denominator x is a polynomial of degree 1. The domain of this function is $\mathbb{R}-\{0\}$. Since the numerator is non-zero, this function can never assume the value 0, and 0 must therefore be excluded from the range. The range therefore is also $\mathbb{R}-\{0\}$.

The graph of the reciprocal function is shown here:

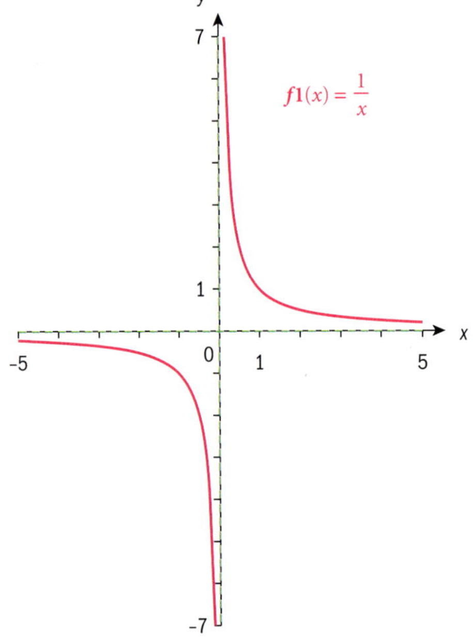

HINT

The reciprocal of a real number a is $\frac{1}{a}$, $a \neq 0$.

The reciprocal function is $y = \frac{1}{x}$, $x \neq 0$, and is the reciprocal of the function $y = x$.

HINT

This shape in a conic section is called a rectangular hyperbola.

You can clearly see from the graph that the values of the function increase rapidly as you take values close to $x = 0$, eg when $x = \frac{1}{100}$, $y = \frac{1}{\left(\frac{1}{100}\right)} = 100$. Similarly, when $x = -0.01$, $y = -100$. As x gets close in value to 0, the values of the function grow without bound. As the y-values get closer to 0, the x-values grow without bound. The lines $x = 0$ and $y = 0$, the y- and x-axis respectively, are called the **asymptotes** of the graph of the function.

The asymptote $x = 0$ is the vertical asymptote, and $y = 0$ is the horizontal asymptote. When producing a sketch of the graph of a function, asymptotes are represented as dotted lines.

HINT
The term "asymptote" will be defined carefully in Chapter 4. For now, think of an asymptote of the graph of a function as a line. The distance between the graph and the line decreases and gets closer to zero the further the graph and the line are extended.

Example 7

Determine the domain and range of the rational function $y = \frac{2}{1-x}$. Confirm your answer graphically, and state the equations of any asymptotes.

$1 - x = 0 \Rightarrow x = 1$, hence domain $= \{x \mid x \neq 1\}$.	Exclude from the domain the value of x that would make the denominator 0.
Since the numerator is non-zero, $y \neq 0$ and the range $= \{y \mid y \neq 0\}$. 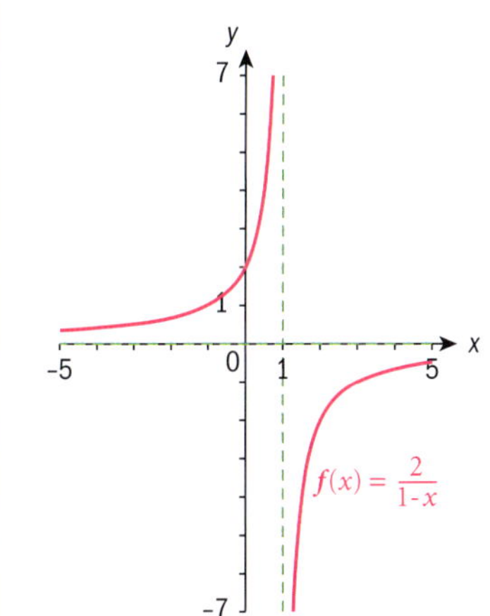 Vertical asymptote: $x = 1$ Horizontal asymptote: $y = 0$	From the graph you can see that as the x-values decrease and approach $x = 1$, the y-values increase without bound. As the x-values increase, the y-values decrease and get closer to $y = 0$.

In rational functions consisting of two linear polynomials, $y = \frac{ax+b}{cx+d}$, you can find the restriction on the domain by setting the denominator equal to zero, $cx + d = 0 \Rightarrow x = -\frac{d}{c}$. The domain must exclude the value $-\frac{d}{c}$, and therefore the vertical asymptote is $x = -\frac{d}{c}$.

International-mindedness

The development of functions bridged many countries including France (René Descartes), Germany (Gottfried Wilhelm Leibniz), and Switzerland (Leonhard Euler).

In order to find any restrictions on y, ie the horizontal asymptote, you need to solve the function for x. This is shown in the following example.

Example 8

Determine the domain and range of the function $y = \frac{2x-1}{1-3x}$ and write down the equations of the asymptotes. Confirm your answer graphically.

$1-3x=0 \Rightarrow x=\frac{1}{3}$ Domain: $\left\{x \mid x \neq \frac{1}{3}\right\}$ Vertical asymptote: $x=\frac{1}{3}$.	Set the denominator equal to 0 and solve. Exclude this x-value from the domain.
Solving for x, $y=\frac{2x-1}{1-3x} \Rightarrow y(1-3x)=2x-1$ $\Rightarrow y-3xy=2x-1$ $\Rightarrow -3xy-2x=-1-y$ $\Rightarrow x(3y+2)=1+y$ $\Rightarrow x=\frac{1+y}{3y+2}$ $3y+2=0 \Rightarrow y=-\frac{2}{3}$ Range: $\left\{y \mid y \neq -\frac{2}{3}\right\}$ Horizontal asymptote: $y=-\frac{2}{3}$. 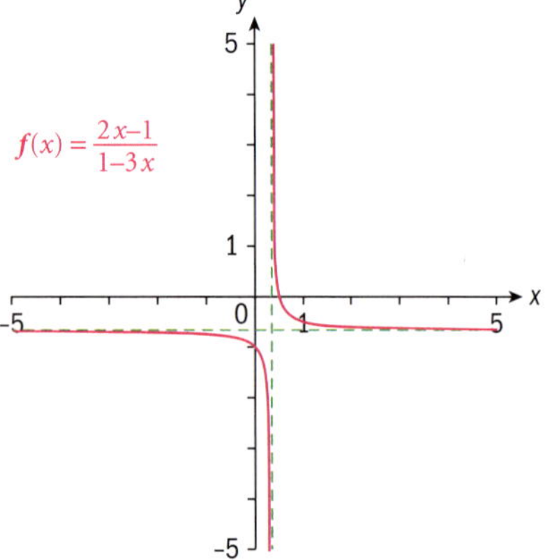	Solve for x and eliminate the y-value that makes the denominator 0.

Reflect How do you think you can check your answer to the horizontal asymptote by using values of x?

Exercise 2C

Find the domain and range of the following functions and state the equations of any asymptotes.

1 $y = \frac{3}{4-2x}$

2 $y = -\frac{1}{3-6x}$

3 $y = \frac{x}{2-4x}$

4 $y = \frac{1+x}{1-x}$

5 $y = \frac{1-2x}{1+2x}$

6 $y = -\frac{2x-3}{2-3x}$

Radical functions

Investigation 6

1 Create a table of values for the function $f(x) = 1 + \sqrt{4x-8}$ for $2 \le x \le 6$, $x \in \mathbb{Z}$.

2 Try to evaluate $f(0)$ and $f(-2)$ and write down what you notice.

3 What values must be excluded from the domain of this function, and why?

4 Define the restricted domain of this function, and the range.

Did you know?

For radical functions the positive root is assumed, unless otherwise indicated.

A function of the type $f(x) = \sqrt{ax+b}$ is a square root function whose radicand (the expression in the square root) is linear. The radicand must be non-negative, hence $ax + b \ge 0$ and the restricted domain is $\left\{x \middle| x \ge -\frac{b}{a},\ a \ne 0\right\}$. The range is the set of non-negative real numbers, ie $\{y | y \ge 0\}$.

Example 9

Determine the domain and range of $y = 2 - \sqrt{2x+3}$ and confirm your answer graphically.

$2x + 3 \ge 0 \Rightarrow x \ge -\frac{3}{2}$	For the function to be real the radicand must be non-negative.
Domain: $\left\{x \middle| x \ge -\frac{3}{2}\right\}$	This is the restricted domain.
$\sqrt{2x+3} \ge 0 \Rightarrow -\sqrt{2x+3} \le 0$ $2 - \sqrt{2x+3} \le 2$ Range: $\{y \mid y \le 2\}$	Since the smallest value that $\sqrt{2x+3}$ can assume is 0, the largest value y can assume is 2.
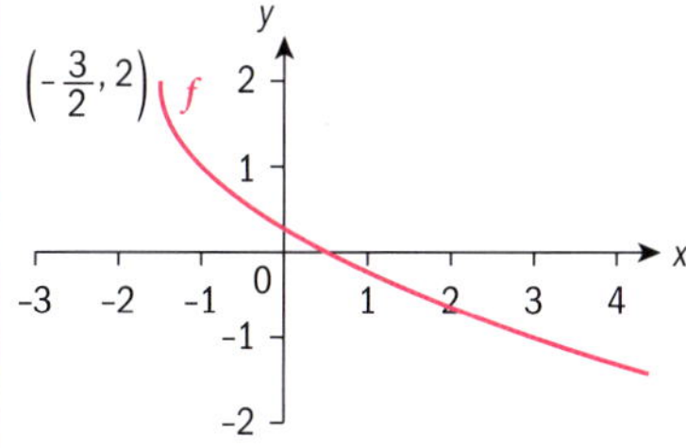	

Investigation 7

1 Given $y^2 = x$, find two solutions for y, and graph both.

2 Is $y^2 = x$ a function? Explain.

3 Is $y = \sqrt{x}$ a function? Explain.

Exercise 2D

1 Determine the domain and range of these functions.

a $y = \sqrt{x-2}$

b $y = \sqrt{3x-2}$

c $y = 1 + \sqrt{2-4x}$

d $y = 3 - \sqrt{2x+1}$

e $y = -2\sqrt{x-1}$

f $y = 1 - 3\sqrt{2-x}$

In the rational functions so far, you have investigated those whose numerator and denominator are linear functions. You will now work with rational functions when the denominator is a quadratic function.

Investigation 8

Given the function $y = \dfrac{1}{x^2 - 3x + 2}$:

a Find the x-values that make the denominator 0.

b State the domain.

c State a value that the rational expression will never assume.

d State the equations of the asymptotes.

e Check your answers by graphing the function, and comment on the shape of the graph.

Example 10

Determine the domain of $f(x) = -\dfrac{2}{2x^2 - 3x - 9}$ and state the equations of any asymptotes.

$2x^2 - 3x - 9 = 0 \Rightarrow (2x+3)(x-3) = 0 \Rightarrow x = -\dfrac{3}{2}; x = 3$	Set the denominator equal to 0 and solve.
Domain: $\left\{x \mid x \neq -\dfrac{3}{2}; x \neq 3\right\}$.	
Vertical asymptotes: $x = -\dfrac{3}{2}; x = 3$	
Horizontal asymptote: $y = 0$	Since the numerator is non-zero, the rational expression will never equal 0.

You will now work with quotient functions that contain a mixture of different types of functions.

International-mindedness

The word *asymptote* is derived from the Greek *asymptotos*, which means "not falling together".

Example 11

Find the domain and range of $y = \frac{1}{\sqrt{x+1}}$ and state the equation of the asymptotes. Confirm your answers graphically.

$\sqrt{x+1} = 0 \Rightarrow x = -1$	Set the denominator equal to 0 and solve.
$x + 1 > 0 \Rightarrow x > -1$ Domain: $\{x \mid x > -1\}$ Vertical asymptote: $x = -1$	Set the radicand to be positive, and solve.
Range: $\{y \mid y > 0\}$ Horizontal asymptote: $y = 0$	Since the numerator is non-zero, the rational expression will never equal 0. Both numerator and denominator are positive, hence the range is all positive real numbers.

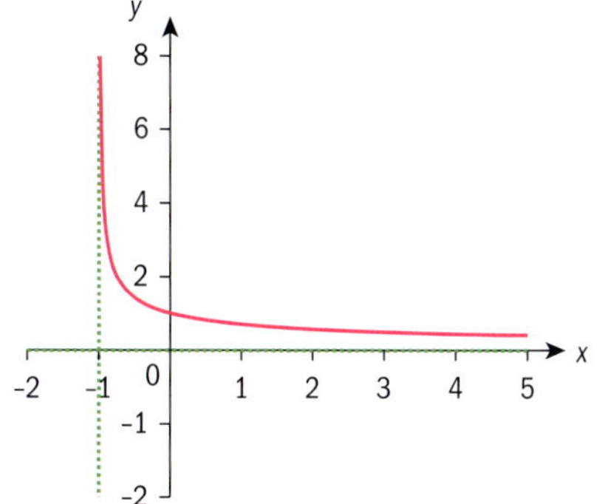

Exercise 2E

Find the domain of these functions and state the equations of any asymptotes. Confirm your answers graphically.

1 $y = \frac{4}{x^2 - 3x}$

2 $y = \frac{1}{x^2 - 9}$

3 $y = -\frac{1}{x^2 + 2x - 3}$

4 $y = \frac{2}{(x+2)^2}$

5 $y = -\frac{1}{2x^2 + 9x - 18}$

6 $y = \frac{2}{\sqrt{3x+6}}$

7 $y = \frac{1}{\sqrt{2x^2 - 3x - 2}}$

8 $y = -\frac{2}{\sqrt{4x^2 - 25}}$

Partial fraction decomposition

In Investigation 8, you sketched the rational function $y = \frac{1}{x^2 - 3x + 2}; x \neq 1, 2$ by exploring key features of its graph. This function is a proper algebraic fraction, since the numerator is a constant term, and the denominator has degree 2. This algebraic fraction can be obtained from the addition of two fractions. How do you "decompose" this algebraic fraction into the sum (or difference) of two algebraic fractions?

Did you know?

A proper rational algebraic fraction is one where the degree of the numerator is less than or equal to the degree of the denominator.

The first step, if possible, is to factorize the denominator, as this will give you the denominators of the fractions you are looking for. Therefore, you can write that

$$\frac{1}{x^2-3x+2} \equiv \frac{A}{x-2} + \frac{B}{x-1}.$$

Combining the fractions on the right-hand side,

$$\frac{1}{x^2-3x+2} \equiv \frac{A}{x-2} + \frac{B}{x-1} \equiv \frac{A(x-1)+B(x-2)}{(x-2)(x-1)}.$$

You now need to find A and B such that $A + B = A(x-1) + B(x-2) = 1$. One way to solve for A and B is to substitute a value of x that will cause either A or B to cancel; for example, $x = 2$ will cause B to cancel. Then,

$A(2-1) + B(2-2) = 1 \Rightarrow A = 1$. Then, letting $x = 1$ in order to cancel A,

$A(1-1) + B(1-2) = 1 \Rightarrow -B = 1$

$\Rightarrow B = -1$

Hence, $\frac{1}{x^2-3x+2} \equiv \frac{1}{x-2} - \frac{1}{x-1}$

TOK

How is mathematics related to reality?

HINT

The use of the triple bar $\equiv$, or identity symbol, means that you want to find A and B such that the LHS and RHS are true for **all** values of x in the domain not just some values of x, as is the case when solving an equation.

To check whether you have decomposed the original fraction correctly, you can combine your fractions algebraically to see whether it gives you the desired result. On the other hand, you can graph both sides and see that they are equivalent.

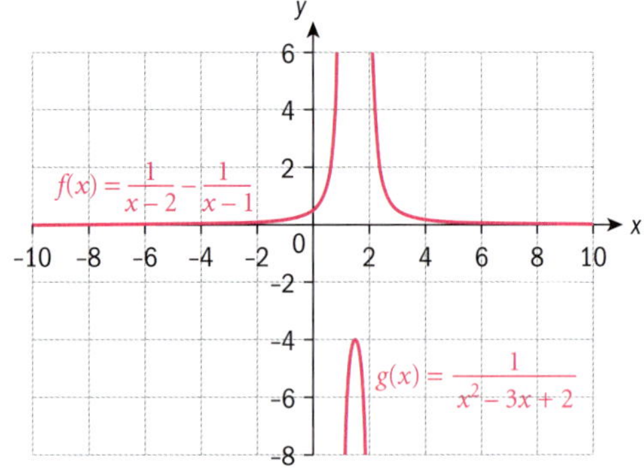

Example 12

Express $\frac{2x-5}{x^2+x-2}$ in partial fractions.

$\frac{2x-5}{x^2+x-2} \equiv \frac{A}{(x+2)} + \frac{B}{(x-1)}$ $\Rightarrow 2x-5 = A(x-1) + B(x+2)$	Since the degree of the numerator is less than the degree of the denominator, this is a proper fraction that can be decomposed into two fractions with linear denominators, where the numerators are constants.
Let $x = 1$: $-3 = 3B \Rightarrow B = -1$	Let $x = 1$ to eliminate A.
Let $x = -2$: $-9 = -3A \Rightarrow A = 3$ $\frac{2x-5}{x^2+x-2} = \frac{3}{(x+2)} - \frac{1}{(x-1)}$	Let $x = -2$ to eliminate B.

Check:

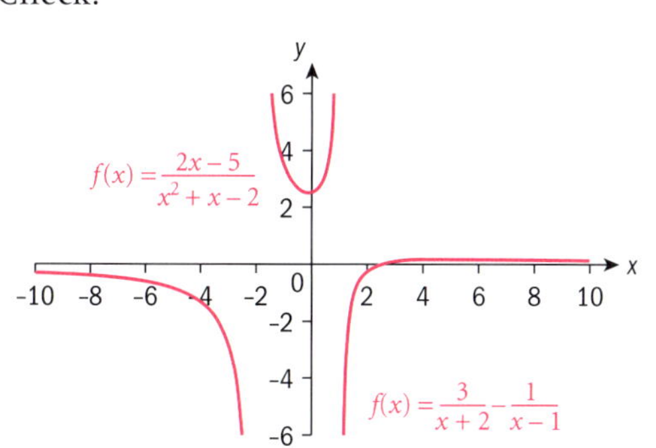

Combine the RHS to make sure that it is equivalent to the LHS, or graph both sides.

Exercise 2F

Express in partial fractions:

1 $\frac{1}{x^2+5x+6}$

2 $\frac{4-x}{x^2+x-2}$

3 $\frac{4x-9}{x^2-3x}$

4 $\frac{x}{x^2-1}$

5 $\frac{5}{-x^2-x+6}$

6 $\frac{10x-1}{8x^2+2x-1}$

7 $\frac{11+3x}{6x^2+5x-6}$

Absolute value functions and equations

You are familiar with the concept of the "absolute value" of a number. It is defined as

$|a| = \begin{cases} a, & a \geq 0 \\ -a, & a < 0 \end{cases}$. The absolute value of a real number is non-negative.

Geometrically, the absolute value is the distance between the point representing a number on the real number line and the origin of the number line. More generally, the absolute value of the difference of two numbers is the distance between the points that represent them on the real line.

From the definition of absolute value and its geometrical meaning, these useful fundamental properties follow, for a, a real number.

Properties of absolute value:

1 $|a| \geq 0$

2 $|-a| = |a|$

3 $|a| = 0 \Rightarrow a = 0$

4 $|a-b| = 0 \Rightarrow a = b$

5 $|ab| = |a||b|$; $\left|\frac{a}{b}\right| = \frac{|a|}{|b|}$, $b \neq 0$

6 $|a+b| \leq |a| + |b|$

7 $|a-b| \geq |a| - |b|$

8 $|a| \leq b \Rightarrow -b \leq a \leq b$; $|a| \geq b \Rightarrow a \leq -b$ or $a \geq b$

Did you know?

Another term for absolute value is the **modulus** of a number. It comes from the French word *module*, which means "unit of measure", and has been used by mathematicians since the early 1800s. Karl Weierstrass first used the vertical bar notation in 1841.

The definition of an absolute value function follows from the definition of the absolute value of a number. It has different definitions within disjoint subsets of its domain.

$$f(x) = |x| = \begin{cases} x, & x \geq 0 \\ -x, & x < 0 \end{cases}.$$

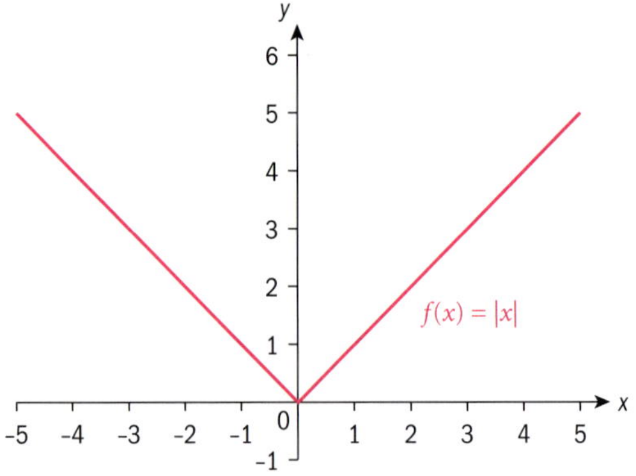

As you can see, the domain is the set of real numbers, and the range is the set of non-negative real numbers since $(0, 0)$ is a minimum point.

Compare the graph of the absolute value function with the graphs of $y = x$ and $y = -x$.

What do they have in common; what is different?

Investigation 9

Consider absolute value functions of the form $y = |x - h| + k;\ h, k \in \mathbb{R}$.

1 Graph functions like these for different values of h and k.

2 What effect do the parameters have on the graph of $y = |x|$?

3 Using different values for a, how does this parameter affect the shape of $y = a|x - h| + k$?

4 What are the coordinates of the maximum or minimum point of $y = a|x - h| + k$, and the condition for being a maximum or minimum point?

5 What are the domain and range of the function in terms of the given parameters?

Example 13

Determine the domain and range of $y = -2|x - 4| + 1$ and confirm graphically.

Domain: $\{x | x \in \mathbb{R}\}$; range: $\{y | y \leq 1\}$.

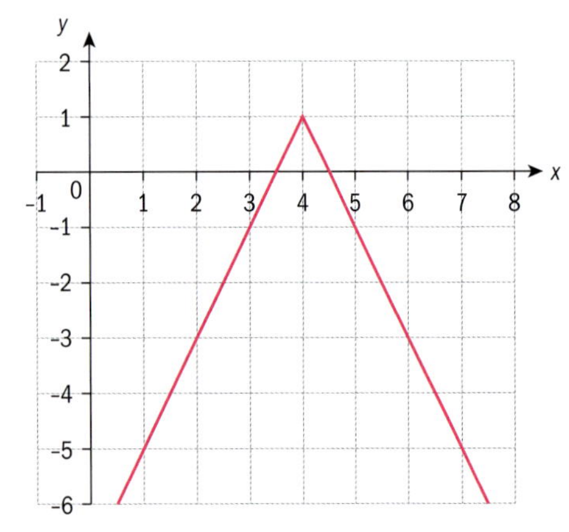

Since $a < 0$, $(4, 1)$ is a maximum vertex.

Example 14

Determine the domain and range of $y = |2 - 2x| + 1$ and sketch the function.

$|2 - 2x| = |-2(x - 1)| = |-2||x - 1| = 2|x - 1|$ $\Rightarrow y = 2|x - 1| + 1$ Domain: $\{x \mid x \in \mathbb{R}\}$; range: $\{y \mid y \geq 1\}$. 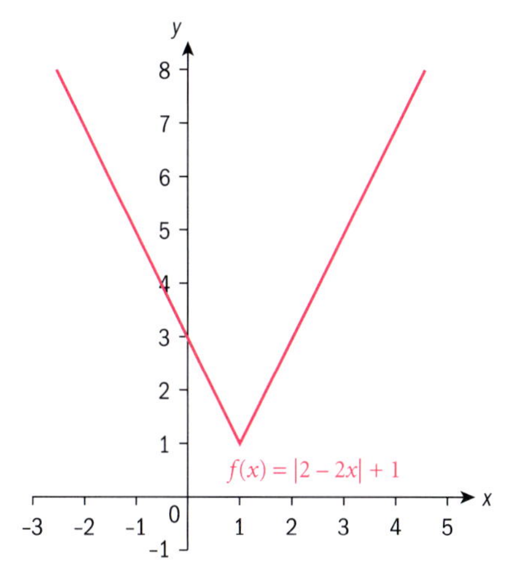	Change to the form $y = a|x - h| + k$ and use absolute value property 5.

TOK

"The object of mathematical rigour is to sanction and legitimize the conquests of intuition" – Jacques Hadamard

Do you think that studying the graph of a function contains the same level of mathematical rigour as studying the function algebraically?

Exercise 2G

Determine the domain and range of these functions, and sketch their graphs.

1 $y = -|x| + 3$ **2** $y = |2x| - 1$

3 $y = |x - 3| - 4$ **4** $y = -|x - 2| + 1$

5 $y = \frac{1}{2}|x + 2| - 1$ **6** $y = -2|3x + 6| + 2$

7 $y = -3|1 - x| + 2$ **8** $y = -|3 - 3x| - 2$

9 $y = |2x^2 + x - 3|$ **10** $y = |6 - 13x - 5x^2| + 2$

You will now consider ways of solving absolute value functions for particular values. For example, using **3** from the previous exercise, find x such that $f(x) = 1$. Graphically, you can see that there are two solutions, $x = -2$ or $x = 8$.

You need to return to the definition of absolute value in order to solve $|x - 3| - 4 = 1$ algebraically.

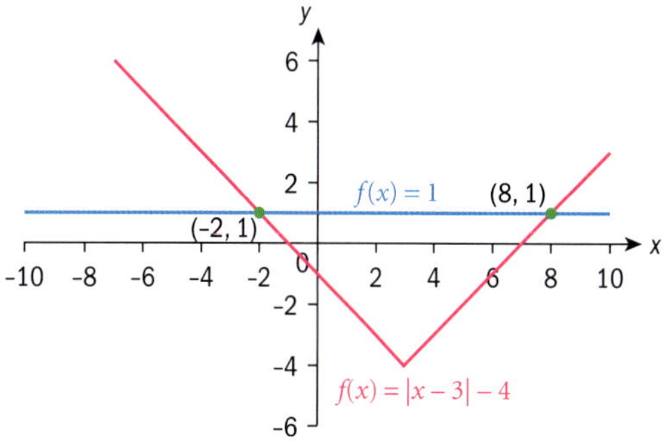

$|x - 3| - 4 = 1$ reduces to $|x - 3| = 5$. Since the distance between the point representing the expression $(x - 3)$ and the origin is 5, this means that either $(x - 3) = 5$ or $(x - 3) = -5$. In the former case, $x = 8$, and in the latter case, $x = -2$. Alternatively, $(x - 3)$ is either positive or negative, hence the equation can also be written as $\pm(x - 3) = 5$, and each case solved separately. Using either method you arrive at the same answers.

Example 15

Solve $-2|3x+6|+2=-10$, and check your answer(s).

$-2\|3x+6\|+2=-10 \Rightarrow \|3x+6\|=6$	Simplify the equation.
$(3x+6)=\pm 6 \Rightarrow 3x=0$ or $3x=-12$.	Use the definition of absolute value.
$x=0; x=-4$	
Check: $-2\|3\times 0+6\|+2=-2\times 6+2=-10$ $-2\|3\times -4+6\|+2=-2\times 6+2=-10$	Check all your solutions in the original equations.

Example 16

Solve $|x+1|=-2x-5$, and check your answer(s).

$\pm(x+1)=-2x-5$	Use the definition of absolute value.
$x+1=-2x-5 \Rightarrow x=-2$ $-(x+1)=-2x-5 \Rightarrow x+1=2x+5$ $\Rightarrow x=-4$	Solve each case separately.
Check: $\|-2+1\|=1; -2(-2)-5=-1$. Since LHS $\neq$ RHS, reject the solution $x=-1$. $\|-4+1\|=3; -2(-4)-5=3$. $x=-4$ is the only solution.	Check your solutions, and discard the answer that does not check.

In this last example, you have to reject one of the two answers because it does not satisfy the original equation. Letting $y_1=|x+1|$ and $y_2=-2x-5$ you can see that the two functions only intersect at $x=-4$.

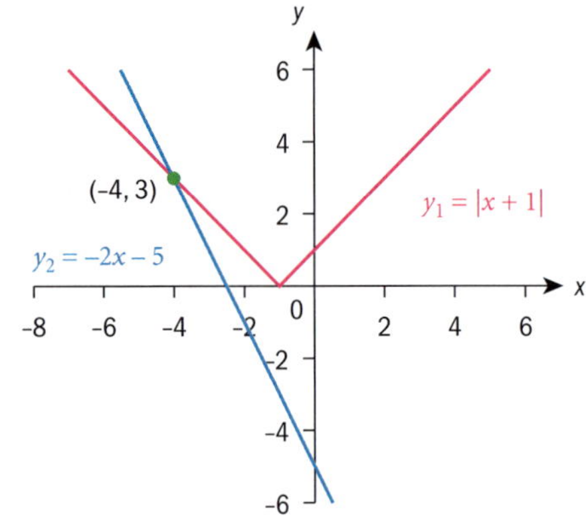

Example 17

Solve $|3x - 4| = |2x + 3|$ and check your answer(s) both numerically and graphically.

Method 1	
$(3x - 4) = (2x + 3) \Rightarrow x = 7$	Let both sides of the equation be positive, and solve.
$-(3x-4) = (2x+3) \Rightarrow x = \frac{1}{5}$	Let one side be negative and the other side positive, and solve.
Method 2	
$(3x - 4)^2 = (2x + 3)^2$ $\Rightarrow 9x^2 - 24x + 16 = 4x^2 + 12x + 9$ $\Rightarrow 5x^2 - 36x + 7 = 0$ $\Rightarrow (5x-1)(x-7) = 0$ $\Rightarrow x = \frac{1}{5}; x = 7$	Since both sides of the equation are positive, square both sides, and solve.
Check: $x = \frac{1}{5}$: LHS: $\left\|3 \times \frac{1}{5} - 4\right\| = \left\|-3\frac{2}{5}\right\| = 3\frac{2}{5}$ RHS: $\left\|2 \times \frac{1}{5} + 3\right\| = 3\frac{2}{5}$ LHS = RHS $x = 7$:	Check your answers.
$\|3 \times 7 - 4\| = 17$; $\|2 \times 7 + 3\| = 17$	Both answers are solutions to the equation.
LHS = RHS 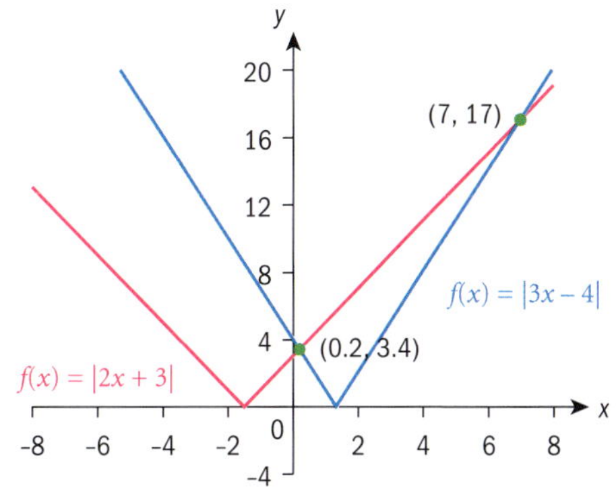 	The graphs intersect at $x = 0.2$ and $x = 7$, therefore the algebraic solution is confirmed.

Exercise 2H

1 Solve the equations and check your answers either numerically or graphically.

a $10 - |3x + 2| = 7$ **b** $8|x + 7| - 3 = 5$

c $|x - 2| = 2x + 1$ **d** $|4x + 3| = 3 - x$

e $|4x + 9| = |2x - 1|$ **f** $|5x + 3| - |2x - 1| = 0$

g $\left|\frac{2x-5}{3}\right| = \left|\frac{3x+4}{2}\right|$

Absolute value property 8 is very useful in solving modulus inequalities, as shown in Example 18.

$|a| \le b \Rightarrow -b \le a \le b$; $|a| \ge b \Rightarrow a \le -b$ or $a \ge b$

Example 18

Solve, and check your answer graphically.

a $|2x - 4| < 4$ **b** $|3 - 2x| \ge 1$

a $|2x - 4| < 4 \Rightarrow -4 < 2x - 4 < 4 \Rightarrow 0 < x < 4$

Use $|a| \le b \Rightarrow -b \le a \le b$

Check:

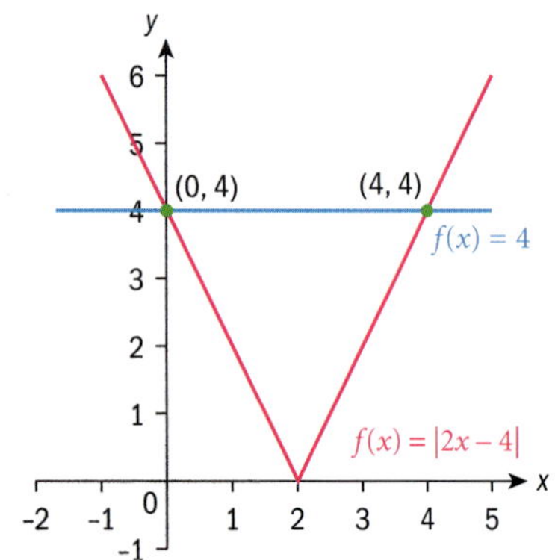

You can see that the graph of $y = |2x - 4|$ is beneath the graph of $y = 4$ in the interval $0 < x < 4$.

b $|3 - 2x| \ge 1 \Rightarrow 3 - 2x \le -1$ or $3 - 2x \ge 1$

$\Rightarrow x \ge 2$ or $x \le 1$

Use $|a| \ge b \Rightarrow a \le -b$ or $a \ge b$

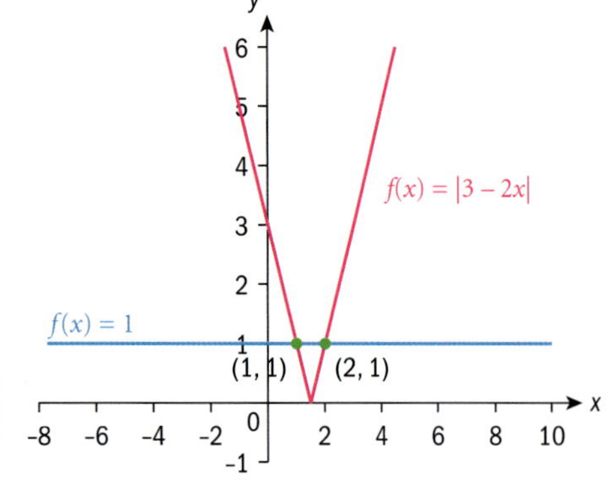

You can see that the graph of $y = |3 - 2x|$ is above the graph of $y = 1$ in the intervals $x \le 1$ and $x \ge 2$.

Example 19

Solve $|x^2 - 5x| < 6$.

$\|x^2 - 5x\| < 6 \Rightarrow -6 < x^2 - 5x < 6$	Rewrite the absolute value as an inequality.
$-6 < x^2 - 5x \Rightarrow 0 < x^2 - 5x + 6$	Work both sides of the inequality separately.
$\Rightarrow 0 < (x - 3)(x - 2)$	
$\Rightarrow x < 2$ or $x > 3$.	The vertex of a quadratic that is concave up and has zeros is below the x-axis. Therefore it is positive to the left and right of the zeros, and negative between the zeros.
$x^2 - 5x < 6 \Rightarrow x^2 - 5x - 6 < 0$	
$\Rightarrow (x - 6)(x + 1) < 0$	
$\Rightarrow -1 < x < 6$.	
$-1 < x < 2,\ 3 < x < 6$.	Taking the solution in line 4 together with the solution in line 7: The intersection of the sets of values $x < 2$ and $-1 < x < 6$ is $-1 < x < 2$. The intersection of the sets of values $x > 3$ and $-1 < x < 6$ is $3 < x < 6$.
Confirm graphically. 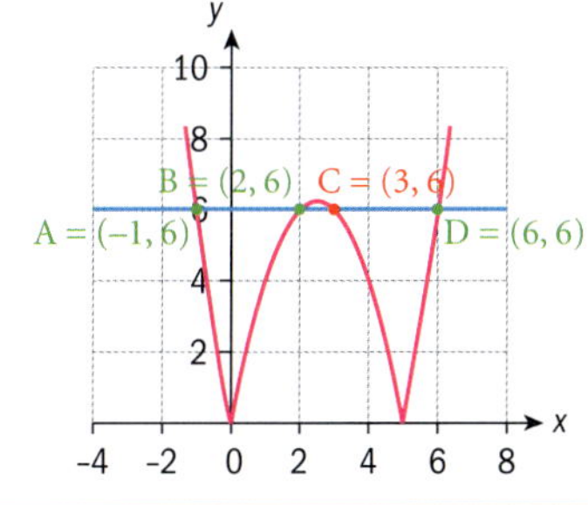 	From the graph you can see that the graph of the absolute value function is less than the graph of $y = 6$ at the stated intervals.

Example 20

Solve $\left|\dfrac{1}{2x-1}\right| < 1$ graphically.

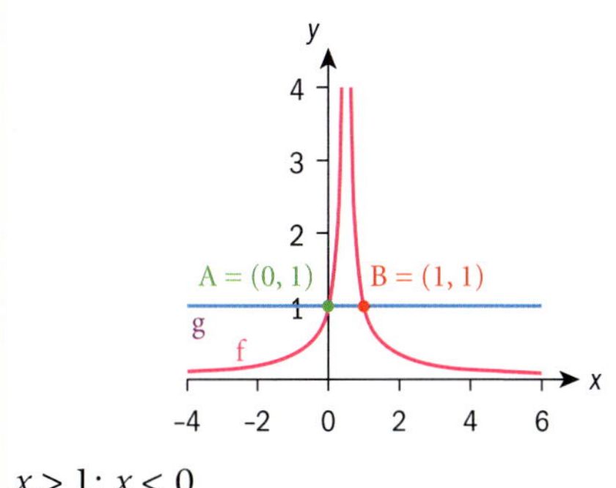 $x > 1;\ x < 0$	Sketch the graphs of $y = \left\|\dfrac{1}{2x-1}\right\|$ and $y = 1$. Find the points of intersection of the LHS and RHS functions. Look to see where the LHS graph is below the RHS graph.

Reflect How would you solve Example 20 algebraically?

TOK

Is zero the same as nothing?

Exercise 2I

1 Solve the inequalities and check your answers graphically.

a $|2x + 3| < 6$ b $|2x - 3| \geq 5$

c $|3 - 2x| < 5$ d $|1 - 3x| \geq 5$

e $|2x + 3| > |x + 3|$ f $x + 6 > |3x + 2|$

2 Solve the inequalities graphically:

a $\left|\frac{x+1}{x-2}\right| > 2$ b $\frac{1}{|x-1|} \geq 1$

c $2|x + 2| - |x + 5| \leq 4$ d $|x^2 + 3x - 4| < 3$

Piecewise-defined functions

You have already experienced these kinds of functions, since the absolute value function is a piecewise-defined function.

$$f(x) = |x| = \begin{cases} x, \ x \geq 0 \\ -x, \ x < 0 \end{cases}$$

For the first branch of the function, the domain is $[0, \infty[$. For the second branch of the function, the domain is $]-\infty, 0[$. The domain therefore is $]-\infty, 0[\cup [0, \infty[$, which is the entire set of real numbers. The range is the set of non-negative real numbers.

You will now graph piecewise functions whose "pieces" are the functions you have studied in this chapter so far.

Example 21

Consider the function $f(x) = \begin{cases} x - 4; \ x \geq 3 \\ -(x-2)^2; \ x < 3 \end{cases}$

a Find $f(-1)$, $f(0)$, $f(3)$, $f(5)$. b Sketch $f(x)$. c Write down the domain and range of f.

a $f(-1) = -(-1 - 2)^2 = -9$ $f(0) = -(0 - 2)^2 = -4$ $f(3) = 3 - 4 = -1$ $f(5) = 5 - 4 = 1$	Substitute the values of x into the function at the appropriate intervals.
b 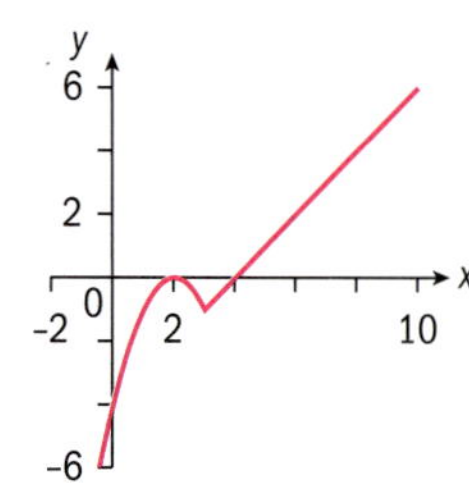	Sketch both functions on their respective domains.
c Domain and range are the set of real numbers.	

Investigation 10

The picture is an attempt at creating a model of a bicycle. So far it consists of linear, quadratic and radical functions, as well as semi-circles.

1 Match the names of the functions on the graph to their equations.

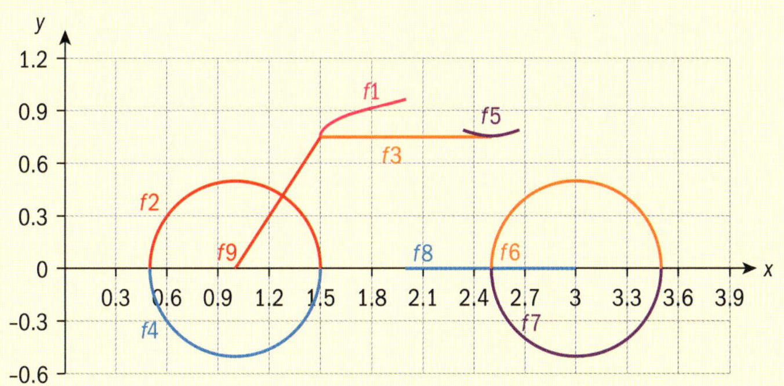

$g(x) = 0.75,\ 1.5 < x < 2.5$

$h(x) = -\sqrt{0.25-(x-3)^2}$

$j(x) = (x-2.5)^2 + 0.76,\ 2.34 < x < 2.66$

$k(x) = 0.3\sqrt{x-1.5} + 0.76,\ 1.5 < x < 3$

$p(x) = -\sqrt{0.25-(x-1)^2}$

$r(x) = 0,\ 0.2 < x < 3$

$s(x) = 1.5x - 1.5,\ 1 < x < 1.5$

$t(x) = \sqrt{0.25-(x-3)^2}$

$v(x) = \sqrt{0.25-(x-1)^2}$

2 Finish the model of the bicycle by creating graphs of different functions from those you have learned so far in this chapter. You may modify those that are given in order to produce a better model of a bicycle.

Exercise 2J

1 $y = \begin{cases} -1, & x > 0 \\ 1, & x \le 0 \end{cases}$

a Find $f(-9)$, $f(0)$, $f(\pi)$, $f(99)$.

b Sketch the function.

c Write down the domain and range of the function.

2 $y = \begin{cases} 4 - x^2, & x > 0 \\ x^2, & x \le 0 \end{cases}$

a Find $f(-4)$, $f(0)$, $f(1)$.

b Sketch the function.

c Write down the domain and range of the function.

3 $y=\begin{cases}\sqrt{x+1}, & x>0\\ 3x-1, & x\le 0\end{cases}$

a Find $f(-1)$, $f(0)$, $f(8)$.

b Sketch the function.

c Write down the domain and range.

4 $f(x)=\begin{cases}|x+1|, & x<-1\\ \sqrt{x+1}, & x\ge -1\end{cases}$

a Find $f(-1)$, $f(0)$, $f(-4)$, $f(8)$.

b Sketch the function.

c State the domain and range.

5 Define the piecewise functions and their intervals shown in this graph.

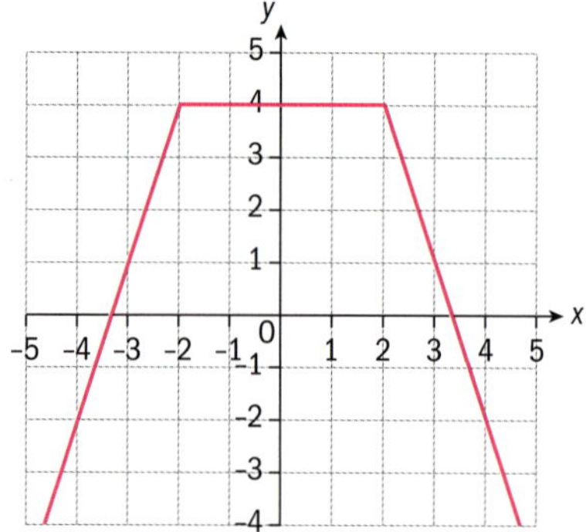

6 Rewrite as a piecewise-defined function.

a $f(x) = |2x + 4|$ b $f(x) = |3x - 9| + 2$

Developing inquiry skills

Let's return to the opening problem. How do the ideas in this section help you?

2.3 Classification of functions

Investigation 11

These mappings represent four different types of relations.

1 Match the mapping with the best description of its type:

a 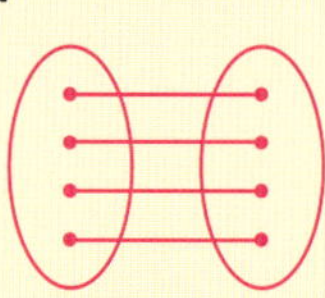

b 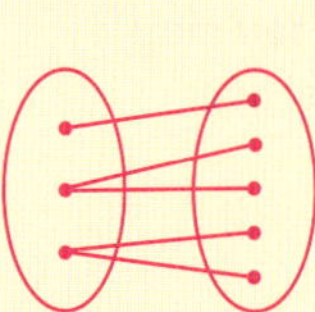

c 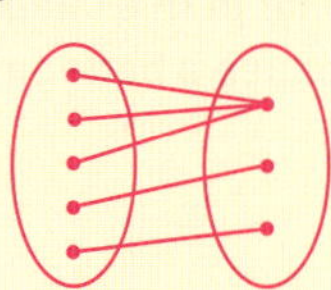

d 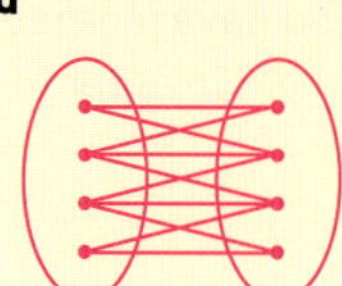

- one-to-one mapping
- many-to-one mapping
- one-to-many mapping
- many-to-many mapping.

2 Which mappings represent functions?

3 Create an algebraic function for each of the mappings that represent functions.

4 Graph your functions, and imagine horizontal lines drawn through the graphs of the functions.

5 **Conceptual** How does the horizontal line test help you classify a function?

A function $f: A \to B$ is **one-to-one** if every element in the range is the image of exactly one element in the domain, ie for any $a \in A$, $b \in B$, $f(a) = f(b) \Rightarrow a = b$. This means that every element in the domain has a different element in the range.

A function is **many-to-one** if more than one element of the domain has the same image.

Example 22

Classify each function as one-to-one or many-to-one.

a $\{(-1.5, 1), (0, 1), (1, 1), (3, 2), (\pi, 1)\}$

b $\{x | x \in \mathbb{R}, x \neq 0\}$ and $f: x \mapsto \frac{1}{x}$.

c

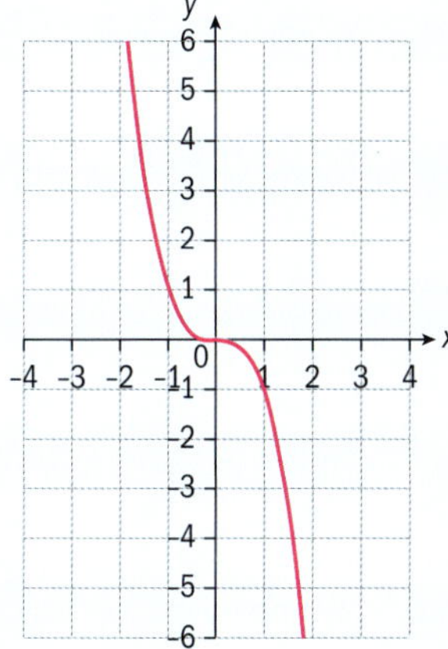

d

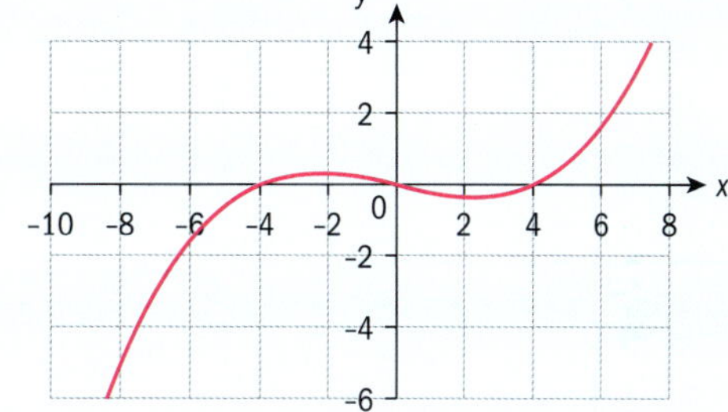

a many-to-one	More than one x-value maps to $y = 1$.
b one-to-one	Each x-value maps to one y-value.
c one-to-one	Use horizontal line test.
d many-to-one	Use horizontal line test.

Onto functions

Another type of mapping is called an **onto** mapping.

This mapping diagram shows a function $f: A \to B$. Each element in the domain maps to one, and only one, element in the range. The element $s \in B$, however, is not the image of any element in A. Therefore $f: A \to B$ is **not** an onto function.

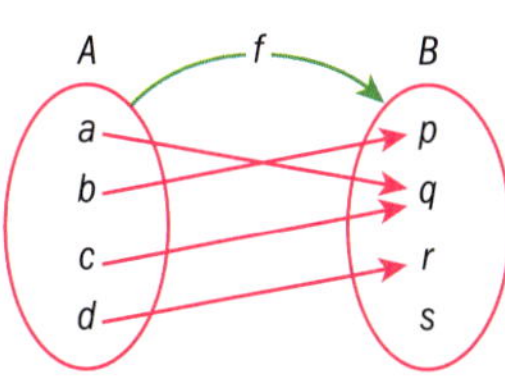

A function $f: A \to B$ is **onto** if every element in B is an image of an element in A, ie for all $b \in B$ there exists an $a \in A$ such that $f(a) = b$.

Investigation 12

1 Is the function $y = 3x - 4$, whose domain and range are the set of real numbers, an onto function?

2 Do all graphs of straight lines represent onto functions? Explain.

3 Is $f: \mathbb{R} \to \mathbb{R}$; $f(x) = x^2$ an onto function? Explain.

4 In question **3**, how can you adjust the set the domain is mapped onto to change your answer?

5 Observe the following graphs of functions, **taking careful consideration of the given mapping of** f. Which graphs represent onto functions? Explain.

a $f: \mathbb{R} \to \mathbb{R}$

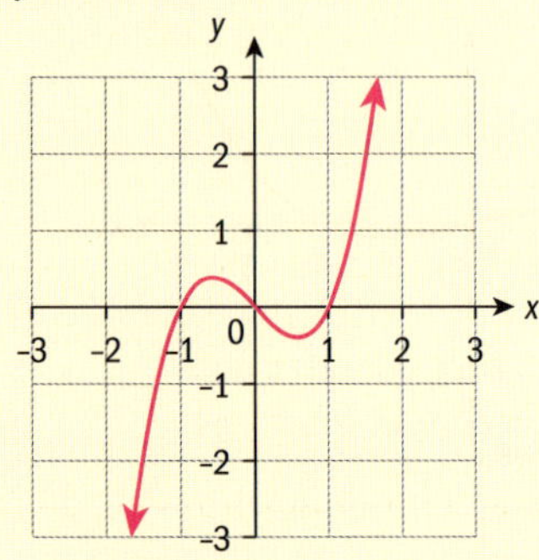

b $f: \mathbb{R}^+ \to \mathbb{R}$

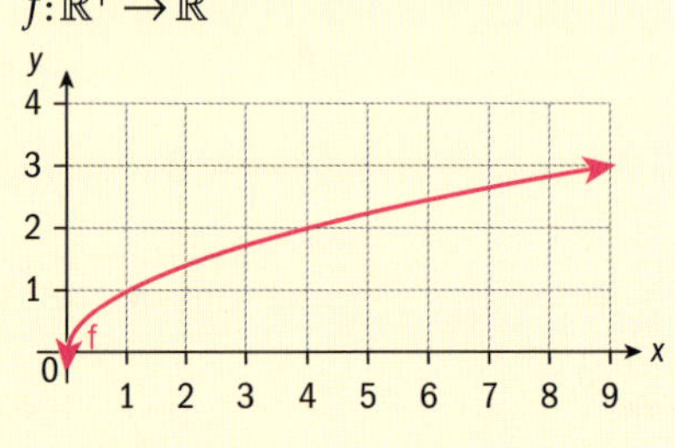

c $f: \mathbb{R} \to \mathbb{R}$

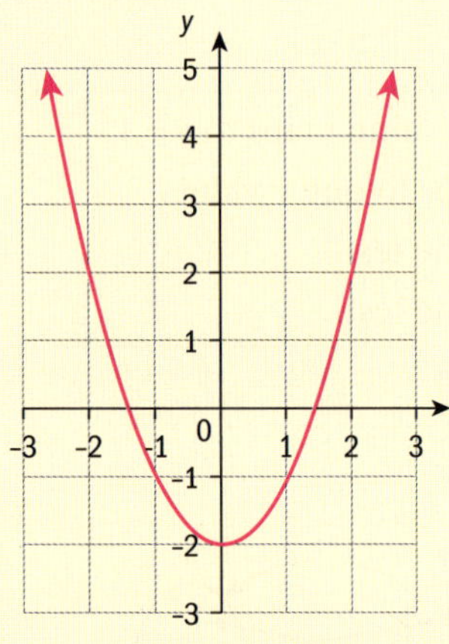

d $f: \mathbb{R} \to \{y \mid y \le 2\}$

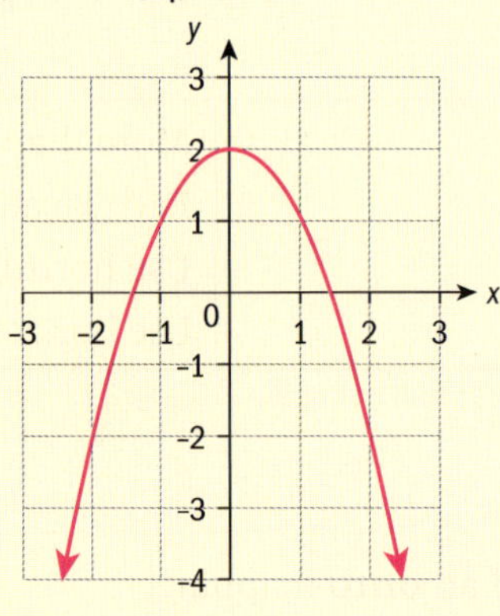

6 Imagine drawing horizontal lines through the **entire** grid of the graphs above. What do the horizontal lines have in common, in relation to the graph, in those functions that are onto/not onto?

7 How can you change the set that the domain is mapped onto for those functions that you have labelled as not onto, in order to make them onto?

8 **Conceptual** Why is the horizontal line test useful for onto functions?

Did you know?

The set of actual images of the domain is called the range. The set of all possible values, ie the set that the range is a subset of, is called the co-domain.

Example 23

Determine whether these functions are onto, one-to-one, neither or both.

a $f:\mathbb{R} \to \mathbb{R}$

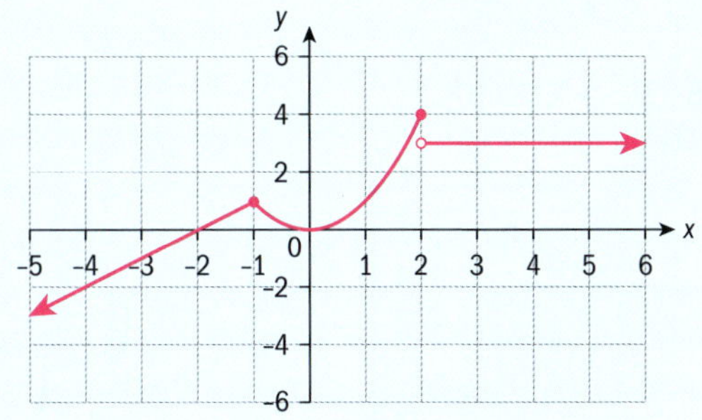

b $f:\mathbb{R} \to y \geq -3$

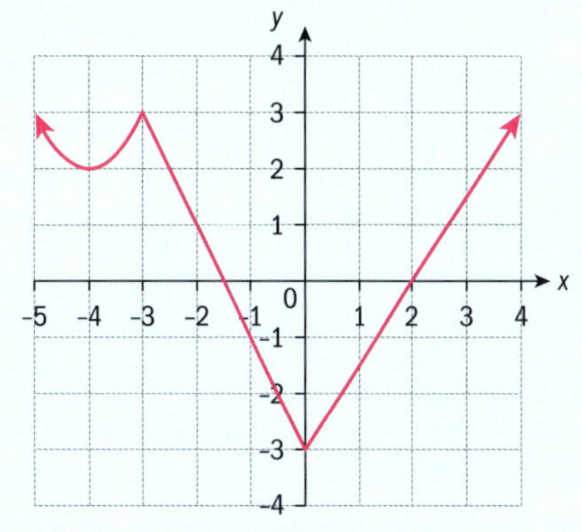

a Neither, as the graph does not pass the horizontal line test for onto nor for one-to-one.	The mapping is to the entire set of real numbers, but since the graph does not extend above $y = 4$, any horizontal lines drawn above this value do not intersect the graph.
b The graph passes the horizontal test for an onto function, but not for a one-to-one function.	Since the indicated mapping is indeed the range of the function, it is onto.

Exercise 2K

Determine whether these functions are onto, one-to-one, neither or both.

1 $f: A \to B,\ A: -5 \leq x < 5;\ B: -3 \leq y < 4$

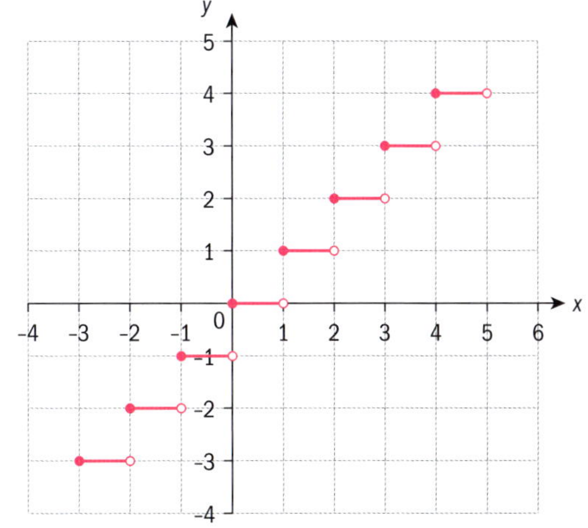

2 $f:\mathbb{R} \to \mathbb{R}$

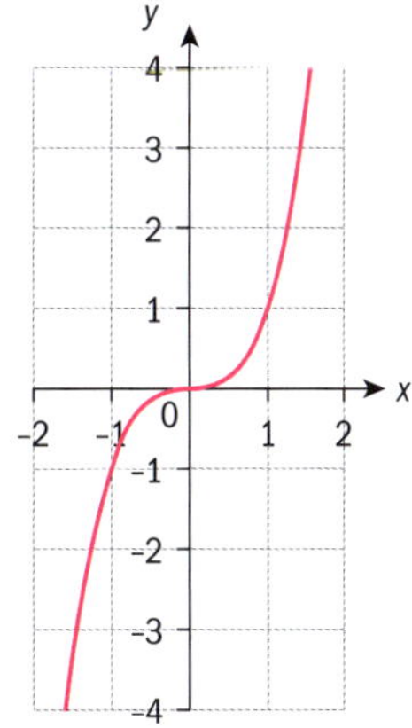

3 $f: \{x \in \mathbb{R}, -3 \leq x \leq 5 \rightarrow \mathbb{R}\}$

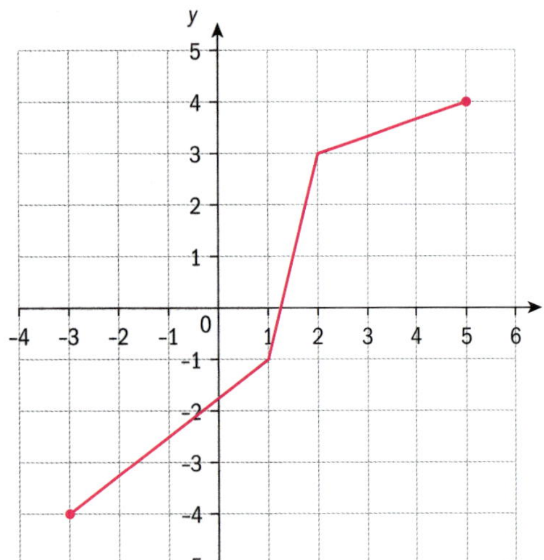

4

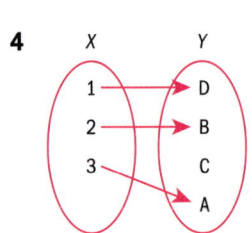

X Y
1 → D
2 → B
3 → A
C

5

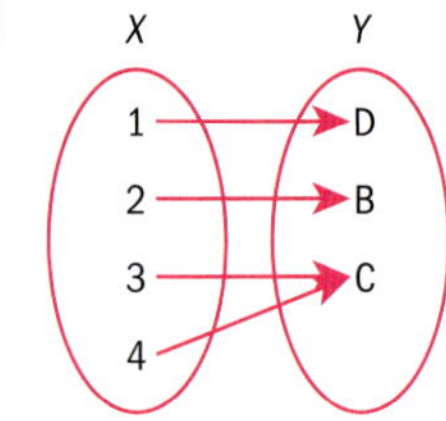

6 $\mathbb{R} \rightarrow \mathbb{R}^+ \cup \{0\}$

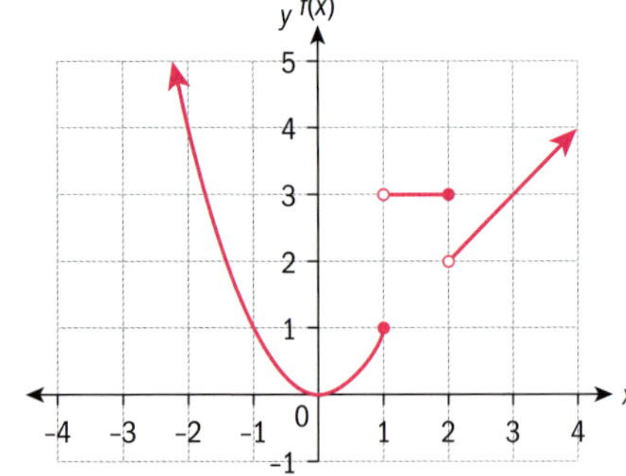

7 Create your own functions that are:

a onto but not one-to-one

b one-to-one but not onto

c neither onto nor one-to-one

d both onto and one-to-one.

Investigation 13

1 Draw two many-to-one graphs symmetrical about the y-axis, eg $y = x^2, y = |x|$.

2 Substitute values for x such that $x = \pm a, a \in \mathbb{R}$ and comment on the relationship between $f(a)$ and $f(-a)$.

3 Draw two graphs that have rotational symmetry about the origin of $180°$, eg $y = x^3, y = x(x-2)(x+2)$.

4 Substitute values for x and evaluate $f(-x)$ and $-f(x)$. Comment on the results.

The functions in **1** are called even functions. The functions in **3** are called odd functions.

5 **Factual** How do you determine graphically whether a function is even or odd?

6 **Conceptual** How do you distinguish whether a function is even or odd?

A function is even if, for all x in the domain, $-x$ is in the domain and $f(x) = f(-x)$. The graph of an even function is symmetrical about the y-axis.

A function is odd if, for all x in the domain, $-x$ is in the domain and $f(-x) = -f(x)$. The graph of an odd function has rotational symmetry of $180°$ about the origin.

Example 24

Determine algebraically whether the following functions are even, odd or neither. Confirm your answers graphically.

a $f(x) = 3 - 2x^2$

b $g(x) = \begin{cases} x - 6, -6 \le x < 0 \\ -x + 6, 0 \le x < 6 \end{cases}$

c $h(x) = \begin{cases} (x+6)^2, -6 \le x < 0 \\ -(x-6)^2, 0 \le x < 6 \end{cases}$

a $f(-x) = 3 - 2(-x)^2 = 3 - 2x^2 = f(x)$

$f(x)$ is even.

Evaluate $f(-x)$ and compare to $f(x)$.

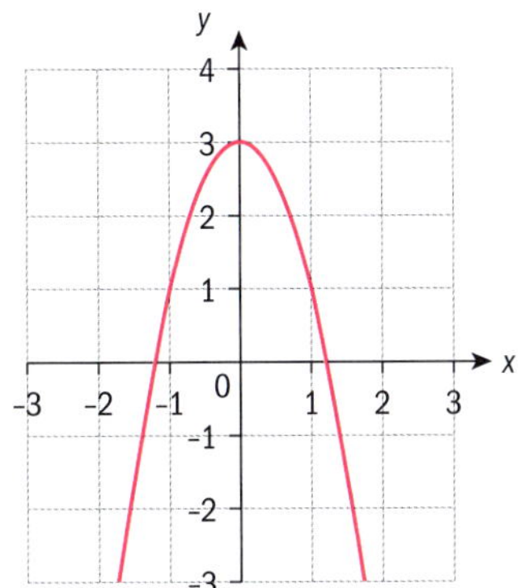

The graph of $f(x)$ is symmetrical about the y-axis.

b $g(-x) = -x - 6$

$g(x) \neq -x - 6 \neq -g(x)$

$g(x)$ is neither even nor odd.

Evaluate $g(-x)$ and compare to $g(x)$ and $-g(x)$.

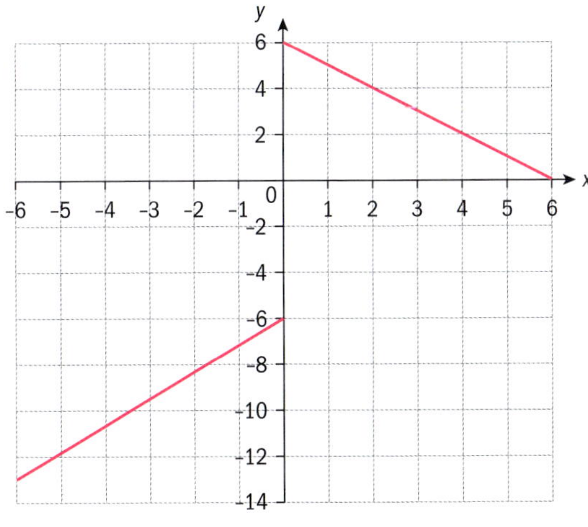

c For $0 \le x < 6$

$h(-x) = -(-x - 6)^2$

$= (x + 6)^2$

$= -h(x)$

Evaluate $h(-x)$ and compare with $-h(x)$.

Continued on next page

$h(x)$ is an odd function.

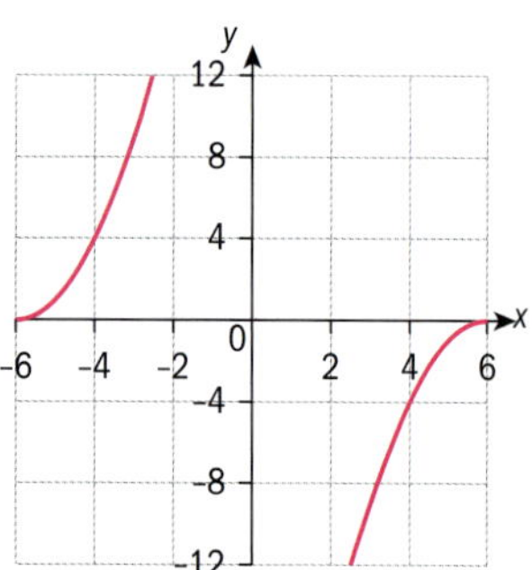

Investigation 14

By testing different examples of even and odd functions, determine whether the following are even, odd or neither, and justify your answers algebraically.

1. The sum (or difference) of:
   - i two even functions
   - ii two odd functions
   - iii an odd and an even function.
2. The product of:
   - i two even functions
   - ii two odd functions
   - iii an odd and an even function.

Exercise 2L

1 Determine algebraically whether the given functions are even, odd or neither. Confirm your answers graphically, and state whether the function is many-to-one or one-to-one.

a $f(x) = 2 - x^2$

b $g(x) = 3x + x^3$

c $h(x) = -\frac{1}{2x}$

d $p(x) = 2(x - 3)^2$

e $f(x) = \begin{cases} -1, & 0 \le x < \pi \\ 1, & \pi \le x < 2\pi \\ -1, & 2\pi \le x < 3\pi \end{cases}$

f $f(x) = x - 2x^3 + x^5$

2 Find a function that is **both** even and odd.

2.4 Operations with functions

Investigation 15

Given the functions $f(x) = 2 - x$ and $g(x) = \frac{1}{x}$, let:

$h(x) = f(x) + g(x)$

$j(x) = f(x) - g(x)$

$k(x) = f(x) \times g(x)$

$$p(x) = \frac{f(x)}{g(x)}$$

$$q(x) = \frac{g(x)}{f(x)}$$

1 State the domain of f and the domain of g.

2 Write out the functions h, j, k, p and q, graph the functions, and state their domains.

3 State the domains of h, j, k, p and q in terms of the domains of f and g.

4 **Conceptual** How can you obtain the domain of the sum, difference, product or quotient of two or more functions?

Function composition

Another operation that combines two functions to form a new function is function composition. In this operation, the output of one function becomes the input of another function.

If f and g are functions, then the notation for the composition of functions f and g is $(f \circ g)(x)$, or $f(g(x))$. The domain of $f \circ g$ is the set of all real numbers x such that:

- x is an element of the domain of g
- $g(x)$ is an element of the domain of f.

The range of $f \circ g$ is a subset of the range of f.

For example, if $f(x) = 2x - 1$ and $g(x) = x^2$, then:

- The domain of f and g is the set of real numbers.
- The range of f is the set of real numbers.
- The range of g is the set of non-negative numbers.
- The range of $(f \circ g)(x)$ is a subset of the range of f.

Hence $(f \circ g)(x)$ is well-defined.

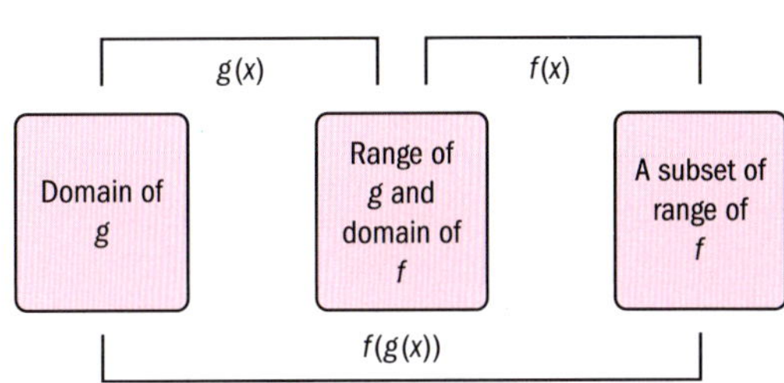

You can obtain specific values of the composite function by two methods.

For example, $f(x) = 2x - 1$; $g(x) = x^2$. Find $f(g(1))$.

Method 1: Finding the composite function first:

$f(g(x)) = f(x^2) = 2x^2 - 1$

$f(g(1)) = 2(1)^2 - 1 = 1$

Method 2: Finding $g(x)$ first:

$g(1) = 1$; $f(1) = 2(1) - 1 = 1$

Example 25

Find $(g \circ f)(x)$ if $f(x) = 2x - 1$ and $g(x) = x^2$. State the domain and range of f, g and $(g \circ f)$. Confirm your answers for $(g \circ f)$ graphically.

$(g \circ f)(x) = g(f(x)) = g(2x - 1) = (2x - 1)^2$ $D_f = \mathbb{R}$; $D_g = \mathbb{R}$; $D_{f \circ g} = \mathbb{R}$ $R_f = \mathbb{R}$; $R_g = \mathbb{R}^+ \cup \{0\}$; $R_{f \circ g} = \mathbb{R}^+ \cup \{0\}$ 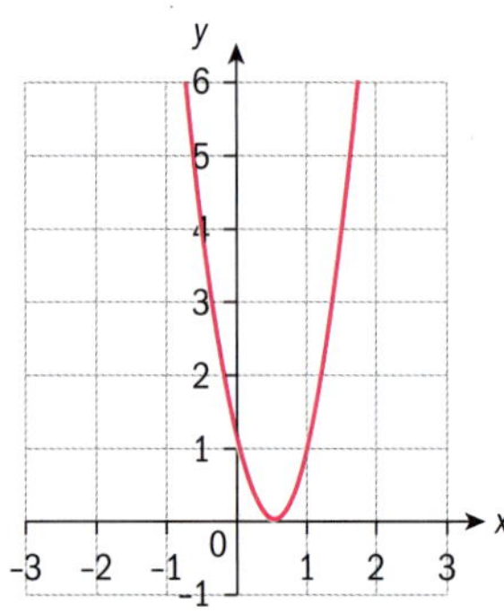	Let g be the argument of f. The range of f is the domain of g. The range of $g(f(x))$ is a subset of the range of g, in this case the range of g itself.

Composition of functions is not limited to two functions, as the following example illustrates.

Example 26

If $f(x) = \frac{1}{x}$; $g(x) = \sqrt{x}$; $h(x) = x^2 - 1$:

a Find:

i the domain of f, g and h

ii $f \circ (g \circ h)$ and state the domain of $(g \circ h)$ and $f \circ (g \circ h)$

iii the range of f, g, h, $(g \circ h)$ and $f \circ (g \circ h)$.

b Determine whether $f \circ (g \circ h)$ is a function.

a i $D_f = \mathbb{R} - \{0\}$; $D_g = \mathbb{R}^+ + \{0\}$; $D_h = \mathbb{R}$ **ii** $(g \circ h)(x) = g(x^2 - 1) = \sqrt{x^2 - 1}$ $f\left(\sqrt{x^2 - 1}\right) = \frac{1}{\sqrt{x^2 - 1}}$ $D_{g \circ h} : x \geq 1; x \leq -1$ $D_{f \circ (g \circ h)} : x < -1; x > 1$ **iii** $R_f = \mathbb{R} - \{0\}$; $R_g = y \geq 0$; $R_h = y \geq -1$ $R_{g \circ h} : y \geq 1; x \leq -1$; $R_{f \circ (g \circ h)} : y > 0$	Define the domains and obtain the composite function.
b The domain of $f \circ (g \circ h)$ is the domain of $g \circ h$, ie $x \geq 1; x \leq -1$. For $f \circ (g \circ h)$ to be a function however, its domain would have to be restricted to exclude ± 1.	For f to be a function every element of the domain must have a corresponding image in the range.

Investigation 16

1 Using the functions in Example 26, and creating your own functions, determine whether function composition is:

a commutative, ie $(f \circ g) = (g \circ f)$

b associative, ie $(f \circ g) \circ h = f \circ (g \circ h)$.

2 Using your own examples for functions, determine whether the following compositions are even, odd or neither:

a an even function composed with an even function

b an even function composed with an odd function

c an odd function composed with an even function

d an odd function composed with an odd function.

3 What kind of a function results (even or odd) when any function is composed with an even function?

4 Attempt to justify your conclusions to questions **2** and **3** algebraically.

TOK

Which do you think is superior: the Bourbaki group analytical approach or the Mandelbrot visual approach to mathematics?

Investigation 17

1 Find functions f and g such that $f(g(x)) = (x-1)^2 + 3$.

2 Expand and simplify $(f \circ g)$. How would you find the functions when the composition is simplified?

3 Find functions h and k such that $(h \circ k)(x) = (x-1)^2 + (x-1) + 3$.

4 Simplify $(h \circ k)$ and now find functions h and k. Are the functions the same as you found in **3**? Explain.

5 Does every composite function have a unique decomposition?

Exercise 2M

1 For each function find, if it exists:

i $g(f(1))$ **ii** $f(g(2))$ **iii** $f(g(x))$ **iv** $g(f(x))$

a $f(x) = 3x; g(x) = \sqrt{x}$

b $f(x) = 5 - 3x; g(x) = x^2 + 4$

c $f(x) = x + 1; g(x) = \sqrt{2x-1}$

2 **a** State the domain and range of functions f and g.

b Find $f \circ g$ and $g \circ f$ and state their domain and range.

i $f(x) = x^2 + x; g(x) = 2 - 3x$

ii $f(x) = |x+1|; g(x) = \sqrt{x^2 - 4}$

3 Let $f(x) = 1 - 2x, g(x) = x^2 - 1, h(x) = \sqrt{2x+4}$.

a Find **i** $f(h(x))$ **ii** $h(g(x))$ **iii** $h(h(x))$ **iv** $f(g(h(x)))$

b State the domain and range of the functions in part **a**.

c Find $h(h(0))$ and $g(g(-1))$.

4 If $f(x) = 3x + a$ and $g(x) = \dfrac{x-4}{3}$ find the value of a such that $f(g(x)) = g(f(x))$.

5 Without using $y = x$ for any of the functions, find expressions for $f(x)$ and $g(x)$ such that

a $f(g(x)) = x^2 - 2x + 3$ **b** $f(g(x)) = \sqrt{2x-3}$

c $f(g(x)) = \sqrt[3]{4-3x}$ **d** $f(g(x)) = \dfrac{3}{(x-2)^2}$

6 The number of bacteria b in food refrigerated at a certain temperature t (Celsius) can be modelled by the function $b(t) = 20t^2 - 80t + 500$. When the food is taken out of the refrigerator, the food temperature can be modelled by the function $t(h) = 4h + 2$, where h is hours not refrigerated.

- **a** Find the composite function $(b \circ t)(h)$ and explain what it means in the context of the problem.
- **b** Find how long it takes for the bacteria count to be 10 000.

7 The speed of a car v (km h^{-1}) at any time t hours can be modelled using the function $v(t) = 40 + 3t + t^2$. The rate of fuel consumption r (litres km^{-1}) at a speed of v can be modelled by the function $r(v) = \left(\frac{v}{500} - 0.1\right)^2 + 0.2$. Find the composite function relating the rate of fuel consumption to the time a trip lasts, and find how long a trip lasted if the rate of fuel consumption was 0.2 litres km^{-1}.

Identity function

The **identity function** is a function f that when composed with g leaves g unchanged. In other words, $(f \circ g)(x) = (g \circ f)(x) = g(x)$. The identity function for function composition is $y = x$.

For example, consider $g(x) = 3x - 2$. Then, if $f(x) = x$, $(f \circ g)(x) = 3x - 2 = g(x)$ and $(g \circ f)(x) = 3x - 2 = g(x)$.

Inverse of a function

A function h that, when composed with g, results in the identity function $f(x) = x$ is called an **inverse function**, ie $(g \circ h)(x) = (h \circ g)(x) = f(x) = x$.

The function $g(x) = 3x - 2$ maps x into $3x - 2$. Reversing the process will map y into x. In other words, the domain of g becomes the range of the inverse of g, and range of g becomes the domain of the inverse of g. This reverse process is called finding the inverse of a function, if the inverse exists.

To find the inverse of g, therefore, interchange the x and y, ie, $x = 3y - 2$, and solve for y, $y = \frac{x+2}{3}$. Let $h(x) = \frac{x+2}{3}$, and perform function composition in order to justify your result.

$$g(h(x)) = 3\left(\frac{x+2}{3}\right) - 2 = x \text{ and } h(g(x)) = \frac{(3x-2)+2}{3} = x.$$

Since $(g \circ h)(x) = (h \circ g)(x) = f(x) = x$, h and g are inverses of each other.

The notation for the inverse of a function g is $g^{-1}(x)$.

- The inverse of a function $f: A \to B$ is the relation formed when the domain and range are interchanged.
- If the inverse relation is also a function, then it is the inverse function of f, $f^{-1}: B \to A$, and we say that f is invertible.

Example 27

Find the inverse of $f(x) = x^2 + 3$ and justify your answer algebraically.

$x = y^2 + 3 \Rightarrow y = \sqrt{x-3}$ Let $g(x) = \sqrt{x-3}$ then $f(g(x)) = \left(\sqrt{x-3}\right)^2 + 3 = x$, and $g(f(x)) = \sqrt{x^2+3-3} = x$.	Interchange the x and y, and solve for y. Form the composite functions fg and gf, and assert that they are equal to x.
Since $g(f(x) = f(g(x)) = x$, $f^{-1}(x) = \sqrt{x-3}$.	Use proper inverse function notation.

Example 28

Find the inverse of the function $f(x) = \dfrac{1}{x-2} + 3$, $x \neq 2$, and graphically determine whether the inverse is also a function. State the domain and range of the function and its inverse.

$x = \dfrac{1}{y-2} + 3 \Rightarrow x - 3 = \dfrac{1}{y-2}$	Interchange the x and y.
$\Rightarrow (x-3)(y-2) = 1$ $\Rightarrow xy - 2x - 3y - 6 = 1$ $\Rightarrow xy - 3y = 7 + 2x$ $\Rightarrow y(x-3) = 7 + 2x$ $\Rightarrow y = \dfrac{7+2x}{x-3}, x \neq 3$	Solve for y.
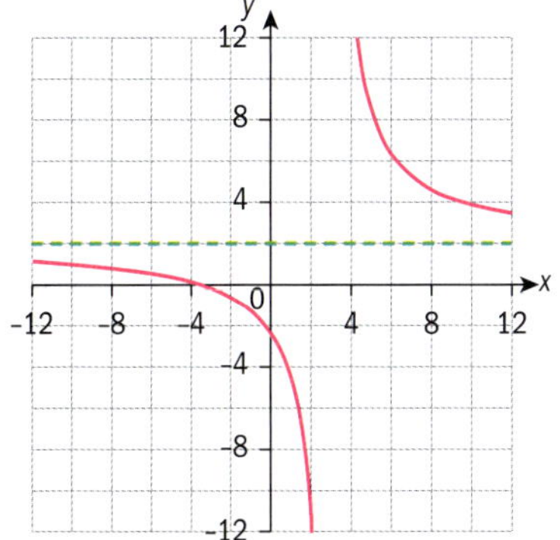	It is a function as its graph passes the vertical line test. Therefore use the notation for inverse function, f^{-1}.
$\Rightarrow f^{-1}(x) = \dfrac{7+2x}{x-3}, x \neq 3$ $D_f = \mathbb{R} - \{2\}$; $R_f = \mathbb{R} - \{3\}$ $D_{f^{-1}} = \mathbb{R} - \{3\}$; $R_{f^{-1}} = \mathbb{R} - \{2\}$	Since $\dfrac{1}{x-2}$ will never equal 0, f will never equal 3. From the graph, $y \neq 2$.

Investigation 18

1 Graph the functions g and h, $g(x) = 3x - 2$ and $h(x) = \frac{x+2}{3}$, that are inverses of each other, on the same set of axes. They are reflections of each other in a certain line. Write the equation of this line.

2 Graph the mutually inverse functions $k(x) = \frac{x}{2} + 5$ and $p(x) = 2(x - 5)$ on the same set of axes, and write the equation of the line about which their graphs have reflective symmetry.

3 Does this graphical test work for all functions that are inverses of each other? To answer this question, find the gradient of the line segment joining (x_1, y_1) and (y_1, x_1) and describe the relationship that the reflection line has to this line segment.

4 **Conceptual** How would you justify graphically that two functions are inverses of each other?

5 Classify the functions in **1** and **2** and their inverses as either one-to-one or many-to-one.

If two functions are inverses of each other, their graphs have reflection symmetry about the line $y = x$.

Investigation 19

The graphs of $f(x) = x^2$ and $y = \pm\sqrt{x}$ are shown. The graphs have reflection symmetry about the line $y = x$, therefore they are mutual inverses.

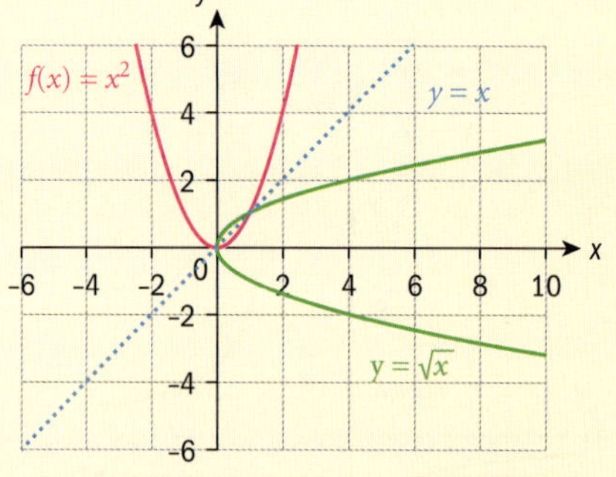

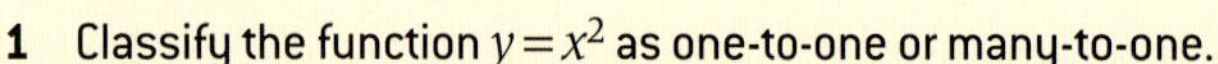

1 Classify the function $y = x^2$ as one-to-one or many-to-one.

2 Is the function an onto function over its natural domain and range?

3 The inverse of the function $y = x^2$ is a relation, but not a function. Explain.

4 How can you restrict the domain of $y = x^2$ in order for its inverse to be a function?

5 Draw the graphs of both the function and the inverse function on this restricted domain. State the domain and range of the inverse function.

6 Choose a different restricted domain, and graph both the function and the inverse function on this restricted domain. State the domain and range of both functions.

7 Which point on the graph of $y = x^2$ allows you to select two different restricted domains of the function such that the inverses are functions?

8 **Conceptual** What kind of functions (one-to-one, many-to-one, onto) have inverses that are also functions? Why?

Only functions that are one-to-one and onto have inverse functions. We say that the functions have a one-to-one correspondence.

Example 29

a Find the inverse relation of $y = x^2 + 1$, and graph both the function and its inverse relation on the same set of axes.

b State two different domain restrictions of the function, and the corresponding ranges, in order that its inverse is a function, and for each, state the domain and range of the inverse function.

c State the two functions, with their restricted domains, and their corresponding inverse functions.

a $x = y^2 + 1 \Rightarrow y = \pm\sqrt{x-1}$

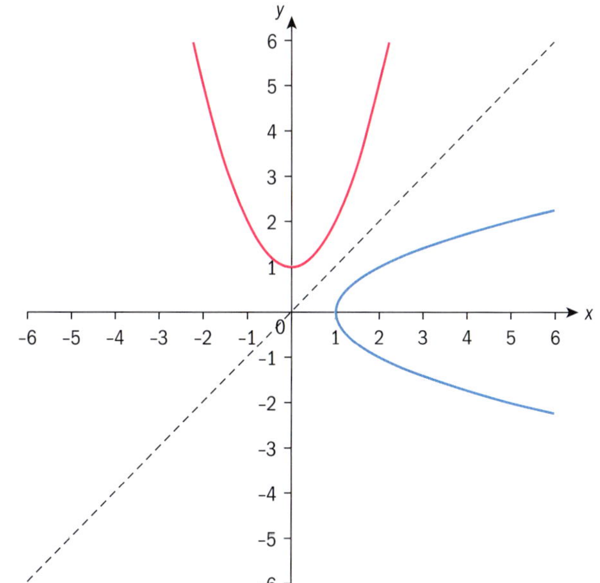

b Function: domain $x \geq 0$; range $y \geq 1$.

Inverse: $x \geq 1$; $y \geq 0$.

or

Function: domain $x \leq 0$; range $y \geq 1$.

Inverse: domain: $x \geq 1$; range: $y \leq 0$.

c $f(x) = x^2 + 1; x \geq 0; f^{-1}(x) = \sqrt{x-1}$

$f(x) = x^2 + 1; x \leq 0; f^{-1}(x) = -\sqrt{x-1}$

The function $f(x) = \dfrac{1}{x}$ has special properties:

- It is an odd function.
- The axes are the asymptotes of its graph.

- It has no x- or y-intercepts.
- It is its own inverse.

To prove the last point algebraically, swap the x and y so that $x=\frac{1}{y}$. Then solving for y, $y=\frac{1}{x}$, hence $f^{-1}(x)=\frac{1}{x}$, and $(f\circ f^{-1})(x)=x=(f^{-1}\circ f)(x)$.

A function f is **self-inverse** if $f^{-1}(x)=f(x)$.

Investigation 20

You have seen that $f(x)=\frac{1}{x}$ is self-inverse.

1 Determine whether functions of the form $y=\frac{k}{x}, x\neq 0, k\in\mathbb{R}$ are self-inverse.

2 Explore functions of the form $y=k-x; k\in\mathbb{R}$, and confirm algebraically and graphically whether they are self-inverse.

3 Determine whether functions of the form $y=\frac{x-k}{x-1}, x\neq 1, k\in\mathbb{R}$ are self-inverse for all k.

4 Can you find other functions that are self-inverses?

Exercise 2N

1 State whether or not the inverse exists, and find the inverse when it does exist.

a $f(x)=\{(4,2),(0,2),(-2,2),(2,2)\}$

b $g(x)=\{(1,3),(-6,2),(-3,-4),(0,0),(-5,-5),(-2,-3)\}$

c $h(x)=\{(-1,-1),(-3,3),(-2,-5),(-4,-4),(1,1),(-5,3),(0,-2)\}$

2 Find the inverse of each function, if it exists. Justify your answer algebraically, and confirm graphically.

a $y=5x-1$ **b** $y=\frac{x-2}{3}$

c $y=x^2-3; x\geq 0$ **d** $y=\frac{2}{x-3}; x\neq 3$

e $y=x^3+1$ **f** $y=\frac{x+1}{x-1}; x\neq 1$

3 Find the inverse of each function. If the inverse is not a function, restrict the domain of the original function so that its inverse is also a function. State the (restricted) domain, and the range of the original function, and the domain and range of the inverse function.

a $y=(x-2)^2$ **b** $y=\frac{2x+1}{x+1}, x\neq -1$

c $y=4x^2+1$

4 Given that $f(x)=(x-1)^2$ and $g(x)=2x+1$, $x\geq 0$, show that $(f^{-1}\circ g^{-1})(x)=(g\circ f)^{-1}$.

5 Justify algebraically that these pairs of functions are inverses of each other.

a $f(x)=-4x+4; g(x)=1-\frac{1}{4}x$

b $f(x)=\frac{x-5}{x-3}, x\neq 3; g(x)=\frac{-2}{x-1}+3, x\neq 1$

c $f(x)=\frac{(2x+3)^3}{2}; g(x)=\frac{\sqrt[3]{2x}-3}{2}$

2.5 Function transformations

Investigation 21

1 a On the same set of axes, graph:

i $y = 2x - 1$ ii $y = |2x - 1|$.

b State the domain and range of both functions.

2 Repeat **1** for the following pairs of functions:

i $y = x^2 - 3$ and $y = |x^2 - 3|$

ii $y = x^3 - x$ and $y = |x^3 - x|$

iii $y = \frac{1}{x-1}$ and $y = \left|\frac{1}{x-1}\right|$; $x \neq 1$.

3 **Conceptual** How does the graph of $y = |f(x)|$ change compared to the graph of $y = f(x)$ over the range of $y = f(x)$?

4 a On the same set of axes, graph:

i $y = 2x - 1$ ii $y = 2|x| - 1$.

b State the domain and range of both functions.

5 Repeat **4** for the following pairs of functions:

i $y = x^2 - 3$ and $y = |x|^2 - 3$

ii $y = x^3 - x$ and $y = |x|^3 - |x|$

iii $y = \frac{1}{x-1}$; $x \neq 1$ and $y = \frac{1}{|x|-1}$; $x \neq 1$.

6 **Conceptual** How does the graph of $y = f(|x|)$ change compared to the graph of $y = f(x)$ over the domain of $y = f(x)$?

TOK

When you see a familiar equation with a transformation, you will often get a "gut feeling" about what the function looks like.

Respond to this question.

Is intuition helpful or harmful in mathematics?

Functions

To transform $y = f(x)$ to $y = |f(x)|$, the graph is unchanged for $y \geq 0$, and reflected in the x-axis for $y \leq 0$.

To transform $y = f(x)$ to $y = f(|x|)$, the graph is unchanged where $x \geq 0$. Where $x < 0$, the part of the graph for $x \geq 0$ is reflected in the y-axis.

Example 30

From the graph of $y = x^3 - 4x$ sketch the graphs of:

a $y = |x^3 - 4x|$ **b** $|x|^3 - 4|x|$

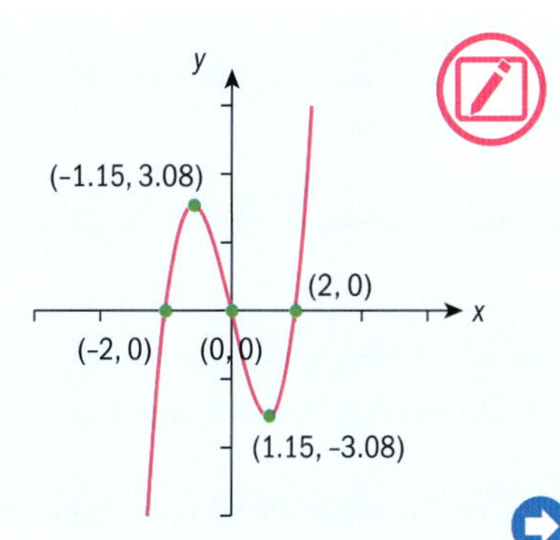

Continued on next page

a

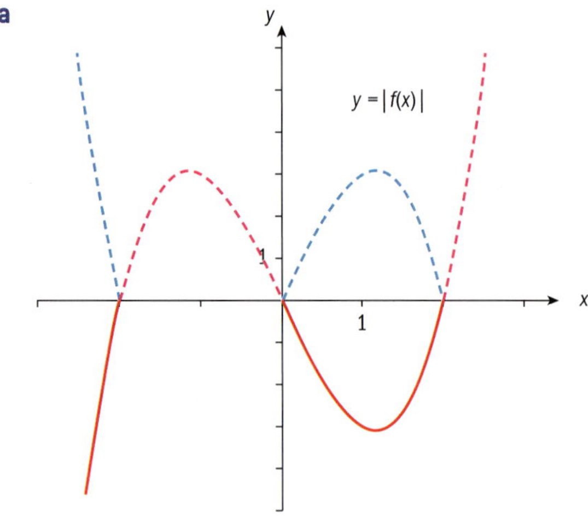

The dotted line is the graph of $y = |x^3 - 4x|$. Wherever $f(x) < 0$, reflect those parts of the graph in the x-axis.

b

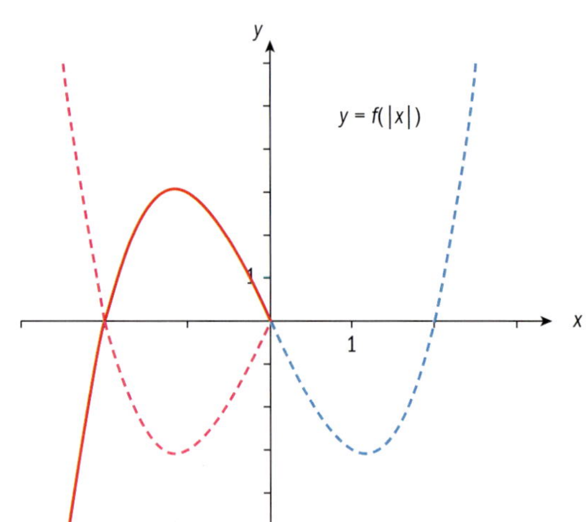

The dotted line is the graph of $y = |x|^3 - 4|x|$. Wherever $x \geq 0$ the graph of $f(x)$ is unchanged. Wherever $x < 0$, the graph of $f(x)$ is reflected in the y-axis.

Exercise 20

For each function $y = f(x)$, draw **a** $y = |f(x)|$ **b** $y = f(|x|)$. Draw each part on the same set of axes as $y = f(x)$.

1 $y = 2 - 5x$

2 $y = x^2 - 4x$

3 $y = 1 - x + 3x^2$

4 $y = 2 - 3x - x^3$

5 $y = \dfrac{2}{3x-1}$

6 $y = 4 - \sqrt{3x+1}$

Investigation 22

1 Graph $f(x) = x^2 - 4$, and state its domain and range.

2 On the same set of axes, graph its reciprocal, $g(x) = \dfrac{1}{x^2-4}$, and state its domain and range.

3 State the relationship between the x-intercepts of f and the graph of g.

4 State the relationship between the y-intercept of f and the y-intercept of g.

5 State the relationship between the minimum of f and the local maximum point of g.

6 Fill in the blanks:

a Where $f(x) > 0$, $\frac{1}{f(x)}$ _______ 0, and where $f(x) < 0$, $\frac{1}{f(x)}$ _______ 0.

b When $f(x)$ approaches $\pm\infty$, $\frac{1}{f(x)}$ approaches _________. When $f(x)$ approaches 0, $\frac{1}{f(x)}$ approaches _________.

7 Summarize how you would transform the graph of any function $y = f(x)$ to its reciprocal function $y = \frac{1}{f(x)}$.

To graph the reciprocal of $y = f(x)$:

- Where they exist, the zeros of $y = f(x)$ are the vertical asymptotes of $y = \frac{1}{f(x)}$.
- If $y = b$ is the y-intercept of $y = f(x)$, then $y = \frac{1}{b}$ is the y-intercept of $y = \frac{1}{f(x)}$.
- The minimum value of $y = f(x)$ occurs at the same value of x as the maximum of $y = \frac{1}{f(x)}$ and vice versa.
- When $f(x) > 0$, $\frac{1}{f(x)} > 0$; when $f(x) < 0$, $\frac{1}{f(x)} < 0$.
- When $y = f(x)$ approaches 0, $\frac{1}{f(x)}$ will approach $\pm\infty$, and vice versa.

Example 31

Draw the graph of $y = x(x-4)$. On the same set of axes, sketch the graph of its reciprocal, $y = \frac{1}{x(x-4)}$. For both graphs, label any intercepts, zeros, extrema and asymptotes.

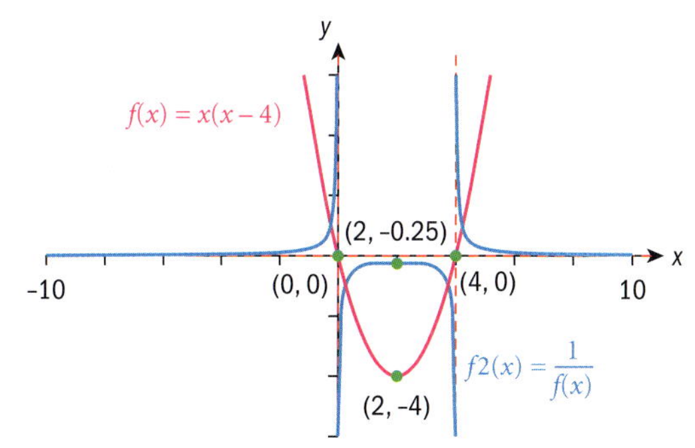

Zeros of f: $(0, 0)$ and $(4, 0)$, therefore $\frac{1}{f}$ has vertical asymptotes at $x = 0$ and $x = 4$.

y-intercept of f is 0, therefore at $y = 0$, $\frac{1}{f}$ has a horizontal asymptote. Minimum of f is $(2, -4)$, therefore maximum of $\frac{1}{f}$ is $\left(2, -\frac{1}{4}\right)$.

Exercise 2P

1 For $y = f(x)$ sketch $y = \dfrac{1}{f(x)}$ on the same set of axes. Make sure to label all intercepts, asymptotes and any maximum/minimum points.

a $y = 3x - 2$ **b** $y = 1 - 4x$

c $y = \dfrac{1}{2x - 3}$ **d** $y = x(2 - x)$

e $y = \dfrac{1}{x^2 - 4}$

2 Copy the graph of $y = f(x)$ and sketch the graph of $y = \dfrac{1}{f(x)}$ on the same set of axes.

a

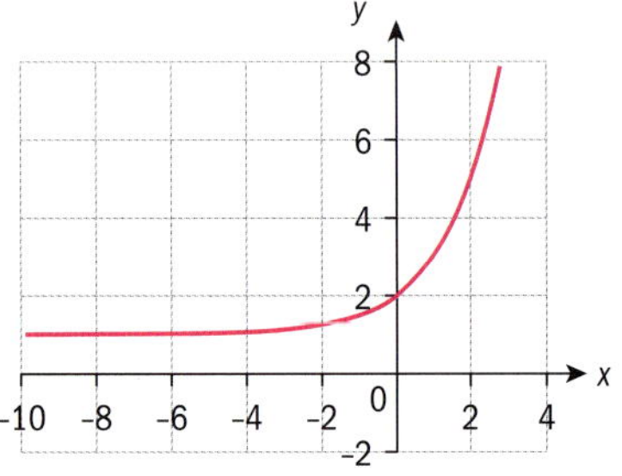

b

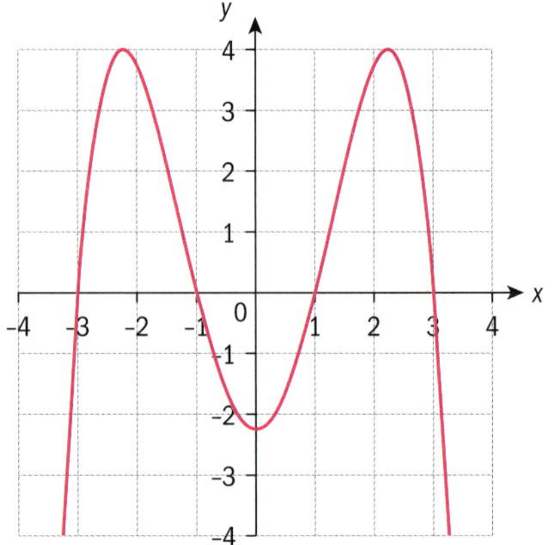

Investigation 23

1 Draw the graph of the function $f(x) = 2x - 4$ and label all intercepts.

2 State $f(-x)$, graph this function on the same set of axes and label all intercepts.

3 What kind of symmetry do the graphs of $f(x)$ and $f(-x)$ have?

4 State $-f(x)$ and graph both f and $-f$ on the same set of axes.

5 What kind of symmetry do the graphs of f and $-f$ have?

The graphs of $y = f(x)$ and $y = f(-x)$ are reflections of each other in the y-axis.

The graphs of $y = f(x)$ and $y = -f(x)$ are reflections of each other in the x-axis.

Example 32

Sketch the graph of $f(x) = (x + 3)(x - 1)$. Clearly label any intercepts and extrema. On the same set of axes, reflect f in the **a** y-axis **b** x-axis and write the functions for each reflection.

a

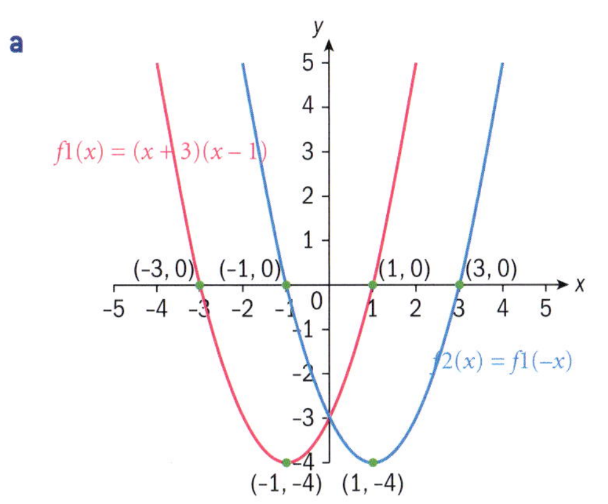

Reflect f in the y-axis.

Label the intercepts and the minimum.

$f(-x) = (-x + 3)(-x - 1) = x^2 - 2x - 3$

Substitute $-x$ for x in the function.

b

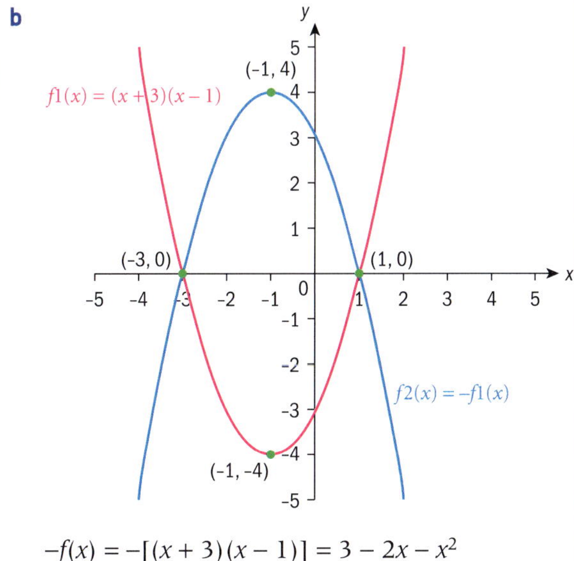

Reflect f in the x-axis. Zeros remain the same as f, and the maximum is reflected in the x-axis.

$-f(x) = -[(x + 3)(x - 1)] = 3 - 2x - x^2$

Change the sign of each term in the function.

Translations

You have worked through investigations considering the effects of the parameters h and k when transforming the graph of $y = x^2$ into the graph of $y = (x - h)^2 + k$, and the graph of $y = |x|$ into the graph $y = |x - h| + k$. The following summary highlights the role of these parameters in the transformation of functions.

- If $h > 0$ then the graph of $y = f(x - h)$ is the translation of the graph of $y = f(x)$, h units in the positive x-direction.
- If $h < 0$ then the graph of $y = f(x - h)$ is the translation of the graph of $y = f(x)$, h units in the negative x-direction.
- If $k > 0$ then the graph of $y = f(x) + k$ is the translation of the graph of $y = f(x)$, k units in the positive y-direction.
- If $k < 0$ then the graph of $y = f(x) + k$ is the translation of the graph of $y = f(x)$, k units in the negative y-direction.

These horizontal and vertical translations are denoted with a column vector $\begin{pmatrix} h \\ k \end{pmatrix}$.

TOK

Is mathematics independent of culture?

Example 33

Describe the transformations necessary to obtain the graph of $y = (x + 3)^2 + 1$ from the graph of $y = x^2$. State the coordinates of the image of the vertex under this translation. Sketch both graphs on the same set of axes.

The graph of $y = x^2$ is translated 3 units in the negative x-direction and 1 unit in the positive y-direction, represented by $\begin{pmatrix} -3 \\ 1 \end{pmatrix}$.

$(x + 3)^2 = (x - (-3))^2$, hence $h = -3$.

The vertex of the transformation is therefore $(-3, 1)$.

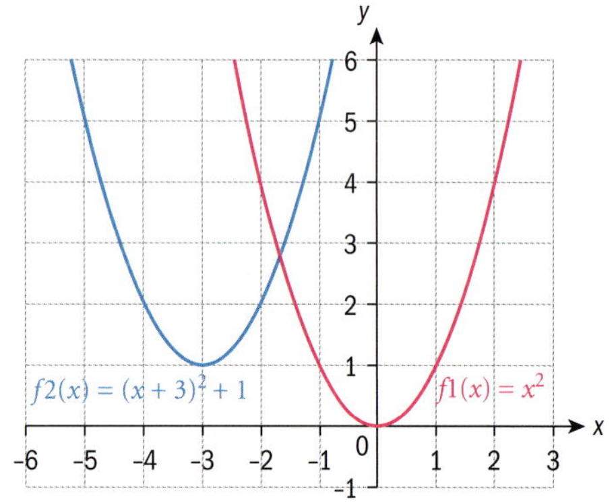

Investigation 24

1. Graph $y = ax^2$ for different values of a, $a > 0$.
2. Describe the effect of the parameter a when **i** $a > 1$ **ii** $0 < a < 1$ on the graph of $y = ax^2$.
3. Graph $y = (ax)^2$ for different positive values of a.
4. Describe the effect of the parameter a when **i** $a > 1$ **ii** $0 < a < 1$ on the graph of $y = (ax)^2$.
5. What is the difference between the transformations in **1** and **3**?

Vertical dilation

For $y = af(x)$, $a > 0$, the graph of $y = f(x)$ is vertically stretched or compressed by a factor of a, parallel to the y-axis.

- If $a > 1$ the graph of $y = f(x)$ is stretched vertically by a factor of a, ie it moves further from the x-axis.
- If $0 < a < 1$ the graph of $y = f(x)$ is compressed vertically by a factor of a, ie it moves closer to the x-axis.

Horizontal dilation

For $y = f(ax)$, $a > 0$, the graph of $y = f(x)$ is horizontally stretched or compressed by a factor of $\frac{1}{a}$, parallel to the x-axis.

- If $a > 1$ the graph of $y = f(x)$ is compressed by a factor of $\frac{1}{a}$, ie the graph moves closer to the y-axis.
- If $0 < a < 1$ the graph of $y = f(x)$ is stretched by a factor of $\frac{1}{a}$, ie the graph moves further from the y-axis.

Did you know?

Dilations are referred to as non-rigid transformations, as they change the shape of the graph of the function. Translations, for example, are rigid transformations as they do not change the shape of the function.

In general, to transform a function $y = f(x)$ to $y = af(b(x + c)) + d$,
a is the vertical dilation
b is the horizontal dilation
c is the horizontal translation
d is the vertical translation.

Example 34

Graph the piecewise function $y = \begin{cases} -(2x+8), & -4 \le x < -3 \\ 2x+4, & -3 < x \le -1 \\ -2x, & -1 \le x < 0 \\ x, & 0 \le x \le 4 \end{cases}$. Clearly label the zeros, intercepts and any maximum or minimum points with their coordinates. Graph the transformations and indicate the coordinates of all the labelled points of the original graph.

a $y = 2f(x)$ **b** $y = f(2x)$

a

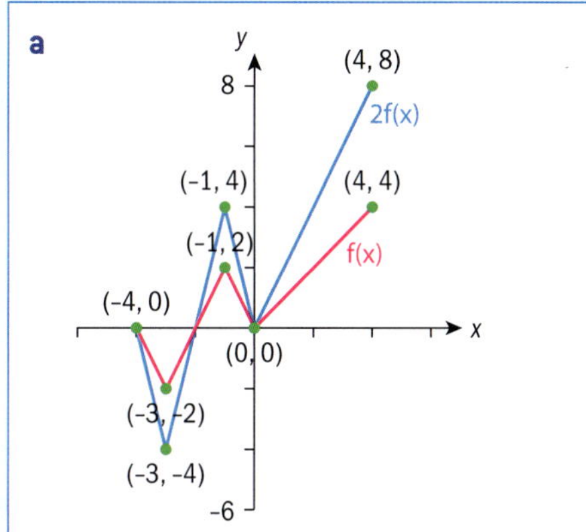

$y = 2f(x)$ stretches the graph of $y = f(x)$ vertically by a factor of 2, hence all y-coordinates are multiplied by 2, and all x-coordinates are unchanged.

Continued on next page

b

$y = f(2x)$ compresses the graph horizontally by a factor of $\frac{1}{2}$, hence each x-coordinate is $\frac{1}{2}$ the x-coordinate of $y = f(x)$, and the y-coordinates are unchanged.

Example 35

Consider the graph of the function $y = f(x)$, where $-2 \leq x \leq 6$.

Sketch the graph of:

a $y = 2f(x)$ **b** $y = f(-x)$

c $y = f(2x)$ **d** $y = f(x - 1) - 3$

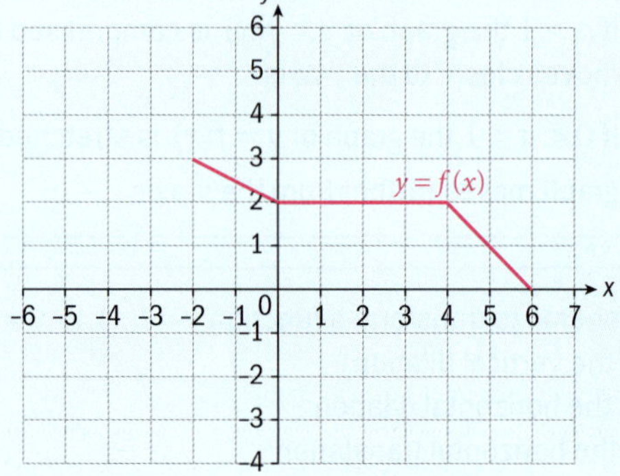

a

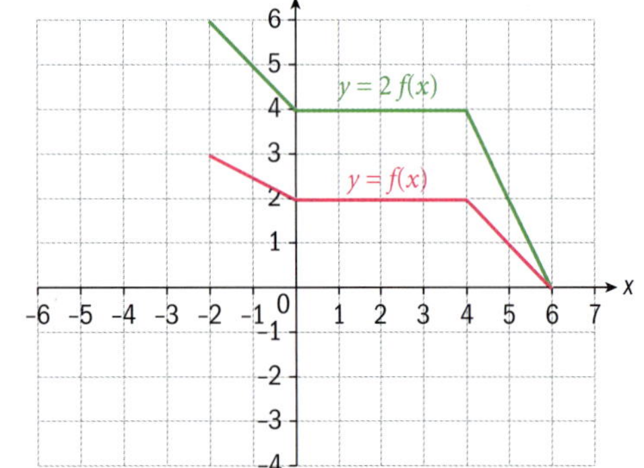

$y = 2f(x)$

$a = 2 \Rightarrow$ vertical stretch with scale factor 2.

Each point on the graph of $y = 2f(x)$ is twice the distance to the x-axis as the corresponding point on the graph of $y = f(x)$.

b

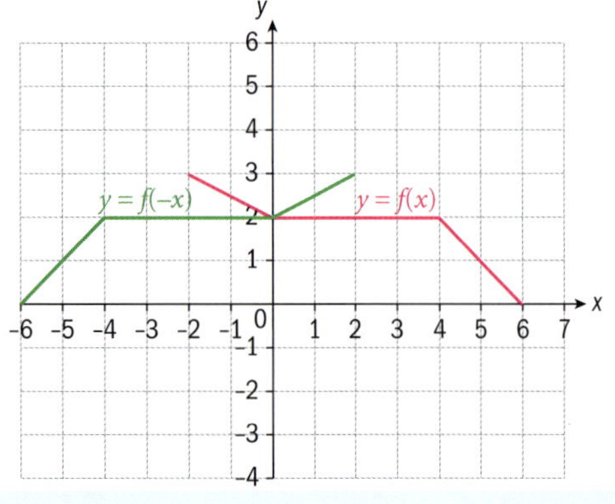

$y = f(-x)$

Reflection of $y = f(x)$ in the y-axis.

c

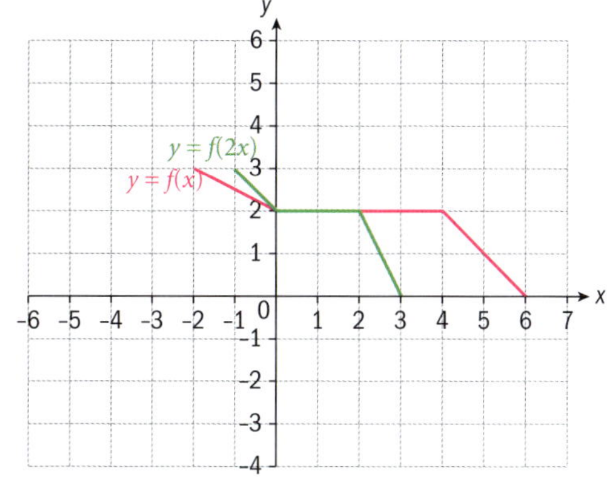

$y = f(2x)$

$a = 2 \Rightarrow$ horizontal compression with scale factor $\frac{1}{2}$.

Each point on the graph of $y = f(2x)$ is half the distance to the y-axis as the corresponding point on the graph of $y = f(x)$.

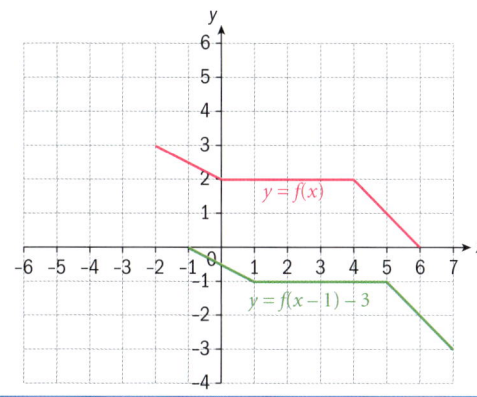

d

y = f(x)

y = f(x − 1) − 3

$y = f(x - 1) - 3$

$h = 1 \Rightarrow$ horizontal translation right 1 unit

$h = -3 \Rightarrow$ and vertical translation down 3 units.

Example 36

Functions g, r and s are transformations of the graph of f. Write the functions g, r and s in terms of f.

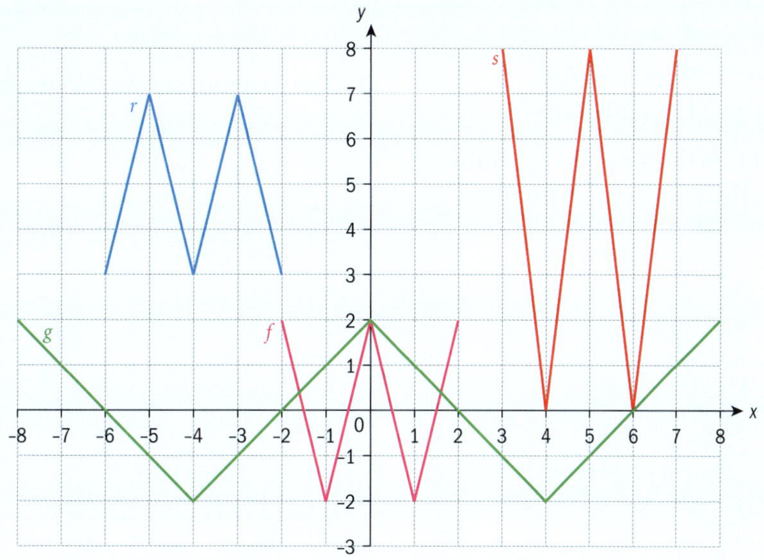

Continued on next page

$g(x) = f\left(\frac{1}{4}x\right)$	The graph of g is a horizontal stretch of the graph of f with scale factor 4. $\frac{1}{q} = 4 \Rightarrow q = \frac{1}{4}$
$r(x) = -f(x + 4) + 5$	The graph of r can be obtained by reflecting the graph of f in the x-axis and then translating $y = -f(x)$ left 4 units and up 5 units. $h = -4$ and $k = 5$
$s(x) = 2f(x - 5) + 4$	The graph of g is a vertical stretch of the graph of f with scale factor 2 and a translation right 5 units and up 4 units. $a = 2 \quad h = 5 \quad k = 4$

Exercise 2Q

1 The graph of $y = f(x)$, where $-3 \le x \le 6$, is shown. Copy the graph of f and draw these functions on the same axes.

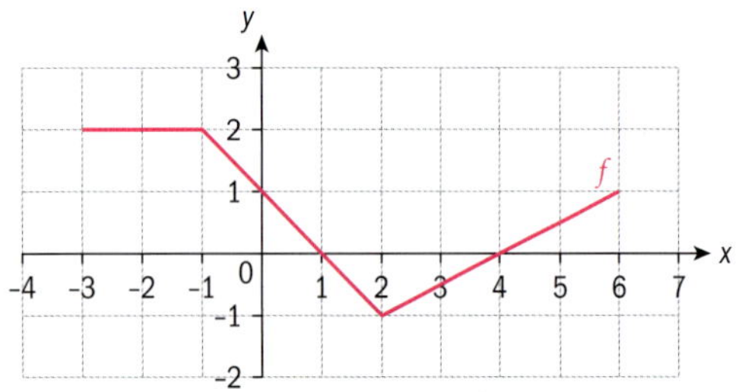

a $g(x) = f(-x)$ **b** $g(x) = -f(x)$

c $g(x) = f(2x)$ **d** $g(x) = 3f(x)$

e $g(x) = f(x + 6)$ **f** $g(x) = f(x) - 3$

2 The graphs of functions r and s are transformations of the graph of f. Write the functions r and s in terms of f.

a

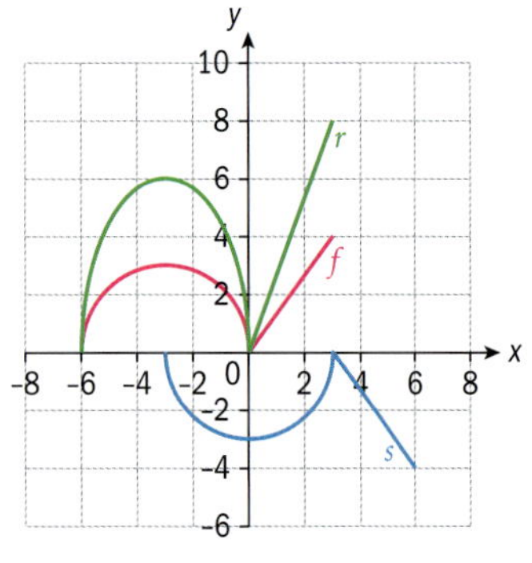

b

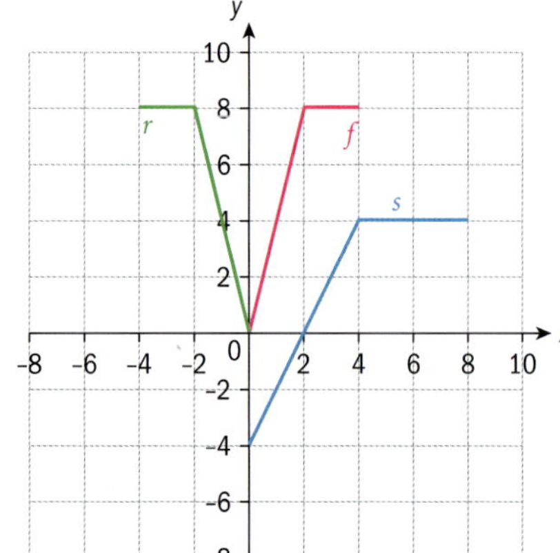

3 The diagram shows the graph of $y = f(x)$, for $2 \le x \le 8$.

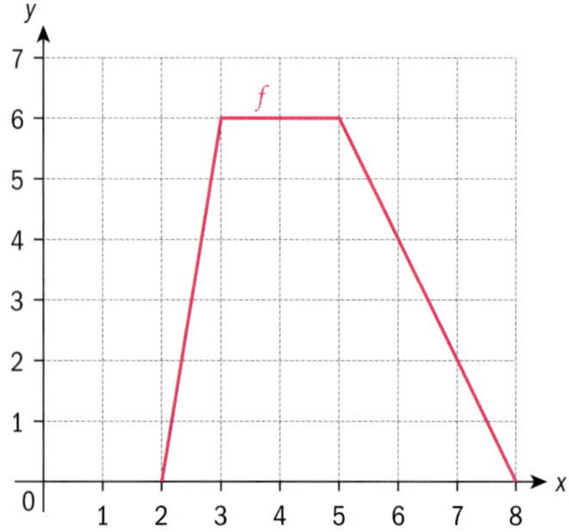

a Write down the range of f.

Let $g(x) = f(-x)$.

b Sketch the graph of g.

c Write down the domain of g.

The graph of h can be obtained by a vertical translation of the graph of g. The range of h is $-4 \leq y \leq 2$.

d Write the equation for h in terms of g.

e Write the equation for h in terms of f.

4 Describe each transformation from $f(x)$ (red) to $g(x)$ (green) in terms of x.

a

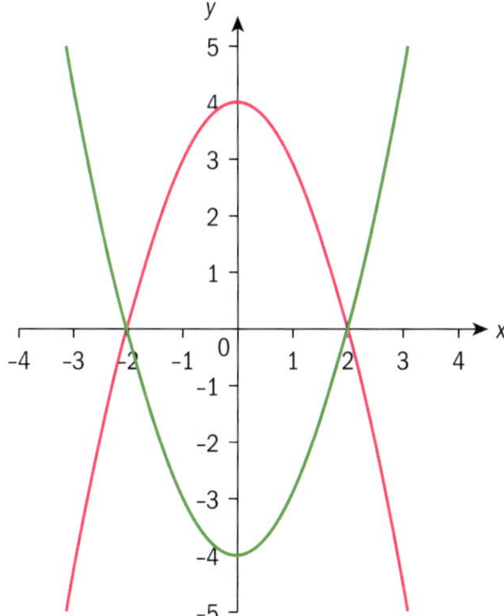

b

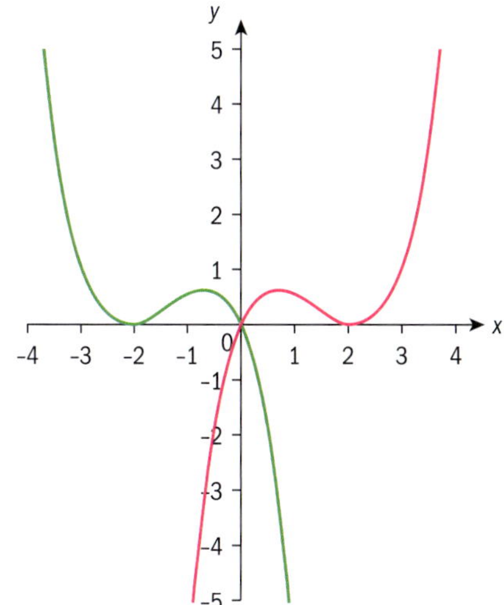

c

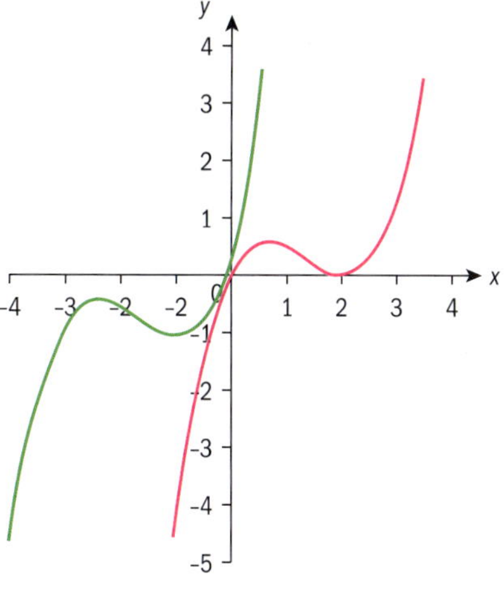

d

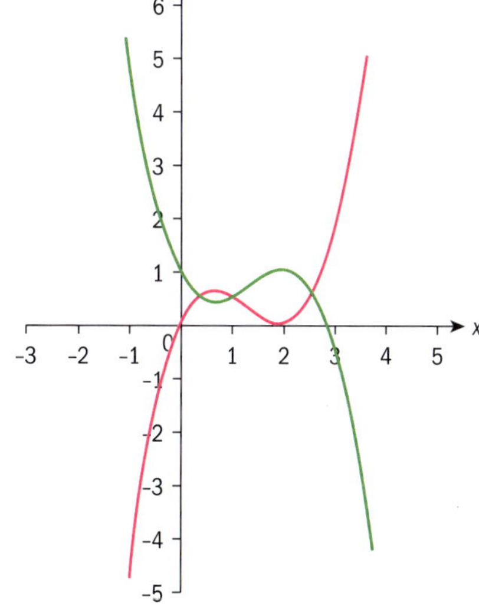

e

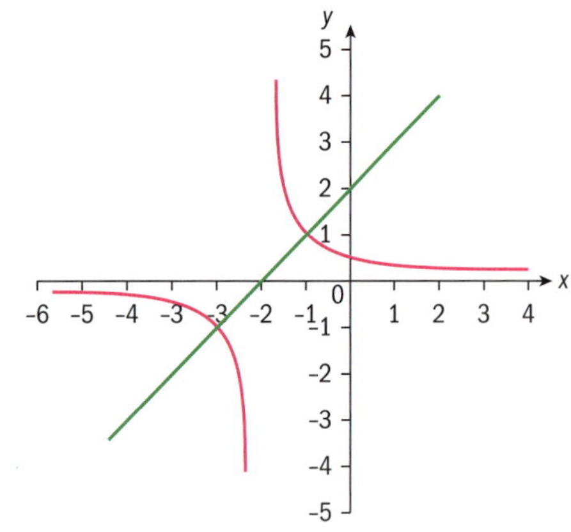

f

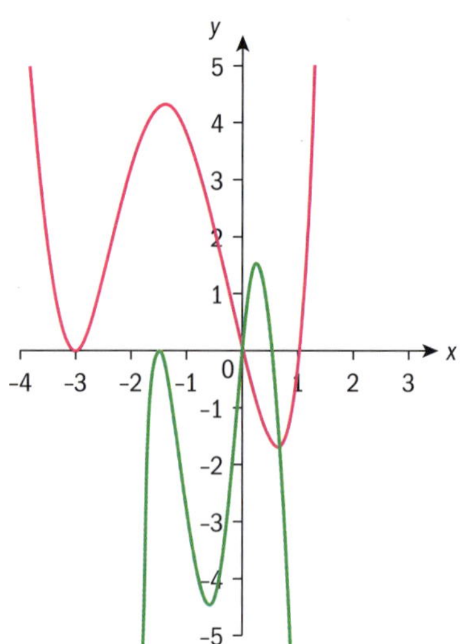

Investigation 25

You can obtain a rational function by performing certain transformations on the reciprocal function $f(x)=\frac{1}{x}$. At each stage of the transformation, graph the new function.

1. Write out the new function g that would represent the transformation on the argument x, that is, $f(x)=f(3x)$, and describe its effect on the graph of f.
2. Now, write out the function h that would represent a vertical stretch factor of 2 on g.
3. Write out the new function k that would represent a horizontal translation of 2 units in the positive x-direction.
4. Now, write out the new function p that would represent a vertical translation of k of 1 unit in the negative direction.
5. Change p into the rational function form $y=\frac{ax+b}{cx+d}; x \neq -\frac{d}{c}$.
6. In order to arrive at the transformed function p, you performed horizontal and vertical dilations, and then horizontal and vertical translations. Change the order in which the transformations are performed. Do you always arrive at the function k?
7. Which transformations are interchangeable?
8. Suppose that the first transformation you want to perform on f is the horizontal translation. What would $g(x)$ be?
9. What would then $h(x)$ be in order to arrive at a denominator of $3(x-2)$?
10. **Conceptual** How is the order in which transformations on a function are performed important to the outcome?
11. Suggest an order for performing transformations when both dilations and translations are involved.

The rational function $y = \frac{ax+b}{cx+d}; x \neq -\frac{d}{c}$ is the result of multiple transformations on the graph of the reciprocal function $y = \frac{1}{x}$. The rational function can also be written in the form $y = \frac{A}{B(x-h)} + k, x \neq h$, where A is the vertical dilation factor, B is the reciprocal of the horizontal dilation factor, h is the horizontal translation and k is the vertical translation. The domain is $\{x \mid x \in \mathbb{R}, x \neq h\}$ and the range is $\{y \mid y \in \mathbb{R}, x \neq k\}$.

Example 37

Graph at each stage the rational function when $y = \frac{1}{x}$ is transformed by a horizontal stretch factor of $\frac{1}{4}$, followed by a vertical stretch factor of 3, then a translation of $\begin{pmatrix} -2 \\ 3 \end{pmatrix}$. State the transformed function, and its domain and range.

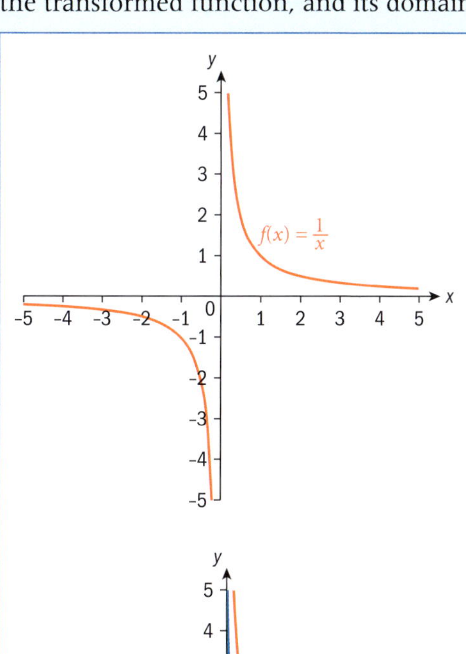

Sketch the graph of $f(x) = \frac{1}{x}$

$g(x) = \frac{1}{4x}$

$f(x) = \frac{1}{x}$

$g(x) = f(4x) = \frac{1}{4x}$, since a horizontal dilation factor a is $f\left(\frac{1}{a}\right)$.

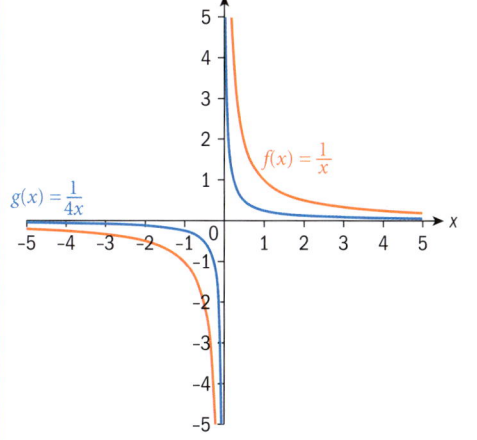

Continued on next page

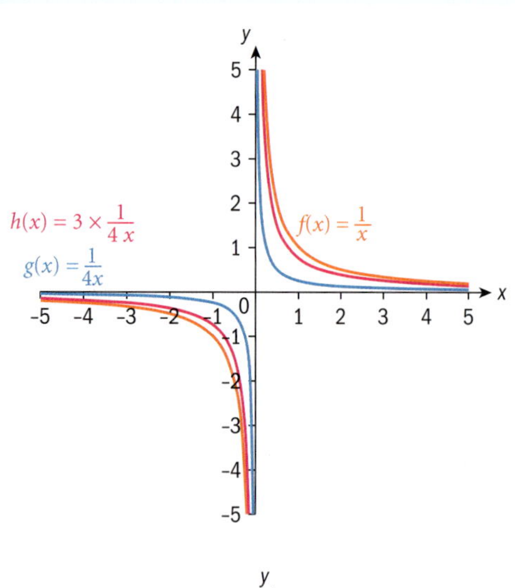

$h(x) = 3g(x) = \frac{3}{4x}$, since a vertical dilation factor a is $af(x)$.

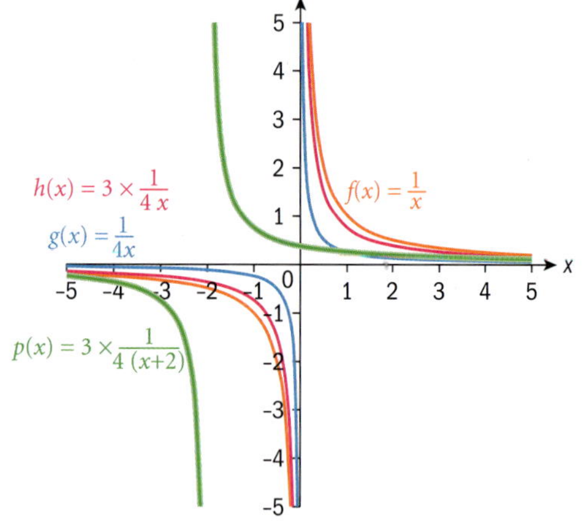

$p(x) = h(x+2) = \frac{3}{4(x+2)}$, since a horizontal translation of -2 units is $f(x+2)$.

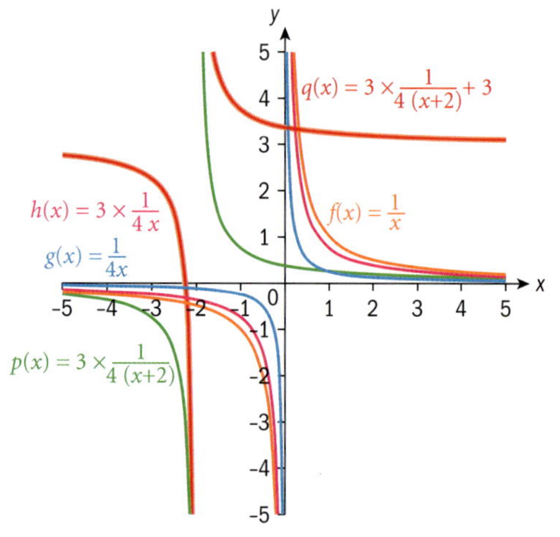

$q(x) = p(x) + 3$, since a vertical translation of $+3$ is $f(x) + 3$.

$y = \frac{3}{4(x+2)} + 3$; D: $x \neq -2$; R: $y \neq 3$

Example 38

Consider the function $y = \frac{-2}{5x+10} + 1$.

a State the domain and range of the function.

b State the transformations, in an appropriate order, that are performed on the graph of $f(x) = \frac{1}{x}$ to obtain the graph of the given function, and describe the transformation.

Graph the transformation at each stage, and check your final graph against the graph of the given function.

a D: $x \neq -2$, R: $y \neq 1$	The asymptotes are $x = -2$ and $y = 0$. Therefore, these must be excluded from the domain and range.
b $g(x) = f(5x) = \frac{1}{5x}$; g represents a horizontal stretch of factor $\frac{1}{5}$.	Make a sketch of $y = \frac{1}{x}$. Start with the argument and work your way outward. Perform the horizontal dilation.
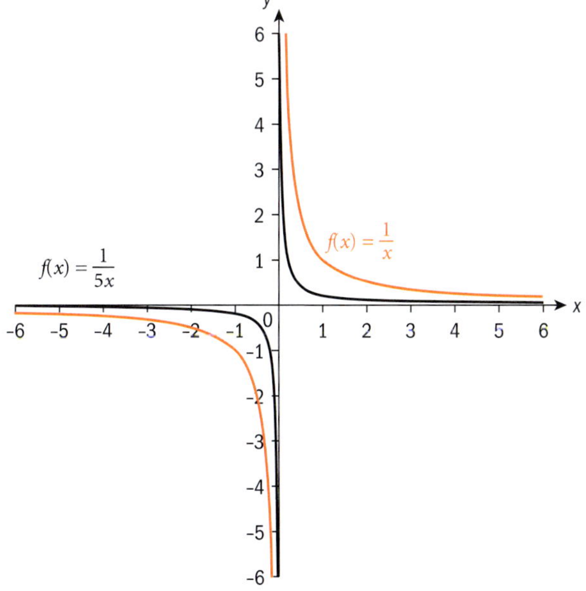	
$5x + 10 = 5(x + 2) \Rightarrow h(x) = g(x + 2)$; h represents a horizontal translation of 2 units in the negative x-direction.	Perform the horizontal translation.

Continued on next page

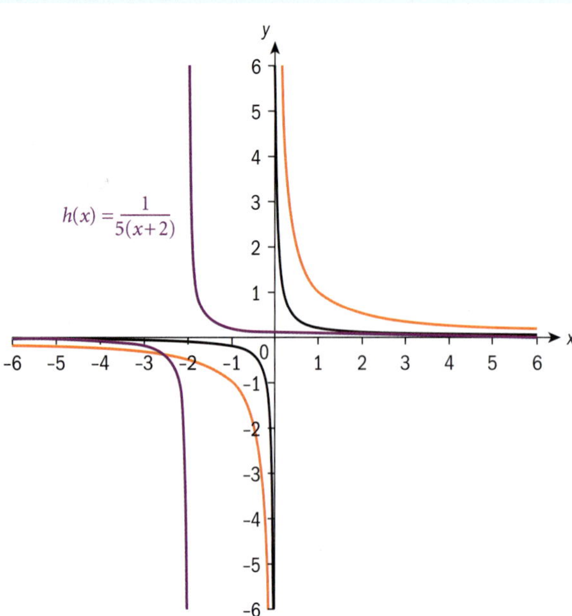

$p(x) = 2h(x) \Rightarrow p(x) = \dfrac{2}{5(x+2)}$. k is a vertical dilation stretch factor 2.

Perform the vertical dilation.

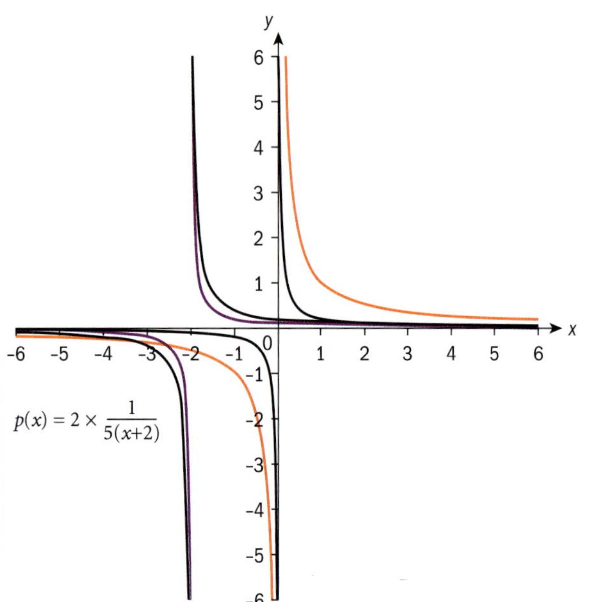

$q(x) = -p(x) \Rightarrow q(x) = -\dfrac{2}{5(x+2)}$	Reflect k in the x-axis.
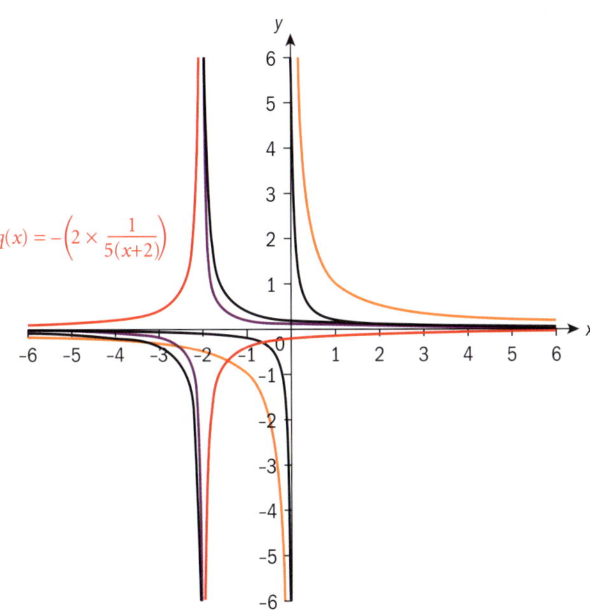	
$r(x) = q(x) + 1 \qquad r(x) = -\dfrac{2}{5(x+2)} + 1$	Perform the vertical translation.
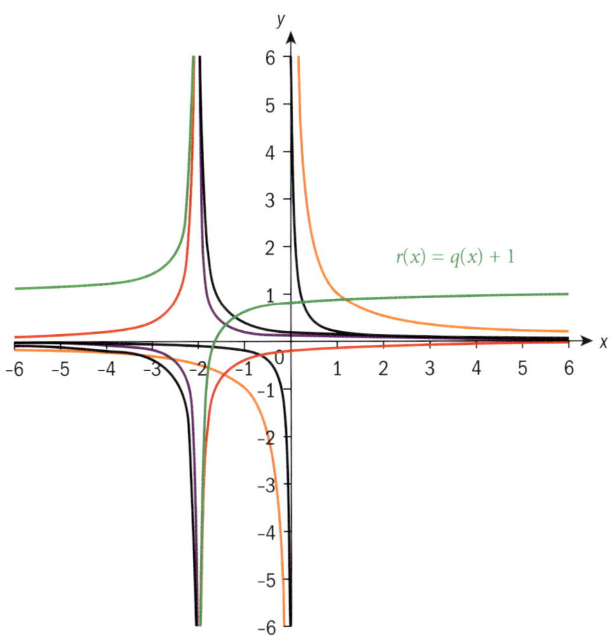	

Example 39

Explain the transformations on the graph of $f(x)=\frac{1}{x}$ to obtain the graph of the function $y=\frac{x+2}{x-3}$.

$\frac{x+2}{x-3}=\frac{(x-3)+5}{(x-3)}=1+\frac{5}{x-3}$	Rewrite in the form $y=\frac{A}{B(x-h)}+k$ by rewriting the numerator to contain the denominator.
Shift the graph of $f(x)=\frac{1}{x}$ horizontally 3 units to the right. Stretch vertically by a factor of 5. Shift vertically 1 unit in the positive y-direction.	h is the horizontal shift, A is the vertical dilation, B is the horizontal dilation, and k is the vertical translation.

HINT

You have seen in the examples that if you want to transform a function $y=f(x)$ to $y=f(ax+b)$, factor out a to get $f\left(a\left(x+\frac{b}{a}\right)\right)$, dilate horizontally by a factor of $\frac{1}{a}$, and translate horizontally $\frac{b}{a}$ units.

TOK

Do you think that mathematics is just the manipulation of symbols under a set of rules?

Exercise 2R

1 Find the rational function when $y=\frac{1}{x}$ is transformed by a vertical stretch of 2, then stretched horizontally by a factor of $\frac{1}{3}$, followed by a translation of $\begin{pmatrix}-2\\3\end{pmatrix}$. Find the domain and range of the new function.

2 Consider the function $y=\frac{2}{3(x-4)}+1$.

a Name the transformations in an appropriate order that are performed on the graph of $y=\frac{1}{x}$ to obtain the given function.

b Write the transformations all together as a function in x.

3 Explain the transformations on the graph of $f(x)=\frac{1}{x}$ to obtain the graphs of the following functions.

a $y=\frac{x-3}{x+5}$ **b** $y=\frac{4x+5}{2x+1}$

c $y=\frac{2x+4}{x+1}$

You will now follow the order of operations for transforming functions using other functions you have learned in this chapter.

Example 40

Explain each transformation necessary in order to obtain the graph of $y = -2|2x + 1| - 2$ from the graph of $f(x) = |x|$, and draw the graph at each stage.

$g(x) = f(2x)$ is a horizontal compression of factor $\frac{1}{2}$.

The graph moves closer to the y-axis.

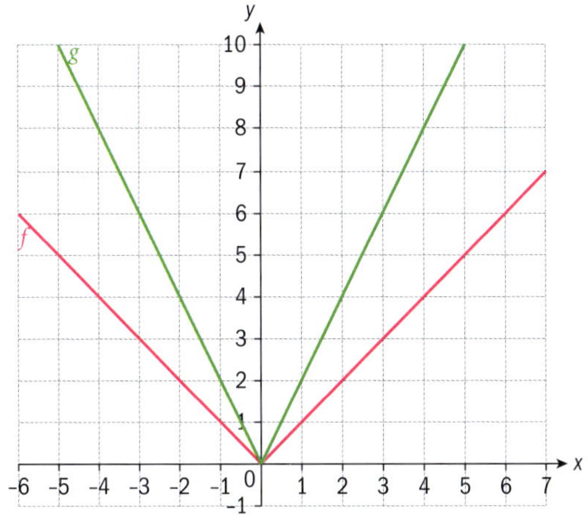

$h(x) = g\left(x + \frac{1}{2}\right)$ is a horizontal shift of the graph of g to the left 0.5 units.

Horizontal translation.

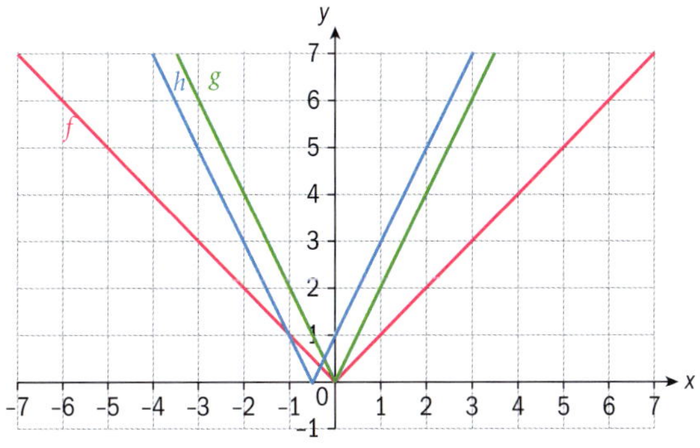

$p(x) = 2h(x)$

Vertical dilation.

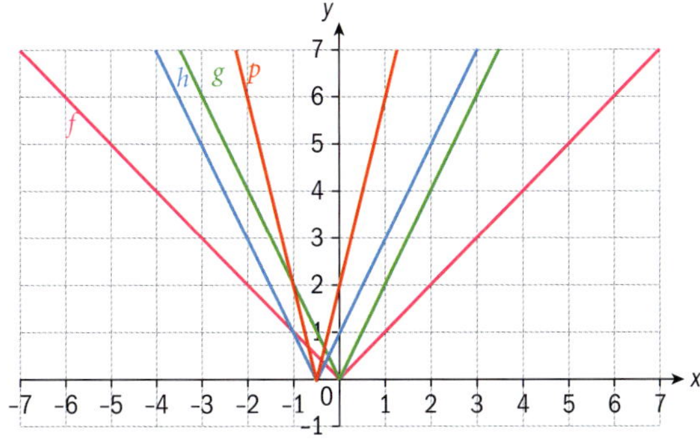

Continued on next page

$q(x) = -p(x)$, a reflection in the y-axis. 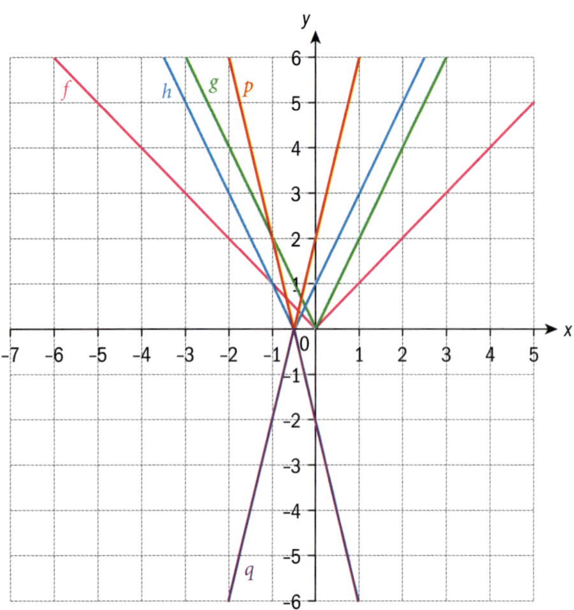	Reflection in the y-axis.
$r(x) = q(x) - 2$, vertical shift of two units in the negative y-direction. 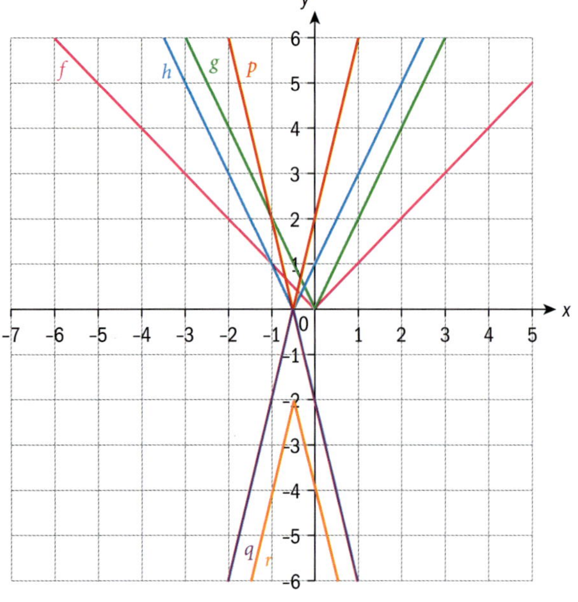	Vertical translation.

Reflect How do the transformations change if you do the horizontal shift before the horizontal dilation?

Developing your toolkit

Now do the Modelling and investigation activity on pages 144–145.

Example 41

Explain each transformation necessary in the correct order to obtain the graph of $y = -\sqrt{2-x}+3$ from the graph of $y = \sqrt{x}$, and show graphically each stage. Confirm graphically that your final graph is correct.

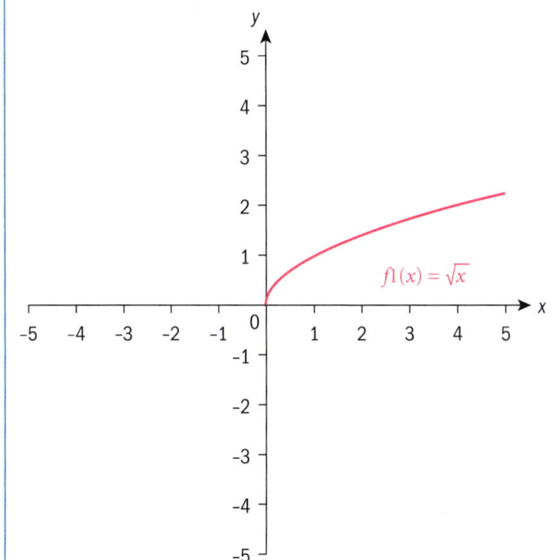

Draw the graph of $y = \sqrt{x}$.

$f2(x) = f1(x+2)$

$f1(x) = \sqrt{x}$

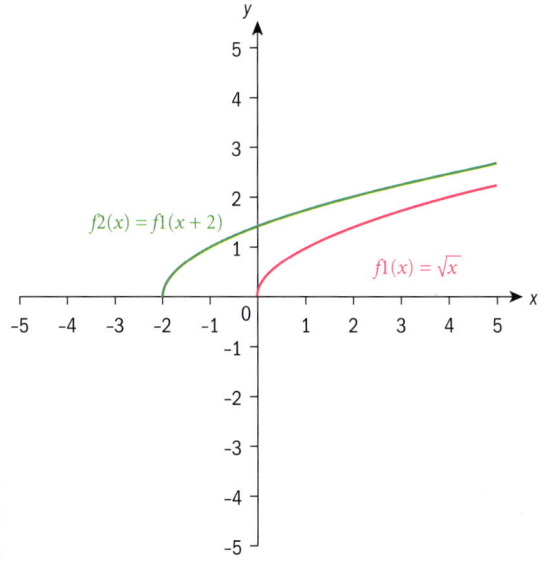

Translate the graph two units in the negative x-direction, since the radical is equivalent to $\sqrt{-x+2}$.

Continued on next page

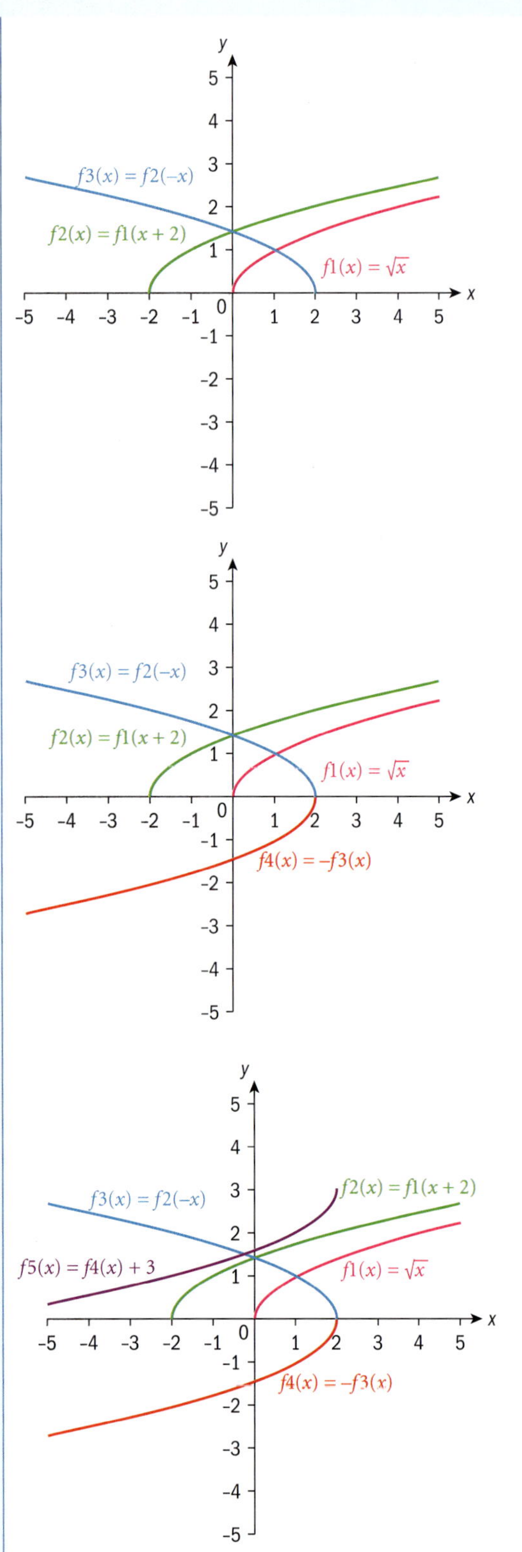

Reflect the graph in the y-axis, since this is $f(-x)$.

Reflect the graph in the x-axis, $-f(x)$.

Translate the graph 3 units in the positive y-direction.

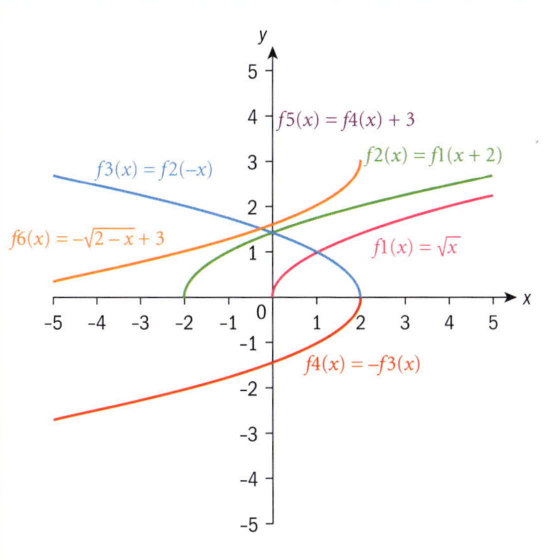

Confirm your final graph by also graphing $y = \sqrt{2 - x} + 3$.

Exercise 2S

On a separate diagram for each pair, transform the first function to the second function, showing the graphs at each stage.

1 $f(x) = x^2$; $y = -2f(2x - 2) + 2$

2 $f(x) = |x|; y = \frac{1}{2} f\left(\frac{x}{2} + 3\right) - 2$

3 $f(x) = \sqrt{x}; y = -f(3 - x) + 1$

4 $f(x) = \frac{1}{x}; -3f(-2x + 3) - 1$

5 $f(x) = x^3$; $-f(2x + 1) - 2$

Chapter summary

- A relation R is a set of ordered pairs (x, y) such that $x \in A, y \in B$ and sets A and B are non-empty. A function f is a relation in which every x-value has a unique y-value.
- A relation R is a function f if:
  - f acts on **all** elements of the domain.
  - f is well-defined, ie it pairs each element of the domain with one and only element of the range. Therefore, if f contains two ordered pairs (a, b_1) and (a, b_2), then $b_1 = b_2$.

  In general, if y is a function of x, you can write $y = f(x)$, or $x \mapsto f(x)$.
- A quadratic function can be expressed in different forms. In all forms, all parameters are real numbers, and $a \neq 0$.

  Standard form: $y = ax^2 + bx + c$.

  Vertex form: $y = a(x - h)^2 + k$, where (h, k) is the vertex.

  Intercept form: $y = a(x - p)(x - q)$, where p and q are the x-intercepts, or zeros of the function.

Continued on next page

- A **rational function** is any function which can be expressed as a quotient $y = \frac{f(x)}{g(x)}$, $g(x) \neq 0$, where $f(x)$ and $g(x)$ are polynomial functions.
- A function of the type $f(x) = \sqrt{ax+b}$ is a square root function whose radicand (the expression in the square root) is linear. The radicand must be non-negative, hence $ax+b \geq 0$ and the restricted domain is $\left\{x \middle| x \geq -\frac{b}{a}, a \neq 0\right\}$. The range is the set of non-negative real numbers, ie $\{y | y \geq 0\}$.
- A function $f: A \rightarrow B$ is **one-to-one** if every element in the range is the image of exactly one element in the domain, ie for any $a \in A$, $b \in B$, $f(a) = f(b) \Rightarrow a = b$. This means that every element in the domain has a different element in the range.

  A function is **many-to-one** if more than one element of the domain has the same image.
- A function $f: A \rightarrow B$ is **onto** if every element in B is an image of an element in A, ie for all $b \in B$ there exists an $a \in A$ such that $f(a) = b$.
- A function is even if, for all x in the domain, $-x$ is in the domain and $f(x) = f(-x)$. The graph of an even function is symmetrical about the y-axis.

  A function is odd if, for all x in the domain, $-x$ is in the domain and $f(-x) = -f(x)$.

  The graph of an odd function has rotational symmetry of $180°$ about the origin.
- If f and g are functions, then the notation for the composition of functions f and g is $(f \circ g)(x)$, or $f(g(x))$. The domain of $f \circ g$ is the set of all real numbers x such that:
  - x is an element of the domain of g
  - $g(x)$ is an element of the domain of f.

  The range of $f \circ g$ is a subset of the range of f.
- The **identity function** is a function f that when composed with g leaves g unchanged. In other words, $(f \circ g)(x) = (g \circ f)(x) = g(x)$. The identity function for function composition is $y = x$.
- A function h that, when composed with g, results in the identity function $f(x) = x$ is called an **inverse function**, ie $(g \circ h)(x) = (h \circ g)(x) = f(x) = x$.
- The notation for the inverse of a function g is $g^{-1}(x)$.
- The inverse of a function $f: A \rightarrow B$ is the relation formed when the domain and range are interchanged.

  If the inverse relation is also a function, then it is the inverse function of f, $f^{-1}: B \rightarrow A$, and we say that f is invertible.
- Only functions that are one-to-one and onto have inverse functions. We say that the functions have a one-to-one correspondence.
- A function f is **self-inverse** if $f^{-1}(x) = f(x)$.
- To transform $y = f(x)$ to $y = |f(x)|$, the graph is unchanged for $y \geq 0$, and reflected in the x-axis for $y \leq 0$.

  To transform $y = f(x)$ to $y = f(|x|)$, the graph is unchanged where $x \geq 0$. Where $x < 0$, the part of the graph for $x \geq 0$ is reflected in the y-axis.
- To graph the reciprocal of $y = f(x)$:
  - Where they exist, the zeros of $y = f(x)$ are the vertical asymptotes of $y = \frac{1}{f(x)}$.
  - If $y = b$ is the y-intercept of $y = f(x)$, then $y = \frac{1}{b}$ is the y-intercept of $y = \frac{1}{f(x)}$.

- The minimum value of $y = f(x)$ occurs at the same value of x as the maximum of $y = \frac{1}{f(x)}$ and vice versa.
- When $f(x) > 0$, $\frac{1}{f(x)} > 0$; when $f(x) < 0$, $\frac{1}{f(x)} < 0$.
- When $y = f(x)$ approaches 0, $\frac{1}{f(x)}$ will approach $\pm\infty$, and vice versa.

- The graphs of $y = f(x)$ and $y = f(-x)$ are reflections of each other in the y-axis.
  The graphs of $y = f(x)$ and $y = -f(x)$ are reflections of each other in the x-axis.
- If $h > 0$ then the graph of $y = f(x-h)$ is the translation of the graph of $y = f(x)$, h units in the positive x-direction.
- If $h < 0$ then the graph of $y = f(x-h)$ is the translation of the graph of $y = f(x)$, h units in the negative x-direction.
- If $k > 0$ then the graph of $y = f(x) + k$ is the translation of the graph of $y = f(x)$, k units in the positive y-direction.
- If $k < 0$ then the graph of $y = f(x) + k$ is the translation of the graph of $y = f(x)$, k units in the negative y-direction.

  These horizontal and vertical translations are denoted with a column vector $\begin{pmatrix} h \\ k \end{pmatrix}$.
- In general, to transform a function $y = f(x)$ to $y = af(b(x + c)) + d$,
  - a is the vertical dilation
  - b is the horizontal dilation
  - c is the horizontal translation
  - d is the vertical translation.
- The rational function $y = \frac{ax+b}{cx+d}; x \neq -\frac{d}{c}$ is the result of multiple transformations on the graph of the reciprocal function $y = \frac{1}{x}$. The rational function can also be written in the form $y = \frac{A}{B(x-h)} + k$, $x \neq h$, where A is the vertical dilation factor, B is the reciprocal of the horizontal dilation factor, h is the horizontal translation and k is the vertical translation. The domain is $\{x | x \in \mathbb{R}, x \neq h\}$ and the range is $\{y | y \in \mathbb{R}, x \neq k\}$.

Developing inquiry skills

Return to chapter opening problem. Can the graphs of the supply and demand functions each be transformed to the other using the function transformations you have learned in this section?

Chapter review

1 Determine which relations represent functions. For those that are functions, state the domain and range.

a

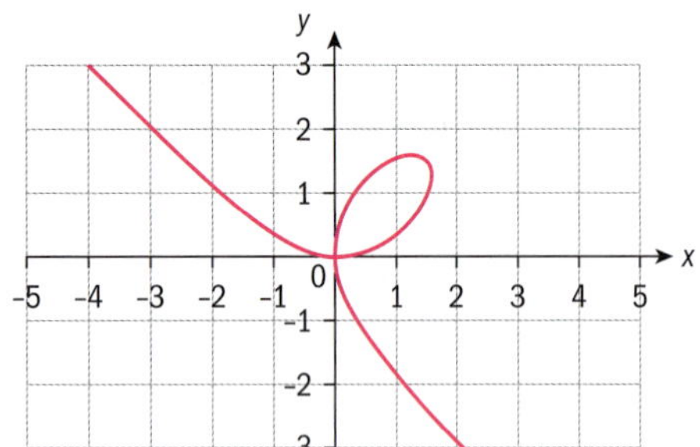

b

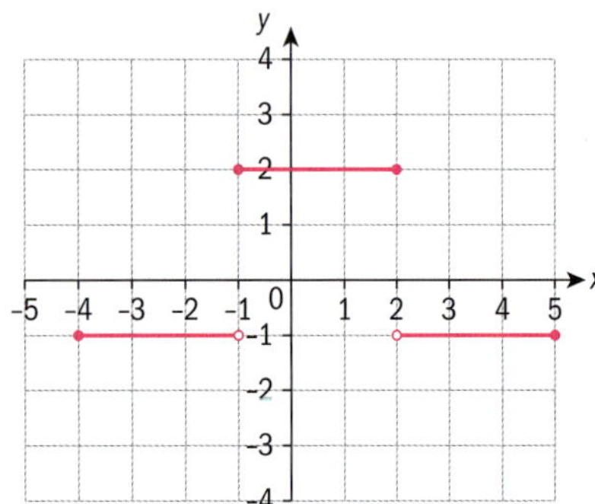

c

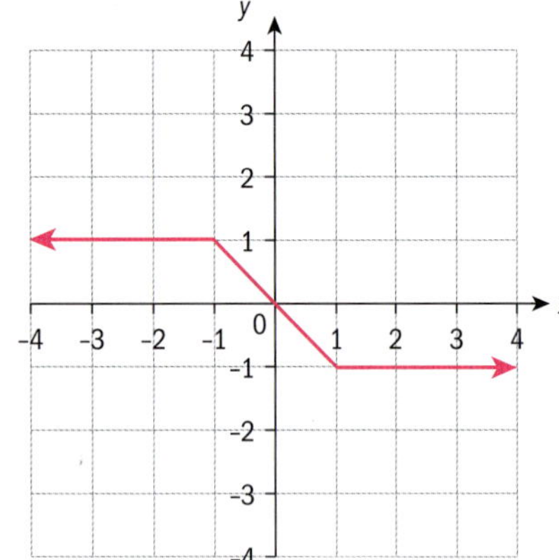

d

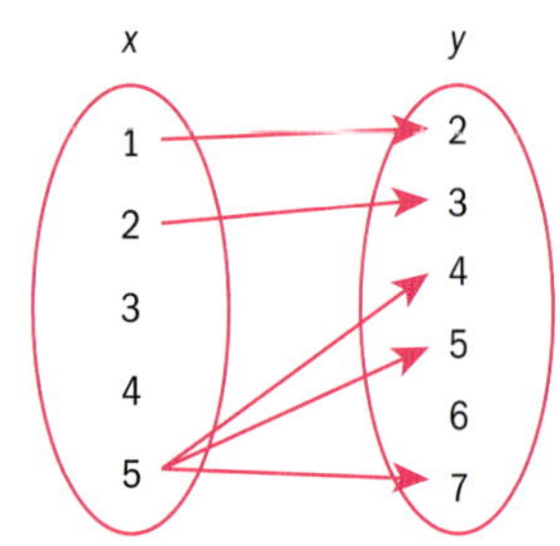

e

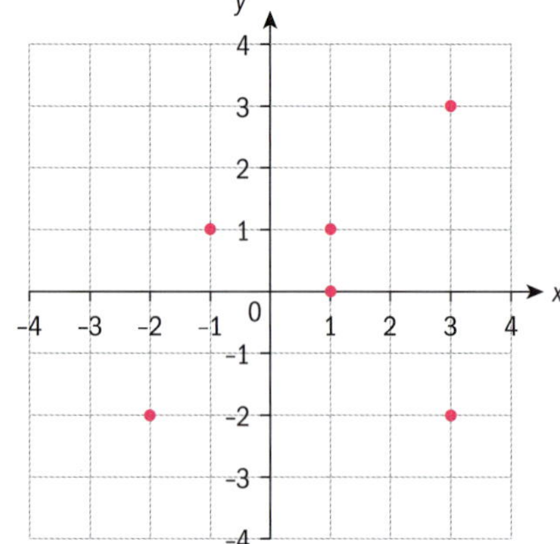

f $\{(-1, \pi), (0, \pi), (3, \pi), (\pi, \pi)\}$

2 If $f(1) = 2$; $g(3) = 1$; $h(2) = 3$, find:

a $f(g(h(2)))$

b If the inverses of f, g and h exist, find $h^{-1}(g^{-1}(f^{-1}(2)))$.

3 Find the inverse of each function. If the inverse is not a function, state the domain restriction that would make the inverse a function.

a $f(x) = \dfrac{x-2}{5}$ **b** $g(x) = \sqrt{1-x}$

c $h(x) = \dfrac{3x}{2-x}; x \neq 2$

4 List the transformations, in order, that transform the graph of $f(x) = \dfrac{1}{x}$ to the graph of $g(x) = \dfrac{2}{3-x} + 4$.

5 Solve the inequality $|x-2| \geq |2x+1|$.

6 Find $g(x)$, the inverse of $f(x) = \dfrac{1}{1+x^2}$ in the domain $[0, 1]$, and prove that f and g are mutual inverses.

7 Determine algebraically whether the function is odd, even or neither.

a $y = \dfrac{x^2+1}{|x|}; x \neq 0$ **b** $y = \dfrac{x}{x^2+1}$

c $y = \dfrac{\sqrt{x}}{x}; x \neq 0$

8 Express $\frac{2}{x^2+5x+6}$ as the sum of two rational expressions with linear denominators.

Exam-style questions

9 **P1:** Consider the equation $f(x)=2kx^2+6x+k,\ x\in\mathbb{R}$.

a If $f(x)$ has no real roots, find the possible values of k. (4 marks)

b If the equation of the line of symmetry of the curve $y = f(x)$ is $x + 1 = 0$, find the value of k. (3 marks)

10 **P1:** The graph of $f(x)=2x^2-4x+7$ is translated using the vector $\begin{pmatrix}2\\-1\end{pmatrix}$.

a Describe the effect of the translation on the graph of f. (2 marks)

b Find the equation of the translated graph, giving your answer in the form $y=ax^2+bx+c$.

11 **P1:** The function f is defined as $f(x)=\frac{3x-4}{x+2},\ x\neq -2$.

a Find an expression for $f^{-1}(x)$.

b Write down the domain of f^{-1}.

12 **P1:** Consider the function $f(x)=\frac{k}{x-1}+1,\ x>1,\ x\in\mathbb{R},\ k\in\mathbb{R}$.

a Show that $f(x)$ is a self-inverse function. (3 marks)

b State the range of f. (2 marks)

c Sketch the graph of $y=f(x)$ for any k, showing clearly any asymptotes. (2 marks)

13 **P1:** The function $f(x)$ is one-to-one and defined such that $f(x)=x^2-6x+13,\ x\geq k,\ x\in\mathbb{R},\ k\in\mathbb{R}$.

a Find the least possible value for k. (3 marks)

b Find an expression for the inverse function $f^{-1}(x)$. (3 marks)

c State the domain and the range of $f^{-1}(x)$. (2 marks)

14 **P1:** The function f is defined by $f(x)=\frac{17-10x}{2x-1},\ x\in\mathbb{R},\ x\neq\frac{1}{2}$.

a Show that f can be written in the form $f(x)=\frac{12}{2x-1}-5$ (4 marks)

b Hence, state

i the equation of any vertical asymptote

ii the equation of any horizontal asymptote. (2 marks)

c Sketch the graph of $y = f(x)$, clearly showing any asymptotes on your sketch. (3 marks)

15 **P2:** The number of butterflies kept in a private conservatory may be modelled by the formula $P=\frac{18(1+0.82t)}{3+0.034t}$, where P is the population of butterflies after time t months.

a Write down the initial butterfly population. (1 mark)

b Use the formula to estimate the number of butterflies in the conservatory after one year. (2 marks)

c Calculate the number of months that will have passed when the butterfly population reaches 100. (2 marks)

d Show that, when governed by this model, the number of butterflies in the conservatory cannot exceed 434. (3 marks)

16 **P1:** Given that $f(x)=x-3$ and $gf(x) = 2x^2 - 18$, find an expression for the function $g(x)$. (4 marks)

To Infinity and ... !

Approaches to learning: Thinking skills, Communication, Research, Collaboration
Exploration criteria: Mathematical communication (B), Personal engagement (C), Use of mathematics (E)
IB topic: Linking different areas

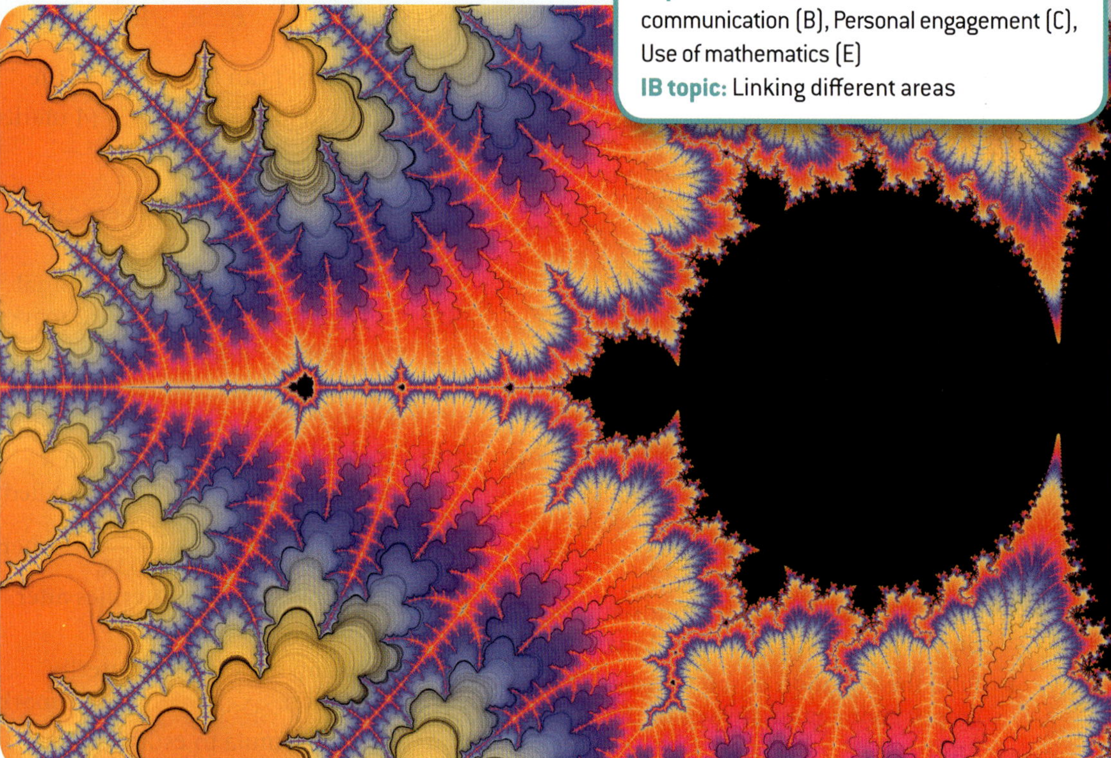

What is infinity?

Discuss these questions with a partner and then share your ideas with the whole class:

1 What is your current understanding of 'Infinity' in mathematics?

2 Do you think an understanding of infinity is important? interesting? necessary?

3 Explain your reasoning.

Processes which go on forever are quite common in mathematics.

Where have you already met the concept of infinity in this course?

Write down any other examples of where you have met the concept of infinity in mathematics, in your other academic studies or outside of your academic studies.

Let's think further about infinity

Game 1

The winner is the person who names the biggest positive natural number.

The first person names the biggest natural number (positive whole number) they can.

The next person names a natural number that is bigger than the previous number if they can.

This continues around the class.

Who will win the game?

Game 2

The winner is the person who names the closest rational number to 0.

The first person names the closest rational number to 0 they can.

The next person names a rational number that is closer to 0 than the previous number if they can.

This continues around the class.

Who will win the game?

Game 3

The winner is the person who names the closest real number to 1.

The first person names the closest real number to 1 they can.

The next person names a real number closer to 1 than the previous number if they can.

This continues around the class.

Who will win the game?

These three games are based on the idea of infinity.

However, they look at infinity in slightly different ways.

What is the difference in the meaning of infinity in these three games?

To consider this question, discuss these three questions:

In game 1, if you counted forever would you miss any numbers out?

In game 2, is it possible to count all of the rational numbers?

In game 3, what is the difficulty in counting the set of numbers?

Conclusion

Use your above reasoning to answer these questions:

Are two sets that contain an infinite number of numbers necessarily the same size?

How can one infinity be larger than another?

Extension presentation

Research one of these concepts, historical developments, applications or paradoxes that result from the existence of infinity. They are all conceptually difficult.

You can use both online sources and/or printed resources.

Present your ideas to your class.

Cantors orders of infinity	Hilbert's paradox of the grand hotel	Cantors Diagonal Proof
0.999999… = 1	The Infinitude of primes	Arithmetic properties of Infinity
Division by 0	The representation of infinity in art	Fractals

Expanding the number system: complex numbers

So far you have only worked with real numbers. In this chapter, you will begin to work with complex numbers: a new type of number that complies with all the properties of real numbers, but expands into another dimension. When complex numbers were first discovered they were called imaginary numbers since mathematicians could not find a real application for them. At a later stage, these numbers were put to use in many practical areas, and their name was changed to complex numbers. You will discover that many equations that we could not find a solution to within the set of real numbers actually have complex solutions. In this chapter you will also learn about one of the most important theorems in mathematics: the fundamental theorem of algebra.

Concepts

- Systems
- Patterns

Microconcepts

- Quadratic function and graph
- Quadratic equations and inequalities
- Discriminant
- Complex numbers
- Modulus of a complex number
- Operations with complex numbers
- Powers and roots of complex numbers
- Polynomial functions and their graphs
- Operations on polynomials
- Linear combination of two polynomials
- Factor and remainder theorem
- The fundamental theorem of algebra
- Polynomial equations
- Sum and product of the roots of polynomial equations
- Polynomial inequalities
- Simultaneous equations

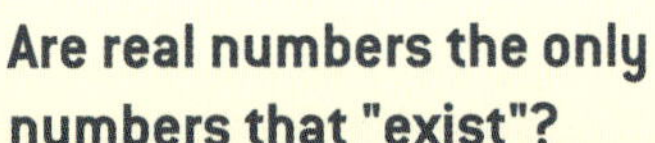

Are real numbers the only numbers that "exist"?

How do physicists at the Large Hadron Collider at CERN model the smashing of particles?

Your city council has asked you to prepare a plan of a roller coaster for a new amusement park. There are some constraints regarding the height of the roller coaster.

The designer would like to model the track using functions. What kind of features are impossible to have, given this restriction?

What kinds of functions are suitable for modelling the roller coaster and why?

Is it possible to model the entire track by using one function only?

If it is not possible, what conditions must the piecewise functions have at their joining points?

How many points are necessary to determine the equation of a curved part of the track?

Developing inquiry skills

Write down any similar inquiry questions you might ask to model another path—for example, the path of a character in a 2D computer game. What different questions might you need to ask?

Think about the questions in this opening problem and answer any you can. As you work through the chapter, you will gain mathematical knowledge and skills that will help you to answer them all.

Before you start

You should know how to:

1 Solve simple quadratic equations, eg

a $x^2 = 225 \Rightarrow x = \pm\sqrt{225} = \pm 15$

b $3x^2 - 72 = 0 \Rightarrow x^2 = 24$

$\Rightarrow x = \pm\sqrt{24} = \pm 2\sqrt{6}$

2 Use a GDC to graph linear and quadratic functions, and to solve equations, eg

Given the functions $f(x) = x^2 - 2x + 2$ and $g(x) = -2x + 3$ find the values of x such that $f(x) = g(x)$.

Graph both functions and find the points of intersection.

Skills check

1 Solve the following quadratic equations.

a $x^2 = 169$

b $7x^2 = 196$

c $11x^2 + 13 = 1102$

2 For the functions f and g, find the values of x such that $f(x) = g(x)$.

a $f(x) = 2x^2 - x - 1$ and $g(x) = -3x + 3$

b $f(x) = 4x^2 - x - 1$ and $g(x) = x^2 + x$

Continued on next page

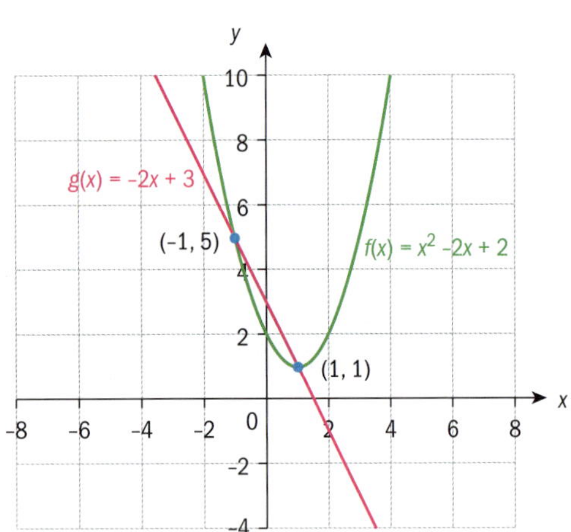

$x = -1$ or $x = 1$

3 Use a GDC to solve linear or quadratic inequalities, eg
Find the values of x such that $x^2 - 3x \le 10$.

Rewrite the inequality as a quadratic function and find all the values of x for which the graph is on or below the x-axis.

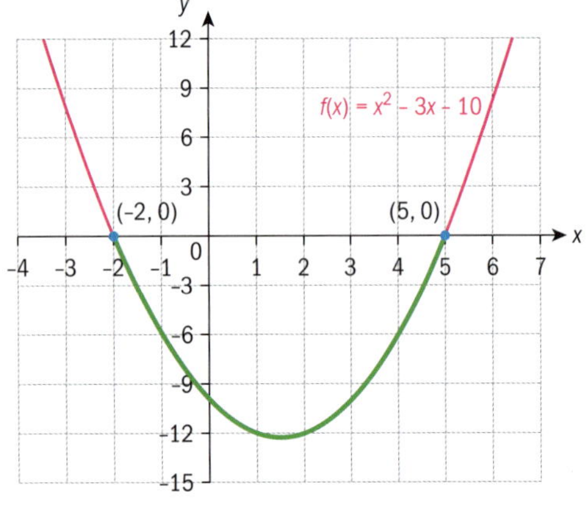

$x \in [-2, 5]$.

4 Solve simultaneous equations, eg

$$\begin{cases} 3x + 4y = 5 \\ 2x + 3y = -1 \end{cases}$$

Use one of the methods to solve, eg the method of elimination.

$$\begin{cases} 3x + 4y = 5 \\ 2x + 3y = -1 \end{cases} \Rightarrow \begin{cases} 6x + 8y = 10 \\ -6x - 9y = 3 \end{cases}$$

$-y = 13 \Rightarrow y = -13 \Rightarrow 2x + 3 \times (-13) = -1$

$\Rightarrow 2x = 38 \Rightarrow x = 19 \Rightarrow (x, y) = (19, -13)$

3 Find the values of x such that

a $2x - 1 > 3x + 2$

b $x^2 + 2x \le 8$

c $3x^2 - 2x \le 2 + 5x - x^2$

4 Solve the following simultaneous equations by a method of your choice.

a $\begin{cases} 7x - 4y = 1 \\ 5x - 3y = 2 \end{cases}$

b $\begin{cases} 2x + 3y = 1 \\ 6x = 3 - 9y \end{cases}$

c $\begin{cases} 2x - y = 4 \\ 2x + 3y = 1 \end{cases}$

d $\begin{cases} x = 5y + 1 \\ 2x - 3 = 10y \end{cases}$

3.1 Quadratic equations and inequalities

Solving quadratic equations by factorization

When a quadratic equation can be expressed as the product of two linear factors, you can factorize the equation and solve it.

International-mindedness

The Babylonians (2000–1600 BC) used quadratics to find the area of fields for agriculture and taxation.

Example 1

Solve the following quadratic equations by factorization.

a $x^2 + 3x - 40 = 0$ **b** $2x^2 - 7x - 4 = 0$

a $x^2 + 3x - 40 = 0 \Rightarrow x^2 + 8x - 5x - 40 = 0$ $\Rightarrow x(x+8) - 5(x+8) = 0$ $\Rightarrow (x+8)(x-5) = 0$	Split the middle term, so that you can factorize a common factor from the first two terms, and a common factor from the last two terms. Use distribution to factorize the expression.
$x + 8 = 0$ or $x - 5 = 0 \Rightarrow x = -8$ or $x = 5$	Use the zero product theorem to find the solutions.
b $2x^2 - 7x - 4 = 0 \Rightarrow 2x^2 - 8x + x - 4 = 0$	Again, split the middle term and factorize in pairs.
$2x(x-4) + (x-4) = 0 \Rightarrow (2x+1)(x-4) = 0$	Use distribution to factorize the expression.
$2x + 1 = 0$ or $x - 4 = 0 \Rightarrow x = -\frac{1}{2}$ or $x = 4$	Use the zero product theorem to find the solutions.

Exercise 3A

Solve the following equations by factorization.

1 $x^2 + 8x + 15 = 0$

2 $x^2 + 5x - 14 = 0$

3 $3x^2 - 7x + 2 = 0$

4 $4x^2 - 20x + 25 = 0$

5 $5x^2 - 4x - 12 = 0$

Solving equations by completing the square

Most quadratic equations that you will come across do not involve a perfect square. However, they can easily be transformed into an expression which *does* involve a perfect square by a process called **completing the square**.

You already know how to solve a quadratic equation by factorization. However, it is not always possible to factorize a quadratic equation. Completing the square is another method that you can use to solve a quadratic equation. You can use this method whenever the equation has real solutions.

First you will find the value of c that can be added to $x^2 + 6x$ to form a perfect square.

The following diagrams will show you how to do this.

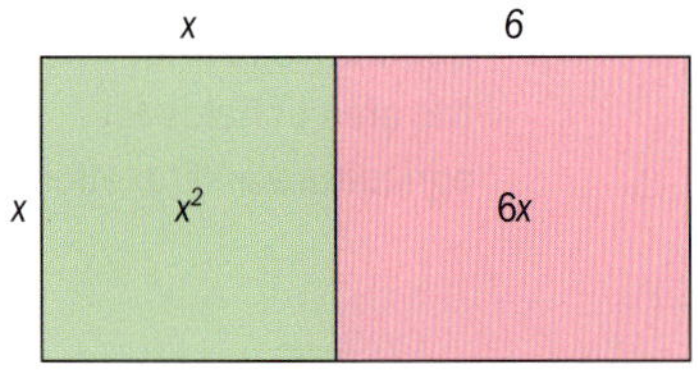	The area of the region shaded in green is x^2 and the area of the region shaded in pink is $6x$. Thus, the area of the whole rectangle is $x^2 + 6x$.
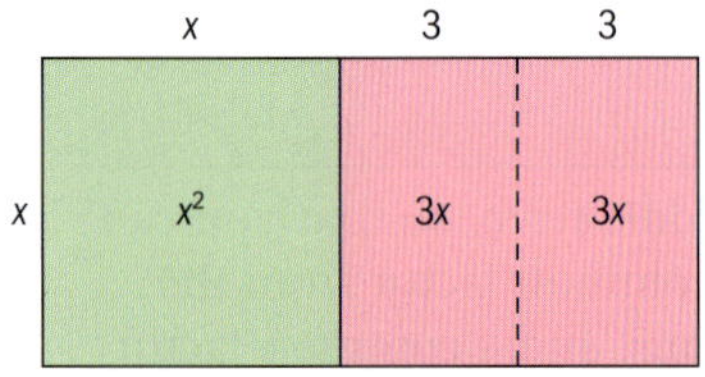	Divide the pink region into two equal parts.
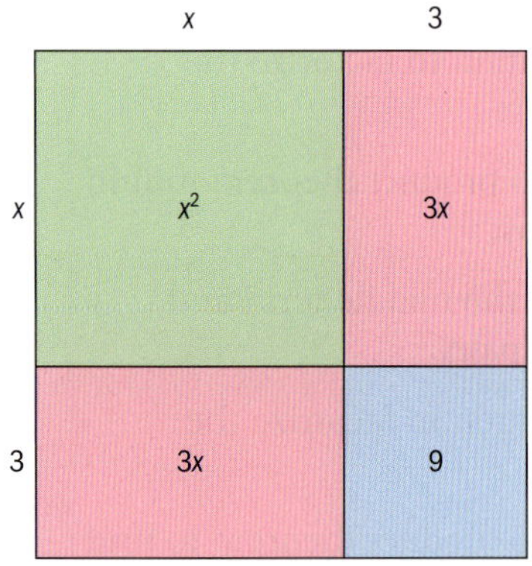	Rearrange the parts. The large square formed has area $(x + 3)^2$ So the expression $x^2 + 6x$ becomes a perfect square when you add 9.

Investigation 1

For each of the following diagrams, find the value of c which makes the expression into a perfect square, and then write the expression as a perfect square.

1 $x^2 + 4x + c$ 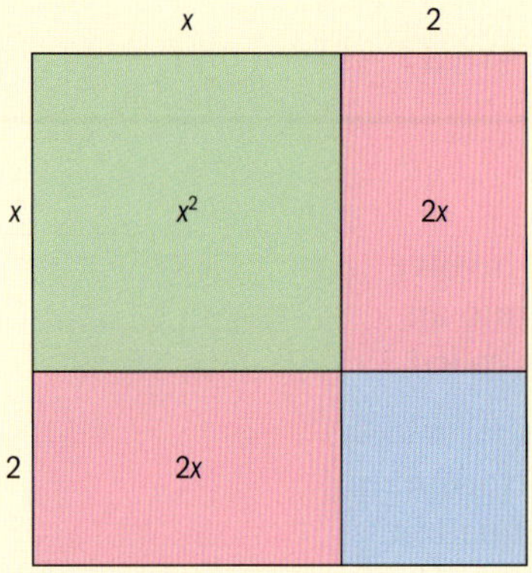	**2** $x^2 + 10x + c$ 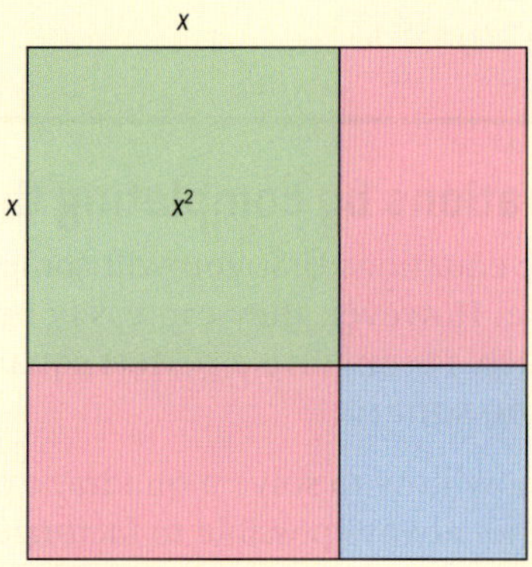

3 $x^2 + 3x + c$

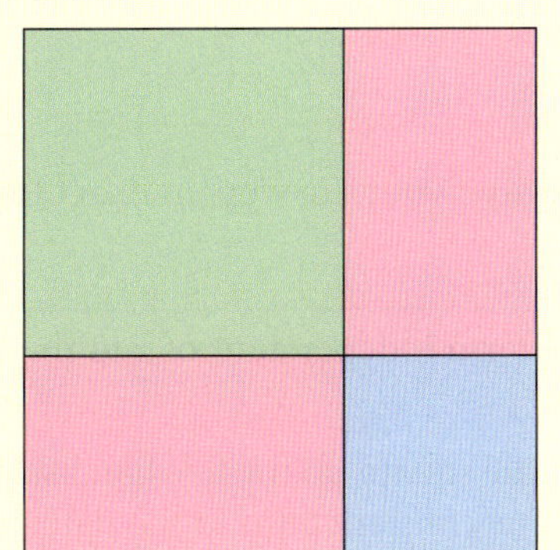

4 $x^2 + bx + c$

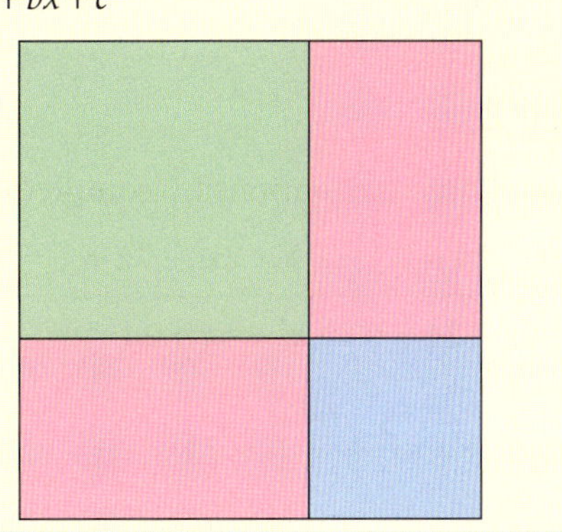

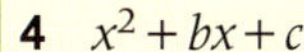

5 Fill in the blank: To complete the square for $x^2 + bx$, add ____________.

6 **Factual** What are the steps of the completing the square method?

7 **Conceptual** How does the area model help you understand the process of completing the square and why is the process called completing the square?

Use your findings from the investigation to algebraically solve the following examples.

Example 2

For each quadratic expression, find the value of k which makes the expression a perfect square. Hence, write each in a factorized form.

a $x^2 + 8x + k$ **b** $x^2 - 7x + k$ **c** $9x^2 + 3x + k$

a $x^2 + 8x + k = x^2 + \overbrace{2 \times x \times 4}^{8x} + \overbrace{4^2}^{k} = (x+4)^2$ $\Rightarrow k = 4^2 = 16$ **b** $x^2 - 7x + k = x^2 \overbrace{-2 \times x \times \frac{7}{2}}^{-7x} + \overbrace{\left(\frac{7}{2}\right)^2}^{k}$ $= \left(x - \frac{7}{2}\right)^2 \Rightarrow k = \left(\frac{7}{2}\right)^2 = \frac{49}{4}$ **c** $9x^2 + 3x + k = (3x)^2 + \overbrace{2 \times (3x) \times \frac{1}{2}}^{3x} + \overbrace{\left(\frac{1}{2}\right)^2}^{k}$ $= \left(3x + \frac{1}{2}\right)^2 \Rightarrow k = \left(\frac{1}{2}\right)^2 = \frac{1}{4}$	Use the binomial expansion and express the term in x as twice the product of x with a constant term. Use this constant term to find the value of k.

Once you have transformed a quadratic equation to a form which involves a perfect square, you can easily solve the equation.

Example 3

Solve the following equations by completing the square. Give your answers in exact form.

a $x^2 + 5x - 24 = 0$ **b** $2x^2 - 7x + 1 = 0$

a $x^2 + 5x - 24 = 0 \Rightarrow x^2 + 5x = 24$	Move the term independent of x to the right-hand side.
$\Rightarrow x^2 + 2 \times x \times \frac{5}{2} + \left(\frac{5}{2}\right)^2 = 24 + \left(\frac{5}{2}\right)^2$	Complete the square on the left-hand side, and add the same constant term to the right-hand side so that you don't change the equation.
$\Rightarrow \left(x + \frac{5}{2}\right)^2 = \frac{121}{4} \Rightarrow x + \frac{5}{2} = \pm\sqrt{\frac{121}{4}}$	Factorize the complete square and solve the equation.
$\Rightarrow x = -\frac{5}{2} \pm \frac{11}{2} \Rightarrow x = -8$ or $x = 3$	
b $2x^2 - 7x + 1 = 0 \Rightarrow x^2 - \frac{7}{2}x + \frac{1}{2} = 0$	Since the coefficient of x^2 is not 1, divide the whole equation by 2.
$\Rightarrow x^2 - 2 \times x \times \frac{7}{4} + \left(\frac{7}{4}\right)^2 = -\frac{1}{2} + \left(\frac{7}{4}\right)^2$	Move the term independent of x to the right-hand side. Complete the square on the left-hand side, and add the same constant term to the right-hand side.
$\Rightarrow \left(x - \frac{7}{4}\right)^2 = \frac{41}{16} \Rightarrow x - \frac{7}{4} = \pm\sqrt{\frac{41}{16}}$	
$\Rightarrow x = \frac{7 \pm \sqrt{41}}{4}$	Factorize the expression and solve the equation. There is no need to separate the solutions since they are written in the surd conjugate form.

In Example **3b**, you left your answer in surd form. However, when solving real-life problems, it is much more practical to give your answers in decimal form to an appropriate degree of accuracy. These values can then be used in making measurements, or whatever the context of the problem is.

Exercise 3B

1 Solve the following equations by completing the square. Give your answers in exact form.

a $x^2 + 6x - 7 = 0$

b $x^2 - 7x - 30 = 0$

c $x^2 - x - 1 = 0$

d $3x^2 - 7x + 2 = 0$

e $4x^2 + 12x + 5 = 0$

f $5x^2 - 10x + 2 = 0$

2 Solve the following equations by completing the square. Give your answers correct to 3 significant figures.

a $x^2 + 2x - 1 = 0$
b $x^2 - 3x + 1 = 0$
c $2x^2 - x - 3 = 0$
d $3x^2 + 9x + 5 = 0$

The quadratic formula

Notice that you can generally solve a quadratic equation by factorization when its roots are integers, but you can *always* solve a quadratic equation by completing the square – even when its roots are surds. However, the disadvantage is that completing the square can be long and repetitive. It is, therefore, helpful to have a quicker way to find the solutions to any quadratic equation with real roots.

Investigation 2

In this investigation, you will start with the general form of a quadratic equation, $ax^2 + bx + c = 0$, $a, b, c \in \mathbb{R}$, $a \neq 0$, and use completing the square to find its solutions in terms of a, b and c.

Copy the following table and fill in the missing steps or explanations.

Mathematical working	Explanation
$ax^2 + bx + c = 0$	
1	**1** Multiply both sides of the equation by a.
2	**2** Subtract ac from both sides of the equation.
3 $a^2x^2 + 2 \times ax \times \frac{b}{2} + \left(\frac{b}{2}\right)^2 = -ac + \left(\frac{b}{2}\right)^2$	**3**
4 $\left(ax + \frac{b}{2}\right)^2 = \frac{b^2}{4} - ac$	**4**
5 $\left(ax + \frac{b}{2}\right)^2 = \frac{b^2 - 4ac}{4}$	**5**
6	**6** Take the square root of both sides.
7 $ax = -\frac{b}{2} \pm \frac{\sqrt{b^2 - 4ac}}{2}$	**7**
8	**8** Divide by a to find x.

Factual What do a, b and c stand for in your formula in line 8?

Conceptual How does the quadratic formula generalize the solutions to quadratic equations?

HINT

When solving quadratic equations the solutions are called roots since in the formula you use a square root to find them. When using quadratic functions, the points of intersection with the x-axis have x values that are obtained by solving the corresponding quadratic equation. In this case you call them zeros because the value of the function for those x values is 0.

HINT

You could take the same approach that we used in Example **3b**, by dividing the whole equation by a, since $a \neq 0$. However, this approach will guide you through using a slightly different method which will give the same result.

TOK

How can you deal with the ethical dilemma of using mathematics to cause harm, such as plotting the course of a missile?

The quadratic formula
For a quadratic equation in the form $ax^2+bx+c=0,\ a,\ b,\ c\in\mathbb{R},\ a\neq 0$, the solutions or roots are given by $x=\dfrac{-b\pm\sqrt{b^2-4ac}}{2a}$.

Example 4

Use the quadratic formula to solve the following equations. Leave your answers in exact form.

a $x^2+11x+24=0$ **b** $3x^2+8x+4=0$ **c** $5x^2-2x-1=0$ **d** $2x^2+x+1=0$

a $x^2+11x+24=0 \Rightarrow a=1,\ b=11,\ c=24$ $\Rightarrow x=\dfrac{-11\pm\sqrt{11^2-4\times1\times24}}{2\times1}=\dfrac{-11\pm\sqrt{121-96}}{2}$	Identify the coefficients *a, b* and *c* and apply the formula.
$=\dfrac{-11\pm5}{2}\Rightarrow x=-8$ or $x=-3$	Separate the solutions.
b $3x^2+8x+4=0\Rightarrow a=3,\ b=8,\ c=4$ $\Rightarrow x=\dfrac{-8\pm\sqrt{8^2-4\times3\times4}}{2\times3}=\dfrac{-8\pm\sqrt{64-48}}{6}$	Identify the coefficients *a, b* and c and apply the formula.
$=\dfrac{-8\pm4}{6}\Rightarrow x=\dfrac{-12}{6}=-2$ or $x=\dfrac{-4}{6}=-\dfrac{2}{3}$	Separate the solutions.
c $5x^2-2x-1=0\Rightarrow a=5,\ b=-2,\ c=-1$ $\Rightarrow x=\dfrac{-(-2)\pm\sqrt{(-2)^2-4\times5\times(-1)}}{2\times5}=\dfrac{2\pm\sqrt{4+20}}{10}$	Identify the coefficients *a, b* and *c* and apply the formula.
$=\dfrac{2\pm2\sqrt{6}}{10}\Rightarrow x=\dfrac{1-\sqrt{6}}{5}$ or $x=\dfrac{1+\sqrt{6}}{5}$	Separate the solutions.
d $2x^2+x+1=0\Rightarrow a=2,\ b=1,\ c=1$ $\Rightarrow x=\dfrac{-1\pm\sqrt{1^2-4\times2\times1}}{2\times2}=\dfrac{-1\pm\sqrt{1-8}}{4}$	Identify the coefficients *a, b* and *c* and apply the formula.
$=\dfrac{-1\pm\sqrt{-7}}{4}\Rightarrow x\notin\mathbb{R}$	Since the expression under the square root is negative, there is no real solution.

When either of the coefficients *b* and/or *c* are equal to zero, we have special cases that we solve by using simpler methods.

Case 1

$b = 0, c \neq 0$

$ax^2 + c = 0 \Rightarrow ax^2 = -c \Rightarrow x^2 = -\frac{c}{a} \Rightarrow x = \pm\sqrt{-\frac{c}{a}}$, where

$x \in \mathbb{R}$ if $-\frac{c}{a} \geq 0$ and $x \notin \mathbb{R}$ if $-\frac{c}{a} < 0$.

Case 2

$b \neq 0, c = 0$

$ax^2 + bx = 0 \Rightarrow x(ax+b) = 0 \Rightarrow x = 0$ or $ax + b = 0 \Rightarrow x = -\frac{b}{a}$, so you have two real solutions, one of which is always 0.

If one or more of the coefficients in a quadratic equation is an unknown constant, you can use the quadratic formula to find the solutions in terms of that constant.

Example 5

Solve the quadratic equation $px^2 + 3 = 3px + x$, $p \neq 0$, by using the quadratic formula. Express x in terms of the real parameter p.

$px^2 + 3 = 3px + x \Rightarrow px^2 - (3p+1)x + 3 = 0$	Rewrite the equation in the general quadratic form.
$\Rightarrow a = p,\ b = -(3p+1),\ c = 3$	Identify the coefficients.
$\Rightarrow x = \frac{-(-(3p+1)) \pm \sqrt{(-(3p+1))^2 - 4 \times p \times 3}}{2 \times p}$	Apply the quadratic formula.
$= \frac{3p+1 \pm \sqrt{9p^2 + 6p + 1 - 12p}}{2p}$	Simplify.
$= \frac{3p+1 \pm \sqrt{9p^2 - 6p + 1}}{2p}$	
$= \frac{3p+1 \pm \sqrt{(3p-1)^2}}{2p}$	Factorize the expression under the square root.
$= \frac{3p+1 \pm (3p-1)}{2p} \Rightarrow x = \frac{3p+1+3p-1}{2p} = \frac{6p}{2p} = 3$	Simplify again.
or $x = \frac{3p+1-3p+1}{2p} = \frac{2}{2p} = \frac{1}{p}$	Write both solutions separately.

Exercise 3C

1 Solve the following equations by using the quadratic formula. Give your answers in the exact form.

a $x^2 + 9x + 18 = 0$ **b** $x^2 - x - 30 = 0$

c $x^2 - x - 1 = 0$ **d** $2x^2 - 3x = 2$

e $\sqrt{2}x^2 + \sqrt{3} = 11x$

2 Solve for x the following quadratic equations by using the quadratic formula. Give your answers in terms of the real parameter.

a $x^2 + 3x = ax + 3a$ **b** $2x^2 - b = x - 2bx$

c $x^2 + kx = 2k^2$ **d** $p^2x^2 + 3px = px + 3$

Discriminant of a quadratic equation

Investigation 3

1 Use the quadratic formula to solve the following equations.

a $x^2 - 3x - 10 = 0$ **b** $3x^2 - 2x + 5 = 0$

c $x^2 - 12x + 36 = 0$ **d** $4x^2 - 20x + 25 = 0$

e $6x^2 - 13x + 6 = 0$ **f** $x^2 - 5x + 7 = 0$

2 **Factual** How many real solutions can different quadratic equations have?

3 **Conceptual** By considering the quadratic formula $x = \frac{-b \pm \sqrt{b^2 - 4ac}}{2a}$, explain why quadratic equations have different numbers of solutions.

The expression $b^2 - 4ac$ is called the **discriminant** of a quadratic.

4 Sketch graphs of $y = f(x)$ for each equation given in question **1**. On your sketch, label any zeros of each function.

5 **Factual** What do you notice about the number of zeros of each graph, and the number of solutions to the quadratic equation?

6 **Factual** What does the graph of a quadratic equation tell you about the value of the discriminant?

7 **Conceptual** What can the discriminant be used for?

Given a quadratic equation of the form $ax^2 + bx + c = 0$, $a, b, c \in \mathbb{R}$, $a \neq 0$, the **discriminant** is the expression in the formula that is under the square root and is denoted by the Greek letter Δ, $\Delta = b^2 - 4ac$.

As you found in the investigation, the sign of the discriminant determines the number of roots of the quadratic.

International-mindedness

Frenchman Nicole Oresme was one of the first mathematicians to consider the concept of functions in the 14th century, the term "function" was introduced by the German mathematician Gottfried Wilhelm Leibniz in the 17th century and the notation was coined by Swiss Leonard Euler in the 18th century.

Case 1: $\Delta > 0$

If the discriminant is positive, then $x = \frac{-b \pm \sqrt{b^2 - 4ac}}{2a}$ and there are **two distinct real roots.**

Case 2: $\Delta = 0$

If the discriminant is equal to zero, then $x = \frac{-b}{2a}$. This is regarded as **one repeated real root.**

Case 3: $\Delta < 0$

If the discriminant is less than zero, then $\sqrt{b^2 - 4ac}$ is not real. In this case, there is **no real solution.**

Example 6

Without solving, determine the nature of the roots of each equation.

a $5x^2 + 7x - 1 = 0$ **b** $25x^2 + 49 = 70x$ **c** $2x^2 + \frac{7}{4}x + \frac{1}{2} = 0$

a $5x^2 + 7x - 1 = 0 \Rightarrow a = 5,\ b = 7,\ c = -1$	Identify the coefficients.
$\Delta = 7^2 - 4 \times 5 \times (-1) = 69 > 0$	Calculate the discriminant.
There are two distinct real roots.	Since the discriminant is greater than 0 there are two distinct real roots.
b $25x^2 + 49 = 70x \Rightarrow 25x^2 - 70x + 49 = 0$	Write in the form $ax^2 + bx + c = 0$.
$\Rightarrow a = 25,\ b = -70,\ c = 49$	Identify the coefficients.
$\Delta = (-70)^2 - 4 \times 25 \times 49 = 4900 - 4900 = 0$	Calculate the discriminant.
There is one repeated real root.	Since the discriminant is equal to 0 there is one repeated real root.
c $2x^2 + \frac{7}{4}x + \frac{1}{2} = 0 \Rightarrow a = 2,\ b = \frac{7}{4},\ c = \frac{1}{2}$	Identify the coefficients.
$\Delta = \left(\frac{7}{4}\right)^2 - 4 \times 2 \times \frac{1}{2} = \frac{49}{16} - 4 = -\frac{15}{16} < 0$	Calculate the discriminant.
There are no real roots.	Since the discriminant is less than 0 there are no real roots.

You might also be asked to determine the value of a real parameter so that the equation has different types of roots.

Example 7

Find the value(s) of the real parameter p so that

a $2x^2 - 3x + p = 0$ has two real roots **b** $px^2 + p = 13x$ has one real repeated root

c $(p + 2)x^2 + 2px = 1 - p$ has no real roots.

Continued on next page

a $2x^2 - 3x + p = 0 \Rightarrow \Delta = (-3)^2 - 4 \times 2 \times p$	Find the discriminant.
$\Rightarrow 9 - 8p > 0 \Rightarrow p < \frac{9}{8}$	Set the discriminant to be greater than 0 and solve the inequality.
b $px^2 + p = 13x \Rightarrow \Delta = (-13)^2 - 4 \times p \times p$ $= 169 - 4p^2$	Find the discriminant.
$\Rightarrow 169 - 4p^2 = 0 \Rightarrow p = \pm\frac{13}{2}$	Set the discriminant to be equal to 0 and solve the equation.
c $(p+2)x^2 + 2px = 1 - p$ $\Rightarrow \Delta = (2p)^2 - 4(p+2)(p-1)$ $= 4p^2 - 4p^2 - 4p + 8$ $= 8 - 4p$	Find the discriminant.
$\Rightarrow 8 - 4p < 0 \Rightarrow p > 2$	Set the discriminant to be less than 0 and solve the inequality.

Example 8

Find the values of r for which the equation $x^2 + 3rx + 1 = 0$ has

a two distinct real roots **b** one real repeated root **c** no real roots.

$x^2 + 3rx + 1 = 0 \Rightarrow \Delta = (3r)^2 - 4 \times 1 \times 1$ $\Rightarrow \Delta = 9r^2 - 4$	Find the discriminant and simplify it.
a Two distinct real roots:	When $\Delta > 0$ there are two distinct real roots.
$\Delta > 0 \Rightarrow 9r^2 - 4 > 0 \Rightarrow r^2 > \frac{4}{9} \Rightarrow \lvert r\rvert > \frac{2}{3}$ $\Rightarrow r < -\frac{2}{3}$ or $r > \frac{2}{3}$	Take the square root of both sides, $\sqrt{r^2} = \lvert r\rvert$. Solve the inequality.
b One repeated real root:	When $\Delta = 0$ there is one repeated real root.
$\Delta = 0 \Rightarrow 9r^2 - 4 = 0 \Rightarrow r^2 = \frac{4}{9} \Rightarrow r = \pm\frac{2}{3}$	Solve the equation.
c No real roots:	When $\Delta < 0$ there is no real root.
$\Delta < 0 \Rightarrow 9r^2 - 4 < 0 \Rightarrow r^2 < \frac{4}{9} \Rightarrow \lvert r\rvert < \frac{2}{3}$ $\Rightarrow -\frac{2}{3} < r < \frac{2}{3}$	Take the square root of both sides, $\sqrt{r^2} = \lvert r\rvert$. Solve the inequality.

All of the inequalities can be solved by graphing on your calculator.

Exercise 3D

1 Without solving these equations, determine the nature of their roots.

a $x^2 + 3x - 7 = 0$ **b** $x^2 + x + 2 = 0$

c $x^2 + 2x + 1 = 0$ **d** $5x^2 + \sqrt{3}x + 2 = 0$

e $2x^2 = \pi x - 1$ **f** $2.25x^2 + 49 = 21x$

2 Find the values of the given parameter for which the equation has

i two distinct real roots

ii one real repeated root

iii no real roots.

a $mx^2 + 2x - 5 = 0$ **b** $4x^2 = 3x + 4 - t$

c $(2s + 1)x^2 = s(3x - 1)$

Solving quadratic inequalities

Apart from using a calculator and identifying the part of the parabola that is above or below the x-axis (shown in Before you start), there are two other methods for solving quadratic inequalities.

Method 1: Algebraic method

In order to solve a quadratic inequality that will factorize, you need to first factorize the quadratic expression into linear factors and then consider the sign of the product. You know that the product of two factors is positive if both factors are of the same sign, and that the product is negative if both factors are of different signs.

$$A \times B > 0 \Rightarrow \begin{cases} A > 0 \text{ and } B > 0 \\ \text{or} \\ A < 0 \text{ and } B < 0 \end{cases} \quad \text{or} \quad A \times B < 0 \Rightarrow \begin{cases} A > 0 \text{ and } B < 0 \\ \text{or} \\ A < 0 \text{ and } B > 0 \end{cases}$$

TOK

We have seen the involvement of several nationalities in the development of quadratics.

To what extent do you believe that mathematics is a product of human social collaboration?

Example 9

Solve the quadratic inequality $2x^2 - 5x - 3 \geq 0$.

$2x^2 - 5x - 3 \geq 0 \Rightarrow 2x^2 - 6x + x - 3 \geq 0$ $\Rightarrow 2x(x - 3) + (x - 3) \geq 0 \Rightarrow (x - 3)(2x + 1) \geq 0$	The quadratic expression will factorize, so split it into two linear factors.
$\Rightarrow \begin{cases} x - 3 \geq 0 \text{ and } 2x + 1 \geq 0 \\ \text{or} \\ x - 3 \leq 0 \text{ and } 2x + 1 \leq 0 \end{cases}$	Consider the signs of the product.
$\Rightarrow \begin{cases} x \geq 3 \text{ and } x \geq -\frac{1}{2} \\ \text{or} \\ x \leq 3 \text{ and } x \leq -\frac{1}{2} \end{cases} \Rightarrow \begin{cases} x \geq 3 \\ \text{or} \\ x \leq -\frac{1}{2} \end{cases}$ which can be written as an interval $x \in \left(-\infty, -\frac{1}{2}\right] \cup [3, \infty)$	Find the range of values of x which satisfy the inequality.

Method 2: Using a "sign table"

If a quadratic expression will not factorize, then you cannot split it into two linear terms to compare signs of factors.

If this is the case, it is best to find any zeros of the quadratic by solving the corresponding quadratic equation, and then consider the signs of the quadratic either side of the zeros.

This method will work whether or not the quadratic will factorize.

Example 10

Solve the quadratic inequality $2x^2 - 5x - 3 \geq 0$.

$$2x^2 - 5x - 3 = 0 \Rightarrow x = \frac{5 \pm \sqrt{25+24}}{4} = \frac{5 \pm 7}{4}$$

$$x = 3 \text{ or } x = -\frac{1}{2}$$

Solve the corresponding quadratic equation, eg by using the formula.

	$x < -\frac{1}{2}$	$x = -\frac{1}{2}$	$-\frac{1}{2} < x < 3$	$x = 3$	$x > 3$
$2x^2 - 5x - 3$	+	0	–	0	+

Order the zeros in the table and investigate the sign of the quadratic between the zeros.

$$x \in \left(-\infty, -\frac{1}{2}\right] \cup [3, \infty)$$

The coefficient of x^2 is positive so the parabola is concave up, so between the zeros the expression has a negative value.

Since the inequality is not strict, you should include the zeros in your solution.

Exercise 3E

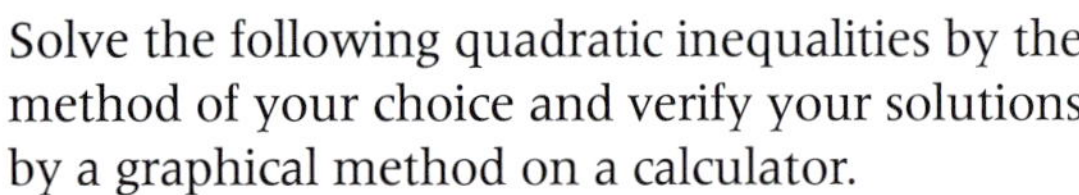

Solve the following quadratic inequalities by the method of your choice and verify your solutions by a graphical method on a calculator.

1 $x^2 + 6x + 8 < 0$

2 $x^2 - 8x + 16 > 0$

3 $x^2 - 13x - 30 \geq 0$

4 $4x^2 - 11x + 6 \leq 0$

5 $8 < 5x^2 - 6x$

6 $12x - 4 \geq 9x^2$

Developing inquiry skills

Return to the opening problem.

Can you use a quadratic equation to model the roller coaster?

3.2 Complex numbers

Algebraic vs. geometric approach towards complex numbers

Investigation 4

Consider the quadratic equation $x^2+4=0$.

1. Can you find real values that are solutions of the equation? Explain why, or why not.

Mathematicians define the imaginary number i such that $i^2=-1$, therefore you can write $i=\sqrt{-1}$.

2. Can you use this fact to solve the quadratic equation, expressing your answers in terms of i?
3. Use the quadratic formula to solve the equation $x^2-2x+5=0$. Give the roots in terms of i.
4. **Conceptual** What does the imaginary number i represent and how is it used in solving quadratics?

TOK

Descartes showed that geometric problems could be solved algebraically and vice versa.

What does this tell us about mathematical representation and mathematical knowledge?

HINT

Recall that

$\sqrt{a\times b}=\sqrt{a}\times\sqrt{b}$

$a, b\geq 0$

The letter i is used to represent the imaginary number $\sqrt{-1}$.

Solutions of any equation $x^2+c=0$, $c>0$, can be written as $x=\pm\sqrt{-c}=\pm \mathrm{i}d$, $d\in\mathbb{R}^+$. We call these **imaginary numbers**.

By defining $i^2=-1$, you have developed a whole new avenue of algebra that you are going to use in this chapter.

International-mindedness

Greek mathematician Heron discussed the root of a negative in the first century BC.

Complex numbers are numbers of the form $z=a+\mathrm{i}b$, where $a, b\in\mathbb{R}$.

a is called the **real part** of z, and we write $\mathrm{Re}(z)=a$.

b is called the **imaginary part** of z, and we write $\mathrm{Im}(z)=b$.

Complex numbers can be written in different forms, as you will see later in the book. The form $z=a+\mathrm{i}b$ is called the **Cartesian form** of a complex number.

Complex numbers were first introduced when mathematicians were trying to solve certain cubic equations. However, as you have seen in the above investigation, complex numbers also allow you to find solutions to any quadratic equation, regardless of the value of its discriminant.

Did you know?

The first person to propose working with the square root of a negative number was Heron of Alexandria (c.10–c.60) when discussing the volume of an impossible frustum of a pyramid.

Later on, in Italy, mathematical tournaments became very popular and solving difficult cubic equations became a way in which they found the winner. Those mathematicians discovered the formula for solutions of cubic equations and introduced complex numbers. Niccolo Fontana Tartaglia (1499–1557) discovered the formula first and shared his knowledge with Gerolamo Cardano (1501–1576).

Cardano introduced complex numbers of the form $a + \sqrt{-b}$, $a \in \mathbb{R}$, $b \in \mathbb{R}^+$. They realized that those two parts cannot be combined and the second part was called the imaginary or even impossible part.

René Descartes (1596–1650) was the first mathematician to establish the term "imaginary part", and John Wallis (1616–1703) made huge progress in giving a geometrical interpretation to $\sqrt{-1}$.

Leonhard Euler (1707–1783) was the mathematician who first introduced the symbol $\mathrm{i} = \sqrt{-1}$ and called it an "imaginary unit".

By looking at the complex number $z = a + \mathrm{i}b$, you might notice that every real number a can be seen as a complex number where the imaginary part is 0, that is, $b = 0$. In a similar way, a complex number that has real part equal to 0, ie $a = 0$, only has an imaginary part and is called a purely imaginary number. The number 0 can be seen as both, $0 = 0 + \mathrm{i} \times 0$.

In this way, you can view the real numbers as a subset of the complex numbers, and hence you can expand the subset definition from Chapter 1: $\mathbb{N} \subset \mathbb{Z} \subset \mathbb{Q} \subset \mathbb{R} \subset \mathbb{C}$.

Example 11

State the real part and the imaginary part of the following complex numbers.

a $2 + 3\mathrm{i}$ **b** $2\mathrm{i} - 5$ **c** $\frac{2}{3} - \sqrt{7}\mathrm{i}$ **d** $\frac{\sqrt{11} + \pi\mathrm{i}}{4}$

a $z = 2 + 3\mathrm{i} \Rightarrow \mathrm{Re}(z) = 2$ and $\mathrm{Im}(z) = 3$ **b** $z = 2\mathrm{i} - 5 \Rightarrow \mathrm{Re}(z) = -5$ and $\mathrm{Im}(z) = 2$ **c** $z = \frac{2}{3} - \sqrt{7}\mathrm{i} \Rightarrow \mathrm{Re}(z) = \frac{2}{3}$ and $\mathrm{Im}(z) = -\sqrt{7}$ **d** $z = \frac{\sqrt{11} + \pi\mathrm{i}}{4} \Rightarrow \mathrm{Re}(z) = \frac{\sqrt{11}}{4}$ and $\mathrm{Im}(z) = \frac{\pi}{4}$	The real part has no imaginary unit and the imaginary part has an imaginary unit.

Geometric approach

Real numbers can be visualized on a number line. Each point on the line represents one real number.

In a similar way, complex numbers can be represented in a two-dimensional coordinate plane, where the horizontal axis represents the real part of the number, and the vertical axis represents the imaginary part of the number.

Each complex number $z = x + \mathrm{i}y$, $x, y \in \mathbb{R}$, is represented by a point $\mathrm{P}(x, y)$ in the plane and the coordinates are the real and imaginary parts of the complex number itself.

Purely real numbers lie on the x-axis, and purely imaginary numbers lie on the y-axis.

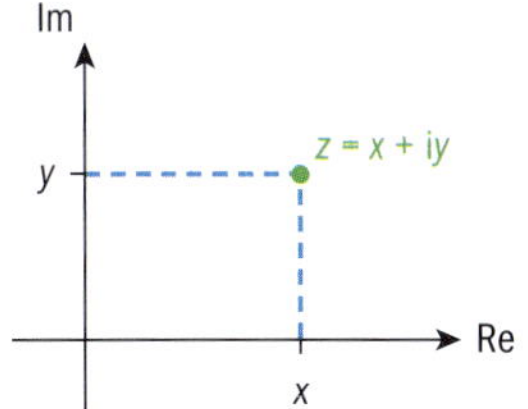

Modulus or absolute value of a complex number

You already know that the modulus or absolute value of a real number is algebraically defined as $|x| = \begin{cases} x, x \geq 0 \\ -x, x < 0 \end{cases}$. Geometrically, the modulus represents the distance along the number line between x and the origin. You can extend this idea into two dimensions in the complex plane where the modulus of a complex number represents the distance between the complex number and the origin.

Did you know?

Jean-Robert Argand (1768–1822) first proposed the idea of representing complex numbers on a two-dimensional coordinate plane. Carl Friedrich Gauss (1777–1855) independently developed and refined the plane model and therefore the plane of complex numbers is known as the **Argand diagram** or **Gaussian plane**. Caspar Wessel (1745–1818) also developed a similar approach using vectors.

Investigation 5

1. Plot the points that represent the following complex numbers on a complex plane. $3+4\mathrm{i}$, $1-2\mathrm{i}$, $-5+4\mathrm{i}$, -2, $3\mathrm{i}$, $-2-3\mathrm{i}$
2. What is the distance between each point you plotted in question **1** and the origin?
3. **Factual** Can you write down a formula for finding the modulus, $|z|$, of a complex number $z = x + \mathrm{i}y$?
4. **Factual** What do you notice about the modulus of a complex number?
5. **Conceptual** What does the modulus of a complex number tell you about that number?

Given the complex number $z = x + \mathrm{i}y$, $x, y \in \mathbb{R}$, the **modulus** of z is given by

$$|z| = |x + \mathrm{i}y| = \sqrt{x^2 + y^2} = \sqrt{(\mathrm{Re}(z))^2 + (\mathrm{Im}(z))^2}.$$

Notice that the modulus of a complex number is always a non-negative real number.

Example 12

Find the modulus of the following complex numbers.

a $5 - 12\text{i}$ **b** $\frac{1}{2} - \frac{\text{i}}{4}$ **c** $\frac{20\text{i} - 21}{29}$

Check your answers with a GDC.

a $\lvert 5-12\text{i}\rvert = \sqrt{5^2+(-12)^2} = \sqrt{169} = 13$	Apply the formula for the modulus.
b $\left\lvert \frac{1}{2} - \frac{\text{i}}{4} \right\rvert = \sqrt{\left(\frac{1}{2}\right)^2 + \left(-\frac{1}{4}\right)^2} = \sqrt{\frac{5}{16}} = \frac{\sqrt{5}}{4}$	Apply the formula for the modulus.
c $\left\lvert \frac{20\text{i}-21}{29} \right\rvert = \sqrt{\left(-\frac{21}{29}\right)^2 + \left(\frac{20}{29}\right)^2}$ $= \sqrt{\frac{841}{841}} = 1$	Apply the formula for the modulus.
Checking: 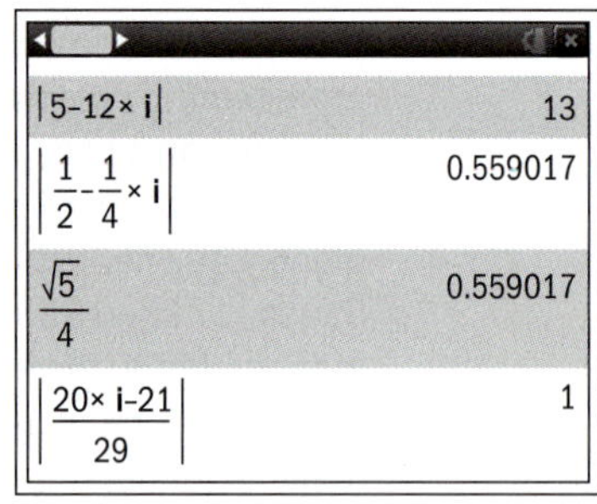	Make sure you know how to use a calculator to find the modulus of a complex number. You can use this to check the results.

Operations with complex numbers

First you need to understand when two complex numbers are equal.

> Two complex numbers $z_1 = a + b\text{i}$ and $z_2 = c + d\text{i}$, $a, b, c, d \in \mathbb{R}$, are equal if and only if their real parts are equal and their imaginary parts are equal, $a = c$ and $b = d$.

Investigation 6

Use your calculator to complete the first row of this table. z_1 and z_2 are complex numbers, and m and n are real constants.

Then add your own examples: for each blank row, define any two complex numbers z_1 and z_2, and any two real constants m and n, and complete the table.

z_1	z_2	$z_1 + z_2$	m	$m \times z_1$	n	$n \times z_2$	$m \times z_1 + n \times z_2$
$z_1 = 2 + \mathrm{i}$	$z_2 = 3 - 2\mathrm{i}$		2		-3		

Use the table you just completed to answer the following questions.

1. **Factual** How do you add two complex numbers?
2. **Factual** How do you subtract two complex numbers?
3. **Conceptual** Do we need two separate operations for adding and subtracting complex numbers?
4. **Factual** How do you multiply (or divide) a complex number by a real number?
5. **Conceptual** How are algebraic operations with complex numbers similar to algebraic operations with polynomials?

When adding complex numbers, or multiplying them by a constant, use the following rules.

$$z_1 + z_2 = (a + b\mathrm{i}) + (c + d\mathrm{i}) = (a + c) + (b + d)\mathrm{i},\ a, b, c, d \in \mathbb{R}$$
$$\lambda z = \lambda(a + b\mathrm{i}) = (\lambda a) + (\lambda b)\mathrm{i},\ \lambda, a, b \in \mathbb{R}$$

International-mindedness

How do you use the Babylonian method of multiplication?

Try 36×14.

Example 13

If $z_1 = 4 - 3\mathrm{i}$ and $z_2 = -2 + 5\mathrm{i}$, calculate the following and check your answers with a calculator.

a $z_1 - z_2$ **b** $\frac{1}{3}z_1 + \frac{2}{5}z_2$

a $z_1 - z_2 = (4 - 3\mathrm{i}) - (-2 + 5\mathrm{i})$
$= (4 + 2) + (-3 - 5)\mathrm{i} = 6 - 8\mathrm{i}$

Use the properties of complex numbers.

b $\frac{1}{3}z_1 + \frac{2}{5}z_2 = \frac{1}{3}(4 - 3\mathrm{i}) + \frac{2}{5}(-2 + 5\mathrm{i})$

$= \frac{4}{3} - \mathrm{i} - \frac{4}{5} + 2\mathrm{i} = \frac{8}{15} + \mathrm{i}$

Multiply both z_1 and z_2 by the given constant, and then add.

Checking:

4-3× i-(-2+5× i)	6-8× i
$\frac{1}{3}$× (4-3× i)+$\frac{2}{5}$× (-2+5× i)	$\frac{8}{15}$+i

Make sure you know how to use a calculator to add and subtract complex numbers and to multiply complex numbers by a constant. You can use this to check the results.

Exercise 3F

1 Find the real and imaginary parts of the following complex numbers.

a $z = -4\text{i}$ **b** $z = 5$

c $z = -24 + 7\text{i}$ **d** $z = \dfrac{5 - 12\text{i}}{13}$

e $z = \dfrac{2\text{i} - 1}{\sqrt{5}}$

2 Find the modulus of each of the complex numbers in question **1**.

3 Given the complex numbers $z_1 = 3 - 4\text{i}$, $z_2 = 5 + \text{i}$ and $z_3 = -1 + 2\text{i}$, calculate the following and check your answers with a calculator.

a $z_1 - z_2 + z_3$ **b** $2z_1 + 3z_2 - 4z_3$

c $-\dfrac{1}{2}z_1 + \dfrac{2}{3}z_2 - \dfrac{1}{4}z_3$ **d** $\dfrac{4z_3 - 5z_1 + 2z_2}{3}$

Investigation 7

Use your calculator and your algebra skills to fill in the following table. The first line has been done for you.

z_1	z_2	$z_1 \times z_2$
$z_1 = 2 + \text{i}$	$z_2 = 1 + 3\text{i}$	$(2+\text{i}) \times (1+3\text{i}) = 2 + \text{i} + 6\text{i} + 3\underbrace{\text{i}^2}_{-1} = -1 + 7\text{i}$
$z_1 = 1 - 2\text{i}$	$z_2 = 3 + \text{i}$	
$z_1 = 3 + 2\text{i}$	$z_2 = 4 - 5\text{i}$	
$z_1 = a + b\text{i}$	$z_2 = 3 + \text{i}$	
$z_1 = 3 + 2\text{i}$	$z_2 = c + d\text{i}$	
$z_1 = a + b\text{i}$	$z_2 = c + d\text{i}$	

1 **Factual** Using words, express how the real part of $z_1 \times z_2$ is related to the real and imaginary parts of both z_1 and z_2.

2 **Factual** In a similar way, express how the imaginary part of $z_1 \times z_2$ is related to the real and imaginary parts of both z_1 and z_2.

3 **Conceptual** How would you compare multiplication of two complex numbers to multiplication of two algebraic linear factors?

To summarize the findings from the investigation, we write the following.

$$z_1 \times z_2 = (a + b\text{i}) \times (c + d\text{i}) = (ac - bd) + (ad + bc)\text{i},\ a, b, c, d \in \mathbb{R}$$

Although the formula in the key point above gives you a general rule for multiplying two complex numbers, in reality this rule does not simplify the process of multiplication much. Therefore, when asked to multiply two complex numbers, it is often simpler to use algebra to find their product.

Example 14

Given the complex numbers $z_1 = 1 - 3i$, $z_2 = 4 + i$ and $z_3 = -2 + 3i$, find the following.

a $z_1 \times z_2 - z_3$ **b** $z_1 \times z_2 \times z_3$ **c** $z^2_1 + 2z_2 \times z_3$

Use technology to check your answer.

a $z_1 \times z_2 - z_3 = (1-3i)(4+i)-(-2+3i)$ $= 4 + i - 12i + 3 + 2 - 3i$ $= 9 - 14i$	Expand the brackets and simplify the expression.
b $z_1 \times z_2 \times z_3 = (1-3i)(4+i)(-2+3i)$ $= (7-11i)(-2+3i)$ $= -14 + 21i + 22i + 33$ $= 19 + 43i$	First multiply the first two factors, then multiply the product with the third factor. Simplify the expression.
c $z^2_1 + 2z_2 \times z_3 = (1-3i)^2 + 2(4+i)(-2+3i)$ $= 1 - 6i - 9 + 2(-8 + 12i - 2i - 3)$ $= -8 - 6i - 22 + 20i$ $= -30 + 14i$	Square z_1 and multiply z_2 by z_3. Expand the expressions and simplify.
Checking: 1-3× i →z1 1-3× i 4+i →z2 4+ i -2+3× i →z3 -2+3× i z1× z2-z3 9-14× i z1× z2×z3 19+43× i z1^2+2× z2× z3 -30+14× i	Make sure you know how to use a calculator to multiply two complex numbers. You can use this to check the results. For ease, store the values of z_1, z_2 and z_3 in the GDC memory and then work with the variables.

Example 15

Find $z \in \mathbb{C}$ that satisfies the equation $z \times (2 - i) = z + 3 + 2i$

Use technology to check your answer.

Let $z = a + bi$	Rewrite z in Cartesian form and substitute it into the equation.
$\Rightarrow (a + bi) \times (2 - i) = (a + bi) + 3 + 2i$ $\Rightarrow (2a + b) + (-a + 2b)i = (a + 3) + (b + 2)i$	Simplify both sides of the equation, and collect real and imaginary parts.
$\Rightarrow \begin{cases} 2a + b = a + 3 \\ -a + 2b = b + 2 \end{cases} \Rightarrow \begin{cases} a + b = 3 \\ -a + b = 2 \end{cases}$	If two complex numbers are equal then both the real parts are equal and the imaginary parts are equal.

Continued on next page

$\Rightarrow (a, b) = \left(\frac{1}{2}, \frac{5}{2}\right) \Rightarrow z = \frac{1}{2} + \frac{5}{2}\text{i}$	Simplify and solve the simultaneous equations.
Checking: $\left(\frac{1}{2}+\frac{5}{2}\times \text{i}\right)\times(2-\text{i})=\frac{1}{2}+\frac{5}{2}\times \text{i}+3+2\times \text{i}$ true	Use your calculator to check the solution.

Conjugate complex numbers

For any complex number $z = a + b\text{i}$, $a, b \in \mathbb{R}$, there is a **conjugate** complex number of the form $z^* = a - b\text{i}$. Their real parts are equal, $\text{Re}(z) = \text{Re}(z^*)$, and their imaginary parts are opposite, $\text{Im}(z) = -\text{Im}(z^*)$.

Notice that when adding two conjugate complex numbers you get **twice the real part**, that is $z + z^* = a + b\text{i} + a - b\text{i} = 2a$.

Also, when subtracting two conjugate complex numbers you get **twice the imaginary part**, multiplied by i, that is $z - z^* = a + b\text{i} - (a - b\text{i}) = 2b\text{i}$.

Investigation 8

Copy and complete the following table. The first row has been done for you. Use a GDC to check your answers.

z	$\lvert z\rvert$	z^*	$z \times z^*$
$z = -1 + \text{i}$	$\lvert z\rvert = \sqrt{(-1)^2 + 1^2} = \sqrt{2}$	$z^* = -1 - \text{i}$	$z \times z^* = 2$
$z = 3 - 4\text{i}$			
$z = -\sqrt{5} + 2\text{i}$			
$z = a + b\text{i}$			

1 **Factual** What is the relationship between the product and the modulus?

2 **Conceptual** Why do a complex number and its conjugate have equal moduli?

3 **Conceptual** Verify that z and z^* are mutually conjugate.

Division of complex numbers

In Investigation 6 and Example 13, you saw how to divide a complex number by a real number. The following example demonstrates how the complex conjugate helps us to divide two complex numbers.

Example 16

Given two complex numbers $z_1 = a + bi$ and $z_2 = c + di$, $a, b, c, d \in \mathbb{R}, z_2 \neq 0$, find $\frac{z_1}{z_2}$.

$\frac{z_1}{z_2} = \frac{a+bi}{c+di} = \frac{a+bi}{c+di} \times \frac{c-di}{c-di}$	Multiply numerator and denominator by the complex conjugate of the denominator.
$= \frac{ac - adi + bci - bd\overbrace{i^2}^{-1}}{c^2 + d^2}$	The denominator is then a real number, and you know how to divide a complex number by a real number.
$= \frac{ac+bd}{c^2+d^2} + \frac{bc-ad}{c^2+d^2}i$	Simplify the numerator, and split into real and imaginary parts.

Notice that, since $z \times z^* = \lvert z \rvert^2$, we can write $\frac{z_1}{z_2} = \frac{z_1 \times z_2^*}{\lvert z_2 \rvert^2}$.

Example 17

If $z_1 = 2 + i$, $z_2 = 2 - 5i$ and $z_3 = -1 + 2i$, find the following:

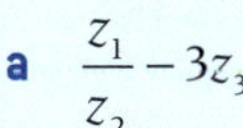

a $\frac{z_1}{z_2} - 3z_3$ **b** $\frac{z_1^{\ 2}}{z_2 \times z_3}$ **c** $\frac{2z_1 - 3z_2}{z_1 \times z_3}$

Check your answers with a calculator.

a $\frac{z_1}{z_2} - 3z_3 = \frac{2+i}{2-5i} \times \frac{2+5i}{2+5i} - 3(-1+2i)$	Multiply both numerator and denominator by the conjugate of the denominator.
$= \frac{4+10i+2i-5}{4+25} + 3 - 6i$	Expand and simplify.
$= \frac{-1+12i+87-174i}{29}$ $= \frac{86-162i}{29}$	Write the expression over a common denominator, and simplify.

Continued on next page

b $\frac{z_1^{\ 2}}{z_2 \times z_3} = \frac{(2+\text{i})^2}{(2-5\text{i})(-1+2\text{i})}$

$= \frac{4+4\text{i}-1}{-2+4\text{i}+5\text{i}+10}$ — Expand the numerator and multiply the factors in the denominator.

$= \frac{3+4\text{i}}{8+9\text{i}}$

$= \frac{(3+4\text{i})(8-9\text{i})}{|8+9\text{i}|^2}$ — You could multiply top and bottom by the conjugate of the bottom. Here, we have used $\frac{z_1}{z_2} = \frac{z_1 \times z_2^*}{|z_2|^2}$.

$= \frac{24-27\text{i}+32\text{i}+36}{64+81}$ — Multiply out both numerator and denominator.

$= \frac{60+5\text{i}}{145} = \frac{12+\text{i}}{29}$ — Simplify the answer.

c $\frac{2z_1 - 3z_2}{z_1 \times z_3} = \frac{2(2+\text{i})-3(2-5\text{i})}{(2+\text{i})(-1+2\text{i})}$

$= \frac{4+2\text{i}-6+15\text{i}}{-2+4\text{i}-\text{i}-2}$ — Expand the numerator and multiply the factors in the denominator.

$= \frac{-2+17\text{i}}{-4+3\text{i}}$ — Simplify.

$= \frac{(-2+17\text{i})(-4-3\text{i})}{|-4+3\text{i}|^2}$ — Using $\frac{z_1}{z_2} = \frac{z_1 \times z_2^*}{|z_2|^2}$

$= \frac{8+6\text{i}-68\text{i}+51}{16+9} = \frac{59-62\text{i}}{25}$ — Multiply the factors in the numerator and simplify the expression.

Checking:

Use a calculator to check the results. For ease, store all three values in the memory and then work with the variables.

Input	Result
$\frac{z1}{z2}-3\times z3$	$\frac{86}{29}-\frac{162}{29}\times i$
$\frac{z1^2}{z2\times z3}$	$\frac{12}{29}+\frac{1}{29}\times i$
$\frac{2\times z1-3\times z2}{z1\times z3}$	$\frac{59}{25}-\frac{62}{25}\times i$

Example 18

Find $z \in \mathbb{C}$ that satisfies the equation $\frac{z+2}{1-\text{i}} = \frac{z-3\text{i}}{2+\text{i}}$

Use your calculator to check the solution.

$\frac{z+2}{1-i} = \frac{z-3i}{2+i}$	Multiply both sides of the equation by $(1-i)(2+i)$.
$\Rightarrow (z+2) \times (2+i) = (z-3i) \times (1-i)$	
$\Rightarrow 2z + zi + 4 + 2i = z - zi - 3i - 3$	Multiply out.
$\Rightarrow 2z - z + zi + zi = -3i - 3 - 4 - 2i$	Isolate the terms in z on one side of the equation.
$\Rightarrow z + 2zi = -7 - 5i$	Simplify.
$\Rightarrow z(1+2i) = -7 - 5i$	Factorize.
$z = \frac{-7-5i}{1+2i}$	Make z the subject.
$= \frac{(-7-5i)(1-2i)}{\lvert 1+2i \rvert^2}$ $= \frac{-7-5i+14i-10}{1+4}$ $= \frac{-17+9i}{5}$	Using $\frac{z_1}{z_2} = \frac{z_1 \times z_2^*}{\lvert z_2 \rvert^2}$
Checking: 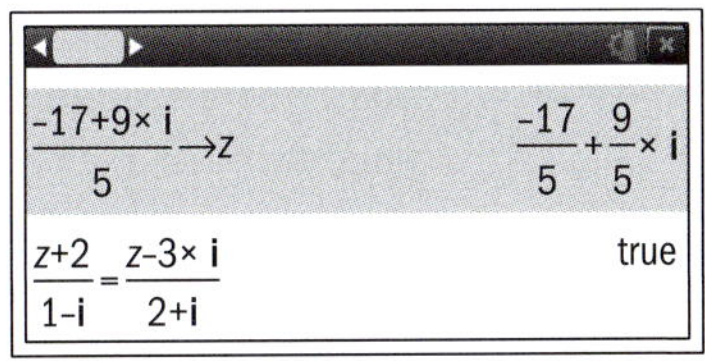	Use your calculator to check the solution.

Exercise 3G

1 Given the complex numbers $z_1 = 1 + i$, $z_2 = 3 - 2i$ and $z_3 = -2 + 3i$, calculate the following and check your answers with a calculator.

a $z_1 \times z_2 + 2z_3^*$ **b** $\frac{z_1}{z_3} - \frac{z_2}{5}$

c $z_1^{\,2} - 3z_2z_3$ **d** $\frac{z_1 z_3}{z_2^*}$

e $\frac{2z_1 - 4z_2^*}{z_3\ z_2^*}$

2 Find the real and imaginary parts of the following complex numbers.

a $\frac{1+2i}{i}$ **b** $\frac{1}{i} + \frac{2i}{1-i}$

c $\frac{1+2i}{1-2i} - \frac{1-2i}{1+2i}$

3 Find the real numbers a and b that satisfy the following:

a $(1+3i)(a+bi) = 5+5i$

b $\frac{a+bi}{1+2i} = -3+i$

4 Find the complex number z that satisfies the following equations.

a $2(z+i) = 3i(z-1)$

b $\frac{z-2}{1-2i} = \frac{z-i}{2+i}$

c $(z+2i)(2+i) = (z-1)(1-i)$

d $\frac{z+1+i}{1-4i} = \frac{z-3i+2}{2i+5}$

5 Given the complex number $z = a + bi$, a, $b \in \mathbb{R}$, find the relationship between a and b such that

a $\frac{z}{2+i} \in \mathbb{R}$ **b** $\frac{1-i}{z^*}$ is purely imaginary.

6 Solve for $z \in \mathbb{C}$.

a $|z| + z = 1$

b $|z| - z^* = i$

c $z^2 + z^* = 2$

7 Prove the following properties of the modulus of complex numbers.

a $|z_1 \times z_2| = |z_1| \times |z_2|$ **b** $\left|\frac{z_1}{z_2}\right| = \frac{|z_1|}{|z_2|}$

c $|z_1 + z_2| \le |z_1| + |z_2|$

8 Prove the following properties of conjugate complex numbers.

a $(z^*)^* = z$

b $(z_1 + z_2)^* = z_1^* + z_2^*$

c $(z_1 \times z_2)^* = z_1^* \times z_2^*$

d $\left(\frac{z_1}{z_2}\right)^* = \frac{z_1^*}{z_2^*}$

e $|z| = |z^*|$

Powers and roots of complex numbers

In order to raise a complex number to a power, you first need to investigate what happens when you raise the imaginary unit, i, to a power.

TOK

Could we ever reach a point where everything important in a mathematical sense is known?

Investigation 9

Fill in the following table by calculating the powers of i.

n	0	1	2	3	4	5	6	7	8	9
i^n	1	i								

1 Describe a pattern that your table shows.

2 Find a general rule for i^n, $n \in \mathbb{N}$.

3 Use your general rule to find i^{2019}.

4 Does your rule also apply for negative powers of i, that is, when $n \in \mathbb{Z}^-$?

5 **Conceptual** Can you generalize the powers of imaginary numbers?

Now you will use your general rule to raise any complex number to an integer power.

6 Let $z = a + bi$. In Chapter 1 you learned the binomial expansion of $(a + b)^n$, $n \in \mathbb{N}$. Use the binomial expansion, and your results about powers of i from question **2**, to find expressions for z^2 and z^3.

7 Use the same method to find $(z^4)^*$ and $(z^*)^4$. What do you notice?

The following example uses some higher powers of a complex number.

Example 19

Given the complex number $z = 2 + \text{i}$, find each of these expressions. Check your answer with a calculator.

a z^3 **b** $(z^*)^5$ **c** $(z^5)^*$

a $z^3 = (2+\text{i})^3$	Use the binomial theorem for $n = 3$.
$= 2^3 + 3 \times 2^2 \times \text{i} + 3 \times 2 \times \text{i}^2 + \text{i}^3$ $= 8 + 12\text{i} - 6 - \text{i} = 2 + 11\text{i}$	Use the powers of i. Simplify the expression.
b $(z^*)^5 = (2-\text{i})^5$	Use the binomial theorem for $n = 5$.
$= 2^5 + 5 \times 2^4(-\text{i}) + 10 \times 2^3(-\text{i})^2 + 10 \times 2^2(-\text{i})^3 + 5 \times 2(-\text{i})^4 + (-\text{i})^5$ $= 32 - 80\text{i} - 80 + 40\text{i} + 10 - \text{i} = -38 - 41\text{i}$	Use the powers of i. Simplify the expression.
c First calculate z^5 and then $(z^5)^*$. $z^5 = (2+\text{i})^5$	Use the binomial theorem for $n = 5$.
$= 2^5 + 5 \times 2^4\text{i} + 10 \times 2^3\text{i}^2 + 10 \times 2^2\text{i}^3 + 5 \times 2\text{i}^4 + \text{i}^5$ $= 32 + 80\text{i} - 80 - 40\text{i} + 10 + \text{i} = -38 + 41\text{i}$	Use the powers of i. Simplify the expression.
$\Rightarrow (z^5)^* = (-38 + 41\text{i})^* = -38 - 41\text{i}$	Find the conjugate complex number.
Checking: 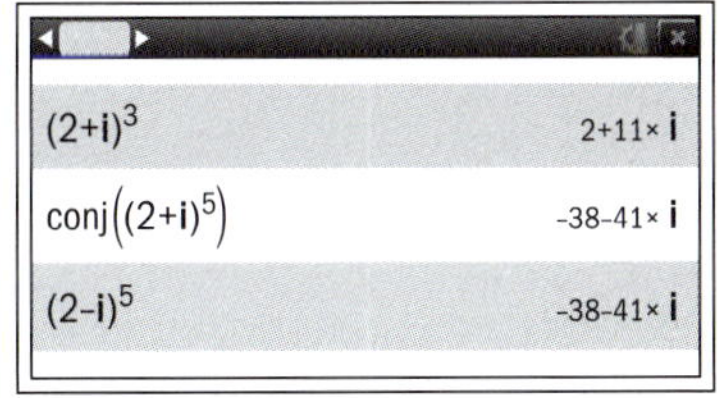	Use a calculator to check the solutions.

You can find square roots of complex numbers algebraically by letting the root be equal to a complex number $a + \text{i}b$. You then square this, and equate real and imaginary parts.

You can use a similar technique for finding higher roots of complex numbers, such as cube roots or fourth roots.

Example 20 will show you how to do this.

Example 20

Evaluate $\sqrt{7-24\mathrm{i}}$.

Let $z = a + b\mathrm{i}, a, b \in \mathbb{R}$	Write z in Cartesian form.
$z = \sqrt{7-24\mathrm{i}} \Rightarrow (a+b\mathrm{i})^2 = 7-24\mathrm{i}$ $\Rightarrow a^2 + 2ab\mathrm{i} - b^2 = 7 - 24\mathrm{i}$	Square both sides and expand the product.
$\Rightarrow \begin{cases} a^2 - b^2 = 7 \\ 2ab = -24 \end{cases} \Rightarrow \begin{cases} a^2 - \frac{144}{a^2} = 7 \\ b = -\frac{12}{a} \end{cases}$	Equate real parts and imaginary parts. Solve simultaneous equations by using the method of substitution.
$\Rightarrow \begin{cases} a^4 - 7a^2 - 144 = 0 \\ b = -\frac{12}{a} \end{cases} \Rightarrow \begin{cases} (a^2+9)(a^2-16) = 0 \\ b = -\frac{12}{a} \end{cases}$	Factorize the expression. Since a is real, we discard $(a^2 + 9) = 0$.
$\Rightarrow \begin{cases} a^2 = 16 \\ b = -\frac{12}{a} \end{cases} \Rightarrow \begin{cases} a = \pm 4 \\ b = -\frac{12}{\pm 4} = \mp 3 \end{cases}$	For two values of a find two values of b.
$z = 4 - 3\mathrm{i}$ or $z = -4 + 3\mathrm{i}$	Find both solutions.

A calculator can also be used to evaluate this square root, but the calculator will only give you one solution.

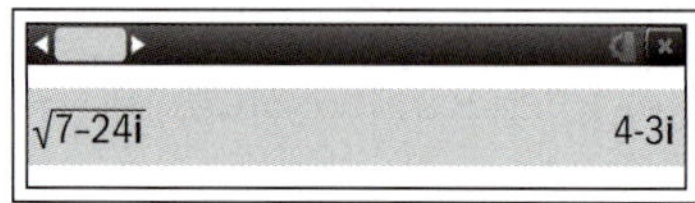

You have to know that the second solution is the opposite number, that is $-4 + 3\mathrm{i}$.

As you saw in Example 20, finding square roots of complex numbers requires algebraic skill in solving simultaneous equations of degree two or higher.

Exercise 3H

1 Calculate

a $\mathrm{i}^7 + \mathrm{i}^{17} + \mathrm{i}^{27} + \mathrm{i}^{37}$

b $\mathrm{i}^{173} - \mathrm{i}^{272} + \mathrm{i}^{351} - \mathrm{i}^{766}$

c $(3 + \mathrm{i}^{77}) \times (1 - 2\mathrm{i}^{93})$

d $\dfrac{3\mathrm{i}^{2018} + 2\mathrm{i}^{2019}}{4\mathrm{i}^{2020} - 3\mathrm{i}^{2021}}$

e $\dfrac{\sum_{k=1}^{2019} \mathrm{i}^k}{\prod_{k=1}^{2019} \mathrm{i}^k}$

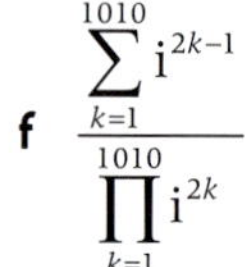

f $\dfrac{\sum_{k=1}^{1010} \mathrm{i}^{2k-1}}{\prod_{k=1}^{1010} \mathrm{i}^{2k}}$

HINT

Similar to Σ-notation for a sum there is a product notation

$$\prod_{k=1}^{n} \mathrm{i}^k = \mathrm{i}^1 \times \mathrm{i}^2 \times \ldots \times \mathrm{i}^n$$

2 Calculate the following and check your answers with a calculator.

a $(3+2i)^3$

b $(1-3i)^4$

c $(1-2i)^4+(1+2i)^4$

d $(1+i)^5+(1-i)^5$

3 Evaluate the following and check your answers with a calculator.

a $\sqrt{i}$ b $\sqrt{-i}$

c $\sqrt{-21+20i}$ d $\sqrt{\frac{5}{36}-\frac{i}{3}}$

4 Show that

a $|z^n|=|z|^n,\ n\in\mathbb{Z}^+$

b $(z^*)^n=(z^n)^*,\ n\in\mathbb{Z}^+$

5 Given that $z=1+i$, find the values of $n\in\mathbb{Z}^+$ such that

a z^n is real

b z^n is purely imaginary.

6 Show that

a $(1-i)^{2n}=(-2i)^n,\ n\in\mathbb{N}$

b $(1-i)^{2n+1}=(1-i)(-2i)^n,\ n\in\mathbb{N}$

3.3 Polynomial equations and inequalities

Investigation 10

1 Draw the graphs of $f_n(x)=x^n,\ n\in\mathbb{N}$, up to at least $n=7$.

2 **Conceptual** How could you classify these graphs by their shapes?

3 **Conceptual** How could you classify these graphs by their symmetrical properties?

Polynomials are functions which map a real variable, often called x, to another real number. We write $f:\ \mathbb{R}\to\mathbb{R}$.

Polynomials are functions of the form $f(x)=a_nx^n+a_{n-1}x^{n-1}+\ldots+a_1x+a_0$, $a_n\neq 0$ where $a_i\in\mathbb{R}$, $i=0,\ \ldots,\ n$ are called the **coefficients.** The highest power (n) of the variable x is called the **degree** of the polynomial, and we write $\deg(f)=n$.

A **linear combination** of two functions f and g is an expression of the form $a\times f(x)+b\times g(x)$, where a and b are some real numbers.

A linear combination of n functions is an expression of the form $\sum_{i=1}^{n}a_i\times f_i(x)$, where f_i are functions and $a_i\in\mathbb{R}$.

In general, you can say that polynomials are linear combinations of the power functions $\{1, x, x^2, x^3, x^4, x^5, \ldots\}$.

By taking a linear combination of such powers, for example $3 \times x^5 - 2 \times x^2 + 8 \times x - 11 \times 1$, you obtain a polynomial $f(x) = 3x^5 - 2x^2 + 8x - 11$.

Polynomial functions are the functions that you have worked with most frequently so far. Here are some examples of polynomials of different degrees that you have learned about.

- Zero polynomial: $\theta(x) = 0$. Its graph is a horizontal line along $y = 0$ (the x-axis).
- Constant polynomial: $f(x) = c$, $c \in \mathbb{R}$, $c \neq 0$. Its graph is the horizontal line $y = c$.
- The degree of a constant polynomial is 0.

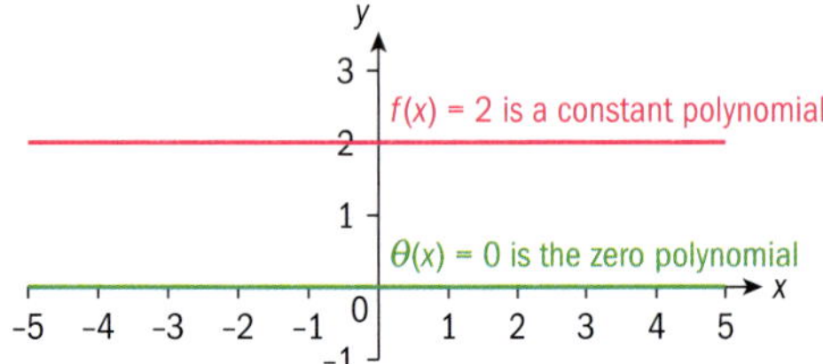

- Linear polynomial: $f(x) = mx + c$, $m \neq 0$, is also called a polynomial of the first degree. The graph is a straight line. By changing the parameters m and c we change the steepness of the line and the y-axis intercept, respectively.

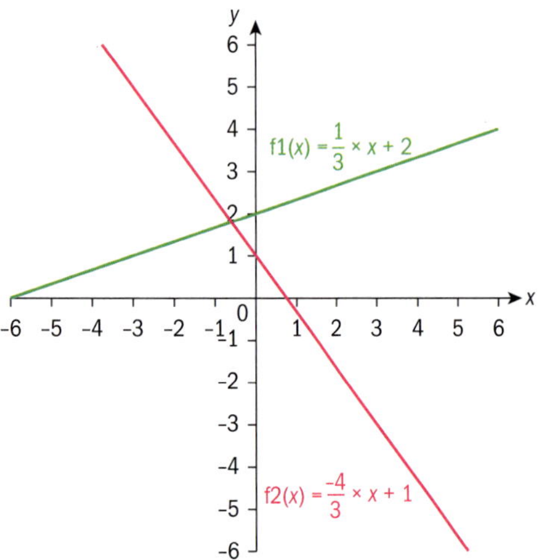

- Quadratic function: $f(x) = ax^2 + bx + c$, $a \neq 0$, is a polynomial of the second degree. Its graph is a parabola whose axis of symmetry is a vertical line which passes through the vertex of the graph. By changing the parameters a, b and c, we change
  - the shape of the graph (wide or narrow)
  - the concavity (open upwards or downwards)
  - the position of the parabola relative to the coordinate axes.

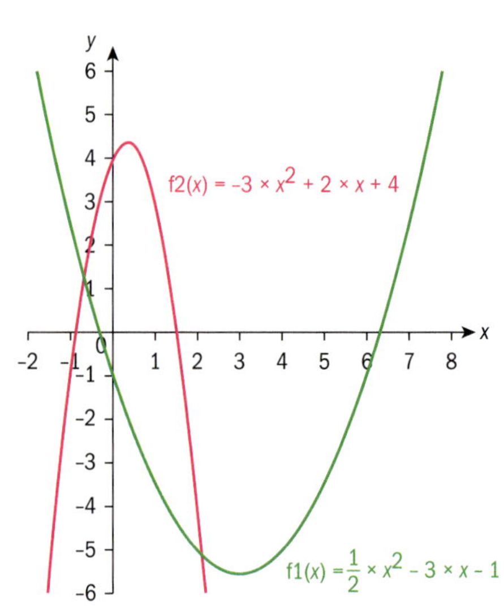

Did you know?

There is still a lot of discussion going on among mathematicians regarding the zero polynomial. Some consider that the zero polynomial is more than just a special case of the constant function. The degree of the zero polynomial is sometimes defined as -1 or even $-\infty$.

HINT

Notice that the notation used for the zero polynomial is $\theta(x) = 0$ so that you can distinguish it from other polynomials. The zero polynomial has an important property, being an additive identity element for the set of polynomials, that is $f(x) + \theta(x) = \theta(x) + f(x) = f(x)$ for all polynomials f.

- Cubic function: $f(x) = ax^3 + bx^2 + cx + d, a \neq 0$, is a polynomial of the third degree. There are two shapes of cubic graphs. One shape looks like a "Flex" shape like those you met in Investigation 10. The second shape is a combination of two U-shapes, one of which is concave up and the other is concave down.

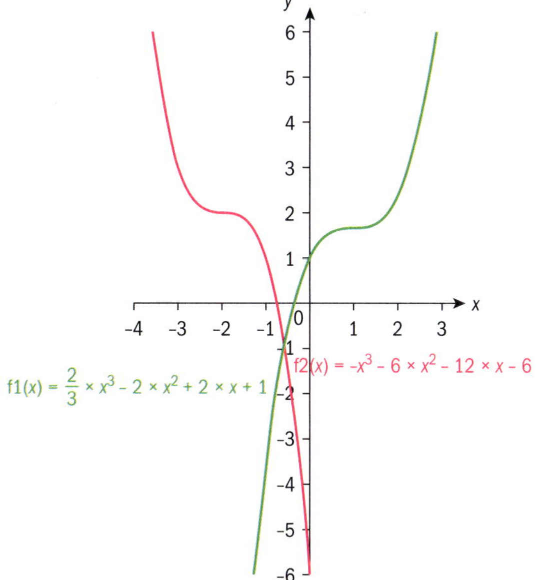

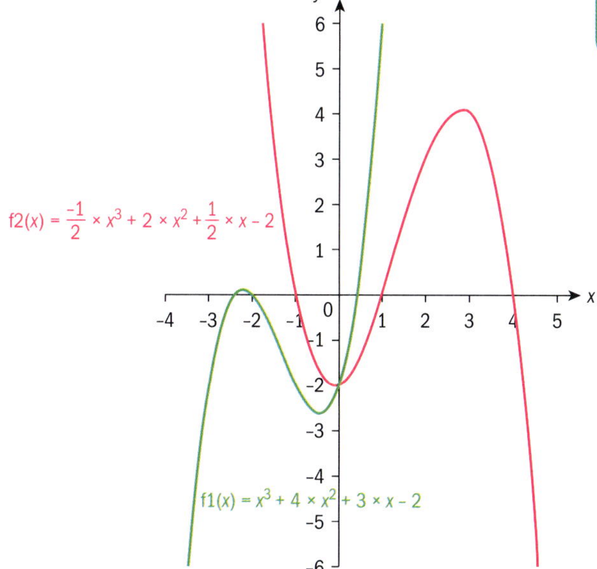

f2(x) = −1/2 × x^3 + 2 × x^2 + 1/2 × x − 2
f1(x) = x^3 + 4 × x^2 + 3 × x − 2

Investigation 11

In Chapter 2, you learned that a quadratic equation $f(x) = ax^2 + bx + c, a \neq 0$, has axis of symmetry $x = -\frac{b}{2a}$ and vertex $\left(-\frac{b}{2a}, f\left(-\frac{b}{2a}\right)\right)$.

In this investigation, you will look at how the parameters of a cubic function affect the graph of the function.

Consider the cubic function $f(x) = ax^3 + bx^2 + cx + d, a \neq 0$.

1 Using graphing software, investigate the effect of changing a. What can you conclude about the effect that a has on the graph?

2 What effect does changing d have on the graph of the function?

3 Investigate the effect of changing the parameter b in the graph of $f(x) = x^3 + bx^2$. How does the shape of the graph change as b changes? Do all graphs of this nature have any common points?

4 Investigate the effect of changing the parameter c in the graph of $f(x) = x^3 + cx$. How does the shape of the graph change as c changes? Do all graphs of this nature have any common points?

5 **Conceptual** How do the parameters in quadratic and cubic functions affect their graphs?

International-mindedness

The Sulba Sutras in ancient India and the Bakhshali manuscript contained an algebraic formula for solving quadratic equations.

HINT

With the help of graphing technology, you might want to investigate how the parameters of a fourth-order polynomial affect the shape, symmetry, vertices and intercepts of the graph.

One interesting feature of polynomials of the same degree with the same leading coefficient is that even though locally they might look very different you can find a larger window in which they look very similar.

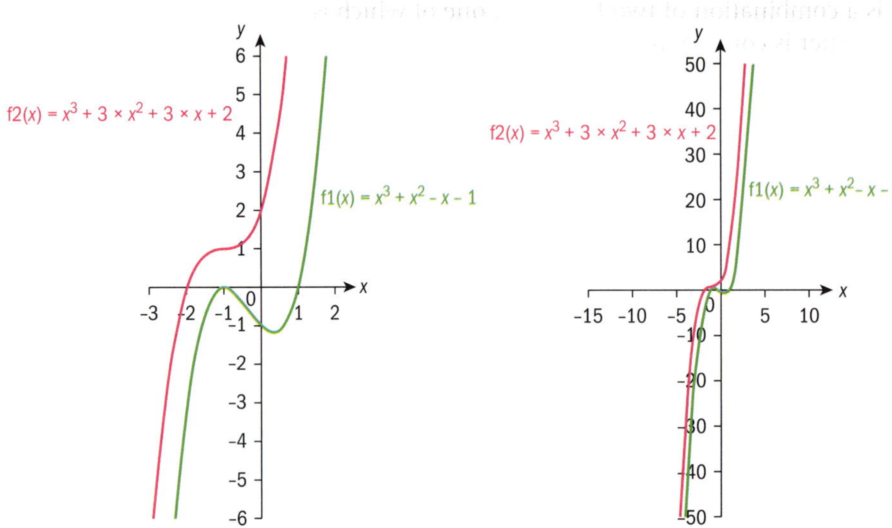

They behave like a polynomial that has only the leading term, since other terms for extremely large values of x are insignificant to the total value and can be neglected. This is the so-called "end behaviour" property of polynomials.

HINT

Polynomials are continuous functions, which means that you can draw their graphs without lifting your pen from the paper. Notice that you always have to proceed in one direction (usually you draw them from left to right). Their graphs are also smooth therefore they have no sharp points.

The factor and remainder theorems

In order to divide two polynomials you need to remind yourself how to divide two numbers by using long division.

Example 21

Use all the steps in the long division of two numbers to divide $3x^4 + 4x^3 + 2x^2 + 3x - 1$ by $x^2 + x - 2$.

			$3x^2$	$+x$	$+7$	Dividing $3x^4$ by x^2 gives $3x^2$.
x^2+x-2	$3x^4$	$+4x^3$	$+2x^2$	$+3x$	-1	
	$3x^4$	$+3x^3$	$-6x^2$			Multiply $x^2 + x - 2$ by $3x^2$ and subtract.
		x^3	$+8x^2$	$+3x$	-1	Dividing x^3 by x^2 gives x.
		x^3	$+x^2$	$-2x$		Multiply $x^2 + x - 2$ by x and subtract.
			$7x^2$	$+5x$	-1	Dividing $7x^2$ by x^2 gives 7.
			$7x^2$	$+7x$	-14	Multiply $x^2 + x - 2$ by 7 and subtract.
				$-2x$	$+13$	Stop when the degree of the remainder is smaller than the degree of the divisor.

You can write:

$$3x^4 + 4x^3 + 2x^2 + 3x - 1 = (x^2 + x - 2)(3x^2 + x + 7) + (-2x + 13)$$

Theorem

For any two polynomials f and g, there are unique polynomials q and r such that $f(x) = g(x) \times q(x) + r(x)$ for all real values of x.

The polynomial q is called the **quotient** and the polynomial r is called the **remainder**. Notice that $\deg(g) > \deg(r)$.

Exercise 3I

1 Use long division to divide polynomial f by polynomial g.

a $f(x) = 2x^3 + 5x^2 - 11x + 4$ and $g(x) = x + 4$

b $f(x) = 3x^5 + x^4 + 6x^3 + 5x^2 + 6$ and $g(x) = x^2 + 2$

c $f(x) = x^6 - x^5 + 2x^4 + x^3 - 5x^2 + 2x - 6$ and $g(x) = x^2 - x + 3$

2 Use long division to find the quotient and remainder when polynomial f is divided by polynomial g.

a $f(x) = 3x^3 - 6x^2 + x - 1$ and $g(x) = x - 1$

b $f(x) = 2x^4 - 3x^3 + 4x^2 + x - 5$ and $g(x) = x^2 + x + 2$

c $f(x) = x^5 + x^4 + 1$ and $g(x) = x^3 + 1$

Polynomial remainder theorem

Given a polynomial $f(x) = a_n x^n + a_{n-1}x^{n-1} + \ldots + a_2x^2 + a_1x + a_0$, $a_i \in \mathbb{R}$, $i = 1, 2, \ldots, n$, $a_n \neq 0$, and a real number p, then the remainder when $f(x)$ is divided by a linear expression $(x - p)$ is $f(p)$.

Proof:

In the unique decomposition of the polynomial $f(x) = (x - p) \times q(x) + r$, where the remainder r is a constant (one degree less than the divisor), we input $x = p \Rightarrow f(p) = \underbrace{(p - p)}_{0}q(p) + r \Rightarrow f(p) = r$.

Factor theorem

A polynomial $f(x) = a_n x^n + a_{n-1}x^{n-1} + \ldots + a_2x^2 + a_1x + a_0$, $a_k \in \mathbb{R}$, $k = 0, 1, 2, \ldots, n$, $a_n \neq 0$, has a factor $(x - p)$, $p \in \mathbb{R}$, if and only if $f(p) = 0$.

The factor theorem is a direct consequence of the remainder theorem and the proof is left as an exercise for you.

Mathematician William George Horner (1786–1837) formalized an algorithm that simplifies the process of finding the value of a polynomial at a given value of x.

If you want to find the value of $f(x) = 3x^3 - 2x^2 - 5x - 1$ when $x = 2$, select the coefficients of all terms, including missing terms, and organize them in tabular form.

Did you know?

The polynomial remainder theorem is also known as "Little Bézout's theorem" as it was proposed by Étienne Bézout (1730–1837). Bézout was inspired by the work of Euler and decided to become a mathematician. In 1763 he was appointed examiner of the Gardes de la Marine (French Naval Academy) with a special task to compose a textbook for teaching mathematics to the students.

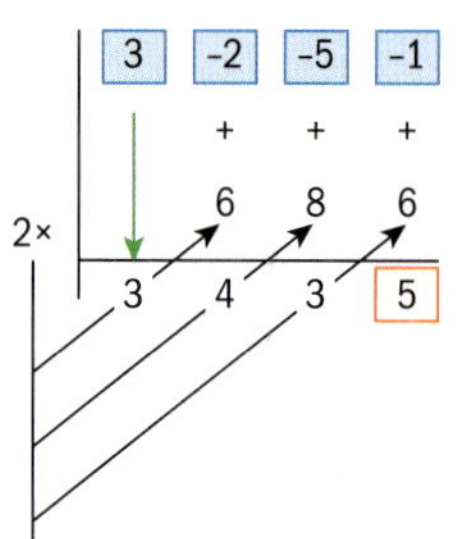

HINT

$f(x) = ((3 \times x - 2) \times (x - 5)) \times x - 1$

$f(2) = ((3 \times 2 - 2)2 - 5) \times 2 - 1$
$= ((6 - 2) \times 2 - 5) \times 2 - 1$
$= (8 - 5) \times 2 - 1$
$= 3 \times 2 - 1$
$= 6 - 1$
$= 5$

Horner's algorithm finds the quotient and remainder when a polynomial $f(x)$ is divided by a linear polynomial $g(x)$. This algorithm is also known as **synthetic division**. Notice that synthetic division only works when you divide a polynomial by a linear polynomial.

Since synthetic division tells us the quotient when f is divided by a linear factor g, it is an effective and quick method to use when searching for factors of f.

Example 22

Use synthetic division to find the remainder when dividing:

a $f(x) = 3x^3 + 11x^2 - 7x - 19$ by $g(x) = x + 2$, **b** $f(x) = 4x^3 - 3x + 8$ by $g(x) = x - 5$.

a 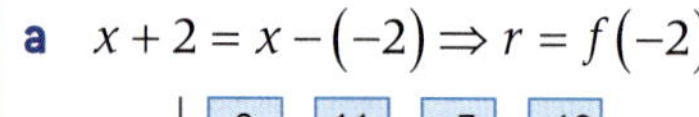 $x + 2 = x - (-2) \Rightarrow r = f(-2)$

Use the remainder theorem.

Use synthetic division to find the remainder.

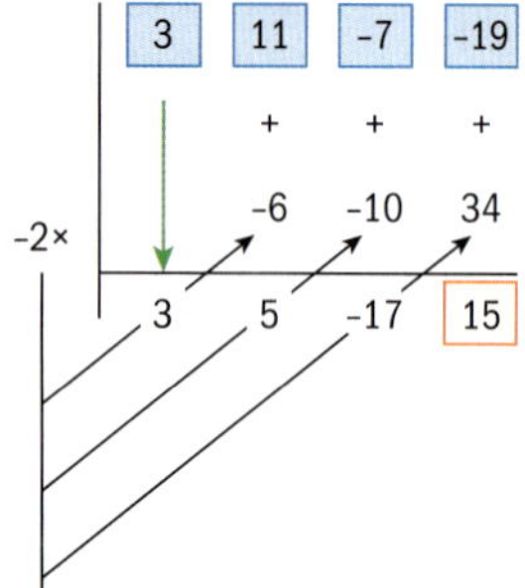

$r = f(-2) = 15$

b $x - 5 \Rightarrow r = f(5)$

Use the remainder theorem.

Use synthetic division to find the remainder.

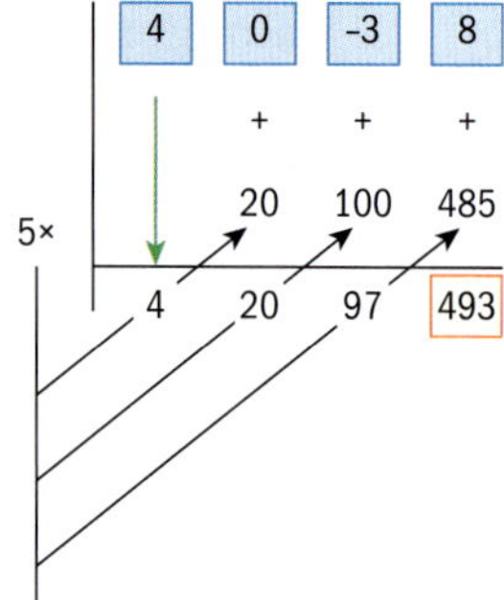

$r = f(5) = 493$

Reflect What do you notice about the quotient polynomial $q(x)$ and the result you obtained from synthetic division in the example above? Can you use the remainder theorem to explain why this is the case?

Example 23

The polynomial $f(x) = 2x^3 + x^2 - 15x - 18$ is given.

a Use synthetic division to show that $(x - 3)$ and $(x + 2)$ are factors of the polynomial.

b Hence fully factorize the polynomial.

a

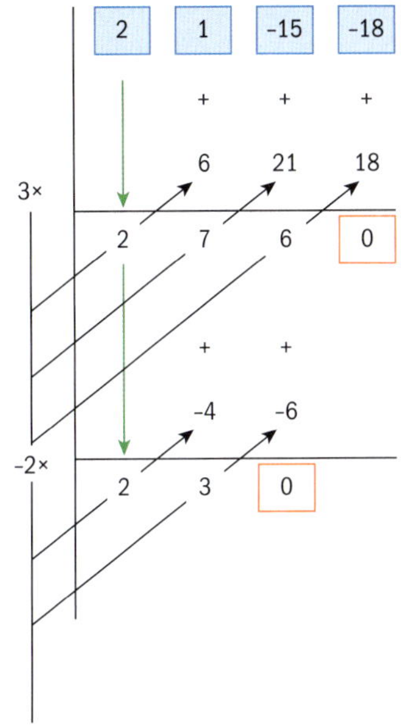

	2	1	-15	-18
		+	+	+
3×		6	21	18
	2	7	6	0
		+	+	
-2×		-4	-6	
	2	3	0	

Use synthetic division twice: first to divide $2x^3 + x^2 - 15x - 18$ by $(x - 3)$, and then to divide the quotient by $(x + 2)$.

Since the remainder is zero in each case, $(x - 3)$ and $(x + 2)$ are factors of the polynomial.

Use the factor theorem.

b $2x^3 + x^2 - 15x - 18$

$= (x - 3)(x + 2)(2x + 3)$

In the synthetic division table, you can see the last factor is the second quotient.

Exercise 3J

1 In each case, use synthetic division to find the quotient and the remainder when polynomial f is divided by g.

a $f(x) = x^3 + 2x^2 - 3x + 1$ and $g(x) = x - 2$

b $f(x) = 2x^3 + 3x^2 - 10x - 2$ and $g(x) = x + 3$

c $f(x) = 2x^4 - 6x^3 - 9x^2 + 7x - 11$ and $g(x) = x - 4$

d $f(x) = 3x^5 + 20x^4 + 10x^3 - 13x^2 + 7x - 3$ and $g(x) = x + 6$

2 The polynomial $2x^4 - x^3 - 32x^2 + 31x + 60$ is given.

a Show that $(x + 1)$ and $(x - 3)$ are factors of the polynomial.

b Hence fully factorize the polynomial.

Corollary

Given a polynomial $f(x) = a_n x^n + a_{n-1}x^{n-1} + \dots + a_2x^2 + a_1x + a_0, a_k \in \mathbb{R}$, $k = 1, 2, \dots, n, a_n \neq 0$, and real numbers a and b, $a \neq 0$, then the remainder when $f(x)$ is divided by a linear expression $(ax - b)$ is $f\left(\frac{b}{a}\right)$.

The proof is left to you as an exercise.

In order to use synthetic division when dividing by a linear expression $(ax - b)$ we need to modify the algorithm.

$f(x) = (ax - b) \times q(x) + r$

$\Rightarrow f(x) = a\left(x - \frac{b}{a}\right) \times q(x) + r$

$\Rightarrow f(x) = \left(x - \frac{b}{a}\right) \times (a \times q(x)) + r$

So, when applying Horner's algorithm, notice that all the coefficients in the quotient will have a common factor of a.

HINT

Use the same technique we used to prove the remainder theorem.

Example 24

Use synthetic division to find the quotient and remainder when dividing $f(x) = 6x^3 - 7x^2 - 15x + 1$ by $g(x) = 3x + 1$.

$g(x) = 3x + 1 = 3\left(x - \left(-\frac{1}{3}\right)\right)$	Rewrite the divisor so that you can use the remainder theorem.
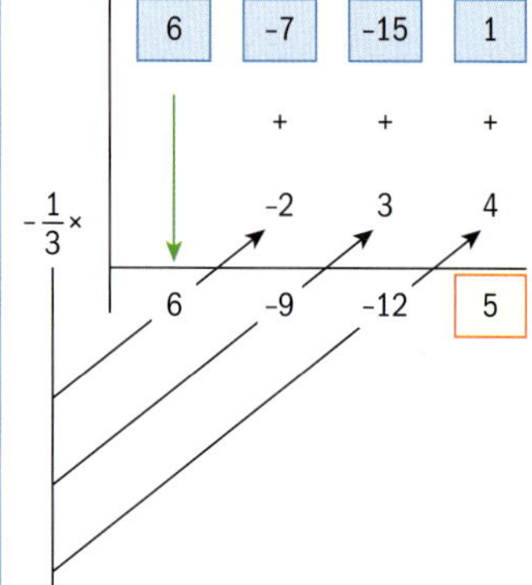	Use Horner's algorithm.
So the quotient is $q(x) = 2x^2 - 3x - 4$ and the remainder is $r(x) = 5$. $6x^3 - 7x^2 - 15x + 1 = (2x^2 - 3x - 4)(3x + 1) + 5$	Divide the coefficients of the remainder in the table by 3.

Example 25

When polynomial $f(x) = x^3 + 3x^2 - ax - 5$ is divided by $(x + 2)$ the remainder is 3. Find the value of a.

$x + 2 \Rightarrow r = f(-2)$	Use the remainder theorem.
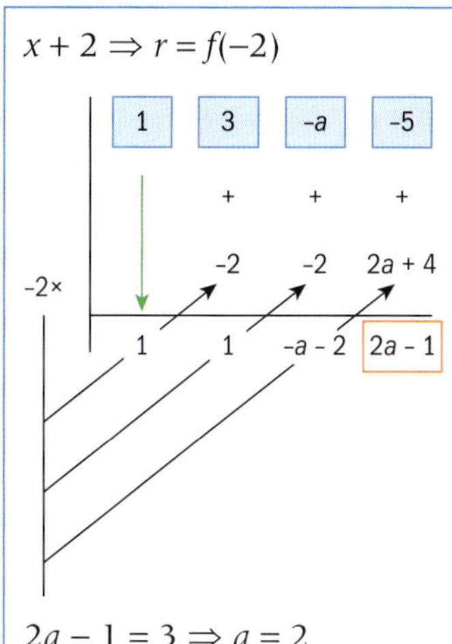 	Use synthetic division to find the remainder in terms of a.
$2a - 1 = 3 \Rightarrow a = 2$	Equate the remainder you obtained in the table with 3, and solve for a.

Example 26

Find the remainder when polynomial $f(x) = 3x^{2019} + 5x^{1019} - 7x + 4$ is divided by $x^2 - 1$.

$f(x) = (x^2 - 1) \times q(x) + \underbrace{ax + b}_{r(x)}$	Write the unique decomposition of the polynomial and notice that the remainder is a linear polynomial.
$x = 1 \Rightarrow f(1) = 3 \times 1^{2019} + 5 \times 1^{1019} - 7 \times 1 + 4 = 5$ $x = -1 \Rightarrow f(-1) = 3 \times (-1)^{2019} + 5 \times (-1)^{1019} - 7 \times (-1) + 4 = 3$	Calculate the value of the polynomial at the zeros of the divisor.
$\begin{cases} f(1) = (1^2 - 1) \times q(1) + a \times 1 + b = 5 \\ f(-1) = ((-1)^2 - 1) \times q(-1) + a \times (-1) + b = 3 \end{cases}$	Use the unique decomposition of the polynomial and form simultaneous equations.
$\begin{cases} a + b = 5 \\ -a + b = 3 \end{cases} \Rightarrow \begin{cases} b = 4 \\ a = 1 \end{cases}$ Therefore the remainder is $r(x) = x + 4$.	Solve the simultaneous equations and write the remainder.

Exercise 3K

1 In each case, use synthetic division to find the quotient and the remainder when polynomial f is divided by g.

a $f(x) = 4x^3 - 5x^2 + 13x - 2$ and $g(x) = 4x - 1$

b $f(x) = 6x^3 + 11x^2 + 5x + 4$ and $g(x) = 2x + 3$

c $f(x) = 2x^4 + 3x^3 - 6x^2 + 4x + 3$ and $g(x) = 2x - 1$

d $f(x) = 3x^5 - 5x^4 + x^3 - 2x^2 + 8x + 2$ and $g(x) = 3x + 1$

2 When the polynomial f is divided by the polynomial $g(x) = 2x^2 - 3x + 1$ you obtain the quotient $q(x) = x^2 + 2$ and the remainder $r(x) = x - 3$. Find the polynomial f.

3 $f(x) = 6x^5 + 17x^4 - 20x^3 - 35x^2 + 44x + a$ is divisible by $(x + 2)$. Find the value of a.

4 $f(x) = 2x^4 + 5x^3 - 4x^2 + bx + 1$ is divisible by $(2x + 1)$. Find the value of b.

5 $f(x) = x^4 + 5x^3 + 5x^2 + ax + b$ is divisible by $(x - 1)$ and $(x + 2)$. Find the values of a and b.

6 $f(x) = 6x^5 + 13x^4 - 30x^3 - 45x^2 + ax + b$ is divisible by $(2x - 1)$ and when divided by $(x - 1)$ the remainder is -40. Find the values of a and b.

7 When polynomial f is divided by $(x + 2)$ the remainder is 4, and when divided by $(x - 5)$ the remainder is -3. Find the remainder when polynomial f is divided by $(x^2 - 3x - 10)$.

8 Find the remainder when $f(x) = x^{2019} + x^{2018} + \ldots + x + 1$ is divided by $(x + 1)$.

9 Show that the polynomial $f(x) = (x + 2)^{2n} + (x + 3)^n - 1$ is divisible by $(x^2 + 5x + 6)$ for all $n \in \mathbb{Z}^+$.

10 Given a polynomial $f(x) = a_n x^n + a_{n-1}x^{n-1} + \ldots + a_2x^2 + a_1x + a_0$, $a_k \in \mathbb{R}$, $k = 1, 2, \ldots, n$, $a_n \neq 0$, and real numbers a and b, $a \neq 0$, show that the remainder when $f(x)$ is divided by the linear expression $(ax - b)$ is $f\left(\frac{b}{a}\right)$.

Developing inquiry skills

Return to the opening problem.

Can you use a polynomial to model the roller coaster?

3.4 The fundamental theorem of algebra

Investigation 12

1 Factorize the following numbers into prime factors. 30, 504, 1155, 35 200

2 Fully factorize the following polynomials.
$x^2 + 10x + 25, x^3 - 2x^2 - 15x, x^3 - 3x^2 + 3x - 1, x^4 - 5x^2 + 4$.

3 **Conceptual** What are the similarities and differences between factorizing a number into prime factors, and factorizing a polynomial expression?

The fundamental theorem of algebra is one of the most important theorems in mathematics. It is an algebraic version of the fundamental theorem of arithmetic (which states that every number is either prime or is the product of a unique combination of prime factors.)

The fundamental theorem of algebra refers to the existence of complex zeros of a polynomial.

The fundamental theorem of algebra (FTA)

A polynomial $f(x) = a_n x^n + a_{n-1} x^{n-1} + \ldots + a_2 x^2 + a_1 x + a_0$, $a_n \neq 0$, with real or complex coefficients has at least one zero. There is an $\omega \in \mathbb{C}$ such that $f(\omega) = 0$.

Gauss proved this theorem, but the proof goes beyond the level of this textbook.

There are a lot of theorems, lemmas and corollaries that are derived from the fundamental theorem of algebra that can help with algebraic manipulation of equations and polynomial functions.

Lemma

Each polynomial $f(x) = a_n x^n + a_{n-1} x^{n-1} + \ldots + a_2 x^2 + a_1 x + a_0$, $a_n \neq 0$, with real coefficients can be written in a factor form, $f(x) = a_n (x - \omega_1)(x - \omega_2)\ldots(x - \omega_n)$, such that $\omega_k \in \mathbb{C}$, $k = 1, \ldots, n$.

Example 27

Express the polynomial $f(x) = x^4 - 6x^3 - 19x^2 + 24x$ as a product of linear factors, and check your answer by using your calculator.

$f(x) = x^4 - 6x^3 - 19x^2 + 24x$

$= x(x^3 - 6x^2 - 19x + 24)$

Use the lemma to factorize the polynomial. All the terms have the common factor x, so you can factorize this first.

	1	-6	-19	24
1×		1	-5	-24
	1	-5	-24	0

Notice that the sum of the coefficients of the remaining cubic expression is 0, therefore apply synthetic division for $x = 1$ to factorize it further. $(x - 1)$ is a factor since the remainder is 0.

$x^2 - 5x - 24 = x^2 \underbrace{-8x + 3x}_{-5x} - 24$

Factorize the remaining quadratic expression by splitting the middle term.

Continued on next page

$= x(x-8) + 3(x-8) = (x-8)(x+3)$	Use all the factors in the factor form.
$f(x) = x(x-1)(x+3)(x-8)$	Make sure you know how to use your calculator to check the zeros.
Checking: polyRoots$(x^4-6\times x^3-19\times x^2+24\times x, x)$ {-3, 0, 1, 8}	

Notice that a calculator gives zeros and you need to use the FTA to write the polynomial in the factor form.

If a certain factor appears more than once we say that the factor has **multiplicity**. So if we have a polynomial f of nth degree with less than n different zeros, then the sum of their multiplicities will add up to n.

$$f(x) = a_n(x-\omega_1)^{p_1}(x-\omega_2)^{p_2}\dots(x-\omega_k)^{p_k},\ k < n,\ \sum_{i=1}^{k} p_k = n$$

Example 28

-1 is a zero of the polynomial $f(x) = 2x^5 + 7x^4 + 3x^3 - 13x^2 - 17x - 6$ with multiplicity 3.

Express $f(x)$ as the product of linear factors.

Check your answers by using a calculator.

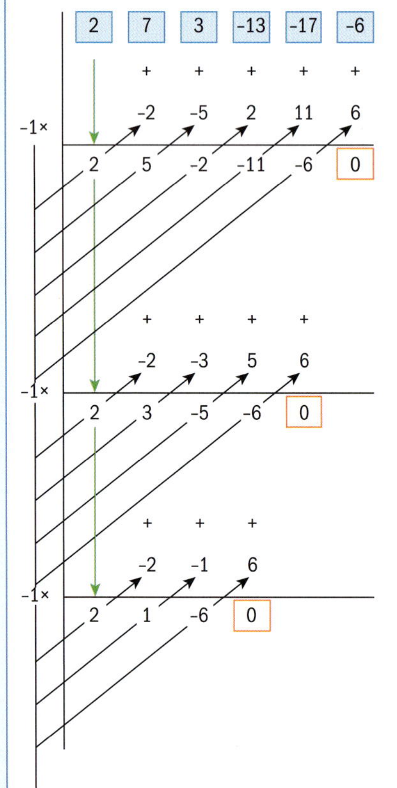

Successively apply synthetic division with respect to the multiplicity of the given zero, in this case three times.

$2x^2 + x - 6 = 2x^2 \underbrace{-3x + 4x}_{x} - 6$

$= x(2x - 3) + 2(2x - 3) = (2x - 3)(x + 2)$

$f(x) = (x + 1)^3(2x - 3)(x + 2)$

Checking:

polyRoots$(2\times x^5 + 7\times x^4 + 3\times x^3 - 13\times x^2 - 17\times x$
$\left\{-2, -1, -1, -1, \frac{3}{2}\right\}$

Factorize the remaining quadratic expression by splitting the middle term.

Use all the factors in the factor form.

Use your calculator to check the zeros.

Notice that the last zero is in fraction form, so you need to multiply it by 2 to obtain the linear factor:

$2\left(x - \frac{3}{2}\right) = (2x - 3)$

Investigation 13

1 Use the remainder theorem and synthetic division to show that the complex number z is a zero of the polynomial f.

- **a** $z = \mathrm{i}, f(x) = 2x^3 - 3x^2 + 2x - 3$
- **b** $z = -4\mathrm{i}, f(x) = 2x^3 - x^2 + 32x - 16$
- **c** $z = -2 + 3\mathrm{i}, f(x) = 3x^3 + 10x^2 + 31x - 26$
- **d** $z = 1 + \mathrm{i}, f(x) = x^4 + x^3 - 2x^2 + 2x + 4$
- **e** $z = -1 + 3\mathrm{i}, f(x) = x^5 + x^4 + 9x^3 - 9x^2 + 8x - 10$

2 Use the remainder theorem and synthetic division to find the values of the polynomial f at z^* for all the complex numbers given in question **1**.

3 **Conceptual** If z is a zero of a polynomial f, what can you say about the complex conjugate z^*?

4 **Factual** By considering the fundamental theorem of algebra and your answer to question **3**, what can you say about the zeros of an odd-degree polynomial?

Did you know?

Due to the imperfection of the calculator's algorithm when finding zeros, you obtain approximations of the multiple zeros, without their multiplicity. This will occur even more times when complex zeros occur.

TOK

Aliens might not be able to speak an Earth language but would they still describe the equation of a straight line in similar terms?

Is mathematics a formal language?

Your findings from Investigation 13 can be summarized by the following theorem.

Conjugate root theorem

Given a polynomial

$f(x) = a_n x^n + a_{n-1}x^{n-1} + \ldots + a_2 x^2 + a_1 x + a_0,\ a_k \in \mathbb{R}, k = 1, 2, \ldots, n, a_n \neq 0,$

that has a complex zero z, then its conjugate z^* is also a zero of the polynomial f.

Example 29

Given that $1 + 3\text{i}$ is a complex zero of the polynomial $f(x) = x^3 - 5x^2 + 16x - 30$, find all the other zeros of f.

Check your answers using a calculator.

Method 1	Use the conjugate zero theorem.
$x_1 = 1 + 3\text{i} \Rightarrow x_2 = 1 - 3\text{i}$ is a zero of f	
$(x - (1 + 3\text{i}))(x - (1 - 3\text{i}))$ $= x^2 - x - 3\text{i}x - x + 3\text{i}x + 1^2 + 3^2 = x^2 - 2x + 10$	Expand and find the quadratic factor.
$$\begin{array}{r\|rrrr} & x & -3 & & \\ \hline x^2-2x+10 & x^3 & -5x^2 & 16x & -30 \\ & -x^3 & -2x^2 & +10x & \\ \hline & 0 & -3x^2 & 6x & -30 \\ & & -3x^2 & +6x & -30 \\ \hline & & 0 & 0 & 0 \end{array}$$	Use long division to find the remaining factor.
$f(x) = (x - (1 + 3\text{i}))(x - (1 - 3\text{i}))(x - 3)$ $x - 3 = 0 \Rightarrow x = 3$	Fully factorize the polynomial and find the remaining zero.
Method 2 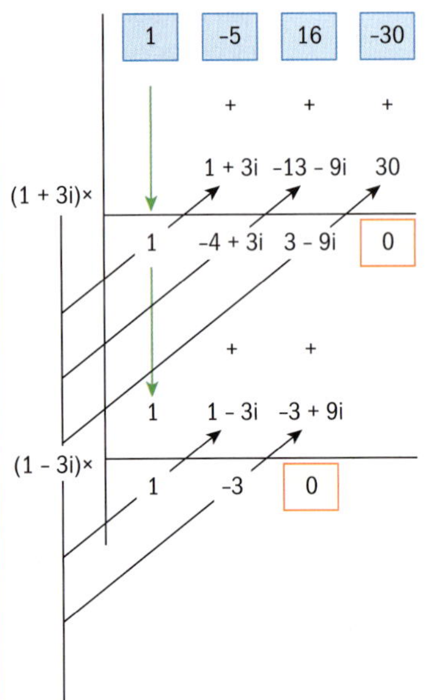	Use synthetic division for complex numbers twice.
$f(x) = (x - (1 + 3\text{i}))(x - (1 - 3\text{i}))(x - 3)$ $x - 3 = 0 \Rightarrow x = 3$	Fully factorize the polynomial and find the remaining zero.
Checking: 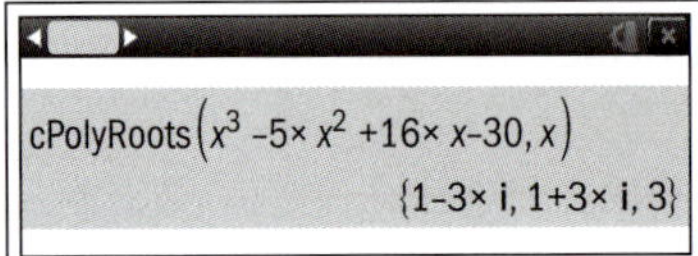	Make sure that you know how to use your calculator to check your answers. Notice that you need to use complex roots of polynomials.

Example 30

The polynomial $f(x) = x^4 - x^3 + 2x^2 + ax - 24, a \in \mathbb{R}$, is given.

a Find the value of a given that $2i$ is a complex zero of f.

b Hence find all the remaining zeros of f, and check your answers using a calculator.

a

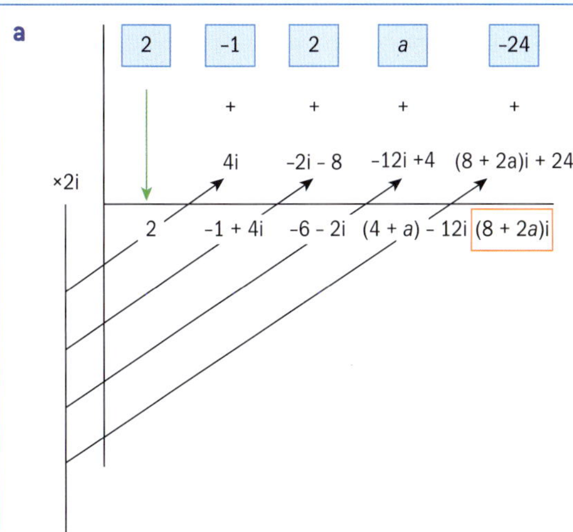

Apply synthetic division for the complex number $2i$.

$f(2i) = (8 + 2a)i = 0 \Rightarrow a = -4$

Use the remainder theorem.

b $x_1 = 2i \Rightarrow x_2 = -2i$ is a root of f

Use the conjugate zero theorem.

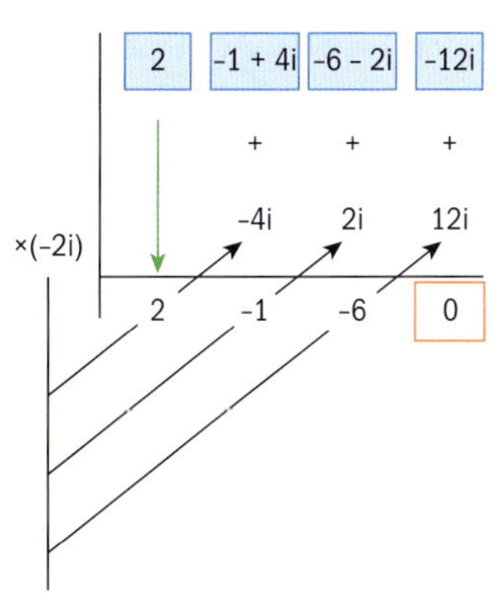

Use $a = -4$ and apply synthetic division a second time.

$q(x) = 2x^2 - x - 6 = 0 \Rightarrow (2x + 3)(x - 2) = 0$

$\Rightarrow 2x + 3 = 0$ or $x - 2 = 0 \Rightarrow x = -\dfrac{3}{2}$ or $x = 2$

Identify the quotient and find its zeros.

Checking:

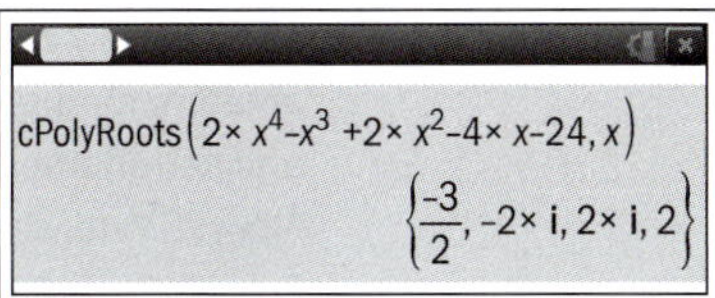

Use your calculator to check your answers.

Exercise 3L

1 Find a polynomial of the smallest degree with integer coefficients whose zeros are

a 0, 2 and 7 **b** −3, −2, 1 and 3

c $-1, -\frac{1}{2}$, 2 and 5.

2 Find a polynomial of the smallest degree with integer coefficients whose zeros include

a $\sqrt{2}$ and 2 **b** $-1, \frac{1}{2}$ and $\sqrt[3]{3}$

c $1-\sqrt{3}$ and $\sqrt[3]{2}$

3 Factorize the following polynomials and check your answers using a calculator.

a $f(x) = x^3 - 3x^2 + 4x - 2$

b $f(x) = 3x^3 - x^2 + 2x + 6$

c $f(x) = 2x^4 - 5x^3 + 11x^2 - 3x - 5$

4 In each case, k is a zero with multiplicity n of the polynomial f.

Factorize f fully and check your answers using a calculator.

a $k = -2, n = 2, f(x) = 3x^3 + 7x^2 - 8x - 20$

b $k = \frac{2}{3}, n = 2, f(x) = 9x^3 + 24x^2 - 44x + 16$

c $k = 1, n = 2, f(x) = x^4 - 2x^3 - 3x^2 + 8x - 4$

d $k = -\frac{1}{2}, n = 3, f(x) = 8x^4 + 20x^3 + 18x^2 + 7x + 1$

e $k = 1, n = 4, f(x) = 5x^5 - 13x^4 + 2x^3 + 22x^2 - 23x + 7$

5 In each case, f is a polynomial with complex zero z. Find the remaining zeros.

a $f(z) = z^3 - 2z^2 + 4z - 8,\ z = 2\text{i}$

b $f(z) = 2z^3 - 13z^2 + 32z - 13,\ z = 3 - 2\text{i}$

c $f(z) = 3z^3 + 4z^2 - 4z + 7,\ z = \frac{1}{2} - \frac{\sqrt{3}}{2}\text{i}$

d $f(z) = z^4 - 2z^3 + 6z^2 - 2z + 5,\ z = -\text{i}$

e $f(z) = z^4 + 5z^3 + 11z^2 + 13z + 10,\ z = -2 - \text{i}$

f $f(z) = 3z^4 + 2z^3 + 19z^2 + 12z + 6,\ z = -\frac{1}{3} + \frac{\sqrt{2}}{3}\text{i}$

6 Given that z is a zero of the polynomial f, find the missing coefficients. Hence find all the remaining zeros of f and check your answers using a calculator.

a $z = 2, f(x) = x^3 + ax^2 + x - 2, a \in \mathbb{R}$

b $z = -5, f(x) = 2x^3 + 10x^2 + ax + 15, a \in \mathbb{R}$

c $z = \sqrt{5}\text{i}, f(x) = x^4 - 2x^3 + ax^2 + bx + 85, a, b \in \mathbb{R}$

d $z = 1 - 2\text{i}, f(x) = 3x^4 - 7x^3 + 18x^2 + ax + b, a, b \in \mathbb{R}$

Sum and product of polynomial roots

Just as you have already shown for quadratic polynomials, François Viète developed formulas that connect the zeros and the coefficients of a polynomial. Viète was the first to investigate this connection but he did it just for positive real zeros, whilst Albert Girard was the first to extend it to complex zeros.

International-mindedness

François Viète (1540–1603) was a French amateur mathematician and astronomer, whilst Albert Girard was a French mathematician and musician.

Investigation 14

1 Use your calculator to find the zeros of the following polynomials.

$f_1(x) = x^3 + x^2 - 4x - 4,\ f_2(x) = x^3 - 6x^2 + 11x - 6,$

$f_3(x) = x^3 - 4x^2 - 7x + 10,\ f_4(x) = x^3 + 27x^2 + 71x - 1155.$

2 Find the sums and the products of all the zeros for each polynomial in question **1**.

3 **Factual** How do the sum and product of the zeros you found in question **2** relate to the coefficients of each polynomial?

4 Use your calculator to find the zeros of the following polynomials.

$f_5(x) = 2x^3 - 11x^2 + 17x - 6, f_6(x) = 3x^3 - 16x^2 + 3x + 10,$
$f_7(x) = 15x^3 + 19x^2 - 4, f_8(x) = 14x^3 + 291x^2 + 1402x - 1155.$

5 Find the sums and the products of all the zeros for each polynomial in question **4**.

6 **Conceptual** How do the sum and product of the zeros of a cubic relate to their coefficients?

Your results from Investigation 14 form part of the following key point.

Given a cubic equation $ax^3 + bx^2 + cx + d = 0$, $a, b, c, d \in \mathbb{R}$, $a \neq 0$, with x_1, x_2 and x_3 as solutions, then

$$\begin{cases} x_1 + x_2 + x_3 = -\dfrac{b}{a} \\ x_1 \times x_2 + x_1 \times x_3 + x_2 \times x_3 = \dfrac{c}{a} \\ x_1 \times x_2 \times x_3 = -\dfrac{d}{a} \end{cases}$$

Example 31

The roots of a cubic equation $3x^3 + 5x^2 + 2x - 4 = 0$ are x_1, x_2 and x_3.

Without solving the equation, find

a $x_1 + x_2 + x_3$ **b** $x_1 \times x_2 \times x_3$ **c** $\dfrac{1}{x_1x_2} + \dfrac{1}{x_1x_3} + \dfrac{1}{x_2x_3}$.

$a = 3,\ b = 5,\ c = 2,\ d = -4$	Identify the coefficients of the cubic polynomial.
a $x_1 + x_2 + x_3 = -\dfrac{5}{3}$	Use the theorem for the sum.
b $x_1 \times x_2 \times x_3 = -\dfrac{-4}{3} = \dfrac{4}{3}$	Use the theorem for the product.

Continued on next page

c $\frac{1}{x_1x_2}+\frac{1}{x_1x_3}+\frac{1}{x_2x_3}=\frac{x_3+x_2+x_1}{x_1x_2x_3}=\frac{-\frac{5}{3}}{\frac{4}{3}}=-\frac{5}{4}$	Rewrite the expression in terms of the sum and product and then use the results from the previous parts. Make sure you know how to use a calculator to check your results. Use the complex roots finder to obtain all the possible zeros. Use the features for sum and product to find the results. Notice that the results obtained in decimal form are the same as the results in fraction form to a certain degree of accuracy.

Investigation 15

1 Use your calculator to find the zeros of the following polynomials.

$f_1(x)=x^4+x^3-7x^2-x+6, f_2(x)=x^4+4x^3-39x^2-86x+280,$

$f_3(x)=6x^4+7x^3-13x^2-4x+4, f_4(x)=35x^4+381x^3-73x^2-309x+110,$

$f_5(x)=4x^4-4x^3+65x^2-64x+16, f_6(x)=6x^4-11x^3+117x^2+397x-145.$

2 Find the sums and the products of all the zeros for each polynomial in question **1**.

3 **Conceptual** How do the sum and product of the zeros of a fourth-degree polynomial relate to its coefficients?

Given a polynomial $f(x)=a_nx^n+a_{n-1}x^{n-1}+\ldots+a_2x^2+a_1x+a_0$ with real or complex coefficients $(a_n\neq 0)$ and zeros $x_1, x_2, \ldots, x_n$ of f, then

$$x_1+x_2+x_3+\ldots+x_n=-\frac{a_{n-1}}{a_n}$$

and $x_1x_2x_3\ldots x_n=(-1)^n\frac{a_0}{a_n}$

The proof comes as a direct consequence of the factor theorem, as you saw when proving the equivalent result for a cubic polynomial. Since every polynomial can be written in the form $f(x)=a_n(x-x_1)(x-x_2)\ldots(x-x_n)$, by expanding this expression we obtain the given formulas.

Example 32

Find the sum and the product of the zeros of the following polynomials.

a $f(x) = x^4 + x^3 - 2x^2 + 3x - 2$

b $f(x) = 2x^5 - 5x^4 + 15x^3 - 511x^2 - 4x + 8$

c $f(x) = -5x^{2019} + 7x^{1457} - 4x^{250} - 15x^{47} + 21$

a $n = 4,\ a_4 = 1,\ a_3 = 1,\ a_0 = -2$ $x_1 + x_2 + x_3 + x_4 = -\frac{a_3}{a_4} = -\frac{1}{1} = -1$ $x_1x_2x_3x_4 = (-1)^4 \frac{a_0}{a_4} = \frac{-2}{1} = -2$	Use the formulas for sum and product from the key point. Notice that for those formulas you need to use only the three coefficients.
b $n = 5,\ a_5 = 2,\ a_4 = -5,\ a_0 = 8$ $\sum_{k=1}^{5} x_k = -\frac{a_4}{a_5} = -\frac{-5}{2} = \frac{5}{2}$ $\prod_{k=1}^{5} x_k = (-1)^5 \frac{a_0}{a_5} = -\frac{8}{2} = -4$	
c $n = 2019,\ a_{2019} = -5,\ a_{2018} = 0,$ $a_0 = 21$ $\sum_{k=1}^{2019} x_k = -\frac{a_{2018}}{a_{2019}} = -\frac{0}{-5} = 0$ $\prod_{k=1}^{2019} x_k = (-1)^{2019} \frac{a_0}{a_{2019}} = -\frac{21}{-5} = \frac{21}{5}$	Use your calculator to check the results. Notice that sometimes, due to the accuracy of the calculator's algorithm, the results are not exact (the complex part is almost 0). In this example the calculator cannot even find the zeros due to the high order, so it is helpful to have an algebraic technique to find their sum and product.

Exercise 3M

1 Find the sum and product of the zeros of these polynomials.

a $f(x) = x^4 - 3x^3 + 2x^2 - 5x + 4$

b $f(x) = 2x^6 - 6x^5 + 11x^4 + 13x^3 - 5x$

c $f(x) = 23x^{17} - \frac{5}{11}x^9 + \sqrt{13} \times x^4 - \frac{\pi}{3}x - 46$

d $f(x) = 3x^{2020} - 4x^{2019} + 17x^{573} + 115x - 8$

2 The cubic equation $4x^3 - 2x^2 + x - 7 = 10$ has roots x_1, x_2 and x_3. Without solving the equation, find

a $x_1 + x_2 + x_3$

b $x_1 \times x_2 \times x_3$

c $10x_1 + 10x_2 + 10x_3$

d $\frac{3}{x_1 \times x_2} + \frac{3}{x_1 \times x_3} + \frac{3}{x_2 \times x_3}$

Check your results using a calculator.

3 The roots of the quartic equation
$6x^4 + 2x^3 - 11x^2 + 13x - 3 = 0$
are x_1, x_2, x_3 and x_4. Without solving the equation, find

a $x_1 + x_2 + x_3 + x_4$

b $x_1 \times x_2 \times x_3 \times x_4$

c $3x_1 + 3x_2 + 3x_3 + 3x_4$

d $\frac{6}{x_1 \times x_2 \times x_3} + \frac{6}{x_1 \times x_2 \times x_4} + \frac{6}{x_1 \times x_3 \times x_4} + \frac{6}{x_2 \times x_3 \times x_4}$

Check your results using a calculator.

3.5 Solving equations and inequalities

Since you have already extensively covered solving quadratic equations in Section 3.1, in this section you are mainly going to be focused on solving polynomial equations of degree 3 or higher.

Historically, a group of Italian mathematicians, del Ferro, Tartaglia, and Cardano (13th–14th century), developed a formula for solving a general cubic equation.

That formula is very complicated and very difficult to use, so you are going to be studying rational solutions that can be found by alternative, simpler methods.

The most common method that you are going to use for solving polynomial equations is factorization. In order to be able to factorize the polynomial you are going to use theorems that are valid for polynomials of all degrees.

Before you start on factorization you are going to look at how many real zeros you may expect for a given polynomial.

Did you know?

Babylonians were able to solve some cases of cubic equations (2000–1600 BC). They had tables with perfect squares and perfect cubes and their sums. By using those tables, they were solving equations of the form $ax^3 + bx = c$.

Solving polynomial equations

In this section you are going to solve polynomial equations by using algebraic methods and then use a calculator to verify the solutions.

Example 33

Solve the equation $x^3 - 7x + 6 = 0$ by using algebra.

$x^3 - 7x + 6 = \underbrace{x^3 - x}\underbrace{-6x + 6}$	Split the linear term so that you can factorize in pairs.
$= x(x^2 - 1) - 6(x - 1) = x(x - 1)(x + 1) - 6(x - 1)$	Factorize the difference of two squares and then factorize the expression.
$= (x - 1)(x(x + 1) - 6)$	Factorize the quadratic factor by inspection.
$= (x - 1)(x^2 + x - 6) = (x - 1)(x - 2)(x + 3)$	
$\Rightarrow (x - 1)(x - 2)(x + 3) = 0$	Use zero product theorem.
$\Rightarrow x_1 = 1, x_2 = 2, x_3 = -3$	
	Make sure you know how to check the result by finding the zeros of the polynomial using a calculator.

On a calculator you can also use the graphical method for finding zeros of the polynomial.

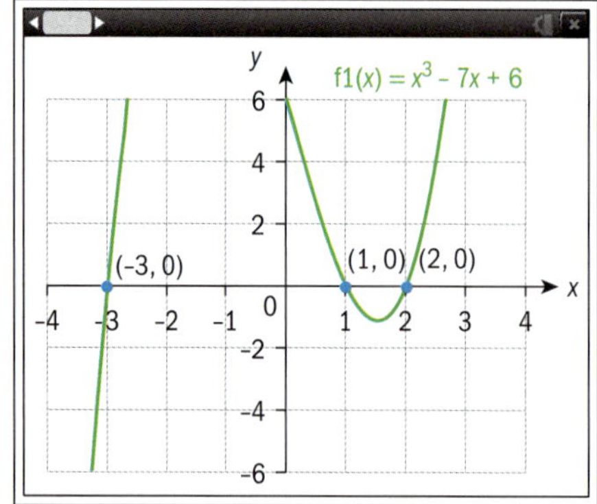

Apart from quadratics, you haven't drawn many graphs of higher-order polynomials. You are going to graph more polynomials once calculus is introduced.

Exercise 3N

1 Solve the following equations. Give all the real solutions, and check your answers using a calculator.

a $x^3 + x^2 - 4x - 4 = 0$

b $x^3 + 2x^2 - 9x - 18 = 0$

c $x^3 - 3x^2 + 3x - 2 = 0$

d $x^4 + x^3 - 3x^2 - x + 2 = 0$

2 The equation $x^3 + x^2 + ax - 4 = 0$ has one root equal to -2.

a Find the value of a.

b Find the remaining two roots.

3 The equation $2x^3 + ax^2 + bx + 9 = 0$ has one repeated root equal to 3.

a Find the values of a and b.

b Find the remaining root.

4 $f(x) = x^3 + ax^2 + bx + c,\ a, b, c \in \mathbb{R}$.

Two zeros of this polynomial are opposite numbers.

a Show that $ab = c$.

b Find the third zero.

Solving polynomial inequalities

When solving polynomial inequalities by an algebraic method, you need to fully factorize the polynomial and then investigate the signs of the factors on the whole set of real numbers by constructing a sign table. You should identify for which values of x the polynomial is either positive or negative, and use this to solve the inequality.

When solving polynomial inequalities by a graphical method, use a calculator to graph the polynomial and again identify for which values of x the polynomial is either positive or negative, and use this to solve the inequality.

Example 34

Use an algebraic method to solve the inequality $x^3 + 4x^2 + x - 6 > 0$ and check your answer using a calculator.

Factorization:

$x^3 + 4x^2 + x - 6 = x^3 + 4x^2 + 4x - 3x - 6$

$= x(x^2 + 4x + 4) - 3(x + 2) = x(x + 2)^2 - 3(x + 2)$

$= (x + 2)(x(x + 2) - 3) = (x + 2)(x^2 + 2x - 3)$

$= (x + 2)(x - 1)(x + 3)$

Factorize the expression by splitting the linear term.

Use the distributive property to factorize the expression.

Factorize the remaining quadratic factor.

x	$]-\infty, -3[$	-3	$]-3, -2[$	-2	$]-2, 1[$	1	$]1, \infty[$
$x+3$	–	0	+	+	+	+	+
$x+2$	–	–	–	0	+	+	+
$x-1$	–	–	–	–	–	0	+
x^3+4x^2+x-6	–	0	+	0	–	0	+

Construct the sign table and investigate the sign.

$x \in]-3, -2[\cup]4, \infty[$

Identify the intervals where the sign is positive.

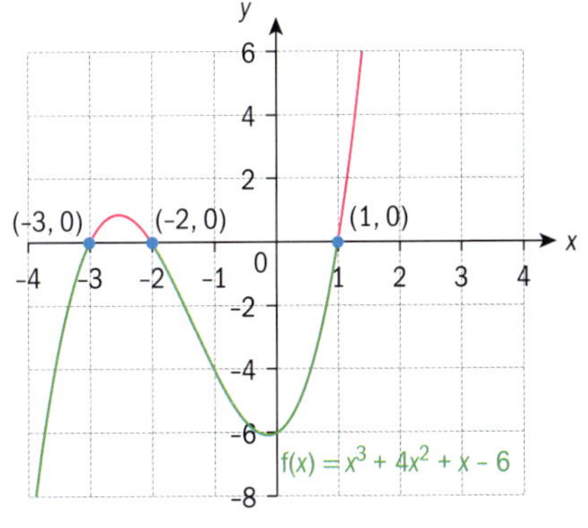

Draw the graph of the polynomial and find all the zeros. Identify the values of x for which the graph is above the x-axis.

Notice that when the inequality is not strict, you need to include the boundaries in the solution.

Example 35

Given the polynomials $f(x) = 3x^3 + 2x^2 - 3$ and $g(x) = x^3 - x^2 + 3x - 1$, find all the values of x such that $f(x) \geq g(x)$ by using a graphical method on a calculator.

Method 1 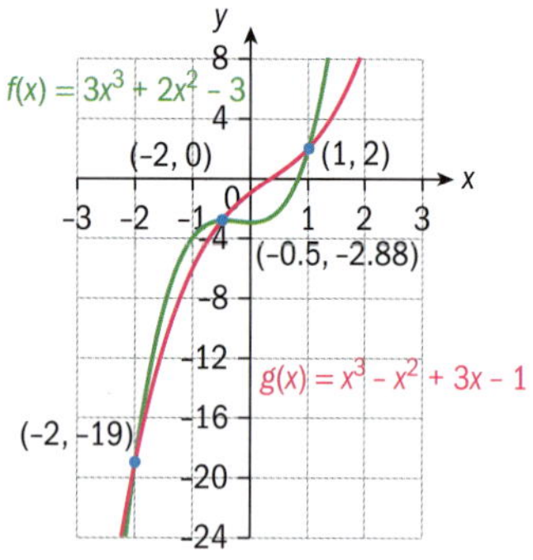	Graph both polynomials and find the points of intersection.
$x \in [-2, -0.5] \cup [1, \infty[$	Identify the intervals on the x-axis where the graph of f is above the graph of g.
Method 2 $3x^3 + 2x^2 - 3 \geq x^3 - x^2 + 3x - 1$ $2x^3 + 3x^2 - 3x - 2 \geq 0$ 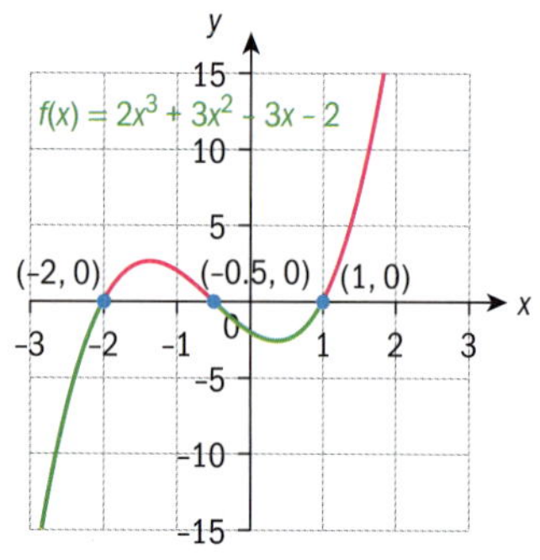 $x \in [-2, -0.5] \cup [1, \infty[$	Rewrite the inequality as one polynomial and find its zeros. Then find the intervals where the polynomial is above or on the x-axis.

HINT

In the case of Method 1 you may have needed to consider changing the scale on your GDC to find all the points of intersection between the graphs that might not be visible in the original window. Also, it can be difficult to read from the graph which function is the upper one and which the lower one. This is even more difficult on calculators with lower resolution.

Method 2 is more suitable because you don't need to find the points of intersection but only find zeros, which are always on the x-axis. This saves time as you don't need to change scales in the window, etc.

Sometimes polynomial equations or inequalities can be easily broken to simpler polynomials that you can sketch and find the solution by inspection.

Example 36

Use a simple polynomial graph to solve the inequality $x^3 - x^2 - 4 \leq 0$.

<table>
<tr><td>$x^3 - x^2 - 4 \leq 0 \Rightarrow \underbrace{x^3}_{f(x)} \leq \underbrace{x^2 + 4}_{g(x)}$
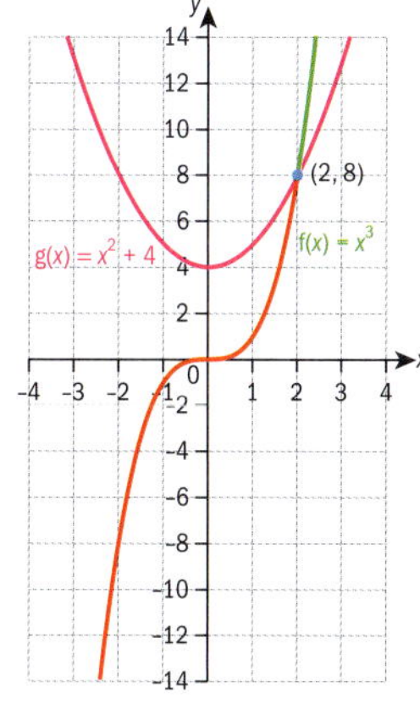</td><td>Rewrite the inequality by using two simple polynomials.
Or
Simply enter the original inequality into the GDC.

Graph those two simple polynomials and find by inspection the point of intersection.</td></tr>
<tr><td>$x \in \mathbb{R}, x \leq 2$</td><td>Identify the interval where the graph of f is below the graph of g.</td></tr>
</table>

Notice that an equivalent inequality $x^3 - 4 \leq x^2$ would also be an example of simple polynomials to graph and find the solution.

Algebraic methods for solving polynomials are restricted to the third degree in this course. A GDC may be used to solve polynomial equations and inequalities of a higher degree.

Example 37

Use a GDC to solve the inequality $4x^2 < 2x^7 + 1$.

<table>
<tr><td>$-2x^7 + 4x^2 - 1 < 0$
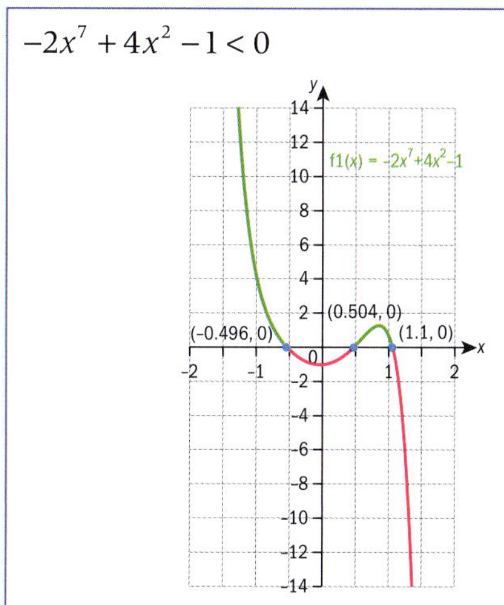</td><td>Rewrite the inequality as one polynomial.</td></tr>
<tr><td>$-0.496 < x < 0.504$ or $x > 1.10$</td><td>Graph the polynomial and find the zeros. Identify which part of the graph is below the x-axis.</td></tr>
</table>

Notice that the equivalent inequality $2x^7 - 4x^2 + 1 > 0$ gives the same result.

Exercise 30

1 Solve the following inequalities in the set of real numbers by the method of your choice.

a $x^3 + x^2 - 4x - 4 < 0$

b $x^3 + 2x^2 - 9x - 18 > 0$

c $x^3 - 3x^2 + 3x - 2 \le 0$

d $4x^3 + 8x^2 + x - 3 \ge 0$

e $3x^3 + 4x^2 + 7x + 2 > 0$

f $12x^3 - 16x^2 - 81x - 35 < 0$

g $x^4 + x^3 - 3x^2 - x + 2 \le 0$

h $2x^4 - x^3 + x^2 - x - 1 < 0$

2 Given the polynomials $f(x) = 3x^3 + 5x^2 - 6x - 9$ and $g(x) = x^3 + 2x^2 - 4x - 4$, find all the values of x such that $f(x) \le g(x)$.

3 Use simple polynomial graphs to find the solutions of the following inequalities.

a $x^3 + x + 2 > 0$ **b** $-2x^3 + x^2 + 1 \ge 0$

c $x^4 + x^2 - 2 \le 0$

4 Use a calculator to solve the following inequalities.

a $x^5 - 4x^3 + 2x + 1 \ge 0$

b $x^{13} + 4 < 3x^8 + 5x$

c $x^{15} + 2x^{14} + 5x^8 > 4x^2 - 1$

3.6 Solving systems of linear equations

Systems of two linear equations with two unknowns

Up to this point in your studies, you have only been asked to solve systems of linear equations (such as a pair of simultaneous equations) where there is a unique solution: a single pair of real numbers x and y that satisfy both equations.

In this section, you will learn about the possible types of solutions that can occur when solving systems of linear equations.

Investigation 16

1 Without using a calculator, solve the following simultaneous equations.

a $\begin{cases} 4x + 3y = 18 \\ 7x - 4y = 13 \end{cases}$ **b** $\begin{cases} 2x - 5y = 4 \\ -6x + 15y = 3 \end{cases}$ **c** $\begin{cases} 10x - 4y = 3 \\ -2x + \frac{4}{5}y = -\frac{3}{5} \end{cases}$

2 For each set of equations **a–c**, sketch both lines on the same axes. Where possible, show on your sketch the solution to the pair of equations.

3 **Factual** What are the different types of solution to a system of two linear equations with two unknowns?

4 **Conceptual** Explain the geometrical significance of each type.

When solving systems of two linear equations with two unknowns there are three possible types of solutions that can occur. Each type has a different geometrical interpretation.

1 When there is **a unique pair** of numbers that satisfy both equations, the lines intersect at one point.

2 When there are **no real numbers** that satisfy both equations, the lines are parallel and distinct.

3 When there are **infinitely many pairs** of real numbers that satisfy both equations, the lines coincide.

TOK

If we can find solutions in higher dimensions can we reason that these spaces exist beyond our sense of perception?

All the results of solving systems of linear equations can be checked on a calculator. Make sure you know how to do this using your GDC.

HINT

In type 3, infinitely many pairs of real numbers (x, y) satisfy this system where x can be **any value**, whilst y is a value **dependent** on x. Instead of expressing x in terms of x as usual, a calculator may provide an answer where x is expressed in terms of y.

Example 38

Find the value of a real parameter m such that the system

$\begin{cases} 4x - y = 1 \\ 3x + my = 2 \end{cases}$ has no solution.

$\begin{cases} 4x - 1 = y \\ 3x + m(4x - 1) = 2 \end{cases} \Rightarrow \begin{cases} 4x - 1 = y \\ (3 + 4m)x = 2 + m \end{cases}$ $x = \dfrac{2+m}{3+4m}$	Apply the method of substitution by expressing y as the subject in the first equation, and substituting it in the second equation.
$3 + 4m = 0 \Rightarrow m = -\dfrac{3}{4}$	The system has no solution when $3 + 4m = 0$, since then you would be dividing by zero.

The system of equations in Example 38 can either have no solutions (in the case where $m = -\frac{3}{4}$), or a unique solution (in the case where $m \neq -\frac{3}{4}$).

It is not possible for this particular system to have infinitely many solutions since the coefficients of the variable x and the constant coefficients are not proportional, ie $\frac{4}{3} \neq \frac{1}{2}$.

TOK

How accurate is a visual representation of a mathematical concept?

Example 39

Consider the system of equations $\begin{cases} 2x + (p-1)y = 1 \\ px + 3y = 2 \end{cases}$.

a Find the values of the real parameter p such that the system has a unique solution.

b Find the unique solution in terms of p.

a $\begin{cases} 2x + (p-1)y = 1 \,\vert \times p,\ p \neq 0 \\ px + 3y = 2 \,\vert \times 2 \end{cases}$	Use the method of elimination; multiply the first equation by p and the second by 2.
$\Rightarrow \begin{cases} 2px + p(p-1)y = p \\ 2px + 6y = 4 \end{cases} \Rightarrow (p^2 - p - 6)y = p - 4$	Subtract the two equations to eliminate x.
$\Rightarrow p^2 - p - 6 \neq 0 \Rightarrow (p-3)(p+2) \neq 0$ $p \neq 3$ and $p \neq -2$	In order to have a unique solution the coefficient next to y must not be equal to 0.
b $(p^2 - p - 6)y = p - 4 \Rightarrow y = \frac{p-4}{p^2 - p - 6}$	Use the elimination in part **a** to express y in terms of p.
$px = -3y + 2 \Rightarrow px = \frac{-3(p-4)}{p^2 - p - 6} + 2$	Substitute your value of y into the second equation.
$\Rightarrow px = \frac{-3p + 12 + 2p^2 - 2p - 12}{p^2 - p - 6}$ $\Rightarrow x = \frac{2p^2 - 5p}{p(p^2 - p - 6)} \Rightarrow x = \frac{2p - 5}{p^2 - p - 6}$ $(x, y) = \left(\frac{2p-5}{p^2 - p - 6}, \frac{p-4}{p^2 - p - 6}\right), p \neq -2, 3$	Solve to find x in terms of p.

Notice that the case when $p = 0$ is covered by the unique solution $(x, y) = \left(\frac{5}{6}, \frac{2}{3}\right)$.

Exercise 3P

1 In each case, find the value of the real parameter m such that the system of equations has no solution.

a $\begin{cases} mx + 6y = -2 \\ 7x + 3y = 5 \end{cases}$ **b** $\begin{cases} (m-1)x + 2y = -1 \\ 3x + my = 11 \end{cases}$

2 In each case, find the values of the real parameter p such that the system of equations has infinitely many solutions. Find the solutions.

a $\begin{cases} 6x + py = 8 \\ 3x + y = 4 \end{cases}$

b $\begin{cases} px - y = p \\ (p-4)x + (p-1)y = -2 \end{cases}$

3 In each case, find the values of the real parameter s such that the system of equations has a unique solution. Hence find the solution in terms of s.

a $\begin{cases} 2x - sy = 1 \\ 3x - y = 2 \end{cases}$ **b** $\begin{cases} (s+2)x + y = s \\ (5-2s)x + sy = 4 \end{cases}$

4 The system

$$\begin{cases} a(x+y) + b(x-y) = 1 \\ a(x-y) + b(x+y) = 1 \end{cases}, a, b \in \mathbb{R}, a, b \neq 0$$

is given.

a Find the solution (x, y) in terms of a and b.

b State and explain whether it is possible for the system to have no solution.

Now let's take two linear equations in a general form and use the method of elimination to find a general solution.

Example 40

Find the general formula for a unique solution of $\begin{cases} ax + by = e \\ cx + dy = f \end{cases}, a, b, c, d \neq 0.$

$\begin{cases} ax + by = e \\ cx + dy = f \end{cases} \Rightarrow \left.\begin{cases} adx + bdy = ed \\ bcx + bdy = fb \end{cases}\right] -$	Multiply the first equation by d and the second equation by b to obtain equal coefficients of the variable y.
$\Rightarrow adx - bcx \Rightarrow ed - fb$	Subtract the equations to eliminate the variable y.
$\Rightarrow x(ad - bc) \Rightarrow ed - fb$ $\Rightarrow x = \dfrac{ed - fb}{ad - bc}$	Use distribution to make x the subject.
$a \times \dfrac{ed - fb}{ad - bc} + by = e$	Substitute x in one of the equations, eg the first equation.

Continued on next page

$\Rightarrow by = \dfrac{e(ad-bc)-a(ed-fb)}{ad-bc}$	
$= \dfrac{ead-ebc-aed+afb}{ad-bc}$	
$= \dfrac{afb-ebc}{ad-bc}$	
$\Rightarrow y = \dfrac{b(af-ec)}{ad-bc} \times \dfrac{1}{b}$	Make y the subject.
$\Rightarrow y = \dfrac{af-ec}{ad-bc}.$	
$\Rightarrow (x, y) = \left(\dfrac{ed-fb}{ad-bc}, \dfrac{af-ec}{ad-bc}\right), ad-bc \neq 0$	State the condition that the denominator cannot be equal to 0, which is required for the system to have a unique solution.

In Example 40 you could have first eliminated x instead of y, which would have given the same end result.

Did you know?

Gabriel Cramer (1704–1752) has developed this formula within his work on determinants. The form $\begin{vmatrix} a & b \\ c & d \end{vmatrix}$ is called a determinant. It is a numerical value such that $\begin{vmatrix} a & b \\ c & d \end{vmatrix} = ad - bc.$

All the expressions in the values of x and y in Example 40 can be written as determinants in the following way.

$$\begin{vmatrix} a & b \\ c & d \end{vmatrix} = D, \quad \begin{vmatrix} e & b \\ f & d \end{vmatrix} = D_x, \quad \begin{vmatrix} a & e \\ c & f \end{vmatrix} = D_y$$

Hence, to calculate the values of x and y we need to calculate those three determinants and then we use the formulas $x = \dfrac{D_x}{D}, y = \dfrac{D_y}{D}, D \neq 0.$

Using determinants to calculate unique solutions works very efficiently when the coefficients of the simultaneous linear equations are complex numbers.

Example 41

Use the formula from Example 40 to solve these simultaneous equations with complex coefficients.

$$\begin{cases} 3x + 2iy = 5 + 7i \\ (1 - i)x - y = i \end{cases}$$

Check your answers using a calculator.

$a = 3,\ b = 2i,\ c = 1 - i,\ d = -1,\ e = 5 + 7i,\ f = i$	Identify the coefficients.
$3\times(-1) - 2i\times(1 - i) = -3 - 2i - 2 = -5 - 2i$	Using your result from Example 40, the denominator is $ad - bc$.
$(5 + 7i)\times(-1) - 2i \times i = -5 - 7i + 2 = -3 - 7i$	The numerator for x is $ed - fb$.
$3\times i - (5 + 7i)\times(1 - i) = 3i - 5 + 5i - 7i - 7 = -12 - i$	The numerator for y is $af - ec$.
$x = \frac{-3-7i}{-5-2i} \times \frac{-5+2i}{-5+2i} = \frac{15-6i+35i+14}{29} = 1+i$	$x = \frac{ed - fb}{ad - bc}$. Divide complex numbers by multiplying top and bottom by the conjugate of the denominator.
$y = \frac{-12+i}{-5-2i} \times \frac{-5+2i}{-5+2i} = \frac{60-24i-5i-2}{29} = 2-i$	$y = \frac{af - ec}{ad - bc}$
A GDC can also be used to confirm the result.	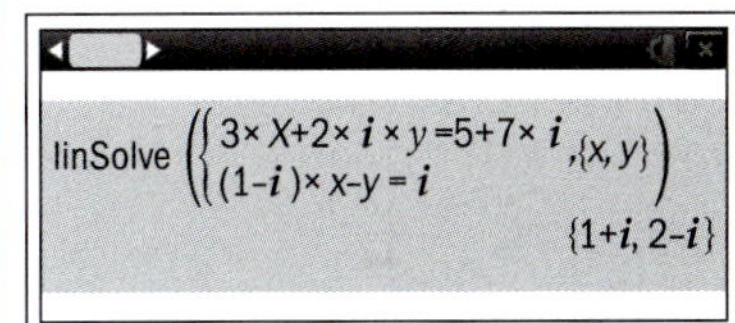

Exercise 3Q

Solve the following simultaneous equations and check your answers with a GDC.

1 $\begin{cases} 3x + (1 + i)y = 4 + 5i \\ (4 - i)x - (1 - i)y = 7 \end{cases}$

2 $\begin{cases} (1 + 4i)x + 3iy = 2 + 4i \\ (3 - 5i)x + (5 + 4i)y = 21 - 27i \end{cases}$

Systems of three linear equations with three unknowns

When you are presented with a system of three equations with three unknowns, you have to reduce them to a system of two equations with two unknowns, which you know how to solve. In this section, you will focus on solving systems algebraically, either by substitution or elimination.

Example 42

Solve the simultaneous equations $\begin{cases} 2x + y - z = 5 \\ x - 2y - 3z = 0 \\ 5x - y + 2z = 1 \end{cases}$ by

a the method of substitution

b the method of elimination.

a $x = 2y + 3z \Rightarrow \begin{cases} 2(2y+3z) + y - z = 5 \\ 5(2y+3z) - y + 2z = 1 \end{cases}$	Isolate one of the variables from one of the equations and substitute it in the remaining two equations. Here we find x in the second equation. Try to target the simplest possible expression.
$\Rightarrow \begin{cases} 5y + 5z = 5 \\ 9y + 17z = 1 \end{cases} \Rightarrow \begin{cases} y + z = 1 \\ 9y + 17z = 1 \end{cases}$	
$y = 1 - z \Rightarrow 9(1 - z) + 17z = 1 \Rightarrow 8z = -8$ $\Rightarrow z = -1 \Rightarrow y = 2 \Rightarrow x = 1$ $\Rightarrow (x, y, z) = (1, 2, -1)$	You can apply the method of substitution again, by expressing y in terms of z, or vice versa. Find z and then the values of y and x respectively.
b $\left.\begin{array}{l} \left.\begin{array}{r} 2x + y - z = 5 \\ 5x - y + 2z = 1 \end{array}\right] + \\ \left.\begin{array}{r} 4x + 2y - 2z = 10 \\ x - 2y - 3z = 0 \end{array}\right] + \end{array}\right\} \Rightarrow \begin{cases} 7x + z = 6 \\ 5x - 5z = 10 \end{cases}$	Add equations 1 and 3. Add 2 × equation 1 to equation 2. This eliminates y to give two equations in two unknowns. When eliminating a variable you must eliminate the same variable, and you must use a combination of all three equations in the process.
$\Rightarrow \left\{ \left.\begin{array}{r} 7x + z = 6 \\ x - z = 2 \end{array}\right] + \right. \Rightarrow 8x = 8 \Rightarrow x = 1$	Add the two equations to eliminate z. Find the value of x.
$\Rightarrow 7 + z = 6 \Rightarrow z = -1$ $\Rightarrow 2 + y - (-1) = 5 \Rightarrow y = 2$ $\Rightarrow (x, y, z) = (1, 2, -1)$ You can again use a calculator to check your solution.	Use this x to find z, and then use x and z to find y.

Notice that once you obtain the system of two equations with two unknowns, you don't have to apply the same method again to solve this system. If you used elimination to find a system of two equations with two unknowns, you could then use substitution to solve them, or vice versa.

When solving equations you can notice that you are just operating on the variables' coefficients and the free coefficients (independent of the variables). Instead of always rewriting everything with the variables you can just organize the coefficients in rows and calculate the elementary operations to obtain the result.

$\left\{\begin{array}{cccc} 2 & 1 & -1 & 5 \\ 1 & -2 & -3 & 0 \\ 5 & -1 & 2 & 1 \end{array}\right. \Rightarrow \left\{\begin{array}{cccc} 1 & -2 & -3 & 0 \\ 2 & 1 & -1 & 5 \\ 5 & -1 & 2 & 1 \end{array}\right.$	Swap the first two rows to simplify the calculations.
$\Rightarrow \left\{\begin{array}{cccc} 1 & -2 & -3 & 0 \\ 0 & 5 & 5 & 5 \\ 0 & 9 & 17 & 1 \end{array}\right. \Rightarrow \left\{\begin{array}{cccc} 1 & -2 & -3 & 0 \\ 0 & 1 & 1 & 1 \\ 0 & 9 & 17 & 1 \end{array}\right.$	Multiply the first row by 2 and subtract it from the second row. Multiply the first row by 5 and subtract it from the third row.
$\Rightarrow \left\{\begin{array}{cccc} 1 & -2 & -3 & 0 \\ 0 & 1 & 1 & 1 \\ 0 & 0 & 8 & -8 \end{array}\right. \Rightarrow \left\{\begin{array}{cccc} 1 & -2 & -3 & 0 \\ 0 & 1 & 1 & 1 \\ 0 & 0 & 1 & -1 \end{array}\right.$	Divide the second row by 5 to simplify the calculations. Multiply the second row by 9 and subtract it from the third row.
$\Rightarrow \left\{\begin{array}{r} x-2y-3z=0 \\ y+z=1 \\ z=-1 \end{array}\right. \Rightarrow \left\{\begin{array}{r} x-2y-3z=0 \\ I+(-1)=1 \\ z=-1 \end{array}\right.$	Divide the third row by 8 to simplify the calculations.
$\Rightarrow \left\{\begin{array}{c} x-2\times 2-3(-1)=0 \\ y=2 \\ z=-1 \end{array}\right. \Rightarrow \left\{\begin{array}{c} x=1 \\ y=2 \\ z=-1 \end{array}\right.$	Now rewrite the coefficients into the equations and find the solutions from the third equation up.

This method is called row reduction or the Gaussian method and it is generally used for matrices.

Investigation 17

1 Use your calculator to solve the following systems of simultaneous equations.

a $\begin{cases} 2x+y-z=5 \\ x-2y-3z=0 \\ 5x-y+2z=1 \end{cases}$

b $\begin{cases} 3x+y-2z=4 \\ 5x-y+z=-3 \\ 6x+2y-4z=5 \end{cases}$

Continued on next page

c $\begin{cases} x-2y+3z=4 \\ 2x-4y+6z=8 \\ -3x+6y-9z=-12 \end{cases}$ **d** $\begin{cases} 7x+2y+z=3 \\ 11x-10y+2z=1 \\ 23x+18y+3z=-3 \end{cases}$

e $\begin{cases} x+y-2z=-12 \\ 3x-3y+2z=22 \\ 7x-2y-2z=3 \end{cases}$ **f** $\begin{cases} 6x+15y-21z=4 \\ 8x+2y-11z=15 \\ -2x-5y+7z=17 \end{cases}$

2 **Conceptual** How can you classify the types of solutions to systems of three linear equations with three unknowns?

3 **Factual** What is the difference between the sets of infinitely many solutions to the equations in parts **c** and **e**?

4 **Factual** What is the relationship between the coefficients of the three equations in part **c**?

When solving systems of three simultaneous linear equations with three unknowns there are again three possible types of solution that can occur.

1 There is **a unique triplet** of numbers that satisfy all three equations.

2 There is **no triplet** of real numbers that satisfy all the equations.

3 There are **infinitely many triplets** of real numbers that satisfy all the equations.

Example 43

Discuss all possible types of solution of the system $\begin{cases} x-y+mz=2 \\ 2x+y-z=3 \\ x+y-2z=5 \end{cases}$ with respect to the real parameter m.

$\begin{cases} x-y+mz=2 \\ 2x+y-z=3 \\ x+y-2z=5 \end{cases} \Rightarrow \begin{cases} 2x+(m-2)z=7 \\ x+z=-2 \end{cases}$	Eliminate the variable y by adding the first and third equations and subtracting the second and third equations.
$\Rightarrow \begin{cases} \left.\begin{matrix} 2x+(m-2)z=7 \\ 2x+2z=-4 \end{matrix}\right] - \Rightarrow (m-4)z=11 \end{cases}$	Eliminate x by multiplying the second equation by 2 and subtracting it from the first equation.

$\Rightarrow z = \frac{11}{m-4}, m \neq 4$	If a real solution exists, you need to divide by the expression $(m-4)$ which cannot be equal to 0.
$\Rightarrow x = -2 - \frac{11}{m-4} = \frac{-2m-3}{m-4}$	Find x in terms of m.
$\Rightarrow y = 5 - \left(\frac{-2m-3}{m-4}\right) + 2\left(\frac{11}{m-4}\right) = \frac{7m+5}{m-4}$	Find y in terms of m.
When $m \neq 4$ there is a unique solution $\Rightarrow (x, y, z) = \left(\frac{-2m-3}{m-4}, \frac{7m+5}{m-4}, \frac{11}{m-4}\right), m \neq 4$	
$m = 4 \Rightarrow 0 \times z = 11 \Rightarrow 0 = 11$, which is false. When $m = 4$ the system has no solution.	When $m = 4$ from the equation you obtain a **false** statement, therefore the system has no solution.

Determining the equation of a polynomial from a system of linear equations

Suppose you are told the coordinates of two points which lie on the graph of the linear function $f(x) = ax + b$. You can use these points to formulate two equations in two unknowns, a and b. By solving this system of equations, you can determine the equation of the function.

In a similar way, when you know three points through which the graph of a quadratic function $f(x) = ax^2 + bx + c$ passes, you can formulate three equations in three unknowns a, b and c. Solving this system of equations gives the equation of the quadratic.

Example 44

Find the equation of the quadratic function $f(x) = ax^2 + bx + c$ that passes through the points $(-1, 10)$, $(2, -2)$ and $(4, 0)$.

$(-1, 10): a(-1)^2 + b(-1) + c = 10$ $\Rightarrow a - b + c = 10 \quad (1)$ $(2, -2): a \times 2^2 + b \times 2 + c = -2$ $\Rightarrow 4a + 2b + c = -2 \quad (2)$ $(4, 0): a \times 4^2 + b \times 4 + c = 0$ $\Rightarrow 16a + 4b + c = 0 \quad (3)$	Input the coordinates of each point on the parabola into the equation.

Continued on next page

$\begin{matrix}(2)-(1)\\(3)-(1)\end{matrix} \Rightarrow \begin{cases}3a+3b=-12\\15a+5b=-10\end{cases} \Rightarrow \begin{cases}a+b=-4\\3a+b=-2\end{cases}$	Eliminate variable c by subtracting the first equation from the second and third equations.
$2a=2 \Rightarrow a=1 \Rightarrow 1+b=-4 \Rightarrow b=-5$	Subtract the first equation from the second to eliminate b.
$1+5+c=10 \Rightarrow c=4$	Find a, b and c.
$\Rightarrow f(x)=x^2-5x+4$	Write the equation of the function using your values of a, b and c.

HINT

You can use your GDC in two different ways to solve this type of problem.

- You can formulate the three linear equations and use your GDC to solve them.
- You can use the statistical feature for finding the regression equation of a quadratic curve which passes through three specified points.

Exercise 3R

1 Find the value(s) of the real parameter m so that the system has no unique solution.

a $\begin{cases}2x+y-z=4\\x-y+mz=2\\3x+2y-z=1\end{cases}$ **b** $\begin{cases}x+my+z=1\\3x+y+mz=-2\\x+2y+z=3\end{cases}$

2 Find the value(s) of the real parameter k so that the system has infinitely many solutions. Find the solutions.

a $\begin{cases}2x+3y-z=1\\kx+9y-3z=3\\3x+7y+5z=2\end{cases}$

b $\begin{cases}x+ky+z=0\\4x+6y+(k+2)z=-6\\2x+3y+2z=-3\end{cases}$

3 Find the values of the real parameter m so that the system $\begin{cases}2x+y+z=m\\x+my+z=m+1\\x+y+mz=-3\end{cases}$ has a unique solution.

4 Find the equation of a quadratic function $f(x)=ax^2+bx+c$ that passes through the points

a $(-3, 1)$, $(-2, -5)$ and $(1, 4)$

b $(-1, 1)$, $(1, -9)$ and $(2, 8)$

Developing your toolkit

Now do the Modelling and investigation activity on page 216.

Chapter summary

- **The quadratic formula**

  For a quadratic equation in the form $ax^2+bx+c=0$, $a, b, c\in\mathbb{R}$, $a\neq 0$, the solutions or roots are $x=\dfrac{-b\pm\sqrt{b^2-4ac}}{2a}$.

- Given a quadratic equation of the form $ax^2+bx+c=0$, $a, b, c\in\mathbb{R}$, $a\neq 0$, the **discriminant** is the expression in the formula that is under the square root and is denoted by Greek letter Δ, $\Delta=b^2-4ac$.
  - **Case 1:** $\Delta>0$

    If the discriminant is positive, then $x=\dfrac{-b\pm\sqrt{b^2-4ac}}{2a}$ and there are **two distinct real roots**.
  - **Case 2:** $\Delta=0$

    If the discriminant is equal to zero, then $x=\dfrac{-b}{2a}$. This is regarded as **one repeated root**.
  - **Case 3:** $\Delta<0$

    If the discriminant is less than zero, then $\sqrt{b^2-4ac}$ is not real. In this case, there is **no real solution**.
- Complex numbers are numbers of the form $z=a+\mathrm{i}b$, where $a, b\in\mathbb{R}$.

  a is called the **real part** of z, and we write $\mathrm{Re}(z)=a$.

  b is called the **imaginary part** of z, and we write $\mathrm{Im}(z)=b$.

  The set of complex numbers is denoted by $\mathbb{C}$.
- Given the complex number $z=x+\mathrm{i}y$, $x, y\in\mathbb{R}$, the **modulus** of z is given by

  $|z|=|x+\mathrm{i}y|=\sqrt{x^2+y^2}=\sqrt{(\mathrm{Re}(z))^2+(\mathrm{Im}(z))^2}$.
- Two complex numbers $z_1=a+b\mathrm{i}$ and $z_2=c+d\mathrm{i}$, $a, b, c, d\in\mathbb{R}$, are equal if and only if their real parts are equal and their imaginary parts are equal, $a=c$ and $b=d$.
- Addition and multiplication by a scalar:

  $z_1+z_2=(a+b\mathrm{i})+(c+d\mathrm{i})=(a+c)+(b+d)\mathrm{i},\quad a, b, c, d\in\mathbb{R}$

  $\lambda z=\lambda(a+b\mathrm{i})=(\lambda a)+(\lambda b)\mathrm{i},\quad \lambda, a, b\in\mathbb{R}$
- Multiplication of complex numbers:

  $z_1\times z_2=(a+b\mathrm{i})\times(c+d\mathrm{i})=(ac-bd)+(ad+bc)\mathrm{i},\quad a, b, c, d\in\mathbb{R}$
- For any complex number $z=a+b\mathrm{i}$, $a, b\in\mathbb{R}$, there is a conjugate complex number of the form $z^*=a-b\mathrm{i}$. Their real parts are equal, $\mathrm{Re}(z)=\mathrm{Re}(z^*)$, and their imaginary parts are opposite, $\mathrm{Im}(z)=-\mathrm{Im}(z^*)$.
- Division of complex numbers:

  $\dfrac{z_1}{z_2}=\dfrac{z_1\times z_2^*}{|z_2|^2}$
- Powers of i: $\mathrm{i}^n=\begin{cases}1, n=4k\\ \mathrm{i}, n=4k+1\\ -1, n=4k+2\\ -\mathrm{i}, n=4k+3\end{cases}, k\in\mathbb{Z}$

Continued on next page

- Polynomials are functions which map a real variable, often called x, to another real number. We write f: $\mathbb{R} \to \mathbb{R}$.
- Polynomials are functions of the form $f(x) = a_n x^n + a_{n-1}x^{n-1} + \ldots + a_1 x + a_0$, where $a_i \in \mathbb{R}$, $i = 0, \ldots, n$ are called the **coefficients.** The highest power (n) of the variable x is called the **degree** of the polynomial, and we write
$\deg(f) = n$.
- A **linear combination** of two functions f and g is an expression of the form $a \times f(x) + b \times g(x)$, where a and b are real numbers.

  A linear combination of n functions is an expression of the form $\sum_{i=1}^{n} a_i \times f_i(x)$, where f_i are functions and $a_i \in \mathbb{R}$.
- For any two polynomials f and g there are unique polynomials q and r such that $f(x) = g(x) \times q(x) + r(x)$ for all real values of x.

  The polynomial q is called the **quotient** and the polynomial r is called the **remainder**. Notice that $\deg(g) > \deg(r)$.
- **Polynomial remainder theorem**

  Given a polynomial $f(x) = a_n x^n + a_{n-1}x^{n-1} + \ldots + a_2 x^2 + a_1 x + a_0$, $a_i \in \mathbb{R}$, $i = 0, 1, 2, \ldots, n$, $a_n \neq 0$, and a real number p, then the remainder when $f(x)$ is divided by the linear expression $(x - p)$ is $f(p)$.
- **Factor theorem**

  A polynomial $f(x) = a_n x^n + a_{n-1}x^{n-1} + \ldots + a_2 x^2 + a_1 x + a_0$, $a_k \in \mathbb{R}$, $k = 0, 1, 2, \ldots, n$, $a_n \neq 0$, has a factor $(x - p)$, $p \in \mathbb{R}$ if and only if $f(p) = 0$.
  - **Corollary**

    Given a polynomial $f(x) = a_n x^n + a_{n-1}x^{n-1} + \ldots + a_2 x^2 + a_1 x + a_0$, $a_k \in \mathbb{R}$, $k = 1, 2, \ldots, n$, $a_n \neq 0$, and real numbers a and b, $a \neq 0$, then the remainder when $f(x)$ is divided by the linear expression $(ax - b)$ is $f\left(\frac{b}{a}\right)$.
- **The fundamental theorem of algebra (FTA)**

  A polynomial $f(x) = a_n x^n + a_{n-1}x^{n-1} + \ldots + a_2 x^2 + a_1 x + a_0$, $a_n \neq 0$, with real or complex coefficients has at least one zero. There is an $\omega \in \mathbb{C}$ such that $f(\omega) = 0$.
- **Lemma**

  Each polynomial $f(x) = a_n x^n + a_{n-1}x^{n-1} + \ldots + a_2 x^2 + a_1 x + a_0$, $a_n \neq 0$, with real coefficients can be written in a factor form $f(x) = a_n(x - \omega_1)(x - \omega_2)\ldots(x - \omega_n)$ such that $\omega_k \in \mathbb{C}$, $k = 1, \ldots, n$.
- **Conjugate root theorem**

  Given a polynomial $f(x) = a_n x^n + a_{n-1}x^{n-1} + \ldots + a_2 x^2 + a_1 x + a_0$, $a_k \in \mathbb{R}$, $k = 1, 2, \ldots, n$, $a_n \neq 0$, that has a complex zero z, then its conjugate z^* is also a zero of the polynomial f.

- **Conjugate root theorem**

  Given a polynomial $f(x) = a_n x^n + a_{n-1}x^{n-1} + \ldots + a_2x^2 + a_1x + a_0$, $a_k \in \mathbb{R}, k = 1, 2, \ldots, n, a_n \neq 0$, that has a complex zero z, then its conjugate z^* is also a zero of the polynomial f.

- If a polynomial $f(x) = a_n x^n + a_{n-1}x^{n-1} + \ldots + a_2x^2 + a_1x + a_0$ with real or complex coefficients $(a_n \neq 0)$ has zeros $x_1, x_2, \ldots, x_n$ then $\sum_{1\le i_1 < i_2 < \ldots < i_k \le n} \left(x_{i_1}x_{i_2}\ldots x_{i_k}\right) = (-1)^k \frac{a_{n-k}}{a_n}$, $1 \le k \le n$; specifically, for the sum and the product,

  $$\begin{cases} x_1 + x_2 + x_3 + \ldots + x_n = -\dfrac{a_{n-1}}{a_n} \\ x_1x_2x_3\ldots x_n = (-1)^n \dfrac{a_0}{a_n} \end{cases}$$

- When solving systems of two linear equations with two unknowns there are three possible types of solutions that can occur. Each type has a different geometrical interpretation.
  - When there is a **unique pair** of numbers that satisfy both equations, the lines intersect at one point.
  - When there are **no real numbers** that satisfy both equations, the lines are parallel and distinct.
  - When there are **infinitely many pairs** of real numbers that satisfy both equations, the lines coincide.
- When solving systems of three simultaneous linear equations with three unknowns there are again three possible types of solutions that can occur.
  - There is a **unique triplet** of numbers that satisfy all three equations.
  - There is **no triplet** of real numbers that satisfy all the equations.
  - There are **infinitely many triplets** of real numbers that satisfy all the equations.
- **Finding polynomials by points**

  When you know three points through which a quadratic function $f(x) = ax^2 + bx + c$ passes, you can formulate three equations in three unknowns a, b and c. Solving these gives the equation of the quadratic.

Developing inquiry skills

Return to the opening problem. Your city council has asked you to prepare a plan of a roller coaster for a new amusement park.

How would you go about this task using what you have learned in this chapter?

Chapter review

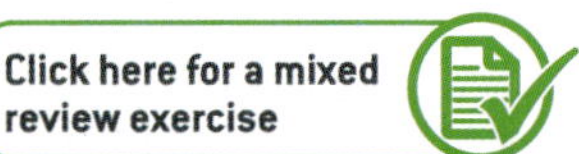

1 Show that the equation $x(x - a - b) = 1 - ab$ always has real solutions for all values of a, $b \in \mathbb{R}$.

2 The polynomial $f(x) = ax^3 + bx^2 + cx + d$, a, b, c, $d \in \mathbb{Z}$, has two zeros, z and ω.

a Show that $z = (1 - i)^3 + (1 + i)^3 = -4$.

b Find $\omega = \dfrac{i^{64} \times i^{63} \times \ldots \times i^{18}}{(1-2i)}$ in the form $a + bi$, a, $b \in \mathbb{R}$.

c Hence find the polynomial f that has integer coefficients which are relatively prime.

3 The system of simultaneous equations $\begin{cases} ax + y = 2 \\ x + ay = -2 \end{cases}$, $a \in \mathbb{R}$, is given.

a Find the values of a for which the system has a unique solution and find the solution in terms of a.

b For which value of a does the system have:

i no solution

ii infinitely many solutions. Write the equation of the line.

4 a Prove that $\dfrac{(1+i)^{2019}}{(1-i)^{2017}}$ is an integer and find its value.

b What must $n \in \mathbb{Z}$ satisfy so that $\dfrac{(1+i)^{n+2}}{(1-i)^{n}}$ is an imaginary number?

5 Use your GDC to solve the inequality $x^5 + 3x^4 - 2x^2 + 2x + 4 \geq x^4 + 3x^3 + 2x^2 + 3$.

6 Solve the simultaneous equations.

$$\begin{cases} 3x + z = 2y + 1 \\ 6x + 8y = 3z + 6 \\ 4y - 7z = 12x + 4 \end{cases}$$

7 The polynomial f is such that $(f(x))^2 = 9x^4 + 12x^3 - 26x^2 - 20x + 25$.

a What is the degree of the polynomial f?

b Given that α and β are the zeros of the polynomial f, find $\dfrac{1}{\alpha} + \dfrac{1}{\beta}$.

8 Find the complex numbers z and ω such that $\begin{cases} z^* - i\omega = 3 \\ \omega^* + iz = 6 - i \end{cases}$

9 Given that $z^2 - z + 1 = 0$, show that $z^{2019} = -1$.

Exam-style questions

10 P1: a Show that the solutions to the equation $x^2 - 6x - 43 = 0$ may be written in the form $x = p \pm q\sqrt{13}$, where p and q are positive integers. (4 marks)

b Hence, or otherwise, solve the inequality $x^2 - 6x - 43 \leq 0$. (2 marks)

11 P1: a Solve the equation $8x^2 + 6x - 5 = 0$ by factorization. (4 marks)

b Determine the range of values of k for which $8x^2 + 6x - 5 = k$ has no real solutions. (3 marks)

12 P1: Find the range of values of k for which the equation $3kx^2 + k\sqrt{3}x + 3 = 0$ has two real roots. (6 marks)

13 P1: a Solve the inequality $(x + 4)(3 - x) > 0$. (3 marks)

b Solve the inequality $2x^2 - 11x + 9 < 0$. (4 marks)

c Hence, find the range of values of x which satisfy both the inequalities $(x + 4)(3 - x) > 0$ and $2x^2 - 11x + 9 < 0$. (1 mark)

14 **P1:** Find the possible values of $\omega \in \mathbb{C}$ which satisfy $\omega^2 = 77 - 36\text{i}$. (9 marks)

15 **P1:** The polynomial $2x^3 + ax^2 - 10x + b$ has a factor of $(x - 1)$, and has a remainder of 15 when divided by $(x + 2)$. Find the values of a and b $(a, b \in \mathbb{R})$. (7 marks)

16 **P1:** $f(z) = z^4 - 8z^3 + 48z^2 - 176z + 260,\ z \in \mathbb{C}$

Given $(3 - \text{i})$ is a zero of f, find the remaining zeros. (9 marks)

17 **P1:** Find the sum and product of the roots of each function.

a $f(\text{x}) = 5\text{x}^4 - 4x^3 + 3x^2 + 2x - 1$ (4 marks)

b $g(x) = 5x^5 - 4x^4 + 3x^3 + 2x^2 + -7x + 10$ (4 marks)

18 **P1:** Prove that $\omega z^* - z\omega^*$ is purely imaginary for all $\omega, z \in \mathbb{C}$. (5 marks)

19 **P2:** Find the equation of a quadratic function $f(x) = ax^2 + bx + c$ passing through the points $(-1, -5)$, $(3, -1)$, and $(10, -71)$. (7 marks)

Making a Mandelbrot

Approaches to learning: Critical thinking, Communication

Exploration criteria: Mathematical communication (B), Personal engagement (C), Use of mathematics (E)

IB topic: Complex numbers

Fractals

You may have heard about fractals.

The image above is from the Mandelbrot set, one of the most famous examples of a fractal.

This is not only a beautiful image in its own right. The Mandelbrot set as a whole is an object of great interest to mathematicians. However, as yet, no practical applications have been found.

This image appears to be very complicated, but is in fact created using a remarkably simple rule.

Exploring an iterative equation

Consider this iterative equation:

$z_{n+1} = (z_1)^2 + c$

Consider a value of $c = 0.5$, so you have $z_{n+1} = (z_1)^2 + 0.5$.

Given that $z_1 = 0$, find the value of z_2.

Now find the value of z_3.

Repeat for a few more iterations.

What do you think is happening to the values found in this calculation?

A different iterative equation

Now consider a value of $c = -0.5$, so you have $z_{n+1} = (z_1)^2 - 0.5$.

Again start with $z_1 = 0$.

What happens this time?

The Mandelbrot set

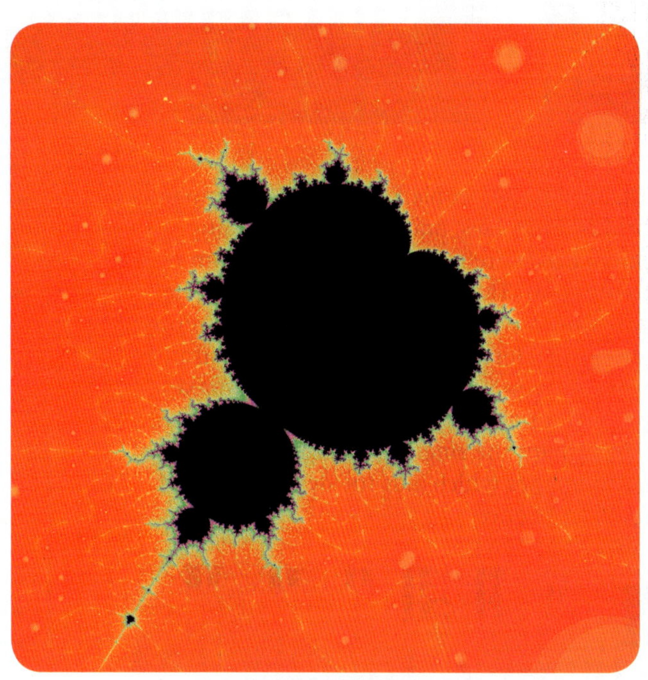

The Mandelbrot set consists of all those values of c for which the sequence starting at $z_1 = 0$ does **not** escape to infinity (those values of c where the calculations zoom off to infinity).

That is only part of the story however.

The value of c in the function does not need to be real.

It could be complex of the form $a + b\text{i}$.

If you pick a complex number for c, is it in the Mandelbrot set or not?

Consider starting with a value of $z_1 = 0 + 0\text{i}$, rather than just "0".

Consider, for example, a value of c of $1 + \text{i}$, so you have $z_{n+1} = (z_1)^2 + (1 + \text{i})$.

$z_1 = 0 + 0\text{i}$

Find z_2.

Repeat to find z_3, z_4, etc.

Does this diverge or converge? (Does it zoom off to infinity or not?)

HINT

One way to check is to plot the different values of z_1, z_2, z_3, z_4, etc, on an Argand diagram and to see whether the points remain within a boundary square. Or you could calculate the modulus of each value and see whether this is increasing.

Repeat for a few more values of c.

Try, for example:

a $c = 0.2 - 0.7\text{i}$ **b** $c = -0.25 + 0.5\text{i}$

What happens if $c = \text{i}$?

Try some values of your own.

You may find a GDC will help with these calculations as it can be used for complex numbers. When the number gets very big, the calculator will not be able to give a value. You may also notice that some numbers get "big" a lot quicker than others.

It is clearly a time-consuming process to try all values!

There are programmes that can be used to calculate the output after several iterations when you input a number for c. This will indicate which values of c belong to the Mandelbrot set and which don't.

The Mandelbrot set diagram is created by colouring points on an Argand diagram of all those values of c that escape to infinity in one colour (say black) and all those that remain bounded in another colour (say red).

Extension

- Try to construct your own spreadsheet or write a code that could do this calculation a number of times for different chosen values of c.
- Explore what happens as you zoom in to the edges of the Mandelbrot set.
- What is the relationship between Julia sets and the Mandelbrot set?
- What is the connection to Chaos Theory?
- What are Multibrot sets?
- How could you find the area or perimeter of the Mandelbrot set?

4 Measuring change: differentiation

The motion of people and objects, a company's profits and losses, and how rates of change of two phenomena are related may change from moment to moment. This chapter is about differential calculus: a way of measuring instantaneous change. Differential calculus provides a framework for you to model, interpret and make predictions about real-life problems. As such, it is perhaps the single most important piece of mathematics which aids the development of modern science.

Concepts

- Change
- Relationships

Microconcepts

- Limit of a function at a point
- Continuity of a function
- Differentiation from first principles
- Tangents and normals to curves
- Differentiation rules
- Methods of differentiation: chain, product and quotient rules
- Maxima, minima and points of inflexion
- Kinematic problems
- Optimization
- Implicit differentiation

How is a runner's speed measured with respect to time?

How can a company maximize its profits and minimize its losses?

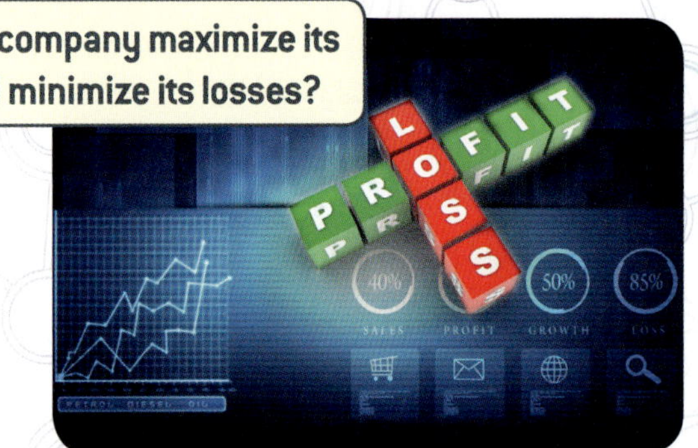

At which instant in time must a rocket make its re-entry into the Earth's atmosphere, or risk being caught in the Earth's orbit?

What are the dimensions of a cylindrical object for maximum volume using minimum material?

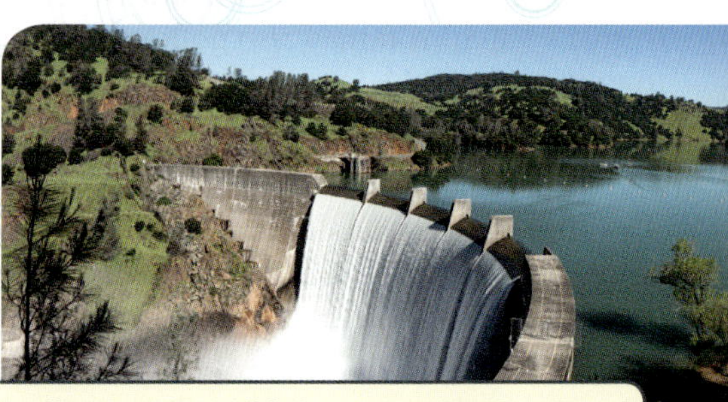

How does the rate at which water flows out of a reservoir affect the rate of change of the water depth in the reservoir?

At the 2009 World Athletics Championships in Berlin, Germany, the Jamaican sprinter Usain Bolt posted a time of 9.58 s in the men's 100m sprint. After this record-breaking time, Bolt was hailed as the fastest man in the world. But what does this mean?

- Was Bolt running at the same speed throughout the whole race?
- How could you determine Bolt's speed at any particular moment in the race?
- How can you determine Bolt's fastest speed in the race?

To answer the above questions, think about the following:

- How is "fastest speed" measured?
- What is the relationship between "fastest speed" and "average speed"?

Developing inquiry skills

List some sports in which the direction of something changes rapidly – for example, a tennis ball after it is hit. What different inquiry questions might you ask in this situation? For example, how can you determine the speed of the tennis ball before and after it changes direction? How can you determine the time at which the ball changes direction?

Think about the questions in this opening problem and answer any you can. As you work through the chapter, you will gain mathematical knowledge and skills that will help you to answer them all.

Before you start

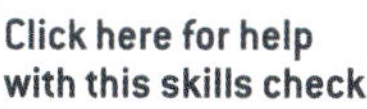

You should know how to:

1 Use rational exponents to rewrite expressions in the form $c \times x^n$.

eg $\frac{2}{x^3} = 2x^{-3}$; $\sqrt[4]{x^3} = x^{\frac{3}{4}}$

2 Sketch graphs of functions, clearly labelling any intercepts and asymptotes.

eg $y = \frac{1}{x-2} + 1$

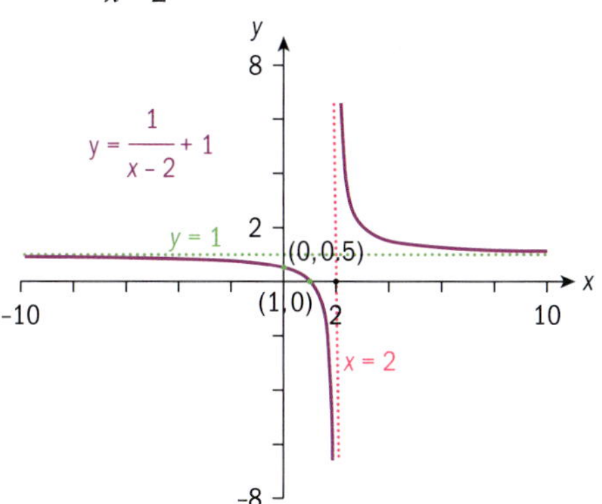

3 Find the sum of an infinite geometric series.

eg $\sum_{n=0}^{\infty}\left(\frac{1}{3}\right)^n$

Since $|r| < 1$, $S_\infty = \frac{1}{1-\frac{1}{3}} = \frac{3}{2}$

Skills check

1 Use rational exponents to rewrite these expressions in the form $c \times x^n$:

$7\sqrt{x}$; $\frac{2}{x^3}$; $\frac{8}{5\sqrt[3]{x^2}}$

2 Sketch the graph of f clearly labelling any intercepts and asymptotes: $f(x) = \frac{2}{x+3}$

3 Find $\sum_{n=0}^{\infty} 5\left(\frac{1}{2}\right)^n$.

4.1 Limits, continuity and convergence

TOK

"This statement is false."

Should paradoxes change the way that mathematics is viewed as an area of knowledge?

Investigation 1

About 2500 years ago, Zeno of Elea—a philosopher and logician—posed the following problem: Achilles and a tortoise were running a race. Achilles allowed the tortoise a head start of 10 metres. Both started running at a constant speed: Achilles at 10 ms^{-1}, and the tortoise at 1 ms^{-1}. After how many metres would Achilles overtake the tortoise?

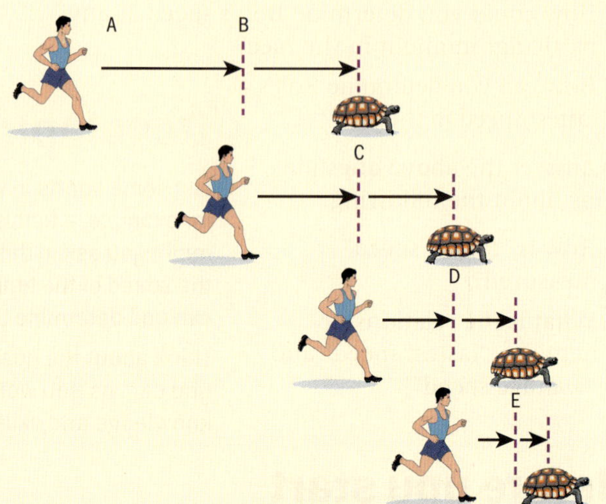

Zeno analysed the problem this way:

- Achilles starts at point A and runs to the tortoise's starting point, B, which is 10 m from A.
- When Achilles reaches B, because the tortoise runs at $\frac{1}{10}$th the speed of Achilles, the tortoise is now at C, which is 1 m from B.
- When Achilles reaches C, the tortoise has run to D, which is 0.1 m from C.

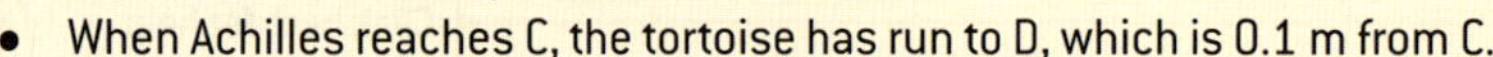

1 List the distance that Achilles covers as he runs from:
a A to B **b** B to C **c** C to D.

Since this pattern continues, whenever Achilles reaches any point where the tortoise has been, he still has to run further to catch up, because the tortoise has advanced to the next point. Zeno's conclusion therefore was that Achilles would never overtake the tortoise!

Since Zeno knew intuitively that Achilles *would* overtake the tortoise, he concluded that the problem needed to be analysed in a different way. Here, you will use a geometric series to analyse the problem.

Fill in the blanks in questions **2** and **3**.

2 When Achilles reaches point D, the tortoise has run to point E, which is __________ m from D.

3 When Achilles reaches point E, the tortoise has run to point F, which is __________ m from E.

4 Using the previous information, represent the distances that Achilles has run using a geometric series. What are the first term and the common ratio?

5 Using what you learned in Chapter 1, state a necessary condition for the sum of an infinite geometric series to exist.

6 Can you find the sum to infinity of this series?

7 Does Achilles overtake the tortoise?

It had taken several millennia for mathematicians to arrive at the language and concepts needed to solve this paradox satisfactorily. In this section you will learn some of the mathematics developed by 17th- and 18th-century mathematicians in an attempt to deal with the concepts of time and infinity.

Limit of a function

Investigation 2

1 Use your GDC to plot the graph of the function $f(x) = \dfrac{x^2 - 1}{x - 1}$.

2 Copy this table and, by tracing along your graph beginning to the left of $x = 1$, write down the value of $f(x)$ at each x-value in the table for $0.6 \leq x < 1$.

Then, beginning at $x = 1.4$, trace along the graph until you reach $x = 1$ and write down the value of $f(x)$ at each x-value in a copy of the table for $1 < x \leq 1.4$.

x	0.6	0.7	0.8	0.9	1	1.1	1.2	1.3	1.4
$f(x) = \dfrac{x^2 - 1}{x - 1}$									

3 Write down the value that $f(x)$ is approaching as you trace along your graph from the left and the right of $x = 1$.

In **2** and **3** you have attempted to find the limit of the function at $x = 1$.

4 **Conceptual** What is meant by the limit of a function at a particular point?

5 Summarize your observations about the values of the function as it approaches $x = 1$ from the left and from the right, and state the value of the function at $x = 1$.

6 **Conceptual** What condition is placed on the left and right limits of f at $x = a$ for the limit of f to exist at a?

7 Plot the graph of $g(x) = x + 1$ on your GDC. Evaluate $g(x)$ at each of the points where you evaluated $f(x)$ (in question **2**).

8 **Conceptual** Does a function need to be defined at $x = c$ in order to have a limit at $x = c$?

9 **Factual** Write down graphical similarities and differences between f and g, and explain. Can you see the graphical difference between f and g on your GDC or graphing software? If not, explain why not.

In Investigation 2, f is *not* defined at $x = 1$ and g *is* defined at $x = 1$. However, as x approaches 1 from the left and the right, both f and g approach 2. This means that the limit of the functions as x approaches 1 *both* from the left and from the right is 2.

We can write this result using the following notation:

$$\lim_{x \to 1} \frac{x^2 - 1}{x - 1} = 2 \text{ and } \lim_{x \to 1}(x + 1) = 2$$

$\lim\limits_{x \to a^-} f(x)$ means the limit of f as x approaches a from the left. $\lim\limits_{x \to a^+} f(x)$ means the limit of f as x approaches a from the right. Leaving out the sign on the a indicates that the limit is the same from both directions.

Example 1

Using your GDC:

a sketch the graph of $y = \frac{3^x - 1}{x}, x \neq 0$

b find $\lim_{x\to 0^-} \frac{3^x - 1}{x}$ and $\lim_{x\to 0^+} \frac{3^x - 1}{x}$ numerically, giving your answer to 1 decimal place.

a

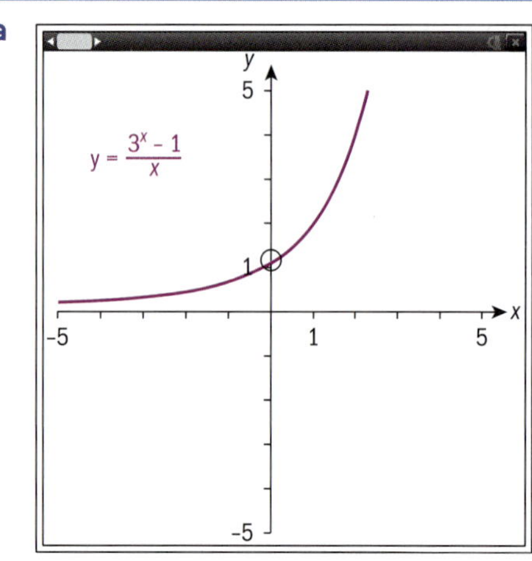

b From the table of values on your GDC, since the limits from both the left and right are the same to 1 dp, $\lim_{x\to 0} \frac{3^x - 1}{x} \approx 1.1$

Example 2

a Use your GDC to plot a graph of the function $f(x) = \begin{cases} 2x+1, & x > 1 \\ 2x-2, & x < 1 \end{cases}$.

b Find $\lim_{x\to 1^-} f(x)$ and $\lim_{x\to 1^+} f(x)$.

a

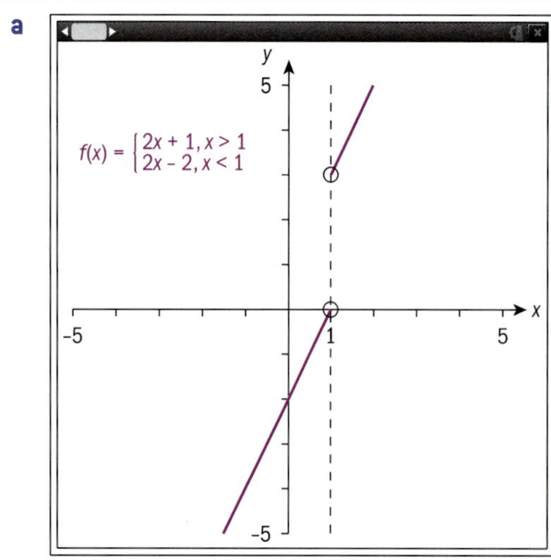

There are open circles on the values of the function at $x = 1$, since the piecewise function is not defined at $x = 1$.

b $\lim_{x\to 1^-} f(x) = 0; \lim_{x\to 1^+} f(x) = 3$	When x approaches 1 from the left, $f(x)$ approaches 0. When x approaches 1 from the right, $f(x)$ approaches 3.

In Example 1, the limits were the same whether approaching from the left or from the right. In Example 2, the limits were different when approaching from the left and from the right. We say, therefore, that

the function $f(x) = \begin{cases} 2x+1, & x > 1 \\ 2x-2, & x < 1 \end{cases}$ has no limit at $x = 1$.

For the limit of a function $f(x)$ to exist as x approaches a certain value c, it is not necessary that the function be defined at $x = c$; however it *is* necessary that the value that $f(x)$ approaches from the left and from the right must be the same.

The notation used to say that the limit, L, of a function f exists as x approaches the same real value c from the left and from the right is:

$$\left(\lim_{x\to c} = L\right) \Leftrightarrow \left(\lim_{x\to c^+} f(x) = L \text{ and } \lim_{x\to c^-} f(x) = L\right),\ L \in \mathbb{R}$$

HINT

The double arrow $\Leftrightarrow$ is read "if and only if".

TOK

What value does the knowledge of limits have?

Exercise 4A

Find the limit of each function, if it exists.

1 $\lim_{x\to -2} \frac{x^2-4}{x+2}$

2 $\lim_{x\to 3} \frac{x^2-9}{x-3}$

3 $\lim_{x\to 2} \begin{cases} x-3, & x < 2 \\ x+1, & x > 2 \end{cases}$

4 $\lim_{x\to 1} \frac{x^3-1}{x-1}$

5 $\lim_{x\to 1} \begin{cases} x+1, & x < 1 \\ x^2+1, & x \ge 1 \end{cases}$

6 $\lim_{x\to 6} \left((x-6)^{\frac{2}{3}}\right)$

7 $\lim_{x\to 2} [x]$

HINT

$y = [x]$ or $y = \text{int}(x)$ is the step function or floor function. It is defined as the "largest integer less than or equal to x". If this function is used in examinations, it will be defined for you.

Limits and continuity

Investigation 3

The following functions are considered continuous on their domains, and continuous everywhere.

a
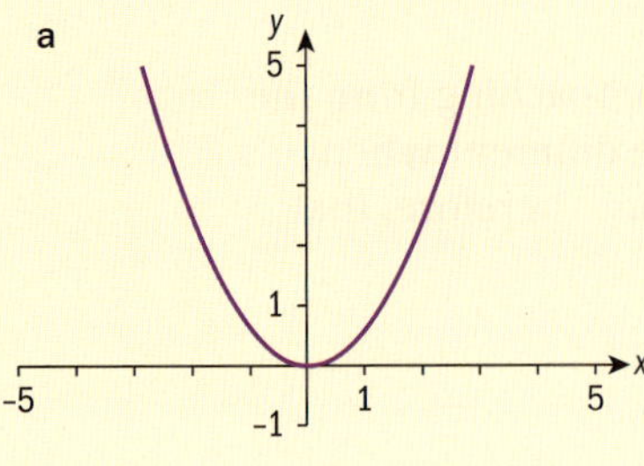

b
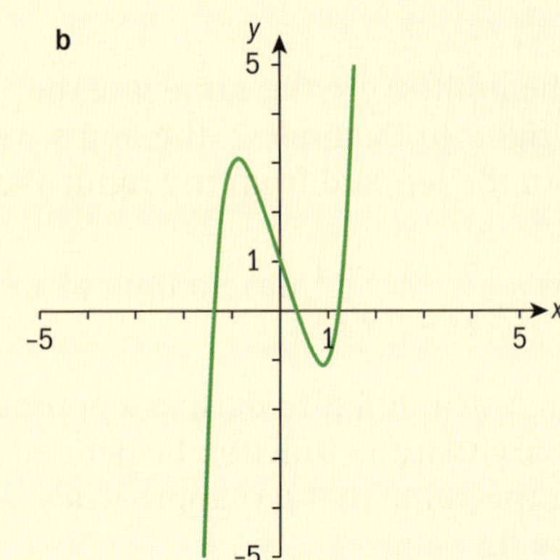

c
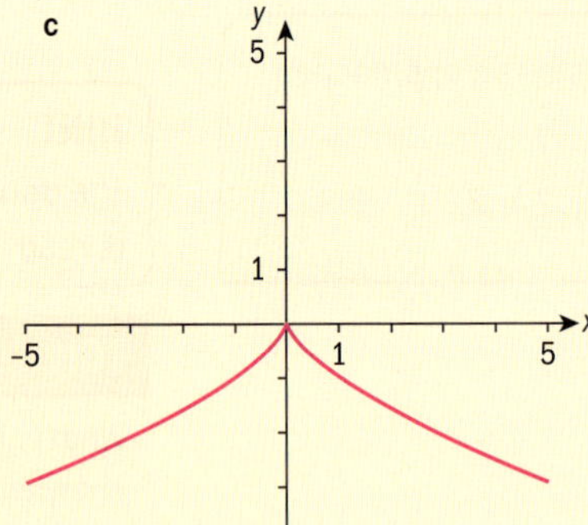

d
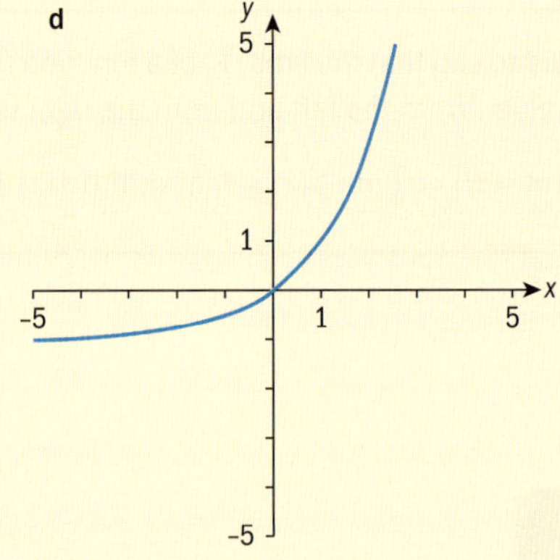

1 Based on the graphs of functions, how would you define a continuous function?

2 Consider any one of the graphs of continuous functions, for example, **b**, whose function is $f(x) = x^5 - 3x + 1$.

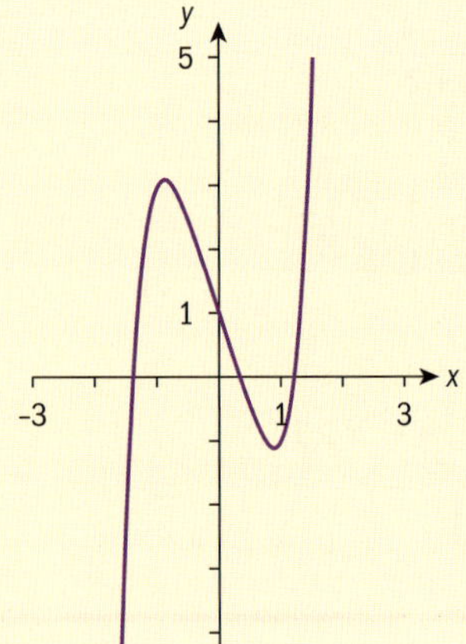

Find:

a $\lim_{x \to 0} f(x)$ and $f(0)$ **b** $\lim_{x \to -1} f(x)$ and $f(-1)$.

Write down what you notice about the limit of the function and the value of the function at that point.

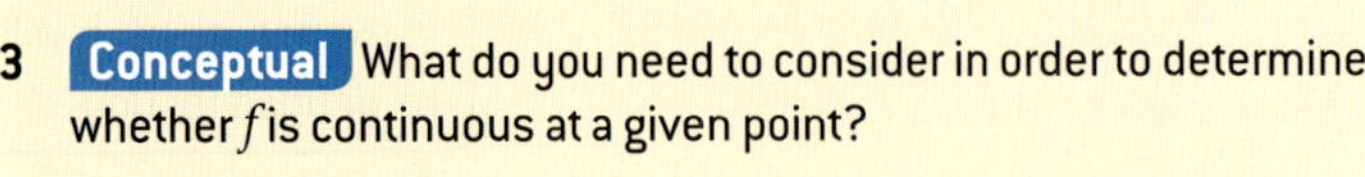

3 **Conceptual** What do you need to consider in order to determine whether f is continuous at a given point?

4 Given the function $f(x) = x^5 - 3x + 1$ and any real value a, explain whether or not:

a $f(a)$ exists

b $\lim_{x \to a} f(x)$ exists.

5 **Conceptual** Using your results from **4a** and **4b** what generalization can you make about any continuous function f?

6 The function $\lim_{x\to 1}\frac{x^2-1}{x-1}$ is continuous in its domain, since its natural domain excludes 1. It is not, however, continuous everywhere, since it is not continuous at $x = 1$. Find

a $\lim_{x\to 1}\frac{x^2-1}{x-1}$ and **b** $f(1)$, and write down what you notice.

7 The graph on the right shows $f(x)=\begin{cases}2^x, & x<2\\ 1, & x>2\end{cases}$, which is not continuous at $x = 2$. Explain whether or not the following exist:

a $\lim_{x\to 2} f(x)$

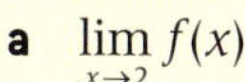

b $f(2)$

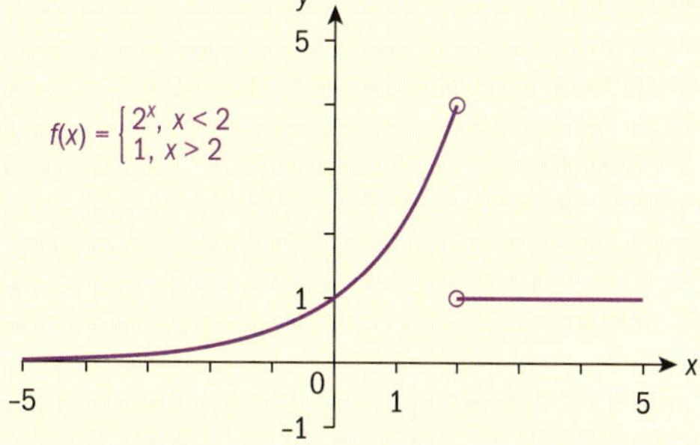

8 **Conceptual** Consider **all** your answers to the questions and write down **three** conditions necessary for a function to be continuous at $x = a$.

In working through the investigation above, you will have arrived at the following definitions.

Three necessary conditions for f to be continuous at $x = a$ are:

1 f is defined at a, ie a is an element of the domain of f.

2 The limit of f at a exists, that is, $\lim_{x\to a} f(x) = L$.

3 The limit of f at a is equal to the value of the function at a, that is, $f(a) = \lim_{x\to a} f(x)$.

A function that is not continuous at a point $x = a$ is said to be **discontinuous** at $x = a$.

A function is said to be continuous on an open interval I if it is continuous at every point in the interval.

A function is continuous if it is continuous at every point in its domain.

A function is continuous everywhere if it is continuous throughout the real numbers.

A function that is not continuous is said to be discontinuous.

Note that the reciprocal function $y=\frac{1}{x}, x\neq 0$, is a continuous function because it is continuous at every point in its domain, since the domain excludes $x = 0$.

However, it is not continuous on an open interval containing 0, for example, $x\in\,]-1, 1[$. That is, it has a point of discontinuity at $x = 0$ because it is not defined at this point.

Reflect Of the types of functions you studied in Chapter 2—polynomial functions, rational functions, modulus functions and radical functions—which are continuous functions, and which functions have discontinuities at certain points?

Example 3

a Use your GDC to plot the graph of:

$$f(x)=\begin{cases}-2x, & -1<x<1\\ 2, & \text{otherwise}\end{cases}$$

b Find the limits, if they exist, as x approaches -1, 0 and 1.

c Determine whether f is continuous at $x=-1$, $x=0$ and $x=1$.

a

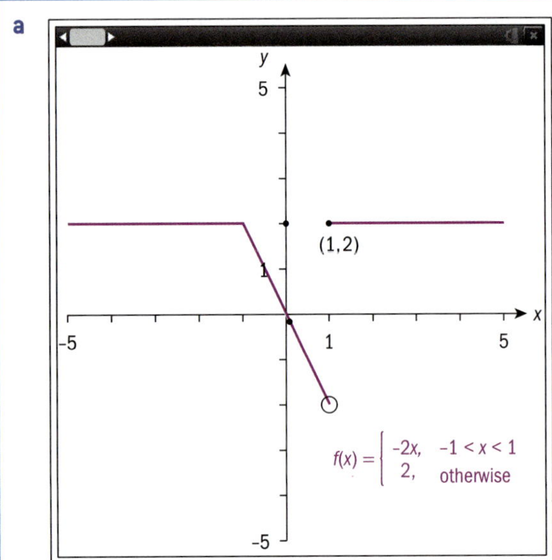

b $\lim_{x\to -1} f(x)=2$

$\lim_{x\to 0} f(x)=0$

$\left(\lim_{x\to 1^-} f(x)=-2, \lim_{x\to 1^+} f(x)=2\right)$

$\Rightarrow \lim_{x\to 1} f(x)$ is undefined

The left and right limits are equal at both $x=-1$ and $x=0$.

As x approaches 1 from the left, $f(x)$ approaches -2, and as x approaches 1 from the right, f approaches 2, hence the limit as x approaches 1 does not exist.

c $\left(\lim_{x\to -1} f(x)=2; f(-1)=2\right)\Rightarrow f$ is continuous at -1.

$\left(\lim_{x\to 0} f(x)=0; f(0)=0\right)\Rightarrow f$ is continuous at 0.

$\left(\lim_{x\to 1} f(x) \text{ is undefined and } f(1)=2\right)\Rightarrow$

f is not continuous at $x=1$.

Definition of continuity at a point.

Example 4

$$f(x)=\begin{cases}\dfrac{x^2-3x^2+5x-3}{x-1}, & x\neq 1\\ k, & x=1\end{cases}$$

Determine the value of k such that $f(x)$ is continuous at $x = 1$.

$\dfrac{x^2-3x^2+5x-3}{x-1}=\dfrac{(x-1)(x^2-2x+3)}{x-1}=(x^2-2x+3)$	Since $x \neq 1$, divide through by $(x-1)$, since $(x-1)\neq 0$.
$\lim\limits_{x\to 1}\dfrac{x^2-3x^2+5x-3}{x-1}=\lim\limits_{x\to 1}(x^2-2x+3)=2$	$\lim\limits_{x\to 1} f(x)$ exists.
For f to be continuous at $x = 1$, $k = f(1) = \lim\limits_{x\to 1} f(x) = 2$ Hence, when $k = 2$, $f(x)$ is continuous at $x = 1$.	For f to be continuous at $x = 1$, $f(1) = \lim\limits_{x\to 1} f(x)$

International-mindedness

Maria Agnessi, an 18th-century Italian mathematician, published a text on calculus and also studied the form $y = \dfrac{8a^3}{x^2+4a^2}$, known as the witch of Agnesi.

Exercise 4B

1 Determine whether $f(x)=\begin{cases}2x+1, & x<3\\ 3x-2, & x\geq 3\end{cases}$ is continuous at $x = 3$.

2 Determine whether $f(x)=\begin{cases}x^2-1, & x\leq 2\\ 2x-1, & x>2\end{cases}$ is continuous at $x = 2$.

3 Determine whether $f(x)=\begin{cases}\dfrac{x-2}{|x-2|}, & x\neq 2\\ 0, & x=2\end{cases}$ is continuous at $x = 2$.

4 Determine whether $f(x)=\begin{cases}\dfrac{x^2-a^2}{x-a}, & x\neq a\\ 2a, & x=a\end{cases}$ is continuous at $x = a$.

5 Determine whether or not

$$f(x)=\begin{cases}\dfrac{x^2-1}{x^2+x-2}, & x\neq 1, x\neq -2\\ \dfrac{2}{3}, & x=1\\ 4, & x=-2\end{cases}$$

is continuous at $x = 1$ and $x = -2$.

6 Find the value of k such that $f(x)=\begin{cases}x^2-2, & x\leq -2\\ 3kx, & x>-2\end{cases}$ is continuous at $x = -2$.

7 Find the value of k such that $f(x)=\begin{cases}kx^2-k, & x\geq 3\\ 4, & x<3\end{cases}$ is continuous for all real values of x.

8 Determine whether each function is continuous on the set of real numbers. If the function is not continuous, state the value(s) of x at which the function is discontinuous.

a $f(x)=\dfrac{x^2+9}{x^2-9}$ **b** $f(x)=\dfrac{x+1}{1-x^2}$

c $f(x)=\dfrac{x}{x^2+4}$ **d** $f(x)=\dfrac{x^2+4x+5}{x^2+4x-5}$

e $f(x)=\dfrac{x^2}{x^3-1}$ **f** $f(x)=\dfrac{x+1}{\sqrt{x^2+1}}$

Limits and infinity

You are already familiar with **infinite discontinuities**, for example the graph of $f(x) = \frac{1}{x^2}$, $x \neq 0$.

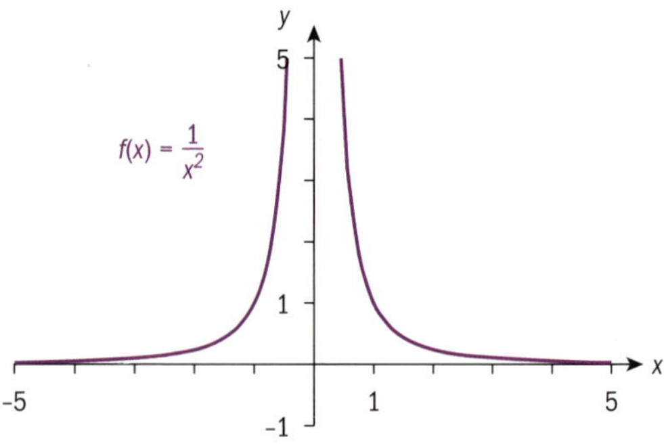

You can see that as x approaches 0 from the left and from the right, the values of the function increase without bound, and approach positive infinity. The limit therefore does not exist, since the limit is not a real number.

The line $x = 0$ is the vertical asymptote of this function. We can now define the vertical asymptote of a function using limits.

> The line $x = c$ is a vertical asymptote of the graph of a function $y = f(x)$ if either $\lim_{x \to c^+} f(x) = \pm\infty$ or $\lim_{x \to c^-} f(x) = \pm\infty$.

Did you know?

The word asymptote comes from the Greek *asymptotes*, meaning "not falling together".

You will also notice in the graph of the function $f(x) = \frac{1}{x^2}$ that as x increases without bound both in the positive and negative directions, the function gets closer and closer to $y = 0$, but is never equal to 0, since there is no real number that satisfies the equation $\frac{1}{x^2} = 0$. The line $y = 0$ is therefore a horizontal asymptote.

> The line $y = k$ is a horizontal asymptote of $f(x)$ if either $\lim_{x \to -\infty} f(x) = k$ or $\lim_{x \to +\infty} f(x) = k$.

Did you know?

The horizontal asymptote tells you the behaviour of the function for very large values of x. However, unlike the vertical asymptote, the function can assume the value of the horizontal asymptote for small values of x. For example,

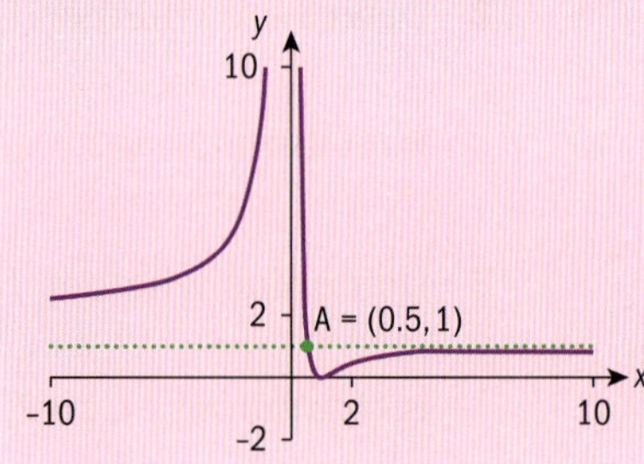

Here you see that $\lim_{x \to \infty} f(x) = 1$ so $y = 1$ is the horizontal asymptote, yet $f\left(\frac{1}{2}\right) = 1$.

Example 5

Sketch the graph of $f(x) = \frac{x+2}{x+1}$, and state the equation of any asymptotes.

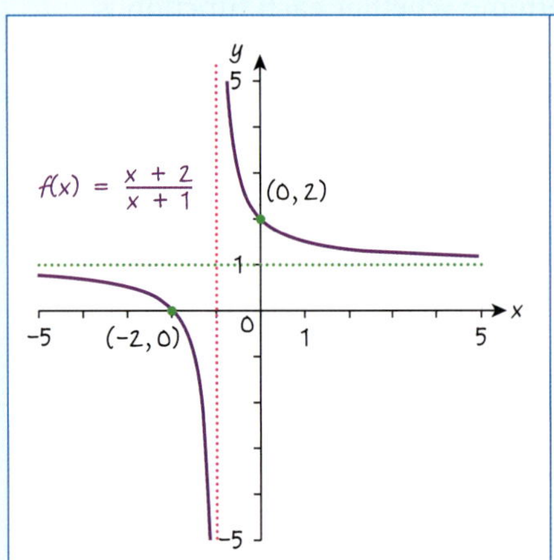	A sketch requires the labelling of intercepts and turning points, and dotted lines for the asymptotes.
Asymptotes: $x = -1$; $y = 1$	Vertical asymptote $\lim_{x \to -1^-} = -\infty$; $\lim_{x \to -1^+} = \infty$ Horizontal asymptote $\lim_{x \to \infty} f(x) = 1$

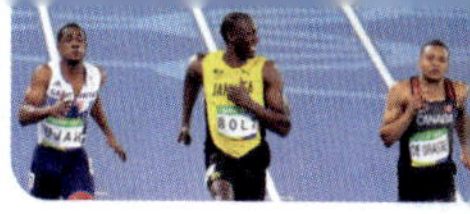

Investigation 4

1 Use your GDC to plot the graph of $f(x) = \frac{x^2 + 2x + 1}{x}$, $x \neq 0$, and investigate the values of the function as x approaches $\pm\infty$. Write down what you notice.

2 Conjecture a relationship between the x and y values as x approaches $\pm\infty$.

3 Simplify $f(x)$ and find the limit of f as x approaches $\pm\infty$.

4 **Conceptual** What special property of a graph can you find by examining the limit of the function as x tends to infinity?

You will have noticed that there is a *slant*, or *oblique*, asymptote that passes between the local minimum and maximum points of the function. As x approaches $\pm\infty$ the function resembles ever more closely the straight line $y = x + 2$.

On the graph, the line $y = x + 2$ is an asymptote to f, which simplified is equal to $f(x) = x + 2 + \frac{1}{x}$.

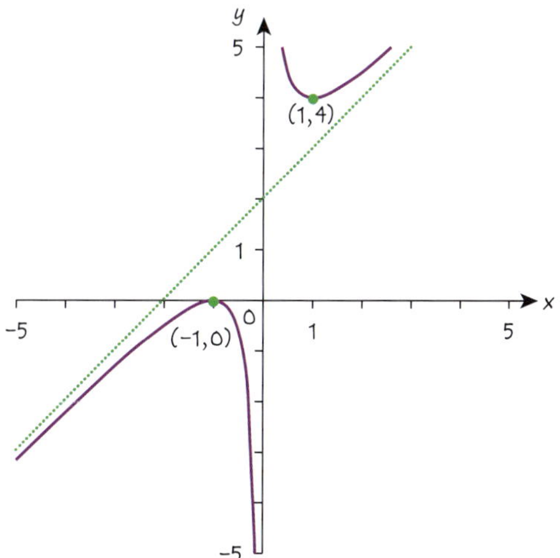

An asymptotic line that is neither vertical nor horizontal but is at an inclination to the x-axis is called a slant or oblique asymptote.

A rational function $P(x) = \frac{f(x)}{g(x)}$, $g(x) \neq 0$, where $f(x)$ and $g(x)$ are both polynomial functions, has an oblique asymptote if the degree of f is exactly one more than the degree of g. Dividing $f(x)$ by $g(x)$, $P(x) = L(x) + \frac{r(x)}{g(x)}$, $r(x)$ is the remainder. The oblique asymptote is $y = L(x)$.

Example 6

Sketch the graph of $f(x)=\dfrac{x^2+1}{x}$, $x \neq 0$, showing clearly any asymptotes. State the equation of the asymptote.

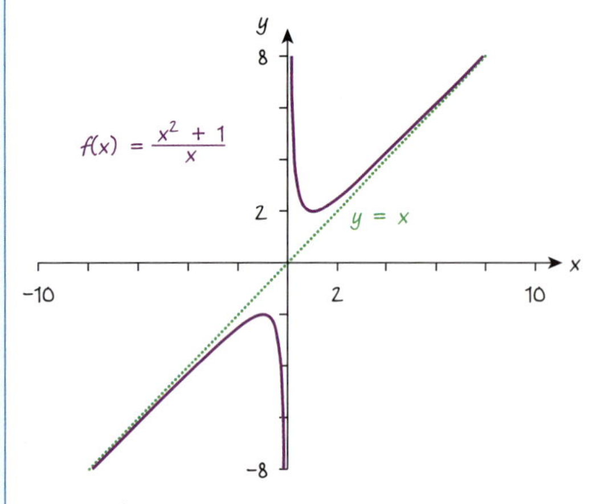

$f(x)=\dfrac{x^2+1}{x}=x+\dfrac{1}{x}$.

The oblique asymptote is $y=x$.

The degree of the numerator is 2 and the degree of the denominator is 1, hence the oblique asymptote is $y=x$.

Properties of limits

Up to now you have been finding limits graphically and confirming your results numerically. You can find some limits algebraically using the properties of limits.

Properties of limits as $x \to \pm\infty$

Let L_1, L_2 and k be real numbers, and let $\lim\limits_{x\to\pm\infty} f(x) = L_1$ and $\lim\limits_{x\to\pm\infty} g(x) = L_2$.

Then:

1. $\lim\limits_{x\to\pm\infty}\left(f(x)\pm g(x)\right) = L_1 + L_2$
2. $\lim\limits_{x\to\pm\infty}\left(f(x)g(x)\right) = L_1L_2$
3. $\lim\limits_{x\to\pm\infty}\left(\dfrac{f(x)}{g(x)}\right) = \dfrac{L_1}{L_2}$, $L_2 \neq 0$
4. $\lim\limits_{x\to\pm\infty} kf(x) = k\lim\limits_{x\to\pm\infty} f(x) = kL_1$
5. $\lim\limits_{x\to\pm\infty}\left[f(x)\right]^{\frac{a}{b}} = \left(L_1\right)^{\frac{a}{b}}$, where $\dfrac{a}{b}\in\mathbb{Q}$ is in simplest form, provided $\left(L_1\right)^{\frac{a}{b}}$ is real.

These properties also hold when finding limits as $x \to c$, $c \in \mathbb{R}$.

Example 7

Find the horizontal asymptote of $f(x) = \dfrac{x+1}{2x+3}$.

$\dfrac{x+1}{2x+3} = \dfrac{\frac{x}{x}+\frac{1}{x}}{\frac{2x}{x}+\frac{3}{x}} = \dfrac{1+\frac{1}{x}}{2+\frac{3}{x}}$	Divide numerator and denominator by the highest power of x.
$\lim\limits_{x\to\infty} f(x) = \lim\limits_{x\to\infty} \dfrac{1+\frac{1}{x}}{2+\frac{3}{x}} = \lim\limits_{x\to\infty}\left(1+\dfrac{1}{x}\right) \div \lim\limits_{x\to\infty}\left(2+\dfrac{3}{x}\right)$	
$\lim\limits_{x\to\infty}\left(1+\dfrac{1}{x}\right) = \lim\limits_{x\to\infty} 1 + \lim\limits_{x\to\infty}\dfrac{1}{x} = 1+0=1$	Apply property 1.
$\lim\limits_{x\to\infty}\left(2+\dfrac{3}{x}\right) = \lim\limits_{x\to\infty} 2 + \lim\limits_{x\to\infty}\dfrac{3}{x} = 2+0=2$	Apply properties 1 and 3.
Hence, $\lim\limits_{x\to\infty}\dfrac{x+1}{2x+3} = \dfrac{1}{2}$, and the horizontal asymptote is $y = \dfrac{1}{2}$.	
The limit can be confirmed graphically and numerically using your GDC. 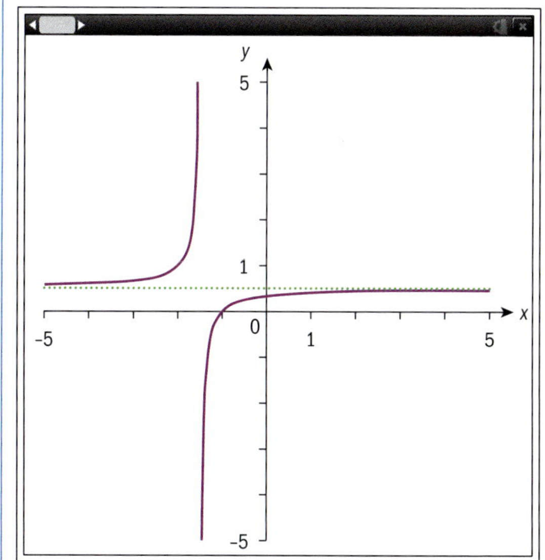	

As shown in Example 7, when finding limits of rational algebraic expressions, it is often useful to divide the numerator and denominator by the largest power of x.

Calculus

Investigation 5

1 For the rational functions $f(x)=\dfrac{p(x)}{q(x)}$, $q(x)\neq 0, p(x)$ and $q(x)$ polynomial functions, find algebraically the horizontal asymptotes, if they exist, by finding the limits of the functions as $x \to \infty$.

a $y=\dfrac{-2x+3}{x+1}$ **b** $y=\dfrac{x+2}{-3x^2+3x-1}$ **c** $y=\dfrac{x^3-25}{x^2+25}$

d $y=\dfrac{3x+2}{x^2-1}$ **e** $y=\dfrac{x^3+x-4}{4x^3+3x^2+x}$ **f** $y=\dfrac{x^2+1}{x-1}$

g $y=\dfrac{x^2+3x}{x^5-1}$ **h** $y=\dfrac{x^5-2x}{x^3-x+1}$

2 Summarize your findings for the functions in question **1** in a copy of the following table. The first one has been done for you.

	Degree $(p(x))$	Leading coefficient	Degree $(q(x))$	Leading coefficient	Horizontal asymptote
a	1	−2	1	1	$y=-2$
b					
c					
d					
e					
f					
g					
h					
i					

3 **Conceptual** Is there a rule for finding the horizontal asymptotes for some types of rational functions?

Exercise 4C

1 Find the required limit, if it exists.

a $\lim\limits_{x\to 3}\dfrac{x+2}{x-2}$ **b** $\lim\limits_{x\to 1}\dfrac{x^2+x-2}{x-1}$

c $\lim\limits_{x\to \sqrt[3]{3}}\dfrac{x^6-9}{x^3-3}$ **d** $\lim\limits_{x\to 2}\dfrac{x^2-4}{x^2-2x}$

e $\lim\limits_{x\to 0}\dfrac{x^2-4}{x^2-2x}$ **f** $\lim\limits_{x\to 2}\dfrac{2}{2+\dfrac{2}{2-x}}$

g $\lim\limits_{x\to 0}\dfrac{(2+3x)^2-4(1+x)^2}{6x}$ **h** $\lim\limits_{x\to a}\dfrac{a^2x^2-b^2}{ax-b}$

2 Find $\lim\limits_{x\to\infty} f(x)$, if it exists.

a $f(x)=\dfrac{3x}{x+3}$ **b** $f(x)=\dfrac{-2x^2}{x^2-1}$

c $f(x)=\dfrac{-x^2+x+2}{3x^2+2x-1}$ **d** $f(x)=\dfrac{5x^2-2}{2x+3}$

e $f(x)=\dfrac{2x+3}{5x^2-2}$ **f** $f(x)=\dfrac{x-1}{x^2+3x-2}$

g $f(x)=\dfrac{(x+2)(x-1)}{2x}$ **h** $f(x)=\dfrac{\sqrt{4x-1}+2\sqrt{x+3}}{\sqrt{x}}$

3 Find any vertical and horizontal asymptotes of these functions.

a $f(x)=\frac{3x}{6x-1}$ b $g(x)=\frac{x^2+3}{3-x^2}$

c $h(x)=\frac{1-x-x^3}{x^3-1}$ d $k(x)=-\frac{5x}{x^2-2}$

e $p(x)=\frac{3x^3-x}{x^2}$ f $r(x)=\frac{x^2}{2x^2-3x+1}$

Limits of sequences

Investigation 6

1 Use your GDC to plot a graph of the function $f(x)=\frac{1}{x}$, $x>0$, and determine $\lim\limits_{x\to\infty}\frac{1}{x}$.

2 Using your GDC or graphing software, on the same set of axes as **1**, graph the sequence $\{a_n\}$ where $a_n=\frac{1}{n}$, $n\in\mathbb{Z}^+$.

3 Determine graphically $\lim\limits_{n\to\infty}\frac{1}{n}$, $n\in\mathbb{Z}^+$.

4 Write down the first few terms of the sequence, and state the relationship of $\frac{1}{m}$ and $\frac{1}{n}$ when $m, n\in\mathbb{Z}^+$, and $m<n$.

5 Use your answer to question **4** to justify your answer to question **3**.

6 **Conceptual** When does a sequence converge?

In Investigation 6 you have seen that $\lim\limits_{n\to\infty}\frac{1}{n}=0$, $n\in\mathbb{Z}^+$, that is, the sequence converges to 0.

If $\lim\limits_{n\to\infty}\{a_n\}=L$, $L\in\mathbb{R}$, then the sequence is said to **converge**; otherwise it **diverges**.

Example 8

Determine whether each sequence is convergent or divergent.

If a sequence is convergent, give the limit of the sequence.

a $1, -1, 1, -1, 1, \ldots$ b $\frac{1}{10}, \frac{1}{100}, \frac{1}{1000}, \frac{1}{10\,000}, \ldots$

c $2, 4, 6, 8, \ldots$ d $2.39, 2.399, 2.3999, \ldots$

Continued on next page

Calculus

a Divergent	The terms of the sequence are oscillating and not approaching a fixed value.
b Convergent; $\lim_{n\to\infty} u_n = 0$	Geometric sequence with $u_n = \left(\frac{1}{10}\right)^n$.
c Divergent	The terms of the sequence are increasing and not approaching a limit.
d Convergent; $\lim_{n\to\infty} u_n = 2.4$	The values of the sequence are approaching 2.4.

The properties of limits of sequences are the same as those for limits of functions.

Example 9

Find $\lim_{n\to\infty}\frac{n^3+4n}{2n^3-1}$. Confirm your answer graphically.

$\lim_{n\to\infty}\frac{n^3+4n}{2n^3-1} = \lim_{n\to\infty}\frac{1+\frac{4}{n^2}}{2-\frac{1}{n^3}} = \frac{1}{2}$ The sequence converges to $\frac{1}{2}$. 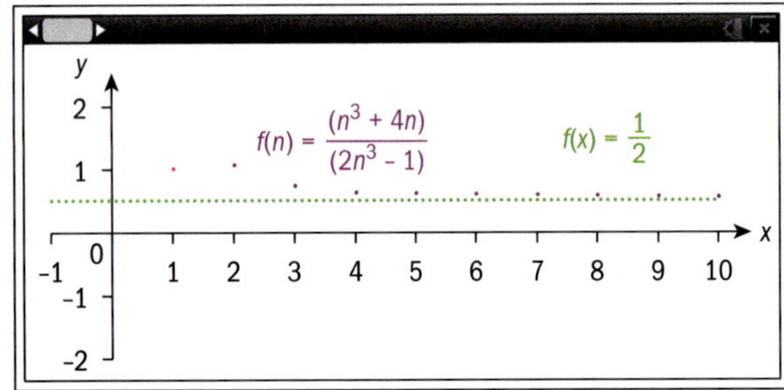 	Divide both numerator and denominator by n^3, and use the property of limits.

In Chapter 1 you learned that if a geometric series has a finite sum it converges to its sum. Recall the formula for finding the sum of a finite geometric series, $S_n = \frac{u_1(1-r^n)}{1-r}$. When $\|r\| < 1$ and $r \neq 0$ then $\lim_{n\to\infty}\left(\frac{u_1(1-r^n)}{1-r}\right) = \frac{u_1}{1-r}$.

> For an infinite geometric series, when $|r| < 1$ and $r \neq 0$ the series converges to its sum, $S_\infty = \frac{u_1}{1-r}$. If the series does not have a finite sum, the series diverges.

Example 10

Determine whether $\sum_{n=0}^{\infty} \frac{2^n + 3^n}{5^n}$ converges.

$\sum_{n=0}^{\infty} \frac{2^n + 3^n}{5^n} = \sum_{n=0}^{\infty} \frac{2^n}{5^n} + \sum_{n=0}^{\infty} \frac{3^n}{5^n} = \sum_{n=0}^{\infty} \left(\frac{2}{5}\right)^n + \sum_{n=0}^{\infty} \left(\frac{3}{5}\right)^n$	Separate the two series.
$\sum_{n=0}^{\infty} \left(\frac{2}{5}\right)^n = \frac{1}{1-\frac{2}{5}} = \frac{5}{3};\ \sum_{n=0}^{\infty} \left(\frac{3}{5}\right)^n = \frac{1}{1-\frac{3}{5}} = \frac{5}{2}$	Each is a geometric series with $\|r\| < 1$, so each converges to a finite limit.
Hence $\sum_{n=0}^{\infty} \frac{2^n + 3^n}{5^n} = \frac{5}{3} + \frac{5}{2} = \frac{25}{6} = 4\frac{1}{6}$	Find the sum of each, and add.

Exercise 4D

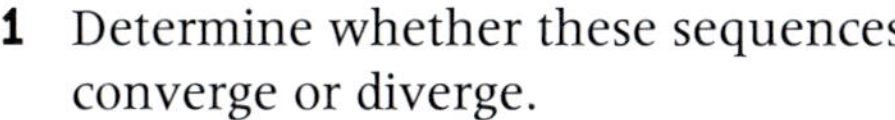

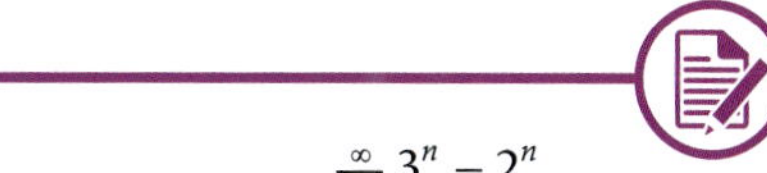

1 Determine whether these sequences converge or diverge.

a 3, 1, 3, 1, 3, 1, ... **b** 1, 1, 2, 3, 5, 8, ...

c $1, \frac{1}{2}, \frac{1}{3}, \frac{1}{4}, \ldots$ **d** $\frac{1}{3}, \frac{1}{9}, \frac{1}{27}, \frac{1}{81}, \ldots$

2 Determine whether these sequences converge. If the sequence converges, state the number it converges to.

a $\lim_{n\to\infty} \frac{n+2}{n}$ **b** $\lim_{n\to\infty} \frac{n+2}{2n+3}$

c $\lim_{n\to\infty} \frac{n^2 - n}{2n^2 + \sqrt{n}}$ **d** $\lim_{n\to\infty} \frac{1-2n^3}{2n^2+1}$

e $\lim_{n\to\infty} \frac{2n^2+1}{1-2n^3}$ **f** $\lim_{n\to\infty} \frac{1+n^2}{1-n^3}$

3 Determine whether each series converges. If it converges, determine its sum.

a $\sum_{n=0}^{\infty} \frac{(-1)^n}{3^n}$ **b** $\sum_{n=1}^{\infty} 3\left(\frac{1}{2}\right)^n$

c $\sum_{n=0}^{\infty} \frac{2}{10^n}$ **d** $\sum_{n=1}^{\infty} \frac{3^n - 2^n}{5^n}$

e $\sum_{n=0}^{\infty} \left(\frac{e}{\pi}\right)^n$ **f** $\sum_{n=1}^{\infty} \left(\frac{\pi}{3.14}\right)^n$

4 A geometric series has $u_1 = 42$ and $r = 2^x$.

a Find the values of x for which the series is convergent.

b Find the value of x for which the series converges to 48.

5 Find the set of values for which $\sum_{n=0}^{\infty} \left(\frac{3x}{x+1}\right)^n$ converges.

6 Prove that the series $\sum_{n=0}^{\infty} \frac{1+2^n}{3^{n-1}}$ converges, and find its sum.

4.2 The derivative of a function

Investigation 7

The Jamaican sprinter Usain Bolt was the first person, since speed times were fully automated, to hold both the 100 m and 200 m world records simultaneously.

The data below shows the time, in seconds, which had elapsed when Usain Bolt passed every 10 m mark in the 100 m final for both the Olympic Games in Beijing (2008) and the World Championships in Berlin (2009).

The graph shows Bolt's time (plotted on the x-axis) against the distance he had run (plotted on the y-axis).

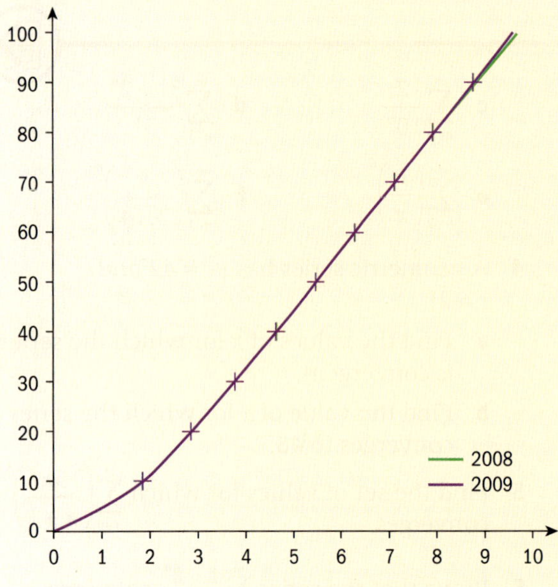

Distance (m)	2008	2009
10	1.83	1.89
20	2.87	2.88
30	3.78	3.78
40	4.65	4.64
50	5.5	5.47
60	6.32	6.29
70	7.14	7.10
80	7.96	7.92
90	8.79	8.75
100	9.69	9.58

1 Write down what you notice about the time Bolt took to cover each 10 m distance in 2008 compared to 2009.

2 For each of the two events, write down Bolt's average running speed over the 100 m.

3 Find Bolt's fastest average speed over any 10 m interval in either race, and state the 10 m interval over which he achieved this.

4 **Factual** Compare Bolt's fastest speed with his average speed in that race. If you were to examine Bolt's speed at *any* point in that race, which of the two speeds you found would be likely to be closest to the speed that he was running at that instant?

5 **Factual** From the given information, is it possible to find his fastest speed at any particular moment in time, eg at 3.25 s?

6 **Conceptual** What is an average rate of change between output and input values of a function?

In Investigation 7, the distance Usain Bolt had covered was a function of time. You found his average speed over any particular 10 m interval by dividing the change in the distance he had covered—that is, 10 m—by the corresponding change in time.

For any two points (t_1, d_1) and (t_2, d_2) on the graph, Bolt's average speed between those two points is given by $\frac{d_2 - d_1}{t_2 - t_1}$.

You may have noticed that this is the same as the formula for the gradient of a straight line joining the points (t_1, d_1) and (t_2, d_2).

Therefore, the average rate of change of a function f between the points x_1 and x_2 is the gradient of the line segment joining $(x_1, f(x_1))$ and $(x_2, f(x_2))$.

The average rate of change of a function f between two values x_1 and x_2 is given by $\frac{\Delta y}{\Delta x} = \frac{f(x_2) - f(x_1)}{x_2 - x_1}$.

(Note that $\frac{\Delta y}{\Delta x}$ means "the change in y divided by the change in x", where Δ is the Greek letter "delta".)

If a function is linear, the gradient between any two points is the same, and hence the rate of change of the function is always constant.

This is not the case, however, for non-linear functions. In order to explore the gradient or rate of change of a non-linear function, you first need to learn some essential definitions.

A **secant** is a line that intersects a curve.

A **tangent** to a curve at a specific point is a straight line that *touches* the curve at that point.

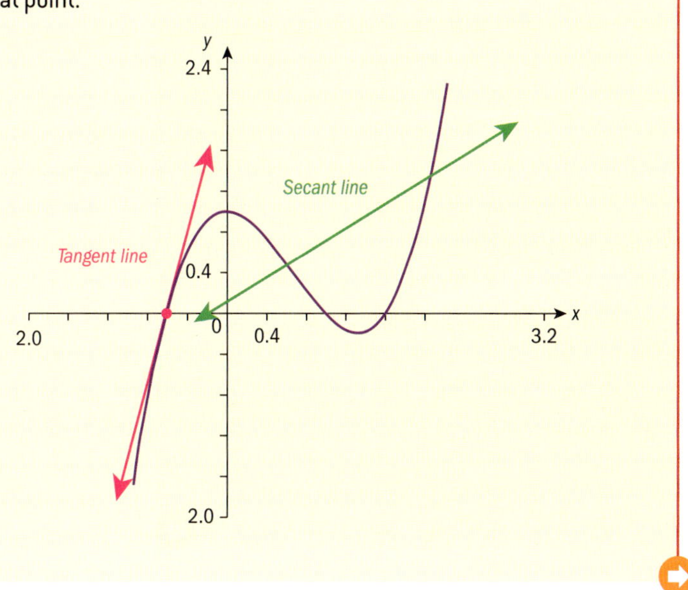

Continued on next page

Calculus

The green line is tangent to the curve at the red marked point. It also intersects the curve at another point.

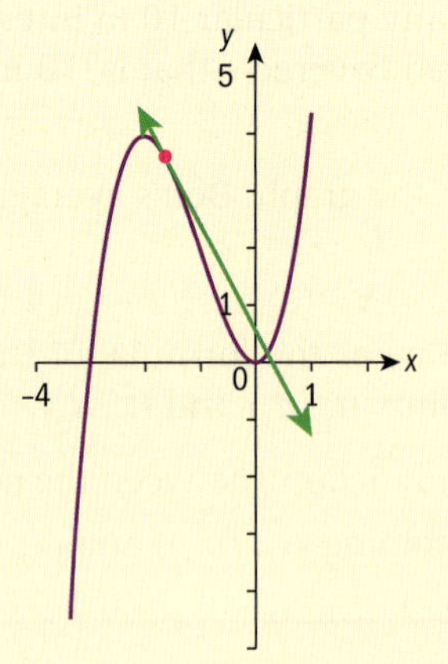

Investigation 8

Let the function f be defined by $f(x) = x^2$.

$\mathrm{A}(1, 1)$ is a point which lies on the graph of $y = f(x)$, and B is any other arbitrary point (x, x^2) on the curve.

Let [AB] represent a secant line joining points A and B.

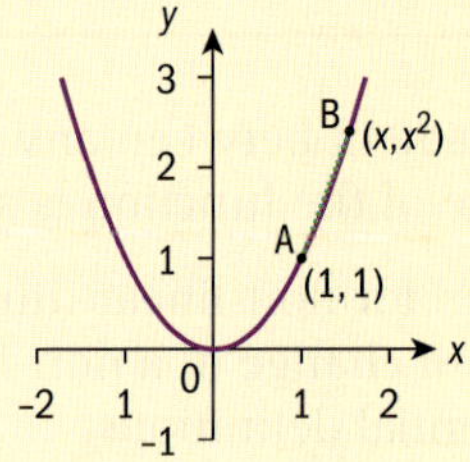

1 Calculate the gradient of [AB] for each of the different values of x given in the table. The first one has been done for you.

x	$\mathrm{B}(x, f(x))$	Gradient of [AB]
2	$(2, 4)$	$\frac{4-1}{2-1} = 3$
1.5		
1.1		
1.01		
1.001		

In the table, the x-values of B approach 1 from the right (ie for values of x greater than 1); which is the x-coordinate of A. Hence, the point $\mathrm{B}\left(x, f(x)\right)$ gets closer and closer to point A.

2 Write down what you notice about the gradient of [AB] as the value of x approaches 1 from the right.

3 Repeat question **1**, but this time let x approach 1 from the left of point A, using x values: 0, 0.5, 0.9, 0.999.

Write down what you notice about the gradient of [AB] as the value of x approaches 1 from the left.

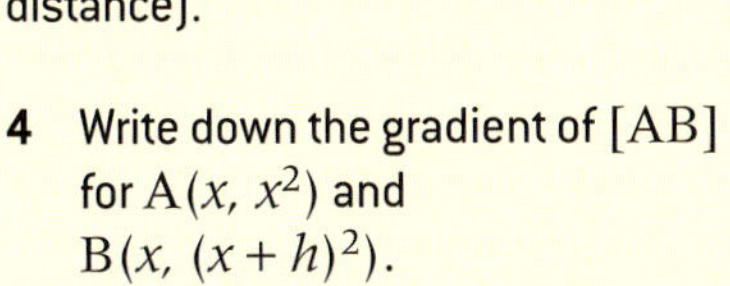

Now consider a fixed point $A(x, x^2)$ on the curve, and a second point B with x coordinate h units away from point A (where h is a very small distance).

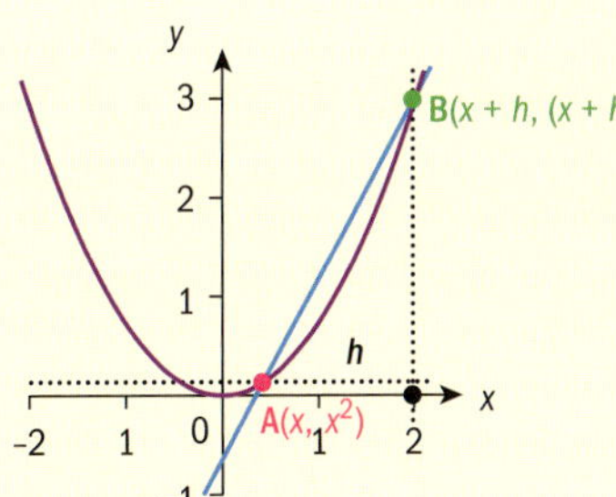

4 Write down the gradient of [AB] for $A(x, x^2)$ and $B(x, (x+h)^2)$.

5 **Conceptual** How is the gradient of a tangent to the curve at a point related to the gradient of a secant line passing through the same point?

6 Letting h tend to 0 and using limits, deduce the gradient of the tangent to $f(x) = x^2$ at any point x.

7 Using your answer to question **6**, find the gradient of the curve at $A(1, 1)$.

8 **Conceptual** What does the gradient of the tangent to the curve $y = f(x)$ at any point tell you about the instantaneous rate of change between x and y at that point?

You saw in Investigation 8 that if point B approaches point A, the gradient of the secant line [AB] approaches the gradient of the tangent to the curve at point A.

The gradient of the secant line [AB] is $\frac{f(x+h)-f(x)}{h}$.

This gradient is a measure of the **average rate of change** of $f(x)$ with respect to x.

As B approaches A, the gradient of [AB] approaches the gradient of the tangent line to the curve at A.

So, the gradient of the tangent to the curve at A is the limiting value of $\frac{f(x+h)-f(x)}{h}$ as h approaches 0.

The gradient of the tangent to the curve at A is the instantaneous rate of change of $f(x)$ at A.

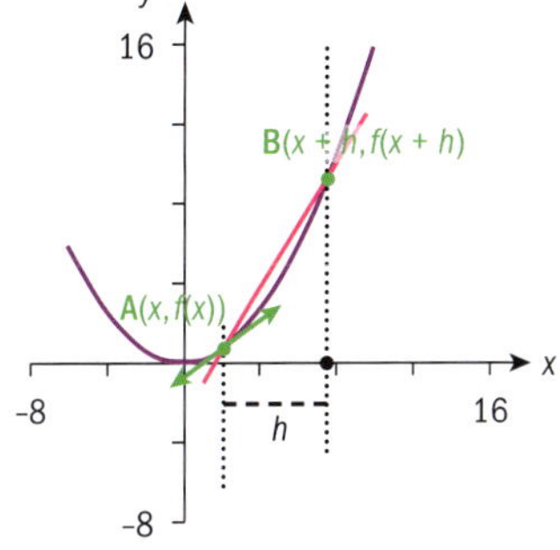

The gradient of a curve $y = f(x)$ at any point $(a, f(a))$ on the curve is $\lim_{h \to 0} \frac{f(a+h)-f(a)}{h}$, provided this limit exists.

In Investigation 7 (Usain Bolt), finding the speed over shorter intervals of time should give a more accurate approximation of his speed at any given time in the race. The examples will show how an interval of time can be made infinitesimally small by using limits.

Example 11

Find the gradient of the curve $y = x^2$ at the point $(-2, 4)$.

Let $f(x) = x^2$. Then the gradient of the curve $y = f(x)$ at $(-2, 4)$ is	
$\lim\limits_{h \to 0} \frac{f(-2+h) - f(-2)}{h}$	Definition of the gradient of a curve at a point.
$= \lim\limits_{h \to 0} \frac{(-2+h)^2 - (-2)^2}{h}$	Substituting $f(x) = x^2$.
$= \lim\limits_{h \to 0} \frac{4 - 4h + h^2 - 4}{h}$	
$= \lim\limits_{h \to 0} \frac{-4h + h^2}{h} = \lim\limits_{h \to 0} \frac{h(-4+h)}{h}$	
$= \lim\limits_{h \to 0} (-4+h) = -4$	Simplify, and evaluate the limit.

Example 12

Find the points on the graph of $y = \frac{1}{x}$ where the gradient is $-\frac{1}{4}$.

Any point on the curve is $\left(a, \frac{1}{a}\right)$ and a neighbouring point is $\left(a+h, \frac{1}{a+h}\right)$.	Express the points in terms of a.
Gradient at $\left(a, \frac{1}{a}\right)$ is	
$= \lim\limits_{h \to 0} \frac{\left(\frac{1}{a+h} - \frac{1}{a}\right)}{h}$ $= \lim\limits_{h \to 0} \frac{\left(\frac{a-(a+h)}{a(a+h)}\right)}{h}$ $= \lim\limits_{h \to 0} \frac{\left(\frac{-h}{a^2+ah}\right)}{h}$	Use the definition for the gradient of a curve at a point, and simplify.

$= \lim_{h\to 0}\left(\frac{-1}{a^2+ah} = -\frac{1}{a^2}\right)$	
$-\frac{1}{a^2} = -\frac{1}{4} \Rightarrow a = \pm 2$	Set the gradient equal to $-\frac{1}{4}$, and solve for a.
At $x = \pm 2$, $y = \pm\frac{1}{2}$	Find y at each value of a.
The points are $\left(2, \frac{1}{2}\right)$ and $\left(-2, -\frac{1}{2}\right)$.	State the coordinates of the points where the gradient of $y = \frac{1}{x}$ is $-\frac{1}{4}$.

Exercise 4E

1 Find the gradient of the graph of the function at the given value of x.

a $y = 2x^2 + 1$; $x = -1$ **b** $y = 1 - 3x^2$; $x = 1$

c $y = \frac{2}{x}$; $x = -1$ **d** $y = -x^2$; $x = -1$

e $y = x^3$; $x = -1$ **f** $y = x^2 + x - 1$; $x = 0$

g $y = \frac{1}{x^2}$; $x = 2$ **h** $y = \frac{x}{x+1}$; $x = 0$

2 A and B are two points on the graph of $f(x) = x^2 + 3$. The x-coordinate of A is 1 and the x-coordinate of B is $1 + h$.

Find, in terms of h, the gradient of the secant line AB. Hence, deduce the gradient of the tangent to $f(x)$ at A.

3 a By using the fact that the gradient of a curve $y = f(x)$ at any point $(a, f(a))$ on the curve is $\lim_{h\to 0}\frac{f(a+h)-f(a)}{h}$, find the gradient function of $y = 3x^2 + 2x - 1$.

b Hence, find the coordinates of the point on the curve where the gradient is -4.

4 Find the coordinates of the point on the graph of $y = \frac{1}{x^2}$ where the gradient is $-\frac{1}{4}$.

The **gradient function** is the same as the **derivative** function.

The **gradient function** $f'(x) = \lim_{h\to 0}\frac{f(x+h)-f(x)}{h}$ is called the **derivative** of f with respect to x.

We sometimes use different notation for the derivative function:

- $f'(x)$ (this is called "f dash of x" or "f prime of x"); or
- $\frac{dy}{dx}$ (meaning "the derivative of y with respect to x")

If f' exists, we say that f has a derivative at x, or is **differentiable** at x.

A function f is said to be **differentiable everywhere** if the derivative exists for all x in the domain of f.

TOK

Who do you think should be considered the discoverer of calculus?

Reflect If you have two real-life physical quantities x and y which change in relation to one another, what does the derivative function tell you about these quantities?

If you plot a graph of x against y, what geometrical property of the graph would the derivative function tell you?

Example 13

Given $f(x)=\sqrt{x}$, find $f'(x)$ at $x=9$.

$f'(x)=\lim\limits_{h\to 0}\dfrac{\sqrt{x+h}-\sqrt{x}}{h}$	
$=\lim\limits_{h\to 0}\dfrac{\sqrt{x+h}-\sqrt{x}}{h}\times\dfrac{\sqrt{x+h}+\sqrt{x}}{\sqrt{x+h}+\sqrt{x}}$	Rationalize the numerator.
$=\lim\limits_{h\to 0}\dfrac{(x+h)-x}{h\left(\sqrt{x+h}+\sqrt{x}\right)}$	Simplify the numerator using the "difference of two squares".
$=\lim\limits_{h\to 0}\dfrac{h}{h\left(\sqrt{x+h}+\sqrt{x}\right)}$	Simplify.
$=\lim\limits_{h\to 0}\dfrac{1}{\sqrt{x+h}+\sqrt{x}}$	Divide through by h.
$f'(x)=\dfrac{1}{2\sqrt{x}}$	Let $h\to 0$.
$f'(9)=\dfrac{1}{2\sqrt{9}}=\dfrac{1}{6}$	Evaluate $f'(x)$ at $x=9$.

Example 14

A particle moves in a straight line so that its position from its starting point at any time t, in seconds, is given by $s=3t^2$ where s is in metres. The particle passes through a point P when $t=a$ and passes though point Q when $t=a+h$.

a Find the average speed of the particle between the points P and Q.

b Let $h\to 0$ and hence deduce the speed of the particle at the instant it passes through P.

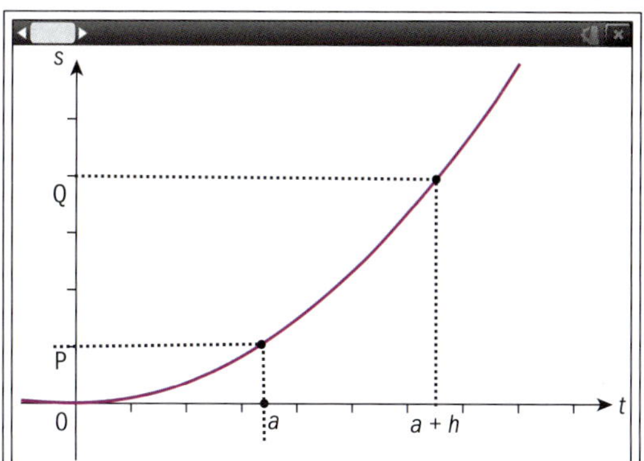

Sketching a quick graph of the function, using your GDC to help you, can often be a good way of visualizing the shape of the function you're working with.

a P is the point $(a, 3a^2)$ and Q is the point $(a + h, 3(a + h)^2)$.

Substitute the values of t at P and Q into $s = 3t^2$ to find the corresponding values of s.

Average speed between P and Q

$$= \frac{3(a+h)^2 - 3a^2}{(a+h) - a}$$

Average speed between P and Q $= \frac{\text{distance travelled from P to Q}}{\text{time taken}}$

$$= \frac{3(a^2 + 2ah + h^2) - 3a^2}{h}$$

Multiply out the numerator and simplify the denominator.

$$= \frac{3h^2 + 6ah}{h}$$

Simplify.

$$= \frac{h(3h + 6a)}{h}$$

Factorize h from the numerator.

$= (3h + 6a)$ m s^{-1}

Divide through by h.

b Letting $h \to 0$,

$$\lim_{h \to 0}(3h + 6a) = 6a$$

hence speed at P is $6a$ m s^{-1}

Use the limit to find the instantaneous speed at P.

If you are asked to find the derivative of a function **from first principles**, you will need to use the definition that $f'(x) = \lim_{h \to 0} \frac{f(x+h) - f(x)}{(x+h) - x}$ and find the derivative by letting $h \to 0$.

Exercise 4F

1 For the following functions, find the derivative from first principles and find the value of the derivative at the given x value.

a $f(x) = x^2 + x - 2$; $f'(0)$

b $f(x) = 2 - x + 3x^2$; $f'(-1)$

c $f(x) = -\frac{2}{x}$; $f'(1)$

d $f(x) = \sqrt{x+1}$; $f'(3)$

e $f(x) = \frac{1}{\sqrt{x}}$; $f'(9)$

f $f(x) = x^3 - 1$; $f'(1)$

2 A particle moves in a straight line so that its position in metres (in relation to its starting point) after t seconds is $8 + 2t^2$ m.

If the particle moves through point A when $t = a$, and through point B when $t = a + h$, find:

a the average velocity (speed) of the particle as it moves from A to B

b the instantaneous velocity (speed) of the particle as it passes through A.

3 A particle moves along a straight line such that, after t seconds, its displacement from a fixed point is s metres.

If $s = 10t^2 - t^3$, find:

a an expression for the velocity of the particle after t seconds

b the velocity of the particle after 1 second and after 10 seconds, and comment on what you think the signs of the velocity signify.

Differentiability and continuity

Some functions do not have a derivative at every point in their domain. However, differentiability at a point implies continuity at the point.

> Given a function $f(x)$ and a point c in the domain of f, if f is differentiable at $x = c$ then it is also continuous at $x = c$.

You can use ideas you've learned about the definition of the derivative function in order to prove that this key point holds true.

To prove that $f(x)$ is continuous at $x = c$, we need to show that $f(c)$ exists and $\lim_{x \to c} f(x) = f(c)$, or equivalently that $\lim_{x \to c}[f(x) - f(c)] = 0$.

Proof:

$$\lim_{x \to c}[f(x) - f(c)] = \lim_{x \to c}\left[\frac{f(x) - f(c)}{x - c} \times (x - c)\right]$$ multiplying numerator and denominator by $(x - c)$

$$= \lim_{x \to c}\frac{f(x) - f(x)}{x - c} \times \lim_{x \to c}(x - c)$$ by properties of limits from Section 4.1

$$= f'(c) \times \lim_{x \to c}(x - c)$$ because the derivative of f at c exists

$$= f'(c) \times 0$$ $\lim_{x \to c}(x - c) = 0$

$$= 0$$

So $\lim_{x \to c}[f(x) - f(c)] = 0$ and hence f is continuous at c.

This theorem therefore says that **differentiability at a point implies continuity at the point**.

Is the converse of this theorem true? That is, if a function is continuous at a point, is it differentiable at the point?

It turns out that this is false. To demonstrate this, you need only find one counterexample: a function that is continuous at $x = c$, but whose left and right limits as x approaches c are either different, or do not exist.

You are already familiar with one such function: the **modulus** or **absolute value** function, $f(x) = |x|$.

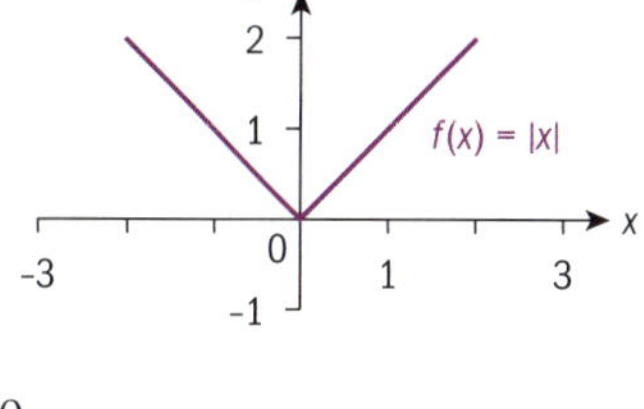

Consider the point where $x = 0$.

$f(x) = |x|$ is continuous at $x = 0$ because:

1 $\lim\limits_{x\to 0^+} f(x) = \lim\limits_{x\to 0^-} f(x) = f(0) = 0.$

So the left and right limits are equal to the value of the function at $x = 0$, and hence f is continuous at 0.

Is f differentiable at $x = 0$?

You can clearly see that for all $x > 0$, $f(x) = x$ and hence the derivative of f at all points where $x > 0$ is equal to the gradient of the line $y = x$, ie $f'(x) = 1$.

For all $x < 0$, $f(x) = -x$ and hence the derivative of f at all points where $x < 0$ is equal to the gradient of the line $y = -x$, ie $f'(x) = -1$.

However, if you try to find the derivative at $x = 0$, you'll get two different values for the derivative depending on whether you let $h \to 0$ from the left or right of zero:

From the right: $f'(x) = \lim\limits_{h\to 0^+} \dfrac{|x+h| - |x|}{h} = 1$

From the left: $f'(x) = \lim\limits_{x\to 0^-} \dfrac{|x+h| - |x|}{h} = -1$

Since the left and right limits are not equal, the function does not have a derivative at $x = 0$.

> If a function is continuous at $x = c$ it *might not* be differentiable at $x = c$.

Investigation 9

1 Use your GDC to plot the graph of the function $f(x) = x^2$. Zoom in at a point on the function that is differentiable, for example at $x = 0$. Write down what happens to the shape of the graph at this point as you zoom further and further in.

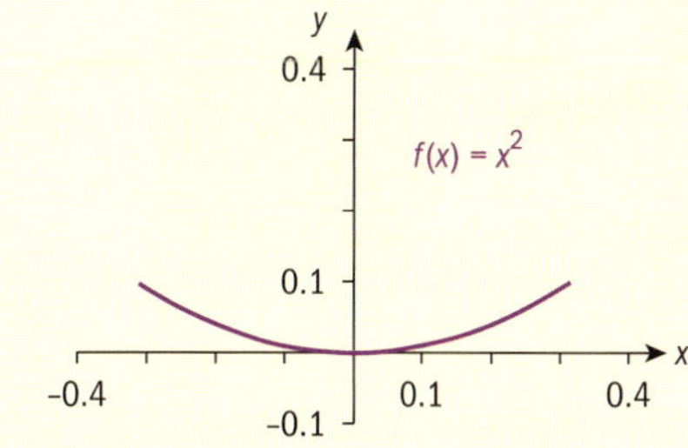

2 Graph $g(x) = |x|$ and zoom in at the point that is not differentiable; $x = 0$. Write down what happens to the shape of the graph at this point as you zoom further and further in.

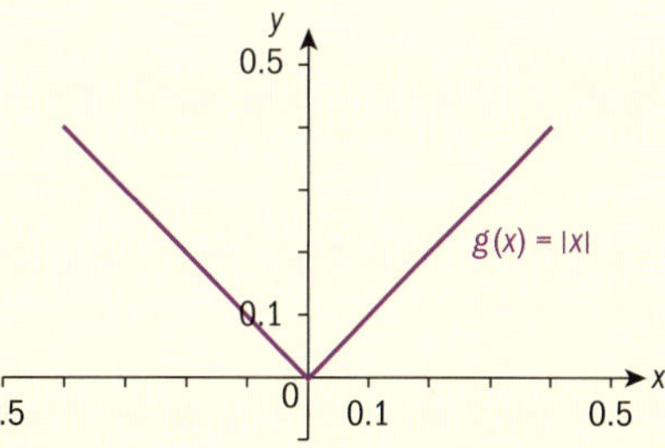

3 **Conceptual** How does zooming in at a point on the graph of a function help you determine the continuity and differentiability of the function at a given point?

TOK

Mathematics and the real world: the seemingly abstract concept of calculus allows us to create mathematical models that permit human feats, such as getting a man on the Moon. What does this tell us about the links between mathematical models and physical reality?

Investigation 10

1 The functions given below are all continuous everywhere on the real numbers. Use your GDC to plot the graphs of these functions, and write down the coordinates of the point on each graph where the function is not differentiable. By investigating the right and left limits at the point, explain why the function has no derivative at this point.

a $f(x) = |x-2|+3$ **b** $f(x) = x^{\frac{2}{3}}$ **c** $f(x) = \sqrt[3]{x}$

2 **Conceptual** Are all functions that are continuous throughout their domain also differentiable throughout their domain?

3 **Conceptual** What generalizations could you make about the nature of points where functions are continuous but not differentiable?

Points at which functions are continuous but not differentiable can be classified as cusp points, corner points or vertical tangent points.

4 **Factual** Identify which of the three functions in question **1** has which type of point. Describe what happens to the derivative function at each of these three points.

Tangents and normals

How much can a cyclist lean inwards, in order to negotiate a bend, without falling off?

In order for a cyclist to move in a circular path, a certain force is required to keep them from shooting away from the circle at a tangent. This force is **perpendicular** to the tangent to the curve of the circular path.

The **normal line** at a point on a curve is the line perpendicular to the curve's tangent at that point.

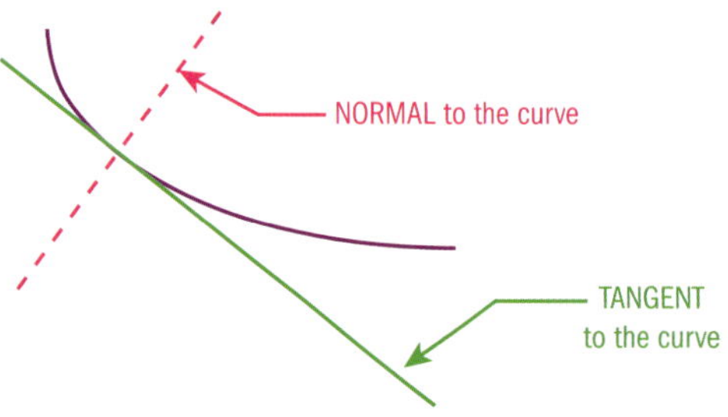

Did you know?

The "Weierstrass function" is a pathological function; meaning its properties are considered "atypical" or "counterintuitive". This function is continuous everywhere, but is not differentiable anywhere. Every point on the function is either a cusp, a corner, or a vertical tangent. It is nicknamed the "blancmange function" because it looks like the pudding with the same name!

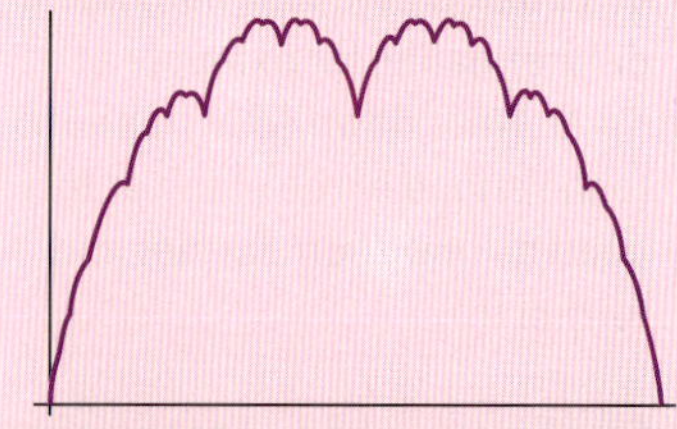

Reflect Can you find other examples of how tangent and normal lines relate to real-life problems?

$f(x)$ is a non-linear function and (x_1, y_1) is a point which lies on the curve $y = f(x)$. If $f'(x_1) = m$, where m is a constant, then m is the gradient of the tangent to $y = f(x)$ at (x_1, y_1), and the equation of the tangent line is $(y - y_1) = m(x - x_1)$.

Recall the different forms of the equation of a straight line:
$y = mx + c$ (gradient–intercept form)
$ax + by + d = 0$ (general form)
$(y - y_1) = m(x - x_1)$ (point–gradient form)

You can also find the equation of the normal to the function at a particular point. As you saw in the example above where the cyclist travels around a bend, the normal to the curve at any point is perpendicular to the tangent at that point.

Recall that for perpendicular lines with gradients m_1 and m_2, then $m_1m_2 = -1$.

If a tangent line has gradient m, then the gradient of the normal to the tangent line is $\frac{-1}{m}$.

Example 15

Given $f(x) = 2 - 3x^2$, find:

a the gradient of the tangent to the curve at $x = 1$

b the equation of the tangent to the curve at $x = 1$

c the equation of the normal to the tangent at $x = 1$ in the form $ax + by + c = 0$ where $a, b, c \in \mathbb{Z}$.

a $f'(x) = \lim_{h \to 0} \frac{\left[2 - 3(x+h)^2\right] - (2 - 3x^2)}{h}$	Find the gradient function.
$= \lim_{h \to 0} \frac{h(-6x - 3h)}{h}$	
$= -6x$	
$f'(1) = -6$	Evaluate the derivative function at $x = 1$.
Hence $m = -6$	This is the gradient of the tangent to the curve at $x = 1$.

Continued on next page

b $f(1) = 2 - 3(1)^2 = -1$ $m = -6$; curve passes through $(1, -1)$ $y - -1 = -6(x - 1)$ $y + 1 = -6x + 6$ $y = -6x + 5$ is the equation of the tangent	The gradient of the tangent is -6, and it goes through $(1, -1)$. Using $(y - y_1) = m(x - x_1)$ for a straight line.
c Gradient of the normal is $\frac{1}{6}$. $y = mx + c \Rightarrow c = y - mx$ $= -1 - \left(\frac{1}{6}\right)(1)$ $= -1\frac{1}{6} = -\frac{7}{6}$ $y = \frac{1}{6}x - \frac{7}{6}$ is the normal.	$m_1 m_2 = -1$ for perpendicular lines. Using $y = mx + c$ as the normal is a straight line with gradient $\frac{1}{6}$ passing through $(1, -1)$.
$x - 6y - 7 = 0$ is the equation of the normal in standard form.	Multiply through by 6, and rearrange.

Exercise 4G

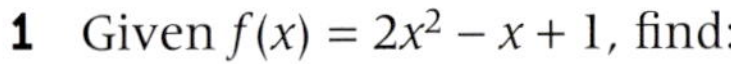

1 Given $f(x) = 2x^2 - x + 1$, find:

- **a** the gradient of the tangent to the curve at $x = 1$
- **b** the equation of the tangent to the curve at $x = 1$
- **c** the equation of the normal to the curve at $x = 1$ in the form $ax + by + c = 0$ where $a, b, c \in \mathbb{Z}$.

2 Find the points on the curve $y = \frac{1}{2 - x}$ where the derivative is 1, and find the equations of the tangents through these points.

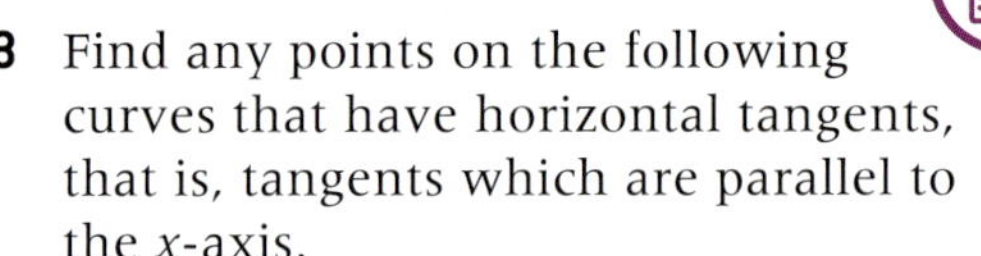

3 Find any points on the following curves that have horizontal tangents, that is, tangents which are parallel to the x-axis.

- **a** $y = 2x^2 - 1$
- **b** $y = 2 - 3x - x^2$
- **c** $y = x^3 - 1$
- **d** $y = x^3 - 3x$

4 Find the equations of the tangent and normal to the curve $y = x + \frac{1}{x}$ at $x = 1$.

Developing inquiry skills

Refer back to the opening problem in which a sprinter was running a race. How does the concept of limits help you analyse the problem of finding a runner's speed at a specific instant in time?

4.3 Differentiation rules

Investigation 11

1 **Factual** What is the *gradient* of a straight line parallel to the x-axis?

2 Write down the derivative of:

a $f(x) = -7$ **b** $f(x) = 3$ **c** $f(x) = 0$ **d** $f(x) = \pi$

3 **Conceptual** What is the *derivative* of a constant function?

4 **Factual** For a straight line with equation $y = mx + c$, which parameter tells you the gradient of the line?

5 Find the derivative of:

a $y = -x + 2$ **b** $4x - 2y + 1 = 0$ **c** $f(x) = \frac{3-2x}{4}$

6 **Conceptual** How is the derivative of a linear function related to the gradient of its graph?

If $f(x) = c$ where $c \in \mathbb{R}$, then $f'(x) = 0$.
If $f(x) = mx + c$ where $m, c \in \mathbb{R}$, then $f'(x) = m$.

TOK

Mathematics—invented or discovered?

If mathematics is created by people, why do we sometimes feel that mathematical truths are objective facts about the world rather than something constructed by human beings?

Derivative of a sum or difference of functions

What is the derivative of the sum or difference of functions?

In Exercise 4F question **1a**, you worked out that the derivative of $f(x) = x^2 + x - 2$ is $f'(x) = 2x + 1$.

Notice that $\frac{\mathrm{d}}{\mathrm{d}x}\left(x^2 + x - 2\right) = \frac{\mathrm{d}}{\mathrm{d}x}\left(x^2\right) + \frac{\mathrm{d}}{\mathrm{d}x}(x) + \frac{\mathrm{d}}{\mathrm{d}x}(-2)$

$$= \quad 2x \quad + \quad 1 \quad + \quad 0$$

In other words, the **derivative of the sum** of functions is equal to the **sum of the derivatives** of the individual functions. Can you prove that this always true, for any differentiable functions?

Let $f(x) = u(x) + v(x)$ where $u(x)$ and $v(x)$ are two functions in x whose derivatives exist. Differentiating $f(x)$ from first principles gives

$$f'(x) = \lim_{h \to 0} \frac{\left[u(x+h) + v(x+h)\right] - \left[u(x) + v(x)\right]}{h}$$

$$= \lim_{h \to 0} \left[\frac{u(x+h) - u(x)}{h} + \frac{v(x+h) - v(x)}{h}\right]$$

$$= \lim_{h \to 0} \frac{u(x+h) - u(x)}{h} + \lim_{h \to 0} \frac{v(x+h) - v(x)}{h}$$ using properties of limits

$$= u'(x) + v'(x)$$

Reflect Suppose $f(x) = u(x) - v(x)$. Can you prove that $f'(x) = u'(x) - v'(x)$?

Calculus

Derivative of the sum and difference of functions

If $f(x) = u(x) \pm v(x)$, then $f'(x) = u'(x) \pm v'(x)$.

Reflect Suppose f is a linear multiple of a differentiable function g, for example $f(x) = c(g(x))$ where c is a constant. Can you prove that $f'(x) = c(g'(x))$?

For $c \in \mathbb{R}$, $(cf)'(x) = cf'(x)$, provided $f'(x)$ exists.

Investigation 12

1 a From first principles, find the derivative of the function $y = x^2$.

b Copy and complete the table of values by entering the value of the derivative of $y = x^2$ at each different value of x.

x	−2	−1	0	1	2	3
$\frac{dy}{dx}$						

2 a From first principles, find the derivative of the function $y = x^3$.

b Copy and complete the table of values by entering the value of the derivative of $y = x^3$ at each different value of x.

x	−2	−1	0	1	2	3
$\frac{dy}{dx}$						

3 Using technology, find the derivative of $y = x^4$ for the same values as in questions **1** and **2**.

4 Based on your findings in questions **1**, **2** and **3**, can you conjecture a rule for the derivative of $y = x^n$? Can you find a way to justify that this rule will apply for any $n \in \mathbb{Z}$?

5 **Conceptual** What is the relationship between the degree of a polynomial function and the degree of its derivative?

6 Write an expression for the derivative of $y = x^n$, $n \in \mathbb{Z}^+$, using the limit definition.

7 Prove the result from question **4** using the algebraic identity $a^n - b^n = (a - b)(a^{n-1} + a^{n-2}b + a^{n-3}b^2 + \ldots + ab^{n-2} + b^{n-1})$ from Chapter 1.

The rule you have proved is called the power rule for differentiation.

8 **Conceptual** How is the power rule useful in differentiating a polynomial function?

HINT

Expand the expression for the derivative of $y = x^n$, $n \in \mathbb{Z}^+$ from question **6**, with $a = x + h$ and $b = x$.

If n is a positive integer and $f(x) = x^n$, then $f'(x) = nx^{n-1}$.

Proving the power rule for negative integer values of n is left for you as an exercise. This result is extended to **all real values of n**, although the proof is not part of the syllabus.

Reflect What is the result when $n = 0$?

Example 16

Differentiate $y = \frac{1}{2}x^4 - 2x^3 + 3x^2 - x + 3$ with respect to x.

$\frac{dy}{dx} = \frac{1}{2} \times 4x^{4-1} - 2 \times 3x^{3-1} + 3 \times 2x^{2-1} - x^{1-1} + 0$ $= 2x^3 - 6x^2 + 6x - 1$	The derivative of a sum is the sum of the derivatives. The first four terms use the power and constant multiple rules, and the last term uses the constant function rule.

When the function is given as $y = \ldots$, write the derivative as $\frac{dy}{dx}$.

Example 17

Find $f'(x)$ if $f(x) = \frac{2x^5 + 3x^3 - 1}{x^2}$.

$f(x) = \frac{2x^5}{x^2} + \frac{3x^3}{x^2} - \frac{1}{x^2}$	Write f as a sum of three individual terms.
$= 2x^3 + 3x - x^{-2}$	Simplify.
$f'(x) = 6x^2 + 3 + 2x^{-3} = 6x^2 + 3 + \frac{2}{x^3}$	Differentiate each term.

Example 18

Find the equation of the normal to the curve $f(x) = -1 - x - 2x^5$ at $x = 0$.

$f'(x) = -1 - 10x^4$; $f'(0) = -1$	Differentiate $f(x)$ and evaluate it at $x = 0$.
Gradient of the normal is 1.	$m_{\text{normal}} = -\frac{1}{m_{\text{tangent}}}$.
Hence, normal is $y = x + c$.	General equation of a straight line.

Continued on next page

$f(0) = 1$, hence tangent passes through $(0, 1)$.	Find a point which the tangent passes through.
$0 = 1 + c \Rightarrow c = -1$	Evaluate the constant and state the equation of the normal.
Normal is $y = x - 1$.	

Exercise 4H

Use the power rule for differentiation throughout this exercise.

1 Find $\frac{dy}{dx}$ for each function.

a $y = (3x-1)^2$

b $y = 3x^5 - 4x^2 + 2x - 1$

c $y = \frac{1}{4}x^2 + \frac{2}{3}x - 1$

d $y = 5x^5 - 4x^3 + x - \frac{1}{4x^3} - \frac{1}{5x^5}$

e $y = \frac{3 - 2x^3 + x^4}{x}$

f $y = \sqrt{x}$ **g** $y = \frac{1}{\sqrt{x}}$

h $y = \sqrt[5]{x^2}$ **i** $y = \frac{2}{\sqrt[3]{x}} - \frac{3}{\sqrt{x^5}}$

j $y = \left(1 + \sqrt{x}\right)\left(3 - \sqrt[3]{x}\right)$

2 Find the equation of the tangent to the curve $y = -2(x^2 + 3x)$ at $x = 1$.

3 Find the equation of the normal to the curve $y = \frac{x-3}{x}$ at $x = -1$.

4 Find the equation of the tangents to the curve $f(x) = 5x^3 + 12x^2 - 7x$ at $x = \pm 1$.

5 Find the equation of the tangents of the function $f(x) = x^3 - 5x^2 + 5x - 4$ whose gradients are 2.

6 Find the coordinates of the point where the normal to the graph of $f(x) = x^2 - 3x + 1$ at $x = 1$ intersects the curve again.

7 Find the equations the tangents of $f(x) = x^3 + x^2 + x - 1$ that are normal to the line $x + 2y + 5 = 0$.

To differentiate **1a** in Exercise 4H, $y = (3x-1)^2$, you expanded the brackets and used the sum and difference rules. This function can also be differentiated using a different rule—the **chain rule**—which you will now discover.

Investigation 13

1 Expand $y = (1 + x)^2$ and hence find $\frac{dy}{dx}$. Give your answer in factorized form.

2 Repeat question **1** for the following functions:

a $y = (1 - 2x)^2$ **b** $y = (3x - 1)^2$ **c** $y = (1 + ax)^2$

3 Write each of the functions in question **2** as the composition of two functions. For example, in part **a**, let $g(x) = (1 - 2x)$ and $f(x) = x^2$. Then $y = f(g(x))$.

4 Using your answers to questions **2** and **3** to help you, find an expression for the derivative of a composite function $y = f(g(x))$.

5 Use your expression from question **4** to find the derivative of $y = (1 - 2x)^4$. Check your answer is correct by expanding $y = (1 - 2x)^4$ and differentiating each term separately.

6 Find the derivative of $y = (1 - 2x)^{15}$.

Questions **7** and **8** will guide you in finding a rigorous definition for the derivative of a composite function.

7 Consider the function $y = (3x - 1)^2$.

a Write this as a composite function $y = f(u)$ in a new variable u, where $u = g(x)$.

b Find $\frac{dy}{du}$, the derivative of y with respect to the new variable u.

c Since u is itself a function of the variable x, find $\frac{du}{dx}$, the derivative of u with respect to x.

d For this function, show that $\frac{dy}{dx} = \frac{dy}{du} \times \frac{du}{dx}$.

In the relationship $\frac{dy}{dx} = \frac{dy}{du} \times \frac{du}{dx}$, $\frac{dy}{du}$ and $\frac{du}{dx}$ are not fractions.

Therefore, $\frac{dy}{dx}$ is not arrived at by cancelling du top and bottom. Since, however, these are rates of change, you can intuitively see that if, for example, y changes twice as fast as u and u changes twice as fast as x, then y would change four times as fast as x.

If f is differentiable at $u = g(x)$, and g is differentiable at x, then the composite function $(f \circ g)(x)$ is differentiable at x. Furthermore, if $y = f(u)$ and $u = g(x)$, then $\frac{dy}{dx} = \frac{dy}{du} \times \frac{du}{dx}$.

Another definition for the chain rule is $(f \circ g)'(x) = f'(g(x)) \times g'(x)$.

Although expanding $y = (3x - 1)^2$ and then using the sum and difference rules to differentiate is very accessible, there are other situations when expanding can be quite tedious. This is an example where the chain rule is very helpful.

Example 19

Differentiate $y = (2 - 11x)^{10}$ with respect to x.

Letting $u = 2 - 11x$ gives $\frac{du}{dx} = -11$.	Define u and find $\frac{du}{dx}$.
$y = u^{10}$ and $\frac{dy}{du} = 10u^9$.	Write y in terms of u and find $\frac{dy}{du}$.
$\frac{dy}{dx} = \frac{dy}{du} \times \frac{du}{dx} = 10u^9 \times (-11) = -110u^9$	Use the chain rule.
$\frac{dy}{dx} = -110(2-11x)^9$	Substitute for u.

Example 20

Differentiate $\sqrt{4-3x^2}$.

$(f \circ g)(x) = \sqrt{4-3x^2}$ $f(x) = \sqrt{x};\ g(x) = 4-3x^2$ $(f \circ g)'(x) = f'(g(x)) \times g'(x)$	Write the expression as the composition of two functions, f and g.
$f'(g(x)) = \frac{1}{2}(g(x))^{-\frac{1}{2}} = \frac{1}{2}(4-3x^2)^{-\frac{1}{2}}$ $g'(x) = -6x$	Differentiate f with respect to $g(x)$ and differentiate g with respect to x.
$(f \circ g)'(x) = \left[\frac{1}{2}(4-3x^2)^{-\frac{1}{2}}\right] \times [-6x]$ $= -3x(4-3x^2)^{-\frac{1}{2}}$ $= -\frac{3x}{\sqrt{4-3x^2}}$	Use the chain rule.

Exercise 4I

1 Find $\frac{dy}{dx}$ for each function.

a $y = (4x-3)^5$

b $y = \sqrt{1-4x}$

c $y = \frac{2-x^2-3x^5}{x}$

d $y = -\frac{2}{\sqrt{1-3x^2}}$

e $y = \left(\frac{1-\sqrt{x}}{x}\right)^3$

f $y = \sqrt[3]{2x^2-4}$

2 Find the equation of the tangent to the curve $y = 3x^2 - 4x^3$ at $x = 1$.

3 Find the equation of the normal to the curve $y = \frac{x-2}{x}$ at $x = -1$.

4 Find $\frac{dy}{dx}$ if $y = \sqrt{2-\sqrt{x}}$.

5 $f(x) = \frac{-20}{x} + 1$ for $x > 0$ and $g(x) = 5x + 3$ for $x \in \mathbb{R}$. Find the values of x for which the graphs of f and g have the same gradient.

6 Let $f(x) = 5ax^3 - 2bx^2 + 4cx$.

a Find $f'(x)$.

b Given that $f'(x) \geq 0$ and $a > 0$, show that $b^2 \leq 15ac$.

7 Use the chain rule to prove that

a the derivative of an even function is an odd function

b the derivative of an odd function is an even function.

You have seen that the derivative of a sum or difference of functions is equal to the sum or difference of the derivatives of the functions. Does the same hold true for the product of functions?

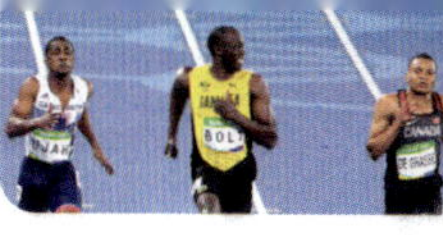

Investigation 14

1 Write down $\frac{\mathrm{d}(x^3)}{\mathrm{d}x}$.

2 Determine whether $\frac{\mathrm{d}(x^3)}{\mathrm{d}x} = \frac{\mathrm{d}(x^2)}{\mathrm{d}x} \times \frac{\mathrm{d}(x)}{\mathrm{d}x}$.

3 Find some other functions $f(x)$ and $g(x)$, and determine whether

$$\frac{\mathrm{d}}{\mathrm{d}x}(f(x) \times g(x)) = \frac{\mathrm{d}}{\mathrm{d}x}(f(x)) \times \frac{\mathrm{d}}{\mathrm{d}x}(g(x))$$

4 **Conceptual** What can you conclude about the derivative of a product of functions?

5 Let $f(x) = u(x) \times v(x)$, where u and v are differentiable functions.

Follow the steps below to help you find an expression for $f'(x)$ in terms of $u(x)$ and $v(x)$.

a Write an expression for the derivative of f using the limit definition from first principles.

b Add $u(x+h)v(x) - u(x+h)v(x)$ to the numerator of your expression in part **a**. Can you explain why this does not change the value of your expression?

c Factorize and rearrange your expression to show that:

$$f'(x) = \lim_{h \to 0}\left[u(x+h)\frac{v(x+h)-v(x)}{h} + v(x)\frac{u(x+h)-u(x)}{h}\right]$$

d Let $h \to 0$ and write down an expression for $f'(x)$ which does not involve limits.

This expression is called the **product rule** for differentiation.

6 **Conceptual** How can the product rule help you find the derivative of a product of functions?

If $f(x) = u(x) \times v(x)$ where u and v are differentiable functions, then $f'(x) = u(x)v'(x) + v(x)u'(x)$.

Another way of writing this is, if $y = uv$ where u and v are differentiable functions in x, then $\frac{\mathrm{d}y}{\mathrm{d}x} = u\frac{\mathrm{d}v}{\mathrm{d}x} + v\frac{\mathrm{d}u}{\mathrm{d}x}$.

Reflect What would the product rule be if you had three differentiable functions in x?

Example 21

Use the product rule to find $f'(x)$ if $f(x) = (3 - 2x)(3x - 1)$.

Let $u(x) = 3 - 2x$, then $u'(x) = -2$. Let $v(x) = 3x - 1$, then $v'(x) = 3$.	Define u and v, and find u' and v'.
$f'(x) = 3(3-2x) - 2(3x-1)$ $= 11 - 12x$	Use the product rule.

TOK

What is the difference between inductive and deductive reasoning?

Calculus

Example 22

Find the equation of the normal to the curve $y = \frac{3x^2 - 1}{x + 2}$, $x \neq -2$, at (0, 1).

$y = \frac{3x^2 - 1}{x + 2} = (3x^2 - 1)(x + 2)^{-1}$	Change the quotient to a product.
Let $u = 3x^2 - 1$, then $u' = 6x$. Let $v = (x + 2)^{-1}$, then $v' = -(x + 2)^{-2}$.	Define u and v, and find u' and v'.
$\frac{dy}{dx} = (3x^2 - 1) \times -(x + 2)^{-2} + 6x(x + 2)^{-1}$	Use the product rule.
$= (x + 2)^{-2}\left[-(3x^2 - 1) + 6x(x + 2)\right]$ $= (x + 2)^{-2}(3x^2 + 12x + 1)$ $= \frac{3x^2 + 12x + 1}{(x + 2)^2}$	Factorize $(x + 2)^{-1}$.
$x = 0 \Rightarrow \frac{dy}{dx} = \frac{1}{4}$	Evaluate $\frac{dy}{dx}$ at $x = 0$.
Gradient of the normal is -4.	
$y - 1 = -4(x - 0)$ $y = -4x + 1$	Using $y - y_1 = m(x - x_1)$ for a straight line.

Exercise 4J

Differentiate these functions with respect to x:

1 $y = (2x - 3)(x + 3)^3$

2 $y = (2x + 3)^2(3 - x)^3$

3 $y = \frac{x - 1}{x + 1}$, $x \neq -1$

4 $y = x\sqrt{2 - 3x}$

5 $y = \frac{1}{x^3 - 2x^2 + 3x + 1}$

6 $y = (x + 1)^4 \sqrt[3]{(2 - 3x)^2}$

7 $y = \frac{(2x - 1)^3}{(4 - x)^2}$, $x \neq 4$

8 $y = \frac{1 - 2x}{\sqrt{3x^2 + 2}}$

9 Find the equation of the tangent and the equation of the normal to the curve $y = \frac{x^2 + 1}{x^2 - 3}$, $x \neq \pm\sqrt{3}$, at (1, 0).

10 Consider the function $y = \sqrt{x + 1}(3 - x)^2$.

a Show that $y' = \frac{-(3 - x)(5x + 1)}{2\sqrt{x + 1}}$.

b Find the x-coordinates of all points on the curve $y = \sqrt{x + 1}(3 - x)^2$ where the tangent is parallel to the x-axis.

The quotient rule

Given any rational function of the form $f(x)=\frac{u(x)}{v(x)}$, where u and v are differentiable functions, you can find the derivative of f by changing it to a product $f(x) = (u(x)) \times (v(x))^{-1}$ and then using the product rule.

However, there are times when having a rule for finding the derivative of a quotient *without* changing it to a product is helpful.

If $f(x)=\frac{u(x)}{v(x)}$, where u and v are differentiable functions and $v(x) \neq 0$,

then:

$$f'(x)=\lim_{h\to 0}\frac{\frac{u(x+h)}{v(x+h)}-\frac{u(x)}{v(x)}}{h}$$

$$=\lim_{x\to\infty}\frac{v(x)u(x+h)-u(x)v(x+h)}{hv(x+h)v(x)}$$

$$=\lim_{h\to 0}\frac{v(x)u(x+h)-v(x)u(x)+v(x)u(x)-u(x)v(x+h)}{hv(x+h)v(x)}$$

Here we have subtracted $v(x)u(x)$ and then added it back on again, so the value of the numerator is unchanged.

$$=\lim_{h\to 0}\frac{v(x)\frac{u(x+h)-u(x)}{h}-u(x)\frac{v(x+h)-v(x)}{h}}{v(x+h)v(x)}$$

$$=\frac{\lim_{h\to 0}v(x)\times\lim_{h\to 0}\frac{u(x+h)-u(x)}{h}-\lim_{h\to 0}(x)\times\lim_{h\to 0}\frac{v(x+h)-v(x)}{h}}{\lim_{h\to 0}\left[v(x+h)v(x)\right]}$$

This uses the distributive properties of limits.

$$=\frac{v(x)u'(x)-u(x)v'(x)}{(v(x))^2}$$

> If $f(x)=\frac{u(x)}{v(x)}$, $v(x)\neq 0$, then $f'(x)=\frac{v(x)u'(x)-u(x)v'(x)}{[v(x)]^2}$.
>
> Alternatively, if $y=\frac{u}{v}$, then $\frac{dy}{dx}=\frac{v\frac{du}{dx}-u\frac{dv}{dx}}{v^2}$, where u and v are differentiable functions in x, and $v\neq 0$.

Example 23

Use the quotient rule to differentiate $y=\frac{3x^2-1}{x+2}$, $x\neq -2$.

(In Example 22, you saw how this function could be differentiated using the product rule. In this example, you will see that the quotient rule gives the same result.)

Continued on next page

Calculus

Let $u = 3x^2 - 1$ and $v = x + 2$ Then $\frac{du}{dx} = 6x$; $\frac{dv}{dx} = 1$	Define u and v, and find u' and v'.
Hence, $\frac{dy}{dx} = \frac{6x(x+2)-(3x^2-1)}{(x+2)^2}$ $= \frac{3x^2+12x+1}{(x+2)^2}$	Use the quotient rule.

Exercise 4K

1 Differentiate the functions using the quotient rule:

a $y = \frac{1+3x}{5-x}$ **b** $y = \frac{\sqrt{x}}{2-x}$

c $y = \frac{1+2x}{\sqrt{1-x^2}}$ **d** $y = \frac{1+3x}{x^2+1}$

2 Differentiate the functions with respect to x.

a $y = \frac{x^2-2}{x^3-1}$ **b** $y = \frac{x^2+2x}{\sqrt{x^3+1}}$

c $y = \frac{1}{x^2\sqrt{1-x+2x^2}}$ **d** $y = \frac{1-\sqrt{x}}{1+\sqrt{x}}$

e $y = \frac{1+3x}{\sqrt{\sqrt{x}-1}}$ **f** $y = \left(1 - \frac{1}{2+x}\right)^{\frac{2}{3}}$

3 Find the equation of the normal to the curve known as "Newton's Serpentine", $y = \frac{4x}{x^2+1}$, at $x = 0$.

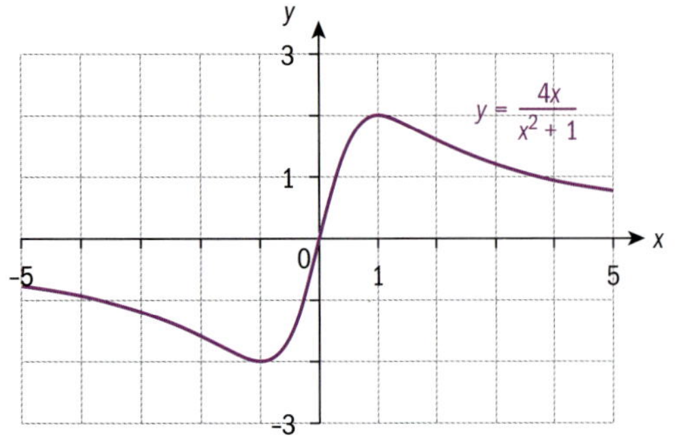

4 Find the equation of the normal to the curve $y = \frac{8}{4+x^2}$ at $x = 1$. This curve is known as the "witch of Agnesi".

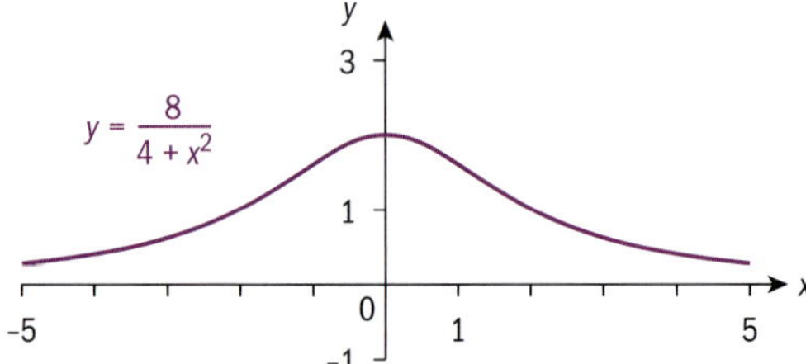

5 The curve $y = \frac{x^3+x^2+x+1}{x}$ is known as the "Trident of Newton".

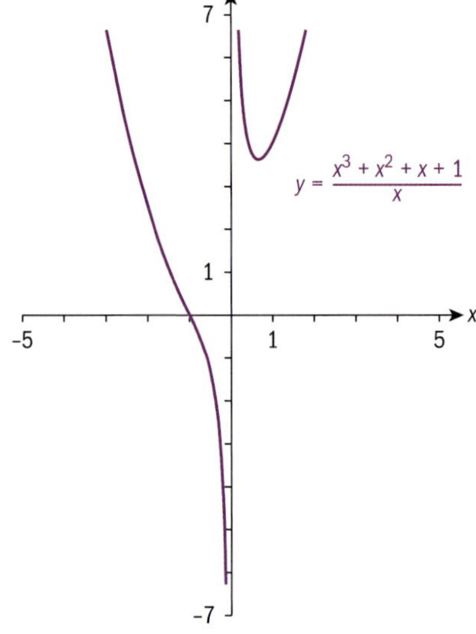

Find the coordinates of the point on this curve where the tangent to the curve is parallel to $y = 2x$.

Higher derivatives

If $f(x)$ is a differentiable function, then $f'(x)$ is the derivative of $f(x)$. Similarly, if $f'(x)$ is a differentiable function, then $f''(x)$ is the derivative of $f'(x)$. Since multiple dash or prime notation begins to lose its efficiency after about the third derivative, for higher derivatives we write $f^{(n)}(x)$.

You can also use $\frac{dy}{dx}$ notation.

$$f'(x)=\frac{dy}{dx};\ f''(x)=\frac{d^2y}{dx^2};\ f'''(x)=\frac{d^3y}{dx^3};\ f^{(n)}(x)=\frac{d^ny}{dx^n}.$$

Did you know?

Throughout history famous curves have often been given special names, such as those in questions **3** to **5** in Exercise 4K. You might want to research applications of these, and other famous curves, which have been used to model real-life situations.

Investigation 15

1 Consider $f(x) = x^4$.

- **a** Find $f'(x)$, $f''(x)$, $f'''(x)$ and $f^{(4)}(x)$.
- **b** Find $f^{(5)}(x)$.

Consider $f(x) = x^k$, $k \in \mathbb{Z}^+$.

- **c** Find expressions for $f'(x)$, $f''(x)$, $f'''(x)$ and $f^{(4)}(x)$.
- **d** Use ${}^kC_p = \binom{k}{p} = \frac{k!}{p!(k-p)!}$ where p is the pth derivative of $f(x)$ to compare the coefficients of the variable x in the expressions found in part **c**.
- **e** Analyse the answers you found in parts **c** and **d** to find a pattern that will help you conjecture a formula for $f^{(p)}(x)$ where $p \le k$, and $f(x) = x^k$.

2 $f(x) = uv$, where u and v are differentiable functions in x.

- **a** Write down an expression for $f'(x)$ using the product rule.
- **b** Find expressions for $f''(x)$ and $f'''(x)$.
- **c** Analyse your answers to find a pattern that will help you conjecture a formula for $f^{(n)}(x)$.
- **d** Use your formula to find $f^{(4)}(x)$.

The general case for $f^{(n)}(x)$ is called **Leibniz's formula**.

3 **Conceptual** Why is finding a pattern for higher-order derivatives of a function useful?

HINT

Recall the binomial theorem.

Example 24

a Find the first three derivatives of $f(x) = 2x^4 - x^3 + 3x^2 + 1$.

b Find x when $f''(x) = 0$ for $f(x) = \frac{x}{x^2+2}$.

c Given $f(x) = -16x^2 - \frac{1}{x}$, find $f''(1)$.

a $f'(x) = 8x^3 - 3x^2 + 6x$ $f''(x) = 24x^2 - 6x + 6$ $f'''(x) = 48x - 6$	Differentiate each function.
b $u = x \quad v = x^2 + 2$ $u' = 1 \quad v' = 2x$ $f'(x) = \frac{x^2+2-2x^2}{(x^2+2)^2} = \frac{2-x^2}{(x^2+2)^2}$	Use quotient rule.
$u = 2 - x^2 \quad v = (x^2+2)^2$ $u' = -2x \quad v' = 2(x^2+2)(2x)$ $= 4x(x^2+2)$ $f''(x) = \frac{-2x(x^2+2)^2 - 4x(x^2+2)(2-x^2)}{(x^2+2)^4}$ $f''(x) = \frac{-2x(x^2+2)(x^2+2+4-2x^2)}{(x^2+2)^4}$ $= \frac{-2x(x^2+2)(-x^2+6)}{(x^2+2)^4}$	Use quotient rule a second time.
$f''(x) = 0 \Rightarrow x = 0,\ x = \pm\sqrt{6}$	Solve $f''(x) = 0$.
c $f(x) = -16x^2 - \frac{1}{x}$	
$f'(x) = -32x + \frac{1}{x^2}$	Find the first derivative.
$f''(x) = -32 - \frac{2}{x^3}$	Find the second derivative.
$f''(1) = -32 - 2$ $f''(1) = -34$	Evaluate the second derivative when $x = 1$.

Example 25

A particle moves in a straight line so that its position from a fixed point after t seconds is given by $s(t) = 4t + 3t^2 - t^3$, where s is in cm.

a Find the velocity of the particle at $t = 1.5$ s if the velocity is the derivative of the position of the function.

b If the acceleration is the derivative of the velocity, find the acceleration of the particle at $t = 1.5$ s.

a $s'(t) = 4 + 6t - 3t^2$ $s'(1.5) = 6.25$ cm s^{-1}	Differentiate $s(t)$ and evaluate at $t = 1.5$ s.
b $s''(t) = 6 - 6t$ $s''(1.5) = -3$ cm s^{-2}	Differentiate $s'(t)$ and evaluate at $t = 1.5$ s.

Exercise 4L

1 If $f(x) = 1 - 4x - \frac{1}{x}$, find $f''(x)$.

2 If $f(x) = 3x^5 - 2x^2 + 1$, find $f^{(4)}(1)$.

3 Given that $y = (1 - ax)^3$, find a such that $\frac{d^3y}{dx^3} = 162$.

4 Given $f(x) = \frac{x^4}{12} - 2x^2 + 5x + 1$, find the values of x for which $f''(x) = 0$.

5 If $y = x^4 - 4x^3 + 16x - 16$, find x such that $\frac{d^2y}{dx^2} = \frac{d^3y}{dx^3}$.

6 $f(x) = x^4 + px^2 + qx + r$ passes through the point $(-1, 16)$, and at this point, $f''(x) = -f'(x) = 16$. Find the values of p, q and r.

7 A particle moves in a straight line such that its position at any t is $s(t) = (t - 2)^3(4 - 2t)^2$ metres. Find:

a the velocity of the particle after 3 seconds

b the acceleration of the particle after 3 seconds

c the jerk (derivative of the acceleration) of the particle after 1 second.

8 Given $f(x) = \frac{1}{x}$, find the first five derivatives, and hence conjecture an expression for $f^{(n)}(x)$. Prove your conjecture using the method of mathematical induction.

9 Use Leibniz's theorem to find $f^{(4)}(x)$ at $x = 1$ if $f(x) = (x - 1)^4[(x + 1)^4(1 - 2x^3)^3]$.

4.4 Graphical interpretation of the derivatives

Investigation 16

Consider the graph of a quadratic function whose leading coefficient is positive, ie $a > 0$.

For $a > 0$, the vertex is a minimum point.

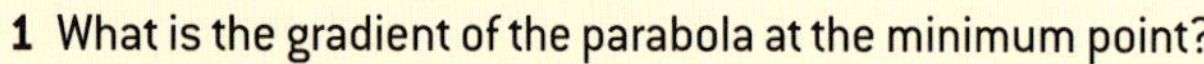

1 What is the gradient of the parabola at the minimum point?

2 What sign do the gradients left of the minimum point on the parabola have?

3 What sign do the gradients right of the minimum point on the parabola have?

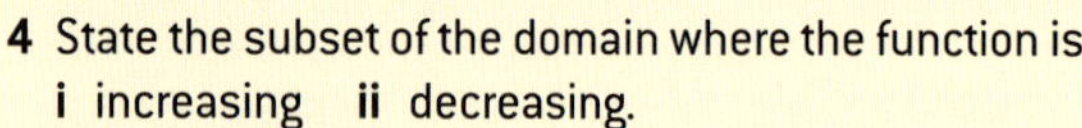

4 State the subset of the domain where the function is
   i increasing **ii** decreasing.

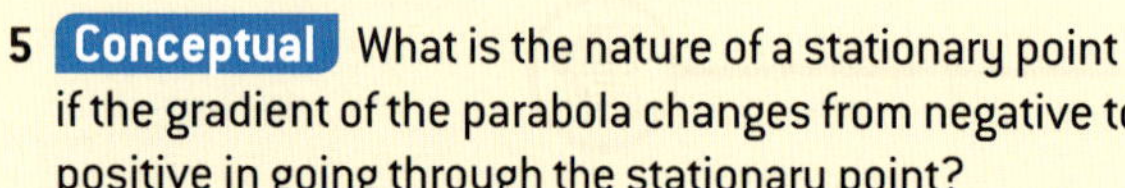

5 **Conceptual** What is the nature of a stationary point if the gradient of the parabola changes from negative to positive in going through the stationary point?

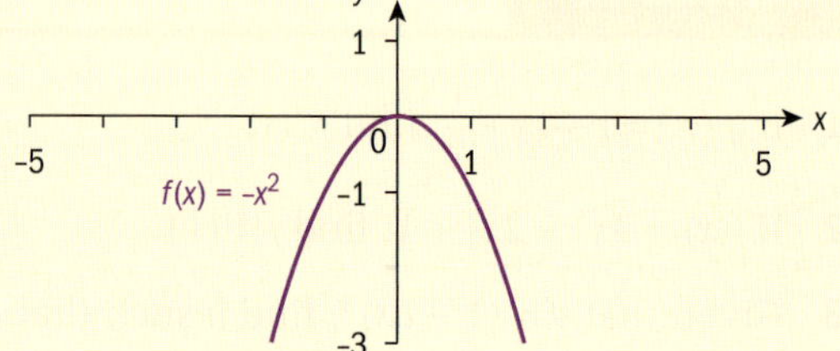

Consider the graph of a quadratic function whose leading coefficient is negative, ie $a < 0$.

For $a < 0$, the vertex is a maximum point.

6 What is the gradient of the parabola at the maximum point?

7 What sign do the gradients left of the maximum point on the parabola have?

8 What sign do the gradients right of the maximum point on the parabola have?

9 **Conceptual** What is the nature of a stationary point if the gradient of the parabola changes from positive to negative in going through the stationary point?

10 **Factual** In the interval where the gradients are negative, is the function increasing or decreasing?

11 **Factual** In the interval where the gradients are positive, is the function increasing or decreasing?

12 State the subset of the domain where the function is **i** increasing **ii** decreasing.

13 **Conceptual** How can you identify the intervals on which a function is increasing/decreasing using the first derivative test?

14 **Conceptual** How does the first derivative test help in classifying local extrema and identifying intervals where a function is increasing/decreasing?

Maximum and minimum points are referred to as extrema, turning points or stationary points.

As you can see in the diagram, the y-values of all points of the function are less than or equal to the y-value at C. A is a local maximum, as the y-value of A is less than the y-value of C. B is a local minimum, as the global minimum of the function is $-\infty$.

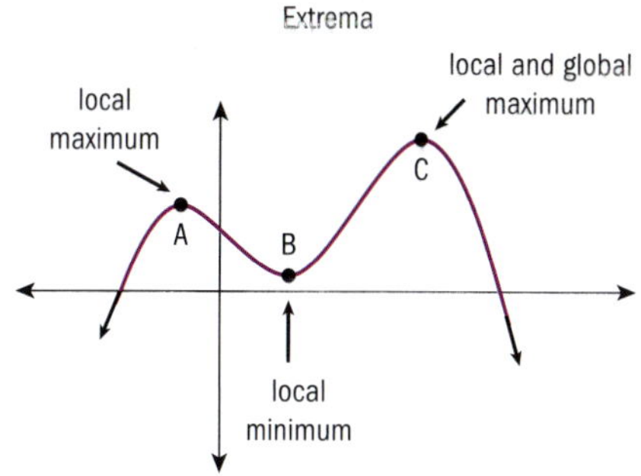

The global maximum (or minimum) of a function is the greatest (or least) value the function attains in its domain.

First derivative test for continuous and differentiable functions on an open interval $]a, b[$

- f has a turning point at $x = c$, $a < c < b$, if and only if $f'(c) = 0$.
- If $f' > 0$ for $x < c$, and $f' < 0$ for $x > c$, then f has a local maximum at $x = c$.
- If $f' < 0$ for $x < c$ and $f' > 0$ for $x > c$, then f has a local minimum at $x = c$.

If $f(x)$ is continuous and differentiable on $[a, b]$:

- If $f'(x) > 0$ for all $x \in\ ]a, b[$, then f' increases on $[a, b]$.
- If $f'(x) < 0$ for all $x \in\ ]a, b[$, then f' decreases on $[a, b]$.

Example 26

If $f(x) = 1 - 4x^2 + 2x^4$:

a Find any turning points.

b Determine the nature of the points and justify your answers.

c State the intervals in which the function increases/decreases.

d Confirm your answers graphically, and state whether the points found in part **a** are local or global extrema.

a $f'(x) = -8x + 8x^3$	
$-8x + 8x^3 = 0$	Set the first derivative equal to 0, and solve.
$\Rightarrow -8x\left(1 - x^2\right) = 0$	
$\Rightarrow x = 0;\ x = \pm 1$	
Turning points: $(0, 1)$, $(1, -1)$, $(-1, -1)$	Evaluate $f(x)$ at 0, -1 and 1 to find the y-coordinates of the turning points.
b	

x	$x < -1$	$-1 < x < 0$	$0 < x < 1$	$x > 1$
$f'(x) = -8x + 8x^3$	$f'(-2) < 0$	$f'(-0.5) > 0$	$f'(0.5) < 0$	$f'(2) > 0$
Sign of $f'(x)$	$-$	$+$	$-$	$+$

Continued on next page

Calculus

Since the gradients go from negative to positive through $x = -1$ and $x = 1$, the minimum points are $(-1, -1)$ and $(1, -1)$.

Since the gradients go from positive to negative through $x = 0$, there is a local maximum at $(0, 1)$.

Use $\frac{dy}{dx}$ to test values to the left and right of the turning points.

c Since $f(x) < 0$ when $x < -1$ and $0 < x < 1$, f is decreasing on $]-\infty, -1[\cup]0, 1[$.

Since $f(x) > 0$ when $-1 < x < 0$ and $x > 1$, f is increasing on $]-1, 0[\cup]1, \infty[$.

d

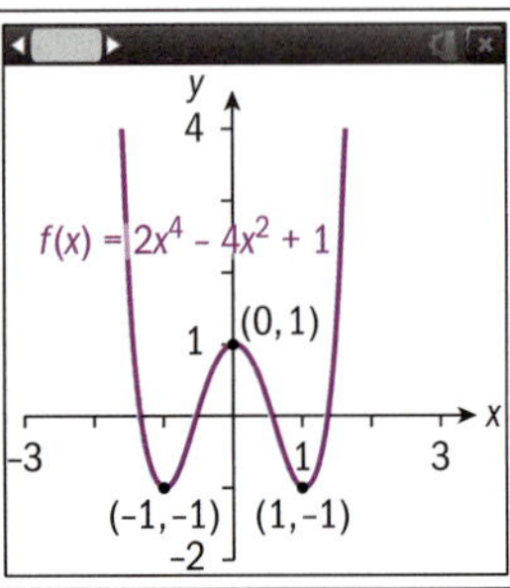

From the graph, $(0, 1)$ is a local maximum point.

$(-1, -1)$ and $(1, -1)$ are global minimum points.

Since the power of the polynomial function is even and the leading coefficient is positive, $f(x) \to +\infty$ as $x \to \pm\infty$.

Exercise 4M

1 The graph of $y = f(x)$ is shown below.

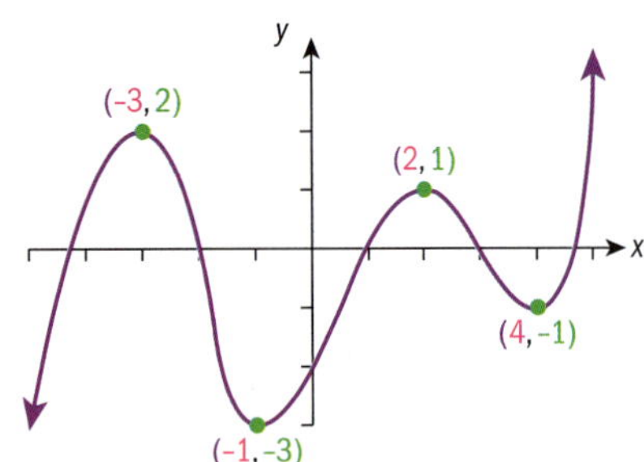

State the values of x where:

a f' is 0 **b** f' is positive

c f' is negative **d** f is increasing

e f' is decreasing.

2 Use these graphs of f' to determine:

i the intervals where f is increasing

ii the intervals where f is decreasing

iii the x values of the turning points of f.

a

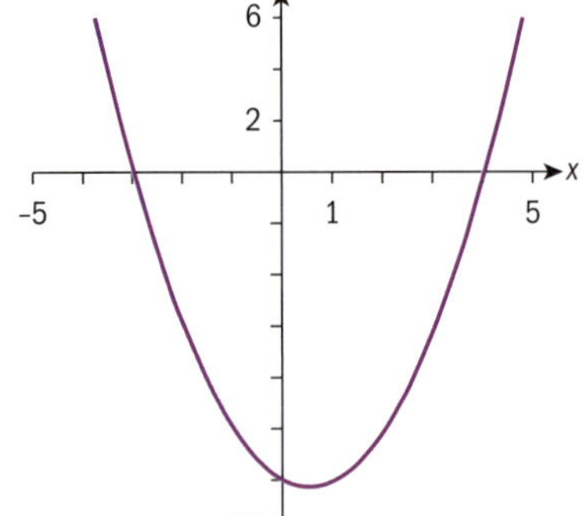

b

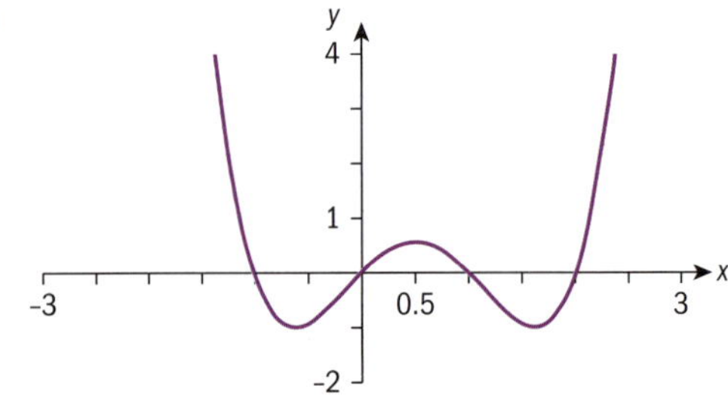

3 For each function:

i Find any stationary points and determine the nature of the points.

ii State the intervals in which the function increases.

iii State the intervals in which the function decreases.

iv Confirm your answers graphically, and state whether the points found in part **i** are local or global extrema.

a $y = 2x^3 - 6x^2 + 3$ **b** $y = -3x^4 + 2x^3 + 3x^2 - 4$

c $y = \dfrac{2}{3 - 2x - x^2}$ **d** $y = \dfrac{3x + 3}{3x - x^2}$

e $y = x + \dfrac{1}{x - 1}$

4 Find the greatest and least values of the function $f(x) = 4 + 5x - x^2 - x^3$ in the interval $-3 \le x \le 3$.

5 Find the global extrema of $f(x) = x^2 + \dfrac{4}{x}$ in the interval $[-1, 2]$.

6 The graph of the function $y = x^3 + ax^2 + b$ has a minimum point at $(4, -11)$. Find the coordinates of the maximum point, and the value of b.

7 The graph of the function $y = x^3 + ax^2 + bx + c$ passes through $(1, 1)$. It has turning points at $x = -1$ and $x = 3$. Find the values of a, b and c.

Investigation 17

Use the graph of $f(x) = x^2$ and the graph of its derivative to answer the following questions.

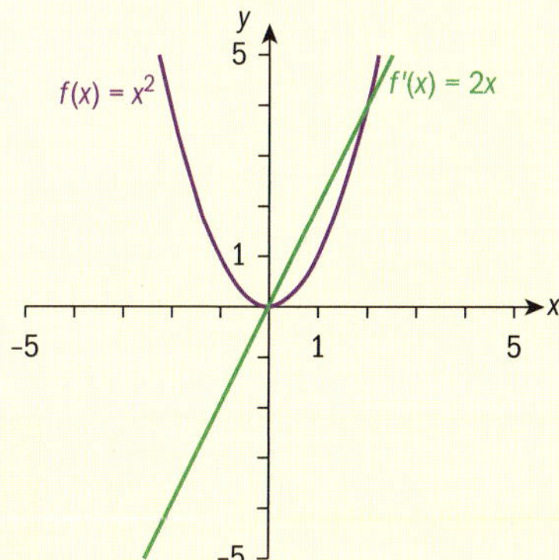

1 State the change in the signs of the gradients of f in going from the points left of the minimum point, through the minimum, then right of the minimum point. Explain how your answer is shown in the graph of $f'(x)$.

2 This graph also shows $f''(x)$, the derivative of $f'(x)$.

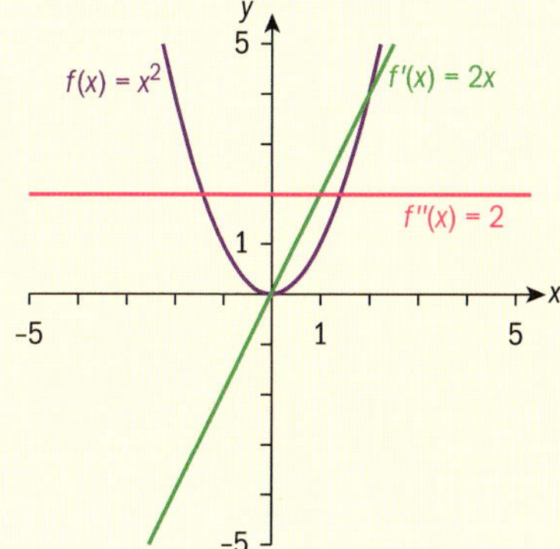

Continued on next page

Calculus

What is the sign of the gradient of $f'(x)$ at any value of x? In particular, what is the sign of the second derivative of the minimum point of $f(x)$?

3 When the second derivative of a function is positive, is the *gradient* of $f'(x)$ increasing or decreasing as you go from left to right?

4 Sketch the graph of $f(x) = -x^2$ together with its first two derivative functions. Answer questions **1** and **2** for this function.

5 When the second derivative of a function is negative, is the *gradient* of $f'(x)$ increasing or decreasing as you go from left to right?

6 Summarize your findings by stating what you think the sign of the second derivative of a function is at a minimum point, and at a maximum point.

7 Analyse the graph of $f(x) = 3 + x + \frac{1}{x}$ and its first two derivative functions. Do your findings from **6** still hold true?

In this investigation, you have discovered the second derivative test for classifying local extrema.

8 **Conceptual** How is the second derivative useful in classifying local extrema?

9 Consider now the graph of $f(x) = x^4$.

Graph its first two derivatives. Does your conjecture work for this function? Why, or why not?

10 **Conceptual** How can you determine the nature of a turning point when the second derivative test is inconclusive?

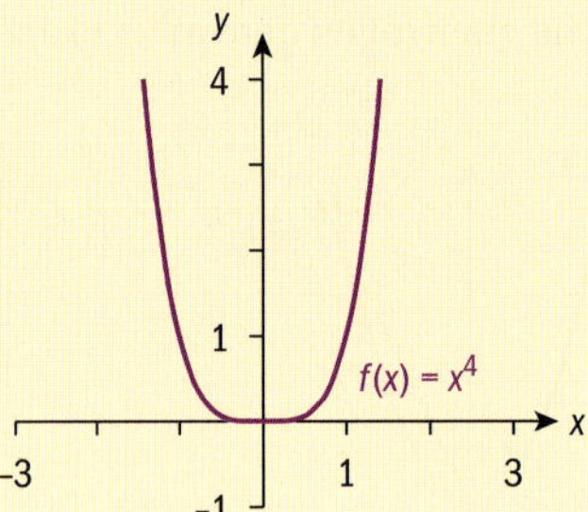

Second derivative test for maxima and minima

If $f'(c) = 0$ and $f''(c) < 0$ then $f(x)$ has a local maximum at $x = c$.

If $f'(c) = 0$ and $f''(c) > 0$ then $f(x)$ has a local minimum at $x = c$.

If $f'(c) = 0$ and $f''(c) = 0$ then the second derivative test is inconclusive. If the second derivative test is inconclusive, revert to the first derivative test for local extrema.

International-mindedness

The Greeks' mistrust of zero meant that Archimedes' work did not lead to calculus.

Example 27

Use the second derivative test to find the greatest and least value of $y = x^3 - 5x^2 + 7x$ in the interval $[0, 4]$.

$\frac{dy}{dx} = 3x^2 - 10x + 7$ $3x^2 - 10x + 7 = 0 \Rightarrow (3x - 7)(x - 1) = 0$ $x = 1; \frac{7}{3}$	Find $\frac{dy}{dx}$ and set equal to zero to find the turning points.

$\frac{d^2y}{dx^2}=6x-10$ $\frac{d^2y}{dx^2}$ at $x=1$ is $-4<0$, hence at $x=1$ f has a local maximum.	Find $\frac{d^2y}{dx^2}$ and evaluate it at each of the turning points.
$y(1)=3$, so $(1, 3)$ is a local maximum. $\frac{d^2y}{dx^2}$ at $x=\frac{7}{3}$ is $4>0$, hence at $x=\frac{7}{3}$ f has a local minimum.	Evaluate y at the local maximum.
$y\left(\frac{7}{3}\right)=\frac{49}{27}$, so $\left(\frac{7}{3}, \frac{49}{27}\right)$ is a local minimum. Testing the endpoints for global extrema: At $x=0$, $y=0$, and at $x=4$, $y=12$. For $x\in[0,4]$, the least value of y is 0 and the greatest value of y is 12.	Evaluate y at the local minimum. Since $x=1$ and $x=\frac{7}{3}$ are in the interval $[0,4]$, they are local extrema. You need to test the endpoints of the interval to ensure you find the greatest and least values of the function.

Example 28

Find any turning points of $y=\frac{x-5}{x}$, $x\neq 0$, and state the interval(s) where y is increasing/decreasing. Sketch the graph.

$\frac{dy}{dx}=\frac{5}{x^2}$. Since $\frac{5}{x^2}\neq 0$ for all x, y has no stationary points. Since $\frac{5}{x^2}>0$ for all x, y is increasing throughout its domain.	Write $y=1-\frac{5}{x}$ and differentiate, or use the quotient rule.
$y=\frac{x-5}{x}$ has a vertical asymptote at $x=0$ and a horizontal asymptote at $y=1$. 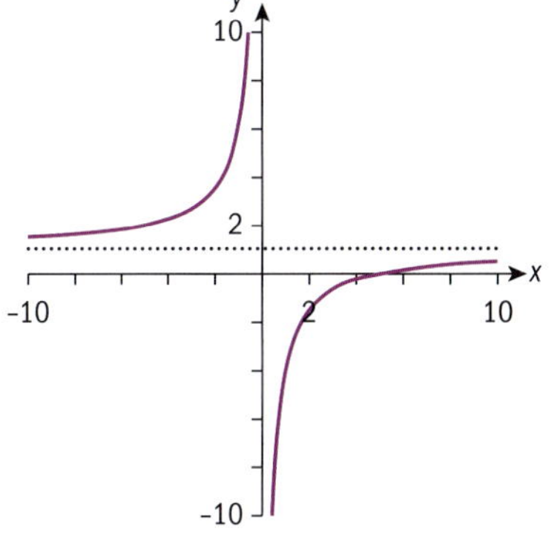 	A vertical asymptote occurs at a point where the function is not defined; here where the denominator is zero. A horizontal asymptote occurs at the limit of the function as $x\to\pm\infty$.

Exercise 4N

1 Find and classify any turning points of the functions:

   a $y = x^7 - 7x$

   b $y = 5x^4 - x^5$

   c $y = 4x + 1 + \frac{1}{x}$

2 The function $f(x) = x^3 - bx$, $b > 0$, has a turning point at $x = 1$.

   a Find the value of b.

   b Find the coordinates of all the turning points.

   c Sketch the function.

3 The quadratic function $f(x) = ax^2 + bx + c$, $a \neq 0$, has one turning point.

   a Find the x-coordinate of its single turning point.

   b State the condition that determines whether the turning point is a local maximum or minimum, and justify your answer.

4 The function $f(x) = x^3 + bx + c$ has a turning point at $(1, 4)$.

   a Find b and c.

   b Find any other turning points.

Investigation 18

Sketch the function $f(x) = x^3 - 3x + 1$.

1 Find the first and second derivatives of f and sketch their graphs.

There is a point on f that corresponds to a stationary point on f'.

2 Write down the x-coordinate of this point and state the gradient of f' at this point.

3 Do the signs of the gradient of the function change in going through this point?

4 State the value of f'' at this point.

5 Describe the concavity of f to the left and to the right of this point.

6 State the signs of f'' to the left and to the right of this point on f.

7 Repeat steps **1–6** for $y = -f(x)$.

A point where the graph of a function is continuous and where the concavity changes is a **point of inflexion**.

8 **Conceptual** Why does the sign of a function's second derivative at any point indicate the concavity of the function at that point?

9 For $f(x) = x^3 - 3x + 1$, find $f''(c)$, where $(c, f(c))$ is a point of inflexion of f.

10 Conjecture the value of the second derivative of a function at a point of inflexion.

11 Does the converse hold? In other words, if the second derivative of a function at point c is the value you obtained in question **10**, is c a point of inflexion of the function? To test this, use the function $f(x) = x^4$.

12 **Conceptual** Analyse the concavity of $f(x) = x^3 - 3x + 1$ on both sides of the point of inflexion, and the concavity of $f(x) = x^4$ on both sides of its stationary point. Can you determine the additional condition necessary for f to have a point of inflexion at $x = c$?

If f is continuous on an open interval $]a, b[$, then if f changes concavity at $x = c$, $c \in]a, b[$, $(c, f(c))$ is a **point of inflexion**.

If f has a point of inflexion at $x = c$, then $f''(c) = 0$.

If $f''(c) = 0$ and f changes concavity through $x = c$, then f has a point of inflexion at $x = c$.

f is concave down in an open interval if, for all x in the interval, $f''(c) < 0$.

f is concave up in an open interval if, for all x in the interval, $f''(c) > 0$.

Example 29

Consider the function $f(x) = 2x^4 - 4x^2 + 1$.

a Find all turning points, and determine their nature (you should justify your answers).

b Find the intervals where the function is:
i decreasing **ii** increasing.

c Find the intervals where the function is:
i concave up **ii** concave down.

d Sketch the function, indicating any maxima, minima and points of inflexion.

a $f'(x) = 8x^3 - 8x$

$8x^3 - 8x = 8x(x^2 - 1) = 0 \Rightarrow x = 0, \pm 1$

Set $f'(x) = 0$ to find any turning points.

$f''(x) = 24x^2 - 8$

$f''(0) = -8 \Rightarrow f$ has a maximum at $x = 0$.

$f''(-1) = 16 \Rightarrow f$ has a minimum at $x = -1$.

$f''(1) = 16 \Rightarrow f$ has a minimum at $x = 1$.

Use the second derivative test to find the nature of the turning points.

Turning points are $(0, 1)$, $(-1, -1)$, and $(1, -1)$.

b Sign diagram

x	$x < -1$	$-1 < x < 0$	$0 < x < 1$	$x > 1$
Sign of f'	−	+	−	+
Behaviour of f	decreasing	increasing	decreasing	increasing

Hence, **i** f is decreasing in the intervals $]-\infty, -1[\cup]0, 1[$ and **ii** f is increasing in the intervals $]-1, 0[\cup]1, \infty[$.

c $24x^2 - 8 = 0 \Rightarrow x = \pm\frac{1}{\sqrt{3}}$

Set $f''(x) = 0$ to find any possible points of inflexion.

x	$x < -\frac{1}{\sqrt{3}}$	$-\frac{1}{\sqrt{3}} < x < \frac{1}{\sqrt{3}}$	$x > \frac{1}{\sqrt{3}}$
Sign of f''	+	−	+
Concavity of f	concave up	concave down	concave up

Continued on next page

i f is concave up on the interval $\left]-\infty, -\frac{1}{\sqrt{3}}\right[\cup \left]\frac{1}{\sqrt{3}}, \infty\right[$.

ii f is concave down on the interval $\left]-\frac{1}{\sqrt{3}}, \frac{1}{\sqrt{3}}\right[$.

d

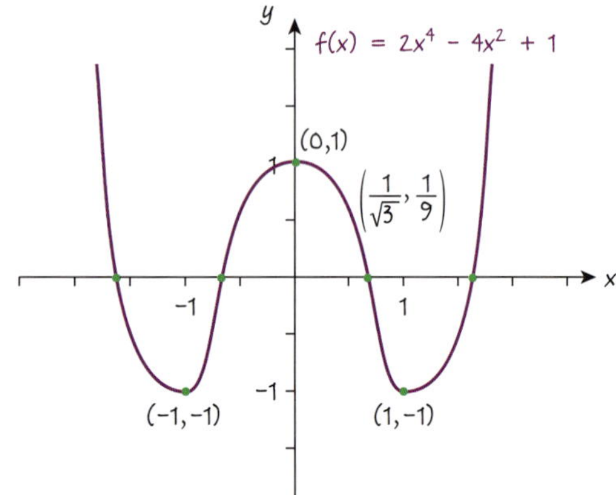

Investigation 19

Consider the function $y = x^3$.

1 Explain why the graph has a point of inflexion at $(0, 0)$.

2 What is the first derivative of the function at $(0, 0)$?

3 What are the signs of the gradients of $f(x)$ to the left and right of $(0, 0)$?

4 How does your answer to **3** differ from the first derivative of local extrema of a function?

5 What is the nature of a stationary point if the gradients of the points of f do not change sign in going through the point, and whose second derivative is 0 at this point?

This kind of inflexion point is called a horizontal point of inflexion.

6 **Conceptual** What three conditions are necessary for a function $f(x)$ to have a horizontal point of inflexion at $x = c$?

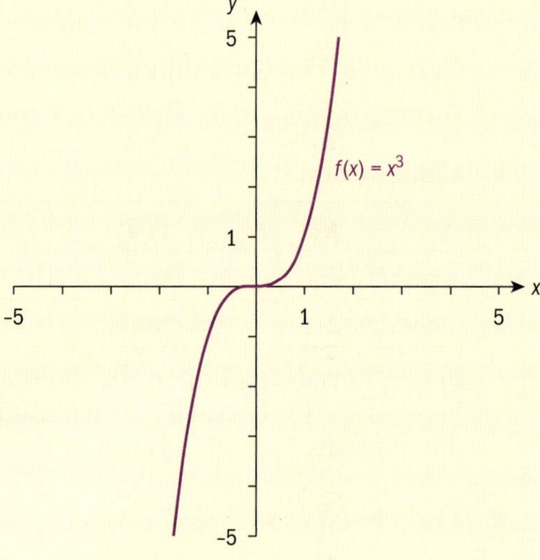

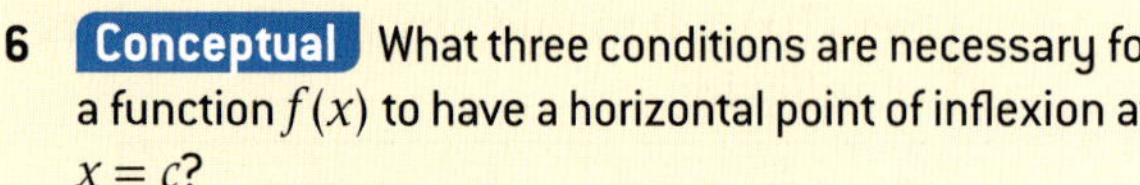

If f has a point of inflexion at $x = c$, and $f'(c) = 0$, the point is called a **horizontal point of inflexion** since its gradient is parallel to the x-axis.

Example 30

Find any turning points and points of inflexion of $y = (x+1)(x-3)^3$ and justify your answers. Confirm your answers graphically.

$\frac{dy}{dx} = 4x(x-3)^2$ $\frac{dy}{dx} = 0 \Rightarrow x = 0, x = 3$	Differentiate using the product and chain rules (or by multiplying out), set $\frac{dy}{dx} = 0$, and solve.
$y(0) = -27$, $y(3) = 0$	Find y.
$\frac{d^2y}{dx^2} = 12(x-3)(x-1)$	Again, use chain and product rules.
$\frac{d^2y}{dx^2} = 36 > 0$ at $x = 0$, hence f has a minimum at $(0, -27)$.	Apply second derivative test.
$\frac{d^2y}{dx^2} = 0$ at $x = 3$, hence test is inconclusive.	Test for a change in concavity of f in going through $x = 3$.
$\frac{d^2y}{dx^2} = -0.0039 < 0$ at $x = 2.9$ $\frac{d^2y}{dx^2} = 0.0041 > 0$ at $x = 3.1$	$\frac{d^2y}{dx^2}$ is negative for small values to the left of $x = 3$ and positive for small values to the right of $x = 3$.
Since f changes concavity at $x = 3$ and $f'(3) = f''(3) = 0$, $(3, 0)$ is a horizontal point of inflexion. 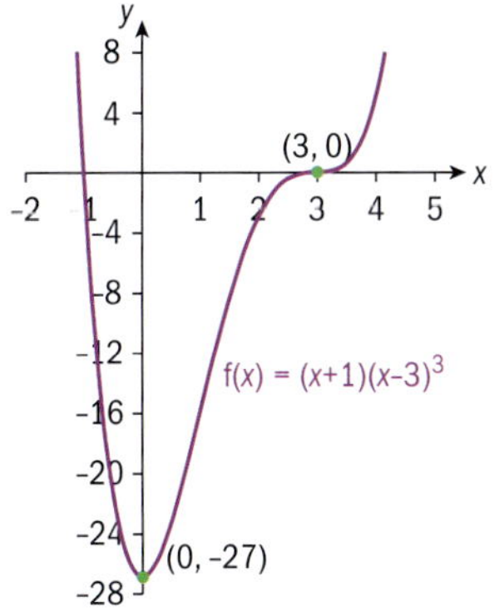 	

Example 31

$f(x) = 2x^3 + x^4$

a Find all turning points and points of inflexion; determine their nature and justify your answers.

b Find the intervals where the function is: **i** concave up **ii** concave down.

c Sketch the function, indicating any maxima, minima and points of inflexion.

Continued on next page

a $f'(x) = 6x^2 + 4x^3$

$6x^2 + 4x^3 = 0 \Rightarrow x = 0; x = -\frac{3}{2}$

Set the first derivative equal to 0 and solve to determine the stationary points.

$f''(x) = 12x + 12x^2$

$f''\left(-\frac{3}{2}\right) = 9 > 0 \Rightarrow f$ has a local minimum at $x = -\frac{3}{2}$. Minimum $\left(-\frac{3}{2}, \frac{153}{16}\right)$.

Evaluate the second derivative at both stationary points.

$f''(0) = 0 \Rightarrow$ at $x = 0$, test is inconclusive.

Use the first derivative test to determine that $x = 0$ is not an extremum.

x	$x < 0$	$x > 0$
$f'(x)$	+	+

Since there is no sign change in going through $x = 0$, $(0, 0)$ is a possible point of inflexion.

To be sure it is a point of inflexion, we need to check that the concavity of f changes as x passes through 0.

b $f''(x) = 12x + 12x^2$

$12x + 12x^2 = 0 \Rightarrow x = 0, -1$

Consider the sign of $f''(x)$ in order to determine concavity change.

x	$x < -1$	$-1 < x < 0$	$x > 0$
Sign of f''	+	−	+
Concavity of f	up	down	up

i f is concave up on $]-\infty, -1[$ and $]0, \infty[$;
ii it is concave down on $]-1, 0[$.

Since $f''(0) = 0$, and concavity changes at $(0, 0)$, and from part **a** $f'(0) = 0$, $(0, 0)$ is a horizontal point of inflexion.

c

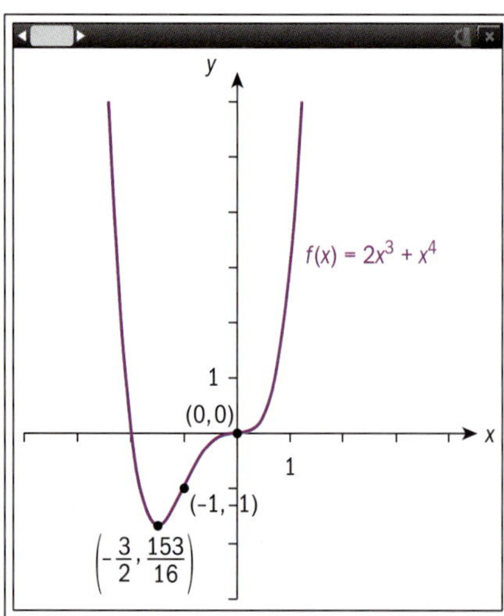

Exercise 4O

For questions **1–8**, find

a any points of inflexion

b the intervals where the function is concave up

c the intervals where the function is concave down.

Justify your answers.

1 $y = x^3 - x$

2 $y = x^4 - 3x + 2$

3 $y = x^3 - 6x^2 - 12x + 2$

4 $y = x^3 + x^2 - 1$

5 $y = 4x^3 - x^4$

6 $y = x^3 - 3x^2 + 3x - 1$

7 $y = 2x^4 + x^3 + 1$

8 $f(x) = x^4 - 4x^3 + 16x - 16$

9 The cubic function $f(x) = ax^3 + bx^2 + cx + d$ has two distinct stationary points, at x_1 and x_2. Show that:

a $b^2 > 3ac$

b The x-coordinate of the point of inflexion is $\frac{x_1 + x_2}{2}$.

TOK

The nature of mathematics: Does the fact that Leibniz and Newton came across calculus at similar times support the argument of Platonists over Constructivists?

Investigation 20

1 Copy the graph of $y = f(x)$ shown on the right.

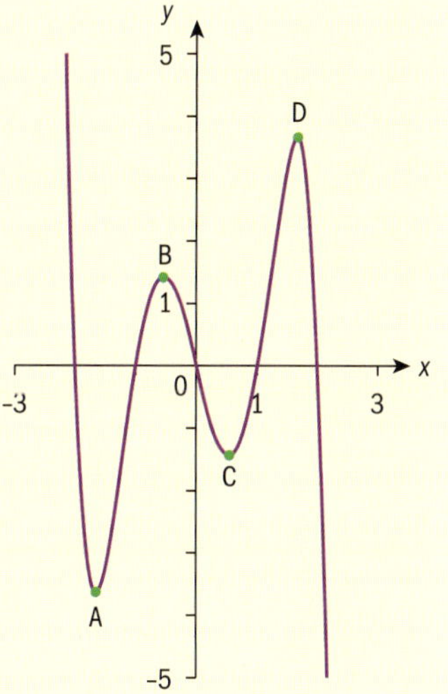

2 On the same axes, mark points which show the value of $f'(x)$ at points A, B, C and D, labelling them A′, B′, C′ and D′. They should have the same x- value as $f(x)$.

3 What are the intervals in which f is: **i** increasing **ii** decreasing? What does this tell you about the sign of f' in each of these intervals?

4 Use your answer to question **3** to complete your sketch of f'.

5 What special points of f do the maximum and minimum points of f' correspond to? Label these points as E, F and G going from left to right on your sketch of f.

6 What are the intervals in which f is: **i** concave up **ii** concave down? What does this tell you about the values of f'' in these intervals?

7 On the same axes, mark points which show the value of $f''(x)$ at points E, F and G. Label these points E′, F′ and G′. Complete a sketch of the graph of f''.

8 **Conceptual** Describe how you can sketch the graphs of $y = f'(x)$ and $y = f''(x)$ from the graph of $y = f(x)$.

Example 32

Copy this graph and sketch the first and second derivatives of the function the graph represents.

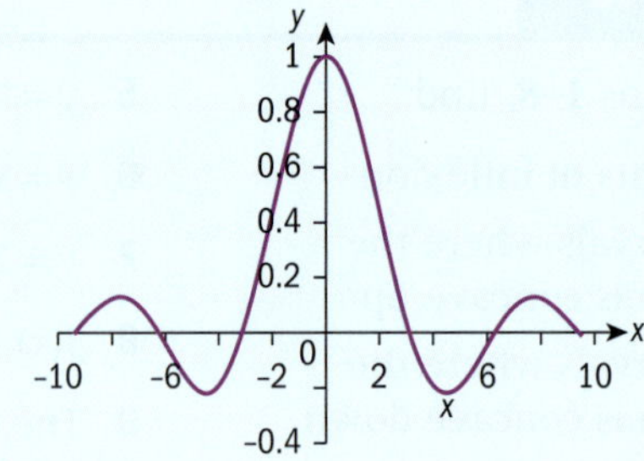

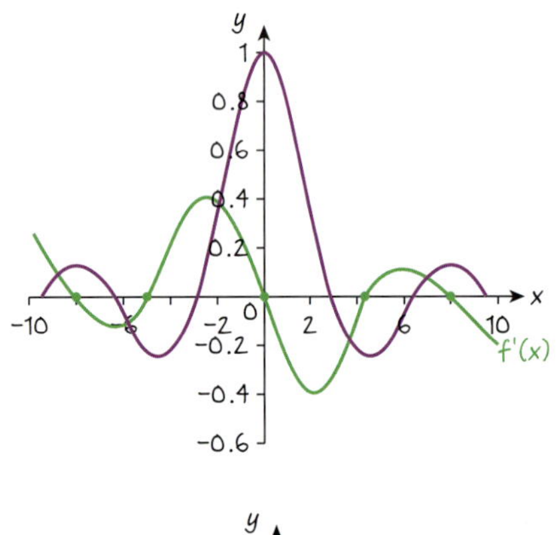

- The turning points of f are the zeros of f'.
- The intervals in which f is increasing/decreasing tell you where the graph of f' is positive/negative.
- The points of inflexion of f correspond to the turning points of f'.

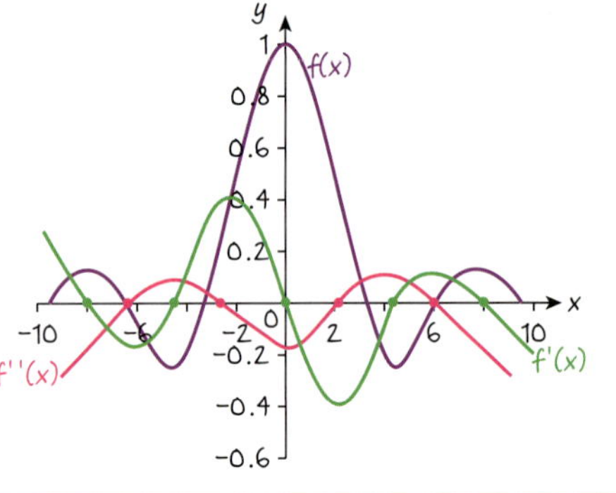

- The turning points of f' are the zeros of f''.
- Where f is concave up, f'' is positive; where f is concave down, f'' is negative.

Exercise 4P

1 Sketch the graphs of these functions, and on the same graph, sketch the first two derivatives of the function.

a

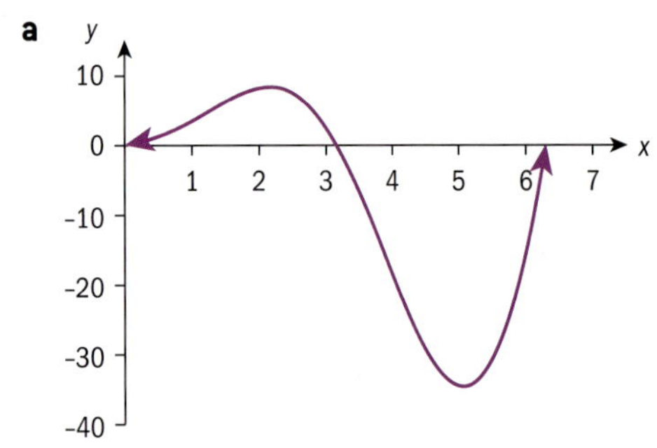

b

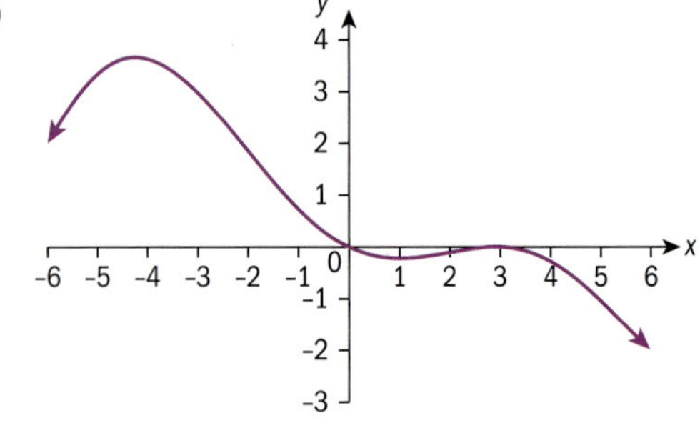

c

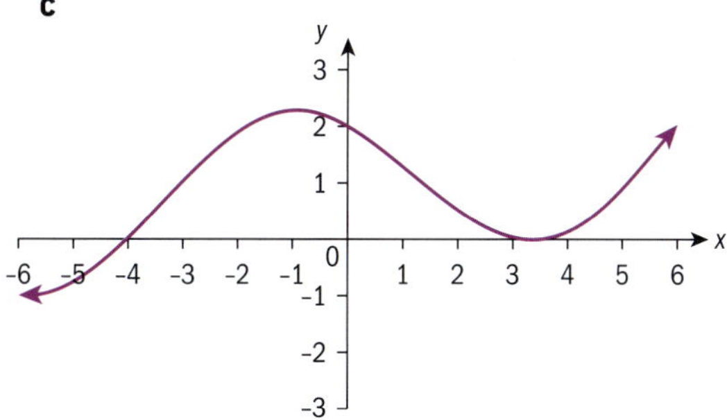

d

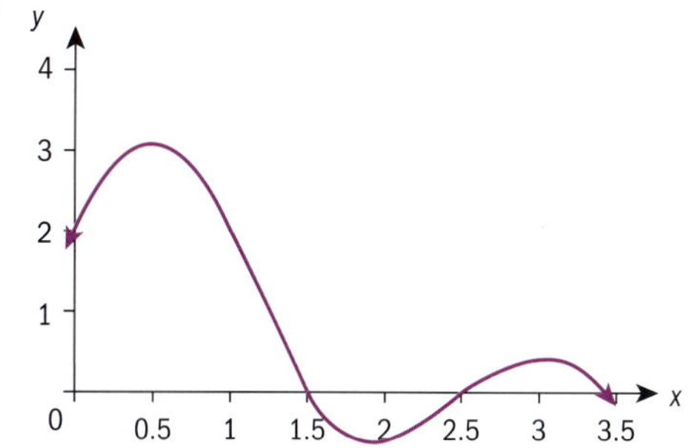

You have worked on sketching the first two derivatives of a function from the graph of the function. How do you sketch the function from its derivatives?

Investigation 21

1 Sketch the graph of $f'(x) = 4x^3 - 12x^2$.

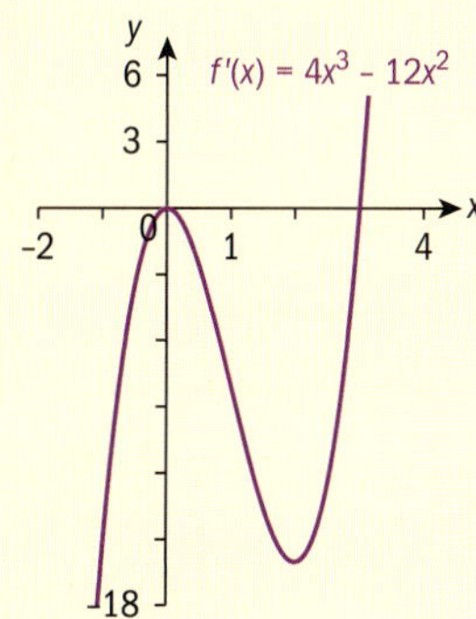

Using the graph of $y = f'(x)$:

2 Describe how you can find the x values of any turning points or horizontal points of inflexion of f, and find these points.

3 Describe how you can find the intervals on which f is: **i** increasing **ii** decreasing, and find these intervals.

4 Describe how you can find the intervals on which f is: **i** concave up **ii** concave down.

5 Sketch a possible graph for $f(x)$ on the same graph.

6 On the same graph, sketch two other functions for f whose derivative is f'. How are the different possible graphs of f related to one another?

7 **Conceptual** Given the graph $y = f'(x)$, describe how you can sketch a possible graph of $y = f(x)$.

Example 33

Here is the graph of $y = f'(x)$ for a function f.

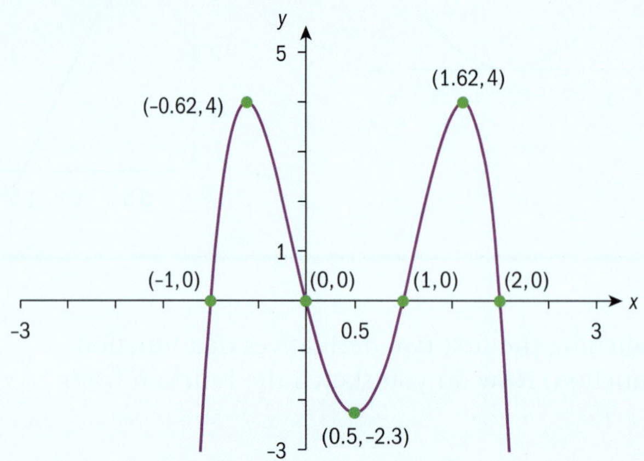

From the graph, indicate:

a the x-coordinate of any points where f has turning points, and determine the nature of these points

b the intervals where f is: **i** increasing **ii** decreasing

c the x-coordinate of the points of inflexion of f

d the intervals where f is: **i** concave up **ii** concave down.

e Sketch a possible graph of f using your answers to parts **a**, **b** and **c**.

HINT

For parts **c** and **d** it helps to make a sketch of f''.

a $x = -1, 0, 1, 2.$	The zeros of $f'(x)$ are the stationary points of f.
At $x = -1$, f has a local minimum, at $x = 0$ f has a local maximum, at $x = 1$ f has a local minimum, at $x = 2$ f has a local maximum.	At $x = -1$ and $x = 1$, f' goes from $-$ to $+$, hence these points are minima. At $x = 0$ and $x = 2$, f' goes from $+$ to $-$, hence these points are maxima.
b **i** $-1 < x < 0$; $1 < x < 2$ **ii** $x < -1$; $0 < x < 1$; $x > 2$	f is increasing where $f'(x) > 0$ and decreasing where $f'(x) < 0$.
c $x = -0.62$, $x = 0.5$, $x = 1.62$	f has points of inflexion where it has a local extremum and there is no sign change in going through the extremum.
d **i** $x < -0.62$; $0.5 < x < 1.62$; **ii** $-0.62 < x < 0.5$; $x > 1.62$	f is concave up when $f''(x) > 0$.
e 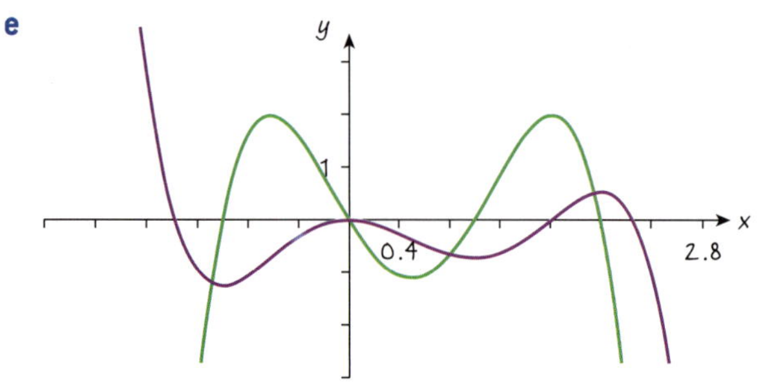	f is concave down when $f''(x) < 0$.

Investigation 22

The graph of f'' is given.

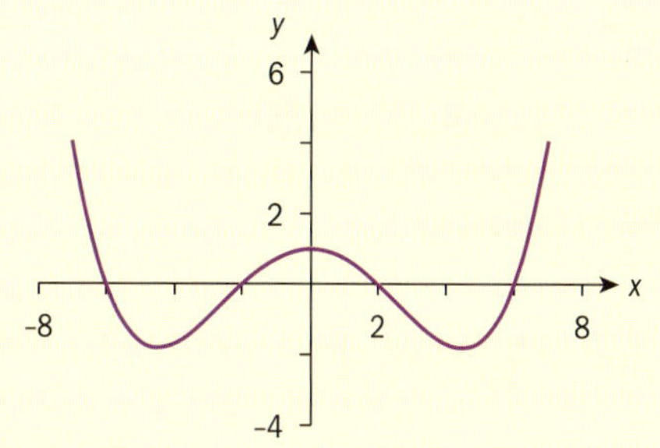

1 Using the graph, how do you identify:

a where f has a point(s) of inflexion?

b the intervals where i f is concave up ii f is concave down?

2 **Conceptual** What can you deduce about the zeros, maximums and minimums of $f'(x)$ from the graph of $y = f''(x)$.

3 Copy the graph of f'' and sketch f' on the same graph.

4 Sketch the graph of f using all your answers.

Exercise 4Q

1 Draw these graphs of f' and, on the same graph, sketch f.

a

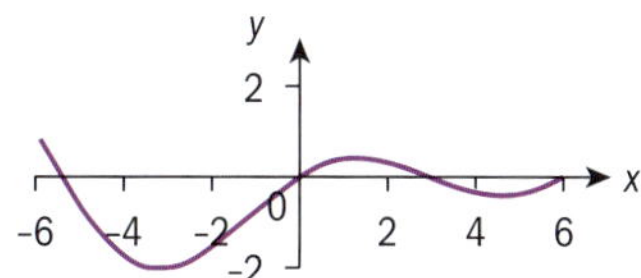

b

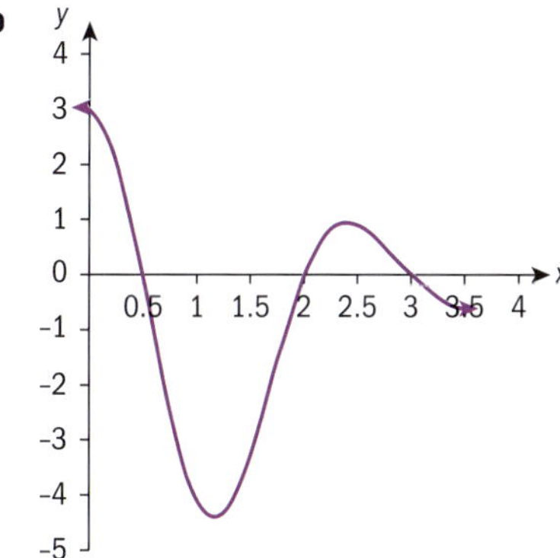

c

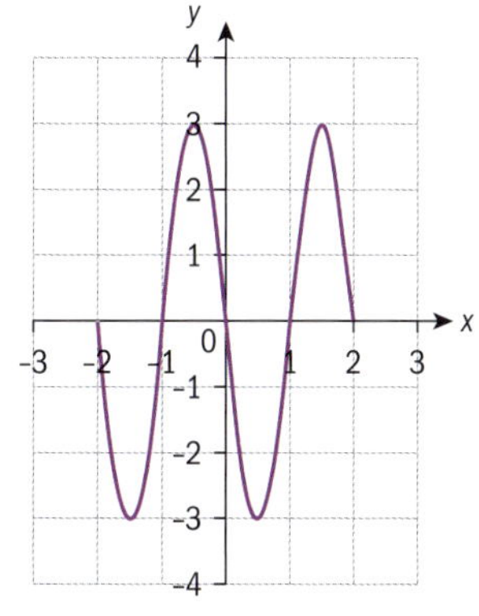

d

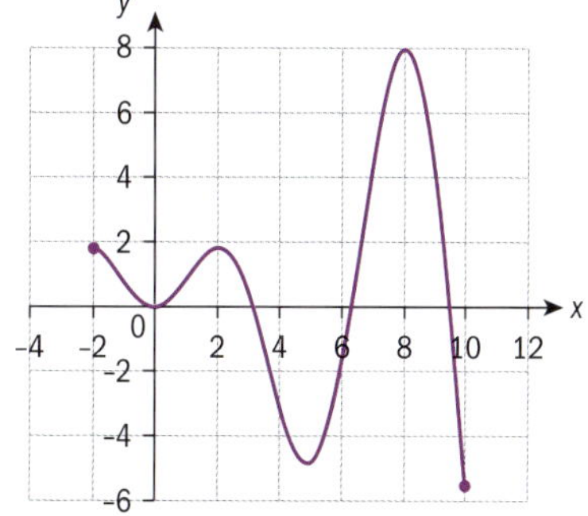

2 Draw these graphs of f'' and, on the same graph, sketch both f and f'.

a

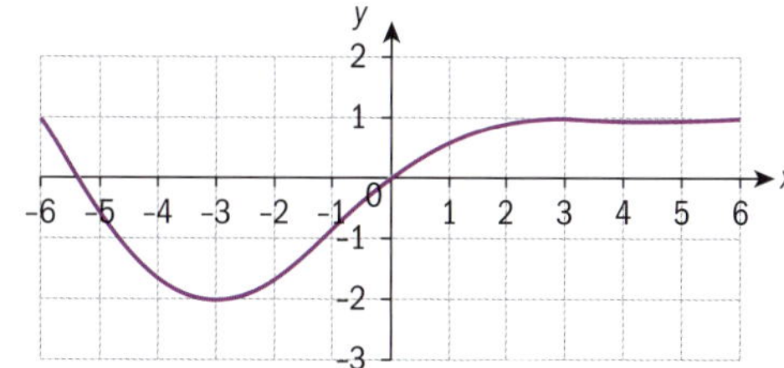

b

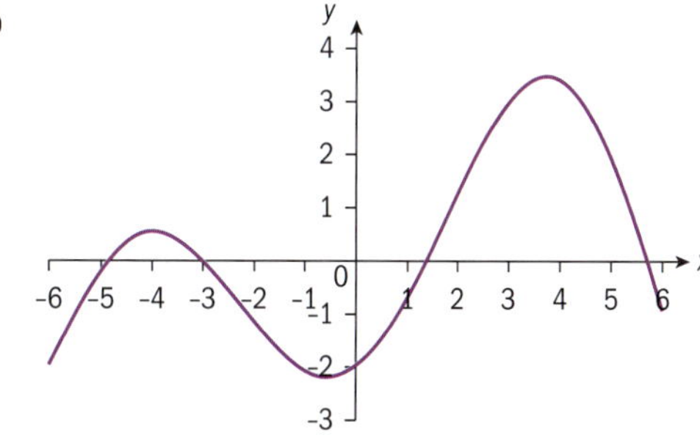

Calculus

4.5 Applications of differential calculus

Optimization

Investigation 23

Soft drinks are often packaged in cylindrical cans made from aluminium. The standard volume of such a can is 330 ml (this may vary among countries). Companies are always interested in minimizing production costs in order to maximize profits. Are companies using the least amount of aluminium in order to produce the 330 ml cans?

Your task is to find the height and diameter of a 330 ml can that will minimize the can's surface area.

1 Write down an expression for the surface area A of a cylinder in terms of the radius and height. This expression should contain two variables. Which variables are these?

2 Write down an expression, containing the same two variables as used in question **1**, for the volume V of a cylinder. Equate your expression to 330 cm^3, since the volume of drink in a can must be 330 cm^3.

3 Use the expression for V (from question **2**) to rewrite the expression for A (from question **1**) in terms of r only, and state a reasonable domain for r.

4 Differentiate your expression for A with respect to r. Hence, find the values of r and h which minimize the surface area of the can, and find the surface area in this case.

5 Check your answers graphically.

6 **Conceptual** What is optimization in calculus?

7 Research actual dimensions and surface areas of your favourite drinks, and deduce whether or not companies are using the least amount of material in order to have the desired volume of 330 ml.

8 If companies are using more than the necessary amount of aluminium, give some reasons as to why this might be so. What other considerations might they need to make in the production of the size and shape of the can?

TOK

How can you justify a rise in tax for plastic containers, eg plastic bags, plastic bottles, etc, using optimization?

When mathematics is applied to real-life situations, its primary purpose is to investigate, explain and solve real-life problems. This process is called mathematical modelling, and in Investigation 23 you used mathematical modelling to investigate the optimum dimensions of a drinks can. Some of the steps involved in creating and using a mathematical model are the same, no matter what the problem is:

- Draw a diagram to represent the problem visually. Label all the known elements, as well as the unknown elements that you need to find. (For the drinks can, you were asked to find the height and radius of a can with volume 330 cm^3 which would minimize the surface area.)
- Identify the independent variable(s) and their constraints. (For the drinks can, this was the radius of the can, which had domain $r > 0$.)
- Identify all other constraints on the problem. (For the drinks can, you were told that the volume must be 330 cm^3.)
- Translate the real-life problem into a mathematical function(s). (You used expressions for the area and volume of a cylinder.)
- Carry out the mathematics necessary in order to solve the problem. (You needed to differentiate the expression for A with respect to r, and find the value of r which minimized A.)
- Reflect on the reasonableness of your results. (Using your GDC to check your results can be helpful. You then compared your theoretical results against the actual dimensions of a drinks can.)
- Apply your methods to other similar problems. (What other problems could the mathematical model for the surface area of a drinks can be applied to?)

In particular, optimization problems deal with finding the most efficient and effective solutions to real-life problems.

Investigation 24

You have 20 m of fencing to enclose part of your garden as an outdoor, rectangular-shaped area for your rabbits. You want to enclose the largest possible area with this amount of fencing.

1 Draw a suitable diagram to represent this scenario. Letting x represent the width of the enclosure, express the sides of the rectangle in terms of x.

2 Write down an expression for the area of the enclosure in terms of x. State the domain of x and explain why it has this domain.

3 Find the dimensions of the enclosure which maximize its area, and find the maximum area.

4 Comment on the geometric significance of the dimensions of the largest enclosure.

5 Show that for any rectangular enclosure with fixed perimeter P, the maximum area will always be the same geometric shape as that which you found in question **3**.

6 Repeat questions **1–3** to find the maximum area that can be enclosed by an isosceles triangle with a perimeter of 20 m. What can you say about the three sides of the triangle?

7 Find the area that can be enclosed by: **i** a regular pentagon **ii** a regular hexagon, using 20 m of fencing.

8 Express in terms of n the area enclosed by 20 m of fencing when in the shape of a regular polygon with n sides.

Continued on next page

9 Reflect on your answers for the areas of a regular triangle, quadrilateral, pentagon and hexagon, and conjecture the geometric figure that you think has the maximum area using 20 m of border. Explain your answer.

10 State, in terms of x, the maximum area of a regular polygon of n sides that can be made with x m of border.

11 Comment on the applicability of your results to real-life problems.

12 **Conceptual** How does finding a mathematical model for a general case help you apply solutions to particular cases?

Example 34

A piece of cardboard, measuring 100 cm by 200 cm, is to be made into a box by cutting out small squares, each with side length x, as shown in the diagram. a is the length between the squares on the longer side of the cardboard, and b is the length between the squares on the shorter side of the cardboard.

i Find expressions for a and b in terms of x, and state the constraints on the lengths of x, a and b.

ii Find the value of x (in cm) which maximizes the volume of the box, and find the maximum volume of the box. Check your answers graphically.

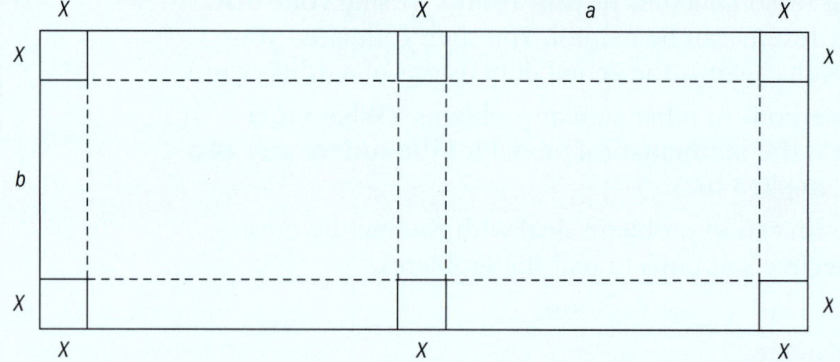

i Length: $3x + 2a = 200 \Rightarrow a = \dfrac{200-3x}{2}$ Width: $2x + b = 100 \Rightarrow b = 100 - 2x$ $a > 0;\ b > 0 \Rightarrow 2x < 100 \Rightarrow x < 50;\ 3x < 200 \Rightarrow x < 66\frac{2}{3}$ $x < 50$ and $x < 66\frac{2}{3} \Rightarrow x < 50$	Express length and width in terms of x. Identify the constraints on the variables.
ii $V = x \times (100-2x)\left(\dfrac{200-3x}{2}\right) = 3x^3 - 350x^2 + 10\,000x$	Volume $= x \times b \times a$
$\dfrac{dV}{dt} = 9x^2 - 700x + 10\,000$	Differentiate, and set equal to 0.
$9x^2 - 700x + 10000 = 0 \Rightarrow x = 18.858\ldots;\ x = 58.9197\ldots$	
Since x must be less than 50, reject the second answer.	Compare your answer with the constraint on x.
$\dfrac{d^2V}{dx^2} = 18x - 700$. When $x = 18.858\ldots$, $\dfrac{d^2V}{dx^2} < 0$, hence $V_{max} = 84230.6\ldots \approx 84200\text{ cm}^3$	Use second derivative test to determine which point is a maximum.

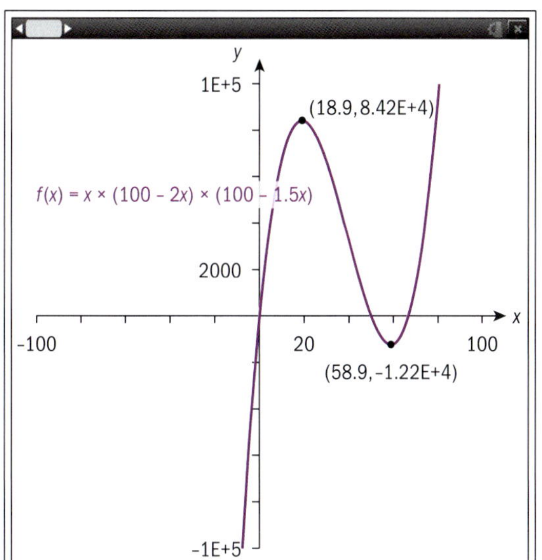

From the graph, you can see that the volume function is maximized at $x = 18.9$ cm. Since $V < 0$ when $x = 58.9$, reject this answer.

Exercise 4R

1. A rectangular plot of land is bounded on one side by a wall. Determine the largest area that can be enclosed using the wall and 800 m of fencing.
2. You wish to enclose a rectangular area of 200 m^2. On one side of the enclosure is a stream, and no fencing is needed along that side. Find the least possible length of fencing required for the enclosure.
3. In a church building, a stained-glass window is to be constructed in the shape of a semicircle whose straight edge sits on top of a rectangle. The diameter of the semicircle is the same length as the horizontal sides of the rectangle. If the perimeter of the window is to be 12 m, determine the radius of the semicircle and dimensions of the rectangle that will allow the maximum amount of light through.
4. Two vertical poles, with heights 6 m and 14 m, stand 20 m apart on horizontal ground. A wire is attached to the tops of the two poles, and is anchored to the ground at a point between the poles. Determine how far from the base of the smaller pole it must be anchored so that the minimum length of wire is used.
5. An open rectangular box is to be made from a piece of cardboard which measures 24 cm by 45 cm. The box is made by cutting out congruent squares from the corners of the cardboard and then folding up the sides of the remaining cardboard. Determine the dimensions of the box of largest volume, and find the maximum volume.
6. A cylindrical can with an open top has a surface area of 3π m^2. Find the radius and height of the can that will maximize the volume, and find the maximum volume.
7. A cylindrical can is to be designed to hold 1 litre of car engine oil. Find the surface area of the can that would allow the costs of material to be minimized.
8. Find the volume of the largest right-circular cone that can be inscribed in a sphere whose radius is 10 cm.
9. Find the shortest distance between the point (1.5, 0) and the curve $y = \sqrt{x}$.
10. A piece of wire 80 cm in length is cut into three parts: two equal circles and a square. Find the radius of the circles if the sum of the three areas is to be minimized.

11 You are designing a poster to have a printed area of 320 cm^2. The margins at the top and bottom of the poster are to be 5 cm each, and the side margins are to be 4 cm each. Find the dimensions of the poster with the smallest total area.

12 You are 2 km offshore in a boat. The coastline can be modelled as a straight line, and you want to reach a town that is 6 km directly along the coast from the point on the shore that is opposite your boat.

You can row at 2 km h^{-1} and can jog at 5 km h^{-1}. Determine where you should land your boat in order to reach the town in the smallest amount of time possible.

Optimization techniques are used throughout industry, since the goal is to maximize profits. A company's profit is the amount left over after the costs of production have been discounted from the company's total income (sometimes called *revenue*). In other words, profit is the difference between revenues and costs. If profit is represented by $p(x)$, revenues by $r(x)$ and costs by $c(x)$, then $p(x) = r(x) - c(x)$.

The break-even point occurs when the production costs and the total revenues are equal. This diagram shows a graphical representation of this.

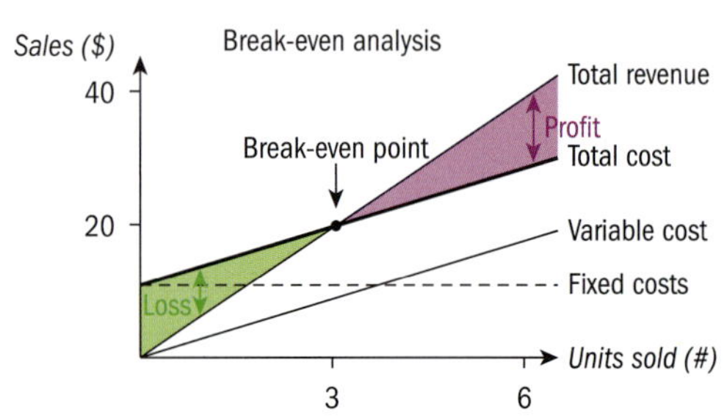

Example 35

The cost of manufacturing fishing poles can be modelled by the function $c(x) = 7x + 3$, where x is the number of batches (each containing 1000 poles) manufactured. Revenue is modelled by $r(x) = x^3 - 10x^2 + 20x$. A small company has enough workers to produce a maximum of 1200 fishing poles.

a State the domain of both the cost and revenue functions.

b Find the number of fishing poles that the company should manufacture to maximize profits.

The company introduces a new process that allows them to produce 6000 poles.

c Find the number of poles that would cause the company to minimize profits (or maximize losses).

d By graphing the cost and revenue functions, find the number of poles that should be manufactured if the company is to just break even (that is, the production level at which the costs and revenues are equal).

e It would not maximize the company's profits to produce as many poles as workers are capable of producing. Use your graph to explain why.

a For both functions, $0 \le x \le 1.2$.	The company can produce a maximum of 1200 poles, which is 1.2 units.
b $p(x) = (x^3 - 10x^2 + 20x) - (7x + 3)$ $= x^3 - 10x^2 + 13x - 3$	$p(x) = r(x) - c(x)$
$p'(x) = 3x^2 - 20x + 13$ $3x^2 - 20x + 13 = 0 \Rightarrow x_1 = 0.730;\ x_2 = 5.94$	Find the stationary points on the curve of the profit function.
Since $x \le 1.2$, reject $x_2 = 5.94$.	Check your answers against the constraints of x.

$p''(x) = 6x - 20$ $p''(0.730) < 0 \Rightarrow x = 0.730$ is a maximum.	Use the second derivative to determine the nature of the stationary points.
Since x is in thousands of poles, the production level necessary to maximize profits is 730 fishing poles.	Interpret your answer in the context of the problem.
c $p''(5.94) > 0 \Rightarrow x = 5.94$ is a minimum, so at $x = 5.94$ the profit function p has an absolute minimum in its domain. The production level that would maximize losses is 5940 fishing poles.	Consider the nature of the other stationary point.
d Break even at 296 fishing poles and at 1190 fishing poles.	The company breaks even when costs equal revenues. From the graph that is at $x = 0.296$ (ie 296 fishing poles), or at $x = 1.19$ (ie 1190 fishing poles).

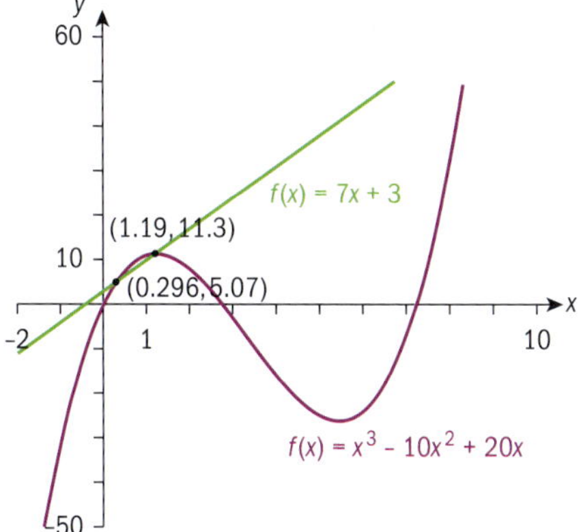

e The company's revenues decrease, despite the costs continuing to increase. This may be because of lack of demand for so many poles, which would mean they cannot sell them all.

Exercise 4S

1 A company manufactures scientific calculators. The cost of manufacturing one calculator in a batch of size x can be modelled by the function $C(x) = 0.01x^3 - 10x + 150$. The company wants to minimize the cost of producing the calculators. Find the size of the batch for which the cost per calculator is minimized.

2 A pharmaceutical company's research department proposes a mathematical model which connects the amount of a blood pressure drug that is injected into a patient (x cm^3) with the decrease in the patient's systolic pressure (the highest pressure when the heart beats, $f(x)$ mmHg)

The model is $f(x) = \frac{3}{4}x^2 - \frac{1}{8}x^3$.

Determine the amount of medication that produces the largest drop in the systolic pressure, and the size of this systolic drop.

3 The production costs of x USB flash drives made by a company can be modelled by the function $c(x) = 500 + 3x$. The projected selling price is modelled by $p(x) = 7 - 0.002x$. The projected revenues therefore can be modelled by the function $r(x) = x(7 - 0.002x)$.

a Find the domain of the price function.

b Find the break-even point(s) of the production of the flash drives.

c State the flash drive production levels, in the form $a < x < b$, that must be met in order for the company to make a profit.

d Find the maximum profit that the company can make from sales of flash drives.

4 The cost, in euros, of producing x ski-suits is $c(x) = 400 + 20x - 0.2x^2 + 0.0004x^3$. The revenue function is modelled by $r(x) = 35x - 3$.

a Find the number of ski-suits that should be produced to minimize costs per unit.

b Find the number of ski-suits that should be produced to maximize profits.

c Interpret your answers to parts **a** and **b**.

5 A gardener plans to enclose three sides of a 1000 m^2 rectangular plot of land with shrubs. The shrubs cost €15 per metre, and the fencing for the fourth side costs €3 per metre. Determine the approximate dimensions of the plot that would minimize the costs for creating the plot, and the costs.

6 The airfare between Vienna and Dubai is approximately €500. A particular airline has a passenger capacity of 300, but at certain times of the year they only fly with an average of 180 passengers. After doing some market research, the airline finds out that for each reduction of €10 in ticket price below €500 they can attract two more passengers on each flight. Determine the best ticket price, and the corresponding number of passengers per flight, that would maximize the airline's revenues on this route.

Kinematics

Kinematics is the study of how objects move. A particle moving in a straight line is the simplest type of motion. To describe simple linear motion, you need a starting point, a direction and a distance.

- A **path** is the set of points along which an object travels.
- The **distance travelled** by a particle is the *length* of the path between the particle and its start point.
- The **displacement** of the particle is the *difference* between the particle's position and its starting point.

If you run a complete lap around an athletics track, the *distance* you will have travelled is 400 m. However, after you have completed the lap, your displacement would be 0 m, since your initial and end positions are the same.

Displacement is a vector quantity (vectors are covered in Chapter 9) as it tells you the *shortest distance* of the object from its starting point. Distance, however, is a scalar quantity as it only tells you the *length of the path* the object has travelled.

In the diagram on the right, the *path* a particle travels along is the curved line joining A to B. The length of this path would tell you the *distance* travelled by the particle. The *displacement* of the particle from point A is the length of the vector $\overrightarrow{AB}$. Hence, the displacement is always less than or equal to the distance travelled.

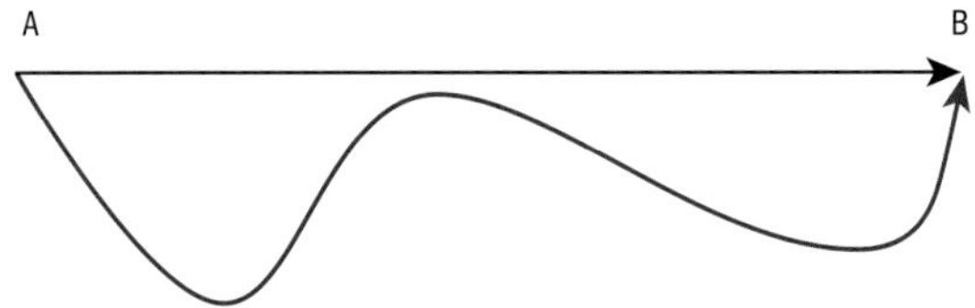

- Velocity is the first derivative of the displacement.
- Acceleration is the second derivative of the displacement, or the first derivative of the velocity.

Investigation 25

Ben runs from point A to point B along a straight line. When he reaches B, he then turns and runs back to point A. When he reaches A, he turns once again and runs back to point B.

Ben's position, s, at any time t after he first left point A (where $0 \le t \le 4$) can be modelled by the function $s(t) = t^3 - 6t^2 + 9t - 2$.

1 Use your GDC to plot the graph of Ben's displacement, $s(t)$, against t.

2 Use your GDC to plot the graph of Ben's velocity, $s'(t)$, against t.

Write down the time intervals where Ben's velocity is:

a positive **b** negative.

Which direction is Ben moving in each time?

3 **Conceptual** What does a positive or negative value for velocity represent?

4 **Conceptual** What is the connection between speed and velocity?

5 Find the times at which Ben's velocity is zero. What is Ben doing at each of these times?

6 At which two times did Ben's maximum velocity occur? Describe Ben's motion at each of these two times.

TOK

Read these statements:

- Mathematical truths are practical simplifications based on a number of experiences.
- Mathematics is analytical, it is true by definition.
- Mathematics gives us knowledge independent of experience.

Which do you agree with the most? Why?

Investigation 26

A particle moves along a line so that its position at any time t, in seconds, is $s(t) = t^3 - 7t^2 + 11t - 2.5$.

1 Find the velocity and acceleration of the particle in terms of t.

2 Find the times at which:

a the particle is at rest

b the particle is speeding up

c the particle is slowing down.

Justify your answers.

3 **Conceptual** What must be true about the signs of the velocity and acceleration in order for a particle to speed up, and in order for a particle to slow down?

4 At what values of t does the particle change direction? How can you tell this from the graph of acceleration against time?

Velocity, $v(t)$, is the derivative of the displacement function $s(t)$: $v(t) = s'(t)$

Acceleration, $a(t)$, is the derivative of the velocity: $a(t) = v'(t) = s''(t)$

When v and a have the same sign, the particle is speeding up (accelerating).

When v and a have different signs, the particle is slowing down (decelerating).

Example 36

A particle moves in a horizontal line so that its position from a fixed point after t seconds, where $t \geq 0$, is s metres, where $s(t) = 5t^2 - t^4$.

a Find the velocity and acceleration of the particle after 1 second.

b Determine whether the particle is speeding up or slowing down at $t = 1$.

c Find the values of t when the particle is at rest.

d Find the time intervals on which the particle is speeding up, and the intervals on which it is slowing down.

a $v(t) = s'(t) = 10t - 4t^3 \text{ m s}^{-1}$

$v(1) = 6 \text{ m s}^{-1}$

Differentiate s to get an expression for v, and evaluate at $t = 1$.

$a(t) = v'(t) = 10 - 12t^2 \text{ m s}^{-2}$

$a(1) = -2 \text{ m s}^{-2}$

Differentiate v to get an expression for a, and evaluate at $t = 1$.

b Since v and a have opposite signs at $t = 1$, the particle is slowing down.

c $10t - 4t^3 = 0$

The particle is at rest when $v = 0$.

$2t(5 - 2t^2) = 0$

$t = 0,\ t = \sqrt{\frac{5}{2}} \approx 1.58 \text{ s.}$

Reject the negative value of the square root, since $t \geq 0$.

d $10 - 12t^2 = 0$

$t = \sqrt{\frac{5}{6}} \approx 0.91 \text{ s.}$

Find the places at which the acceleration changes sign. Again, ignore the negative value for t.

t	$0 < t < \sqrt{\frac{5}{6}}$	$\sqrt{\frac{5}{6}} < t < \sqrt{\frac{5}{2}}$	$t > \sqrt{\frac{5}{2}}$
Sign of v	+	+	−
Sign of a	+	−	−

Compare the signs of velocity and acceleration each time at each interval on which one of them changes.

The particle is speeding up when $0 \text{ s} < t < 0.91 \text{ s}$; $t > 1.58 \text{ s}$.

Both v and a have the same sign on these intervals.

The particle is slowing down when $0.91 \text{ s} < t < 1.58 \text{ s}$.

v and a have different signs on this interval.

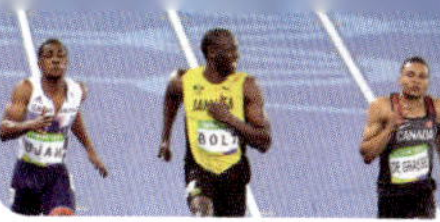

Exercise 4T

1 The height, h metres, of an object that is moving along a vertical path is modelled by the function $h = 112 + 96t - 16t^2$, where t is the time, in seconds. Find:

a the object's velocity at $t = 0$

b the maximum height of the object and the time at which it reaches maximum height

c the object's velocity when its height is 0 m.

2 A person jumps from a diving board above a swimming pool. At time t seconds after leaving the board, the person's height above the surface of the pool, s, can be modelled by the function $s(t) = 10 + 5t - 5t^2$. Find:

a the height of the diving board above the surface of the pool

b the time between the person leaving the board and hitting the water

c the velocity and acceleration of the diver upon impact with the water. Interpret your answers in the context of this problem.

3 A child's toy launches toy rockets into the air. The height, h m, of a toy rocket which is launched into the air with an initial velocity of v_0 and initial height of h_0 can be modelled by the function $h(t) = h_0 + v_0 t - 4.9t^2$, where t is time in seconds that have passed since the rocket was launched.

A toy rocket is launched from the ground with an initial velocity of $50\ \text{m s}^{-1}$. Find the maximum height the rocket reaches, and the time that passes before it hits the ground again.

4 You begin walking eastward through a park, and your velocity graph is shown below.

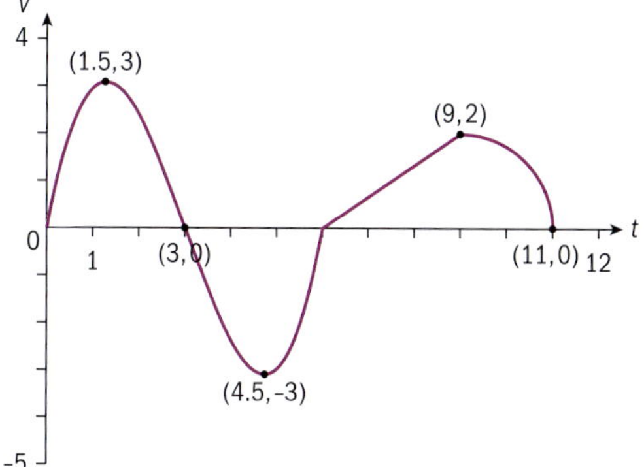

Find:

a the times when you are standing still

b the time intervals when you are moving: **i** eastward **ii** westward

c the time intervals when you are moving most quickly: **i** eastward **ii** westward

d the time(s) when you are moving most quickly

e the time intervals when you are:
i speeding up **ii** slowing down.

5 The position, s cm, of a particle moving along a line can be modelled by the function $s(t) = -t^3 - 3t^2 + 4t + 3$ where t is the time in seconds. Find the velocity and acceleration of the particle at any time t, and describe the motion of the particle.

6 A particle moves in a straight line such that its displacement from a fixed point after t seconds is s metres where $s(t) = 10t^2 - t^3$.

a Find the average velocity of the particle in the first 3 seconds.

b Find the velocity and acceleration of the particle at $t = 3$ s.

c Determine whether the particle is speeding up or slowing down at $t = 3$ s.

d Find the value of t when the particle changes direction.

7 A particle moves along a line so its position at any time t in seconds is

$s(t) = t^3 - 7t^2 + 11t - 2.5$.

a Find the velocity and acceleration of the particle at any time t.

b Find the times when the particle is at rest.

c Find the times when the particle is:
i speeding up
ii slowing down.

d Find the values of t when the particle changes direction.

e Find the total distance travelled in the first 3 seconds.

4.6 Implicit differentiation and related rates

Investigation 27

An **implicit relation** is one where neither of the variables is the subject of the equation. For example, a circle centred on the origin with radius r is defined as the set of points (x, y) such that $x^2 + y^2 = r^2$. This is an example of an implicitly defined relation, as neither x nor y is the subject of the equation. If you rearrange to make y the subject, then you will have an explicitly defined relation.

1 Write down the equation of a circle, centred on the origin, whose radius is 1.

2 Can you rearrange this equation to find an explicit expression for y? Differentiate your explicit relation to find $\frac{dy}{dx}$.

3 Evaluate $\frac{dy}{dx}$ at $x = 0.5$. You should have two answers. Explain why this is the case.

4 Using your answer to question **3**, state and explain whether the explicit equation for a circle is a function.

5 Solve for y, if possible, and state which relations are functions.

a $3x + 2y = 4$ **b** $4x + 3y + 3xy = 1$

c $x + y^2 = 2$ **d** $x^3 + y^3 - 9xy = 0$

In your work so far, you have always changed an implicitly defined relation to an explicit one in order to find its derivative.

6 **Conceptual** Is it always possible to change an implicitly defined relation into an explicitly defined relation? What do you need in order to find the derivative of any implicitly defined relation?

7 The expression in **5d** is called the "folium of Descartes". It is not a function, but rather can be seen as three distinct piecewise functions joined together to form a graph. From the graph, identify the domains of three possible functions that make up the graph.

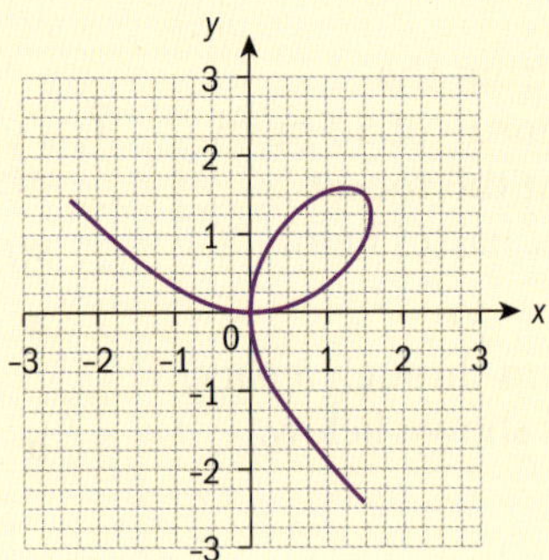

Since it is not always possible to change an implicitly defined relation into an explicit one, it is necessary to have a technique to differentiate implicit functions. This technique is called **implicit differentiation**.

Consider the implicit function $2x^2 + y^3 = 3x$. In implicit differentiation, we differentiate each term separately with respect to the variable x:

Term 1: $2x^2$. This is itself a function in x, so you can differentiate it as normal with respect to the variable x, that is $\frac{d(2x^2)}{dx} = 4x$.

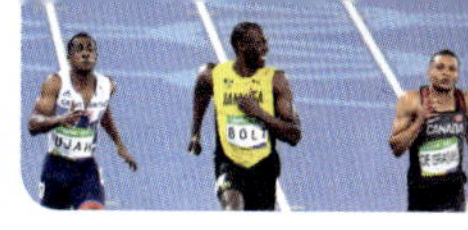

Term 3: $3x$. As with term 1, this is itself a function in x, so you can differentiate it as normal with respect to the variable x, that is $\frac{d(3x)}{dx} = 3$.

Term 2: y^3. This is more tricky, since the variable y is itself a function in x. To differentiate this, you need to let $u = y^3$ and use the chain rule:

$$\frac{du}{dx} = \frac{du}{dy} \times \frac{dy}{dx}$$

Now $\frac{du}{dy} = 3y^2$. Since you do not know y explicitly in terms of x, you cannot find $\frac{dy}{dx}$, so you write $\frac{du}{dx} = 3y^2 \times \frac{dy}{dx}$

and hence $\frac{d(y^3)}{dx} = 3y^2 \times \frac{dy}{dx}$

Now put the derivatives of the three terms back together:

$$4x + 3y^2 \times \frac{dy}{dx} = 3$$

This is an expression which you can rearrange to make $\frac{dy}{dx}$ the subject:

$$\frac{dy}{dx} = \frac{3-4x}{3y^2}$$

At this point, you are nearly complete. However, it's best to express $\frac{dy}{dx}$ in terms of the variable x only, if possible. To do this, you should find an expression for $3y^2$ in terms of x.

Solving for y^2 in the original equation, $y = \sqrt[3]{(3x-2x^2)} \Rightarrow y^2 = (3x-2x^2)^{\frac{2}{3}}$.

Hence, $\frac{dy}{dx} = \frac{3-4x}{3(3x-2x^2)^{\frac{2}{3}}}$

Note that, in this instance, you can also find $\frac{dy}{dx}$ by finding an explicit expression for y from the original equation, and then differentiating this directly:

Solving for y, $y = \sqrt[3]{3x-2x^2}$.

Then, using the power and chain rules,

$$\frac{dy}{dx} = \frac{1}{3}(3x-2x^2)^{-\frac{2}{3}} \times (3-4x) = \frac{3-4x}{3(3x-2x)^{\frac{2}{3}}}.$$

This gives the same expression for $\frac{dy}{dx}$ as you found by differentiating implicitly.

When differentiating a term $u(y)$ with respect to a *different* variable x, the chain rule $\frac{d(u(y))}{dx} = u'(y) \times \frac{dy}{dx}$ allows you to differentiate implicitly.

The answers obtained in differentiating implicitly and explicitly are equivalent. In the case of the function $2x^2 + y^3 = 3x$, you are able to check your answer to the implicit differentiation by defining the relation explicitly, and differentiating.

TOK

Mathematics and knowledge claims: Euler was able to make important advances in mathematical analysis before calculus had been put on a solid theoretical foundation by Cauchy and others. However, some work was not possible until after Cauchy's work.

What does this suggest regarding intuition and imagination in mathematics?

You will now see an example where it is not easily possible to define the relation explicitly.

Example 37

a Differentiate, with respect to x, the folium of Descartes: $x^3 + y^3 - 9xy = 0$.

b Find the equation of the tangent to the curve at the point $(2, 4)$ in the form $ax + by + c = 0$.

a $\frac{d(x^3)}{dx} = 3x^2$	
$\frac{d(y^3)}{dx} = 3y^2\frac{dy}{dx}$	Use the chain rule.
$\frac{d(9xy)}{dx} = 9\left(\frac{d(xy)}{dx}\right) = 9\left(y + x\frac{dy}{dx}\right)$	Use the product rule to differentiate xy.
$3x^2 + 3y^2\frac{dy}{dx} - 9\left(y + x\frac{dy}{dx}\right) = 0$	Put the differentiated terms back together in the equation.
$3x^2 + 3y^2\frac{dy}{dx} - 9y - 9x\frac{dy}{dx} = 0$	Expand and simplify.
$\frac{dy}{dx}(3y^2 - 9x) = 9y - 3x^2$	Factorize $\frac{dy}{dx}$.
$\frac{dy}{dx} = \frac{9y - 3x^2}{3y^2 - 9x}$	Solve for $\frac{dy}{dx}$.
b $\frac{dy}{dx} = \frac{9y - 3x^2}{3y^2 - 9x} = \frac{9 \times 4 - 3(2)^2}{3(4)^2 - 9(2)} = \frac{4}{5}$	Substitute $(2, 4)$ into the expression for $\frac{dy}{dx}$.
$m = \frac{4}{5}$; $(2, 4)$ $y - 4 = \frac{4}{5}(x - 2)$	Use the point–intercept form of the straight line.
$-4x + 5y - 12 = 0$	Multiply through by the denominator, and rearrange.

In this last example, do not substitute for y since the equation cannot easily be solved explicitly.

Reflect Why is it necessary to have a method for implicit differentiation?

Exercise 4U

1 Find $\frac{dy}{dx}$ by differentiating implicitly with respect to x.

a $2y^2 - 3x^2 = 1$ **b** $y^3 = x^4 - 1$

c $2x^2 + 4y^2 - 3x + 4y = 0$

d $x^2 - 3xy + 2y^2 = 5$

e $(x + y)^2 = 2 - 3y$ **f** $2x^2 = \frac{x + y}{x - y}$

g $x = \sqrt{2x^2 - 6y^3}$

2 Find the equations of the tangent and the normal to the curve $\frac{1}{x^3}+\frac{1}{y^3}=2$ at the point $(1, 1)$.

3 Find the equations of the tangent and the normal to the curve $\frac{1}{x+1}+\frac{1}{y+1}=1$ at the point $(1, 1)$.

4 Find the coordinates of the point(s) on the curve $x^2+y^2=6x+8y$ where the gradient is 0.

5 Given the curve $x+y=x^2-2xy+y^2$:

a Find $\frac{\mathrm{d}y}{\mathrm{d}x}$.

b Show that $1-\frac{\mathrm{d}y}{\mathrm{d}x}=\frac{2}{2x-2y+1}$.

c Show that $\frac{\mathrm{d}^2y}{\mathrm{d}x^2}=\left(1-\frac{\mathrm{d}y}{\mathrm{d}x}\right)^3$.

The technique of implicit differentiation can also be used to differentiate a function with respect to an extraneous variable, for example, time. Imagine filling a cone-shaped paper cup with water. As you fill the cup, the volume of water is changing with respect to time. The height of the water in the cup and the radius of the circular surface of the water are also both changing with time.

Implicit differentiation can sometimes provide an equation which links rates of change that are related to one another; for example, the rates of change of volume, height and radius are linked in the example below.

Example 38

A cone-shaped paper cup is being filled with water. The volume of a cone is given as $V=\frac{1}{3}\pi r^2h$.

Use implicit differentiation to find an expression which links $\frac{\mathrm{d}V}{\mathrm{d}t}$ (the rate of change of the water's volume), $\frac{\mathrm{d}r}{\mathrm{d}t}$ (the rate of change of the water's radius) and $\frac{\mathrm{d}h}{\mathrm{d}t}$ (the rate of change of the height of the water).

Interpret your answer in the context of the problem.

$\frac{\mathrm{d}V}{\mathrm{d}t}=\frac{\pi}{3}\frac{\mathrm{d}(r^2h)}{\mathrm{d}t}$	Differentiate the volume formula with respect to t.
Now $\frac{\mathrm{d}(r^2h)}{\mathrm{d}t}=2rh\frac{\mathrm{d}r}{\mathrm{d}t}+r^2\frac{\mathrm{d}h}{\mathrm{d}t}$	Use the product rule.
So $\frac{\mathrm{d}V}{\mathrm{d}t}=\frac{\pi}{3}\left(2rh\frac{\mathrm{d}r}{\mathrm{d}t}+r^2\frac{\mathrm{d}h}{\mathrm{d}t}\right)$	

The rate at which the volume of a cone changes is related to the rates at which its radius and height change. As the volume increases, the height of the water level and the length of the radius of the water level also increase.

TOK

How do we choose the axioms underlying mathematics?

Is this an act of faith?

"Mathematics is the language with which God wrote the Universe" – Galileo

Investigation 28

A cone-shaped paper cup has a height of 10 cm and a radius of 3 cm. The cup is being filled with water. The volume of the water in the cup is changing as it is being filled. (Volume of a cone: $V = \frac{1}{3}\pi r^2 h$.)

1 Make a sketch of the cone. Label an arbitrary height of liquid as h and the height of the cone as 10 cm. Label the radius at the top of the cone as 3 cm and the radius of any arbitrary water level as r.

2 Use similar triangles to write an equation expressing r in terms of h.

3 Substitute your expression for r into the formula for volume, and simplify. You will now have the formula for the volume in terms of h only.

Water is now poured into the cup so that the rate of change of the volume with respect to time (ie $\frac{dV}{dt}$) is 2 cm^3 per second.

The height of the water level also changes as water is poured into the cone. The rate of change of the height h after t seconds is $\frac{dh}{dt}$.

4 Differentiate your expression for the volume of water in the cone (from question **3**) with respect to time. [**Remember:** you'll need to use implicit differentiation, since you are differentiating with respect to a different variable (ie t) from the one given in the formula, which is h.]

Set your expression for $\frac{dV}{dt}$ equal to 2 cm^3, and make $\frac{dh}{dt}$ the subject.

You now have a expression relating the height of the water level, h, with its rate of change, $\frac{dh}{dt}$.

5 Use your expression to find how fast the water level is rising when the height of the water level is 2 cm.

6 **Factual** Summarize your findings on how the rates of change of the variables in this problem are related.

7 **Conceptual** What do "related rates" problems analyse?

In related rates problems you investigate the effect that a change in a particular rate has on a different rate. There are many examples of this in real life. Here are just a few:

- The rate at which air is pumped into a balloon affects the rate at which the balloon's surface area changes.
- The rate at which water is released from a reservoir affects the rate at which the water level in the reservoir changes.
- The speed of a car (remember, speed is the rate of change of distance with time) affects the rate at which the car consumes fuel.

Reflect Can you think of some other real-life related rates problems?

A method for solving related rates problems, using the steps in Investigation 28, is outlined for you.

- Make a sketch or diagram of the problem, including geometric shapes.
- Label your sketch with the variables and fixed quantities.

- Identify an equation that relates your variables.
- If your equation has more than two variables, identify another equation that will help you express one of the variables in terms of the other. Substitute into the main equation. (For example, V, r and h were all variables in the paper cup investigation, but you expressed r in terms of h.)
- Differentiate your equation implicitly.
- Find the unknown quantity.

Example 39

A 25 m-long industrial ladder is leaning against a wall on a building site. The bottom of the ladder starts to slip along the floor at a rate of 0.2 m s^{-1}. Determine how fast the top of the ladder is moving down the wall at the instant when it is 20 m above the floor. Interpret your answer.

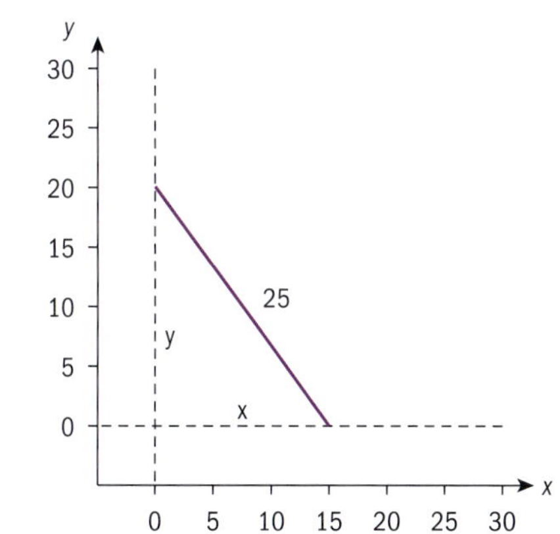

$x^2 + y^2 = 625$	Use the Pythagorean theorem to form an equation connecting x and y.
$2x\dfrac{dx}{dt} + 2y\dfrac{dy}{dt} = 0$	Differentiate your equation implicitly with respect to t.
$y = 20$ m; $\dfrac{dx}{dt} = 0.2$ (1)	Use the information given.
$x^2 + 20^2 = 25^2 \Rightarrow x = 15$ (2)	Use the fact that the ladder is 25 m long to find x when y is 20 m.
$2(15)(0.2) + 2(20)\dfrac{dy}{dt} = 0$	Substitute the values from (1) and (2) into the implicit equation.
$\Rightarrow \dfrac{dy}{dt} = -0.15$ m s^{-1}	Find the unknown quantity.
The top of the ladder is moving down the wall at 0.15 m s^{-1} at the instant it is 20 m above the floor.	Interpret the answer in the context of the problem. Remember that the answer is negative because y is decreasing.

Calculus

Exercise 4V

1 Write an equation relating the rates of change of the variables in these formulae.

   a Area of a circle: $A = \pi r^2$

   b Surface area of a closed cylinder: $A = 2\pi r + 2\pi h$

   c Volume of a cone: $V = \frac{1}{3}\pi r^2 h$

   d Volume of a sphere: $V = \frac{4}{3}\pi r^3$

2 If l, w and h represent the length, width and height of a rectangular box, express the rate of change of the diagonal length of the box in terms of the rates of change of l, w and h.

3 The length of a rectangle is increasing at a rate of 3 cm s^{-1} and its width is decreasing at a rate of 3 cm s^{-1}. If the length and width of the rectangle are initially 24 cm and 7 cm respectively, find the initial rate of change of its **a** area; **b** perimeter; **c** diagonal.

4 A 10 m-long industrial ladder is leaning against a wall on a building site. The top of the ladder starts to slip down the wall at a rate of 0.5 m s^{-1}. Determine how fast the foot of the ladder is moving along the ground when it is 6 m from the wall.

5 The volume of a cube is increasing at a rate of 1.5 $m^3 s^{-1}$. Determine the rate at which the surface area of the cube is changing when the cube has a volume of 27 cm^3.

6 A closed, right-circular cylinder's radius is decreasing at a rate of 3 cm s^{-1} and its height is increasing at twice the rate of the radius. Determine how fast the cylinder's surface area is changing when its height is 10 cm and its radius is 12 cm.

7 Side a of a right-angled triangle is increasing at 5 cm s^{-1}, whilst side b of the triangle is decreasing at a rate of 4 cm s^{-1}. Determine how fast the length of the hypotenuse h is changing when $a = 15$ cm and $b = 20$ cm.

8 Air is being pumped into a spherical balloon at the rate of 7 $cm^3 s^{-1}$. Find the rate of change of the balloon's radius, in terms of π, when the volume of air in the balloon is 36π cm^3.

9 Water is poured into a conical tank at the rate of 3 m^3 per minute. The tank stands with the point downward. Determine how fast the water level in the tank is rising when the depth is 2 m and the radius of the water surface is 1.5 m.

10 A spark from a fireplace burns a hole in a newspaper. The initial radius of the hole is 1 cm, and its area is increasing at a rate of 2 $cm^2 s^{-1}$. Find the rate of change of the radius when the radius is 5 cm.

Developing your toolkit

Now do the Modelling and investigation activity on page 302.

Chapter summary

Limits

- The notation used to say that the limit, L, of a function f exists as x approaches the same real value c from the left and from the right is $\left(\lim_{x\to c} = L\right) \Leftrightarrow \left(\lim_{x\to c^+} f(x) = L \text{ and } \lim_{x\to c^-} f(x) = L\right)$, $L \in \mathbb{R}$.
- A function $y = f(x)$ is said to be continuous at $x = a$ if
  1 f is defined at a, ie a is an element of the domain of f
  2 the limit of f at a exists, that is, $\lim_{x\to a} f(x) = L$
  3 the limit of f at a is equal to the value of the function at a, that is, $f(a) = \lim_{x\to a} f(x)$.
- A function that is not continuous at a point $x = a$ is said to be discontinuous at $x = a$.
- A function is said to be continuous on an open interval I if it is continuous at every point in the interval.
  - A function is continuous if it is continuous at every point in its domain.
  - A function is continuous everywhere if it is continuous throughout $(-\infty, \infty)$.
- A function that is not continuous is said to be discontinuous.
- The line $x = c$ is a vertical asymptote of the graph of a function $y = f(x)$ if either $\lim_{x\to c^+} f(x) = \pm\infty$ or $\lim_{x\to c^-} f(x) = \pm\infty$.
- The line $y = k$ is a horizontal asymptote of $f(x)$ if either $\lim_{x\to -\infty} f(x) = k$ or $\lim_{x\to +\infty} f(x) = k$.
- An asymptotic line that is neither vertical nor horizontal but is at an inclination to the x-axis is called a slant or oblique asymptote. A rational function $P(x) = \frac{f(x)}{g(x)}$, $g(x) \neq 0$, where $f(x)$ and $g(x)$ are both polynomial functions, has an oblique asymptote if the degree of f is exactly one more than the degree of g. Dividing $f(x)$ by $g(x)$, $P(x) = L(x) + \frac{r(x)}{g(x)}$, $r(x)$ is the remainder. The oblique asymptote is $y = L(x)$.
- Let L_1, L_2 and k be real numbers, and let $\lim_{x\to\pm\infty} f(x) = L_1$ and $\lim_{x\to\pm\infty} g(x) = L_2$. Then:
  1 $\lim_{x\to\pm\infty}\left(f(x) \pm g(x)\right) = L_1 \pm L_2$
  2 $\lim_{x\to\pm\infty}\left(f(x)g(x)\right) = L_1 L_2$
  3 $\lim_{x\to\pm\infty}\left(\frac{f(x)}{g(x)}\right) = \frac{L_1}{L_2}$, $L_2 \neq 0, g(x) \neq 0$
  4 $\lim_{x\to\pm\infty} kf(x) = k\lim_{x\to\pm\infty} f(x) = kL_1$
  5 $\lim_{x\to\pm\infty}\left[f(x)\right]^{\frac{a}{b}} = \left(L_1\right)^{\frac{a}{b}}$, where $\frac{a}{b} \in \mathbb{Q}$ is in simplest form, provided $\left[L_1\right]^{\frac{a}{b}}$ is real.
- If $\lim_{n\to\infty}\{a_n\} = L$, $L \in \mathbb{R}$, then the sequence is said to converge; otherwise it diverges.
- For an infinite geometric series, when $|r| < 1$, $r \neq 0$, the series converges to its sum, $S_\infty = \frac{u_1}{1-r}$. If the series does not have a finite sum, the series diverges.

Continued on next page

The derivative of a function

- The average rate of change of a function f between two values x_1 and x_2 is given by:
$\frac{\Delta y}{\Delta x} = \frac{f(x_2) - f(x_1)}{x_2 - x_1}$
- The gradient of a curve $y = f(x)$ at any point $(a, f(a))$ on the curve is $\lim_{h \to 0} \frac{f(a+h) - f(a)}{h}$, provided this limit exists.
- The gradient function $f'(x) = \lim_{h \to 0} \frac{f(x+h) - f(x)}{h}$ is called the **derivative** of f with respect to x.
- We sometimes use different notation for the derivative function:
    - $f'(x)$ (this is called "f dash of x" or "f prime of x"); or
    - $\frac{dy}{dx}$ (meaning "the derivative of y with respect to x").
- If f' exists, we say that f has a derivative at x, or is **differentiable** at x.
- A function f is said to be **differentiable everywhere** if the derivative exists for all x in the domain of f.
- Given a function $f(x)$ and a point c in the domain of f, if f is differentiable at $x = c$ then it is also continuous at $x = c$, but the converse is not always true.

Differentiation rules

- If $f(x) = c$ where $c \in \mathbb{R}$, then $f'(x) = 0$.
- If $f(x) = mx + c$ where $m, c \in \mathbb{R}$, then $f'(x) = m$.
- If $f(x) = u(x) \pm v(x)$, then $f'(x) = u'(x) \pm v'(x)$.
- For $c \in \mathbb{R}$, $(cf)'(x) = cf'(x)$, provided $f'(x)$ exists.
- If n is a positive integer and $f(x) = x^n$, then $f'(x) = nx^{n-1}$.
- **Chain rule**: If f is differentiable at $u = g(x)$, and g is differentiable *at* x, then the composite function $(f \circ g)(x)$ is differentiable at x. Furthermore, if $y = f(u)$ and $u = g(x)$, then $\frac{dy}{dx} = \frac{dy}{du} \times \frac{du}{dx}$.
- Another definition for the chain rule is $(f \circ g)'(x) = f'(g(x)) \times g'(x)$
- **Product rule**: If $f(x) = u(x) \times v(x)$ where u and v are differentiable functions, then:
$f'(x) = u(x)v'(x) + v(x)u'(x)$.
- Another way of writing this is, if $y = uv$ where u and v are differentiable functions in x, then $\frac{dy}{dx} = u\frac{dv}{dx} + v\frac{du}{dx}$.
- **Quotient rule**: If $f(x) = \frac{u(x)}{v(x)}, v(x) \neq 0$, then $f'(x) = \frac{v(x)u'(x) - u(x)v'(x)}{[v(x)]^2}$.
- Alternatively, if $y = \frac{u}{v}$, then $\frac{dy}{dx} = \frac{v\frac{du}{dx} - u\frac{dv}{dx}}{v^2}$, where u and v are differentiable functions in x, and $v \neq 0$.

Graphical interpretation of the derivatives

- Maximum and minimum points are referred to as extrema, turning points or stationary points.
- The global maximum (or minimum) of a function is the greatest (or least) value the function attains in its domain.

- First derivative test for continuous and differentiable functions on an open interval $]a, b[$:
  - f has a stationary point at $x = c$ if and only if $f'(c) = 0$.
  - If $f' > 0$ for $x < c$, and $f' < 0$ for $x > c$, then f has a local maximum at $x = c$.
  - If $f' < 0$ for $x < c$ and $f' > 0$ for $x > c$, then f has a local minimum at $x = c$.
- If $f(x)$ is continuous and differentiable on $[a, b]$:
  - If $f'(x) > 0$ for all $x \in]a, b[$, then f' increases on $[a, b]$.
  - If $f'(x) < 0$ for all $x \in]a, b[$, then f' decreases on $[a, b]$.
- Second derivative test for maxima and minima:
  - If $f'(c) = 0$ and $f''(c) < 0$ then $f(x)$ has a local maximum at $x = c$.
  - If $f'(c) = 0$ and $f''(c) > 0$ then $f(x)$ has a local minimum at $x = c$.
  - If $f'(c) = 0$ and $f''(c) = 0$ then the second derivative test is inconclusive. If the second derivative test is inconclusive, revert to the first derivative test for local extrema.
- A point on the graph of f whose concavity changes in going through the point is a **point of inflexion**.
  - If f has a point of inflexion at $x = c$, then $f''(c) = 0$.
  - If $f''(c) = 0$ **and** f changes concavity through $x = c$, then f has a point of inflexion at $x = c$.
  - f is concave down in an open interval if, for all x in the interval, $f''(c) < 0$.
  - f is concave up in an open interval if, for all x in the interval, $f''(c) > 0$.
- If f has a point of inflexion at $x = c$, and $f'(c) = 0$, the point is called a **horizontal point of inflexion** since its gradient is parallel to the x-axis. Since $f'(c) = 0$, $x = c$ is also a stationary point.

Applications of differential calculus

- Velocity, $v(t)$, is the derivative of the displacement function $s(t)$: $v(t) = s'(t)$
- Acceleration, $a(t)$, is the derivative of the velocity: $a(t) = v'(t) = s''(t)$
- When v and a have the same sign, the particle is speeding up (accelerating).
- When v and a have different signs, the particle is slowing down (decelerating).

Implicit differentiation and related rates

- When differentiating a term $u(y)$ with respect to a *different* variable x, the chain rule $\frac{\mathrm{d}(u(y))}{\mathrm{d}x} = u'(y) \times \frac{\mathrm{d}y}{\mathrm{d}x}$ allows you to differentiate implicitly.
- Differentiating a variable y with respect to time, t, $\frac{\mathrm{d}(y)}{\mathrm{d}t} = \frac{\mathrm{d}y}{\mathrm{d}x} \times \frac{\mathrm{d}y}{\mathrm{d}t}$ or alternatively, $\frac{\mathrm{d}(y)}{\mathrm{d}t} = y' \frac{\mathrm{d}y}{\mathrm{d}t}$

Developing inquiry skills

The distance a sprinter has travelled at various times is given by the function $s(t) = -0.102t^3 + 1.98t^2 + 0.70t$ for $0 \le t \le 10$ seconds.

1 What is the fastest instantaneous speed of the sprinter in the race?

2 Draw the graph of $v(t)$ and state whether or not the sprinter is running at a constant speed throughout the race.

3 How could you determine the sprinter's speed at a particular instant $t = c$?

Chapter review

Click here for a mixed review exercise

1 The graph of a function $y = f(x)$ is shown below.

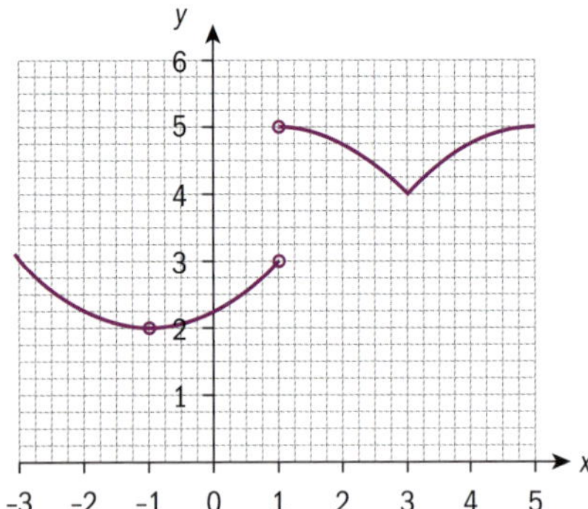

Use the graph to find the following limits, if they exist. If a limit does not exist, explain why.

a $\lim\limits_{x \to 1} f(x)$ **b** $\lim\limits_{x \to -1} f(x)$ **c** $\lim\limits_{x \to 0} f(x)$

d $\lim\limits_{x \to 3} f(x)$

2 Find the following limits, if they exist.

a $\lim\limits_{x \to 2}\left(-x^2 + 5x - 2\right)$ **b** $\lim\limits_{x \to 2} \dfrac{x+3}{x+6}$

c $\lim\limits_{x \to 0}(8 - 2x)^{\frac{1}{3}}$ **d** $\lim\limits_{x \to 3} \dfrac{x^2 - 4x + 3}{x - 3}$

e $\lim\limits_{x \to \infty} \dfrac{\sqrt{x^2 - 2}}{x}$ **f** $\lim\limits_{x \to -\infty} \dfrac{2}{x^3 + 1}$

3 For $f(x) = \begin{cases} x^2 + 2x, & x \le -2 \\ x^3 - 6x, & x > -2 \end{cases}$, determine if $f(x)$ is continuous at -2.

4 Find the value of a for which the given function is continuous.

a $f(x) = \begin{cases} 2x^2 + a, & x \ge -1 \\ -x^3, & x < -1 \end{cases}$

b $f(x) = \begin{cases} 2x^2 + 4x, & x \ge 1 \\ a - x, & x < 1 \end{cases}$

5 Determine whether each sequence converges as $n \to \infty$. If the sequence converges, find the limit.

a $\dfrac{n+3}{3n-4}$ **b** $\dfrac{4-2n^2}{n^3-1}$ **c** $\dfrac{n^2-2n+3}{n-1}$

6 Determine whether the series $\sum\limits_{n=0}^{\infty} 2\left(\dfrac{(-1)^n}{3^n}\right)$ converges and, if it does, find its sum.

7 Find the values of n for which the series $n^2 + \dfrac{n^2}{n^2+1} + \dfrac{n^2}{(n^2+1)^2} + \ldots$ is convergent, and find its sum.

8 Find all asymptotes of the given functions.

a $f(x) = \dfrac{6x^2+7}{x^2-9}$ **b** $f(x) = \dfrac{4x}{3x^3+81}$

c $f(x) = \dfrac{x^3+x^2-4}{x^2}$

9 Differentiate each of these with respect to x.

a $y = (1-2x)^5(3x-2)^6$ **b** $y = \dfrac{x^2-1}{\sqrt{x-1}}$

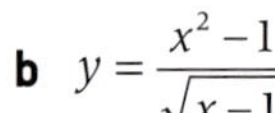

c $y = \sqrt{2x + \sqrt{x^3+1}}$

10 Given the function $y = \dfrac{x}{x^2-1}$:

a Find the vertical and horizontal asymptotes.

b Show that the function is an odd function.

c Show that $\dfrac{dy}{dx} < 0$ for all x.

d Sketch the function.

11 The graph of $y = 3 - x^2$ is shown.

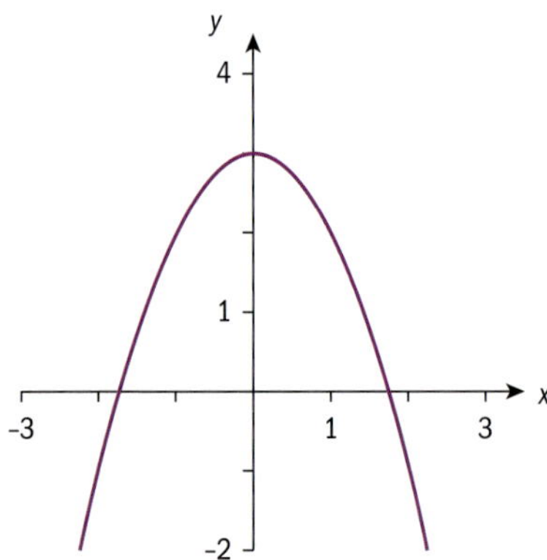

A point P on the graph has x-coordinate a.

a Show that $|OP| = \sqrt{a^4 - 5a^2 + 9}$.

b Find the values of a for which the curve is closest to $(0, 0)$.

12 Given the function $f(x) = \dfrac{x^2 - 3x + 2}{x^2 + 3x + 2}$:

a Find the equations of all asymptotes.

b Find the x- and y-intercepts.

c Find the x-coordinate of the extrema of the function and justify your answer.

13 Given that $f(x) = \dfrac{9(x+1)}{x^2}$; $f'(x) = -\dfrac{9(x+2)}{x^3}$; $f''(x) = \dfrac{18(x+3)}{x^4}$:

a Find the x-intercept(s) of $f(x)$.

b Find the equations of the asymptote(s) of $f(x)$.

c Find the coordinates of the local minimum of $f(x)$, and justify that it is a minimum.

d Find the interval(s) where $f(x)$ is concave up.

e Sketch the graph of $f(x)$, clearly showing and labelling the features you found in **a–d**.

14 $f(x) = x - b\sqrt{x}; x \geq 0, b \in \mathbb{R}^+$. Find, in terms of b:

a the zeros of f

b the intervals where f is:
i increasing **ii** decreasing

c the range of f

d the concavity of f.

15 Copy the following graph and, given that $f(0) = 0$, sketch the graph of $f(x)$. You should clearly indicate any stationary points, and state their nature.

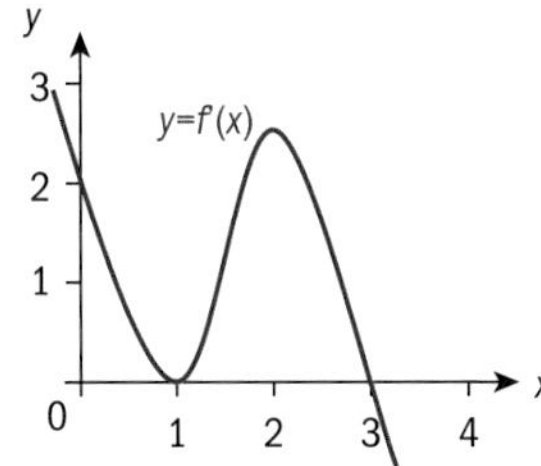

16 Find the gradient of the tangent to the curve $4x^2 + y^2 = 4$ at the point $\left(\dfrac{2}{\sqrt{5}}, \dfrac{2}{\sqrt{5}}\right)$.

17 Find the equation of the normal to the curve $2x^2y + 3xy^2 = 1$ at the point $(1, 1)$. Give your answer in the form $ax + by + c = 0$; $a, b, c \in \mathbb{Z}$.

18 The point A with coordinates (a, b) lies on the curve $\sqrt{x} + \sqrt{y} = \sqrt{p}$, $p > 0$. The tangent to the curve at A intersects the axes at $(0, r)$ and $(s, 0)$. Show that $p = r + s$.

19 A drone is rising vertically at a rate of 3 m s^{-1} when it is 8 m above its launch pad. Determine the rate of change of the distance between the drone and an observer who is standing 6 m from the launch pad and on the same horizontal level.

20 A particle travels in a straight line and its velocity at any time t is $v(t) = 1 + t - \sqrt{4t + 9}$ m s^{-1}. Find:

a the particle's initial velocity

b the time when the particle comes to rest

c the particle's acceleration at this time

d the interval(s) when the particle is: **i** slowing down **ii** speeding up.

21 Two concentric circles are expanding in size. At time t, the radius of the outer circle is 3 cm and is expanding at a rate of 0.8 cm s^{-1}; the radius of the inner circle is 1 cm and is expanding at a rate of 1.2 cm s^{-1}. Find the rate of change of the area of the ring between the two circles at any time t.

22 The volume of a spherical balloon is increasing at a rate of 3 cm^3 s^{-1}. Find the rate at which the surface area of the balloon is increasing when its radius is 1 cm.

23 PQRST is a pentagon whose perimeter is p cm. PQT is an equilateral triangle, and QRST is a rectangle. If $|PQ| = x$ cm, find the ratio of p to x for which the area of the pentagon is maximum.

Exam-style questions

24 P1: Consider the function defined for all $x \in \mathbb{R}$ by

$$f(x) = \begin{cases} -\dfrac{x^2-4}{x^2-5x+6}, & \text{for } x < 2 \\ a^2+3a+6, & \text{for } x = 2, a \in \mathbb{R} \\ e^{2-x}+x+1, & \text{for } x > 2 \end{cases}$$

a Show that $\lim_{x\to 2} f(x)$ exists and state its value. (4 marks)

b Hence, find possible value(s) for a such that f is continuous at $x = 2$. (3 marks)

25 P2: Consider the rational function defined by $f(x) = \dfrac{x^3-8}{x^2+x-2}$.

a Find the largest possible domain of f. (2 marks)

b Sketch the graph of $y = f(x)$. (3 marks)

c Hence

i state the equations of the vertical asymptotes to the graph of f

ii justify that $y = x - 1$ is a slant asymptote to the graph of f. (4 marks)

26 P1: **a** Find an expression for the derivative of $g(x) = \dfrac{\sqrt{x}}{x^2+1}, x \geq 0$ (3 marks)

b Hence find an equation for the normal to the graph of g at $x = 1$. (3 marks)

c Explain why there is no tangent to the graph of g at $x = 0$. (2 marks)

27 P1: Consider the curve with equation $x^2 = \dfrac{x+y}{x-y}$, where $y \neq x$.

a Show that $\dfrac{dy}{dx} = (x-y)^2 + \dfrac{y}{x}$ (5 marks)

b Hence find the equations of the tangent to the curve at the points where $y = 0$. (4 marks)

28 P2: The following table shows the annual profits (in thousands of dollars) for a company during a five-year period.

Year	2000	2001	2002	2003	2004
Profit	98.5	100.9	101.1	101.2	102.3

a Find

i the average rate at which the profits varied between 2000 and 2002

ii the average rate at which the profits varied between 2002 and 2004. (3 marks)

b Compare the values obtained in part **a** and state their meaning in the context of the question. (2 marks)

29 P1: Consider the function defined by $f(x) = 2x^2+3x - 4$.

a By differentiating from first principles, show that $f'(x) = 4x + 3$. (5 marks)

b Hence, find an equation for the tangent to the graph of f at $x = 1$. (3 marks)

30 P2: A projectile is launched vertically from the ground. After t seconds its height is given by $h(t) = 112t - 4.9t^2$ metres.

a Determine $h(4)$ and $h(5)$. (2 marks)

b Find an expression for the velocity of the projectile in terms of t. (2 marks)

c Find the instant in time, t, for which the height of the projectile reaches a maximum. (2 marks)

d Determine how long it takes for the projectile to hit the ground. (2 marks)

e Sketch the graph of $h = h(t)$, stating clearly the domain of validity of the model and its maximum point. (3 marks)

f Find the velocity of the projectile at the instant it hits the ground. (2 marks)

g Show that the acceleration of the projectile is constant, stating its value. (2 marks)

31 P1: Consider the information about functions f and g, given in the following table.

x	$f(x)$	$f'(x)$	$g(x)$	$g'(x)$
0	1	−2	12	−12
1	2	4	6	−3
2	9	10	4	$-\frac{4}{3}$
3	22	16	3	$-\frac{3}{4}$

a Find

i $\left(\frac{f}{g}\right)'(2)$

ii $(g \circ f)'(1)$ (5 marks)

b State whether the following statements are true or false. Justify your answers with a reason.

i f is an increasing function.

ii The tangents to the graph of g at the points $x = 2$ and $x = 3$ are perpendicular lines. (4 marks)

32 P1: A crude oil well W lies 75 km offshore. An oil company is planning to construct a pipeline in two linear sections:

The first section will run underground, and take the oil from the well to a point A on the shore.

The second section will run above ground, and will take oil from point A to a harbour H, where it is loaded into tankers.

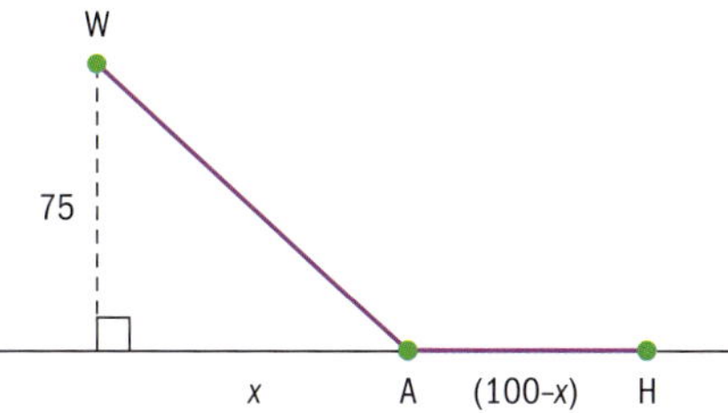

All distances are shown in kilometres.

a Find an expression, in terms of x, for the total length of the pipeline. (2 marks)

It costs the company three times as much to construct a pipeline below ground than it does to construct it above ground.

b Determine the value of x that results in minimum construction costs. (7 marks)

c Hence, show that the total length of the pipeline is $\frac{75}{2}\sqrt{2}+100$. (4 marks)

33 P2: Consider the function defined by $f(x)=\frac{ax-4}{x-a}$, $x \neq a$, $a \geq 0$.

a Write down the equations of the vertical and horizontal asymptotes of the graph of f. (2 marks)

b Show that $f'(x)=\frac{4-a^2}{(x-a)^2}$. (3 marks)

c Show that the graph of f has no turning points. (4 marks)

d Find the equation, in terms of a, of the normal to the graph of f at $x = 1$. (4 marks)

The horizontal and vertical asymptotes of f meet at the point X. The normal to the graph of f at $x = 1$ passes through X for certain values of a.

e Show that these values of a satisfy the equation $4a^3 - 14a^2 + 4a + 15 = 0$. (4 marks)

f Hence, find the values of a for which the normal to the graph of f at $x = 1$ passes through X. (3 marks)

Click here for further exam practice

River crossing

Approaches to learning: Thinking skills: Evaluate, Critiquing, Applying
Exploration criteria: Personal engagement (C), Reflection (D), Use of mathematics (E)
IB topic: Differentiation, Optimization

Modelling and investigation activity

The problem

You are standing at the edge of a slow-moving river which is one kilometre wide. You want to return to your campground on the opposite side of the river. You can swim at 3 km h^{-1} and run at 8 km h^{-1}. You must first swim across the river to any point on the opposite bank. From there you must run to the campground, which is 2 km from the point directly across the river from where you start your swim.

What route will take the least amount of time?

Visualize the problem

Here is a diagram of this situation:

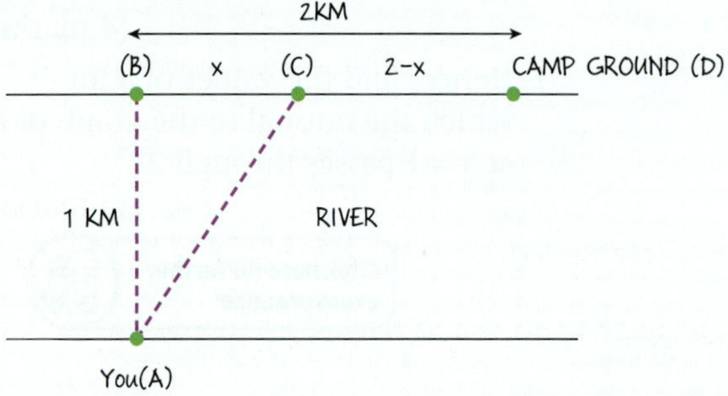

Discuss what each label in the diagram represents.

Solve the problem

What is the length of AC in terms of x?

Using the formula for time taken for travel at a constant rate of speed from this chapter, write down an expression in terms of x for:

1 the time taken to swim from A to C

2 the time taken to run from C to D.

Hence write down an expression for the total time taken, T, to travel from A to D in terms of x.

You want to minimize this expression (find the minimum time taken).

Find $\frac{dT}{dx}$.

Now solve $\frac{dT}{dx} = 0$ to determine the value of x that minimizes the time taken.

How do you know this is a valid value?

Use the second derivative test to show that the value you found is a minimum value.

For this value of x, find the minimum time possible and describe the route.

Assumptions made in the problem

The problem is perhaps more accurately stated as:

You are standing at the edge of a river. You want to return to your campground which you can see further down the river on the other side. You must first swim across the river to any point on the opposite bank. From there you must run to the campground.

What route will take the least amount of time?

Look back at the original problem.

- What additional assumptions have been made in the original question?
- What information in the question are you unlikely to know when you are standing at the edge of the river?
- What additional information would you need to know to determine the shortest time possible?

The original problem is a simplified version of a real-life situation. Criticize the original problem, and the information given, as much as possible.

Extension

In an exploration it is important to reflect critically on any assumptions made and the subsequent significance and limitations of the results.

Consider the box problem on page 280, the rectangular region problem on page 279, the cylindrical can problem on page 278 and/or some of the questions in Exercise 4R and/or Exercise 4S.

If you were writing an exploration using these problems as the basis or inspiration of that exploration, then:

What assumptions have been made in the question?

What information in the question are you unlikely to know in real-life?

How could you find this missing information?

What additional information would you need to know?

Criticize the questions as much as possible!

Paper 3 question and comments

Paper 3 in the IBDP Mathematics Higher Level course consists of two extended, closed questions, each of which should take approximately 30 minutes to complete.

Usually the questions will provide an opportunity for you to apply the mathematics you have learned to solve a "real-life problem".

Often you will be using mathematics in situations you have not encountered before, and certainly some of the later parts of the questions will have a "problem solving" aspect where you have to choose the best method – out of possibly several different approaches – available to you.

But because paper 3 questions are "closed" problems, there will always be a solution for you to find.

The question will be structured so that the easier parts will normally be at the beginning of the question and, as the question progresses, less and less direction will be provided by the questions.

Paper 3 questions do require a particular set of skills. It's important to make use of the five-minute reading time so you can best assess all that the question is asking of you: You need to be familiar with the overall shape of the question. You need to look at which parts you are confident you can do, and which will require more thought.

In these questions, not being able to do an earlier part should not prevent you answering some of the later parts. Always read through the question carefully and check which parts you can do even if you have not completed all the previous parts. A question part which asks you to "show that..." is often an indicator that the result will be needed in a later part of the question.

The question that follows is typical of the style of a paper 3 question. The notes given are general exam hints, and do not contain any instructions with regard to how to do this particular question.

Give yourself 30 minutes and see how much you can do in that time. If you are not used to extended questions you might struggle to complete it, but with practice you will quickly be able to maximize the marks you can attain.

1 The aim of this question is to discover and prove properties of Pascal's triangle and the binomial coefficients ${}^nC_r$.

HINT

Most paper 3 questions will begin with a statement of the task.

a Write down the first 4 rows of Pascal's triangle, with the first row for $n = 1$ being 1. (2 marks)

b Simplify **i** ${}^nC_0$ **ii** ${}^nC_1$. (2 marks)

c Write down **i** ${}^6C_2$ **ii** ${}^6C_4$. (2 marks)

HINT

Easier marks are awarded to get you started as you familiarize yourself with Pascal's triangle.

HINT

"Write down" means exactly that; no working is required. You can use your GDC. Here you look at a numerical example of the general case, which is introduced in part **d**.

d **i** Using the formula ${}^nC_r = \dfrac{n!}{(n-r)!r!}$, show that ${}^nC_{n-r} = {}^nC_r$.

ii State what property of Pascal's triangle your answer to **i** explains. (3 marks)

HINT
Here, the question begins to test your insight and understanding of the mathematics you're using.

e Find **i** ${}^5C_2 + {}^5C_3$ **ii** ${}^6C_3$. (2 marks)

HINT
As in part **c**, here you are generating a numerical example of the general case which follows in the next part.

f Using the formula ${}^nC_r = \dfrac{n!}{(n-r)!r!}$ and putting fractions over a common denominator, show that ${}^nC_{r-1} + {}^nC_r = {}^{n+1}C_r$. (3 marks)

g **i** Calculate $\sum_{r=0}^{3} {}^3C_r$. **ii** Calculate $\sum_{r=0}^{4} {}^4C_r$.

iii Suggest a formula for $\sum_{r=0}^{n} {}^nC_r$. (4 marks)

h **i** Write down the binomial expansion for $(a + b)^n$.

ii Substitute $a = b = 1$ into your expansion in **i**, simplify the result, and deduce what result you have just proved using this substitution.

HINT
Even if you cannot deduce the final result, you will be awarded marks for following the instructions you are given.

iii Substitute $a = 1$, $b = -1$ into your expansion in **i**, simplify the result, and deduce what result you have just proved using this substitution. (8 marks)

j By considering the coefficient of x^n in the expansion of $(1 + x)^{2n}$, show that ${}^{2n}C_n = \left({}^nC_0\right)^2 + \left({}^nC_1\right)^2 + \left({}^nC_2\right)^2 + \ldots + \left({}^nC_n\right)^2$.

(You might find it helpful to use $(1 + x)^{2n} = (1 + x)^n\,(1 + x)^n$, $x^n = x^r\,x^{n-r}$, and your answer to part **d**.) (4 marks)

Total 30 marks

HINT
"Show that" is similar to "prove". You have to use the definition you are given and logically deduce the required result. The answer cannot be assumed.

HINT
"Show that..." in a paper 3 question is often an indication that you will need to use this result in a later question part. In this case, you will use this result in part **j**.

HINT
Take note of the hint, which should help you with the algebraic manipulation.

HINT
Here, the command word "suggest" means you will have to look back at your answers to previous question part(s) and identify a pattern.

HINT
The insight and algebra that this question part requires of you is quite difficult, so hints are sometimes given.

5 Analysing data and quantifying randomness: statistics and probability

Univariate data analysis involves handling data which has only one variable ("uni" means "one"). For example, the heights of all students in your class is univariate data. The main purpose of univariate analysis is to describe the data and find patterns that lead to generalizations regarding certain populations. You can draw charts, find averages and analyse this data in many other ways.

Univariate analysis *doesn't* deal with relationships between variables, such as the relationship between height and weight for students in your class. This is called bivariate analysis, which you will study later in this chapter.

Concepts

- Quantity
- Validity

Microconcepts

- Population, sample, discrete and continuous data
- Sampling: convenience, simple random, systematic, quota, stratified
- Frequency distributions (tables)
- Grouped data
- Histogram
- Central tendency: mean, mode, median
- Spread (or dispersion): cumulative frequency, cumulative frequency graphs
- Median, quartiles, percentiles
- Range, interquartile range, outliers
- Box-and-whisker diagrams
- Skew
- Standard deviation and variance

How could you choose a sample of people from your school to survey which would reliably tell you about the eating habits of students in the whole school?

How can you use different charts to tell you different information about a data set?

What does it mean to be average?

The study of statistics allows mathematicians to **collect** a set of data, like this one, and **organize** it, **analyse** it, **represent** it and **interpret** it.

A group of 32 students took a test with maximum mark 10. Their scores are listed below.

2, 7, 5, 8, 3, 7, 0, 5, 7, 4, 8, 6, 1, 7, 5, 1, 9, 4, 7, 5, 2, 6, 10, 6, 2, 4, 7, 5, 8, 3, 6, 8

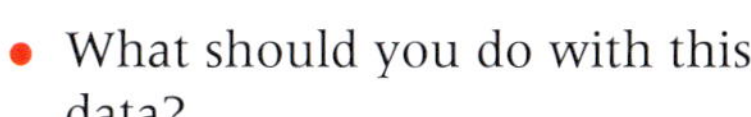

- What should you do with this data?
- How can you organize the data to give a better picture of the scores?
- How should you display the scores?

Can you draw any conclusions from the data?

Developing inquiry skills

How would you collect, organize, analyse, represent and interpret the data if you had two data sets?

Think about the questions in this opening problem and answer any you can. As you work through the chapter, you will gain mathematical knowledge and skills that will help you to answer them all.

Before you start

Click here for help with this skills check

You should know how to:

1 Find the mean, median and mode for sets of data, both non-grouped and grouped.
eg Find the mean, median and mode for the non-grouped data set:

5, 6, 7, 8, 10, 11, 13, 14, 14, 18, 19

mean = 11.4 (3 s.f.)   median = 11
mode = 14

Find the mean, median and mode for the grouped data set:

0–4	1
5–9	3
10–14	5
15–19	7

mean = 12.7 (3 s.f.)
The median lies in the 10–14 interval.
The mode lies in the 15–19 interval.

2 Find the range, quartiles and interquartile range for non-grouped data sets.
eg Find the range, quartiles and interquartile range for the non-grouped data set above.

range = max − min = 14

lower quartile (Q_1) = 7

upper quartile (Q_3) = 14

interquartile range = $Q_3 - Q_1$ = 7

Skills check

1 The masses in kg of 10 cows in a milking parlour are listed below.
720, 750, 690, 975, 700, 710, 720, 680, 695, 645

a Find the mean, mode and median.

b Find the range, quartiles and interquartile range.

2 The amount of milk produced in litres from all four milkings in a time period is given in the frequency table.

$0 < x \le 5$	5
$5 < x \le 10$	2
$10 < x \le 15$	6
$15 < x \le 20$	8
$20 < x \le 25$	4
$25 < x \le 30$	5
$30 < x \le 35$	8

Find the mean, mode and median.

5.1 Sampling

Investigation 1

A group of IB students wish to investigate the mean average time that each student in their school spends on homework per night. They would like to ask each of the 1500 students in their school, but realize that they do not have enough time to collect and analyse this amount of data, so they decide to ask a subset (sample) of the student population in the hope that the sample will give a good approximation for students in the whole school.

To decide how to choose the sample:

- Beth suggests they should just interview their friends
- Emily suggests they should interview two people from each year group
- Natasha suggests they should pick 10 boys and 10 girls
- Amanda suggests they should assign a number to each student in the school, and use a random number generator to choose a sample
- Greg suggests they obtain a list of all students in the school, organized alphabetically by surname, and choose every 20th person in the list.

1 In a group, discuss the advantages and disadvantages of each of the five methods. For each suggestion, you should consider the following.
   - How easy would it be to obtain a sample?
   - What sample size would it give? Is this sample size big enough?
   - Would it give results which are representative of all students in the school?

Record your results in a table like this.

Sampling technique	Advantages	Disadvantages
Beth's suggestion: interview their friends		
Emily's suggestion: interview two people from each year group		

2 How could you combine the suggestions of two or more of the students in order to get results which are better representative of all students in the school?

3 Can you think of any other techniques to obtain a sample in this scenario?

4 **Conceptual** Why must you consider the context of a scenario in order to choose an appropriate sampling technique?

International-mindedness

The Kinsey Reports – famous sampling issues.

The Kinsey Reports were translated into thirteen languages, sold over three quarter of a million copies and have been called some of the most influential scientific books of the 20th century.

The data you collect can be **qualitative** or **quantitative**.

Qualitative data (sometimes called categorical data) is described in words. In answering such questions as, "How are you feeling today?" the responses may include "happy", "sad", "down", "excited".

Quantitative data are always numbers. Quantitative data describes information that can be counted. For example, "How many people live in your house?" (eg 2, 3, 4) or "How long does it take you to get home after school?" (eg 6 minutes, 20 minutes, 45 minutes).

Quantitative data can be **discrete** or **continuous.**

Continuous data describes quantities that are measured, such as height, weight or age.

Discrete data describes quantities that are counted, such as number of t-shirts owned, number of DP subjects studied or number of throws needed to hit a bullseye in darts.

Even though these variables will be presented correct to a certain degree of accuracy, they could take any real value within their range. For example, a length can be measured to different degrees of accuracy: 5 cm could actually be 5.1 cm, or 5.08 cm, or 5.08236 cm.

Prices are usually thought of as continuous, even though in reality most currencies are not subdivided below 0.01 units.

Reflect Is the taxiing speed of an airplane discrete or continuous? Is the number of airplanes waiting to take off discrete or continuous?

Could data ever be classified as both discrete *and* continuous? Why is it important to consider the nature of the variable, rather than just the data values themselves, when classifying whether data are discrete or continuous?

What is the difference between a population and a sample?

Suppose you want to know the number of people in your city who drive a car to work and take one or more passengers with them. Here, the **population** is everyone who lives in the city and drives a car to work.

In this scenario it is very impractical to ask every single person from the population about the number of passengers they take, and so you will probably ask a small selection of drivers. This selection of drivers is called a **sample**; it is a subset of the population.

The **population** consists of every member in the group that you want to find out about.

A **sample** is a subset of the population that will give you information about the population as a whole.

HINT

The aim of this chapter is to introduce basic statistical concepts. It is expected that most of the calculations required will be done using technology, but explanations of calculations by hand may enhance understanding and full working is beneficial in your internal assessment.

It is not just about you getting the numbers, but it is about you understanding what those numbers mean. The emphasis is on understanding and interpreting the results obtained, in context.

International-mindedness

Ronald Fisher (1890-1962) lived in the UK and Australia and has been described as "a genius who almost single-handedly created the foundations for modern statistical science". He used statistics to analyse problems in medicine, agriculture and social sciences.

Choosing a sample which gives a good representation of the population

There are a number of ways in which you can draw a sample from the population. You should always try to choose a method which results in the sample giving the best approximation for the population as a whole.

In Investigation 1, you learned about choosing a sample that would give a good representation of the population as a whole. Statisticians have defined the following key words to clarify how a sample is taken:

- The **target population** is the population from which you want to take a sample.
- The **sampling frame** is a list of the items or people (within the target population) from which you can take your sample.
- A **sampling unit** is a single member (such as an item or person) from the sampling frame that is chosen to be sampled.
- The **sampling variable** is the variable under investigation. This is the characteristic that you want to measure from each sampling unit.
- The **sampling values** are the possible values which the sampling variable can take.

Reflect Define each of these in terms of the context presented in Investigation 1.

Example 1

You must carry out a survey to estimate the number of children in the families that live in an apartment block.

Given this context, define the:

a target population
b sampling unit
c sampling frame
d sampling variable
e sampling values.

a The target population is "all the apartments in the block". **b** The sampling unit is "each apartment". **c** The sampling frame is "an itemized list of individual apartments" **d** The sample variable is "the number of children who live in the apartment". **e** The sampling values are "0, 1, 2, 3, 4, 5, …"	Use the definitions to carefully describe the context.

Exercise 5A

For each of questions **1–4**, define:

- **a** the target population
- **b** the sampling unit
- **c** the sampling frame
- **d** the sampling variable
- **e** the sampling value.

1 An investigation into the lengths of celery sticks grown in a certain US state.

2 The weight of ball bearings manufactured by a company.

3 The volume, to the nearest ml, that a soft drink factory fills its 1 litre bottles of soda drink.

4 The weight of crates of oranges, where each crate contains 50 oranges.

Sampling techniques

As you saw in Investigation 1, there are a number of different sampling methods. Here you will study a few of the most well known methods.

Simple random sampling (SRS)

- Each member of the population has an equal chance of being selected. A sample is chosen by drawing names from a hat, or assigning numbers to the population and using a random number generator.
- **Example**: To conduct a simple random sample to find the mean length of time spent doing homework, you might put the names of every student in the school into a hat and draw out the names of 100 students to form a sample.

HINT

Today, the word random is used to mean haphazard or unpredicted. In statistics it means that the probability of each member being selected is equal.

The sum of these probabilities equals 1.

Systematic sampling

- Here, you list the members of the population and select a sample according to a random starting point and a fixed interval.
- **Example**: If you wanted to create a systematic sample of 100 students at a school with an enrolled population of 1000, you would choose every tenth person from the list of all students.

Stratified sampling

- This involves dividing the population into non-overlapping smaller groups known as *strata*. The strata are formed based on members' shared characteristics. You then choose a random sample from each strata, and put them together to form your sample.
- **Example**: In a high school of 1000 students, you could choose 25 students from each of the four year groups to form a sample of 100 students.

Quota sampling

- This is like stratified sampling, but involves taking a sample size from each strata in proportion to the size of the population.
- **Example:** In a high school of 1000 students where 60% are female and 40% are male, your sample should also have 60% female and 40% male.

Convenience sampling

- This is the easiest method by which you can generate a sample. You select those members of the population who are most easily accessible or readily available.
- **Example**: To conduct a convenience sample to find the mean length of time spent doing homework, you might survey those students who are in the same class as you.

HINT

Most calculators have a random generator.

Search in the menu on your GDC.

Investigation 2

A sample of five distinct digits can be formed from the digits 0–9, eg {1, 2, 3, 4, 5}, {2, 4, 6, 3, 5}, {0, 3, 7, 8, 9}, etc. A digit is not repeated in the same sample.

1 Construct a few samples yourself using a random number generator or other method.

2 Ben uses a simple random sample to choose five digits. Is each digit equally likely to be chosen? Explain why.

3 Define the population in Ben's sample, and calculate the mean of the population.

4 Calculate the mean of each of the three samples given at the start of the investigation.

5 Randomly generate your own samples and calculate the mean of these samples.

6 When the mean of a sample is not an integer, you may have noticed that the first decimal digit (the "tenths" digit) is always an even number. Explain why this is the case.

7 In Chapter 1, while studying the binomial theorem, you learned about the number of ways to choose r objects out of a total of n objects. Use this to find the number of different five-digit samples you can take from the digits 0–9.

8 A frequency table showing the values of the means of all possible samples is given below. Complete this table with the relative frequencies: that is, the frequency of obtaining a sample mean in that interval divided by the total number of sample means.

Interval	1.8–2.4	2.6–3.2	3.4	4.0	4.2–4.8	5.0–6.4	6.6–7.2	Total
Frequency	4	24	59	78	59	24	4	
Relative frequency								

9 What does the relative frequency tell you about the probability of obtaining a sample mean in that interval?

10 Calculate the mean *of the sample means*. How does this relate to the population mean?

11 **Conceptual** By considering your answer to question **10**, why can the mean of a simple random sample provide an estimate for the mean of a population?

Provided the sample is a random sample, the mean of the sample means is approximately equal to the population mean. Therefore, the sample mean can be used as an estimator for the population mean.

TOK

The nature of knowing: is there a difference between information and data?

Example 2

A researcher wishes to understand the target market for a new variant of Bluetooth headphones. It has been established that the target market falls into the following age ranges.

Age	
16–25	58%
26–35	25%
36–45	12%
46–55	5%

a Describe how you could use quota sampling to produce informed results on a sample of 200 people.

b Discuss the advantages and disadvantages of this technique.

a Set up the specific quota for the 200 people

Age	
16–25	116
26–35	50
36–45	24
46–55	10

Complete the survey.

The researcher is expected to evaluate the proportion in which the subgroups exist in the population. This proportion has to be maintained in the sample selected using this sampling method.

For example, if 58% of the people who are interested in purchasing the Bluetooth headphones are between the age group of 16–25 years, the subgroups also should have the same percentages of people belonging to the respective age group.

b Advantages:

This sampling process is quick, simple and convenient

Disadvantages:

In practice, it may be difficult to find the appropriate amount of people and asking their ages may be a sensitive subject.

Effective representation of a population may be achieved using quota sampling.

This sampling technique helps researchers in studying a population using specific quotas.

Exercise 5B

1 Describe how you could choose a systematic sample of 40 books from a library containing 2000 books. Discuss the advantages and disadvantages of using this technique.

2 A non-governmental organisation (NGO) would like to take a sample from its various worldwide bases to give some information about its results globally. Below is a list of the number of bases the NGO has in each continent.

Europe	Africa	Antarctica	Asia	Oceania	North America	South America
57	35	2	35	57	85	35

a Describe how you could use this information to conduct a stratified sample.

b Discuss the advantages and disadvantages of using this technique.

3 A population consists of five digits 0, 1, 2, 3, 4.

A sample consists of three non-repeating digits.

a Show that there are 10 possible samples of three digits.

b List all the possible samples.

c Find the mean of each sample.

d Show that the mean of the sample means is equal to the population mean.

4 A bank prepares statements for its clients at the end of each month. A machine seals the envelopes. A batch of 50 envelopes is inspected to see whether the machine is sealing properly.

Identify the variable, sample and population. State whether the variable is discrete or continuous.

5 A company produces 1000 batteries. The manufacturer claims they have a life of 4000 years.

Explain why testing the population would not be possible. Suggest a sampling technique which may be beneficial to help test these batteries.

6 The government would like to gather information on a new standardized test for all 16–19 year olds in the USA. A random sample of 15 schools in each state is generated and each student aged 16–19 in these schools takes the test.

Describe which type of sampling this is.

7 A high school has the following number of students in each year group.

Grade	Number of students
9	80
10	100
11	150
12	170

Explain how you could use stratified sampling to obtain a sample of 50 students.

Dealing with bias in sampling

Bias in a sample can be defined as **concluding some incorrect statistical information.**

This tends to give a skewed interpretation of the general population.

Investigation 3

In this investigation you will compare the estimates of the mean length of a population of pencils using different methods of sampling.

Equipment required: A selection of 50 coloured pencils (preferably of five different colours) of varying lengths.

1 Choose four different samples of five pencils by using the following sampling techniques:

A Simple random sample

B Systematic sample

C Stratified sample

D Quota sample

2 Find the mean length in each of the four samples. Compile frequency tables for the mean obtained by each method.

3 Calculate the mean for each sample.

4 Compare each sample mean with the mean of the population lengths.

HINT

- Simple random: allocate each pencil a number and use a random number generator.
- Systematic: line the pencils up in height order and take each 10th pencil.
- Stratified: randomly choose one pencil of each colour.
- Quota: split the population into colours, and choose a sample from each colour which is proportional to the number of pencils of that colour.

5 **Factual** Which sampling technique produced the closest estimate to the population mean?

6 **Conceptual** Which sampling technique would minimize the bias and best represent the population?

7 **Conceptual** How do you reduce bias?

The purpose of sampling is to gather information which represents the general population by using an efficient and effective method.

It is essential that each sample is representative of the population and bias has been minimized or eliminated.

Sources of bias can include:

i Some members of the population being excluded from the sampling frame

Example: A survey is going to be conducted by phoning members of the population. A possible sample may be a systematic sample of each kth person down the list. However, this excludes people who either do not have a fixed phone line, or have not registered their number in the directory.

ii Non-response

Example: A university sent out a survey to people by mail, and the respondents were asked to fill in the questionnaire and send it back to the university. It may be that only a certain demographic of people *do* respond (such as those over 50, or with first language English), which means that the results of the survey could be biased.

iii Bad design

It is important that the questionnaire is clear, unambiguous and not leading, otherwise it may lead to respondents giving misleading information.

iv Bias by the respondent

This is the hardest bias to eliminate as often people do not like to tell the truth if it is negative. Personal questions about health, weight or income are the most common topics for respondents to give untruthful answers.

Reliability of data

Data is **reliable** if you can repeat the data collection process and obtain similar results. For example, could you repeat a survey and obtain similar findings?

Data is **sufficient** if there is enough data available to support your conclusions.

When collecting data, you might ask "How many data items should I have?" There is no fixed value, but you need to ensure that you collect enough data so that your results are repeatable, and are **representative** – that is, they represent the whole population well.

Reflect How many students would you need to survey to find a good estimate of the average time students at your school spend doing homework?

Two factors which can cause unreliable data are:

1 **Missing data**. This is a common problem in nearly all types of research. Missing data might be caused by

  a a lack of response to questionnaires or surveys

  b it not being possible to record the data (for example, if you were surveying the number of cars on a road at different times of day, it might not be possible to record data at night as you would probably be asleep).

Missing data reduce the validity of a sample and can therefore distort inferences about the population. Missing values are automatically excluded from analysis. A few data values missing, like 10 in a sample of 1000, are not a problem, but if they make up 20% or more of the sample it is a serious issue.

In a survey:

- One way to minimize missing data in surveys is to avoid asking questions where "not available" or "don't know" answers are available for selection.
- You might be missing data because certain questions only applied to *some* of the people interviewed. For instance, if a question was only applicable for students who took a certain test, then data might appear to be missing for other students in the school.

2 **Errors in handling data**. Data might be entered incorrectly, or columns within a table might get muddled up; all of which affects the final results that you obtain from the data.

To avoid these issues, data collection should be closely monitored and checked to minimize errors.

Developing inquiry skills

In the opening problem for the chapter, you were given the test scores, out of 10, of 32 students.

- Are the test scores an example of discrete or continuous data?
- Before marking every student's test paper, the teacher wishes to choose a sample of eight that will give her an estimate of the mean average mark for the class. Describe a suitable sampling method the teacher could use.

5.2 Descriptive statistics

Measures of central tendency

The measures of central tendency describe what is occurring in the middle of the data set.

The three most common measures of central tendency are the mode, the mean and the median.

Arithmetic mean

The arithmetic mean is the quotient of the sum of all the data values and the number of data values in the population.

To calculate the arithmetic mean from grouped data, the mid-value of each class is taken.

In a sample of data the mean is written as $\overline{x}$:

$\overline{x} = \frac{\sum_{i=1}^{k} f_i x_i}{n}$, where $\sum_{i=1}^{k} f_i x_i$ is the sum of the set of data and n is the number of values in the data set.

If we are dealing with the whole population of data, we write the mean as μ, "mu":

$$\mu = \frac{\sum_{i=1}^{k} f_i x_i}{n}$$

Mode

The mode is the value (or values) which occurs most frequently in a set of data.

If there is one value (or interval in grouped data) that occurs most frequently, then that value is the mode and the frequency is the modal value.

In a frequency table this is the value or interval taking the highest frequency.

Median

The median is the value in the middle position when the data is arranged in order of size.

If the number of data values n is odd, the median is the $\left(\frac{n+1}{2}\right)$ th term.

If the number of data values is even, the median is the mean average of the $\left(\frac{n}{2}\right)$th and $\left(\frac{n}{2}+1\right)$ th terms.

International-mindedness

"Lies, damned lies and statistics" is a popularized quote often attributed to Benjamin Disraeli (1804–81) and possibly developed from a court expression stating that there are "liars, damned liars and expert witnesses."

Over the years statistics have been used to promote beliefs and opinions by carefully choosing the analysis and representation of the data.

International-mindedness

The 19th century German psychologist Gustav Fechner popularized the median although French mathematician Pierre Laplace had used it earlier.

Measures of dispersion

The measures of dispersion describe what is happening to the spread of the data set.

The three most common measures of dispersion are the range, inter-quartile range and the standard deviation.

Range

The range is the difference between the maximum and minimum values of the data set.

$$\text{Range} = x_{\max} - x_{\min}$$

The range is a simple measure of how spread out the data is (the **dispersion**).

In the opening problem, you studied the test results of 32 students:

5, 8, 6, 3, 7, 4, 6, 2, 5, 7, 0,
6, 8, 1, 4, 7, 2, 7, 5, 8, 1, 4,
6, 9, 7, 3, 8, 5, 10, 2, 5, 7

Here, the minimum test score is 0 and the maximum score is 10.

Therefore, the range is 10 – 0 = 10.

In a similar way to the mean, the range can be affected by one very large or small value.

Suppose the test scores of ten students from one class were:

2, 10, 3, 1, 4, 2, 0, 2, 3, 1

Here, the range is still 10 – 0 = 10, but the data is much more compact than the range suggests. The value 10 influences the range so that it is not representative of the spread of the whole data.

Knowing whether the data is widely spread out or bunched around the mean provides you with a better understanding of the distribution of the whole data. You need to consider more values than just the smallest and the largest.

TOK

Do different measures of central tendency express different properties of the data?

How reliable are mathematical measures?

International-mindedness

What are the benefits of sharing and analysing data from different countries?

Quartiles

Quartiles are values that divide your data into quarters.

- The first quartile (called the **lower quartile** or Q_1) has 25% of the data below it.
- The second quartile is the median; the middle value in the data set. 50% of the data lies below the median, and 50% lies above it.
- The third quartile (the **upper quartile** or Q_3) has 75% of the data below it.

Q_1 is the median of the lower 50% of the data, Q_3 is the median of the upper 50% of the data, and n is the number of data points.

$$Q_1 = \left(\frac{n+1}{4}\right)\text{th} \quad \text{and} \quad Q_3 = 3\left(\frac{n+1}{4}\right)\text{th}$$

The interquartile range is the difference between Q_3 and Q_1.

Reflect The interquartile range is sometimes called the range of the "middle half" of the data. Explain why this is the case.

The important advantage of the interquartile range is that it can be used to give a more accurate measure of the data spread if extreme values distort the range.

The interquartile range is the measure of spread associated with the median.

$$\text{IQR} = Q_3 - Q_1$$

Histograms

When you have a lot of data, it is often convenient to organize it in a **grouped frequency table**. If the data is continuous, you can draw a **histogram**.

A histogram looks similar to a bar chart, but a bar chart is only suitable for discrete data. When you have continuous data, you should use a histogram.

The main differences between a histogram and a bar chart are:

- A histogram has no gaps between the bars because it shows continuous data.
- The horizontal axis on a histogram has a continuous scale, but the horizontal axis on a bar chart has discrete categories.
- For a bar chart, the width is always the same and is not relevant to the x-scale. For a histogram, the width of each rectangle should span the width of the data class it represents.

Steps for drawing an equal-class-size histogram by hand:

- Write down the class intervals on the horizontal axis.
- Place the *frequency* on the vertical axis.
- Carefully assess the scale so that the histogram fits on the page.
- Draw the top line of each class interval at the corresponding frequency.
- Draw the rectangles with a ruler to represent the data. Ensure there are no spaces.

HINT

In order to generate a good histogram, it is usual to have between 5 and 15 classes. Find the minimum and maximum numbers first and decide class widths accordingly.

The *class width* is the difference between the maximum and the minimum possible values in each class interval.

For example, the class $18 < x \leq 20$ has a class width of 2, because $20 - 18 = 2$.

Example 3

The weights of 50 baby chicks hatched in one week were recorded in the table below.

a Organize the data in a table by grouping the data into classes of equal width.

b Draw a frequency histogram to represent the data.

22.4	20.0	23.3	25.2	24.1	23.6	19.7	21.3	24.0	24.7
23.6	22.3	21.4	22.4	23.9	25.0	22.2	23.5	21.9	23.0
24.3	21.6	19.9	21.2	23.5	24.6	23.5	19.9	20.2	20.3
23.0	24.6	23.5	22.5	23.0	24.5	24.9	21.3	20.5	20.4
21.4	22.4	23.6	24.5	23.4	22.9	23.8	24.6	21.0	21.3

a

Class interval	Frequency
$19.5 < x \leq 20$	3
$20 < x \leq 20.5$	5
$20.5 < x \leq 21$	1
$21 < x \leq 21.5$	6
$21.5 < x \leq 22$	2
$22 < x \leq 22.5$	6
$22.5 < x \leq 23$	4
$23 < x \leq 23.5$	6
$23.5 < x \leq 24$	5
$24 < x \leq 24.5$	5
$24.5 < x \leq 25$	6
$25 < x \leq 25.5$	1

Start by finding the minimum, maximum and range:

- minimum = 19.7
- maximum = 25.2
- range = 5.5

Since the range is 5.5, and we want between 5 and 15 categories, a class width of 0.5 seems suitable, since $\frac{5.5}{0.5} = 11$.

It will be easier and neater if we start at 19.5 rather than starting at 19.7 as the minimum value.

b

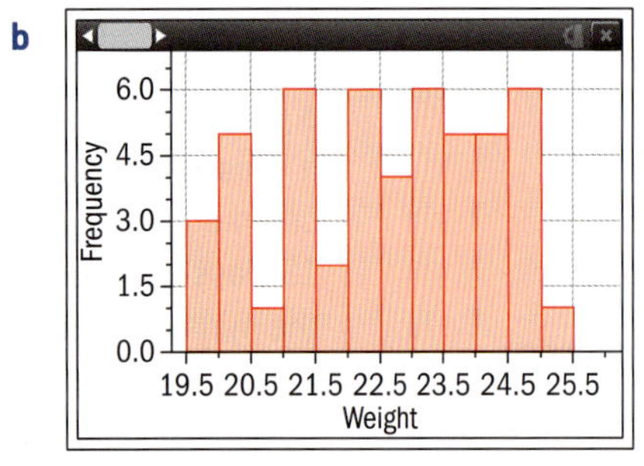

Use your frequency table to draw the histogram.

HINT

The weights of chicks are continuous data, but for measuring purposes the data have been rounded to one decimal place so that the continuous data takes discrete values. The data is still continuous and therefore we use a histogram.

Example 4

Mr Neave examined the eggs that he was trying to hatch into chicks. He had 25 trays of 24 eggs which he placed in the incubator. The number of hatched chicks per tray is shown in the table below.

12	18	17	13	9
21	2	5	12	1
5	14	11	13	15
18	21	14	15	9
12	14	15	16	24

He decided to group the numbers of eggs into the following class intervals to make a frequency table.

$0.5 \le x < 3.5$ $3.5 \le x < 6.5$ $6.5 \le x < 9.5$ $9.5 \le x < 12.5$

$12.5 \le x < 15.5$ $15.5 \le x < 18.5$ $18.5 \le x < 21.5$ $21.5 \le x < 24.5$

a Complete the frequency table for the data.

b Classify the data as discrete or continuous.

c Classify the data as grouped or not grouped.

d Determine whether you should use a bar chart or histogram and draw an appropriate representation.

a

Number of chicks hatched from 24 eggs	Frequency
$0.5 \le x < 3.5$	2
$3.5 \le x < 6.5$	2
$6.5 \le x < 9.5$	2
$9.5 \le x < 12.5$	4
$12.5 \le x < 15.5$	8
$15.5 \le x < 18.5$	4
$18.5 \le x < 21.5$	2
$21.5 \le x < 24.5$	1

You can complete the frequency table for the data by counting the appropriate entries in the data table.

b The number of chicks hatched is a discrete variable.

The data is discrete because we are counting, not measuring.

Continued on next page

c This is grouped data.	The data is grouped because a number of different possible values have been joined to make a single category.
d A histogram is the appropriate representation because the data is grouped and quantitative.	

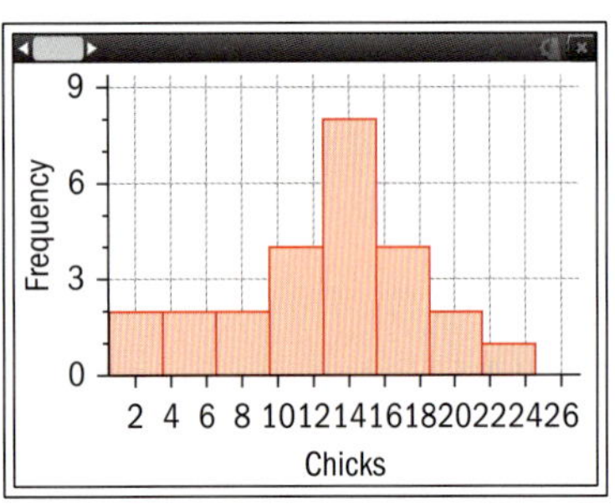

Frequency histograms for discrete data

Number of siblings in a class of 100 students	
Number of siblings	**Frequency**
0	48
1	32
2	17
3	3

When working with discrete data such as that given in the table above, it is quite common to use a vertical line graph to represent the data, as illustrated on the right. However, this becomes problematic if the data is grouped. If you wish to represent the category "1–3 eggs", for example, drawing a single vertical line would be misleading. Similarly, drawing separate lines at 1, 2 and 3 could lead to confusion.

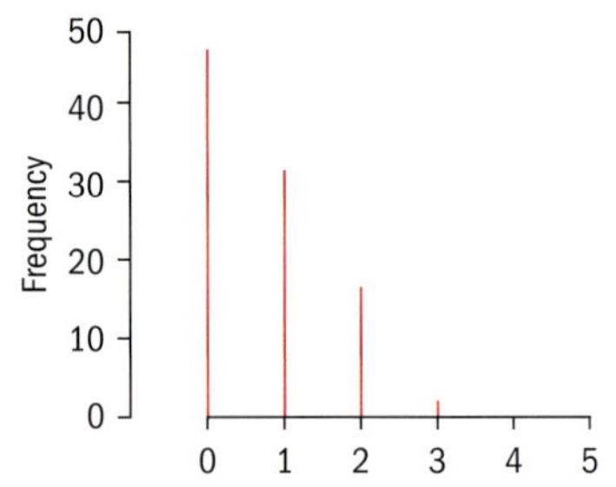

For this reason, we would draw the rectangle for the 1–3 eggs so that it starts at 0.5 and ends at 3.5.

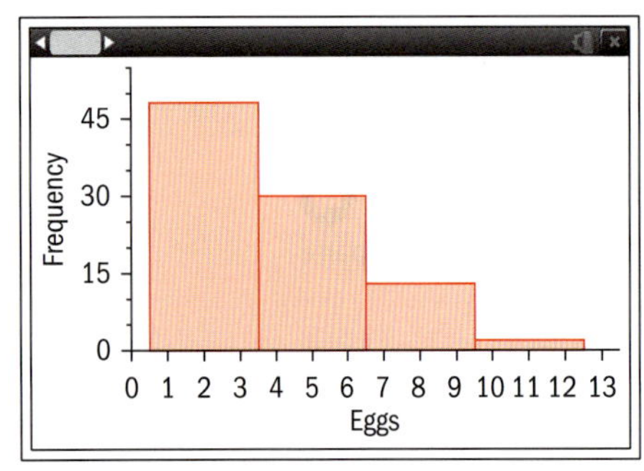

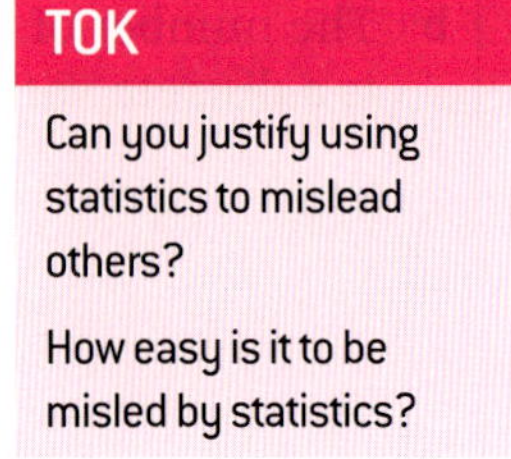

TOK

Can you justify using statistics to mislead others?

How easy is it to be misled by statistics?

Exercise 5C

In each of the following practice questions, decide whether you are working with continuous or discrete data.

Draw a frequency histogram accordingly.

1 The number of people travelling in each of 33 cars was counted, and the results are shown in the table below.

Number of people	1	2	3	4	5	6
Frequency	8	11	6	4	2	2

a Find the limits for each category.

b Complete the following frequency table.

Number of people	Frequency	Interval on histogram
1	8	$0.5 < x \leq 1.5$
2		
3		
4		
5		
6		

c Draw a frequency histogram to represent the data.

HINT

Just as with grouped data, we need to draw the bars as if they were rounding to 1, 2, 3, 4, 5, or 6. The first bar should go from 0.5 to 1.5.

2 The lengths of 30 Swiss cheese plant leaves were measured and the information grouped as shown. Measurements are taken to the nearest cm.

Length of leaf (to the nearest cm)	$10 \leq x < 15$	$15 \leq x < 20$	$20 \leq x < 25$	$25 \leq x < 30$
Frequency	3	8	12	7

Draw a frequency histogram to illustrate the data.

3 The ages of 100 users of a social media app were surveyed and their reported ages are shown below.

Age	$15 \leq x < 25$	$25 \leq x < 35$	$35 \leq x < 45$	$45 \leq x < 55$	$55 \leq x < 65$
Number of app users	41	30	15	8	6

Draw a frequency histogram to illustrate the data.

HINT

Ages are rounded differently to most data. If we say a tree is 10 metres tall, we usually mean that it is 10 metres correct to the nearest metre. So the height h satisfies $9.5 \leq h < 10.5$. If we say a person is 10 years old, we mean their age a satisfies $10 \leq a < 11$, where a is a continuous quantity. The graphs that you draw must reflect this fact.

4 The histogram below has lost its title and axis labels.

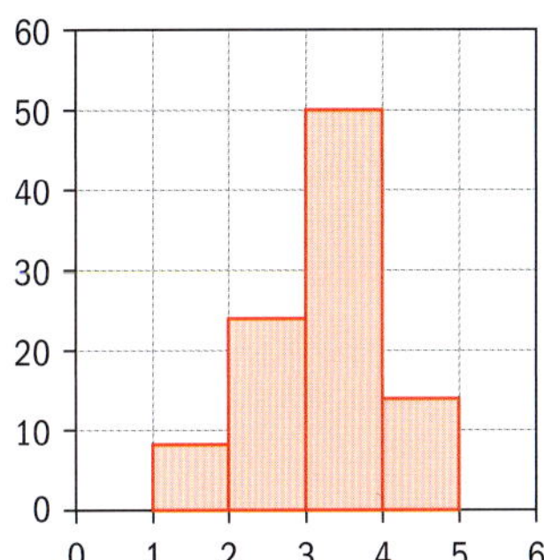

a Determine whether it is displaying discrete or continuous data.

b Suggest what it might be displaying.

5 The number of hours that Macey spends doing homework each day in the month of June is represented in the following frequency histogram.

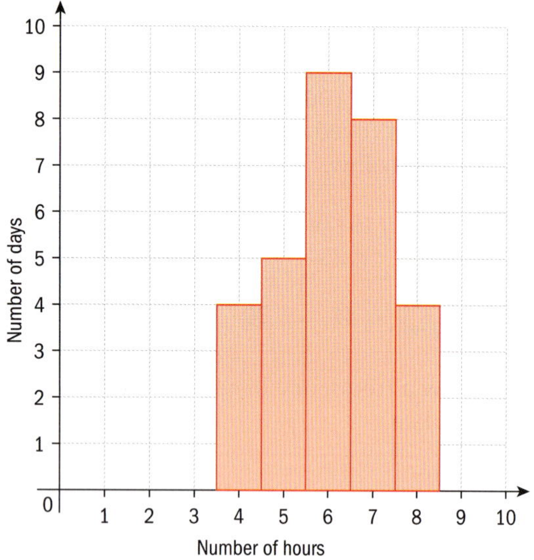

a From the frequency histogram construct a frequency table.

b Use your frequency table to calculate the average number of hours Macey spends doing homework each day.

Reflect 1 How do you analyse the distribution of data using a histogram?

2 How do the different class intervals change the shape of the distribution?

3 Why do we sometimes use relative frequencies?

When examining a histogram you can start to visualize certain characteristics of the set of data from which the histogram originated.

Commenting on the **distribution of the data** (as you are asked to do in question 1 of the "Reflect" box above) means outlining some of the most basic features of how the data is arranged. There are four key points that you should always try to include. The acronym CSOS is a way for you to remember these:

Centre

Look for the mean, median and mode.

Spread

Look for the range, maximum and minimum.

Outliers

In the histogram there may be data which looks out of place. An outlier is defined as a data item which is more than 1.5 × Interquartile range (IQR) above the upper quartile, or less than 1.5 × Interquartile range (IQR) below the lower quartile.

Shape

Examine whether it is unimodal, bimodal or multimodal, whether it is symmetric or skewed.

Did you known?

The word *histogram* originates from Ancient Greek.

Histos can be translated as "stands upright" and *gram* can mean "a drawing". The English mathematician Karl Pearson was the first to use this word.

Unimodal

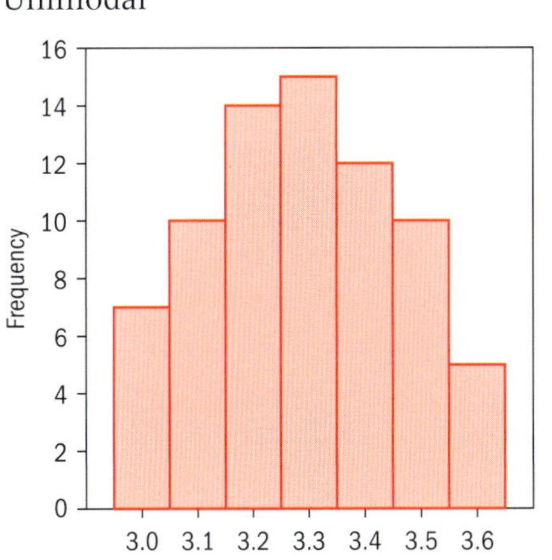

Bimodal

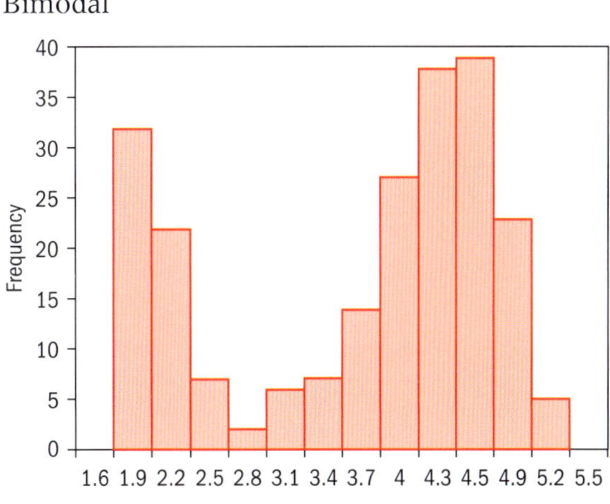

Multimodal

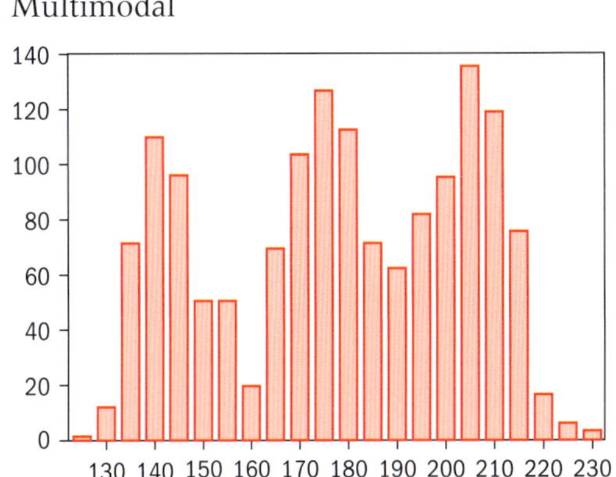

Skewed (not symmetrical)

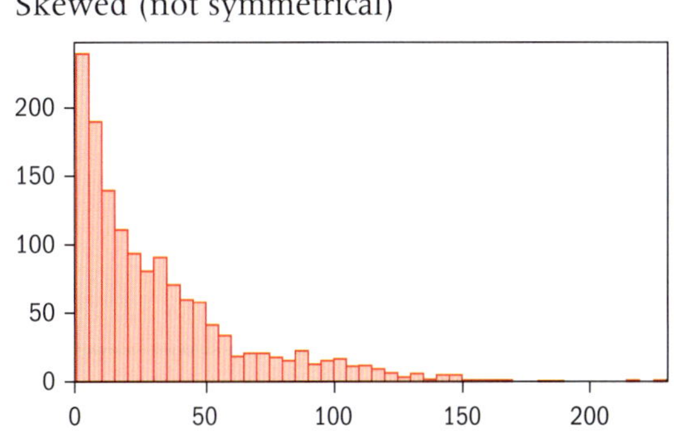

When you are comparing two distributions it is important to be concise and clear in your description of the distribution. Use words such as "wider", "narrower", "more varied", "less varied".

You can make estimates for the mean, median or mode if they cannot be calculated directly. In order to calculate an estimate of the mean for grouped data, it would be necessary to place the data into a frequency table.

Example 5

Consider the following frequency histogram showing the length (x cm) of 51 fish caught in the River Avon.

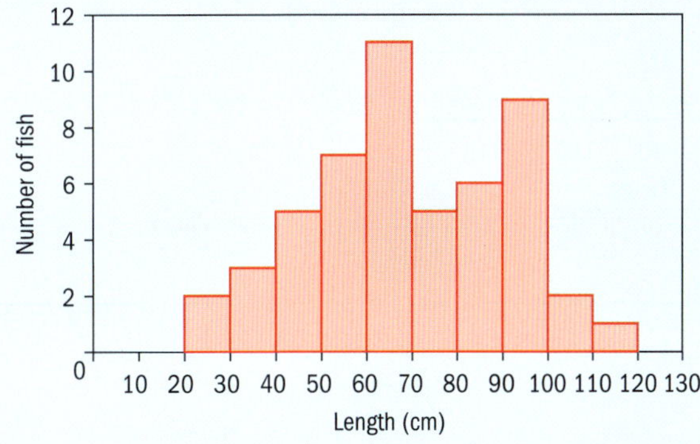

Continued on next page

a State the median class.

b State the range.

c Comment on the distribution of the data.

d Estimate the mean.

a $60 \le x < 70$.

Since there are 51 pieces of data, the median class would contain the 26th data point.

b Range is 100.

The maximum length of the fish caught is 120 cm and the minimum length is 20 cm. Range $= x_{max} - x_{min} = 120 - 20 = 100$.

c **C**entre: The midpoint would fall in the $60 \le x < 70$ class interval.

Spread: The length of fish in the river varies from 20 cm to 120 cm.

Outliers: There are no outliers.

There are no outliers in this sample.

Shape: the distribution is bimodal, most fish caught had either a length of $60 \le x < 70$ cm or $90 \le x < 100$ cm.

From the shape of the distribution, you can say that because there are two peaks with almost identical height, the data is bimodal.

d

	A length	B freq	C	D
=				=OneVar(
2	35	3	x̄	69.5098
3	45	5	Σx	3545.
4	55	7	Σx²	270675.
5	65	11	sx := sn-...	22.0285
6	75	5	σx := σn...	21.8115
7	85	6	n	51.
8	95	9	MinX	25.
9	105	2	Q₁X	55.
10	115	1	MedianX...	65.
11			Q₃X	85.
12			MaxX	115.

D7 =51.

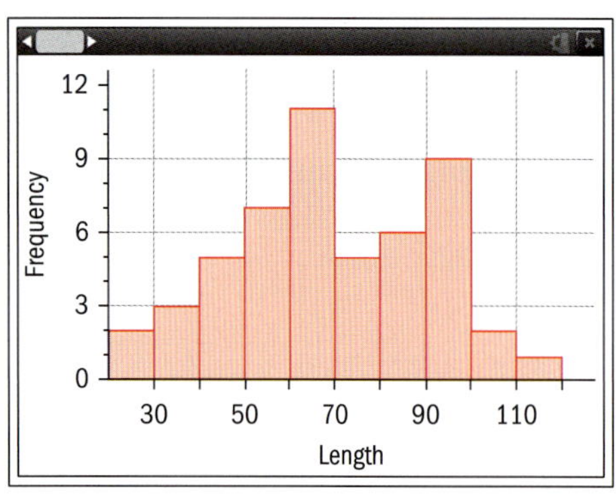

$\bar{x} = 69.5$ cm

Use your GDC to estimate the mean. Make sure you know how to do this.

Remember that you use the mid-interval value when working with grouped data. For the group $20 \le x < 30$, you would write 25.

Analysis of the distribution of a histogram is very precise and can change depending on the class intervals. You can examine the effect of changing the class interval.

Below is the data for the weight of 50 eggs one week before they hatch (measured to the nearest gram).

63	48	51	50	61	56	42	60	38	48
53	55	40	50	54	33	50	42	35	49
51	39	40	65	33	33	53	39	48	65
53	52	52	44	55	44	61	52	70	52
54	51	53	50	55	45	59	71	41	70

First, consider the histogram drawn by taking each integer to be a discrete value.

This would look like a multimodal distribution.

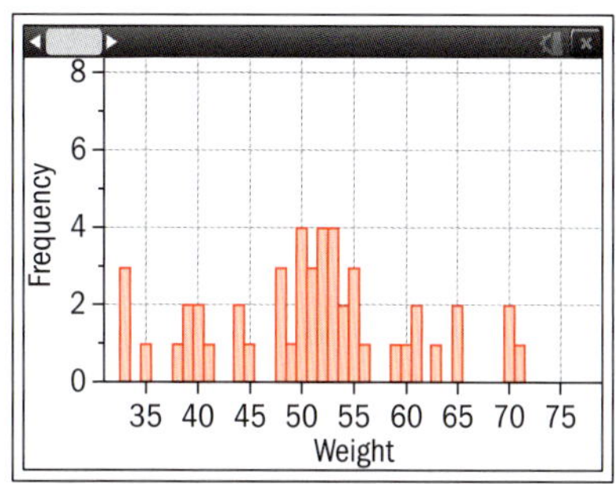

If you made four equal class widths then you would obtain the following frequency distribution table and histogram:

Weight (g)	Frequency
$33 \le w < 43$	12
$43 \le w < 53$	18
$53 \le w < 63$	14
$63 \le w < 73$	6

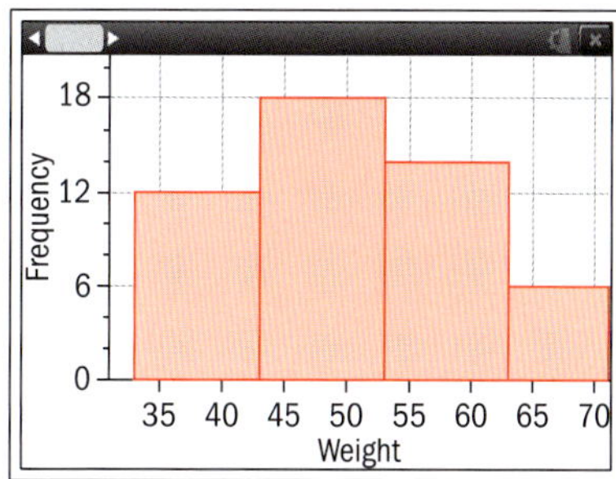

This histogram shows a different distribution. It is unimodal at $43 \le w < 53$ and does not appear to be symmetrical. The distribution therefore appears to be skewed.

However, if you use 10 class intervals,

Class interval (cm)	Frequency
$33 \le w < 37$	4
$37 \le w < 41$	5
$41 \le w < 45$	5
$45 \le w < 49$	3
$49 \le w < 53$	12
$53 \le w < 57$	10
$57 \le w < 61$	2
$61 \le w < 65$	3
$65 \le w < 69$	2
$69 \le w < 73$	3

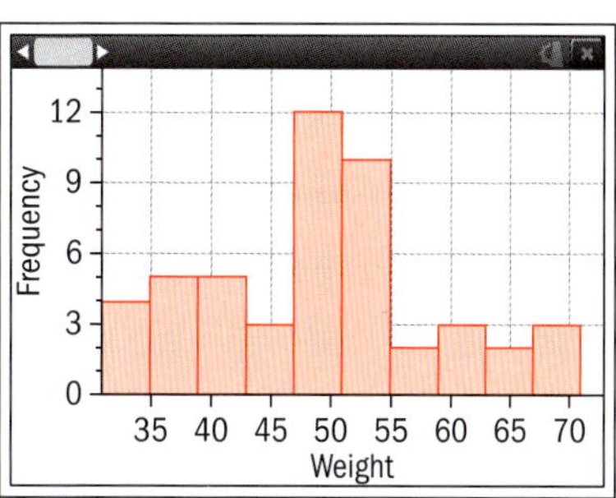

this histogram appears to be more symmetrical with the modal class as $49 \le w < 53$.

Statistics and probability

You can be asked to compare the distributions of two data sets using histograms. Consider again the frequency histogram showing the length of fish caught in the River Avon from Example 5. Imagine you are given a second histogram for a sample of fish taken from the river Blyth, and are asked to compare the distribution.

Investigation 4

Histogram for fish in River Avon

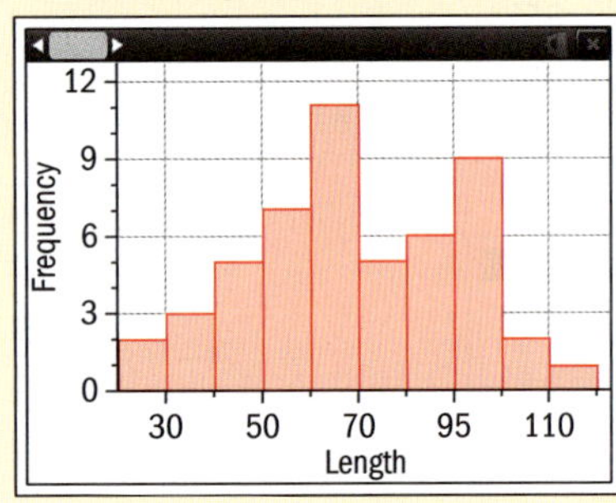

Histogram for fish in River Blyth

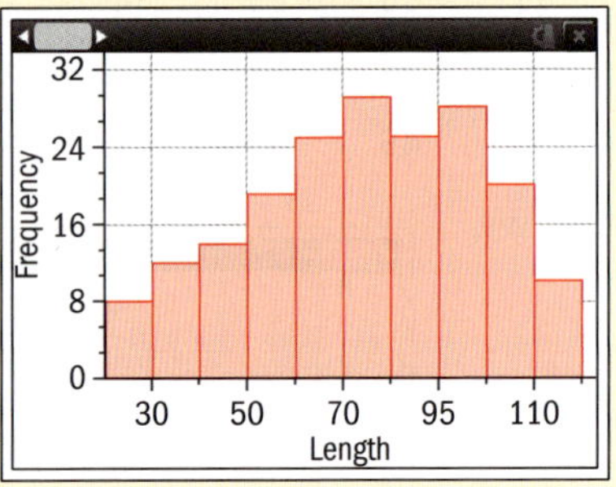

Compare the two graphs.

1 What are the visual differences between the graphs?

2 What does this difference tell you about the samples that were taken?

3 Why might you consider this not to be an important comparison about the distribution of the fish lengths in each river?

TOK

Why have mathematics and statistics sometimes been treated as separate subjects?

It is very difficult to make fair comparisons between the data sets if the number of data points in each set is different.

In order to make fairer comparisons you can use a *relative frequency histogram*. This uses the same raw data as the histogram; however, on the y-axis the *relative frequency* is plotted. It is a percentage of the total frequency and leads to more accurate comparisons.

Relative frequency is the proportion (or percentage) of the data set belonging in the given class interval.

For a data set with n members, a category with frequency f has relative frequency $\frac{f}{n}$.

The relative frequency can be calculated by finding the frequency as a percentage of the total number of elements (see Example 6).

Example 6

The following frequency tables show the lengths of samples of fish caught in the River Chippin and the River Dere.

River Chippin

Length (cm)	Frequency
$10 < l \le 20$	2
$20 < l \le 30$	52
$30 < l \le 40$	80
$40 < l \le 50$	30
$50 < l \le 60$	25
$60 < l \le 70$	33
$70 < l \le 80$	30
$80 < l \le 90$	30
$90 < l \le 100$	40
$100 < l \le 110$	38
$110 < l \le 120$	30
$120 < l \le 130$	5
$130 < l \le 140$	5

River Dere

Length (cm)	Frequency
$10 < l \le 20$	5
$20 < l \le 30$	8
$30 < l \le 40$	21
$40 < l \le 50$	40
$50 < l \le 60$	24
$60 < l \le 70$	23
$70 < l \le 80$	20
$80 < l \le 90$	17
$90 < l \le 100$	16
$100 < l \le 110$	10
$110 < l \le 120$	9
$120 < l \le 130$	5
$130 < l \le 140$	2

a Calculate the relative frequency for each category.

b Draw comparative relative frequency histograms for each river, where the frequency ranges from 0 to 1 on each vertical axis.

c Compare the distribution of fish lengths in the two rivers.

Continued on next page

a

River Chippin	River Dere
0.005	0.025
0.13	0.04
0.2	0.105
0.075	0.2
0.0625	0.12
0.0825	0.115
0.075	0.1
0.075	0.085
0.1	0.08
0.095	0.05
0.075	0.045
0.0125	0.025
0.0125	0.01
1	**1**

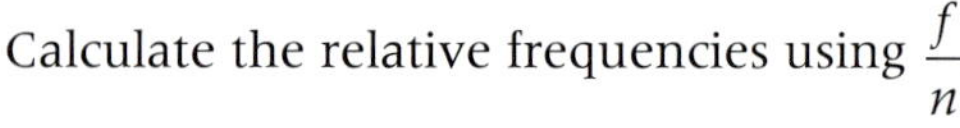

Calculate the relative frequencies using $\frac{f}{n}$.

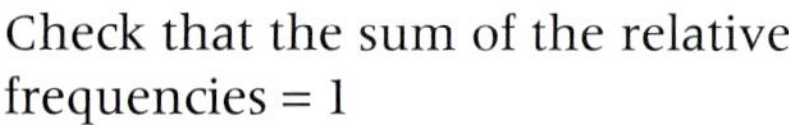

Check that the sum of the relative frequencies = 1

b

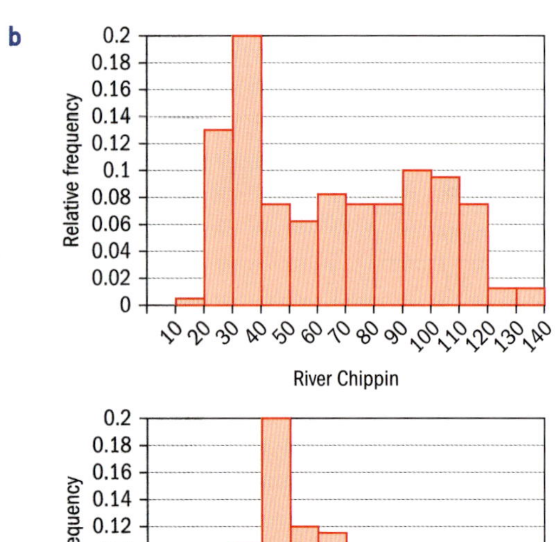

River Chippin

River Dere

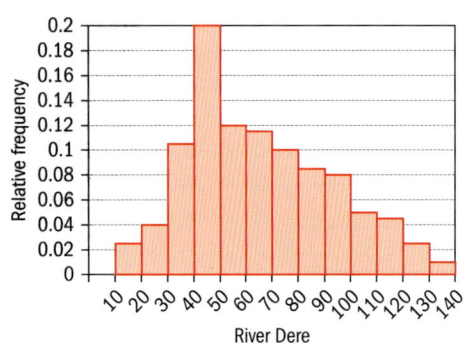

c From the relative frequency histograms it is evident to see that River Chippin is more uniformly distributed.

Both Rivers have a maximum relative frequency of 0.2.

Exercise 5D

1 The data given has been produced from the masses of 50 Koalas (kg) in an Australian nature reserve.

33	19	24	35	36	24	29	29	29	34
38	35	35	35	36	60	35	50	34	48
41	41	51	42	35	36	32	61	30	40
41	19	33	34	17	35	35	38	35	42
20	29	50	33	37	28	49	58	45	40

Construct a frequency histogram of this distribution. Describe the distribution, commenting on the shape, centre and spread.

2 You ask everyone in your class to give you the day of the week on which they were born.

a Explain why you should use a bar chart to show the results.

b Sketch the bar chart to predict what the distribution would look like.

3 You are given the following frequency tables on the length of time males and females spend on their mobile phones during a period of one day.

Time spent per day in minutes (male)	Frequency
$0 \le x < 15$	5
$15 \le x < 30$	8
$30 \le x < 45$	10
$45 \le x < 60$	5
$60 \le x < 75$	2

Time spent per day in minutes (female)	Frequency
$0 \le x < 15$	4
$15 \le x < 30$	5
$30 \le x < 45$	7
$45 \le x < 60$	14
$60 \le x < 75$	2

a Explain why it is necessary to use a relative frequency histogram to compare the data in this context.

b By adding an additional column to the table, write down the relative frequencies.

c Draw relative frequency histograms for the two sets of data.

d Analyse each distribution.

e Compare the length of time per day spent on the phone by the male and female subjects.

4 The histogram below shows the amount of money, given to the nearest \$10, spent on food each week by the families living in a particular neighbourhood.

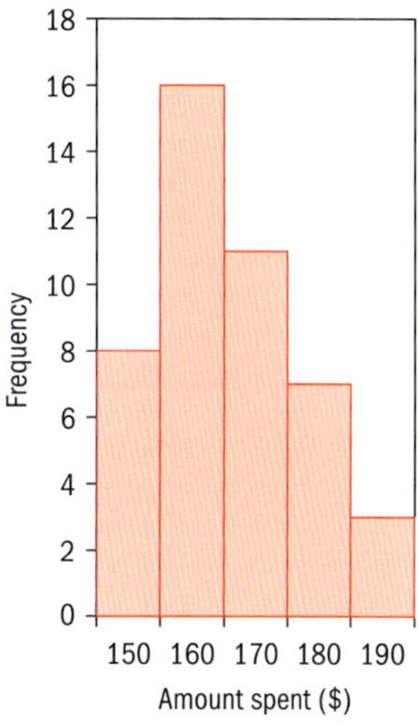

a Write down the number of families that were interviewed.

b By converting the histogram into a frequency table, calculate the mean amount spent on food by the families.

c Comment on the distribution of the data.

5 The histogram below shows the frequency of items sold for a certain amount of dollars at a market.

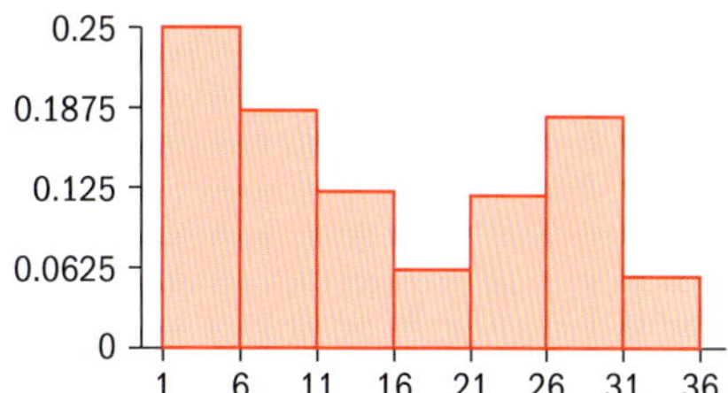

There were 32 items in total.
Calculate how many items cost less than 16 dollars.

6 Determine whether each of the following histograms:

a are symmetric or skewed

b are unimodal, bimodal or multimodal

c contain outliers.

i

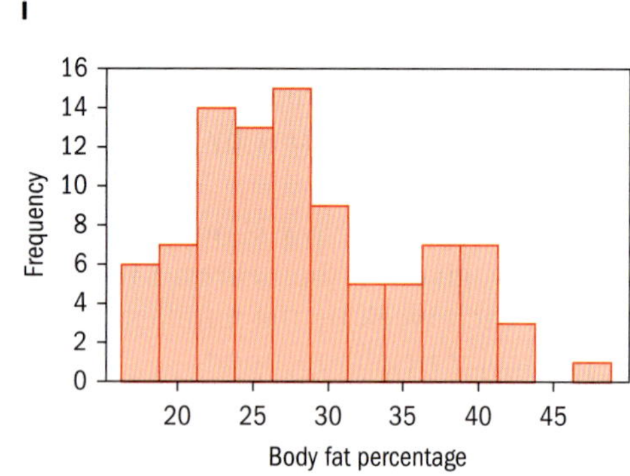

ii

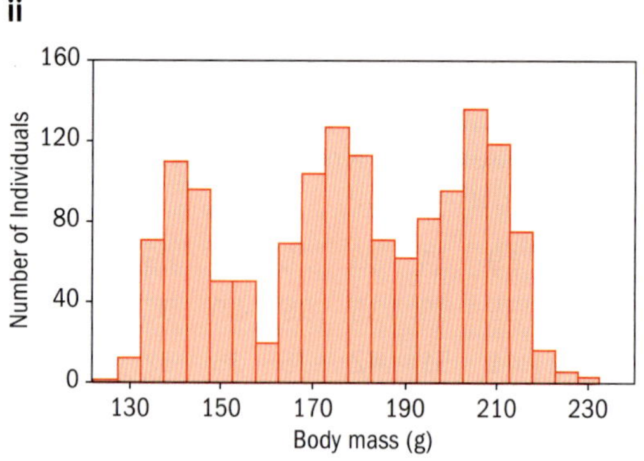

iii

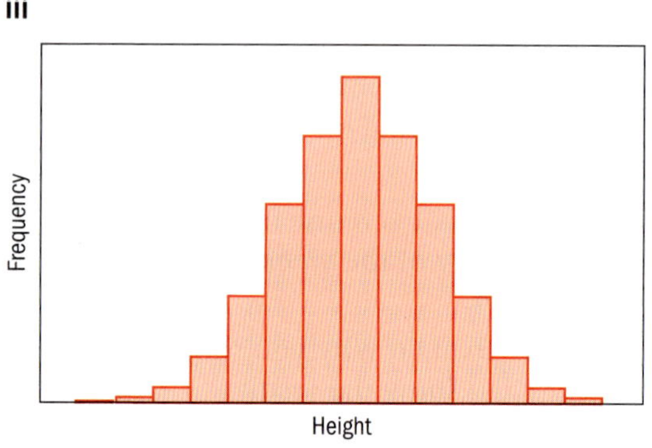

Histograms with unequal class widths

Consider the following set of data for the weights of the fish caught in a river (measured to the nearest kg):

Stem	Leaf
0	8 8 9
1	
2	1 2 2 2 3 4 8
3	2 3 5 5 5 6 7 7
4	1 3 3 3 3 3 3 3 3 3 3 3 3 4 4 4 5 6 6 6 6 7 7 7 7 8 8 8 9 9 9 9 9
5	1 1 1 1 2 2 3 3 4 4 4 4 5 5 5 5 5 5 6 6 6 6 6 7 7 7 7 7 7 7 8 8 8 8 8 8 8 8 9 9 9 9 9 9 9 9
6	1 1 1 2 2 2 3 3 3 4 4 4 4 4 5 5 6 6 7 7 8 8 8 8 9
7	
8	
9	0 0 1 2 4 5 5 5 7 8

Key: 2|5 means the weight of the fish is 25 kg.

HINT

The data have been represented as a stem-and-leaf diagram, where the "stem" (the left-hand column) represents the category figure and the "leaves" (the right-hand figures) represent the final digit.

Why does using a small number of wide classes make it difficult to comment on the distribution?

The graph shown here is an accurate representations of the data but could be more useful. You have seen that wide classes are appropriate when data is very spread out, and narrow classes help when the data is concentrated together.

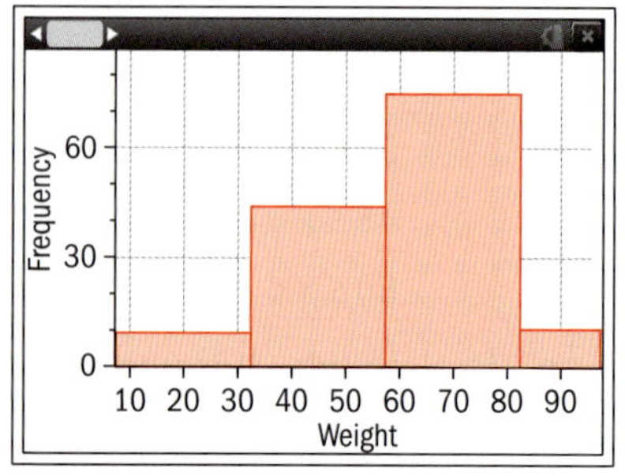

These ideas can be combined to create a better system using wider classes at the edges, where the data is sparse (spread out), and narrow classes in the middle where the data is dense (concentrated together).

Weight (kg)	Frequency
$0 < w \le 30$	10
$30 < w \le 40$	8
$40 < w \le 45$	20
$45 < w \le 50$	16
$50 < w \le 55$	12
$55 < w \le 57$	11
$57 < w \le 59$	15
$59 < w \le 61$	9
$61 < w \le 65$	14
$65 < w \le 70$	11
$70 < w \le 100$	10

Suppose now that you draw a frequency histogram to represent this frequency table. It would look like this:

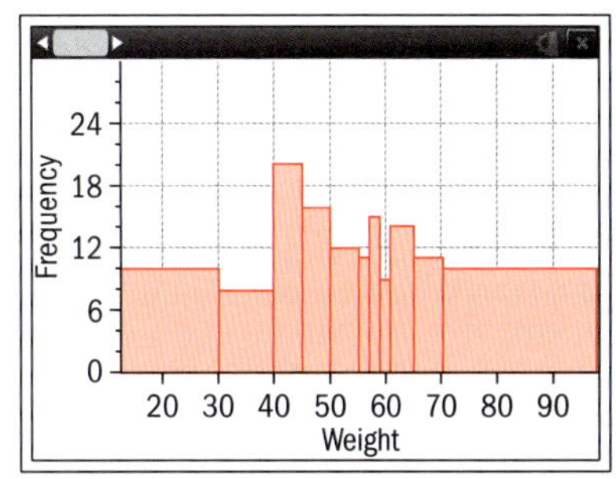

- Do you think this graph represents the data well?
- Does it obey the principle that, in a histogram, the area is proportional to the frequency?

We can see here that because we have fixed the heights of the rectangles to the frequency, wider categories have ended up looking much too large.

Exercise 5E

1 a A group of 25 females from Mexico were asked how many children they each had. The results are shown in the frequency histogram below.

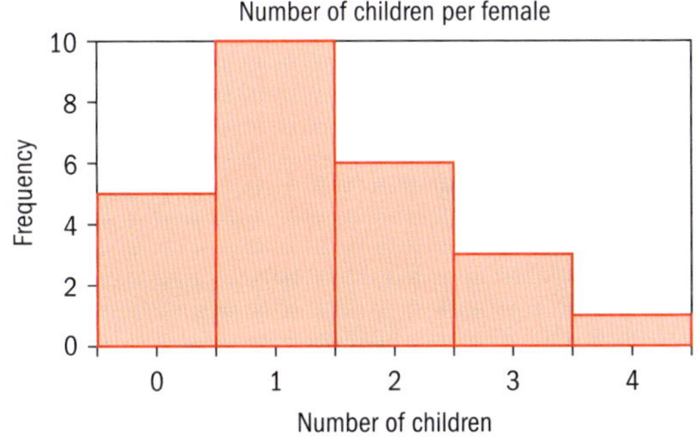

Show that the mean number of children per female is 1.4.

b A group of 25 females from Australia were surveyed and the results are given below.

Number of children	0	1	2	3	4	5
Frequency	4	8	5	4	5	2

Use the data from parts **a** and **b** to compare and describe the distribution between the number of children the two groups of females have.

Statistics and probability

2 The following table shows the age distribution of teachers who have moved schools.

Ages	Number of schools
$20 \le x < 30$	5
$30 \le x < 40$	4
$40 \le x < 50$	3
$50 \le x < 60$	2
$60 \le x < 70$	3

a Calculate an estimate of the mean age of the teachers.

b Construct a histogram to represent this data.

c Describe the distribution of this histogram.

3 The heights of 14-year-old students from Sri Lanka and Peru were recorded in the frequency tables below.

Heights of students in Sri Lanka (measured to the nearest cm)	Frequency
120–130	3
130–140	11
140–150	14
150–160	8
160–170	6

Heights of students in Peru (measured to the nearest cm)	Frequency
120–130	22
130–140	41
140–150	32
150–160	6
160–170	6

a Explain why it is necessary to construct a relative frequency histogram for a comparison of these data sets.

b Construct a relative frequency histogram for the comparison.

c Comment on the distributions of the heights of 14-year-old students in Sri Lanka and Peru.

4 A survey of 90 mothers was taken in New Zealand to inquire about their age when giving birth to their first child.

Age, A (Years)	Frequency
$15 < A \le 20$	5
$20 < A \le 23$	15
$23 < A \le 25$	20
$25 < A \le 30$	20
$30 < A \le 40$	30

a Determine which type of histogram would give the best representation.

b Construct the histogram.

c Calculate the mean and determine the modal class.

Developing inquiry skills

1 Find the mean, mode and median of the class test scores from the start of this chapter.

2 Which of mean, median or mode gives the best indication of the "average" score? Are any two averages roughly equal?

5.3 The justification of statistical techniques

Box-and-whisker diagrams

It is possible to get a sense of the distribution of a data set by examining summary statistics. These include:

1. the minimum value
2. the first quartile
3. the median (or second quartile)
4. the third quartile
5. the maximum.

A five-number summary can be represented graphically as a **box-and-whisker diagram** (sometimes called a box plot).

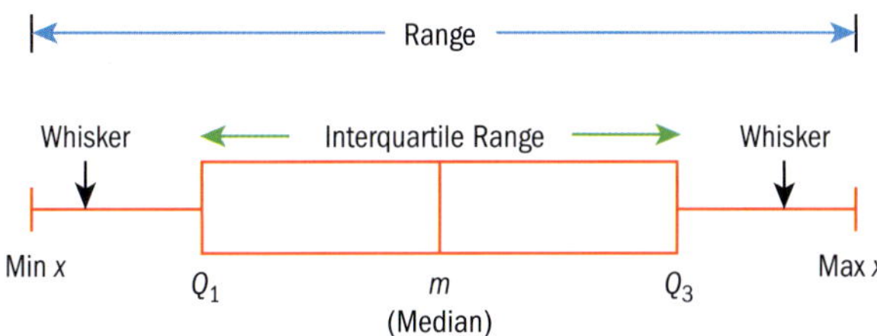

The first and third quartiles are at the ends of the rectangle (box); the median is indicated with a vertical line in the interior of the box, and the maximum and minimum points are at the ends of the horizontal lines (whiskers). The diagram should be drawn to scale along a single, horizontal axis. The test data from the class scores at the start of this chapter are displayed in this box-and-whisker diagram.

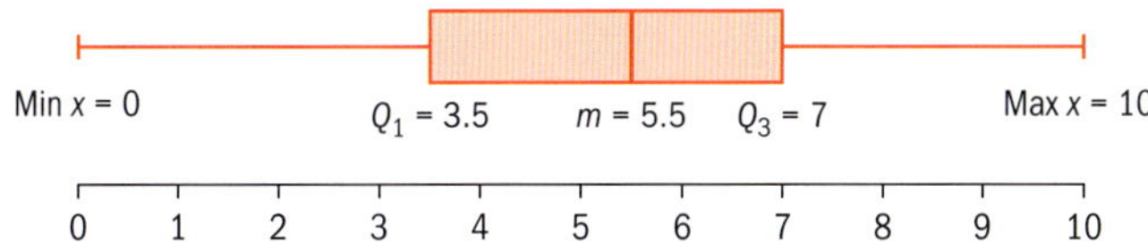

Outliers

Outliers can occur in a data set. This can be due to error in data collection or naturally occurring anomalies in the population. One method of identifying outliers is using the theory of the IQR. This theory states that:

An outlier is defined as a data item which is more than 1.5 × interquartile range (IQR) below the lower quartile or above the upper quartile.

HINT

Make sure you know how to use your GDC to find summary statistics and construct a box-and-whisker diagram.

TOK

To what extent can we rely on technology to produce our results?

Variance and standard deviation

The range and quartiles are good measures of how spread out a data set is around the median, but they do not use all the data values to give an indication of the data's spread. The **variance** and standard deviation, on the other hand, combine *all* the data values in a data set to give a measure of a data's spread about its mean. In particular, the standard deviation is a measure of how far, on average, each data value differs from the mean.

Investigation 5

The temperature (°C) has been recorded on the first day of the first five months of the year in two separate towns, Whitby and Mullion Cove.

	Whitby	Mullion Cove
January	10	18
February	16	10
March	14	13.5
April	12	14
May	18	14.5

1 Calculate the three measures of central tendency (mean, mode and median) and the range.

2 Compare these answers.

You can now explore to see if there is a further statistical measure that can help distinguish between the two groups.

For the Whitby data:

3 Calculate by how much each individual value differs from the mean.

4 What do you notice about the sign (positive or negative) of these values?

5 How can you ensure that they all have the same sign?

6 Square each of the values from part **3**.

7 What is your unit at this stage?

8 Find the mean of these values.

9 In order to get your unit back to the original unit quantity, which mathematical operation must you perform?

10 Calculate this final value as the mean of the deviations from the mean.

11 **Conceptual** How does this technique find and describe the dispersion of the data?

12 Can you generalize this algorithm into a mathematical formula?

HINT

Squaring the difference or taking the absolute value would both make each term positive so that the values above the mean do not cancel out the values below the mean.

However, squaring the differences adds more weighting to the smaller and larger differences. In many cases this extra weighting is important because the data points further away from the mean may be more significant.

Investigation 5 has led you to find the **standard deviation**, which is the square root of the **variance**.

Standard deviation gives a *mean average* of the *distance* between each data point and the mean.

$$\sigma = \text{Population standard deviation} = \sqrt{\frac{\sum_{i=1}^{n}(x-\mu)^2}{n}}$$

Variance gives a *mean average* of the *squared distance* between each data point and the mean.

$$\sigma^2 = \text{Population variance} = \frac{\sum_{i=1}^{n}(x-\mu)^2}{n}$$

If the data set is close together, the standard deviation will be small. If the data is spread out, the standard deviation will be large. The units of standard deviation are the same as the units of the original data.

HINT

It is expected that a GDC will be used to calculate the population standard deviation and variance.

Example 7

A group of 40 students were asked how many times they visited the dentist in the last year.

Their responses were:

3, 0, 2, 5, 7, 6, 8, 0, 4, 1, 6, 3, 0, 5, 6, 5, 3, 6, 2, 7, 6, 0, 4, 4, 6, 6, 5, 7, 0, 1, 2, 5, 8, 0, 4, 3, 4, 6, 7, 5.

Calculate the mean and standard deviation for this data.

$\bar{x} = \frac{\Sigma fx}{\Sigma f} = \frac{160}{40} = 4$

Calculate the mean.

x	f	fx	$(x-\bar{x})$	$\left(x-\bar{x}\right)^2$	$\left(x-\bar{x}\right)^2 \times f$
0	6	0	−4	16	96
1	2	2	−3	9	18
2	3	6	−2	4	12
3	4	12	−1	1	4
4	5	20	0	0	0
5	7	35	1	1	7
6	8	48	2	4	32
7	3	21	3	9	27
8	2	16	4	16	32
	40	160	5		228

Find $(x-\bar{x})$ for each data value x. This is the distance between x and the mean $\bar{x}$.

Square each value of $(x-\bar{x})$.

Find $(x-\bar{x})^2 \times f$. This is the frequency of x, multiplied by the squared distance from the mean for that x.

$\sigma = \sqrt{\frac{\Sigma(x-\bar{x})^2 \times f}{\Sigma f}} = \sqrt{\frac{228}{40}} = 2.39$

Find the standard deviation.

Continued on next page

Or

Use technology to calculate the values:

	A visits	B freq	C	D	E	F
=						=OneVar(
1	0	6			Title	One-Va...
2	1	2			$\bar{x}$	4.
3	2	3			Σx	160.
4	3	4			Σx²	868.
5	4	5			sx := sn-...	2.41788
6	5	7			σx := σn...	2.38747
7	6	8			n	40.
8	7	3			MinX	0.
9	8	2			Q₁X	2.

$\bar{x} = 4$; $\sigma = 2.39$

Your GDC will allow you to find the mean and standard deviation by entering the data and the frequency. Make sure you know how to do this. To find the variance, simply square the standard deviation.

Investigation 6

What happens to the mean and standard deviation when changes are made to the whole data set?

Consider the following data sets:

a 1, 1, 1, 1, 1 **b** 1, 2, 3, 4, 5 **c** 2, 6, 7, 5, 4

1 Find the mean.

2 Find the standard deviation.

Each value in each data set is increased by 3.

3 Find the new mean and standard deviation and comment on your findings.

Each value in each data set is multiplied by three.

4 Find the new mean and standard deviation and comment on your findings.

5 **Conceptual** What happens to the mean and standard deviation when we add a constant to a set of data and multiply a set of data by a constant?

Example 8

The number of text messages sent by a group of 15 students on a one-week residential trip were: 36, 40, 12, 0, 15, 25, 25, 78, 45 ,28, 18, 3, 15, 19, 20,

a Find the mean and median number of text messages.

b Find the interquartile range.

c Determine if any of the data values can be considered as outliers.

a mean = 25.3 (3 s.f) median = 20	Calculate the mean using $\sum_{i=1}^{k} \frac{f_i x_i}{n}$. The median is in the 8th position.
b Interquartile range = 36 − 15 = 21	The lower quartile (Q_1) value lies in the $\frac{n+1}{4}$th position when arranged in order. Therefore, 4th position = 15 The upper quartile (Q_3) value lies in the $\frac{3(n+1)}{4}$th position when arranged in order. Therefore, 12th position = 36
c $1.5 \times \text{IQR} = 31.5$ $15 - 31.5 = -16.5$; not possible $36 + 31.5 = 67.5$ Therefore, the data point 78 can be considered an outlier.	A data point is an outlier if it is less than $1.5 \times \text{IQR}$ below Q_1, or greater than $1.5 \times \text{IQR}$ above Q_3.

Exercise 5F

1 Find the mean and the standard deviation of the following data.

a 2, 3, 3, 4, 4, 5, 5, 6, 6, 6

b 21 kg, 21 kg, 24 kg, 25 kg, 27 kg, 29 kg

c

x	f
3	2
4	3
5	2

d

Interval	Frequency
1–5	2
6–10	4
11–15	4
16–20	5
21–25	2

2 For a particular data set, the following summary statistics were observed.

$$\Sigma f = 20$$
$$\Sigma fx = 563$$
$$\Sigma fx^2 = 16\,143$$

Find the values of the mean and the standard deviation.

3 In a biscuit factory a sample of 10 packets of biscuits were weighed.

Mass (g)	196	197	199	200	200	200	202	203	203	205

a Calculate the mean and standard deviation of this data.

4 Samples of water were taken from near a chemical plant to see if the lead content was too high.

The following data was collected in μg per litre.

6.3	9.6	12.2	12.3	10.3
12.1	10.3	8.4	9.2	4.3

a Use this data to predict the mean and standard deviation for this sample.

It is said that over 10 μg per litre of lead in water is deemed a dangerous amount.

b Comment on whether there are grounds for further investigation.

Investigation 7

1 Recall the formula for the population standard deviation: $\sigma = \sqrt{\frac{\Sigma(x-\mu)^2}{n}}$

2 Expand the numerator: $\sigma = \sqrt{\frac{\Sigma(\ldots\ldots\ldots\ldots\ldots\ldots\ldots).}{n}}$

3 Distribute the sigma notation: $\sigma = \sqrt{\frac{\Sigma\ldots.+\Sigma\ldots\ldots+\Sigma\ldots..}{n}}$

4 Distribute the n: $\sigma = \sqrt{\frac{\Sigma\ldots.}{n}+\frac{\Sigma\ldots.}{n}+\frac{\Sigma\ldots.}{n}}$

5 Substitute the mean of the population; $\mu = \frac{\Sigma x}{n}$ and $\frac{\Sigma\mu^2}{n} = \mu^2$

6 Collect like terms.

7 Resubstitute the mean of the population; $\mu = \frac{\Sigma x}{n}$

8 Your result should be $\sigma = \sqrt{\frac{\Sigma x^2}{n} - \left(\frac{\Sigma x}{n}\right)^2}$

This is your alternate representation of the formula for the standard deviation which can be derived from summary statistics, as opposed to the actual data.

9 **Conceptual** Why is the alternative form of the standard deviation formula useful?

TOK

Is standard deviation a mathematical discovery or a creation of the human mind?

When calculating the standard deviation, you must decide whether your sample is representative of the entire population. In order to make predictions about the standard deviation of the whole population we have to take into account that the population is much larger than the sample and therefore there is more likely to be a larger spread.

For example, a sample of the weights of 25 apples (in grams) is shown below.

132	122	132	125	134
129	130	131	133	129
126	132	133	133	131
133	138	135	135	134
142	140	136	132	135

You have been given the raw data and therefore the mean and standard deviation is calculated with the formulae:

$$\bar{x} = \frac{\Sigma x}{n} = 132.48 \text{ g}$$

$$s_n = \sqrt{\frac{\Sigma(x-\bar{x})^2}{n}} = 4.28 \text{ g}$$

You don't have all the data available and you will have to estimate the mean and standard deviation of the population from this sample.

In order to make an *estimate* for the population mean, you can use the sample mean, an unbiased estimator of the population mean.

The notation we use to show the estimator for the population mean is μ.

There is only one difference in the calculation for the estimator of the population standard deviation. When you calculate the population standard deviation, you divide the sum of squared deviations from the mean by the number of items in the population (dividing by 25 in this case).

When you calculate an estimate of the population standard deviation from a sample, you divide the sum of squared deviations from the mean by the number of items in the sample less one. In this case, you would divide by 24 (that is, by 25 − 1).

HINT

By dividing by the sample size lowered by one, we compensate for the fact that we are working only with a sample rather than with the whole population.

As a result, the calculated sample standard deviation will be slightly higher than if you had used the population standard deviation formula. The purpose of this difference is to get a better and unbiased estimate of the population's standard deviation by taking into account the fact that the sample may not exactly represent the whole range of the population.

The unbiased estimator for the population mean is the same value as the sample mean.

$$x = \mu$$

The unbiased estimator for the population standard deviation is

$$s_{n-1} = \sqrt{\frac{\Sigma(x-\mu)^2}{n-1}}$$

If you have a sample and you want to make predictions about the population, use these values and call them unbiased estimators.

It is always necessary to use $n - 1$ as the denominator in the formula. When we calculate the population standard deviation we need to use the sample mean and predict each deviation from the sample mean. If we did not have one of these measurements (either the sample mean or a deviation) we could calculate it from the other information.

So if you have n data points only $n - 1$ of them are free to vary. Hence $n - 1$ is called our "degrees of freedom" and we replace n by $n - 1$.

Reflect Is it sufficient to always subtract one from the sample size to make this estimate?

This is known as Bessel's correction.

Example 9

A machine produces 1000 marbles each day with a mean diameter of 1 cm. A sample of eight marbles was taken from the production line and the diameters measured. The results in centimetres were

1.0 1.1 1.0 0.8 1.4 1.3 0.9 1.1

Determine what the standard deviation of the machine would be for the whole day.

$$\bar{x} = \frac{1.0+1.1+1.0+0.8+1.4+1.3+0.9+1.1}{8}$$

$\bar{x} = 1.075$

x	$(x-\bar{x})$	$(x-\bar{x})^2$
1.0	(1.0 – 1.075)	0.005625
1.1	(1.1 – 1.075)	0.000625
1.0	(1.0 – 1.075)	0.005625
0.8	(0.8 – 1.075)	0.075625
1.4	(1.4 – 1.075)	0.105625
1.3	(1.3 – 1.075)	0.050625
0.9	(0.9 – 1.075)	0.030625
1.1	(1.1 – 1.075)	0.000625
		$\Sigma(x-\bar{x})^2 = 0.275$

$$s_{n-1} = \sqrt{\frac{\Sigma(x-\bar{x})^2}{n-1}}$$

$$= \sqrt{\frac{0.275}{7}}$$

$$= 0.1982\text{cm (4 dp)}$$

You require the estimate of standard deviation formula:

$$s_{n-1} = \sqrt{\frac{\Sigma(x-\mu)^2}{n-1}}$$

HINT

You can also calculate standard deviation directly using a GDC.

Exercise 5G

1 Clancy recorded the number of minutes late the bus was over a 10-day period. Find an estimate of the population standard deviation for the time in minutes the bus would be late for his whole town.

10	12	5	0	14	2	5	8	9	6

2 There are 25 rabbits born in one week on a farm. Their weights (in grams) are recorded below.

450	453	452	480	501
462	475	460	470	430
485	435	425	465	456
475	435	466	482	455
462	435	462	478	455

From this information, make a prediction about the mean and standard deviations of the weights for the whole year.

HINT

The alternate formula for calculating an unbiased estimate of the population standard deviation from summary statistics is

$$s_{n-1} = \sqrt{\frac{\Sigma x^2}{n-1} - \frac{n}{n-1}\left(\frac{\Sigma x}{n}\right)^2}$$

3 The summary statistics for a sample of the life of batteries is given in the table. Use the alternate formula above to calculate the mean and standard deviation of the batteries in the factory.

n	25
Σx	38 750
Σx^2	60 100 000

4 Sylvain recorded the number of passengers in a carriage on his train each day for 50 days.

The results are shown in the table.

1	7	6	7	7	6	7	6	7	8
8	2	10	6	10	10	5	12	5	8
6	8	10	8	9	3	7	12	9	5
8	6	7	5	9	11	12	4	9	6
7	8	9	7	9	11	7	13	14	15

a Construct an appropriate diagram to represent this data.

b Comment on the shape of the distribution of the representation.

c Estimate the mean and standard deviation of the whole population.

d Infer some information for Sylvain to report on his findings.

Cumulative frequency

Investigation 8

This is the number of times that 50 students have lost a pencil this week.

5, 9, 10, 5, 9, 9, 8, 4, 9, 8, 5, 7, 3, 10, 7, 7, 8, 7, 6, 6, 9, 6, 4, 4, 10, 5, 6, 6, 3, 8, 7, 8, 3, 4, 6, 6, 5, 7, 5, 4, 3, 5, 2, 4, 2, 8, 1, 0, 3, 5

1 Construct a frequency table for this data.

2 Copy and complete this table.

Pencils	0	1	2	3	4	5	6	7	8	9	10
Number of students who lost this many or less (cumulative frequency)	1	2	4	9							

The **cumulative frequency** for x pencils is found by adding up the frequency of students who have lost "x pencils or less".

3 Plot the number of pencils on the x-axis, and cumulative frequency on the y-axis. Join up the points to form a smooth curve.

4 **Factual** Explain what this graph tells you about how the number of students losing x pencils changes as x increases? Discuss your answer with a classmate.

5 Explain why the cumulative frequency line cannot turn down or to the left.

6 **Conceptual** How would you use the cumulative frequency curve to find the median and quartiles?

When you have raw data (eg a list of numbers), you can use the formula median $= \left(\frac{n+1}{2}\right)$th value to find the median and quartiles when the data is arranged in ascending order.

However, when you have grouped data, it is hard to tell accurately what the median or quartile might be if the $\left(\frac{n+1}{2}\right)$th value lies in the middle of a group.

Cumulative frequency curves enable you to find the median and quartiles from a set of grouped data. To find the median, you draw a horizontal line across from the frequency axis to the curve at the $\left(\frac{n+1}{2}\right)$th value, and then down to the *x*-axis to determine the median data point. Finding the quartiles is similar.

Example 10

The cumulative frequency graph represents the weight in grams of 80 peaches picked from a particular peach tree.

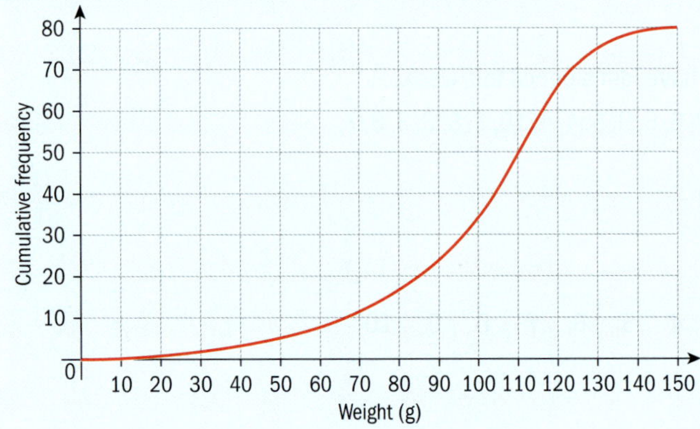

a Estimate the

i median weight of the peaches

ii 40th percentile of the weight of the peaches.

b Estimate the number of peaches that weigh more than 100 grams.

a i There are 80 peaches in total, so the median lies in the 40th position.	Use the graph to draw a horizontal line at 40 and a corresponding vertical line.

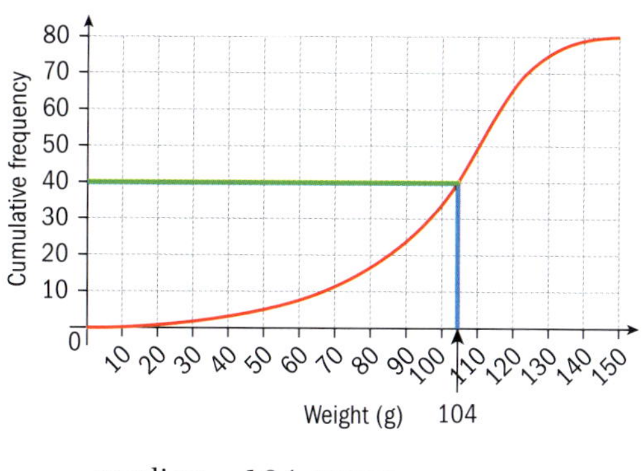

median = 104 grams	Read the value where the vertical line intercepts the x-axis.
	The 40th percentile implies 40% of the peaches are less than or equal to this weight. There are 80 peaches therefore 40% of 80 = 32 Use the graph to draw a horizontal line at 32 and a corresponding vertical line.
ii 40th percentile = 98 grams	Read the value where the vertical line intercepts the x-axis.
	Draw a vertical line from the weight 100 and draw the corresponding horizontal line.
b 80 − 34 = 46 peaches	Ensure to read the question carefully to see whether the quantity above or below this value is required.

Cumulative frequency distributions

It is useful to add a cumulative frequency column to calculate the class interval which contains the median.

The representation of the cumulative frequency curve can also be used to calculate the five-point summary from grouped data and other percentiles.

You can use the five-point summary from the cumulative frequency curve and represent it on the same diagram above or below the x-axis as a box-and-whisker diagram.

The actual cumulative frequency curve is referred to as an **ogive**. Different frequency distributions give rise to different-shaped ogives.

Key points for constructing a cumulative frequency graph:

- The cumulative frequency figures are expressed on the y-axis and the variate is expressed on the x-axis.

HINT

When dealing with grouped data, finding percentiles and quartiles as estimates is simplified by using a cumulative frequency curve.

HINT

Outliers on the box-and-whisker diagram should be indicated by a cross.

- The initial point is obtained by plotting 0 (cumulative frequency figure) against the lower boundary of the lowest class, as all the values must lie above this.
- Each cumulative frequency figure is plotted against the upper boundary of the corresponding class.
- The cumulative frequency figure associated with the highest class must be the same as the total frequency.
- All points are joined with a smooth curve.

International-mindedness

Ogive originates from the French language as a pointed or gothic arch. Galton was the first statistician to use it to mean a cumulative frequency curve.

Exercise 5H

1 A quiz marked out of 100 is written by 800 students. The cumulative frequency graph for the marks is given below.

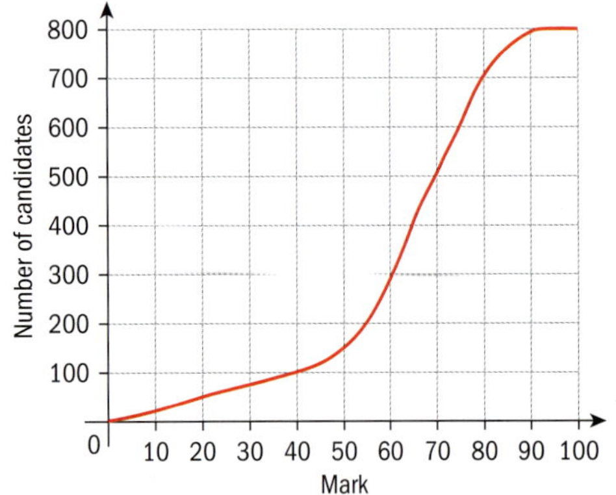

a Estimate the number of students who scored 25 marks or less on the quiz.

b The middle 50% of the quiz results lie between marks a and b, where $a < b$.

Find the values of a and b.

2 Consider the data set $\{k - 3, k, k + 2, k + 5\}$, where $k \in \mathbb{R}$.

a Find the mean of this data set in terms of k.

Each number in the above data set is now *decreased* by 3.

b Find the mean of this *new* data set in terms of k.

3 200 vehicles are tested for their air pollution efficiency. The results are given in the cumulative frequency graph

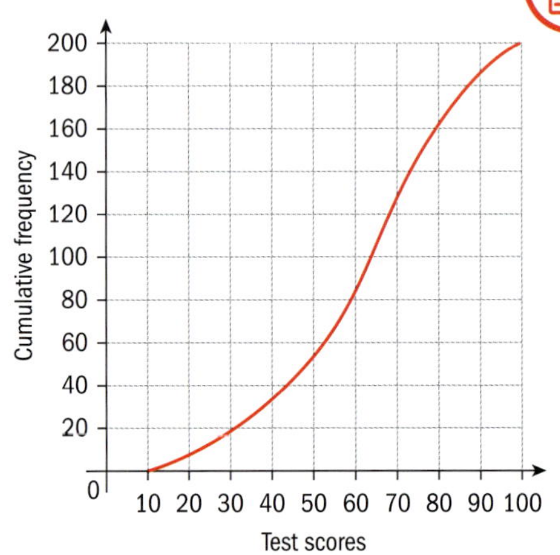

a Estimate the median test score.

b The top 10% of the vehicles receive a lower insurance premium price A and the next best 20% of the vehicles receive a price B. Estimate

i the minimum score required to obtain price A

ii the minimum score required to obtain a price B.

4 The heights in metres of a random sample of 80 cows in a field was measured and the following cumulative frequency graph obtained.

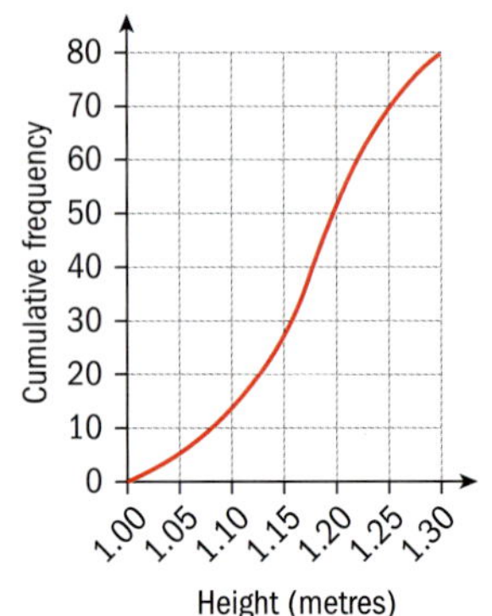

a **i** Estimate the median of these data.

ii Estimate the interquartile range for this data.

b Draw a frequency table for the heights of the cows, using a class width of 0.05 metres.

c A cow is selected at random.

i Find the probability that its height is less than or equal to 1.15 metres.

ii Given that its height is less than or equal to 1.15 metres, find the probability that its height is less than or equal to 1.12 metres.

5 **a** Consider the set of numbers $a, 2a, 3a, \ldots, na$ where a and n are positive integers.

i Show that the expression for the mean of this set is $\frac{a(n+1)}{2}$.

ii Let $a = 4$. Find the minimum value of n for which the sum of these numbers exceeds its mean by more than 100.

b Consider now the set of numbers $x_1, \ldots, x_m, y_1, \ldots, y_n$ where $x_i = 0$ for $i = 1, \ldots, m$ and $y_i = 1$ for $i = 1, \ldots, n$.

i Show that the mean M of this set is given by $\frac{n}{m+n}$ and the standard deviation S by $\frac{\sqrt{mn}}{m+n}$.

ii Given that $M = S$, find the value of the median.

Developing inquiry skills

Return to the opening problem.

1 Draw a box-and-whisker diagram for the data.

2 What does this tell you about the performance of the class?

5.4 Correlation, causation and linear regression

Bivariate data

Many of the methods which have been developed in this chapter are applied when dealing with univariate data. There are often situations when several different characteristics can be measured on each member of the data set and it can be investigated whether the variates are interrelated. This is known as bivariate data.

Scatter diagrams

A scatter diagram shows the relationship between two quantitative variables from an individual in the data set. The values of one variable appear on the x-axis and the values of the other variable lie on the y-axis.

Each individual in the data set appears as a single point.

Unlike univariate graphs, bivariate graphs often form a cloud or "scatter" of points which are subject to considerable variability, and even though there may be a strong underlying linear relationship between the variables the actual data points may not lie on the specific linear equation.

When you describe a scatter diagram you should be looking for patterns and it is important to classify your observations as:

1 **Association**. If the pattern moves up from left to right it has **positive** association; if it moves down from left to right it has **negative** association.
2 **Form**. This refers to the general shape of the pattern (often **linear**).
3 **Strength**. This is determined by how closely the points follow the specified association and form.

HINT

Always plot the independent variable on the x-axis and the dependent variable on the y-axis.

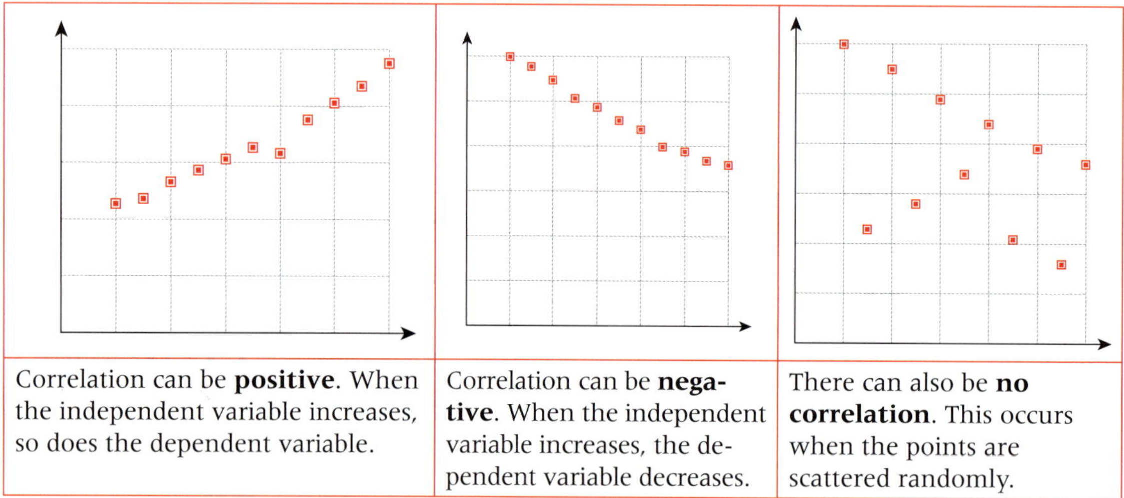

Correlation can be **positive**. When the independent variable increases, so does the dependent variable.

Correlation can be **negative**. When the independent variable increases, the dependent variable decreases.

There can also be **no correlation**. This occurs when the points are scattered randomly.

Correlation can also be described as strong, moderate or weak.

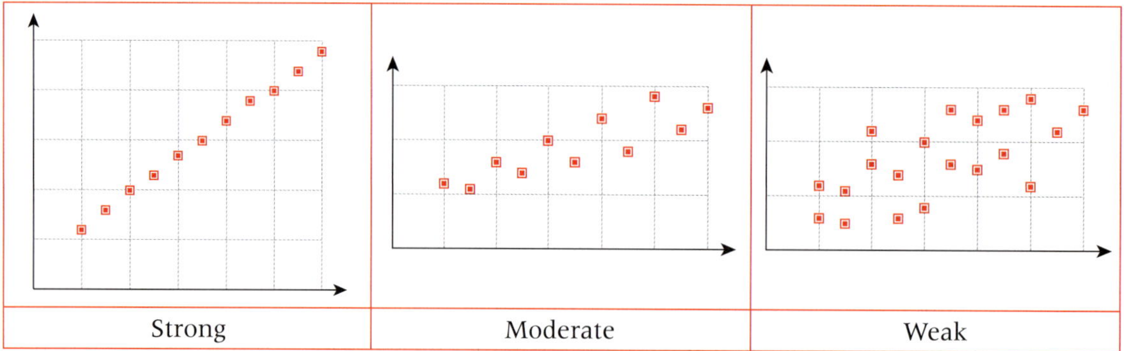

Strong | Moderate | Weak

The correlation r measures the direction of the association and the strength of the linear relationship between two variables. It can be given as a verbal explanation or a numerical quantity.

Linear correlation of bivariate data including product moment correlation coefficient

From any scatter diagram there are multiple approaches to ascertain the correlation.

Method 1: correlation from line of best fit

Method 2: correlation by product moment correlation coefficient (PMCC)

In the following investigation, you will use method 1 to find the line of best fit by eye.

International-mindedness

In 1956 Australian statistician Oliver Lancaster made the first convincing case for a link between exposure to sunlight and skin cancer using statistical tools including correlation and regression.

Investigation 9

A group of 12 students took part in a ski race and the positions were decided on the combined times of two runs. One student wished to determine whether there was a relationship between the times of the two runs, and recorded the times. The results are shown in the table.

Pupil	A	B	C	D	E	F	G	H	I	J	K	L	Total	Mean
Run 1	53.3	56.7	53.8	54	61.3	62.5	56.7	58.9	61.0	58.7	70.1	56.8		
Run 2	54.3	57.6	53.9	55.6	67.5	63.4	55.1	57.8	68.9	DNF	66.6	57.7		

1. Using either 1 mm or 2 mm graph paper, plot the points (Run1, Run2).
2. Find the mean of each run, and plot the point this point on the graph.
3. Draw a line that goes through the point you found in part **2**, and passes as closely to as many data points as possible.
4. Find the equation of your line.
5. Describe the strength of the association between the two variables.

Linear regression

A line of best fit is often an inaccurate measure. Therefore, a method called **linear regression** is used to find the equation of the line that best fits the data.

Correlation is measured in direction and strength between two quantitative variables. When a scatterplot shows a linear relationship, then we emphasize this relationship by drawing a line on the scatterplot.

This regression line should only be drawn when one of the variables helps to explain or predict the values of the other variable.

Regression requires an explanatory variable and a response variable.

In order to draw the most precise line of best fit it is necessary to use a method called **least squares regression**. This method ensures that the line is drawn through the points in such a way that the sum of the squares of the deviations from the points to the line is a minimum.

If the vertical deviations are considered then the line becomes the regression line y on x, where x is the explanatory variable and y is the response variable. This is the most frequent regression line.

HINT

You can produce a scatter plot of the data in investigation 9 on your GDC.

The GDC will also give you the regression equation, or equation of the line of best fit. In this case, the line of best fit is given as $y = 4.72 + 0.940x$. Compare this with the equation you found in **4** of investigation 9.

Statistics and probability

If the horizontal deviations are minimized the line becomes *x* on *y*.

You have seen that the graphical representations of two variables can help to give a general idea about the information contained in a set of data, but for precision you need a numerical measure.

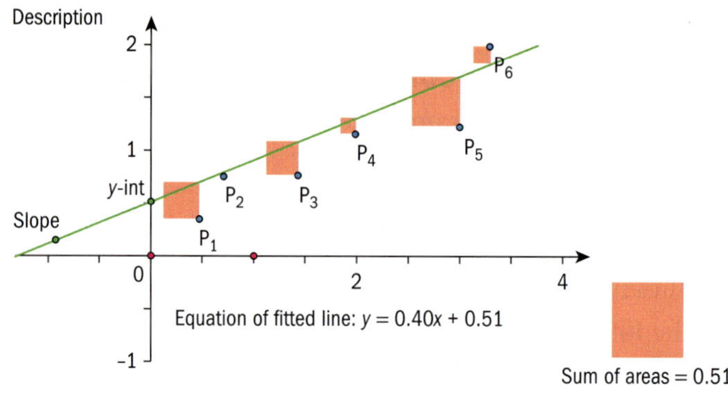

A calculator can identify each of the data points. It calculates *x*, *y* and places the line through this point; it then measures the vertical distance from each point to the line and forms the corresponding square.

The line is then pivoted until the sum of the areas of the squares is the least possible value.

Through this procedure the line of best fit is generated.

Once the line has been generated the equation of the line can be calculated to use as a model.

TOK

How can causal relationships be established in mathematics?

Example 11

The following table show the percentages of 10 students in English and Humanities exams.

English	92	18	88	20	30	80	60	54	46	40
Hums	100	42	80	54	60	66	80	68	62	54

a Use your GDC to find the regression equation.

b Generate a scatter plot on your GDC and insert the equation line from part **a**. Describe the strength of the association between exam results in English and exam results in Humanities.

a Enter the data into your GDC and use the appropriate commands to generate the regression equation.

The regression equation is $y = 0.528x + 38.7$

Notice that the GDC has given you the value of $r = 0.863$ (to 3 s.f.). This demonstrates a moderate positive correlation between test results in English and test results in Humanities.

b

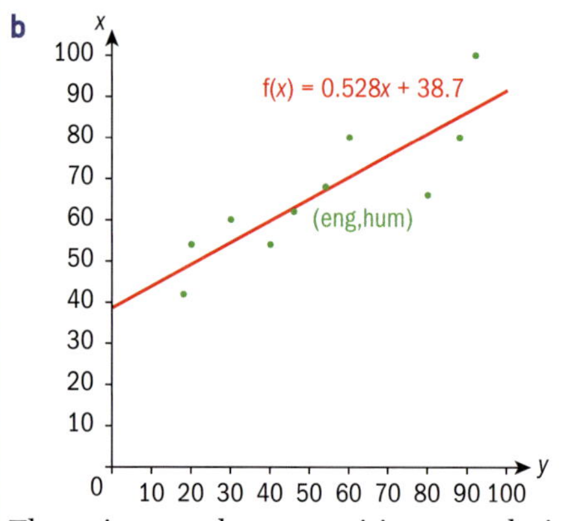

There is a moderate positive correlation between the exam results.

Express the correlation in your own words.

HINT

At this stage we define the correlation coefficient as r and we use the table below to determine the level of correlation. The values of r range from −1 to 1.

Strong negative	Moderate negative	Weak negative	Very weak negative	Zero	Very weak positive	Weak positive	Moderate positive	Strong positive
−1	−0.75	−0.5	−0.25	0	0.25	0.5	0.75	1

Example 12

Using the data from Example 11, find the product moment correlation coefficient of the data and compare it to the r value the GDC gives you for this data.

x	y	x^2	y^2	xy
92	100	8464	10000	9200
18	42	324	1764	756
88	80	7744	6400	7040
20	54	400	2916	1080
30	60	900	3600	1800
80	66	6400	4356	5280
60	80	3600	6400	4800
54	68	2916	4624	3672
46	62	2116	3844	2852
40	54	1600	2916	2160
$\sum x = 528$	$\sum y = 666$	$\sum x^2 = 34\,464$	$\sum y^2 = 46\,820$	$\sum xy = 38\,640$

Produce a table of values for x^2, y^2, and xy.

There are 10 students in the data set. Therefore, $n = 10$

$$\bar{x} = \frac{\sum x}{n} = \frac{528}{10} = 52.8 \qquad \bar{y} = \frac{\sum y}{n} = \frac{666}{10} = 66.6$$

$$s_{xy} = \frac{\sum xy}{n} - \bar{x}\,\bar{y} = \frac{38\,640}{10} - (52.8)(66.6) = 347.52$$

$$s_x^{\ 2} = \frac{\sum x^2}{n} - \bar{x}^2 = \frac{34\,464}{10} - (52.8)^2 = 658.56$$

$$s_y^{\ 2} = \frac{\sum y^2}{n} - \bar{y}^2 = \frac{46\,820}{10} - (66.6)^2 = 246.44$$

$$r = \frac{s_{xy}}{s_x\, s_y} = \frac{347.52}{\sqrt{658.56}\sqrt{246.44}} = 0.8626\ldots$$

Therefore the PMCC is 0.863 (3 dp), indicating a moderate positive linear correlation.

Notice the PMCC of 0.863 is the same r value the GDC shows in Example 11.

Example 13

a A student obtained a mark of 75 on an English exam, but was absent for the Humanities exam. Justify using the data and the regression line equation from Example 11 to predict the grade he would obtain on the Humanities exam, and find his exam grade.

b If another student was absent for the English exam, and his grade on the English exam was 20, state with reason whether it is justifiable to predict his English grade using the regression equation.

a Since 75 is within the given English exam data range of 18 to 92, and the data showed a moderate positive correlation, the regression equation may be used to predict his Humanities grade.	Notice that $75 \in \{x \mid 18 \le x \le 92\}$, hence the regression equation may be used to interpolate.
$y = 0.528x + 38.7$ $\Rightarrow y = 0.528\ldots(75) + 38.7\ldots$ $= 78.3$ The student is predicted to obtain a grade of 78 on his Humanities exam.	Substitute $x = 75$ into the regression equation.
b Since 20 is outside the range of data for the Humanities exam, $42 \le y \le 100$, it would not be reliable to predict his English grade from his Humanities grade.	If the data used to make a prediction is outside the given range, then extrapolation is not usually reliable to make a prediction.

HINT

Descriptive statistics for both sets of data can be found using your GDC.

You can find the mean, median, standard deviation, IQR, etc, for **both** sets of data.

Investigation 10

A chemical fertilizer company wishes to determine the extent of correlation between the quantity of compound x used and the lawn growth per day.

Find Pearson's correlation coefficient between the two variables:

Lawn	Compound x (g)	Lawn growth y (mm)
A	1	3
B	2	4
C	4	6
D	5	8

1 Complete the table.

x	y	xy	x^2	y^2
1	3			
2	4			
4	6			
5	8			

International-mindedness

Karl Pearson (1857–1936) was an English lawyer and mathematician. His contributions to statistics include the product-moment correlation coefficient and the chi-squared test.

He founded the world's first university statistics department at the University College of London in 1911.

$n =$

$\bar{x} =$

$\bar{y} =$

$\Sigma xy =$

$\Sigma x^2 =$

$\Sigma y^2 =$

2 **Factual** What is the formula for standard deviation?

3 **Conceptual** What does product moment correlation coefficient represent?

Interpolation and extrapolation

A value predicted from within the smallest and largest values is called **interpolation**.

A value predicted from outside the smallest and largest values is called **extrapolation**.

The **accuracy of interpolating** depends on how linear the data is. This can be gauged by the correlation coefficient.

The **accuracy of extrapolating** depends not only on the original data but also assumes that the relationship extends outside the data that we have collected. This depends on the situation under investigation and is **unlikely to be reliable.**

TOK

Is extrapolation knowledge gained using intuition and, possibly, emotion?

If so, how would you describe interpolation in terms of ways of knowing?

Correlation fallacies

International-mindedness

On 28th January 1986, millions of people across the world watched the space shuttle *Challenger* break apart just 73 seconds after being launched.

All seven of the crew members aboard perished in the crash, among them was Christa McAuliffe, the first teacher to be invited into an astronaut team.

The cause of this disaster was the failure of an O-ring, which was meant to prevent the hot gases escaping. It is now known that a leading factor for the O-ring failure was the exceptionally low temperature at the time of the launch.

The consideration of the correlation between the external temperature and the erosion of O-rings is now considered on subsequent space launches.

Below are the definitions of six common fallacies about correlation. Investigate which of these fallacies are predominant in the following examples (there may be more than one for each).

- **Correlation vs causation**

A scatter graph often shows a relationship between two variates of a data set (correlation). However, this does not imply that a change in one variable causes a change in the other (causation). For example, it is not a good explanation of the relation between run 1 and run 2 of the ski race to say that the time in run 2 was slow because the time in run 1 was slow.

- **Correlation is only linear**

Using the PMCC as a measure of correlation only allows the recording of linear correlation. Technology allows non-linear functions of best fit to be allocated to the scatter graph.

- **Theory of the third variable**

This fallacy is very common. The theory of the third variable states that there exists a third underlying variable to which the two variables observed may be related. Time is a common third variable. For example, during the years 2000–15, the number of detox juice bars opening in London and the frequency of knife crime in London were recorded. The data showed a high correlation value. Clearly the third variable time played an important part in the correlation study.

- **Spurious mathematical relationships (ratios, fractions, additions)**

It is essential to ascertain whether there exists a mathematical relationship between the variables which may give rise to an invalid correlation. For example, if a relationship is required between a student's Maths HL grade and total Diploma points, because the Maths grade forms part of the total Diploma grade, then careful examinations must be made before correlation analysis can be applied.

- **Combinations from separate populations**

It is essential that the data set arises from one homogeneous population rather than two distinct groups.

- **Incorrect sampling techniques**

As in the previous section the way that the data has been collected plays an important part in the correlation analysis.

TOK

What is the difference between correlation and causation?

To what extent do these different processes affect the validity of the knowledge obtained?

Investigation 11

A group of marine biologists were investigating the effects of torrential rainfall on the coral in the Great Barrier Reef. They collected the following data. Each marine biologist supplied three data points.

Researcher	1	1	1	2	2	2	3	3	3	4	4	4
Depth of rainfall (mm)	72	70	70	73	68	65	66	70	75	59	59	62
% of new bleaching on coral	0.63%	0.58%	0.56%	0.67%	0.54%	0.53%	0.052%	0.048%	0.061%	0.45%	0.46%	0.55%

1 The Australian government predict that the average rainfall on a nearby reef in the storm season is 78 mm. Using the above data, show that the estimate of percentage of coral bleaching on the new reef would be 0.376%.

2 It is then realized that researcher 3 had not calculated the percentages properly and therefore all of his results were inaccurate and could be classified as unreliable data. They were removed from the data.

Calculate the new prediction for bleaching if 78mm of rain has fallen.

3 **Conceptual** How can you determine if a value is an outlier or unreliable data?

Exercise 5I

1 Consider the following scatter plots and suggest explanations for the *form, strength* and *association*.

A Grade in Maths vs grade in Art

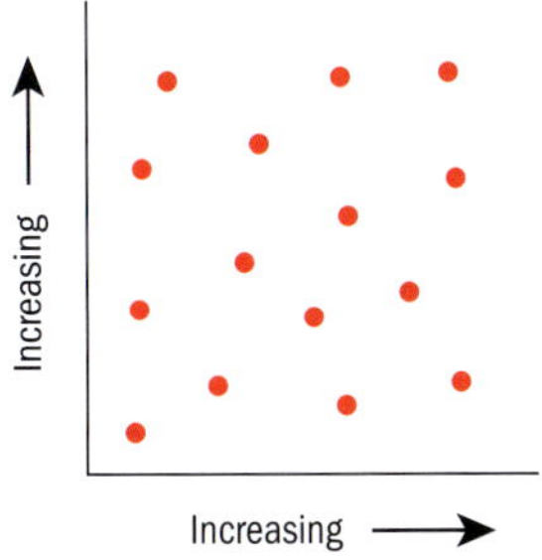

B Height vs shoe size

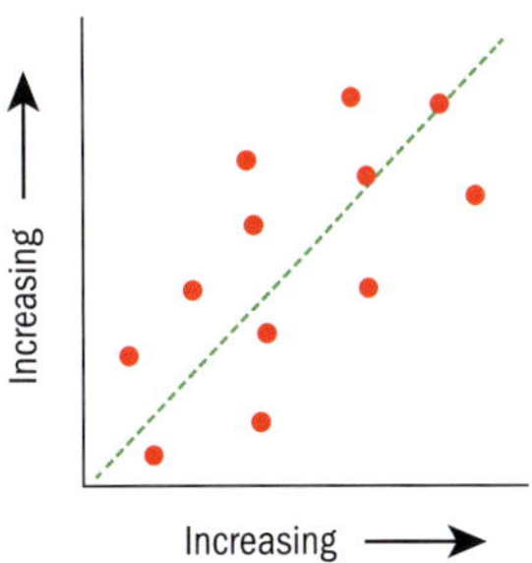

C Amount of revision vs score in test

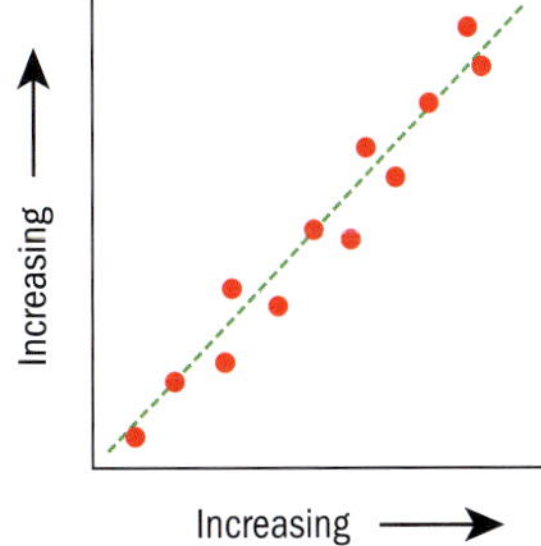

D Gas mileage vs weight of car

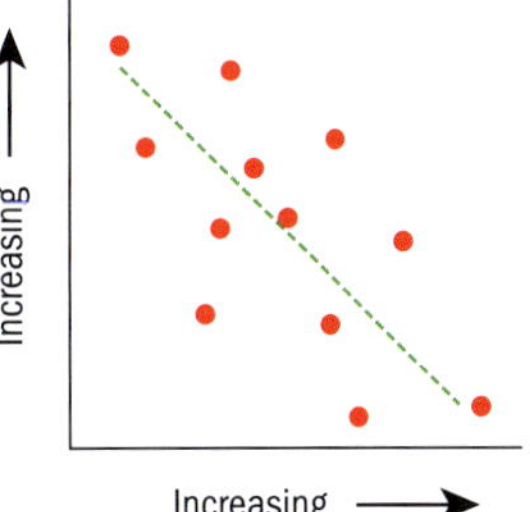

E Temperature vs heating bill

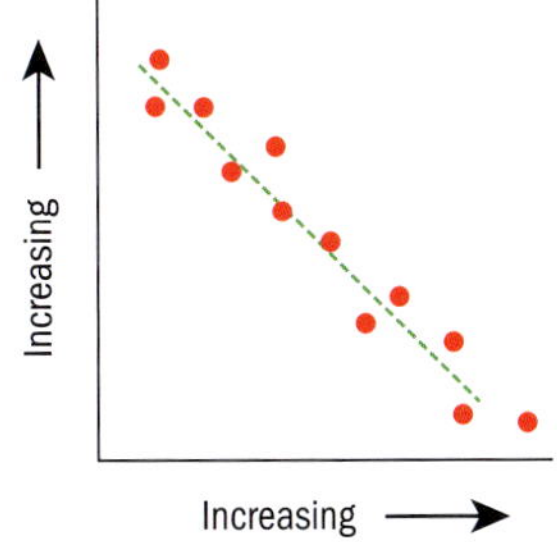

2 You conduct a series of timed experiments on 10 members of a gym in two circuit training events and record the following information.

a 20 squat jumps **b** 20 pull ups **c** weight **d** height

The results are shown below.

	1	2	3	4	5	6	7	8	9	10
a (seconds)	43	52	34	53	43	34	57	32	52	34
b (seconds)	44	49	33	50	45	36	61	42	67	33
c (kg)	68	63	56	73	63	50	75	57	77	55
d (m)	1.64	1.50	1.67	1.75	1.45	1.35	1.87	1.67	1.56	1.70

Draw separate scatter diagrams to represent the following information:

a height vs weight **b** weight vs squat jumps **c** squat jumps vs pull ups.

Comment on the direction, association, form and strength of the correlation.

i Draw a line of best fit on each of the scatter diagrams.

ii Use the line of best fit to estimate the weight of a person measuring 1.68 m.

iii Use the line of best fit to estimate the time taken for a person to do 20 squat jumps given that they took 45 seconds to do 20 pull ups.

iv Determine which graphs would give the most accurate predictions.

3 Cars which are bought new immediately begin to lose their value once they begin to be used. Can you predict the value of a car, given that you know the mileage on the odometer?

Distance driven (km)	4895	75 256	8563	24 495	68 562	58 200	34 011	70 568
Value ($)	37 900	27 495	32 595	38 995	33 895	29 495	34 995	21 000

a Draw a scatter plot to represent the data.

b Determine the strength of the correlation. Use the regression line to predict the price of a car that has driven 50 000 miles.

c State any other information about each car which is important to know when determining its value.

4 The following table gives the heights and weights of 14 race horses.

Height (m) X	1.48	1.51	1.23	1.57	1.29	1.30	1.37	1.17	1.2	1.34	1.42	1.42	1.37	1.44
Weight (kg) Y	329	314	185	356	228	230	257	171	185	214	315	271	242	285

a Use technology to draw a scatter diagram.

b Calculate the line of best fit.

c Comment on the correlation coefficient in the context of the question.

d Use the equation generated by your calculator to predict the weight of a race horse with height of 1.38 m.

Developing your toolkit

Now do the Modelling and investigation activity on pages 364–365.

Chapter summary

- Using statistics:
  - Selection of a relevant aim (the purpose for inquiry)
  - Collection of data (action)
  - Organization of data (action)
  - Analysis of data (action)
  - Reflection on the original aim.
- **Qualitative** variates (data) describe a certain non-numerical characteristic using words.
- **Quantitative** variates (data) have a numerical value. There are two types:
  - **Discrete** variates (data) can be counted (eg number of goals scored) or can only take certain values (eg shoe size).
  - **Continuous** variates (data) can be measured (eg weight and height) and can take any numerical value within a given range.
- The **target population** is the population under investigation.
- The **sampling unit** is each specific member of the population.
- The **sampling frame** is the list of the elements in the population.
- The **sampling variable** is the variable under investigation.
- The **sampling values** are the possible values for the variable.
- A **simple random sample** is each member of the population having an equal chance of being selected.

  Examples of this might include giving each member of the sampling frame a number and using a random number generator to choose the sample.
- Obtaining a simple random sample from a large population can be time-consuming. Often it is better to arrange the population with a system, for example alphabetical order, size order or completion order, and then take the kth member from the list. This is called **systematic sampling**.
- **Stratified sampling** can be used when the population includes clearly defined sections or strata, which possess their own characteristics and are different from each other. The strata must be exhaustive (make up the whole of the population). Separate random samples are taken from each of the individual strata.
- In **quota sampling** the population is divided into certain groups depending on a specific characteristic, eg gender, age, income level. The sample is then obtained by asking as many people as possible until each quota is fulfilled.
- **Cluster sampling** is when the groups are naturally formed and then only that cluster is sampled to ascertain the information on the whole population.
- **Convenience sampling** is when the first/easiest sampling units are approached and their information is recorded.
- If we have a sample of data we write the **arithmetic mean** as $\bar{x}$:

$$\bar{x} = \frac{\sum_{i=1}^{k} f_i x_i}{n}$$

where $\sum_{i=1}^{k} f_i x_i$ is the sum of the set of data and n is the number of values in the data set.

Continued on next page

Statistics and probability

- If we are dealing with the whole population of data, we call the **population mean** μ:

$$\mu = \frac{\sum_{i=1}^{k} f_i x_i}{n}$$

- The **mode** is the value (or values) which occur the most frequently in a set of data.
  - If there is one value (or interval in grouped data) which occurs the most frequently, then that value is defined as the mode and the frequency is the modal value.
  - In a frequency table this is evident as the value taking the highest frequency.
- The **median** is the value which lies exactly in the middle of a set of data.

  If the number of values in the data set is even then the median will lie between these two values.
- The **range** of a data set is a single numerical value (unlike the range of a function) and is calculated by the maximum value in the data set – the minimum value in the data set.
- The median is the middle value when the data is arranged in order of size. The **quartiles** are the middle values on each side of the median and separate the data into quarters:
  - the **lower quartile** or Q_1 represents the value below which 25% of the data set lies
  - the **upper quartile** or Q_3 represents the value above which 25% of the data lies.
- A **bar chart** is used for qualitative data.
  - All rectangles have an equal width.
  - The frequency is recorded on the vertical axis.
  - There are spaces between the rectangles.
- A **histogram** is used for quantitative discrete data and all continuous data.
  - The area of the rectangle is proportional to the frequency.
  - The vertical axis can record *frequency*, *relative frequency* or *frequency density*.
  - There are no spaces between the rectangles.
- One formula used for the **standard deviation** of the population is

$$\sigma = \sqrt{\frac{\Sigma(x-\mu)^2}{n}}$$

- The **standard deviation** shows the variation in a set of data. If the data set is close together, the standard deviation will be small. If the data is spread out, the standard deviation will be large. The units of the standard deviation are the same as the units of the original data.
- A value in the data set is an **outlier** if and only if it is either less than 1.5 x IQR below the lower quartile or more than 1.5 x IQR above the upper quartile.
- If we have a sample and we want to make predictions about the population we use these values and call them unbiased estimators.
  - The unbiased estimator for the population mean is the same value as the sample mean:

$$x = \mu$$

  - The unbiased estimator for the population standard deviation is

$$s_{n-1} = \sqrt{\frac{\Sigma(x-\mu)^2}{n-1}}$$

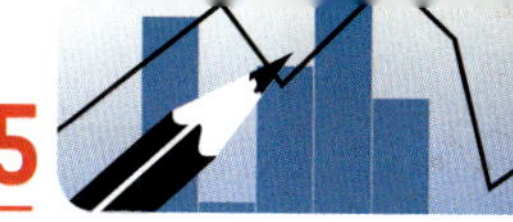

- A **scatter diagram** shows the relationship between two quantitative variables from an individual in the data set. The values of one variable appear on the x-axis and the values of the other variable lie on the y-axis.
  - Each individual in the data set appears as a single point.
  - Unlike univariate graphs, bivariate graphs often form a cloud or "scatter" of points which are subject to considerable variability, and even though there may be a strong underlying linear relationship between the variables the actual data points do not lie on the specific linear equation.
- When you describe a scatter diagram you should be looking for patterns and it is important to classify your observations as:
  - **Association**. If the pattern moves up from left to right you call it ***positive association***: if it moves down from left to right you call it ***negative*** association.
  - **Form**. This refers to the general shape of the pattern (often *linear*).
  - **Strength**. This is determined by how closely the points follow the specified form: **strong, moderate, weak, zero.**
- The correlation r measures the direction of the association and the strength of the linear relationship between two variables. It can be given as a verbal explanation or a numerical quantity.
- The covariance is simply an extension of the formula for variance, specifically looking at the relationship between x and y:

$$s_{xy} = \frac{1}{n}\sum(x-\bar{x})(y-\bar{y}) \quad \text{or} \quad s_{xy} = \frac{\sum xy}{n} = \bar{x}\,\bar{y}$$

Strong negative	Moderate negative	Weak negative	Very weak negative	zero	Very weak positive	Weak positive	Moderate positive	Strong positive
−1	−0.75	−0.5	−0.25	0	0.25	0.5	0.75	1

- In order to draw the most precise line of best fit it is necessary to use a method called **least squares regression**. This method ensures that the line is drawn through the points in such a way that the sum of the squares of the deviations from the points to the line is a minimum.

Statistics and probability

Developing inquiry skills

Return to the opening problem.

1. Represent the scores on a cumulative frequency graph.
2. Write a five-number summary for the data.
3. What would happen to the mean, median and standard deviation if the teacher decided to multiply all of the scores by 10 to show them as a percentage?

Chapter review

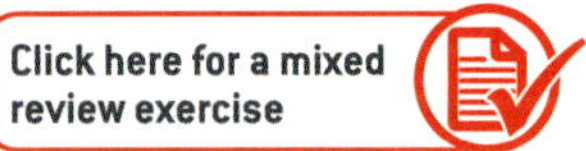

1 A co-educational school comprises 15 classes of 30 students (MYP 1–5) and 150 DP students. There are equal numbers of girls and boys throughout the school.

Describe how you would obtain a stratified sample size of 40 students.

2 A medical school wishes to decide whether to publish a monthly magazine. They decide to select a committee of 12 students to look at the advantages and disadvantages.

The medical school has a seven-year programme and consists of six hundred students in total, 150 in the first year, 100 in the 2nd, 3rd and 4th and 50 in the 5th, 6th and 7th years.

Describe how you would select a committee by

a random sampling

b stratified sampling

c quota sampling

d convenience sampling.

3 Give three rules to be followed when designing a questionnaire.

4 Name a feature of a magazine which is:

a a discrete variable

b a continuous variable.

5 Determine which of the following are discrete or continuous and quantitative or qualitative variables

a taste

b temperature

c cinema attendance

d height.

6 Sketch a frequency distribution where the mode is larger than the mean.

7 The English department wished to perform statistical anaylsis on the first 450 sentences of one of their set literature texts. They counted the number of words in each of the first 450 sentences.

The following frequency table was generated.

Number of words per sentence	Number of sentences
1–10	13
11–20	16
21–30	146
31–40	139
41–50	84
51–60	32
61–70	20

a Estimate the mean, median and modal group.

b Estimate the percentage of sentences containing more than 40 words.

c Construct a cumulative frequency table and cumulative frequency curve.

d Extract the five-point summary from the cumulative frequency curve.

e Draw both box-and-whisker diagrams.

f State the assumption upon which the estimates are made.

g Discuss the differences between the box-and-whisker diagram from the table and from the curve.

8 The number of siblings of a class of 60 students is given below.

1	0	0	1	1	0	2	0	2	1
2	1	2	0	3	1	3	1	1	2
1	3	1	1	5	0	1	2	1	0
2	3	1	1	1	0	1	0	1	1
0	5	2	3	2	2	0	1	1	0
1	1	1	0	1	0	1	1	1	2

a Tabulate the results and describe the frequency distribution.

b Calculate the mean number of siblings and the standard deviation.

c Draw a box-and-whisker diagram to show your findings.

d A further 32 students' data was then given. The mean number of siblings for all 92 students was 1.25. Of the further 32 students, calculate the number with two siblings given that the remaining students had 27 siblings between them.

9 In an experiment the following values were obtained for x and y.

x	0	0.5	1	1.5	2	2.5	3	3.5
y	0.6	0..45	0.8	0.85	1.4	1.65	2.4	2.85

a Calculate y^2.

b Draw a scatter diagram to illustrate this data for y^2 against x and draw a regression line.

c Hence find an equation for y^2 in terms of x and therefore y in terms of x.

10 A regression line passes through the point (1, 8) and has a gradient of −0.5.

Estimate the value of y when $x = 7$.

Exam-style questions

11 P1: Consider the following set of data:
3, 6, 1, 5, a, b, where $a > b$.

The mode of this data is 5. The median of this data is 4.5.

a Find the value of a and the value of b. (5 marks)

b Find the mean of this data. (2 marks)

12 P1: a A set of 10 student friends have a mean mass of 70 kg. A new student called Steve joins this friendship group. The new mean mass of the 11 students is now 72 kg. Find Steve's mass. (4 marks)

b The new lower quartile and upper quartile for the masses of the 11 students are 66 and 76 respectively. Determine whether Steve's mass is an outlier, and justify your answer. (3 marks)

13 P2: Sue collects continuous data on the heights of flowers, and displays it in the table below.

Height, h (cm)	$0 < h \le 10$	$10 < h \le 20$	$20 < h \le 30$	$30 < h \le 40$	$40 \le h \le 50$
Frequency	40	45	50	60	5

a State how many flowers Sue measured the heights of. (1 mark)

b Find the mid-point of the modal interval. (1 mark)

c Calculate an estimate for the **i** mean **ii** standard deviation. (5 marks)

d Sue's calculator states that the median is 25. Find a better estimate than this by considering the median's position within the interval that it belongs to. Give your answer to the nearest integer. (3 marks)

14 P2: Sally and Rob both teach different IB Maths classes. In a test, Sally's students scored

1, 1, 4, 7, 8, 8, 10, 10

In the same test, Rob's students scored
4, 4, 4, 5, 6, 6, 10, 10, 10, 10

a For the data from Sally's class, find **i** the median **ii** the mean. (3 marks)

b For the data from Rob's class find **i** the median **ii** the mean. (3 marks)

c Give a reason why Sally could claim that her class performed better than Rob's class. (1 mark)

d Give a reason why Rob could claim that his class performed better than Sally's class. (1 mark)

15 P2: Discrete data showing the marks of 76 students who each took the same IB exam is given in the table below.

Grade	1	2	3	4	5	6	7
Frequency	4	8	16	20	16	8	4

a Sketch a bar chart to represent this data. (3 marks)

b For this data find:

i the mode

ii the median

iii the mean. (3 marks)

c Explain any similarities between the answers to parts **b ii** and **iii** by referring to a geometrical property of the bar chart drawn in part **a**. (2 marks)

16 P2: Paired, bivariate data (x, y) that is strongly correlated has a y on x line of best fit of the form $y = mx + c$. When $x = 70$ an estimate for y is 100. When $x = 100$ an estimate for y is 140.

a Find the values of **i** m **ii** c. (3 marks)

b State if there is positive or negative correlation. (1 mark)

c The value of $\bar{x}$ is 90. Find the value of $\bar{y}$. (3 marks)

d When $x = 60$ find an estimate for the value of y, given that this is interpolation. (2 marks)

17 P2: Ten teenagers were asked their age and the number of brothers and sisters they have. The data is shown in the scatter diagram below, where x represents the teenager's age and y represents the number of brothers and sisters that they have.

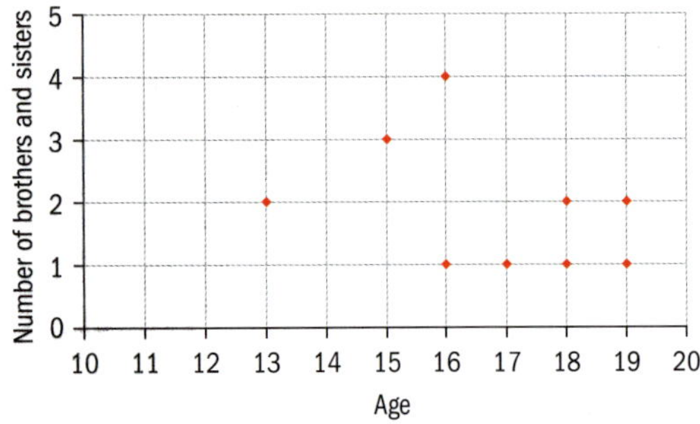

a Use the scatter diagram to copy and complete the following table.

x	13	14	15	16	16	17	18	18	19	19
y					4			2	1	

(3 marks)

b Calculate the Pearson product moment correlation coefficient for this data. (2 marks)

c Give two reasons why it would not be valid to use this scatter diagram to estimate the number of brothers and sisters a 25-year-old might have. (2 marks)

18 P1: A set of bivariate data has a Pearson product-moment correlation coefficient of $r = 0.870$ for 25 pairs (x, y). The y on x line of best fit is given by $y = 15x + 11$. Consider each of the scatter diagram, the value of r and line of best fit – together with their definitions – to answer the following.

a All the original x values are increased by adding 5 and all the original y values are decreased by subtracting 4.

i State the new value of r.

ii State the new value for the gradient of the y on x line of best fit.

iii Give a reason for your answers to **i** and **ii**.

iv Describe in two words the linear correlation that exists for this new data. (5 marks)

b All the original y values are altered by multiplying by 2 and all the original x values remain unchanged.

i State the new value of r.

ii State the new value for the gradient of the y on x line of best fit.

iii Give a reason for your answers to **i** and **ii**. (3 marks)

c All the original x values are altered by multiplying by -3 and all the original y values remain unchanged.

i State the new value of r.

ii State the new value for the gradient of the y on x line of best fit.

iii Give a reason for your answers to **i** and **ii**.

iv Describe in two words the linear correlation that exists for the new data. (6 marks)

19 **P2:** Paired bivariate data (x, y) is given in the table below. The data represents the heights (x metres) and lengths (y metres) of a rare type of animal found on a small island.

x	2.4	3.6	2.8	1.8	2.0	2.2	3.0	3.4
y	3.0	4.0	3.0	1.7	2.0	2.3	3.1	2.7

a **i** Calculate the Pearson product moment correlation coefficient for this data.

ii In two words, describe the linear correlation that is exhibited by this data.

iii Calculate the y on x line of best fit. (6 marks)

Another four examples of this rare animal are found on a nearby smaller island. This extra data is given in the table below.

x	2.3	2.7	3.0	3.5
y	4.1	1.5	4.2	1.5

b **i** Calculate the Pearson product moment correlation coefficient for the combined data of all 12 animals.

ii In two words, describe the linear correlation that is exhibited by the combined data.

iii Suggest a reason why it would not be particularly valid to calculate the y on x line of best fit for the combined data.

(5 marks)

20 **P2:** Ten pairs of twins take an intelligence test. For each pair of twins, one twin is female and the other is male. The bivariate data obtained is given in the table below.

Female	100	110	95	90	103	120	97	105	89	111
Male	98	107	95	89	100	112	99	101	89	109

a Find the Pearson product moment correlation coefficient, r. (2 marks)

b State, in two words, a description for this linear correlation. (2 marks)

c Letting the male score be represented by x and the female score by y, find the equation of the

i y on x line of best fit

ii x on y line of best fit. (4 marks)

d Another pair of female/male twins are discovered. The male scored 105 on the test but the female was too ill to take it. Estimate the score that she would have obtained, giving your answer to the nearest integer. (1 mark)

e Yet another pair of female/male twins are discovered. The female scored 95 on the test but the male refused to take it. Estimate the score that he would have obtained, giving your answer to the nearest integer. (1 mark)

f If, for a further pair of male/female twins, the male scored 140 on the test, explain why it would be unreliable to use a line of best fit to estimate the female's score.

(1 mark)

Rank my maths!

Approaches to learning: Collaboration, Communication
Exploration criteria: Personal engagement (C), Use of mathematics (E)
IB topic: Bivariate data, Correlation, Spearman's rank

Pearson product moment correlation coefficient

As you have seen in this chapter, the Pearson product moment correlation coefficient (PMCC) evaluates the strength of linear correlation between two variables.

What happens if there appears to be a relationship, but the correlation does not appear to be linear?

You can represent these data on a scatter diagram as shown:

x	4	5	6	7	8	9	10	11	12	13	14
y	4	4.1	4.2	4.4	4.7	5.1	5.5	6	7	9	20

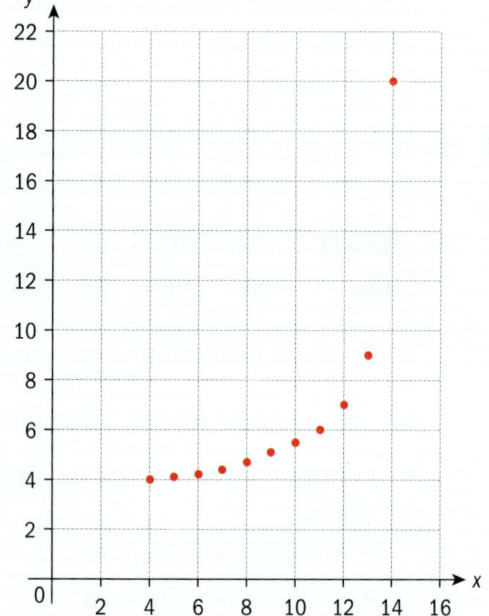

Do you think there is a relationship between the variables x and y?

Is the relationship strong? Describe the relationship.

Find the Pearson's correlation coefficient, r, for this data.

What does this tell you about the correlation for this data?

Explain why the relationship is very strong but the correlation coefficient does not reflect this.

Spearman rank correlation

You can also use **Spearman rank correlation** to analyse the relationship:

The Spearman correlation evaluates the relationship between variables which is not necessarily linear. The Spearman correlation coefficient is based on the **ranked values** for each variable rather than on the **raw data**.

What are the main differences between the Pearson correlation and the Spearman correlation?

The formula for Spearman's correlation, r_s, is:

$$r_s = 1 - \frac{6\sum D^2}{n\left(n^2 - 1\right)}$$

where D is the difference between a pair of scores and n is the number of pairs of ranks.

Research how to interpret Spearman's rank correlation.

What is the value of Spearman's rank for the above data?

What does this value of r_s tell you about these data?

Country	GDP/Capita in US$(2017)	Number of children per woman(2017)
Afghanistan	1981	5.12
Argentina	20787	2.26
Australia	47047	1.77
Barbados	18640	1.68
Brazil	15484	1.75
Cambodia	4002	2.52
China	16807	1.60
Gambia	1715	3.52
Germany	50715	1.45
Haiti	1815	2.72
India	7056	2.43
Japan	43876	1.41
Kenya	3286	2.98
Malawi	1202	5.49
Moldova	5698	1.57
Nepal	2682	2.12
Niger	1017	6.49
Paraguay	9691	1.90
Peru	13434	2.12
Philippines	8342	3.02
Poland	29291	1.35
Qatar	128378	1.90
Russia	25533	1.61
Serbia	15090	1.44
Spain	38091	1.50
Tunisia	11911	2.23
United Kingdom	43877	1.88
United States	59532	1.87
Uzbekistan	6865	1.76
Zimbabwe	2086	3.98

Activity 1

Monotonic

Monotonic

Non-Monotonic

Describe the relationship between GDP/Capita and Number of children per woman.

Why is Pearson's correlation not appropriate for these data?

Calculate the Spearman's rank correlation coefficient for this data.

Describe the relationship between GDP/Capita and Number of children per woman.

Activity 2

Your teacher will instruct you, in groups, on how to create a selection of different pieces of music.

Individually rank the pieces of music you have selected by number, with 1 being the favourite.

Record the rankings of everyone in your group. Do **not** collaborate or communicate with each other.

Find the Spearman rank correlation between each pair of students in your group.

Do any of the pairs of students display strong correlations?

Write a conclusion for this experiment based on the results you have found.

Extension

What else could you compare the ranking of using the same process as in Activity 2?

Brainstorm some possible ideas within your groups.

Design and run an experiment for one of your ideas.

Think carefully about what you would need to be aware of as you run this experiment to ensure that your results are accurate and unbiased.

6 Relationships in space: geometry and trigonometry

Geometry and trigonometry allow us to quantify the physical world, enhancing our spatial awareness in two and in three dimensions. This topic provides you with the tools for analysis, measurement and transformation of quantities, movements and relationships.

In this chapter you will investigate how the properties of shapes depend on the space they occupy in both two and three dimensions, and how formulae can be used to make specific calculations. Leading on from this you will identify 2D relationships extended into 3D shapes and the importance of right-angled triangles. Finally, you will look at how to solve trigonometric equations and use differential calculus to find related rates involving trigonometric expressions.

Concepts

- Space
- Relationships

Microconcepts

- Angle between two lines/line and a plane
- Distance between/midpoint of two points
- Volume and surface area of 3D solids
- The circle: length of arc, area of sector
- Radian measure of angles
- Trigonometric ratios
- Exact values of trigonometric ratios of special angles
- Sine rule, including ambiguous case. Cosine rule.
- Area of a triangle and applications
- Pythagorean identities
- Compound angle and double angle identities
- Circular functions (domain, range, periodicity)
- Graphs of trigonometric functions including general form
- Relationships between trigonometric functions and symmetry properties of their graphs
- Inverse trigonometric functions
- Solving trigonometric equations in a finite interval both graphically and analytically
- Equations leading to quadratic equations in $\sin x$, $\cos x$ and $\tan x$
- Derivatives of trigonometric functions

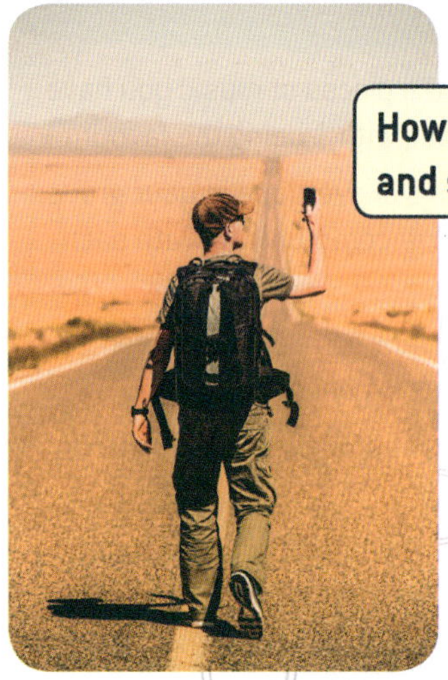

How far can I travel and still be in range?

Will this hold enough water for a month?

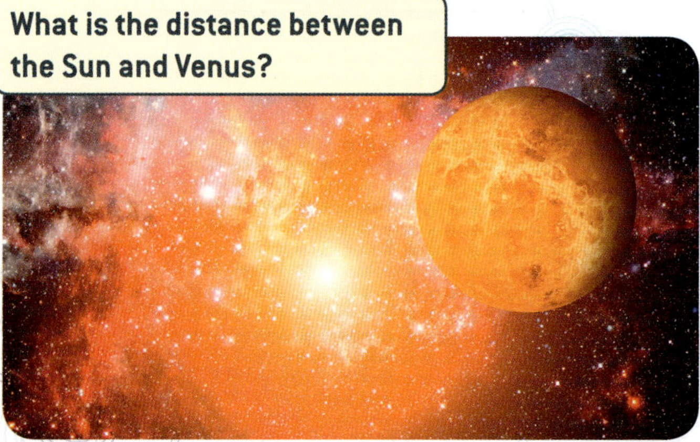

What is the distance between the Sun and Venus?

The Ain Dubai, or Dubai Eye, is set to be the largest Ferris wheel in the world. It will be over 200 metres tall, with a capacity of 1400 passengers. In planning, some of the following questions need careful consideration.

The London Eye has 32 pods, each with a maximum capacity of 25 people. The total expected capacity of the Ain Dubai is 1400. Presuming the pods are the same size, how many pods will the Ain Dubai have?

The boarding platform is 15 metres above the ground.

- Find the relevant dimensions of the Ferris wheel.
- Find the volume of each pod.
- Find the arc distance between each pod.

Given that one revolution takes 30 minutes, find the angle of rotation per minute and find the speed in km per hour. Do you think this constant speed would allow passengers to embark and alight without the wheel stopping?

- What are the limitations of this model?

Determine the vertical height from the boarding platform at each of the following times:

a after 14 minutes

b three minutes after the pod is at its highest point.

In order to see the best views, the pod must be 170 m from the floor. Write a mathematical inequality to determine the maximum time the pod is above 170 m. Determine the distance travelled within that time.

Developing inquiry skills

Write down any similar inquiry questions you might ask if you were building another structure. What different questions might you need to ask?

Think about the questions in this opening problem and answer any you can. As you work through the chapter, you will gain mathematical knowledge and skills that will help you to answer them all.

Before you start

You should know how to:

1 Use Pythagoras' theorem.

eg In the diagram AC and BD are perpendicular to AB. Point P can move along AB. If [AC] = 3 cm, [BD] = 6 cm and [AB] = 8 cm, find the distance of P from A for which DP + PC is a minimum.

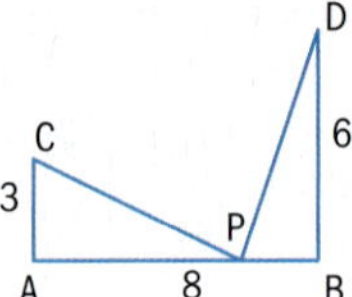

Let AP = x

In triangle APC, PC = $\sqrt{9+x^2}$

In triangle PBD, PD = $\sqrt{36+(8-x^2)}$

PC + PD = $\sqrt{9+x^2} + \sqrt{100-(16x+x^2)}$

Using a GDC, minimum occurs when $x = 2.67$cm.

2 Work with similar triangles.

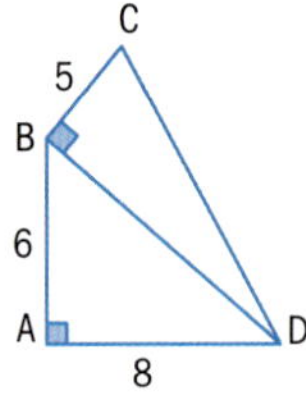

eg the diagram shows two right-angled triangles ABD and BCD. Use similar triangles to find the height of C above AD.

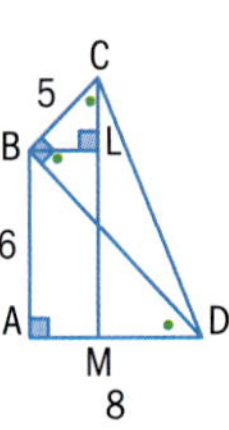

Construct a perpendicular from C to AD and a horizontal line through B.

In triangle ABD, BD = $\sqrt{6^2+8^2} = 10$

$A\hat{D}B = D\hat{B}L = B\hat{C}L$

Therefore triangles BCL and BAD are similar.

$\frac{CL}{AD} = \frac{CB}{BD}$

$\frac{CL}{8} = \frac{5}{10}$

CL = 4

LM = AB = 6

CM = 6 + 4 = 10

Skills check

Click here for help with this skills check

1 The length of a rectangle is twice the width of the rectangle. The length of a diagonal of the rectangle is 25 cm. Calculate the area of the rectangle.

2 ACQ and BCP are straight lines.

AB is parallel to PQ.

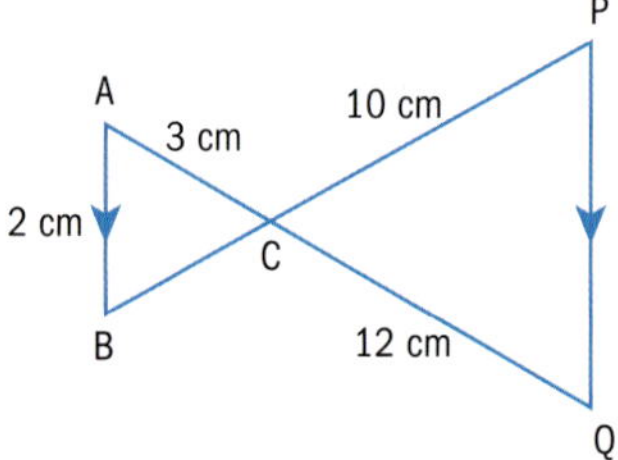

Prove that triangle ABC is similar to triangle CPQ.

6.1 The properties of three-dimensional space

In 2D space the x-axis is horizontal and the y-axis is vertical. When moving to 3D space the x–y plane can be considered the horizontal plane and the z-axis is now vertical.

For mathematical consistency, the "right-hand orientation" for the three axes is generally used. The positive x-axis is pointing in the direction of the first finger. The positive y-axis pointing in the direction of the second finger and the positive z-axis pointing in the direction of the thumb.

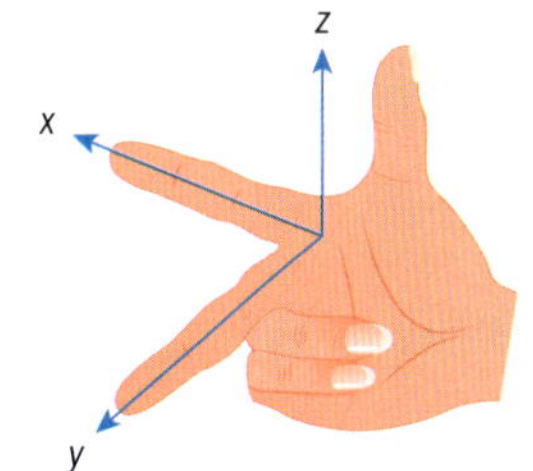

Investigation 1

Conduct this investigation in pairs.

On a piece of A4 paper, draw any two-dimensional net of a cube (there are 11 possibilities), with sides 4 cm, and cut it out.

Reflect Why are there precisely 11 nets of a cube?

Fold along the lines to make a 3D space. Tape the cube together so that it keeps its shape.

Label one vertex as the origin, O.

Label points ABCDEFG as the vertices of the cube, such that OA indicates the x-axis, OC indicates the y-axis and OD indicates the z-axis

Mark the points 1, 2, 3, 4 on the x-, y- and z-axes at 1 cm intervals.

You now have cube OABCDEFG with sides 4 cm.

The rule for the game is to start at O, and move in the x-, y- or z-direction along the outside of your cube.

From O move 4 units along the x-axis to get to A. Therefore, A has coordinates (4, 0, 0).

1 Find the 3D coordinates of B, C, D, E, F and G.

2 Place a dot on one of the faces. Ask your partner to find the coordinates. Change roles and try this a few times.

3 Place a dot on one of the edges. Ask your partner to find the coordinates. Change roles and try this a few times.

4 **Conceptual** How do 3D coordinates differ from 2D coordinates?

5 Find the coordinate of the point at the centre of the cube.

6 Find the length of OB. Find the length of OF.

7 Take turns to calculate the length between any two points from step 2 and step 3 of the game.

8 **Conceptual** How does understanding 2D geometry enhance the understanding of 3D geometry?

In order to generalize relationships in 2D and 3D space, it is important to create formulae.

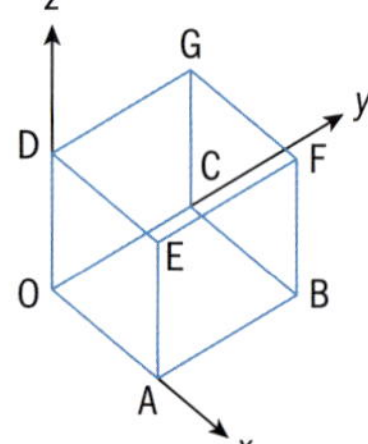

The midpoint and distance formulae

The coordinates of the midpoint of a line segment are the averages of the x-coordinates, the y-coordinates and the z-coordinates of the corresponding endpoints.

The midpoint of (x_1, y_1, z_1) and (x_2, y_2, z_2) is $\left(\frac{x_1 + x_2}{2}, \frac{y_1 + y_2}{2}, \frac{z_1 + z_2}{2}\right)$.

In the cube from Investigation 1, the distance OF can be calculated using Pythagoras' theorem. When calculating with 3D spaces, it is easier to show the appropriate individual 2D space.

In triangle OAB:

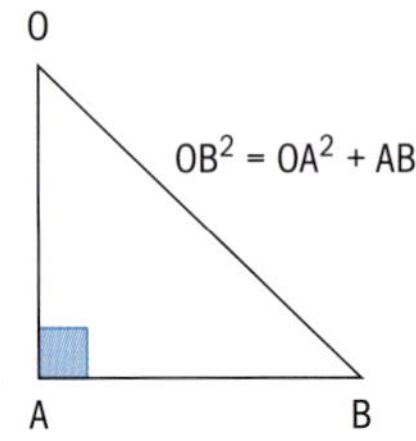

In triangle OBF:

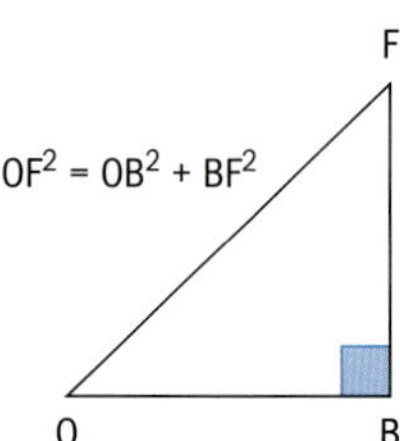

$OF^2 = (OA^2 + AB^2) + BF^2$

$OF = \sqrt{OA^2 + AB^2 + BF^2}$

Since OA is the change in x, $(x_2 - x_1)$, AB is the change in y, $(y_2 - y_1)$ and BF is the change in z, $(z_2 - z_1)$.

The distance (d) from (x_1, y_1, z_1) to (x_2, y_2, z_2) is

$$d = \sqrt{(x_2 - x_1)^2 + (y_2 - y_1)^2 + (z_2 - z_1)^2}$$

HINT

The squares of the expressions are always positive, therefore the points are interchangeable.

Example 1

Find the midpoint and distance of AB when A is (1, 2, 5) and B is (3, 3, 7).

Midpoint of AB: $\left(\frac{x_1+x_2}{2}, \frac{y_1+y_2}{2}, \frac{z_1+z_2}{2}\right)$ $= \left(\frac{1+3}{2}, \frac{2+3}{2}, \frac{5+7}{2}\right)$ $= (2, 2.5, 6)$	Apply the formula.
Distance AB: $d = \sqrt{(x_2-x_1)^2+(y_2-y_1)^2+(z_2-z_1)^2}$ $d = \sqrt{(3-1)^2+(3-2)^2+(7-5)^2}$ $d = \sqrt{4+1+4}$ $d = \sqrt{9}$ $d = 3$	

EXAM HINT

One of the best ways to show working and obtain full marks in DP exams is to write down a formula, substitute and evaluate. Good mathematical layout is to keep the equals signs aligned as you move down the page.

Relationships in triangular space

Consider the triangle ABC with a right angle at B and A placed on the origin.

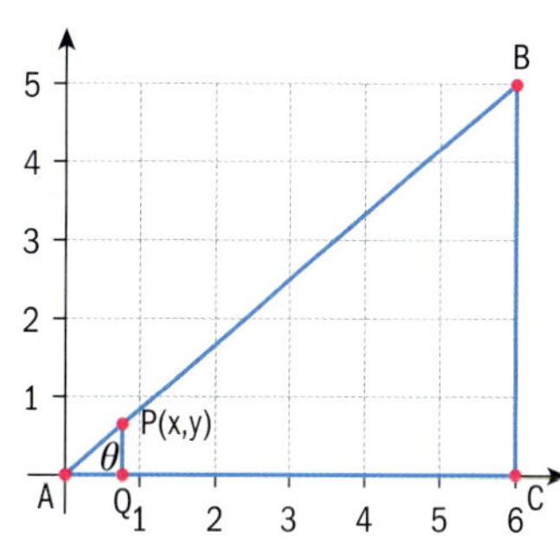

Triangles ABC and AQP are similar since they have equal angles.

Therefore, it follows that $\frac{PQ}{AP} = \frac{BC}{AC}$.

This ratio is $\frac{\text{opposite side}}{\text{hypotenuse}}$.

Due to the relative position of the sides and the acute angle θ this ratio is called sine θ.

Similarly, $\frac{AQ}{AP} = \frac{AB}{AC} = \cos\theta$ and $\frac{PQ}{AQ} = \frac{BC}{AB} = \tan\theta$.

Since you can do this for any size of right-angled triangle ABC you can also say that the trigonometric ratios for angle θ in triangle ABC are:

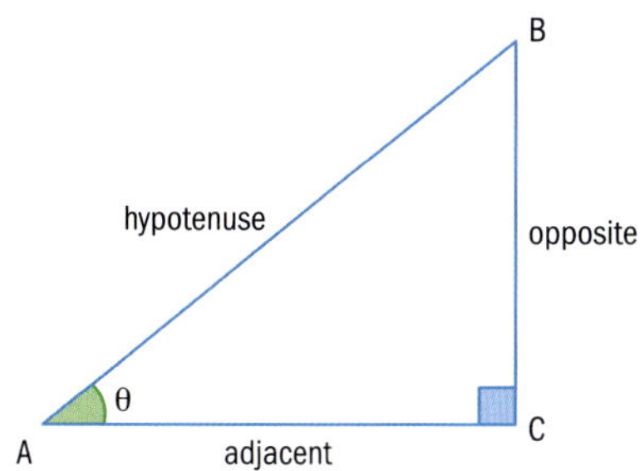

$$\sin\theta = \frac{BC}{AC} = \frac{\text{opposite}}{\text{hypotenuse}}$$

$$\cos\theta = \frac{AB}{AC} = \frac{\text{adjacent}}{\text{hypotenuse}}$$

$$\tan\theta = \frac{BC}{AB} = \frac{\text{opposite}}{\text{adjacent}}$$

HINT

You have to refer to the sides of the triangle as opposite, adjacent, and hypotenuse because by our definition of the trigonometric ratios, you need the sides and their relative position to the angle θ.

With these definitions and a calculator, right-angled triangles can be solved:

- Given any acute angle and the length of one side you can find the third angle and the lengths of the other two sides.
- Given any two sides of the triangle you can find the length of the third side and the sizes of the two unknown angles.

HINT

A GDC can calculate trigonometric ratios of any angle measured in either degrees or radians. You must ensure that the mode is set to the angle measure you are using.

If you want to find the angle θ whose trigonometric ratio you know, for example, $\sin\theta = \frac{5}{7}$, you use the notation $\theta = \arcsin\frac{5}{7}$. On a GDC this is denoted by $\sin^{-1}$. Therefore if $\sin\theta = \frac{p}{q}$, then $\theta = \sin^{-1}\left(\frac{p}{q}\right) = \arcsin\left(\frac{p}{q}\right) \neq \frac{1}{\sin\theta}$.

This will be discussed in more detail later on in the chapter.

Example 2

Solve for the unknown angles and sides of the following triangles:

a

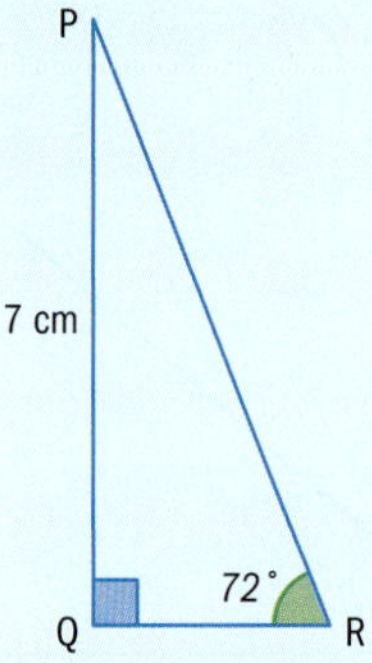

b

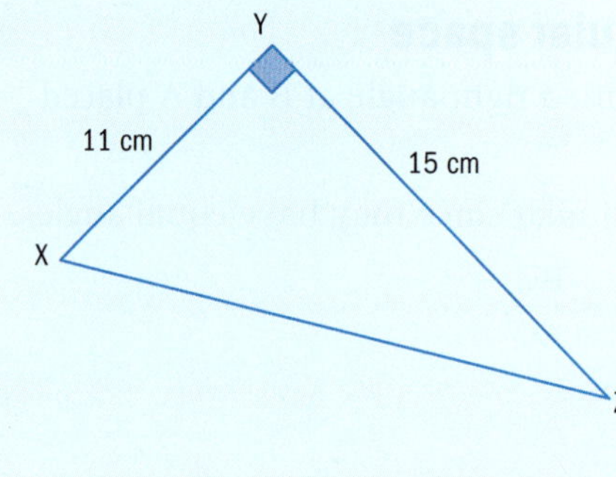

a In triangle PQR, $\text{Q}\hat{\text{P}}\text{R} = 90° - 72° = 18°$	
$\frac{\text{PQ}}{\text{PR}} = \sin 72° \Rightarrow \text{PR} = \frac{7}{\sin 72} = 7.36\text{ cm}$	When using an answer in a new calculation you must make sure that you use the unrounded answer, not the approximate answer to 3 sf.
$\text{QR} = \sqrt{(7.36..)^2 - 7^2} = 2.27\text{ cm}$	Using Pythagoras' theorem.
b In triangle XYZ, $\text{XZ} = \sqrt{11^2 + 15^2} = 18.6\text{cm}$	
$\tan \text{Y}\hat{\text{X}}\text{Z} = \frac{15}{11}$	
$\Rightarrow \text{Y}\hat{\text{X}}\text{Z} = \arctan\left(\frac{15}{11}\right) = 53.7°$	
$\text{Y}\hat{\text{X}}\text{Z} = 90° - 53.7° = 36.3°$	

Angles in three-dimensional space

In the cuboid, ABCD is the horizontal plane. The angle between the diagonal CE and the plane ABCD is labelled θ (theta). As the 3D shape is a cuboid, right-angle trigonometry can be used.

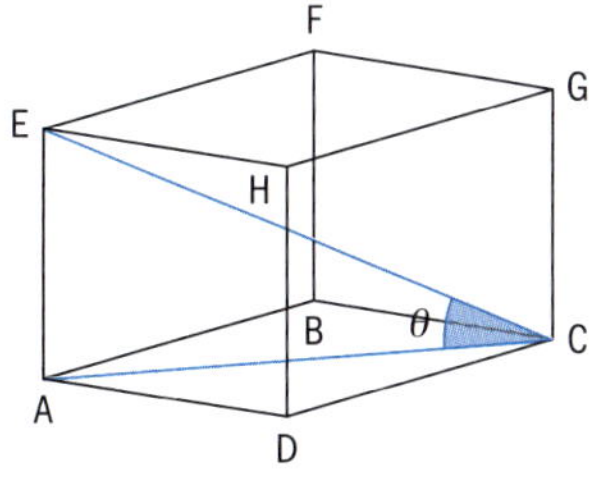

HINT

All the angles in the cuboid can be found with this method by rotating the cuboid.

In right-angled triangles there exists a specific ratio between the sides and the angles.

$$\sin\theta = \frac{\text{opposite side}}{\text{hypotenuse}}$$

$$\cos\theta = \frac{\text{adjacent side}}{\text{hypotenuse}}$$

$$\tan\theta = \frac{\text{opposite side}}{\text{adjacent side}}$$

International-mindedness

The word "trigonometry" is derived from the Greek words "trigon", meaning triangle and "metria", meaning measure.

Example 3

Find the angle between the diagonal CE and the plane ABCD in this cuboid. Give your answer to the nearest degree.

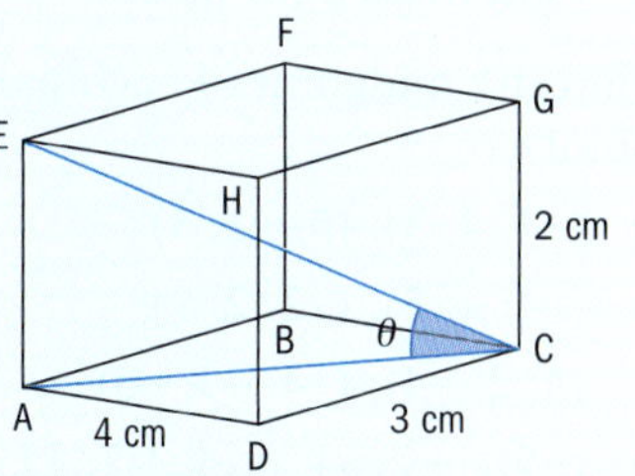

$AC^2 = 3^2 + 4^2$ $AC^2 = 25$ $AC = 5$	Use Pythagoras for the base ABCD.
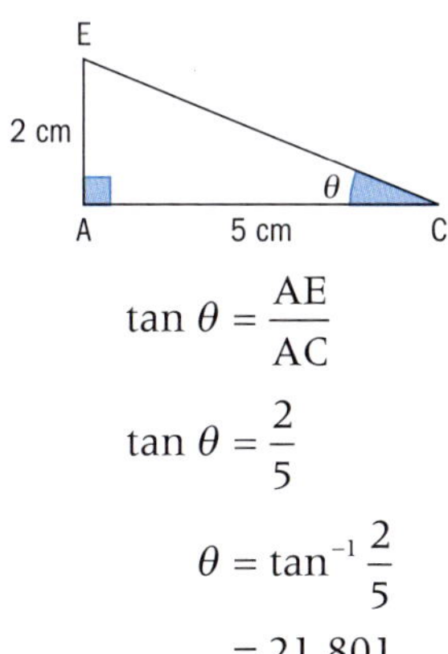	Construct a new 2D representation of the 3D space.
$\tan\theta = \frac{AE}{AC}$ $\tan\theta = \frac{2}{5}$ $\theta = \tan^{-1}\frac{2}{5}$ $= 21.801\ldots$	Use right-angled trigonometry.
$\theta = 22°$	Re-read the question for the desired degree of accuracy.

Exercise 6A

1 The cuboid, OABCDEFG, is such that the length of OA is 3 units, the length of OC is 4 units and the length of OD is 2 units.

A lies on the x-axis, C lies on the y-axis and D lies on the z-axis.

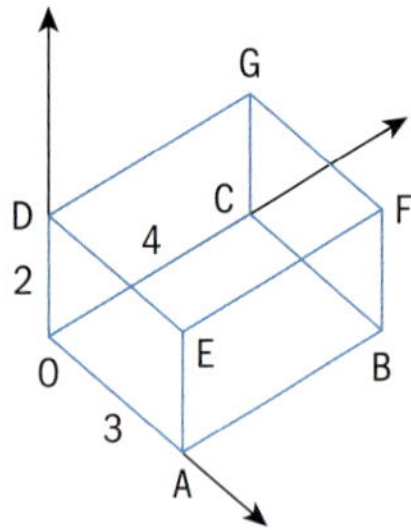

a Write down the coordinates of

i A **ii** B **iii** E **iv** F

b Find the midpoint of OF.

c Find the length of OF.

2 Find the midpoint of each pair of points.

a (−4, 4, 3), (5, −1, 3)

b (−4, 4, 5), (−2, 2, 9)

c (5, 2, −4), (−4, −3, −8)

d (−5.1, −2, 9), (1.4, 1.7, 11).

3 Find the distance between each pair of points.

a (2, 3, 5) and (4, 3, 1)

b (−3, 7, 2) and (2, 4, −1)

c (−1, 3, − 4) and (1, −3, 4)

d (2, −1, 3) and (−2, 1, 3).

4 ABCDE is a square-based pyramid, where A is the apex and BCDE are the vertices of the square. The lengths of the sides of the square are 20 metres and the vertical height is 15 metres.

a Draw a labeled sketch of the pyramid.

Add the point M, the midpoint of BC.

b Calculate the angle between the face ABC and the base BCDE.

c Calculate the angle between the slant edge AB and the base BCDE.

5 Find the angle between the diagonal FD and the base ABCD in this cuboid. Give your answer to the nearest degree.

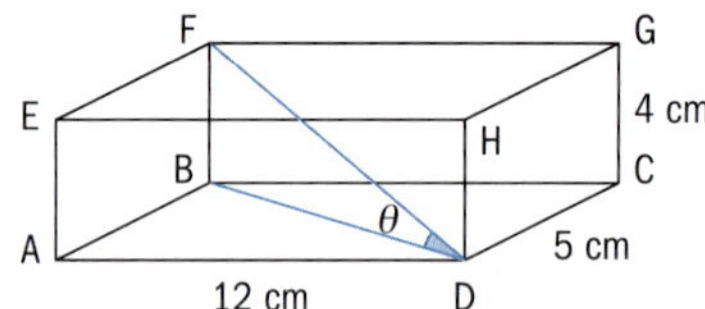

6 ABCDEFGH is a cuboid with AB 4 cm, BC 4 cm and CG 8 cm.

a Find the length of AG.

b Find the angle between the diagonal AG and the plane ABCD to the nearest degree.

c Hence find the angle between AG and the plane EFGH.

d Find the angle between the diagonal AG and the plane ADHE to the nearest degree.

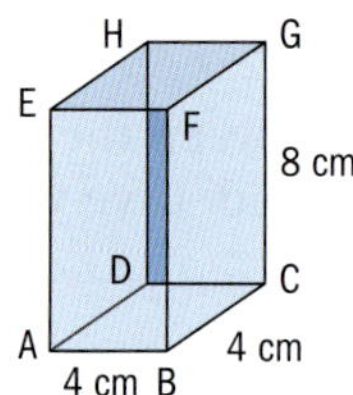

7 ABCDE is a square-based right pyramid with AB 8 cm. The point E is 8 cm directly above the point M.

a Find CA.

b Find AM.

c Find EA.

d Find the angle between EA and the base.

Surface area of 3D shapes using differential calculus

Relationship between 3D, 2D and 1D

When you first studied circles, the most important one-dimensional measurements were the circumference and diameter. From the ratio of these measurements, the constant π was introduced.

Area is the measure of space inside a 2D shape. Once this had been defined, an important relationship between the area and the radius and circumference could be inferred:

$$\text{circumference} = \frac{\text{change of area}}{\text{change in radius}}$$

Using differential calculus this can be expressed as:

$$\text{circumference} = \frac{\mathrm{d}A}{\mathrm{d}r}$$

In 3D shapes, the surface area can be expressed as the change in volume with respect to the change in radius:

$$\text{surface area} = \frac{\mathrm{d}V}{\mathrm{d}r}$$

International-mindedness

Area can be measured in square units like cm^2 or m^2, but also "packets" like hectares or acres.

Which countries use bigha, mou, feddan rai, tsubo?

Investigation 2

1 Given that the area of a circle is πr^2, use calculus to find the circumference of a circle.

2 Given that the volume of a sphere is $\frac{4}{3}\pi r^3$, use calculus to find the surface area.

3 **Conceptual** How can these releationships be extended to squares and cubes?

Consider a square in the following way:

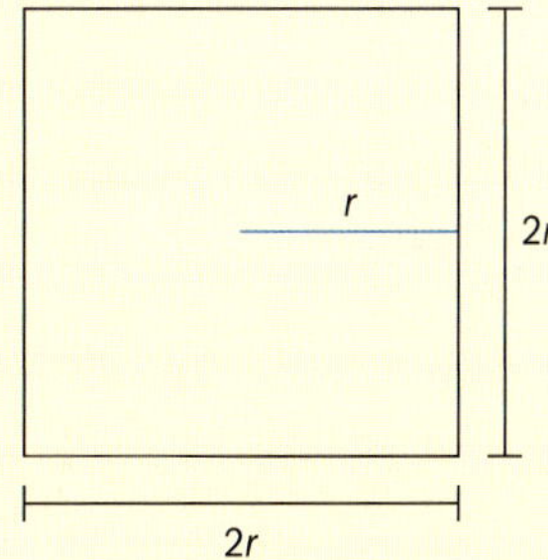

4 Find the area of this square and the corresponding volume of the cube.

5 Hence, find the perimeter of the square and the surface area of the cube.

6 **Conceptual** What is the relationship between circumference and area in a 2D shape? What is the relationship between surface area and volume in a 3D shape?

Geometry and trigonometry

Volumes and surface area of other 3D shapes

- The surface area of a regular pyramid: $S = A_{base} + nA_{triangle}$ where n is the number of sides of the regular polygon that makes up the base.

Pythagoras' theorem is used to find missing lengths in triangles in order to find the areas of each triangular face

- Surface area of a cone: $S = \pi r^2 + \pi rs$ where r is the radius of the circle at the base of the cone and s is the slant height of the cone.

TOK

What are the platonic solids and why are they an important part of the language of mathematics?

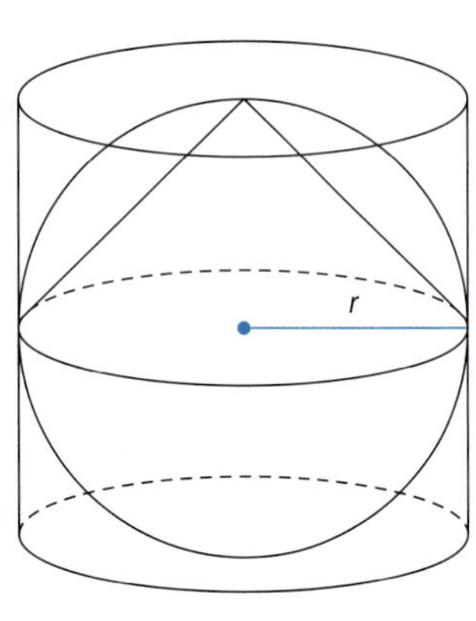

Use in the formulae:

- Volume of a pyramid: $V = \frac{1}{3}A_{base}h$ where h is the vertical height of the pyramid.
- Surface area of a cone: $S = \pi r^2 + \pi rs$ where r is the radius of the circle at the base of the cone and s is the slant height of the cone.
- Volume of a cone: $V = \frac{1}{3}\pi r^2 h$
- Surface area of a sphere: $S = 4\pi r^2$
- Volume of a sphere: $V = \frac{4}{3}\pi r^3$

Example 4

A cone has a base with circumference 40 cm and has a capacity of 4.8 litres.

a Calculate the exact ratio between the height and the radius.

b Find the capacity of a similar cone which has a radius of 15 cm.

a $r = \frac{C}{2\pi}$	Calculate the radius.
$r = \frac{40}{2\pi}$	
$V = \frac{1}{3}\pi r^2 h$	Use the volume formula to calculate the height.
$4800 = \frac{1}{3}\pi\left(\frac{40}{2\pi}\right)^2 h$	Recall 1 litre = 1000 cm^3
$4800 = \frac{1}{3}\pi\left(\frac{1600}{4\pi^2}\right) h$	
$4800 = \frac{1600}{12\pi}h$	

$\dfrac{4800(12\pi)}{1600} = h$	
$36\pi = h$	
$\dfrac{h}{r} = \dfrac{9\pi^2}{5}$	Calculate the exact ratio between the height and the radius.
b $h = 15 \times \dfrac{9\pi^2}{5} = 27\pi^2$	Calculate the height of a similar cone
$V = \dfrac{1}{3}\pi r^2 h$	
$V = \dfrac{1}{3}\pi 15^2 27\pi^2$	
$V = 2025\pi^3 = 62.8$ litres	

Exercise 6B

1 Find the volume and surface area of each 3D shape.

a

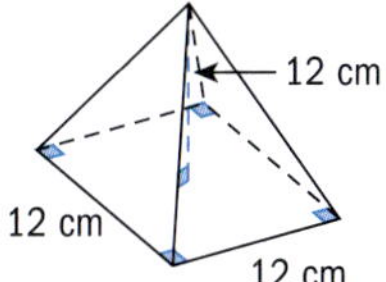

b

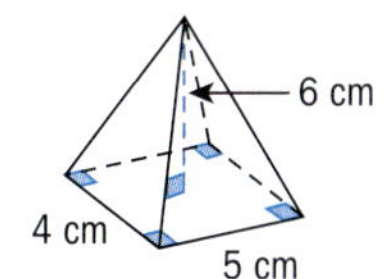

c

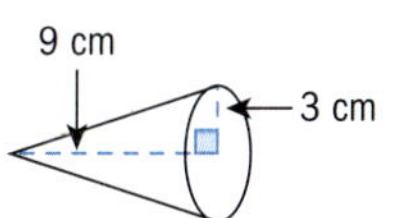

d

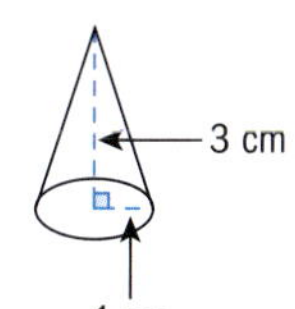

e

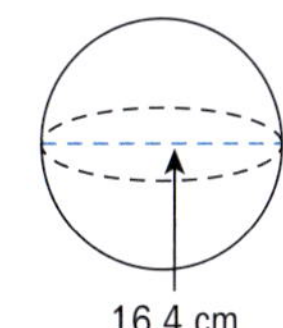

f

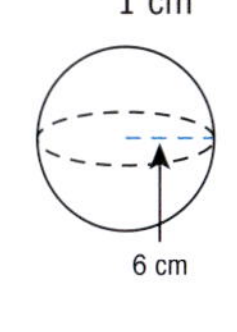

2 A hemisphere has diameter 5.6 cm. Find the volume of the solid.

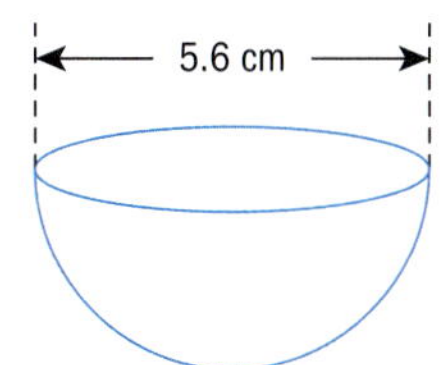

3 An ornamental structure is made from a cylinder and a cone. Find the volume of the solid.

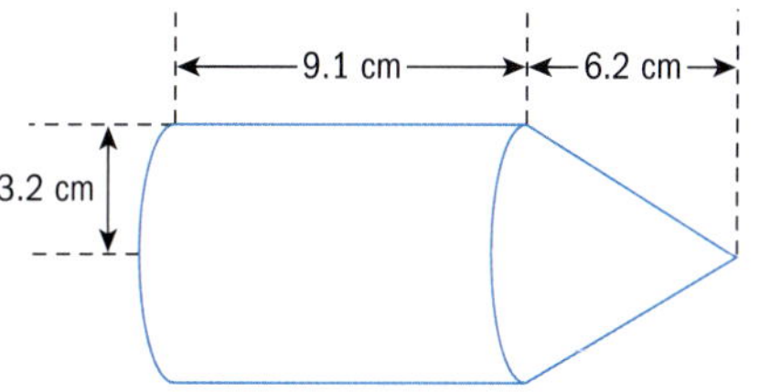

4 A plant pot is made out of clay. The pot has the shape of a hemisphere, with total radius 9 cm and a hollow hemisphere of 8 cm. Assuming no clay is wasted, find the volume of clay required to make the plant pot.

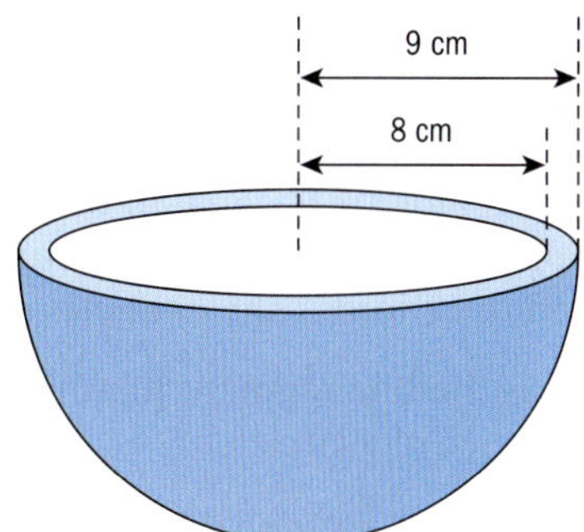

5 Twelve spheres each with a radius of 3 cm are fully immersed in a cylinder of water with radius 10 cm. Find the rise in the water level.

6 The diagram represents a large cone of height 45 cm and base 15 cm. It is constructed by placing a small cone A of height 15 cm and base of diameter 5 cm on top of a frustum. Find the volume of the frustum.

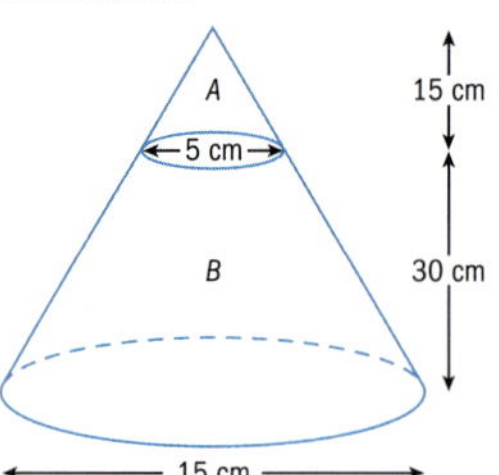

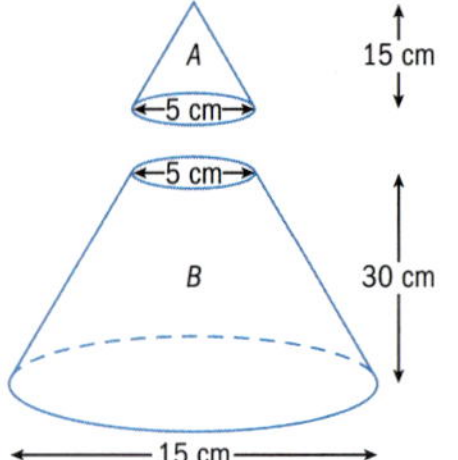

Developing inquiry skills

Return to the opening problem. The pods constructed for the passengers on the Ferris wheel are in the shape of an ellipsoid.

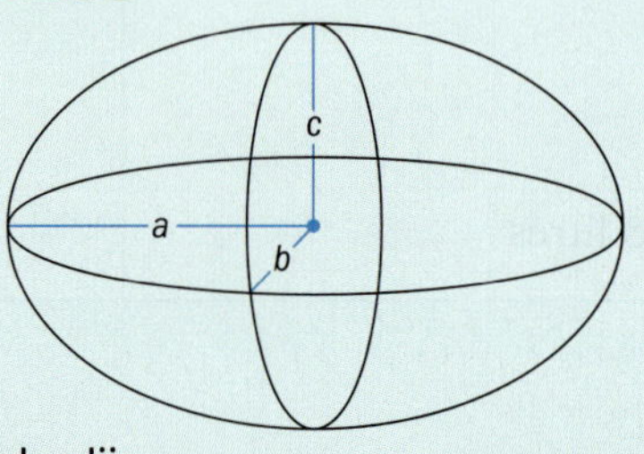

The formula for the volume of an ellipsoid is very similar to the formula for a sphere:

$\frac{4}{3}\pi bc$ where a, b and c are the cross sectional radii

In the case of Ferris wheel pods, the shapes are called spheroids as the vertical cross section is circular.

$$\text{Volume} = \frac{4}{3}\pi r^2 a$$

If the pods have a horizontal diameter of 8 m and a vertical diameter of 6 m, find the volume of a pod.

6.2 Angles of measure

Arcs, secants and chords

An **arc** is part of the circumference of a circle. The minor arc AB is the shortest distance between two points A and B on the circumference of a circle. The major arc AB is the longest distance around the circumference from A to B. Unless otherwise indicated in a question, assume arc AB refers to the minor arc.

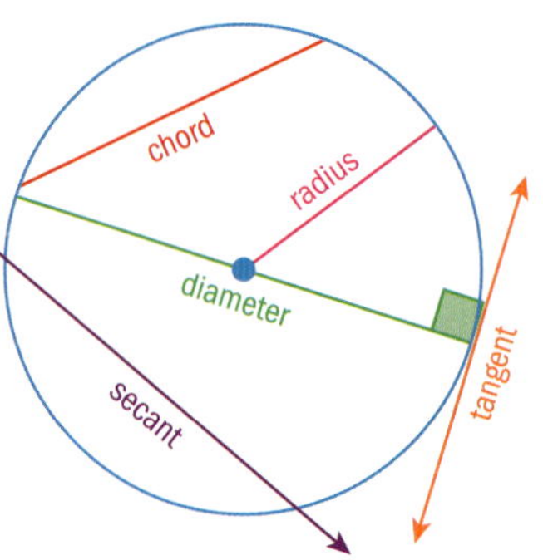

When the major arc is equal to the minor arc, the points are opposite each other on the circumference of the circle and the chord is the diameter of the circle.

A **chord** is the line segment joining two points on the circumference of a circle.

A **secant** is a straight line that intersects the circumference.

Investigation 3

1 Draw a circle with its centre at the origin of a coordinate plane and a radius of 1 arbitrary unit. Label the point A (1, 0) on your diagram.

2 Find the circumference of this circle in terms of π.

3 Complete the following table by finding the length of the following arcs on this unit circle, in terms of π:

Central angle in degrees	Length of corresponding arc
360° (circumference)	
180°	
90°	
60°	
45°	
30°	

The values added in the second column are the **radian measures** of the angles given in degrees in the first column.

4 **Conceptual** How do you explain what a radian is?

5 Consider a circle that has a radius r of any length and an arc of length s along the circle. Complete the table from question **3** again by finding the length of the arcs on this circle, in terms of r and π.

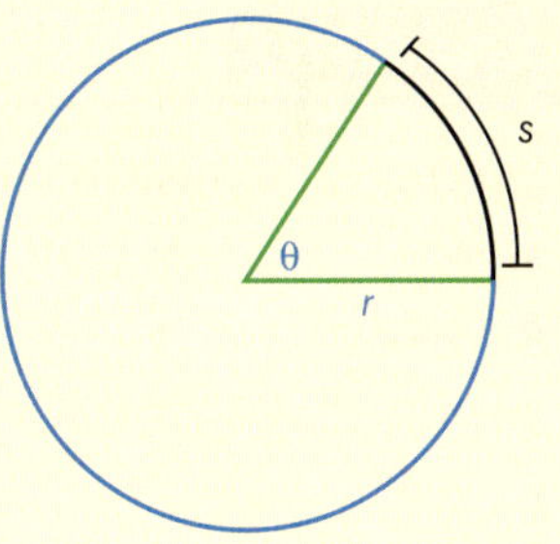

6 Use your findings to suggest a formula that describes the relationship between the radius r and arc length s of the corresponding arc.

7 **Conceptual** Explain why radians are a *unitless* measure. You might find rearranging the formula obtained in question **6** helpful.

Radians

If a circular arc is drawn so that it is equal in length to the radius of the circle, then the angle is called one radian. The definition does not depend on the size of the circle.

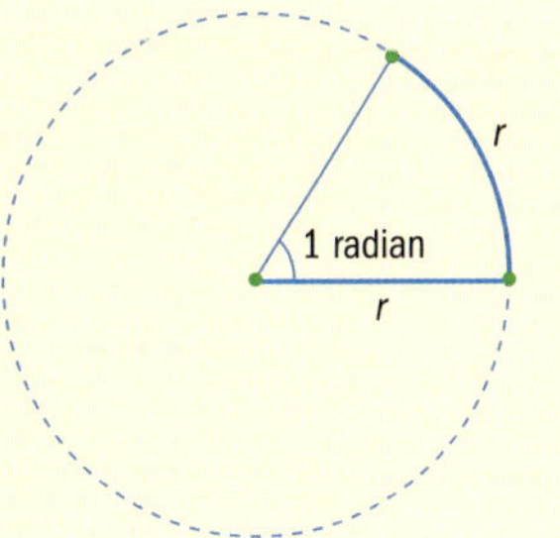

HINT

Make sure you know how to use your GDC when working with both degrees and radians.

HINT

The symbol "c" or the abbreviation "rad" are used for radians. However, they are often omitted, particularly when π is present. The symbol for degrees is never omitted.

Converting between degrees and radians

In order to easily convert between degrees and radians, familiarize yourself with the concept of equivalence:

$$360° \equiv 2\pi \text{ radians}$$

Example 5

a Convert 20° to radians. **b** Convert 56.5° to radians.

c Convert $\frac{4\pi}{3}$ radians to degrees.

a $360° \equiv 2\pi$ $20° \equiv \frac{2\pi}{18} = \frac{\pi}{9}$	Divide both sides by 18.
b $360° \equiv 2\pi$ $56.5° = 0.986$ rad	Divide by 360 and multiply by 56.5.
c $\frac{4\pi}{3}\text{ rad} = \frac{4\pi}{3} \times \frac{180°}{\pi} = 240°$	The π cancels. Do not forget the degree symbol.

Exercise 6C

1 Express each angle in radians.

a 45° **b** 60° **c** 270°
d 360° **e** 18° **f** 225°
g 80° **h** 200° **i** 120°
j 135°

2 Express each angle in degrees.

a $\frac{\pi}{6}$ **b** $\frac{\pi}{10}$ **c** $\frac{5\pi}{6}$
d 3π **e** $\frac{7\pi}{20}$ **f** $\frac{4\pi}{5}$
g $\frac{7\pi}{4}$ **h** $\frac{14\pi}{9}$ **i** $\frac{5\pi}{3}$
j $\frac{13\pi}{4}$

3 Express each angle in radians, correct to 3 significant figures.

a 10° **b** 40° **c** 25°
d 300° **e** 110° **f** 75°
g 85° **h** 12.8° **i** 37.5°
j 1°

4 Express each angle in degrees, correct to 3 significant figures.

a 1 rad **b** 2 rad **c** 0.63^c
d 1.41^c **e** 1.55^c **f** 3^c
g 0.36^c **h** 1.28^c **i** 0.01^c
j 2.15^c

5 **a** Find an exact value for the base angles in an isosceles triangle when the apex angle is 1 radian.

b Find the exact values of the vertex angle V in an isosceles triangle when the base angles are both 1 radian.

Developing inquiry skills

Return to the opening problem. The Ain Dubai has a diameter of 210 m. There will be 48 luxury pods evenly dispersed around the circumference.

Find the angle from the centre between each pod in radians and degrees.

Reflect

Discuss your answer with a partner then share your ideas with your class.

- Who "invented" the symbol for pi?
- Write a sentence stating who you think it could have been.
- Research the correct answer.
- Write a few sentences to explain why this symbol for pi was chosen.

Sectors and segments

A **sector** is the region enclosed by two radii and an arc. Unless otherwise indicated in a question, the minor arc refers to the minor sector and the major arc refers to the major sector.

A **segment** is the region between a chord and an arc.

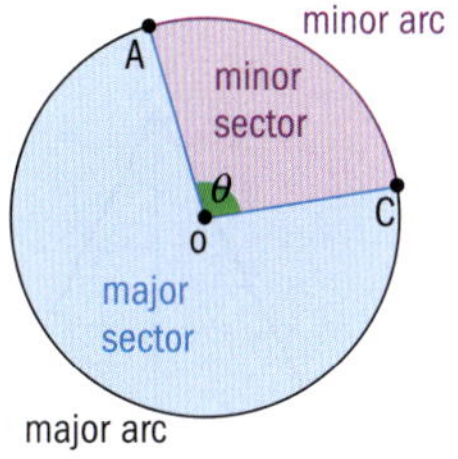

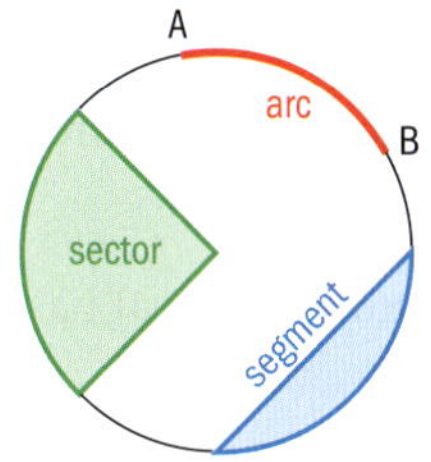

The length of arc, l, can be found by using the circumference formula for a circle.

$$l = \frac{\theta}{2\pi} \times 2\pi r = r\theta$$

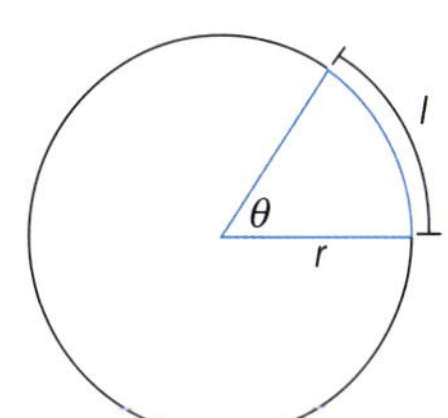

The area of a sector, A, can be found by using the area formula for a circle.

$$A = \frac{\theta}{2\pi} \times \pi r^2 = \frac{1}{2} r^2 \theta$$

The length of an arc is given by

$$l = r\theta$$

and the area of a sector by

$$A = \frac{1}{2} r^2 \theta$$

where θ is measured in radians.

Example 6

Find

a the minor arc length AB **b** the area of the minor sector AOB.

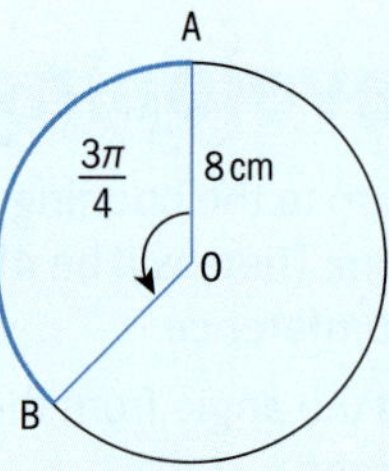

a $l = r\theta = 8 \times \frac{3\pi}{4} = 6\pi\,\text{cm}$	The exact answer would leave your arc length in terms of π. You can also round to 3sf to give 18.8 cm.
b $A = \frac{1}{2}r^2\theta = \frac{1}{2}(8)^2\left(\frac{3\pi}{4}\right) = 24\pi\,\text{cm}^2$	You can also round to 3sf to give 75.4 cm^2.

Example 7

The solid is made up of a right cone and a hemisphere.

a Find the vertical height of the solid.

b Show that the surface area of the hemispherical base is equal to the curved surface area of the cone.

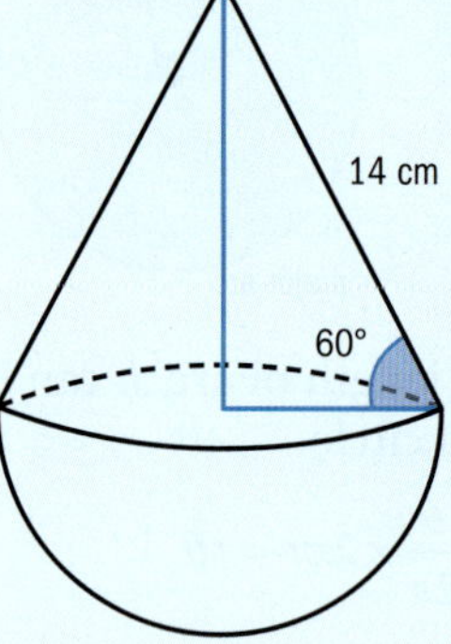

a $\frac{h}{14} = \sin\frac{\pi}{3}$ $h = 14\left(\frac{\sqrt{3}}{2}\right)$ $h = 7\sqrt{3}$ $\frac{r}{14} = \cos\frac{\pi}{3}$ $r = 14\left(\frac{1}{2}\right)$ $r = 7$ height of object $= 7 + 7\sqrt{3} = 19.1\,\text{cm}$	Convert to radians for the entirety of the question. Apply trigonometry in the right-angled triangles.
b length of arc $L = 2\pi r = 14\pi$ $L = s\theta = 14\theta$ Therefore, $14\pi = 14\theta$ $\pi = \theta$	Length of arc is equal to the circumference of the base of the cone. Where s is the slant height of the cone.

Area of sector $= \frac{1}{2}\varphi s^2 = 307.87\ \text{cm}^2$

Curved surface area of hemispherical base $= 2\pi r^2 = 308\ \text{cm}^2$

A cone opens up into a sector of a circle when cut along a slanting edge.

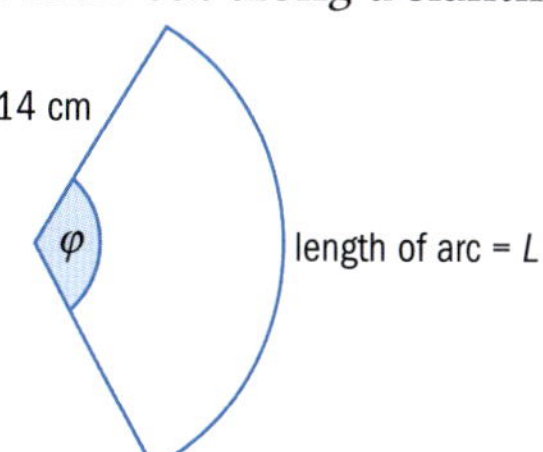

Exercise 6D

1 For each circle, find

a the arc length

b the area of the sector.

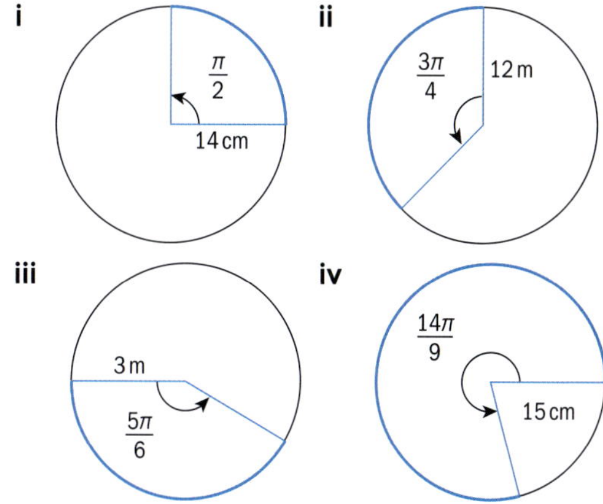

2 The sector formed by a central angle of $\frac{\pi}{12}$ has an area of $3\pi\,\text{cm}^2$. Find the radius of a circle.

3 A sector has a radius of 12 m and area of $36\pi\ \text{m}^2$. Find

a the angle in the sector between the radii

b the perimeter of the sector.

4 Find the shaded area when $\theta = 1.5$ rad.

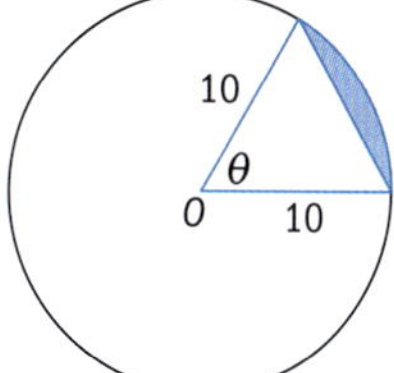

5 The pendulum of a large grandfather clock swings from side to side once every second. The length of the pendulum is 4 m and the angle through which it swings is $\frac{\pi}{12}$. Find the total distance travelled in 1 minute by the tip of the pendulum.

6 Verify that the radius of a sector with perimeter p is given by $r = \frac{p}{2+\theta}$

7 A quadrant (quarter) of a circle whose diameter is 20 cm is reshaped into a semi-circle.

Given that the area is constant, find the perimeter of the semi-circle and find the change in the radius of the circle.

Developing inquiry skills

Return to the opening problem. It has been established that there are 56 pods on the Ferris wheel.

You are in Pod 1 and your friend is in Pod 22. Find the distance between you and your friend.

6.3 Ratios and identities

Investigation 4

Part 1

Draw a unit circle with its centre at the origin of a coordinate plane and a radius of 1 abritrary unit.

Label the point $A(1, 0)$ on your diagram.

Draw the point $S(x, y)$ on the circle such that the angle θ is 45° or $\frac{\pi}{4}$.

Draw the radius OS and draw a vertical line from S down to the x-axis, as shown in the diagram.
Label the point on the x-axis $Q(x, 0)$.

1 Use Pythagoras' theorem to determine the relationship between the x- and y-coordinates of S. Hence, find the values of x and y. Leave your answers as simplified radicals.

2 Take an arbitrary point on the circumference of the circle such that the acute angle is θ. Label this point P. Draw the radius OP and draw a vertical line from P down to the x-axis.

3 Use the trigonometric ratios to write down the values $\sin\theta$, $\cos\theta$ and $\tan\theta$ in terms of x and y. Remember, the radius of the circle is 1.

4 Reflect point S in the y-axis so that it appears in the 2nd quadrant. Label this point K.

5 Find the angle $K\hat{O}A$ in degrees and in radians.

6 Write down the coordinates of K in terms of $\sin\theta$ and $\cos\theta$ as simplified radicals. Find the exact value of $\tan\theta$.

7 Reflect point K in the x-axis so that it appears in the 3rd quadrant. Label this point L.

8 Find the reflex angle $L\hat{O}A$ in degrees and in radians.

9 Write down the coordinates of L in terms of $\sin\theta$ and $\cos\theta$ as simplified radicals. Find the exact value of $\tan\theta$.

10 Reflect point L in the y-axis so that it appears in the 4th quadrant. Label this point M.

11 Find the reflex angle $M\hat{O}A$ in degrees and in radians.

12 Write down the coordinates of M in terms of $\sin\theta$ and $\cos\theta$ as simplified radicals. Find the exact value of $\tan\theta$.

45°
1
y
45°
x

International-mindedness

The stones used to make the pyramids in Ancient Egypt were constructed with a 13 knot rope replicating the shape of a 3-4-5 triangle to ensure sides meet at right angles.

Part 2

Draw an equilateral triangle with a side length of 2 units.

Bisect any vertex to the midpoint of the opposite side, dividing your equilateral triangle into two triangles.

Use Pythagoras' theorem and trigonometry to calculate the side lengths and angles in each of the two triangles.

13 Use your diagram to find, in exact form, the values of

a $\cos 30°$ **b** $\sin 30°$ **c** $\tan 30°$

d $\cos 60°$ **e** $\sin 60°$ **f** $\tan 60°$.

14 Sketch a new unit circle with its centre as the origin of a coordinate plane and a radius of 1 arbitrary unit. Sketch two radii that make a 30° angle and a 60° angle from the x-axis. Label the points on the circle P1 and P2 respectively.

15 Write down the coordinates of P1 and P2.

16 Repeat steps for both points P1 and P2: reflect both points in the x- and y-axis to find the corresponding sin and cos values in quadrants II, III and IV.

17 Summarize your findings in the table below.

Angle θ in degrees	Angle θ in radians	$\sin\theta$	$\cos\theta$	$\tan\theta$
0°				
30°				
45°				
60°				
90°				undefined
120°				
135°				
150°				
180°				
210°				
225°				
240°				
270°				undefined
300°				
315°				
330°				
360°				

18 Explain why some values in the diagram are undefined.

Conceptual Why are the angles 45, 30 and 60 called special angles?

Conceptual Why do different angles lead to the same ratios?

There are constant ratios for angles since all the similar triangles have proportional sides, therefore those ratios are constant.

This gives the opportunity to find trigonometrical ratios of any angle.

Reflect Discuss your answers with a friend, then share your ideas with your class.

- What facts do you know about the unit circle?
- What can the unit circle help you to find?
- Research why we use radians instead of degrees.

The four quadrants

The coordinate axes divide the plane into four quadrants: First (I), Second (II), Third (III) and Fourth (IV), as shown.

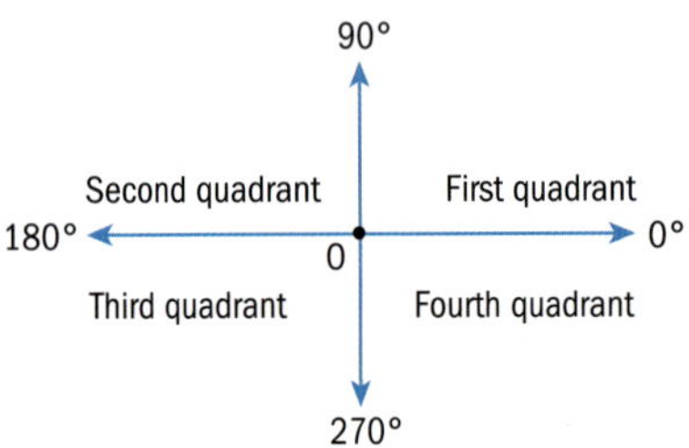

HINT

$\sin\theta = y$-coordinate

$\cos\theta = x$-coordinate

A good way to remember this is by using the alphabet:

x comes before y

$\cos\theta$ comes before $\sin\theta$.

Angles in QI are acute.

Angles in QII are obtuse.

Angles in QIII and QIV are reflex.

Using the unit circle it follows:

$$\sin\theta = \frac{y}{OP} = \frac{y}{1}$$

$$\cos\theta = \frac{x}{OP} = \frac{x}{1}$$

$$\tan\theta = \frac{y}{x}, x \neq 0$$

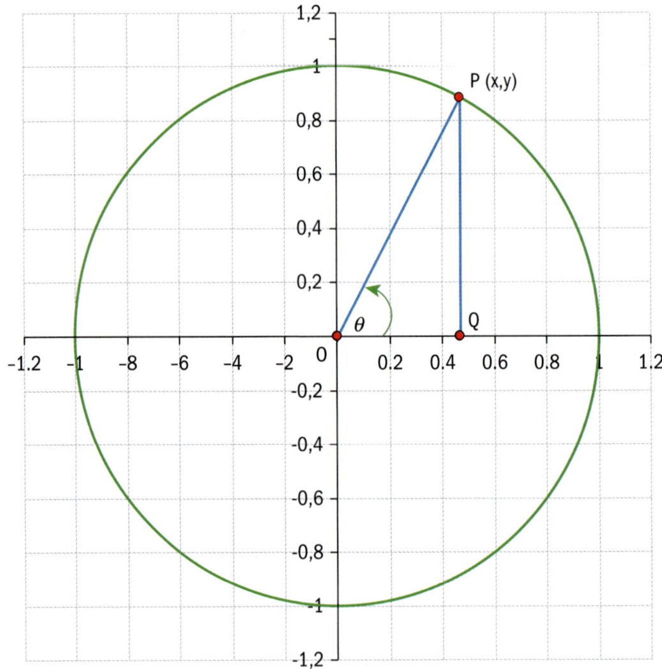

This information gives rise to the first trigonometric identity:

$$\sin\theta = \frac{y}{1}$$

$$\cos\theta = \frac{x}{1}$$

$$\tan\theta = \frac{y}{x}$$

$$\text{then } \tan\theta = \frac{\sin\theta}{\cos\theta}$$

Since the coordinates of P lie on the unit circle, they can also take negative values depending on the position of P relative to the axes. The hypotenuse OP is always equal to 1. When P is in the second quadrant, θ is obtuse, the x-coordinate is negative and the y-coordinate is positive, so $\sin\theta$ is the only positive ratio in this quadrant. If you consider the signs of the coordinates of P in each quadrant, you can therefore assign signs to the three ratios. This is summarized in the diagram below.

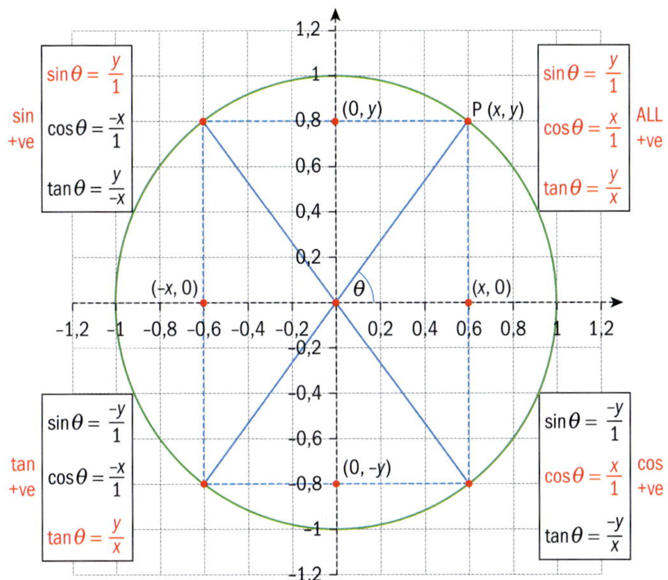

From the symmetrical properties of a circle you also obtain the following identities for $0 \leq \theta \leq \frac{\pi}{2}$:

$\sin(\pi - \theta) = \sin\theta$	$\sin(\pi + \theta) = -\sin\theta$	$\sin(2\pi - \theta) = -\sin\theta$
$\cos(\pi - \theta) = -\cos\theta$	$\cos(\pi + \theta) = -\cos\theta$	$\cos(2\pi - \theta) = \cos\theta$
$\tan(\pi - \theta) = -\tan\theta$	$\tan(\pi + \theta) = \tan\theta$	$\tan(2\pi - \theta) = -\tan\theta$

Angles in the unit circle are measured starting from the point (1, 0) and turning anticlockwise (positive angles) or clockwise (negative angles).

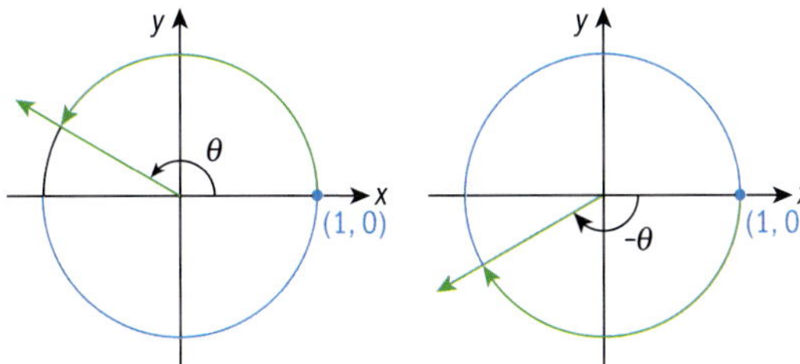

Looking at your results, the x- and y-coordinates of any point that lies in each of the four quadrants, you can identify the sign of each of the trigonometric ratios in a given quadrant.

TOK

To what extent do instinct and reason create knowledge?

Do different geometries (Euclidean and non-Euclidean) refer to or describe different worlds?

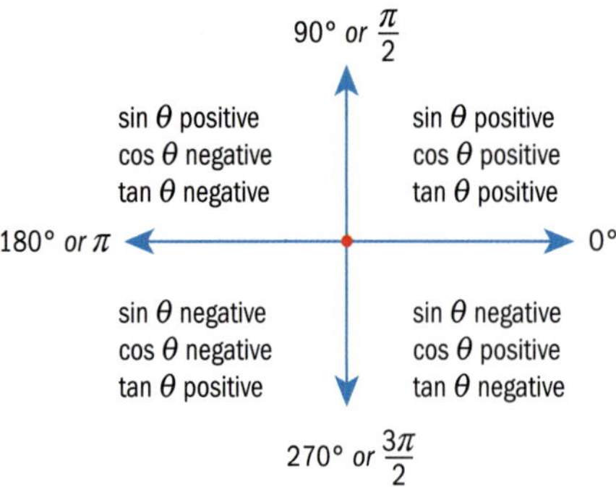

To help remember the signs of the three trigonometric ratios in the quadrants, you can see that all of the ratios are positive in the first quadrant, then only one of the ratios is **positive** in each subsequent quadrant.

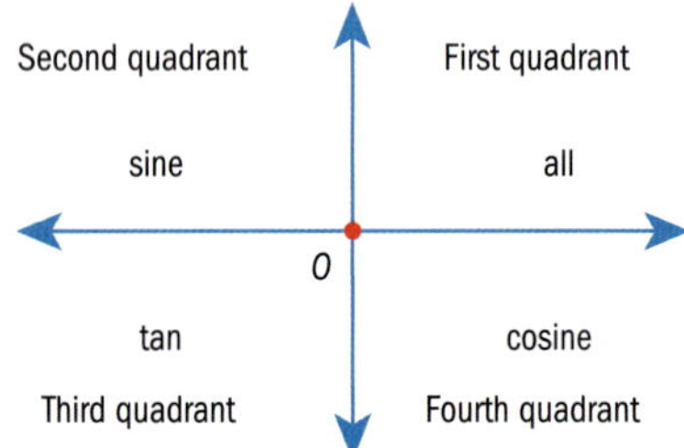

HINT

The mnemonics "**A**dd **S**ugar **T**o **C**offee," "**A**ll **S**tudents **T**ake **C**alculus," or just **ASTC** are good ways to remember this.

The **associated acute angle** of an angle is the position of an angle with its vertex at the origin of a unit circle in the first quadrant.

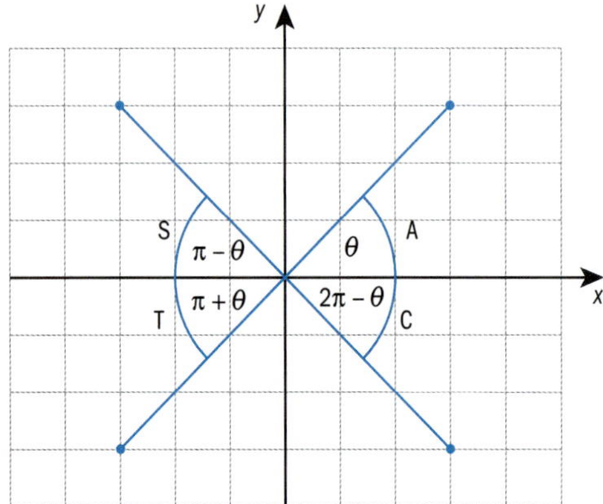

International-mindedness

The word "sine" started out as a totally different word and passed through Sanskrit, Arabic and Latin before becoming the word that we use today.

Angles with the same y-value have equal sines.

$\sin 60° = \sin 120°$

$\sin 240° = \sin 300°$

Angles with the same x-values have equal cosines.

$\cos 60° = \cos 300°$

$\cos 120° = \cos 240°$

Angles directly opposite each other on a line drawn through the origin have the same tangents.

$\tan 60° = \tan 240°$

$\tan 120° = \tan 300°$

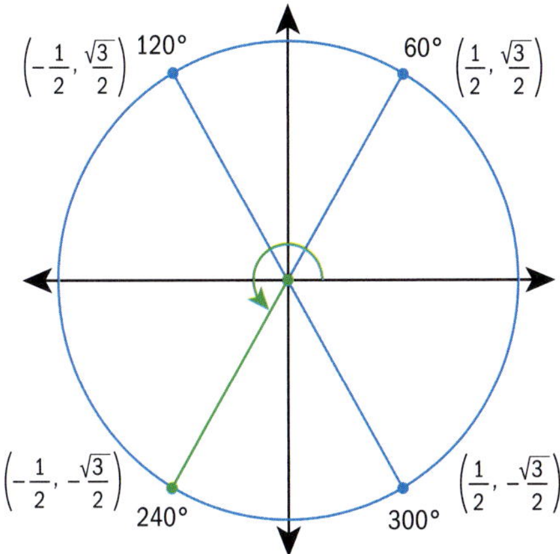

Example 8

Find the sine, cosine and tangent of 225°.

$\theta = 225°$	
θ lies in QIII	Identify the quadrant.
$\alpha = \theta - 225° = 45°$	Identify the corresponding acute angle, α.
hence $\sin 225 = -\sin 45$ $= -\frac{\sqrt{2}}{2}$	
$\cos 225 = -\cos 45 = -\frac{\sqrt{2}}{2}$	
$\tan 225 = \tan 45 = 1$	In QIII only tan is positive.

Example 9

If $\cos\theta = 0.5$ and $\tan\theta$ is negative, find $\sin\theta$ and θ.

If $\cos\theta = 0.5$ then $\alpha = 60°$.	Find the acute angle.
If cosine is positive and tan is negative, then θ must lie in the 4th quadrant, hence $\sin\theta$ is negative. Therefore, $\sin\theta = -\frac{\sqrt{3}}{2}$ and $\theta = 300°$.	Identify the quadrant.

Exercise 6E

1 Write down the quadrant and the associated acute angle when θ is:

a $-215°$ **b** $\frac{-5\pi}{4}$ **c** $\frac{7\pi}{2}$ **d** $\frac{11\pi}{6}$

e $564°$ **f** $-22°$ **g** $\frac{8\pi}{3}$

2 Complete the following table:

	$\sin\theta$	$\cos\theta$	$\tan\theta$	θ
A	0.5	–0.866		
B				$\frac{11\pi}{6}$
C	0.3907		–0.4245	
D		$\frac{\sqrt{2}}{2}$	1	
E		$-\frac{\sqrt{2}}{2}$	–1	

3 Given θ lies between 0° and 360°, find all the values of θ such that:

a $\sin\theta = 0.6$

b $\cos\theta = -2.6$

c $\tan\theta = -2.36$

4 Given that θ is reflex and $\cos\theta = \frac{3}{5}$, find $\sin\theta$ and $\tan\theta$.

You are now able to find the values of sine, cosine and tangent of any angle. This leads to solving trigonometric equations.

International-mindedness

This feature of periodicity is used in wave theory, oscillations and many other real-life applications.

Example 10

1 Find the sine of each angle and state an obtuse angle that has the same sine.

a 70° **b** $\frac{\pi}{4}$

2 Find the cosine of each angle and a reflex angle less than 360° or 2π that has the same cosine.

a 50° **b** $\frac{2\pi}{3}$

1 a $\sin 70° = \sin(180° - 70°)$ $= \sin 110° = 0.94$ **b** $\sin\frac{\pi}{4} = \sin\left(\pi - \frac{\pi}{4}\right) = \sin\frac{3\pi}{4}$ $= 0.71$	Two angles have the same sine if the angles add up to 180° or π radians.
2 a $\cos 50° = \cos(360° - 50°)$ $= \cos 310° = 0.64$ **b** $\cos\frac{2\pi}{3} = \cos\left(2\pi - \frac{2\pi}{3}\right)$ $= \cos\frac{4\pi}{3} = -0.5$	Two angles have the same cosine if the angles add up to 360° or 2π radians.

Exercise 6F

For each of the following, find the value of θ for $0 \le \theta \le 2\pi$:

1 $\cos\theta = 0.45$

2 $\tan\theta = -0.56$

3 $\sin\theta = 0.23$

4 $\cos\theta = 2.15$

5 $\cos^2\theta + \cos\theta = 0$

6 $2\cos^2\theta - 3\cos\theta = -1$

7 $4\sin^2\theta = 1$

8 $\sin\frac{3\theta}{2} = -0.62$

9 $\tan 2\theta = -0.4555$

The cosine rule proof

In the triangle shown below, the perpendicular height has been drawn from B to AC. The perpendicular has length h and the lengths of the sides AB, AC and BC are c, b and a, respectively.

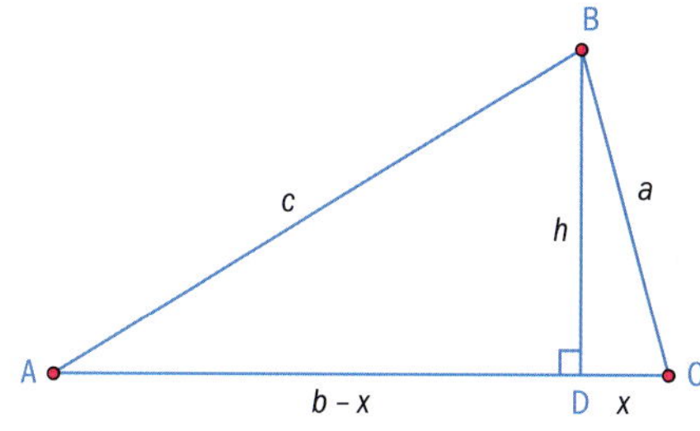

In ΔABD, $h^2 = c^2 - (b - x)^2$

In ΔBDC, $h^2 = a^2 - x^2$

By simplifying:

$c^2 - (b^2 - 2bx + x^2) = a^2 - x^2$

$\Rightarrow c^2 - b^2 + 2bx = a^2$

$\Rightarrow c^2 = a^2 + b^2 - 2bx$

$\Rightarrow c^2 = a^2 + b^2 - 2ba\cos C$

By rearranging:

$$\cos C = \frac{a^2 + b^2 - c^2}{2ab}$$

HINT

In ΔBDC, $\cos C = \frac{x}{a} \Rightarrow x = a\cos C$

Other methods for establishing the cosine rule can be obtained by constructing perpendiculars from A to BC and from C to AB.

The **cosine rule** states that for any triangle *ABC* with corresponding lengths of the sides *a*, *b* and *c*:

$$a^2 = b^2 + c^2 - 2bc\cos A \Leftrightarrow \cos A = \frac{b^2 + c^2 - a^2}{2bc}$$

$$b^2 = a^2 + c^2 - 2ac\cos B \Leftrightarrow \cos B = \frac{a^2 + c^2 - b^2}{2ac}$$

$$c^2 = a^2 + b^2 - 2ab\cos C \Leftrightarrow \cos C = \frac{a^2 + b^2 - c^2}{2ab}$$

The cosine rule is used to solve triangle problems in the following cases.

- Given three sides to find the angles.
- Given two sides and an included angle to find the third side and the remaining angles.

The cosine rule generalizes Pythagoras' theorem by considering acute and obtuse angles and can be used to solve a triangle given the measures of all three sides or the measures of two sides and the included angle.

TOK

How certain is the shared knowledge of mathematics?

Example 11

In triangle PQR, PQ = 9 cm, QR = 16 cm and PR = 11 cm. Find the smallest angle in the triangle to the nearest degree.

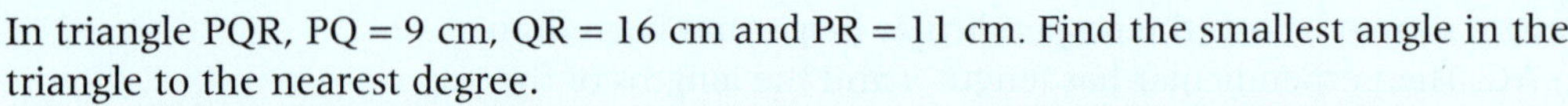

The smallest angle is opposite the smallest side: angle R.

Using the cosine rule:

$$\cos \text{R} = \frac{p^2 + q^2 - r^2}{2pq}$$

$$\Rightarrow \text{P}\hat{\text{R}}\text{Q} = \arccos\left(\frac{16^2 + 11^2 - 9^2}{2 \times 16 \times 11}\right)$$

$$\Rightarrow \text{P}\hat{\text{R}}\text{Q} = 33°$$

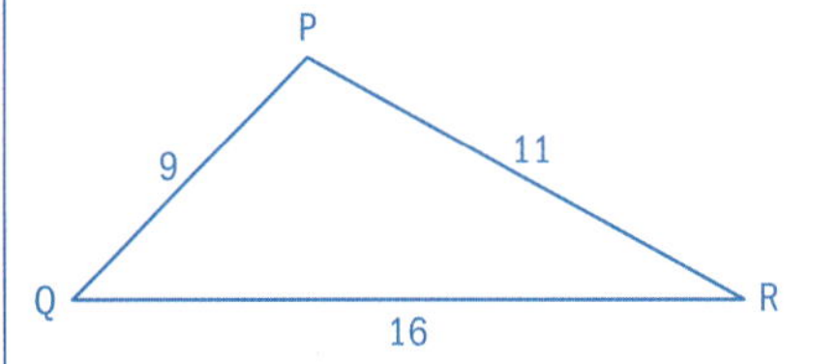

Example 12

In triangle ABC, the lengths of the sides a, b and c are in the ratio 2 : 5 : 6 respectively. Find the largest angle of the triangle.

$a : b : c = 2 : 5 : 6$

This means that triangle ABC is similar to triangle XYZ with sides 2, 5 and 6 units long. Since similar triangles are equiangular we can therefore solve for triangle ABC.

The largest angle is opposite the largest side.

$$\cos Z = \frac{x^2 + y^2 - z^2}{2xy}$$

$$\Rightarrow \text{X}\hat{\text{Z}}\text{Y} = \arccos\left(\frac{2^2 + 5^2 - 6^2}{2 \times 2 \times 5}\right) = 110.5°$$

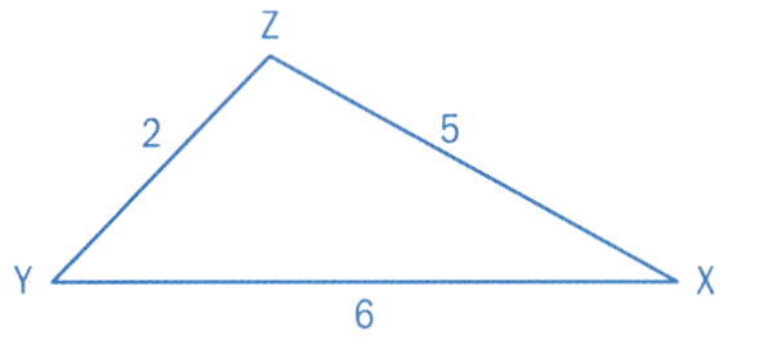

Exercise 6G

1 Use the cosine rule to find the missing angles and sides of the following triangles:

a

b

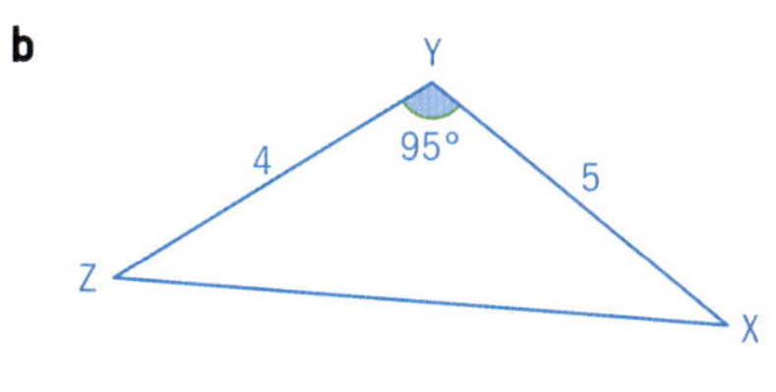

c

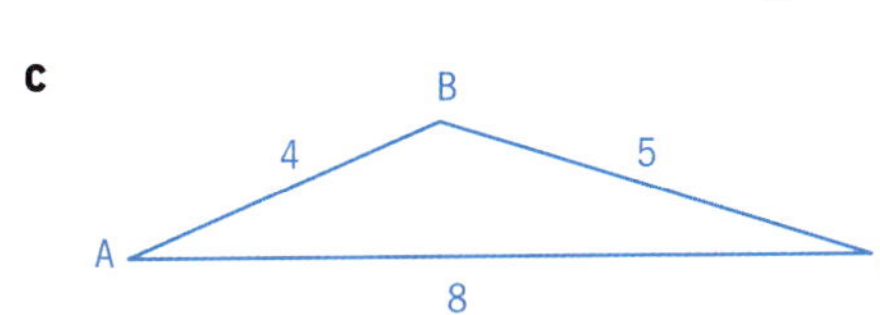

2 Find the largest angle in ΔABC given that $a = 4.5$ cm, $b = 3.9$ cm and $c = 2.3$ cm.

3 In triangle PQR the sides PQ, QR and RP are in the ratio 3 : 2 : 4 respectively. Find the smallest angle of the triangle.

4 Triangle ABC has sides of length 5, x and $(2x - 1)$. Given that $\hat{BAC} = 60°$ find x, and hence calculate the other two angles of the triangle.

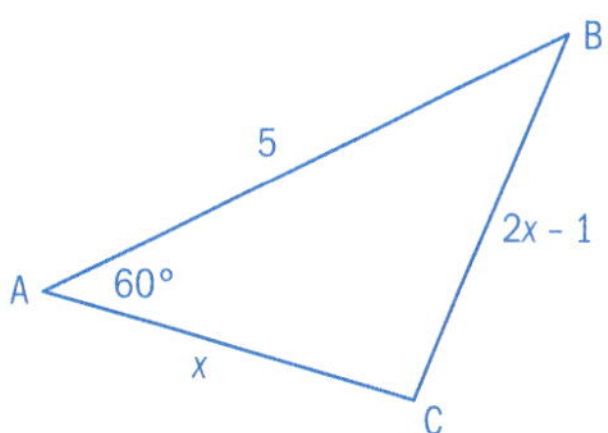

5 A parallelogram ABCD has sides AB and AD of length a and b respectively. The diagonals AC and BD have lengths p and q as shown in the diagram. Show that
$p^2 + q^2 = 2(a^2 + b^2)$

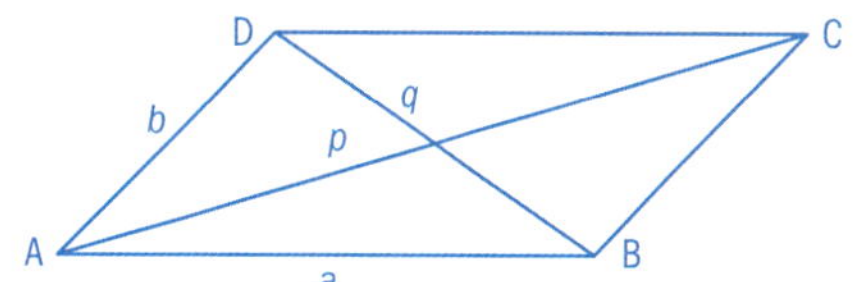

The sine rule proof

In the triangle shown, a perpendicular has been drawn from B to AC. The height of the perpendicular is h and the lengths of the sides AB, AC and BC are c, b and a respectively.

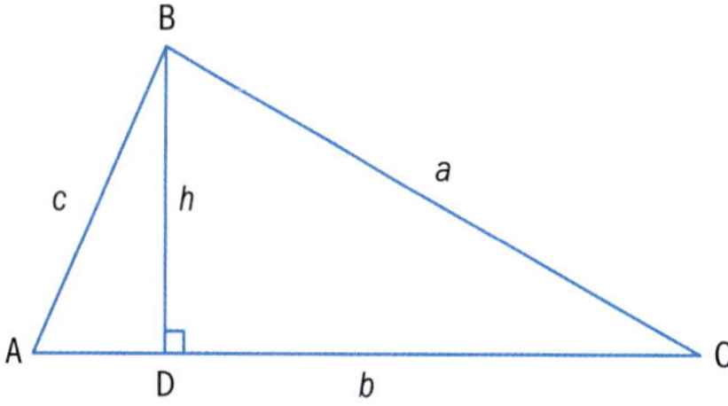

In ΔABD $\quad \sin A = \frac{h}{c} \Rightarrow h = c\sin A$

In ΔCBD $\quad \sin C = \frac{h}{a} \Rightarrow h = a\sin C$

$$\Rightarrow c\sin A = a\sin C \Rightarrow \frac{c}{\sin C} = \frac{a}{\sin A}$$

Now consider the same triangle this time with a perpendicular of height H drawn from C to AB.

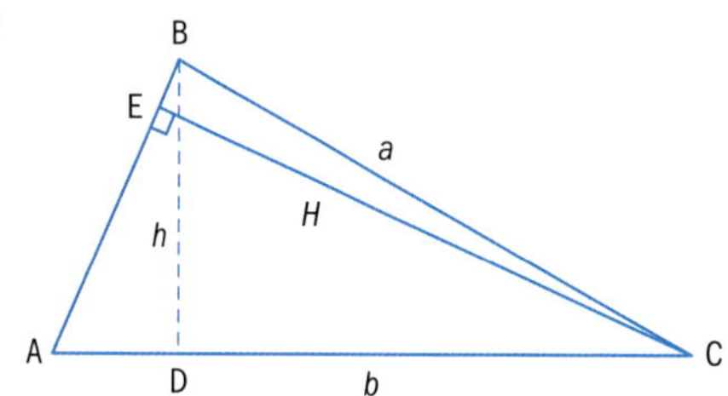

In ΔBCE $\sin B = \frac{H}{a} \Rightarrow H = a\sin B$

In ΔACE $\sin A = \frac{H}{b} \Rightarrow H = b\sin A$

$\Rightarrow a\sin B = b\sin A \Rightarrow \frac{a}{\sin A} = \frac{b}{\sin B}$

Combining the two results: $\frac{a}{\sin A} = \frac{b}{\sin B} = \frac{c}{\sin C}$

International-mindedness

Diagrams of Pythagoras' theorem occur in early Chinese and Indian manuscripts. The earliest references to trigonometry are in Indian mathematics.

The **sine rule** states that in a triangle *ABC* with side lengths *a*, *b* and *c*

$$\frac{a}{\sin A} = \frac{b}{\sin B} = \frac{c}{\sin C}.$$

HINT

The sine rule can also be written as

$$\frac{\sin A}{a} = \frac{\sin B}{b} = \frac{\sin C}{c}$$

This alternate representation is preferred for finding unknown angles, as the angles are the numerators.

The sine rule is used to solve triangle problems in the following cases.

- Given two angles and the length of one side of a triangle to find missing lengths.
- Given the length of two sides and an angle not **included** between the two given sides to find the missing side and angles.

The ambiguous case of the sine rule

The ambiguous case of the sine rule occurs when two triangles can be constructed from given information about two sides and the non-included angle in a triangle.

Given the length of sides b and a as well as angle A, two triangles ABC can be constructed. The angles B are supplementary.

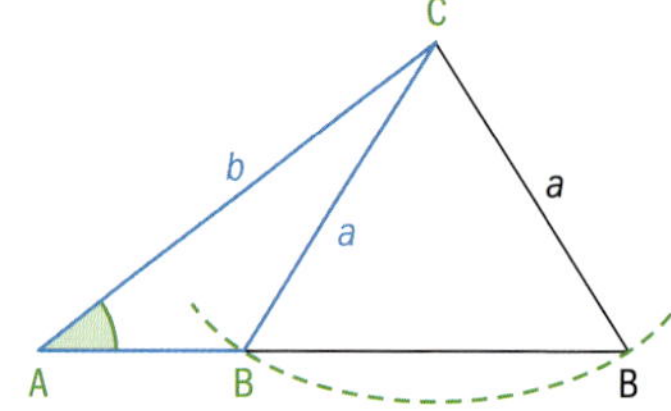

HINT

When the question says, "Find the values…," this generally involves the ambiguous case of the sine rule.

Example 13

Consider the triangle ABC, where AB = 10, BC = 7 and CAB = 30°.

a Find the two possible values of the angle at C.

b Hence, find the angle at B, given that it is acute.

a If possible try to draw the case diagram above and label accordingly.

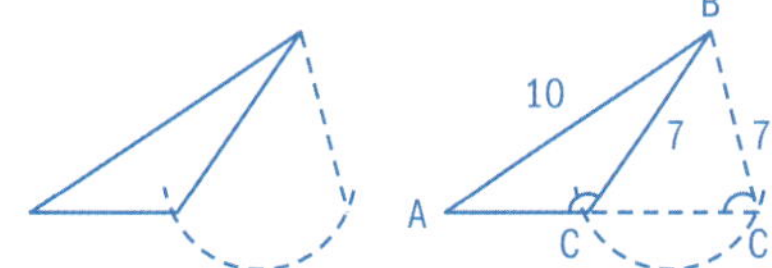

As you know two sides and an angle that is not between these two sides, you use the sine rule.

$$\frac{\sin 30°}{7} = \frac{\sin ACB}{10}$$

$$\sin ACB = \sin^{-1}\left(\frac{10}{14}\right) = 45.6°$$

or

$$\sin ACB = 180 - \sin^{-1}\left(\frac{10}{14}\right) = 134.4°$$

Example 14

The illustration to the right shows the angles of elevation of the highest point of the Great Pyramid of Giza, measured from two observation points A and B. The angle of elevation at A is 46° and the angle of elevation at B is 40°. Given that A and B are 35 m apart, find the height of the pyramid, h.

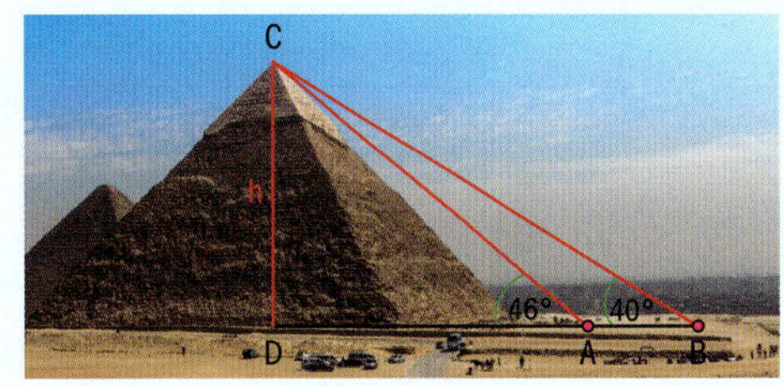

Apply the sine rule to triangle ABC:

$$\frac{CB}{\sin 134°} = \frac{AB}{\sin 6°}$$

$C\hat{A}B = 180° - 46° = 134°$

$A\hat{C}B = 180° - (134° + 40°) = 6°$

$$CB = \frac{35\sin 134°}{\sin 6°} \approx 241 \text{ m}$$

In triangle BCD:

$$\sin 40° = \frac{h}{BC} \Rightarrow h \approx 241\sin 40°$$

$\therefore h \approx 154.9$ m (to 3 s.f.)

Example 15

The diagram to the right shows a river with a 5 m long fence AB, built at an angle of 34° to the riverside.

Farmer Brown wants to fence off an area in the shape of a triangle ABC (as shown in the diagram) for his three goats. He has 3 m of fencing left. Find the angles ACB and ABC.

Continued on next page

Geometry and trigonometry

Using the sine rule in triangle ABC:

$$\frac{\sin C}{5} = \frac{\sin 34°}{3} \Rightarrow \sin C = \left(\frac{5\sin 34°}{3}\right) = 0.9319...$$

$$C = \arcsin\left(\frac{5\sin 34°}{3}\right) = 68.7°$$

$A\hat{C}B = 68.7°$ or $111.3°$

When

$A\hat{C}B = 68.7°,\ A\hat{B}C = 180° - (34 + 68.7)° = 77.3°$

When

$A\hat{C}B = 111.3°,\ A\hat{B}C = 180° - (34 + 111.3)° = 34.7°$

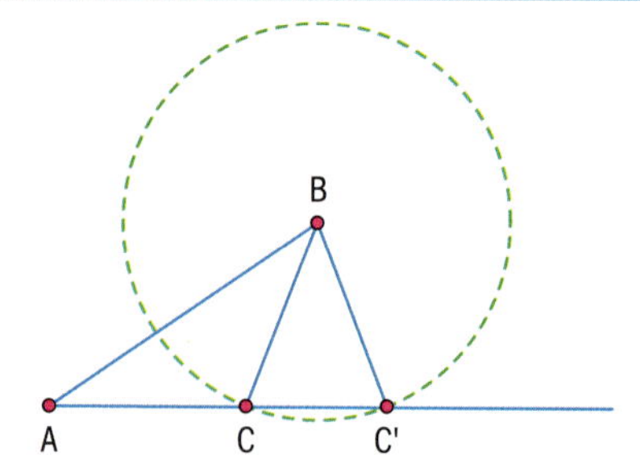

AB = 5.00 cm

CB = 3.00 cm

$B\hat{C}A = 111.38°$

$B\widehat{C'}A = 68.62°$

However, we know that

$\sin(180 - 68.7)° = 0.9319$

Exercise 6H

1 Find the unknown angles and sides in the following triangles.

a

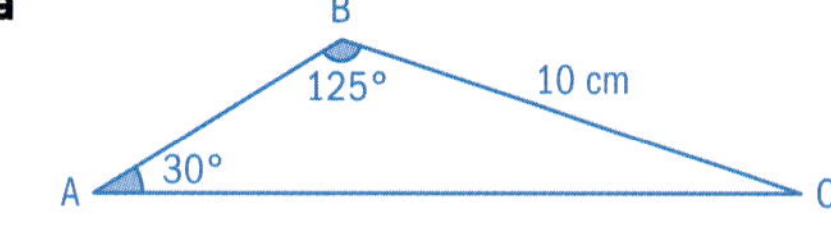

b

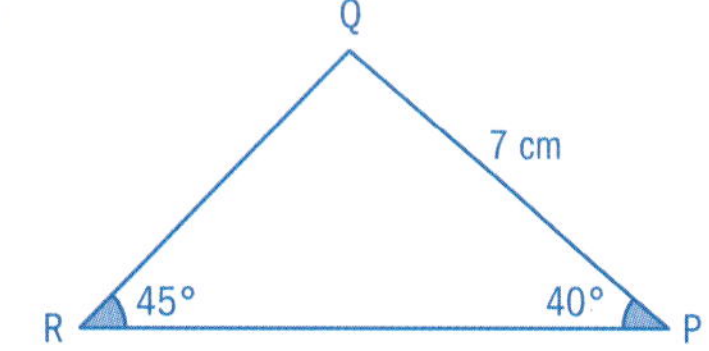

c

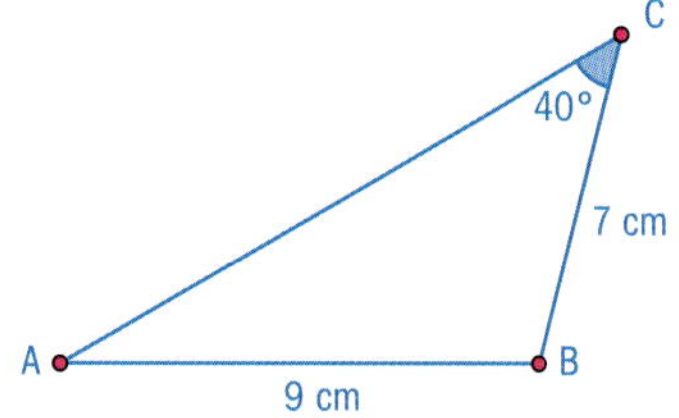

2 A plane flying from P to Q followed a course from P that had a 15° error, as shown in the diagram below. After travelling for 80 km, the pilot corrected the course by changing direction at point R and flew a further 150 km to reach Q. Assuming that the plane flew at a constant speed, $400\,\text{km}\,\text{h}^{-1}$, how much time (to the nearest second) was lost due to the error?

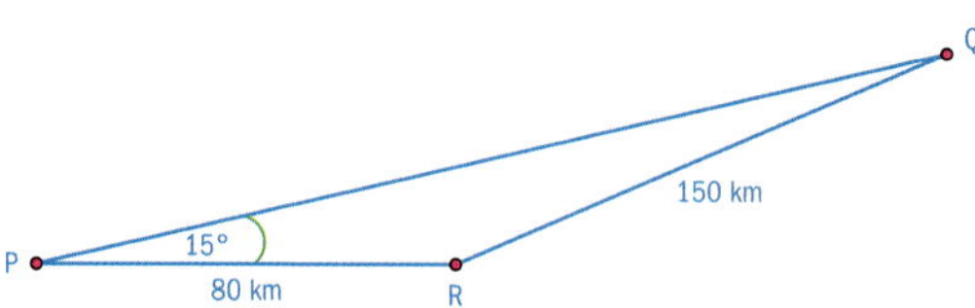

3 From a hot air balloon the angles of depression to each end of the lake are 68° and 32°. Given that the balloon is 250 m above the ground, find the length of the lake. Give your answer to the nearest metre.

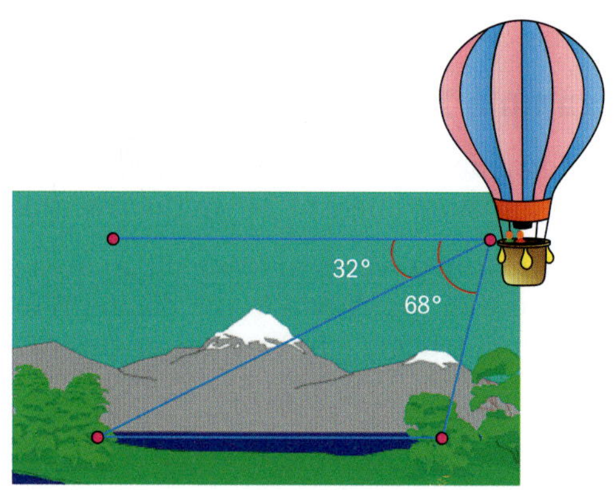

4 The diagram below shows three points ABC on level ground. A vertical mast MA stands at A. The top of the mast is supported by wires on the ground. Two such wires are fastened at B and at C, with $M\hat{B}A = 64°$ and $M\hat{C}A = 23°$. Given that B and C are 15 m apart, find the length of both wires and the height of the mast.

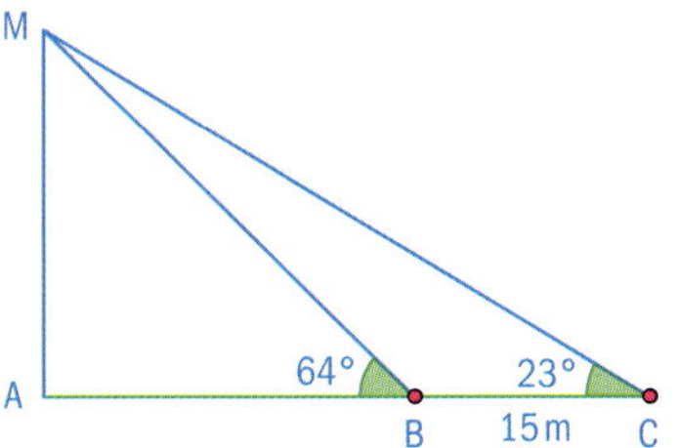

5 Show that two triangles, ABC, can be drawn such that AB = 31 cm, AC = 27 cm and $A\hat{B}C = 55°$. Find the size of the angles of each triangle. Give your answer to the nearest degree.

Area of a triangle

Consider triangle ABC shown in the diagram to the right.

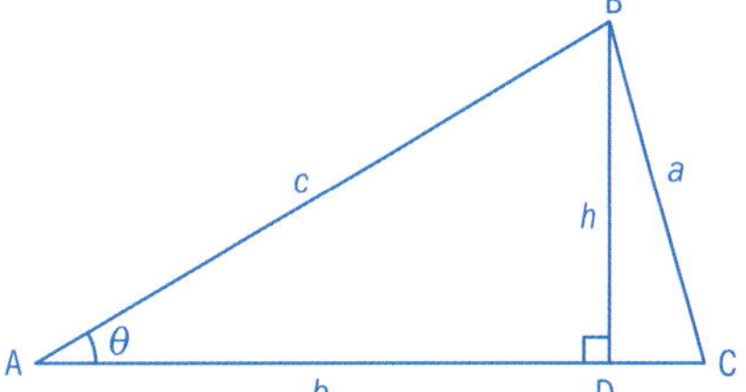

Area of triangle ABC $= \frac{1}{2}bh$

In triangle ABD $\sin\theta = \frac{h}{c} \Rightarrow h = c\sin\theta$

Substituting for h we obtain:

Area of $\Delta ABC = \frac{1}{2}bc\sin A$

Example 16

Find the area of the quadrilateral ABCD shown on the right.

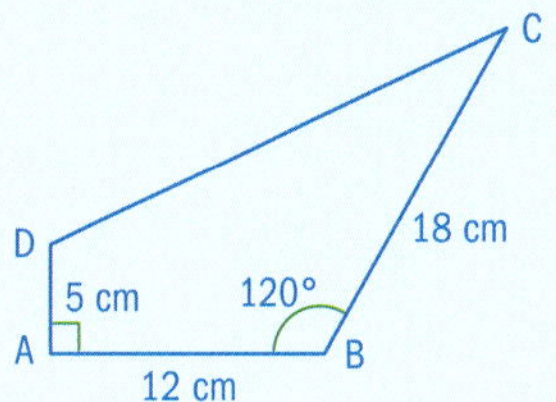

By joining BD you obtain two triangles whose areas you can add.

Area of triangle ABD $= \frac{1}{2}\times 12\times 5 = 30\,\text{cm}^2$

$\theta = \arctan\left(\frac{5}{12}\right) = 22.6°$

$\therefore D\hat{B}C = 120° - 22.6° = 97.4°$

$\tan\theta = \frac{AD}{AB} = \frac{5}{12}$

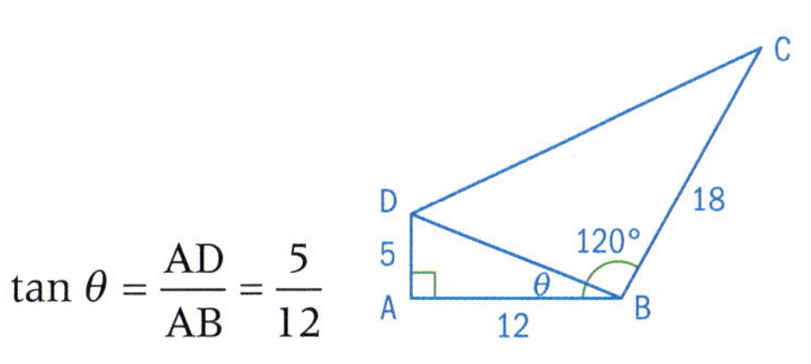

Continued on next page

$BD = \sqrt{5^2 + 12^2} = 13$	Using Pythagoras' theorem.
Area of triangle BDC $= \frac{1}{2} \times 13 \times 18 \times \sin 97.4°$ $= 116\text{ cm}^2$	Area of a triangle $= \frac{1}{2}bc\sin A$
Area of quadrilateral $= 30 + 116 = 146\text{ cm}^2$	

Exercise 6I

1 Find the area of quadrilateral PQRS shown in the diagram below.

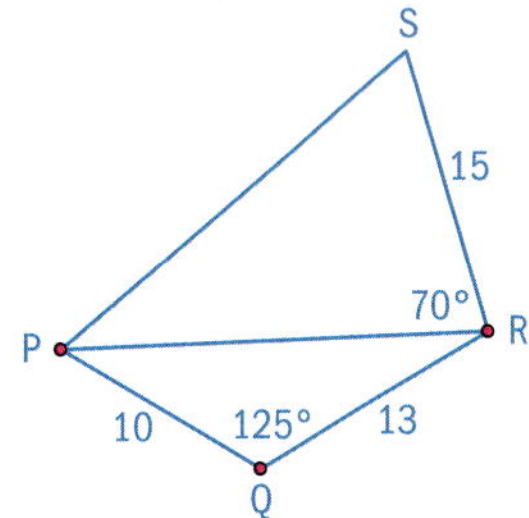

2 Find the difference between the areas of two possible triangles ABC in which $B\hat{A}C = 20°$, BC = 52 cm and AC = 2BC.

3 The diagram below shows two chords XY and YZ drawn on a circle with centre O and radius 5 cm. Given that XY = 3 cm and YZ = 7 cm, find the area of quadrilateral OXYZ.

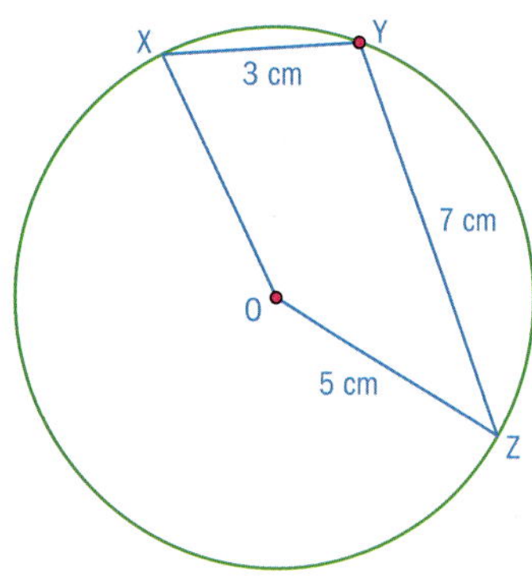

4 The diagram below shows a mast AB which is 12 m long. Points C and D lie on the ground such that the angle of elevation from C to B is 60° and the angle of elevation from D to B is 55°. Given that the distance between C and D is 15 m, find the angle CAD and hence find the area of the triangle CAD.

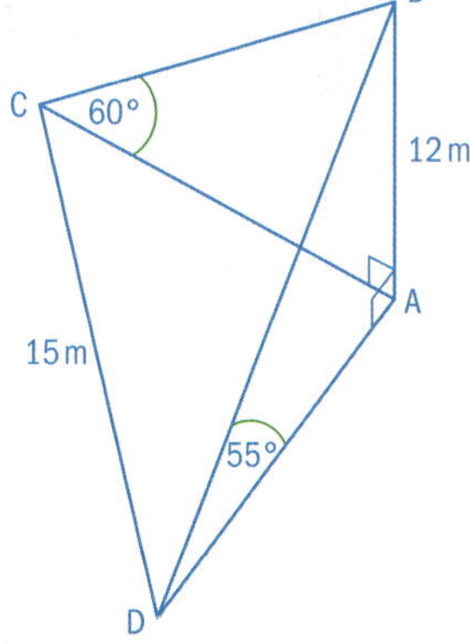

5 In the diagram below, O is the centre of a circle with radius r. $P\hat{O}R = Q\hat{O}S = \frac{\pi}{4}$ and $R\hat{O}S = \frac{\pi}{6}$.

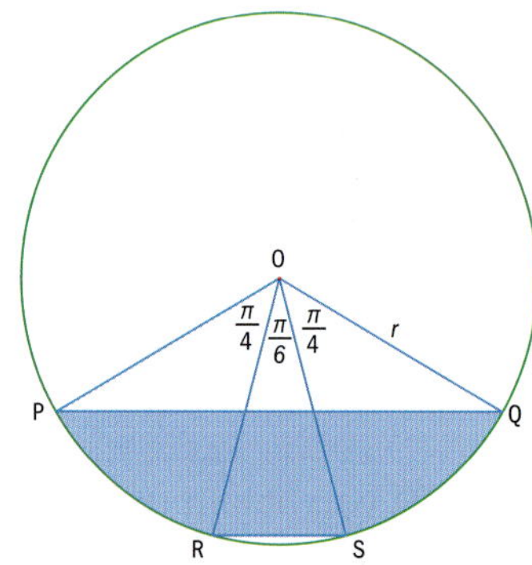

a Find the area of triangles POQ and ROS.

b Find the area of the minor segment formed by the chord PQ.

c Find the area of the minor segment formed by the chord RS.

d Show that the shaded area is equal to $\frac{r^2}{4}\left(\pi + 1 - \sqrt{3}\right)$.

6 A yacht and a catamaran leave the harbour at 0900 hours. The yacht sails at 24 knots on course 037 and the catamaran at 15 knots on 139. Find the bearing and the distance in nautical miles of the yacht from the catamaran at 1400 hours. (1 knot is a speed of 1 nautical mile per hour.)

Reciprocal trigonometric functions

For any acute angle in a right-angled triangle:

$$\sin\theta = \frac{\text{opposite}}{\text{hypotenuse}},\ \cos\theta = \frac{\text{adjacent}}{\text{hypotenuse}} \text{ and } \tan\theta = \frac{\text{opposite}}{\text{adjacent}}.$$

Three further trigonometric ratios are defined by:

$$\operatorname{cosec}\theta = \frac{\text{hypotenuse}}{\text{opposite}},\ \sec\theta = \frac{\text{hypotenuse}}{\text{adjacent}} \text{ and } \cot\theta = \frac{\text{adjacent}}{\text{opposite}}.$$

HINT

A quick way to remember these is to look at the third letter of the reciprocal trigonometric ratio.

$$\text{co}\underline{\text{s}}\text{ec}\ \theta = \frac{1}{\sin\theta},\ \text{se}\underline{\text{c}}\ \theta = \frac{1}{\cos\theta} \text{ and } \text{co}\underline{\text{t}}\ \theta = \frac{1}{\tan\theta}$$

As it follows that $\operatorname{cosec}\theta = \frac{1}{y}$, $\sec\theta = \frac{1}{x}$ and $\cot\theta = \frac{x}{y}$, then it also follows that these reciprocal ratios take the same signs as their reciprocal functions in the four quadrants respectively.

Example 17

Given that θ is a reflex angle and $\cos\theta = \frac{2}{11}$, find $\sin\theta$, $\tan\theta$, $\sec\theta$, $\operatorname{cosec}\theta$ and $\cot\theta$.

As $\cos\theta$ is positive and θ is reflex, the angle must be in QIV.	Identify the relevant quadrant.
The associated acute angle would be 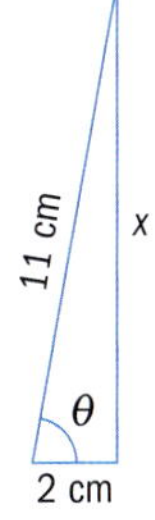	Sketch a diagram.

Continued on next page

$x = \sqrt{117}$

$\sin\theta = -\frac{\sqrt{117}}{11}$

$\tan\theta = -\frac{\sqrt{117}}{2}$

$\sec\theta = \frac{11}{2}$

$\operatorname{cosec}\theta = -\frac{11}{\sqrt{117}}$

and $\cot\theta = -\frac{2}{\sqrt{117}}$

Exercise 6J

Evaluate exactly:

1 $\sec\frac{\pi}{6} - \cot\frac{\pi}{3}$

2 $\operatorname{cosec}\left(\frac{-2\pi}{3}\right) + 2\tan\frac{7\pi}{6}$

3 Simplify each of the following expressions to give a single trigonometric function.

a $\cos\theta\tan\theta$ **b** $\cot\theta\sec\theta$

c $\operatorname{cosec}\theta\tan\theta$ **d** $\cos\theta\sec^2\theta\sin\theta$

e $\frac{\tan\theta\cot\theta}{\sin\theta}$ **f** $\tan\theta\operatorname{cosec}\theta$

4 Given that $\operatorname{cosec}\theta = \frac{13}{5}$ and $\frac{\pi}{2} < \theta < \pi$, find the values of $\cos\theta$ and $\cot\theta$.

5 Given that $\sec\theta = -\frac{5}{4}$ and $\pi < \theta < 2\pi$, find the values of $\cos\theta$, $\tan\theta$ and $\sin\theta$.

Trigonometric identities

It is evident that right-angled triangles and Pythagoras' theorem are useful in analysing the relationships between the trigonometric ratios.

From the unit circle it can be seen that:

- the hypotenuse has a of length of one,
- the two shorter sides have a length of $\sin\theta$ and $\cos\theta$ respectively.

Pythagoras' theorem can be applied to obtain the result:

$\sin^2\theta + \cos^2\theta \equiv 1$

This is called the Pythagorean identity.

In the same unit circle diagram, $\tan\theta \equiv \frac{\text{opposite side}}{\text{adjacent side}} \equiv \frac{\sin\theta}{\cos\theta}$

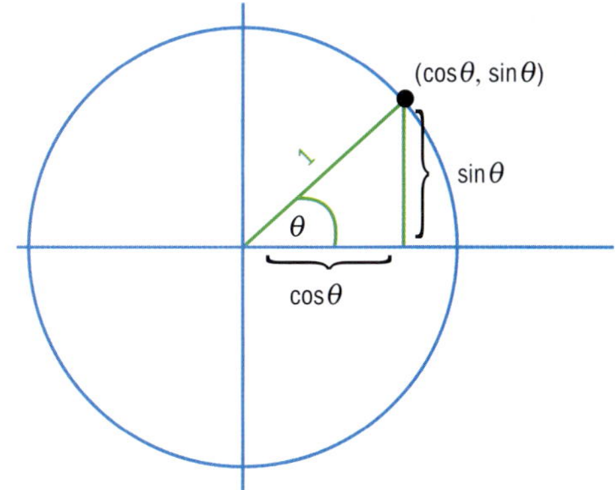

$$\sin^2\theta + \cos^2\theta \equiv 1$$

$$\tan\theta \equiv \frac{\sin\theta}{\cos\theta}$$

HINT

Notice that $(\sin\theta)^2$ is written $\sin^2\theta$.

The ≡ symbol is used to show that this relationship is an equivalence relationship and therefore holds for all values of θ.

Example 18

Without the use of a calculator, find $\cos\theta$, when $\sin\theta = \frac{3}{5}$ and θ is acute.

$\sin^2\theta + \cos^2\theta = 1$ $\left(\frac{3}{5}\right)^2 + \cos^2\theta = 1$ $\cos^2\theta = 1 - \left(\frac{3}{5}\right)^2 = \frac{16}{25}$	The second trigonometric identity allows for a second method in solving problems. Use the Pythagorean identity.
$\cos\theta = \frac{4}{5}$	Use the positive value as the angle is acute.

The original trigonometric identity $\sin^2\theta + \cos^2\theta \equiv 1$ is a very important identity. From it, two similar identities can be deduced:

Dividing through by $\cos^2\theta$ (because $\cos^2\theta$ is not equal to 0):

$$1 + \frac{\sin^2\theta}{\cos^2\theta} \equiv \frac{1}{\cos^2\theta}$$

but, as $\tan\theta = \frac{\sin\theta}{\cos\theta}$, it follows that $\tan^2\theta = \frac{\sin^2\theta}{\cos^2\theta}$

and, as $\sec\theta = \frac{1}{\cos\theta}$, it follows that $\sec^2\theta = \frac{1}{\cos^2\theta}$.

Therefore, the second trigonometric identity is

$1 + \tan^2\theta \equiv \sec^2\theta$

If the original trigonometric identity is divided by $\sin^2\theta$:

$$\frac{\cos^2\theta}{\sin^2\theta} + 1 \equiv \frac{1}{\sin^2\theta}$$

TOK

Is it ethical that Pythagoras gave his name to a theorem that may not have been his own creation?

Geometry and trigonometry

but, as $\cot\theta = \dfrac{\cos\theta}{\sin\theta}$, it follows that $\cot^2\theta = \dfrac{\cos^2\theta}{\sin^2\theta}$

and, as $\operatorname{cosec}\theta = \dfrac{1}{\sin\theta}$, it follows that $\operatorname{cosec}^2\theta = \dfrac{1}{\sin^2\theta}$.

Therefore, the third identity is

$\cot^2\theta + 1 \equiv \operatorname{cosec}^2\theta$

Example 19

Solve $1 + \cos\theta = 2\sin^2\theta$, for all values of θ between 0 and 2π.

$1 + \cos\theta = 2\sin^2\theta$	The square on the RHS indicates a quadratic, hence be prepared for two solutions. $\sin^2\theta + \cos^2\theta \equiv 1$ Hence, $\sin^2\theta \equiv 1 - \cos^2\theta$
$1 + \cos\theta = 2(1 - \cos^2\theta)$ $2\cos^2\theta + \cos\theta - 1 = 0$ $(2\cos\theta - 1)(\cos\theta + 1) = 0$ $\cos\theta = \frac{1}{2}$ or $\cos\theta = -1$ If $\cos\theta = \frac{1}{2}$, then $\theta = \frac{\pi}{3}$ or $\frac{5\pi}{3}$ If $\cos\theta = -1$, then $\theta = \pi$ Therefore, the roots of the equation between 0 and 2π are $\frac{\pi}{3}, \pi, \frac{5\pi}{3}$.	This quadratic can be solved by factorizing.

Example 20

Given that $x = a\sin\theta$ and $y = b\tan\theta$, find the relationship between x and y.

If $x = a\sin\theta$, then $\frac{x}{a} = \sin\theta$ or $\frac{a}{x} = \operatorname{cosec}\theta$ And if $y = b\tan\theta$, then $\frac{y}{b} = \tan\theta$ or $\frac{b}{y} = \cot\theta$ $\cot^2\theta + 1 \equiv \operatorname{cosec}^2\theta$ $\frac{b^2}{y^2} + 1 \ldots \frac{a^2}{x^2}$	Since $\sin\theta$ and $\tan\theta$ are the reciprocals of $\operatorname{cosec}\theta$ and $\cot\theta$, we use the third Pythagorean identity: $\cot^2\theta + 1 \equiv \operatorname{cosec}^2\theta$

Exercise 6K

1 Given $s = \sin\theta$, $c = \cos\theta$, $t = \tan\theta$, simplify the following expressions:

a $\sqrt{1-s^2}$ **b** $\dfrac{s}{\sqrt{1-s^2}}$ **c** $\dfrac{\sqrt{1-c^2}}{c}$

d $\dfrac{c}{1-c^2}$ **e** $\sqrt{1+t^2}$ **f** $\dfrac{t}{\sqrt{1+t^2}}$

2 Given $x = a\sin\theta$, $y = b\cot\theta$ and $z = a\sec\theta$, simplify the following expressions:

a $x^2 - a^2$ **b** $y\sqrt{b^2+y^2}$

c $\dfrac{b}{\sqrt{b^2+y^2}}$ **d** $\dfrac{\sqrt{z^2+a^2}}{z}$

3 Solve the following equations, for all values of θ between 0 and 2π.

a $3 - 3\cos\theta = 2\sin^2\theta$ **b** $\sec\theta = 2$

c $\cos^2\theta + \sin\theta + 1 = 0$ **d** $\sec^2\theta = 1 + \tan\theta$

e $3\tan^2\theta - 5\sec\theta + 1 = 0$

f $2\cot\theta\cos\theta + 7 = 7\operatorname{cosec}\theta$

4 Simplify:

a $(\operatorname{cosec}\theta + 1)(\operatorname{cosec}\theta - 1)$ **b** $\dfrac{1+2\sin\theta\cos\theta}{\sin\theta+\cos\theta}$

c $\dfrac{2\tan\theta + \sec^2\theta}{(\sin\theta+\cos\theta)^2}$

Ratios of compound angles

Given A and B are angles, there are often advantages in expressing the trigonometrical ratios of $(A + B)$ and $(A - B)$ in terms of the trigonometric ratios of A and of B.

In Investigation 5, the non-distributive properties of trigonometric ratios and the proof of the compound angle identities should become apparent.

Investigation 5

Are trigonometrical ratios distributive?

1 Complete the table using your knowledge of special triangles.

2 Can the angles in trigonometrical ratios be distributed?

3 Find an expression for $\sin 90°$ using combinations from $\sin 30°$, $\sin 60°$, $\cos 30°$ and $\cos 60°$.

sin 30°	
cos 30°	
tan 30°	
sin 45°	
cos 45°	
tan 45°	
sin 60°	
cos 60°	
tan 60°	
sin 90°	
cos 90°	
tan 90°	

Conceptual Do the trigonometric functions of compound angles follow the distributive property? Explain your answer.

$\sin(A + B) = \sin A \cos B + \cos A \sin B$

The proof is shown here:

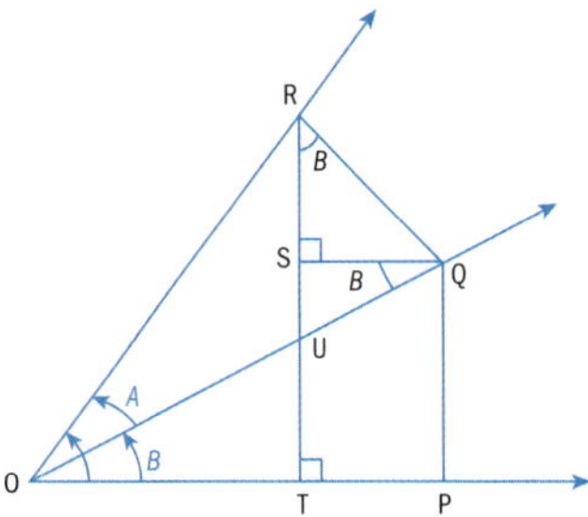

The right-angled triangles OQR and OPQ contain angles A and B respectively.

The line RT is a construction line, such that $\text{O}\hat{\text{U}}\text{T} = 90 - B$, therefore $\text{U}\hat{\text{R}}\text{Q} = B$

$$\sin(A + B) \equiv \frac{\text{TR}}{\text{OR}} \equiv \frac{\text{TS}+\text{SR}}{\text{OR}} \equiv \frac{\text{PQ}+\text{SR}}{\text{OR}}$$

$$\equiv \frac{\text{PQ}}{\text{OR}} + \frac{\text{SR}}{\text{OR}}$$

$$\equiv \frac{\text{PQ}}{\text{OQ}}\frac{\text{OQ}}{\text{OR}} + \frac{\text{SR}}{\text{QR}}\frac{\text{QR}}{\text{OR}}$$

$$\equiv \sin A \cos B + \cos A \sin B$$

In the following exercise, the complete set of compound angle identities will be discovered.

> **HINT**
>
> $\frac{\text{PQ}}{\text{OR}}$ does not help with trig ratios.
>
> However, we require combinations of PQ with OQ and OQ with OR. Hence if we introduce
>
> $\frac{\text{PQ}}{\text{OR}} = \frac{\text{PQ}}{\text{OQ}}\frac{\text{OQ}}{\text{OR}}$
>
> and $\frac{\text{SR}}{\text{OR}} = \frac{\text{SR}}{\text{QR}}\frac{\text{QR}}{\text{OR}}$
>
> this solves the problem.

Exercise 6L

1 Given that $\sin(A + B) \equiv \sin A \cos B + \cos A \sin B$, replace B with $-B$ to find the identity for $\sin(A - B)$.

2 **a** Using the same diagram as above, prove that $\cos(A + B) = \cos A \cos B - \sin A \sin B$.

> **HINT**
>
> Introduce OQ and RQ respectively to make the expressions relevant.

b Replace B with $-B$ to show the identity $\cos(A - B) \equiv \cos A \cos B + \sin A \sin B$

3 **a** Given the identity $\tan\theta \equiv \frac{\sin\theta}{\cos\theta}$ implies $\tan(A + B) \equiv \frac{\sin(A + B)}{\cos(A + B)}$, show that $\tan(A + B) \equiv \frac{\tan A + \tan B}{1 - \tan A \tan B}$

b Replace B with $-B$ to find the identity $\tan(A - B)$.

4 Evaluate

a $\sin 75°$

b $\tan 105°$

c $\sin 33°\cos 3° - \cos 33°\sin 3°$

d $\cos 75°\cos 15° + \sin 75°\sin 15°$

5 Given A is obtuse and $\sin A = \frac{3}{5}$, B is acute and $\sin B = \frac{5}{13}$, calculate directly

a $\cos(A - B)$ **b** $\tan(A + B)$.

6 Prove the following identities:

a $\cot(A + B) \equiv \frac{\cot A \cot B - 1}{\cot A + \cot B}$

b $\frac{\sin(A + B)}{\cos A \cos B} \equiv \tan A + \tan B$

c $\sec A + \tan A = \frac{1 + \sin A}{\cos A}$

d $\tan A + \cot A = \sec A \operatorname{cosec} A$

e $\sec^2\theta + \text{cosec}^2\theta = \sec^2\theta\,\text{cosec}^2\theta$

f $\dfrac{\text{cosec}\,\theta - \cot\theta}{1-\cos\theta} = \text{cosec}\,\theta$

g $\text{cosec}\,x - \sin x = \cos x \cot x$

h $1+\cos^4 x - \sin^4 x = 2\cos^4 x$

i $\sec\theta + \tan\theta = \dfrac{\cos\theta}{1-\sin\theta}$

j $\dfrac{\sin A\tan A}{1-\cos A} = 1+\sec A$

7 a Express $\tan(A+B+C)$ in terms of $\tan A$, $\tan B$ and $\tan C$.

b Given A, B, C are acute and $\tan A = \frac{1}{2}$, $\tan B = \frac{1}{5}$ and $\tan C = \frac{1}{8}$, find $A+B+C$.

c If A, B, C are the angles of a triangle, show that the sum of their tangents is equal to the product.

Double angle identities for sine and cosine

Now it has been established that $\sin(A+B) \equiv \sin A\cos B + \cos A\sin B$, there exist special cases of this identity when $A = B$. These are sometimes even more useful than the compound angle identities.

$\sin(A+B) \equiv \sin A\cos B + \cos A\sin B$

$\sin(A+A) \equiv \sin A\cos A + \cos A\sin A$

$\sin 2A \equiv 2\sin A\cos A$

It also follows that $\cos 2A = \cos^2 A - \sin^2 A$

$\tan 2A = \dfrac{2\tan A}{1-\tan^2 A}$

The reason the double angle formula for cosine is stated in three alternate forms depends on the question which has been given. They all are deduced from the same identity.

$\cos(A+B) \equiv \cos A\cos B - \sin A\sin B$

by substituting $B = A$ in to the identity

$\cos 2A = \cos^2 A - \sin^2 A$

But $\cos^2 A + \sin^2 A = 1$

Hence $\cos^2 A = 1 - \sin^2 A$

$\cos 2A = (1-\sin^2 A) - \sin^2 A$

$\cos 2A = 1 - 2\sin^2 A$

$1 - \cos^2 A$

But $\cos^2 A + \sin^2 A = 1$

Hence $\sin^2 A = 1 - \cos^2 A$

$\cos 2A = \cos^2 A - (1-\cos^2 A)$

$\cos 2A = 2\cos^2 A - 1$

Alternatively, $\cos^2 A = \frac{1}{2}(1+\cos 2A)$

$\sin^2 A = \frac{1}{2}(1-\cos 2A)$

TOK

Trigonometry was developed by successive civilizations and cultures.

To what extent is mathematics a product of human social interaction?

Double Angle formulae:

$\sin 2A \equiv 2\sin A \cos A$

$\cos 2A = \cos^2 A - \sin^2 A$

$= 2\cos 2A - 1$

$= 1 - 2\sin 2A$

Example 21

Solve the equation $3\cos 2\theta - \sin \theta = -1$ for all values from 0 to 360 degrees.

$3\cos 2\theta - \sin \theta = -1$	Use $\cos 2\theta = 1 - 2\sin^2 \theta$.
$3(1 - 2\sin^2 \theta) - \sin \theta = -1$	Simplify.
$0 = -3 + 6\sin^2 \theta + \sin \theta + 1$	
$0 = 6\sin^2 \theta + \sin \theta - 2$	
$0 = 6\sin^2 \theta + 4\sin \theta - 3\sin \theta - 2$	
$0 = 2\sin \theta(3\sin \theta + 2) - 1(3\sin \theta + 2)$	Factorize and solve for $\sin\theta$.
$0 = (2\sin \theta - 1)(3\sin \theta + 2)$	
$\sin \theta = \frac{1}{2}$ $\quad$ $\sin \theta = -\frac{2}{3}$	
Basic $\theta = 30°$	sin +ve therefore QI and QII
Basic $\theta = 41.8°$	sin −ve therefore QIII and QIV
30°, 150°, 221.8°, 318.2°	

Example 22

Evaluate $2\sin 15 \cos 15$ without a calculator.

$2\sin 15 \cos 15$	Recognize $\sin 2A \equiv 2\sin A \cos B$
$= \sin(2 \times 15)$	
$= \sin(30)$	
$= \frac{1}{2}$	

Example 23

If $\sin A = \frac{4}{5}$ and $\cos B = \frac{8}{17}$, where A is obtuse and B is acute, find the exact value of $\sin(A + B)$.

$\sin A = \frac{4}{5}$, $\cos A = -\frac{3}{5}$	If A is obtuse, it lies in quadrant II and therefore cosine is negative.
$\cos B = \frac{8}{17}$, $\sin B = \frac{15}{17}$	If B is acute, it lies in quadrant I and therefore sine is positive.
$\sin(A+B) = \frac{4}{5} \times \frac{8}{17} + \left(-\frac{3}{5}\right) \times \frac{15}{17}$ $= \frac{-13}{85}$	$\sin(A+B) \equiv \sin A \cos B + \cos A \sin B$

Proving trigonometric identities

An important skill is using compound angles, double angles, and other trigonometric identities in order to manipulate trigonometric expressions, to verify equivalence. There are many ways of approaching this; ultimately both sides have to be equal, to prove equivalence.

Example 24

Prove the trigonometric identity $\frac{\sin 2A}{1+\cos 2A} = \tan A$.

$\frac{\sin 2A}{1+\cos 2A}$	As the RHS is already a single trigonometric expression it is best to manipulate the LHS.
$= \frac{2\sin A \cos A}{1+2\cos^2 A - 1}$	Use the double angle formula for $\cos 2A$ to eliminate the "1".
$= \frac{2\sin A \cos A}{2\cos^2 A}$	Divide by $2\cos A$.
$= \tan A = \text{RHS}$	Use the first trigonometric identity.

Example 25

Use your GDC to solve $\sin 3x + \cos 2x = 1$ for $0° \le x \le 180°$.

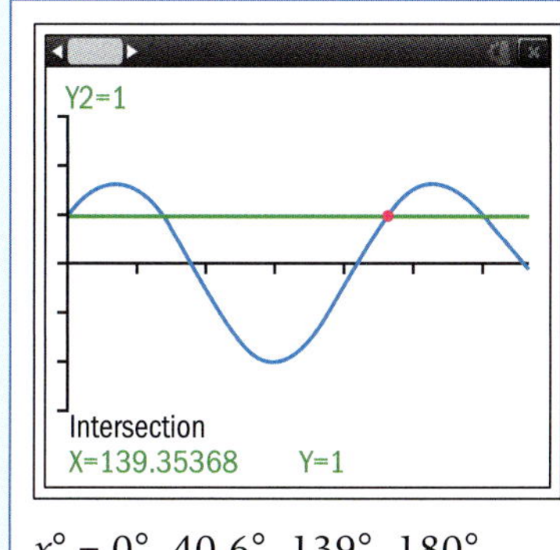 $x° = 0°, 40.6°, 139°, 180°$	

Exercise 6M

1 Prove the following identities.

a $\tan A + \cot A = 2\operatorname{cosec} 2A$

b $\frac{\sin 2A + \cos 2A + 1}{\sin 2A - \cos 2A + 1} = \cot A$

c $\frac{\cos 3X - \sin 3X}{1 - 2\sin 2X} = \cos X + \sin X$

d $\cot x - \operatorname{cosec} 2x = \cot 2x$

2 Use the fact that $4A = 2 \times 2A$ to prove $\frac{\sin 4A}{\sin A} = 8\cos^3 A - 4\cos A$

3 If $\tan A = \frac{3}{4}$, find $\sin 2A$ $\sec 2A$ and $\tan 2A$.

4 Express $\sin 3X$ and $\cos 3X$ in terms of $\sin X$ and $\cos X$, respectively.

5 Express $\cos 4A$ in terms of $\cos A$.

6 Express $\tan 4A$ in terms of $\tan A$.

7 Find the values of x between 0 and 2π such that:

a $\sin 3x = \sin^2 x$

b $\cot 2x = 2 + \cot x$

c $\cos 3x - 3\cos x = \cos 2x + 1$.

Investigation 6

You will need: spaghetti (or drinking straws), protractor, string, colourful ball of wool, scissors, glue, poster paper.

Work in groups of two or three.

1 At one end of the paper, construct a circle on a set of axes, which has a radius of one spaghetti length.

2 Use radian measure to measure and mark every $\frac{\pi}{6}$ around the circle.

3 To the right of the circle draw a set of axes with an x-axis that is about 2π spaghetti lengths long and 2 spaghetti lengths tall.

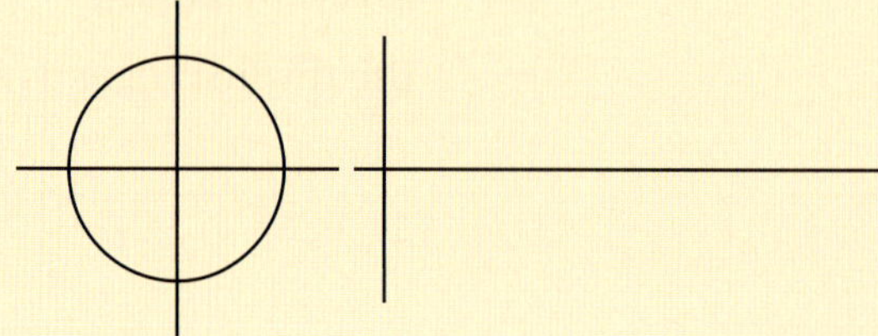

4 Place the string around the circle with the end at zero. Mark the $\frac{\pi}{6}$ marks onto the string. Then put the string on the x-axis of the right-hand set of axes and transfer the marks labelling every $\frac{\pi}{6}$.

5 Place a piece of spaghetti from the origin to the $\frac{\pi}{6}$ mark on the circle. Take another piece of spaghetti and measure the vertical distance from the point P to the x-axis. Glue this piece of spaghetti at $\frac{\pi}{6}$ on the other set of axes.

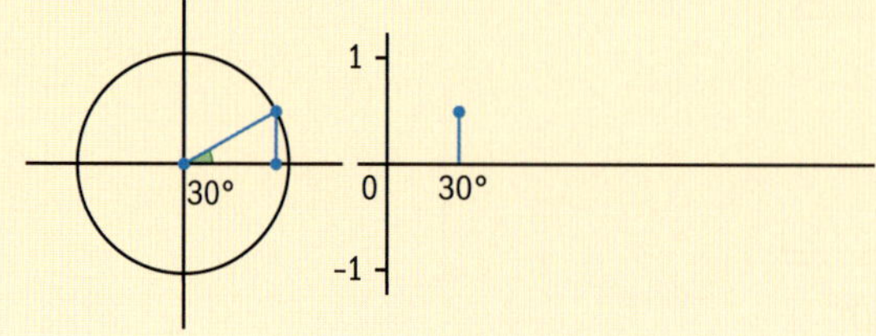

6 Now repeat for 45°, 60°, 90° and so on, until you have gone completely around the circle to 360°.

7 Take the wool and glue it to your poster from zero degrees, along the top of the spaghetti pieces to form a smooth curve.

8 The sine curve has been produced.

Go to your device and research periodic functions. Is the sine function periodic? Why?

Factual What is a periodic function?

What shape is the graph of a periodic function?

9 Use your calculator to graph $y = \sin x$. Does it look like your spaghetti graph?

Conceptual How does the shape of a periodic function show there will always be multiple x-values that give the same values of y?

10 Write an explanation about why $\sin 60° = \sin 120°$.

Conceptual How is a sine curve produced?

Developing inquiry skills

Discuss your answers with your group and then **share** your ideas with your class.

- What is the radius of the circle in spaghetti units?
- What is the circumference in spaghetti units?
- What happens after 2π?
- Why is this the graph of the sine function?
- What are the zeros of your sine function?
- Where is the highest point above the x-axis? At what angle is this?
- Where is the lowest point above the x-axis? At what angle is this?
- How would you do this investigation differently for a cosine curve?

Learning the graphs of trigonometrical functions in terms of both degrees and radians is very important.

Reflect Using a graphical software package, explore the properties of the graphs of $\sin\theta$ and $\cos\theta$. As the angle θ increases from 0 to 2π describe the values of $\sin\theta$, $\cos\theta$ and $\tan\theta$.

There is a unique value for each trigonometric ratio of any specific angle. Therefore, the mapping $\theta \to \sin\theta$ is a function and the graph showing this periodic nature can be found.

The graph is continuous with range $-1 < \sin\theta < 1$ and period 2π.

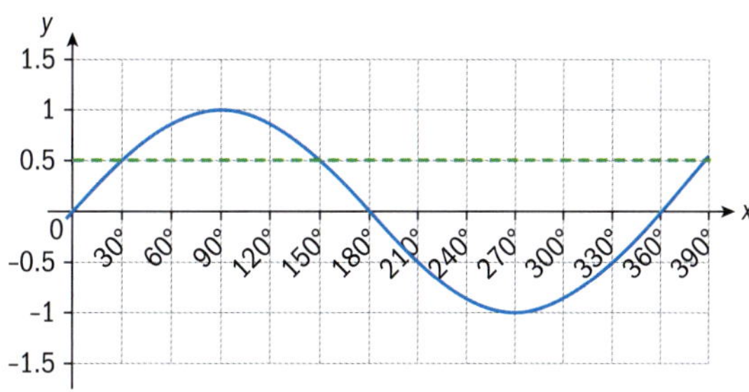

Geometry and trigonometry

Example 26

Find the values of θ between 0° and 360° for which

a $\sin\theta = 0.38$ **b** $\cos\theta = -0.62$.

a $\sin\theta = 0.38$ $\theta = 22.3°$ $22.3°, 180° - 22.3°$	You know there are two solutions by observing the graph. 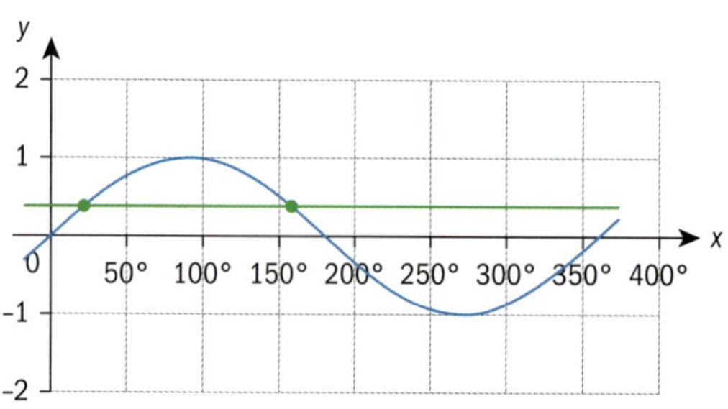
$22.3°, 157.7°$	Because the sine of the angle is positive, the solutions must come in QI and QII.
b $\cos\theta = -0.62$ if $\cos\alpha = 0.62$ $\alpha = 51.68°$ $\theta = 180° + \alpha, \theta = 180° - \alpha$ $231.7°, 128.3°$	Because the cosine is negative the solutions are in QII and III. 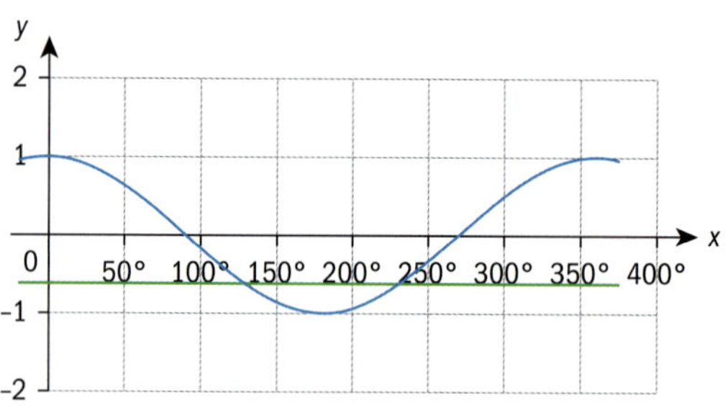

6.4 Trigonometric functions

Recall the definition of trigonometric ratios using the unit circle. The diagram shows an angle θ made by the radius OP

P has coordinates (x, y).

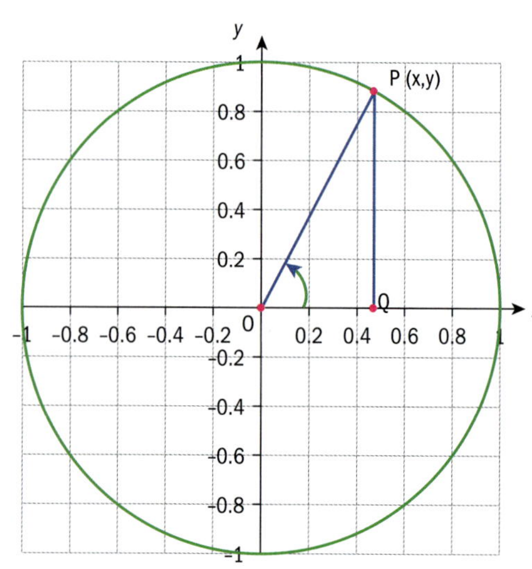

Take several angles on the unit circle and for each one consider $\sin\theta$, which is represented by the y-coordinate on the unit circle (imagine a number of vertical lines from points on the circumference of the circle to the x-axis). These values are represented on the graph below.

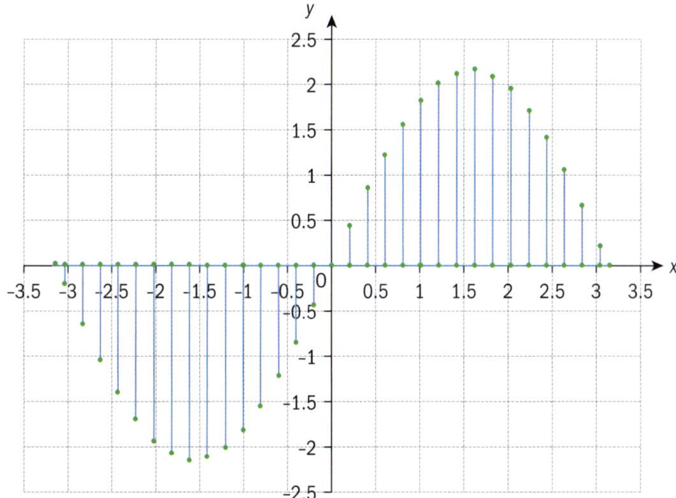

Now consider $\cos\theta = \dfrac{x}{\text{OP}} = x$ and take several angles on the unit circle.

Imagine a number of horizontal lines from points on the circumference of the circle to the y-axis. Representing these lines on a graph:

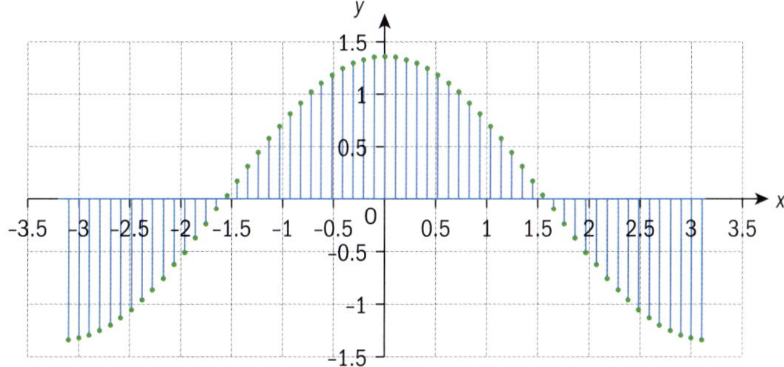

The tips of the vertical line segments lie on the sine and cosine curves and represent the functions $f(\theta) = \sin\theta$ and $f(\theta) = \cos\theta$.

Note that since $\theta \in \mathbb{R}$ the domain of both functions will be all real values of θ.

The functions are periodic with period 2π.

The range of the functions will be $-1 \le f(\theta) \le 1$.

In order to explore $f(\theta) = \tan\theta$, where $\tan\theta = \dfrac{\sin\theta}{\cos\theta} = \dfrac{y}{x}$, $x \ne 0$. Therefore, when the value of $\cos\theta$ is zero, the graph is undefined which is when $\theta = \dfrac{(2n+1)\pi}{2}$, $n \in \mathbb{Z}$. The graph of $f(\theta) = \tan\theta$ will have vertical asymptotes for all these values of θ.

The domain of the tangent function is $\theta \in \mathbb{R}$, $\theta \ne \dfrac{(2n+1)\pi}{2}$, $n \in \mathbb{Z}$.

The range of the tangent function will be $f(\theta) \in \mathbb{R}$

In order to visualize the shape of the tangent function, consider the diagram to the right which shows the unit circle and a tangent drawn at the point (1, 0).

Triangles OPQ and OLM are similar and hence $\frac{\text{LM}}{\text{OM}} = \frac{\text{PQ}}{\text{OQ}} \Rightarrow \text{LM} = \frac{y}{x} = \tan\theta$. Therefore, the value of $\tan\theta$ is obtained by extending OP to meet this tangent line at L. The y-coordinate of the point L is then equal to $\tan\theta$. If we do this for several values of θ we get the following graph.

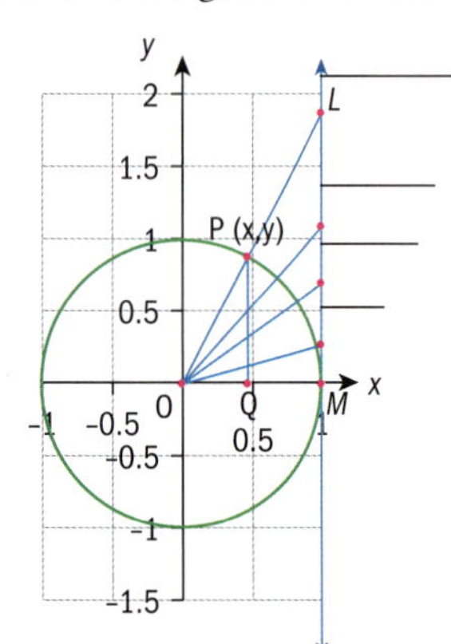

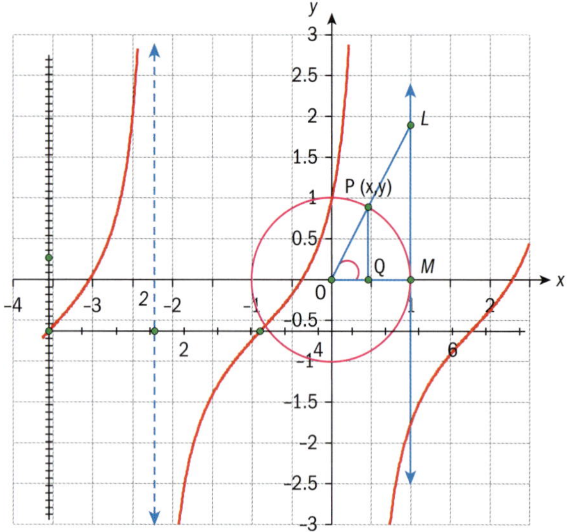

Due to the symmetrical property of the circle, two points on a diameter represent two angles with the same tangent value.

Note that for $f(\theta) = \tan\theta$ the period of the function is π.

We can summarize these results as follows

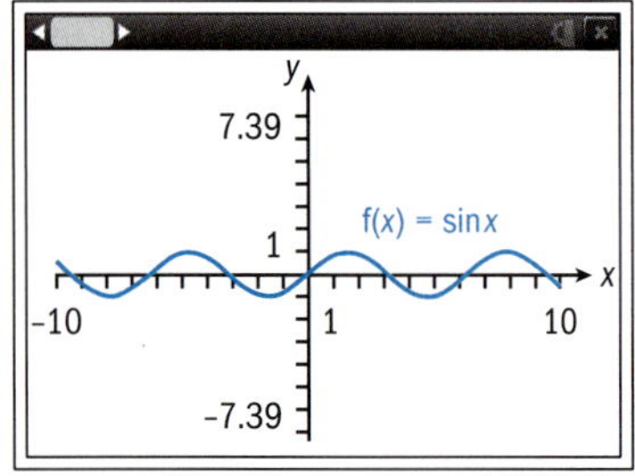

Domain: $x \in \mathbb{R}$

Range: $y \in \mathbb{R}, -1 \le y \le 1$

We also say that the amplitude of y is 1.

Period $= 2\pi$

$\Rightarrow \sin(x + 2n\pi) = \sin x$

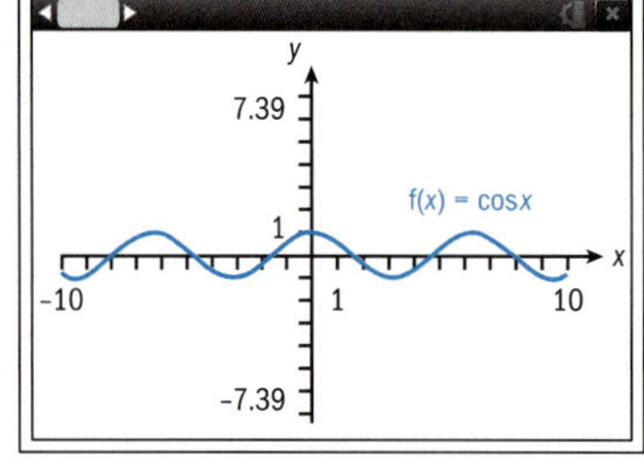

Domain: $x \in \mathbb{R}$

Range: $y \in \mathbb{R}, -1 \le y \le 1$

We also say that the amplitude of y is 1.

Period $= 2\pi$

$\Rightarrow \cos(x + 2n\pi) = \cos x$

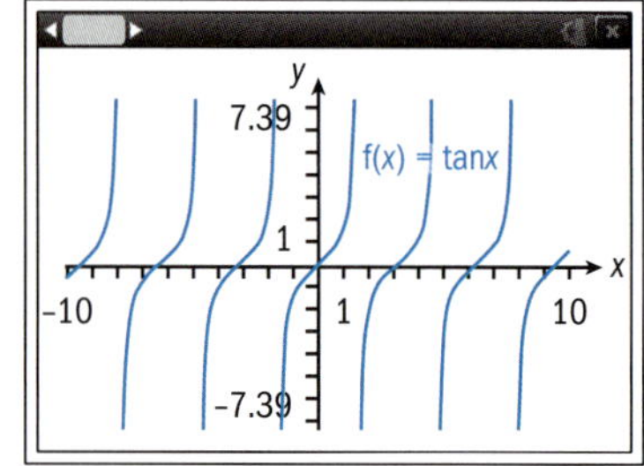

Domain: $x \in \mathbb{R},\ x \neq \frac{(2n+1)\pi}{2}$

Range: $y \in \mathbb{R}$

Period $= \pi$

$\Rightarrow \tan(x + n\pi) = \tan x$

HINT

The amplitude of the wave is the vertical distance between the peak (maximum) and the horizontal axis of the wave.

Example 27

Use the graph $f(\theta) = \cos\theta$ to deduce the graph and properties of $g(\theta) = \sec\theta$.

Since $g(\theta) = \dfrac{1}{f(\theta)}$, you can obtain the graph using the following reasoning:

- $\sec\theta = 1$ when $\cos\theta = 1$ at $\theta = \pm\, 2n\pi,\ n \in \mathbb{Z}$
- $\sec\theta = -1$ when $\cos\theta = -1$ at $\theta = \pm\, (2n+1)\pi,\ n \in \mathbb{Z}$
- $g(\theta)$ is undefined when $f(x) = 0 \Rightarrow$ vertical asymptotes at $\theta = \pm(2n+1)\dfrac{\pi}{2}$
- $\sec\theta \to \infty$ as $\cos\theta \to 0^+$
- $\sec\theta \to -\infty$ as $\cos\theta \to 0^-$

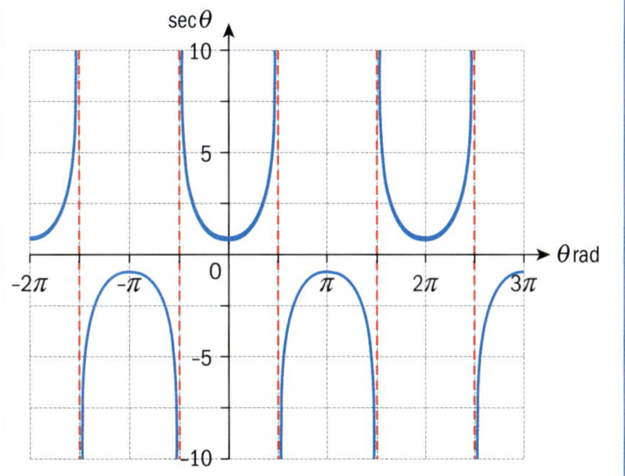

Relationships between trigonometric functions and the symmetry properties of their graphs

HINT

In mathematics certain symbolic vocabulary is introduced to simplify sentences.

iff or $\Leftrightarrow$ mean “if and only if.”

$\forall$ is the notation for “for all values”

Example 28

Use the graph of $y = \sin x$ to sketch the graph of $y = 3\sin 4x$.

HINT

A vertical stretch with a scale factor between 0 and 1 is called a compression.

Let $f(x) = \sin x$ and $g(x) = 3\sin 4x$ Then $g(x) = 3f(4x)$ The amplitude of $g(x)$ is 3.	Apply a vertical stretch of scale factor 3 parallel to the y-axis. Stretching the function of $y = 3\sin x$ by a scale factor of $\frac{1}{4}$ parallel to the x-axis.
The period of $g(x)$ is $\frac{\pi}{2}$.	The period of the new function is therefore $\frac{2\pi}{4} = \frac{\pi}{2}$.

Continued on next page

Geometry and trigonometry

Below is a sketch of the function $g(x) = 3\sin 4x$

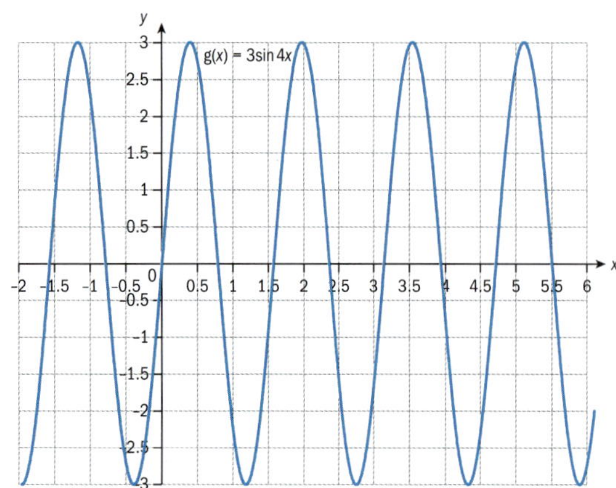

Exercise 6N

1 Use the graph $f(\theta) = \sin\theta$ to deduce the graph and properties of $g(\theta) = \operatorname{cosec}\theta$.

2 Use the graph $f(\theta) = \tan\theta$ to deduce the graph and properties of $g(\theta) = \cot\theta$.

3 Use transformations of $y = \sin x$ to graph each of the following functions:

a $y = 3\sin 4x$

b $y = \sin\left(x + \frac{3\pi}{2}\right)$

c $y = -2\sin(\pi x)$

d $y = 2\sin(4x + \pi) - 1$

4 Use transformations of $y = \cos x$ to graph each of the following functions:

a $y = 2\cos x + 2$

b $y = \cos 3x - 1$

c $y = -2\cos 3x$

d $y = 3(1 + \cos 2x)$

5 On the same axes sketch the graphs of $f(x) = \cos 2x$ and $g(x) = 2\cos x$ for $0 \le x \le \pi$.

Find how many solutions there are to the equation $f(x) = g(x)$ in the interval $0 \le x \le \pi$.

Investigation 7

Consider the function $f(x) = a\sin[b(x+c)] + d,\ b,\ c \in \mathbb{R}, b > 0$.

By considering different values of a, b, c and d, show that

1 the amplitude of the function is given by $|a|$

2 the function has a period of $\frac{2\pi}{b}$

3 the line $y = d$ is a principal axis

4 $d = \frac{\text{maximum value} + \text{minimum value}}{2}$

5 the function can be obtained by shifting the graph of $y = a\sin(bx)$ by c units to the left and d units vertically upwards when $c, d > 0$.

Conceptual How do we describe the transformation of a trigonometric function?

TOK

Sine curves model musical notes and the ratios of octaves. Does this mean that music is mathematical?

HINT

The number c is called the phase shift of $f(x)$.

Steps for fitting data to a sine or cosine function

For $f(x) = a\sin[b(x + c)] + d$:

- Calculate the amplitude $a = \dfrac{\text{maximum value} - \text{minimum value}}{2}$
- Calculate the vertical shift $d = \dfrac{\text{maximum value} + \text{minimum value}}{2}$
- Find the value of $b = \dfrac{2\pi}{\text{period}}$
- Find the phase shift c.
- Calculate the horizontal shift by choosing given coordinates of a data point.

Example 29

Find the amplitude, period, phase shift and vertical shift of the function $f(x) = 2\sin(4x + \pi) + 5$.

Hence sketch the graph.

We can rewrite the function as follows $f(x) = 2\sin\left[4\left(x + \dfrac{\pi}{4}\right)\right] + 5$ Amplitude = 2 Period = $\dfrac{2\pi}{4} = \dfrac{\pi}{2}$ Phase shift = $\dfrac{\pi}{4}$ Vertical shift = 5	Using $b = \dfrac{2\pi}{\text{period}}$ 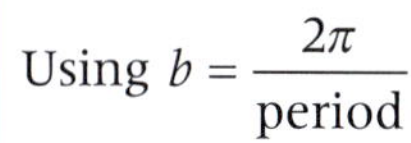 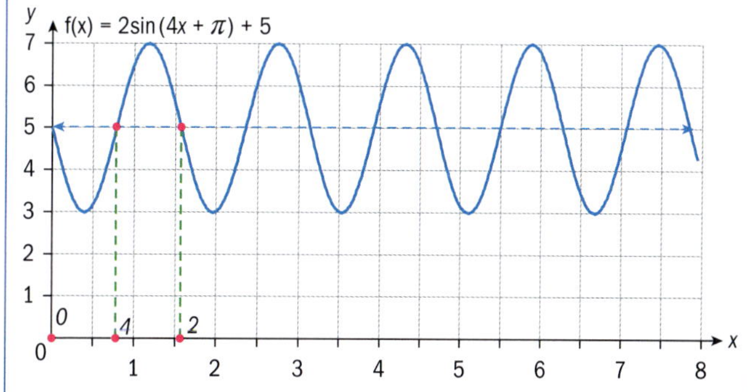

Example 30

The amount of sunlight in Reykjavik, the northernmost capital city in Europe, on the shortest day of the year is 5.82 h and on the longest day is 18.92 h. Assume that the hours of sunlight over a year follow a function of the form $f(x) = a\sin[b(x + c)] + d$. Find the values of a, b, c and d. Use a graph of this function to find the number of hours of sunshine on 1st March. (Assume that this is the 60th day of the year, the shortest day falls on the 21st December and that there are 365 days in a year.)

$-6.55 = 6.55\sin\left[\dfrac{2\pi}{365}(355 + c)\right]$ Let $f(x) = a\sin[b(x + c)] + d$	

Continued on next page

The amplitude of the function is given by $\frac{(18.92-5.82)}{2}=6.55=a$	Calculate the amplitude of wave by $\left(\frac{\text{maximum}-\text{minimum}}{2}\right)$.
The vertical shift is given by $\frac{(18.92+5.82)}{2}=12.37=d$	Calculate the vertical shift.
The period of the function is $\frac{2\pi}{b}=365 \Rightarrow b=\frac{2\pi}{365}$	The data repeats itself every 365 days.
$f(x)=6.55\sin\left[\frac{2\pi}{365}(x+c)\right]+12.37$	
$x = 365 - 10 = 355$ and $f(x) = 5.82$.	The shortest day occurs on 21st December.
$5.82=6.55\sin\left[\frac{2\pi}{365}(355+c)\right]+12.37$	Substitute in values.
$-1=\sin\left[\frac{2\pi}{365}(355+c)\right]$	
$\left[\frac{2\pi}{365}(355+c)\right]=\arcsin -1=-\frac{\pi}{2}$	
$355+c=-\frac{\pi}{2}\times\frac{365}{2\pi}$	
$c=-\frac{\pi}{2}\times\frac{365}{2\pi}-355=-263.75$	
$f(x)=6.55\sin\left[\frac{2\pi}{365}(x-263.75)\right]+12.37$	
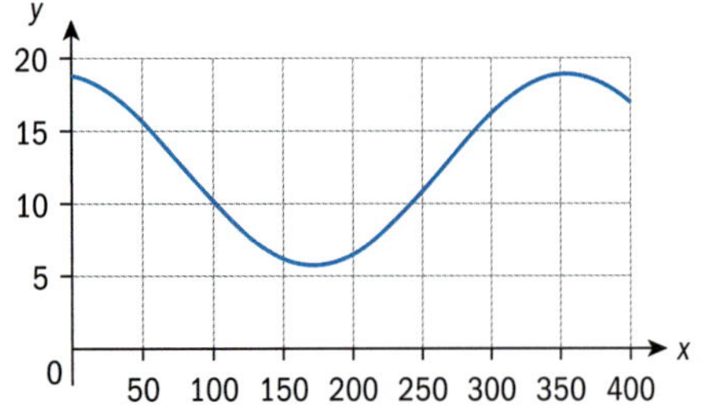	
$f(60)=6.55\sin\left[\frac{2\pi}{365}(60-263.75)\right]+12.37$	

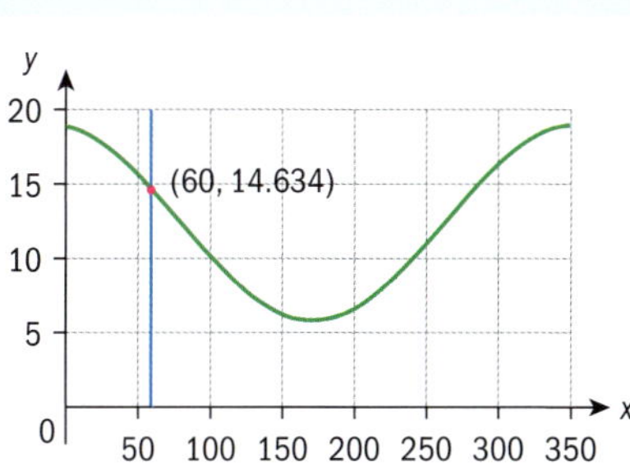

From the graph there are 14.63 hours of sunshine on 1st March.

Exercise 60

1 Find the amplitude, period and phase shift of each of the following functions. Calculate the minimum and maximum values and sketch each function:

a $f(x) = 5\sin\left[3\left(x + \dfrac{\pi}{12}\right)\right] + 1$

b $g(x) = -2\cos\left(3x + \dfrac{5\pi}{2}\right) - 2$

2 The voltage V produced by an AC generator is given by $V(t) = 220\sin(120\pi t)$.

a What is the maximum voltage produced?

b What is the minimum voltage produced?

c What is the amplitude of the function V?

d What is the period of the function V?

e Sketch the graph of V over two periods starting at $t = 0$.

3 Water tides can be modelled using the function $h(t) = a\sin[b(t + c)] + d$ where $h(t)$ is the height of water at time t measured in hours after midnight. The time between consecutive high tides is 12 hours. The height of the water at high tide is 14.4 m and the height of the water at low tide is 1.2 m.

On a particular day the first high tide occurs at 08:15 h.

a Use the information given to find the values a, b, c and d.

b Plot the graph of the function and calculate the time of the first low tide.

c A lifeboat is only allowed to leave or enter the harbour if the height of the water is at least 5 m. Find the time intervals during which a boat could enter or leave the harbour on that particular day.

4 Find equations for the following graphs.

a

b

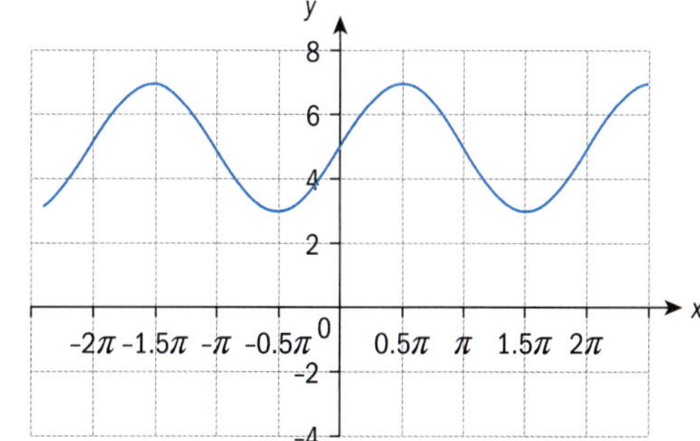

c

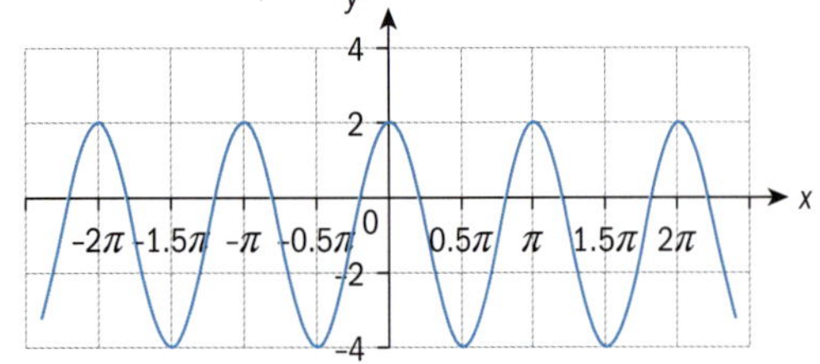

e

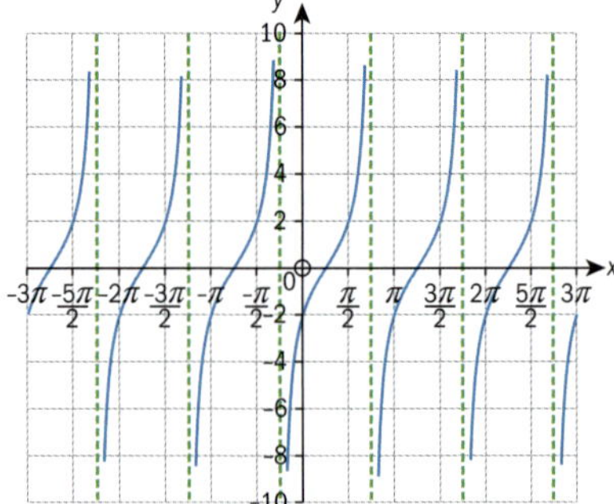

d

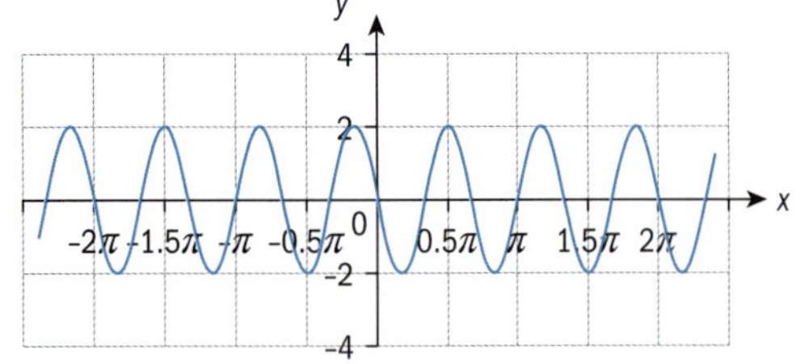

f

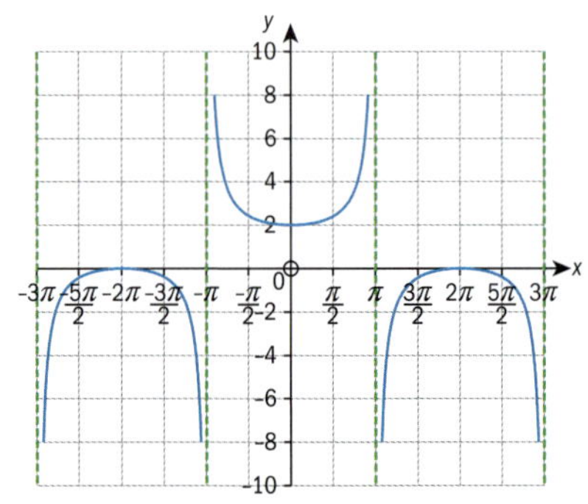

Inverse trigonometric functions

Equations of the type $\sin x = \frac{1}{2}$ arise frequently in mathematics. In order to isolate the variable, it is necessary to perform equivalent operations to both sides.

The inverse operation to $\sin x$ is therefore required. This is known as arcsin (or $\sin^{-1}$ on calculator buttons).

$$\sin x = \frac{1}{2} \quad \arcsin(\sin x) = \arcsin \frac{1}{2}$$

$$\arcsin(\sin x) = \sin x = \frac{1}{2}$$

$$x = \frac{\pi}{6} + 2n\pi, \ \frac{5\pi}{6} + 2n\pi$$

Recall from Chapter 2 that the inverse of a function occurs if $f(x)$ is a one-to-one function

If $\sin x = y$, then $x = \arcsin y$. However, as $\sin x$ is a periodic and not a one-to-one function, it is necessary to add a domain restriction onto the original function to obtain a valid inverse function

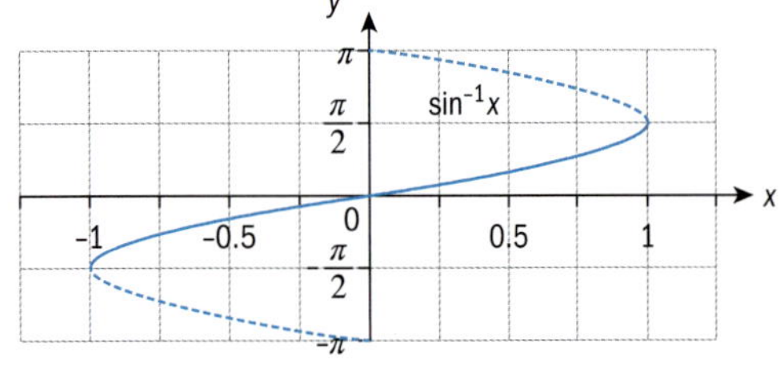

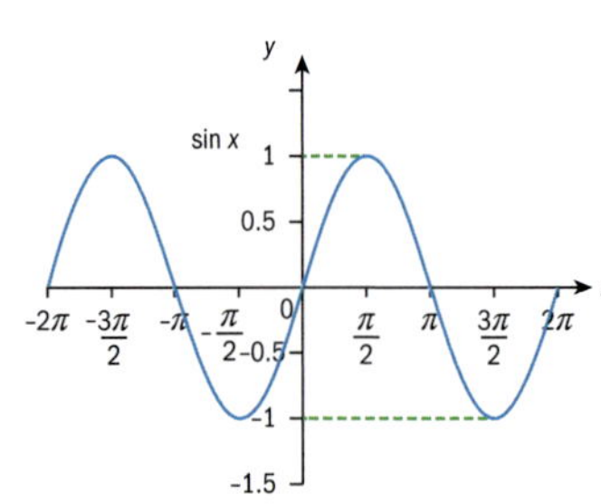

Example 31

a Sketch the graph of $f(x) = 2\cos\left(\frac{x}{2}\right)$.

b Determine the domain for a one-to-one function to occur.

c Sketch the inverse function, $f^{-1}(x)$.

d Use the graph to find the value of x for which $f(x) = f^{-1}(x)$.

a

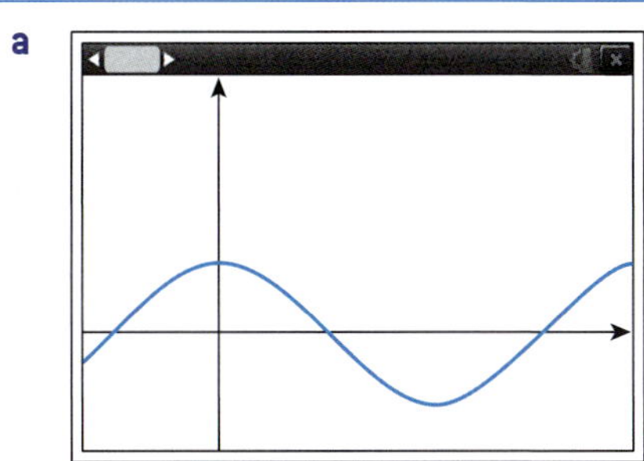

Use technology to draw the curve

$$f(x) = 2\cos\left(\frac{x}{2}\right)$$

b

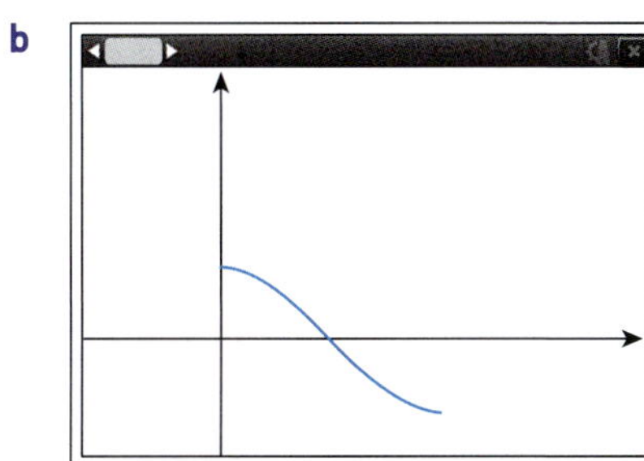

Restrict the domain to $0 \le x \le 2\pi$ to ensure a one-to-one function.

c

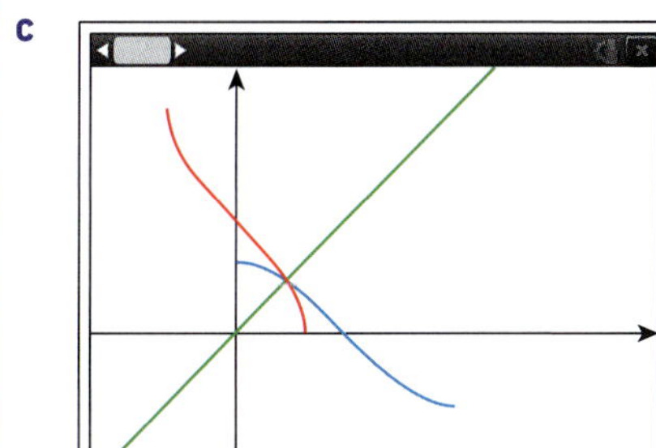

Draw the line $y = x$ and the inverse function.

d

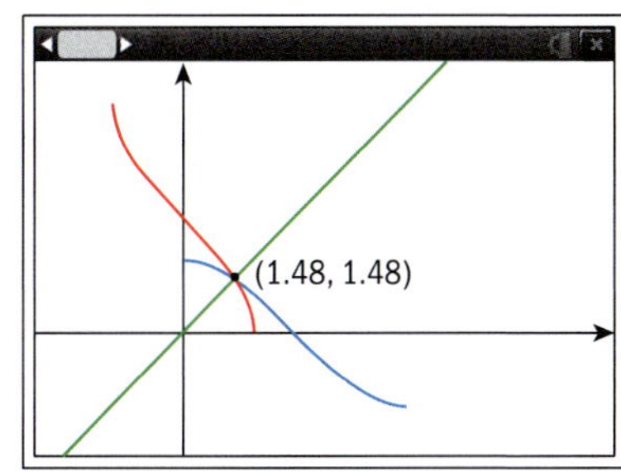

Find the point of intersection of either the two curves of one of the curves and the line $y = x$. $x = 1.48$.

Geometry and trigonometry

Exercise 6P

1 Write down the values of the following angles

a $\arcsin \frac{1}{2}$ **b** $\arctan 1$

c $\arccos -1$ **d** $\arctan \frac{1}{\sqrt{3}}$

e $\arcsin 0$ **f** $\arccos \frac{1}{\sqrt{2}}$

g $\arccos (-\frac{1}{2})$

2 If $\sin^{-1} x = \frac{2\pi}{9}$, find the exact value of $\cos^{-1}x$.

3 Draw appropriate graphs and hence solve the inequalities of x in the given range $0 < x < 2\pi$.

a $\tan x < 0$ **b** $\sec 2x < 0$ **c** $\sin 4x > 3$

4 Two functions $f(x) = \sin 2x$ and $g(x) = \frac{1}{2}x$ are defined on the interval $0 \le x \le \pi$.

a Determine whether $f^{-1}(x)$ and $g^{-1}(x)$ exist.

b The domain is now restricted to $0 \le x \le \frac{\pi}{2}$. Hence find $g^{-1}(\frac{1}{2})$ and $f^{-1}g\left(\frac{\pi}{6}\right)$.

5 For the graph of $y = 6\sin 5x$, find:

a the period

b the amplitude

c the equation of the line of symmetry.

d Describe the series of transformations which map $y = \sin x$ onto the graph.

6.5 Trigonometric equations

Throughout this chapter, you have seen how there are often multiple methods for solving the same equation.

Most equations in algebra have a finite number of roots (zeros). However, due to the periodic characteristic of the trigonometric functions, trigonometric equations can have an infinite number of roots. In the exam, a specific domain will often be given in order to ensure precision.

TOK

Solving an equation has given you an answer in mathematics but, how can an equation have an infinite number of solutions?

Example 32

Solve the equation $4\cos\theta - 3\sec\theta = 2\tan\theta$ for $-180° \le x \le 180°$.

Method 1	
$4\cos\theta - 3\sec\theta = 2\tan\theta$ for $-180° \le x \le 180°$	
$4\cos\theta - 3\sec\theta = 2\tan\theta$	
$4\cos\theta - \frac{3}{\cos\theta} = 2\frac{\sin\theta}{\cos\theta}$	
$4\cos^2\theta - 3 = 2\sin\theta$	Multiply by $\cos\theta$.
$4(1-\sin^2\theta) - 3 = 2\sin\theta$	Use trigonometric identity.
$4\sin^2\theta + 2\sin\theta - 1 = 0$	
$-126°, -54°, 18°, 162°$	Solve the quadratic equation.

sine is positive in the 1st and 2nd quadrants and negative in the 3rd and 4th quadrants. Therefore, the roots of the equation in the required range are $-126°, -54°, 18°, 162°$	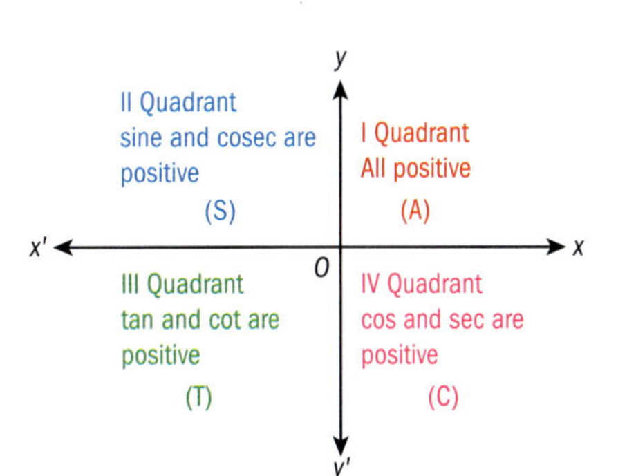
Method 2 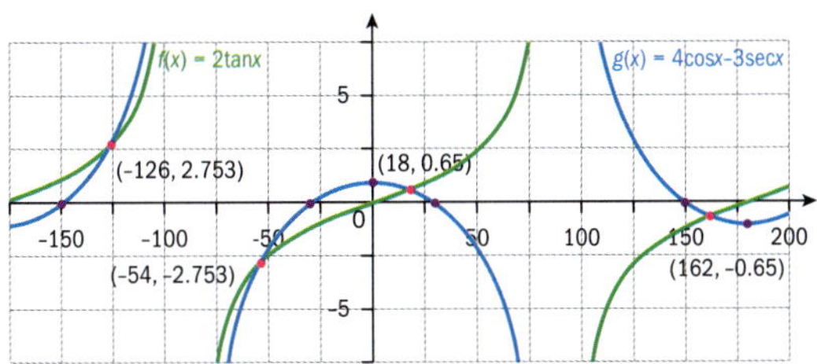 Therefore, the roots of the equation in the required range are $-126°, -54°, 18°, 162°$	Using a graphical approach.

Example 33

Solve the equation $\sin 3x + \sin x = 0$ for values of $\theta°$ from 0 to 2π.

$\sin 3x + \sin x = 0$	
$\sin 3x + \sin x = 0$ for values of $\theta°$ from 0 to 2π	
$\sin(2x + x) + \sin x = 0$	Split $3x = 2x + x$
$\sin 2x \cos x + \cos 2x \sin x + \sin x = 0$	Use compound angle identities.
$2\sin x \cos x \cos x + (1 - 2\sin^2 x)\sin x + \sin x = 0$	Use double angle formula.
$\sin x(2\cos^2 x + 1 - 2\sin^2 x + 1) = 0$	
$\sin x(2\cos^2 x - 2\sin^2 x + 2) = 0$	
$\sin x(2\cos 2x + 2) = 0$	
$\sin x = 0$ or $2\cos 2x = -2$ $2\cos 2x = -2$ $\cos 2x = -1$	
$x = 0, \pi, 2\pi$ or $2x = \pi, 3\pi$ $x = \frac{\pi}{2}, \frac{3\pi}{2}$	

Exercise 6Q

1 Find the roots of the following equations both algebraically and graphically.

a $\text{cosec}^2\theta = 3\cot\theta - 1$

b $2\tan\theta = 3 + 5\cot\theta$

c $2\sec^2\theta - 3 + \tan\theta = 0$

d $5\,\text{cosec}\,\theta + \cot\theta = 2\tan\theta$

Differentiation of trigonometric functions from first principles

By definition, it has been established that if $y = f(x)$ then

$$\frac{dy}{dx} = \lim_{h\to 0}\frac{f(x+h)-f(x)}{h}$$

In order to use this definition to find the derivatives of trigonometric functions it is also important to include two important results:

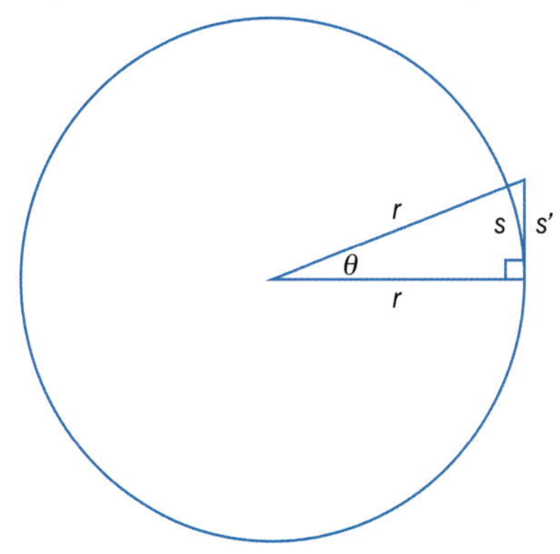

Using small angle theory:

Consider the diagram of a right-angled triangle where one of the shorter sides is tangent to a circle, as shown.

Let r be the radius of the circle, s be the arc length and s' be the opposite side to angle θ.

As θ tends to 0, s tends to s'.

From the sine rule:

$$\frac{s'}{\sin\theta} = \frac{r}{\sin\frac{\pi}{2}}$$

$s' = r\sin\theta$

From the formula for arc length $s = r\theta$:

if $s \approx s'$, then for all small angles $\sin\theta \approx \theta$

Using the double angle identity:

$\cos 2x = 1 - 2\sin^2 x$ with the substitution of $x = \frac{\theta}{2}$ and substituting $\sin\theta \approx \theta$

$$\cos 2\frac{\theta}{2} = 1 - 2\sin^2\frac{\theta}{2}$$

$$\cos\theta = 1 - \frac{\theta^2}{2}$$

As θ is small $\frac{\theta^2}{2}$ tends to 0, hence:

$\cos\theta \approx 1$

International-mindedness

Road signs often show slopes or gradients as percentages or ratios. How do they show slopes in your country? What about other countries?

If $f(x) = \sin x$,

then $f(x + h) = \sin(x + h)$

and

$f(x + h) - f(x) = \sin(x + h) - \sin x$

Example 34

Find the derivative of $f(x) = \sin x$ from first principles.

$f'(x) = \lim_{h\to 0}\frac{f(x+h)-f(x)}{h}$	Use the definition of the derivative.
$= \lim_{h\to 0}\frac{\sin(x+h)-\sin(x)}{h}$	If $f(x) = \sin x$, then $f(x + h) = \sin(x + h)$
$= \lim_{h\to 0}\frac{\sin x\cos h+\cos x\sin h-\sin x}{h}$	Use compound angle identities.
$= \lim_{h\to 0}\left(\cos x\frac{\sin h}{h}+\sin x\frac{\cos h-1}{h}\right)$	Rewrite the expression.
$= \cos x\times\lim_{h\to 0}\frac{\sin h}{h}+\sin x\times\lim_{h\to 0}\frac{\cos h-1}{h}$	Use the properties of limits.
$= \cos x \times 1 + \sin x \times 0 = \cos x$	Use the values of found limits.

Exercise 6R

1 Prove the following statements by differentiating from first principles:

a $\frac{\mathrm{d}(\cos x)}{\mathrm{d}x} = -\sin x$

b $\frac{\mathrm{d}(\sin 2x)}{\mathrm{d}x} = 2\cos 2x$

c $\frac{\mathrm{d}\left(\sin\frac{x}{3}\right)}{\mathrm{d}x} = \frac{1}{3}\cos\frac{x}{3}$

d $\frac{\mathrm{d}(\sin(2x+3))}{\mathrm{d}x} = 2\cos(2x+3)$

2 Use trigonometric identities and properties of limits and differentiation from first principles to show that the following statements are true:

a $\frac{\mathrm{d}(\tan x)}{\mathrm{d}x} = \sec^2 x$

b $\frac{\mathrm{d}(\cot x)}{\mathrm{d}x} = -\mathrm{cosec}\, x^2$

c $\frac{\mathrm{d}(\tan 3x)}{\mathrm{d}x} = 3\sec^2 3x$

The derivatives will be found by using the properties of differentiation and simple derivative results. Since we know the derivatives of sine and cosine functions, we can find the derivative of the tangent function by using the quotient rule.

Function	Derivative
$y = \sin x$	$\frac{dy}{dx} = \cos x$
$y = \cos x$	$\frac{dy}{dx} = -\sin x$
$y = \tan x$	$\frac{dy}{dx} = \sec^2 x$
$y = \operatorname{cosec} x$	$\frac{dy}{dx} = -\operatorname{cosec} x \cot x$
$y = \sec x$	$\frac{dy}{dx} = \sec x \tan x$
$y = \cot x$	$\frac{dy}{dx} = -\operatorname{cosec}^2 x$

HINT

Note that all "co-" functions give a negative function upon differentiation.

Example 35

Find the derivative of $f(x) = \tan x$ by using the quotient rule.

$f(x) = \tan x = \frac{\sin x}{\cos x} \Rightarrow$	Rewrite tangent as a quotient of sine and cosine.
$f'(x) = \frac{\frac{d(\sin x)}{dx} \times \cos x - \sin x \times \frac{d(\cos x)}{dx}}{(\cos x)^2}$	Apply the quotient rule.
$= \frac{\cos x \times \cos x - \sin x \times (-\sin x)}{\cos^2 x}$ $= \frac{\cos^2 x + \sin^2 x}{\cos^2 x}$	Use the found derivatives and simplify the expression.
$= \frac{1}{\cos^2 x} = \sec^2 x$	Use the fundamental trigonometric identity.

Example 36

Find the derivative of the function $f(x) = \sin x \times \cos x$.

Solution 1	
$f'(x) = \frac{d(\sin x)}{dx} \times \cos x + \sin x \times \frac{d(\cos x)}{dx}$	Use the product rule.

$= \cos x \times \cos x + \sin x \times (-\sin x)$	Use the found derivatives.
$= \cos^2 x - \sin^2 x$	Simplify the expression.
Solution 2	
$f(x) = \sin x \times \cos x = \frac{1}{2}\sin 2x \Rightarrow$	Use $\sin 2\theta = 2\sin\theta\cos\theta$ to rewrite the product.
$f'(x) = \frac{1}{2} \times \frac{d(\sin 2x)}{dx}$	Use the product by a constant rule.
$= \frac{1}{2} \times \cos 2x \times 2$	Use the chain rule.
$= \cos 2x$	Simplify the expression.

HINT

Notice that those two results are equivalent since the trigonometric formula for a cosine of a double angle is $\cos 2\theta = \cos^2\theta - \sin^2\theta$.

Other trigonometric functions:

Example 37

Find the derivative of $f(x) = \sec x$.

In this case we are going to use the fact that the reciprocal functions can be written as composite functions and then we are going to apply the chain rule.	
$f(x) = \sec x$	
$f(x) = \frac{1}{\cos x}$ $= (\cos x)^{-1} \Rightarrow$	Secant is reciprocal cosine.
$f'(x) = -1(\cos x)^{-2}(-\sin x)$	Use the chain rule.
$= \frac{\sin x}{\cos^2 x}$ $= \sec x \tan x$	Simplify and rewrite.

Geometry and trigonometry

Exercise 6S

1 Differentiate with respect to x:

a $y = \cot x$ **b** $y = \operatorname{cosec} x$

2 **a** $y = \sin 2x$ **b** $y = \cos(2x + 1)$

c $y = \cos(8 - 3x)$ **d** $y = \cot\left(\frac{7 - 2x}{13}\right)$

3 Find the derivative of the following functions:

a $f(x) = \cos(x^5 - 3)$

b $f(x) = \operatorname{cosec}(x^2 + 1)$

c $f(x) = \sec(4x^3 - 2x^2 + 7x + 17)$

d $f(x) = \sec\left(\sqrt{\cos x + 1}\right)$

e $f(x) = \sin(\cos(\tan x))$

Example 38

Find the derivative with respect to variable x.

$y = x^2\sin 2x$

$y = x^2\sin 2x \Rightarrow y' = 2x\sin 2x + x^2\cos 2x \times 2$ $= 2x\sin 2x + 2x^2\cos 2x$	Use the product rule. Simplify the expression.

Once you know how to differentiate a trigonometric function you can calculate the gradient of a curve at the given point.

Example 39

Find the gradient of the curve $y = 3x\cos(2x)$ at the point $\left(\frac{5\pi}{6}, \frac{5\pi}{4}\right)$.

$y = 3x\cos(2x) \Rightarrow y' = 3\cos(2x) + 3x(-\sin(2x)\times 2)$ $= 3(\cos(2x) - 2x\sin(2x))$	Use product rule. Simplify expression.
$y'\left(\frac{5\pi}{6}\right) = 3\left(\cos\left(2\times\frac{5\pi}{6}\right) - 2\times\frac{5\pi}{6}\times\sin\left(2\times\frac{5\pi}{6}\right)\right)$	Substitute the x-value in the expression.
$= 3\left(\underbrace{\cos\left(\frac{5\pi}{3}\right)}_{\frac{1}{2}} - \frac{5\pi}{3}\underbrace{\sin\left(\frac{5\pi}{3}\right)}_{-\frac{\sqrt{3}}{2}}\right) = \frac{3}{2} + \frac{5\pi\sqrt{3}}{2}$ This result can be obtained from a GDC.	Calculate and simplify. Sometimes it is easier to rewrite and simplify the trigonometric expression and then to differentiate it rather than going into differentiation immediately.

Exercise 6T

1 Use product and quotient rules to differentiate with respect to x:

a $y = \sin x(2x + 1)$

b $y = \cos 2x(x + x^2)$

c $y = \frac{\cos x}{x}$

d $y = \frac{2x+3}{\sin 2x}$

e $y = \frac{\tan x}{\sqrt{2-x}}$

2 Find the gradient of the curve at the given point.

a $y = \sin 3x$, at $x = \frac{\pi}{3}$

b $y = \cos 2x$, at $x = \frac{5\pi}{4}$

c $y = (x - 2)\sin x$, at $x = 0$

d $y = -3x\cos x$, at $x = \frac{\pi}{2}$

e $y = x^3\tan x$, at $x = \frac{3\pi}{4}$

3 Use technology to find the gradient at the indicated value:

a $y = \sin^2\alpha + \cos^2\alpha,\ \alpha$

b $y = \dfrac{\tan\beta}{\sin\beta},\ \beta$

c $y = \dfrac{\sin\rho + \sin 2\rho}{\cos\rho + \cos 2\rho},\ \rho$

Derivatives of inverse trigonometric functions

In order to differentiate the inverse trigonometric functions, arcsin, arccos and arctan, we have to be familiar with many manipulations.

Investigation 8

1 If $f(x) = \sin x$, find $f^{-1}(x)$.

2 For any composition of functions what is $f(f^{-1}(x))$?

3 If $y = \arcsin x$, rewrite x in terms of y.

4 Implicitly differentiate both sides with respect to x.

5 Show that this can be rearranged to $\dfrac{dy}{dx} = \dfrac{1}{\cos y}$.

6 Using a trigonometric identity show that $\cos y$ can be written as $\sqrt{1-x^2}$.

HINT

if $\sin y = x$ then $\sin^2 y = x^2$. Use $\sin^2 y + \cos^2 y = 1$.

7 Therefore write the derivative of $\arcsin x$ in terms of x.

8 **Conceptual** Could this logical reasoning work for arccos$x$ and arctan$x$.

9 Complete this procedure for arccos and arctan.

10 Notice that the function $\arcsin x : [-1,\ 1] \to \left[-\dfrac{\pi}{2}, \dfrac{\pi}{2}\right]$. Therefore, the cosine value will always be positive so we don't have to consider the positive or negative square root.

Conceptual How can we use the concepts of inverse and composite functions to find the derivative of an inverse trigonometric function?

TOK

Which is a better measure of angle: radian or degree? What are the "best" criteria by which to decide?

Geometry and trigonometry

The investigation showed the derivatives of the inverse trigonometric functions.

For $y = \arcsin x$ then $\dfrac{dy}{dx} = \dfrac{1}{\sqrt{1-x^2}}$

For $y = \arccos x$ then $\dfrac{dy}{dx} = \dfrac{-1}{\sqrt{1-x^2}}$

For $y = \arctan x$ then $\dfrac{dy}{dx} = \dfrac{1}{1+x^2}$

We can now consider:

$y = \arcsin\frac{x}{a}$

$\frac{x}{a} = \sin y$

$x = a\sin y$

Differentiating both sides with respect to x:

$1 = a\cos y\frac{\mathrm{d}y}{\mathrm{d}x}$

$\frac{\mathrm{d}y}{\mathrm{d}x} = \frac{1}{a\cos y}$

$$\frac{\mathrm{d}y}{\mathrm{d}x} = \frac{1}{a\sqrt{1-\frac{x^2}{a^2}}}$$

$$= \frac{\sqrt{a^2}}{a\sqrt{a^2-x^2}}$$

$$= \frac{1}{\sqrt{a^2-x^2}}$$

HINT

Because $\sin y = \frac{x}{a}$

$\sin^2 y = \frac{x^2}{a^2}$

$1-\cos^2 y = \frac{x^2}{a^2}$

Similar results follow for:

For $y = \arccos\frac{x}{a}$ then $\frac{\mathrm{d}y}{\mathrm{d}x} = \frac{-1}{\sqrt{a-x^2}}$

For $y = \arctan\frac{x}{a}$ then $\frac{\mathrm{d}y}{\mathrm{d}x} = \frac{a}{a^2+x^2}$

Example 40

Differentiate $\arctan\sqrt{x}$.

$\frac{\mathrm{d}y}{\mathrm{d}x} = \frac{1}{1+(\sqrt{x})^2}\frac{1}{2}x^{-\frac{1}{2}}$ $\frac{\mathrm{d}y}{\mathrm{d}x} = \frac{1}{2\sqrt{x}(1+x)}$	Use the chain rule.

Exercise 6U

1 Find the derivative of the following:

a $f(x) = \arccos 2x$

b $f(x) = \arcsin\frac{3}{2}x$

c $f(x) = \arctan(2x+1)$

2 Find $\frac{dy}{dx}$:

a $y = 2x \arccos x$

b $y = \frac{\arccos x}{2x}$

c $y = (x^2 - 1)\arctan 3x$

3 Show that the following identities are valid and explain why:

a $\frac{d(\arcsin x + \arccos x)}{dx} = 0$

b $\frac{d(\arctan x + \arctan(-x))}{dx} = 0$

Tangents and normals

Given the equation of a curve $y = f(x)$, the equations of the normal and the tangent at a specified point (x_1, y_1) are given by:

Equation of tangent $y = f'(x_1)(x - x_1) + y_1$

Equation of the normal $y = -\frac{1}{f'(x_1)}(x - x_1) + y_1$

This can also be applied to trigonometric functions.

Example 41

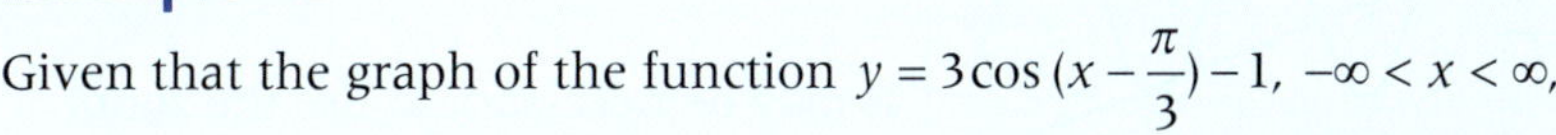

Given that the graph of the function $y = 3\cos(x - \frac{\pi}{3}) - 1$, $-\infty < x < \infty$, intersects the y-axis at point P.

a Find the equation of the tangent at P.

b Find the equation of the normal at P.

c Verify using your calculator.

$y = 3\cos\left(x - \frac{\pi}{3}\right) - 1$ $y = 3\cos\left(0 - \frac{\pi}{3}\right) - 1$ $y = 0.5$	Calculate the y-intercept; the y-intercept occurs when $x = 0$. P(0, 0.5)
a $f'(x) = -3\sin\left(x - \frac{\pi}{3}\right)$ $f'(0) = -3\sin\left(0 - \frac{\pi}{3}\right)$ $= -3\sin\left(-\frac{\pi}{3}\right) = 2.60$ (3 s.f.)	Equation of the tangent. Find the derivative at the point x_1.
Therefore equation of tangent is $y = 2.60x + 0.5$	Using $y = f'(x_1)(x - x_1) + y_1$ at the point (0, 0.5).

Continued on next page

b $-\frac{1}{f'(x_1)} = -\frac{1}{-3\sin\left(x - \frac{\pi}{3}\right)}$	The gradient of the normal is the negative reciprocal of the gradient of the tangent.
$\frac{1}{f'(0)} = -\frac{1}{-3\sin\left(0 - \frac{\pi}{3}\right)}$	
$= -0.385$ (3 s.f.)	
Therefore the equation of the normal is	
$y = -0.385(x - x_1) + y_1$	Using $y = -\frac{1}{f'(x_1)}(x - x_1) + y_1$ at (0.0.5).
$y = -0.385(x - 0) + 0.5$	
$y = -0.385x + 0.5$	
c 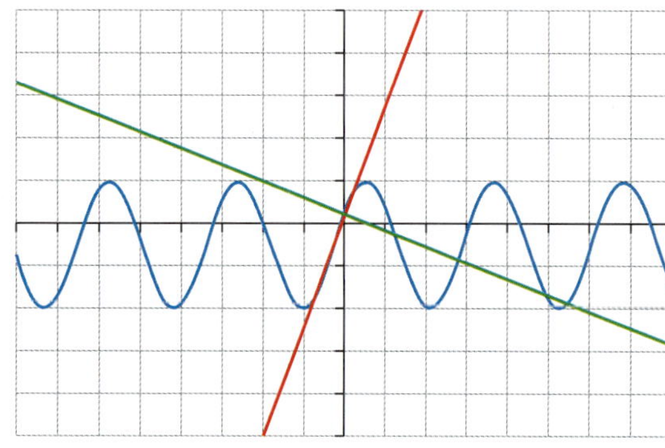	It may be necessary to use the zoom function on your calculator to ensure the normal and the tangent appear to be perpendicular.

Example 42

Find the area of the triangle enclosed by the tangent and normal to the curve $y = 2\sin(3x + \pi) + 2$, at the point P(1, P) and the y-axis.

$y = 2\sin(3x + \pi) + 2$	Find the coordinates of point P.
$y = 2\sin(3(1) + \pi) + 2$	P(1, 1.72)
1.72 (3 s.f.)	Find the equations of the tangent and normal.
$y = f'(x_1)(x - x_1) + y_1$	Equation of tangent $y = f'(x_1)(x - x_1) + y_1$
	Equation of the normal
	$y = -\frac{1}{f'(x_1)}(x - x_1) + y_1$
Using GDC	Find the gradient at point P.
$f'(x_1) = 5.94$	
$y = 5.94(x - 1) + 1.72$	

$y = 5.94x - 4.22$	Equation of tangent.
$y = -0.168x + 1.89$	Equation of normal.
$A = \frac{1}{2}bh$	Calculate area of triangle.
$A = \frac{1}{2}(1.89 + 4.22)(1)$	
$3.055\ \text{cm}^2$	

Exercise 6V

1 Given a function $f(x)$ and the point P, calculate the equations of the tangent and the normal at point P.

a $f(x) = \sin(3x + \pi)$ at $P\left(\frac{\pi}{3}, 0\right)$

b $f(x) = \arccos 2x$ at $P(0.05, y)$

c $f(x) = x\sin 2x$ at $P(-0.5, y)$

2 Consider the curves $y = x\cos 2x$ and $y = \tan(3x + \pi) - 1$.

a Find the point of intersection between the two curves in the first quadrant.

b Find the equations of normals to both curves at this point of intersection.

c Find the acute angle between the two normals.

3 The curve C is given by the equation

$$y = \frac{x\cos x}{x + \cos x},\ x \geq 0.$$

a Find the derivative of C.

b Find the equations of the tangent and the normal at point $\left(\frac{\pi}{2}, 0\right)$.

4 The curve D is given by the equation $y = (4x^2 + 1)\arctan 2x$.

a Show that the derivative is $8x\arctan\left(\frac{4x^2+1}{x^2+1}\right)$.

b Hence find the equation of the tangent at $(0.5, y)$.

Related rates of change with trigonometric expressions

Example 43

A ladder of length 10 m is leaning against a wall. It starts to slip down the wall at a rate of $0.5\,\text{m}\,\text{s}^{-1}$. How fast is the angle between the ladder and the ground changing when the vertical height of the ladder is 8 m?

Continued on next page

Geometry and trigonometry

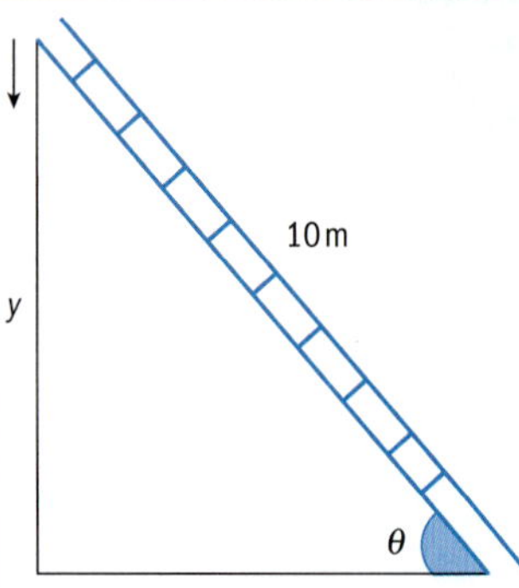

	Sketch a diagram representing the given information.
Given that $\frac{dy}{dt} = -0.5$ and $y = 8$ we need to find $\frac{d\theta}{dt}$.	Write down the given information and what you are asked to find.
$\sin\theta = \frac{y}{10}$	Identify the relationship between the height of the ladder and the angle.
$\cos\theta \times \frac{d\theta}{dt} = \frac{1}{10}\frac{dy}{dt}$	Differentiate as an implicit function with respect to time.
$y = 8 \Rightarrow \sin\theta = \frac{8}{10} = \frac{4}{5}$ $\Rightarrow \cos\theta = \sqrt{1-\left(\frac{4}{5}\right)^2} = \frac{3}{5}$	Find any missing information necessary to solve for the required rate.
$\frac{3}{5} \times \frac{d\theta}{dt} = \frac{1}{10} \times \left(-\frac{1}{2}\right) \Rightarrow \frac{d\theta}{dt} = -\frac{1}{12}$ (radians) $\sec^{-1}$ So the angle is decreasing at a rate of $\frac{1}{12}$ (radians) $\sec^{-1}$ or $4.77^\circ \sec^{-1}$.	Substitute and solve.

Exercise 6W

1 The beacon on a lighthouse 100 m from a straight shoreline rotates six times per minute.

a Find how fast the beam is moving along the shoreline at the moment when the light beam and the shoreline are at right-angles.

b Find how fast the beam is moving along the shoreline when the beam hits the shoreline 50 m from the point on the shoreline closest to the lighthouse.

c What is happening to the velocity of the light beam when the ray is parallel to the shoreline?

2 A tourist is filming from a spot that is 30 metres from the tree, following birds that are moving at a speed of $90\,\text{km}\,\text{h}^{-1}$. The birds are moving perpendicularly to the line joining the tree and the spot. Find how fast the tourist must rotate the camera:

a when the bird is directly in front of the camera

b one second after the bird is directly in front of the camera.

3 An isosceles triangle with sides 6 cm, 5 cm and 5 cm is going through a transformation in which the longest side is decreasing at a rate of $0.1\,\text{cm}\,\text{s}^{-1}$.

a Find the rate of change of the angle opposite to the decreasing side at the beginning.

b Find the rate of change of the angle opposite to the decreasing side when the triangle is equilateral.

4 A balloon has a spherical shape. There is a hole in the balloon and the air is leaking at $3\,\text{cm}^3\,\text{min}^{-1}$.

a Find the rate at which the radius is decreasing when $r = 10\,\text{cm}$.

b Find the rate at which the surface area is decreasing when $r = 4.5\,\text{cm}$.

5 A radar is located on the ground vertically below the path of a plane. The plane is flying at the constant height of 10 000 m and is maintaining a speed of $1025\,\text{km}\,\text{h}^{-1}$. Find the rate, in degrees per second, of the rotating radar when:

a the horizontal distance of the plane is 8 km from the radar

b the plane is directly above the radar.

6 A train is moving along a straight track at $75\,\text{km}\,\text{h}^{-1}$ heading east. A camera is positioned 2 km west of the train track and is focused on the train.

a Find the rate of change of the distance between the camera and the train when the train is 4 km from the camera.

b Find the rate at which the camera is rotating when the train is 4 km from the camera. Give your answer in degrees per second correct to the nearest tenth of a degree.

7 A 3 m long ladder is leaning against a wall. It starts to slip horizontally along the ground at a rate of $6\,\text{cm}\,\text{s}^{-1}$. Find how fast the angle between the ladder and the ground is changing when the bottom of the ladder is 1m away from the wall.

Developing your toolkit

Now do the Modelling and investigation activity on pages 440–441.

Chapter summary

- The midpoint of (x_1, y_1, z_1) and (x_2, y_2, z_2) is $\left(\frac{x_1 + x_2}{2}, \frac{y_1 + y_2}{2}, \frac{z_1 + z_2}{2}\right)$
- The distance (d) from (x_1, y_1, z_1) to (x_2, y_2, z_2) is

$$d = \sqrt{(x_2 - x_1)^2 + (y_2 - y_1)^2 + (z_2 - z_1)^2}$$

- In right-angled trigonometry:

$$\sin\theta = \frac{\text{opposite side}}{\text{hypotenuse side}}$$

$$\cos\theta = \frac{\text{adjacent side}}{\text{hypotenuse side}}$$

$$\tan\theta = \frac{\text{opposite side}}{\text{adjacent side}}$$

Continued on next page

- Volume of a cuboid: $V_c = l \times w \times h$
- Volume of corresponding pyramid: $V_p = \frac{V_c}{3} = \frac{l \times w \times h}{3}$
- The volume of a pyramid is $\frac{1}{3}$ the volume of its corresponding prism, which is found by first calculating the area of the base, then multiplying it by the height. $V = \frac{1}{3} A_{base} h$
- For a cone, the area of the base is the area of a circle πr^2.
  The volume of a cone is $\frac{1}{3}$ the volume of the cylinder with the same measurements. $V = \frac{1}{3}\pi r^2 h$
- Surface area of a pyramid: $S = A_{base} + nA_{triangle}$ where n is the number of sides of the regular polygon that makes up the base.
- The Pythagorean theorem is used to find missing lengths in triangles in order to find the areas of each triangular face.
- Surface area of a cone: $S = \pi r^2 + \pi rs$, where r is the radius of the circle at the base of the cone and s is the slant height of the cone.
- Surface area of a sphere: $S = 4\pi r^2$
- Volume of a sphere: $V = \frac{4}{3}\pi r^3$
- If a circular arc is drawn so that it is equal in length to the radius of the circle, then the angle is called one radian. The definition does not depend on the size of the circle. Familiarizing the concept of equivalence $360° \equiv 2\pi$ enables easy conversions.

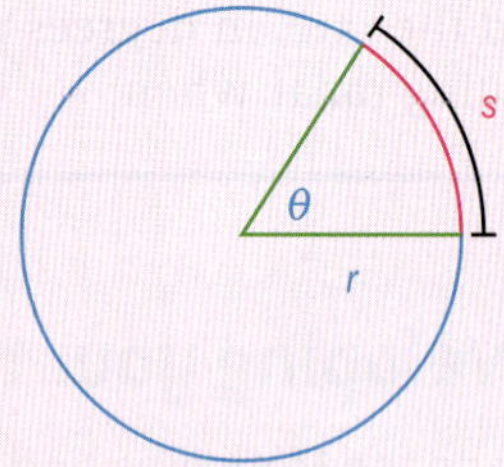

- The length of an arc is given by: $s = r\theta$
- The area of a sector is given by: $A = \frac{1}{2}r^2\theta$

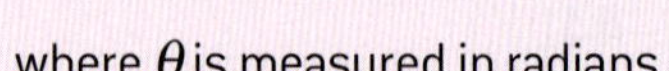

where θ is measured in radians.

$\sin\theta = y\text{-coordinate}$

$\cos\theta = x\text{-coordinate}$

$\sin(\pi - \theta) = \sin\theta$
$\cos(\pi - \theta) = -\cos\theta$
$\tan(\pi - \theta) = -\tan\theta$

$\sin(\pi + \theta) = -\sin\theta$
$\cos(\pi + \theta) = -\cos\theta$
$\tan(\pi + \theta) = \tan\theta$

$\sin(2\pi - \theta) = -\sin\theta$
$\cos(2\pi - \theta) = \cos\theta$
$\tan(2\pi - \theta) = -\tan\theta$

- The cosine rule states that for any triangle ABC:

  $a^2 = b^2 + c^2 - 2bc\cos A \Rightarrow \cos A = \dfrac{b^2 + c^2 - a^2}{2bc}$

- The sine rule states that for any triangle ABC:

  $\dfrac{a}{\sin A} = \dfrac{b}{\sin B} = \dfrac{c}{\sin C}$

- The sine rule can also be written as:

  $\dfrac{\sin A}{a} = \dfrac{\sin B}{b} = \dfrac{\sin C}{c}$

  This alternate representation is preferred for finding unknown angles, as the angles are the numerator.

- $\sin^2\theta + \cos^2\theta \equiv 1$
- $\tan\theta \equiv \dfrac{\sin\theta}{\cos\theta}$
- $\cos^2\theta + \sin^2\theta = 1$
- $1 + \tan^2\theta = \sec^2\theta$
- $1 \cot^2\theta = \mathrm{cosec}^2\theta$
- Reciprocal trigonometric functions

  $\sec\theta = \dfrac{1}{\cos\theta}$

  $\mathrm{cosec}\,\theta = \dfrac{1}{\sin\theta}$

  $\cot\theta = \dfrac{1}{\tan\theta}$

- Compound angle identities

  $\sin(A \pm B) = \sin A\cos B \pm \cos A\sin B$

  $\cos(A \pm B) = \cos A\cos B \mp \sin A\sin B$

  $\tan(A \pm B) = \dfrac{\tan A \pm \tan B}{1 \mp \tan A\tan B}$

- Double angle identities

  $\sin 2\theta = 2\sin\theta\cos\theta$

  $\cos 2\theta = \cos^2\theta - \sin^2\theta = 2\cos^2\theta - 1 = 1 - 2\sin^2\theta$

  $\tan 2\theta = \dfrac{2\tan\theta}{1 - \tan^2\theta}$

- Derivatives of trigonometric functions

  $f(x) = \sin x \Rightarrow f'(x) = \cos x$

  $f(x) = \cos x \Rightarrow f'(x) = -\sin x$

  $f(x) = \tan x \Rightarrow f'(x) = \sec^2 x$

Continued on next page

- Derivatives of inverse trigonometric functions

  $f(x) = \arcsin x \Rightarrow f'(x) = \dfrac{1}{\sqrt{1-x^2}}$

  $f(x) = \arccos x \Rightarrow f'(x) = -\dfrac{1}{\sqrt{1-x^2}}$

  $f(x) = \arctan x \Rightarrow f'(x) = \dfrac{1}{1+x^2}$

- Derivatives of reciprocal trigonometric functions

  $f(x) = \sec x \Rightarrow f'(x) = \sec x \tan x$

  $f(x) = \csc x \Rightarrow f'(x) = -\text{cosec}\, x \cot x$

  $f(x) = \cot x \Rightarrow f'(x) = -\text{cosec}^2 x$

- $\sin(A+B) \equiv \sin A \cos B + \cos A \sin B$
- $\sin 2A \equiv 2\sin A \cos B$
- $\cos 2A = \cos^2 A - \sin^2 A$
- $\tan 2A = \dfrac{2\tan A}{1-\tan^2 A}$
- If $f(x) = a\sin[b(x+c)] + d$, then:

  To calculate the amplitude: $a = \dfrac{\text{maximum value} - \text{minimum value}}{2}$

  To calculate the vertical shift: $d = \dfrac{\text{maximum value} + \text{minimum value}}{2}$

- For a curve $y = f(x)$ at a specified point (x_1, y_1):
  Equation of the tangent $y = f'(x_1)(x - x_1) + y_1$
  Equation of the normal $y = \dfrac{-1}{f'(x_1)} \times (x - x_1) + y_1$

Developing inquiry skills

A Ferris wheel with a diameter of 50 metres makes two revolutions per 30 minutes. Assume that the wheel is tangent to the ground and let P be the point of tangency. Find the rate at which the distance between P and a rider R is changing when they are five metres above the ground and travelling upwards.

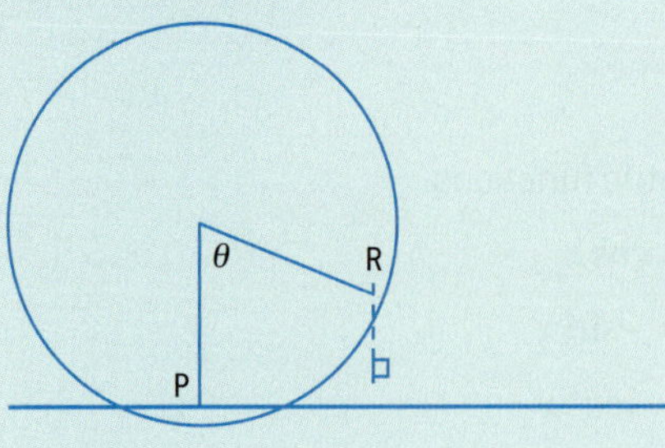

Chapter review

1 A solid metal cube, with side length 4 cm, is melted down and recast as a sphere.

Find the radius of the sphere.

2 A costume hat is formed by attaching a cone to a hemisphere.

The radius of the hemisphere is 6 cm and the total height of the hat is 14 cm.

Find:

a the volume

b the surface area.

3 The diagram shows a triangle ABC where AB = 5 cm, AC = BC = 3 cm. A is the centre of a 3 cm circle which intersects AB at X and B is the centre of a 3 cm circle which cuts AB at Y.

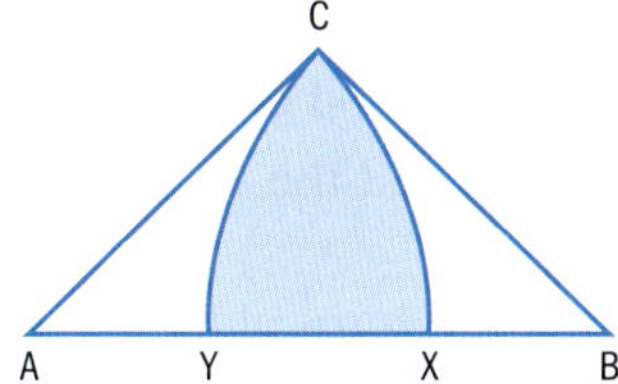

a Find the size of the angle CAB, giving your answer in degrees to the nearest decimal place and radians to 3 s.f.

b The shaded region, R, is bounded by the arcs CX, CY and the straight line XY.

Find:

i the length of the perimeter of R

ii the area of the sector ACX

iii the area of the sector R.

4 a Find the values for $\cos x$ which satisfy the equation $6\sin^2 x = 5 + \cos x$.

b Find all the values of x in the interval $180° \le x \le 540°$. Show your answer graphically.

5 Solve the equation $4\tan^2 x + 12\sec x + 1 = 0$ giving your answers in degrees to the nearest degree, where $-180° \le x \le 180°$.

6 Express $\tan 3A$ in terms of $\tan A$.

7 Find all the solutions to the equation $\cos 4\theta = \cos 3\theta$. Hence or otherwise, show that the roots of the equation $8x^3 + 4x^2 - 4x - 1 = 0$ are $\cos\frac{2\pi}{7}, \cos\frac{4\pi}{7}, \cos\frac{6\pi}{7}$.

Therefore, prove that

$$\sec\frac{2\pi}{7} + \sec\frac{4\pi}{7} + \sec\frac{6\pi}{7} = -4$$

8 Solve the following equations in the interval $0° \le x \le 360°$.

a $\sin(x + 60°) = \cos x$

b $\tan(A - x) = \frac{2}{3}$ and $\tan A = 3$

9 Prove that $\sin 3A \equiv 3\sin A - 4\sin^3 A$

10 Find x if $\arcsin x + \arccos\frac{x}{2} = \frac{5\pi}{6}$

11 Show that for any triangle ABC

$$\tan\frac{A}{2}\tan\frac{B-C}{2} = \frac{b-c}{b+c}$$

12 Three towns A, B and C are at sea level. The bearings of towns B and C from A are 36 degrees and 247 degrees respectively. If B is 120 km from A and A is 234 km from C, calculate the distance and bearing of town B from C.

13 A particle is moving along a straight line. The distance x m from a fixed point O on the line is given by the equation $x = a(1 + \cos^2 t)$. Find:

a the acceleration of the particle at time t

b the values of x at the point where the particle is at rest

c the values of x where the velocity of the particle is at a maximum.

Exam-style questions

14 P1: The diagram below shows a quadrilateral ABCD such that AB = 10 cm, AD = 5 cm, BC = 13 cm, $B\hat{C}D = 45°$ and $B\hat{A}D = 30°$.

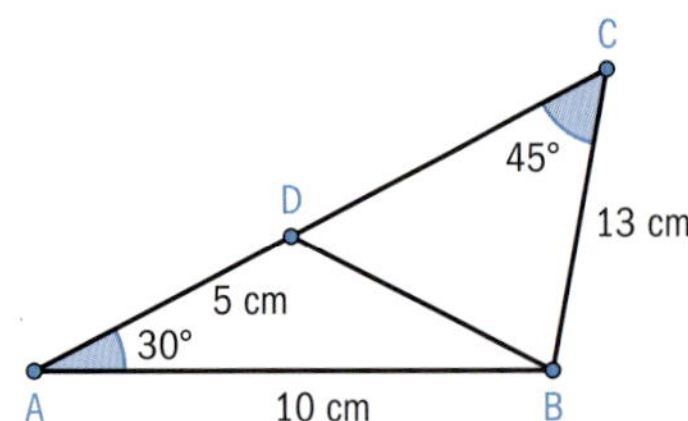

a Find the exact area of the triangle ABD. (2 marks)

b Show that $BD = 5\sqrt{5-2\sqrt{3}}$. (4 marks)

c Find the exact value of $\sin C\hat{D}B$. (3 marks)

d Hence explain why there are two possible values for the size of the angle $C\hat{B}D$, stating the relationship between the two possible values. (2 marks)

15 P2: A spinner is made of two identical cones with heights 5 cm and radii 3 cm, as illustrated in the diagram.

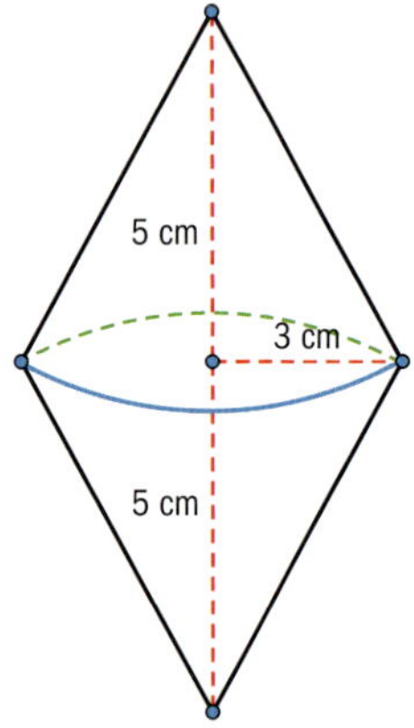

a Calculate the spinner's surface area. (4 marks)

b If the spinner is placed in a cylinder-shaped container with height 10.1 cm and diameter 6.1 cm, determine the percentage of the volume of the container occupied by the spinner. (2 marks)

16 P1: Solve the equation $2\cos^2 x = \sin 2x$ for $0 \le t \le 2\pi$, giving your answer in terms of π. (5 marks)

17 P1: A function $f(x)$ is defined by $f(x) = a\cos bx + c$ for $-1 \le x \le 3$, where a, b and c are constants.

The graph of $y = f(x)$, shown below, passes through the points $(1, 3)$ and $(2, -1)$.

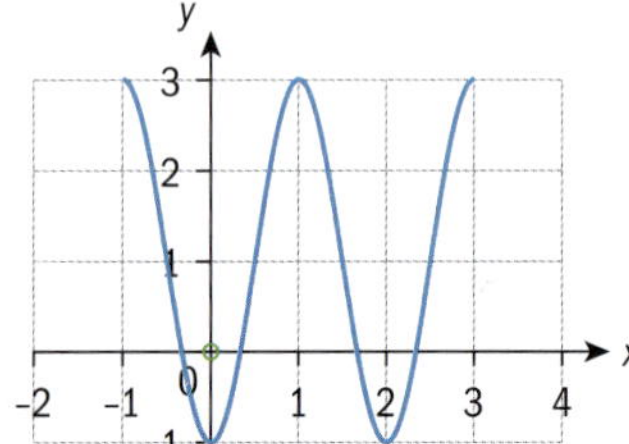

a State:

i the range of f

ii the period of f. (2 marks)

b Hence, find the values of a, b and c. (4 marks)

c Find the zeros of f. (3 marks)

18 P1: a Prove the identity

$$\frac{\cos x}{1-\sin x} - \tan x = \sec x \quad \text{(5 marks)}$$

b Hence, or otherwise, solve the equation $\frac{\cos 2x}{1-\sin 2x} - \tan 2x = \sqrt{2}$ for $0 \le x < 2\pi$. (6 marks)

19 P1: Find $\frac{dy}{dx}$ when

a $y = \arctan\left(\frac{1}{x}\right)$ (3 marks)

b $y = x^2 e^{\arctan x}$ (4 marks)

20 P1: Find the equation of the normal to the curve defined by the equation $\sin y = \tan(\cos x)$ at the point $\left(\frac{\pi}{2}, 0\right)$. (9 marks)

21 **P1:** Suppose $S(x) = (\sin 2x + \cos 2x)^2$.

a Show that $S(x) = \sin(4x) + 1$, for all $x \in \mathbb{R}$. (3 marks)

b Hence sketch the graph of S for $0 \le x \le \pi$. (3 marks)

c State:

i the period of S;

ii the range of S. (2 marks)

d Sketch the graph of the function C defined by $C(x) = \cos(2x) - 1$, for $0 \le x \le 2\pi$. (3 marks)

The graph of C can be obtained from the graph of S under a horizontal stretch of scale factor k, followed by a translation by the vector $\begin{pmatrix} p \\ q \end{pmatrix}$.

e Write down:

i the value of k.

ii a possible vector $\begin{pmatrix} p \\ q \end{pmatrix}$. (3 marks)

22 **P2:** Over the course of a single December day in Limassol, Cyprus, the highest temperature was found to be 22°C. The lowest temperature was 12°C, which occurred at 03:00 hours.

a If t is the number of hours since midnight, the temperature T may be modelled by the equation $T = A\sin(B(t - C)) + D$, where A, B, C and D are positive integers.

b Find the smallest possible values of A, B, C and D. (9 marks)

c Using your answer to part **a**, along with your GDC, find the number of hours during which the temperature was above 20°C. (4 marks)

23 **P1:** A right circular cone with fixed height 50 cm is increasing in volume at a rate of $2\,\text{cm}^3\,\text{min}^{-1}$. Find the rate at which the radius of the base r is increasing when $r = 0.4$ cm. Give your answer in exact form. (6 marks)

Click here for further exam practice

The sound of mathematics

Approaches to learning: Research, Critical thinking, Using technology

Exploration criteria: Mathematical communication (B), Personal engagement (C), Use of mathematics (E)

IB topic: Trigonometric functions

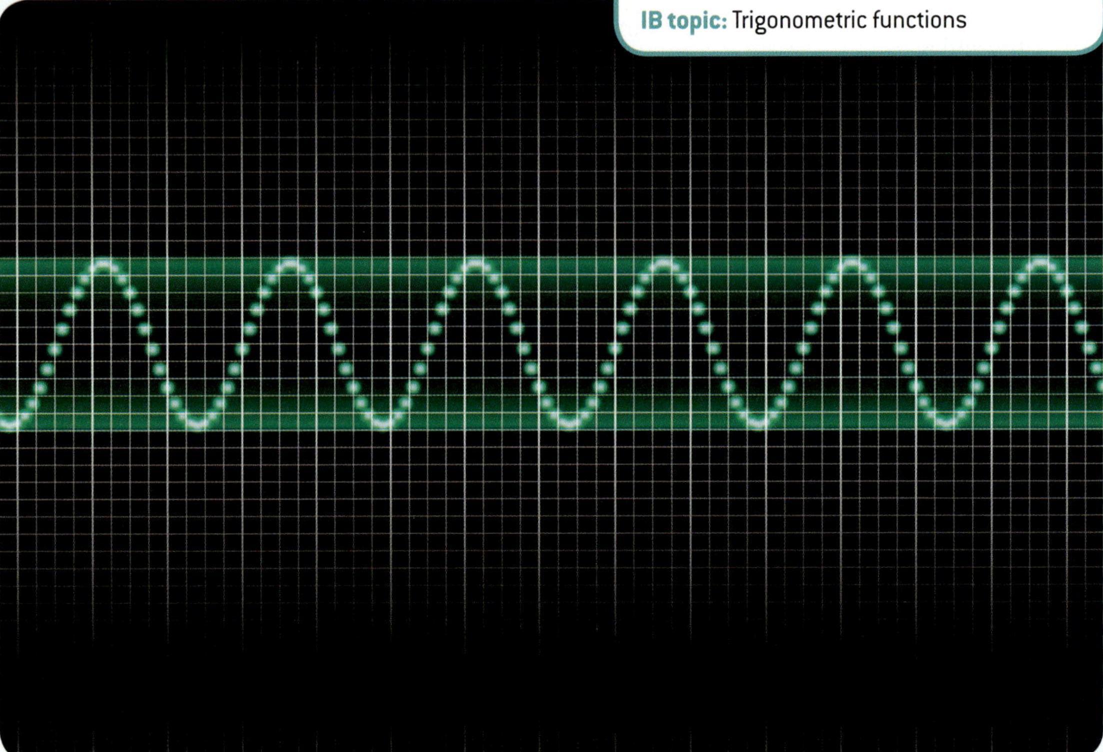

Brainstorm

In small groups, brainstorm some ideas that link music and mathematics.

Construct a **mind map** from your discussion with the topic "MUSIC" in the centre.

Share your mindmaps with the whole class and discuss.

This task concentrates on the relationship between sound waves and trigonometric functions.

Research

Research how sound waves and trigonometric functions are linked.

Think about:

How do vibrations cause sound waves?

How do sound waves travel?

What curve can be used to model a sound wave?

The fundamental properties of a basic sound wave are its **frequency** and its **amplitude**.

What is the frequency of a sound?

What units is it measured in?

What is the amplitude of a sound?

Use what you studied in this chapter to answer these questions:

If you have a sine wave with the basic form $y = a \sin(bt)$, where t is measured in seconds, how do you determine its period, frequency and its amplitude?

What do the values of a and b represent? What does y represent in this function?

With this information, determine the equivalent sine wave for a sound of 440Hz.

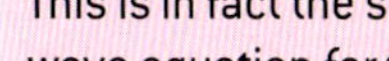

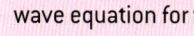

Did you know?

This is in fact the sine wave equation for the note A.

Technology

There are a large number of useful programmes that you can use to consider sound waves.

Using these programmes it is possible to record or generate a sound and view a graphical representation of the soundwave with respect to time.

If you have a music department in your school and/or access to such a programme, you could try this.

Design an investigation

Using the available technology and the information provided here, what could you investigate and explore further?

What experiment could you design regarding sounds?

Discuss your ideas with your group.

What exactly would be the aim of each investigation/exploration/experiment that your group thought of?

Select one of the ideas in your group and plan further.

What will you need to think about as you conduct this exploration?

How will you ensure that your results are reliable?

How will you know that you have completed the exploration and answered the aim?

Extension

Trigonometric functions also occur in many other areas that aid your understanding of the physical world.

Some examples are:

- Temperature modelling
- Tidal measurements
- The motion of springs and pendulums
- The electromagnetic spectrum

They can be thought of in similar terms of waves with different frequencies, periods, amplitudes and phase shifts too.

Think about the task that you have completed here and consider how you could collect other data that can be modelled using a trigonometric function. You could use one of the examples here or research your own idea.

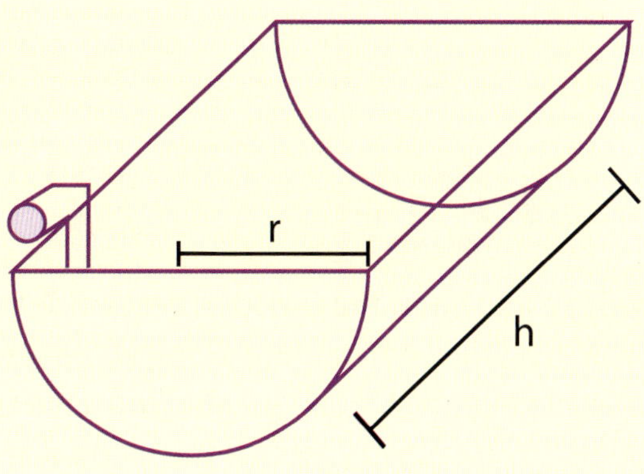

7 Generalizing relationships: exponents, logarithms and integration

Concepts

- Relationships
- Generalization

Microconcepts

- Integration as antidifferentiation
- Analytic approach to areas under curves—Riemann sums
- Integration of trigonometric functions, polynomial, radical and rational functions
- Laws of exponents
- Laws of logarithms
- Exponential and logarithmic functions and their graphs
- Solution of exponential equations using logarithms
- Derivatives of exponential and logarithmic functions
- Including tangents, normals and optimization
- Indefinite integrals of exponential functions and $\frac{1}{x}$
- The composites of these with the linear function $(ax + b)$
- Integration by inspection, by substitution, by parts
- Repeated integration by parts

Mathematical models are representations of real-life situations using variables, and various forms that show relations between these variables. Forms may include expressions, relations as well as visual representation in the form of graphs. When creating mathematical models, some problems are simplified because of constraints or assumptions made.

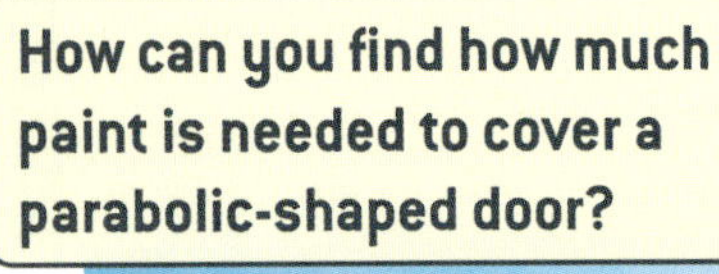

How can you find how much paint is needed to cover a parabolic-shaped door?

Suppose you know the function that models the rate at which a given item moves along an assembly line in metres per second. How can you find a function that models the number of metres the manual is from the start of the assembly line t seconds after it begins moving?

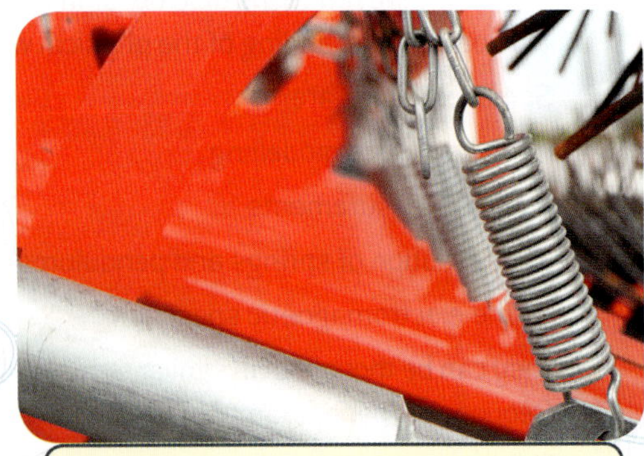

How much work is needed to stretch a spring a fixed amount?

Piyakan wants to plant a flowering shrub in her garden but does not want the shrub to grow higher than 1.5 m. Her best friend, who is a mathematician, says that a hydrangea would be ideal and gives her the following reasons: When a hydrangea is planted its height during the first 20 days is modelled by the function $h(t) = \frac{1}{5}e^{\frac{t}{10} - 0.9}$, $0 \le t \le 20$, where h is in metres and t is measured in days. After 20 days the rate of growth is modelled by the function $g(t) = \frac{1}{50}e^{-\frac{t}{10} + 3.09}$, where t is measured in days since the shrub was planted.

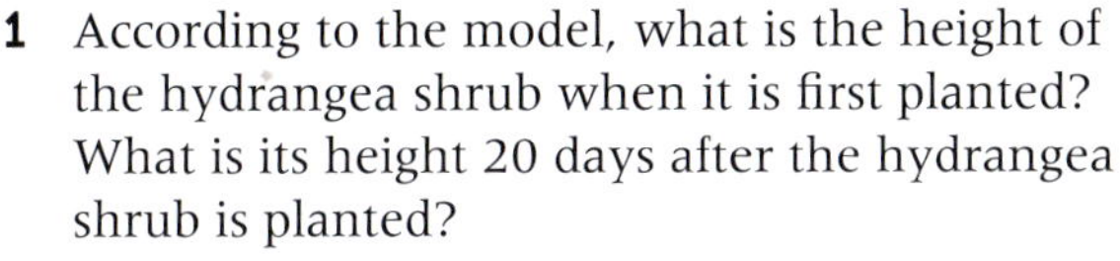

1 According to the model, what is the height of the hydrangea shrub when it is first planted? What is its height 20 days after the hydrangea shrub is planted?

2 When is the plant growing fastest during the first 20 days? What is the rate of growth at this time? Explain why the height model has a restricted domain.

3 Find a model for the height of the hydrangea for $t > 20$.

4 Explain why the model for $t > 20$ confirms that the hydrangea will not grow indefinitely and find the maximum height reached.

Developing inquiry skills

Think about the questions in this opening problem and answer any you can. As you work through the chapter, you will gain mathematical knowledge and skills that will help you to answer them all.

Before you start

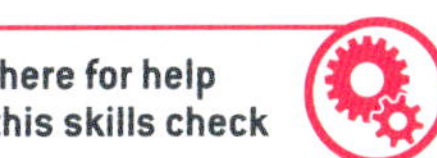

Click here for help with this skills check

You should know how to:	Skills check
Find derivatives using different techniques. eg	
1 Differentiate $x^2 \cos 3x$. $\frac{dy}{dx} = -3x^2 \sin 3x + 2x \cos 3x$	**1** Differentiate $y = \frac{x^3 - 5x}{\cos x}$.
2 Determine $\frac{dy}{dx}$ in terms of x and y given that $x \tan y - x^2 y = \sin^2 y$. $x\sec^2 y \frac{dy}{dx} + \tan y - x^2 \frac{dy}{dx} - 2xy = 2\sin y \cos y \frac{dy}{dx}$ $\Rightarrow (x\sec^2 y - x^2 - \sin 2y)\frac{dy}{dx} = 2xy - \tan y$ $\Rightarrow \frac{dy}{dx} = \frac{2xy - \tan y}{x\sec^2 y - x^2 - \sin 2y}$	**2** Determine $\frac{dy}{dx}$ in terms of x and y given that $xy^3 - y\sin^2 x = y\cos x$.

7.1 Integration as antidifferentiation and definite integrals

Investigation 1

The graph shows the derivative function $y = f'(x)$ of the function $f(x)$.

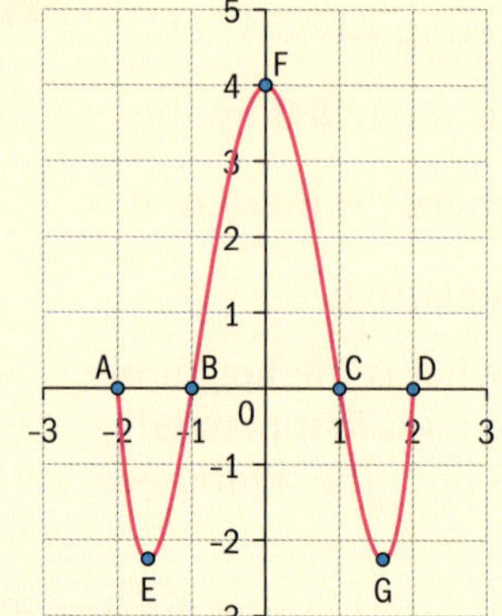

1 Over what intervals is $f(x)$ an increasing function?

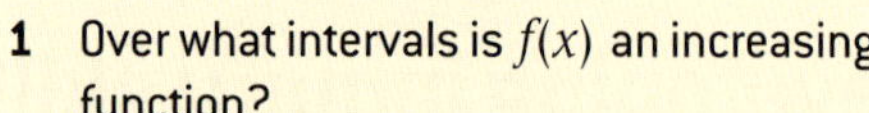

2 Over what intervals is $f(x)$ a decreasing function?

3 Describe the nature of each of the points A to G on the graph of $f(x)$. Explain your answers.

4 Over which intervals is $f(x)$ concave up?

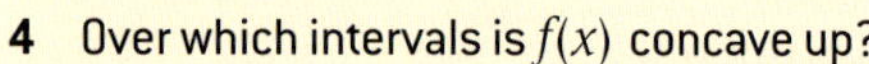

5 Over which intervals is $f(x)$ concave down?

6 How accurately can you sketch $y = f(x)$? Give reasons for your answer.

7 How many possible correct sketches of $y = f(x)$ can be drawn? What do these curves have in common? Give examples.

8 **Conceptual** To what extent does the derivative function limit our knowledge about the original function?

In Investigation 1, you sketched a possible curve for $y = f(x)$ from the graph of its derivative, $y = f'(x)$.

Integration, or **antidifferentiation**, is the process of finding a function, $f(x)$, when you are given the derivative function, $f'(x)$.

Suppose that $f'(x) = 2x$. You saw in Investigation 1 that there are an infinite number of functions which $f(x)$ could be, and they are all of the form $f(x) = x^2 + c,\ c \in \mathbb{R}$. This is because any function of the form $f(x) = x^2 + c,\ c \in \mathbb{R}$ has derivative $f'(x) = 2x$.

The **indefinite integral** of a function is the set of all antiderivatives of that function. In the example where $f'(x) = 2x$, we write

$\int 2x\,dx = x^2 + c,\ c \in \mathbb{R}$.

In this case, $2x$ is called the integrand, and the real number c is called the **constant of integration**.

In general, we say that $\int f(x)\,dx = F(x) + c,\ c \in \mathbb{R}$ where $F'(x) = f(x)$.

HINT

When writing down an integral, you need to include:

- the integral sign $\int$
- the integrand (the expression which is being integrated)
- "dx" which indicates that you are integrating "with respect to the variable x".

Investigation 2

1 Copy and complete the following table.

$\frac{d}{dx}(x^2) = 2x$	$\int 2x\,dx = x^2 + c$	$\int x\,dx =$
$\frac{d}{dx}(x) = 1$		
	$\int 3x^2\,dx =$	$\int x^2\,dx =$
$\frac{d}{dx}(x^4) = 4x^3$		$\int x^3\,dx =$
		$\int x^4\,dx =$
		$\int x^5\,dx =$
$\frac{d}{dx}\left(x^{\frac{3}{2}}\right) = \frac{3}{2}x^{\frac{1}{2}}$		$\int x^{\frac{1}{2}}\,dx =$
		$\int x^n\,dx =$
$\frac{d}{dx}(5x^4) = 20x^3$		$\int 20x^3 dx = 5\int x^3 dx =$
$\frac{d}{dx}(-x^2) = -2x$	$\int -2x\,dx =$	
	$\int -\frac{2}{3}x^3\,dx =$	
	$\int kx^n\,dx =$ where $k \in \mathbb{R}$	
$\frac{d}{dx}(2x^3 - x^2 + 4x) =$	$\int (6x^2 - 2x + 4)\,dx =$	

2 Explain in words how you would find the integral of x^n.

3 What is the rule for integrating kx^n, $k \in \mathbb{R}$?

4 **Conceptual** What can you say about the integral of a polynomial function?

You could summarize your results from Investigation 2 as follows:

- $\int x^n\,dx = \frac{x^{n+1}}{n+1} + c$
- $\int kx^n\,dx = k\int x^n\,dx = \frac{kx^{n+1}}{n+1} + c$
- $\int (a_n x^n + a_{n-1}x^{n-1} + \ldots + a_1 x + a_0)dx = \int a_n x^n dx + \int a_{n-1}x^{n-1}dx + \ldots + \int a_1 x\,dx + \int a_0\,dx$

Functions

Calculus

There is another rule that is useful in integrating functions. This is the same as the sum and difference differentiation rule:

$$\int (f(x) \pm g(x))\,dx = \int f(x)\,dx \pm \int g(x)\,dx$$

Using the results for derivatives of trigonometric functions which you met in Chapter 6, the following antiderivatives also follow:

$\frac{d}{dx}(\sin x) = \cos x$	$\int \cos x\,dx = \sin x + c,\ c \in \mathbb{R}$
$\frac{d}{dx}(\cos x) = -\sin x$	$\int \sin x\,dx = -\cos x + c,\ c \in \mathbb{R}$
$\frac{d}{dx}(\tan x) = \sec^2 x$	$\int \sec^2 x\,dx = \tan x + c,\ c \in \mathbb{R}$

Example 1

Find the following indefinite integrals.

a $\int \frac{3}{x^5}\,dx$ **b** $\int (3x^2 + 5x + 2)\,dx$

a $\int \frac{3}{x^5}\,dx = 3\int \frac{1}{x^5}\,dx$ $= 3\int x^{-5}\,dx$ $= 3 \times \frac{x^{-4}}{-4} + c$ $= -\frac{3}{4x^4} + c$	Using $\int kx^n\,dx = k\int x^n\,dx$. Write integrand as an integer power of x. Using $\int x^n\,dx = \frac{x^{n+1}}{n+1} + c$.
b $\int (3x^2 + 5x + 2)\,dx$ $= 3\int x^2\,dx + 5\int x\,dx + 2\int 1\,dx$ $= x^3 + \frac{5x^2}{2} + 2x + c$	Using $\int (f(x) \pm g(x))\,dx = \int f(x)\,dx \pm \int g(x)\,dx$ and $\int kx^n\,dx = k\int x^n\,dx$. Integrate using $\int x^n\,dx = \frac{x^{n+1}}{n+1} + c$.

Differentiate your answers to check your results.

Example 2

Integrate the following functions.

a $f(x) = x^{\frac{2}{5}}$ **b** $f(x) = x^3 + 3 - x^{-3}$

a $\int x^{\frac{2}{5}}\,dx = \dfrac{x^{\frac{7}{5}}}{\frac{7}{5}} + c = \dfrac{5}{7}x^{\frac{7}{5}} + c$	Using $\int kx^n\,dx = k\int x^n\,dx$.
b $\int\left(x^3 + 3 - x^{-3}\right)dx = \int x^3\,dx + 3\int 1\,dx - \int x^{-3}dx$ $= \dfrac{x^4}{4} + 3x - \dfrac{x^{-2}}{(-2)} + c$ $= \dfrac{x^4}{4} + 3x + \dfrac{x^{-2}}{2} + c$	Using $\int(f(x) \pm g(x))dx = \int f(x)dx \pm \int g(x)\,dx$ and $\int kx^n\,dx = k\int x^n\,dx$.

Differentiate your results to check that your answers are correct. Do not forget the constant of integration.

Example 3

Determine the following indefinite integrals.

a $\int(x + \sin x)dx$ **b** $\int\left(\theta^2 - 2\tan^2\theta\right)d\theta$

a $\int(x + \sin x)dx = \int x\,dx + \int \sin x\,dx$	Separate integrals.
$= \dfrac{x^2}{2} - \cos x + c$	Use $\int \sin x\,dx = -\cos x + c$.
b $\int\left(\theta^2 - 2\tan^2\theta\right)d\theta$	Separate integrals.
$= \int\theta^2\,d\theta - 2\int\tan^2\theta\,d\theta$ $= \int\theta^2\,d\theta - 2\int\left(\sec^2\theta - 1\right)d\theta$	Use identity $\tan^2\theta + 1 = \sec^2\theta$.
$= \int\theta^2\,d\theta - 2\int\sec^2\theta\,d\theta + 2\int 1\,d\theta$ $= \dfrac{\theta^3}{3} - 2\tan\theta + 2\theta + c$	Integrate with respect to θ. Use $\int\sec^2 x\,dx = \tan x + c$.
$= \dfrac{\theta^3}{3} + 2\theta - 2\tan\theta + c$	Simplify.

When certain information about the function is given, you can use this to find the value of the constant c. In this case, you can find a unique solution for the integral of a given function. In Example 4, you can find the specific curve that passes through a given point.

Example 4

$$\frac{dy}{dx}=\left(1-x-x^2\right)^2.$$

Find y, given that the graph of the function passes through the point $(1,0)$.

$(1-x-x^2)^2=((1-x)-x^2)^2$ $=(1-x)^2-2(1-x)x^2+x^4$ $=1-2x+x^2-2x^2+2x^3+x^4$	Expand.
$\Rightarrow \frac{dy}{dx}=1-2x-x^2+2x^3+x^4$	Simplify.
$y=\int 1dx-2\int x\,dx-\int x^2dx+2\int x^3dx+\int x^4dx$ $\Rightarrow y=x-x^2-\frac{x^3}{3}+\frac{x^4}{2}+\frac{x^5}{5}+c$	Integrate term by term.
At $(1,0)$, $0=1-1-\frac{1}{3}+\frac{1}{2}+\frac{1}{5}+c$	Substitute $(1,0)$ into the equation to find c.
$\Rightarrow c=-\frac{11}{30}$ $\therefore y=x-x^2-\frac{x^3}{3}+\frac{x^4}{2}+\frac{x^5}{5}-\frac{11}{30}$	Rewrite y with the value of c found above.

Integrating functions of the form $f(ax+b)$

Investigation 3

Consider the function $y=ax+b$.

1 **a** Find the integral of the function by integrating term by term.

b Expand and simplify $\frac{y^2}{2a}$.

c How can you combine your results to parts **a** and **b** in order to integrate $y=ax+b$ without integrating each term separately?

2 **a** Expand y^2 and hence find $\int (ax+b)^2 \, dx$.

b Expand and simplify $\frac{y^3}{3a}$.

c How can you combine your results to parts **a** and **b** in order to integrate y^2 without integrating each term separately?

3 **a** Expand y^3 and hence find $\int (ax+b)^3 \, dx$.

b Expand and simplify $\frac{y^4}{4a}$.

c How can you combine your results to parts **a** and **b** in order to integrate y^3 without integrating each term separately?

4 Based on your results above, make a conjecture about the value of the integral $\int (ax+b)^n \, dx$.

5 Find the derivatives of the following functions.

a $\sin(ax+b)$

b $\cos(ax+b)$

c $\tan(ax+b)$

6 Use your results to question **5** to find the following integrals.

a $\int \cos(ax+b)\, dx$

b $\int \sin(ax+b)\, dx$

c $\int \sec^2(ax+b)\, dx$

7 **Factual** What differentiation rule have you used the reverse of in this investigation?

8 **Conceptual** Why is knowledge of derivatives useful for integration?

As you saw in Investigation 3, the reverse of the chain rule allows you to state the following results.

- $\int (ax+b)^n \, dx = \frac{(ax+b)^{n+1}}{a(n+1)} + c, \quad c, n \in \mathbb{R}$
- $\int \sin(ax+b)\, dx = -\frac{\cos(ax+b)}{a} + c, \; c \in \mathbb{R}$
- $\int \cos(ax+b)\, dx = \frac{\sin(ax+b)}{a} + c, \; c \in \mathbb{R}$
- $\int \sec^2(ax+b)\, dx = \frac{\tan(ax+b)}{a} + c, \; c \in \mathbb{R}$

Functions

Calculus

Example 5

Integrate the following with respect to x.

a $(3x-4)^8$ **b** $(5-2x)^{-6}$ **c** $\tan^2(5x-1)$

a $\int(3x-4)^8\,dx = \frac{(3x-4)^9}{3(9)}+c$ $= \frac{(3x-4)^9}{27}+c$	Apply the rule $\int(ax+b)^n\,dx = \frac{(ax+b)^{n+1}}{a(n+1)}+c$ and simplify.
b $\int(5-2x)^{-6}\,dx = \frac{(5-2x)^{-5}}{-2(-5)}+c$ $= \frac{(5-2x)^{-5}}{10}+c$ $= \frac{1}{10(5-2x)^5}+c$	
c $\int\tan^2(5x-1)dx = \int(\sec^2(5x-1)-1)dx$ $= \int\sec^2(5x-1)\,dx - \int 1dx$ $= \frac{\tan(5x-1)}{5}-x+c$	Use $\tan^2\theta+1=\sec^2\theta$. Use $\int\sec^2(ax+b)dx = \frac{\tan(ax+b)}{a}+c$.

Exercise 7A

1 Find indefinite integrals, with respect to x, of the following functions.

a $-\frac{2}{3}x$ **b** $\frac{5}{4}x^3$ **c** $4x^{\frac{3}{2}}$

d $\frac{7}{2}x^{-\frac{1}{2}}$ **e** $2\sin\frac{x}{2}\cos\frac{x}{2}$

2 Integrate the following functions with respect to x.

a x^3+3x^2-4x+3

b $x^4+4x^3-3x^{-3}+x^{-4}$

c $(x^2-x+2)^2$ **d** $(2-x)^3$

e $x+\cos x-\frac{1}{\cos^2 x}$

3 Find $y=f(x)$ given that

$\frac{dy}{dx}=3x^2-4$ and $f(-1)=2$.

4 If $f'(t)=t-2+t^{-\frac{1}{2}}$, find $f(t)$ given that the graph of the function passes through the point (4, 4).

5 Given that $y=2$ when $x=-1$ and $\frac{dy}{dx}=(2x-1)^3$, find y in terms of x.

6 Use trigonometric identities to find the indefinite integral $\int\left(\sin\frac{\theta}{2}+\cos\frac{\theta}{2}\right)^2 d\theta$.

7 Given that $f'(\theta)=(2-3\sin\theta)$, find the function $f(\theta)$ given that $f(0)=-2$.

8 In parts **a**–**d**, find $f(x)$ with the given conditions.

a $f'(x)=3-\cos x$ and $f\left(\frac{\pi}{6}\right)=\frac{\pi}{2}$

b $f'(x)=\frac{2}{\cos^2 x}-3\sin x$ and $f(0)=4$

c $f''(x)=2\sin x$, $f'\left(\frac{\pi}{4}\right)=0$ and $f(0)=1$

d $f''(x)=2+3\cos x$, $f'(0)=4$ and $f(0)=5$

9 A particle moves in a straight line such that at time t seconds, its acceleration $a(t) = 18t - 2$. When $t = 0$ its velocity is $1\,\text{m}\,\text{s}^{-1}$, and when $t = 1$ its displacement from the origin is 3 m.

Find expressions for the velocity and the displacement of the particle at any given time t.

10 A particle moves in a straight line such that, at time t seconds, its acceleration is $a(t) = 6\cos t$.

When $t = 0$ its velocity is $0\,\text{m}\,\text{s}^{-1}$, and when $t = \frac{\pi}{3}$ its displacement from the origin is 2 m.

Find expressions for the velocity and the displacement of the particle at any given time t.

11 Integrate the following with respect to x.

a $(1 - 3x)^6$ **b** $3(4 - x)^{\frac{1}{2}}$

c $3\cos(5x + 2) + 4\sin(5x + 2)$

d $2(2 - 3x)^{-\frac{1}{3}} + (1 + 2x)^{\frac{1}{3}}$

12 Given that $f'(\theta) = 4\sin\left(\theta + \frac{\pi}{4}\right)\cos\left(\theta + \frac{\pi}{4}\right)$, find the function $f(\theta)$ given that $f(0) = 1$.

Did you know?

Finding the area bounded by a curve has always intrigued mathematicians. Seeking a method to do this helped to develop the differential calculus.

However, before the formulation of calculus, Archimedes devised a way to find the area bounded by a parabola and a chord using the method of exhaustion. Investigation 4 leads you through a simplified version of the work of Archimedes.

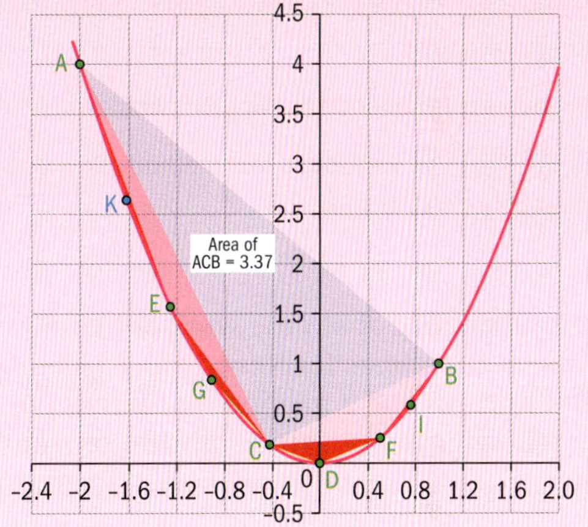

Investigation 4

In this investigation you are going to approximate the area bounded by the parabola $y = 4 - x^2$ and the x-axis using the method of exhaustion.

The figure on the right shows the parabola $y = 4 - x^2$ and triangle ABC.

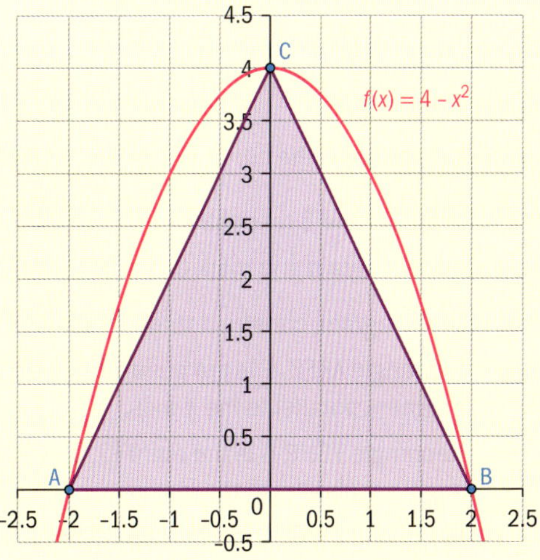

1 Find the area of triangle ABC.

2 Write an inequality relating the area bounded by the parabola and the x-axis and the area of the triangle.

Continued on next page

The figure on the right shows two other triangles constructed with D and E being points on the parabola having the x-coordinate at the midpoint of OA and OB respectively.

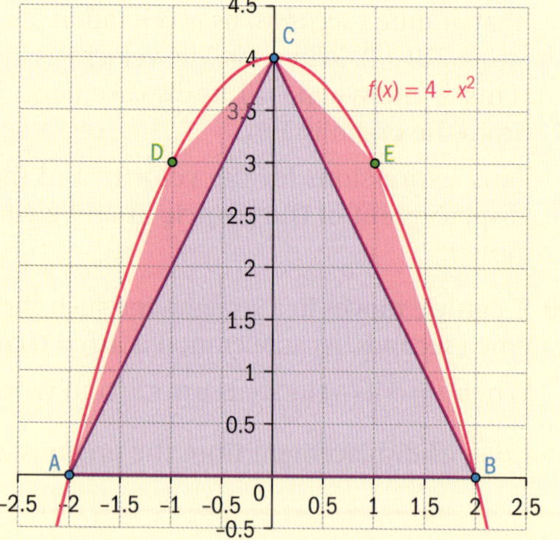

3 Show that the area of each of the triangles ADC and CEB is equal to 1 unit².

4 Extend your inequality from question **2** to relate the area under the curve to the sum of the areas under all three triangles.

For the following questions you may want to use technology.

5 Construct points P, Q, R and S on the parabola which have x-coordinates at $-1.5, -0.5, 0.5, 1.5$ respectively.

6 Calculate the areas of triangles APD, DQC, CRE and ESB.

7 Extend your inequality from question **2** to relate the area under the curve to the sum of the areas under all the triangles.

8 If you were to continue drawing further triangles in this way, what could you say about the sum of their areas?

9 **Conceptual** How does this method allow you to find a good approximation of the area bounded by the parabola and the x-axis?

What is fascinating about the work done by Archimedes is that it took about 2000 years for Newton and Leibnitz to improve on it and develop calculus and a method for finding the area under curves. In the following investigation you will see that the concept used in calculus is not too far removed from Archimedes' method.

Investigation 5

In this investigation you will approximate the area bounded by the curve $f(x) = \frac{1}{x}$, the lines $x = 1$, $x = 2$, and the x-axis.

The diagram shows the graph of $f(x) = \frac{1}{x}$ with the required area ACDB shaded.

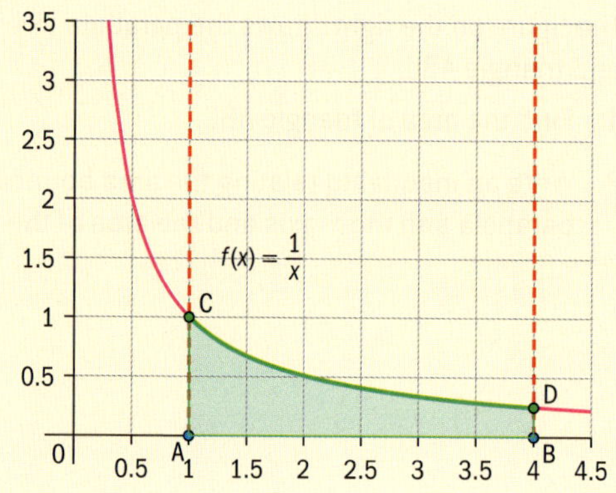

Points P and Q lie on the curve $f(x) = \frac{1}{x}$ and have x-coordinates 2 and 3 respectively.

1 Copy the diagram and mark on points P and Q.

2 On your diagram, construct three rectangles as follows.

- Rectangle AEFC. E is the point (2, 0) and FC is a line segment parallel to the x-axis.
- Rectangle EGHP. G is the point (3, 0) and PH is a line segment parallel to the x-axis.
- Rectangle GBIQ. IQ is a line segment parallel to the x-axis.

3 Calculate the area of each of the three rectangles.

The sum of the areas of these three rectangles is called an *upper sum* for the area under the curve.

4 On the same diagram, construct three more rectangles as follows.

- Rectangle AEPR. PR is a line segment parallel to the x-axis.
- Rectangle EGQS. QS is a line segment parallel to the x-axis.
- Rectangle GBDT. DT is a line segment parallel to the x-axis.

5 Calculate the area of each of these three rectangles.

The sum of the areas of these three rectangles is called a *lower sum* for the area under the curve.

6 Write an inequality relating both sums of the areas of rectangles (from question **3** and question **5**) to the area ACDB.

For the next questions you will need to use technology.

7 Repeat method described in questions **2** to **6**, but this time construct a set of six rectangles which lie below the curve, and another set of six rectangles which lie above the curve. (Hint: The bases of each of the six rectangles should be determined by the points (1, 0), (1.5, 0), (2, 0), (2.5, 0), (3, 0), (3.5, 0), (4, 0).)

8 What happens if you were to repeat the above using two sets of 12 rectangles?

9 Write an inequality relating the areas of two sets of n rectangles, each with base width Δx, with the area ACDB.

10 **Conceptual** How does the concept of limits lead to an approximation of area under a curve?

In Investigation 5, you saw that the area bounded by the curve $f(x)=\frac{1}{x}$, the lines $x = 1$ and $x = 4$ is given by $\lim_{n\to\infty}\sum_{i=1}^{n}\frac{1}{x_i}\times\Delta x$. This is equal to $\lim_{n\to\infty}\sum_{i=1}^{n}\frac{1}{(x_i+\Delta x)}\times\Delta x$ as $\Delta x\to 0$, since $f(x)=\frac{1}{x}$ is continuous over this domain and there are no "breaks" or "jumps" in its graph.

HINT

We will look at this function in more detail later on in this chapter.

This result can be generalized for all functions that are continuous over an interval $[a, b]$ as follows:

The red curve is the graph of a function $f(x)$ which is continuous over the interval $[a, b]$. The aim is to find the area bounded by the curve, the lines $x = a$ and $x = b$, and the x-axis. The upper sum has been drawn with 20 rectangles which gives an approximation of the area required. Let the width of each rectangle be Δx. Then for a general rectangle with vertex at $x = x_i$, the rectangle area is given by $f(x_i) \times \Delta x$.

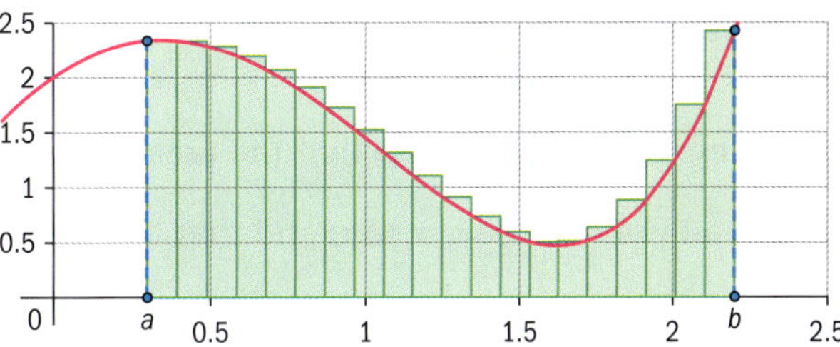

The total shaded area is therefore $\sum_{i=1}^{20} f(x_i)\times\Delta x$.

International-mindedness

A Riemann sum, named after nineteenth-century German mathematician Bernhard Riemann, approximates the area of a region, obtained by adding up the areas of multiple simplified slices of the region.

Functions

Calculus

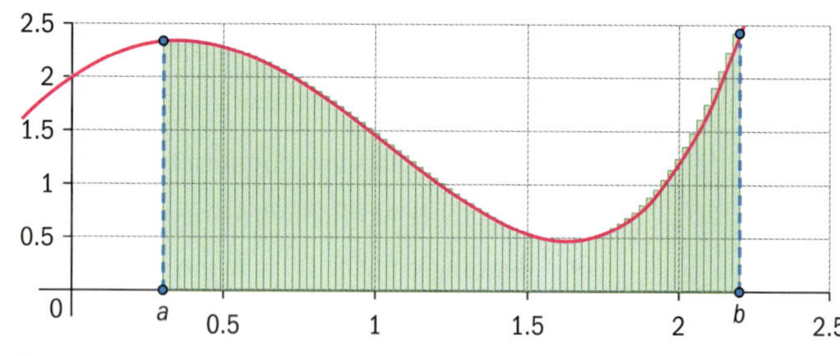

In this next figure, the upper sum is shown with 80 rectangles and you can see that the upper sum of these rectangles, given by $\sum_{i=1}^{80} f(x_i) \times \Delta x$, is a more accurate approximation of the area required.

Suppose now that you divide the interval $[a, b]$ into n equal intervals of width Δx. As you let $n \to \infty$, this upper sum of the areas tends towards the area bounded by the curve and the x-axis over the interval $[a, b]$.

Hence $A = \lim_{n\to\infty} \sum_{i=1}^{n} f(x_i) \Delta x$.

What happens if the lower sum is taken instead of the upper sum?

The sums discussed are called Riemann sums after the mathematician Georg Riemann. He was the person who gave the first rigorous definition of a definite integral.

Investigation 6

1 Draw the graph of the function $f(x) = 16 - x^2$ for $0 \le x \le 4$.

2 Draw a Riemann upper sum with four rectangles and find this sum.

3 Use technology to increase the number of rectangles to 20, 40, 50 and 80. Calculate the respective Riemann upper sums.

4 Find the value of the indefinite integral $\int (16 - x^2) \mathrm{d}x$.

5 a Substitute $x = 4$ in your answer.

b Now substitute $x = 0$ in your answer.

c Calculate the difference of these two answers.

6 Repeat questions **1** to **5** for the following functions.

a $f(x) = 8 - x^3, 0 \le x \le 2$. Substitute $x = 2$ and $x = 0$ into the indefinite integral, and find the difference.

b $f(x) = \cos x, 0 \le x \le \frac{\pi}{2}$. Substitute $x = \frac{\pi}{2}$ and $x = 0$ into the indefinite integral, and find the difference.

7 What do you notice about your results?

8 How would you expect your answer to question **7** to be different if you had been asked to find Riemann lower sums for each function?

9 **Conceptual** How does your understanding of limits and integration lead to an accurate measure of area under a curve?

In Investigation 6, you saw that the area under the curve of $y = f(x)$ is given by $A = \lim_{n\to\infty} \sum_{i=1}^{n} f(x_i) \Delta x$, and that this limit is related to the integral of the function $f(x)$. In fact, it can be proved that the limit at infinity of the Riemann sum is the *exact* area under the graph over an interval $[a, b]$. The proof, however, is beyond the scope of this course.

The **definite integral** is equal to the limit at infinity of the Riemann sum, and hence gives the exact area under the curve between $x = a$ and $x = b$.

Suppose $f(x)$ is continuous over the interval $[a, b]$.
The definite integral of $f(x)$ from a to b is equal to the number given by $\int_a^b f(x)\,\mathrm{d}x$.

Note that the definite integral always gives a numerical answer. Here is how the definite integral is worked out for the examples in Investigation 6.

Upper limit — Evaluate at upper limit — Evaluate at lower limit

$$\int_0^4 (16 - x^2)\,\mathrm{d}x = \left[16x - \frac{x^3}{3}\right]_0^4 = \left(64 - \frac{64}{3}\right) - (0 - 0) = 42.67$$

Lower limit

Write the integral in square brackets, with upper and lower limits as shown. Note that c always cancels out, so you don't need to include it.

$$\int_0^2 (8 - x^3)\,\mathrm{d}x = \left[8x - \frac{x^4}{4}\right]_0^2 = (16 - 4) - (0 - 0) = 12$$

$$\int_0^{\frac{\pi}{2}} 5\cos x\,\mathrm{d}x = [5\sin x]_0^{\frac{\pi}{2}} = \left(5\sin\frac{\pi}{2}\right) - (5\sin 0) = 5$$

Properties of definite integrals

If the integral of $f(x)$ with respect to x in the interval $[a, b]$ exists, then:

- $\int_a^b f(x)\,\mathrm{d}x = -\int_b^a f(x)\,\mathrm{d}x$
- $\int_a^a f(x)\,\mathrm{d}x = 0$
- $\int_a^b kf(x)\,\mathrm{d}x = k\int_a^b f(x)\,\mathrm{d}x$
- $\int_a^b (f(x) \pm g(x))\,\mathrm{d}x = \int_a^b f(x)\,\mathrm{d}x \pm \int_a^b g(x)\,\mathrm{d}x$
- $\int_a^c f(x)\,\mathrm{d}x + \int_c^b f(x)\,\mathrm{d}x = \int_a^b f(x)\,\mathrm{d}x,\ a < c < b$

HINT

You can test these properties by taking combinations of the three definite integrals introduced in Investigation 6.

Functions

Calculus

Example 6

Evaluate the following definite integrals.

a $\int_0^1 (3x^2 - 4x + 7)\,\mathrm{d}x$ **b** $\int_0^\pi \cos\frac{x}{3}\,\mathrm{d}x$ **c** $\int_{-1}^0 5(1 - 2x)^3\,\mathrm{d}x$

Continued on next page

a $\int_0^1 (3x^2 - 4x + 7)\,dx$

$= [x^3 - 2x^2 + 7x]_0^1$

$= (1 - 2 + 7) - (0)$

$= 6$

You can confirm these results using technology.

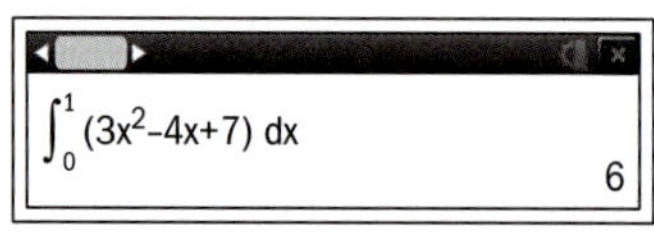

b $\int_0^{\pi} \cos\frac{x}{3}\,dx = \left[\frac{\sin\frac{x}{3}}{\frac{1}{3}}\right]_0^{\pi}$

$= \left(3\sin\frac{\pi}{3}\right) - (3\sin 0)$

$= \frac{3\sqrt{3}}{2}$

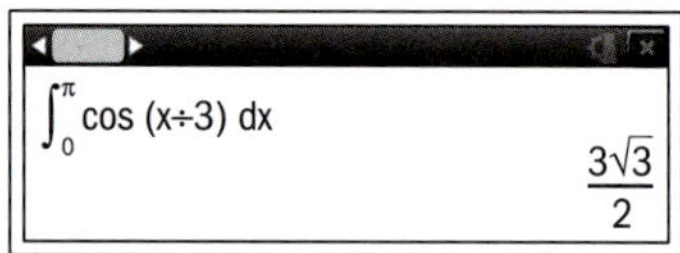

c $\int_{-1}^{0} 5(1-2x)^3\,dx$

$= 5\left[\frac{(1-2x)^4}{-2\times 4}\right]_{-1}^{0}$

$= 5\left(-\frac{1}{8}\right) - 5\left(-\frac{(1+2)^4}{8}\right)$

$= -\frac{5}{8} + \frac{405}{8} = 50$

$\int_{-1}^{0} 5(1-2x)^3\,dx$ 50

Example 7

Find the area of the triangle formed by the x-axis and the lines $f(x) = 2x + 4$ and $g(x) = 4 - x$ by using:

a the formula for area of triangle **b** definite integration.

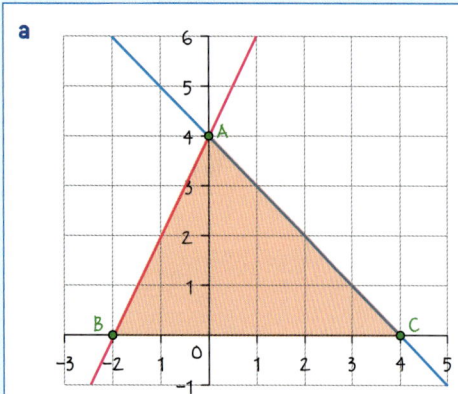

a

Sketch the graph and find the points of intersection with the x-axis.

The two lines intersect on the y-axis.

Area $= \frac{1}{2}bh$.

Area $= \frac{1}{2} \times 6 \times 4 = 12$ units^2

b Area of triangle ABC

$= \int_{-2}^{0}(2x+4)\,dx + \int_{0}^{4}(4-x)\,dx$

$= \left[x^2+4x\right]_{-2}^{0} + \left[4x - \frac{x^2}{2}\right]_{0}^{4}$

$= (0) - (4 + (-8)) + (16 - 8) - (0)$

$= 4 + 8 = 12$ units^2

Divide triangle ABC into two areas to be found by integration.

Investigation 7

Students should only use a scientific calculator.

1 Copy and complete the following table.

Definite integral $\int_a^b f(x)\,dx$	**Numerical answer**	**Sketch of** $f(x)$
$\int_{-2}^{2}(1-x^3)\,dx$		
$\int_{-2}^{0}(1-x^3)\,dx$		
$\int_{0}^{2}(1-x^3)\,dx$		
$\int_{-\frac{\pi}{4}}^{\frac{\pi}{2}}\sin x\,dx$		
$\int_{-\frac{\pi}{4}}^{0}\sin x\,dx$		
$\int_{0}^{\frac{\pi}{2}}\sin x\,dx$		
$\int_{-\frac{\pi}{2}}^{\frac{\pi}{2}}\sin x\,dx$		
$\int_{-2}^{2}(2+3x-x^2)\,dx$		
$\int_{-2}^{0}(2+3x-x^2)\,dx$		
$\int_{0}^{2}(2+3x-x^2)\,dx$		

Continued on next page

2 **Factual** Why do some of the answers to question **1** not give the total area bounded by the graph, the x-axis and the upper and lower limits?

3 **Conceptual** Explain how you could you use integration to calculate the total area of a function which is partly above and partly below the x-axis.

When the function $f(x)$ is negative for $x \in [a, b]$, then the area bounded by the curve, the x-axis and the lines $x = a$ and $x = b$ is $\left|\int_a^b f(x)\,dx\right|$.

TOK

We are trying to find a method to evaluate the area under a curve.

"The main reason knowledge is produced is to solve problems." To what extent do you agree with this statement?

Example 8

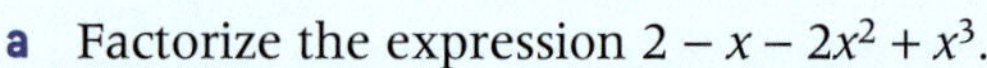

a Factorize the expression $2 - x - 2x^2 + x^3$.

b Hence sketch the graph $f(x) = 2 - x - 2x^2 + x^3$.

c Find the area of the region bounded by the graph $f(x) = 2 - x - 2x^2 + x^3$ and the x-axis.

a $2 - x - 2x^2 + x^3 = (2 - x) - x^2(2 - x)$ $= (2 - x)(1 - x^2)$ $= (2 - x)(1 - x)(1 + x)$	This function is easy to factorize by inspection. You could also use the factor theorem to factorize.
b 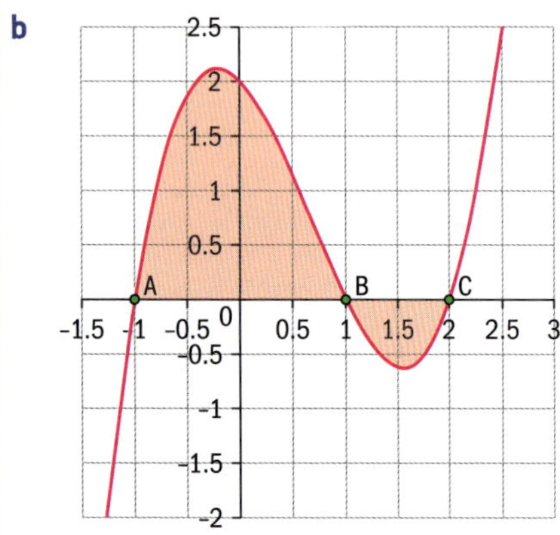	Since the coefficient of x^3 is positive you know that this is the shape of the curve. Note you only need the general shape and the x-intercepts to determine the parts of the curve which are below and above the x-axis.
c $A = \int_{-1}^{1}(2 - x - 2x^2 + x^3)\,dx + \left|\int_{1}^{2}(2 - x - 2x^2 + x^3)\,dx\right|$ $= \left[2x - \frac{x^2}{2} - \frac{2x^3}{3} + \frac{x^4}{4}\right]_{-1}^{1} + \left|\left[2x - \frac{x^2}{2} - \frac{2x^3}{3} + \frac{x^4}{4}\right]_{1}^{2}\right|$ $= \frac{8}{3} + \left|-\frac{5}{12}\right| = \frac{37}{12}$ units^2	Integrate the sections of the curve above and below the x-axis separately. Find the absolute value of the section below the x-axis.

Note that if you sketch $|(2 - x - 2x^2 + x^3)|$ you would obtain this graph, so that the required area becomes

$$\int_{-1}^{2} \left|\left(2 - x - 2x^2 + x^3\right)\right| dx$$

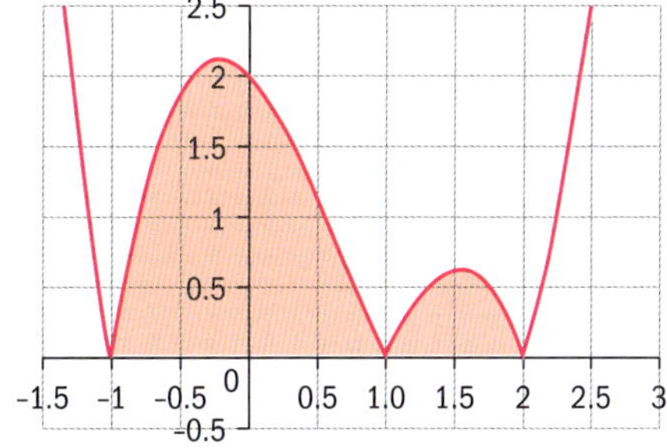

In general, when the graph of a function $f(x)$ is partly above and partly below the x-axis, the total area bounded by the graph and the x-axis for $x \in [a, b]$ is given by $A = \int_a^b |f(x)| dx$.

HINT

When using your GDC to find the total area bounded by a graph which is partly above and partly below the x-axis, you need to find $\int_a^b |f(x)| dx$ in order to obtain the correct numerical result.

You are now ready to understand one of the most astonishing and important formalizations of Newton's and Leibnitz's work, which connects differentiation and integration. The theorem justifies the procedures for evaluating definite integrals and is still regarded as one of the most significant developments of modern-day mathematics.

The fundamental theorem of calculus

If f is continuous on $[a, b]$ and if F is any antiderivative of f on $[a, b]$, then $\int_a^b f(x) dx = F(b) - F(a)$.

International-mindedness

The **fundamental theorem of calculus** shows the relationship between the derivative and the integral and was developed in the seventeenth century by Gottfried Wilhelm Leibniz and Isaac Newton.

Exercise 7B

For questions **1** to **4**, use definite integration to find the area of the region bounded by the graph of the function, the x-axis, and the given lines. Verify your result by finding the area of a geometric shape.

1 $y = 5, x = 0, x = 4$

2 $y = 2x - 5, x = 0, x = 10$

3 $y = 3 - |x|, x = -3, x = 3$

4 $y = \begin{cases} 4, & x \le 0 \\ 4 - x, & x > 0 \end{cases}, x = -3, x = 4$

5 Find the area bounded by the lines $y = x + 3$, $y = 6 - 2x$ and the x-axis using:

a appropriate definite integrals

b geometry.

For questions **6** to **8**, find the areas bounded by the function, the x-axis, and the given lines. Verify your results using technology.

6 $y = x - x^3$, $x = -1$ and $x = 1$

7 $y = x^2 - 5x - 6$, $x = -1$ and $x = 6$

8 $y = x^2 - 5x + 6$, $x = 1$ and $x = 4$

9 Show that $4\sin\left(\frac{x}{3}+\frac{\pi}{2}\right)=0$ when $\frac{x}{3}=n\pi-\frac{\pi}{2}, n\in\mathbb{Z}$.

By taking appropriate values of $n\in\mathbb{Z}$, find the first two:

a positive x-intercepts of $y=4\sin\left(\frac{x}{3}+\frac{\pi}{2}\right)$

b negative x-intercepts of $y=4\sin\left(\frac{x}{3}+\frac{\pi}{2}\right)$.

Hence, find the total area bounded by the curve $y=4\sin\left(\frac{x}{3}+\frac{\pi}{2}\right)$ and the x-axis between $x=-\frac{3\pi}{2}$ and $x=\frac{9\pi}{2}$.

Use technology to verify your answer.

10 Sketch the graph of the function

$$f(x)=\begin{cases}4-x^2, -2\le x\le 0\\ 4-x, 0\le x\le 4\end{cases}$$

Calculate the area bounded by the graph of $f(x)$ and the x-axis.

Developing inquiry skills

In the opening scenario for this chapter, the function $g(t)=\frac{1}{50}e^{-\frac{t}{10}+3.09}$ models the rate of growth of the hydrangea shrub in meters per day for $t\ge 20$ days. Find the antiderivatives of the function g.

7.2 Exponents and logarithms

Let $m\in\mathbb{Z}^+$ and $a\in\mathbb{R}$.

Then $a^m=\underbrace{a\times a\times a\times\ldots\times a}_{m}$ where there are m factors of a.

a is called the base.

m is called the **exponent** (or **index**, or **power**) of a.

International-mindedness

The term *power* was used by the Greek mathematician Euclid for the square of a line.

Investigation 8

You have actually been using the properties of exponents in previous chapters (and, indeed, before beginning your Higher Level course). However, in this investigation you will derive these properties of exponents, and reflect on why some conditions are tied to these properties.

In all the cases below, $m, n\in\mathbb{Z}^+$ and $m\ge n$.

1 Use the definition that $a^m=\underbrace{a\times a\times a\times\ldots\times a}_{m}$ to show the following properties are true.

a $a^m\times a^n=a^{m+n}$ **b** $a^m\div a^n=a^{m-n}$ **c** $(a^m)^n=a^{mn}$

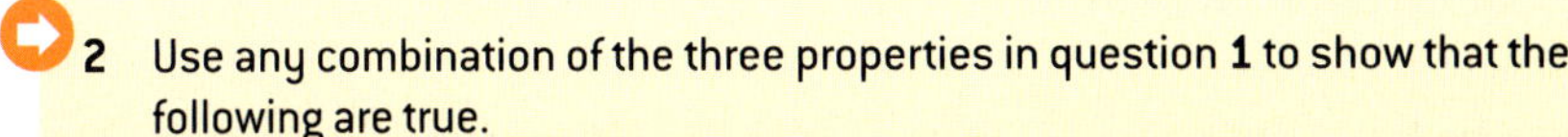

2 Use any combination of the three properties in question **1** to show that the following are true.

a $a^0 = 1$ **b** $a^{-n} = \frac{1}{a^n}, a \neq 0$

c $a^{\frac{1}{m}} = \sqrt[m]{a}, a > 0$ **d** $a^{\frac{m}{n}} = \sqrt[n]{a^m} = \left(\sqrt[n]{a}\right)^m, a > 0$

3 Explain why $a \neq 0$ is a necessary condition for property **2b** to be true.

4 Give an example to show that if $a \leq 0$, inconsistencies may arise for property **2c**.

5 Explain why $a^x > 0$ for all $a, x \in \mathbb{R}, a > 0$.

6 **Factual** What is an axiom?

7 **Conceptual** Why is it important to know that mathematics is axiomatic?

International-mindedness

French mathematician Nicolas Chuquet created his own notation for exponentials and might have been the first to recognize zero and negative exponents.

Example 9

Evaluate the following:

a $\left(\frac{64}{27}\right)^{\frac{2}{3}}$ **b** $121^{-\frac{1}{2}}$ **c** $-32^{-\frac{1}{5}}$

a $\left(\frac{64}{27}\right)^{\frac{2}{3}} = \left(\sqrt[3]{\frac{64}{27}}\right)^2 = \left(\frac{4}{3}\right)^2 = \frac{16}{9}$	Using $a^{\frac{m}{n}} = \sqrt[n]{a^m} = \left(\sqrt[n]{a}\right)^m, a > 0$.
b $121^{-\frac{1}{2}} = \frac{1}{121^{\frac{1}{2}}} = \frac{1}{\sqrt{121}} = \frac{1}{11}$	Using $a^{-n} = \frac{1}{a^n}, a \neq 0$ and $a^{\frac{1}{m}} = \sqrt[m]{a}, a > 0$.
c $-32^{-\frac{1}{5}} = \frac{1}{-32^{\frac{1}{5}}} = \frac{1}{\sqrt[5]{-32}} = -\frac{1}{2}$	Using $a^{-n} = \frac{1}{a^n}, a \neq 0$ and $a^{\frac{1}{m}} = \sqrt[m]{a}$, which you can do because $m = 5$ which is odd.

Example 10

Simplify the expression $(15 \times 4^{2n+1}) \div (25 \times 2^{4n})$.

$\left(15 \times 4^{2n+1}\right) \div \left(25 \times 2^{4n}\right) = \frac{15 \times \left(2^2\right)^{2n+1}}{25 \times 2^{4n}}$	Change the numerator to an expression involving powers of 2.
$= \frac{15 \times 2^{4n+2}}{25 \times 2^{4n}}$	Using $(a^m)^n = a^{mn}$.
$= \frac{15 \times 4 \times 2^{4n}}{25 \times 2^{4n}}$	Using $a^m \times a^n = a^{m+n}$.
$= \frac{12}{5}$	Divide numerator and denominator by 2^{4n}.

Example 11

Solve the equation $2^x - 2^{2-x} = 3$.

$2^x - 2^{2-x} = 3$	
$\Rightarrow 2^x - 2^2 \times 2^{-x} = 3$	Using $a^m \times a^n = a^{m+n}$.
$\Rightarrow 2^x - \dfrac{2^2}{2^x} = 3$	Using $a^{-n} = \dfrac{1}{a^n}$, $a \neq 0$.
$\Rightarrow (2^x)^2 - 2^2 = 3 \times 2^x$	Multiply each term by 2^x.
$\Rightarrow (2^x)^2 - 3 \times 2^x - 4 = 0$	Rearrange. You now have a quadratic equation in 2^x.
Let $2^x = y$ $\Rightarrow y^2 - 3y - 4 = 0$ $\Rightarrow (y - 4)(y + 1) = 0$ $\Rightarrow y = 4$ or $y = -1$	Make a substitution and solve the quadratic equation.
$2^x = 4 \Rightarrow x = 2$	Substitute $2^x = y$ back in
$2^x \neq -1$	$2^x > 0$ for all real values of x.
$x = 2$	

There are many applications of exponents in financial matters, including a scenario that you met when working on geometric sequences and series in Chapter 1. The following examples illustrate how the laws of exponents are used to solve financial problems.

International-mindedness

The word *exponent* was first used in the 16th century by German monk and mathematician Michael Stifel. Englishman Samuel Jeake introduced the term indices in 1696.

Example 12

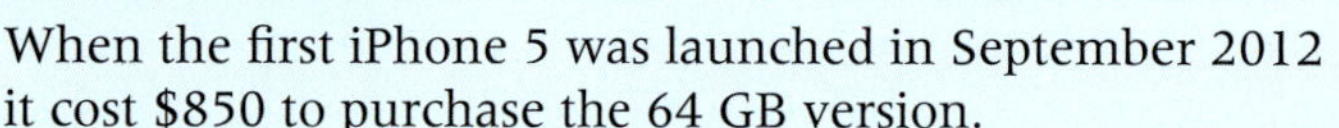

When the first iPhone 5 was launched in September 2012 it cost \$850 to purchase the 64 GB version.

In September 2017 the same iPhone 5 could be purchased, brand new, for \$80.

Use the laws of exponents to calculate the average depreciation per month of this model of phone.

Let V_0 represent the value when it was launched, and V_n represent the value n months after.	

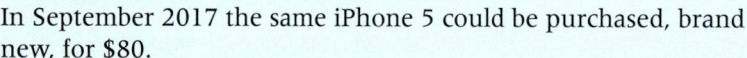

Let r be the depreciation rate per month. $V_1 = V_0(1-r)$ $V_2 = V_1(1-r) = V_0(1-r)^2$ . . . $V_n = V_{n-1}(1-r) = V_0(1-r)^n$	Assuming that the depreciation rate remains constant. These terms form a geometric sequence with first term V_0 and common ratio $(1-r)$.
After 5 years $V_{60} = V_0(1-r)^{60}$ $\Rightarrow 80 = 850(1-r)^{60}$	There are 60 months in 5 years.
$\Rightarrow 1-r = \sqrt[60]{\frac{8}{85}} = 0.9613...$ $\Rightarrow r = 1-0.9613... \approx 0.038$	Substitute $V_{60} = 80$ and $V_0 = 850$ and solve for r.
So the average depreciation rate per month is 3.8%.	

Example 13

The value of a sailing boat depreciates at the rate of 15% per year for the first three years. After that, the rate of depreciation r% remains constant. A new boat costing €60 000 is worth one fifth of its original value after 15 years. Find:

a the value of the boat, to the nearest euro, after three years

b the rate of depreciation after the first three years of purchase.

a $V_3 = 60\,000(1-0.15)^3 = €36\,848$	Depreciation over first three years forms a geometric series with initial value 60 000 and ratio (1 – 0.15).
b 12 years after the initial 3 years, the value depreciates to €12 000.	Since the boat is worth one-fifth of its value after 15 years.
$V_{12} = 12000 = 36\,848 \times (1-r)^{12}$ $\Rightarrow 1-r = \sqrt[12]{\frac{12\,000}{36\,848}}$	This is another geometric series with initial value 36 848 and ratio r.
$\Rightarrow r = 1-\sqrt[12]{\frac{12\,000}{36\,848}} = 0.08925$	Solve for r.
After 3 years the value of the boat depreciates at an average rate of 8.93% (3 s.f.)	

Exercise 7C

1 Find the value of these expressions.

a $(81)^{\frac{3}{4}}$ **b** $\left(\frac{8}{125}\right)^{-\frac{1}{3}}$ **c** $\left(\frac{32}{243}\right)^{\frac{2}{5}}$

2 Show that each of these equalities are true.

a $\left(\frac{a^{12}y^{-3}}{16y}\right)^{\frac{3}{4}} = \frac{a^9}{8y}$

b $\frac{a^{-2} + 2a^{-1} + 1}{a^{-3}} = a(a+1)^2$

c $\frac{b^4 \times b^{-11}}{b^{-7}} = 1$

3 Simplify $\sqrt{9y^3} \div \sqrt[3]{8y^2}$ and use your answer to show that when $y = 64$, $\sqrt{9y^3} \div \sqrt[3]{8y^2} = 48$.

4 Show that $\frac{(a^5b^2c^{-3}) \times \sqrt{a^{-3}b^3c}}{\sqrt{abc}} = \left(\frac{ab}{c}\right)^3$

5 Simplify the following.

a $21 \times 3^{4n-1} - 7 \times 9^{2n}$

b $48 \times 4^{2n-1} + 6 \times 2^{4n+1}$

6 Solve the following equations.

a $2^{3x} - 2^{7x-2} = 0$ **b** $9^{1-2x} = \frac{1}{27^{x-4}}$

c $9^x + 9 = 10 \times 3^x$ **d** $2^{x+2} + 7 = \frac{1}{2^{x-1}}$

7 A house bought for €300 000 in 2000 was sold for €500 000 in 2010. Determine the annual rate of appreciation of the house. Give your answer to the nearest percent.

8 Tensions in the Middle East cause oil prices to change. The table shows the cost price for a barrel of oil at various points in time. You may assume that all data represents prices at the end of each month.

December 2015	\$37.21
June 2016	\$48.76
December 2016	\$51.97
February 2017	\$53.47
March 2017	\$49.33

a **i** Determine the average percentage monthly price rise in oil price between December 2015 and June 2016.

ii Determine the average percentage monthly price rise in oil price between June 2016 and December 2016.

iii Comment on how your answers to parts **i** and **ii** compare to the average monthly percentage price rise between December 2015 and December 2016, when considering only data from these two months.

b State how your results from part **a** compare to the monthly average increase in oil price between December 2016 and February 2017.

c Determine the average daily decrease in oil price between February 2017 and March 2017.

9 Paloma invests \$2000 at Superior Bank which offers 6% interest compounded quarterly provided she leaves the money in the bank for 10 years. Her sister Concita invests \$1500 at Best Bank which offers 6% interest compounded annually and \$500 at Better Bank which offers 3% interest compounded monthly provided the money is fixed for 10 years. Calculate the value of each sister's investment (to the nearest dollar) at the end of 10 years.

10 Sureepan borrows 40 000 Bhat to buy a new computer. The bank charges 7.5% interest compounded yearly. Sureepan agrees to pay a half of the amount still owing at the end of each year after interest has been compounded. Determine how much Sureepan would have paid back at the end of four years.

Investigation 9

1 Use table 1 to calculate the following:

a 729×27

b $243 \div 27$

c $\frac{27}{729} - \frac{1}{27}$

2 Use the table to verify the following statements:

a $\frac{81}{2187} + \frac{486}{6561} = \frac{1}{9}$

b $\frac{45}{1458} - \frac{243}{13122} = \frac{1}{81}$

3 Copy and complete table 2 using powers of 2.

4 Use your table to create some statements similar to the ones in questions **1** and **2**.

5 **Conceptual** How has the use of exponents helped in calculations?

n	3^n
−5	$\frac{1}{243}$
−4	$\frac{1}{81}$
−3	$\frac{1}{27}$
−2	$\frac{1}{9}$
−1	$\frac{1}{3}$
0	1
1	3
2	9
3	27
4	81
5	243
6	729
7	2187
8	6561
9	19 683

Table 1

n	2^n
−5	
−4	
−3	
−2	
−1	
0	
1	
2	
3	
4	
5	
6	
7	
8	
9	

Table 2

Logarithms

Logarithms are defined as follows.

$a = b^x \Leftrightarrow x = \log_b a$, where $a, b \in \mathbb{R}^+$ and $b \neq 1$

The restrictions on a and b are required to obtain sensible results and you should always check that they are satisfied.

International-mindedness

Logarithms do not have units but many measure ments use a log scale such as earthquakes, the pH scale and human hearing.

Example 14

Write the following identities in logarithmic form.

a $3^4 = 81$ **b** $8^{\frac{1}{3}} = 2$

a $3^4 = 81 \Rightarrow \log_3 81 = 4$	The base is 3 and the power is 4. So the logarithm to base 3 of 81 is 4.
b $8^{\frac{1}{3}} = 2 \Rightarrow \log_8 2 = \frac{1}{3}$	The base is 8 and the power is $\frac{1}{3}$. So the logarithm to base 8 of 2 is $\frac{1}{3}$.

Example 15

Write the following identities in exponent form.

a $\log_2 32 = 5$ **b** $\log_9 3 = \frac{1}{2}$

a $\log_2 32 = 5 \Rightarrow 2^5 = 32$	The base is 2 and the logarithm is 5. So 2 to the power of 5 is 32.
b $\log_9 3 = \frac{1}{2} \Rightarrow 9^{\frac{1}{2}} = 3$	The base is 9 and the logarithm is $\frac{1}{2}$. So 9 to the power of $\frac{1}{2}$ is 3.

Example 16

Solve the following equations for x.

a $\log_x 81 = 2$ **b** $x = \log_4 16$ **c** $\log_5 x = 3$ **d** $\log_a 1 = x$

a $\log_x 81 = 2 \Rightarrow x^2 = 81$ $\Rightarrow x^2 = 9^2$ $\Rightarrow x = 9$	Use the definition of logarithms to write each equation in exponent form, and then solve. x cannot be negative.
b $x = \log_4 16 \Rightarrow 4^x = 16$ $\Rightarrow 4^x = 4^2$ $\Rightarrow x = 2$	
c $\log_5 x = 3 \Rightarrow 5^3 = x$ $\Rightarrow x = 125$	
d $\log_a 1 = x \Rightarrow a^x = 1$ $\Rightarrow x = 0$	$a^0 = 1$ for any value of a.

Investigation 10

1 Use the definition of logarithms and the properties of exponents to verify the following:

a $\log_a x + \log_a y = \log_a xy$ **b** $\log_a x - \log_a y = \log_a \frac{x}{y}$

c $\log_a x^n = n \log_a x$

2 Show that:

a $\log_a 1 = 0$ **b** $\log_a a = 1$ **c** $-\log_a x = \log_a \frac{1}{x}$

3 For which values of a, x and y are these properties valid?

4 Show that $\log_a x = \frac{\log_b x}{\log_b a}$, hence show that $\log_a b = \frac{1}{\log_b a}$

5 **Conceptual** What can you deduce about logarithms and exponents from your results to this investigation?

HINT

This is called the change of base rule.

In Investigation 10 you derived the following **properties of logarithms:**

1. $\log_a x + \log_a y = \log_a xy$
2. $\log_a x - \log_a y = \log_a \frac{x}{y}$
3. $\log_a x^n = n \log_a x$
4. $\log_a 1 = 0$
5. $\log_a a = 1$
6. $-\log_a x = \log_a \frac{1}{x}$
7. $\log_a x = \frac{\log_b x}{\log_b a}$
8. $\log_a b = \frac{1}{\log_b a}$

HINT

Since 10 is a very common base, instead of writing $\log_{10} a$ we often just write $\log a$.

The following examples will show you how these properties can be used to solve problems.

Example 17

Express each of the following as a single logarithm.

a $\log 2 + 2\log 3 - \log 6$ **b** $\log_a p - 2\log_a q + \frac{1}{2}\log_a r$ **c** $1 + \log_b ab$

a $\log 2 + 2\log 3 - \log 6$	
$= \log 2 + \log 3^2 - \log 6$	Use property 3.
$= \log \frac{2 \times 9}{6} = \log 3$	Use properties 1 and 2.
or	
$\log 2 + 2\log 3 - \log 6$	
$= \log 2 + 2\log 3 - (\log 2 + \log 3)$	
$= \log 3$	
b $\log_a p - 2\log_a q + \frac{1}{2}\log_a r$	
$= \log_a p - \log_a q^2 + \log_a r^{\frac{1}{2}}$	Use property 3.
$= \log_a \frac{p \times \sqrt{r}}{q^2}$	Use properties 1 and 2.
c $1 + \log_b ab$	
$= \log_b b + \log_b ab$	Use property 5.
$= \log_b (b \times ab) = \log_b ab^2$	Use property 1.

Functions

Calculus

Example 18

Without the use of a calculator, find the value of the following.

a $\log_3 7 \times \log_7 3$ **b** $\log_2 27 \times \log_3 64$

a $\log_3 7 \times \log_7 3$	
$= \log_3 7 \times \frac{\log_3 3}{\log_3 7}$	Use property 7 to change logarithms to base 3 and simplify.
$= \log_3 3$	
$= 1$	By property 5.
b $\log_2 27 \times \log_3 64$	
$= 3\log_2 3 \times \frac{\log_2 64}{\log_2 3}$	Use properties 3 and 7.
$= 3 \times 6\log_2 2$	
$= 18$	

You can use the properties of exponents and logarithms to solve equations, as shown in the following examples.

Example 19

Solve for x.

a $5^x = 36$ **b** $4^{2x+1} = 10$ **c** $2(3^x)^2 = 7$

a $5^x = 36$	
$\Rightarrow x\log 5 = \log 36$	Take logarithms of both sides.
$\Rightarrow x = \frac{\log 5}{\log 36} = 0.449$	Solve for x.
b $4^{2x+1} = 10$	
$\Rightarrow (2x-1)\log 4 = \log 10$	Take logarithms of both sides.
$\Rightarrow 2x - 1 = \frac{\log 10}{\log 4}$	
$\Rightarrow 2x = 1 + \frac{\log 10}{\log 4}$	Solve for x.
$\Rightarrow x = \frac{1}{2}\left(1 + \frac{\log 10}{\log 4}\right) = 1.33$	
c $2(3^x)^2 = 7$	
$\Rightarrow 3^{2x} = \frac{7}{2}$	Use properties of exponents.
$\Rightarrow 2x\log 3 = \log 3.5$	Take logarithms of both sides.
$\Rightarrow x = \frac{\log 3.5}{2\log 3} = 0.570$	Solve for x.

Example 20

Solve the equations:

a $2^{3-x} = 8^x$ **b** $3\sqrt{27} = 9^{x-1}$

a $2^{3-x} = 8^x$	
$\Rightarrow 2^{3-x} = \left(2^3\right)^x$	Use laws of exponents to write both sides to base 2.
$\Rightarrow 2^{3-x} = 2^{3x}$	
$\Rightarrow 3 - x = 3x$	Since both sides are to the same base, you can equate exponents.
$\Rightarrow 4x = 3$	
$\Rightarrow x = \frac{3}{4}$	
b $3\sqrt{27} = 9^{x-1}$	
$\Rightarrow 3 \times 3^{\frac{3}{2}} = \left(3^2\right)^{x-1}$	Use laws of exponents to write both sides to base 3.
$\Rightarrow 3^{\frac{5}{2}} = 3^{2x-2}$	Combine terms on LHS and simplify RHS.
$\Rightarrow \frac{5}{2} = 2x - 2$	Equate the exponents.
$\Rightarrow 2x = \frac{5}{2} + 2$	Solve for x.
$\Rightarrow 2x = \frac{9}{2}$	
$\Rightarrow x = \frac{9}{4}$	

Example 21

Solve the following equations.

a $\log_{15} x + \log_{15}(2x - 1) = 1,\ x > 0$ **b** $\log_4 x + \log_4(x - 6) = 2,\ x > 0$

a $\log_{15} x + \log_{15}(2x - 1) = 1$	
$\Rightarrow \log_{15} x(2x - 1) = 1$	Use property 1.
$\Rightarrow \log_{15} x(2x - 1) = \log_{15} 15$	Use property 5.
$\Rightarrow x(2x - 1) = 15$	Since both sides are log in the same base, you can equate the terms.
$\Rightarrow 2x^2 - x - 15 = 0$	
$\Rightarrow (2x + 5)(x - 3) = 0$	
$\Rightarrow x = 3,\ x \neq -\frac{5}{2}$ since $x > 0$	

Continued on next page

b $\log_4 x + \log_4 (x-6) = 2$	
$\Rightarrow \log_4 x(x-6) = 2$	Use property 1.
$\Rightarrow \log_4 x(x-6) = 2\log_4 4$	By property 5, $1 = \log_4 4$.
$\Rightarrow \log_4 x(x-6) = \log_4 16$	By property 3, $2\log_4 4 = \log_4 4^2$.
$\Rightarrow x(x-6) = 16$	
$\Rightarrow x^2 - 6x - 16 = 0$	
$\Rightarrow (x-8)(x+2) = 0$	
$\Rightarrow x = 8,\ x \neq -2$ since $x > 0$	

In Chapter 1 you used tables to solve questions which involved finding the value of an exponent. Such questions can be solved using logarithms, as shown in Example 22.

Example 22

Find the number of terms in each of these geometric sequences.

a 0.15, 0.45, 1.35, ..., 12.15 **b** 440, 110, 27.5, ..., 0.4296875

a $u_1 = 0.15,\ r = \dfrac{0.45}{0.15} = 3$	Find the first term and the common ratio.
$\Rightarrow u_n = 0.15 \times 3^{n-1} = 12.15$	Express 12.5 as the *n*th term of the sequence.
$\Rightarrow \log 3^{n-1} = \log \dfrac{12.15}{0.15}$	Divide both sides by 0.15 and take logarithms.
$\Rightarrow (n-1)\log 3 = \log \dfrac{12.15}{0.15}$	Use property 3 of logs to bring the exponent down.
$\Rightarrow n = 1 + \dfrac{\log \dfrac{12.15}{0.15}}{\log 3} = 5$	Use technology to find the answer.
This sequence has 5 terms.	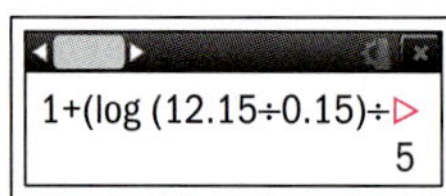
b $u_1 = 440,\ r = \dfrac{110}{440} = 0.25$	
$u_n = 440 \times 0.25^{n-1} = 0.4296875$	Express 0.4296875 as the *n*th term of the sequence.
$\Rightarrow (n-1)\log 0.25 = \log \dfrac{0.4296875}{440}$	Divide both sides by 440 and take logarithms
$\Rightarrow n = 1 + \dfrac{\log \dfrac{0.4296875}{440}}{\log 0.25}$	Use technology to find the answer.
$\Rightarrow n = 6$	
This sequence has 6 terms.	

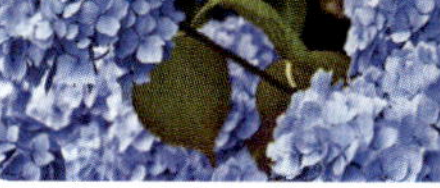

A more straightforward method of finding a solution is to use a GDC with Nsolve (SolveN, depending on the GDC type) by entering Nsolve $\left(\log 3^{n-1} = \log\left(\frac{12.5}{0.15}\right)\right)$.

Example 23

Calculate the number of terms that are required for the sum of the geometric series given by $\sum_{i=1}^{n} 3 \times 2^i$ to exceed 1000.

$u_1 = 6,\ r = \frac{u_2}{u_1} = 2$	Identify the first term and common ratio.
$\frac{6(1-2^n)}{1-2} > 1000$ $\Rightarrow 2^n - 1 > \frac{1000}{6}$	Use $S_n = \frac{u_1(1-r^n)}{1-r}$ for the sum to n terms of a geometric series.
$\Rightarrow 2^n > 167.\dot{6}$ $\Rightarrow n\log 2 > \log 167.\dot{6}$ $\Rightarrow n > \frac{\log 167.\dot{6}}{\log 2} = 7.38$	Use technology to find the answer.  log 167.666÷log 2 7.38944635
Therefore $n = 8$. When 8 or more terms are added the sum exceeds 1000.	You need a whole number of terms.

Example 24

A company provides Mr Awadi with a new car worth €30 000 on 1 January.

On the 1 January of each subsequent year, the value of the car depreciates by 15% of its value from the previous year.

a Find the value of the car three years after it was purchased. Give your answer to the nearest euro.

b The company has a policy of replacing cars when their value falls to €2500. Determine how many years it takes for Mr Awadi to receive a new company car.

a After three years the car would have depreciated to $30\,000 \times (1 - 0.15)^3 = €18\,424$	Use technology to find the answer. 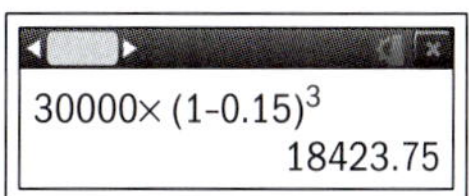 30000× (1-0.15)^3 18423.75

Continued on next page

b $30\,000\times(0.85)^n = 2500$

$\Rightarrow n\log 0.85 = \log\dfrac{2500}{30\,000}$

$\Rightarrow n = \dfrac{\log\dfrac{2500}{30\,000}}{\log 0.85} = 15.29\text{ years}$

Mr Awadi will have to wait for more than 15 years to obtain a new company car.

Let n be the number of years that have passed when the value of the car reaches 2500.

Divide by 30 000, take logs, and bring the n down.

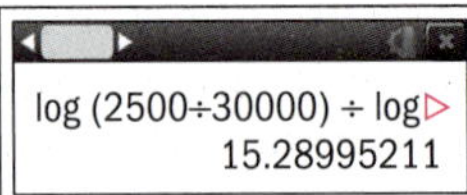

Exercise 7D

1 Express each of the following in logarithmic form.

a $3^5 = 243$ **b** $16^{\frac{1}{4}} = 2$ **c** $p = q^5$

d $10^{-4} = 0.0001$ **e** $x^y = 11$

2 Write each of the following in exponent form.

a $\log_5 625 = 4$ **b** $\log_{64} 8 = \dfrac{1}{2}$

c $\log_m n = p$ **d** $\log_b 1 = 0$

e $\log 0.01 = -2$

3 Solve for x.

a $\log_x 2 = 128$ **b** $\log_4 x = 3$

c $\log_x 8 = \dfrac{3}{4}$ **d** $\log_9 x = \dfrac{3}{2}$

e $\log_x 49 = 2$

4 Express the following in terms of $\log_a m$ and $\log_a n$.

a $\log_a \dfrac{m}{n^2}$ **b** $\log_a\left(\dfrac{\sqrt{m}}{n^3}\right)^{\frac{2}{3}}$

5 Express each of the following as a single logarithm.

a $\log 6 - 3\log 2 + \log 40$ **b** $\log_3 36 - 2$

c $\dfrac{1}{4}\log_a m + \dfrac{3}{4}\log_a mn^2$

6 Express each of the following as a rational number.

a $\log_2 12 - \log_2 48 - 3$

b $\dfrac{1}{4}\log_2 81 + \log_2 48 - \dfrac{2}{3}\log_2 27$

7 Express a in terms of b.

a $4\log a = 3\log b$

b $\log a = \log b - \log 2$

c $\log b = 1 - 4\log a$

8 Evaluate the following.

a $\log_2 3 \times \log_3 2$ **b** $\log_6 10 \times \log 36$

c $\log_4 3 \times \log_3 8$ **d** $\log_5 8 \div \log_{25} 8$

e $\dfrac{1}{\log_3 6} + \dfrac{1}{\log_2 6}$ **f** $\log_5 40 - \dfrac{1}{2\log_{64} 5}$

9 Show that

a $x^{\log y} = y^{\log x}$ **b** $\dfrac{1}{\log_x xy} + \dfrac{1}{\log_y xy} = 1$

10 Let $p = \log_a x$ and $q = \log_a y$. Express $\log_x a$ in terms of p and $\log_y a$ in terms of q. Show that:

a $\log_{xy} a = \dfrac{1}{p+q}$ **b** $\log_{\frac{x}{y}} a = \dfrac{1}{p-q}$

11 Use technology to solve the following equations.

a $3^x = 10$

b $5^{3x-1} = 12$

c $2^x \times 5^{x-1} = 0.01$

12 Solve the following equations.

a $\log_5 x = 9\log_x 5$

b $3\log_7 x + \log_7 49 = 8$

c $\log_4 x + \log_x 4 = 2$

13 Solve the equation $25^x - 6\times 5^x - 16 = 0$.

14 Solve the equation $\log_2 x + \log_4 x = 9$.

15 Solve the following equations. Give your answers to 3 significant figures.

a $\log_5 x + 12\log_x 5 = 7$ **b** $5 \times 7^x + \frac{4}{7^x} = 21$

16 Solve the equation $4 \times 9^x + 3 \times 4^x = 13 \times 6^x$ Give your answer to 3 significant figures.

17 Solve each pair of simultaneous equations.

a $2\log_5 x + 6\log_9 y = 16$; $5\log_5 x - \log_3 y = 6$

b $3\log_a b = 1$; $ab = 16$

c $2m\log_4 16 = n$; $81^m + 3^n = 54$

d $\log_4 x = y = \log_{16}(6x - 9)$

18 A geometric series has first term 6 and common ratio $\frac{2}{3}$. Given that the sum to k terms of the series is greater than 17.8, find the least possible value of k.

19 A rubber ball is dropped from a height of 5 metres. Each time it hits the ground it bounces up to a height of $\frac{7}{8}$ of the height of the previous bounce.

a Determine how far the ball has travelled just before it hits the ground for the third time.

b Write an expression for the distance travelled by the ball when it hits the ground for the kth time.

c Calculate the maximum number of times that the ball hits the ground before the total distance travelled exceeds 39.5 m.

20 Paula is carrying out a series of experiments using increasing amounts of a certain chemical in each attempt. She has 300 g of the chemical to experiment with. In the first experiment she uses 5 g of the chemical and in the second experiment she uses 5.5 g. Given that the amount of chemical used increases according to a geometrical progression, determine how many experiments she is able to carry out.

Euler's number and exponential functions

Investigation 11

1 If €1000 is invested in a bank that offers an interest rate of 2% compounded annually how much is the investment worth after m years?

2 How much is the investment worth after m years if the interest were compounded:

a every six months? **b** monthly?

3 For each of the three cases above evaluate the value of the investment after exactly one year.

4 Investigate the growth of €1 invested for one year at 100% interest, compounded at n different intervals over the year.

a Write down the general formula to obtain the growth of this investment after one year.

b Use technology to draw up a table with the value of the investment for different values of n.

c Plot these values and comment on your results.

In Investigation 11 you saw that the value of $\left(1+\frac{1}{n}\right)^n$ seems to be approaching a limit as the value of n increases. In fact, it can be shown that $\lim_{n\to\infty}\left(1+\frac{1}{n}\right)^n \approx 2.718$ (3 dp).

The proof that this limit exists and finding its value are beyond the scope of this course.

This limit is called Euler's number and is denoted by $e \approx 2.718281828459045...$

TOK

Mathematics is all around us in patterns, shapes, time and space.

What does this tell you about mathematical knowledge?

Investigation 12

1 Sketch the function $f(x) = 2^x$, for $-6 \le x \le 6$, $-6 \le y \le 6$.

2 From your graph comment on the following:

 a stationary points

 b intercepts

 c asymptotes

 d concavity.

3 Using the same scale and axes, sketch the functions $g(x) = a^x$ for different values of $a \in \mathbb{R}^+$.

4 What do the graphs have in common?

5 With reference to the graphs sketched in question **3**, answer the following questions.

 a What do the graphs of $g(x) = a^x$, $a > 1$ have in common?

 b Why do we refer to these graphs as representations of exponential growth?

 c How are the graphs of $g(x) = a^x$, $0 < a < 1$ different? Why do they represent exponential decay?

 d Why do we put the following restrictions on a, $a > 0$, $a \ne 1$?

 e **Conceptual** Compare the graphs of $g(x) = a^x$ and $h(x) = x^a$ for $a \in \{2, 3, 5, 9\}$ to help you answer the question: How does exponential growth compare to polynomial growth?

6 Use technology to copy and complete the table below by graphing the function $g(x) = a^x$ for $a \in \{1, 1.5, 2, 2.5, 3, 3.5, 4\}$ and finding the derivative of each function at $x = 0$.

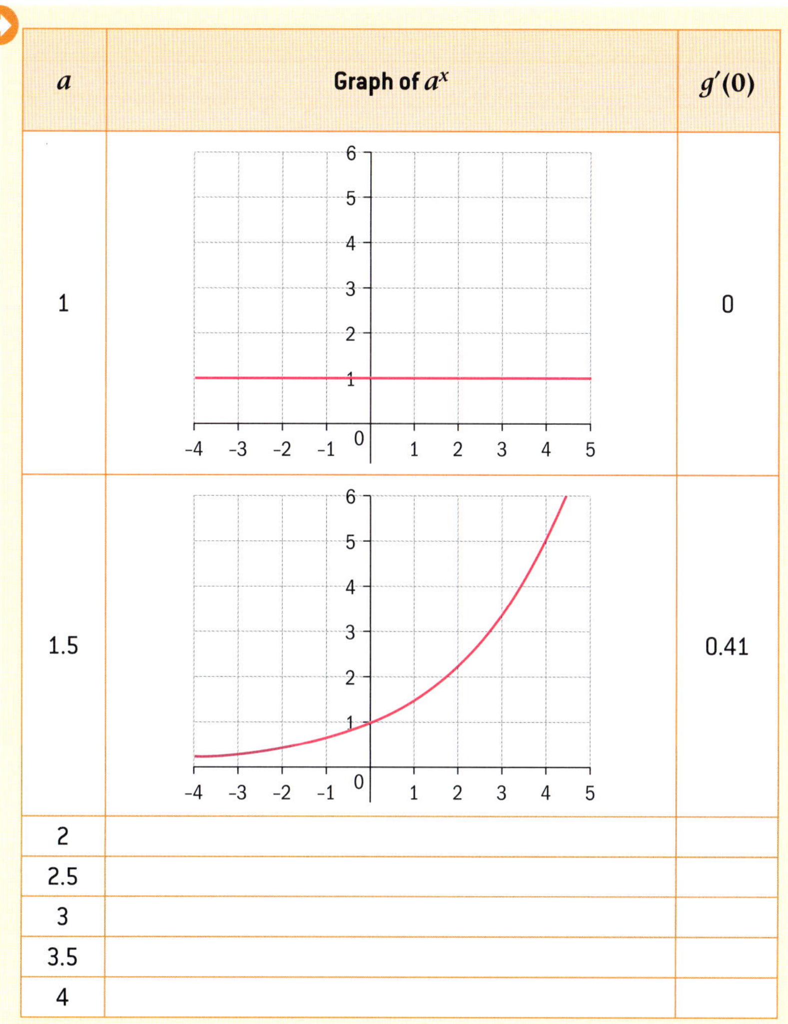

a	Graph of a^x	$g'(0)$
1		0
1.5		0.41
2		
2.5		
3		
3.5		
4		

7 **Conceptual** What happens to the value of $g'(0)$ as a increases?

HINT

In some books you will see the notation $\left.\frac{\mathrm{d}y}{\mathrm{d}x}\right|_{x=0}$ which is equivalent to the notation $g'(0)$

TOK

The phrase "exponential growth" is used popularly to describe a number of phenomena.

Do you think that using mathematical language can distort understanding?

You should now be able to answer questions **1** and **2** of the shrub problem at the beginning of the chapter.

The family of curves $g(x) = ax$ *for* $a \in \mathbb{R}^+$, $a > 1$ are known as **exponential growth functions and** have the following properties:

- They have no stationary points or points of inflection.
- They are increasing.
- The x-axis is a horizontal asymptote.
- There are no vertical asymptotes.
- They are one-to-one functions.
- They all pass through the point $(0, 1)$.
- The gradient of $g(x)$ at $(0, 1)$ increases as the value of a increases.

HINT

A function is injective or one-to-one if it maps every point in its domain to a unique point in its range. A function such as $f(x) = x^2$ is not injective as $f(a) = f(-a) = a^2$; ie it would fail the horizontal test.

It is left as an exercise for you to verify the following properties of exponential decay functions.

The family of curves $g(x) = a^x$ for $a \in \mathbb{R}^+$, $0 < a < 1$ are known as **exponential decay functions** and have the following properties:

- They have no turning points.
- They are decreasing.
- The x-axis is a horizontal asymptote.
- There are no vertical asymptotes.
- They are one-to-one functions.
- They all pass through the point (0, 1).
- The gradient of $g(x)$ at (0, 1) decreases as the value of a increases.

In Investigation 12 you saw that as the family of curves $g(x) = a^x$ for $a \in \mathbb{R}^+$, $a > 1$ moves from $a = 2$ to $a = 3$, the gradient of $g(x)$ at the y-intercept moves from a value less than 1 to a value greater than 1.

It is possible to find a value of a such that $\left.\frac{\mathrm{d}}{\mathrm{d}x}(a^x)\right|_{x=0} = 1$.

In fact, when $a = \mathrm{e}$, then $\left.\frac{\mathrm{d}}{\mathrm{d}x}(\mathrm{e}^x)\right|_{x=0} = 1$. Here is a simplified way of showing this:

Find the derivative of $y = a^x$ from first principles.

$$\frac{\mathrm{d}(a^x)}{\mathrm{d}x} = \lim_{h\to 0}\frac{a^{x+h} - a^x}{h}$$

$$= \lim_{h\to 0}\frac{a^x(a^h - 1)}{h} \qquad \text{since } a^{x+h} = a^x a^h$$

$$= a^x \lim_{h\to 0}\frac{(a^h - 1)}{h} \qquad \text{taking } a^x \text{ outside the limit as it is independent of } h$$

HINT

In a rigorous proof we would need to show that $\lim_{h\to 0}\frac{(a^h - 1)}{h}$ exists, however this is beyond the scope of this course so you can accept that it does exist.

Let $k = \lim_{h\to 0}\frac{(a^h - 1)}{h}$. Then $\frac{\mathrm{d}}{\mathrm{d}x}(a^x) = ka^x$.

This implies that the derivative of any exponential function is proportional to the function itself, and that $\left.\frac{\mathrm{d}}{\mathrm{d}x}(a^x)\right|_{x=0} = k$.

So the number a such that $\left.\frac{\mathrm{d}}{\mathrm{d}x}(a^x)\right|_{x=0} = 1$ is actually the number for which $k = 1$.

Recall that $k = \lim_{h\to 0}\frac{(a^h - 1)}{h}$ so you want to find the value of a which gives $\lim_{h\to 0}\frac{(a^h - 1)}{h} = 1$.

Suppose that $\lim_{h\to 0}\frac{(a^h-1)}{h}=1$. Then for a very small value of h:

$$\frac{(a^h-1)}{h}\approx 1$$

$$\Rightarrow a^h-1\approx h$$

$$\Rightarrow a^h\approx h+1$$

$$\Rightarrow a\approx (h+1)^{\frac{1}{h}}$$

Now let $h=\frac{1}{m}$, so that a small value of h corresponds to a large value of m. Then:

$$a=\lim_{h\to 0}\frac{(a^h-1)}{h}=\lim_{m\to\infty}\left(1+\frac{1}{m}\right)^m=\mathrm{e}$$

HINT

You saw in Investigation 11 that $\lim_{m\to\infty}\left(1+\frac{1}{m}\right)^m=\mathrm{e}$

We say that the exponential function is **invariant under differentiation**, that is, $\frac{\mathrm{d}}{\mathrm{d}x}\left(\mathrm{e}^x\right)=1\times\mathrm{e}^x$.

HINT

Note that $y=a^x$ is called *an* exponential function but $y=\mathrm{e}^x$ is the exponential function.

Investigation 13

1 **Factual** Explain why an exponential function $f(x)=a^x, a>0$ has an inverse.

2 **Factual** Explain how you would obtain the graph of the inverse of $f(x)=a^x$.

3 On the same axes, sketch the graphs of $f(x)=a^x$ and its inverse.

4 Find the equation of the inverse function $f^{-1}(x)$ of the function $f(x)=a^x$. State the domain and range of the inverse.

5 **Conceptual** What do your answers tell you about the relationship between exponential and logarithmic functions?

The inverse of an exponential function $f(x)=a^x$ is $f^{-1}(x)=\log_a x,\ x>0$.

Hence the inverse of $y=10^x$ is given by $y=\log x$.

The inverse of the exponential function $y=\mathrm{e}^x$ is given by $y=\log_{\mathrm{e}} x$, which is often written as $y=\ln x$. This is called the **natural logarithm function**.

HINT

Notation: $\log_{\mathrm{e}} x=\ln x$

Example 25

The table shows two values that lie on the graph of a function of the form $y=ba^x,\ a>0$. Find the values of a and b, and sketch the graph of y.

x	1	3
y	4.5	40.5

Continued on next page

Functions

Calculus

$x = 1 \Rightarrow y = ba = 4.5 \quad (1)$ $x = 3 \Rightarrow y = ba^3 = 40.5 \quad (2)$ Therefore	Form two equations in a and b using the values (1, 4.5) and (3, 40.5) which lie on the graph.
$(2) \div (1) \Rightarrow a^2 = \frac{40.5}{4.5} = 9$ $\Rightarrow a = 3$ since a is bigger than 0 $(1) \Rightarrow ab = 4.5 \Rightarrow b = \frac{4.5}{3} = 1.5$	Solve simultaneously. Here, we divide the two equations.
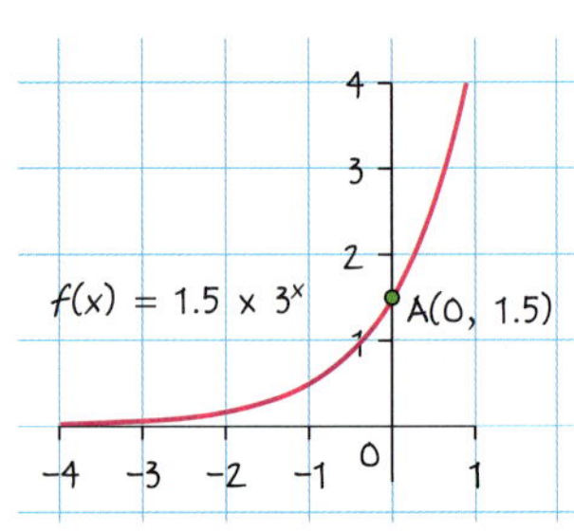	Sketch the graph of $y = 1.5 \times 3^x$. On your sketch, mark the point where the graph crosses the y-axis.

Example 26

The graph of $f(x) = 3^x$ is shown in red.

$g(x)$ is the reflection of $f(x)$ in the y-axis, and $h(x)$ is the reflection of $f(x)$ in the x-axis.

a Find the functions $g(x)$ and $h(x)$.

b On separate axes, sketch the graph of $f(|x|)$

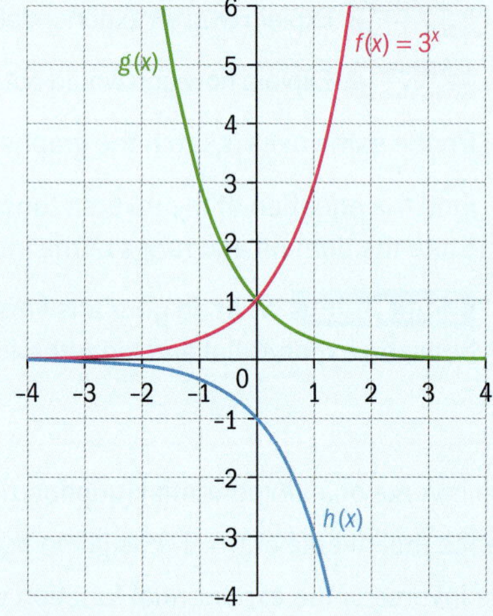

a $g(x) = f(-x)$ $\therefore g(x) = 3^{-x} = (3^{-1})^x = \left(\frac{1}{3}\right)^x$	$g(x)$ is a reflection of $f(x)$ in the y-axis, so all the x-values are reversed.
$h(x) = -f(x) = -(3^x)$	$h(x)$ is a reflection of $f(x)$ in the x-axis, so all the y-values are reversed.

b $f(\lvert x \rvert) = \begin{cases} 3^x, & x \geq 0 \\ 3^{-x}, & x < 0 \end{cases}$ 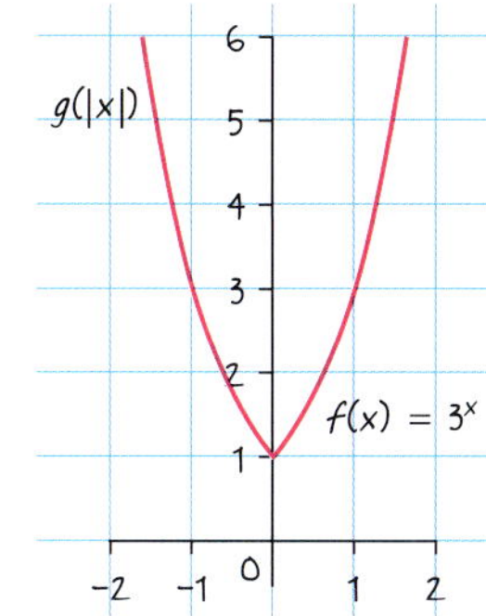	$f(\lvert x \rvert)$ is the graph of $f(x)$ when $x \geq 0$. $f(\lvert x \rvert)$ is the graph of $g(x)$ when $x < 0$.

Example 27

Show that $\log x = (\log \mathrm{e}) \times (\ln x)$.

$\ln x = \log_{\mathrm{e}} x$ $= \dfrac{\log_{10} x}{\log_{10} \mathrm{e}}$ $= \dfrac{\log x}{\log \mathrm{e}}$ $\Rightarrow \log x = (\log \mathrm{e}) \times (\ln x)$	Write the bases in to avoid confusion. Change from base e to base 10.

Example 28

The graph of $y = \log_a x + b$ is shown.

Find b in terms of a.

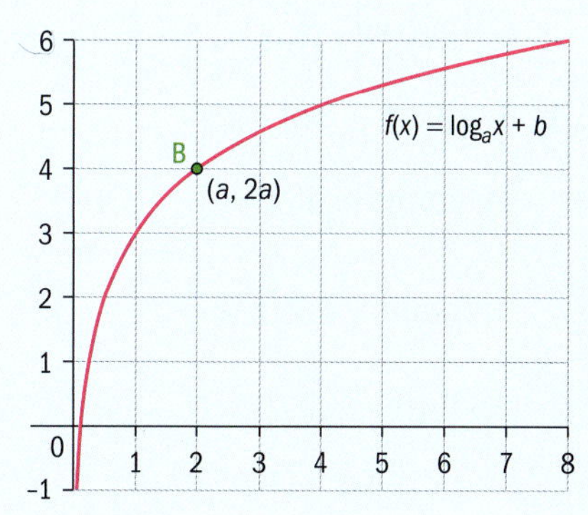

$(a, 2a) \Rightarrow 2a = \log_a a + b$ $\Rightarrow b = 2a - 1$	$(a, 2a)$ is on the graph. $\log_a a = 1$

Functions

Calculus

Example 29

For each of the following functions, state **i** the domain; **ii** any asymptotes; **iii** coordinates of any axes intercepts.

Hence sketch the graphs.

a $y = \log_3(x+1) - 2$ **b** $y = |\ln(1-2x)| + 3$

a Asymptote: $x = -1$ is a vertical asymptote

log 0 is undefined, so there is an asymptote where $x + 1 = 0$, that is at $x = -1$.

Domain: $\{x | x \in \mathbb{R},\ x > -1\}$

The log functions must be applied to positive values.

$$y = 0$$
$$\log_3(x+1) = 2$$

x-intercept occurs at $y = 0$.

$$x + 1 = 3^3$$

Convert to exponent form.

$$x = 8$$

x-intercept $(8, 0)$

$x = 0 \Rightarrow y = \log_3 1 - 2$

y-intercept occurs at $x = 0$.

$y = 0 - 2 = -2$

$\log 1 = 0$ in any base.

y-intercept $(0, 2)$

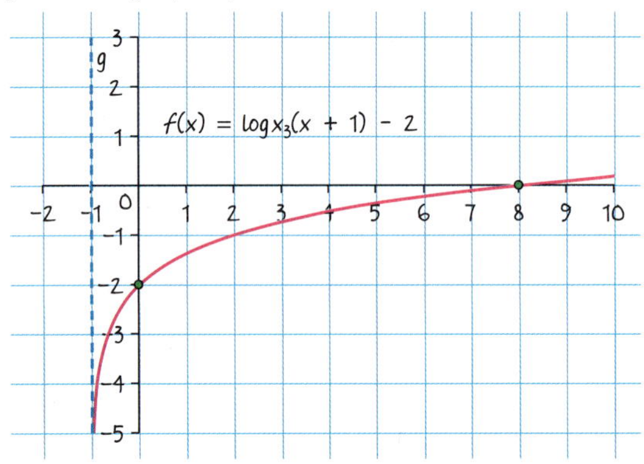

b $y = |\ln(1-2x)| + 3$

Asymptote at $x = \frac{1}{2}$

log 0 is undefined.

Domain $x \in \mathbb{R}, x < \frac{1}{2}$

The log functions must be applied to positive values.

$\Rightarrow |\ln(1-2x)| + 3 = 0$

For *x*-intercept $y = 0$

$|\ln(1-2x)| \neq -3$

$|\ln(1-2x)| \geq 0$ for all *x*.

$\therefore$ no *x*-intercept

$y = |\ln(1)| + 3$

For *y*-intercept $x = 0$

$y = 0 + 3 = 3$

$\log 1 = 0$ in any base.

Therefore *y*-intercept at $(0, 3)$.

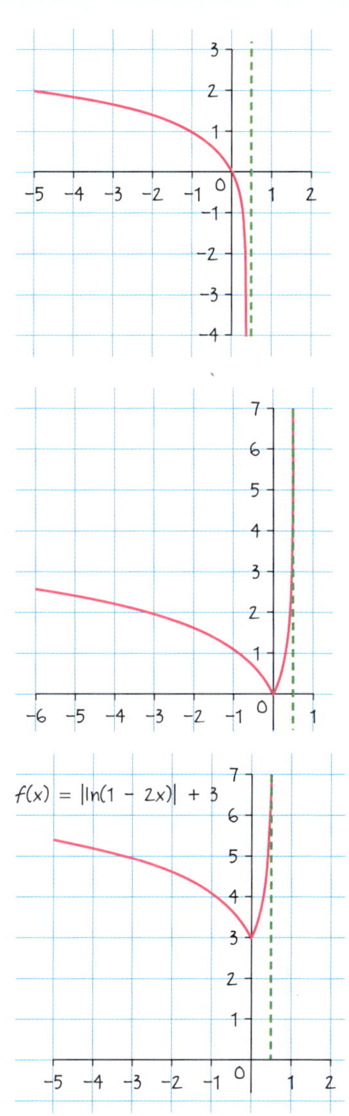

First, sketch the graph of $\ln(1-2x)$.

Reflect the negative portion in the x-axis to give $\ln|(1-2x)|$.

Translate $y = \ln|(1-2x)|$ by 3 units in the positive y direction to get $y = \ln|(1-2x)| + 3$.

Exercise 7E

1 The diagram shows graphs of exponential functions of the form $y = a^x$. For each of the functions determine the value of a.

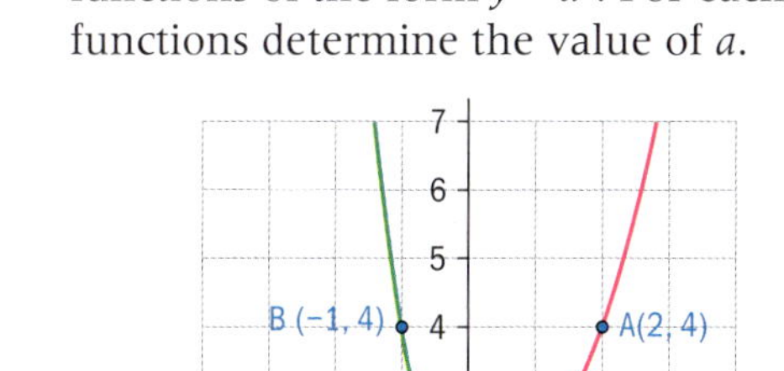

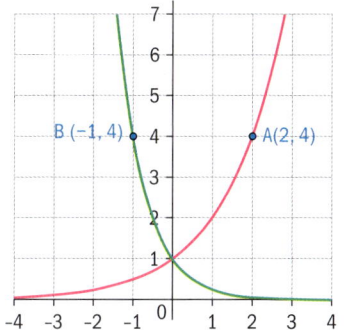

2 Given that $f(x) = ka^x$, $k, a \in \mathbb{R}$, $a > 0$,

a Show that $f(x+1) = af(x)$.

b Find $f(x+2)$ in terms of $f(x)$.

c Find $f(x-1)$ in terms of $f(x)$.

d Make a conjecture for expressing $f(x+n)$, $n \in \mathbb{Z}$ and prove it.

3 On the same set of axes, sketch the graphs of $f(x) = a^x$ and $g(x) = a^{|x|}$. State the domain and range of each function.

4 Solve the equation $e^x + e^{-x} = 2$.

5 The three graphs shown are geometric transformations of the function $f(x) = e^x$. State the transformations and give the new functions in terms of x.

a

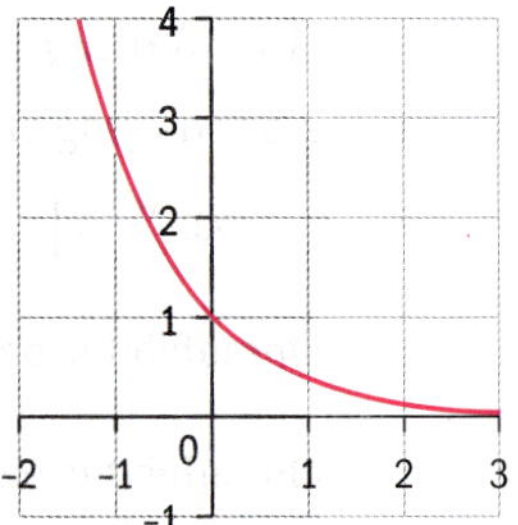

b

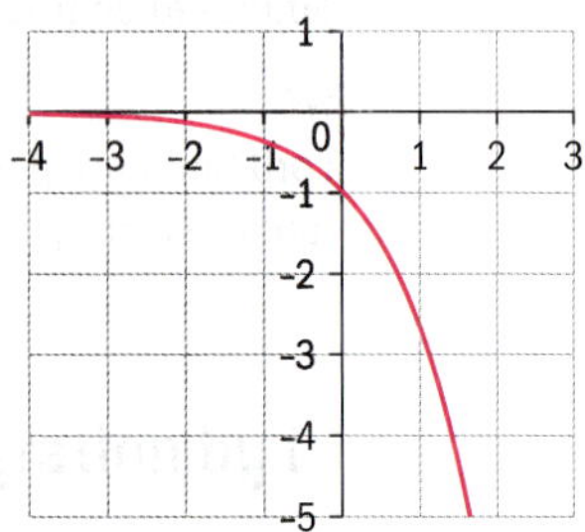

c

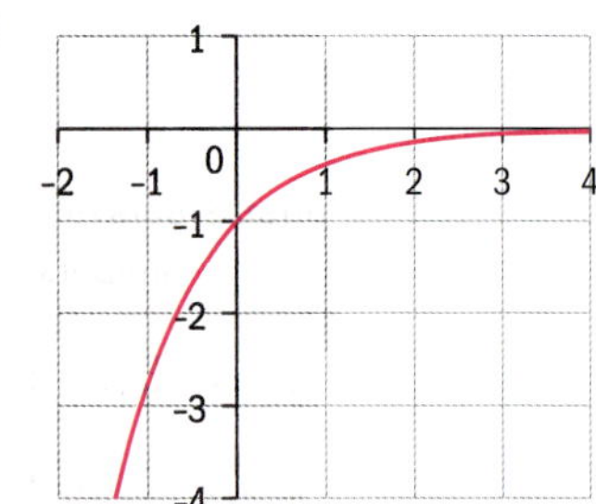

6 On the same set of axes, sketch the graphs of $f(x) = e^x$ and $f^{-1}(x) = \ln x$. State the domain and range of each function.

> **HINT**
>
> The answer to questions **6** and **7** are important key facts.

7 Given that $f(x) = a^x$, find $f^{-1}(x)$ and $f \circ f^{-1}(x)$. Hence show that $a^{\log_a x} = x$.

8 The diagrams show the graphs $f(x)$, $g(x)$ and $h(x)$. Given that each function is a transformation of $y = \ln x$, write each function in terms of x.

a

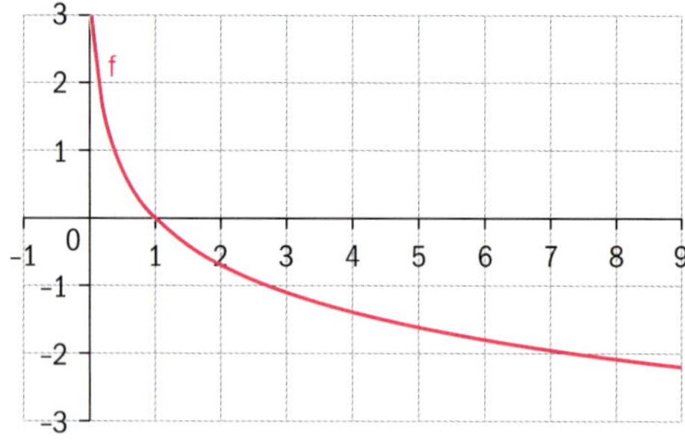

$f(x)$

b

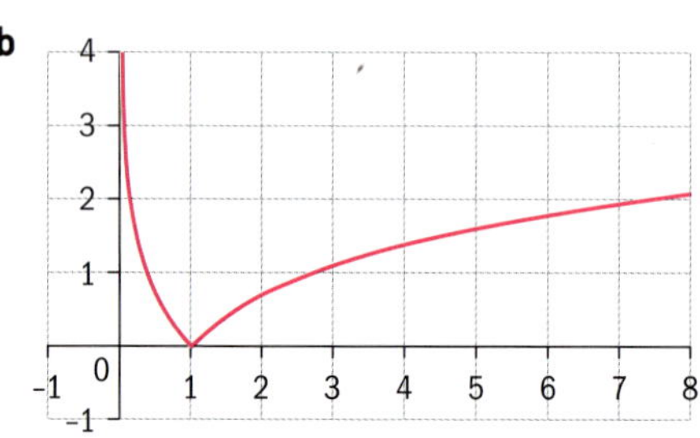

$g(x)$

c

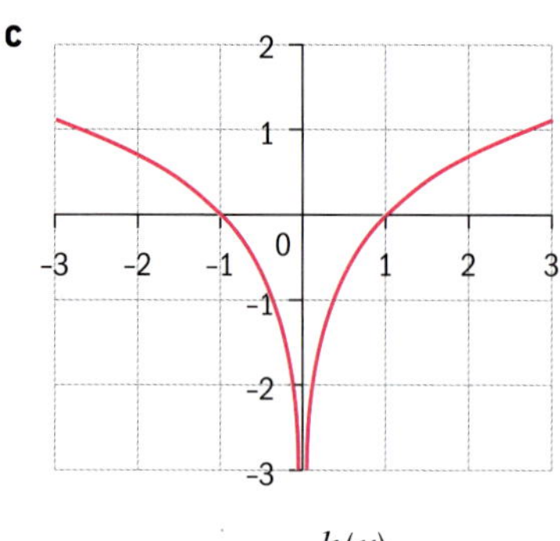

$h(x)$

9 On separate axes, draw the graphs of $y = \log_2(x - 2)$ and $y = (\log_2 x) - 2$. Identify the geometrical transformations of $y = \log_2 x$ that produce these functions.

10 Identify the domain, any asymptotes and axes intercepts of the graphs of the following functions. Hence, sketch their graphs.

a $y = \ln(x + 1) - 2$

b $y = \log_2(4 + 2x) - 5$

7.3 Derivatives of exponential and logarithmic functions; tangents and normals

Investigation 14

In the previous section you have already established that the exponential function is invariant under differentiation; ie $\Rightarrow \frac{\mathrm{d}}{\mathrm{d}x}(\mathrm{e}^x) = \mathrm{e}^x$.

1 Write $y = a^x$ in an equivalent form using a logarithm.

2 Write $\ln y = x$ in an equivalent form using an exponential.

3 Explain why exponential and logarithmic functions are differentiable over their domain.

4 Use your answer to question **2** to find the derivative of the function $y = \ln x$ by using implicit differentiation.

5 Determine the derivative of the function $y = \log_a x$.

6 In section 7.2 you used differentiation from first principles to show that $\frac{\mathrm{d}}{\mathrm{d}x}(a^x) = ka^x$, where $k = \lim_{h\to 0}\frac{a^h - 1}{h}$. Use you result from question **4** to find the value of k.

7 What do your results imply about the derivative of:

a exponential functions at the y-intercept

b the exponential function $y = \mathrm{e}^x$ at the y-intercept?

8 **Conceptual** From what you have learned in this investigation, how is it useful to be able to switch between exponential and logarithmic form?

In Investigation 14 you learned the following important points.

- $\frac{\mathrm{d}}{\mathrm{d}x}(\mathrm{e}^x) = \mathrm{e}^x$
- $\frac{\mathrm{d}}{\mathrm{d}x}(a^x) = (\ln a)a^x$
- $\frac{\mathrm{d}}{\mathrm{d}x}(\ln x) = \frac{1}{x}$
- $\frac{\mathrm{d}}{\mathrm{d}x}(\log_a x) = \frac{1}{x \ln a}$

TOK

"One reason why mathematics enjoys special esteem, above all other sciences, is that its propositions are absolutely certain and indisputable" – Albert Einstein.

How can mathematics, being after all a product of human thought which is independent of experience, be appropriate to the objects of the real world?

Functions

Calculus

Now that you know how to differentiate exponential and logarithmic functions, you can apply all the rules of differentiation to more complicated compound functions as shown in the following examples.

Example 30

Find the x-intercept of the implicitly defined function $e^{\sin y} = x^3$.

Differentiate the function with respect to x and hence find the equation of the normal line which passes through the x-intercept. Give your answer in the form $ax + by = c$.

$y = 0 \Rightarrow e^{\sin 0} = x^3$ $\Rightarrow x^3 = 1$ $\therefore x-\text{intercept } (1, 0)$	On the x-axis, $y = 0$. $x > 0$ since $e^y > 0$ for all values of $y \in \mathbb{R}$.
$e^{\sin y} = x^3 \Rightarrow \cos y\left(e^{\sin y}\right)\frac{dy}{dx} = 3x^2$	Differentiating implicitly with respect to x.
$\Rightarrow \frac{dy}{dx} = \frac{3x^2}{\cos y\left(e^{\sin y}\right)}$ $\Rightarrow \left.\frac{dy}{dx}\right\rvert_{(1,0)} = \frac{3}{1} = 3 = m_T$	Substitute $(1, 0)$ to find the gradient of the tangent m_T.
The gradient of the normal m_N at $(1, 0)$ is $-\frac{1}{3}$. Equation of normal is given by $y = -\frac{1}{3}x + c$ $(1, 0) \quad 0 = -\frac{1}{3} + c \quad c = \frac{1}{3}$ Equation of normal is $3y + x = 1$.	$m_N = -\frac{1}{m_T}$

Example 31

Given that $\sin\frac{y}{2} = \ln(xy)$ show that $\frac{dy}{dx} = \frac{2y}{xy\cos\frac{y}{2} - 2x}$.

$\sin\frac{y}{2} = \ln x + \ln y$	Use law of logs.
$\Rightarrow \frac{1}{2}\cos\frac{y}{2} \times \frac{dy}{dx} = \frac{1}{x} + \frac{1}{y}\frac{dy}{dx}$	Differentiate implicitly.

$\Rightarrow \frac{dy}{dx}\left(\frac{1}{2}\cos\frac{y}{2} - \frac{1}{y}\right) = \frac{1}{x}$ $\Rightarrow \frac{dy}{dx}\left(\frac{y\cos\frac{y}{2} - 2}{2y}\right) = \frac{1}{x}$ $\Rightarrow \frac{dy}{dx} = \frac{2y}{xy\cos\frac{y}{2} - 2x}$	Rearrange and simplify.

Example 32

Given that $y = xe^{3x}$ show that $\frac{d^2y}{dx^2} - 3\frac{dy}{dx} = \frac{3y}{x}$.

Either	
$y = xe^{3x} \Rightarrow \frac{dy}{dx} = 3xe^{3x} + e^{3x}$	Use product rule and chain rule.
$\frac{d^2y}{dx^2} = 9xe^{3x} + 3e^{3x} + 3e^{3x}$ $= 9xe^{3x} + 6e^{3x}$	Find second derivative.
$\frac{d^2y}{dx^2} - 3\frac{dy}{dx} = 9xe^{3x} + 6e^{3x} - 3\left(3xe^{3x} + e^{3x}\right)$ $= 3e^{3x}$ $= \frac{3y}{x}$	Substitute your expressions into $\frac{d^2y}{dx^2} - 3\frac{dy}{dx}$.
Therefore $\frac{d^2y}{dx^2} - 3\frac{dy}{dx} = \frac{3y}{x}$	
Or	
$y = xe^{3x} \Rightarrow \frac{dy}{dx} = 3xe^{3x} + e^{3x}$	Differentiate with respect to x.
$\Rightarrow \frac{dy}{dx} = 3y + e^{3x}$	Substitute y for xe^{3x}.
$\Rightarrow \frac{d^2y}{dx^2} = 3\frac{dy}{dx} + 3e^{3x}$	Differentiate again with respect to x.
$\Rightarrow \frac{d^2y}{dx^2} - 3\frac{dy}{dx} = \frac{3y}{x}$	

Functions

Calculus

Example 33

Given that $f(x) = 3xe^{-x}$ for $-3 \le x \le 3$:

a Show that $f(x)$ has one turning point and one point of inflection.

b Identify the nature of the turning point and state its coordinates.

c Find the coordinates of the point of inflection and hence sketch the graph of $f(x)$.

d Find the equation of the tangent to the curve at the point of inflection.

e Determine the coordinates of the point where this tangent meets the x-axis.

f Calculate the area bounded by this tangent and the x- and y-axes.

a $f(x) = 3xe^{-x} \Rightarrow f'(x) = -3xe^{-x} + 3e^{-x} = 3e^{-x}(1 - x)$

$f''(x) = -3e^{-x} - 3(1 - x)e^{-x} = 3xe^{-x} - 6e^{-x} = 3e^{-x}(x - 2)$

Turning point at $f'(x) = 0 \Rightarrow 3e^{-x}(1 - x) = 0 \Rightarrow x = 1$

Inflection point at $f''(x) = 0 \Rightarrow 3e^{-x}(x - 2) = 0 \Rightarrow x = 2$

x	$x < 2$	$x > 2$
Sign of f''	$-$	$+$
Concavity of f	down	up

Since $f''(2) = 0$ and the function changes concavity at $x = 2$, there is a point of inflection at $x = 2$. (Note: this is not a horizontal point of inflection since $f'(2) \ne 0$.)

b $f(1) = 3e^{-1} = \frac{3}{e} \Rightarrow$ turning point at $\left(1, \frac{3}{e}\right)$

$f''(1) = 3e^{-1}(1-2) < 0 \Rightarrow \left(1, \frac{3}{e}\right)$ is a maximum

c $f(2) = 6e^{-2} \Rightarrow$ point of inflection at $\left(2, \frac{6}{e^2}\right)$

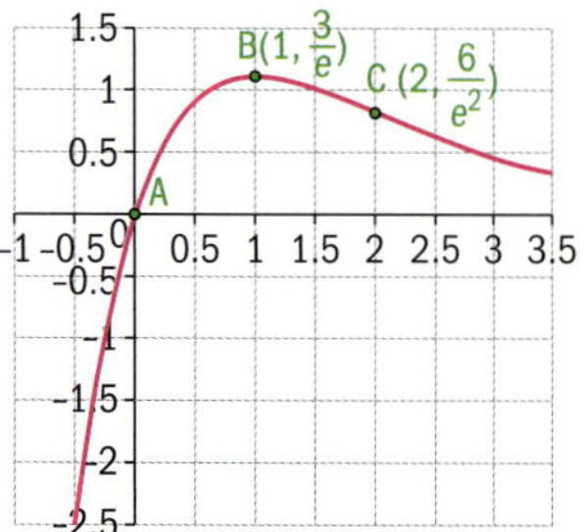

d Gradient of tangent at

$x = 2$ is given by $f'(2) = 3e^{-2}(1-2) = -\frac{3}{e^2}$

$\therefore$ tangent is $y = -\frac{3}{e^2}x + c$

Substituting the point $\left(2, \frac{6}{e^2}\right)$ into the equation of

tangent: $\frac{6}{e^2} = -\frac{3}{e^2} \times 2 + c \Rightarrow c = \frac{12}{e^2}$

Therefore equation of tangent is $y = -\frac{3}{e^2}x + \frac{12}{e^2}$

e $y = 0 \Rightarrow x = 4$

f $x = 0 \Rightarrow y = \frac{12}{e^2}$

Required area of triangle ADE

$= \frac{1}{2} \times 4 \times \frac{12}{e^2} = \frac{24}{e^2}$

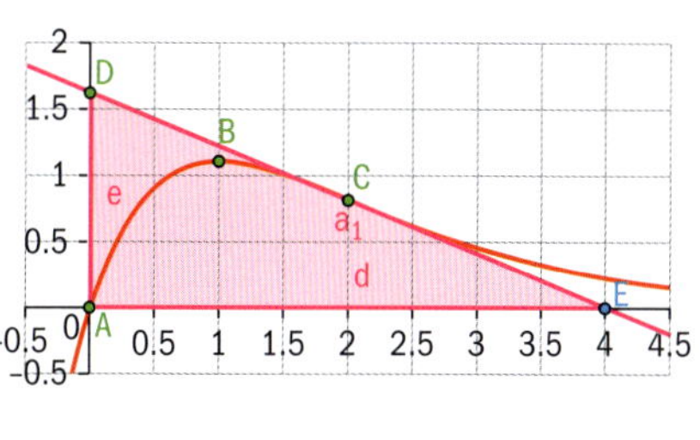

Exercise 7F

1 Differentiate the following functions with respect to x.

a $y = 3e^{5x+4}$ **b** $y = e^{3x^2}$ **c** $y = 5^{4x}$

d $y = e^{\cos x}$ **e** $y = \ln\sqrt{x}$ **f** $y = 3 - 5\ln x$

g $y = \ln(5x + 4)$ **h** $y = \ln(f(x))$

2 Use the product rule to find the derivatives of these functions with respect to x.

a $y = 3xe^{2x+4}$ **b** $y = \sqrt{x}e^{\sqrt[3]{x}}$ **c** $y = x^3 \ln(2x + 1)$

d $y = \frac{3\sin x}{e^x}$ **e** $y = e^{2x}\tan 3x$

3 Use the quotient rule to find the derivatives of these functions with respect to x.

a $y = \frac{\sqrt{x}}{e^{\sqrt{x}}}$ **b** $y = \frac{\sqrt{x}}{\ln x}$ **c** $y = \frac{1-e^x}{1+e^x}$

d $y = \frac{\ln x}{1+x}$ **e** $y = \frac{x+e^x}{e^{-x}}$

4 Find the derivatives of the following functions.

a $y = e^{\sqrt{x}} + \frac{1}{\ln x}$ **b** $y = x^x$

5 Find the derivatives of the following functions at the points indicated.

a $y = \sqrt{1+e^x}$ at the point $\left(0, \sqrt{2}\right)$

b $y = \ln(1 + x)^2$ at the point $\left(\frac{1}{2}, \left(\ln\frac{3}{2}\right)^2\right)$

6 $f(x) = \ln x - x$

a Show that the function has only one turning point.

b Find the coordinates of the turning point and classify its nature.

c State the domain and range of the function.

d Write the equation of any asymptotes.

e Sketch the curve of $f(x) = \ln x - x$.

7 $f(x) = e^x - x$

- **a** Show that the function has only one turning point.
- **b** Find the coordinates of the turning point and classify its nature.
- **c** Explain why the function has no points of inflection.
- **d** State the domain and range of the function.
- **e** Give the equations of any asymptotes to the function.
- **f** Sketch the curve of $f(x) = e^x - x$.

8 $f(x) = \begin{cases} x\ln|x| & \text{for } -2 \le x \le 2, x \ne 0 \\ 0 & \text{for } x = 0 \end{cases}$

- **a** Explain why the function $g(x) = x \ln |x|$ is not defined for $x = 0$.
- **b** Find the x-intercepts of the function $f(x)$.
- **c** Show that the function $f(x)$ has two turning points.
- **d** Identify the nature of the two turning points and find their exact positions.
- **e** Sketch the function $f(x)$.

9 $f(x) = x^2e^x$

- **a** Show that $f(x)$ has two turning points and two points of inflection.
- **b** Locate and identify the nature of the turning points.
- **c** Give the x-coordinates of the points of inflection.
- **d** Give the equations of any asymptotes to the function.
- **e** Sketch the function $f(x) = x^2e^x$.
- **f** Find the equation of the tangent to the curve at the point on the curve where $x = 1$.
- **g** Find the area of the triangle formed by this tangent and the x- and y-axes.

10 $f(x) = e^x \sin x$ for $-\frac{3\pi}{2} < x < \frac{3\pi}{2}$.

- **a** Show that $f(x)$ has two turning points and two points of inflection.
- **b** Identify the nature of the turning points and the value of the x-coordinate at each turning point.
- **c** Find the coordinates of the points of inflection and hence sketch the graph of $f(x)$.
- **d** Show that the normal at one point of inflection is parallel to the tangent at the other point of inflection.

7.4 Integration techniques

Investigation 15

Part A

Consider the function $f(x) = e^x$.

1 Use antidifferentiation to evaluate $\int e^x dx$.

2 Find the value of $A_n = \int_0^{na} e^x dx$.

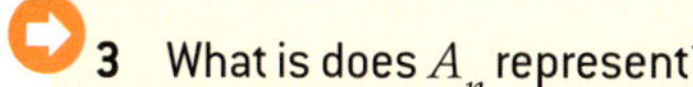

3 What is does A_n represent?

4 Let $I_n = \int_{(n-1)a}^{na} e^x dx$. Copy and complete the following table.

n	$I_n = \int_{(n-1)a}^{na} e^x dx$
1	
2	
3	
4	
5	

5 Show that the integrals $I_n = \int_{(n-1)a}^{na} e^x dx$ form a geometric sequence and state the first term and the common ratio.

6 Evaluate $\sum_{r=1}^{n} I_r$.

7 Write a general comment to summarize your results from part A.

Part B

8 Use antidifferentiation to find $\int \frac{1}{x} dx$. Comment on any limitations.

9 Find the value of $B_n = \int_1^{na} \frac{1}{x} dx,\ n \in \mathbb{Z}^+, n > 1$.

10 What is does B_n represent?

11 Let $H_1 = \int_1^a \frac{1}{x} dx$ and $H_n = \int_{(n-1)a}^{na} e^x dx$ for $a \in \mathbb{R}^+,\ n \in \mathbb{Z}^+, n > 1$.

Copy and complete the following table:

n	H_n
1	$H_1 = \int_1^a \frac{1}{x} dx =$
2	
3	
4	
5	

12 Find the value of $\sum_{r=1}^{n} H_r$.

13 Write a general comment to summarize your results from part B.

14 **Factual** What two areas of mathematics have you combined in answering this investigation?

15 **Conceptual** Why are different areas of mathematics combined in some problems?

Functions

Calculus

Investigation 16

1 Find the derivative of:

a $y = e^{ax+b}$

b $y = \ln(ax+b)$

2 Use your answers to question **1** to find the following indefinite integrals.

a $\int e^{ax+b}\,dx$ **b** $\int \frac{1}{(ax+b)}\,dx$

3 Use antidifferentiation to determine $\int a^x\,dx$.

4 Use antidifferentiation to determine $\int \frac{f'(x)}{f(x)}\,dx$. (Hint: Think about the composite function which would be differentiated to give this form.)

5 **Conceptual** What techniques have you used in this investigation to help you find the integrals of composite exponential functions and reciprocal functions?

Integration by inspection

Investigation 16 will have shown you that the following results are true.

- To evaluate definite integrals, apply the fundamental theorem of calculus:

$$\int f(x)\,dx = F(c) + c \Rightarrow \int_a^b f(x)\,dx = F(b) - F(a)$$

- $\int e^x dx = e^x + c$
- $\int \frac{1}{x} dx = \ln|x| + c$
- $\int e^{ax+b}\,dx = \frac{1}{a}e^{ax+b} + c$
- $\int \frac{1}{(ax+b)}\,dx = \frac{1}{a}\ln|ax+b| + c$
- $\int a^x dx = \frac{a^x}{\ln a} + c$
- $\int \frac{f'(x)}{f(x)}\,dx = \ln(f(x)) + c$

Earlier in this chapter you also obtained the following results:

- $\int (ax+b)^n \, dx = \frac{(ax+b)^{n+1}}{a(n+1)} + c, \quad c, n \in \mathbb{R}$
- $\int \sin(ax+b) \, dx = -\frac{\cos(ax+b)}{a} + c, \; c \in \mathbb{R}$
- $\int \cos(ax+b) \, dx = \frac{\sin(ax+b)}{a} + c, \; c \in \mathbb{R}$
- $\int \sec^2(ax+b) \, dx = \frac{\tan(ax+b)}{a} + c, \; c \in \mathbb{R}$

We can now generalize these results as **the reverse chain rule**.

You should recall from Chapter 4 how to use the chain rule to differentiate a composite function:

If $y = f(g(x))$ then $y = f(u)$ where $u = g(x)$ and

$\frac{dy}{dx} = \frac{dy}{du} \times \frac{du}{dx} = f'(g(x)) \times g'(x)$.

When the function that you want to integrate has the form of a derivative of a composite function $f'(g(x)) \times g'(x)$, you can integrate it by using the chain rule in reverse as follows.

$$\frac{d}{dx}\left(f\left(g(x)\right)\right) = f'\left(g(x)\right) \times g'(x) \Rightarrow \int f'\left(g(x)\right) \times g'(x) \, dx = f\left(g(x)\right) + c$$

In the key points which followed Investigation 16, the "inner" function, $g(x)$, of the composite function was a linear function of the form $g(x) = ax + b$. Hence, $g'(x)$ was simply a.

The following examples will help you understand how to integrate a composite function by the inverse chain rule. This process is often called "integrating by inspection", since you "inspect" the integrand to find suitable functions f and g for which the integrand is equal to $f'(g(x)) \times g'(x)$. You can check your integration is correct by differentiating your answer.

A special case of the inverse chain rule is when you are asked to integrate an expression of the form $\frac{f'(x)}{f(x)}$.

You know that $\frac{d}{dx}(\ln x) = \frac{1}{x}$.

Hence, by the chain rule, $\frac{d}{dx}\left(\ln f(x)\right) = \frac{f'(x)}{f(x)}$.

Therefore $\int \frac{f'(x)}{f(x)} \, dx = \ln\left(f(x)\right) + c$.

HINT

$f'(g(x))$ is another way of writing $\frac{dy}{du}$ and $g'(x)$ is another way of writing $\frac{du}{dx}$.

TOK

Where does mathematics come from?

Galileo said that the universe is a grand book written in the language of mathematics.

Does it start in our brains or is it part of the universe?

Example 34

Find the following integrals by inspection.

a $\int 12(3x+5)^3\,dx$ **b** $\int \sin\frac{\pi x}{2}\,dx$ **c** $\int x e^{x^2+3}\,dx$

a $\int 12(3x+5)^3\,dx = (3x+5)^4 + c$	You can see that if $f(g(x)) = (3x+4)^4$ then $f'(g(x)) = 4(3x+4)^3$ and $g'(x) = 3$ so $f'(g(x)) \times g'(x) = 12(3x+5)^3$.
b $\int \sin\frac{\pi x}{2}\,dx = -\frac{2}{\pi}\cos\frac{\pi x}{2} + c$	By inspection, letting $f(u) = -\frac{2}{\pi}\cos u$ and $g(x) = \frac{\pi}{2}x$. Then $f(g(x)) = -\frac{2}{\pi}\cos\frac{\pi}{2}x$ and $f'(g(x)) \times g'(x) = \left(-\frac{2}{\pi}\right) \times -\frac{\pi}{2}\sin\frac{\pi}{2}x = \sin\frac{\pi}{2}x$.
c $\int x e^{x^2+3}\,dx = \frac{1}{2}\int 2x e^{x^2+3}\,dx = \frac{1}{2}e^{x^2+3} + c$	By inspection, if you let $f(u) = e^u$ and $g(x) = x^2 + 3$ then $f(g(x)) = e^{x^2+3} \Rightarrow f'(x) = 2xe^{x^2+3}$ so multiply the integrand by $\frac{1}{2}$.

Example 35

Find the value of these definite integrals.

a $\int_{-\frac{\pi}{4}}^{\frac{\pi}{4}} \left(1 - 2\sin 2x + \frac{x^2}{2}\right)dx$ **b** $\int_0^{\frac{\pi}{3}} \left(16e^{4x} + 2\sin 3x\right)dx$

a $\int_{-\frac{\pi}{4}}^{\frac{\pi}{4}} \left(1 - 2\sin 2x + \frac{x^2}{2}\right)dx$	
$= \left[x + \cos 2x + \frac{x^3}{6}\right]_{-\frac{\pi}{4}}^{\frac{\pi}{4}}$	Integrate term by term.
$= \left(\frac{\pi}{4} + \cos\frac{\pi}{2} + \frac{\left(\frac{\pi}{4}\right)^3}{6}\right) - \left(-\frac{\pi}{4} + \cos\left(-\frac{\pi}{2}\right) + \frac{\left(-\frac{\pi}{4}\right)^3}{6}\right)$	Substitute the upper and lower limits into the integrated expression.
$= \frac{\pi}{2} + \frac{\pi^3}{192}$	

b $\int_0^{\frac{\pi}{3}}\left(16e^{4x}+2\sin 3x\right)dx$	Integrate term by term by inspection.
$=\left[4e^{4x}-\frac{2}{3}\cos 3x\right]_0^{\frac{\pi}{3}}$	
$=\left(4e^{\frac{4\pi}{3}}-\frac{2}{3}\cos\pi\right)-\left(4e^{0}-\frac{2}{3}\cos 0\right)$	Substitute the upper and lower limits into the integrated expression.
$=4e^{\frac{4\pi}{3}}+\frac{2}{3}-1+\frac{2}{3}=4e^{\frac{4\pi}{3}}+\frac{1}{3}$	Simplify.

Check your results by finding the derivative of your answers.

Example 36

Work out these integrals.

a $\int 10x\left(5x^2+3\right)^3 dx$ **b** $\int \frac{1}{2\sqrt{x}}\sec^2\sqrt{x}\,dx$ **c** $\int \sin x\cos^4 x\,dx$

a $\int 10x\left(5x^2+3\right)^3 dx$ $=\frac{\left(5x^2+3\right)^4}{4}+c$	By inspection, suppose $f(u)=\frac{1}{4}u^4$ and $g(x)=5x^2+3$. Then $10x(5x^2+3)^3=f'(g(x))\times g'(x)$ So by the inverse chain rule, $\int f'(g(x))\times g'(x)\,dx=f(g(x))$.
b $\int \frac{1}{2\sqrt{x}}\sec^2\sqrt{x}\,dx$ $=\tan\sqrt{x}+c$	Let $f(u)=\tan u$ and $g(x)=x^{\frac{1}{2}}$. Then $\frac{1}{2\sqrt{x}}\sec^2\sqrt{x}=f'(g(x))\times g'(x)$ $f'(g(x))=\sec^2\sqrt{x}$.
c $\int \sin x\cos^4 x\,dx$ $=-\frac{1}{5}\int -5\sin x\cos^4 x\,dx$ $=-\frac{1}{5}\cos^5 x+c$	Let $f(u)=u^5$ and $g(x)=\cos x$ Then $f'(g(x))\times g'(x)=\left(5\cos^4 x\right)\times(-\sin x)$ $=-5\sin x\cos^4 x$ $f'(g(x))=(\cos x)^4$ and $g(x)=\cos x\Rightarrow g'(x)=-\sin x$

Example 37

Find the function $f(x)$ that satisfies the conditions, $f''(x) = e^x + \sin 3x$, $f'(0) = f(0) = 1$.

$f''(x) = e^x + \sin 3x$ $\Rightarrow f'(x) = e^x - \frac{1}{3}\cos 3x + c$	Integrate the second derivative function to obtain the first derivative.
$\Rightarrow f'(0) = e^0 - \frac{1}{3}\cos 0 + c = 1$ $\Rightarrow 1 - \frac{1}{3} + c = 1$ $\Rightarrow c = \frac{1}{3}$	Substitute $x = 0$ to obtain a value for c.
$f'(x) = e^x - \frac{1}{3}\cos 3x + \frac{1}{3}$ $\Rightarrow f(x) = e^x - \frac{1}{9}\sin 3x + \frac{x}{3} + c_1$	Integrate the first derivative to find $f(x)$.
$\Rightarrow f(0) = e^0 - \frac{1}{9}\sin 0 + \frac{0}{3} + c_1 = 1$ $\Rightarrow 1 - 0 + 0 + c_1 = 1$ $\Rightarrow c_1 = 0$ $\Rightarrow f(x) = e^x - \frac{1}{9}\sin 3x + \frac{x}{3}$	Use the condition given for $f(x)$ to find the value of c_1.

Example 38

Integrate the following expressions.

a $\int\left(\frac{x}{x^2-1}\right)dx$ **b** $\int\left(\frac{x+1}{x^2+2x+1}\right)dx$ **c** $\int \tan x \, dx$

a $\int\left(\frac{x}{x^2-1}\right)dx = \frac{1}{2}\int\left(\frac{2x}{x^2-1}\right)dx$ $= \ln\lvert x^2 - 1\rvert + c$	Note that if you multiply the numerator by 2 it becomes the derivative of the denominator, so you can use $\int\frac{f'(x)}{f(x)}dx = \ln(f(x)) + c$.
b $\int\left(\frac{x+1}{x^2+2x+1}\right)dx = \frac{1}{2}\int\left(\frac{2x+2}{x^2+2x+1}\right)dx$ $= \ln(x+1)^2 + c$	Again, use $\int\frac{f'(x)}{f(x)}dx = \ln(f(x)) + c$. The denominator is a perfect square so there is no need for absolute value.

c $\int \tan x \, dx = \int \left(\frac{\sin x}{\cos x}\right) dx$	Write $\tan x$ as $\frac{\sin x}{\cos x}$.
$= -\int \left(\frac{-\sin x}{\cos x}\right) dx$	The derivative of $\cos x$ is $-\sin x$.
$= -\ln\lvert\cos x\rvert$	Use $\int \frac{f'(x)}{f(x)} dx = \ln(f(x)) + c$.
$= \ln\left\lvert\left(\frac{1}{\cos x}\right)\right\rvert + c$	Use rules of logarithms.
$= \ln\lvert\sec x\rvert + c$	$\sec x = \frac{1}{\cos x}$

Exercise 7G

1 Find the following integrals.

a $\int (x^3 - \sec^2 x) dx$ **b** $\int \left(3e^x + \frac{1}{2x} + \sin x\right) dx$

c $\int 2\sin x \cos x \, dx$ **d** $\int (\tan^2 3x + 1) dx$

e $\int \left(\frac{3}{2x} + 3^x\right) dx$ **f** $\int \left(\frac{3}{1-3x}\right) dx$

g $\int \left(\frac{x}{x-1} + \cos x\right) dx$

h $\int \left((\cos x)e^{\sin x} - \frac{2}{x}\right) dx$

i $\int \left(\frac{e^{\sqrt{x}}}{2\sqrt{x}} - 4\sin x \cos x\right) dx$

2 Find $f(x)$, given the extra conditions.

a $f'(x) = (2x - \sin 4x)$, $f(0) = 0$

b $f'(x) = x^2 + \frac{1}{e^x} + \sec^2 x$, $f(0) = -1$

c $f'(x) = \frac{3}{2x-5} - 3x^2 - 3e^{x-3}$, $f(3) = -25$

3 Find $f(x)$, given the extra condition.

a $f''(x) = 2\cos^2 x - 1$, $f'\left(\frac{\pi}{2}\right) = 1$, $f(0) = \frac{3}{4}$

b $f''(x) = e^{2x-1} + \sin(1-2x)$,

$f'\left(\frac{1}{2}\right) = f\left(\frac{1}{2}\right) = \frac{1}{2}$

4 Evaluate the following definite integrals and check your answers using technology.

a $\int_{-\frac{\pi}{4}}^{\frac{\pi}{4}} (x - \sec^2 x) dx$ **b** $\int_0^{\frac{\pi}{3}} (e^{3x-2} + \sin x) dx$

c $\int_1^2 \left(\frac{1}{2^x} + x^2 - e^{2x}\right) dx$ **d** $\int_0^1 \left(3^x - \frac{6}{1-3x}\right) dx$

e $\int_0^{\frac{\pi}{2}} \left(\sec^2\left(\frac{x}{2}\right) + xe^{x^2}\right) dx$ **f** $\int_2^4 \frac{2x}{(x-1)} dx$

g $\int_{\frac{\pi}{6}}^{\frac{\pi}{3}} \left(\frac{\cos 6x}{\cos 3x + \sin 3x}\right) dx$ **h** $\int_0^{\frac{\pi}{4}} \left(\frac{\sin 2x}{\cos x}\right) dx$

i $\int_0^{\frac{\pi}{6}} (2\sin^2 x \cot x) dx$ **j** $\int_0^{\frac{1}{2}} \left(2^x + \frac{2}{1-x}\right) dx$

5 Given that $f'(x) = \tan x$ and $f(x)$ is defined on the interval $\left[0, \frac{\pi}{2}\right]$ with $f(0) = 0$, find $f(x)$ and state its range.

Find the equations of the tangent and normal to $f(x)$ at $x = \frac{\pi}{4}$.

Show that the area of the triangle formed by these two lines and the y-axis is $\frac{\pi^2}{16}$.

Functions

Calculus

6 Reduce $y = \dfrac{x+9}{2x^2+x-3}$ into partial fractions. Hence find $\displaystyle\int \frac{x+9}{2x^2+x-3}\,dx$.

7 Use partial fractions to find $\displaystyle\int \left(\frac{1}{x^2-1}\right) dx$.

8 Express $\dfrac{5x+9}{x^2-9}$ as partial fractions. Hence, find $\displaystyle\int \left(\frac{5x+9}{x^2-9}\right) dx$.

9 Use partial fractions to show that the area bounded by the curve $y = \dfrac{1-2x}{x+x^2}$, the x-axis, and the lines $x = \dfrac{1}{2}$ and $x = 1$ is equal to $\ln\left(\dfrac{27}{32}\right)$.

10 Show that $\displaystyle\int_0^{\frac{1}{2}} \left(\frac{2+3x-x^2}{1-x^2}\right) dx = \frac{1}{2} + \ln\left(\frac{8}{3}\right)$.

Integration by substitution

Sometimes it's too complicated to integrate a composite function by inspection using the inverse chain rule. In such cases, you will need another technique. One of these techniques is to use a substitution. Here, you'll study integration by substitution using one of the results that you have already obtained by inspection.

You saw previously that $\displaystyle\int (5x-4)^7 dx = \frac{1}{40}(5x-4)^8 + c$.

The benefit of using a substitution is that it helps to simplify the function that you are trying to integrate. In this case, you would proceed as follows.

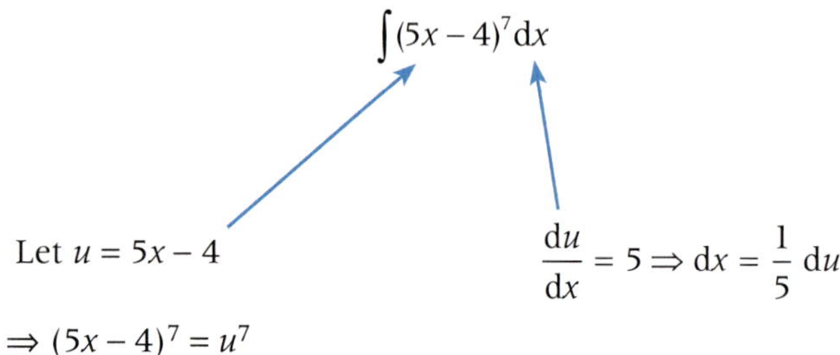

$\int (5x-4)^7 dx$
Let $u = 5x - 4$
$\dfrac{du}{dx} = 5 \Rightarrow dx = \dfrac{1}{5}\,du$
$\Rightarrow (5x-4)^7 = u^7$

HINT

It is important to know that when you substitute for dx, although it looks like you can cross multiply, there is more into it than this. It is beyond the course to explain why this can be done so you will have to do this bearing in mind that it is not really multiplication or division.

You cannot integrate a function of u with respect to x so you need to find $\dfrac{du}{dx}$ and hence substitute dx with an expression involving du.

$\displaystyle\int (5x-4)^7 dx = \int u^7 \times \frac{1}{5}\,du$ Substitute x for u

$\displaystyle = \frac{1}{5}\int u^7\,du$

$\displaystyle = \frac{1}{40}u^8 + c$ Integrate with respect to u

Now that integration has been done you can substitute x back into the integrated expression. If the initial integral is in terms of x, your answer must also be in terms of x.

$$\int (5x-4)^7 \mathrm{d}x = \frac{1}{40}(5x-4)^8 + c$$

Example 39

Find the following integrals using an appropriate substitution.

a $\int 3(6x-1)\mathrm{e}^{3x^2-x}\mathrm{d}x$

b $\int \cot x \, \mathrm{d}x$

a Let $u = 3x^2 - x$.	By choosing this substitution you simplify the given function.
Then $\frac{\mathrm{d}u}{\mathrm{d}x} = 6x - 1 \Rightarrow (6x-1)\,\mathrm{d}x = \mathrm{d}u$	Find du in terms of dx.
$\int 3(6x-1)\mathrm{e}^{3x^2-x}\mathrm{d}x$ $= 3\int \mathrm{e}^{3x^2-x}(6x-1)\mathrm{d}x$ $= 3\int \mathrm{e}^u \mathrm{d}u$	Substitute $u = 3x^2 - x$ and $(6x-1)\mathrm{d}x = \mathrm{d}u$.
$= 3\mathrm{e}^u + c$	Integrate the simplified function with respect to u
$= 3\mathrm{e}^{3x^2-x} + c$	Substitute back for x.
b $\int \cot x \, \mathrm{d}x = \int \frac{\cos x}{\sin x}\mathrm{d}x$	Rewrite in terms of sine and cosine.
Let $u = \sin x$. Then $\frac{\mathrm{d}u}{\mathrm{d}x} = \cos x \Rightarrow \mathrm{d}u = \cos x \, \mathrm{d}x$.	When you have to integrate a quotient, you should generally let u be the denominator. Integrating this will give an expression in ln of the denominator.
$\int \cot x \, \mathrm{d}x = \int \frac{\cos x}{\sin x}\mathrm{d}x$ $= \int \frac{1}{\sin x} \times \cos x \mathrm{d}x$	Use the substitution to find the integral.
$= \int \frac{1}{u}\mathrm{d}u$	Substitute $u = \sin x$ and $\mathrm{d}u = \cos x \, \mathrm{d}x$.
$= \ln\lvert u \rvert + c$	Integrate with respect to u.
$= \ln\lvert \sin x \rvert + c$	Substitute back for x.

You could also find the two integrals in Example 39 by inspection.

In part **a**, $f(u) = e^u$ and $g(x) = 3x^2 - 2$. Then

$$\begin{aligned}\int e^{3x^2-x}(6x-1)\,dx &= \int f'(g(x)) \times g'(x)\,dx \\ &= f(g(x)) + c \\ &= e^{3x^2-x} + c\end{aligned}$$

In part **b**, letting $f(x) = \sin x$ gives $\int \frac{\cos x}{\sin x}dx = \int \frac{f'(x)}{f(x)}dx$

$$\begin{aligned} &= \ln(f(x)) + c \\ &= \ln|\sin x| + c\end{aligned}$$

However, when it's very difficult to find suitable functions f and g for which the integrand is $f'(g(x)) \times g'(x)$, then integration by substitution is the best way to proceed.

TOK

Is mathematics invented or discovered?

Example 40

Find $\int x\sqrt{x+1}\,dx$.

Let $u = x + 1 \Rightarrow x = u - 1$ $\Rightarrow dx = du$	This substitution is chosen to avoid a composite function involving square root.
$\int x\sqrt{x+1}\,dx = \int (u-1) \times u^{\frac{1}{2}}du$ $= \int \left(u^{\frac{3}{2}} - u^{\frac{1}{2}}\right)du$	Substitute for u and simplify.
$= \frac{2}{5}u^{\frac{5}{2}} - \frac{2}{3}u^{\frac{3}{2}} + c$	Integrate term by term.
$= 2u^{\frac{3}{2}}\left(\frac{1}{5}u - \frac{1}{3}\right) + c$	Simplify by factorizing.
$= 2(x+1)^{\frac{3}{2}}\left(\frac{(x+1)}{5} - \frac{1}{3}\right) + c$	Substitute back for x.
$= 2(x+1)^{\frac{3}{2}}\left(\frac{3x+3-5}{15}\right) + c$ $= \frac{2}{15}(x+1)^{\frac{3}{2}}(3x-2) + c$	Simplify.

The next example is a standard integral which is given in the data booklet. However, it is important to understand how substitution is used to simplify the integral.

Example 41

Find $\int \frac{1}{\sqrt{1-x^2}}\,dx$.

Let $x = \sin u \Rightarrow 1 = \cos u \frac{du}{dx}$	Differentiate implicitly with respect to x.
$\Rightarrow dx = \cos u\, du$	
$\int \frac{1}{\sqrt{1-x^2}}\,dx = \int \frac{1}{\sqrt{1-\sin^2 u}} \cos u\, du$	Substitute for x and dx.
$= \int \frac{1}{\cos u} \cos u\, du$	Use $\sin^2 u + \cos^2 u = 1$.
$= \int du$	Simplify and integrate.
$= u + c$	Substitute back for x.
$= \arcsin x + c$	$x = \sin u \Rightarrow u = \arcsin x$

Example 42

Find $\int 3^x\,dx$.

$3^x = \left(e^{\ln 3}\right)^x = e^{(\ln 3)x}$	Remember that $(e^{\ln k}) = k$.
$\int 3^x\,dx = \int e^{(\ln 3)x}\,dx$	
Let $u = (\ln 3)x \Rightarrow du = \ln 3\,dx$	Use the substitution $u = (\ln 3)x$.
$\int 3^x\,dx = \frac{1}{\ln 3}\int e^u\,du$	Write the integral in terms of u.
$= \frac{e^u}{\ln 3} + c$	Integrate.
$= \frac{e^{(\ln 3)x}}{\ln 3} + c$	Substitute back to x.
$= \frac{3^x}{\ln 3} + c$	Rewrite $(e^{\ln 3})$as 3.

Exercise 7H

Find the integrals using an appropriate substitution.

1 $\int 6x\sqrt{3x^2+4}\,dx$ **2** $\int 3x^2\cos x^3\,dx$

3 $\int (1-2x)e^{2+x-x^2}\,dx$ **4** $\int 2\sin 2x e^{\cos 2x}\,dx$

5 $\int 3^x \ln 3 \sin 3^x\,dx$ **6** $\int \frac{3}{2}\sqrt{x}\,e^{\sqrt{x^3}}\,dx$

7 $\int 2x\sqrt{x-1}\,dx$ **8** $\int (1-x)\sqrt{1+x}\,dx$

9 $\int \left(x^2+x\right)\sec^2\left(x^3-\frac{3}{2}x^2\right)dx$

10 $\int \sin 2x\left(2^{\cos 2x}\right)dx$

11 $\int x\sqrt{1-x^2}\,dx$

16 $\int e^{5-x}\sqrt{e^{5-x}}\,dx$

17 $\int\left(\frac{x}{3-x}\right)^2 dx$

12 $\int x^2\sqrt{1-x}\,dx$

13 $\int \frac{x^2}{\sqrt{1-x}}dx$

18 $\int(1-x)\sqrt{2x-3}\,dx$

19 $\int \frac{3(x-4)}{(1-x)^2}dx$

14 $\int x(1+x)^5 dx$

15 $\int\left(\frac{\sin x}{1-\cos x}\right)dx$

20 $\int\left(\frac{\sec^2 x}{2+\tan x}\right)dx$

Investigation 17

1 Use an appropriate u-substitution to work out the indefinite integral $I = \int \frac{1}{\sqrt{4-x^2}}dx.$

2 Hence find the value of the definite integral $I_D = \int_0^1 \frac{1}{\sqrt{4-x^2}}dx.$

3 What is the value of the substituted variable u at the lower and upper bounds of x?

4 Re-write I_D, including the upper and lower bounds, purely in terms of the variable u. Find the value of this definite integral.

5 What do you notice about your answers to question **2** and question **4**?

6 Repeat questions **1** to **5** for $I = \int \frac{1}{\sqrt{4-9x^2}}dx$ and $I_D = \int_0^{\frac{1}{3}} \frac{1}{\sqrt{4-9x^2}}dx.$

7 Now repeat questions **1** to **5** for the general integrals

$I = \int \frac{1}{\sqrt{a^2-b^2x^2}}dx$ and $I_D = \int_0^{\frac{a}{2b}} \frac{1}{\sqrt{a^2-b^2x^2}}dx$

8 Find an appropriate u-substitution for the indefinite integral $I = \int \frac{1}{a^2+b^2x^2}dx$ and find I in terms of x. (Hint: Remember the identity $1+\tan^2 x = \sec^2 x$.)

9 How could you rearrange the denominator in the integrals $\int \frac{1}{x^2-2x+5}dx$ in order to find an appropriate trigonometric u-substitution?

10 Evaluate $\int_3^{1+2\sqrt{3}} \frac{1}{x^2-2x+5}dx$

11 **Conceptual** How does manipulating the form in which an integral is presented allow you to integrate more easily?

HINT

Refer to Example 41 to find an appropriate substitution.

HINT

Think of the identity $1+\tan^2 x = \sec^2 x$.

Definite integrals and integration by substitution: Trigonometric substitutions

When the integrand contains a quadratic radical expression in the denominator, use one of the following trigonometric substitutions to transform the integral.

- If the form is $\frac{1}{\sqrt{a^2 - x^2}}$ use the substitution $x = a \sin u$.
- If the form is $\frac{1}{\sqrt{x^2 - a^2}}$ use the substitution $x = a \sec u$.
- If the form is $\frac{1}{\sqrt{a^2 + x^2}}$ use the substitution $x = a \tan u$.

You obtain the same value for the definite integral, whether you

- Work out the indefinite integral in terms of the substituted variable u, then substitute back to the original variable x and evaluate the limits.

OR

- Change the limits so that they are given in terms of u, then perform the whole definite integration in terms of u.

TOK

Is imagination more important than knowledge?

Example 43

Find the value of the following definite integrals.

a $\int_0^{\frac{\sqrt{3}}{2}} 2x\sqrt{4x^2+1}\,dx$ **b** $\int_0^{\frac{\pi}{3}} \sin x \cos^2 x\,dx$

a Method 1

Let

$$\left.\begin{aligned} 4x^2+1 &= u \\ 8x\,dx &= du \end{aligned}\right\} \Rightarrow$$

When there is a root, it generally works to let u be the expression within the root.

Differentiate to give $8x\,dx = du$.

$$\int_0^{\frac{\sqrt{3}}{2}} 2x\sqrt{4x^2+1}\,dx = \frac{1}{4}\int_0^{\frac{\sqrt{3}}{2}} \sqrt{4x^2+1}\times(8x\,dx)$$

$$= \frac{1}{4}\int_{x=0}^{x=\frac{\sqrt{3}}{2}} \sqrt{u}\,du$$

Write the integral in terms of u and du, but keep the limits in terms of x. Make sure you write them as $x = \ldots$

$$= \frac{1}{4}\left[\frac{2u^{\frac{3}{2}}}{3}\right]_{x=0}^{x=\frac{\sqrt{3}}{2}}$$

Integrate with respect to u.

$$= \left[\frac{(4x^2+1)^{\frac{3}{2}}}{6}\right]_0^{\frac{\sqrt{3}}{2}}$$

Substitute back to obtain the answer in terms of x.

Continued on next page

$$= \left(\frac{(3+1)^{\frac{3}{2}}}{6}\right) - \left(\frac{(1)^{\frac{3}{2}}}{6}\right)$$

$$= \frac{8}{6} - \frac{1}{6} = \frac{7}{6}$$

Substitute your limits in, now that the integrated expression is written in terms of x.

Method 2

Let $4x^2 + 1 = u \Rightarrow 8x\,\mathrm{d}x = \mathrm{d}u$

$$\text{Limits} \Rightarrow \begin{cases} x = \frac{\sqrt{3}}{2} \Rightarrow u = 3 + 1 = 4 \\ x = 0 \quad \Rightarrow u = 0 + 1 = 1 \end{cases}$$

Identify the u-substitution and calculate the new limits in terms of u.

$$\int_0^{\frac{\sqrt{3}}{2}} 2x\sqrt{4x^2+1}\,\mathrm{d}x = \frac{1}{4}\int_1^4 \sqrt{u}\,\mathrm{d}u$$

Write the integral in terms of u and du, and write the limits in terms of u.

$$= \frac{1}{4}\left[\frac{2u^{\frac{3}{2}}}{3}\right]_1^4$$

Integrate with respect to u.

$$= \left(\frac{1}{6} \times 4^{\frac{3}{2}}\right) - \left(\frac{1}{6} \times 1^{\frac{3}{2}}\right)$$

$$= \frac{8}{6} - \frac{1}{6} = \frac{7}{6}$$

Do not switch back to x, but calculate the value of the integrated expression using your limits in u.

b Method 1

Let $\cos x = u \Rightarrow -\sin x\,\mathrm{d}x = \mathrm{d}u$

Identify the u-substitution and find $\frac{\mathrm{d}u}{\mathrm{d}x}$.

$$\int_0^{\frac{\pi}{3}} \sin x \cos^2 x\,\mathrm{d}x = \int_{x=0}^{x=\frac{\pi}{3}} -u^2\,\mathrm{d}u$$

Write the integral in terms of u and du, but keep the limits in terms of x. Make sure you write them as $x = \ldots$

$$= \left[-\frac{u^3}{3}\right]_{x=0}^{x=\frac{\pi}{3}}$$

Integrate with respect to u.

$$= \left[-\frac{\cos^3 x}{3}\right]_{x=0}^{x=\frac{\pi}{3}}$$

Substitute back to obtain the answer in terms of x.

$$= \left(-\frac{1}{3} \times \left(\frac{1}{2}\right)^3\right) - \left(-\frac{1}{3} \times (1)^3\right)$$

$$= -\frac{1}{24} + \frac{1}{3} = \frac{7}{24}$$

Substitute your limits in, now that the integrated expression is written in terms of x.

Method 2

Let $\cos x = u \Rightarrow -\sin x\,\mathrm{d}x = \mathrm{d}u$

$$\text{Limits} \Rightarrow \begin{cases} x = \frac{\pi}{3} \Rightarrow u = \frac{1}{2} \\ x = 0 \Rightarrow u = 1 \end{cases}$$

Identify the u-substitution and calculate the new limits in terms of u.

$\int_0^{\frac{\pi}{3}} \sin x \cos^2 x \,dx = \int_1^{\frac{1}{2}} -u^2 \,du$	Write the integral in terms of u and du, and write the limits in terms of u.
$= \int_{\frac{1}{2}}^{1} u^2 \,du$	Changing the sign of the integrand has the same effect as reversing the limits.
$= \left[\frac{u^3}{3}\right]_{\frac{1}{2}}^{1}$	Integrate with respect to u.
$= \left(\frac{1}{3}\times(1)^3\right) - \left(\frac{1}{3}\times\left(\frac{1}{2}\right)^3\right)$ $= \frac{1}{3} - \frac{1}{24} = \frac{7}{24}$	Do not switch back to x, but calculate the value of the integrated expression using your limits in u.

Exercise 7I

For questions **1** to **10**, find the value of the definite integral using an appropriate substitution. Give your answers in exact form. You may then check your answers using technology.

1 $\int_2^5 \frac{x}{\sqrt{x-1}} \,dx$

2 $\int_3^4 \left(\frac{x}{2-x}\right)^2 dx$

3 $\int_0^{\frac{\pi}{2}} \left(\frac{\sin x}{1+\cos x}\right) dx$

4 $\int_0^{\frac{3}{2}} \left(\frac{1}{\sqrt{9-4x^2}}\right) dx$

5 $\int_1^2 \frac{16x^4}{(2x-1)^2} \,dx$

6 $\int_1^{\frac{3}{2}} \frac{(1-2x)\sqrt{1+x-x^2}}{1+x-x^2} \,dx$

7 $\int_{\frac{\pi}{4}}^{\frac{\pi}{3}} \frac{\cos x}{\sin^3 x} \,dx$

8 $\int_0^{\frac{\pi}{4}} \sec^2 x \tan^3 x \,dx$

9 $\int_{\frac{\pi}{12}}^{\frac{\pi}{6}} \frac{\sin x \cos x}{\cos^3 2x} \,dx$

10 $\int_0^2 3^x \sqrt{3^x} \,dx$

11 Show that $\tan^3 x \equiv \sec^2 x \tan x - \tan x$. Hence, show that $\int_0^{\frac{\pi}{4}} \tan^3 x \,dx = \frac{1}{2}(1-\ln 2)$.

12 Given that $\int_0^{\frac{\pi}{3}} \sin kx \cos^3 kx \,dx = \frac{3}{16k}$, find the value of k.

13 Write $3x^2 + 12x + 16$ in the form $a(x+h)^2 + k$. Hence, using an appropriate substitution, show that $\int_{-2}^{\frac{2}{\sqrt{3}}-2} \left(\frac{1}{3x^2+12x+16}\right) dx = \frac{\pi}{8\sqrt{3}}$.

14 Show that $e^x + e^{-x} = \frac{e^{2x}+1}{e^x}$.

Hence, use an appropriate substitution to show that $\int_0^{-\ln\sqrt{3}} \frac{1}{e^x + e^{-x}} \,dx = -\frac{\pi}{12}$.

15 **a** Given that $t = \tan\left(\frac{x}{2}\right)$, show that $\sin x = \frac{2t}{1+t^2}$.

b Show that $\frac{dx}{dt} = \frac{2}{1+t^2}$.

c Hence, show that $\int_0^{\frac{\pi}{2}} \left(\frac{1}{1+\sin x}\right) dx = 1$.

Investigation 18

The diagram represents a one-to-one function $u(v)$. Point A has coordinates (v_1, u_1) and point B has coordinates (v_2, u_2).

1 How would you represent area 1 using integrals?

2 If the axes are interchanged the function can be interpreted as $v(u)$. How would you represent area 2 using integrals?

3 What is the sum of your results?

4 What is the area of rectangle OFBD?

5 What is the area of rectangle OEAC?

6 Express area 1 in terms of the areas of rectangles OFBD and OEAC and your expression for area 2 from question **2**.

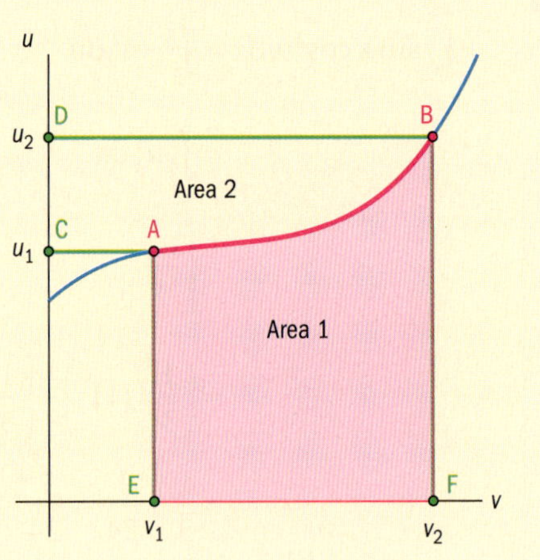

The result you obtained in Investigation 18 is a non-rigorous visualization of the technique of integration called **integration by parts**. It is related to the product rule for differentiation where

$$\frac{\mathrm{d}(u \times v)}{\mathrm{d}x} = v \times \frac{\mathrm{d}u}{\mathrm{d}x} + u \times \frac{\mathrm{d}v}{\mathrm{d}x}$$

If you integrate this identity with respect to x you obtain

$$\int \frac{\mathrm{d}(u \times v)}{\mathrm{d}x}\,\mathrm{d}x = \int v \times \frac{\mathrm{d}u}{\mathrm{d}x}\,\mathrm{d}x + \int u \times \frac{\mathrm{d}v}{\mathrm{d}x}\,\mathrm{d}x$$

which reduces to $\int u \times \frac{\mathrm{d}v}{\mathrm{d}x}\,\mathrm{d}x = uv - \int v \times \frac{\mathrm{d}u}{\mathrm{d}x}\,\mathrm{d}x$ where u and v are functions of x.

> Given functions $u(x)$ and $v(x)$, integration by parts gives
>
> $$\int u\frac{\mathrm{d}v}{\mathrm{d}x}\,\mathrm{d}x = uv - \int v\frac{\mathrm{d}u}{\mathrm{d}x}\,\mathrm{d}x$$

Integration by parts allows you to convert an integral of the product of two functions into another integral that is often simpler.

Typically, integration by parts is used to integrate the logarithmic function and inverse trigonometric functions. It is also useful when you want to integrate products of functions such as:

- $x \ln x$
- $x \arcsin x$
- $x \cos x$
- $x\mathrm{e}^x$
- $\mathrm{e}^x \sin x$

The following examples will illustrate some of these integrals.

Example 44

Find the following integrals:

a $\int xe^x \, dx$ **b** $\int x \ln x \, dx$

a Let $u = x \Rightarrow \frac{du}{dx} = 1$ $\frac{dv}{dx} = e^x \Rightarrow v = \int e^x \, dx = e^x$	Allocate u and $\frac{dv}{dx}$ to appropriate functions in the product, and find $\frac{du}{dx}$ and v.
$\int xe^x \, dx = xe^x - \int e^x \, dx$	Use $\int u \frac{dv}{dx} \, dx = uv - \int v \frac{du}{dx} \, dx$.
$= xe^x - e^x + c$ $= e^x(x-1) + c$	Integrate and simplify.
b Let $u = \ln x \Rightarrow \frac{du}{dx} = \frac{1}{x}$ $\frac{dv}{dx} = x \Rightarrow v = \int x \, dx = \frac{x^2}{2}$	Allocate u and $\frac{dv}{dx}$ to appropriate functions in the product, and find $\frac{du}{dx}$ and v.
$\int x \ln x \, dx = \frac{x^2}{2} \ln x - \int \frac{x^2}{2} \times \frac{1}{x} dx$	Use $\int u \frac{dv}{dx} \, dx = uv - \int v \frac{du}{dx} \, dx$.
$= \frac{x^2}{2} \ln x - \frac{x^2}{4} + c$ $= \frac{x^2}{2}\left(\ln x - \frac{1}{2}\right) + c$	Integrate and simplify.

HINT

When integrating to obtain v you do not add a constant of integration. This is added to the final answer.

In cases where you need to integrate the product of a polynomial function with the exponential function, you should always differentiate the polynomial.

Reflect What happens if you integrate "x" in $\int xe^x \, dx$?

In cases where you need to integrate the product of a polynomial function (of degree ≥ 1) with the logarithmic function, you should always differentiate the logarithmic function.

Reflect What happens if you differentiate "x" in $\int x \ln x \, dx$?

TOK

To what extent can shared knowledge be distorted and misleading?

Functions

Calculus

Example 45

Find the integral $\int \ln x \, dx$.

$\int \ln x \, dx = \int 1 \times \ln x \, dx$	
Let $u = \ln x \Rightarrow \frac{du}{dx} = \frac{1}{x}$ $\frac{dv}{dx} = 1 \Rightarrow v = \int 1 \, dx = x$	You can integrate 1 and differentiate ln x.
$\int \ln x \, dx = x \ln x - \int x \times \frac{1}{x} \, dx$	Use $\int u \frac{dv}{dx} \, dx = uv - \int v \frac{du}{dx} \, dx$.
$= x \ln x - x + c$ $= x(\ln x - 1) + c$	Integrate and simplify.

Example 46

Use integration by parts to find $\int \arccos x \, dx$.

$\int \arccos x \, dx = \int 1 \times \arccos x \, dx$	You can integrate 1 and differentiate arccos x.
Let $u = \arccos x \Rightarrow \frac{du}{dx} = -\frac{1}{\sqrt{1-x^2}}$	To find the derivative let $y = \arccos x$ $\Rightarrow x = \cos y$ and differentiate implicitly.
$\frac{dv}{dx} = 1 \Rightarrow v = \int 1 \, dx = x$	
$\int \arccos x \, dx = x \arccos x - \int x\left(-\frac{1}{\sqrt{1-x^2}}\right) dx$ $= x \arccos x + \int \left(\frac{x}{\sqrt{1-x^2}}\right) dx$ $= x \arccos x + \int \left(-\frac{1}{2\sqrt{t}}\right) dt$ $= x \arccos x - \sqrt{1-x^2} + c$	Use the substitution $t = 1 - x^2$ to find the integral. Integrate and write answer in terms of x.

Exercise 7J

Use integration by parts to find the following indefinite integrals.

1 $\int 2xe^x\,dx$ **2** $\int 3x\sin x\,dx$

3 $\int(1-2x)e^x\,dx$ **4** $\int(2-x)\sin(2-x)\,dx$

5 $\int\left(\frac{1+2x}{3}\right)\sec^2\frac{x}{2}\,dx$ **6** $\int x2^{x+1}\,dx$

7 $\int\frac{x}{3^x}\,dx$ **8** $\int x^3\ln x\,dx$

9 $\int(2-5x)\ln\frac{x}{3}\,dx$ **10** $\int\arcsin x\,dx$

11 $\int(1+3x-x^2)\ln 4x\,dx$ **12** $\int\log_a x\,dx$

13 $\int x\arccos x\,dx$ **14** $\int 4x\arctan x\,dx$

15 $\int x^2\arccos x\,dx$

Find the value of the following definite integrals and check your answers using technology.

16 $\int_0^{\sqrt{3}}\arctan x\,dx$ **17** $\int_1^3\left(\frac{\ln x}{x^5}\right)dx$

18 $\int_0^{\frac{\pi}{4}}\left(\frac{x}{\cos^2 x}\right)dx$ **19** $\int_1^e(1-2x+x^2)\ln 3x\,dx$

20 $\int_0^{\frac{\pi}{4}}x\sin x\,dx$

You will encounter examples where you have to apply integration by parts more than once before you reach the final answer. This often happens when the integral involves a product of a trigonometric or exponential function with a polynomial of order > 1.

Example 47

Show that $\int_0^{\frac{\pi}{2}}x^2\cos x\,dx=\frac{1}{4}(\pi^2-8)$.

Integrating by parts

Let $u=x^2\Rightarrow\frac{du}{dx}=2x$

$\frac{dv}{dx}=\cos x\Rightarrow v=\int\cos x\,dx=\sin x$

$$\int_0^{\frac{\pi}{2}}x^2\cos x\,dx=\left[x^2\sin x\right]_0^{\frac{\pi}{2}}-\int_0^{\frac{\pi}{2}}2x\sin x\,dx$$

To deal with the term $\int_0^{\frac{\pi}{2}}2x\sin x\,dx$, integrate by parts again.

$u=2x\Rightarrow\frac{du}{dx}=2$

$\frac{dv}{dx}=\sin x\Rightarrow v=\int\sin x\,dx=-\cos x$

Continued on next page

$$\int_0^{\frac{\pi}{2}} 2x\sin x \, dx = [-2x\cos x]_0^{\frac{\pi}{2}} - \int_0^{\frac{\pi}{2}} -2\cos x \, dx$$

$$= [-2x\cos x]_0^{\frac{\pi}{2}} + 2\int_0^{\frac{\pi}{2}} \cos x \, dx$$

$$= [-2x\cos x]_0^{\frac{\pi}{2}} + 2[\sin x]_0^{\frac{\pi}{2}}$$

Therefore

$$\int_0^{\frac{\pi}{2}} x^2 \cos x \, dx = \left[x^2 \sin x\right]_0^{\frac{\pi}{2}} + [-2x\cos x]_0^{\frac{\pi}{2}} + 2[\sin x]_0^{\frac{\pi}{2}}$$

$$= \left(\frac{\pi^2}{4}\times 1 - 0\right) + (-\pi\times 0 - 0) + 2(1-0)$$

$$= \frac{\pi^2}{4} + 2$$

$$= \frac{1}{4}\left(\pi^2 + 8\right)$$

Exercise 7K

Integrate the following indefinite integrals.

1 $\int 2x^2 e^{2x} \, dx$

2 $\int x^3 \sin x \, dx$

3 $\int \left(x - x^2\right)\cos x \, dx$

4 $\int x^2 \sin\left(\frac{x}{4}\right) dx$

5 $\int x^3 e^{\frac{x}{3}} \, dx$

Integrate the following definite integrals.

6 $\int_0^2 x^2 e^x \, dx$

7 $\int_0^{\frac{\pi}{2}} x^2 \sin x \, dx$

8 $\int_0^{\frac{\pi}{2}} \left(1 + x^2\right)\cos x \, dx$

9 $\int_0^1 \frac{x^2}{3} e^{3x} \, dx$

10 $\int_0^1 \frac{2x^2}{e^{2x}} \, dx$

Check your answers to questions **6** to **10** using technology.

Investigation 19

Given that $I = \int e^x \sin x \, dx$:

1 Choosing either function to differentiate and the other to integrate, use integration by parts once on I. (Note: your integrated expression will involve another integral which you won't be immediately able to integrate.)

2 Use integration by parts again on the integral which remains in each of the results from question **1**. Each time, make sure you differentiate the same function as before (ie if you differentiated the exponential function the first time, differentiate that again; if you differentiated the trigonometric function the first time, differentiate the trigonometric function again).

3 **Factual** What do you notice about each of the expressions you obtained for I in question **2**?

4 **Conceptual** How can you obtain an expression for I which does not contain an integral?

5 Use the method above to find $I = \int e^x \cos x \, dx$.

6 **Conceptual** What properties of exponential and trigonometric functions allowed you to integrate the product of an exponential function with a trigonometric function?

TOK

How do "believing that" and "believing in" differ?

How does belief differ from knowledge?

In Investigation 19 you were able to use **cyclic integration by parts**. This is a method that is used when the product in the integrand is made up of a combination of exponential function, a sine or cosine function.

Here, you integrate the expression, I, by parts twice to give an expression involving I again. This allows you to isolate I.

Exercise 7L

Use cyclic integration by parts to find the following.

1 $\int \tan x \sec^2 x \, dx$

2 $\int \sin x \cos x \, dx$

3 $\int \sin 2x \cos 3x \, dx$

4 $\int e^{3x} \cos 2x \, dx$

5 $\int \sin^2 x \, dx$

Functions

Calculus

Developing your toolkit

Now do the Modelling and investigation activity on page 516.

Chapter summary

Integration

- $\int f(x)\,dx = F(x) + c,\ c \in \mathbb{R}$
- $\int x^n\,dx = \dfrac{x^{n+1}}{n+1} + c,\ \ n \neq -1,\ c \in \mathbb{R}$
- $\int kx^n\,dx = k\int x^n\,dx = \dfrac{kx^{n+1}}{n+1} + c$
- $\int (a_n x^n + a_{n-1}x^{n-1} + \ldots + a_1 x + a_0)\,dx = \int a_n x^n dx + \int a_{n-1}x^{n-1} dx + \ldots$ $+ \int a_1 x\,dx + \int a_0\,dx$
- $\int \left(f(x) \pm g(x)\right) dx = \int f(x)\,dx \pm \int g(x)\,dx$
- $\int \cos x\,dx = \sin x + c,\ c \in \mathbb{R}$
- $\int \sin x\,dx = -\cos x + c,\ c \in \mathbb{R}$
- $\int \sec^2 x\,dx = \tan x + c,\ c \in \mathbb{R}$
- $\int (ax+b)^n\,dx = \dfrac{(ax+b)^{n+1}}{a(n+1)} + c,\ \ c, n \in \mathbb{R}$
- $\int \sin(ax+b)\,dx = -\dfrac{\cos(ax+b)}{a} + c,\ c \in \mathbb{R}$
- $\int \cos(ax+b)\,dx = \dfrac{\sin(ax+b)}{a} + c,\ c \in \mathbb{R}$
- $\int \sec^2(ax+b)\,dx = \dfrac{\tan(ax+b)}{a} + c,\ c \in \mathbb{R}$

Definite integration

- If $f(x)$ is continuous over the interval $[a, b]$ then the definite integral of $f(x)$ from a to b is equal to the number given by $\int_a^b f(x)\,dx$.
- If the integral of $f(x)$ with respect to x in the interval $[a, b]$ exists, then:
- $\int_a^b f(x)\,dx = -\int_b^a f(x)\,dx$

- $\int_a^a f(x)\,dx = 0$
- $\int_a^b kf(x)\,dx = k\int_a^b f(x)\,dx$
- $\int_a^b (f(x) \pm g(x))\,dx = \int_a^b f(x)\,dx \pm \int_a^b g(x)\,dx$
- $\int_a^c f(x)\,dx + \int_c^b f(x)\,dx = \int_a^b f(x)\,dx,\ a < c < b$
- When the function $f(x)$ is negative for $x \in [a, b]$, then the area bounded by the curve, the x-axis and the lines $x = a$ and $x = b$ is $\left|\int_a^b f(x)\,dx\right|$.
- The total area bounded by $f(x)$ and the x-axis for $x \in [a, b]$ where the graph is partly above and partly below the x-axis is given by $A = \int_a^b |f(x)|\,dx$.

The fundamental theorem of calculus

- If f is continuous in $[a, b]$ and if F is any antiderivative of f on $[a, b]$ then $\int_a^b f(x)\,dx = F(b) - F(a)$.

Exponents and logarithms

- Properties of exponents
- $a^m \times a^n = a^{m+n}$
- $a^m \div a^n = a^{m-n}$
- $(a^m)^n = a^{mn}$
- $a^0 = 1$
- $a^{-n} = \frac{1}{a^n},\ a \neq 0$
- $a^{\frac{1}{m}} = \sqrt[m]{a},\ a > 0$
- $a^{\frac{m}{n}} = \sqrt[n]{a^m} = (\sqrt[n]{a})^m,\ a > 0$
- Logarithms and their properties
- $a = b^x \Leftrightarrow x = \log_b a,\ a, b \in \mathbb{R}^+$ and $b \neq 1$
- $\log_a x + \log_a y = \log_a xy$
- $\log_a x - \log_a y = \log_a \frac{x}{y}$
- $\log_a x^n = n\log_a x$
- $\log_a 1 = 0$
- $\log_a a = 1$

Continued on next page

Functions

Calculus

- $-\log_a x = \log_a \frac{1}{x}$
- $\log_a b = \frac{1}{\log_b a}$
- The family of curves $g(x) = a^x$ for a $\in \mathbb{R}^+$, $a > 1$ are known as exponential growth functions and have the following properties:
    - They have no stationary points.
    - They are increasing.
    - The x-axis is a horizontal asymptote.
    - There are no vertical asymptotes.
    - They are one-to-one functions.
    - They all pass through the point $(0, 1)$.
    - The gradient of at $(0, 1)$ increases as the value of a increases.
- The family of curves $g(x) = a^x$ for $a \in \mathbb{R}^+$, $0 < a < 1$ are known as exponential decay functions and have the following properties:
    - They have no turning points.
    - They are decreasing.
    - The x-axis is a horizontal asymptote.
    - There are no vertical asymptotes.
    - They are one-to-one functions.
    - They all pass through the point $(0, 1)$.
    - The gradient of at $(0, 1)$ decreases as the value of a increases.
- The Euler number is denoted by $\mathrm{e} \approx 2.718281828459045...$ and the exponential function is given by $f(x) = \mathrm{e}^x$.
- The inverse of an exponential function $f(x) = a^x$ is $f^{-1}(x) = \log_a x$, $x > 0$. The inverse of the exponential function $y = \mathrm{e}^x$ is given by $y = \ln x$, which is also called the natural logarithm function.
- $\frac{\mathrm{d}}{\mathrm{d}x}(\mathrm{e}^x) = \mathrm{e}^x$
- $\frac{\mathrm{d}}{\mathrm{d}x}(a^x) = (\ln a)a^x$
- $\frac{\mathrm{d}}{\mathrm{d}x}(\ln x) = \frac{1}{x}$
- $\frac{\mathrm{d}}{\mathrm{d}x}(\log_a x) = \frac{1}{x \ln a}$
- $\int \mathrm{e}^x \mathrm{d}x = \mathrm{e}^x + c$
- $\int \frac{1}{x} \mathrm{d}x = \ln|x| + c$

- $\int e^{ax+b}\,dx = \frac{1}{a}e^{ax+b} + c$
- $\int \frac{1}{(ax+b)}\,dx = \frac{1}{a}\ln|ax+b| + c$
- $\int a^x dx = \frac{a^x}{\ln a} + c$
- $\int \frac{f'(x)}{f(x)}\,dx = \ln(f(x)) + c$
- $\frac{d}{dx}(f(g(x))) = f'(g(x)) \times g'(x) \Rightarrow \int f'(g(x)) \times g'(x)\,dx$
  $= f(g(x)) + c$

Definite integrals and integration by substitution:

Trigonometric substitutions

- When the integrand contains a quadratic radical expression use one of the following trigonometric substitutions to transform the integral:
- If the form is $\sqrt{a^2 - x^2}$ use the substitution $x = a\sin u$.
- If the form is $\sqrt{x^2 - a^2}$ use the substitution $x = a\sec u$.
- If the form is $\sqrt{a^2 + x^2}$ use the substitution $x = a\tan u$.

Integration by parts

- $\int u \times \frac{dv}{dx}\,dx = uv - \int v \times \frac{du}{dx}\,dx$

Developing inquiry skills

Return to the opening problem. You should now be able to solve the problem.

How have the skills you have learned in this chapter helped you?

Functions

Calculus

Chapter review

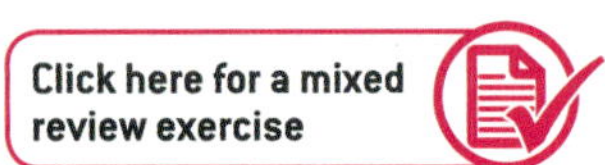

1 Solve the following equations.

a $4^{x-1} - 3(2^x) + 8 = 0$

b $2(5^x) - 9 = \frac{1}{5^{x-1}}$

2 Solve the following logarithmic equations.

a $\log_2 x + \log_2 3 - \log_2 5 = \log_2 6$

b $\log_9 x - \log_9 7 = \frac{3}{2}$

3 Solve for x:

a $\log_2 x - \frac{3}{\log_x 2} = 4$

b $\log_7 x - 4 \log_x 7 = 0$

4 Given that $y = 2e^{3x} - 7e^{-3x}$, show that $\frac{d^2y}{dx^2} = 9y$.

5 Given that $\ln x + \ln x^2 + \ln x^3 + \ldots + \ln x^k = 2k(k+1)$, show that $x = e^4$.

6 Determine the values of $\ln x$ given that $\ln x - \frac{1}{\ln x} = 4$.

7 Find the coordinates of the maximum point of the function $f(x) = \frac{3e^x}{1+2e^{2x}}$.

8 Given that $y = |\ln x|$, show that $\frac{dy}{dx} = \frac{\ln x}{x|\ln x|}$.

9 The function $f(x)$ is defined as $f(x) = (\ln x)^3, x \in \mathbb{R}, 0 < x \le 4$.

a Show that there is only one point of inflection and find its coordinates.

b Find the equations of the tangent and normal to the curve at the point $x = e$.

c Calculate the area of the triangle formed by these normal and tangent lines, and the y-axis.

10 Find the value of $\int_1^3 \frac{(2x-3)\sqrt{x^2-3x+3}}{x^2-3x+3}\,dx$.

11 Find a such that $\int_a^{\frac{1}{2}} \frac{1}{\sqrt{1-4x^2}}\,dx = \frac{\pi}{24}$.

12 Find the equation of a tangent to the curve $f(x) = e^{\frac{x}{2}-1}$ that passes through the origin.

13 Use the substitution $3x = 2\cos u$ to determine $\int \sqrt{4-9x^2}\,dx$.

14 Determine $\int x^2 e^{\frac{x}{2}}\,dx$.

15 Use integration by parts to find $\int 3^x \sin x\,dx$.

Exam-style questions

16 **P1: a** Solve the equation $3e^{2x} + 11e^x - 4 = 0$ giving your answer(s) in the form $a\ln b$, where a and b are integers. (5 marks)

b Solve the equation $(\ln x)^2 - 5\ln x - 36 = 0$. (4 marks)

17 **P1:** Solve each of the following equations, giving your answers in terms of logarithms.

a $3 \times 10^{5x-1} = 45$ (4 marks)

b $3^{2-x} = 7^{\frac{x}{2}}$ (4 marks)

18 **P1: a** Describe a sequence of transformations that map the graph of the function $f(x) = \log_{10} x$ onto the graph of $y = \log_{10}(2x+100)^3$. (5 marks)

b Find an expression for $\frac{dy}{dx}$. (5 marks)

19 **P1: a** Show that $\log_{16} 4 = \frac{1}{2}$. (3 marks)

b Hence or otherwise, solve the equation $\log_{16}(x-4) - \log_{16}(x-12) = \frac{1}{2}$.

Give your answer in exact form. (4 marks)

20 P1: Consider the curve given by the graph of $y = x^3e^{-x}$.

a Prove that the point $(3,\ 27e^{-3})$ is a maximum point on the curve. (7 marks)

b Find the equation of the tangent to the curve at the point where $x = 1$, giving your answer in the form $ax + by + c = 0$, where a, b and c are constants. (5 marks)

21 P1: Given that $\int_1^5 \frac{5x}{2x^2 + 3x - 2}\,dx = \ln p$, determine the value of p. (10 marks)

22 P2: a By using a suitable substitution, find an exact value for the integral $\int_{\frac{1}{6}}^{\frac{1}{3}} \frac{dx}{\sqrt{1-9x^2}}$. (8 marks)

b Use technology to determine the value of the integral $\int_{\frac{1}{6}}^{\frac{1}{3}} \frac{dx}{\sqrt{1-9x^2}}$ correct to 3 significant figures. (1 mark)

23 P1: a Find $\int \frac{x}{\sin^2 x}\,dx$. (6 marks)

b The area bounded by the curve $y = \frac{x}{\sin^2 x}$, the x – axis and the lines $x = \frac{\pi}{4}$ and $x = \frac{\pi}{2}$ is denoted by A.

Show that $A = \frac{1}{4}(\pi + \ln k)$, where k is a constant to be determined. (7 marks)

24 P2: The first four terms in an arithmetic series are given by $\log_2 343$, $\log_2 x$, $\log_2 y$, $\log_2 1331$

Find the value of x. (9 marks)

25 P1: Find $\int e^{-x}\sin 3x\,dx$. (9 marks)

Functions

Calculus

A passing fad?

Approaches to learning: Communication, Research

Exploration criteria: Mathematical presentation, Reflection (D), Use of mathematics (E)

IB topic: Exponentials and Logarithms

Look at the data

Fortnite was released by Epic games in July 2017 and quickly grew in popularity.

Here are some data for the total number of registered players worldwide from August 2017 to June 2018:

Date	August 2017	November 2017	December 2017	January 2018	June 2018
Months since launch, t	1	4	5	6	11
Number of registered players, P (million)	1	20	30	45	125

The data is taken from the press releases of the developers, Epic Games.

- Are these data reliable?
- Are there any potential problems with the data that has been collected?
- What other data might be useful?

Plotting these data on your GDC or other graphing software gives this graph:

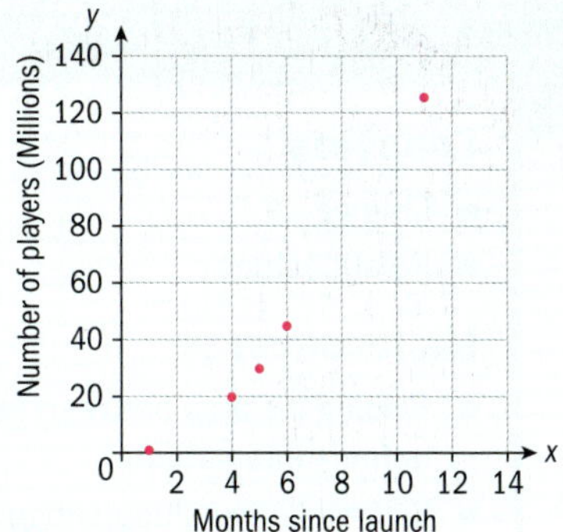

Describe the shape of the graph.

What could be some possible explanations for this growth?

Model the data

Why might it be useful to find a model that links the number of players of Fortnite against the number of months since the launch of the game?

Who might this be useful for?

Let t be the number of months since July 2017.

Let P be the number of players in millions.

Assume that the data is modelled by an exponential function of the form $P = a - b^t$ where a and b are constants to be found.

Use the techniques from the chapter or previous tasks to help you.

Consider different models **by hand** and **using technology**.

How do these models differ? Why?

Which one is preferable? Why?

What alterations could be made to the model?

Use the model to predict the number of users for the current month.

Do you think this is likely to be a reliable prediction?

How reliable do you think the models are at predicting how many Fortnite players there are now? Justify your answer.

Research the number of players there are in this current month who play Fortnite.

Compare this figure with your prediction based on your model above. How big is the error?

What does this tell you about the reliability of your previous model?

Plot a new graph with the updated data and try to fit another function to this data.

Will a modified exponential model be a good fit?

If not, what other function would be a better model that could be used to predict the number of users now?

Extension

Think of another example of data that you think may currently display a similar exponential trend (or exponential decay).

Can you collect reliable and relevant data for your example?

Find data and present it in a table and a graph.

Develop a model or models for the data (ensure that your notation is consistent and your variables are defined) - you could use technology or calculations by hand.

For how long do you think your model will be useful for making predictions?

Explain.

8 Modelling change: more calculus

If you know the value of a quantity and you wish to predict its value in the future, you need to understand how the quantity has changed, and is changing, over time. This is an important skill in many real-life scenarios: geologists model the phases of the earth's cooling and heating; biologists aim to understand and predict the spread of diseases; bankers analyse the changes in profit and loss in the financial markets; and so forth. For all of these problems, building models of change is essential. In this chapter you will study applications of integration such as areas between curves and volumes of revolution, limits, Maclaurin series, and differential equations. These mathematical concepts help you to analyse and understand the science of change.

Concepts

- Change
- Relationships

Microconcepts

- Finding the area between two curves using definite integration
- Volumes of revolution about the x- and y-axes
- Displacement as the integral of the velocity function
- Total distance as the integral of the absolute value of the velocity function
- Separable differential equations
- Homogeneous differential equations
- Integrating factors
- Euler's method for solving differential equations
- L'Hopital's Rule
- Maclaurin polynomials

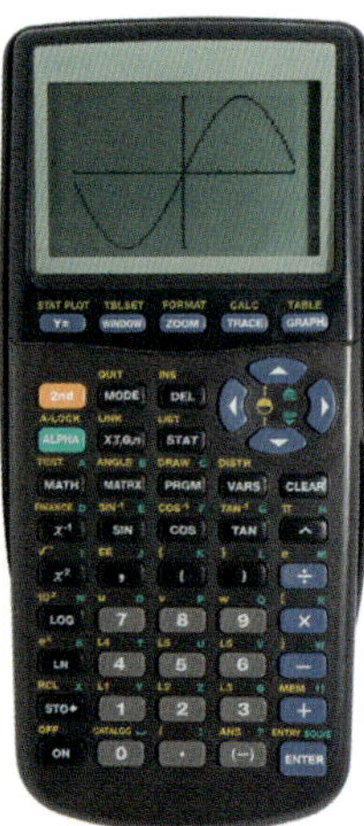

How does your calculator approximate the values of functions such as sine and cosine?

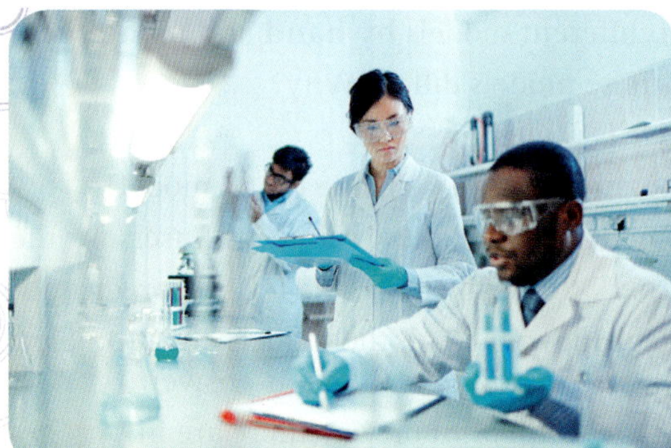

How do forensic scientists figure out the time a person was murdered when there are no witnesses to the crime?

How do medical researchers model and predict the spread of disease?

A mathematician on the staff of a glass making company has been asked to design a new series of glasses for different drinks, eg water, white wine, red wine, champagne, etc. In modelling the design for this series of glasses, the mathematician creates a profile of a glass on the x-axis, similar to this one.

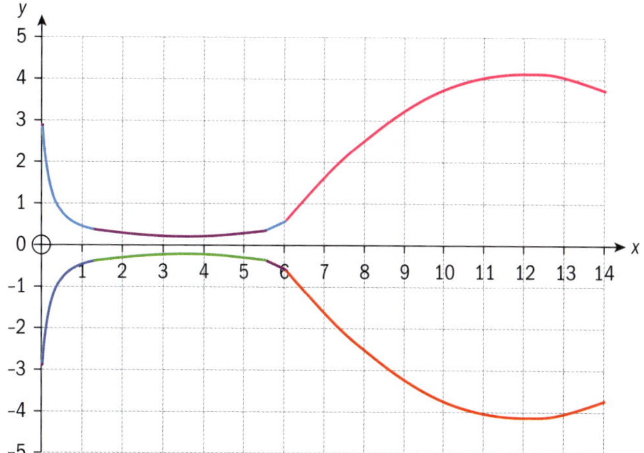

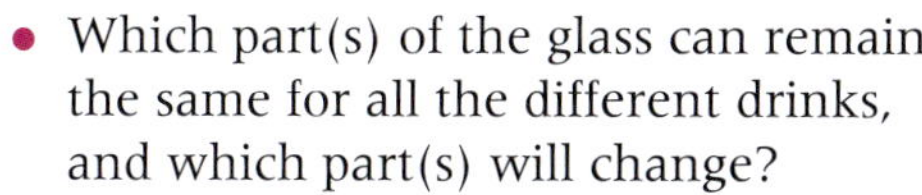

- Which part(s) of the glass can remain the same for all the different drinks, and which part(s) will change?
- What kinds of functions can model the different parts of the design?
- How do you find the surface area between the curves of the bowl of the glass?
- How do you model the volume of the glass?
- How does the rate of change of the volumes for the different drinks affect the rate of change of the surface areas and lengths of the bowls of the glass?
- Can polynomial functions be used to model all parts of the design?

Developing inquiry skills

Think about the questions in this opening problem and answer any you can. As you work through the chapter, you will gain mathematical knowledge and skills that will help you to answer them all.

Before you start

<table>
<tr>
<th>You should know how to:</th>
<th>Skills check 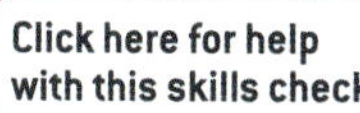 
Click here for help with this skills check</th>
</tr>
<tr>
<td>1 Find limits algebraically.
eg, $\lim_{x\to 0}(x^2 - x + 1) = \lim_{x\to 0}(0^2 - 0 + 1) = 1$</td>
<td>1 Find the limits, if they exist, of
a $\lim_{x\to -4}\frac{x^2-16}{x+4}$ b $\lim_{x\to 0}\frac{3x^3+x^2}{x^2}$
c $\lim_{x\to 0}\frac{1}{x^2+1}$</td>
</tr>
<tr>
<td>2 Find horizontal asymptotes algebraically.
eg, find the horizontal asymptote of
$y = \frac{x^3 - x^2 + 1}{2 - x - x^3}$.
$\lim_{x\to\infty}\left(\frac{x^3-x^2+1}{2-x-x^3}\right) = \lim_{x\to\infty}\left(\frac{\frac{x^3}{x^3}-\frac{x^2}{x^3}+\frac{1}{x^3}}{\frac{2}{x^3}-\frac{x}{x^3}-\frac{x^3}{x^3}}\right)$
Horizontal asymptote is $y = -1$</td>
<td>2 Find any horizontal asymptotes of
a $y = \frac{2x^4-3}{2+3x^4}$ b $y = \frac{3x}{x^3-1}$
c $y = \frac{(x-5)^2}{x^2-5}$</td>
</tr>
<tr>
<td>3 Direct integration, integration by substitution, and integration by parts.
eg, $\int xe^x dx = (x-1)e^x$.</td>
<td>3 Use an appropriate method to integrate:
a $\int 2x\sqrt{x^2-1}dx$ b $\int(\sin x\cos x)dx$
c $\int(4x+5)\ln x dx$</td>
</tr>
</table>

8.1 Areas and volumes

Investigation 1

1 Sketch the graph of $y = \sqrt{x}$, and shade the area between the curve, the x-axis, and the line $x = 1$. Find the shaded area.

2 On the same axes, draw the graph of $y = x^2$ and shade the area between the curve, the x-axis, and the lines $x = 0$ and $x = 1$.

3 Find the area between the curves $y = \sqrt{x}$ and $y = x^2$, and explain how you found this area.

4 The two functions $y = \sqrt{x}$ and $y = x^2$ have graphs that are continuous curves in the interval between their points of intersection. Letting one of these be f and the other g, then in this interval $f(x) \geq g(x)$. Identify which function is f and which is g.

Find the value of the definite integral $\int_a^b [f(x) - g(x)]\,\mathrm{d}x$, where $x = a$ and $x = b$ are the points of intersection of the curves.

Compare your answer to the value you found in question **3** for the area between the curves. What do you notice?

5 **a** Find the area bounded by the curve $y = 2x^2 + 10$, the x-axis, and the lines $x = -1$ and $x = 3$.

b Find the area bounded by the line $y = 4x + 16$, the x-axis, and the lines $x = -1$ and $x = 3$.

c Using your answers to parts **a** and **b**, find the area between the curve $y = 2x^2 + 10$ and the line $y = 4x + 16$.

6 As with the functions you worked with in question **4**, these two functions are also continuous in the interval between their intersection and $f(x) \geq g(x)$. Identify which of these two functions is f and which is g.

Find the value of the definite integral $\int_a^b [f(x) - g(x)]\,\mathrm{d}x$, where $x = a$ and $x = b$ are the points of intersection of the curves.

Compare your answer to the value you found in question **5c** for the area between the line and the curve. What do you notice?

7 If you mistakenly interchange the two functions in the integral, how would your answer change?

8 How could you change the integral so that the placement of the functions is irrelevant?

9 **Conceptual** How can you use a single integral to find the area between two curves whose functions are f and g, that intersect at $x = a$ and $x = b$, and are continuous in the interval $[a, b]$?

If functions f and g are continuous in the interval $[a, b]$ and $f(x) \geq g(x)$, $x \in [a, b]$ then the area between the graphs of f and g is

$$A = \int_a^b f(x)\,\mathrm{d}x - \int_a^b g(x)\,\mathrm{d}x = \int_a^b [f(x) - g(x)]\,\mathrm{d}x$$

International-mindedness

Egyptian mathematician Ibn al-Haytham is credited with calculating the integral of a function in the 10th century.

Example 1

Graph the functions $f(x) = 2\cos x$ and $g(x) = x^2 - 1$ and find the area of the region enclosed by the curves.

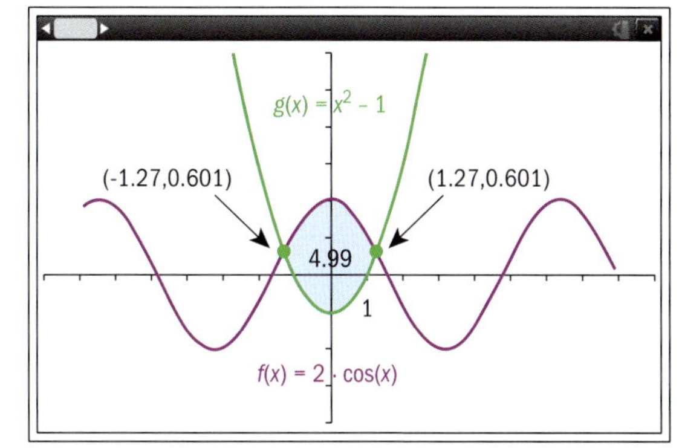

Using technology, graph both functions and find their points of intersection. Then find the area of the region enclosed by the curves.

$$A = \int_{-1.27}^{1.27}\left[2\cos x - \left(x^2 - 1\right)\right]dx \approx 4.99 \text{ square units}$$

Even when using technology, correct mathematical communication is expected.

Example 2

Find the total area enclosed by the graph $f(x) = x^3 - 2x + 1$ and its reflection in the x-axis.

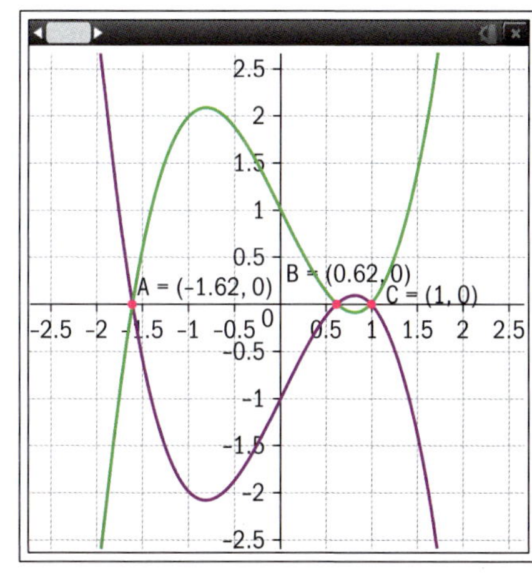

Graph $y = f(x)$ and $y = -f(x)$.

Find their points of intersection.

Method 1

$$A = \int_{-1.62}^{0.62}\left[\left(x^3 - 2x + 1\right) - \left(-x^3 + 2x - 1\right)\right]dx + \int_{0.62}^{1}\left[\left(-x^3 + 2x - 1\right) - \left(x^3 - 2x + 1\right)\right]dx$$

Write the integrals representing both regions, and add the results.

Continued on next page

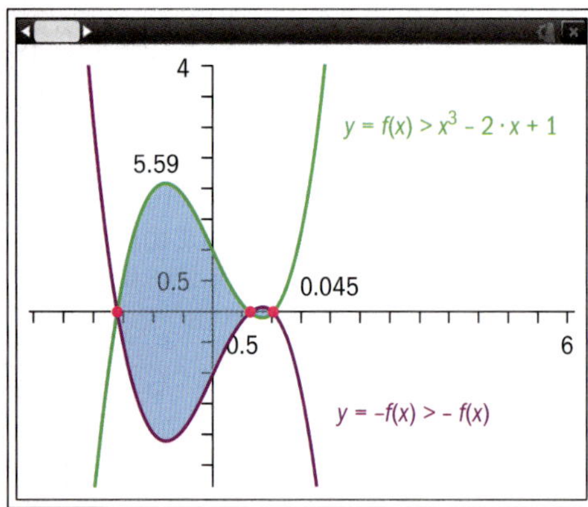

$A = 5.59 + 0.045 \approx 5.64$ units squared.

Method 2

$\int_{-1.62}^{1} \left|(x^3 - 2x + 1) - (-x^3 + 2x - 1)\right| dx = 5.64$

Find the definite integral of the absolute value of the difference of the functions over the entire interval.

ie $\int_{-1.62}^{1} \left|f(x) - (-f(x))\right| dx$

Exercise 8A

1 In each case, find the area of the region enclosed by the graphs of the two curves.

a $y = \frac{1}{2}x^3 + 2x^2 + 2x - \frac{1}{2}$ and $y = -\frac{1}{2} + 3x + 2x^2 - \frac{1}{2}x^3$.

b $y = |x|; y = x^{\frac{2}{3}}$ **c** $y = 3x - x^2; y = 2x^3 - x^2 - 5x$

d $y = 8\cos^2 x, -1 \le x \le 1; y = \sec^2 x$

e $y = \tan\left(\frac{x}{2}\right); y = 1$

2 Find the area of the region enclosed by the graphs of the curves

a $y = (x-2)^4 - 4; y = \frac{1}{1+x}$ for $x > 0$.

b $y = e^{1-x} - 1; y = \sqrt{x}; x = 4$

c $y = 2\sin x, x > 0; y = e^{\frac{x}{2} - 4} + 1$

3 Find the area of the finite regions enclosed by the graphs of the functions $y = \cos 2x$ and $y = \ln(x - 1)$.

A region may not be entirely enclosed between two functions. In this case, the region may need to be partitioned into sub-regions in order to find the required area.

Investigation 2 shows an example of partitioning areas.

TOK

Consider $f(x) = \frac{1}{x}$, $1 \le x \le \infty$.

An infinite area sweeps out a finite volume. How does this compare to our intuition?

Investigation 2

1. Draw the graphs of the functions $y = \sqrt{x}$ and $y = x - 2$. Label R the region in the first quadrant which is bounded by the functions and the x-axis.
2. Why is it not possible to use the techniques you learned in Investigation 1 to find the area of the region R?
3. Divide R into two separate regions whose areas are possible to find with the integration techniques you have learned so far, and label them as R_1 and R_2.
4. Find the areas of R_1 and R_2, and the area of the entire region.

Example 3

Find the area of the region bounded by the graphs of $y = \frac{1}{x}$, $y = x^{\frac{2}{3}}$, the x-axis, and the line $x = 3$.

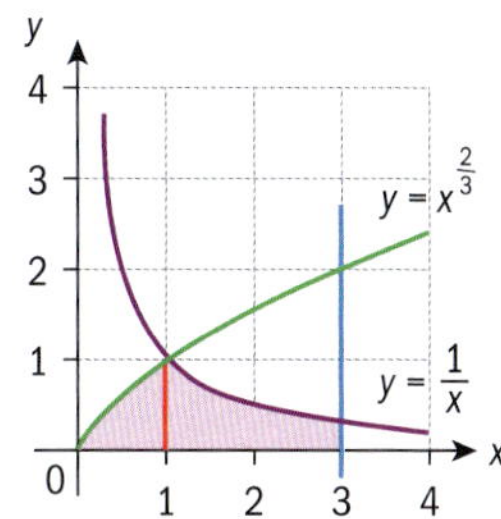	Draw the graphs of the functions and lines. Shade the desired region. Partition the region into sub-regions whose areas can be found.
$A = \int_0^1 x^{\frac{2}{3}}\text{d}x + \int_1^3 \frac{1}{x}\text{d}x$ $= \left[\frac{3x^{\frac{5}{3}}}{5}\right]_0^1 + \left[\ln\lvert x\rvert\right]_1^3$ $= \frac{3}{5} + \ln 3 \approx 1.70$ square units	Write out the definition of the area of the region, and solve.

Exercise 8B

1. Find the area enclosed by the graphs of the functions $y = \frac{4}{x^2}; y = x - \frac{1}{4}x^2; y = x^2 + 3x$.
2. Find the area bounded by the graphs of $y = 2^x; y = 4^x; y = \frac{1}{x}$.
3. Find the sum of the areas of the regions bounded by the graphs of $f(x) = 1 + x + e^{x^2-2x}$ and $g(x) = x^4 - 7x^2 + 6x + 2$.
4. Consider the region in the first quadrant bounded by the graphs of $y = x^3$ and $y = 4x$.

   Find

   a The value of k such that the line $x = k$ divides the region into two regions of equal area.

   b The value of m such that the line $y = m$ divides the region into two regions of equal area.

Volumes of revolution

Investigation 3

1 Sketch the graph of $y = 2x$ between $x = 0$ and $x = 2$.

2 Imagine rotating the line $y = 2x$ about the x-axis 2π radians so that a cone is formed.

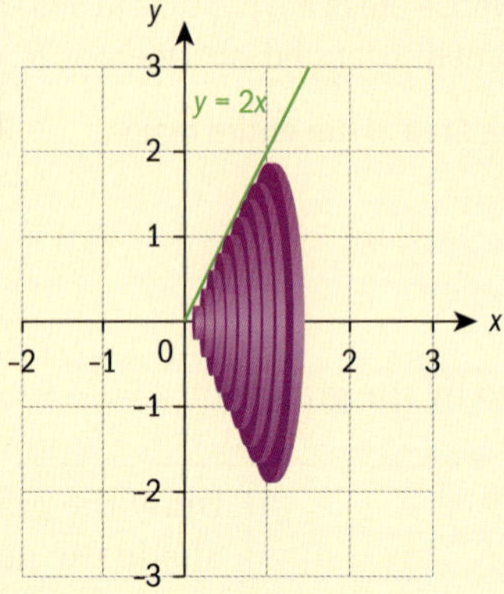

The vertical cross-sections of the cone are thin, upright cylinders, which are parallel to the y-axis.

The radius of each thin cylinder is the function y, and the height of each cylinder is designated as dx.

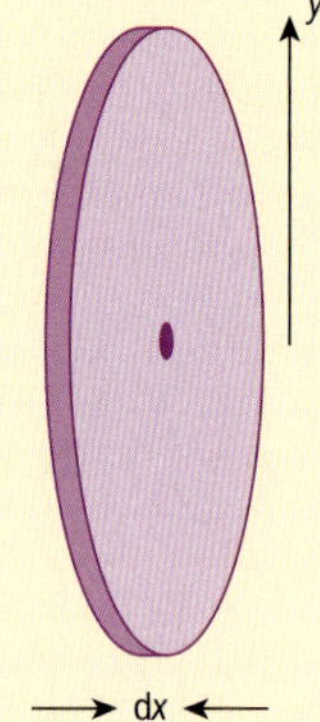

3 State the volume of one of the cylinders, in terms of y.

4 Write the volume of the entire cone as the summation of the cylindrical cross-sections, in terms of y.

Based on your answer to questions **1–4**, when dx is infinitesimally small, the summation of all the cylindrical cross-sections in an interval $[a, b]$ can be written as $\int_a^b \pi y^2 \, dx$.

5 Find the volume of the cone obtained by rotating the line $y = 2x$ in the interval [0,2] through 2π radians about the x-axis.

6 Compare your answer for the volume of the cone with the answer you would get using the formula for the volume of a cone, $V = \dfrac{\pi r^2 h}{3}$, where the radius of the base of the cone is 4, and the height is 2.

Now rotate the line $y = 2x$, $0 \le x \le 2$, through 2π radians about the y-axis. An interval of [0,2] on the x-axis corresponds to [0,4] on the y-axis.

Here, the cross-sections of the cone are thin, horizontal cylinders, which are parallel to the x-axis.

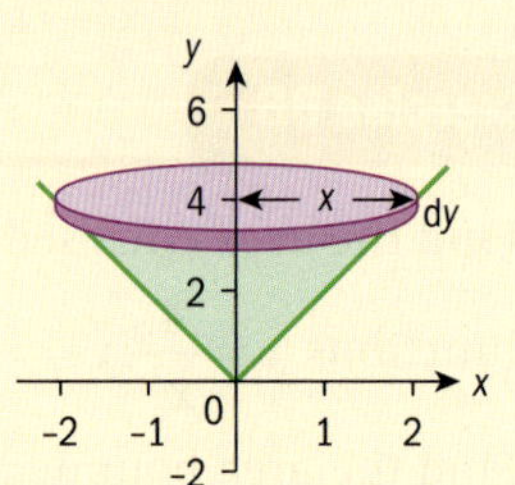

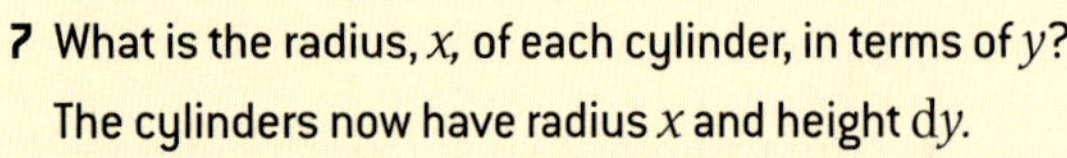

7 What is the radius, x, of each cylinder, in terms of y?

The cylinders now have radius x and height dy.

8 What is the volume of one of the cylindrical cross-sections?

9 Write the volume of the entire cone as the summation of the cylindrical cross-sections, in terms of x.

Based on your findings in questions **7** and **8**, when dy is infinitesimally small, the summation of all the cylindrical cross-sections in an interval $[c,d]$ (where $f(a)=c$ and $f(b)=d$) can be written as $\int_c^d \pi x^2 dy$.

10 Find the volume of the cone obtained by rotating the line $x=\frac{y}{2}$ in the interval $0 \le y \le 4$ through 2π radians about the y-axis.

11 Compare your answer for the volume of the cone with the answer you would get using the formula for the volume of a cone.

12 **Factual** What is a solid of revolution?

13 **Conceptual** Describe how you can calculate the volume of a solid of revolution.

- The volume of a solid of revolution formed when $y=f(x)$, which is continuous in the interval $[a, b]$, is rotated 2π radians about the x-axis is $V=\pi\int_a^b y^2 dx$.
- The volume of a solid of revolution formed when $y=f(x)$, which is continuous in the interval $y=c$ to $y=d$, is rotated 2π radians about the y-axis is $V=\pi\int_c^d x^2 dy$.

Example 4

Find the volume of the solid formed when the graph of the curve $y=e^{1-x}$ in the interval $[0, 1]$ is rotated 2π radians about the x-axis.

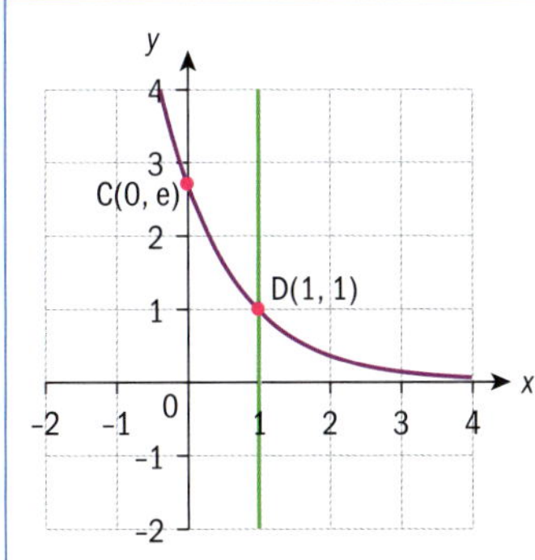	Sketch the curve. It is helpful to draw vertical lines at $x=0$ and $x=1$ to mark where the limits of the solid would be. C and D are the points of intersection of the curve and the lines.
$V=\pi\int_0^1 (e^{1-x})^2 dx=\pi\int_0^1 e^{2(1-x)} dx$ $=-\frac{\pi}{2}\left[e^{2(1-x)}\right]_0^1=\frac{\pi}{2}(e^2-1)$ cubic units ≈ 10.0 cubic units	Use the formula $V=\pi\int_a^b y^2 dx$

Calculus

Example 5

Find the volume of the solid of revolution formed by rotating the region enclosed by the graph of $y = \frac{x^2+1}{2}$ and the lines $y = 0.5$ and $y = 3$, by 2π radians about the y-axis.

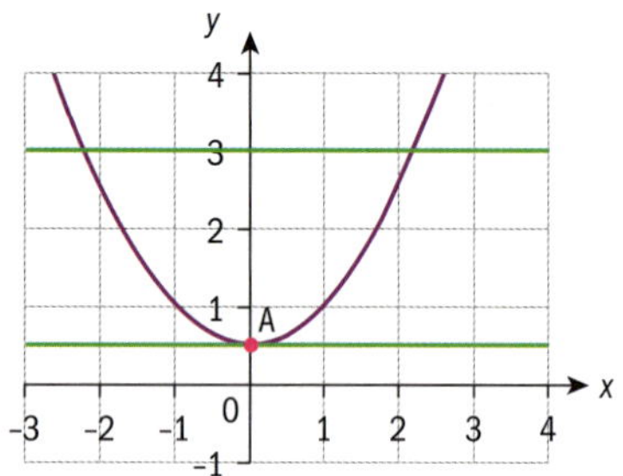	Sketch the curve and the lines.
$y = \frac{x^2+1}{2} \Rightarrow x = \sqrt{2y-1}$	Solve for x.
$V = \pi\int_{0.5}^{3}(2y-1)\mathrm{d}y = \pi\left[y^2 - y\right]_{0.5}^{3} = 6.25\pi$ cubic units	Use the formula $V = \pi\int_c^d x^2\mathrm{d}y$

You will now see how a volume of a solid formed by the region between two curves is calculated.

Example 6

Find the volume of the solid formed when the region between the two curves $f(x) = \sqrt{\frac{x}{2}}$ and $g(x) = \frac{x^2}{4}$ is rotated about

a the x-axis **b** the y-axis.

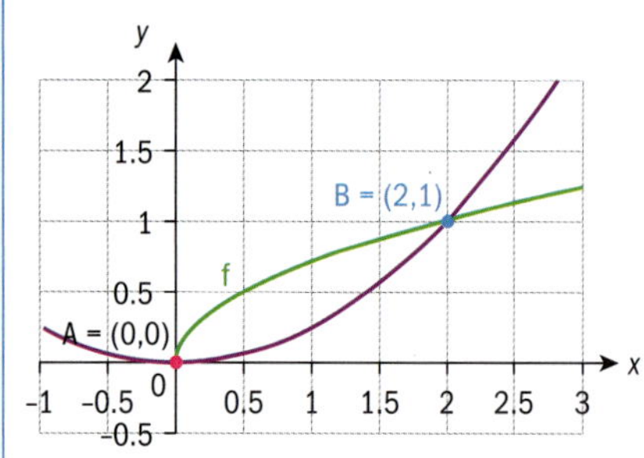	Sketch the graphs and find the points of intersection.

a $V = \pi\int_0^2\left(\left[f(x)\right]^2 - \left[g(x)\right]^2\right)\mathrm{d}x$

$V = \pi\int_0^2\left(\left[\sqrt{\frac{x}{2}}\right]^2 - \left[\frac{x^2}{4}\right]^2\right)\mathrm{d}x$

$V = \pi\int_0^2\left(\frac{x}{2} - \frac{x^4}{16}\right)\mathrm{d}x$

$V = \pi\left[\frac{x^2}{4} - \frac{x^5}{80}\right]_0^2 = \pi\left(1 - \frac{2}{5}\right)$

$= \frac{3\pi}{5}$ cubic units

In the interval $[0,2]$, $f(x) \geq g(x)$.

Geometrically, the volume of the region between the two curves is the difference in the volume of the solids formed by the curves and the x-axis.

b $y = \sqrt{\frac{x}{2}}$ $\quad x = 2y^2; y = \frac{x^2}{4} \quad x = 2\sqrt{y}$

$y = 0$, $y = 1$ are the boundaries

$V = \pi\int_0^1\left[\left(2\sqrt{y}\right)^2 - \left(2y^2\right)^2\right]\mathrm{d}y$

$V = \pi\int_0^1\left[4y - 4y^4\right]\mathrm{d}y$

$\Rightarrow V = 4\pi\left[\frac{y^2}{2} - \frac{y^5}{5}\right]_0^1$

$\Rightarrow V = \frac{6\pi}{5}$ cubic units

In the y-interval $[0,1]$, $g(x) \geq f(x)$.

- If $f(x) \geq g(x)$ for all x in the interval $[a, b]$, then the volume of revolution formed when rotating the region between the two curves 2π radians about the x-axis is $V = \pi\int_a^b\left(\left[f(x)\right]^2 - \left[g(x)\right]^2\right)\mathrm{d}x$.
- If x_1 and x_2 are relations in y such that $x_1 \geq x_2$ for all y in the interval $[c, d]$, then the volume formed by rotating the region between the two curves 2π radians about the y-axis is $V = \pi\int_c^d\left(x_1^2 - x_2^2\right)\mathrm{d}y$.

Exercise 8C

1 Find the volume of the solid formed when the graph of the curve $y = \sqrt{x}$ in the interval $[1,4]$ is rotated 2π radians about

a the x-axis

b the y-axis.

2 Find the volume of the solid formed by rotating the region enclosed by the graph of the function and the x-axis through 2π radians about the x-axis in the given interval.

a $y = x^2 + x$; $[0, 1]$ **b** $y = 1 - \sqrt{x}$; $[1, 4]$

c $y = \sqrt{\cos x \sin x}$; $\left[0, \frac{\pi}{2}\right]$

d $y = \tan x$; $\left[0, \frac{\pi}{4}\right]$

e $y = e^{-x}$; $[0, 1]$ **f** $y = \frac{1}{\sqrt{x+1}}$; $[0, 3]$

3 A bottle stopper is modelled by the function $y = \frac{x}{12}\sqrt{36 - x^2}$. Find the volume of the stopper when it is rotated 2π radians about the x-axis between $x = 0$ and $x = 6$.

4 Find the volume of the solid of revolution when the graph of the given function is rotated 2π radians about the y-axis between the given lines which are parallel to the x-axis.

a $x = 2\sin y$; $y = 0$ and $y = 2$.

b $x = \tan\left(\frac{\pi}{4}y\right)$; $y = 0$ and $y = 1$.

5 Find the volume of the solid of revolution generated by rotating the region bounded by the graph of $y = \arcsin x$ and the lines $x = 1$ and $x = 0$ through 2π radians about the y-axis.

6 Find the volume of a solid that is obtained by rotating the curve $y = \arccos x$, $0 \le x \le 1$, through 2π radians about the y-axis.

7 Find the volume of the solid generated by rotating the region enclosed by the curves $y = e^{\frac{x}{3}} - 1$ and $y = \arctan x$ through 2π radians about

a the x-axis **b** the y-axis.

8 Find the volume of a solid that is obtained by rotating the finite region enclosed by the curves $y = 1 + \ln x$ and $y = \tan\frac{x}{2}$ through 2π radians about the x-axis.

9 Find the volume of revolution of the solid generated by rotating the region in the first quadrant bounded by the x-axis and the curves $y = e^x - 1$ and $y = \cos x$ through 2π radians about the x-axis.

Developing inquiry skills

Return to the opening problem for the chapter. You should have been able to answer the first two questions before working on this chapter. How does what you have learned in this section enable you to now solve the next two questions?

8.2 Kinematics

Investigation 4

A particle moves along a straight line. It's velocity at time t seconds is modelled by the function $v(t) = t^2 - \frac{4}{(t+1)^2}$ cm s^{-1}, where $0 \le t \le 6$.

1 Sketch the graph of the particle's velocity for $0 \le t \le 6$.

2 Find and interpret the initial velocity of the particle.

3 Describe the motion of the particle during $t \in [0, 6]$.

TOK

Does the inclusion of kinematics as core mathematics reflect a particular cultural heritage?

Since the velocity is the derivative of the displacement, performing anti-differentiation on the velocity of the particle will give you an expression for its displacement.

4 Find the displacement of the particle from $t = 0$ to $t = 1$.

5 The initial position of the particle is $s(0) = 5$ cm. Use the displacement from $t = 0$ to $t = 1$, and the initial position of the particle to find the position of the particle after the 1st second.

6 Find the particle's final position.

7 Using the initial position and the expression for $v(t)$, find an expression for $s(t)$, the particle's displacement at any time t.

8 Use the displacement function you found in question **7** to find the position of the particle at $t = 1$ and $t = 6$. Compare your answers to those you found in questions **5** and **6**.

9 **Factual** Describe the changes in displacement which occur over the 6 seconds of motion.

10 Explain how you find the *total distance* the particle actually travelled in the six seconds, and find this distance.

11 **Factual** What is the difference between displacement and distance?

12 **Conceptual** How can you integrate the velocity function to find the displacement of a particle over a given time interval? How can you adapt this method to instead find the total distance travelled?

If $v(t)$ is the velocity function in terms of t, then

- the displacement of the particle between t_1 and t_2 is $\int_{t_1}^{t_2} v(t)\,dt$
- the total distance travelled by the particle between t_1 and t_2 is $\int_{t_1}^{t_2} |v(t)|\,dt$.

Example 7

A particle moves in a straight line such that its velocity at any time t can be modelled by $v(t) = t - t^3$ m s^{-1}. Find

a the displacement of the particle in the first two seconds, and interpret your answer

b the total distance travelled by the particle in the first two seconds.

a $\int_0^2 (t - t^3)dt = \left[\frac{t^2}{2} - \frac{t^4}{4}\right]_0^2 = -2$	Since the value of the definite integral is negative, the final displacement is to the left of the starting position.

Continued on next page

b Method 1: Analytical solution	Sketch the function to see if it is entirely above or below the t-axis, or if part of the graph is above and part is below.

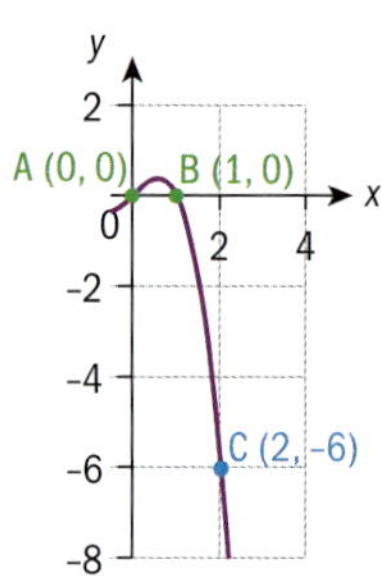

$s(t)=\int_0^1 (t-t^3)\mathrm{d}t+\left\lvert\int_1^2 (t-t^3)\mathrm{d}t\right\rvert$	Integrate the parts above and below the axis separately.
$=\left[\frac{t^2}{2}-\frac{t^4}{4}\right]_0^1+\left\lvert\left[\frac{t^2}{2}-\frac{t^4}{4}\right]_1^2\right\rvert$	
$=\left(\frac{1}{2}-\frac{1}{4}\right)+\left\lvert\left(\frac{4}{2}-\frac{16}{4}\right)-\left(\frac{1}{2}-\frac{1}{4}\right)\right\rvert=2.5\text{ m}$	Use the formula for total distance.

Method 2: Using technology

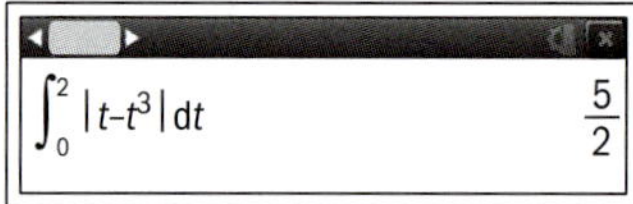

$$\int_0^2 |t-t^3|\,\mathrm{d}t \qquad \frac{5}{2}$$

Example 8

The velocity of a particle moving in a straight line can be modelled by $v(t) = 5\sin^2 t\cos t$ for $0 \le t \le 2\pi$, where v is measured in cm s^{-1} and t is measured in seconds.

a Find the times when the particle is at rest, and describe its movement over the entire interval.

b Find the displacement of the particle in the given time interval, and interpret your answer.

c Find the total distance covered by the particle.

a

Method 1: Analytical solution

$5\sin^2 t\cos t = 0 \Rightarrow \sin t = 0,\ \cos t = 0$

$\Rightarrow$ at $t = 0, \frac{\pi}{2}, \pi, \frac{3\pi}{2}, 2\pi$ seconds the particle is at rest.

Find the zeros of the velocity function to determine when the particle is at rest.

t	$0 < x < \frac{\pi}{2}$	$\frac{\pi}{2} < x < \pi$	$\pi < x < \frac{3\pi}{2}$	$\frac{3\pi}{2} < x < \pi$
$v(t)$	+	−	−	+

Find the sign of the function between each of the zeros.

From $t = 0$ s to $t = \frac{\pi}{2}$ s the particle is moving to the right. It stops at $t = \frac{\pi}{2}$ s, then moves to the left until $t = \pi$ s and stops. It continues moving to the left until $t = \frac{3\pi}{2}$ s, then stops, and then moves to the right again until $t = 2\pi$ s.

Moving right when the velocity function is positive.

Moving left when the velocity function is negative.

Method 2: Using a graph

Graph the velocity function and identify its zeros.

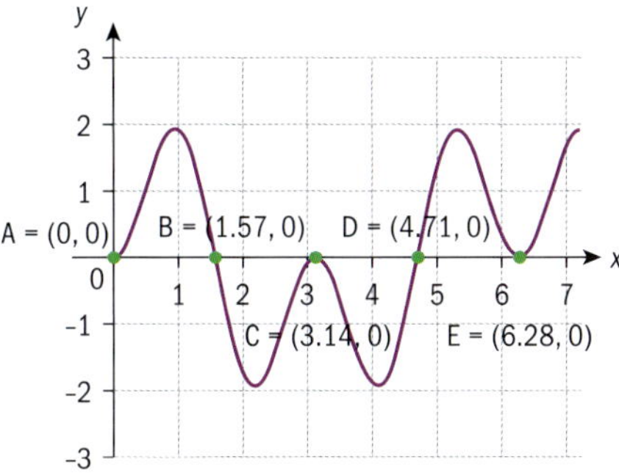

Interpret the graph. Notice that here, you can only give approximate, decimal values for t and not exact values.

From $t = 0$ s to $t = 1.57$ s the particle is moving to the right. It stops at $t = 1.57$ s, then moves to the left until $t = 3.14$ s and stops. It then continues moving to the left until $t = 4.71$ s, then stops, and then moves to the right until $t = 6.28$ s.

b displacement $= \int_0^{2\pi} (5\sin^2 t \cos t)\mathrm{d}t$

$$= \left[\frac{5\sin^3 t}{3}\right]_0^{2\pi} = 0\,\text{cm}$$

The total displacement is zero.

The particle starts and stops at the same place.

c total distance travelled $= \int_0^{2\pi} \left|5\sin^2 t \cos t\right| \mathrm{d}t$

$$= 4\int_0^{\frac{\pi}{2}} 5\sin^2 t \cos t \,\mathrm{d}t$$

$$= 4\left[\frac{5\sin^3 t}{3}\right]_0^{\frac{\pi}{2}}$$

$$= \frac{20}{3}\,\text{cm}$$

By looking at the symmetry of the graph, you can see that the distance travelled in each interval of $\frac{\pi}{2}$ is the same.

Exercise 8D

1 A particle travels with velocity $v(t) = (1 - t)\sin 2t$, $0 \le t \le 2$, where v is in metres per second and t is in seconds.

a Find the particle's displacement after 2 seconds.

b Find the total distance travelled by the particle in the 2 seconds.

In questions **2 – 6**, you are given a velocity function and a corresponding time interval. For each function, find

a the times in the given interval when the particle

i stops

ii moves to the left

iii moves to the right

b the particle's displacement at the end of the time interval

c the total distance the particle has travelled in the time interval.

2 $v(t) = 3\cos t;\ 0 \le t \le 2\pi$

3 $v(t) = \sin 2t;\ 0 \le t \le \frac{\pi}{2}$

4 $v(t) = \sqrt{2-t};\ 0 \le t \le 2$

5 $v(t) = e^{\cos t}\sin t;\ 0 \le t \le 2\pi$

6 $v(t) = \dfrac{t}{1+t^2};\ 0 \le t \le 3$

7 The velocity of an object, in m s^{-1}, is given by $v(t) = 3e^{-\frac{t}{3}}$ where t is in seconds.

Find the distance travelled by the object in the first t seconds.

8 A particle moves along the x-axis so that its initial position is 8 cm to the right of the origin. Given that

$$\int_0^{t_1} v(t)\mathrm{d}t = -2; \int_{t_1}^{t_2} v(t)\mathrm{d}t = 16; \int_{t_2}^{t_3} v(t)\mathrm{d}t = -7,$$

a Find the particle's displacement between 0 and t_1.

b Find the total distance travelled between t_1 and t_2.

c Find the position of the particle at times t_1, t_2, and t_3.

9 The graph shows the velocity, in cm s^{-1}, of a particle moving along the x-axis from $t = 0$ s to $t = 7$ s. At $t = 0$ the particle is 2 cm to the left of the origin.

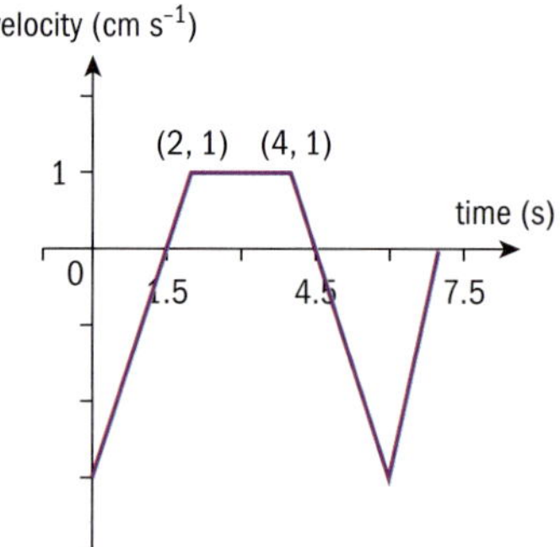

a Find the particle's position at the end of the 7 s.

b Find the total distance the particle travelled in the seven seconds.

Developing your toolkit

Now do the Modelling and investigation activity on page 568.

8.3 Ordinary differential equations (ODEs)

Investigation 5

When astronauts from the Apollo 15 spacecraft landed on the moon, one of the experiments they conducted was to test Galileo's theory that if two objects of different masses are in free fall in a vacuum, they will land at the same time, regardless of their weight. The moon was an ideal place to test this theory, since its gravitational pull is minimal and there is no air resistance. So, one of the astronauts let a hammer and a feather fall from his hands at about 1.22 metres from the ground, and indeed they landed at the same time!

The acceleration due to gravity on the moon is about 1.625 m s^{-2}. This is far less than that on earth, which is about 9.8 m s^{-2}.

1 Write an equation relating an object's acceleration due to gravity on the moon with the rate of change of the object's velocity, v.

Hence, find a formula for v using the initial condition that at $t = 0$, $v = 0$.

2 Write an equation relating the object's velocity, which you found in question **1**, with its rate of change of height above the ground, s.

Use the initial height of the feather and hammer to find a formula for their height above the ground at any time t.

3 How long did it take the two objects to reach the ground after the astronaut let them drop?

In Investigation 5 you saw equations that contain a derivative. Such equations are called **differential equations**. This family of equations is used to describe dynamic processes, as they connect variables together with rates of change. The value of a differential equation at a particular point is called the initial value or condition. When you find the function that satisfies the differential equation, you have solved the differential equation, and when you find the particular solution that takes into the account the initial condition, you have solved the initial value problem.

International-mindedness

French mathematician Jean d'Alembert's analysis of vibrating strings using differential equations, plays an important role in modern theoretical physics.

Students who study physics will be familiar with Newton's laws of motion. Newton's 2nd law states that force is the product of mass and acceleration, that is $F = ma$.

Since acceleration is the derivative of velocity, you can rewrite his law as $F = m\frac{\mathrm{d}v}{\mathrm{d}t}$.

Since velocity is the derivative of displacement, then acceleration is the 2nd derivative of displacement. Hence, Newton's second law can also be written as $F = m\frac{\mathrm{d}^2s}{\mathrm{d}t^2}$.

All of these laws are expressed in terms of rates of change, and are therefore differential equations.

The highest derivative in the differential equation is the **order of the equation**. Newton's Law expressed as $F = m\frac{\mathrm{d}v}{\mathrm{d}t}$ is a differential equation of order one. Newton's Law expressed as $F = m\frac{\mathrm{d}^2s}{\mathrm{d}t^2}$ is a differential equation of order two, and so on.

The differential equations you formed in Investigation 5 could be easily solved using the integration techniques you have already studied. Using these same techniques, you will now work on solving differential equations using the method of **separation of variables**.

Example 9

Solve the differential equation for y if $\frac{dy}{dx} = x(1+y)e^x$.

$\frac{dy}{dx} = x(1+y)e^x \Rightarrow \frac{dy}{1+y} = xe^x dx$	Separate the variables. Express all terms in x on one side, and all terms in y on the other.
$\Rightarrow \int \frac{dy}{1+y} = \int xe^x dx$	Integrate both sides with respect to their variable.
Integrating the left-hand side: $\int \frac{dy}{1+y} = \ln \lvert 1+y \rvert$	Integrate by inspection.
Integrating the right-hand side: $u = x; du = dx; v = e^x; dv = e^x dx$ $\Rightarrow \int (xe^x) dx = xe^x - \int e^x dx = xe^x - e^x + k$	Integrate by parts.
LHS = RHS $\therefore \ln \lvert 1+y \rvert = xe^x - e^x + d$	Combine the constants of integration into one constant, d.
$\Rightarrow y+1 = e^{xe^x - e^x + d} = Ae^{xe^x - e^x}$ $y = Ae^{xe^x - e^x} - 1$	Solve for y, and since e^d is also a constant, you can represent it using A.

A differential equation $\frac{dy}{dx} = f(x,y)$ is separable if f can be expressed as a product of a function in x and a function in y, $\frac{dy}{dx} = g(x)h(y)$.

If $h(y) \neq 0$, the variables can be separated to give

$$\frac{dy}{h(y)} = g(x)dx \Rightarrow \int \frac{dy}{h(y)} = \int g(x)dx.$$

Example 10

Solve the differential equation $(1+x^2)\frac{dy}{dx} = 1+y^2$ for y which satisfies the initial condition $y(0) = 2$.

$(1+x^2)\frac{dy}{dx} = 1+y^2 \Rightarrow \frac{dy}{1+y^2} = \frac{dx}{1+x^2}$ $\Rightarrow \int \frac{dy}{1+y^2} = \int \frac{dx}{1+x^2}$	Separate the variables.

$\Rightarrow \arctan y = \arctan x + c$ $\Rightarrow y = \tan(\arctan x + c)$	Integrate and combine the constants of integration.
$y(0) = 2 \Rightarrow \tan c = 2$	Substitute the initial condition.
$\Rightarrow y = \tan(\arctan x + \arctan 2)$ $\Rightarrow y = \dfrac{x+2}{1-2x}$	Use compound angle formula for $\tan(\alpha + \beta)$.

Exercise 8E

1 Solve each differential equation to find an expression for y

a $\dfrac{dy}{dx} = \dfrac{2x^2}{y^2}$ **b** $\dfrac{dy}{dx} = e^{y-x}$

c $e^{-2x}\dfrac{dy}{dx} = \dfrac{3}{y}$ **d** $\dfrac{dy}{dx} = y\cos x$

e $\dfrac{dy}{dx} = \dfrac{x+\sin x}{5y^4}$ **f** $\left(1+9x^2\right)\dfrac{dy}{dx} = 1$

2 Solve the differential equations for y given an initial value.

a $x^{\frac{1}{2}} + y^{\frac{1}{2}}\dfrac{dy}{dx} = 0;\ y(1) = 4$

b $xe^{x^2} + y\dfrac{dy}{dx} = 0; y(0) = 1$

c $\dfrac{dy}{dx} = e^{x-2y};\ y(0) = 0$

d $\dfrac{dy}{dx} = 2xy\sin x^2; y(0) = 1$

e $e^x\dfrac{dy}{dx} + x = 0; y(0) = 2$

Differential equations are used to model many phenomena. The problems that follow will show you the variety of uses for ODEs. In some problems the model will be given, and in others you may be asked to formulate the differential equation, and then solve.

Example 11

The rate of growth of a colony of bacteria, $\dfrac{dP}{dt}$, is proportional to its current size, P.

At $t = 0$ hours, the population of the bacteria is 1000.

a Write a differential equation to model this population.

After one hour the colony has grown by 20%.

b Find the constant of proportionality.

c Determine how many hours it will take for the population of bacteria to grow to 50 000, and confirm your answer graphically.

Continued on next page

Calculus

a $\frac{dP}{dt} = kP \Rightarrow \frac{dP}{P} = kdt$	The rate of growth is directly proportional to its population. k is the constant of proportionality.
b $\Rightarrow \int \frac{dP}{P} = \int kdt$	Separate the variables.
$\Rightarrow \ln P = kt + c$	Integrate.
$\Rightarrow P = e^{kt+c} = Ae^{kt}$	Let $A = e^c$, since e^c is a constant.
$P(0) = 1000 \Rightarrow 1000 = A$	
$\Rightarrow P = 1000e^{kt}$	Write the function substituting for A.
$P(1) = 1200 \Rightarrow 1200 = 1000e^{k}$	Use $P(1) = 1200$ to solve for k.
$\Rightarrow k = \ln 1.2$	
$\therefore P = 1000e^{t\ln 1.2}$	This is the solution of the differential equation.
c $50\,000 = 1000e^{t\ln 1.2}$	Substitute $P = 50\,000$ and solve for t.
$\Rightarrow 50 = e^{t\ln 1.2} = e^{\ln 1.2^t}$	
$1.2^t = 50 \Rightarrow t = 21.5$ hours	

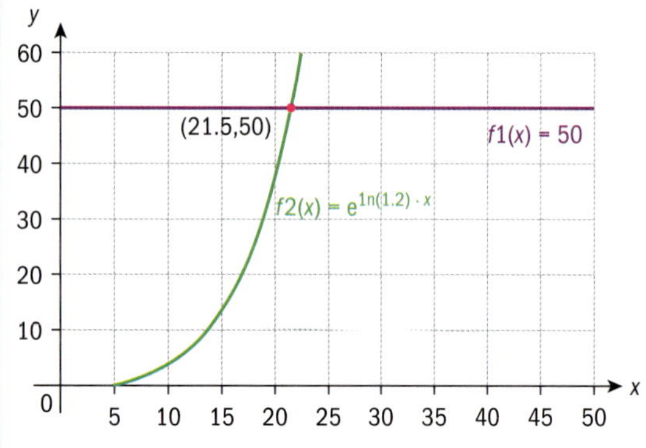

Reflect Do you think that populations grow as in the model shown? That is, does a population grow exponentially for an indefinite period of time?

Example 12

Newton's Law of Cooling (which also applies to warming) states that the rate at which an object's temperature is changing at any given time t is proportional to the difference between its temperature T and the temperature of the surrounding environment, T_0.

Police discover a dead body at 06:00 in the morning in a room maintained at 22°C. When discovered, the body's temperature was 30°C. Three hours later, the body's temperature had dropped to 26°C. The body temperature at the time of death is assumed to be 37°C. Estimate the time at which death occurred.

$\frac{dT}{dt} = -k(T - T_0)$	Let t be the number of hours that have passed since midnight the night before the body was found.
$\Rightarrow \frac{dT}{dt} = -k(T - 22)$	k is the constant of proportionality. The negative rate of change indicates that the body temperature is falling.
$\Rightarrow \frac{dT}{T - 22} = -kdt$ $\Rightarrow \int \frac{dT}{T - 22} = \int -kdt$ $\Rightarrow \ln(T - 22) = -kt + c$	Separate the variables, and integrate.
$\Rightarrow e^{-kt+c} = T - 22$ $\Rightarrow Ae^{-kt} = T - 22$	Since $e^{-kt+c} = e^{-kt} \times e^{c}$, e^{c} can be represented by the constant A.
When $t = 6$, $T = 30$, and when $t = 9$, $T = 26$: $\Rightarrow Ae^{-6k} = 30 - 22 = 8$, $Ae^{-9k} = 26 - 22 = 4$	Use the given information to find two equations in A and k.
$e^{3k} = \frac{8}{4} \Rightarrow k = \frac{\ln 2}{3}$ Using 1st equation, $Ae^{-6\frac{\ln 2}{3}} = 8 \Rightarrow A = 8e^{2\ln 2} = 8 \times 4 = 32$ So $T = 32e^{-\frac{t\ln 2}{3}} + 22$	Solve simultaneously by dividing both equations. It's easiest to leave your answer in exact form until the end.
Person died when $T = 37°C$. $37 = 32e^{-\frac{t\ln 2}{3}} + 22$ $e^{-\frac{t\ln 2}{3}} = \frac{15}{32}$ $e^{\frac{t\ln 2}{3}} = \frac{32}{15}$ $t = \frac{3}{\ln 2}\left(\ln\left(\frac{32}{15}\right)\right) = 3.28$	Substitute $T = 37°C$ and solve for t.
The person died approximately 3.28 hours after midnight, that is 03:17 am.	State your answer in context.

Using an exponential model for population growth (as you did in Example 11) does not take into account certain limitations on population. For a population of people, limitations on growth could be available resources, natural catastrophes, effects of diseases, war, etc; all of which

Calculus

will influence the population's rate of growth. Therefore, the equation that is typically used to model the population size is called the logistic differential equation. In a sense, these equations prevent the indefinite growth (or decay) of an exponential model. Other than population studies, logistic functions are used in medical research, eg in modelling the growth of tumours, in the field of economics, and even in linguistics.

The standard logistic equation is of the form $\frac{\mathrm{d}n}{\mathrm{d}t} = kn(a-n), a, k \in \mathbb{R}$.

t = the time during which a population grows
n = the population after time t
k = relative growth
a = constant (depends on problem being modelled)

*There are other forms of logistic equations. The logistic equation will always be given to you in the problem.

Example 13

The rate of growth of a giraffe population in a national park can be modelled by the logistic equation $\frac{\mathrm{d}P}{\mathrm{d}t} = 0.0002P(200-P)$, where t is in years. If there were 30 giraffes at the time scientists starting studying this population, determine

a how many years it would take for the giraffe population to reach 150

b the maximum the giraffe population can grow to.

a $\frac{\mathrm{d}P}{\mathrm{d}t} = 0.0002P(200-P) \Rightarrow \frac{\mathrm{d}P}{P(200-P)} = 0.0002\mathrm{d}t$	Separate the variables.
$\Rightarrow \int \frac{\mathrm{d}P}{P(200-P)} = \int 0.0002\mathrm{d}t$	Express as integrals.
$\frac{1}{P(200-P)} = \frac{1}{200}\left(\frac{1}{P} + \frac{1}{200-P}\right)$	Partial fraction decomposition.
$\Rightarrow \int \left(\frac{1}{P} + \frac{1}{200-P}\right)\mathrm{d}P = 200\int 0.0002\mathrm{d}t$	
$\Rightarrow \ln P - \ln(200-P) = 0.04t + c$	Integrate.
$\Rightarrow \ln\left(\frac{P}{200-P}\right) = 0.04t + c$	
$\Rightarrow e^{0.04t+c} = \frac{P}{200-P}$	

$Ae^{0.04t} = \frac{P}{200-P} \Rightarrow P = \frac{200Ae^{0.04t}}{1+Ae^{0.04t}}$	Let $e^c = A$ and make P the subject.
$P(0) = 30 \Rightarrow A = 0.1764...$	Use the initial condition to solve for A.
$\Rightarrow 0.1765e^{0.04t} = \frac{P}{200-P}$	
$0.1765e^{0.04t} = \frac{150}{200-150} = 3$	Substitute $P = 150$, and solve for t.
$t = 71$ years	
b $A = 0.1765 \Rightarrow P = \frac{200 \cdot 0.1765e^{0.04t}}{1+0.1765e^{0.04t}}$	Substitute the value you found for A into the expression for P.
$\lim_{t\to\infty} \frac{200 \cdot 0.1765e^{0.04t}}{1+0.1765e^{0.04t}} = 200$	Find the limit of P as t tends to infinity.

Reflect Compare the graph of the exponential model used in Example 11 to the graph of the logistic model shown here, which is used in Example 13. Why is the logistic model superior to the exponential model to model population growth?

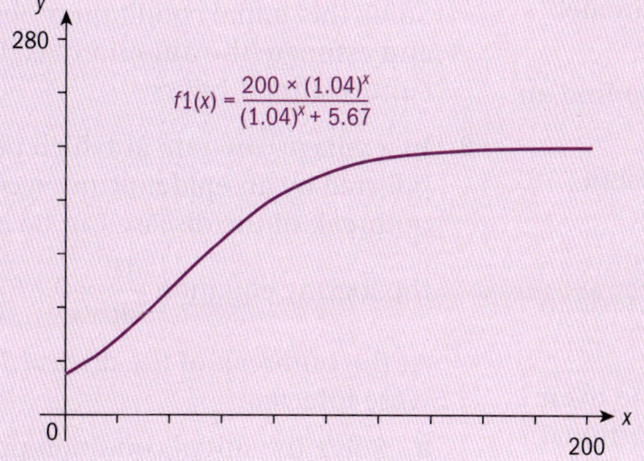

TOK

π vs τ

You have been using radians to measure angles instead of degrees in recent chapters.

Why has this change been necessary? What are its advantages?

Exercise 8F

1 Newton's Law of Cooling states that the rate at which an object's temperature is changing at any given time t is proportional to the difference between its temperature T and the temperature of the surrounding environment T_0. If an object in a laboratory that is kept at a constant temperature cools from 100°C to 80°C in ten minutes, and from 80°C to 65°C after a further ten minutes, find

a a differential equation to model this problem

b the temperature of the laboratory

c the temperature of the object after another 10 minutes has elapsed.

2 An industrial oven is kept at a constant temperature of 180°C. A chemical, initially at 0°C, is placed into the oven.

The rate of increase in temperature of the chemical is proportional to the difference between the oven's temperature and the chemical's temperature, T.

After 5 minutes have passed, the temperature of the chemical is 120°C.

a Find a differential equation that models this problem.

b Determine the temperature of the chemical after a further 5 minutes in the oven.

3 At any point P on a curve in the first quadrant, the tangent to the curve at P intersects the x-axis at point Q. The distance from point P to the origin is equal to the distance from point P to point Q.

Find the equation of the curve, in terms of y, given that the point (2,1) lies on the curve.

4 The rate of growth of a population P is proportional to the size of the population. A culture of bacteria in a lab has an initial population of 500, and after 3 hours, the population has grown to 10 000 bacteria.

a Write a differential equation to model this problem.

b Solve the differential equation to find an expression for P.

c Find the number of bacteria present after 5 hours.

d Determine how many hours are necessary for the number of bacteria to reach 500 000.

5 Once a spherical soufflé is put in the oven, its volume increases at a rate proportional to its radius.

a Show that the radius r cm of the soufflé at time t minutes after it has been put into the oven satisfies the differential equation $\frac{\mathrm{d}r}{\mathrm{d}t} = \frac{k}{r}$.

b The radius of the soufflé is 8 cm when it goes in to the oven, and 12 cm when it is fully cooked, 30 minutes later. Find, to the nearest centimetre, the radius of the soufflé after 15 minutes in the oven.

6 On the first day of school after a break, one student comes to school having been infected with the flu virus. There are 100 students in the school. The rate of the spread of the flu virus in the school can be modelled by $\frac{\mathrm{d}N}{\mathrm{d}t} = kN(100 - N)$, where N is the number of students infected after t days.

a Show that the solution to this differential equation is $Ae^{100tk} = \frac{N}{100 - N}$.

b If on day 1 one student was infected, and on day 2 two students were infected, find A and k.

c Solve the differential equation for N, and hence determine how many students were infected after one week.

7 The cells in a culture in a lab grow according to the logistic equation

$\frac{\mathrm{d}P}{\mathrm{d}t} = 0.02P\left(1 - \frac{P}{2000}\right)$. Initially the population of cells is 300.

Solve this initial condition problem for P, and estimate the amount of cells in the culture after 3 hours.

8 In a village, the rate at which people are infected by an epidemic per week after the outbreak of the disease can be modelled by the logistic equation $\frac{\mathrm{d}P}{\mathrm{d}t} = 0.8P\left(1 - \frac{P}{350}\right)$.

At the outbreak of the disease 7 people were infected.

a Solve the initial condition problem for P.

b Estimate how many people in total will have been infected by the disease.

9 The acceleration, in m s^{-2}, of a particle moving in a straight line at time t seconds is modelled by $a = -\frac{1}{2}v$.

a Express a in terms of v and v'.

b If $v(0) = 20$, find v in terms of t.

10 The acceleration of an object is given in terms of its displacement s as $a = \frac{2s}{s^2+1}$ m s^{-2}.

a If $a = v\frac{dv}{ds}$, find v given that $s = 1$ m when $v = 2$ m s^{-1}.

b Hence, find the velocity when the object has travelled 5 metres.

11 The acceleration of a particle moving in a chemical solution can be expressed as $a = -\left(\frac{1}{50}v^2 + 32\right)$ where a is in m s^{-2}. At $t = 0$ its velocity is 40 m s^{-1}. Solve the initial value problem for v, and find the velocity of the object at $t = 10$ s.

Reflect Do you think that the model derived in question **4** is appropriate? Describe how you might adapt it to better suit the situation.

Investigation 6

Consider the differential equation $x\frac{dy}{dx} = x + y$.

1 Is it possible to solve this differential equation using previous methods you have learned? Explain why, or why not.

2 Show that this equation can be written as $\frac{dy}{dx} = 1 + \frac{y}{x}$.

3 Rewrite the equation for $\frac{dy}{dx}$ in terms of a new variable v, where $v = \frac{y}{x}$ and v is a function in x. Solve for y and find $\frac{dy}{dx}$.

4 Since $v = \frac{y}{x} \Rightarrow y = vx$, find another expression for $\frac{dy}{dx}$ using implicit differentiation.

5 Equate the two expressions you found for $\frac{dy}{dx}$ in questions **3** and **4**. Simplify your answer.

6 Solve the differential equation in question **5**.

7 Substituting $v = \frac{y}{x}$ into the expression you obtained in question **6**, show that $y = x(\ln x + c)$.

8 Using the initial condition that $y = -1$ when $x = 1$, find c and hence solve the differential equation for y.

International-mindedness

Differential equations first became solvable with the invention of calculus by Newton and Leibniz. In Chapter 2 of his 1671 work "Method of Fluxions", Newton displayed 3 types of differential equations.

The type of differential equation you have worked through in Investigation 6 is called a **homogeneous differential equation**. It can be written in the form $\frac{dy}{dx} = f\left(\frac{y}{x}\right)$, and is solved using the substitution $v = \frac{y}{x}$, where v is a function in x.

Example 14

Use the substitution $v = \frac{y}{x}$ to find the general solution of the differential equation $x\frac{dy}{dx} - y = x^2 \sin x$.

$x\frac{dy}{dx} - y = x^2 \sin x$ $\quad \frac{dy}{dx} = x\sin x + \frac{y}{x}$	Make $\frac{dy}{dx}$ the subject.
$\Rightarrow \frac{dy}{dx} = x\sin x + v$	Use the given substitution.
$v = \frac{y}{x} \Rightarrow vx = y$	
$\frac{dy}{dx} = v + x\frac{dv}{dx}$	Use implicit differentiation.
$\Rightarrow v + x\frac{dv}{dx} = x\sin x + v$	Equate both expressions for $\frac{dy}{dx}$.
$\Rightarrow \frac{dv}{dx} = \sin x$	Make $\frac{dv}{dx}$ the subject.
$\Rightarrow v = -\cos x + c$	Integrate both sides.
$v = \frac{y}{x} \Rightarrow y = x(-\cos x + c)$	Solve for y.

HINT

For simplicity's sake, it is normal to write down derivatives in differential equations using 'prime notation'. Newton's laws can therefore be expressed as $F = mv'$ and $F = ms''$.

Example 15

Use the substitution $v = \frac{y}{x}$ to solve the following initial value problem for y.

$xy' = y(\ln x - \ln y)$ and $y(1) = 4$.

$xy' = y(\ln x - \ln y) \Rightarrow y' = \frac{y}{x}\ln\left(\frac{x}{y}\right) = v\ln\left(\frac{1}{v}\right)$ $= -v\ln v$	Substitute v and rearrange for y'.
$y' = v + xv'$	Differentiate vx.
$v + x\frac{dv}{dx} = -v\ln v$	Equate expressions for y'.
$1 + \frac{x}{v}\frac{dv}{dx} = -\ln v$ $\frac{x}{v}\frac{dv}{dx} = -(\ln v + 1)$	Divide through by v.
$\Rightarrow \frac{dv}{v(\ln v + 1)} = -\frac{dx}{x}$	Separate the variables.
$\Rightarrow \ln\left(\ln\left(\frac{y}{x}\right) + 1\right) = c - \ln x$	Integrate LHS by substitution where $u = \ln v + 1$, and substitute back for $v = \frac{y}{x}$.

$y(1) = 4 \Rightarrow c = \ln(\ln 4 + 1)$	Use the initial condition to solve for c.
$\ln\left(\ln\left(\frac{y}{x}\right)+1\right) = \ln(\ln 4+1) - \ln x$	
$\Rightarrow \ln\left(\ln\left(\frac{y}{x}\right)+1\right) = \ln\left(\frac{\ln 4+1}{x}\right)$	
$\Rightarrow \ln\left(\frac{y}{x}\right)+1 = \frac{\ln 4+1}{x}$	Simplify, and solve for y.
$\Rightarrow \ln\left(\frac{y}{x}\right) = \frac{\ln 4+1}{x} - 1 = \frac{\ln 4+1-x}{x}$	
$\Rightarrow y = xe^{\frac{\ln 4+1-x}{x}}$	

Exercise 8G

Use the substitution $v = \frac{y}{x}$ to solve the differential equations in this exercise.

1 **a** $xy' = x^2\cos x + y$ **b** $x^2\frac{dy}{dx} = 3x^2 - xy$

c $x^2\frac{dy}{dx} = y^2 + xy + 4x^2$

2 Solve for y when $x^2y' = y^2 + xy + 4x^2$ and $y(1) = 2$.

3 Solve $y' = \frac{y}{x} + \frac{y^2}{x^2}$ for y, if $y(1) = 2$.

4 Solve the differential equation $x^2y' = y^2 + 3xy + 2x^2$ when $y(1) = -1$. Give your answer in the form $y = f(x)$.

Investigation 7

Consider the differential equation $x^2\frac{dy}{dx} + 2xy = 1$.

1 Rewrite this equation in the form $\frac{d[f(x)g(x)]}{dx} = 1$ for functions f and g.

This is called an exact differential equation.

2 Integrate, with respect to x, both sides of the equation you found in question **1**. Hence find an expression for y in terms of x.

Consider the equation $\frac{dy}{dx} + p(x)y = q(x)$ (*)

The left-hand side is not an exact differential equation. We want to multiply each term by another function, $I(x)$, in order to transform it into an exact differential equation.

3 Multiply each term in (*) by $I(x)$.

Continued on next page

TOK

Is there always a trade-off between accuracy and simplicity?

Calculus

4 What can you say about the first term in your answer to question **3** and the first term in the product rule for $\frac{d}{dx}(Iy) = Iy' + I'y$?

In order for the second term in your answer to question **3** to match $I'y$, you need $I(x)p(x)y = \frac{dI}{dx}y$, and hence $I(x)p(x) = \frac{dI}{dx}$.

5 Solve the differential equation $I(x)p(x) = \frac{dI}{dx}$ for $I(x)$ by separating the variables.

6 Now re-write your answer to question **3** using the function $I(x)$ you found in question **5**.

7 The left-hand side should now be an exact differential equation. Use the reverse of the product rule to write the equation in question **6** with a single term on the left-hand side.

The term $I = e^{\int p(x)\,dx}$ is called the integrating factor.

8 **Factual** What is the standard form for a differential equation?

9 **Conceptual** How does the integrating factor help you solve an ordinary differential equation?

A differential equation of the form $\frac{d}{dx}(y(x)u(x)) = v(x)$ is an **exact differential equation**.

A differential equation of the form $\frac{dy}{dx} + p(x)y = q(x)$ is in **standard form**, and can be made **exact** by multiplying both sides by the integrating factor, $I = e^{\int p(x)dx}$.

Example 16

Consider the first order differential equation $(x+1)\frac{dy}{dx} - 3y = (x+1)^5$.

a Find an integrating factor for this equation.

b Hence, solve the equation.

a $(x+1)\frac{dy}{dx} - 3y = (x+1)^5$ $\Rightarrow \frac{dy}{dx} - \frac{3y}{x+1} = (x+1)^4$	Convert the given equation into the required form: $\frac{dy}{dx} + p(x)y = q(x)$.
$p(x) = -\frac{3}{x+1} \Rightarrow I = e^{\int -\frac{3}{x+1}dx}$	Identify $p(x)$, and find $I = e^{\int p(x)\,dx}$.
$e^{\int -\frac{3}{x+1}dx} = e^{-3\ln(x+1)} = \frac{1}{(x+1)^3}$	Simplify.

b $\frac{1}{(x+1)^3}\left(\frac{dy}{dx}-\frac{3y}{x+1}\right)=\frac{1}{(x+1)^3}(x+1)^4$	Multiply both sides of the converted form of the equation by I.
$\frac{1}{(x+1)^3}\frac{dy}{dx}-\frac{3y}{(1+x)^4}=(x+1)$	Simplify.
$\frac{d}{dx}\left(\frac{y}{(x+1)^3}\right)=x+1$	Since the left-hand side is now exact, write it as a single term using the product rule.
$\frac{y}{(x+1)^3}=\frac{1}{2}x^2+x+c$	Integrate both sides.
$y=(x+1)^3\left(\frac{1}{2}x^2+x+c\right)$	Solve for y.

Example 17

Solve the initial value problem for the first order differential equation $\frac{dy}{dx}-y\tan x=-\sec x$ given that $y(0) = 1$.

$I=e^{\int -\tan x\,dx}=e^{\ln(\cos x)}=\cos x$	If the equation is already in the correct form, find I.
$\cos x\left(\frac{dy}{dx}-y\tan x\right)=\cos x(-\sec x)$	Multiply by I.
$\cos x\frac{dy}{dx}-y\sin x=-1$	Simplify.
$\frac{d}{dx}(y\cos x)=-1$	Write LHS as a single term.
$\Rightarrow y\cos x=-x+c$	Integrate both sides.
$y(0)=1\Rightarrow c=1$	Substitute the initial values.
$\Rightarrow y=\frac{1-x}{\cos x}$	Solve for y.

Exercise 8H

1 Find the general solution, in terms of y, of these differential equations.

a $y' + y = e^x$

b $(x-1)y' = y - x^2y$

c $xy' + y = x^2 + 1$

d $y' + y = \sin(e^x)$

e $y' + xy = xe^{x^2}$

f $x^2y' + 2xy = \cos x$

g $xy' + 2y = \cos x$

2 Solve these initial value problems.

a $xy' = y + x^3 \sin x;\ y(\pi) = 0$

b $xy' - \frac{3}{x} = 2y; y(2) = 5$

c $\frac{dy}{dx} + y\tan x = \sec x; y(0) = 2$

Numerical methods for solving ordinary differential equations

Up until now you have explored various analytical methods for solving differential equations given various forms of the equation. These methods however are still not sufficient for solving all differential equations, so numerical methods are used to solve, for example, a differential equation of the form $\frac{dy}{dx} = f(x, y)$ given an initial condition.

The derivative of any point on a curve $(x_0, y(x_0))$ can be approximated using the gradient of the tangent to the curve at x_0, which is given by $y'(x_0) = \frac{y(x_0 + h) - y(x_0)}{h}$.

Rearranging this formula gives $y(x_0 + h) = y(x_0) + h \times y'(x_0)$.

The smaller the values of h become, the better the formula for the gradient of the tangent approximates the actual derivative of the curve at $(x_0, y(x_0))$.

This process of approximating the gradient at a point by choosing ever smaller values of h is helpful in approximating solutions of differential equations. This method is called the linearization method, or Euler's method, and you can write the gradient formula recursively as $y_{n+1} = y_n + hf(x_n, y_n)$, where $x_{n+1} = x_n + h$ and $f(x_n, y_n) = y'(x_n)$, as given by the differential equation.

Investigation 8

Consider the differential equation $y' = 1 + y$ and the initial condition that $y(0) = 1$.

This is of the form $\frac{dy}{dx} = f(x, y)$, where $f(x, y) = 1 + y$.

In this investigation, you will use the recurrence formulas $x_{n+1} = x_n + h$ and $y_{n+1} = y_n + hf(x_n, y_n)$, with step size $h = 0.5$, to solve the equation numerically.

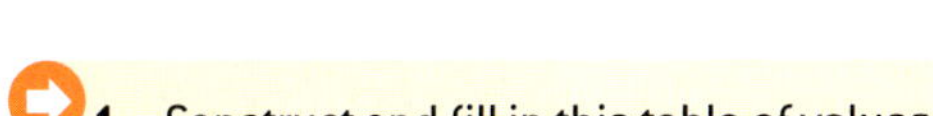

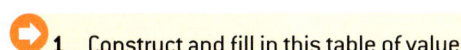

1 Construct and fill in this table of values.

n	x_n	y_n	$f(x_n, y_n) = 1 + y_n$
0	0	1	2
1	0.5	$1 + 2(0.5) = 2$	3
2	1	$2 + 3(0.5) = 3.5$	4.5
3			
4			
5			

2 Now plot the points (x_n, y_n) on the coordinate axes.

3 Select an appropriate analytical method and solve the differential equation $y' = 1 + y$.

4 On the same set of axes you used in question **2**, plot the graph of the function y you obtained by solving $y' = 1 + y$ in question **3**. Comment on your result.

5 To the right is part of the table of values for the function you obtained in question **3**: the analytical solution to $y' = 1 + y$.

Compare these values with the approximate values you obtained in question **1**.

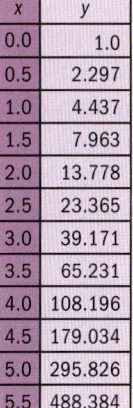

x	y
0.0	1.0
0.5	2.297
1.0	4.437
1.5	7.963
2.0	13.778
2.5	23.365
3.0	39.171
3.5	65.231
4.0	108.196
4.5	179.034
5.0	295.826
5.5	488.384

6 Perform steps **1**, **2** , **4** and **5** with $h = 0.1$. Use this table as your starting point.

n	x_n	y_n	$f(x_n, y_n) = 1 + y$
0	0	1	2
1	0.1	$1 + (2)(0.1) = 1.2$	2.2
2	0.2	$1.2 + (2.2)(0.1) = 1.42$	2.42
3			
4			
5			
6			
7			
8			
9			
10			

7 How do your results show that varying the value of h in the numerical approximation of the solution to the differential equation helps to better approximate the solution of the differential equation at $x = 1$?

8 **Conceptual** How can a numerical approximation to the solution of an ODE be made stronger?

Calculus

Euler's Method for solving differential equations numerically:

Given a first order differential equation in the form $\frac{dy}{dx} = f(x, y)$:

- select the initial point (x_0, y_0)
- use the recurrence formulas $x_{n+1} = x_n + h$ and $y_{n+1} = y_n + hf(x_n, y_n)$ to generate as many points as instructed
- plot the points (x_n, y_n) and connect with a smooth curve.

International-mindedness

The Euler method is named after Swiss mathematician Leonhard Euler (pronounced "oiler"), who proposed this in his book "Institutionum calculi integralis" in 1768.

Example 18

Using Euler's method with step size $h = 0.1$, approximate the solution to the initial value problem $\frac{dy}{dx} = \sin(x + y) - e^x$, $y(0) = 4$.

a Find the coordinates of five points which lie on the approximate solution to the differential equation, and sketch the graph of an approximate solution to $y = y(x)$.

b Find the approximate solution of the differential equation when $x = 0.5$.

c Determine if it's possible to solve this differential equation analytically with the methods you have learned.

a

n	x_n	y_n	$\frac{dy}{dx} = f(x_n, y_n) = \sin(x_n + y_n) - e^{x_n}$
0	0	4	−1.7568...
1	0.1	4+(0.1)(−1.7568) = 3.8243...	−1.8103...
2	0.2	3.8243+(0.1)(−1.8104) = 3.6432...	−1.8668...
3	0.3	3.6433 +(0.1) (−1.8669) = 3.4566...	−1.9268...
4	0.4	3.4566 +(0.1) (−1.9268) = 3.2639...	−1.9907...
5	0.5	3.2639+(0.1) (−1.9907) = 3.0648	−2.0594...

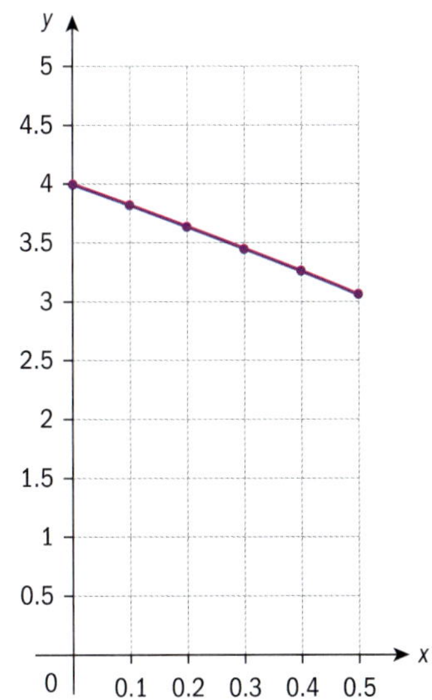

b $y = 3.06$

Draw a line from $x = 0.5$ up to the graph, and then across to the y-axis.

c This differential equation cannot be transformed into any of the forms necessary for an analytical solution.

Exercise 8I

1 Apply Euler's method with step size $h = 0.1$ to approximate the solution to the initial value problem $y' + 2y = 2 - e^{-4x}$; $y(0) = 1$, at $x = 0.5$.

2 Use Euler's method with a step size of 0.4 to approximate the solution of the differential equation $y' = 2xy$; $y(1) = 2$, at $x = 3$.

3 Use Euler's method with a step size of 0.1 to find an approximate value of y when $x = 0.4$ that satisfies the differential equation $y' = x^2 + y^2$ with the initial condition $y(0) = 1$. Explain whether your approximate value is greater than or less than the actual value.

4 Use Euler's method with a step size of 0.1 to find an approximation for the value of y when $x = 0.3$ given that $\frac{dy}{dx} = e^x + 2y^2$ and $y(0) = 1$.

5 (1,2) is a point on the curve defined by $y' = 2x(1 + x^2 - y)$.

a Use Euler's method with $h = 0.1$ to find y when $x = 1.3$, and explain how a more accurate answer can be obtained using this same method.

b Solve the differential equation using an appropriate analytical method, and find y when $x = 1.3$.

Calculus

Developing inquiry skills

Return to the opening problem for the chapter. How can you answer the 5th question with what you have learned in this section?

8.4 Limits revisited

Investigation 9

1 What do you obtain when you try to solve $\lim_{x\to 0}\frac{\sin x}{x}$ algebraically, that is, by substituting $x=0$ into the given function?

2 Let $f(x)=\sin x$ and $g(x)=x$. Graph the function $y=\frac{f(x)}{g(x)}$, and find graphically $\lim_{x\to 0}\frac{\sin x}{x}$. Justify your answer.

3 Find $\lim_{x\to 0}\frac{f'(x)}{g'(x)}$, and compare your answer with $\lim_{x\to 0}\frac{\sin x}{x}$ from question **2**.

Now let $f(x)=1+\cos x$ and $g(x)=x^2\sin x$.

4 Repeat question **1** to show that you cannot algebraically find $\lim_{x\to \pi}\frac{1+\cos x}{x^2\sin x}$, and repeat question **2** to find this limit graphically.

5 Find $\lim_{x\to 0}\frac{f'(x)}{g'(x)}$ algebraically, and compare it with $\lim_{x\to 0}\frac{f(x)}{g(x)}$ which you found graphically.

6 Since $\lim_{x\to 0}\frac{f'(x)}{g'(x)}$ is equivalent to $\frac{f'(0)}{g'(0)}$, write $\frac{f'(0)}{g'(0)}$ using the formal definition of the derivative of a function at $x=a$,

$f'(a)=\lim_{x\to a}\frac{f(x)-f(a)}{x-a}$.

Simplify your expression to show that $\frac{f'(0)}{g'(0)}=\lim_{x\to 0}\frac{f(x)}{g(x)}$.

7 What do you obtain when you try to find $\lim_{x\to\infty}\frac{x}{e^x}$?

8 Graph the function in question **7** and write down its limit as $x\to\infty$.

9 If $h(x)=x$ and $k(x)=e^x$, find $\lim_{x\to\infty}\frac{h'(x)}{k'(x)}$.

The limits in questions **1**, **4**, and **7** are called 'indeterminate forms'.

10 **Factual** What is an indeterminate form?

11 **Conceptual** How can you find an analytical solution to $\lim_{x\to a}\frac{f(x)}{g(x)}$ when substituting for $x=a$ results in the indeterminate forms $\frac{0}{0}$ or $\frac{\pm\infty}{\pm\infty}$?

International-mindedness

Antiphon, Democritus and Leucippus contributed to the Greek method of exhaustion which was put together by Eudoxus about 370 BC to approximate areas that would now be dealt with by means of limits.

L'Hopital's Rule

If $\lim\limits_{x \to a} \frac{f(x)}{g(x)} = \frac{f(a)}{g(a)} = \frac{0}{0}$, and if $\lim\limits_{x \to a} \frac{f'(x)}{g'(x)}$ exists, then $\lim\limits_{x \to a} \frac{f(x)}{g(x)}$ also exists, and $\lim\limits_{x \to a} \frac{f(x)}{g(x)} = \lim\limits_{x \to a} \frac{f'(x)}{g'(x)}$.

This rule can be used when the limit of a quotient results in the indeterminate forms $\frac{0}{0}$ or $\frac{\pm\infty}{\pm\infty}$.

TOK

Is it ethically fair to name this theorem L'Hopital's Rule?

Example 19

Use L'Hopital's Rule, if possible, to find the following limits, and confirm your answers graphically.

a $\lim\limits_{x \to 0} \frac{1 - \cos x}{x^2 + x}$ **b** $\lim\limits_{x \to \frac{\pi}{2}} \frac{\sec x}{1 + \tan x}$ **c** $\lim\limits_{x \to 0} \frac{\cos x}{x}$

a $\lim\limits_{x \to 0} \frac{1 - \cos x}{x^2 + x} = \frac{0}{0}$

If the limit is one of the two indeterminate forms, use L'Hopital's Rule.

$\lim\limits_{x \to 0} \frac{1 - \cos x}{x^2 + x} = \lim\limits_{x \to 0} \frac{\sin x}{2x + 1} = \frac{0}{1} = 0$

Differentiate the numerator and denominator, and evaluate each derivative at $x = 0$.

Confirm graphically.

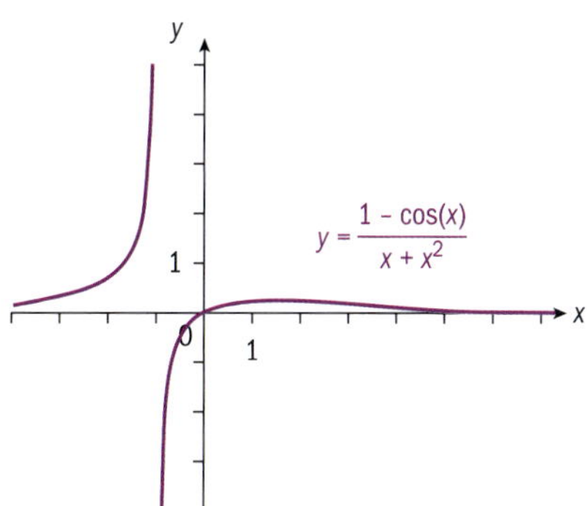

b $\lim\limits_{x \to \frac{\pi}{2}} \frac{\sec x}{1 + \tan x} = \frac{\infty}{\infty}$

If the limit is one of the two indeterminate forms, use L'Hopital's Rule.

$\lim\limits_{x \to \frac{\pi}{2}} \frac{\sec x}{1 + \tan x} = \lim\limits_{x \to \frac{\pi}{2}} \frac{\sec x \tan x}{\sec^2 x} = \lim\limits_{x \to \frac{\pi}{2}} (\sin x) = 1$

Differentiate the numerator and denominator, simplify, and evaluate at $x = \frac{\pi}{2}$.

Confirm graphically.

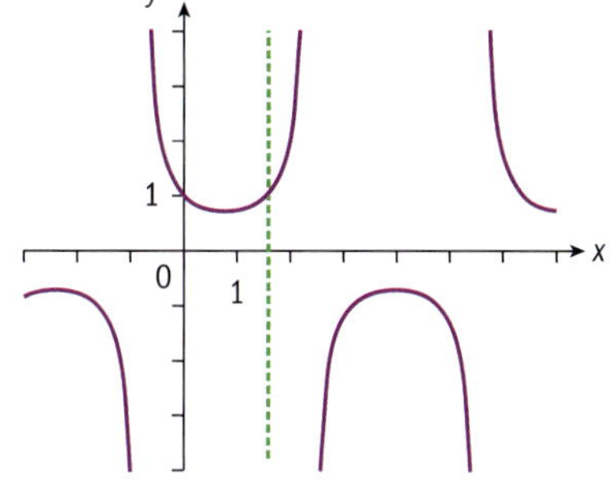

Here, note that the right- and left-hand limits are both 1, but the right-hand indeterminate form is $\frac{-\infty}{-\infty}$.

Continued on next page

Calculus

c $\lim_{x\to 0}\frac{\cos x}{x}=\frac{1}{0}\Rightarrow$ L'Hopital's Rule cannot be used since $\lim_{x\to 0}\frac{f(x)}{g(x)}$ is not one of the two indeterminate forms.	This indeterminate form does not apply.

Did you know?

In Chapter 6 you found the derivative of $y=\sin(x)$ from first principles. In doing this, you calculated $\lim_{h\to 0}\frac{\sin(x+h)-\sin x}{(x+h)-x}$, and saw graphically that this limit is 1. You can now prove that this is true analytically using L'Hopital's Rule.

There are times when you will need to apply L'Hopital's Rule more than once in the same problem. For example, find $\lim_{x\to 0}\frac{\cos x-1}{x^2}$.

Substituting $x=0$ in the function, $\lim_{x\to 0}\frac{\cos x-1}{x^2}=\frac{0}{0}$, so you can apply L'Hopital's Rule. Therefore, $\lim_{x\to 0}\frac{\cos x-1}{x^2}=\lim_{x\to 0}\frac{-\sin x}{2x}$.

Substituting again $x=0$, $\lim_{x\to 0}\frac{\cos x-1}{x^2}=\lim_{x\to 0}\frac{-\sin x}{2x}=\frac{0}{0}$, so you can apply L'Hopital's Rule again.

$\lim_{x\to 0}\frac{\cos x-1}{x^2}=\lim_{x\to 0}\frac{-\sin x}{2x}=\lim_{x\to 0}\frac{-\cos x}{2}$. Now substituting $x=0$,

$\lim_{x\to 0}\frac{-\cos x}{2}=-\frac{1}{2}$.

Therefore, $\lim_{x\to 0}\frac{\cos x-1}{x^2}=-\frac{1}{2}$.

TOK

Does mathematics have a prescribed method of its own?

Example 20

Find $\lim_{x\to 0}\frac{1-\cos 3x}{x^2}$.

$\lim_{x\to 0}\frac{1-\cos(3\times 0)}{0}=\frac{0}{0}$	Since the limit is one of the two indeterminate forms, apply L'Hopital's Rule.
$\lim_{x\to 0}\frac{3\sin 3x}{2x}=\frac{0}{0}$	Apply L'Hopital's Rule again.
$\lim_{x\to 0}\frac{9\cos 3x}{2}=\frac{9}{2}$	

L'Hopital's Rule is also useful in finding horizontal asymptotes, as the next example illustrates.

Example 21

Find the horizontal asymptote of the function $y = \dfrac{x^2-1}{x^2+1}$. Verify your answer graphically.

$\lim\limits_{x\to\infty}\dfrac{x^2-1}{x^2+1} = \dfrac{\infty}{\infty}$	The horizontal asymptote of $y = f(x)$ is $\lim\limits_{x\to\infty} f(x)$.
Applying L'Hopital's Rule, $\lim\limits_{x\to\infty}\dfrac{x^2-1}{x^2+1} = \lim\limits_{x\to\infty}\dfrac{2x}{2x} = \lim\limits_{x\to\infty} 1 = 1$	L'Hopital's Rule is applicable since substituting $x = \infty$ gives the indeterminate form $\dfrac{\infty}{\infty}$.
Asymptote is $y = 1$.	Plotting the graph confirms that the horizontal asymptote is $y = 1$.

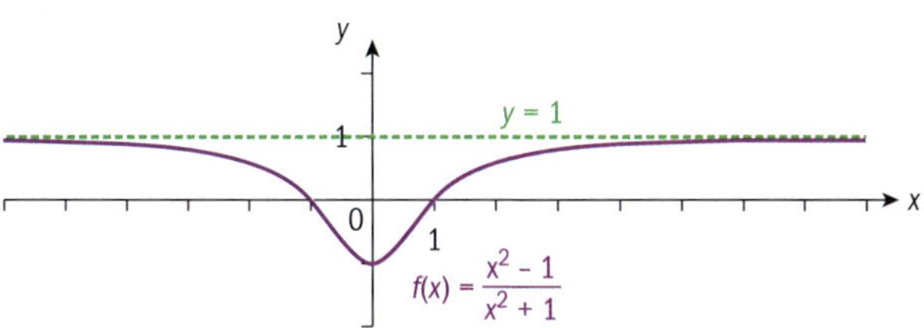

Exercise 8J

1 Use L'Hopital's Rule to find the following limits, if possible.

a $\lim\limits_{x\to 0}\dfrac{\sin x}{x}$

b $\lim\limits_{x\to 0}\dfrac{\tan 3x}{\tan 4x}$

c $\lim\limits_{x\to 0}\dfrac{1-\cos x}{x}$

d $\lim\limits_{x\to 3}\dfrac{e^{x^2}-e^9}{x-3}$

e $\lim\limits_{x\to e}\dfrac{1-\ln x}{\frac{x}{e}-1}$

f $\lim\limits_{x\to 1}\dfrac{\arctan x-\frac{\pi}{4}}{\tan\frac{\pi}{4}x-1}$

g $\lim\limits_{x\to 0}\dfrac{\cos x-\cos 2x}{x-\cos x}$

h $\lim\limits_{x\to 0}\dfrac{\sin(x^2)}{\ln(\cos x)}$

i $\lim\limits_{x\to 0}\dfrac{\ln(1+x)-x}{\cos x-1}$

j $\lim\limits_{x\to\infty}\dfrac{x^2}{e^{1-x}}$

k $\lim\limits_{x\to\infty}\dfrac{x^2}{e^x\ln x}$

l $\lim\limits_{x\to 0}\dfrac{\cos x-\cos 2x}{x^2}$

2 Use L'Hopital's Rule, if possible, to find any horizontal asymptotes of the following functions.

a $y = \dfrac{2x^3-x^2}{2-x-x^4}$

b $y = \dfrac{3x^3-2x^2+1}{2-x-x^3}$

c $y = \dfrac{3x^5-7x}{2x^2+4}$

d $y = \dfrac{2^x}{x^2}$

Constructing polynomials

Investigation 10

Consider the polynomial of order 5 given by $P(x) = a_0 + a_1x + a_2x^2 + a_3x^3 + a_4x^4 + a_5x^5$ and the function $f(x) = \ln(1 + x)$. You will examine the behaviour of both functions at $x = 0$.

1. If $f(0) = P(0)$, find a_0.
2. Find and equate the derivatives $f^{(n)}(x)$ and $P^{(n)}(x)$ for $n = 1, 2, 3, 4, 5$. Each time, solve $f^{(n)}(0) = P^{(n)}(0)$ and find the value of a_n. Do not simplify your answers.
3. Find an expression for each coefficient a_n of the polynomial in terms of $f^{(n)}(0)$ and n.
4. Now, simplify the coefficients and write out the polynomial.
5. Graph $f(x)$ and $P(x)$ on the same set of axes, and describe the behaviour of the functions around an interval having $x = 0$ as its centre.
6. **Conceptual** How can a finite number of terms of an infinite series approximate a function?

What you have just done in Investigation 10 was to construct the 5th order **Maclaurin polynomial** for the function $f(x) = \ln(1 + x)$ at $x = 0$. It is quite astounding that you can construct such a polynomial just by knowing its behaviour around a single point, $x = 0$. You can see that the polynomial approximates the function in the interval $]-1, 1[$, whose centre is $x = 0$.

If $f(x)$ has n derivatives at $x = 0$, then $P(x)$, the **Maclaurin polynomial** of degree n for $f(x)$ centred at $x = 0$, is the unique polynomial of degree n which satisfies the following conditions:

- $f(0) = P(0)$
- $f^{(n)}(0) = P^{(n)}(0)$
- $a_1 = \frac{f'(0)}{1!}; a_2 = \frac{f''(0)}{2!}; a_3 = \frac{f'''(0)}{3!}; a_4 = \frac{f^{(4)}(0)}{4!}; a_5 = \frac{f^{(5)}(0)}{5!}$
- $P(x) = f(0) + \frac{f'(0)}{1!}x + \frac{f''(0)}{2!}x^2 + \ldots + \frac{f^{(n)}(0)}{n!}x^n = \sum_{k=0}^{n} \frac{f^{(k)}(0)}{k!}x^k$

TOK

The Maclaurin series is named after the Scottish mathematician Colin Maclaurin (1678–1746).

Example 22

Compute the 6th order Maclaurin polynomial for the given functions. Graph the polynomial and f on the same set of axes, and comment on how accurately the polynomial approximates the function in the region around $x = 0$. State how you could improve the approximation.

a $f(x) = e^x$ **b** $f(x) = \cos x$ **c** $f(x) = x\cos 2x$ Hint: use your result from **b**.

a $f(x) = \mathrm{e}^x; f(0) = \mathrm{e}^0 = 1 \Rightarrow a_0 = 1; f^{(n)}(x) = \mathrm{e}^x; f^{(n)}(0) = \mathrm{e}^0 = 1$ for all n.

Find successive derivatives of $f(x)$ and evaluate them at $x = 0$.

$$\mathrm{e}^x \cup 1 + x + \frac{x^2}{2!} + \frac{x^3}{3!} + \frac{x^4}{4!} + \frac{x^5}{5!} + \frac{x^6}{6!}$$

Use the general form of a Maclaurin polynomial.

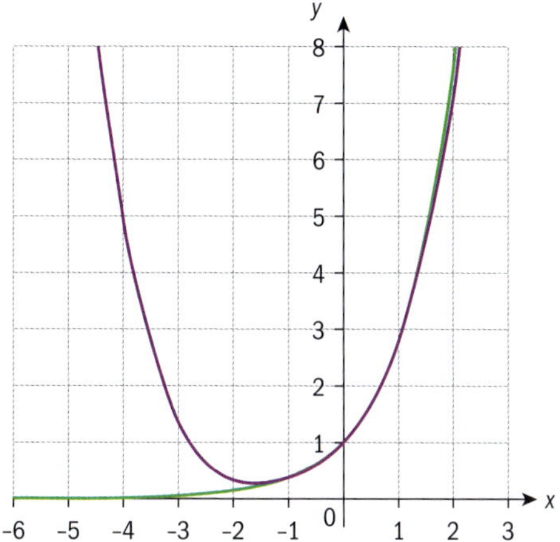

The terms of the polynomial show a good approximation of the function in the interval]−1, 1[. Generating more terms would improve the approximation.

b $f(x) = \cos x; f(0) = 1$

$f'(x) = -\sin x; f'(0) = 0$

$f''(x) = -\cos x; f''(0) = -1$

$f'''(x) = \sin x; f'''(0) = 0$

$f^{(4)}(x) = \cos x; f^{(4)}(0) = 1$

$f^{(5)}(x) = -\sin x; f^{(5)}(0) = 0$

Find successive derivatives of $f(x)$ and evaluate them at $x = 0$.

Observe that $f^{(n)}(0)$ repeats in cycles of 4, ie 1, 0, −1, 0, 1, 0, −1, 0,…

$$\cos x \approx 1 - \frac{x^2}{2!} + \frac{x^4}{4!} - \frac{x^6}{6!}$$

Use the general form of a Maclaurin polynomial.

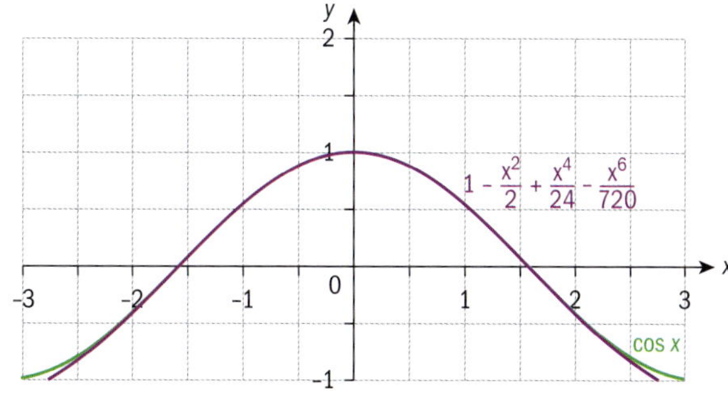

The polynomial shows a good approximation of the function in the interval]−1, 1[. Generating more terms would improve the approximation.

Continued on next page

Calculus

c $\cos x \approx 1 - \frac{x^2}{2!} + \frac{x^4}{4!} - \frac{x^6}{6!} \Rightarrow \cos 2x \approx 1 - \frac{(2x)^2}{2!} + \frac{(2x)^4}{4!} - \frac{(2x)^6}{6!}$

Substitute $2x$ for x in the Maclaurin polynomial for $\cos(x)$.

$x\cos 2x \approx x\left(1 - \frac{(2x)^2}{2!} + \frac{(2x)^4}{4!} - \frac{(2x)^6}{6!}\right) = x - \frac{4x^3}{2!} + \frac{16x^5}{4!} - \frac{64x^7}{6!}$

Multiply the polynomial by x.

$x\cos 2x \approx x - 2x^3 + \frac{2}{3}x^5 - \frac{4x^7}{45}$

Simplify.

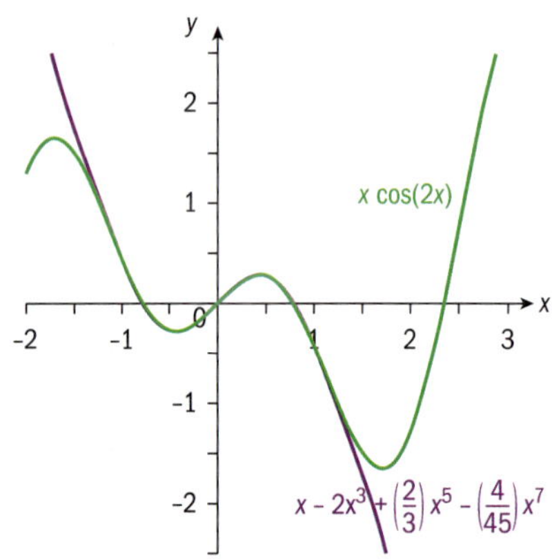

Alternatively, you could have used the product rule to differentiate $f(x) = x\cos 2x$.

Graph both functions.

In the interval]−1, 1[the Maclaurin polynomial is a good approximation of the function. Generating more terms would improve the approximation.

Exercise 8K

1 Compute the 4th order Maclaurin polynomial to approximate the following functions around $x = 0$, and confirm your answer graphically.

a $f(x) = xe^x$ **b** $f(x) = e^{-x}$

c $f(x) = \sin x$ **d** $f(x) = \cos^2 x$

e $f(x) = \frac{1}{1-x}$ **f** $f(x) = \frac{1}{1+x^2}$

g $f(x) = \frac{x}{1+x^2}$

2 Using the results of Investigation 10, determine how many terms of the Maclaurin polynomial are needed to approximate the function $f(x) = \ln(1 + x)$ exactly to four decimal places in the interval]−1, 1[.

There is nothing special about the first six derivatives of the polynomial in the given examples. You could continue and construct the polynomial for any $P^{(n)}(0)$, since the functions have derivatives of all orders at $x = 0$. The approximation of the polynomial to the function improves with each term that is added, until finally you obtain an infinite series.

If $f(x)$ has derivatives of all orders throughout an open interval I such that $0 \in I$, then the Maclaurin series generated by f at $x = 0$ is

$$P(x) = f(0) + \frac{f'(0)}{1!}x + \frac{f''(0)}{2!}x^2 + \frac{f'''(0)}{3!}x^3 + \ldots = \sum_{n=0}^{\infty} \frac{f^{(n)}(0)}{n!}x^n$$

Example 23

Generate the Maclaurin series for sin x and use the first four terms to approximate sin 3° to 6 d.p. Compare your answer to that given by your GDC.

$f(x) = \sin x;\ f(0) = 0$ $f'(x) = \cos x;\ f'(0) = 1$ $f''(x) = -\sin x;\ f''(0) = 0$ $f'''(x) = -\cos x;\ f'''(0) = -1$ $f^{(4)}(x) = \sin x;\ f^{(4)}(0) = 0$ $f^{(5)}(x) = \cos x;\ f^{(5)}(0) = 1$ ...	Find successive derivatives of $f(x)$ and evaluate them at $x = 0$.
$P(x) = x - \frac{x^3}{3!} + \frac{x^5}{5!} - \frac{x^7}{7!} + \ldots = \sum_{n=0}^{\infty}(-1)^n \frac{x^{2n+1}}{(2n+1)!}$	Write out the terms of the Maclaurin series, and its general term.
$3^\circ = \frac{\pi}{60} \Rightarrow P\left(\frac{\pi}{60}\right) = \frac{\pi}{60} - \frac{\left(\frac{\pi}{60}\right)^3}{3!} + \frac{\left(\frac{\pi}{60}\right)^5}{5!} - \frac{\left(\frac{\pi}{60}\right)^7}{7!}$ $= 0.052336$	Substitute 3° in radians for x in the series, since radian is a number, and degree is a magnitude.
Using the GDC, sin 3° = 0.052336 To 6 d.p., the Maclaurin series and the GDC have the same answer.	Find sin 3° using your GDC.

Example 24

Generate the Maclaurin series for $(1+x)^p,\ p \in \mathbb{R}$.

$f(x) = (1+x)^p;\ f(0) = 1$ $f'(x) = p(1+x)^{p-1};\ f'(0) = p$ $f''(x) = p(p-1)(1+x)^{p-2};\ f''(0) = p(p-1)$ $f'''(x) = p(p-1)(p-2)(1+x)^{p-3};\ f'''(0) = p(p-1)(p-2)$ ...	Find successive derivatives of $f(x)$ and determine their value at $x = 0$.

Continued on next page

Calculus

$f^{(n)}(x) = p(p-1)(p-2)\ldots(p-(n-1))(1+x)^{p-n}$ $f^{(n)}(0) = p(p-1)(p-2)\ldots(p-(n-1))$ $P(x) = 1 + px + \frac{p(p-1)}{2!}x^2 + \frac{p(p-1)(p-2)}{3!}x^3 + \ldots$ $= \sum_{n=0}^{\infty} \frac{p(p-1)\ldots(p-n+1)}{n!}x^n$ $= \sum_{n=0}^{\infty} \binom{p}{n} x^n$	Write out the terms of the Maclaurin series, and its general term.

As you have seen in Chapter 1, the series in Example 24 is called the **Binomial series**, and is the generalization of the Binomial Theorem when p is not an integer. When p is a positive integer and $n \le p$ the expression $\binom{p}{n}$ contains a factor $(p - p)$ so $\binom{p}{n} = 0$.

This means that the Binomial series terminates, and reduces to

$$(1+x)^p = \sum_{n=0}^{p} \binom{p}{n} x^n,\ 1 \le n \le p,\ \binom{p}{n} = \frac{p!}{n!(p-n)!} = \frac{p(p-1)(p-2)\ldots(p-(n-1))}{n!}.$$

The Binomial series is the Maclaurin expansion for $f(x) = (1 + x)^p$ and is equal to

$$1 + px + \frac{p(p-1)}{2!}x^2 + \frac{p(p-1)(p-2)}{3!}x^3 + \ldots + \frac{p(p-1)\ldots(p-n+1)}{n!}x^n + \ldots$$

TOK

What does it mean to say that mathematics is an axiomatic system?

Example 25

Use the Binomial series to find the Maclaurin series for

a $f(x) = \frac{1}{(1+x)^2}$ **b** $f(x) = \frac{1}{\sqrt{2-x}}$

a $f(x) = \frac{1}{(1+x)^2} = (1+x)^{-2}$	
$\binom{-2}{n} = \frac{(-2)(-3)(-4)\ldots(-2-n+1)}{n!}$	Use the binomial series with $p = -2$.
$= (-1)^n \frac{2 \times 3 \times 4 \times \ldots \times n(n+1)}{n!} = (-1)^n (n+1)$	Factor out -1.

$$\frac{1}{(1+x)^2} = \sum_{n=0}^{\infty}\binom{-2}{n}x^n = \sum_{n=0}^{\infty}(-1)^n(n+1)x^n$$

$$= 1 - 2x + 3x^2 - 4x^3 + \ldots + (-1)^n(n+1)x^n + \ldots$$

Substitute the expression above, and generalize.

b $$\frac{1}{\sqrt{2-x}} = \frac{1}{\sqrt{2\left(1-\frac{x}{2}\right)}} = \frac{1}{\sqrt{2}\sqrt{1-\frac{x}{2}}} = \frac{1}{\sqrt{2}}\left(1-\frac{x}{2}\right)^{-\frac{1}{2}}$$

Rewrite the function in a form that allows you to use the Binomial series.

$$= \frac{1}{\sqrt{2}}\sum_{n=0}^{\infty}\binom{-\frac{1}{2}}{n}\left(-\frac{x}{2}\right)^n$$

Use the Binomial series with $p = -\frac{1}{2}$ and x replaced by $-\frac{x}{2}$.

As you saw in Example 22c, if you know how to find the Maclaurin series for certain standard functions, you can find the Maclaurin series for more complicated functions, such as those shown in the following example.

Example 26

Find the Maclaurin series for

a $f(x) = e^{x^2}$ **b** $f(x) = \ln\frac{(1+x)}{(1-x)}$ **c** $f(x) = \frac{x}{(1+x)^2}$

a $$e^x = 1 + x + \frac{x^2}{2!} + \frac{x^3}{3!} + \ldots + \frac{x^n}{n!} + \ldots$$

Write down the Maclaurin series for $f(x) = e^x$.

Substituting x^2 for x in the series,

$$f(x) = e^{x^2} = 1 + x^2 + \frac{x^4}{2!} + \frac{x^6}{3!} + \ldots + \frac{x^{2n}}{n!}$$

$$= \sum_{n=0}^{\infty}\frac{x^{2n}}{n!}$$

Write out the new series with x^2 instead of x.

b $$f(x) = \ln\frac{(1+x)}{(1-x)} = \ln(1+x) - \ln(1-x)$$

$$\ln(1+x) = x - \frac{1}{2}x^2 + \frac{1}{3}x^3 - \frac{1}{4}x^4 + \ldots + (-1)^{n+1}\frac{x^n}{n} + \ldots$$

Use the results of Investigation 10 to write the infinite series for $f(x) = \ln(1+x)$.

$$\Rightarrow \ln(1-x) = -x - \frac{x^2}{2} - \frac{x^3}{3} - \ldots - \frac{x^n}{n} - \ldots$$

Substitute $-x$ for x in the Maclaurin series for $\ln(1+x)$.

$$\ln(1+x) - \ln(1-x) = 2\left(x + \frac{x^3}{3} + \frac{x^5}{5} + \ldots\right) = 2\sum_{n=1}^{\infty}\frac{x^{2n-1}}{2n-1}$$

Combine and write the general term.

Continued on next page

c $f(x)=\dfrac{x}{(1+x)^2}=x\left(1-2x+3x^2-4x^3+\ldots+(-1)^n(n+1)x^n+\ldots\right)$ $=x\sum_{n=0}^{\infty}(-1)^n(n+1)x^n=\sum_{n=0}^{\infty}(-1)^n(n+1)x^{n+1}$	Use the Binomial series found in Example 25a and multiply by x.

Some limits can be evaluated using Maclaurin series, as the following example shows.

Example 27

Use the Maclaurin expansion for $y=e^{x^2}$ and $y=\cos x$ to find $\lim\limits_{x\to0}\dfrac{1-e^{x^2}}{1-\cos x}$.

$\lim\limits_{x\to0}\dfrac{1-e^{x^2}}{1-\cos x}=\lim\limits_{x\to0}\dfrac{1-\left(1+x^2+\dfrac{x^4}{2!}+\dfrac{x^6}{3!}+\ldots\right)}{1-\left(1-\dfrac{x^2}{2!}+\dfrac{x^4}{4!}-\ldots\right)}$	Use the Maclaurin expansion for e^{x^2} and $\cos x$.
$=\lim\limits_{x\to0}\dfrac{-x^2-\dfrac{x^4}{2!}-\dfrac{x^6}{3!}-\ldots}{\dfrac{x^2}{2!}-\dfrac{x^4}{4!}+\ldots}$	Simplify. Consider only the smallest powers of x, as higher powers will go to zero much quicker.
$=\dfrac{-1}{\frac{1}{2}}=-2$	Evaluate the limit at $x=0$.

Maclaurin series can be used to make the solution of certain types of differential equations more accessible. A Maclaurin series represents a function on an open interval, and successive differentiations of the series generates series for f', f'', etc. The following example will illustrate how a Maclaurin series can be developed from a differential equation.

International-mindedness

Indian mathematicians (500–1000AD) sought to explain division by zero.

Example 28

Use the first six terms of a Maclaurin series to approximate the solution of $y'=y^2-x$ on an open interval centred at $x=0$ if $y(0)=1$.

$P(x)=f(0)+\dfrac{f'(0)}{1!}x+\dfrac{f''(0)}{2!}x^2+\dfrac{f'''(0)}{3!}x^3+\ldots=\sum_{n=0}^{\infty}\dfrac{f^{(n)}(0)}{n!}x^n$	General form of a Maclaurin series.

	$y(0)=1$
$y'=y^2-x$	$y'(0)=1$
$y''=2yy'-1$	$y''(0)=1$
$y'''=2yy''+2(y')^2$	$y'''(0)=4$
$y^{(4)}=2yy'''+6y'y''$	$y^{(4)}(0)=14$
$y^{(5)}=2yy^{(4)}+8y'y'''+6(y'')^2$	$y^{(5)}(0)=66$

$$P(x)=1+x+\frac{1}{2}x^2+\frac{4}{3!}x^3+\frac{14}{4!}x^4+\frac{66}{5!}x^5$$

Differentiate and use the initial condition.

Finding successive derivatives of y allows you to generate a Maclaurin series for y.

Substitute the table values into the general Maclaurin Series.

Example **24** shows the Maclaurin series (Binomial series) for the function $f(x)=(1+x)^p, p\in\mathbb{Q}$

You can actually prove that the series is equal to the function using differential equations, as seen in the next example.

Example 29

Prove that function $f(x)=(1+x)^p, p\in\mathbb{R}$ is equal to its Binomial series using the initial condition $y(0)=1$.

$$y=1+px+\frac{p(p-1)}{2!}x^2+\frac{p(p-1)(p-2)}{3!}x^3+\dots$$
$$\dots+\frac{p(p-1)\dots(p-n+1)}{n!}x^n+\dots$$

Write out the Binomial series.

$$\Rightarrow y'=p+p(p-1)x+\frac{p(p-1)(p-2)}{2!}x^2+\dots$$

Differentiate term by term.

$$\Rightarrow xy'=px+p(p-1)x^2+\frac{p(p-1)(p-2)}{2!}x^3+\dots$$

Multiply by x.

$$\Rightarrow y'+xy'=p+[p(p-1)+p]x+\left[\frac{p(p-1)(p-2)}{2!}+p(p-1)\right]x^2+\dots$$

Add y'.

$$=p+p^2x+\frac{p^2(p-1)}{2!}x^2+\dots$$
$$=p\left(1+px+\frac{p(p-1)}{2!}x^2+\dots\right)$$
$$=py$$

Simplify.

$$\Rightarrow y'-\frac{p}{1+x}y=0$$

$$I=\int e^{-\frac{p}{1+x}dx}\Rightarrow y=A(1+x)^p$$

$$y(0)=1\Rightarrow y=(1+x)^p$$

Find the integrating factor and use the initial condition.

Calculus

Here is a list of Maclaurin series for special functions. The general term of the series is included.

Maclaurin series
$e^x = 1 + x + \frac{x^2}{2!} + \ldots + \frac{x^n}{n!} + \ldots = \sum_{n=0}^{\infty} \frac{x^n}{n!}$
$\sin x = x - \frac{x^3}{3!} + \frac{x^5}{5!} + \ldots + (-1)^n \frac{x^{2n+1}}{(2n+1)!} + \ldots = \sum_{n=0}^{\infty} (-1)^n \frac{x^{2n+1}}{(2n+1)!}$
$\cos x = 1 - \frac{x^2}{2!} + \frac{x^4}{4!} + \ldots + (-1)^n \frac{x^{2n}}{(2n)!} + \ldots = \sum_{n=0}^{\infty} (-1)^n \frac{x^{2n}}{(2n)!}$
$\ln(1+x) = x - \frac{x^2}{2} + \frac{x^3}{3} - \ldots + \frac{x^n}{n} + \ldots = \sum_{n=0}^{\infty} \frac{x^n}{n}$
$\arctan x = x - \frac{x^3}{3} + \frac{x^5}{5} - \ldots + (-1)^n \frac{x^{2n+1}}{2n+1} + \ldots = \sum_{n=0}^{\infty} (-1)^n \frac{x^{2n+1}}{2n+1}$
Binomial series
$(1+x)^p = 1 + px + \frac{p(p-1)}{2!} x^2 + \ldots + \frac{p(p-1)\ldots(p-n+1)}{n!} x^n + \ldots = \sum_{n=0}^{\infty} \frac{p(p-1)\ldots(p-n+1)}{n!} x^n$

Exercise 8L

1 Find the Maclaurin series for the following functions.

a $y = e^{3x}$ **b** $f(x) = \frac{1}{1+x}$

c $f(x) = \frac{1}{1+2x}$ **d** $y = \arctan x^2$

e $y = \sin^2 x$

2 Find the Maclaurin series for $\sin^2 x$ by using the identity $\sin^2 x = \frac{1}{2} - \frac{\cos 2x}{2}$ and compare it to your answer for **1e**.

3 Use a partial fraction decomposition to express $f(x) = \frac{7x-2}{(x+1)(x-2)}$ as the sum of two rational functions, and find the Maclaurin series for f.

4 Find the following limits using Maclaurin series.

a $\lim_{x \to 0} \frac{\sin x - x}{x^3}$ **b** $\lim_{x \to 0} \frac{e^x - e^{-x}}{x}$

c $\lim_{x \to 0} \frac{(2 - 2\cos x)^3}{x^6}$ **d** $\lim_{x \to 0} \frac{x^2 - \sin^2 x}{x^2 \sin^2 x}$

5 Use the Binomial series to find a Maclaurin series for

a $f(x) = \sqrt{1-x}$ **b** $f(x) = \frac{1}{(1+x)^3}$

c $f(x) = \frac{1}{(1-4x^2)^2}$ **d** $f(x) = \frac{1}{\sqrt[4]{1+2x^3}}$

6 Find the Maclaurin series of $f(x) = \sqrt[3]{1+x}$ up to and including the term in x^3, and use it to approximate $\sqrt[3]{1.2}$to 5 d.p.

7 Find the first three terms of a Maclaurin series to approximate the solution of $y' - y \tan x = \cos x$ if $y(0) = -\frac{\pi}{2}$.

8 Find a Maclaurin series to approximate the solution of $y' = y^2 - x$ if $y(0) = 1$. Use its first six terms to approximate the value of y to 4 d.p. when $x = 0.2$.

9 Use Maclaurin series to show that the solution of the initial value problem $y' = y^2 + 1$, $y(0) = 0$ is $y = \tan x$.

Chapter summary

Areas and volumes

- If functions f and g are continuous in the interval $[a, b]$ and $f(x) \geq g(x)$, $x \in [a, b]$ then the area between the graphs of f and g is $A = \int_a^b f(x)\mathrm{d}x - \int_a^b g(x)\mathrm{d}x = \int_a^b [f(x) - g(x)]\mathrm{d}x$.
- A region may not be entirely enclosed between two functions. In this case, the region may need to be partitioned into sub-regions in order to find the required area.
- Volume of revolution under a single curve.
    - The volume of a solid of revolution formed when $y = f(x)$, continuous in the interval $[a, b]$, is rotated 2π radians about the x-axis is $V = \pi\int_a^b y^2\mathrm{d}x$.
    - The volume of a solid of revolution formed when $y = f(x)$ in the interval $y = c$ to $y = d$ is rotated 2π radians about the y-axis is $V = \pi\int_c^d x^2\mathrm{d}y$.
- Volume of revolution contained between two curves.
    - If $f(x) \geq g(x)$ for all x in the interval $[a, b]$, then the volume of revolution formed when rotating the region between the two curves 2π radians about the x-axis is $V = \pi\int_a^b \left([f(x)]^2 - [g(x)]^2\right)\mathrm{d}x$.
    - If x_1 and x_2 are relations in y such that $x_1 \geq x_2$ for all y in the interval $[c, d]$, then the volume formed by rotating the region between the two curves 2π radians about the y-axis is $V = \pi\int_c^d \left(x_1^2 - x_2^2\right)\mathrm{d}y$.

Kinematics

- If $v(t)$ is the velocity function in terms of t, then the following are true.
    - The displacement of the particle between t_1 and t_2 is $\int_{t_1}^{t_2} v(t)\,\mathrm{d}t$.
    - The total distance travelled by the particle between t_1 and t_2 is $\int_{t_1}^{t_2} |v(t)|\,\mathrm{d}t$.

Ordinary differential equations (ODEs)

- A differential equation $\frac{\mathrm{d}y}{\mathrm{d}x} = f(x, y)$ is separable if f can be expressed as a product of a function in x and a function in y, $\frac{\mathrm{d}y}{\mathrm{d}x} = g(x)h(y)$.
- If $h(y) \neq 0$, the variables can be separated to give $\frac{\mathrm{d}y}{h(y)} = g(x)\mathrm{d}x \Rightarrow \int\frac{\mathrm{d}y}{h(y)} = \int g(x)\mathrm{d}x$.

Continued on next page

Calculus

- When modelling population growth, the standard logistic equation is of the form $\frac{\mathrm{d}n}{\mathrm{d}t} = kn(a-n), a, k, \in \mathbb{R}$.

  t = the time during which a population grows

  n = the population after time t

  k = relative growth

  a = constant (depends on problem being modelled)

- A homogeneous differential equation can be written in the form $\frac{\mathrm{d}y}{\mathrm{d}x} = f\left(\frac{y}{x}\right)$ and solved using the substitution $v = \frac{y}{x}$, where v is a function in x.
- A differential equation of the form $\frac{\mathrm{d}}{\mathrm{d}x}\left(y(x)u(x)\right) = v(x)$ is an **exact differential equation.**

  A differential equation of the form $\frac{\mathrm{d}y}{\mathrm{d}x} + p(x)y = q(x)$ is in **standard form**, and can be made exact by multiplying both sides by the integrating factor, $I = \mathrm{e}^{\int p(x)\mathrm{d}x}$.
- Euler's Method for solving differential equations numerically:

  Given a first order differential equation in the form $\frac{\mathrm{d}y}{\mathrm{d}x} = f(x, y)$
  - select the initial point (x_0, y_0)
  - use the recurrence formulas $x_{n+1} = x_n + h$ and $y_{n+1} = y_n + hf(x_n, y_n)$ to generate as many points as instructed
  - plot the points (x_n, y_n) and connect with a smooth curve.

Limits revisited

- **L'Hopital's Rule**
  - If $\lim\limits_{x\to a}\frac{f(x)}{g(x)} = \frac{f(a)}{g(a)} = \frac{0}{0}$, and if $\lim\limits_{x\to a}\frac{f'(x)}{g'(x)}$ exists, then $\lim\limits_{x\to a}\frac{f(x)}{g(x)}$ also exists, and $\lim\limits_{x\to a}\frac{f(x)}{g(x)} = \lim\limits_{x\to a}\frac{f'(x)}{g'(x)}$.
  - This rule can be used when the limit of a quotient results in the indeterminate forms $\frac{0}{0}$ or $\frac{\pm\infty}{\pm\infty}$.
- If $f(x)$ has n derivatives at $x = 0$, then $P(x)$, the Maclaurin polynomial of degree n for $f(x)$ centred at $x = 0$, is the unique polynomial of degree n which satisfies the following conditions:
  - $f(0) = P(0)$
  - $f^{(n)}(0) = P^{(n)}(0)$
  - $a_1 = \frac{f'(0)}{1!}; a_2 = \frac{f''(0)}{2!}; a_3 = \frac{f'''(0)}{3!}; a_4 = \frac{f^{(4)}(0)}{4!}; a_5 = \frac{f^{(5)}(0)}{5!}$

- $P(x) = f(0) + \frac{f'(0)}{1!}x + \frac{f''(0)}{2!}x^2 + \ldots + \frac{f^{(n)}(0)}{n!}x^n = \sum_{k=0}^{n} \frac{f^{(k)}(0)}{k!}x^k.$
- If $f(x)$ has derivatives of all orders throughout an open interval I such that $0 \in I$, then the Maclaurin series generated by f at $x = 0$ is

  $$P(x) = f(0) + \frac{f'(0)}{1!}x + \frac{f''(0)}{2!}x^2 + \frac{f'''(0)}{3!}x^3 + \ldots = \sum_{n=0}^{\infty} \frac{f^{(n)}(0)}{n!}x^n$$
- The Binomial series is the Maclaurin expansion for $f(x) = (1 + x)^p$ and is equal to

  $$1 + px + \frac{p(p-1)}{2!}x^2 + \frac{p(p-1)(p-2)}{3!}x^3 + \ldots + \frac{p(p-1)\ldots(p-n+1)}{n!}x^n + \ldots.$$

Developing inquiry skills

Return to the opening problem for the chapter. Are you able to approximate any of the functions used to model the glass using a Maclaurin series?

Chapter review

1 Find the area of one of the regions enclosed by the graphs of

a $y = \sin x$ and $y = \cos x$.

b $y = \sin 2x$ and $y = 2\sin x$

2 Find the area of the region enclosed by the graph of the curves

a $y = x^2$; $y = x^3$

b $y = |x|$; $y = x^{\frac{2}{3}}$

c $y = x^4 - 2x^2$; $y = 2x^2$

d $y = |x^2 - 4|$; $y = \frac{x^2}{2} + 4$

e $y = \sqrt{|x|}$; $y = \frac{x+6}{5}$

f $y = \frac{8}{4+x^2}$; $y = \frac{x^2}{4}$

3 Find the area enclosed by the graphs of $y = \frac{2x}{x^2+1}$ and $y = x^3$.

4 In each case, find the general solution to the differential equation in terms of y.

a $\frac{xy}{x+1} = \frac{dy}{dx}$

b $\frac{dy}{ydx} = \frac{x}{x^2+1}$

c $1 + xy' = y^2$

d $y' = \frac{1}{xy + y}$

e $\frac{dy}{dx} + \frac{2}{x}y + x^2 = \sin 2x$

f $y' - y\tan x = 1$

g $\frac{dy}{dx} - \frac{1}{2}y = \frac{1}{2}e^{\frac{1}{2}x}$

5 The radioactive rate of decay of a substance, $\frac{dy}{dt}$, is proportional to the amount of radioactive substance present, y. The half-life of the radioactive substance is approximately 5500 years.

a Express the decay as a differential equation.

b Solve the equation for y.

c Determine the age of a fossil, to the nearest thousand years, if its radioactive substance has decayed to 20% of the original amount.

6 Given the differential equation $\frac{dy}{dx} = 1 - \frac{xy}{4 - x^2}$ and $y(0) = 1$,

a use Euler's method with $h = 0.25$ to find an approximate value of y when $x = 1$

b solve the differential equation using an appropriate analytical method, and find the value of y when $x = 1$

c sketch the values you found in **a** and the graph of the function you found in **b** on the same coordinate axes, and use your sketch to explain why your answer in **a** is greater than or less than the value in **b**.

7 Find the first five terms of the Maclaurin series for $f(x) = \ln(1 + \sin x)$.

8 Find $\lim\limits_{x \to 0} \frac{\sin x - x}{x \sin x}$

9 Find $\lim\limits_{x \to 1} \frac{\ln x}{\sin 2\pi x}$

Exam-style questions

10 **P1: a** On the same axes, sketch the graphs of the functions $f(x) = x^2$ and $g(x) = |x|$. On your sketch, mark the co-ordinates of the points of intersection of f and g. (3 marks)

b Find the sum of the areas of the regions enclosed by the graphs of f and g. (4 marks)

11 **P2:** Two particles P_1 and P_2 are moving along the same straight line. Relative to the origin, their positions after t seconds are given by $s_1(t) = \sin 2t$ and $s_2(t) = \sin(t - 0.24)$ respectively, for $0 \le t \le 3$.

a Find an expression for the distance between the particles at time t. (2 marks)

b Hence determine the furthest distance between the two particles, and the time at which this occurs. (2 marks)

c Determine whether the particles will collide in the first 3 seconds of the movement and, if so, state the time at which collision occurs. (2 mark)

12 **P2:** The diagram shows the graph of the function f defined by $f(x) = 3 - x^2$ for $-2 \le x \le 2$.

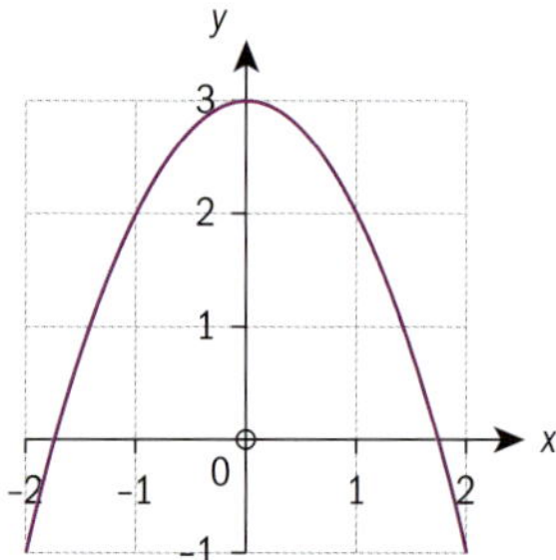

Consider the function g defined by $g(x) = \sin(e^x)$ for $-2 \le x \le 2$.

a On the same axes, sketch the graph of g. (2 marks)

b Solve $f(x) = g(x)$. (2 marks)

c Hence state the solutions to $f(x) > g(x)$. (1 mark)

d Find the area of the region enclosed by the graphs of f and g. (2 marks)

13 **P2:** Pesticides control the population of insects in a specific area around a river. The rate of decrease of the number of insects is proportional to the number of insects at any time t. The initial population of insects is 500 000, and this decreases to 400 000 after five years.

Find the time (in years) it takes for the population of insects to decrease to half its initial size. (8 marks)

14 **P1:** Given that $\frac{dy}{dx} = \cos x \cos^2 y$ and $y = \frac{\pi}{4}$ when $x = \pi$,

a Show that the solution of the differential equation is $y = \arctan(1 + \sin x)$. (5 marks)

b Determine the value of the constant k for which the following limit exists, and evaluate the limit:

$$\lim_{x \to \frac{\pi}{2}} \frac{\arctan(1 + \sin x) - k}{\left(x - \frac{\pi}{2}\right)^2}.$$ (7 marks)

15 **P1: a** By finding the Maclaurin series up to and including the term in x^4, show that $\ln(1 + \sin x) = x - \frac{1}{2}x^2 + \frac{1}{6}x^3 - \frac{1}{12}x^4 + \ldots$ (8 marks)

b Hence, find the Maclaurin series up to and including the term in x^4 for each of the following functions.

i $y = \ln(1 - \sin x)$

ii $y = \ln(\cos x)$

iii $y = \tan x$ (10 marks)

c Calculate $\lim_{x \to 0} \frac{\tan(x^2)}{\ln(\cos x)}$. (4 marks)

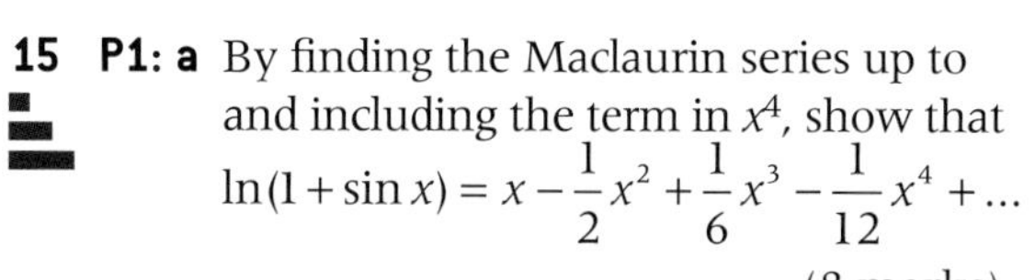

Calculus

Be the particle!

Approaches to learning: Collaboration, Communication
Exploration criteria: Personal engagement (C), Use of mathematics (E)
IB topic: Calculus, Kinematics–motion in a straight line

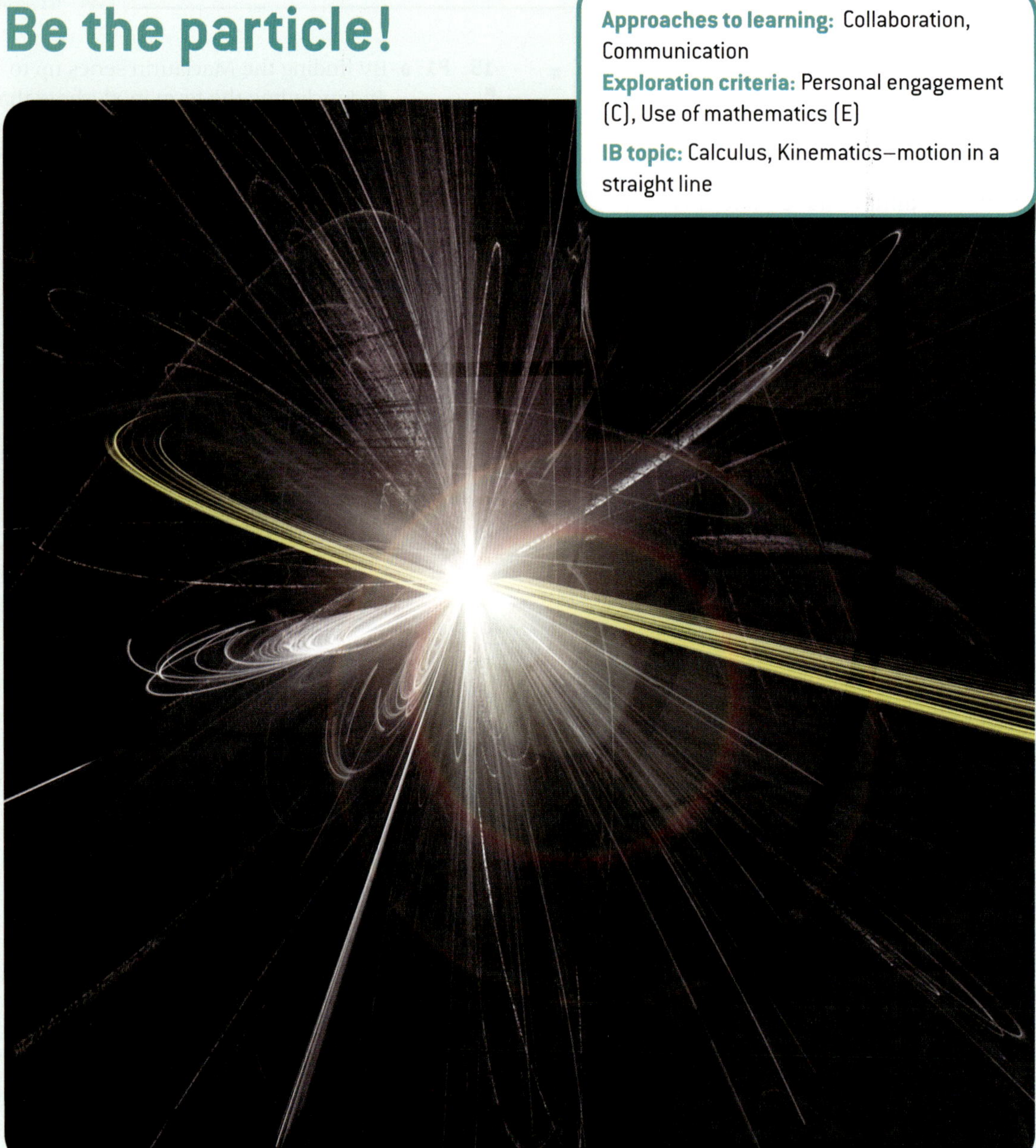

Motion of a particle in a straight line

In this task all times are in seconds and all distances are in 'units'.

A particle, P, moves with **velocity function** $v(t) = t^2 - 7t + 10,\ 0 \le t \le 6$.

In your groups discuss:

- What is the initial speed of the particle?
- When is the particle stationary?
- At what velocity is the particle moving at the end of the journey?

The **displacement function** is $s(t)$.

In your groups discuss:

- At time $t = 0$, $s(t) = -5$. What does $s(t) < 0$ mean?
- What does $s(1)$ mean?
- What is the displacement function $s(t)$ of the particle?
- How far is the particle away from its starting point at the end of the journey?
- What is the particle's displacement at each second of the journey?

Find the **acceleration function** $a(t)$ of the particle.

In your groups discuss:

- What is the initial acceleration?
- When does the particle change direction?
- Find total distance travelled by the particle in the 6 second 'journey'.

Produce graphs of the displacement, velocity and acceleration against time for the particle's journey.

Be the particle for 6 seconds!

How could you walk to model the journey of the particle?

Discuss.

One member of your group should walk the 'path of the particle' using the scale given on the board.

The other members should use a timer to advise the walker when to change direction, etc.

How could you check whether this is an accurate attempt?

Use a motion detector and/or a graphing progamme to fit a cubic curve to the displacement graph.

Compare this to the actual cubic curve for the displacement.

How similar is it to the cubic you produced previously?

What can you do to improve the model found?

Extension

Repeat the above experiment with another velocity function and initial displacement.

Perhaps try $v(t) = 2\sin\left(\frac{1}{2}t\right) + 1,\ 0 \le t \le 10$ and $s(0) = -2$.

Ensure here that all calculations are completed in radians.

You could also devise your own problem similar to the one in this task but try to consider a real-life situation.

For example,

how could you model the movement (displacement, velocity and acceleration) of an elevator or of a 100m runner?

9 Modelling 3D space: Vectors

Nowadays vectors are used in many areas, not just academically but in technology and the arts. Today it is almost impossible to produce a film without the use of vector geometry. Many films are shot in front of blank screens (blue or green) and then moving scenery is developed using vector geometry. This technique is predominantly used in films that combine real actors with animated fiction characters like Avatar, Sméagol and the Hulk.

How can a 3D character be mapped out and created?

Concepts

- Space
- Modelling

Microconcepts

- Vector represented by directed line segments
- Position vector
- Direction vector
- Magnitude of a vector
- Unit vector
- Base vectors i, j and k
- Components of a vector
- Addition of vectors
- Zero vector
- Multiplication of a vector by a scalar
- Opposite vector
- Scalar product of two vectors
- Angle between two vectors
- Perpendicular and parallel vectors
- Vector product of two vectors
- Volume of a parallelepiped
- Vector equations of lines and planes
- Intersections between lines and planes

How can a fingerprint scanner calculate whether this is the right finger?

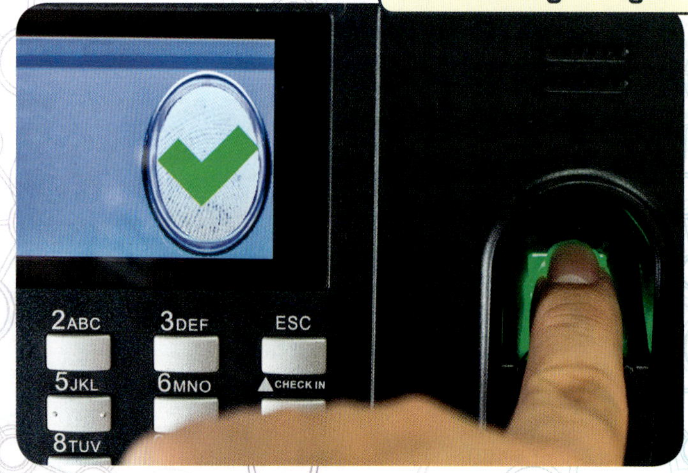

On a ski-slope there are many forces that influence the movement of a skier.

- What are the forces, represented by the arrows, that you can identify on this picture?
- If the lengths of the arrows represent the strength of the force, what will happen if the red and yellow arrows are equal?
- Why is it disadvantageous if the yellow arrow is longer than the red arrow?
- Why are the blue and green arrows equal in length but pointing in opposite directions?
- What will cause the blue and brown arrows to coincide?
- Why is the blue arrow perpendicular to the red and yellow arrows?
- How does the steepness of the slope affect the angles between the brown and blue, and the brown and red arrows?
- What can you say about the forces when the skier is at rest on the slope?
- What other factors can influence the movement of a skier?

Developing inquiry skills

Think about the questions in this opening problem and answer any you can. As you work through the chapter, you will gain mathematical knowledge and skills that will help you to answer them all.

Before you start

You should know how to:

1 Apply Pythagoras theorem.

eg In the right-angled triangle on the diagram below find the missing side.

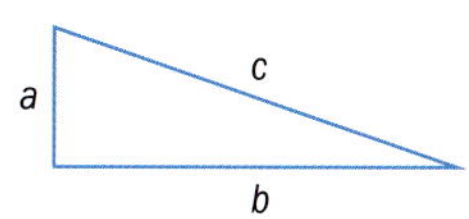

$a = 5$, $b = ?$, $c = 13$

$b^2 = c^2 - a^2 \Rightarrow b = \sqrt{13^2 - 5^2}$

$\Rightarrow b = \sqrt{169 - 25} = 12$

Skills check

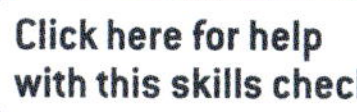

1 In the right-angled triangle find the missing sides.

a $a = 21$, $b = 20$, $c = ?$

b $a = ?$, $b = 7$, $c = 13$

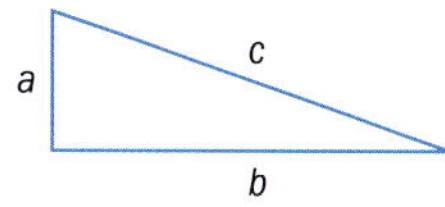

Continued on next page

2 Calculate the distance between two points in the Cartesian plane.

e.g. Find the distance between the points A(1, 2) and B(−3, 4).

$$AB = \sqrt{(x_1 - x_2)^2 + (y_1 - y_2)^2}$$

$$= \sqrt{(1-(-3))^2 + (2-4)^2} = \sqrt{16+4}$$

$$= \sqrt{20} = 2\sqrt{5}$$

3 Recognize parallel and perpendicular lines by their equations.

eg Identify which lines are parallel and which lines are perpendicular:

a $3x + 4y = -3$ **b** $3x + 5y = 0$

c $y = \frac{5}{3}x + 2$ **d** $y = -\frac{3}{4}x$

Two lines are

i parallel if they have equal gradients

ii perpendicular if the product of their gradients is −1.

$$y = -\frac{3}{4}x - \frac{3}{4} \Rightarrow m_a = -\frac{3}{4}$$

$$y = -\frac{3}{5}x \Rightarrow m_b = -\frac{3}{5}$$

$$m_c = \frac{5}{3} \qquad m_d = -\frac{3}{4}$$

$$m_a = m_d \Rightarrow a \parallel d$$

$$m_b \times m_c = -1 \Rightarrow b \perp c$$

4 Solve simultaneous equations.

eg Solve the simultaneous equations.

$$\begin{cases} 2x + 4y = 3 \\ 5x - 3y = 2 \end{cases}$$

eg Use the method of elimination.

$$\begin{cases} 2x + 4y = 3 \\ 5x - 3y = 2 \end{cases} \Rightarrow \begin{cases} 6x + 12y = 9 \\ 20x - 12y = 8 \end{cases}$$

$$26x = 17 \Rightarrow x = \frac{17}{26} \Rightarrow 2 \times \frac{17}{26} + 4y = 3$$

$$\Rightarrow 4y = 3 - \frac{17}{13} \Rightarrow y = \frac{11}{26} \Rightarrow \left(\frac{17}{26}, \frac{11}{26}\right)$$

2 Find the distance between the points A(−2, 3) and B(4, −5).

3 Identify which lines are parallel and which lines are perpendicular:

a $2x + 3y = 4$ **b** $y = \frac{3}{2}x + 1$

c $y = -\frac{2}{3}x$ **d** $3x - 2y - 1 = 0$

e $y = \frac{2}{3}x - 3$

4 Solve the simultaneous equations.

$$\begin{cases} 4x - 3y = 1 \\ 5x - 4y = 2 \end{cases}$$

9.1 Geometrical representation of vectors

Investigation 1

1 Why is it not possible for a triangle to have parallel sides?

2 How many sides must a polygon have in order to have at least one pair of parallel sides?

3 What are the simplest polygons that have parallel sides that have equal lengths?

4 What regular polygons have parallel sides?

In geometry, you have some intuitively defined concepts that you start from, eg **point**, **line**, **plane** and **space**. All the other terms are defined by using those four concepts. Here are some geometrical definitions to better understand the concept of a vector.

A **line segment** is a part of a straight line bounded by two points.

A line segment has no direction since the endpoints are not distinct. The line segment does not start from or end at any specified point, for example, line segments [AB] = [BA].

A **directed line segment** is a line segment that has a direction, i.e. there is a starting point and an end point. The notation is $\overrightarrow{AB}$ where A is the starting point and B is the end point.

Unlike for line segments we distinguish endpoints, therefore $\overrightarrow{AB} \neq \overrightarrow{BA}$.

Two directed line segments $\overrightarrow{AB}$ and $\overrightarrow{CD}$ are called **equivalent**, we write $\overrightarrow{AB} \cong \overrightarrow{CD}$, if and only if line segments [AD] and [BC] have a common midpoint.

International-mindedness

Greek philosopher and mathematician, Aristotle, calculated the combined effect of two or more forces called the Parallelogram law.

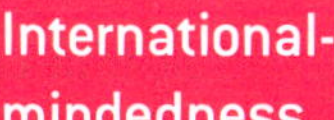

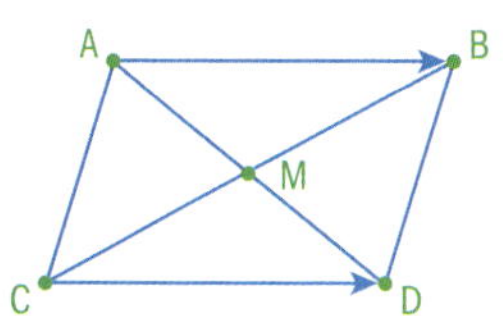

Geometry and trigonometry

Example 1

Given the parallelogram PQRS, find all possible pairs of equivalent directed line segments.

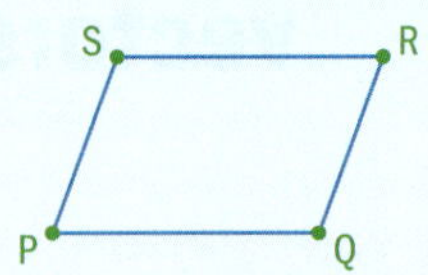

	PQRS form a parallelogram therefore the diagonals bisect each other:
$\overrightarrow{PQ} \cong \overrightarrow{SR}$	[PR] and [QS] have a common midpoint.
$\overrightarrow{PS} \cong \overrightarrow{QR}$	[PR] and [SQ] have a common midpoint.
$\overrightarrow{QP} \cong \overrightarrow{RS}$	[QS] and [PR] have a common midpoint.
$\overrightarrow{SP} \cong \overrightarrow{RQ}$	[SQ] and [PR] have a common midpoint.

Points are said to be collinear if they lie on the same line.

Points or lines are said to be coplanar if they lie in the same plane.

If two directed line segments $\overrightarrow{AB}$ and $\overrightarrow{CD}$ are equivalent, and if the points A, B, C and D are not collinear, then it follows that ABDC forms a parallelogram.

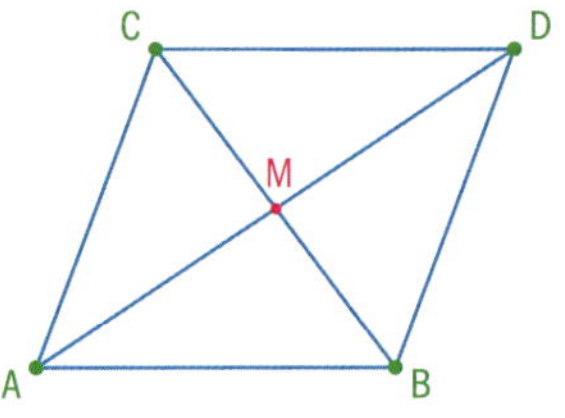

You can see that for example $\overrightarrow{AB}$ and $\overrightarrow{CD}$ have the same direction and the same length.

Notice that $\overrightarrow{AC}$ and $\overrightarrow{DB}$ do not have the same direction and they are not equivalent since [AB] and [CD] do not intersect; therefore, they cannot have a common midpoint.

You can deduce the same for $\overrightarrow{AB}$ and $\overrightarrow{DC}$, that they are not equivalent since [AC] and [BD] do not intersect.

If the points A, B, C and D are collinear as shown in the diagram, you can see that $\overrightarrow{AB}$ and $\overrightarrow{CD}$ have the same direction and you also notice that their lengths are equal, AB = CD.

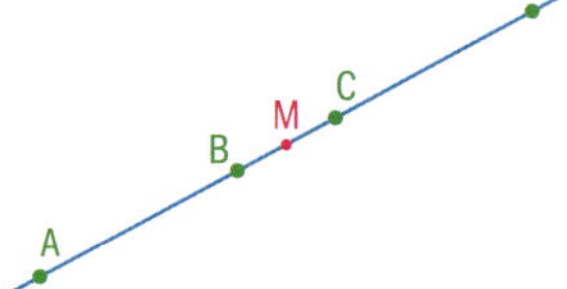

Investigation 2

The diagrams show a square and a regular pentagon inscribed in a circle with the centres marked.

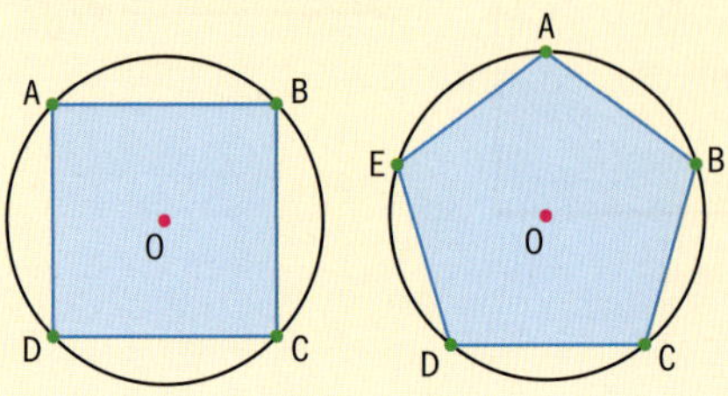

1 Draw the radii of the circle to the vertices on both polygons.

2 Join the vertices (draw diagonals) of both polygons. What do you notice?

3 Draw regular polygons with 6 to 9 sides and draw all the radii and only the longest diagonals for each polygon. What do you notice?

4 In each of your diagrams mark all the equivalent directed line segments.

5 What do you notice about pairs of equivalent directed line segments in your diagrams?

6 **Conceptual** Can you make a conjecture connecting your observations to equivalent directed line segments?

Research the meaning of these properties. Illustrate the properties of the equivalence relation by drawing appropriate diagrams. Determine whether the class of equivalent directed line segments satisfy these properties. This could be a topic for exploration.

HINT

Curiosity-Equivalence relations

The relation $\cong$ is an **equivalence relation** if and only if it has the following properties:

1 *Reflexive*

2 *Symmetrical*

3 *Transitive*

A **vector** is a **collection** of **all equivalent directed line segments**.

The figure shows **equivalent directed line segments** that represent **the same vector**.

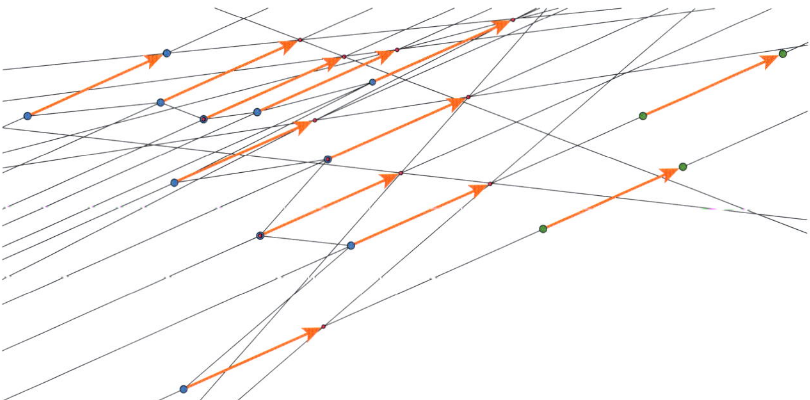

To visualize geometrical terms you use drawings, but the actual drawings are just the representations of these terms.

How do you represent a vector?

The notation used for vectors is a small bold letter or a small letter with an arrow above, **a** or $\vec{a}$.

There are special types of vectors that are represented by particular directed line segments. They are called displacement vectors.

Did you know?

In 2007, Flatland the movie was produced based on the book Flatland published in 1884 by Edwin A. Abbott. It is an award-winning animation that brings to life the properties of 0, 1, 2 and 3-dimensional spaces. The movie can be accessed on YouTube.

A **displacement vector** is represented by **AB** or $\overrightarrow{AB}$, where A is called the **initial or starting point** and B is called the **terminal or end point**.

Notation for displacement vectors is the same as for directed line segments.

Notice that vectors represented by the directed line segments $\overrightarrow{AB}$ and $\overrightarrow{BA}$ have the same magnitude but not the same direction. They lie on the same line but they have the opposite direction. Such vectors are called **opposite vectors** and you write $\overrightarrow{BA} = -\overrightarrow{AB}$.

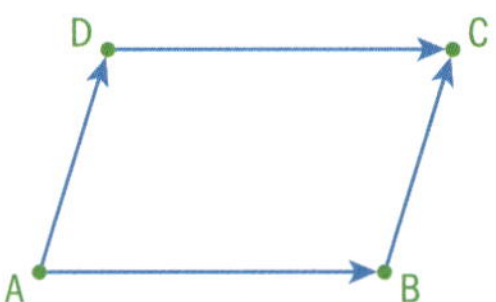

In the parallelogram ABCD the direction vector written as $\overrightarrow{AB}$ is a collection of all the directed line segments that are equivalent to the directed line segment $\overrightarrow{AB}$, therefore $\overrightarrow{AB} = \overrightarrow{DC}$, also $\overrightarrow{AD} = \overrightarrow{BC}$, etc.

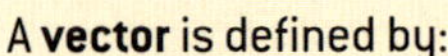

A **vector** is defined by:

i direction

ii magnitude

Direction of a vector is represented by a family of parallel lines that carry all the equivalent directed line segments. The direction of a vector is determined by the starting point and end point.

Magnitude of a vector is the length of the vector. The magnitude of the displacement vector $\overrightarrow{AB}$ is simply the length of AB. The magnitude will be denoted by |**a**| or $|\overrightarrow{AB}|$.

HINT

You will sometimes see that a vector can be defined by direction, orientation and magnitude. Some authors make a distinction between direction and orientation where direction is just the inclination from the horizontal position of a line segment, while orientation is the direction from the initial to the terminal point.

Operations with vectors

Investigation 3

Addition of vectors

In the parallelogram ABCD

$\overrightarrow{AB} + \overrightarrow{AD} = \overrightarrow{AC}$

This is called the **parallelogram law** of vector addition.

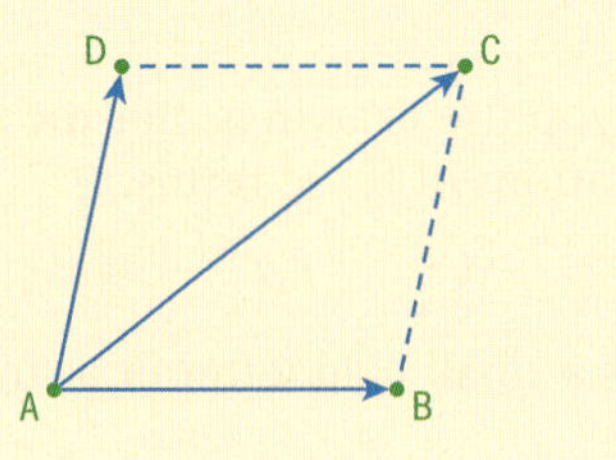

HINT

You will sometimes see the notation for the magnitude of a vector **a** as ||**a**|| or $||\overrightarrow{AB}||$ to distinguish between the absolute value of a number $|a|$ and the magnitude of a vector ||**a**||.

In the triangle ABC

$\overrightarrow{AB} + \overrightarrow{BC} = \overrightarrow{AC}$

This is called the **triangle law** of vector addition.

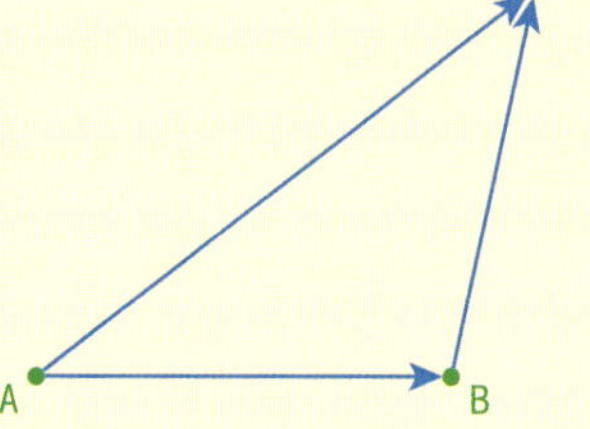

1 **Conceptual** Are the parallelogram and triangle laws equivalent?

2 **Factual** Can you think of any situation where

a the parallelogram law cannot be used;

b the triangle law cannot be used?

3 How would you use these two laws to find $\overrightarrow{AB} - \overrightarrow{AD}$?

4 Explain how you arrived at your result.

5 In the diagram below, use each law separately to show that

$$\overrightarrow{AB} + \overrightarrow{BC} + \overrightarrow{CD} + \overrightarrow{DE} + \overrightarrow{EF} = \overrightarrow{AF}$$

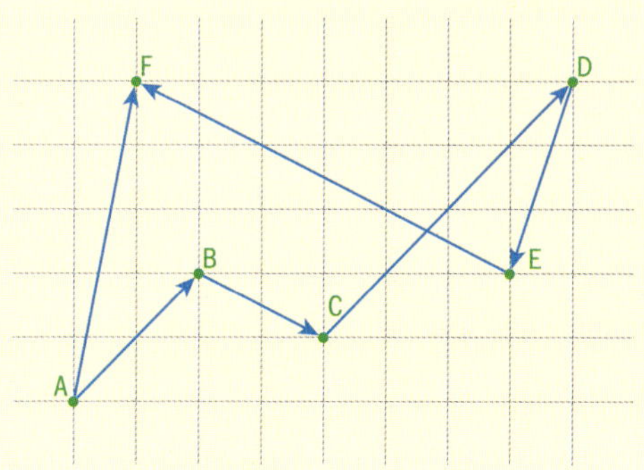

TOK

The $25 billion dollar eigenvector.

Google's success derives in large part from its PageRank algorithm, which ranks the importance of webpages according to an eigenvector of a weighted link matrix.

How ethical is it to create mathematics for financial gain?

Unlike the parallelogram law the triangle law can be used for collinear points.

$\overrightarrow{AB} + \overrightarrow{BC} = \overrightarrow{AC}$

Example 2

Given the diagram on the right, draw the vector

$\overrightarrow{AB} + \overrightarrow{CD} + \overrightarrow{EF}$

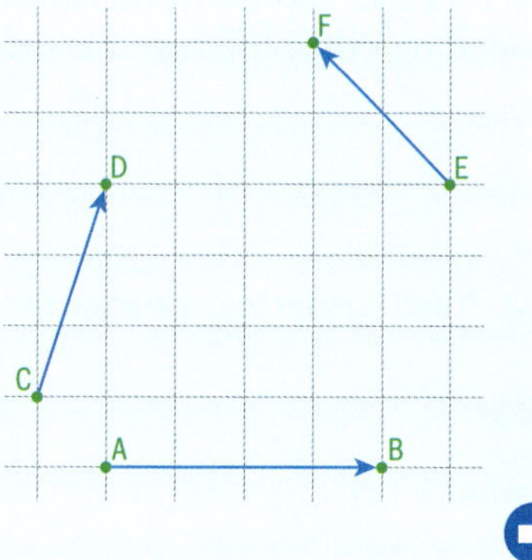

Continued on next page

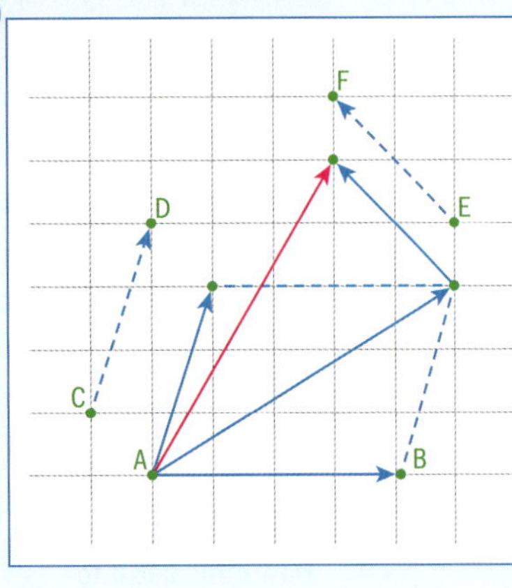

You can either use the parallelogram law or the triangle law.

The **zero vector** is the vector whose magnitude is equal to 0 and whose direction is not defined (or it is better to say it has all the directions).

The **zero vector** can be represented by a direction vector with the same initial and terminal point. You write $\overrightarrow{AA} = \vec{0} = \mathbf{0}$.

Example 3

Show that in a triangle ABC

$\overrightarrow{AB} + \overrightarrow{BC} + \overrightarrow{CA} = \vec{0}$

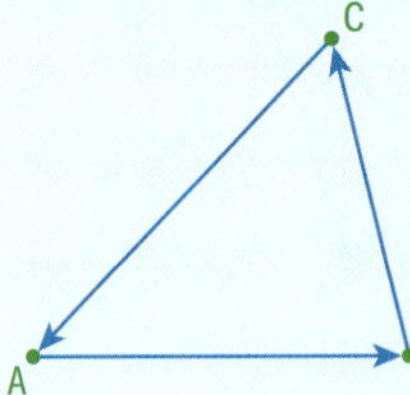

$\overrightarrow{AB} + \overrightarrow{BC} + \overrightarrow{CA} = \overrightarrow{AC} + \overrightarrow{CA}$	Add the first two vectors by using the triangle law.
$= \overrightarrow{AA}$	Add the third vector to the sum of the first two vectors by using the triangle law.
$= \vec{0}$	

Notice that if the direction vectors that represent the sides of a triangle in one particular order are added together the result is the same, a zero vector.

Investigation 4

Recall that for all real numbers a, b and c under the operation of addition, the following properties hold:

i **Commutative** $a + b = b + a$

ii **Associative** $(a + b) + c = a + (b + c)$

iii Identity element 0 is the identity element for addition, $a+0=0+a=a$ for all a.

iv Opposite (inverse) element For all real a, there is the opposite element $-a$ such that $a+(-a)=-a+a=0$.

By drawing appropriate diagrams, determine whether vector addition satisfies the same properties as real numbers for the operation of addition.

Investigation 5

1 For vectors $\mathbf{a}$ and $\mathbf{b}$ on the diagram below,

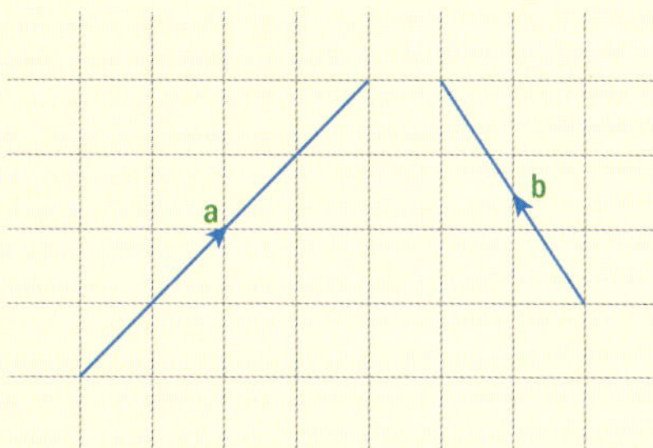

draw diagrams illustrating $\mathbf{a}+\mathbf{a}+\mathbf{a}$ and $\mathbf{b}+\mathbf{b}+\mathbf{b}+\mathbf{b}+\mathbf{b}$.

2 Comment on the magnitude and direction of the vector sums.

3 Draw diagrams illustrating $-\mathbf{a}+(-\mathbf{a})$ and $-\mathbf{b}+(-\mathbf{b})+(-\mathbf{b})+(-\mathbf{b})$.

4 Comment on the magnitude and direction of these vector sums.

5 Represent each vector addition in questions **1** and **3** as a single vector in terms of $\mathbf{a}$ and $\mathbf{b}$.

6 Use the diagrams to illustrate

i $\frac{1}{2}\mathbf{a}$

ii $-\frac{5}{4}\mathbf{a}$

iii $\frac{4}{3}\mathbf{b}$

iv $-\frac{2}{5}\mathbf{b}$

7 Comment on the magnitude and direction of your vector results.

8 How can you represent the vector $\mathbf{a}+(-\mathbf{a})$

i algebraically ii geometrically?

Multiplying vector $\mathbf{a}$ by a non-zero scalar k you obtain a new vector $\mathbf{b}=k\mathbf{a}$ such that

i $\mathbf{b}$ has the **same direction** as $\mathbf{a}$ when k is positive $(k>0)$, or the **opposite direction** when k is negative $(k<0)$.

ii the **magnitude** of $\mathbf{b}$ is the product of the absolute value of k and magnitude of the vector $\mathbf{a}$, $|\mathbf{b}|=|k||\mathbf{a}|$.

A vector and the vector obtained by multiplying that vector by a scalar have the same or opposite directions and are said to be **collinear** or **parallel**.

HINT

Points are collinear if they lie on the same line. Vectors are collinear if you can find their representatives (directed line segments) that lie on the same line.

Linear combination of vectors

Given two non-collinear vectors **a** and **b**, we say that a linear combination of those two vectors is a new vector $\mathbf{c} = \lambda\mathbf{a} + \mu\mathbf{b}$, $\lambda, \mu \in \mathbb{R}$.

Example 4

Given the vectors **a** and **b**, draw the following linear combinations:

a $3\mathbf{a} + 2\mathbf{b}$

b $\frac{1}{2}\mathbf{a} - \frac{2}{3}\mathbf{b}$

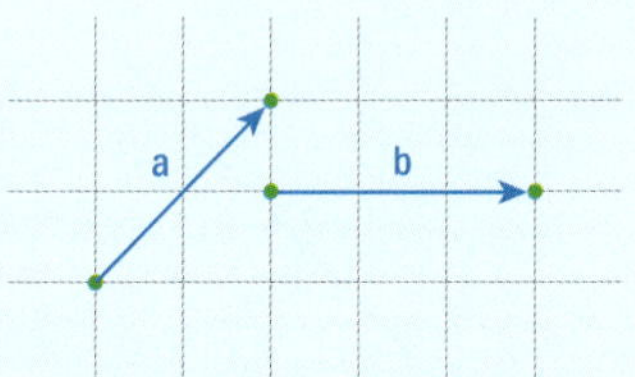

a 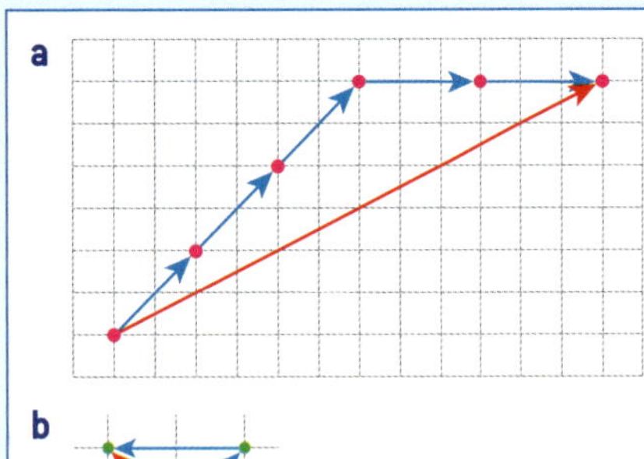	Use the triangle law to add the vectors $3\mathbf{a}$ and $2\mathbf{b}$.
b	Use the triangle law to add the vectors $\frac{1}{2}\mathbf{a}$ and $-\frac{2}{3}\mathbf{b}$.

The operation **subtraction** of two vectors is **not defined**, since it can be seen as a special case of a linear combination.

$\mathbf{a} - \mathbf{b} = 1 \cdot \mathbf{a} + (-1) \cdot \mathbf{b}$, $\lambda = 1, \mu = -1$

The vectors **a**, **b** and **c** are said to be **coplanar** if they lie in the same plane. Vectors **a** and **b** define a sort of coordinate system, and every vector coplanar with them can be shown as a linear combination of **a** and **b**.

International-mindedness

Rene Descartes used a (x, y, z) to represent points in space in the 17th century.

In the 19th century, Arthur Cayley reasoned that we might go further than three values.

Using an analogy in 3D space, the linear combination of three non-coplanar vectors **a, b** and **c**, is a new vector $\mathbf{d} = \lambda\mathbf{a} + \mu\mathbf{b} + \nu\mathbf{c}$, $\lambda, \mu, \nu \in \mathbb{R}$. Also, any three non-coplanar vectors in 3D space define a coordinate system and any other vector in the 3D space can be represented as a linear combination of these three vectors.

Example 5

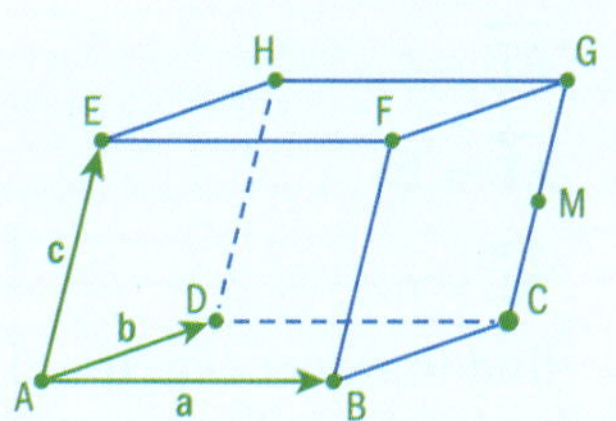

Given the vectors **a**, **b** and **c** on a parallelepiped ABCDEFGH , find as a linear combination of **a**, **b** and **c** the following vectors

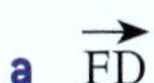

a $\overrightarrow{FD}$ **b** $\overrightarrow{MA}$, where M is the midpoint of the edge [CG].

a $\overrightarrow{FD} = \overrightarrow{FB} + \overrightarrow{BA} + \overrightarrow{AD} = -\mathbf{c} - \mathbf{a} + \mathbf{b}$	Triangle law.
b $\overrightarrow{MA} = \overrightarrow{MC} + \overrightarrow{CD} + \overrightarrow{CD}$ $= -\frac{1}{2}\mathbf{c} - \mathbf{a} - \mathbf{b}$	Triangle law. M is the midpoint of [CG].

Notice that you can obtain the same result by using a different combination, eg $\overrightarrow{MC} + \overrightarrow{CB} + \overrightarrow{BA}$. Can you find more linear combinations that will obtain the same result, not necessarily with only three vectors?

Exercise 9A

1 Given the vectors in the diagram, find:

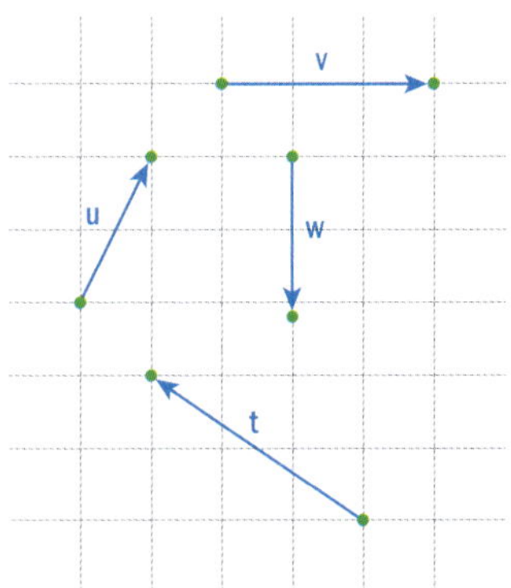

a $\mathbf{t} + \mathbf{u}$

b $\mathbf{v} + \mathbf{w}$

c $\mathbf{t} - 2\mathbf{w}$

d $2\mathbf{v} - 3\mathbf{w}$

e $\mathbf{w} - 3\mathbf{u}$

f $\frac{1}{3}\mathbf{v} - \frac{1}{2}\mathbf{w}$

g $\frac{1}{2}\mathbf{t} - \frac{1}{3}\mathbf{v} + \frac{1}{2}\mathbf{u}$

h $\frac{4}{3}\mathbf{t} - \frac{2}{3}\mathbf{v} - \frac{3}{2}\mathbf{w}$

2 A regular hexagon ABCDEF with the centre O of the circumscribed circle is given. The points M, N and P are the midpoints of the sides [EF], [DE] and [CD] respectively.

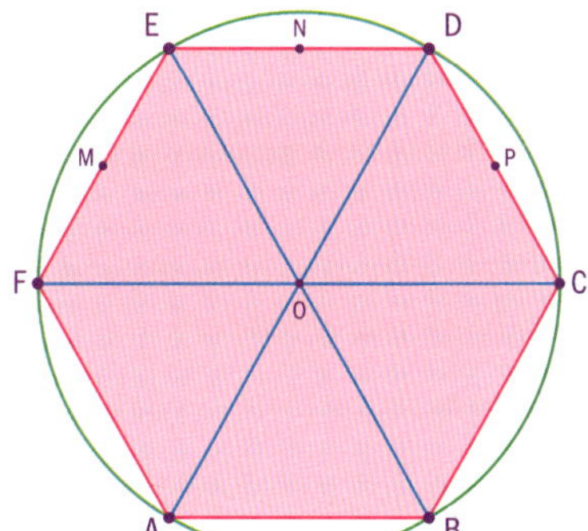

If the vectors $\overrightarrow{AB} = \mathbf{a}$ and $\overrightarrow{AF} = \mathbf{b}$, find in terms of **a** and **b**:

a $\overrightarrow{AO}$ **b** $\overrightarrow{CB}$ **c** $\overrightarrow{CE}$

d $\overrightarrow{DF}$ **e** $\overrightarrow{PA}$ **f** $\overrightarrow{AN}$

g $\overrightarrow{CM}$ **h** $\overrightarrow{PN}$

3 Given a cuboid ABCDEFGH and the vectors

$\overrightarrow{AB} = \mathbf{a}$

$\overrightarrow{AD} = \mathbf{b}$

$\overrightarrow{AE} = \mathbf{c}$

find in terms of **a**, **b** and **c** the following vectors:

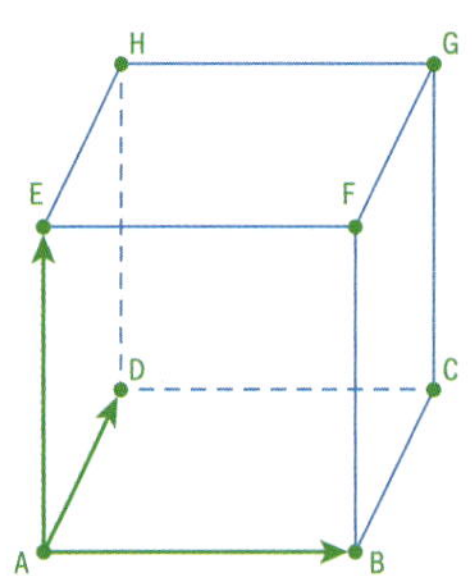

a $\overrightarrow{AG}$ **b** $\overrightarrow{CE}$

c $\overrightarrow{DF}$ **d** $\overrightarrow{MN}$

where M and N are midpoints of edges [BF] and [GH] respectively.

4 Use the definition of multiplication of a vector by a scalar to show the following properties:

i $\lambda(\mu\mathbf{a}) = (\lambda\mu)\mathbf{a} = \mu(\lambda\mathbf{a})$

ii $\lambda(\mathbf{a} + \mathbf{b}) = \lambda\mathbf{a} + \lambda\mathbf{b}$

iii $(\lambda + \mu)\mathbf{a} = \lambda\mathbf{a} + \mu\mathbf{a}$

iv $1 \cdot \mathbf{a} = \mathbf{a}$

v $0 \cdot \mathbf{a} = \mathbf{0}$

for all real parameters λ and μ and for all vectors **a** and **b**.

Use of vectors in geometry

The use of vectors and their properties in geometry provides another method for geometrical proof. In that case, you are simultaneously proving the fact that two line segments are parallel and that they have equal lengths.

For example, in a rectangle ABCD, if $\overrightarrow{AB} = \overrightarrow{DC}$, two facts can be deduced:

i the sides [AB] and [DC] are parallel [AB] ∥ [DC]

ii their lengths are equal AB = DC

You can make the same conclusions if you write $\overrightarrow{AB} = -\overrightarrow{CD}$.

Example 6

In triangle ABC, M and N are midpoints of the sides [AC] and [BC] respectively.

Show that the line segment [MN] is parallel to the side [AB] and that $MN = \frac{1}{2}AB$.

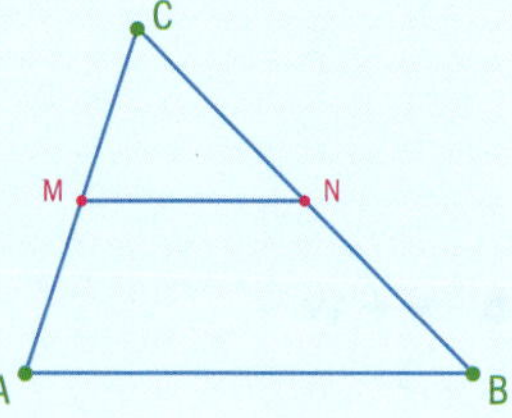

Method 1	
$\overrightarrow{MN} = \overrightarrow{MC} + \overrightarrow{CN}$	Triangle law.
$= \frac{1}{2}\overrightarrow{AC} + \frac{1}{2}\overrightarrow{CB}$	M and N are midpoints of the sides.

HINT

[AB] represents a line segment.

AB represents the distance between the points A and B.

(AB) represents the line that passes through the points A and B.

$= \frac{1}{2}(\overrightarrow{AC} + \overrightarrow{CB})$	Property of multiplication by a scalar.
$= \frac{1}{2}\overrightarrow{AB}$	Triangle law.
So the vectors $\overrightarrow{MN}$ and $\overrightarrow{AB}$ are parallel therefore [MN] is parallel to the side [AB]. Also since the scalar is $\frac{1}{2}$, the magnitudes of the vectors are in the same relation, so $MN = \frac{1}{2}AB$.	
Method 2	
$\overrightarrow{MN} = \overrightarrow{MA} + \overrightarrow{AB} + \overrightarrow{BN}$	The triangle law twice.
$= \frac{1}{2}\overrightarrow{CA} + \overrightarrow{AB} + \frac{1}{2}\overrightarrow{BC}$	M and N are midpoints of the sides.
$= \frac{1}{2}(\overrightarrow{BC} + \overrightarrow{CA}) + \overrightarrow{AB} = \frac{1}{2}\overrightarrow{BA} + \overrightarrow{AB}$	Use the property of multiplication by a scalar.
$= -\frac{1}{2}\overrightarrow{AB} + \overrightarrow{AB} = \frac{1}{2}\overrightarrow{AB}$	The opposite vector and triangle law.
Conclusion is the same as for Method 1.	

Example 7

Given a regular hexagon ABCDEF and midpoints M and N of the sides [EF] and [CD] respectively, use vectors to show that the lines [MN] and [AB] are parallel and that $MN = \frac{3}{2}AB$.

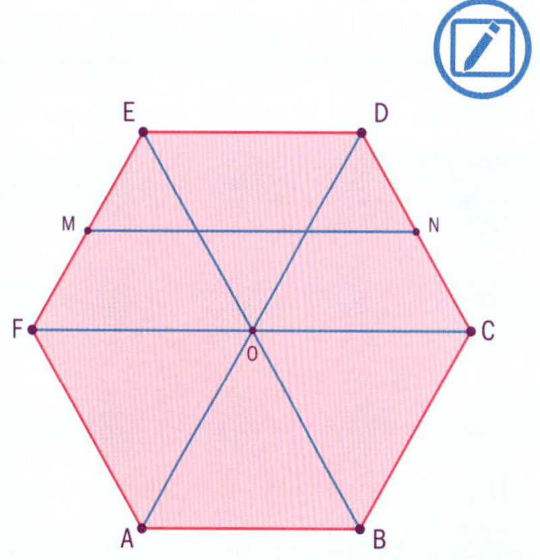

Continued on next page

You need to show that $\overrightarrow{MN} = \frac{3}{2}\overrightarrow{AB}$.	When using vectors the statement to the right is enough to prove both statements.
$\overrightarrow{MN} = \overrightarrow{ME} + \overrightarrow{ED} + \overrightarrow{DN}$	Triangle law to rewrite the vector $\overrightarrow{MN}$.
$= \frac{1}{2}\overrightarrow{FE} + \overrightarrow{AB} + \frac{1}{2}\overrightarrow{DC}$	M and N are the midpoints.
$= \frac{1}{2}\overrightarrow{OD} + \frac{1}{2}\overrightarrow{DC} + \overrightarrow{AB}$	Find equal vectors in the hexagon.
$= \frac{1}{2}(\overrightarrow{OD} + \overrightarrow{DC}) + \overrightarrow{AB}$	
$= \frac{1}{2}\overrightarrow{OC} + \overrightarrow{AB} = \frac{1}{2}\overrightarrow{AB} + \overrightarrow{AB}$	Add the vectors by using the triangle law and simplify the expression.
$= \frac{3}{2}\overrightarrow{AB}$	
So the vectors $\overrightarrow{MN}$ and $\overrightarrow{AB}$ are parallel therefore the line [MN] is parallel to the side [AB].	
Also since the scalar is $\frac{3}{2}$, the magnitudes of the vectors are in the same ratio, so $MN = \frac{3}{2}AB$.	

Example 8

In a parallelogram PQRS the points M and N are on the sides [PQ] and [PS] respectively. Given that the point M divides the side [PQ] in the ratio 5:2, find in which ratio point N has to divide side [SP] so that [MN] is parallel to [QS].

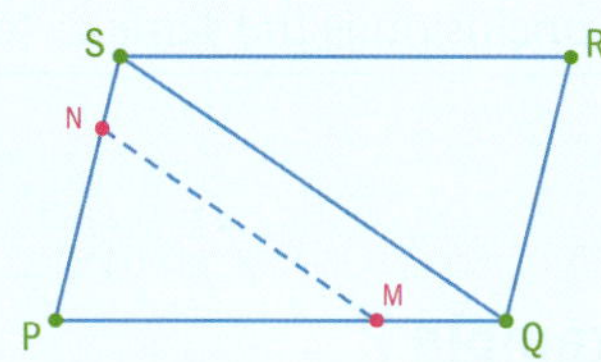

You need to show that vectors $\overrightarrow{MN}$ and $\overrightarrow{QS}$ are parallel.	
$\overrightarrow{MN} = \overrightarrow{MP} + \overrightarrow{PN}$	Use triangle law to rewrite the vector.
$= \frac{5}{7}\overrightarrow{QP} + \frac{k}{k+l}\overrightarrow{PS}$	Use that M divides [PQ] in the ratio 5 : 2, and that N divides [SP] in an unknown ratio $k : l$.
If [MN] and [QS] are parallel then $\overrightarrow{MN} = \lambda\overrightarrow{QS} = \lambda(\overrightarrow{QP} + \overrightarrow{PS})$	Use triangle law, $\overrightarrow{QS} = \overrightarrow{QP} + \overrightarrow{PS}$.
$= \lambda\overrightarrow{QP} + \lambda\overrightarrow{PS}$	Use the distributive property of multiplication of vectors by a scalar.

$\frac{5}{7}\overrightarrow{QP} + \frac{k}{k+l}\overrightarrow{PS} = \lambda\overrightarrow{QP} + \lambda\overrightarrow{PS} \Rightarrow \begin{cases} \lambda = \frac{5}{7} \\ \lambda = \frac{k}{k+l} \end{cases}$ $\Rightarrow \frac{k}{k+l} = \frac{5}{7} \Rightarrow k : l = 5 : 2$ So N divides [SP] in the ratio $l : k$, therefore $2 : 5$.	Use the uniqueness of the linear combination decomposition.

Example 9

In a triangle ABC the midpoints L, M and N of the sides [BC], [AC] and [AB] respectively are given.

Show that the medians [AL], [BM] and [CN] can form a triangle.

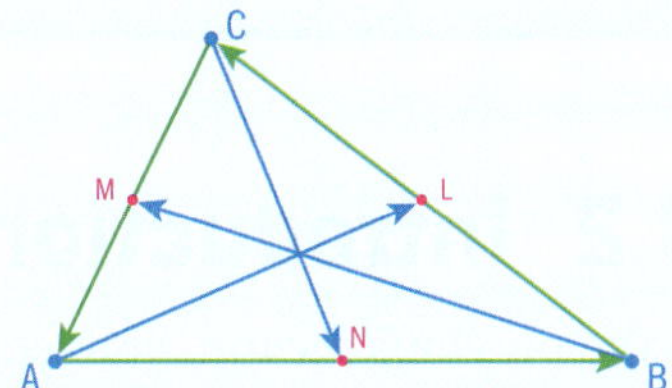

In a triangle ABC the direction vectors of the sides satisfy $\overrightarrow{AB} = \overrightarrow{BC} + \overrightarrow{CA} = \mathbf{0}$	Use the triangle law.
$\overrightarrow{AL} = \overrightarrow{AB} + \frac{1}{2}\overrightarrow{BC}$ $\overrightarrow{BM} = \overrightarrow{BC} + \frac{1}{2}\overrightarrow{CA}$ $\overrightarrow{CN} = \overrightarrow{CA} + \frac{1}{2}\overrightarrow{AB}$	Express all the medians in the terms of the direction vectors of the sides.
$\overrightarrow{AL} = \overrightarrow{BM} + \overrightarrow{CN} = \frac{3}{2}\overrightarrow{AB} + \frac{3}{2}\overrightarrow{BC} + \frac{3}{2}\overrightarrow{CA}$ $= \frac{3}{2}(\overrightarrow{AB} + \overrightarrow{BC} + \overrightarrow{CA}) = \frac{3}{2}\mathbf{0} = \mathbf{0}$	By adding all three equations you obtain the zero vector, therefore the medians can form a triangle.

Exercise 9B

1 A rectangle ABCD is given. The points P and Q are the midpoints of the sides [BC] and [CD] respectively. Show that [PQ] is parallel to the diagonal [BD] and that $PQ = \frac{1}{2}BD$.

2 A square ABCD is given. The points P and Q are the midpoints of the sides [BC] and [CD] respectively. Show that [PQ] is perpendicular to the diagonal [AC] and that $PQ = \frac{1}{2}AC$.

3 In a trapezium ABCD, where the sides [AB] and [CD] are parallel, the points M and N are the midpoints of [BC] and [AD] respectively. Show that [MN] is parallel to both [AB] and [CD] and that MN is the average of AB and CD.

4 In a rectangle ABCD the points P and Q are on the sides [BC] and [CD] respectively. Given that the point P divides the side [BC] in the ratio 2 : 3, find in which ratio point Q has to divide side [CD] so that [PQ] is parallel to [BD].

5 A regular octagon ABCDEFGH is given. Show that:

a [HC] ∥ [AB] and that $HC = \left(1+\sqrt{2}\right)AB$

b [MN] ∥ [AB] and that $MN = \left(1+\frac{\sqrt{2}}{2}\right)AB$,

where M and N are the midpoints of [DE] and [FG] respectively.

6 Given a quadrilateral ABCD and the midpoints K, L, M and N of the sides [AB], [BC], [CD] and [DA] respectively, use direction vectors to show that

KLMN forms a parallelogram.

9.2 Introduction to vector algebra

In the previous section, you learned that every vector in a plane can be written as a linear combination of two non-parallel vectors. When you look at the coordinate plane you can select these two vectors to be unit vectors (magnitude of 1 unit) in the direction of the positive *x*-axis and the positive *y*-axis, and call them **i** and **j** respectively.

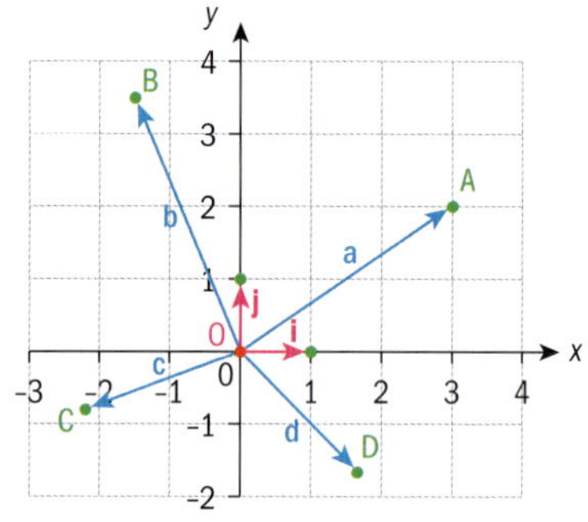

Each vector on the diagram to the right can be written as a linear combination of vectors **i** and **j**. Even vectors **i** and **j** can be written as **i** = 1 × **i** + 0 × **j** and **j** = 0 × **i** + 1× **j**. When you look at the scalars of the linear combination

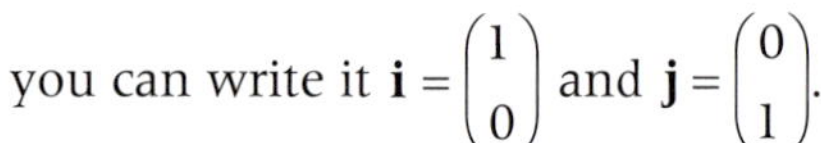

you can write it $\mathbf{i} = \begin{pmatrix}1\\0\end{pmatrix}$ and $\mathbf{j} = \begin{pmatrix}0\\1\end{pmatrix}$.

Notice that the notations used below are equivalent and you can use the one that is the most suited to the given problem.

The vectors can be written as:

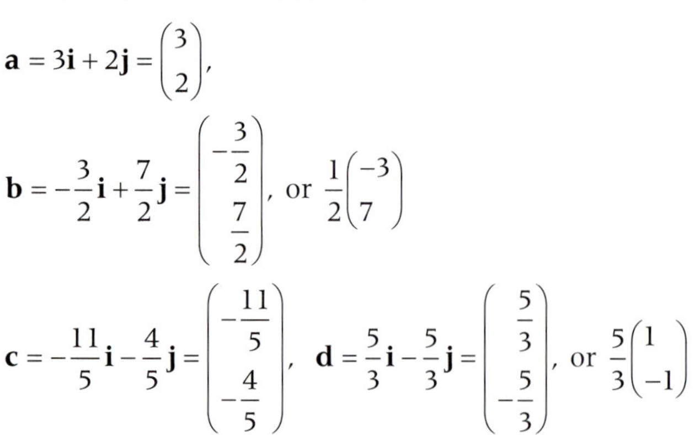

$$\mathbf{a} = 3\mathbf{i} + 2\mathbf{j} = \begin{pmatrix}3\\2\end{pmatrix},$$

$$\mathbf{b} = -\frac{3}{2}\mathbf{i} + \frac{7}{2}\mathbf{j} = \begin{pmatrix}-\frac{3}{2}\\ \frac{7}{2}\end{pmatrix}, \text{ or } \frac{1}{2}\begin{pmatrix}-3\\7\end{pmatrix}$$

$$\mathbf{c} = -\frac{11}{5}\mathbf{i} - \frac{4}{5}\mathbf{j} = \begin{pmatrix}-\frac{11}{5}\\ -\frac{4}{5}\end{pmatrix}, \quad \mathbf{d} = \frac{5}{3}\mathbf{i} - \frac{5}{3}\mathbf{j} = \begin{pmatrix}\frac{5}{3}\\ -\frac{5}{3}\end{pmatrix}, \text{ or } \frac{5}{3}\begin{pmatrix}1\\-1\end{pmatrix}$$

These two forms are called the algebraic form and the column vector form.

International-mindedness

Different ways of representing vectors appear around the world such as row vectors being shown as <a,b>.

A vector that has a starting point at the origin (0,0) and an end point $P(x,y)$ at any point in the plane is called a **position vector** of the point P.

We write $\mathbf{p} = \overrightarrow{OP} = x\mathbf{i} + y\mathbf{j} = \begin{pmatrix} x \\ y \end{pmatrix}$, where x and y are called the horizontal and vertical components of the vector $\mathbf{p}$ respectively.

Notice that the horizontal and vertical components of a position vector correspond to the x and y coordinates of the end point of the position vector.

In the previous section, vector algebra was only discussed from a geometrical point of view. The position vectors of the points A and B are given by their components $\mathbf{a} = a_1\mathbf{i} + a_2\mathbf{j} = \begin{pmatrix} a_1 \\ a_2 \end{pmatrix}$, $\mathbf{b} = b_1\mathbf{i} + b_2\mathbf{j} = \begin{pmatrix} b_1 \\ b_2 \end{pmatrix}$,

and now you need to investigate what the implications on the components are when performing operations.

First, it is necessary to state that two vectors are equal if and only if their corresponding components are equal.

$$\mathbf{a} = \mathbf{b} \Leftrightarrow \begin{pmatrix} a_1 \\ a_2 \end{pmatrix} = \begin{pmatrix} b_1 \\ b_2 \end{pmatrix} \Leftrightarrow \begin{cases} a_1 = b_1 \\ a_2 = b_2 \end{cases}$$

Investigation 6

Take two position vectors in the coordinate system and draw a diagram representing their sum. Use any rule you like. Organize your working in the following table.

Remember to try vectors with negative components.

$\mathbf{a}$	$\mathbf{b}$	$\mathbf{a}+\mathbf{b}$
$\begin{pmatrix} 2 \\ 1 \end{pmatrix}$	$\begin{pmatrix} 3 \\ 4 \end{pmatrix}$	

How do you obtain the components of the sum?

TOK

Do you think that one form of symbolic representation is preferable to another?

You can summarize your findings in the following definition.

Vector addition

If two vectors $\mathbf{a}$ and $\mathbf{b}$ are given by their components, then

$$\mathbf{a} + \mathbf{b} = (a_1\mathbf{i} + a_2\mathbf{j}) + (b_1\mathbf{i} + b_2\mathbf{j}) = (a_1 + b_1)\mathbf{i} + (a_2 + b_2)\mathbf{j}$$

or by using the column vector form

$$\mathbf{a} + \mathbf{b} = \begin{pmatrix} a_1 \\ a_2 \end{pmatrix} + \begin{pmatrix} b_1 \\ b_2 \end{pmatrix} = \begin{pmatrix} a_1 + b_1 \\ a_2 + b_2 \end{pmatrix}$$

Investigation 7

Take a position vector **a** and a scalar λ. In the coordinate system, draw a diagram representing the vector $\lambda\mathbf{a}$. Use any rule you like. Organize your working in the following table. Remember to try vectors with negative components, and also scalars with negative values.

a	λ	$\lambda\mathbf{a}$
$\begin{pmatrix}1\\3\end{pmatrix}$	2	

How do you obtain the components of the vector $\lambda\mathbf{a}$?

You can summarize your findings in the following definition.

Multiplication of a vector by a scalar (real number)

$$\lambda\mathbf{a} = \lambda(a_1\mathbf{i} + a_2\mathbf{j}) = (\lambda a_1)\mathbf{i} + (\lambda a_2)\mathbf{j}$$

or by using the column vector form

$$\lambda\mathbf{a} = \lambda\begin{pmatrix}a_1\\a_2\end{pmatrix} = \begin{pmatrix}\lambda a_1\\\lambda a_2\end{pmatrix}$$

Now you can summarize both rules in the following rule.

Linear combination of two vectors

$$\lambda\mathbf{a} + \mu\mathbf{b} = \lambda(a_1\mathbf{i} + a_2\mathbf{j}) + \mu(b_1\mathbf{i} + b_2\mathbf{j}) = (\lambda a_1 + \mu b_1)\mathbf{i} + (\lambda a_2 + \mu b_2)\mathbf{j}$$

or by using the column vector form

$$\lambda\mathbf{a} + \mu\mathbf{b} = \lambda\begin{pmatrix}a_1\\a_2\end{pmatrix} + \mu\begin{pmatrix}b_1\\b_2\end{pmatrix} = \begin{pmatrix}\lambda a_1 + \mu b_1\\\lambda a_2 + \mu b_2\end{pmatrix}$$

Again, notice that **subtraction** of two vectors doesn't need to be defined as a separate operation since subtraction is just a special case of the linear combination of two vectors.
$\lambda = 1,\ \mu = -1 \Rightarrow \mathbf{a} - \mathbf{b} = 1\cdot\mathbf{a} - 1\cdot\mathbf{b}$.

To solve the following problems, use vector algebra.

International-mindedness

Belgian/Dutch mathematician Simon Stevin used vectors in his theoretical work on falling bodies and his treatise "Principles of the art of weighing" in the 16th century.

Example 10

Given the vectors $\mathbf{a} = 3\mathbf{i} - 2\mathbf{j}$ and $\mathbf{b} = -\mathbf{i} + 4\mathbf{j}$ find the following vectors:

a $\mathbf{a} + \mathbf{b}$ **b** $3\mathbf{a}$ **c** $-2\mathbf{b}$ **d** $5\mathbf{a} + 4\mathbf{b}$

a $\mathbf{a} + \mathbf{b} = (3\mathbf{i} - 2\mathbf{j}) + (-\mathbf{i} + 4\mathbf{j})$ $= (3 - 1)\mathbf{i} + (-2 + 4)\mathbf{j} = 2\mathbf{i} + 2\mathbf{j}$	Add the corresponding components.
b $3\mathbf{a} = 3(3\mathbf{i} - 2\mathbf{j}) = 9\mathbf{i} - 6\mathbf{j}$ **c** $-2\mathbf{b} = -2(-\mathbf{i} + 4\mathbf{j}) = 2\mathbf{i} - 8\mathbf{j}$	Multiply each component by the scalar.
d $5\mathbf{a} + 4\mathbf{b} = 5(3\mathbf{i} - 2\mathbf{j}) + 4(-\mathbf{i} + 4\mathbf{j})$ $= 15\mathbf{i} - 10\mathbf{j} - 4\mathbf{i} + 16\mathbf{j} = 11\mathbf{i} + 6\mathbf{j}$	Multiply each component by the appropriate scalar and then add the corresponding components.

Example 11

Express the vector $\mathbf{c} = 4\mathbf{i} - 7\mathbf{j}$ as a linear combination of $\mathbf{a} = -\mathbf{i} + \mathbf{j}$ and $\mathbf{b} = 2\mathbf{i} - 3\mathbf{j}$.

$\mathbf{c} = \lambda\mathbf{a} + \mu\mathbf{b}$ $4\mathbf{i} - 7\mathbf{j} = \lambda(-\mathbf{i} + \mathbf{j}) + \mu(2\mathbf{i} - 3\mathbf{j})$ $4\mathbf{i} - 7\mathbf{j} = (-\lambda + 2\mu)\mathbf{i} + (\lambda - 3\mu)\mathbf{j}$	Rewrite the linear combination by using the components. Make the corresponding components equal.
$\begin{cases} -\lambda + 2\mu = 4 \\ \lambda - 3\mu = -7 \end{cases} \Rightarrow \mu = 3, \lambda = 2$	Solve the simultaneous equations by any method.
$\mathbf{c} = 2\mathbf{a} + 3\mathbf{b}$	Write the linear combination.

Now you can find the components of any direction vector $\overrightarrow{AB}$.

The position vectors of the points A and B are given by their components:

$$\mathbf{a} = \begin{pmatrix} a_1 \\ a_2 \end{pmatrix}, \mathbf{b} = \begin{pmatrix} b_1 \\ b_2 \end{pmatrix}$$

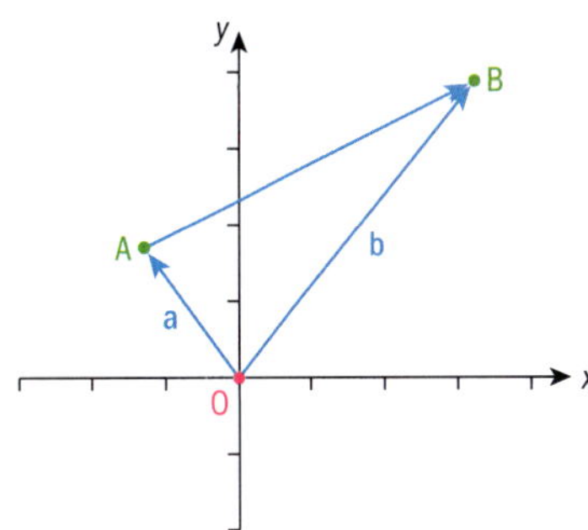

Use the triangle law and properties of vector addition.

$$\overrightarrow{OA} + \overrightarrow{AB} = \overrightarrow{OB} \Rightarrow$$

$$\overrightarrow{AB} = \overrightarrow{OB} - \overrightarrow{OA} \Rightarrow$$

$$\overrightarrow{AB} = \mathbf{b} - \mathbf{a} = \begin{pmatrix} b_1 \\ b_2 \end{pmatrix} - \begin{pmatrix} a_1 \\ a_2 \end{pmatrix} = \begin{pmatrix} b_1 - a_1 \\ b_2 - a_2 \end{pmatrix}$$

Notice that the subtraction is always performed in the order "terminal point minus initial point".

Geometry and trigonometry

Example 12

Given a parallelogram ABCD, with the points A(−2,−1), B(3,0) and D(0,2), find the coordinates of the point C.

Let point C(x,y). 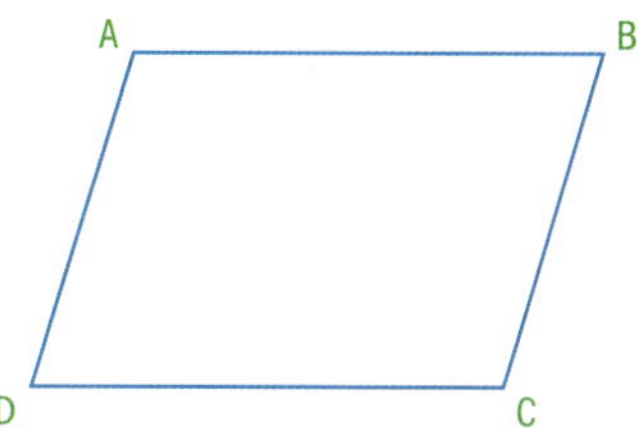	
Method 1	
$\overrightarrow{AB} = \mathbf{b} - \mathbf{a} = \begin{pmatrix} 3-(-2) \\ 0-(-1) \end{pmatrix} = \begin{pmatrix} 5 \\ 1 \end{pmatrix}$	Use the formula to find direction vector $\overrightarrow{AB}$.
$\overrightarrow{DC} = \mathbf{c} - \mathbf{d} = \begin{pmatrix} x-0 \\ y-2 \end{pmatrix} = \begin{pmatrix} 5 \\ 1 \end{pmatrix} \Rightarrow$	Since ABCD is a parallelogram $\overrightarrow{AB} = \overrightarrow{DC}$. Use the formula again to find the coordinates of C.
$\begin{cases} x = 5 \\ y-2 = 1 \end{cases} \Rightarrow \begin{cases} x = 5 \\ y = 3 \end{cases} \Rightarrow \text{C}(5,3)$	Two vectors are equal if their corresponding components are equal.
Method 2	
$\overrightarrow{AD} = \mathbf{d} - \mathbf{a} = \begin{pmatrix} 0-(-2) \\ 2-(-1) \end{pmatrix} = \begin{pmatrix} 2 \\ 3 \end{pmatrix}$	Use the formula to find direction vector $\overrightarrow{AD}$.
$\overrightarrow{BC} = \mathbf{c} - \mathbf{b} = \begin{pmatrix} x-3 \\ y-0 \end{pmatrix} = \begin{pmatrix} 2 \\ 3 \end{pmatrix} \Rightarrow$	Since ABCD is a parallelogram $\overrightarrow{AD} = \overrightarrow{BC}$. Use the formula again to find the coordinates of C.
$\begin{cases} x-3 = 2 \\ y = 3 \end{cases} \Rightarrow \begin{cases} x = 5 \\ y = 3 \end{cases} \Rightarrow \text{C}(5,3)$	Two vectors are equal if their corresponding components are equal.

When extending geometry into a 3-dimensional space you add an additional component that is moving away from a 2-dimensional plane.

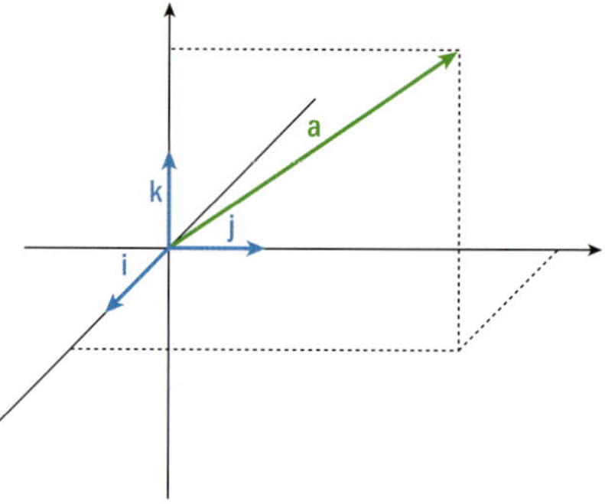

In a similar way you are going to replace the 2-dimensional coordinate system by a 3-dimensional coordinate system by adding another axis, z, that is perpendicular to both x and y axes. For vectors you are going to add another vector, **k**, that will also be a unit vector in the direction of the positive z-axis. So now any vector **a** in the 3-dimensional space can be written as a linear combination of vectors **i**, **j** and **k**.

Performing operations on 3–dimensional vectors is the same as performing operations on 2–dimensional vectors just with the operations on the additional component.

The following problem is in three dimensions.

Example 13

Let ABCDEFGH be a parallelepiped, with the vertices A(0,0,0), B(2,3,0), D(–1,4,–2) and E(1,–2,6). Find the coordinates of the points:

a C **b** G

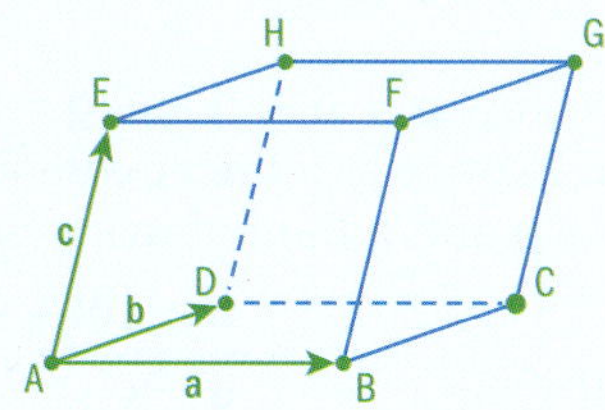

a Take the point $C(x,y,z)$.	
$\overrightarrow{AB} = \overrightarrow{OB} - \overrightarrow{OA} = \begin{pmatrix} 2-0 \\ 3-0 \\ 0-0 \end{pmatrix} = \begin{pmatrix} 2 \\ 3 \\ 0 \end{pmatrix}$	Find direction vector $\overrightarrow{AB}$.
$\overrightarrow{DC} = \overrightarrow{OC} - \overrightarrow{OD} = \begin{pmatrix} x-(-1) \\ y-4 \\ z-(-2) \end{pmatrix} = \begin{pmatrix} 2 \\ 3 \\ 0 \end{pmatrix} \Rightarrow$	Since ABCD is a parallelogram $\overrightarrow{AB} = \overrightarrow{DC}$.
$\begin{cases} x+1=2 \\ y-4=3 \\ z+2=0 \end{cases} \Rightarrow \begin{cases} x=1 \\ y=7 \\ z=-2 \end{cases} \Rightarrow C(1,7,-2)$	Two vectors are equal if their corresponding components are equal. Find the coordinates of C.
b Take the point $G(x,y,z)$.	Point G is above point C so you need to find the vector $\overrightarrow{CG}$.
$\overrightarrow{AE} = \overrightarrow{OE} - \overrightarrow{OA} = \begin{pmatrix} 1\ \ 0 \\ -2-0 \\ 6-0 \end{pmatrix} = \begin{pmatrix} 1 \\ -2 \\ 6 \end{pmatrix}$	
$\overrightarrow{CG} = \overrightarrow{OG} - \overrightarrow{OG} = \begin{pmatrix} x-1 \\ y-7 \\ z-(-2) \end{pmatrix} = \begin{pmatrix} 1 \\ -2 \\ 6 \end{pmatrix} \Rightarrow$	Since ACGE is a parallelogram $\overrightarrow{AE} = \overrightarrow{CG}$.
$\begin{cases} x-1=1 \\ y-7=-2 \\ z+2=6 \end{cases} \Rightarrow \begin{cases} x=2 \\ y=5 \\ z=4 \end{cases} \Rightarrow G(2,5,4)$	Two vectors are equal if their corresponding components are equal. Find the coordinates of G.

Exercise 9C

1 Given the vectors $\mathbf{a} = 2\mathbf{i} - 5\mathbf{j}$ and $\mathbf{b} = -3\mathbf{i} + 4\mathbf{j}$ find the following vectors:

a $\mathbf{a} + \mathbf{b}$ b $\mathbf{a} - \mathbf{b}$

c $5\mathbf{a} - 6\mathbf{b}$ d $7\mathbf{b} - 4\mathbf{a}$

e $\frac{3}{5}\mathbf{a} + \frac{3}{4}\mathbf{b}$

2 The vectors $\mathbf{p} = 3\mathbf{i} + 2\mathbf{j}$ and $\mathbf{q} = \mathbf{i} + 5\mathbf{j}$ are given. Express as a linear combination of vectors $\mathbf{p}$ and $\mathbf{q}$ the following vectors.

a $5\mathbf{i} - \mathbf{j}$ b $10\mathbf{i} + 9\mathbf{j}$

c $-9\mathbf{i} + 7\mathbf{j}$ d $\mathbf{i}$

e $-\mathbf{j}$ f $-\frac{1}{2}\mathbf{i} + \frac{2}{3}\mathbf{j}$

3 Given a rectangle PQRS, with the points Q(−4,−1), R(−1,−3) and S(3,3), find the coordinates of the point P.

4 A cuboid ABCDEFGH, with the vertices B(3,2,0), C(2,3,1), D(1,1,2) and G(0,3,−1) is given. Find the coordinates of all the remaining vertices.

5 Use the properties of real numbers to show the following properties of vector addition:

i **Commutative** $\mathbf{a} + \mathbf{b} = \mathbf{b} + \mathbf{a}$

ii **Associative** $(\mathbf{a} + \mathbf{b}) + \mathbf{c} = \mathbf{a} + (\mathbf{b} + \mathbf{c})$

iii **Identity element 0** is the identity element for addition, $\mathbf{a} + \mathbf{0} = \mathbf{0} + \mathbf{a} = \mathbf{a}$ for all $\mathbf{a}$.

iv **Opposite (inverse) element** For all vectors $\mathbf{a}$, there is the opposite vector $-\mathbf{a}$ such that $\mathbf{a} + (-\mathbf{a}) = -\mathbf{a} + \mathbf{a} = \mathbf{0}$.

6 Use the properties of real numbers to show the following properties of multiplication of a vector by a scalar:

i $(\lambda\mu)\mathbf{a} = \lambda(\mu\mathbf{a}) = \mu(\lambda\mathbf{a})$

ii $\lambda(\mathbf{a} + \mathbf{b}) = \lambda\mathbf{a} + \lambda\mathbf{b}$

iii $(\lambda + \mu)\mathbf{a} = \lambda\mathbf{a} + \mu\mathbf{a}$

iv $1\mathbf{a} = \mathbf{a}$

v $0\mathbf{a} = \mathbf{0}$ or $\lambda\mathbf{0} = \mathbf{0}$

Investigation 8

Take a point A in the coordinate plane. Draw the position vector $\overrightarrow{OA}$ and find the exact value of its magnitude (length). Vary the point A and find the magnitude of $\overrightarrow{OA}$.

1 What did you use to find the magnitude?

2 **Factual** What is the formula you used to find the magnitude?

3 How would you find the magnitude of a direction vector $\overrightarrow{AB}$?

Now take a point A in a 3-dimensional coordinate system. Sketch an appropriate diagram and find the magnitude of the position vector $\overrightarrow{OA}$. Vary the point A and find the corresponding magnitudes of $\overrightarrow{OA}$.

4 What did you use to find the magnitude?

5 **Factual** What is the formula you used to find the magnitude?

6 **Conceptual** How can representing a vector in 3D enhance your knowledge of vectors?

You can summarize all your findings in the following definitions.

Magnitude of a vector

Given a position vector **in a plane** with the components $\mathbf{a} = \overrightarrow{OA} = \begin{pmatrix} a_1 \\ a_2 \end{pmatrix}$, the magnitude of the position vector **a** is $OA = |\overrightarrow{OA}| = \sqrt{a_1^2 + a_2^2}$.

The magnitude of the direction vector $\overrightarrow{AB}$ is

$$|\overrightarrow{AB}| = \left|\begin{pmatrix} b_1 - a_1 \\ b_2 - a_2 \end{pmatrix}\right| = \sqrt{(b_1 - a_1)^2 + (b_2 - a_2)^2}.$$

Given a position vector **in 3D space** with the components $\mathbf{a} = \overrightarrow{OA} = \begin{pmatrix} a_1 \\ a_2 \\ a_3 \end{pmatrix}$,

the magnitude of the position vector **a** is $OA = |\overrightarrow{OA}| = \sqrt{a_1^2 + a_2^2 + a_3^2}$.

The magnitude of the direction vector $\overrightarrow{AB}$ is given by

$$|\overrightarrow{AB}| = \left|\begin{pmatrix} b_1 - a_1 \\ b_2 - a_2 \\ b_3 - a_3 \end{pmatrix}\right| = \sqrt{(b_1 - a_1)^2 + (b_2 - a_2)^2 + (b_3 - a_3)^2}$$

Notice that the magnitude of vector $\overrightarrow{AB}$ is the distance formula AB in two or in three dimensions.

Geometry and trigonometry

The following problems are about magnitude.

Example 14

Given the vectors $\mathbf{a} = 2\mathbf{i} - 6\mathbf{j}$ and $\mathbf{b} = -3\mathbf{i} - 5\mathbf{j}$, find the magnitudes of the following vectors:

a **a** **b** $2\mathbf{a} - 3\mathbf{b}$

a $\lvert\mathbf{a}\rvert = \sqrt{2^2 + (-6)^2} = \sqrt{4+36} = \sqrt{40} = 2\sqrt{10}$	Use the formula for magnitude.
b $2\mathbf{a} - 3\mathbf{b} = 2(2\mathbf{i} - 6\mathbf{j}) - 3(-3\mathbf{i} - 5\mathbf{j})$ $= 4\mathbf{i} - 12\mathbf{j} + 9\mathbf{i} + 15\mathbf{j} = 13\mathbf{i} + 3\mathbf{j}$	First find the components of the vector.
$\lvert 2\mathbf{a} - 3\mathbf{b}\rvert = \sqrt{13^2 + 3^2} = \sqrt{169+9} = \sqrt{178}$	Then use the formula for magnitude.

Example 15

Let $\mathbf{c} = \begin{pmatrix} -7 \\ 6 \end{pmatrix}$ and $\mathbf{d} = \begin{pmatrix} k \\ 2k \end{pmatrix}, k \in \mathbb{R}$. Find the values of k such that $|\mathbf{c}| = 2|\mathbf{d}|$.

Continued on next page

$\lvert\mathbf{c}\rvert = 2\lvert\mathbf{d}\rvert$ $\sqrt{(-7)^2 + 6^2} = 2\sqrt{k^2 + (2k)^2}$ $49 + 36 = 4(k^2 + 4k^2) \Rightarrow 85 = 20k^2$ $k^2 = \frac{17}{4} \quad k = \pm\frac{\sqrt{17}}{2}$	Use the formula for magnitude. Square both sides. Make k the subject. Find both values of k.

Can you make a geometrical interpretation of these two solutions?

For vectors in three dimensions, you include another component.

Example 16

Show that the points A(1, −1, −2), B(0, 4, 1) and C(2, 2, 2) are vertices of a right-angled triangle.

$\overrightarrow{AB} = \overrightarrow{OB} - \overrightarrow{OA}$ $\lvert\overrightarrow{AB}\rvert = \left\lvert\begin{pmatrix}0-1\\4-(-1)\\1-(-2)\end{pmatrix}\right\rvert = \sqrt{(-1)^2+5^2+3^2} = \sqrt{35}$ $\overrightarrow{BC} = \overrightarrow{OC} - \overrightarrow{OB}$ $\lvert\overrightarrow{BC}\rvert = \left\lvert\begin{pmatrix}2-0\\2-4\\2-1\end{pmatrix}\right\rvert = \sqrt{2^2+(-2)^2+1^2} = 3$ $\overrightarrow{AC} = \overrightarrow{OC} - \overrightarrow{OA}$ $\lvert\overrightarrow{AC}\rvert = \left\lvert\begin{pmatrix}2-1\\2-(-1)\\2-(-2)\end{pmatrix}\right\rvert = \sqrt{1^2+3^2+4^2} = \sqrt{26}$	Find the direction vectors of all three sides and calculate the magnitudes of those vectors.
$(\sqrt{26})^2 + 3^2 = (\sqrt{35})^2$ Magnitudes satisfy Pythagoras' theorem, so the triangle ABC is a right-angled triangle.	Verify that the magnitudes satisfy Pythagoras' theorem.

International-mindedness

Vectors developed quickly in the first two decades of the 19th century with Danish-Norwegian Caspar Wessel, Swiss Jean Robert Argand and German Carl Friedrich Gauss.

Investigation 9

Draw in the coordinate plane the position vector $\mathbf{a} = 3\mathbf{i} - 4\mathbf{j}$, then draw other position vectors with the same magnitude.

1 What can you conclude about the positions of the end points of your vectors?

2 What shape do all the end points of the position vectors with the same magnitude form in two dimensions?

3 Can you write an equation for the set of points?

4 What shape do all the end points of position vectors with the same magnitude form in three dimensions?

Unit vectors

Unit vectors are vectors with a magnitude of 1 unit in any direction. So far you have only been using vectors **i** and **j** in a 2-dimensional plane, and **i**, **j** and **k** in a 3-dimensional space as unit vectors. They have directions in the positive parts of the coordinate axes.

Unit vector **u** in the direction of a non-zero vector **a** is collinear with the vector **a**, therefore you can write $\mathbf{u} = \lambda\mathbf{a}$, $\lambda \in \mathbb{R}$.

To find this vector, you need to find the scalar λ. In the previous section, it was stated that the magnitude of vector **u** satisfies the following.

$$|\mathbf{u}| = |\lambda||\mathbf{a}| \Rightarrow |\lambda| = \frac{|\mathbf{u}|}{|\mathbf{a}|} = \frac{1}{\sqrt{a_1^2 + a_2^2}} \Rightarrow \lambda = \pm\frac{1}{\sqrt{a_1^2 + a_2^2}}$$

The unit vector **u** will have the same direction as **a** if $\lambda > 0$ and the opposite direction if $\lambda < 0$.

To summarize the finding and by using an analogy you can find the following formulas.

$$\mathbf{a} = \begin{pmatrix} a_1 \\ a_2 \end{pmatrix} \Rightarrow \mathbf{u} = \frac{1}{\sqrt{a_1^2 + a_2^2}}\begin{pmatrix} a_1 \\ a_2 \end{pmatrix} \text{ for a 2D vector}$$

$$\mathbf{a} = \begin{pmatrix} a_1 \\ a_2 \\ a_3 \end{pmatrix} \Rightarrow \mathbf{u} = \frac{1}{\sqrt{a_1^2 + a_2^2 + a_3^2}}\begin{pmatrix} a_1 \\ a_2 \\ a_3 \end{pmatrix} \text{ for a 3D vector}$$

Example 17

Find the unit vector that has the same direction as the given vector.

a $\mathbf{v} = 5\mathbf{i} - 12\mathbf{j}$ **b** $\mathbf{w} = 2\mathbf{i} + 3\mathbf{j} - 6\mathbf{k}$

a $\hat{v} = \frac{1}{\sqrt{5^2 + (-12)^2}}(5\mathbf{i} - 12\mathbf{j}) = \frac{5}{13}\mathbf{i} - \frac{12}{13}\mathbf{j}$	Use the formula for a 2D unit vector.
b $\widehat{w} = \frac{1}{\sqrt{2^2 + 3^2 + (-6)^2}}(2\mathbf{i} + 3\mathbf{j} - 6\mathbf{k})$ $= \frac{2}{7}\mathbf{i} + \frac{3}{7}\mathbf{j} - \frac{6}{7}\mathbf{k}$	Use the formula for a 3D unit vector.

A calculator has a feature to find a unit vector, but that feature will not be available on the IB exam, not compliant with the test mode.

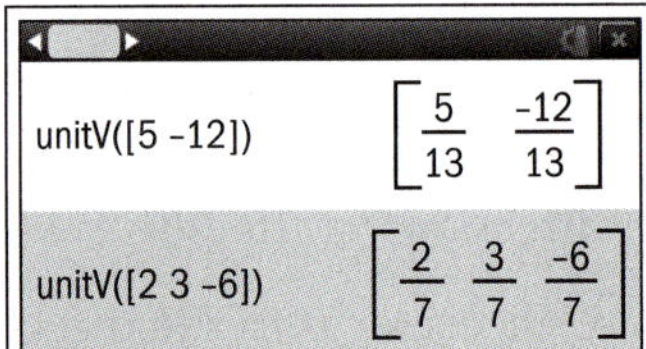

A similar method needs to be used when you are finding a vector of a given magnitude that is collinear or parallel to a particular vector.

Example 18

Find all the vectors that are parallel to $\mathbf{a} = 2\mathbf{i} - \mathbf{j} + 2\mathbf{k}$, and that have a magnitude of 5.

Call the vector **b**.	
$\lvert\mathbf{b}\rvert = \lvert\lambda\rvert\lvert\mathbf{a}\rvert \quad \lvert\lambda\rvert = \frac{\lvert\mathbf{b}\rvert}{\lvert\mathbf{a}\rvert} = \frac{5}{\sqrt{2^2 + (-1)^2 + 2^2}} = \frac{5}{3}$	Apply the formula for magnitudes of parallel vectors to find the scalar.
$\lambda = \pm\frac{5}{3} \quad \mathbf{b} = \pm\frac{5}{3}\mathbf{a}$	
$\Rightarrow \mathbf{b}_1 = \frac{10}{3}\mathbf{i} - \frac{5}{3}\mathbf{j} + \frac{10}{3}\mathbf{k}$ or $\mathbf{b}_2 = -\frac{10}{3}\mathbf{i} + \frac{5}{3}\mathbf{j} - \frac{10}{3}\mathbf{k}$	Use both values for the scalar to find the vectors in the same and the opposite direction.

Exercise 9D

1 Find the unit vectors in the direction of the following vectors:

a $\mathbf{a} = 7\mathbf{i} + 24\mathbf{j}$ **b** $\mathbf{b} = -3\mathbf{i} + 2\mathbf{j}$

c $\mathbf{c} = 4\mathbf{i} - 5\mathbf{j} + 20\mathbf{k}$ **d** $\mathbf{d} = -\mathbf{i} + 3\mathbf{j} + 4\mathbf{k}$

2 Find all the unit vectors that are parallel with the following vectors:

a $\mathbf{a} = 20\mathbf{i} - 21\mathbf{j}$ **b** $\mathbf{b} = \mathbf{i} - 3\mathbf{j}$

c $\mathbf{c} = 5\mathbf{i} + 6\mathbf{j} - 30\mathbf{k}$ **d** $\mathbf{d} = 2\mathbf{i} + \mathbf{j} - 5\mathbf{k}$

3 Given the vectors $\mathbf{a} = 3\mathbf{i} - 2\mathbf{j} + \lambda\mathbf{k}$ and $\mathbf{b} = -\mathbf{i} + 5\mathbf{j} + (\lambda - 5)\mathbf{k}$, find the real value of λ such that $2\lvert\mathbf{a}\rvert = \lvert\mathbf{b}\rvert$.

4 Find all the vectors that are parallel to the given vectors and that have a magnitude of m, given that:

a $\mathbf{a} = 5\mathbf{i} - \mathbf{j}$, $m = 6$

b $\mathbf{b} = -4\mathbf{i} + 5\mathbf{j} + 20\mathbf{k}$, $m = 63$

5 A cuboid ABCDEFGH with vertices B(3,2,0), C(2,3,1), D(1,1,2) and G(0,3,−1) is given. Find:

a the length of a space diagonal

b the volume of the cuboid.

Developing inquiry skills

Return to the opening problem. How might vectors be used to understand the scenario of the skier on the slope?

9.3 Scalar product and its properties

The **scalar product** of two vectors is a **real number (scalar)**. If the vectors are given by their components then the formula to calculate the scalar product is called an **algebraic definition**.

Given the 2-dimensional vectors $\mathbf{a} = \begin{pmatrix} a_1 \\ a_2 \end{pmatrix}$ and $\mathbf{b} = \begin{pmatrix} b_1 \\ b_2 \end{pmatrix}$, the scalar or dot product is defined as $\mathbf{a} \cdot \mathbf{b} = a_1 b_1 + a_2 b_2$.

Example 19

Find the scalar product of vectors **a** and **b** given that $\mathbf{a} = 2\mathbf{i} - \mathbf{j}$ and $\mathbf{b} = -\mathbf{i} - 4\mathbf{j}$.

$\mathbf{a} \times \mathbf{b} = \begin{pmatrix} 2 \\ -1 \end{pmatrix} \cdot \begin{pmatrix} -1 \\ -4 \end{pmatrix} = 2 \cdot (-1) + (-1) \cdot (-4) = -2 + 4 = 2$	Apply the definition given the components of the vectors.

Investigation 10

Take two vectors and draw them in the coordinate system. Find their magnitudes and calculate the scalar product of the two vectors. Draw a different pair of vectors in the coordinate system and vary their magnitudes but keep the angle between the vectors the same as for the initial pair. Fill in the following table.

Continued on next page

TOK

Why do you think that we use this definition of scalar product?

Are different proofs of the same theorem equally valid?

Geometry and trigonometry

a	**b**	$\lvert\mathbf{a}\rvert$	$\lvert\mathbf{b}\rvert$	$\lvert\mathbf{a}\rvert\ \lvert\mathbf{b}\rvert$	**a b**
$\begin{pmatrix}1\\2\end{pmatrix}$	$\begin{pmatrix}-1\\1\end{pmatrix}$				
$\begin{pmatrix}3\\6\end{pmatrix}$	$\begin{pmatrix}-2\\2\end{pmatrix}$				
$\begin{pmatrix}4\\-2\end{pmatrix}$	$\begin{pmatrix}5\\5\end{pmatrix}$				

1. What do you notice?
2. Which terms are proportional?
3. How can you explain this proportionality?

Investigation 11

Take the vector $\mathbf{a} = 5\mathbf{i}$ and draw it in the coordinate system. Take another vector $\mathbf{b}$ with the same magnitude and find the angle θ between them by using trigonometric ratios. Find the scalar product of those two vectors. Now take a different vector $\mathbf{b}$ with the same magnitude. Fill in the following table with more vectors $\mathbf{b}$.

a	**b**	θ	$\mathbf{a}\cdot\mathbf{b}$
$\begin{pmatrix}5\\0\end{pmatrix}$	$\begin{pmatrix}2\sqrt{5}\\\sqrt{5}\end{pmatrix}$		
$\begin{pmatrix}5\\0\end{pmatrix}$	$\begin{pmatrix}3\\4\end{pmatrix}$		
$\begin{pmatrix}5\\0\end{pmatrix}$	$\begin{pmatrix}3\sqrt{2}\\\sqrt{7}\end{pmatrix}$		
$\begin{pmatrix}5\\0\end{pmatrix}$	$\begin{pmatrix}0\\5\end{pmatrix}$		
$\begin{pmatrix}5\\0\end{pmatrix}$	$\begin{pmatrix}-\sqrt{5}\\2\sqrt{5}\end{pmatrix}$		

1 Organize your results in order of the magnitude of the angle $0 < \theta < \pi$. On the x-axis put the angle θ while on the y-axis put the value of the scalar product. Plot the points obtained in the table and connect them by a smooth curve.

2 How can you describe the curve?

3 Can you find an equation of the curve?

To summarize the results in your investigations, the geometrical interpretation of the scalar product is given by another definition.

This definition is called the **geometric definition** of the scalar product.

The scalar or dot product of two non-zero vectors **a** and **b** is given by $\mathbf{a} \cdot \mathbf{b} = |\mathbf{a}||\mathbf{b}|\cos\theta$, where θ is the angle between the two vectors.

Example 20

Find the scalar product of vectors **a** and **b** given that $|\mathbf{a}| = 3, |\mathbf{b}| = 5, \theta = \frac{\pi}{6}$.

$\mathbf{a} \cdot \mathbf{b} = |\mathbf{a}||\mathbf{b}|\cos\theta = 3 \cdot 5 \cdot \cos\left(\frac{\pi}{6}\right) = \frac{15\sqrt{3}}{2}$	Apply the geometric definition since the magnitudes and the angle are given.

There are many ways to show that these two algebraic and geometric definitions are equivalent. Here is one of them.

The position vectors **a** and **b** are given with their components $\mathbf{a} = \begin{pmatrix} a_1 \\ a_2 \end{pmatrix}$ and $\mathbf{b} = \begin{pmatrix} b_1 \\ b_2 \end{pmatrix}$.

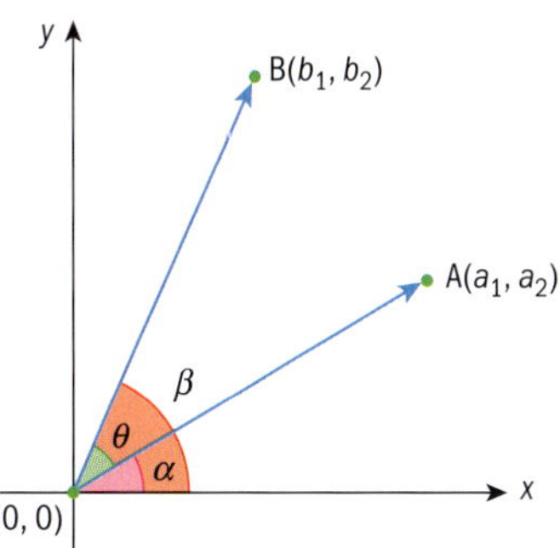

Denote the angles that both vectors enclose with the positive part of the x-axis by α and β respectively. The angle between vectors **a** and **b** is denoted by θ. Notice that $\beta - \alpha = \theta$.

Now start with the geometric definition.

$$\mathbf{a} \cdot \mathbf{b} = |\mathbf{a}||\mathbf{b}|\cos\theta = \sqrt{a_1^2 + a_2^2}\sqrt{b_1^2 + b_2^2}\cos\theta \, (1)$$

The angles α and β can be found by using the lines through the origin and the points A and B respectively, $\tan\alpha = \frac{a_2}{a_1}$ and $\tan\beta = \frac{b_2}{b_1}$.

Now you need to convert the tangent into the cosine and sine by using the formulas $\cos\alpha = \dfrac{1}{\sqrt{1+\tan^2\alpha}}$ and $\sin\alpha = \dfrac{\tan\alpha}{\sqrt{1+\tan^2\alpha}}$, therefore

$$\cos\alpha = \frac{1}{\sqrt{1+\left(\frac{a_2}{a_1}\right)^2}} = \frac{a_1}{\sqrt{a_1^2+a_2^2}} \text{ and } \sin\alpha = \frac{\frac{a_2}{a_1}}{\sqrt{1+\left(\frac{a_2}{a_1}\right)^2}} = \frac{a_2}{\sqrt{a_1^2+a_2^2}}$$

$$\cos\beta = \frac{1}{\sqrt{1+\left(\frac{b_2}{b_1}\right)^2}} = \frac{b_1}{\sqrt{b_1^2+b_2^2}} \text{ and } \sin\beta = \frac{\frac{b_2}{b_1}}{\sqrt{1+\left(\frac{b_2}{b_1}\right)^2}} = \frac{b_2}{\sqrt{b_1^2+b_2^2}}$$

By using the addition formula for cosine you obtain

$\cos\theta = \cos(\beta - \alpha) = \cos\beta\cos\alpha + \sin\beta\sin\alpha$

$$= \frac{b_1}{\sqrt{b_1^2+b_2^2}}\frac{a_1}{\sqrt{a_1^2+a_2^2}} + \frac{b_2}{\sqrt{b_1^2+b_2^2}}\frac{a_2}{\sqrt{a_1^2+a_2^2}} = \frac{a_1b_1+a_2b_2}{\sqrt{\left(a_1^2+a_2^2\right)\left(b_1^2+b_2^2\right)}} \qquad (2)$$

Substituting (2) into equation (1) you obtain the algebraic definition of the scalar product.

$$|\mathbf{a}||\mathbf{b}|\cos\theta = \sqrt{a^2+a_2^2}\sqrt{b_1^2+b_2^2}\cos\theta = \sqrt{a_1^2+a_2^2}\sqrt{b_1^2+b_2^2}\frac{a_1b_1+a_2b_2}{\sqrt{\left(a_1^2+a_2^2\right)\left(b_1^2+b_2^2\right)}} = a_1b_1+a_2b_2$$

Knowing that the two definitions are equivalent, you can investigate some properties of the scalar product by using a definition that you find appropriate.

Combining the two definitions you can obtain the formula for finding the angle enclosed by two vectors.

$$|\mathbf{a}||\mathbf{b}|\cos\theta = a_1b_1+a_2b_2 \Rightarrow \cos\theta = \frac{a_1b_1+a_2b_2}{\sqrt{\left(a_1^2+a_2^2\right)\left(b_1^2+b_2^2\right)}}$$

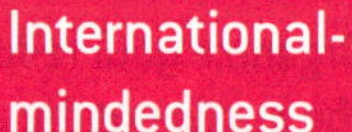

International-mindedness

Vectors are used widely to create movie animations and develop computer games. You can search "The use of mathematics in computer games".

Example 21

Find the angle in radians between the vectors **a** and **b** given that

a $\mathbf{a} = \begin{pmatrix}1\\1\end{pmatrix}$, $\mathbf{b} = \begin{pmatrix}-1\\3\end{pmatrix}$, correct to 3 sf **b** $\mathbf{a} = \begin{pmatrix}-\sqrt{3}\\1\end{pmatrix}$, $\mathbf{b} = \begin{pmatrix}2\\-2\sqrt{3}\end{pmatrix}$, giving the exact value.

a $\cos\theta = \dfrac{1\cdot(-1)+1\cdot 3}{\sqrt{(1^2+1^2)((-1)^2+3^2)}} = \dfrac{2}{2\sqrt{5}} = \dfrac{1}{\sqrt{5}}$ $\theta = \cos^{-1}\left(\dfrac{1}{\sqrt{5}}\right) = 1.11$	Use the given formula and your calculator to find the angle correct to 3 significant figures.
b $\cos\theta = \dfrac{-\sqrt{3}\cdot 2+1\cdot(-2\sqrt{3})}{\sqrt{((-\sqrt{3})^2+1^2)(2^2+(-2\sqrt{3})^2)}} = \dfrac{-4\sqrt{3}}{\sqrt{4\cdot 16}} = -\dfrac{\sqrt{3}}{2}$ $\theta = \cos^{-1}\left(-\dfrac{\sqrt{3}}{2}\right) = \dfrac{5\pi}{6}$	Use the given formula and the special value of trigonometric functions to find the exact value of the angle.

The dimension of vectors can be easily extended to a higher dimension.

Let's apply the algebraic definition in three dimensions.

Given the 3-dimensional vectors $\mathbf{a} = \begin{pmatrix} a_1 \\ a_2 \\ a_3 \end{pmatrix}$ and $\mathbf{b} = \begin{pmatrix} b_1 \\ b_2 \\ b_3 \end{pmatrix}$, the scalar or dot product is given by $\mathbf{a}\cdot\mathbf{b} = a_1b_1 + a_2b_2 + a_3b_3$.

Notice that the geometric definition will not change regardless of the dimension.

The angle between two vectors in three dimensions is given by the following formula.

$$\cos\theta = \frac{a_1b_1 + a_2b_2 + a_3b_3}{\sqrt{(a_1^2+a_2^2+a_3^2)(b_1^2+b_2^2+b_3^2)}}$$

Example 22

Given the vectors **a** and **b** such that **a** = 2**i** + 3**j** − **k** and **b** = **i** + 2**j** + 4**k**, find:

a the scalar product **a** · **b**

b the angle in radians between vectors **a** and **b**, correct to 3 significant figures.

a $\mathbf{a}\cdot\mathbf{b} = 2\cdot 1 + 3\cdot 2 + (-1)\cdot 4 = 4$	Use the algebraic formula to find the scalar product.
b $\cos\theta = \dfrac{4}{\sqrt{2^2+3^2+(-1)^2}\ \sqrt{1^2+2^2+4^2}}$ $= \dfrac{4}{7\sqrt{6}} \Rightarrow \theta = \cos^{-1}\left(\dfrac{4}{7\sqrt{6}}\right) = 1.34$	Use the given formula and your calculator to find the angle correct to 3 significant figures.

Geometry and trigonometry

There are some additional interesting results regarding the geometrical definition of a scalar product.

The fact $\mathbf{a} \cdot \mathbf{b} = 0$ derives several conclusions.

$$\mathbf{a} \cdot \mathbf{b} = |\mathbf{a}||\mathbf{b}|\cos\theta = 0 \Rightarrow \begin{cases} \mathbf{a} = \mathbf{0} \text{ or} \\ \mathbf{b} = \mathbf{0} \text{ or} \\ \theta = \frac{\pi}{2} \end{cases}$$

Therefore, you can conclude that either one of the vectors is a zero vector or that vectors **a** and **b** are orthogonal (perpendicular).

Properties of the scalar product

i $\mathbf{a} \cdot \mathbf{b} = \mathbf{b} \cdot \mathbf{a}$

ii $\mathbf{a} \cdot \mathbf{a} = |\mathbf{a}|^2$

iii $\mathbf{a} \cdot (\mathbf{b} + \mathbf{c}) = \mathbf{a} \cdot \mathbf{b} + \mathbf{a} \cdot \mathbf{c}$

iv $\lambda(\mathbf{a} \cdot \mathbf{b}) = (\lambda\mathbf{a}) \cdot \mathbf{b} = \mathbf{a} \cdot (\lambda\mathbf{b}),\ \lambda \in \mathbb{R}$

These properties are left to you to prove as an exercise.

Example 23

The vectors **a** and **b** are such that $|\mathbf{a}| = 3$ and $|\mathbf{b}| = 2$. Given that $\mathbf{a} + \mathbf{b}$ is perpendicular to $\mathbf{a} - 2\mathbf{b}$ find the angle between the vectors **a** and **b**.

$(\mathbf{a} + \mathbf{b}) \cdot (\mathbf{a} - 2\mathbf{b}) = 0$	The scalar product of two orthogonal vectors is 0.
$\Rightarrow \mathbf{a} \cdot \mathbf{a} - \mathbf{a} \cdot 2\mathbf{b} + \mathbf{b} \cdot \mathbf{a} - \mathbf{b} \cdot 2\mathbf{b} = 0$	Use the properties of the scalar product.
$\Rightarrow |\mathbf{a}|^2 - \mathbf{a} \cdot \mathbf{b} - 2|\mathbf{b}|^2 = 0$	
$\Rightarrow |\mathbf{a}|^2 - 2|\mathbf{b}|^2 = |\mathbf{a}||\mathbf{b}|\cos\theta$	Use the geometrical definition of the scalar product.
$\Rightarrow \cos\theta = \dfrac{|\mathbf{a}|^2 - 2|\mathbf{b}|^2}{|\mathbf{a}||\mathbf{b}|}$	Make $\cos\theta$ the subject.
$\cos\theta = \dfrac{3^2 - 2 \cdot 2^2}{3 \cdot 2} = \dfrac{1}{6} \Rightarrow \theta = \cos^{-1}\left(\dfrac{1}{6}\right) = 1.40$	Use the given magnitudes and find the angle.

If two vectors are collinear or parallel you can conclude that $\mathbf{a} \cdot \mathbf{b} = \pm|\mathbf{a}||\mathbf{b}|$, where the scalar product is positive if the vectors have the same direction and negative if they have the opposite direction.

Investigation 12

Copy and complete the following table and add some 2- and 3-dimensional vectors of your choice.

a	b	a · b	θ
$\begin{pmatrix}1\\2\end{pmatrix}$	$\begin{pmatrix}2\\1\end{pmatrix}$		
$\begin{pmatrix}3\\-1\end{pmatrix}$	$\begin{pmatrix}1\\-4\end{pmatrix}$		
$\begin{pmatrix}-2\\4\end{pmatrix}$	$\begin{pmatrix}2\\1\end{pmatrix}$		
$\begin{pmatrix}1\\2\\3\end{pmatrix}$	$\begin{pmatrix}2\\2\\-1\end{pmatrix}$		
$\begin{pmatrix}4\\0\\3\end{pmatrix}$	$\begin{pmatrix}-6\\3\\2\end{pmatrix}$		

1 What can you say about the values of the scalar product?

2 What can you say about the values of the angle?

3 **Conceptual** What is the relationship between the values of the scalar product and the angle between two vectors?

To summarize your findings from the investigation above you can say that $\mathbf{a} \cdot \mathbf{b} > 0$ if the vectors enclose an **acute angle** and that $\mathbf{a} \cdot \mathbf{b} < 0$ if the vectors enclose an **obtuse angle**.

Example 24

The vectors **a** and **b** are given in component form as $\mathbf{a} = \begin{pmatrix}2\\\lambda\end{pmatrix}, \mathbf{b} = \begin{pmatrix}\mu\\-5\end{pmatrix}, \lambda, \mu \in \mathbb{R}$

a Find the relationship between real parameters λ and μ given that the vectors **a** and **b** are perpendicular.

b Find the two possible vectors **a** in part **a** given that the magnitude of **b** is 13.

Continued on next page

Geometry and trigonometry

a $\mathbf{a}\cdot\mathbf{b}=0 \Rightarrow 2\mu-5\lambda=0$	The scalar product of perpendicular vectors is equal to zero.
b $\lvert\mathbf{b}\rvert=13 \Rightarrow \sqrt{\mu^2+(-5)^2}=13$ $\mu^2=169-25=144$ $\mu_1=12$ or $\mu_2=-12$	Use the formula for magnitude to find μ.
$\lambda=\frac{2}{5}\mu \Rightarrow \lambda_1=\frac{24}{5}$ or $\lambda_2=-\frac{24}{5}$	Use the relationship from part a to find λ. Notice that there are two different values of λ.
$\mathbf{a}_1=\begin{pmatrix}2\\ \frac{24}{5}\end{pmatrix}$ or $\mathbf{a}_2=\begin{pmatrix}2\\ -\frac{24}{5}\end{pmatrix}$	Write down two possible answers for vector **a**.

Exercise 9E

1 Find the scalar product of the vectors **a** and **b**:

a $\lvert\mathbf{a}\rvert=\sqrt{3}, \lvert\mathbf{b}\rvert=4, \theta=30°$

b $\lvert\mathbf{a}\rvert=12, \lvert\mathbf{b}\rvert=8, \theta=115°$

c $\lvert\mathbf{a}\rvert=3, \lvert\mathbf{b}\rvert=5, \theta=\frac{\pi}{7}$

d $\lvert\mathbf{a}\rvert=5\sqrt{2}, \lvert\mathbf{b}\rvert=17, \theta=\frac{3\pi}{4}$

2 Given the vectors $\mathbf{a}=3\mathbf{i}-4\mathbf{j}$ and $\mathbf{b}=6\mathbf{i}+5\mathbf{j}$ find:

a the scalar product $\mathbf{a}\cdot\mathbf{b}$

b the angle between vectors **a** and **b**.

3 Given the vectors $\mathbf{a}=\mathbf{i}+4\mathbf{j}-3\mathbf{k}$ and $\mathbf{b}=-2\mathbf{i}+3\mathbf{j}+\mathbf{k}$ find:

a the scalar product $\mathbf{a}\cdot\mathbf{b}$

b the angle between vectors **a** and **b**.

4 The vectors **a** and **b** are such that $\lvert\mathbf{a}\rvert=2$ and $\lvert\mathbf{b}\rvert=\sqrt{3}$. If $\mathbf{a}-2\mathbf{b}$ and $2\mathbf{a}+\mathbf{b}$ are perpendicular find the angle between the vectors **a** and **b**.

5 Use a suitable definition of the scalar product to deduce the following properties:

i $\mathbf{a}\cdot\mathbf{b}=\mathbf{b}\cdot\mathbf{a}$

ii $\mathbf{a}\cdot\mathbf{a}=\lvert\mathbf{a}\rvert^2$

iii $\mathbf{a}\cdot(\mathbf{b}+\mathbf{c})=\mathbf{a}\cdot\mathbf{b}+\mathbf{a}\cdot\mathbf{c}$

iv $\lambda(\mathbf{a}\cdot\mathbf{b})=(\lambda\mathbf{a})\cdot\mathbf{b}=\mathbf{a}\cdot(\lambda\mathbf{b}), \lambda\in\mathbb{R}$

6 Use the properties in question **5** to show that:

i $(\mathbf{a}+\mathbf{b})\cdot(\mathbf{a}+\mathbf{b})=\lvert\mathbf{a}\rvert^2+\lvert\mathbf{b}\rvert^2+2\lvert\mathbf{a}\rvert\lvert\mathbf{b}\rvert\cos\theta$

ii $(\mathbf{a}-\mathbf{b})\cdot(\mathbf{a}-\mathbf{b})=\lvert\mathbf{a}\rvert^2+\lvert\mathbf{b}\rvert^2-2\lvert\mathbf{a}\rvert\lvert\mathbf{b}\rvert\cos\theta$

Verify the results by using the cosine theorem.

7 The vectors $\mathbf{a}=a_1\mathbf{i}+a_2\mathbf{j}$ and $\mathbf{b}=b_1\mathbf{i}+b_2\mathbf{j}$ are given in component form. The vectors **a**, **b** and $\mathbf{a}-\mathbf{b}$ form a triangle. Use the properties of the scalar product and the cosine theorem to show that the geometric and algebraic definitions are equivalent, $a_1b_1+a_2b_2=\lvert\mathbf{a}\rvert\lvert\mathbf{b}\rvert\cos\theta$.

8 Given the vectors **a** and **b** such that $\mathbf{a}-\mathbf{b}$ and $2\mathbf{a}+\mathbf{b}$ are perpendicular, and $\mathbf{a}-2\mathbf{b}$ and $3\mathbf{a}+\mathbf{b}$ are perpendicular, find the angle between the vectors **a** and **b**.

9.4 Vector equation of a line

Geometrically there is only one line that passes through two distinct points. In the coordinate plane you can find an equation of a line that passes through two points.

$$A(x_1, y_1) \text{ and } B(x_2, y_2) \qquad m = \frac{y_2 - y_1}{x_2 - x_1} \qquad y = \frac{y_2 - y_1}{x_2 - x_1}(x - x_1) + y_1$$

The form $y = mx + k$ is called the slope, y-intercept form. You have also been using the general form $ax + by = c$, $a, b, c \in \mathbb{R}$, where a and b cannot be equal to zero at the same time.

International-mindedness

In 1827, August Ferdinand Mobius published a short book, The Barycentric Calculus where he denoted vectors by letters of the alphabet.

There are other forms of the equation of a line and one such form involves vectors. It is called a **vector form**. The two given points A and B have corresponding position vectors $\overrightarrow{OA}$ and $\overrightarrow{OB}$, respectively.

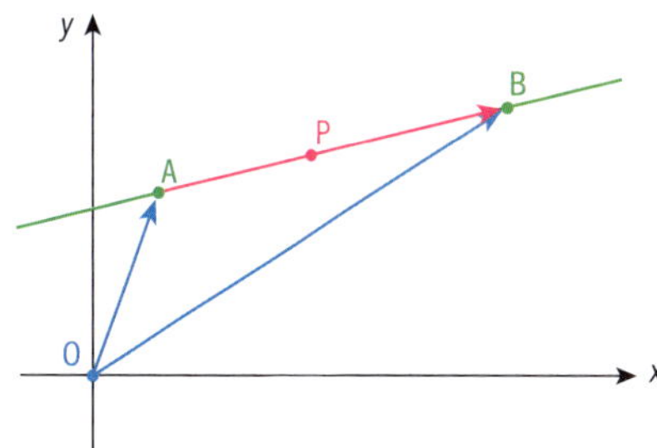

For every point P(x,y) that lies on the line (AB) the vector $\overrightarrow{AP}$ must be collinear or parallel to the vector $\overrightarrow{AB}$.

$\overrightarrow{AP} = k\overrightarrow{AB}$, $k \in \mathbb{R}$

The vector $\overrightarrow{AB}$ is called a **direction vector** of the line, since all the vectors that are parallel to $\overrightarrow{AB}$ can also define the same line. So, you can define the line by using a **direction vector** and **one point**. Notice that any vector parallel to the direction vector will define the same line.

In general, given the direction vector of a line **d** and one point A with position vector **a** that lies on the line, then for every point P with the position vector **p** that lies on the line, $\overrightarrow{AP}$ must be parallel to **d**.

$\overrightarrow{AP} = k\mathbf{d} \Rightarrow \mathbf{p} - \mathbf{a} = k\mathbf{d} \Rightarrow \mathbf{p} = \mathbf{a} + k\mathbf{d}$, $k \in \mathbb{R}$

This is called the **vector equation** of a line. Now rewrite this equation by using the vector components.

$$\begin{pmatrix} x \\ y \end{pmatrix} = \begin{pmatrix} x_1 \\ y_1 \end{pmatrix} + k\begin{pmatrix} d_1 \\ d_2 \end{pmatrix}, k \in \mathbb{R}$$

From this formula you can derive the **parametric form**.

$$\begin{cases} x = x_1 + kd_1 \\ y = y_1 + kd_2 \end{cases}, k \in \mathbb{R}$$

There is another form of the equation of a line that we rarely use in two dimensions, this is called the **Cartesian form**. This form is obtained by eliminating k from both equations.

$$\Rightarrow \begin{cases} \dfrac{x - x_1}{d_1} = k \\ \dfrac{y - y_1}{d_2} = k \end{cases} \Rightarrow \frac{x - x_1}{d_1} = \frac{y - y_1}{d_2}$$

However, in two dimensions it is much easier to algebraically manipulate the slope, y-intercept form or the general form.

The Cartesian form can be rearranged to obtain the slope, y-intercept form as follows.

$$\Rightarrow \frac{d_2(x-x_1)}{d_1} = y - y_1 \Rightarrow y = \frac{d_2}{d_1}(x-x_1)+y_1$$

Now you can see that the slope of the line $m = \frac{d_2}{d_1}$.

This form is equivalent to the form that you saw at the beginning of the section.

$$\overrightarrow{AP} = k\overrightarrow{AB} \Rightarrow \mathbf{p} - \mathbf{a} = k(\mathbf{b} - \mathbf{a})$$
$$\Rightarrow \mathbf{p} = (1-k)\mathbf{a} + k\mathbf{b},\ k \in \mathbb{R}$$

When written using vector components you will obtain the following formula.

$$\begin{pmatrix} x \\ y \end{pmatrix} = (1-k)\begin{pmatrix} x_1 \\ y_1 \end{pmatrix} + k\begin{pmatrix} x_2 \\ y_2 \end{pmatrix}, k \in \mathbb{R}$$

$$\begin{cases} x = x_1 - kx_1 + kx_2 \\ y = y_1 - ky_1 + ky_2 \end{cases} \Rightarrow \begin{cases} x = x_1 + k(x_2 - x_1) \\ y = y_1 + k(y_2 - y_1) \end{cases}, k \in \mathbb{R}.$$

By using the method of substitution you can eliminate the parameter k.

$$\begin{cases} k = \dfrac{x-x_1}{x_2-x_1} \\ y = y_1 + k(y_2 - y_1) \end{cases} \Rightarrow y = y_1 + \frac{x-x_1}{x_2-x_1}(y_2-y_1) \Rightarrow y = \frac{y_2-y_1}{x_2-x_1}(x-x_1)+y_1$$

This is the formula that you have been using in coordinate geometry for the equation of a line through two points.

Example 25

A line is passing through the points A(2,−3) and B(−1,2). Find:

a the vector equation of the line

b the Cartesian equation of the line

c the slope, y-intercept form of the line.

a $\overrightarrow{AB} = (-1-2)\mathbf{i} + (2-(-3))\mathbf{j} = -3\mathbf{i} + 5\mathbf{j}$ So the vector equation of the line is $\begin{pmatrix} x \\ y \end{pmatrix} = \begin{pmatrix} -1 \\ 2 \end{pmatrix} + k\begin{pmatrix} -3 \\ 5 \end{pmatrix}, k \in \mathbb{R}$	Find a direction vector. Take the position vector of one of the points, eg B, to write the vector equation of the line.
b $\begin{cases} x = -1-3k \\ y = 2+5k \end{cases} \Rightarrow \begin{cases} \dfrac{x+1}{-3} = k \\ \dfrac{y-2}{5} = k \end{cases}$ $\dfrac{x+1}{-3} = \dfrac{y-2}{5}$	Use the vector components to obtain the parametric form. Express k in both equations and equate the left sides.

c $\begin{cases} x = -1 - 3k \\ y = 2 + 5k \end{cases} \Rightarrow \begin{cases} 5x = -5 - 15k \\ 3y = 6 + 15k \end{cases}$	Again start with the parametric form. Eliminate the parameter k by multiplying the first equation by 5 and second by 3 and then adding the equations.
$\Rightarrow 5x + 3y = 1 \Rightarrow y = -\frac{5}{3}x + \frac{1}{3}$	Convert the general form to slope, y-intercept form.

Geometrically, there is only one line in the plane that is perpendicular to a given line at a particular point on that line.

By using this fact, you can find the equation of this line involving a normal vector.

A normal vector is perpendicular or orthogonal to any vector on the line so $\mathbf{n} \cdot \mathbf{AP} = 0$

Using the position vectors of the points P and A you obtain the following.

$\mathbf{n} \cdot (\mathbf{p} - \mathbf{a}) = 0 \Rightarrow \mathbf{n} \cdot \mathbf{p} = \mathbf{n} \cdot \mathbf{a}$

If the direction vector $\mathbf{d} = \begin{pmatrix} d_1 \\ d_2 \end{pmatrix}$ then one possible normal vector could be $\mathbf{n} = \begin{pmatrix} d_2 \\ -d_1 \end{pmatrix}$ or any vector parallel to it.

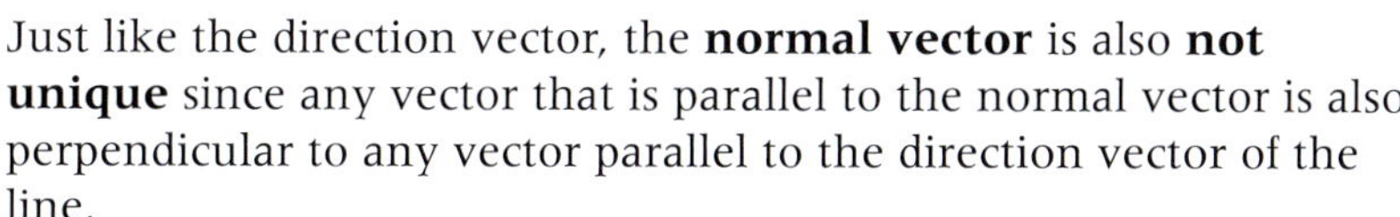

Just like the direction vector, the **normal vector** is also **not unique** since any vector that is parallel to the normal vector is also perpendicular to any vector parallel to the direction vector of the line.

So when rewriting the equation by using vector components you obtain

$$\begin{pmatrix} x \\ y \end{pmatrix} \cdot \begin{pmatrix} d_2 \\ -d_1 \end{pmatrix} = \begin{pmatrix} x_1 \\ y_1 \end{pmatrix} \cdot \begin{pmatrix} d_2 \\ -d_1 \end{pmatrix} \Rightarrow xd_2 - yd_1 = x_1d_2 - y_1d_1 \Rightarrow xd_2 - x_1d_2 + y_1d_1 = yd_1$$

$\Rightarrow y = \frac{d_2}{d_1}(x - x_1) + y_1$, which is the same equation obtained from the vector form.

Example 26

Find the equation of the line in slope, y-intercept form passing through the point A(−3,4) and with:

a direction vector $2\mathbf{i} - \mathbf{j}$

b normal vector $3\mathbf{i} + 5\mathbf{j}$.

Continued on next page

a $\mathbf{p} = \mathbf{a} + \lambda\mathbf{d} \Rightarrow \begin{pmatrix} x \\ y \end{pmatrix} = \begin{pmatrix} -3 \\ 4 \end{pmatrix} + \lambda\begin{pmatrix} 2 \\ -1 \end{pmatrix}$	Write the coordinates of the point and the components of the direction vector in the vector equation.
$\Rightarrow \begin{cases} x = -3 + 2\lambda \\ y = 4 - \lambda \end{cases} \Rightarrow \begin{cases} \dfrac{x+3}{2} = \lambda \\ \dfrac{y-4}{-1} = \lambda \end{cases}$	Rewrite the vector equation in parametric form.
$\Rightarrow \dfrac{x+3}{2} = \dfrac{y-4}{-1} \Rightarrow -\dfrac{1}{2}x - \dfrac{3}{2} = y - 4$	Eliminate the real parameter λ to obtain the Cartesian form.
$y = -\dfrac{1}{2}x + \dfrac{5}{2}$	Rewrite the equation in the slope, y-intercept form.
b $\mathbf{n} \cdot (\mathbf{p} - \mathbf{a}) = 0 \Rightarrow \mathbf{n} \cdot \mathbf{p} = \mathbf{n} \cdot \mathbf{a}$	The normal vector is perpendicular to any vector on the line therefore the scalar product is equal to 0.
$\Rightarrow \begin{pmatrix} 3 \\ 5 \end{pmatrix} \cdot \begin{pmatrix} x \\ y \end{pmatrix} = \begin{pmatrix} 3 \\ 5 \end{pmatrix} \cdot \begin{pmatrix} -3 \\ 4 \end{pmatrix}$	Rewrite the vectors by using the given components.
$\Rightarrow 3x + 5y = -9 + 20 \Rightarrow y = -\dfrac{3}{5}x + \dfrac{11}{5}$	Apply the algebraic definition of the scalar product and rewrite the equation in the slope, y-intercept form.

Different forms of the equation of a line can be used for solving problems with lines.

Example 27

Find the point of intersection between the lines $\mathbf{r} = 2\mathbf{i} - 5\mathbf{j} + \lambda(-\mathbf{i} + 3\mathbf{j})$ and $\dfrac{x+2}{3} = 1 - y$.

$\begin{pmatrix} x \\ y \end{pmatrix} = \begin{pmatrix} 2 \\ -5 \end{pmatrix} + \lambda\begin{pmatrix} -1 \\ 3 \end{pmatrix} \Rightarrow \begin{cases} x = 2 - \lambda \\ y = -5 + 3\lambda \end{cases}$	Rewrite the equation of the first line in the parametric form and substitute the coordinates x and y into the equation of the second line.
$\dfrac{2-\lambda+2}{3} = 1 - (-5 + 3\lambda) \Rightarrow 4 - \lambda = 18 - 9\lambda$ $8\lambda = 14 \Rightarrow \lambda = \dfrac{7}{4}$	Solve the equation for λ.
$\Rightarrow \begin{cases} x = 2 - \dfrac{7}{4} \\ y = -5 + \dfrac{21}{4} \end{cases} \Rightarrow \begin{cases} x = \dfrac{1}{4} \\ y = \dfrac{1}{4} \end{cases}$ $\left(\dfrac{1}{4}, \dfrac{1}{4}\right)$	Use the parametric equations of the first line to find the coordinates of the point of intersection.

Investigation 13

1. Take three distinctive lines in a plane and draw all possible mutual positions they can be in.
2. How many different positions can they form?
3. Which position has a common point of all three lines?

HINT

Lines passing through the same point are called concurrent.

Example 28

Determine whether or not the following lines $L_1: \frac{x-1}{2} = \frac{y+1}{3}$, L_2: $3x + 2y = -3$ and L_3: $\mathbf{r} = (-5\mathbf{i} + 6\mathbf{j}) + \lambda(2\mathbf{i} - 3\mathbf{j})$ are concurrent.

First find the point of intersection between the first two lines.	You need to solve simultaneous equations for two lines to find the point of intersection.
$\begin{cases} \frac{x-1}{2} = \frac{y+1}{3} \\ 3x+2y=-3 \end{cases} \Rightarrow \begin{cases} 3x-3=2y+2 \\ 3x+2y=-3 \end{cases}$	Subtract the first equation from the second equation to eliminate x.
$\Rightarrow 2y + 3 = -5 - 2y \Rightarrow 4y = -8$ $\Rightarrow y = -2 \Rightarrow x = \frac{1}{3}$	Find the value of y and use the second equation to find the value of x.
$\begin{pmatrix} \frac{1}{3} \\ -2 \end{pmatrix} = \begin{pmatrix} -5 \\ 6 \end{pmatrix} + \lambda \begin{pmatrix} 2 \\ -3 \end{pmatrix} \Rightarrow \begin{cases} \frac{1}{3} = -5 + 2\lambda \\ -2 = 6 - 3\lambda \end{cases}$	Check whether the point $\left(\frac{1}{3}, -2\right)$ lies on the third line.
$\Rightarrow \begin{cases} \frac{16}{3} = 2\lambda \\ -8 = -3\lambda \end{cases} \Rightarrow \begin{cases} \lambda = \frac{8}{3} \\ \lambda = \frac{8}{3} \end{cases}$	Since the value of λ is the same, the point is on the line.
The lines pass through the same point.	

When finding lines in vector form you need to focus on direction vectors.

- Parallel lines have collinear direction vectors.
- Perpendicular lines have orthogonal direction vectors, such that the scalar product is equal to 0.

Example 29

The line L: $\mathbf{p} = (1 + \lambda)\mathbf{i} + (\lambda - 2)\mathbf{j}$ and the point A$(-4,1)$ are given. Find the vector equation of a line that is:

a parallel to the line L and passing through the point A;

b perpendicular to the line L and passing through the point A.

Continued on next page

a $\mathbf{p} = (1+\lambda)\mathbf{i} + (\lambda - 2)\mathbf{j}$ $\Rightarrow \mathbf{p} = \mathbf{i} - 2\mathbf{j} + \lambda(\mathbf{i}+\mathbf{j}) \Rightarrow \mathbf{d} = \mathbf{i}+\mathbf{j}$ $\mathbf{r} = -4\mathbf{i} + \mathbf{j} + \mu(\mathbf{i}+\mathbf{j}),\ \mu \in \mathbb{R}$	Rewrite the vector equation of the line to find the direction vector of the line. The parallel line must have a collinear direction vector, so take the same direction vector. Use the position vector of the point A and write the vector equation.
b $\mathbf{d} = \mathbf{i}+\mathbf{j} \Rightarrow \mathbf{n} = \mathbf{i}-\mathbf{j}$ $\mathbf{r} = -4\mathbf{i} + \mathbf{j} + \lambda(\mathbf{i}-\mathbf{j}),\ \lambda \in \mathbb{R}$	The perpendicular line must have a normal vector for its direction vector. Use the position vector of the point A and write the vector equation.

Exercise 9F

1 Find a vector equation of the line passing through the points

a A(0,0) and B(2, 3)

b C(2,1) and D(−1, 3)

c E(−2,−5) and F(3,−6)

d $G\left(\frac{2}{3},-1\right)$ and $H\left(-\frac{1}{2},\frac{3}{4}\right)$

2 Find an equation of the line that is passing through the point P(2,−7) and has:

a direction vector $\mathbf{i}+\mathbf{j}$

b normal vector $2\mathbf{i} - 3\mathbf{j}$.

3 The line $L: \frac{x-3}{2} = \frac{y+1}{-3}$ and the point T(−3, 8) are given. Find a vector equation of the line that is:

a parallel to the line L and passing through the point T

b perpendicular to the line L and passing through the point T

c passing through the point T and the intersection between the line L and the line $\mathbf{r} = (1+2\lambda)\mathbf{i} + (4\lambda - 1)\mathbf{j}$.

4 Find the value of the real parameter a such that the lines $\mathbf{r} = 3\mathbf{i} + 2\mathbf{j} + \lambda(-\mathbf{i} + a\mathbf{j})$ and the line $\mathbf{p} = (1+2\mu)\mathbf{i} + (5\mu - 2)\mathbf{j}$ are:

a parallel **b** perpendicular.

The **vector equation** of a line has the same form regardless of the dimension because two distinctive points always define one line. In a 3D space the formula is:

$$\mathbf{r} = \mathbf{a} + \lambda\mathbf{d} \Rightarrow \begin{pmatrix} x \\ y \\ z \end{pmatrix} = \begin{pmatrix} a_1 \\ a_2 \\ a_3 \end{pmatrix} + \lambda \begin{pmatrix} d_1 \\ d_2 \\ d_3 \end{pmatrix}, \lambda \in \mathbb{R}$$

To find the **Cartesian form** you will need to work through the **parametric form**.

$$\begin{cases} x = a_1 + \lambda d_1 \\ y = a_2 + \lambda d_2, \lambda \in \mathbb{R} \\ z = a_3 + \lambda d_3 \end{cases}$$

Use the method of substitution from each equation to eliminate the parameter λ.

International-mindedness

In 1827, August Ferdinand Mobius published a short book, The Barycentric Calculus where he denoted vectors by letters of the alphabet.

$$\frac{x-a_1}{d_1}=\frac{y-a_2}{d_2}=\frac{x-a_3}{d_3}$$

This is the **Cartesian form** of a line in space.

Notice that the Cartesian form has components of a position vector of a point in the numerator and components of a direction vector in the denominator.

TOK

Why might it be argued that vector equations are superior to Cartesian equations?

Example 30

A line is passing through the points A(1,0,–4) and B(–3,1,–2). Find:

a the vector equation of the line **b** the Cartesian form of the line.

a $\overrightarrow{BA}=(1-(-3))\mathbf{i}+(0-1)\mathbf{j}+(-4-(-2))\mathbf{k}$ $=4\mathbf{i}-\mathbf{j}-2\mathbf{k}$	Find a direction vector.
So the vector equation of the line is $\begin{pmatrix}x\\y\\z\end{pmatrix}=\begin{pmatrix}1\\0\\-4\end{pmatrix}+\lambda\begin{pmatrix}4\\-1\\-2\end{pmatrix},\lambda\in\mathbb{R}$	Take the position vector of one of the points, eg A, to write the vector equation of the line.
b $\frac{x-1}{4}=\frac{y-0}{-1}=\frac{z-(-4)}{-2}$ $\Rightarrow\frac{x-1}{4}=-y=\frac{z+4}{-2}$	Use the Cartesian form and simplify the algebraic expressions.

Example 31

The line L: $\mathbf{r}=(1-\lambda)\mathbf{i}+(2+3\lambda)\mathbf{j}+2\lambda\mathbf{k}$, $\lambda\in\mathbb{R}$ is given.

a Show that the point D(1,2,–3) does not lie on the line.

b Find the vector equation of the line parallel to the line L that passes through the point D.

a $\begin{pmatrix}1\\2\\-3\end{pmatrix}=\begin{pmatrix}1-\lambda\\2+3\lambda\\2\lambda\end{pmatrix}\Rightarrow\begin{cases}1=1-\lambda\\2=2+3\lambda\\-3=2\lambda\end{cases}$	Write the vector equation by using the components. Since the corresponding components are equal solve for λ. Notice that you have obtained three simultaneous equations with one unknown.
$\Rightarrow\lambda=0$ and $\lambda=-\frac{3}{2}$, which is impossible, so the point D does not lie on the line.	By obtaining two different values of λ you can conclude that the system is not consistent, therefore there is no solution.
b $\mathbf{r}=\mathbf{i}+2\mathbf{j}-3\mathbf{k}+\mu(-\mathbf{i}+3\mathbf{j}+2\mathbf{k})$, $\mu\in\mathbb{R}$	Since the lines are parallel they have the same direction vector. Use the formula for a line with the given direction vector and a point.

Two lines

In the 2-dimensional plane, two distinctive lines can either be parallel or they can intersect.

In 3-dimensional space two distinctive lines can have an additional relationship.

i Lines are parallel (L_1 and L_2).

ii Lines intersect at one common point (L_2 and L_3).

iii Lines are skewed (do not intersect and they are not parallel) (eg L_2 and L_4).

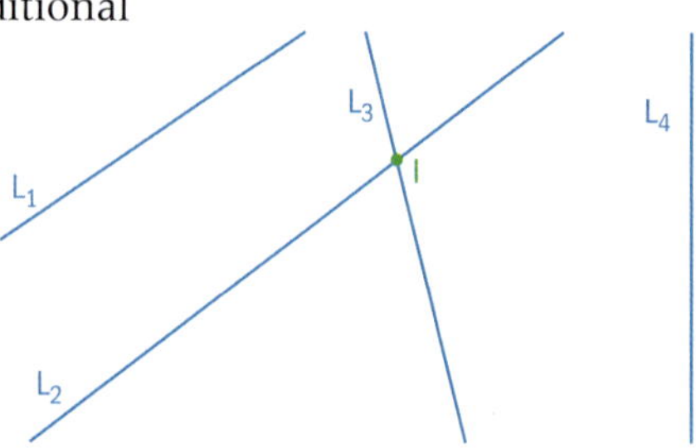

How can you see that two lines are skew?

They have no common point and they are not parallel, which means that they have no equal or collinear direction vectors.

Example 32

Show that the lines $\frac{x-1}{3}=\frac{y+2}{2}=\frac{z}{1}$ and $\frac{x+3}{-2}=\frac{y}{3}=\frac{z-2}{4}$ are skew.

The direction vectors are $\mathbf{d}_1=\begin{pmatrix}3\\2\\1\end{pmatrix}$ and $\mathbf{d}_2=\begin{pmatrix}-2\\3\\4\end{pmatrix}$, they are not parallel.	Since $\mathbf{d}_1$ is not equal nor co-linear to $\mathbf{d}_2$, the vectors are not parallel.
$\begin{cases}x=1+3\lambda\\y=-2+2\lambda\\z=\lambda\end{cases}$ and $\begin{cases}x=-3-2\mu\\y=3\mu\\z=2+4\mu\end{cases}$ $\lambda,\mu\in\mathbb{R}$	Rewrite the equations in parametric form.
$\begin{cases}1+3\lambda=-3-2\mu\\-2+2\lambda=3\mu\\\lambda=2+4\mu\end{cases}\Rightarrow\begin{cases}3(2+4\mu)+2\mu=-4\\2(2+4\mu)-3\mu=2\\\lambda=2+4\mu\end{cases}$	Equate the corresponding coordinates and solve the simultaneous equations for λ and μ. The third equation gives the substitution that you are going to use in the first two equations.
$\Rightarrow\begin{cases}6+12\mu+2\mu=-4\\4+8\mu-3\mu=2\\\lambda=2+4\mu\end{cases}\Rightarrow\begin{cases}14\mu=-10\\5\mu=-2\\\lambda-4\mu=2\end{cases}$	
$\Rightarrow\begin{cases}\mu=-\frac{5}{7}\\\mu=-\frac{2}{5}\\\lambda=2+4\mu\end{cases}$, which is a contradiction. There is no point of intersection and they are not parallel, therefore the lines are skew.	To have a point of intersection the system must be consistent and have only one value for both variables, which is not the case because μ has two different values.

Example 33

Determine the values of a and b so that the following lines $L_1: \frac{x-5}{-4} = \frac{y+1}{3} = \frac{z-1}{2}$,

$L_2: 3-x = \frac{6-y}{2} = \frac{z+1}{2}$ and $L_3: x-a = \frac{y-8}{b} = \frac{z+5}{4}$ are concurrent.

First find the point of intersection between the first two lines. $\begin{cases} 5-4\lambda = 3-\mu \\ -1+3\lambda = 6-2\mu, \lambda, \mu \in \mathbb{R} \\ 1+2\lambda = -1+2\mu \end{cases} \Rightarrow \begin{cases} \mu - 4\lambda = -2 \\ 2\mu + 3\lambda = 7 \\ 2+2\lambda = 2\mu \end{cases}$	Rewrite the equations in parametric form and solve the simultaneous equations to find the point of intersection. Substitute μ from the third equation into the first and second. Find μ. Use the third equation to find λ.
$\Rightarrow \begin{cases} 1+\lambda-4\lambda = -2 \\ 2+2\lambda+3\lambda = 7 \\ 1+\lambda = \mu \end{cases} \Rightarrow \begin{cases} \lambda = 1 \\ \lambda = 1 \\ \mu = 2 \end{cases} \Rightarrow P(1,2,3)$	
$\begin{cases} 1-a = \omega \\ 2-8 = b\omega, \quad \omega \in \mathbb{R} \\ 3+5 = 4\omega \end{cases} \Rightarrow \begin{cases} \omega = 2 \\ 1-2 = a \\ -6 = 2b \end{cases}$ $a=-1,\ b=-3$	Use the coordinates of the point in the parametric form of the third line. In the third equation find the value of parameter ω and use it in the remaining two equations to find the values of a and b.

Exercise 9G

1 Find a vector equation of the line passing through the points:

a A(1,3,−2) and B(4,2,1); **b** C(3,0,−5) and D(5,7,−2).

2 The line $\mathbf{r} = 3\mathbf{i} - \mathbf{j} + 2\mathbf{k} + \lambda(-\mathbf{i} + \mathbf{j} + 3\mathbf{k}), \lambda \in \mathbb{R}$ is given.

a Show that the point P(0,2,5) does not lie on the line.

b Find the equation of the line parallel to it that passes through the point P.

c For what value of $a \in \mathbb{R}$ does the point T(−2,4,a) lie on the line?

3 Two lines $L_1: \frac{x-1}{2} = \frac{y-2}{3} = \frac{z+3}{5}$ and $L_2: x+2 = \frac{y-1}{-2} = \frac{z-2}{4}$ are given.

a Explain why the lines are not parallel.

b Show that the lines are skew.

4 Find any points of intersection between the following lines.

a $L_1: \frac{x-3}{5} = \frac{y+2}{4} = \frac{z-1}{3}$ and $L_2: 7-x = \frac{y-4}{2} = \frac{z+1}{-3}$

b $L_1: x = \frac{y+1}{2} = 3-z$ and $L_2: \frac{x+7}{3} = y = \frac{7-z}{2}$

c $L_1: \frac{x}{2} = \frac{y-2}{5} = \frac{z}{4}$ and $L_2: \frac{x-1}{3} = \frac{y+1}{2} = z+3$

5 Show that the lines $L_1 : \frac{x+5}{2} = y+3 = 5-z$, $L_2 : \frac{x+3}{4} = y+1 = \frac{3-z}{2}$, $L_3 : \frac{x-1}{3} = -\frac{y}{2} = z-2$ are not concurrent.

Developing inquiry skills

Return to the opening problem. How might a vector equation model a skier moving down the mountain?

9.5 Vector product and properties

The vector product is an operation that takes two vectors and results in another **vector**. You will perform this operation only for 3-dimensional vectors by using the following definition.

Given the two vectors and their components, $\mathbf{a} = \begin{pmatrix} a_1 \\ a_2 \\ a_3 \end{pmatrix}$, $\mathbf{b} = \begin{pmatrix} b_1 \\ b_2 \\ b_3 \end{pmatrix}$ then the **vector product** is given by $\mathbf{a} \times \mathbf{b} = \begin{pmatrix} a_2b_3 - a_3b_2 \\ a_3b_1 - a_1b_3 \\ a_1b_2 - a_2b_1 \end{pmatrix}$.

Example 34

Given the vectors $\mathbf{a} = 2\mathbf{i} + 3\mathbf{j} - \mathbf{k}$ and $\mathbf{b} = 4\mathbf{i} - \mathbf{j} + \mathbf{k}$, find $\mathbf{a} \times \mathbf{b}$.

$\mathbf{a} \times \mathbf{b} = \begin{pmatrix} 3\cdot1-(-1)\cdot(-1) \\ (-1)\cdot4-2\cdot1 \\ 2\cdot(-1)-3\cdot4 \end{pmatrix} = \begin{pmatrix} 2 \\ -6 \\ -14 \end{pmatrix}$	Use the given formula by the vector components.

Investigation 14

Take two non-parallel vectors $\mathbf{a}$ and $\mathbf{b}$ and find the vector $\mathbf{c} = \mathbf{a} \times \mathbf{b}$. Find the scalar products between the vectors $\mathbf{a}$ and $\mathbf{c}$ and $\mathbf{b}$ and $\mathbf{c}$.

1. What do you notice?
2. What can you say about the vectors $\mathbf{a}$, $\mathbf{b}$ and $\mathbf{c}$?
3. Is it true for other non-parallel vectors?

Investigation 15

Take two vectors and find their magnitudes. By using the formula from Section 9.3 find the angle θ, between the vectors. Then calculate the vector product of those two vectors and the magnitude of it. Now keep their magnitudes the same but vary the angle between the vectors. Copy and complete the following table.

$\mathbf{a}$	$\mathbf{b}$	$\lvert\mathbf{a}\rvert$	$\lvert\mathbf{b}\rvert$	$\lvert\mathbf{a}\rvert\lvert\mathbf{b}\rvert$	θ	$\mathbf{a} \leftarrow \mathbf{b}$	$\lvert\mathbf{a} \times \mathbf{b}\rvert$	$\mathbf{a} \cdot (\mathbf{a} \times \mathbf{b})$	$\mathbf{b} \cdot (\mathbf{a} \times \mathbf{b})$
$\begin{pmatrix}1\\2\\2\end{pmatrix}$	$\begin{pmatrix}1\\1\\\sqrt{2}\end{pmatrix}$								
$\begin{pmatrix}1\\\sqrt{5}\\\sqrt{3}\end{pmatrix}$	$\begin{pmatrix}1\\1\\\sqrt{2}\end{pmatrix}$								
$\begin{pmatrix}1\\2\\2\end{pmatrix}$	$\begin{pmatrix}2\\0\\0\end{pmatrix}$								
$\begin{pmatrix}1\\\sqrt{6}\\\sqrt{2}\end{pmatrix}$	$\begin{pmatrix}\sqrt{2}\\0\\\sqrt{2}\end{pmatrix}$								
$\begin{pmatrix}1\\-\sqrt{2}\\-\sqrt{6}\end{pmatrix}$	$\begin{pmatrix}\sqrt{2}\\0\\-\sqrt{2}\end{pmatrix}$								

1. What can you say about the ratio between the magnitude of the vector product and the product of the magnitudes?

Continued on next page

Geometry and trigonometry

2 Organize your results in order of the magnitude of the angle $0 < \theta < \pi$. On the x-axis put the angle θ while on the y-axis put the magnitude of the vector product. Plot the points obtained in the table and connect them by a smooth curve.

3 Can you describe the curve?

4 **Conceptual** What is the relationship between the magnitude of the vector product, the magnitudes of the vectors and the angle between the vectors?

The results of this investigation can be summarized as follows:

- The vector product of two vectors is another vector that is perpendicular to both vectors.
- The magnitude of the vector product is given by the formula $|\mathbf{a}\times\mathbf{b}| = |\mathbf{a}||\mathbf{b}|\sin\theta$, where θ is the angle between those two vectors.

Combining those two findings you can conclude that the geometrical definition of the cross product of two vectors is given by the following formula.

Given two vectors **a** and **b** then the **vector product** is given by $\mathbf{a}\times\mathbf{b} = (|\mathbf{a}||\mathbf{b}|\sin\theta)\times\hat{\mathbf{n}}$, where $\hat{\mathbf{n}}$ is the **unit vector** whose direction is given by the right-hand screw rule to both **a** and **b** and the vectors **a**, **b** and **n** follow the right-hand rule.

Consequently the geometrical meaning of the magnitude of the vector product is that it is equal to the area of the parallelogram enclosed by those two vectors.

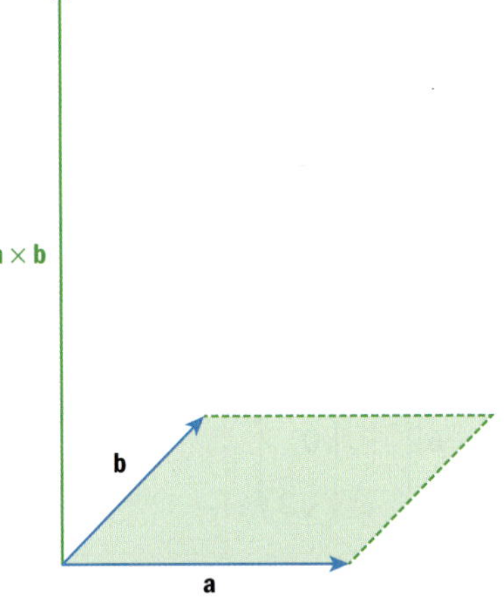

Properties of the vector product

i $\mathbf{a}\times\mathbf{b} = -(\mathbf{b}\times\mathbf{a})$

ii $(\mathbf{a}\times\mathbf{b})\times\mathbf{c} = \mathbf{a}\times(\mathbf{b}\times\mathbf{c})$

iii $\lambda(\mathbf{a}\times\mathbf{b}) = (\lambda\mathbf{a})\times\mathbf{b} = \mathbf{a}\times(\lambda\mathbf{b}),\ \lambda\in\mathbb{R}$

iv $(\mathbf{a}+\mathbf{b})\times\mathbf{c} = (\mathbf{a}\times\mathbf{c})+(\mathbf{b}\times\mathbf{c})$

These properties are left to you to prove as an exercise.

Example 35

Show that two non-zero vectors **a** and **b** are parallel if and only if $\mathbf{a}\times\mathbf{b} = \mathbf{0}$.

If two vectors **a** and **b** are parallel $\mathbf{b} = \lambda\mathbf{a},\ \lambda\in\mathbb{R}$ then the angle between them $\theta = 0$ if $\lambda > 0$ or $\theta = \pi$ if $\lambda < 0$. In both cases $\sin\theta = 0$ therefore	First we need to show one direction. Find the possible values for the angle between those two vectors.

$\lvert \mathbf{a} \times \mathbf{b} \rvert = \lvert \mathbf{a} \rvert \lvert \mathbf{b} \rvert \underbrace{\sin\theta}_{0} = 0$ $\lvert \mathbf{a} \times \mathbf{b} \rvert = 0 \Rightarrow \mathbf{a} \times \mathbf{b} = \mathbf{0}$	Use definition of the magnitude of the vector product. If the magnitude of the vector is 0 then the vector is a zero vector.
$\mathbf{a} \times \mathbf{b} = \mathbf{0} \Rightarrow \lvert \mathbf{a} \times \mathbf{b} \rvert = 0$	Now show the other direction.
$\Rightarrow \lvert \mathbf{a} \rvert \lvert \mathbf{b} \rvert \sin\theta = 0$ $\Rightarrow \sin\theta = 0 \Rightarrow \theta = 0$ or $\theta = \pi$	Vector product is a zero vector therefore the magnitude of it is 0. Since **a** and **b** are non–zero vectors $\sin\theta = 0$.
So the vectors are parallel.	If the angle is 0 then the vectors have the same direction, or if the angle is π then the vectors have opposite directions therefore they are collinear.

Using the algebraic definition of the vector product will also prove the same statement.

There are some interesting properties when combining the vector and scalar product.

Mixed product

An operation with three vectors **a**, **b** and **c** combining both the vector and scalar product is called a **mixed product** $(\mathbf{a} \times \mathbf{b}) \cdot \mathbf{c}$.

HINT

Even though there is no reference to the mixed product in the guide, performing two successive operations with vectors should not be a problem for you to use it for an exploration.

Geometry and trigonometry

Example 36

Given the vectors with their components $\mathbf{a} = \begin{pmatrix} a_1 \\ a_2 \\ a_3 \end{pmatrix}, \mathbf{b} = \begin{pmatrix} b_1 \\ b_2 \\ b_3 \end{pmatrix}$ and $\mathbf{c} = \begin{pmatrix} c_1 \\ c_2 \\ c_3 \end{pmatrix}$, prove that the mixed product is given by the formula $(\mathbf{a} \times \mathbf{b}) \cdot \mathbf{c} = a_1b_2c_3 + a_2b_3c_1 + a_3b_1c_2 - a_1b_3c_2 - a_2b_1c_3 - a_3b_2c_1$

$(\mathbf{a} \times \mathbf{b}) \cdot \mathbf{c} = \begin{pmatrix} a_2b_3 - a_3b_2 \\ a_3b_1 - a_1b_3 \\ a_1b_2 - a_2b_1 \end{pmatrix} \cdot \begin{pmatrix} c_1 \\ c_2 \\ c_3 \end{pmatrix}$	Use the algebraic definition of the vector product.
$= (a_2b_3 - a_3b_2)c_1 + (a_3b_1 - a_1b_3)c_2 + (a_1b_2 - a_2b_1)c_3$ $= a_1b_2c_3 + a_2b_3c_1 + a_3b_1c_2 - a_1b_3c_2 - a_2b_1c_3 - a_3b_2c_1$	Use the algebraic definition of the scalar product and simplify the expressions.

Investigation 16

Take three vectors **a**, **b** and **c** and calculate the following mixed products. Copy and complete the table:

a	**b**	**c**	$(\mathbf{a}\times\mathbf{b})\cdot\mathbf{c}$	$(\mathbf{b}\times\mathbf{c})\cdot\mathbf{a}$	$(\mathbf{c}\times\mathbf{a})\cdot\mathbf{b}$	$(\mathbf{c}\times\mathbf{b})\cdot\mathbf{a}$	$(\mathbf{b}\times\mathbf{a})\cdot\mathbf{c}$	$(\mathbf{a}\times\mathbf{c})\cdot\mathbf{b}$
i	**j**	**k**						
$\begin{pmatrix}1\\2\\1\end{pmatrix}$	$\begin{pmatrix}1\\0\\1\end{pmatrix}$	$\begin{pmatrix}3\\1\\2\end{pmatrix}$						

1 What do you notice about the results?

2 How can you justify these results?

To summarize your findings, you can write the following:

$$(\mathbf{a}\times\mathbf{b})\cdot\mathbf{c}=(\mathbf{b}\times\mathbf{c})\cdot\mathbf{a}=(\mathbf{c}\times\mathbf{a})\cdot\mathbf{b}=-(\mathbf{a}\times\mathbf{c})\cdot\mathbf{b}$$
$$=-(\mathbf{c}\times\mathbf{b})\cdot\mathbf{a}=-(\mathbf{b}\times\mathbf{a})\cdot\mathbf{c}$$

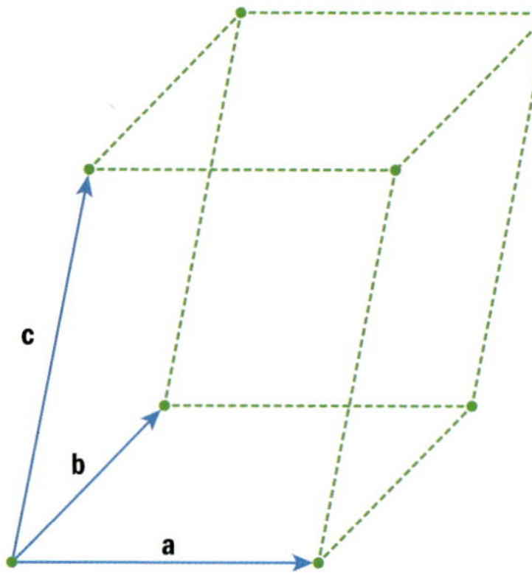

The geometrical meaning of the mixed product of three vectors is summarized in the following.

The volume of a parallelepiped formed by three non-coplanar vectors **a**, **b** and **c** is given by the formula:
$V=|(\mathbf{a}\times\mathbf{b})\cdot\mathbf{c}|$

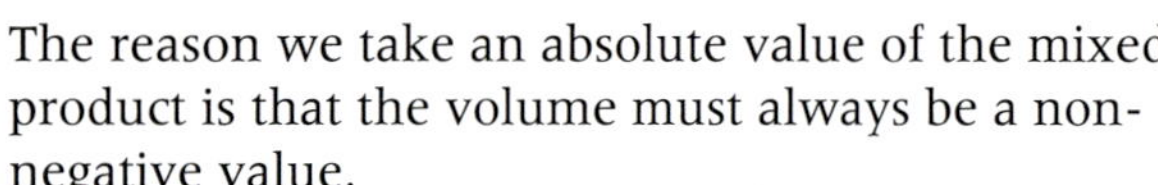

The reason we take an absolute value of the mixed product is that the volume must always be a non-negative value.

To justify this formula you can look at the diagram to the right.

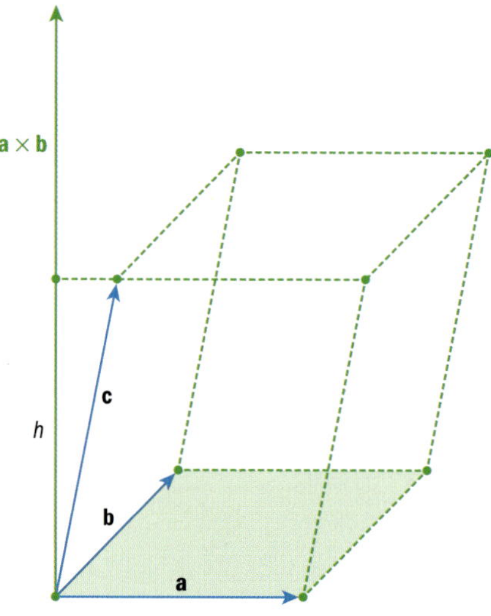

The volume of the parallelepiped is the product of the *Base* that is a parallelogram enclosed by the vectors **a** and **b** and the perpendicular height h.

The vector product has the magnitude that is equal to the area of the parallelogram, and the height h is the perpendicular projection of the vector **c** in the direction of the vector $\mathbf{a}\times\mathbf{b}$.

$$V = Base \cdot h = |\mathbf{a}\times\mathbf{b}||\mathbf{c}||\cos\theta| = |(\mathbf{a}\times\mathbf{b})\cdot\mathbf{c}|$$

Notice that this definition does not depend on the order of the vectors due to the results of the previous investigation.

Example 37

Find the volume of the parallelepiped enclosed by the vectors $\mathbf{a} = -\mathbf{i} + \mathbf{j} - 2\mathbf{k}$, $\mathbf{b} = 2\mathbf{i} + \mathbf{j} - 3\mathbf{k}$ and $\mathbf{c} = \mathbf{i} - 2\mathbf{j} + \mathbf{k}$.

$V = \lvert(\mathbf{a} \times \mathbf{b}) \cdot \mathbf{c}\rvert$ $= \lvert -1 \cdot 1 \cdot 1 + 1 \cdot (-3) \cdot 1 + (-2) \cdot 2 \cdot (-2)$ $-(-1) \cdot (-3) \cdot (-2) - 1 \cdot 2 \cdot 1 - (-2) \cdot 1 \cdot 1 \rvert$ $= 10$	Use the formula for the mixed product.

Three or more vectors are said to be coplanar if they lie in the same plane.

Example 38

Show that the vectors $\mathbf{a} = \mathbf{i} - 2\mathbf{j} - 4\mathbf{k}$, $\mathbf{b} = -\mathbf{i} + \mathbf{j} + \mathbf{k}$ and $\mathbf{c} = 2\mathbf{i} - \mathbf{j} + \mathbf{k}$ are coplanar.

$V = 0 \Rightarrow (\mathbf{a} \times \mathbf{b}) \cdot \mathbf{c} = 0$ $1 \cdot 1 \cdot 1 + (-2) \cdot 1 \cdot 2 + (-4) \cdot (-1) \cdot (-1)$ $-1 \cdot 1 \cdot (-1) - (-2) \cdot (-1) \cdot 1 - (-4) \cdot 1 \cdot 2 = 0$	Coplanar vectors enclose a parallelepiped with the volume 0. Use the algebraic definition of the mixed product.

The mixed product of three vectors can also be used to find the volume of a triangular pyramid. Since the base is not a parallelogram but a triangle, that is half an area of the parallelogram, then the volume of a pyramid is one third of the product of the base and the perpendicular height so you can derive the following formula.

$$V = \frac{1}{3} Base \cdot h = \frac{1}{3} \cdot \frac{1}{2} \lvert \mathbf{a} \times \mathbf{b} \rvert \lvert \mathbf{c} \rvert \lvert \cos\theta \rvert = \frac{1}{6} \lvert (\mathbf{a} \times \mathbf{b}) \cdot \mathbf{c} \rvert$$

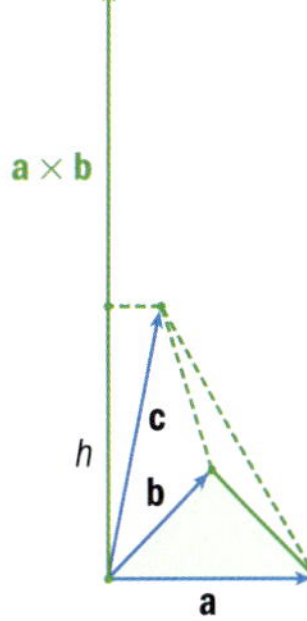

Example 39

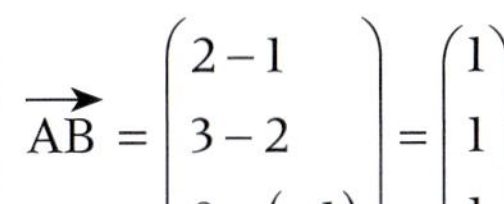

Find the volume of a triangular pyramid with the vertices A(1, 2, −1), B(2, 3, 0), C(−3, 1, 1) and D(0, −4, 5).

$\overrightarrow{AB} = \begin{pmatrix} 2-1 \\ 3-2 \\ 0-(-1) \end{pmatrix} = \begin{pmatrix} 1 \\ 1 \\ 1 \end{pmatrix}$	First you need to find the vectors that enclose the pyramid. Notice that you could have taken any point to be the vertex of the pyramid and the result would have always be the same.

Continued on next page

Geometry and trigonometry

$\overrightarrow{AC} = \begin{pmatrix} -3-1 \\ 1-2 \\ 1-(-1) \end{pmatrix} = \begin{pmatrix} -4 \\ -1 \\ 2 \end{pmatrix}$

$\overrightarrow{AD} = \begin{pmatrix} 0-1 \\ -4-2 \\ 5-(-1) \end{pmatrix} = \begin{pmatrix} -1 \\ -6 \\ 6 \end{pmatrix}$

$V = \frac{1}{6}|(\mathbf{a} \times \mathbf{b}) \cdot \mathbf{c}|$

$= \frac{1}{6}|1 \cdot (-1) \cdot 6 + 1 \cdot 2 \cdot (-1) + 1 \cdot (-4) \cdot (-6)$

$-1 \cdot 2 \cdot (-6) - 1 \cdot (-4) \cdot 6 - 1 \cdot (-1) \cdot (-1)|$

$= \frac{1}{6}|-6 - 2 + 24 + 12 + 24 - 1| = \frac{51}{6} = \frac{17}{2}$

Use the formula for the mixed product.

$a_1b_2c_3 + a_2b_3c_1 + a_3b_1c_2 - a_1b_3c_2 - a_2b_1c_3 - a_3b_2c_1$

Exercise 9H

1 Find the vector product of the following vectors by using the algebraic definition and check your answers by using a calculator:

a $\mathbf{a} = 2\mathbf{i} + 3\mathbf{j} - 5\mathbf{k}$ and $\mathbf{b} = \mathbf{i} - 2\mathbf{j} + 3\mathbf{k}$

b $\mathbf{c} = \mathbf{i} + \mathbf{j}$ and $\mathbf{d} = 3\mathbf{i} - 2\mathbf{k}$

c $\mathbf{m} = 3\mathbf{i} - 4\mathbf{j} - \mathbf{k}$ and $\mathbf{n} = 2\mathbf{i} + \mathbf{j} + 2\mathbf{k}$

d $\mathbf{p} = \frac{1}{2}\mathbf{i} - \frac{3}{4}\mathbf{j} + \mathbf{k}$ and $\mathbf{r} = \mathbf{i} - \frac{2}{3}\mathbf{j} - 2\mathbf{k}$

2 Use the vector product to find the area of the parallelogram enclosed by the vectors $\mathbf{a} = 2\mathbf{i} + 3\mathbf{j} - 6\mathbf{k}$ and $\mathbf{b} = 3\mathbf{i} - \mathbf{j} - 4\mathbf{k}$.

3 A triangle with the vertices A(1, 4, 2), B(−2, 0, 3) and C(−1, 2, 4) is given.

a Find the vectors $\overrightarrow{AB}$ and $\overrightarrow{AC}$.

b Use the vector product to find the area of the triangle ABC.

4 Use either algebraic or geometric definitions to show the following properties of the vector product:

i $\mathbf{a} \times \mathbf{b} = -(\mathbf{b} \times \mathbf{a})$

ii $(\mathbf{a} \times \mathbf{b}) \times \mathbf{c} = (\mathbf{a} \cdot \mathbf{c})\mathbf{b} - (\mathbf{a} \cdot \mathbf{b})\mathbf{c}$

iii $\lambda(\mathbf{a} \times \mathbf{b}) = (\lambda\mathbf{a}) \times \mathbf{b} = \mathbf{a} \times (\lambda\mathbf{b}),\ \lambda \in \mathbb{R}$

iv $(\mathbf{a} + \mathbf{b}) \times \mathbf{c} = (\mathbf{a} \times \mathbf{c}) + (\mathbf{b} \times \mathbf{c})$

5 Show that the points A(1, 1, 1), B(2, −1, 0), C(2, 4, 2) and D(−2, 2, 2) are coplanar by using the mixed product.

6 A parallelepiped ABCDEFGH is given. The vertices of the parallelepiped have the following coordinates: A(1, 2, 1), B(2, −1, 3), C(4, 5, −1) and H(4, 3, 6).

a Find the coordinates of the vertex D.

b Find the vectors that enclose the parallelepiped.

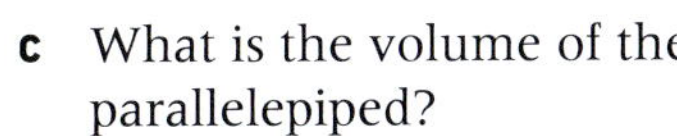

c What is the volume of the parallelepiped?

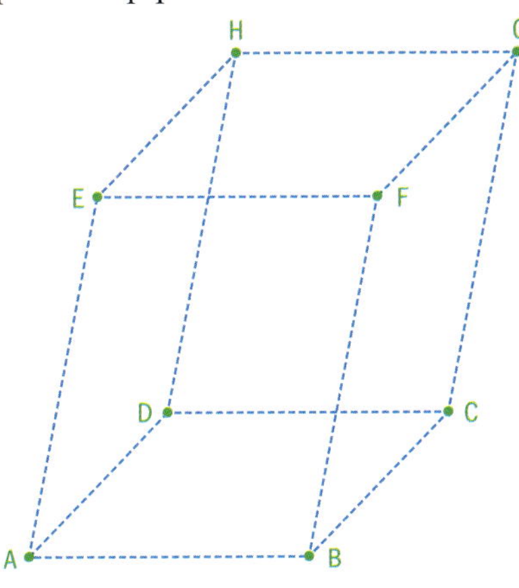

7 A triangular pyramid ABCD is given. If $\overrightarrow{AB} = 2\mathbf{i} + 3\mathbf{j} - 2\mathbf{k}$, $\overrightarrow{BD} = \mathbf{i} - 2\mathbf{j} + 4\mathbf{k}$ and $\overrightarrow{DC} = -2\mathbf{i} + \mathbf{j} + 5\mathbf{k}$, find the volume of the pyramid.

8 Show that $(\mathbf{a}\cdot\mathbf{b})^2 + |\mathbf{a}\times\mathbf{b}|^2 = |\mathbf{a}||\mathbf{b}|^2$

9 Use the algebraic definition of the vector and scalar products to show the following identity, $\mathbf{a}\times(\mathbf{b}\times\mathbf{c}) = (\mathbf{a}\cdot\mathbf{c})\mathbf{b} - (\mathbf{a}\cdot\mathbf{b})\mathbf{c}$.

9.6 Vector equation of a plane

In geometry, there is a unique line passing through two points. Now using an analogy, there is a unique plane containing three non–collinear points.

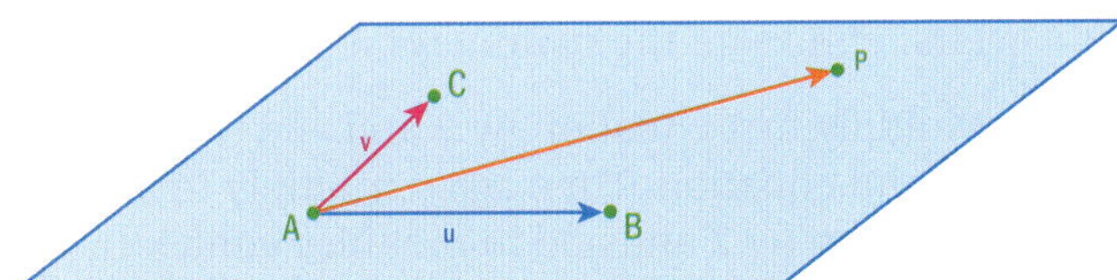

If a point P lies in the same plane as the points A, B and C then the vectors must satisfy the following equation.

$\overrightarrow{AP} = \lambda\overrightarrow{AB} + \mu\overrightarrow{AC},\ \lambda, \mu \in \mathbb{R}$

Rewriting this equation by using the position vectors you obtain the **vector equation of the plane**.

$\mathbf{r} - \mathbf{a} = \lambda(\mathbf{b} - \mathbf{a}) + \mu(\mathbf{c} - \mathbf{a}) \Rightarrow \mathbf{r} = \mathbf{a} + \lambda\mathbf{u} + \mu\mathbf{v},\ \lambda, \mu \in \mathbb{R}$

Using the vector components, you obtain the **parametric form**.

$$\begin{pmatrix} x \\ y \\ z \end{pmatrix} = \begin{pmatrix} a_1 \\ a_2 \\ a_3 \end{pmatrix} + \lambda\begin{pmatrix} u_1 \\ u_2 \\ u_3 \end{pmatrix} + \mu\begin{pmatrix} v_1 \\ v_2 \\ v_3 \end{pmatrix} \Rightarrow \begin{cases} x = a_1 + \lambda u_1 + \mu v_1 \\ y = a_2 + \lambda u_2 + \mu v_2 \quad \lambda, \mu \in \mathbb{R} \\ z = a_3 + \lambda u_3 + \mu v_3 \end{cases}$$

Eliminating the real parameters λ and μ you obtain the **Cartesian form**, $ax + by + cz = d$, where $a, b, c, d \in \mathbb{R}$.

International–mindedness

Benjamin Peirce (1809–1880), a prominent mathematician in the United States, expanded on what he called "this wonderful algebra of space".

Geometry and trigonometry

Example 40

A plane passes through the points A(1, 2, 1), B(2, 0, −1) and C(3, −1, 0). Find

a the vector equation of the plane **b** the parametric equations of the plane

c the Cartesian equation of the plane.

Let point $P(x,y,z)$	
a $\mathbf{u} = \overrightarrow{AB} = (2-1)\mathbf{i} + (0-2)\mathbf{j} + (-1-1)\mathbf{k}$ $= \mathbf{i} - 2\mathbf{j} - 2\mathbf{k}$ $\mathbf{v} = \overrightarrow{AC} = (3-1)\mathbf{i} + (-1-2)\mathbf{j} + (0-1)\mathbf{k}$ $= 2\mathbf{i} - 3\mathbf{j} - \mathbf{k}$	Find a pair of vectors **u** and **v** that define this plane.
So the vector equation of the plane is $\begin{pmatrix} x \\ y \\ z \end{pmatrix} = \begin{pmatrix} 1 \\ 2 \\ 1 \end{pmatrix} + \lambda \begin{pmatrix} 1 \\ -2 \\ -2 \end{pmatrix} + \mu \begin{pmatrix} 2 \\ -3 \\ -1 \end{pmatrix}, \lambda, \mu \in \mathbb{R}$	Use the position vector of one of the three given points in the plane, eg point A.
b $\begin{cases} x = 1 + \lambda + 2\mu \\ y = 2 - 2\lambda - 3\mu \quad \lambda, \mu \in \mathbb{R} \\ z = 1 - 2\lambda - \mu \end{cases}$	Use the vector components to rewrite this equation in parametric form.
c You need to eliminate the parameters λ and μ. $\begin{cases} 2x + y = 4 + \mu \\ 2x + z = 3 + 3\mu \end{cases}$	First eliminate the parameter λ. Multiply first equation by 2 and add to the second equation. Then again multiply first equation by 2 and add to the third equation.
$6x + 3y - 2x - z = 12 + 3\mu - 3 - 3\mu$ $4x + 3y - z = 9$	Now eliminate the parameter μ. Multiply first equation by 3 and subtract the second equation from it.

Two planes

In a 3-dimensional space two distinctive planes can have similar relationships as two distinctive lines in a 2-dimensional plane. The planes can either intersect at a line or they can be parallel.

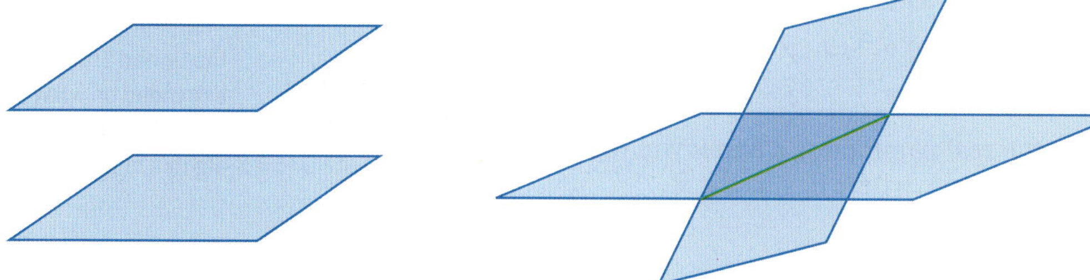

Two planes are parallel if their normal vectors are collinear; otherwise they intersect at a line.

Example 41

Given the three planes $\pi_1 : 2x - 3y + 5z = 1, \pi_2 : x + 2y - z = 0$ and $\pi_3 : 2x + 4y - 2z = 1$, show that:

a $\pi_2 \parallel \pi_3$ **b** π_1 and π_2 intersect and find the equation of the line.

a $\mathbf{d}_2 = \begin{pmatrix} 1 \\ 2 \\ -1 \end{pmatrix}$ and $\mathbf{d}_3 = \begin{pmatrix} 2 \\ 4 \\ -2 \end{pmatrix}$	Identify the normal vectors and check whether they are parallel.
$\mathbf{d}_3 = 2\mathbf{d}_2$ so $\pi_2 \parallel \pi_3$.	Check whether the components are proportional.
	Check for the common points.
$\begin{cases} x + 2y - z = 0 \\ 2x + 4y - 2z = 1 \end{cases} \Leftrightarrow \begin{cases} 2x + 4y - 2z = 0 \\ 2x + 4y - 2z = 1 \end{cases}$ $0 = 1$ Parallel normal vectors and no common point means that the planes are parallel.	Multiply the first equation by 2 and since the left sides are equal, equate the right sides. The remaining statement is false therefore the planes have no common point.
b **Method 1** $\begin{cases} 2x - 3y + 5z = 1 \\ x + 2y - z = 0 \end{cases} \Rightarrow \begin{cases} 2x - 3y + 5(x + 2y) = 1 \\ x + 2y = z \end{cases}$ $\Rightarrow \begin{cases} 7x + 7y = 1 \\ x + 2y = z \end{cases} \Rightarrow \begin{cases} y = \frac{1}{7} - x \\ z = x + 2\left(\frac{1}{7} - x\right) \end{cases}$	Solve simultaneous equations by using the method of substitution and find the Cartesian equation of the line of intersection.
$\Rightarrow \begin{cases} y = \frac{1}{7} - x \\ z = \frac{2}{7} - x \end{cases} \Rightarrow \begin{cases} x = \lambda \\ y = \frac{1}{7} - \lambda, \lambda \in \mathbb{R} \\ z = \frac{2}{7} - \lambda \end{cases}$	Express all the variables in terms of x and then substitute x by a new real parameter λ.
$\frac{x}{1} = \frac{y - \frac{1}{7}}{-1} = \frac{z - \frac{2}{7}}{-1}$	Rewrite the parametric form by eliminating λ into the Cartesian form.

Continued on next page

HINT

You could have taken an arbitrary point from one of the planes and checked whether that point lies in the other plane.

Method 2

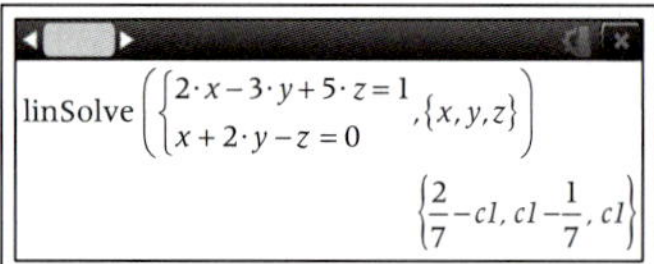

Use a GDC to solve simultaneous equations. Notice that the solution is given in terms of one parameter **c1**.

The line is given in the parametric form.

$$\Rightarrow \begin{cases} x = \frac{2}{7} - \mu \\ y = \mu - \frac{1}{7}, \mu \in \mathbb{R} \\ z = \mu \end{cases}$$

Notice that the direction vector is the opposite, but the point is different. You can check that the point lies on the line obtained by method 1. Also, that the point in the equation obtained by method 1 lies on the line obtained by method 2.

Just like for a line in a plane, there are two directions in space that are perpendicular to a plane and those two directions are mutually opposite. You can again call it the normal but this time to the plane.

Since the normal vector **n** is perpendicular to every vector in the plane, it is perpendicular to the vector $\overrightarrow{AP}$ and the scalar product is equal to 0.

$\mathbf{n} \cdot \overrightarrow{AP} = 0 \Rightarrow \mathbf{n} \cdot (\mathbf{p} - \mathbf{a}) = 0 \Rightarrow \mathbf{n} \cdot \mathbf{p} - \mathbf{n} \cdot \mathbf{a} = 0$

You can see that the equation of a plane by using a normal vector remains the same as the equation of a line in the plane.

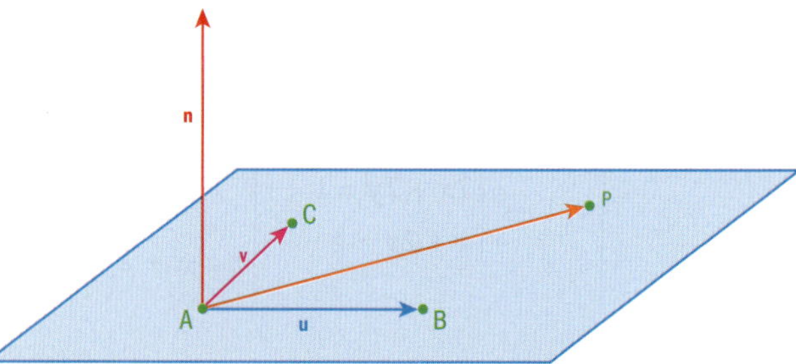

$$\mathbf{n} \cdot \mathbf{p} = \mathbf{n} \cdot \mathbf{a} \Rightarrow \begin{pmatrix} a \\ b \\ c \end{pmatrix} \cdot \begin{pmatrix} x \\ y \\ z \end{pmatrix} = \begin{pmatrix} a \\ b \\ c \end{pmatrix} \cdot \begin{pmatrix} a_1 \\ a_2 \\ a_3 \end{pmatrix} \Rightarrow ax + by + cz = \underbrace{aa_1 + ba_2 + ca_3}_{d}$$

This leaves you with the Cartesian form $ax + by + cz = d$, where $a, b, c, d \in \mathbb{R}$.

We know that the vector product of two vectors is going to define a vector that is perpendicular to the plane defined by those two vectors.

In the following example, you are going to check the result from Example 40.

Example 42

A plane passes through the points A(1,2,1), B(2,0,–1) and C(3,–1,0). Find the Cartesian equation of the plane by using a normal vector.

$\mathbf{u} = \overrightarrow{AB} = \mathbf{i} - 2\mathbf{j} - 2\mathbf{k}$ $\mathbf{v} = \overrightarrow{AC} = 2\mathbf{i} - 3\mathbf{j} - \mathbf{k}$	Take the vectors **u** and **v** that define this plane and find the vector product.
$\begin{pmatrix} 1 \\ -2 \\ -2 \end{pmatrix} \times \begin{pmatrix} 2 \\ -3 \\ -1 \end{pmatrix} = \begin{pmatrix} 2-6 \\ -4+1 \\ -3+4 \end{pmatrix} = \begin{pmatrix} -4 \\ -3 \\ 1 \end{pmatrix}$	Since the normal vector can be taken from any vector parallel with this vector product, it is more convenient to take the opposite $4\mathbf{i} + 3\mathbf{j} - \mathbf{k}$.
Let point $P(x,y,z)$ $\begin{pmatrix} 4 \\ 3 \\ -1 \end{pmatrix} \cdot \begin{pmatrix} x \\ y \\ z \end{pmatrix} = \begin{pmatrix} 4 \\ 3 \\ -1 \end{pmatrix} \cdot \begin{pmatrix} 1 \\ 2 \\ 1 \end{pmatrix}$ $4x + 3y - z = 4 + 6 - 1$ $4x + 3y - z = 9$	Use the position vector of one of the three given points in the plane, eg the point A and then apply the scalar product.

What would happen to the equation of the plane if you take position vector of the point B or C? Check your answer by appropriate working.

Three planes

In Section 9.4 you investigated positions of three lines, now by using an analogy you can investigate positions of three planes.

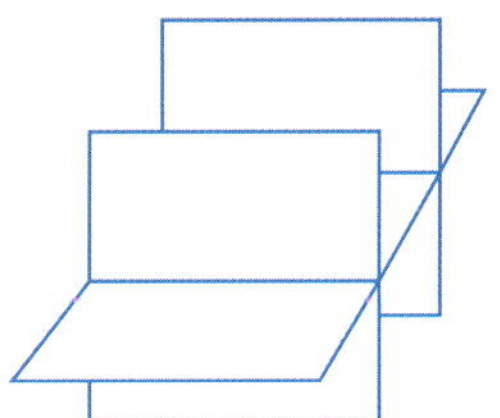

Investigation 17

1 Take three distinctive planes and draw all possible mutual positions they can have in a 3-dimensional space.

2 How many different positions can they form?

3 How can you describe the sets of intersections?

4 **Conceptual** What is the relationship between the sets of intersections of three planes and the solutions of simultaneous equations?

TOK

When is it ethically correct to provide vector locations?

Example 43

Find the point of intersection between the planes $\pi_1 : x + 2y - z = 4, \pi_2 : 2x - 3y + z = -3$ and $\pi_3 : 3x + y - 4z = 1$.

Method 1	
$\begin{cases} x + 2y - z = 4 \\ 2x - 3y + z = -3 \\ 3x + y - 4z = 1 \end{cases}$	The coordinates of the point of intersection satisfy all three equations of the plane. You need to apply a method to solve simultaneous equations. Use for example, the method of substitution.
$\Rightarrow \begin{cases} x = -2y + z + 4 \\ 2(-2y + z + 4) - 3y + z = -3 \\ 3(-2y + z + 4) + y - 4z = 1 \end{cases}$	Express x in the first equation and substitute it in the following two equations.
$\Rightarrow \begin{cases} x = -2y + z + 4 \\ -7y + 3z = -11 \\ -5y - z = -11 \end{cases} \Rightarrow \begin{cases} x = -2y + z + 4 \\ -7y + 3(-5y + 11) = -11 \\ -5y + 11 = z \end{cases}$	Simplify it and use another substitution by expressing z in the third equation and substitute it in the second equation.
$\Rightarrow \begin{cases} x = -2y + z + 4 \\ -22y = -44 \\ z = -5y + 11 \end{cases} \Rightarrow \begin{cases} x = -2 \cdot 2 + 1 + 4 = 1 \\ y = 2 \\ z = -5 \cdot 2 + 11 = 1 \end{cases}$	
So the point is $(1,2,1)$.	
Method 2 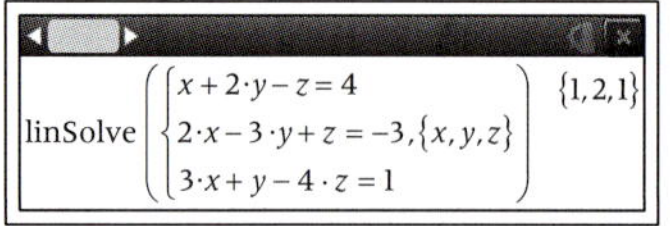 linSolve$\left(\begin{cases} x+2\cdot y-z=4 \\ 2\cdot x-3\cdot y+z=-3 \\ 3\cdot x+y-4\cdot z=1 \end{cases}, \{x,y,z\}\right)$ $\{1,2,1\}$	Use your GDC to solve simultaneous equations.

Example 44

Determine the values of the real parameter m so that the planes $\pi_1 : 2x - 3y - z = 5, \pi_2 : x + 5y + 2z = 6$ and $\pi_3 : 4x + my - 4z = -1$, intersect at a point.

$\begin{cases} 2x - 3y - z = 5 \\ x + 5y + 2z = 6 \end{cases} \Rightarrow \begin{cases} 4x - 6y - 2z = 10 \\ x + 5y + 2z = 6 \end{cases}$	Find the line of intersection between first two planes.
$5x - y = 16 \Rightarrow y = 5x - 16$	Multiply first equation by 2 and add it to the second equation to eliminate z.
$z = 2x - 3(5x - 16) - 5 \Rightarrow z = -13x + 43$	Make y the subject and also express z in terms of x only.
So the line has a vector equation: $\mathbf{r} = -16\mathbf{j} + 43\mathbf{k} + \lambda(\mathbf{i} + 5\mathbf{j} - 13\mathbf{k})$	Find equation of the line.

$\mathbf{d} \cdot \mathbf{n} \neq 0 \Rightarrow \begin{pmatrix} 1 \\ 5 \\ -13 \end{pmatrix} \cdot \begin{pmatrix} 4 \\ m \\ -4 \end{pmatrix} = 4 + 5m + 52 \neq 0$ $\Rightarrow m \in \mathbb{R},\ m \neq -\frac{56}{5}$	The line intersects the plane at one point therefore the scalar product of the direction vector and normal vector cannot be 0.

Example 45

Find possible values of a, b, c and d so that the planes $\pi_1 : 2x + y - 3z = 1$, $\pi_2 : x - 2y + 2z = 4$ and $\pi_3 : ax + by + cz = d$ intersect in one line.

Find the line of intersection between the first two planes. $\begin{cases} 2x + y - 3z = 1 \\ x - 2y + 2z = 4 \end{cases} \Rightarrow \begin{cases} y = 1 - 2x + 3z \\ x - 2(1 - 2x + 3z) + 2z = 4 \end{cases}$ $\Rightarrow \begin{cases} y = 1 - 2x + 3z \\ x - 2 + 4x - 6z + 2z = 4 \end{cases} \Rightarrow \begin{cases} y = 1 - 2x + 3z \\ 5x - 4z = 6 \end{cases}$	You need to apply a method to solve the first two simultaneous equations. Use for example, the method of substitution.
$\Rightarrow \begin{cases} y = 1 - 2\left(\frac{6 + 4z}{5}\right) + 3z \\ x = \frac{6 + 4z}{5} \end{cases} \Rightarrow \begin{cases} x = \frac{6}{5} + \frac{4}{5}\lambda \\ y = -\frac{7}{5} + \frac{7}{5}\lambda, \lambda \in \mathbb{R} \\ z = \lambda \end{cases}$	Find the equation of the line of intersection in parametric form.
$\mathbf{d} = \frac{4}{5}\mathbf{i} + \frac{7}{5}\mathbf{j} + \mathbf{k} = \frac{1}{5}(4\mathbf{i} + 7\mathbf{j} + 5\mathbf{k})$ One possible normal vector can be $\mathbf{n} = 3\mathbf{i} - \mathbf{j} - \mathbf{k} \Rightarrow a = 3,\ b = -1,\ c = -1$	Identify the direction vector of the line. If the line also lies in the third plane then the direction vector of the line must be perpendicular to the normal vector of the plane, scalar product is 0. Use it to find possible values of a, b and c.
When $\lambda = 1$ the point on the line is $(2,0,1)$.	Take an arbitrary point on the line to calculate d.
$3x - y - z = d \Rightarrow 3 \cdot 2 - 0 - 1 = d \Rightarrow d = 5$ So the equation of one possible plane is $3x - y - z = 5$	Use the coordinates of one point on the line to find d.

You can check your result by using a calculator.

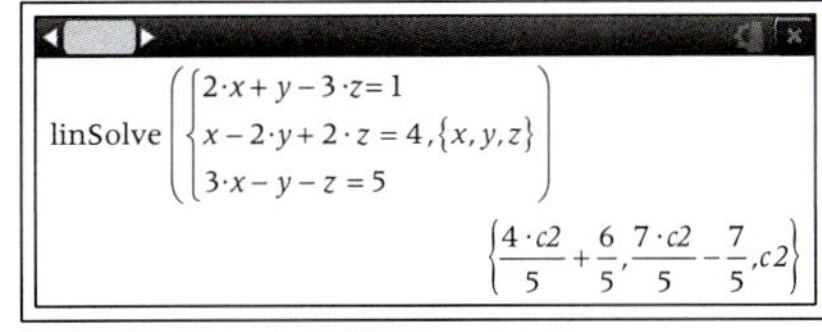

Exercise 9I

1 Find the vector equation of the plane given by two vectors **u** and **v** and one point P:

a $\mathbf{u} = \begin{pmatrix} 2 \\ 1 \\ 4 \end{pmatrix}, \mathbf{v} = \begin{pmatrix} -1 \\ 2 \\ -1 \end{pmatrix}$, $P(0,2,-1)$ **b** $\mathbf{u} = \begin{pmatrix} 0 \\ -3 \\ 2 \end{pmatrix}, \mathbf{v} = \begin{pmatrix} 1 \\ 4 \\ -2 \end{pmatrix}$, $P(1,-2,3)$

c $\mathbf{u} = \begin{pmatrix} -2 \\ 0 \\ 3 \end{pmatrix}, \mathbf{v} = \begin{pmatrix} 2 \\ 1 \\ 5 \end{pmatrix}$, $P(-3,4,2)$

2 A plane is passing through the points $A(0,1,3)$, $B(-1,2,0)$ and $C(3,-2,4)$. Find

a the vector equation of the plane

b the parametric equations of the plane

c the Cartesian equation of the plane.

3 Find the normal vector and the Cartesian equation of the planes in question **1**.

4 The point $A(5,4,-2)$ and the plane $3x - 4y + 2z = 5$ are given.

a Show that the point A is not on the plane.

b Find the equation of the plane that contains the point A and that it is parallel to the given plane.

5 Find the point of intersection between the planes $\pi_1 : x + y - z = 1$, $\pi_2 : 2x - 3y - 9z = 10$ and $\pi_3 : x + 2y - 3z = -4$.

Check your answer by using a calculator.

6 Two planes $x + y - 2z = 3$ and $2x - 3y + z = 1$ intersect in a line.

a Find the Cartesian equation of the line.

b Find the equation of the plane that passes through the point $T(2,-4,1)$ and contains the line from part **a**.

7 Show that two planes are parallel if and only if the vector product of their normal vectors is **0**.

9.7 Lines, planes and angles

When trying to find relationships between lines and/or planes you are trying to find points they have in common. The coordinates of those common points must always satisfy the equations of those lines and/or planes, so you need to solve **simultaneous equations**.

A line and a plane can have three different relationships.

i The line is in the plane, N.

ii The line intersects the plane, M (they have one common point).

iii The line is parallel to the plane, L.

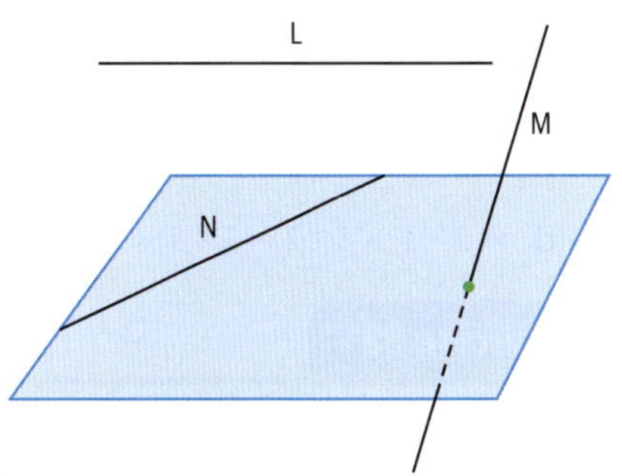

If the line is in the plane or parallel to the plane the normal vector of the plane must be perpendicular to the direction vector of the line. Otherwise the line intersects the plane at one point.

To find the relationship between a line and a plane you need to solve simultaneous equations representing that line and the plane.

Example 46

A plane $2x + y - z = 3$ and three lines $L_1 : \frac{x+1}{2} = \frac{y-3}{-1} = \frac{z+2}{3}$,

$L_2 : \frac{x-2}{1} = \frac{y}{2} = \frac{z+1}{5}$ and $L_3 : \frac{x}{-1} = \frac{y+5}{4} = \frac{z+3}{2}$ are given.

Identify which of the lines is in the plane, parallel to the plane or intersecting the plane at a point, and find the coordinates of the point.

$L_1 : \begin{cases} x = -1 + 2\lambda \\ y = 3 - \lambda \quad , \lambda \in \mathbb{R} \Rightarrow \\ z = -2 + 3\lambda \end{cases}$	All the lines should be rewritten in parametric form so that you can apply the equation of the plane.
$2(-1 + 2\lambda) + (3 - \lambda) - (-2 + 3\lambda) = 3 \Rightarrow$	Solve the equation for λ.
$-2 + 4\lambda + 3 - \lambda + 2 - 3\lambda = 3 \Rightarrow 3 = 3$ $\lambda \in \mathbb{R}$, the line L_1 is in the plane.	The variable is cancelled and the remaining statement is true, so there are infinitely many solutions.
$L_2 : \begin{cases} x = 2 + \lambda \\ y = 2\lambda \quad , \lambda \in \mathbb{R} \Rightarrow \\ z = -1 + 5\lambda \end{cases}$ $2(2 + \lambda) + (2\lambda) - (-1 + 5\lambda) = 3 \Rightarrow$	The second line is substituted into the equation of the plane.
$4 + 2\lambda + 2\lambda + 1 - 5\lambda = 3 \Rightarrow$ $5 - \lambda = 3 \Rightarrow \lambda = 2$	Solve the equation for λ.
There is a unique solution, the point of intersection (4, 4, 9).	Input the value of λ into the parametric form of the line to obtain the coordinates of the point.
$L_3 : \begin{cases} x = -\lambda \\ y = -5 + 4\lambda , \lambda \in \mathbb{R} \Rightarrow \\ z = -3 + 2\lambda \end{cases}$	The third line is substituted into the equation of the plane.
$2(-\lambda) + (-5 + 4\lambda) - (-3 + 2\lambda) = 3 \Rightarrow$	Solve the equation for λ.
$-2\lambda - 5 + 4\lambda + 3 - 2\lambda = 3 \Rightarrow -2 = 3$ $\lambda \in \emptyset$, the line L_3 is parallel to the plane.	The variable is cancelled and the remaining statement is false, so there is no solution.

Notice that the **direction vectors** of the lines L_1 and L_3 are perpendicular to the **normal vector** of the plane and the direction vector of the line L_2 is not.

$\mathbf{n} \cdot \mathbf{d}_1 = 2 \cdot 2 + 1 \cdot (-1) - 1 \cdot 3 = 0$

$\mathbf{n} \cdot \mathbf{d}_3 = 2 \cdot (-1) + 1 \cdot 4 - 1 \cdot 2 = 0$

$\mathbf{n} \cdot \mathbf{d}_2 = 2 \cdot 1 + 1 \cdot 2 - 1 \cdot 5 = -1 \neq 0$

Exercise 9J

1 Find the intersection between the line and the plane:

a $x - 5 = \dfrac{y+1}{2} = \dfrac{1-z}{3}$ and $2x - 4y + z = -3$

b $1 - 2x = \dfrac{y-3}{4} = \dfrac{2z+2}{3}$ and $5x + y - 4z = 3$

c $\dfrac{x-5}{4} = \dfrac{y+2}{-2} = \dfrac{z-3}{3}$ and $2x + y - 2z = 3$

d $\dfrac{1-x}{2} = \dfrac{y+2}{3} = 1 - 3z$ and $2x + y - 3z = -1$

2 Find the value of the real parameter m such that the line $\dfrac{x}{m} = \dfrac{y-1}{2} = \dfrac{z+2}{4}$ is parallel to the plane $2x + my - 3z = -1$.

3 Show that a line is parallel to a plane if the scalar product between the direction vector of the line and normal vector of the plane is 0.

Two planes

To find the angle between two non-parallel planes you need to find the angle between their normal vectors.

A normal is perpendicular to all the vectors in the plane; therefore angles with perpendicular rays are mutually congruent. Since the angle between two vectors can be obtuse but the angle between two planes is always acute you need to use the absolute value of the scalar product.

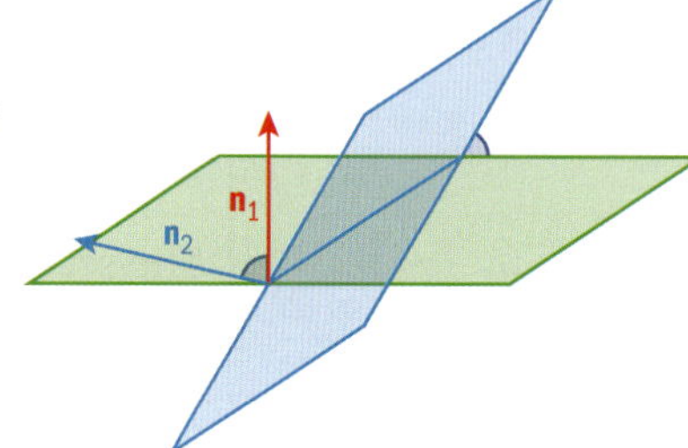

International-mindedness

"As long as algebra and geometry have been separated, their progress have been slow and their uses limited, but when these two sciences have been united, they have lent each mutual force, and have marched together towards perfection." –18th century French mathematician Joseph Lagrange.

Example 47

Find the angle in radians between the planes $\pi_1: x + 2y - 2z = 1$ and $\pi_2: 2x - 3y - 6z = -17$ correct to 3 significant figures.

$\mathbf{n}_1 = \mathbf{i} + 2\mathbf{j} - 2\mathbf{k}$ and $\mathbf{n}_2 = 2\mathbf{i} - 3\mathbf{j} - 6\mathbf{k}$	Identify the normal vectors for both planes.
$\cos\theta = \dfrac{\lvert \mathbf{n}_1 \cdot \mathbf{n}_2 \rvert}{\lvert \mathbf{n}_1 \rvert \lvert \mathbf{n}_2 \rvert} \Rightarrow$	Use the scalar product to find the angle between the two normal vectors.

$\theta = \cos^{-1}\left(\frac{\lvert 1\cdot 2+2\cdot(-3)+(-2)\cdot(-6)\rvert}{\sqrt{1^2+2^2+(-2)^2}\sqrt{2^2+(-3)^2+(-6)^2}}\right)$ $= \cos^{-1}\left(\frac{8}{21}\right) = 1.18$	

Two lines

To find an angle between two non-parallel lines you need to find the angle between their direction vectors. Even though the angle between the two vectors can be obtuse, the angle between two lines is always acute, therefore when calculating the scalar product you need to use the absolute value.

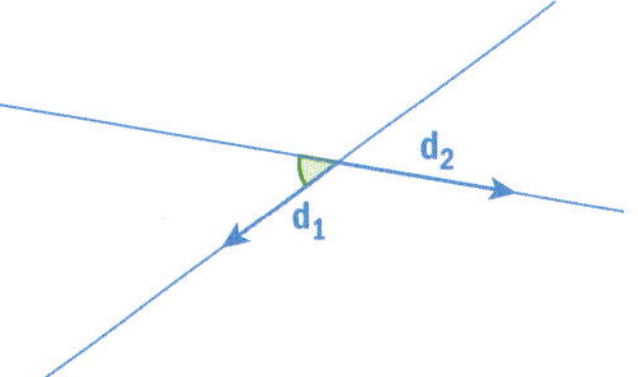

Example 48

Find the angle in radians between two lines, $L_1: \frac{x-5}{3} = \frac{y+1}{4} = \frac{z-4}{12}$ and $L_2: \frac{x-4}{2} = 5 - y = \frac{z-20}{-2}$ correct to 3 significant figures.

$\mathbf{d}_1 = 3\mathbf{i} + 4\mathbf{j} + 12\mathbf{k}$ and $\mathbf{d}_2 = 2\mathbf{i} - \mathbf{j} - 2\mathbf{k}$	Identify the direction vectors of both lines.
$\cos\theta = \frac{\lvert \mathbf{d}_1 \cdot \mathbf{d}_2 \rvert}{\lvert \mathbf{d}_1 \rvert \lvert \mathbf{d}_2 \rvert} \Rightarrow$ $\theta = \cos^{-1}\left(\frac{\lvert 3\cdot 2+4\cdot(-1)+12\cdot(-2)\rvert}{\sqrt{3^2+4^2+12^2}\sqrt{2^2+(-1)^2+(-2)^2}}\right)$ $= \cos^{-1}\left(\frac{22}{39}\right) = 0.971$	Use the scalar product to find the angle between the two direction vectors.

A line and a plane

To find the angle between a line and a plane, the line must not lie in the plane, nor should it be parallel to it. To find the angle, you have to find the angle between the line and its perpendicular projection into the plane. Then you can determine the angle in the same way you did when finding the angle between two lines.

There is also a much easier way by involving the normal vector of the plane. The normal vector of the plane is perpendicular to any line in the plane, therefore the angle between the normal vector of the plane and the direction vector of the line is complementary to the angle between the line and its perpendicular projection to the plane.

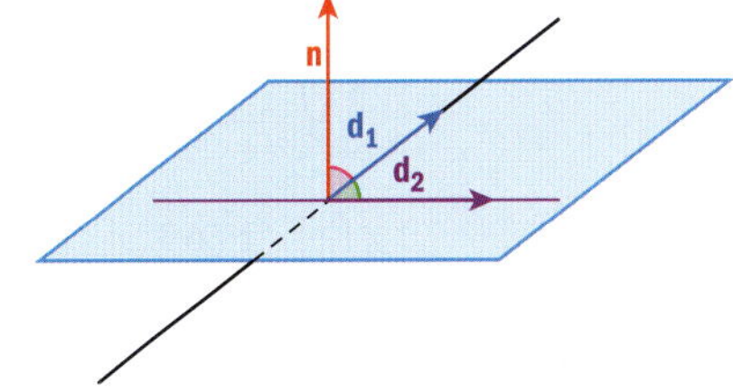

Geometry and trigonometry

Example 49

The line $\frac{x+3}{2}=\frac{y-5}{4}=\frac{3z+3}{2}$ and the plane $5x-y+4z=2$ are given.

a Find the point of intersection between the line and the plane.

b Determine the angle in radians between the line and the plane. Give your answer to 3 significant figures.

a $\begin{cases} x=2\lambda-3 \\ y=4\lambda+5,\ \lambda\in\mathbb{R} \\ z=\frac{2}{3}\lambda-1 \end{cases}$	Rewrite the equation of the line in parametric form and substitute it in the equation of the plane.
$5(2\lambda-3)-(4\lambda+5)+4\left(\frac{2}{3}\lambda-1\right)=2$	Solve the equation for λ.
$10\lambda-15-4\lambda-5+\frac{8}{3}\lambda-4=2\Rightarrow\frac{26}{3}\lambda=26$ $\lambda=3\Rightarrow \text{P}(3,17,1)$	Input the value of λ in the parametric equations of the line to find the point.
b **Method 1** A(−3,5,−1)	Identify another point on the line.
$\begin{cases} x=5\lambda-3 \\ y=-\lambda+5,\lambda\in\mathbb{R} \\ z=4\lambda-1 \end{cases}$ $5(5\lambda-3)-(-\lambda+5)+4(4\lambda-1)=2$	Find the line perpendicular to the plane that passes through A.
$42\lambda=26\Rightarrow\lambda=\frac{13}{21}\Rightarrow \text{F}\left(\frac{2}{21},\frac{92}{21},\frac{31}{21}\right)$	Find the point of intersection F between the perpendicular line and the plane.
$\mathbf{d}=2\mathbf{i}+4\mathbf{j}+\frac{2}{3}\mathbf{k}=\frac{2}{3}(3\mathbf{i}+6\mathbf{j}+\mathbf{k})$ and $\overrightarrow{\text{PF}}=-\frac{61}{21}\mathbf{i}-\frac{265}{21}\mathbf{j}+\frac{10}{21}\mathbf{k}$ $=-\frac{1}{21}(61\mathbf{i}+265\mathbf{j}-10\mathbf{k})$	Identify the direction vector of the line and the vector $\overrightarrow{\text{PF}}$ or vectors collinear with them.
$\cos\theta=\frac{\lvert\mathbf{d}\cdot\overrightarrow{\text{PF}}\rvert}{\lvert\mathbf{d}\rvert\lvert\overrightarrow{\text{PF}}\rvert}\Rightarrow$ $\theta=\cos^{-1}\left(\frac{\lvert3\cdot61+6\cdot265+1\cdot(-10)\rvert}{\sqrt{3^2+6^2+1^2}\sqrt{61^2+265^2+(-10)^2}}\right)$ $=\cos^{-1}\left(\frac{1763}{\sqrt{46\cdot74046}}\right)=0.300$	Find the angle between the two vectors.

Method 2

$\mathbf{d} = 2\mathbf{i} + 4\mathbf{j} + \frac{2}{3}\mathbf{k} = \frac{2}{3}(3\mathbf{i} + 6\mathbf{j} + \mathbf{k})$ and $\mathbf{n} = 5\mathbf{i} - \mathbf{j} + 4\mathbf{k}$	Identify the direction vector of the line or one parallel to it, and identify the normal vector of the plane.
$\sin\theta = \frac{\lvert\mathbf{d}\cdot\mathbf{n}\rvert}{\lvert\mathbf{d}\rvert\lvert\mathbf{n}\rvert} \Rightarrow$ $\theta = \sin^{-1}\left(\frac{\lvert 3\cdot5+6\cdot(-1)+1\cdot4\rvert}{\sqrt{3^2+6^2+1^2}\sqrt{5^2+(-1)^2+4^2}}\right)$	Since the angle between the line and the plane is complementary to the angle between the direction vector of the line and the normal vector of the plane you need to use the sine ratio.
$= \sin^{-1}\left(\frac{13}{2\sqrt{483}}\right) = 0.300$	Find the angle.

Exercise 9K

1 The planes $3x + y - 2z = -1$ and $x - 4y + 2z = 3$ are given.

a Find the equation of the line of intersection.

b Determine the angle between these two planes.

2 Two lines $\frac{x-5}{2} = \frac{y+7}{-5} = 7 - z$ and $-x = \frac{y-5}{2} = \frac{4-z}{5}$ are given.

a Show that the lines do not intersect and find the point of intersection.

b Find the angle between the lines.

3 The line $x - 2 = \frac{1-y}{3} = \frac{z-2}{2}$ and the plane $3x + 2y - z = 1$ are given.

a Find the point of intersection between the line and the plane.

b Determine the angle between the line and the plane.

4 Show that the planes $ax + az = c$ and $bx - by = d$, where $a, b, c, d \in \mathbb{R}$, $a, b \neq 0$, always form an angle of $\frac{\pi}{3}$ or 60°.

9.8 Application of vectors

There are many subjects that use vectors as tools to calculate quantities. Physics is probably the one in which vectors are used the most. In physics, you need to be able to distinguish between vector and scalar quantities.

- **Displacement** is a **vector quantity** since it has a direction and magnitude.
- **Distance** is a **scalar** and it is the magnitude of displacement.
- **Velocity** is a **vector quantity** since it has a direction and magnitude.

TOK

Do you think that there are times when analytical reasoning is easier to use than sense perception when working in three dimensions?

- **Speed** is a **scalar** and it is the magnitude of velocity.
- **Acceleration** is a **vector quantity** that has the same direction as velocity. However, we also use acceleration as a **scalar**, just the magnitude of that vector quantity, so you need to carefully read the context of the problem.

Here are some examples with the use of vectors.

Example 50

On a sunny day two sailing boats "Rab" and "Pag" were spotted from a lighthouse. The courses of the boats "Rab" and "Pag" from the moment they were spotted, can be given by the equations $\mathbf{r} = (-1 + 2t)\mathbf{i} + (10 - 2t)\mathbf{j}$ and $\mathbf{p} = 2s\mathbf{i} + (4 + 3s)\mathbf{j}$ respectively. All the units are given in km and the parameters represent the time in hours.

a Find the positions of the boats when they were spotted and the distance between them.

b Calculate the speed of each sailing boat.

c Find the coordinates of the point where the courses meet.

d Explain whether or not the boats will collide.

a $t = 0 \Rightarrow R(-1 + 2\times 0,\ 10 - 2\times 0) \Rightarrow R(-1, 10)$ $s = 0 \Rightarrow P(2\times 0,\ 4 + 3\times 0) \Rightarrow P(0, 4)$	To find the initial position the parameter is set to 0.
$PR = \sqrt{(-1-0)^2 + (10-4)^2} = \sqrt{37} = 6.08$ The distance between the boats when they were spotted was 6.08 km.	Use the distance formula between two points.
b $\mathbf{d_R} = 2\mathbf{i} - 2\mathbf{j} \Rightarrow \lvert\mathbf{d_R}\rvert = \sqrt{2^2 + (-2)^2} = \sqrt{8} = 2.83$ $\mathbf{d_P} = 2\mathbf{i} + 3\mathbf{j} \Rightarrow \lvert\mathbf{d_P}\rvert = \sqrt{2^2 + 3^2} = \sqrt{13} = 3.61$ The speed of "Rab" is 2.83 km h^{-1} and the speed of "Pag is 3.61 km h^{-1}.	Find the magnitude of each direction vector.
c $\begin{cases} -1 + 2t = 2s \\ 10 - 2t = 4 + 3s \end{cases} \Rightarrow 9 = 4 + 5s \Rightarrow s = 1\text{ h}$ $-1 + 2t = 2\cdot 1 \Rightarrow t = \frac{3}{2}\text{ h}$ The courses meet when $t = \frac{3}{2}$ h and $s = 1$ h, therefore their coordinates are (2,7).	Solve the simultaneous equations by the method of elimination. Add the equations to eliminate t. Substitute s in the first equation to find t.
d Since the values of the parameters (times) are different the sailing boats will not collide.	Use s and t to find the coordinates.

Example 51

Every year Luca swims across the canal from Zadar to Preko. His average speed in a pool is 1.8 m s^{-1}, but there is also a sea current that is directly perpendicular to the route and its speed is 1.2 m s^{-1}.

a Find at what angle to the route Luca should swim to travel directly to Preko. Give your answer correct to 3 significant figures.

b Determine how long it will take if the length of the route is 4.5 km.

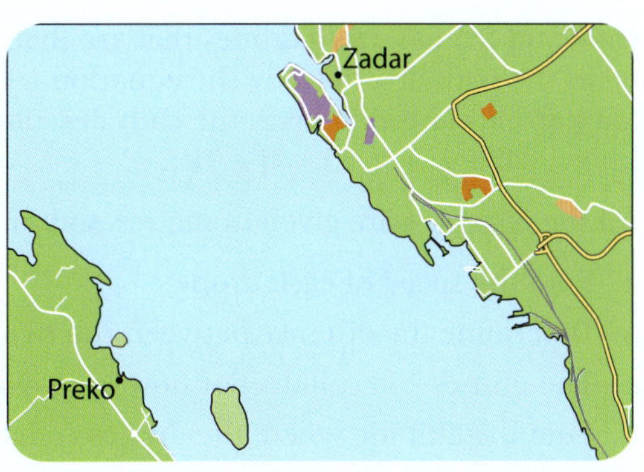

a Take that Luca's route is in the positive direction of the y-axis. 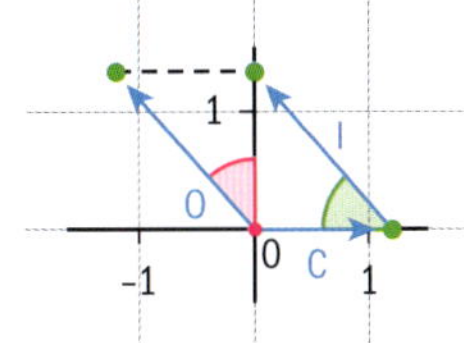	Draw in the coordinate system a vector representing the current and a vector representing Luca's swimming direction.
Method 1 $\sin\theta = \frac{1.2}{1.8} = \frac{2}{3} \Rightarrow \theta = \sin^{-1}\left(\frac{2}{3}\right) = 41.8°$	Use right-angle trigonometry to find the angle.
Method 2 The vector representing the current is $\mathbf{c} = 1.2\mathbf{i}$. The magnitude of vector representing Luca's swimming is $\lvert\mathbf{l}\rvert = 1.8$.	
$\sqrt{1.8^2 - 1.2^2} = 1.34 \Rightarrow \mathbf{l} = -1.2\mathbf{i} + 1.34\mathbf{j}$	Use Pythagoras theorem to find the vertical component of the vector **l**.
$\cos\phi = \frac{\mathbf{c}\cdot\mathbf{l}}{\lvert\mathbf{c}\rvert\cdot\lvert\mathbf{l}\rvert} \Rightarrow \phi = \cos^{-1}\left(\frac{1.2\cdot(-1.2)+0\cdot 1.34}{1.2\cdot 1.8}\right)$ $= \cos^{-1}\left(\frac{-1.2}{1.8}\right) = 131.8° \Rightarrow \theta = \phi - 90° = 41.8°$	Use the scalar product to find the angle between the two vectors.
So, Luca should swim at an angle of 41.8° with respect to the route.	Since the angle is obtuse to find the angle of the route you need to subtract 90°.
b Luca's speed in the sea is approximately 1.34 m s^{-1}. $t = \frac{s}{v} \Rightarrow t = \frac{4500}{\sqrt{1.8^2 - 1.2^2}} = 3354$ s (to 4 sf)	Use the formula for time, distance and velocity.
Luca will swim for approximately 55 minutes and 54 seconds.	Convert seconds into minutes and seconds.

Example 52

Irma and Ida have two drones that are flying at a constant speed. Irma's drone starts first and takes the path described by the equation L_1: $\mathbf{r} = 2\mathbf{i} - 3\mathbf{j} + t(-2\mathbf{i} + 3\mathbf{j} + 4\mathbf{k})$. After 3 seconds Ida's drone starts and takes the path described by the equation L_2: $\mathbf{r} = -4\mathbf{i} + 2\mathbf{j} + s(-2\mathbf{i} + 4\mathbf{j} + 7\mathbf{k})$.

All the distances are given in metres and times in seconds.

a Find the speed of each drone.

b Determine the distance between the two drones just before Ida's drone starts.

c The drones will collide. For how long was Irma's drone flying when they collide?

d Find the altitude where the drones collide.

a $\mathbf{d}_1 = -2\mathbf{i} + 3\mathbf{j} + 4\mathbf{k}$ $\Rightarrow \lvert\mathbf{d}_1\rvert = \sqrt{(-2)^2 + 3^2 + 4^2} = \sqrt{29} = 5.39\,\text{ms}^{-1}$ $\mathbf{d}_2 = -2\mathbf{i} + 4\mathbf{j} + 7\mathbf{k}$ $\Rightarrow \lvert\mathbf{d}_2\rvert = \sqrt{(-2)^2 + 4^2 + 7^2} = \sqrt{69} = 8.31\,\text{ms}^{-1}$	Identify the direction vectors and find their magnitudes.
b $t = 3 \Rightarrow \mathbf{r} = 2\mathbf{i} - 3\mathbf{j} + 3(-2\mathbf{i} + 3\mathbf{j} + 4\mathbf{k})$ $= -4\mathbf{i} + 6\mathbf{j} + 12\mathbf{k} \Rightarrow A(-4, 6, 12)$ $s = 0 \Rightarrow \mathbf{r} = -4\mathbf{i} + 2\mathbf{j} \Rightarrow B(-4, 2, 0)$	Calculate the position of Irma's drone when $t = 3$, point A and Ida's drone when $s = 0$, point B.
$AB = \sqrt{(-4-(-4))^2 + (6-2)^2 + (12-0)^2}$ $= \sqrt{160} = 4\sqrt{10} = 12.6\,\text{m}$	Find the distance between those two points, AB.
c $\begin{cases} 2 - 2t = -4 - 2s \\ -3 + 3t = 2 + 4s \\ 4t = 7s \end{cases} \Rightarrow \begin{cases} \frac{8}{7}t - 2t = -6 \\ 3t - \frac{16}{7}t = 5 \\ s = \frac{4}{7}t \end{cases}$ $t = 7$ and $s = 4$	Solve simultaneous equations by the method of substitution. Express s in terms of t in the third equation and substitute it in the first and second equation. The system is consistent, it has only one solution.
Irma's drone was flying for 7 seconds.	For the flying time of Irma's drone you need to take the parameter t.
d $t = 7 \Rightarrow \mathbf{r} = 2\mathbf{i} - 3\mathbf{j} + 7(-2\mathbf{i} + 3\mathbf{j} + 4\mathbf{k})$ $= -12\mathbf{i} + 18\mathbf{j} + 28\mathbf{k} \Rightarrow C(-12, 18, 28)$	Find the coordinates of the point of collision, C.
The altitude of the collision was 28 m.	Take the third coordinate since it represents the height.

In the following example we are going to revisit the opening problem.

Example 53

Hannah is skiing on a slope at 35° to the horizontal. Given that Hannah has a mass of 30 kg find the following:

a Hannah's weight

b the magnitude of the reaction of the slope on her weight

c Hannah's speed from resting to 50 m down the slope given that the traction vector has a magnitude of 80 N

d the maximum angle of the slope so that Hannah will not slide at all.

a Since the gravitational acceleration is 9.81 m s^{-2} the force has the magnitude $|\mathbf{w}| = m \cdot |\mathbf{g}| = 30 \cdot 9.81 = 294.3$ N

Take an average value of Earth's gravitational acceleration to be 9.81 m s^{-2}. Force is the product of mass and acceleration.

b On the following diagram you can find the forces that act on Hannah.

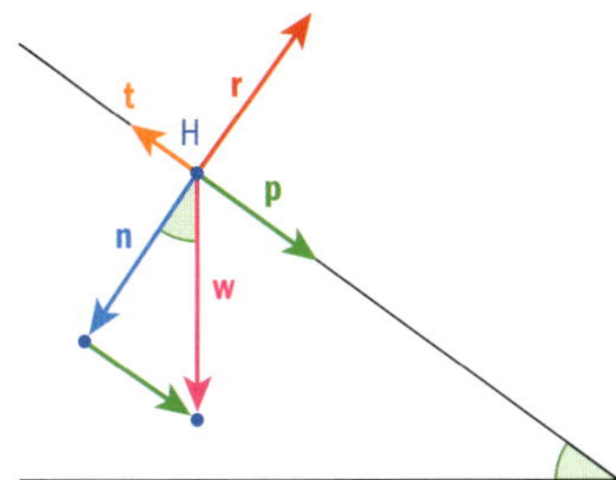

$|\mathbf{n}| = |\mathbf{r}| = |\mathbf{w}|\cos(35°) = 241$ N

The indicated angles on the diagram are congruent since they have perpendicular rays.

The vector **n** is the perpendicular projection of **w** to the normal to the slope, therefore by using the right–angled trigonometry you obtain the formula.

c First you need to find the magnitude of the vector in the direction of the slope.

$|\mathbf{p}| = |\mathbf{w}|\sin(35°) = 169\text{N}$

$\mathbf{d} = \mathbf{p} - \mathbf{t} \Rightarrow |\mathbf{d}| = 169 - 80 = 89\text{N}.$

The vector **p** in the direction of the slope is the perpendicular projection of **w** to the slope, therefore by using the right–angled trigonometry you obtain the formula.

Since the vectors **p** and **t** have opposite directions, to find the magnitude of the sum you need to subtract their magnitudes.

$$|\mathbf{a}| = \frac{|\mathbf{d}|}{m} \Rightarrow |\mathbf{a}| = \frac{89}{30} = 2.96 \text{ m s}^{-2}.$$

Acceleration is a quotient of the force and the mass.

$$|\mathbf{v}|^2 = 2|\mathbf{a}||\mathbf{s}| \Rightarrow |\mathbf{v}| = \sqrt{2 \cdot 2.96 \cdot 50} = 17.2$$

Hannah has a speed of 17.2 m s^{-1}.

Since the initial speed is 0 the velocity is a square root of double the product of the acceleration and the distance.

d $|\mathbf{w}|\sin(\theta) = 80 \Rightarrow \theta = \sin^{-1}\left(\frac{80}{294.3}\right) = 15.8°$

In order for Hannah not to slide, the force parallel to the slope must be opposite to the traction force.

Geometry and trigonometry

Exercise 9L

1 There is a boat on the sea that needs help. Two boats are nearby and they have received the call. If you place the boat in need at the origin of the coordinate plane the other two boats called "Adrianne" and "Lilly" are at the points A(−3,5) and L(7,9) respectively. The coordinates are given in kilometres. They move directly towards the boat in need, "Adrianne" with a speed of 4 m s^{-1} and "Lilly" with a speed of 6 m s^{-1}.

a Show that the courses of these two boats have the following equations t seconds after starting

$$\mathbf{a} = (-3000\mathbf{i} + 5000\mathbf{j}) + t\left(\frac{12}{\sqrt{34}}\mathbf{i} - \frac{20}{\sqrt{34}}\mathbf{j}\right)$$

and

$$\mathbf{l} = (7000\mathbf{i} + 9000\mathbf{j}) + t\left(-\frac{42}{\sqrt{130}}\mathbf{i} - \frac{54}{\sqrt{130}}\mathbf{j}\right).$$

b Determine which boat will arrive first and how much longer it will take the other boat to reach the boat in need.

2 A particle with a constant speed moves along the line $\mathbf{p} = (23 + 2t)\mathbf{i} + (8 - t)\mathbf{j} + (43 + 4t)\mathbf{k}$, where t is the time measured in seconds. All the measurements are given in metres.

a Determine the initial position of the particle.

b Find the speed of the particle.

c Find after how much time the particle will reach the plane $12x - 3y - 5z = -2$.

d Calculate the total distance the particle travelled before reaching the plane.

3 The radar controller Boris spotted two planes in the air. The computer is calculating paths of both planes and these are found to be

$\mathbf{p}_1 = (147 - 8t)\mathbf{i} + (-156 + 9t)\mathbf{j} + (5 + 0.25t)\mathbf{k}$ and
$\mathbf{p}_2 = (-118 + 7t)\mathbf{i} + (189 - 11t)\mathbf{j} + (7 + 0.2t)\mathbf{k}$.

The distance between the points is given in kilometres and the time in the equations of the paths is given in minutes.

a Find the speed of both planes in km h^{-1}.

b Find whether the two paths cross and if yes, find the point of intersection.

c Determine whether the planes will collide. Justify your answer.

Developing your toolkit

Now do the Modelling and investigation activity on page 646.

Chapter summary

- A **directed line segment** is a line segment that has a direction, as there is a starting point and an end point. The notation is $\overrightarrow{AB}$, where A is the starting point and B is the end point.
- Two directed line segments $\overrightarrow{AB}$ and $\overrightarrow{CD}$ are called **equivalent**, you write $\overrightarrow{AB} \cong \overrightarrow{CD}$, if and only if line segments [AD] and [BC] have a common midpoint.
- A **vector** is a collection of all equivalent directed line segments.
- A **displacement vector** is represented by **AB** or $\overrightarrow{AB}$, where **A** is called the **initial or starting point** and **B** is called the **terminal or end point**.
- A **vector** is defined by:

  i **direction** **ii** **magnitude**

- **Direction** of a vector is represented by a family of parallel lines that carry all the equivalent directed line segments. The direction of a vector is determined by the starting and ending point.
- **Magnitude** of a vector is the length of the vector. The magnitude of the displacement vector $\overrightarrow{AB}$ is simply the length AB. The magnitude will be denoted by $|\mathbf{a}|$ or $|\overrightarrow{AB}|$.

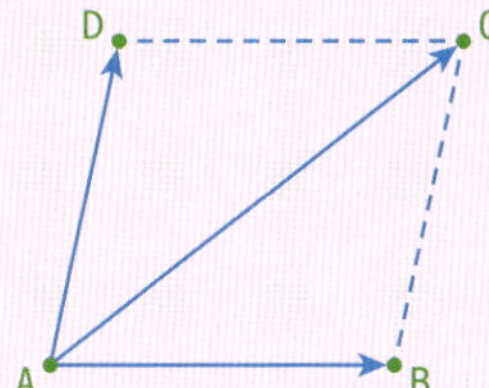

The parallelogram law of vector addition

The triangle law of vector addition

- Multiplying vector $\mathbf{a}$ by a scalar k you obtain a new vector $\mathbf{b} = k\mathbf{a}$ such that
  - i $\mathbf{b}$ has the **same direction** as $\mathbf{a}$ when k is positive $(k > 0)$, or the **opposite direction** when k is negative $(k < 0)$.
  - ii the **magnitude** of $\mathbf{b}$ is the product of the absolute value of k and magnitude of the vector $\mathbf{a}$, $|\mathbf{b}| = |k||\mathbf{a}|$.
- A vector and a vector obtained by multiplication of that vector by a scalar have the same or opposite directions and are said to be **collinear** or **parallel**.
- Given two non-collinear vectors **a** and **b**, a **linear combination** of those two vectors is a new vector,

$$\mathbf{c} = \lambda\mathbf{a} + \mu\mathbf{b} = \lambda\begin{pmatrix} a_1 \\ a_2 \end{pmatrix} + \mu\begin{pmatrix} b_1 \\ b_2 \end{pmatrix} = \begin{pmatrix} \lambda a_1 + \mu b_1 \\ \lambda a_2 + \mu b_2 \end{pmatrix}, \mu, \lambda \in \mathbb{R}$$

- The vectors $\mathbf{a}$, $\mathbf{b}$ and $\mathbf{c}$ are said to be **coplanar** if they lie in the same plane. Vectors $\mathbf{a}$ and $\mathbf{b}$ define a sort of coordinate system and every vector coplanar with them can be shown as a linear combination of $\mathbf{a}$ and $\mathbf{b}$.
- In a 3D space a **linear combination** of three non-coplanar vectors $\mathbf{a}$, $\mathbf{b}$ and $\mathbf{c}$, is a new vector $\mathbf{d} = \lambda\mathbf{a} + \mu\mathbf{b} + v\mathbf{c}$, $\lambda, \mu, v \in \mathbb{R}$.
- A vector that has a starting point at the origin (0,0) and an end point at any point of the plane $P(x, y)$ is called a **position vector** of the point P.

Vector algebra

- Each vector in a coordinate plane can be written as a linear combination of vectors $\mathbf{i} = \begin{pmatrix} 1 \\ 0 \end{pmatrix}$ and $\mathbf{j} = \begin{pmatrix} 0 \\ 1 \end{pmatrix}$, which are unit vectors in the positive direction of the x-axis and y-axis respectively.
- You write $\mathbf{p} = \mathbf{OP} = x\mathbf{i} + y\mathbf{j} = \begin{pmatrix} x \\ y \end{pmatrix}$, where x and y are called the horizontal and vertical component of the vector $\mathbf{p}$ respectively.

Addition

- If two vectors $\mathbf{a}$ and $\mathbf{b}$ are given by their components, then

$$\mathbf{a} + \mathbf{b} = (a_1\mathbf{i} + a_2\mathbf{j}) + (b_1\mathbf{i} + b_2\mathbf{j}) = (a_1 + b_1)\mathbf{i} + (a_2 + b_2)\mathbf{j}$$

Continued on next page

Geometry and trigonometry

or by using the column matrix form

$$\mathbf{a}+\mathbf{b}=\begin{pmatrix}a_1\\a_2\end{pmatrix}+\begin{pmatrix}b_1\\b_2\end{pmatrix}=\begin{pmatrix}a_1+b_1\\a_2+b_2\end{pmatrix}$$

Multiplication by a scalar

- $\lambda\mathbf{a}=\lambda(a_1\mathbf{i}+a_2\mathbf{j})=(\lambda a_1)\mathbf{i}+(\lambda a_2)\mathbf{j}$
  or by using the column matrix form
  $$\lambda\mathbf{a}=\lambda\begin{pmatrix}a_1\\a_2\end{pmatrix}=\begin{pmatrix}\lambda a_1\\\lambda a_2\end{pmatrix}$$

Linear combination of two vectors in a plane

- $\lambda\mathbf{a}+\mu\mathbf{b}=\lambda(a_1\mathbf{i}+a_2\mathbf{j})+\mu(b_1\mathbf{i}+b_2\mathbf{j})=(\lambda a_1+\mu b_1)\mathbf{i}+(\lambda a_2+\mu b_2)\mathbf{j}$
  or by using the column matrix form
  $$\lambda\mathbf{a}+\mu\mathbf{b}=\lambda\begin{pmatrix}a_1\\a_2\end{pmatrix}+\mu\begin{pmatrix}b_1\\b_2\end{pmatrix}=\begin{pmatrix}\lambda a_1+\mu b_1\\\lambda a_2+\mu b_2\end{pmatrix}$$

Magnitude of a vector

- Given a position vector **in a plane** with the components $\mathbf{a}=\overrightarrow{\mathrm{OA}}=\begin{pmatrix}a_1\\a_2\end{pmatrix}$, the magnitude of the position vector **a** is $\mathrm{OA}=|\overrightarrow{\mathrm{OA}}|=\sqrt{a_1^2+a_2^2}$.
- The magnitude of the direction vector is
  $$|\overrightarrow{\mathrm{AB}}|=\left|\begin{pmatrix}b_1-a_1\\b_2-a_2\end{pmatrix}\right|=\sqrt{(b_1-a_1)^2+(b_2-a_2)^2}.$$
- Given a position vector **in a space** with the components $\mathbf{a}=\overrightarrow{\mathrm{OA}}=\begin{pmatrix}a_1\\a_2\\a_3\end{pmatrix}$, the magnitude of the position vector **a** is $\mathrm{OA}=|\overrightarrow{\mathrm{OA}}|=\sqrt{a_1^2+a_2^2+a_3^2}$.
- The magnitude of the direction vector is given by
  $$|\overrightarrow{\mathrm{AB}}|=\left|\begin{pmatrix}b_1-a_1\\b_2-a_2\\b_3-a_3\end{pmatrix}\right|=\sqrt{(b_1-a_1)^2+(b_2-a_2)^2+(b_3-a_3)^2}$$

Unit vectors

- $\mathbf{a}=\begin{pmatrix}a_1\\a_2\end{pmatrix}\Rightarrow\mathbf{u}=\dfrac{1}{\sqrt{a_1^2+a_2^2}}\begin{pmatrix}a_1\\a_2\end{pmatrix}$ for a 2D vector
- $\mathbf{a}=\begin{pmatrix}a_1\\a_2\\a_3\end{pmatrix}\Rightarrow\mathbf{u}=\dfrac{1}{\sqrt{a_1^2+a_2^2+a_3^2}}\begin{pmatrix}a_1\\a_2\\a_3\end{pmatrix}$ for a 3D vector

Scalar product: Algebraic definition

- Given the 2-dimensional vectors $\mathbf{a} = \begin{pmatrix} a_1 \\ a_2 \end{pmatrix}$ and $\mathbf{b} = \begin{pmatrix} b_1 \\ b_2 \end{pmatrix}$ then the scalar product is given by

  $\mathbf{a} \cdot \mathbf{b} = a_1b_1 + a_2b_2$.

- Given the 3-dimensional vectors $\mathbf{a} = \begin{pmatrix} a_1 \\ a_2 \\ a_3 \end{pmatrix}$ and $\mathbf{b} = \begin{pmatrix} b_1 \\ b_2 \\ b_3 \end{pmatrix}$ then the scalar product is given by

  $\mathbf{a} \cdot \mathbf{b} = a_1b_1 + a_2b_2 + a_3b_3$.

Scalar product: Geometric definition

- The scalar product of two non-zero vectors **a** and **b** is given by $\mathbf{a} \cdot \mathbf{b} = |\mathbf{a}||\mathbf{b}|\cos\theta$, where θ is the angle between those two vectors.

Angle between two vectors

- $\cos\theta = \dfrac{a_1b_1 + a_2b_2}{\sqrt{(a_1^2 + a_2^2)(b_1^2 + b_2^2)}}$, for 2D vectors.

- $\cos\theta = \dfrac{a_1b_1 + a_2b_2 + a_3b_3}{\sqrt{(a_1^2 + a_2^2 + a_3^2)(b_1^2 + b_2^2 + b_3^2)}}$, for 3D vectors.

Properties of the scalar product

i $\mathbf{a} \cdot \mathbf{b} = \mathbf{b} \cdot \mathbf{a}$

ii $\mathbf{a} \cdot \mathbf{a} = |\mathbf{a}|^2$

iii $\mathbf{a} \cdot (\mathbf{b} + \mathbf{c}) = \mathbf{a} \cdot \mathbf{b} + \mathbf{a} \cdot \mathbf{c}$

iv $\lambda(\mathbf{a} \cdot \mathbf{b}) = (\lambda\mathbf{a}) \cdot \mathbf{b} = \mathbf{a} \cdot (\lambda\mathbf{b}), \lambda \in \mathbb{R}$

The vector equation of a line

- $\mathbf{p} = \mathbf{a} + k\mathbf{d}, k \in \mathbb{R}$
- $\begin{pmatrix} x \\ y \end{pmatrix} = \begin{pmatrix} x_1 \\ y_1 \end{pmatrix} + k\begin{pmatrix} d_1 \\ d_2 \end{pmatrix}, k \in \mathbb{R}$ in a plane
- $\begin{pmatrix} x \\ y \\ z \end{pmatrix} = \begin{pmatrix} a_1 \\ a_2 \\ a_3 \end{pmatrix} + k\begin{pmatrix} d_1 \\ d_2 \\ d_3 \end{pmatrix}, k \in \mathbb{R}$ in space

Parametric equation of a line

- $\begin{cases} x = x_1 + kd_1 \\ y = y_1 + kd_2 \end{cases}, k \in \mathbb{R}$ in a plane
- $\begin{cases} x = a_1 + kd_1 \\ y = a_2 + kd_2, k \in \mathbb{R} \\ z = a_3 + kd_3 \end{cases}$ in space

Continued on next page

Cartesian equation of a line

- $\frac{x-x_1}{d_1}=\frac{y-y_1}{d_2}$ in a plane
- $\frac{x-a_1}{d_1}=\frac{y-a_2}{d_2}=\frac{x-a_3}{d_3}$ in space

Vector product: Algebraic definition

- Given the two vectors and their components, $\mathbf{a}=\begin{pmatrix}a_1\\a_2\\a_3\end{pmatrix}, \mathbf{b}=\begin{pmatrix}b_1\\b_2\\b_3\end{pmatrix}$ then the **vector product** is given by $\mathbf{a}\times\mathbf{b}=\begin{pmatrix}a_2b_3-a_3b_2\\a_3b_1-a_1b_3\\a_1b_2-a_2b_1\end{pmatrix}$.

Vector product: Geometric definition

- Given two vectors $\mathbf{a}$ and $\mathbf{b}$ then the **vector product** is given by $\mathbf{a}\times\mathbf{b}=(|\mathbf{a}||\mathbf{b}|\sin\theta)\times\hat{\mathbf{n}}$, where $\hat{\mathbf{n}}$ is the **unit vector** orthogonal (perpendicular) to both $\mathbf{a}$ and $\mathbf{b}$ and the vectors $\mathbf{a}$, $\mathbf{b}$ and $\mathbf{n}$ follow the right–hand rule.

Properties of the vector product

i $\mathbf{a}\times\mathbf{b}=-(\mathbf{b}\times\mathbf{a})$

ii $(\mathbf{a}\times\mathbf{b})\times\mathbf{c}=\mathbf{a}\times(\mathbf{b}\times\mathbf{c})$

iii $\lambda(\mathbf{a}\times\mathbf{b})=(\lambda\mathbf{a})\times\mathbf{b}=\mathbf{a}\times(\lambda\mathbf{b}), \lambda\in\mathbb{R}$

iii $(\mathbf{a}+\mathbf{b})\times\mathbf{c}=(\mathbf{a}\times\mathbf{c})+(\mathbf{b}\times\mathbf{c})$

- The **volume of a parallelepiped** formed by three non–coplanar vectors $\mathbf{a}$, $\mathbf{b}$ and $\mathbf{c}$ is given by the formula:

  $V=|(\mathbf{a}\times\mathbf{b})\cdot\mathbf{c}|$
- The **volume of a pyramid** formed by three non–coplanar vectors $\mathbf{a}$, $\mathbf{b}$ and $\mathbf{c}$ is given by the formula

  $V=\frac{1}{6}|(\mathbf{a}\times\mathbf{b})\cdot\mathbf{c}|$

Vector equation of a plane

- $\mathbf{p}=\mathbf{a}+\lambda\mathbf{u}+\mu\mathbf{v}$, $\lambda, \mu\in\mathbb{R}$ where $\mathbf{u}$ and $\mathbf{v}$ are two non–collinear vectors.

Parametric equation of a plane

- $\begin{cases}x=a_1+\lambda u_1+\mu v_1\\y=a_2+\lambda u_2+\mu v_2\\z=a_3+\lambda u_3+\mu v_3\end{cases}$ $\lambda,\mu\in\mathbb{R}$ where $\mathbf{u}$ and $\mathbf{v}$ are two non–collinear vectors.

Cartesian equation of a plane

- $ax+by+cz=d$, where $a, b, c, d\in\mathbb{R}$

Normal equation of a plane

- $\mathbf{n}\cdot\mathbf{p}=\mathbf{n}\cdot\mathbf{a}$, where $\mathbf{n}$ is a normal vector of the plane and $\mathbf{a}$ is the position vector of a point in the plane.
- Two planes can intersect at a line, they can be parallel or coincide.

- Two lines can intersect, be parallel, be skew or coincide.
- A line and a plane can intersect, be parallel or the line can lie in the plane.

Angle between two lines

- It is the angle between their direction vectors if it is acute and it is the supplementary angle if it is obtuse.

Angle between two planes

- It is the angle between their normal vectors if it is acute and it is the supplementary angle if it is obtuse.

Angle between a line and a plane

- It is the complementary angle between the direction vector of the line and the normal vector of the plane if it is acute and it is the complementary angle of its supplementary angle if it is obtuse.

Developing inquiry skills

Return to the chapter opening problem. How has what you have learned in this chapter helped you to understand the problem?

Chapter review

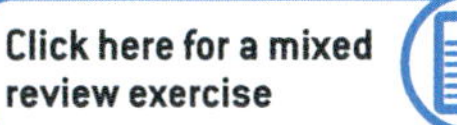

Click here for a mixed review exercise

1 Points A and B in the plane are given with their position vectors **a** and **b** respectively.

a Given that M is the midpoint of the line segment [AB], show that the position vector **m** of the point M is

$$\mathbf{m} = \frac{1}{2}(\mathbf{a} + \mathbf{b}).$$

Points $A(1, -4)$, $B(11, 6)$, $C(6, 5)$ and $D(3, 2)$ form a quadrilateral.

b Use vectors to show that ABCD is a trapezium.

c Show that the midpoints of the sides of the trapezium in part **b** form a parallelogram and determine whether or not that parallelogram is a rhombus.

2 A plane is given by the vector equation $\mathbf{r} = (2 + \lambda - 3\mu)\mathbf{i} + (2\lambda + \mu)\mathbf{j} + (\mu - 1)\mathbf{k}$.

a Show that the Cartesian form of the plane is $2x - y + 7z = -3$.

b The points $A(2, 0, a)$, $B(b, 4, -1)$ and $D(-1, d, 0)$ lie in the plane. Find the values of the real parameters a, b and d.

c Find the coordinates of the point C such that ABCD forms a parallelogram.

d Show that point $E(1, -2, 1)$ doesn't lie in the plane.

e Find the volume of the triangular prism ABCE.

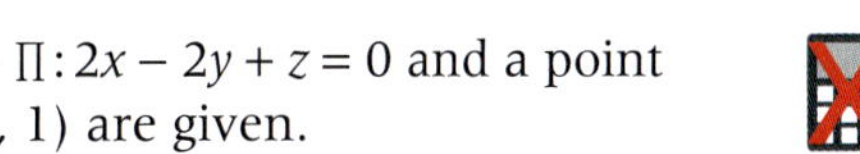

3 A plane $\Pi: 2x - 2y + z = 0$ and a point $A(2, -2, 1)$ are given.

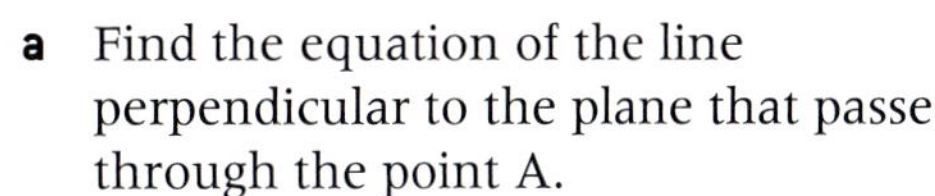

a Find the equation of the line perpendicular to the plane that passes through the point A.

b Find the point of intersection between the line and the plane. Hence find the distance between the point A and the plane.

c The point $P(x_0, y_0, z_0)$ is given.

Show that the distance between the point P and the plane Π is given by the formula $\dfrac{|2x_0 - 2y_0 + z_0|}{3}$.

4 The vectors $\mathbf{a} = \begin{pmatrix} p \\ 2 \\ r \end{pmatrix}$ and $\mathbf{b} = \begin{pmatrix} r \\ 2 \\ p \end{pmatrix}$ have components that form arithmetic sequences with the common differences $\pm d$.

a Given that θ is the angle between the vectors, show that $\cos\theta = \dfrac{6 - d^2}{6 + d^2}$.

b Find the common difference given that the vectors **a** and **b** form an angle of 60°.

5 Two planes $\sin\alpha \cdot x + \cos\alpha \cdot y + z = 3$ and $\cos\alpha \cdot x + \sin\alpha \cdot y - z = 5$, where $\alpha \in \mathbb{R}$, are given.

Show that these two planes are not perpendicular.

6 Find the angle between two unit vectors **u** and **v** such that the vectors $2\mathbf{u} - 3\mathbf{v}$ and $5\mathbf{u} + 2\mathbf{v}$ are perpendicular. Give you answer correct to the nearest degree.

7 A plane Π: $4x - 3y + z = 1$ and a line

$L: \dfrac{x-4}{3} = 1 - y = \dfrac{z-5}{2}$ are given.

a Find the coordinates of the point of intersection between the line and the plane.

b Determine the angle between the line and the plane.

c Find the equation of the line that is symmetrical to the line L with respect to the plane Π.

8 The vectors $\mathbf{a} = 2^x\mathbf{i} + 4^x\mathbf{j} + 5\mathbf{k}$ and $\mathbf{b} = 2^x\mathbf{i} + 0.5^x\mathbf{j} - 4\mathbf{k}$ are given, where $x \in \mathbb{R}$.

a Find the value of x such that the vectors are perpendicular.

b Find the equation of the plane that is given by the vectors in part a and that passes through the point (1, 1, −2).

9 The vector **a** encloses the angles α, β and γ with the unit vectors **i**, **j** and **k** respectively.

a Show that $\cos^2\alpha + \cos^2\beta + \cos^2\gamma = 1$ for any vector **a**.

Given that $\mathbf{a} = 3\mathbf{i} - 6\mathbf{j} + 2\mathbf{k}$:

b find the angles α, β and γ

c show that the plane that passes through the origin can be written in the form $x\cos\alpha + y\cos\beta + z\cos\gamma = 0$.

10 A cuboid OABCDEFG with a square base of edge length 2 and a height of 4 is placed in the coordinate system. The points P and Q are the midpoints of the line segments [EF] and [FG] respectively.

a Show that the plane that passes through the point A, P and Q is given by $4x - 4y + 3z = 8$.

b Find the equation of the line that carries the space diagonal [BG].

c Determine the angle between the plane APQ and the line [BG].

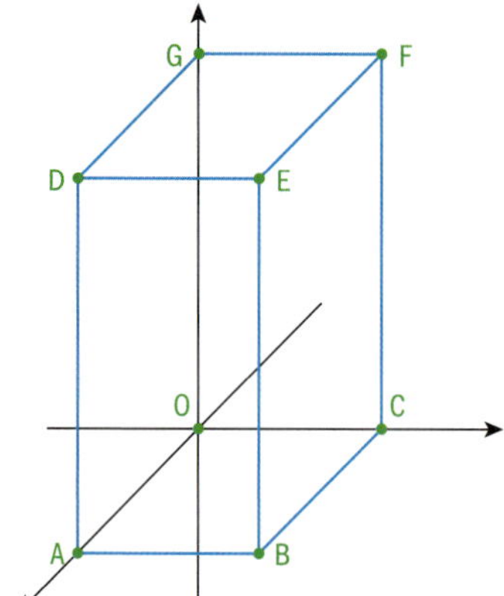

Exam–style questions

11 P1: Consider the points given by the coordinates A(1, 0, 1), B(3, 0, 0), C(4, 2, 3).

a Find the vector $\overrightarrow{AB} \times \overrightarrow{AC}$ (4 marks)

b Find an exact value for the area of triangle ABC. (3 marks)

c Show that the cartesian equation of the plane Π_1 which containing the triangle ABC is $2x-7y+4z=6$ (3 marks)

d A second plane is given by the equation $\Pi_2: 3x-5y+z=1$. Find the line of intersection of the planes Π_1 and Π_2 (5 marks)

12 P1: A tetrahedron has vertices at the points A(1, 0, 1), B(−2, 2, 3), C(0, 4,2), D(3, 1, 3) relative to a fixed point O.

Find the volume of the tetrahedron. (6 marks)

13 P1: In the given semicircle, AB is the diameter and P is a general point on the arc. O is the centre of AB.

It is also given that $\overrightarrow{OA} = \mathbf{a}$, $\overrightarrow{OB} = \mathbf{b}$ and $\overrightarrow{OP} = \mathbf{p}$.

By using the properties of the scalar product, prove that $A\hat{P}B = 90°$. (8 marks)

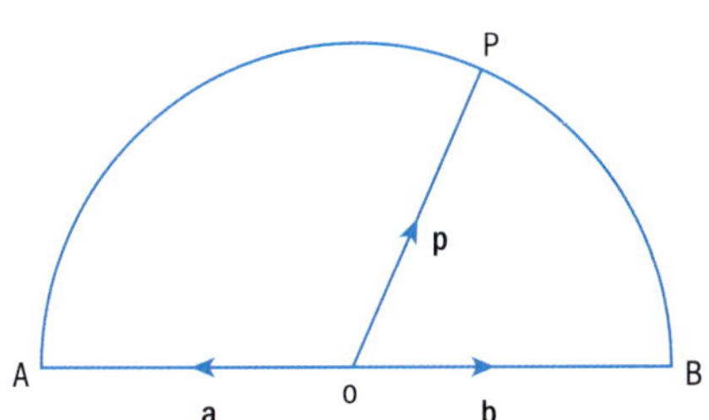

14 P1: Consider a point P(1, 0, 2) and the plane π: $4x-3y+z=19$. Point Q is such that P and Q are equidistant from π, and the line PQ is perpendicular to Π.

a Determine the coordinates of Q. (6 marks)

b Find the exact value of the distance PQ. (3 marks)

15 P1: The plane π has equation $4x+3y-z=14$ and the line L has equation

$$\mathbf{r} = \begin{pmatrix} 1 \\ 5 \\ -3 \end{pmatrix} + \lambda \begin{pmatrix} 6 \\ -2 \\ 2 \end{pmatrix}.$$

a Given that L meets π at the point P, find the coordinates of P. (4 marks)

b Find the shortest distance from the origin O to Π. (4 marks)

16 P2: Points A,B,C,D have coordinates given by A(8,2,0), B(2,0,6), C(4,4,4) and D(12, 3, 0).

a Find a vector equation of the line AB. (3 marks)

b Find a vector equation of the line CD. (3 marks)

c Hence, or otherwise, find the shortest distance between the lines AB and CD. (7 marks)

17 P1: A plane Π contains the point (5,8,0) and the line $\mathbf{r} = \begin{pmatrix} 10 \\ -4 \\ 4 \end{pmatrix} + \lambda \begin{pmatrix} 1 \\ 2 \\ 1 \end{pmatrix}$.

Find the equation of Π in the form $ax+by+cz=1$, where a, b and c are constants. (12 marks)

18 P2: A line is given by the equation $\frac{x-1}{2} = \frac{y}{5} = \frac{z-5}{p}$ and a plane is given by the equation $5x+py+pz=8$, where p is a constant.

Determine the value of p that maximises the angle between the line and the plane, and hence find the maximum acute angle between the line and the plane. Give your answer in degrees, correct to 1 decimal place. (10 marks)

19 P1: Determine whether or not the lines

$L_1: \mathbf{r} = \begin{pmatrix} 1 \\ 0 \\ 2 \end{pmatrix} + \lambda \begin{pmatrix} 3 \\ -1 \\ 1 \end{pmatrix}$ and

$L_2: \mathbf{r} = \begin{pmatrix} 2 \\ 1 \\ 1 \end{pmatrix} + \mu \begin{pmatrix} 1 \\ -1 \\ 1 \end{pmatrix}$ are skew. (9 marks)

Three squares

Approaches to learning: Research, Critical thinking

Exploration criteria: Personal engagement (C), Use of mathematics (E)

IB topic: Proof, Geometry, Trigonometry, Vectors

Modelling and investigation activity

The problem

Three identical squares with length of 1 are adjacent to one another. A line is connected from one corner of the first square to the opposite corner of the same square, another to the opposite corner of the second square and another to the opposite corner of the third square:

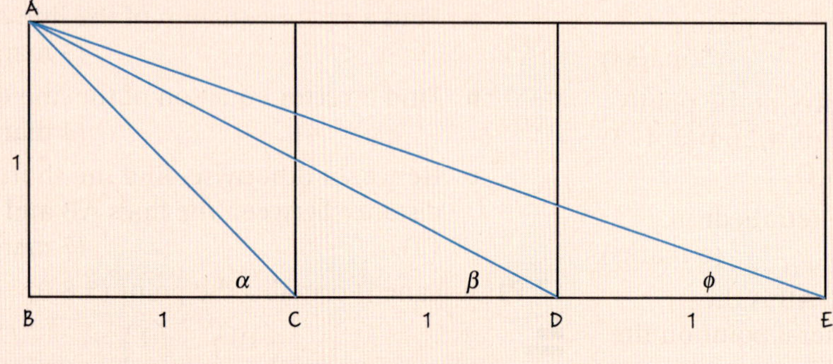

Find the sum of the three angles α, β and ϕ.

Exploring the problem

Look at the diagram

What do you think the answer may be?

Use a protractor if that helps.

How did you come to this conjecture?

Is it convincing?

This is not an accepted mathematical truth. It is a conjecture, based on observation.

You now have the conjecture $\alpha + \beta + \phi = 90°$ to be proved mathematically.

Direct proof

What is the value of α?

Given that $\alpha + \beta + \phi = 90°$, what does this tell you about α and $\beta + \phi$?

What are the lengths of the three hypotenuses of ΔABC, ΔABD and ΔABE?

Hence explain how you know that ΔACD and ΔACE are similar.

What can you therefore conclude about $C\hat{A}D$ and $C\hat{E}A$?

Hence determine why $A\hat{C}B = C\hat{A}D + A\hat{D}C$ and conclude the proof.

Proof using an auxiliary line

An additional diagonal line, CF, is drawn in the second square and the intersection point between CF and AE is labelled G:

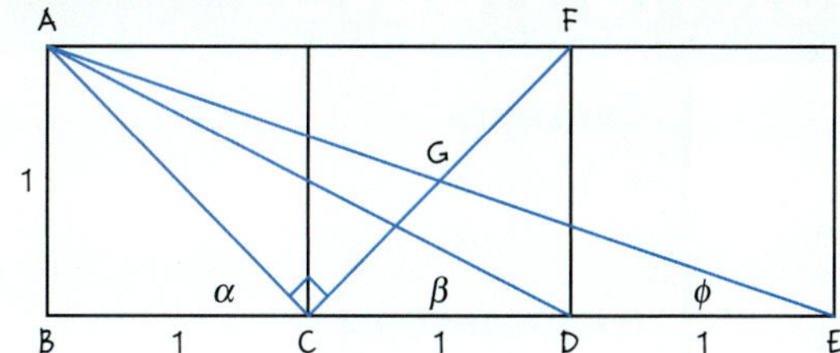

Explain why $B\hat{A}C = \alpha$.

Explain why $E\hat{A}F = \phi$.

If you show that $G\hat{A}C = \beta$, how will this complete the proof?

Explain how you know that ΔGAC and ΔABD are similar.

Hence explain how you know that $G\hat{A}C = B\hat{D}A = \beta$.

Hence complete the proof.

Proof using the cosine rule

The diagram is extended and the additional vertices of the large rectangle are labelled X and Y and the angle is labelled θ:

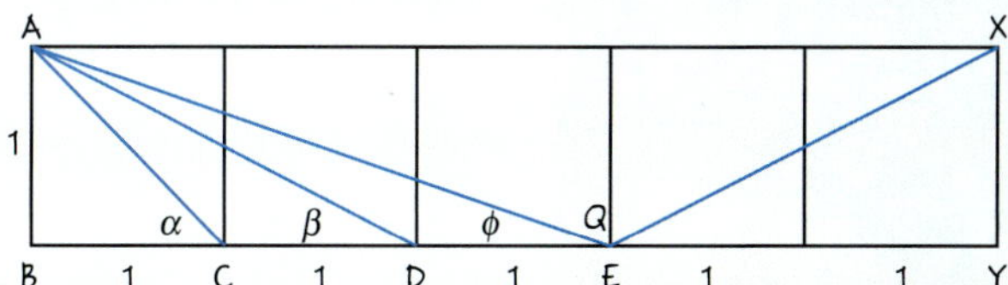

Explain why $X\hat{E}Y = \beta$.

Calculate the lengths AE and AY.

Now calculate $A\hat{E}Y$ (θ) using the cosine rule.

Hence explain how you know that $\beta + \phi = 45°$.

Hence complete the proof.

Proof using vectors

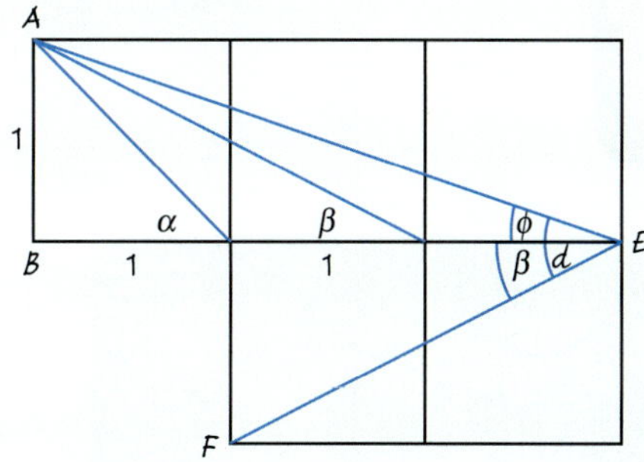

Explain why $B\hat{E}F = \beta$.

Now let $\overrightarrow{EA} = \mathbf{a}$ and $\overrightarrow{EF} = \mathbf{b}$.

Write down vectors for **a** and **b**.

Find the angle d (where $d = \phi + \beta$) using $\cos(d) = \dfrac{a \cdot b}{|a||b|}$.

Hence complete the proof.

Extension

Research other proofs on the Internet.

You could also try to produce a proof yourself.

You do not have to stop working when you have the proof.

What could you do next?

You could use the methods devised in the task in Chapter 5 on Spearman's Rank to rank the proofs and discuss the results.

10 Equivalent systems of representation: more complex numbers

In Chapter 3 you were introduced to the set of complex numbers, and you learned how to graph them on the Argand plane. In this chapter you will learn other forms of complex numbers that facilitate operations with complex numbers. You know how to graph functions with real variables on coordinate axes. Functions of complex variables are used in fields such as engineering and physics, and are graphed on what is called phasor diagrams. A phasor is a complex number that represents a sinusoidal function whose amplitude, angular frequency and initial phase are all invariant with respect to time.

Concepts

- Equivalence
- Systems

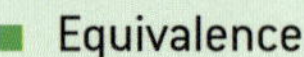

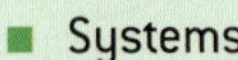

Microconcepts

- The complex plane
- Modulus-argument (polar) form
- Euler's form
- Cartesian form
- De Moivre's theorem
- Sums, products and quotients
- Geometric interpretation
- Rational exponents
- Powers and roots of complex numbers

What kinds of numbers are used in applications such as electrical engineering and film animation?

Is Euler's identity $e^{i\pi} + 1 = 0$ **"beautiful"?**

You know how to perform arithmetic operations on real numbers, and you have learned in Chapter 3 how to compute some operations on complex numbers. In particular, you have seen a very complicated way to find the powers (by using the binomial theorem) and roots (by solving simultaneous equations of higher degree) of complex numbers, but are there simpler ways to compute powers and roots of complex numbers?

- What form(s) of a complex number enable you to efficiently compute powers and roots of the number?
- How do you find different roots of the number 1? That is, how do you find roots of unity?
- When graphing roots of unity, what geometrical properties become evident?
- How can you use Maclaurin's expansion to find a series of polynomial terms for $e^{i\theta}$?

Developing inquiry skills

Think about the questions in this opening problem and answer any you can. As you work through the chapter, you will gain mathematical knowledge and skills that will help you to answer them all.

Before you start

Click here for help with this skills check

You should know how to:

1 Draw the points representing complex numbers in the Argand diagram, eg $z_1 = 1 + 2i$, $z_2 = -3 + i$ and $z_3 = 3 - 4i$

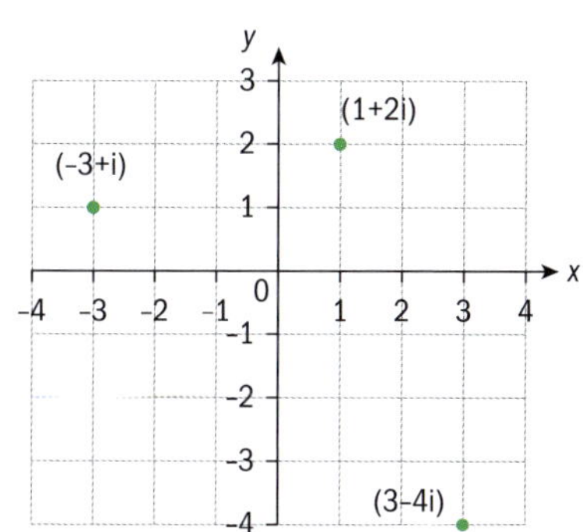

2 Identify the real and imaginary part of a complex number, eg given that

$$z = \frac{2i - 5}{7} \Rightarrow \text{Re}\, z = -\frac{5}{7} \text{ and } \text{Im}\, z = \frac{2}{7}$$

3 Performing operations with complex numbers, eg given that $z_1 = 1 + i$ and $z_2 = 2 - i$ find:

$$5z_1 - 3z_2 + z_1z_2$$
$$= 5(1+i) - 3(2-i) + (1+i)(2-i)$$
$$= 5 + 5i - 6 + 3i + 2 - i + 2i + 1$$
$$= 2 + 9i$$

4 Determine the conjugate, the opposite, the reciprocal and the magnitude of a complex number, eg given that

$z = 3 - i \Rightarrow z^* = 3 + i$, $-z = -3 + i$,

$$\frac{1}{z} = \frac{1}{3-i} \cdot \frac{3+i}{3+i} = \frac{3}{10} + \frac{1}{10}i$$

$$|z| = \sqrt{3^2 + (-1)^2} = \sqrt{10}$$

Skills check

1 Draw the points representing complex numbers in the Argand diagram:

$$z_1 = 2 + \frac{2}{3}i,\ z_2 = -\frac{3}{4} - i \text{ and } z_3 = \frac{1-3i}{2}$$

2 Identify the real and imaginary part of the complex numbers in question **1**.

3 Given that $z_1 = 1 - i$ and $z_2 = 3 + 2i$, find:

a $3z_1 - 4z_2 + 2z_1z_2$ **b** $\dfrac{z_2^2}{z_1^3}$

4 Determine the conjugate, the opposite, the reciprocal and the magnitude of the complex numbers:

a $z = 2 - 3i$ **b** $z = \dfrac{4+3i}{5}$

10.1 Forms of a complex number

In Chapter 3 you looked at some investigations regarding the modulus of a complex number. You found that the set of all the points in the complex plane with the same modulus form a circle. That circle has the centre at the origin and the radius is the modulus itself. To find the exact position of a point, you need another parameter that is the angle that the radius of the circle at the point encloses with the positive part of the x-axis.

In the following investigation you will discover a new system of positioning points in a plane with respect to the origin.

TOK

Why is it called the Argand plane and not the Wessel plane or Gaussian plane?

Investigation 1

1 Plot all of the complex numbers in the Argand plane:

$$\left\{2,\ 1+\mathrm{i},\ 3\mathrm{i},\ -\frac{1}{2}+\frac{\sqrt{3}}{2}\mathrm{i},\ -4,\ -\sqrt{3}-\mathrm{i},\ -2\mathrm{i},\ 6-2\sqrt{3}\mathrm{i}\right\}$$

2 Draw a line segment that represents the modulus of each number and find its value.

3 Find the angle θ that the line segment encloses with the positive part of the x-axis for each complex number.

4 **Conceptual** Given a complex number $z=x+y\mathrm{i}$, $x, y \in \mathbb{R}$, what is the relationship between the angle θ and the rectangular coordinates?

5 **Factual** What is the value of the angle θ if the complex number is:

a real **b** imaginary?

Your investigation leads to the polar form of a complex number.

Every complex number $z=x+y\mathrm{i}$, $x, y \in \mathbb{R}$ represents a point $Z(x, y)$ in the complex plane by using the rectangular coordinates. Each of those points can be represented by two different parameters. One parameter represents the distance to the origin $r=|z|=\sqrt{x^2+y^2}$ or **modulus**. The other parameter represents the angle with the positive part of the x-axis $\theta=\arg z=\arctan\frac{y}{x}+k\pi$, where $k=0, 1, 2$, or **argument**. Therefore $Z(r, \theta)$ represents a new type of coordinates called the polar coordinates, where $r \geq 0$, $0 \leq \theta < 2\pi$.

When plotting points with rectangular coordinates, you can choose whether you will first move from the origin horizontally and then vertically, or vice versa. When plotting points with polar coordinates, you can also choose whether you will move from the origin by first drawing the circle with the radius r and then by drawing the ray representing the angle θ, or vice versa.

Example 1

Draw the complex numbers in the coordinate system given by the modulus and argument:

a $|z| = 3$ and $\arg z = \frac{\pi}{6}$ **b** $|z| = 1$ and $\arg z = \frac{2\pi}{3}$

c $|z| = 2$ and $\arg z = \frac{7\pi}{4}$

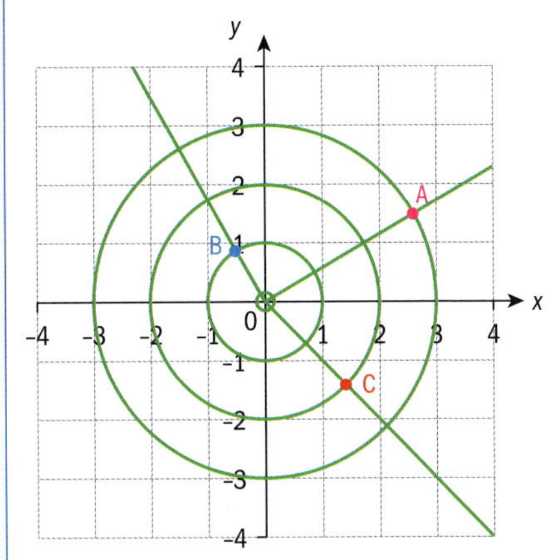

Draw the circles with the given radii.

Draw the rays from the origin that enclose the given angles with the positive part of the *x*-axis.

Find the points of intersection between the corresponding circle and the ray.

International-mindedness

Jean-Robert Argand was an 18th century amateur mathematician born in Geneva, Switzerland.

Investigation 2

1 Plot all of the complex numbers in the Argand plane given by the modulus and the argument:

$|z_1| = 3$ and $\arg z_1 = 0$, $|z_2| = \sqrt{2}$ and $\arg z_2 = \frac{\pi}{4}$, $|z_3| = 1$ and $\arg z_3 = \frac{5\pi}{6}$

$|z_4| = 6$ and $\arg z_4 = -\frac{2\pi}{3}$, $|z_5| = 5$ and $\arg z_5 = -\frac{\pi}{2}$, $|z_6| = 2\sqrt{3}$ and $\arg z_6 = -\frac{\pi}{6}$

2 Find the rectangular coordinates, and therefore the Cartesian form, of all the complex numbers.

3 **Conceptual** What is the relationship between the real and imaginary parts of a complex number and the modulus and argument of the same number?

Your investigation can be summarized into the following formulae:

$x = r\cos\theta$, $y = r\sin\theta$

You therefore obtain the polar form of a complex number.

Given a complex number with the modulus r, $r \in \mathbb{R}$, $r \geq 0$ and the argument θ, $\theta \in \mathbb{R}$ the Cartesian form is obtained by:

$$z = r\,\text{cis}\theta = r(\cos\theta + \text{i}\sin\theta) = r\cos\theta + r\,\text{i}\sin\theta.$$

The form $z = r\,\text{cis}\,\theta$ is the **modulus-argument** or **polar form** of a complex number.

Example 2

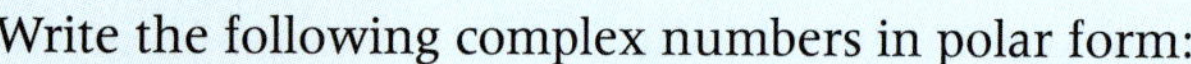

Write the following complex numbers in polar form:

a $4\sqrt{3}+4\text{i}$ **b** $-2+3\text{i}$ **c** $-12-5\text{i}$ **d** $4-2\text{i}$

a $\left\|4\sqrt{3}+4\text{i}\right\| = \sqrt{\left(4\sqrt{3}\right)^2+4^2}$ $= \sqrt{48+16} = \sqrt{64} = 8$	Find the modulus.
$\arctan\dfrac{4}{4\sqrt{3}} = \arctan\dfrac{1}{\sqrt{3}} = \dfrac{\pi}{6}$	Find the argument.
$4\sqrt{3}+4\text{i} = 8\text{ cis }\dfrac{\pi}{6}$	Convert to polar form.
b $\left\|-2+3\text{i}\right\| = \sqrt{(-2)^2+3^2} = \sqrt{13} = 3.61$	Find the modulus.
$\pi + \arctan\dfrac{3}{-2} = 2.16$	Find the argument.
$-2+3\text{i} = 3.61\text{ cis}\,2.16$	Convert to polar form.
c $\left\|-12-5\text{i}\right\| = \sqrt{144+25} = \sqrt{169} = 13$	Find the modulus.
$\pi + \arctan\dfrac{-5}{-12} = \pi + \arctan\dfrac{5}{12} = 3.54$	Find the argument.
$-12-5\text{i} = 13\text{ cis}\,3.54$	Convert to polar form.
d $\left\|4-2\text{i}\right\| = \sqrt{16+4} = \sqrt{20} = 2\sqrt{5} = 4.47$	Find the modulus.
$2\pi + \arctan\dfrac{-2}{4} = 2\pi + \arctan -\dfrac{1}{2} = 5.82$	Find the argument.
$4-2\text{i} = 4.47\text{ cis}\,5.82$	Convert to polar form.

HINT

To choose the value of θ you need to look at the position of the complex number in the Argand diagram, in which quadrant it lies and use the formula:

$\theta = \arctan\left(\dfrac{y}{x}\right) + k\pi$, $k = 0, 1, 2$

So far, you have taken the value of the argument to be $\theta \in [0, 2\pi[$. However, some textbooks and calculators take the argument to be $\theta \in]-\pi, \pi]$. It is acceptable to take both and, if it is not stated beforehand, you can use any in answering your question.

There is another form that is equivalent to polar form, which uses the same parameters r and θ but written as $z = r\text{e}^{\text{i}\theta}$, $r, \theta \in \mathbb{R}$, $r \geq 0$. This is known as **Euler's** or **exponential form** of a complex number.

In Chapter 8 we discussed the infinite Maclaurin series and, in particular, the series for the function $f(x) = \text{e}^x = \sum_{k=0}^{\infty}\dfrac{x^k}{k!}$. If you use a complex variable $x = \text{i}\theta$ instead of a real variable you obtain the following:

$$f(\text{i}\theta) = \text{e}^{\text{i}\theta} = \sum_{k=0}^{\infty}\frac{(\text{i}\theta)^k}{k!} = 1 + \text{i}\theta + \frac{(\text{i}\theta)^2}{2!} + \frac{(\text{i}\theta)^3}{3!} + \frac{(\text{i}\theta)^4}{4!} + \frac{(\text{i}\theta)^5}{5!} + \ldots$$

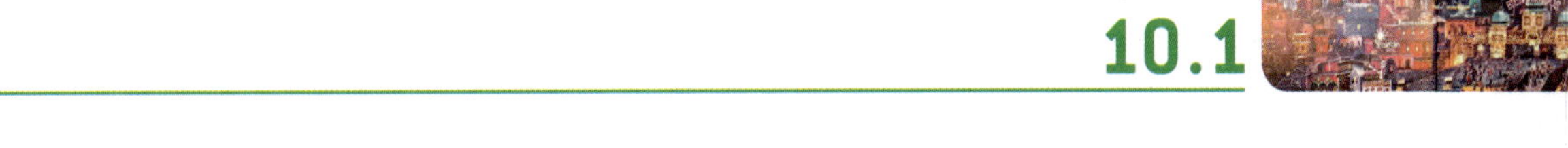

Group like terms, such as the real part and the imaginary part of the complex numbers, using the powers of imaginary unit i:

$$e^{i\theta} = \underbrace{1 - \frac{\theta^2}{2!} + \frac{\theta^4}{4!} - \ldots}_{\cos\theta} + i\left(\underbrace{\theta - \frac{\theta^3}{3!} + \frac{\theta^5}{5!} - \ldots}_{\sin\theta}\right) = \cos\theta + i\sin\theta$$

To obtain the full form, multiply by the modulus, r.

HINT

Since these two forms $r\,\text{cis}\theta = r\,e^{i\theta}$ use the same parameters r and θ, they are sometimes referred to by the same name.

On a GDC, when the document is set in degrees, the polar form will give you the modulus and the argument in degrees. You can see that by checking the results of the previous example.

(4·√3 +4·i)▸Polar	(8. ∠ 30.)
(-2+3·i)▸Polar	(3.60555 ∠ 123.69)
(-12-5·i)▸Polar	(13 ∠ -157.38)
(4-2·i)▸Polar	(4.47214 ∠ -26.5651)

If the document is set in radians, the same feature provides Euler's form.

(4·√3 +4·i)▸Polar	$e^{0.523599\cdot i}\cdot 8.$
(-2+3·i)▸Polar	$e^{2.1588\cdot i}\cdot 3.60555$
(-12-5·i)▸Polar	$e^{-2.7468\cdot i}\cdot 13$
(4-2·i)▸Polar	$e^{-0.463648\cdot i}\cdot 4.47214$

Notice that on the calculator the given angles $\theta \in\,]-\pi, \pi]$.

Using the general formulae for any angle, you can define when two complex numbers in polar form or Euler's form are **equal**.

Given $z_1 = r_1\text{cis}\theta_1 = r_1 e^{i\theta_1}$ and $z_2 = r_2\text{cis}\theta_2 = r_2 e^{i\theta_2}$, if $z_1 = z_2$ then $r_1 = r_2$ and $\theta_1 = \theta_2 + 2k\pi,\ k \in \mathbb{Z}$.

Therefore, two complex numbers in polar form are equal if their moduli are equal and their arguments are equal or differ for a multiple of 2π.

Reflect Why do the arguments differ for a multiple of 2π?

Example 3

Write the following complex numbers in Cartesian form. Check your answer by using technology.

a $2\operatorname{cis}\frac{\pi}{3}$ **b** $5\mathrm{e}^{\mathrm{i}\frac{3\pi}{4}}$ **c** $8\operatorname{cis}\frac{11\pi}{6}$ **d** $\mathrm{e}^{\mathrm{i}\frac{17\pi}{12}}$

a $z = 2\cos\frac{\pi}{3} + 2\mathrm{i}\sin\frac{\pi}{3}$

$= 2\times\frac{1}{2} + 2\mathrm{i}\frac{\sqrt{3}}{2} = 1 + \sqrt{3}\,\mathrm{i}$

Apply the formula: $z = r\cos\theta + r\mathrm{i}\sin\theta$.

b $z = 5\cos\frac{3\pi}{4} + 5\mathrm{i}\sin\frac{3\pi}{4}$

$= 5\times\left(-\frac{\sqrt{2}}{2}\right) + 5\mathrm{i}\frac{\sqrt{2}}{2} = -\frac{5\sqrt{2}}{2} + \frac{5\sqrt{2}}{2}\mathrm{i}$

Apply the formula: $z = r\cos\theta + r\mathrm{i}\sin\theta$.

c $z = 8\cos\frac{11\pi}{6} + 8\mathrm{i}\sin\frac{11\pi}{6}$

$= 8\times\frac{\sqrt{3}}{2} + 8\mathrm{i}\left(-\frac{1}{2}\right) = 4\sqrt{3} - 4\mathrm{i}$

Apply the formula: $z = r\cos\theta + r\mathrm{i}\sin\theta$.

d $z = \cos\frac{17\pi}{12} + \mathrm{i}\sin\frac{17\pi}{12}$

$= -0.259 - 0.966\mathrm{i}$

Apply the formula: $z = r\cos\theta + r\mathrm{i}\sin\theta$.

(2 ∠ 60)	1.+1.73205· *i*
(5 ∠ 135)	-3.53553+3.53553· *i*
(8 ∠ 330)	6.9282-4.· *i*
(1 ∠ 255)	-0.258819-0.965926· *i*

Notice that since the document was set in degree mode, the angles need to be converted to degrees. Also the exact form, if not an integer, is given in decimal form.

Due to the symmetries of the circle it is very easy to determine the polar form of an **opposite**, a **conjugate** and an **opposite-conjugate** complex number.

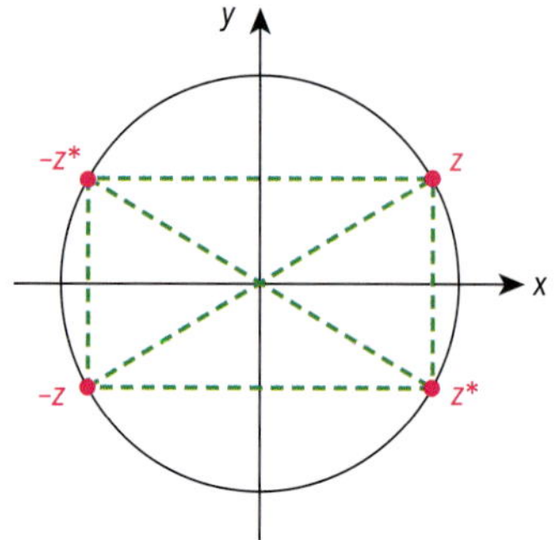

TOK

Imagination is one of the ways of knowing in TOK. How does this relate to imaginary numbers?

So by looking at the symmetries you can see that they all have equal modulus and the angles are related in the following way:

$$z = r\operatorname{cis}\theta \Rightarrow \begin{cases} -z = r\operatorname{cis}(\pi+\theta) \\ z^* = r\operatorname{cis}(\pi-\theta) \\ -z^* = r\operatorname{cis}(2\pi-\theta) \end{cases}$$

Example 4

Given the complex number $z = 3 + 3\text{i}$, find z^* and $-z^*$ in polar form.

$r = \sqrt{3^2+3^2} = \sqrt{18} = 3\sqrt{2}$	First convert to polar form.
$\theta = \arctan\frac{3}{3} = \arctan 1 = \frac{\pi}{4}$	
$\Rightarrow z = 3+3\text{i} = 3\sqrt{2}\operatorname{cis}\frac{\pi}{4}$	
$\Rightarrow z^* = 3\sqrt{2}\operatorname{cis}\left(\pi - \frac{\pi}{4}\right) = 3\sqrt{2}\operatorname{cis}\frac{3\pi}{4}$	Use the formula for conjugate complex number in polar form.
$\Rightarrow -z^* = 3\sqrt{2}\operatorname{cis}\left(2\pi - \frac{\pi}{4}\right) = 3\sqrt{2}\operatorname{cis}\frac{7\pi}{4}$	Use the formula for opposite conjugate complex number in polar form.

Exercise 10A

1 Draw the complex numbers in the coordinate system given by the modulus and argument:

a $|z| = 1$ and $\arg z = \pi$

b $|z| = 2$ and $\arg z = \frac{\pi}{2}$

c $|z| = 5$ and $\arg z = \frac{5\pi}{3}$

d $|z| = 3$ and $\arg z = \frac{5\pi}{6}$

e $|z| = \frac{1}{2}$ and $\arg z = \frac{5\pi}{4}$

f $|z| = \frac{4}{5}$ and $\arg z = \frac{3\pi}{2}$

2 Find the following complex numbers in polar form and check your answers by GDC.

a $2+2\text{i}$ **b** $\frac{3}{2}\text{i}$ **c** $-4-3\text{i}$

d $21-20\text{i}$ **e** $-1+\sqrt{3}\text{i}$ **f** $-\frac{4}{3}\text{i}$

g $\frac{\sqrt{2}}{3} - \frac{\sqrt{2}}{4}\text{i}$

3 Find the Cartesian form of the complex numbers in question **1** and check your answers by GDC.

4 Given the complex number $z = \frac{7}{12}e^{\text{i}\frac{\pi}{9}}$, find in polar form:

a $-z$ **b** z^* **c** $-z^*$

10.2 Operations with complex numbers in polar form

Some operations are simpler to conduct in polar form, but some are not.

Addition

Complex numbers can be seen as vectors in two dimensions. Therefore, addition can be executed through the parallelogram law for adding vectors.

For example, if $z_1 = 4 + 2\text{i}$ and $z_2 = -1 + 3\text{i}$, then the sum is $z_3 = 3 + 5\text{i}$, as seen in the diagram below.

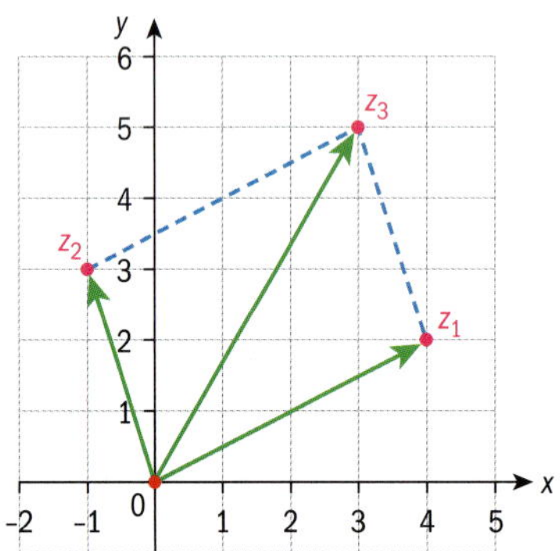

HINT

Some geometrical properties of quadrilaterals in relations to addition of complex numbers in polar form can be a topic for an exploration.

In modulus-argument form, it is very difficult to find the relationship between two moduli and the modulus of the sum. In the example you find:

$r_1 = \sqrt{20}$, $r_2 = \sqrt{10}$ and $r_3 = \sqrt{34}$

The same conclusion works for the arguments:

$\theta_1 = 0.464$, $\theta_2 = 1.89$ and $\theta_3 = 1.03$

Only when two non-collinear vectors form a rectangle (or a square) can you establish some relationships between the moduli and arguments by using the geometrical properties.

Multiplication by a scalar

As seen in Example 3, modulus is a non-negative real number and $\text{cis}\theta$ is a complex number that has a modulus equal to 1.

In Chapter 3 you discovered how to multiply a complex number by a real number, so now you can state the properties of multiplication of a complex number in polar form by a real number:

$$z = r\,\text{cis}\theta \Rightarrow \lambda z = \begin{cases} (\lambda r)\,\text{cis}\,\theta, \lambda > 0 \\ 0, \lambda = 0 \\ (|\lambda| r)\,\text{cis}(\pi + \theta), \lambda < 0 \end{cases}$$

HINT

What is the geometrical transformation?

In combination with addition this could also be a topic for an exploration.

When a complex number is written in polar form it is very simple to describe geometrical transformation of the plane when multiplying by a scalar (real number).

Multiplication of complex numbers in polar form

Investigation 3

1 Copy and complete the following table using your own examples.

z_1	z_2	z_1z_2	$\lvert z_1\rvert$	$\lvert z_2\rvert$	$\lvert z_1z_2\rvert$	$\arg z_1$	$\arg z_2$	$\arg z_1z_2$
-2	$3\mathrm{i}$							
-5	$-2\mathrm{i}$							
$1+\mathrm{i}$	$\sqrt{3}+\mathrm{i}$							
$3-4\mathrm{i}$	$-1-\sqrt{2}\mathrm{i}$							
$-2\sqrt{3}+2\mathrm{i}$	$3\sqrt{2}-3\sqrt{2}\mathrm{i}$							

2 What is the relationship between the moduli of the factors and the modulus of the product?

3 What is the relationship between the arguments of the factors and the argument of the product?

Your investigation can be summarized into the following:

Given $z_1 = r_1\text{cis}\theta_1$ and $z_2 = r_2\text{cis}\theta_2$, if $z_1 \times z_2 = z_3 = r_3\text{cis}\theta_3$ then $r_3 = r_1r_2$ and $\theta_3 = \theta_1 + \theta_2 + 2k\pi,\ k \in \mathbb{Z}$.

TOK

Was the complex plane already there before it was used to represent complex numbers geometrically?

You multiply two complex numbers in polar form by **multiplying the moduli and adding the arguments.**

Example 5

Given that $z_1 = r_1\text{cis}\theta_1$ and $z_2 = r_2\text{cis}\theta_2$, show that $r_3 = r_1r_2$ and $\theta_3 = \theta_1 + \theta_2 + 2k\pi,\ k \in \mathbb{Z}$.

$z_1z_2 = (r_1\text{cis}\theta_1)(r_2\text{cis}\theta_2)$ $= (r_1r_2)(\cos\theta_1 + \mathrm{i}\sin\theta_1)(\cos\theta_2 + \mathrm{i}\sin\theta_2)$	Rewrite the complex part in rectangular form.
$= (r_1r_2)\left(\cos\theta_1\cos\theta_2 + \mathrm{i}\sin\theta_1\cos\theta_2 + \cos\theta_1\mathrm{i}\sin\theta_2 + \underbrace{\mathrm{i}^2}_{-1}\sin\theta_1\sin\theta_2\right)$	Use distribution to multiply the complex parts.
$= (r_1r_2)\left(\underbrace{(\cos\theta_1\cos\theta_2 - \sin\theta_1\sin\theta_2)}_{\cos(\theta_1+\theta_2)} + \mathrm{i}\left(\underbrace{\sin\theta_1\cos\theta_2 + \cos\theta_1\sin\theta_2}_{\sin(\theta_1+\theta_2)}\right)\right)$	Use trigonometric identities for double angles.
$= (r_1r_2)(\cos(\theta_1 + \theta_2) + \mathrm{i}\sin(\theta_1 + \theta_2))$ $r_3 = r_1r_2$ and $\theta_3 = \theta_1 + \theta_2 + 2k\pi,\ k \in \mathbb{Z}$	Use the equality of two complex numbers in polar form.

The proof is even easier if you use Euler's form: $z_1z_2 = \left(r_1\mathrm{e}^{\mathrm{i}\theta_1}\right)\left(r_2\mathrm{e}^{\mathrm{i}\theta_2}\right) = r_1r_2\mathrm{e}^{\mathrm{i}(\theta_1+\theta_2)}$

Example 6

Multiply the following complex numbers in Euler's form and polar form.

a $z_1 = 3\,\mathrm{e}^{\mathrm{i}\frac{2\pi}{3}}$ and $z_2 = 5\,\mathrm{e}^{\mathrm{i}\frac{3\pi}{4}}$ **b** $z_3 = 11\operatorname{cis}210°$ and $z_4 = 23\operatorname{cis}315°$

a $z_1 \times z_2 = 3\,\mathrm{e}^{\mathrm{i}\frac{2\pi}{3}} \times 5\,\mathrm{e}^{\mathrm{i}\frac{3\pi}{4}}$ $= 15\,\mathrm{e}^{\mathrm{i}\left(\frac{2\pi}{3}+\frac{3\pi}{4}\right)} = 15\,\mathrm{e}^{\mathrm{i}\frac{17\pi}{12}}$	Apply the formula for multiplication in Euler's form.
b $z_3 \times z_4 = 11\operatorname{cis}210° \times 23\operatorname{cis}315°$ $= 253\operatorname{cis}(210° + 315°) = 253\operatorname{cis}525°$ $= 253\operatorname{cis}165°$	Apply the formula for multiplication in polar form. Reduce the sum of the arguments by 360 degrees.

With the formula for multiplying two numbers, you can identify two transformations of the plane. If the first number is represented by a point, eg A, and the product is represented by the second point, eg A′, then the second number represents the composition of two transformations:

- enlargement by the scale factor of the modulus of the second number
- rotation about the origin by the angle, argument of the second number.

The multiplication of complex numbers in different forms can sometimes help you to find exact values of some angles.

HINT

This geometrical representation of multiplication can be a topic for an exploration.

Example 7

Given the numbers $z_1 = 4\operatorname{cis}120°$ and $z_2 = 3 + 3\mathrm{i}$, find:

a z_1 in Cartesian form **b** z_2 in polar form **c** $z_1 \times z_2$ in both forms.

Hence, find the exact value of:

d $\cos 165°$ **e** $\tan 165°$.

a $z_1 = 4\operatorname{cis}120°$ $= 4\cos 120° + 4\mathrm{i}\sin 120°$ $= 4\left(-\frac{1}{2}\right) + 4 \times \frac{\sqrt{3}}{2}\mathrm{i} = -2 + 2\sqrt{3}\mathrm{i}$	Use the formula for converting to Cartesian form. Write z_1 in Cartesian form.
b $z_2 = 3 + 3\mathrm{i} \Rightarrow r = \sqrt{3^2 + 3^2} = 3\sqrt{2}$ $\theta = \arctan\frac{3}{3} = \arctan 1 = 45°$ $z_2 = 3\sqrt{2}\operatorname{cis}45°$	First find the modulus and then find the argument. Write z_2 in polar form.

c $z_1 \times z_2 = 4\,\text{cis}\,120° \times 3\sqrt{2}\,\text{cis}\,45°$ $= 12\sqrt{2}\,\text{cis}\,165°$	Find the product in polar form.
$z_1 \times z_2 = \left(-2 + 2\sqrt{3}\text{i}\right)(3 + 3\text{i})$ $= -6 - 6\text{i} + 6\sqrt{3}\text{i} - 6\sqrt{3}$ $= -6\left(1 + \sqrt{3}\right) + 6\left(\sqrt{3} - 1\right)\text{i}$	Find the product in Cartesian form.
d $12\sqrt{2}\cos 165° = -6\left(1 + \sqrt{3}\right)$ $\cos 165° = \frac{-\left(1+\sqrt{3}\right)}{2\sqrt{2}} \times \frac{\sqrt{2}}{\sqrt{2}} = \frac{-\sqrt{2} - \sqrt{6}}{4}$	Use the real part of the polar form and equate it with real part of the Cartesian form. Rationalize the denominator.
e First you need to find the sine value so that you can find the tangent value.	
$12\sqrt{2}\sin 165° = 6\left(\sqrt{3} - 1\right)$ $\sin 165° = \frac{\sqrt{3}-1}{2\sqrt{2}} \times \frac{\sqrt{2}}{\sqrt{2}} = \frac{\sqrt{6} - \sqrt{2}}{4}$	Use the imaginary part of the polar form and equate it with imaginary part of the Cartesian form.
$\tan 165° = \frac{\frac{\sqrt{6}-\sqrt{2}}{4}}{\frac{-\sqrt{6}-\sqrt{2}}{4}} = \frac{\sqrt{2}\left(\sqrt{3}-1\right)}{-\sqrt{2}\left(\sqrt{3}+1\right)} \times \frac{\sqrt{3}-1}{\sqrt{3}-1}$	Find the tangent value. Rationalize the denominator.
$= \frac{\left(\sqrt{3}-1\right)}{-\left(\sqrt{3}+1\right)} \times \frac{\sqrt{3}-1}{\sqrt{3}-1} = -\frac{3 - 2\sqrt{3} + 1}{2} = -2 + \sqrt{3}$	

Exercise 10B

1 Multiply the complex numbers in Euler's or polar form.

a $z_1 = 2\,\text{e}^{\text{i}\frac{\pi}{3}}$ and $z_2 = 4\,\text{e}^{\text{i}\frac{\pi}{4}}$

b $z_3 = 5\,\text{cis}\,90°$ and $z_4 = 6\,\text{cis}\,45°$

c $z_5 = \frac{2}{3}\,\text{e}^{\text{i}\frac{11\pi}{7}}$ and $z_6 = \frac{5}{6}\,\text{e}^{\text{i}\frac{23\pi}{14}}$

d $z_7 = \frac{5}{6}\text{cis}\,220°$ and $z_8 = \frac{6}{5}\,\text{cis}\,275°$

2 Given the numbers

$z_1 = \text{cis}\,\frac{3\pi}{4}$ and $z_2 = -\frac{1}{2} + \frac{\sqrt{3}}{2}\text{i}$, find:

a z_1 in Cartesian form

b z_2 in polar form

c $z_1 \times z_2$ in both forms.

Hence find the exact values of:

d $\sin\frac{17\pi}{12}$ **e** $\tan\frac{17\pi}{12}$.

3 Given the number $z = re^{i\theta}$, find the values of r and θ so that:

a $z\left(\sqrt{3}-i\right)$ is a real number less than 3

b $z(-1 + i)$ is an imaginary number with a modulus greater than 4.

4 Calculate the following product:

$$\left(\sin\frac{\pi}{12}+i\cos\frac{\pi}{12}\right)\left(\sin\frac{\pi}{6}+i\cos\frac{\pi}{6}\right)\times\left(\sin\frac{\pi}{4}+i\cos\frac{\pi}{4}\right)$$

Investigation 4

1 Copy and complete the following table using your own examples, $z \neq 0$:

z	$\frac{1}{z}$	$\lvert z\rvert$	$\left\lvert\frac{1}{z}\right\rvert$	$\arg z$	$\arg\frac{1}{z}$
-3					
$4i$					
$1-i$					
$-3+4i$					
$-3+3\sqrt{3}i$					

2 What is the relationship between the modulus of the number and the modulus of the reciprocal number?

3 What is the relationship between the argument of the number and the argument of the reciprocal number?

Your investigation can be summarized in the following:

Given the complex number in polar form $z = r\operatorname{cis}\theta$, then the reciprocal value in polar form is $\frac{1}{z} = \frac{1}{r}\operatorname{cis}\left(2\pi-\theta\right) = \frac{1}{r}\operatorname{cis}(-\theta)$.

The **reciprocal complex number** of a given complex number in a polar form has the **reciprocal modulus** and **the opposite argument**.

The proof of the polar form of the reciprocal complex number can be done by using the properties of multiplication.

Example 8

Given the complex number in polar form $z = r\operatorname{cis}\theta$ and its reciprocal value in polar form $\frac{1}{z} = \rho\operatorname{cis}\phi$, find the relationship between moduli and arguments.

$z \times \frac{1}{z} = 1 \Rightarrow r \operatorname{cis}\theta \; \rho \operatorname{cis}\phi = 1$	The product of a number and its reciprocal is 1.
$\Rightarrow r\rho \operatorname{cis}(\theta + \phi) = 1 \operatorname{cis} 0$	Use the formula for product in polar form and rewrite 1 in polar form.
$\Rightarrow \begin{cases} r\rho = 1 \\ \theta + \phi = 2k\pi \end{cases} \Rightarrow \begin{cases} \rho = \frac{1}{r} \\ \phi = 2k\pi - \theta, k \in \mathbb{Z} \end{cases}$	Two complex numbers in polar form are equal if their moduli are equal and if their arguments are opposite or add up to a multiple of 2π.

To simplify, you can say that the **reciprocal** value of a **complex number** is a complex number with the **reciprocal moduli** and the **opposite argument**.

Again the proof can be very easily conducted by Euler's form:

$$z = r\,e^{i\theta} \Rightarrow \frac{1}{z} = \frac{1}{r\,e^{i\theta}} = \frac{1}{r}\,e^{-i\theta}$$

TOK

Do the words imaginary and complex make the concepts more difficult than if they had different names?

Example 9

Given the complex number in polar form $z = 2\operatorname{cis}\frac{\pi}{3}$ find its reciprocal value in Cartesian form.

Method 1	
$\frac{1}{z} = \frac{1}{2}\operatorname{cis}\left(-\frac{\pi}{3}\right) = \frac{1}{2}\left(\cos\left(-\frac{\pi}{3}\right) + i\sin\left(-\frac{\pi}{3}\right)\right)$ $= \frac{1}{2}\left(\frac{1}{2} - \frac{\sqrt{3}}{2}i\right) = \frac{1}{4} - \frac{\sqrt{3}}{4}i$	Use the formula for finding the reciprocal number in polar form. Rewrite in Cartesian form.
Method 2	
$z = 2\operatorname{cis}\frac{\pi}{3} = 2\left(\cos\frac{\pi}{3} + i\sin\frac{\pi}{3}\right)$ $= 2\left(\frac{1}{2} + \frac{\sqrt{3}}{2}i\right) = 1 + \sqrt{3}i$	Rewrite the number in Cartesian form.
$\Rightarrow \frac{1}{1+\sqrt{3}i} \times \frac{1-\sqrt{3}i}{1-\sqrt{3}i} = \frac{1-\sqrt{3}i}{4} = \frac{1}{4} - \frac{\sqrt{3}}{4}i$	Find its reciprocal value. Multiply both numerator and denominator by the conjugate.

Division of complex numbers in polar form

Now that you know how to multiply two numbers in polar form and how to find the reciprocal value of a complex number, you do not need to define operation division separately. As for real numbers, division can be defined as multiplication by the reciprocal value.

$$\frac{z_1}{z_2} = z_1 \times \frac{1}{z_2} = r_1 \text{ cis } \theta_1 \times \frac{1}{r_2 \text{ cis}\theta_2} = r_1 \text{ cis } \theta_1 \times \frac{1}{r_2} \text{cis}(-\theta_2) = \frac{r_1}{r_2}\text{cis}(\theta_1 - \theta_2)$$

Given the two complex numbers $z_1 = r_1 \text{ cis } \theta_1$ and $z_2 = r_2 \text{ cis } \theta_2$, $z_2 \neq 0$,

then $\frac{z_1}{z_2} = \frac{r_1}{r_2}\text{cis}(\theta_1 - \theta_2)$, $z_2 \neq 0, r_2 \neq 0$.

Divide two complex numbers in polar form by dividing the moduli and subtracting the arguments.

If the numbers are given in Euler's form you obtain $\frac{z_1}{z_2} = \frac{r_1 e^{i\theta_1}}{r_2 e^{i\theta_2}} = \frac{r_1}{r_2} e^{i(\theta_1 - \theta_2)}$.

Example 10

Given the complex numbers $z_1 = 6\text{cis}\frac{3\pi}{4}$ and $z_2 = 2\text{cis}\frac{2\pi}{3}$, find:

a $\frac{z_1}{z_2}$ **b** $\frac{z_2{}^*}{-z_1}$

a $\frac{z_1}{z_2} = \frac{6\text{cis}\frac{3\pi}{4}}{2\text{cis}\frac{2\pi}{3}} = \frac{6}{2}\text{cis}\left(\frac{3\pi}{4} - \frac{2\pi}{3}\right) = 3\text{cis}\frac{\pi}{12}$	Use the formula for division of two complex numbers in polar form.
b $\frac{z_2{}^*}{-z_1} = \frac{2\text{cis}\left(\pi - \frac{2\pi}{3}\right)}{6\text{cis}\left(\pi + \frac{3\pi}{4}\right)} = \frac{2\text{cis}\left(\frac{\pi}{3}\right)}{6\text{cis}\left(\frac{7\pi}{4}\right)}$	Write the polar form of the conjugate and opposite number.
$= \frac{2}{6}\text{cis}\left(\frac{\pi}{3} - \frac{7\pi}{4}\right) = \frac{1}{3}\text{cis}\left(-\frac{17\pi}{12}\right) = \frac{1}{3}\text{cis}\frac{7\pi}{12}$	Use the formula for division of two complex numbers in polar form and add 2π to the argument.

As with multiplication, division of complex numbers in different forms helps you to find exact values of some angles.

Example 11

Given the numbers $z_1 = e^{i\frac{\pi}{4}}$ and $z_2 = \frac{\sqrt{3}}{2} + \frac{1}{2}i$, find:

a z_1 in Cartesian form **b** z_2 in Euler's form **c** $\frac{z_1}{z_2}$ in both forms.

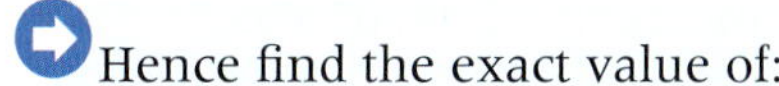

Hence find the exact value of:

d $\sin\frac{\pi}{12}$

a $z_1 = \text{cis}\,\frac{\pi}{4}$	Write the number in polar form and then use the formula to convert to Cartesian form.
$= \cos\frac{\pi}{4} + \sin\frac{\pi}{4}\text{i} = \frac{\sqrt{2}}{2} + \frac{\sqrt{2}}{2}\text{i}$	Write z_1 in Cartesian form.
b $z_2 = \frac{\sqrt{3}}{2} + \frac{1}{2}\text{i} \Rightarrow r = \sqrt{\left(\frac{\sqrt{3}}{2}\right)^2 + \left(\frac{1}{2}\right)^2} = 1$	First find the modulus and then find the argument.
$\theta = \arctan\frac{\frac{1}{2}}{\frac{\sqrt{3}}{2}} = \arctan\frac{1}{\sqrt{3}} = \frac{\pi}{6}$	
$z_2 = \text{e}^{\text{i}\frac{\pi}{6}}$	Write z_2 in Euler's form.
c $\frac{z_1}{z_2} = \frac{\text{e}^{\text{i}\frac{\pi}{4}}}{\text{e}^{\text{i}\frac{\pi}{6}}} = \text{e}^{\text{i}\left(\frac{\pi}{4} - \frac{\pi}{6}\right)} = \text{e}^{\text{i}\frac{\pi}{12}}$	Find the quotient in Euler's form.
$\frac{z_1}{z_2} = \frac{\frac{\sqrt{2}}{2} + \frac{\sqrt{2}}{2}\text{i}}{\frac{\sqrt{3}}{2} + \frac{1}{2}\text{i}} \times \frac{\frac{\sqrt{3}}{2} - \frac{1}{2}\text{i}}{\frac{\sqrt{3}}{2} - \frac{1}{2}\text{i}}$	Find the quotient in Cartesian form.
$= \frac{\sqrt{6} + \sqrt{2} + \left(\sqrt{6} - \sqrt{2}\right)\text{i}}{4}$	
d $\text{e}^{\text{i}\frac{\pi}{12}} = \cos\frac{\pi}{12} + \text{i}\sin\frac{\pi}{12}$ $\Rightarrow \sin\frac{\pi}{12} = \frac{\sqrt{6} - \sqrt{2}}{4}$	Use the imaginary part of the polar form and equate it with imaginary part of the Cartesian form.

You can always check the result by using a calculator.

Exercise 10C

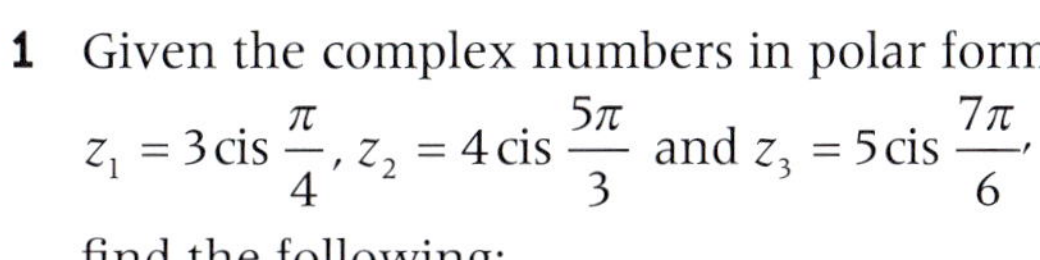

1 Given the complex numbers in polar form $z_1 = 3\,\text{cis}\,\frac{\pi}{4}$, $z_2 = 4\,\text{cis}\,\frac{5\pi}{3}$ and $z_3 = 5\,\text{cis}\,\frac{7\pi}{6}$, find the following:

a $\frac{z_1}{z_2}$ **b** $\frac{z_3{}^*}{-z_2}$ **c** $\frac{z_1}{z_2 z_3}$ **d** $\frac{-z_3{}^*}{(z_1 z_2)^*}$

2 Let $z_1 = 1 + \text{i}$ and $z_2 = 1 - \sqrt{3}\text{i}$. Find the following in Euler's form:

a $\dfrac{z_1}{z_2}$ **b** $\dfrac{-z_2{}^*}{z_1}$ **c** $\dfrac{1}{z_1 z_2}$ **d** $\dfrac{-z_1{}^*}{(z_1 z_2)^*}$

3 Express, in polar form:

a $\dfrac{3}{2+2\text{i}}$ **b** $\dfrac{4-4\text{i}}{-1+\sqrt{3}\text{i}}$ **c** $\dfrac{\sqrt{15}-\sqrt{5}\text{i}}{\sqrt{2}+\sqrt{6}\text{i}}$

4 Given the numbers $z_1 = 5 \operatorname{cis} 60°$ and $z_2 = 3 + 3\text{i}$, find:

a z_1 in Cartesian form **b** z_2 in polar form

c $\dfrac{z_1}{z_2}$ in both forms.

Hence find the exact value of:

d $\cos 15°$ **e** $\sin 15°$ **f** $\tan 15°$

10.3 Powers and roots of complex numbers in polar form

Powers of complex numbers in polar form

Investigation 5

1. Let $z = r \operatorname{cis}\theta$. Use the formula for multiplying two complex numbers in polar form to obtain expressions for: z^2, z^3, z^4 and z^5.
2. Write the general expression for $n \in \mathbb{Z}^+$.
3. **Conceptual** What is the relationship between the modulus and argument of the original number and the modulus and argument of its power?
4. **Factual** Does the formula work for $n = 0$? Justify your answer.
5. **Factual** Does the formula work for $n \in \mathbb{Z}^-$? Justify your answer.

Your investigation can be summarized in the following formula.

De Moivre's theorem

Given a complex number in polar or Euler's form $z = r \operatorname{cis}\theta = r\text{e}^{\text{i}\theta}$, $z \neq 0$, $r \in \mathbb{R}^+$, $\theta \in \mathbb{R}$ then $z^n = r^n \operatorname{cis} n\theta = r^n \text{e}^{\text{i}n\theta}$, $n \in \mathbb{Z}$.

TOK

Why might it be said that $\text{e}^{\text{i}\pi} + 1 = 0$ is beautiful?

International-mindedness

French mathematician Abraham de Moivre (1667–1754) had notable colleagues Isaac Newton and Edmond Halley. He developed a formula that links complex numbers and trigonometry, and also worked on the normal distribution and probability theory.

Example 12

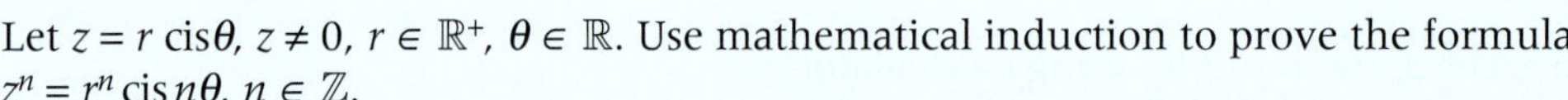

Let $z = r \operatorname{cis}\theta$, $z \neq 0$, $r \in \mathbb{R}^+$, $\theta \in \mathbb{R}$. Use mathematical induction to prove the formula $z^n = r^n \operatorname{cis} n\theta$, $n \in \mathbb{Z}$.

The proof will be conducted in two parts. The first part will be the proof by mathematical induction for all natural numbers and the second part will be a direct proof for all negative integers by using the previously proved statement. **Part I** $P(n)$: $z^n = r^n \operatorname{cis} n\theta$, $n \in \mathbb{N}$	Conduct the proof by mathematical induction in four stages.

$n = 0 \Rightarrow P(0)$: $z^0 = r^0\text{cis}(0 \times \theta)$	**Base**
$\Rightarrow 1 = 1\left(\underbrace{\cos 0}_{1} + \text{i}\underbrace{\sin 0}_{0}\right) \Rightarrow 1 = 1$	Show that the statement is true for the smallest natural number.
Let $P(k)$ be true for some value of $k \geq 0$.	**Assumption**
$n = k \Rightarrow P(k)$: $z^k = r^k\text{cis}\, k\theta$	
When $n = k + 1 \Rightarrow P(k+1)$: $z^{k+1} = z^k \times z$	**Inductive Step**
$= r^k\text{cis}\, k\theta \times r\,\text{cis}\,\theta = r^k \times r\ \text{cis}(k\theta + \theta)$	Use the statement from the assumption.
$= r^{k+1}\ \text{cis}((k+1)\theta)$	The statement is also true for $n = k + 1$.
Since $P(0)$ is true and $P(k) \Rightarrow P(k+1)$, by the principle of mathematical induction $P(n)$ is true for all $n \in \mathbb{N}$.	**Conclusion**
Part II	
$z^n = r^n\ \text{cis}\, n\theta$, $n \in \mathbb{Z}^-$	
Let $n = (-1) \times m$, $m \in \mathbb{Z}^+$.	Every negative integer can be written as a product of −1 and a positive integer.
$z^n = z^{-1\times m} = \left(\frac{1}{z}\right)^m = \left(\frac{1}{r}\text{cis}(-\theta)\right)^m$	Use the polar form of the reciprocal complex number.
$= \left(\frac{1}{r}\right)^m \text{cis}(m \times (-\theta)) = r^{-m}\text{cis}(-m\theta)$	Use the statement from Part I restricting it for $m \in \mathbb{Z}^+$.
$= r^n\ \text{cis}\, n\theta$, $n \in \mathbb{Z}^-$	

It is much simpler to prove this formula if you are using Euler's form and the properties of an exponential function.

$z^n = (r\,\text{e}^{\text{i}\theta})^n = r^n(\text{e}^{\text{i}\theta})^n = r^n\text{e}^{\text{i}n\theta}$

Example 13

Use De Moivre's theorem to calculate $\left(2 - 2\sqrt{3}\text{i}\right)^5$.

$r = \sqrt{2^2 + \left(-2\sqrt{3}\right)^2} = \sqrt{4+12} = 4$	First you need to find z in Euler's form.
$\theta = \arctan\frac{-2\sqrt{3}}{2} = -\frac{\pi}{3}$	
$z = 4\ \text{e}^{-\frac{\pi}{3}\text{i}} \Rightarrow z^5 = 4^5\text{e}^{-\frac{5\pi}{3}\text{i}}$	Apply De Moivre's theorem.
$= 1024\ \text{e}^{\text{i}\frac{\pi}{3}} = 1024\left(\frac{1}{2} + \frac{\sqrt{3}}{2}\text{i}\right)$	Simplify the argument.
$= 512 + 512\sqrt{3}\text{i}$	Write the number in Cartesian form and simplify it.

You can check your result by using a calculator.

Example 14

Let $z = \frac{1}{2} + \frac{\sqrt{3}}{2}\text{i}$.

a Find all the powers z^n where $n \in \mathbb{Z}^+$.

b Draw the complex numbers in the Argand diagram.

a $r = \sqrt{\left(\frac{1}{2}\right)^2 + \left(\frac{\sqrt{3}}{2}\right)^2} = 1$

Rewrite the complex number in polar form.

$$\theta = \arctan \frac{\frac{\sqrt{3}}{2}}{\frac{1}{2}} = \arctan \sqrt{3} = \frac{\pi}{3}$$

$z = \text{cis}\,\frac{\pi}{3} \Rightarrow z^2 = \text{cis}\,\frac{2\pi}{3}$,

Use De Moivre's theorem to find all the powers.

$z^3 = \text{cis}\,\frac{3\pi}{3} = \text{cis}\,\pi = -1$, $z^4 = \text{cis}\,\frac{4\pi}{3}$,

Simplify the argument whenever possible.

$z^5 = \text{cis}\,\frac{5\pi}{3}$, $z^6 = \text{cis}\,\frac{6\pi}{3} = \text{cis}\,0 = 1$

Since you obtained 1, all the powers are going to repeat with the period of 6.

b

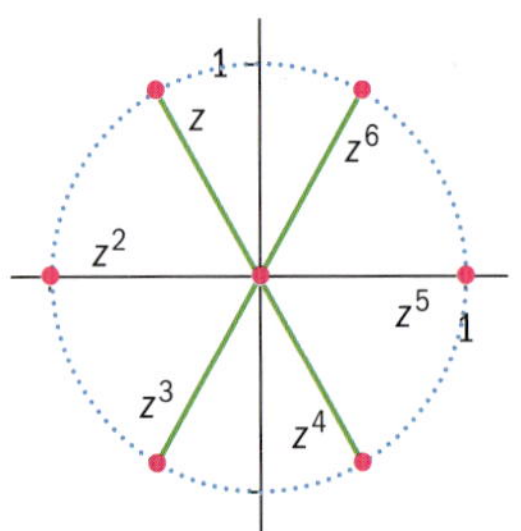

Draw the points in the Argand diagram.

Notice that the powers are forming vertices of a regular hexagon inscribed in the circle with the centre at the origin and the radius 1 (modulus of z).

Example 15

Show that $\left(3\cos\frac{2\pi}{3} - 3\sin\frac{\pi}{3}\text{i}\right)^9 = 3^9$.

$$\left(3\cos\frac{2\pi}{3} - 3\sin\frac{\pi}{3}\text{i}\right)^9$$

$$= \left(3\cos\frac{4\pi}{3} + 3\sin\frac{4\pi}{3}\text{i}\right)^9$$

Rewrite the complex number in polar form by using the properties of trigonometric functions:

$\cos\frac{2\pi}{3} = \cos\frac{4\pi}{3}$ and $-\sin\frac{\pi}{3} = \sin\frac{4\pi}{3}$.

$= \left(3 \text{ cis } \frac{4\pi}{3}\right)^9 = 3^9 \text{cis } \frac{36\pi}{3}$	Apply De Moivre's theorem.
$= 3^9 \text{cis } 12\pi = 3^9 \underbrace{\text{cis} 0}_{1} = 3^9$	Simplify the argument.

You can check your result by using a calculator.

HINT

Notice that the imaginary part, due to the imperfection of the algorithm the calculator is using, is not exactly equal to 0 but very close to it.

Exercise 10D

1 Given the numbers

$z_1 = 1 + \text{i}$ and $z_2 = 2\,\text{e}^{\text{i}\frac{2\pi}{3}}$, find:

a $z_1^3 \times z_2^2$ **b** $\frac{z_1^5}{z_2^3}$

c $\left(z_1^4\right)^* \times \left(z_2^*\right)^5$ **d** $\frac{\left(z_2^*\right)^6}{\left(-z_1^3\right)^*}$

2 Given

$z_1 = \sqrt{2}\left(\cos\frac{\pi}{3} - \cos\frac{5\pi}{6}\text{i}\right)$

and $z_2 = 2\left(\sin\frac{5\pi}{6} - \sin\frac{\pi}{3}\text{i}\right)$,

find $\frac{z_1^3}{z_2^5}$.

3 Find the Cartesian form of

$\left(\frac{\sin\theta + \text{i}\cos\theta}{\cos\theta - \text{i}\sin\theta}\right)^{2019}$.

4 A geometric sequence of complex numbers is given: 2 + i, 1 + 3i, −2 + 4i, ...

a Find the common ratio.

b What is the ninth term of the sequence?

c Find the sum of the first nine terms of the sequence.

TOK

Can we see the beauty of mathematics in fractals?

Investigation 6

Consider the equations $z^2 - 1 = 0$, $z^3 - 1 = 0$ and $z^4 - 1 = 0$.

1 Factorize the expressions and find exact forms of the solution to each equation.

2 Write all the solutions in polar form and plot them on separate Argand diagrams.

3 **Factual** What do solutions of the second and third equation form on the diagrams?

4 Can you predict the shape and write the polar form for the solutions of the equation $z^5 - 1 = 0$?

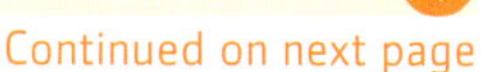

Continued on next page

5 **Conceptual** What do solutions of the equation $z^n - 1 = 0$, $n \geq 3$ form on the Argand diagram?

6 Use the geometrical representation of the solutions to predict the polar form of the solutions of the equation $z^n - 1 = 0$, $n \geq 3$.

7 **Conceptual** How does representing solutions in a different form allow you to solve $z^n - 1 = 0$, $n \geq 3$?

Example 16

Given that the complex number $\omega \neq 1$ is a solution of $z^3 = 1$, show that:

a $\omega^2 + \omega + 1 = 0$

b $(\omega^*)^2 + \omega^* + 1 = 0$

c Hence, calculate $\omega^{2019} + \omega^{2020} + \omega^{2021} + \omega^{2022}$.

a $\omega^3 = 1 \Rightarrow \omega^3 - 1 = 0$ $\Rightarrow (\omega - 1)(\omega^2 + \omega + 1) = 0$ $\Rightarrow \omega^2 + \omega + 1 = 0$	ω must satisfy the equation. Factorize the expression. Since $\omega \neq 1$ then ω must be the root of the quadratic expression.
b $\omega^2 + \omega + 1 = 0 \Rightarrow (\omega^*)^2 + \omega^* + 1 = 0$	Given that ω is the solution of the quadratic equation, then its conjugate is also the solution of the same equation.
c $\omega^{2019} + \omega^{2020} + \omega^{2021} + \omega^{2022}$ $= \omega^{2019}\left(\underbrace{1 + \omega + \omega^2}_{0}\right) + \left(\underbrace{\omega^3}_{1}\right)^{674} = 1$	Factorize the first three terms of the expression and rewrite the fourth term as a power of ω^3. Use the values from the original equation and from part **a**.

Using complex numbers in polar form enables you to factorize polynomials of higher degree into linear (with real zeros) and quadratic (with conjugate complex zeros).

Example 17

Let the complex number $\omega \neq -1$ be a solution of $z^5 + 1 = 0$.

a Find all the solutions of the equation.

b Show that $\omega^4 - \omega^3 + \omega^2 - \omega + 1 = 0$.

c Hence, factorize the expression, giving the coefficients correct to 3 significant figures where appropriate.

a $z^5 + 1 = 0 \Rightarrow z^5 = -1 = \text{cis}\,\pi$	Rewrite −1 in the polar form.
$z_1 = \text{cis}\frac{\pi}{5}, z_2 = \text{cis}\frac{3\pi}{5}, z_3 = \text{cis}\,\pi = -1,$ $z_4 = \text{cis}\frac{7\pi}{5}, z_5 = \text{cis}\frac{9\pi}{5}$	Apply the formula for the roots.
b $\omega^5 + 1 = (\omega + 1)(\omega^4 - \omega^3 + \omega^2 - \omega + 1)$ $\Rightarrow \omega^4 - \omega^3 + \omega^2 - \omega + 1 = 0$	ω must satisfy the equation. Factorize the expression. Since $\omega \neq -1$ then ω must be the root of the quartic expression.
c Since $z_1 = z_5{}^*$ and $z_2 = z_4{}^*$ then $\omega^4 - \omega^3 + \omega^2 - \omega + 1$	Combine conjugate complex solutions for the quadratic expressions.
$= \left(\omega^2 - 2\cos\frac{\pi}{5}\omega + 1\right)\left(\omega^2 - 2\cos\frac{3\pi}{5}\omega + 1\right)$	The linear coefficient is double the real part and the constant coefficient is the modulus, which is 1.
$= (\omega^2 - 1.68\omega + 1)(\omega^2 + 0.681\omega + 1)$	Calculate the linear coefficients.

To find the roots of any complex number you just need to use the value of modulus different to 1.

Let $z = r\,\text{cis}\theta$ and $z^n = w = \rho\ \text{cis}\varphi$. Now use De Moivre's theorem and equate these two complex numbers.

$$z^n = w \Rightarrow r^n\,\text{cis}\,n\theta = \rho\,\text{cis}\,\phi \Rightarrow \begin{cases} r^n = \rho \\ n\theta = \phi + 2k\pi \end{cases} \Rightarrow \begin{cases} r = \sqrt[n]{\rho} \\ \theta = \dfrac{\phi + 2k\pi}{n} \end{cases}, \text{ where,}$$

due to the periodical property of trigonometric functions, $k = 0, 1, 2, \ldots, n - 1$.

This is summarized in the following:

TOK

Why does "i" appear in so many fundamental laws of physics?

Given the complex number $z = r\ \text{cis}\theta,\ r,\ \theta \in \mathbb{R},\ r \geq 0$, then

$$\sqrt[n]{z} = \sqrt[n]{r}\ \text{cis}\,\frac{\theta + 2k\pi}{n},\ k = 0, 1, 2, \ldots, n-1$$

In the set of complex numbers, there are n different complex numbers that are nth roots of a given complex number. They all have **the same modulus,** that is nth root of the original modulus, but ***n* different arguments.** Therefore, they are equally spaced complex numbers on the circle starting with the original argument divided by n. (As seen in Investigation 6.)

Example 18

Solve the equation $z^4 = -8 + 8\sqrt{3}\text{i}$. Draw the solutions on the Argand diagram.

$$r = \sqrt{(-8)^2 + \left(8\sqrt{3}\right)^2} = \sqrt{64 + 192} = 16$$

First rewrite the complex number in polar form.

$$\theta = \arctan\frac{8\sqrt{3}}{-8} = \arctan\left(-\sqrt{3}\right) = \frac{5\pi}{3}$$

$$z = \sqrt[4]{16\text{ cis }\frac{5\pi}{3}}$$

Apply the formula for the roots.

$$= \sqrt[4]{16}\text{ cis}\left(\frac{5\pi + 2k\pi}{12}\right), k = 0, 1, 2, 3$$

Since the $n = 4$ the values of k go from 0 to $n - 1$, that is 3.

$$z_1 = 2\text{ cis }\frac{5\pi}{12}, z_2 = 2\text{ cis }\frac{11\pi}{12},$$

$$z_3 = 2\text{ cis }\frac{17\pi}{12}, z_4 = 2\text{ cis }\frac{23\pi}{12}$$

Write all four solutions.

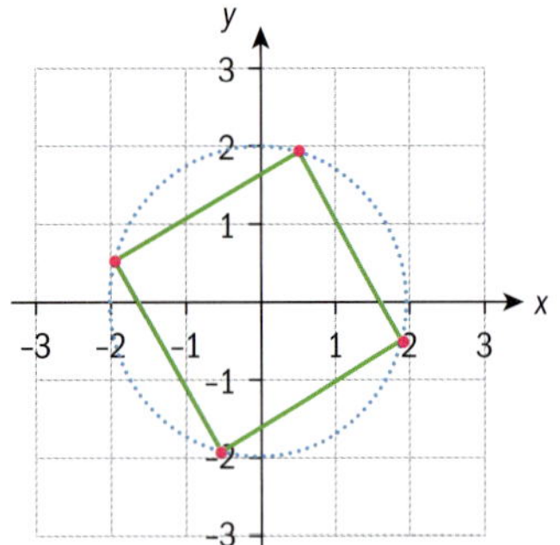

Draw all the solutions in the Argand diagram.

Notice that the vertices form a square.

Sometimes the exact solutions are too complicated to be found. In such cases, you need to use a calculator.

Example 19

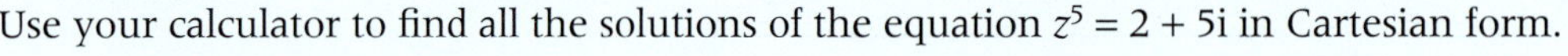
Use your calculator to find all the solutions of the equation $z^5 = 2 + 5\text{i}$ in Cartesian form.

$$z^5 = 2 + 5\text{i} \Rightarrow z = \sqrt[5]{2 + 5\text{i}}$$

$\sqrt[5]{2+5\cdot i}$	1.36087+0.330227·*i*
|1.360867038465+0.33022713557936·*i*|→	1.40036
angle(1.3608670384654+0.33022713557936▸	0.238058

Use your calculator to find the fifth root of the complex number.

Find the modulus and argument and store them into ***r*** and ***a*** respectively for the further calculations.

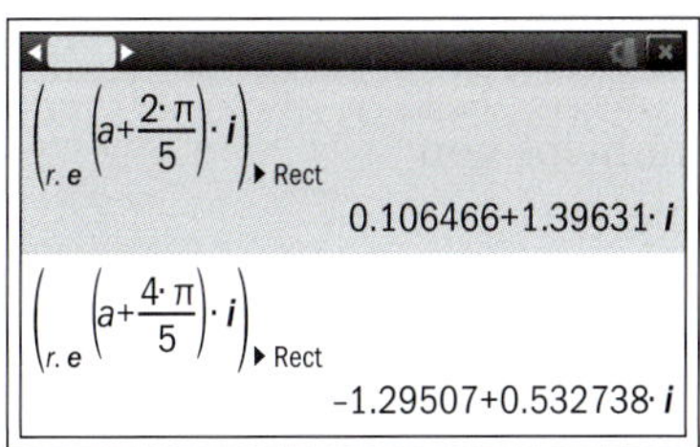

$\left(r \cdot e^{\left(a+\frac{6\cdot\pi}{5}\right)\cdot i}\right)$ ▸Rect

-0.906862-1.06706·i

$\left(r \cdot e^{\left(a+\frac{8\cdot\pi}{5}\right)\cdot i}\right)$ ▸Rect

-0.734596-1.19222·i

$z_1 = 1.36 + 0.330\text{i}$, $z_2 = 0.106 + 1.40\text{i}$,
$z_3 = -1.30 + 0.533\text{i}$, $z_4 = -0.907 - 1.07\text{i}$,
$z_5 = 0.735 - 1.19\text{i}$,

Use the formula for arguments to find all the solutions.

Convert all the solutions from Euler's form to Cartesian form.

Give your answers correct to 3 significant figures.

Notice that the calculator will always give just one solution of the multiple root of a complex number. In such cases, you need to determine the remaining roots.

Exercise 10E

1 Given that the complex number $\omega \neq -1$ is a solution of $z^3 + 1 = 0$, show that:

a $\omega^2 - \omega + 1 = 0$

b $(\omega^*)^2 - \omega^* + 1 = 0$

c Hence, calculate $\omega^{2019} + \omega^{2020} - \omega^{2021} + \omega^{2022}$.

2 Let the complex number $\omega \neq 1$ be a solution of $z^7 - 1 = 0$. Factorize the expression

$\omega^6 + \omega^5 + \omega^4 + \omega^3 + \omega^2 + \omega + 1$.

3 Find all the complex numbers in Euler's form:

a $\sqrt[4]{-625}$ **b** $\sqrt[5]{\sqrt{3} - \text{i}}$

c $\sqrt[6]{-\frac{1}{2} + \frac{\sqrt{3}}{2}\text{i}}$

4 Use your calculator to find all the complex numbers in Cartesian form:

a $\sqrt[4]{1 + 2\text{i}}$ **b** $\sqrt[5]{5 - 2\text{i}}$

c $\sqrt[6]{3 + 8\text{i}}$

5 The complex number $z = -4\sqrt{2} + 4\sqrt{2}\text{i}$ is given.

a Write the number in Euler's form.

b Show that the real part of the complex number $\sqrt[6]{z}$ is $\frac{\sqrt{2\sqrt{2} + 4}}{2}$.

c Find the exact form of the imaginary part of the complex number $\sqrt[6]{z}$.

Trigonometric formulae and polar form of a complex number

Using both the binomial theorem and De Moivre's theorem helps you to find some multiple angle formulae for trigonometric functions.

Let $z = \text{cis}\theta = \cos\theta + \text{i}\sin\theta$.

By using the binomial theorem you obtain:

$$z^2 = \left(\cos\theta + \text{i}\sin\theta\right)^2 = \cos^2\theta + 2\cos\theta\,\text{i}\sin\theta - \sin^2\theta$$
$$= \left(\cos^2\theta - \sin^2\theta\right) + \text{i}\left(2\sin\theta\cos\theta\right)$$

On the other hand, by using De Moivre's theorem you obtain:

$z^2 = \text{cis}2\theta = \cos 2\theta + \text{i}\sin 2\theta$

These two complex numbers are equal. Therefore, their real and imaginary parts are equal as well:

$$\begin{cases}\cos 2\theta = \cos^2\theta - \sin^2\theta \\ \sin 2\theta = 2\sin\theta\cos\theta\end{cases}$$

These double angle formulae have been proved previously by using geometry, so it is very interesting to see a different kind of proof.

This method is very simple and helps you to find more of the formulae for trigonometric functions of multiple angles.

TOK

Has "i" been invented or was it discovered?

Example 20

Let $z = \text{cis}\theta = \cos\theta + \text{i}\sin\theta$.

a Use the binomial theorem to find the real and imaginary part of z^3.

b Use De Moivre's theorem to find the formulae for:

i $\cos 3\theta$ **ii** $\sin 3\theta$ **iii** $\tan 3\theta$

a $z^3 = (\cos\theta + \text{i}\sin\theta)^3$	
$= \cos^3\theta + 3\cos^2\theta\,\text{i}\sin\theta + 3\cos\theta\left(\text{i}\sin\theta\right)^2 + \left(\text{i}\sin\theta\right)^3$	Expand the expression.
$= \cos^3\theta + 3\cos^2\theta\,\text{i}\sin\theta - 3\cos\theta\sin^2\theta - \text{i}\sin^3\theta$	Use the powers of i.
$= \underbrace{\left(\cos^3\theta - 3\cos\theta\sin^2\theta\right)}_{\text{Re}\,z^3} + \underbrace{\left(3\cos^2\theta\sin\theta - \sin^3\theta\right)}_{\text{Im}\,z^3}\text{i}$	
b $z^3 = \text{cis}3\theta = \cos 3\theta + \text{i}\sin 3\theta$	
i $\cos 3\theta = \cos^3\theta - 3\cos\theta\sin^2\theta$	Equate the real parts of z^3.
$= \cos^3\theta - 3\cos\theta\left(1 - \cos^2\theta\right) = 4\cos^3\theta - 3\cos\theta$	Express sine in terms of cosine.
ii $\sin 3\theta = 3\cos^2\theta\sin\theta - \sin^3\theta$	Equate the imaginary parts of z^3.
$= 3\left(1 - \sin^2\theta\right)\sin\theta - \sin^3\theta = 3\sin\theta - 4\sin^3\theta$	Express cosine in terms of sine.

iii $\tan 3\theta = \frac{\sin 3\theta}{\cos 3\theta} = \frac{3\sin\theta - 4\sin^3\theta}{4\cos^3\theta - 3\cos\theta}$	Express tangent in terms of sine and cosine. Use the results from previous parts.
$= \frac{3\frac{\sin\theta}{\cos^3\theta} - 4\frac{\sin^3\theta}{\cos^3\theta}}{4 - 3\frac{\cos\theta}{\cos^3\theta}} = \frac{3\tan\theta\frac{1}{\cos^2\theta} - 4\tan^3\theta}{4 - 3\frac{1}{\cos^2\theta}}$	Divide the numerator and the denominator by $\cos^3\theta$.
$= \frac{3\tan\theta(1+\tan^2\theta) - 4\tan^3\theta}{4 - 3(1+\tan^2\theta)} = \frac{3\tan\theta - \tan^3\theta}{1 - 3\tan^2\theta}$	Use the Pythagorean identity $\frac{1}{\cos^2\theta} = 1 + \tan^2\theta$ and simplify.

Multiple angle formulae for higher values of n can be found in a similar way.

Investigation 7

Let the number $z = \text{cis}\,\theta$.

1 Use De Moivre's theorem to find the following sums

$z + \frac{1}{z}$, $z^2 + \frac{1}{z^2}$, $z^3 + \frac{1}{z^3}$ and $z^4 + \frac{1}{z^4}$ in their simplest form.

2 **Factual** What can you say about the real and the imaginary parts of these sums?

3 **Factual** What is the general formula for $z^n + \frac{1}{z^n}, n \in \mathbb{Z}^+$?

4 Use De Moivre's theorem to find the following differences

$z - \frac{1}{z}$, $z^2 - \frac{1}{z^2}$, $z^3 - \frac{1}{z^3}$ and $z^4 - \frac{1}{z^4}$ in their simplest form.

5 **Factual** What can you say about the real and the imaginary parts of these differences?

6 **Factual** What is the general formula for $z^n - \frac{1}{z^n}, n \in \mathbb{Z}^+$?

7 **Conceptual** What is the relationship between the expressions

$z^n + \frac{1}{z^n}$ and $z^n - \frac{1}{z^n}$ and $z^n, n \in \mathbb{Z}^+$ in polar form?

The formulae for sum and difference of powers and reciprocals of complex numbers are useful when integrating powers of trigonometric functions.

i $z^n + \frac{1}{z^n} = 2\cos n\theta$

ii $z^n - \frac{1}{z^n} = 2\text{i}\sin n\theta$

for n is natural and θ is real.

Example 21

Let the number $z = \text{cis}\theta$.

a Use the binomial theorem to expand $\left(z - \frac{1}{z}\right)^4$.

b Use the formulae $z^n + \frac{1}{z^n} = 2\cos n\theta$ and $z^n - \frac{1}{z^n} = 2\text{i}\sin n\theta$ to find $\sin^4\theta$.

c Hence, find $\int \sin^4 x\,\text{d}x$.

a $\left(z - \frac{1}{z}\right)^4 = z^4 - 4z^3\frac{1}{z} + 6z^2\frac{1}{z^2} - 4z\frac{1}{z^3} + \frac{1}{z^4}$	Use the binomial theorem for $n = 4$ to expand the expression.
$= z^4 - 4z^2 + 6 - 4\frac{1}{z^2} + \frac{1}{z^4}$	Simplify the terms.
b $\left(z - \frac{1}{z}\right)^4 = z^4 + \frac{1}{z^4} - 4\left(z^2 + \frac{1}{z^2}\right) + 6$	Group the terms to obtain the form and
$\Rightarrow (2\text{i}\sin\theta)^4 = 2\cos 4\theta - 4(2\cos 2\theta) + 6$	apply the formulae $z^n - \frac{1}{z^n} = 2\text{i}\sin n\theta$ and
$\Rightarrow 16\sin^4\theta = 2\cos 4\theta - 8\cos 2\theta + 6$	$z^n + \frac{1}{z^n} = 2\cos n\theta$.
$\Rightarrow \sin^4\theta = \frac{1}{8}\cos 4\theta - \frac{1}{2}\cos 2\theta + \frac{3}{8}$	Express $\sin^4\theta$ as a subject.
c $\int \sin^4 x\,\text{d}x$	Use the expression from the previous part to simplify the integration.
$= \int\left(\frac{1}{8}\cos 4x - \frac{1}{2}\cos 2x + \frac{3}{8}\right)\text{d}x$	Apply the compound formula for integration.
$= \frac{1}{8}\left(-\frac{\sin 4x}{4}\right) - \frac{1}{2}\left(-\frac{\sin 2x}{2}\right) + \frac{3}{8}x + c$	
$= -\frac{1}{32}\sin 4x + \frac{1}{4}\sin 2x + \frac{3}{8}x + c, c \in \mathbb{R}$	

Exercise 10F

1 Use mathematical induction to show that $(\text{cis}\theta)^n = \text{cis}\,n\theta$, $n \in \mathbb{Z}^+$.

2 Let $z = \cos\theta + \text{i}\sin\theta$.

a Use the binomial theorem to find the real and imaginary part of z^4.

b Use De Moivre's theorem to find the formulae for:

i $\cos 4\theta$ **ii** $\sin 4\theta$

c Hence, show that

$$\tan 4\alpha = \frac{4\tan\alpha - 4\tan^3\alpha}{1 - 6\tan^2\alpha + \tan^4\alpha}.$$

3 Let the number $z = \text{e}^{\text{i}\theta}$.

a Use the binomial theorem to expand $\left(z + \frac{1}{z}\right)^4$.

b Use the formula $z^n + \frac{1}{z^n} = 2\cos n\theta$ to find $\cos^4\theta$.

c Hence, find $\int \cos^4 x\,\text{d}x$.

4 Given that $\omega \neq \pm 1$ is a solution of the equation $z^6 - 1 = 0$:

a Show that $1 + \omega^2 + \omega^4 = 0$.

b Evaluate $1 + \omega^{102} + \omega^{1004} + \omega^{20008}$.

5 Find the sum $1 + \text{e}^{\text{i}\frac{\pi}{8}} + \text{e}^{\text{i}\frac{\pi}{4}} + \text{e}^{\text{i}\frac{3\pi}{8}} + \ldots + \text{e}^{\text{i}\pi}$.

Developing your toolkit

Now do the Modelling and investigation activity on page 678.

Chapter summary

- **Cartesian form of a complex number**
  $z = x + y\text{i}, \quad x, y \in \mathbb{R}$
- **Modulus-argument or polar form of a complex number**
  Given a complex number with the modulus r, $r \in \mathbb{R}$, $r \geq 0$ and the argument θ, $\theta \in \mathbb{R}$
  $z = r\,\text{cis}\,\theta$.
- **Euler's or exponential form of a complex number**
  Given a complex number with the modulus r, $r \in \mathbb{R}$, $r \geq 0$ and the argument θ, $\theta \in \mathbb{R}$
  $z = r\,\text{e}^{\text{i}\theta}$.
- **Relationship between the rectangular and polar coordinates**
  $$\begin{cases} r = \sqrt{x^2 + y^2} \\ \theta = \arctan \dfrac{y}{x} \end{cases} \Rightarrow \begin{cases} x = r\cos\theta \\ y = r\sin\theta \end{cases}$$
- **Two equal complex numbers**
  Given $z_1 = r_1\text{cis}\theta_1 = r_1\text{e}^{\text{i}\theta_1}$ and $z_2 = r_2\text{cis}\theta_2 = r_2\,\text{e}^{\text{i}\theta_2}$, if $z_1 = z_2$ then $r_1 = r_2$ and $\theta_1 = \theta_2 + 2k\pi$, $k \in \mathbb{Z}$.
- **Conjugate, opposite and conjugate opposite complex numbers**
  $$z = r\,\text{cis}\,\theta \Rightarrow \begin{cases} -z = r\,\text{cis}(\pi + \theta) \\ z^* = r\,\text{cis}(\pi - \theta) \\ -z^* = r\,\text{cis}(2\pi - \theta) \end{cases}$$
- **Multiplication**
  Given $z_1 = r_1\text{cis}\theta_1$ and $z_2 = r_2\text{cis}\theta_2$, if $z_1 \times z_2 = z_3 = r_3\,\text{cis}\theta_3$ then $r_3 = r_1\,r_2$ and $\theta_3 = \theta_1 + \theta_2 + 2k\pi$, $k \in \mathbb{Z}$
  You multiply two complex numbers in polar form by multiplying the moduli and adding the arguments.
- **Reciprocal number**
  Given the complex number in polar form $z = r\,\text{cis}\theta$ then the reciprocal value in polar form is $\dfrac{1}{z} = \dfrac{1}{r}\text{cis}(2\pi - \theta) = \dfrac{1}{r}\text{cis}(-\theta)$.
  The reciprocal complex number of a given complex number in polar form has the reciprocal modulus and the opposite argument.

Continued on next page

- **Division**

  Given the two complex numbers $z_1 = r_1 \operatorname{cis}\theta_1$ and $z_2 = r_2 \operatorname{cis}\theta_2$, $z_2 \neq 0$, then

  $$\frac{z_1}{z_2} = \frac{r_1}{r_2}\operatorname{cis}(\theta_1 - \theta_2), z_2 \neq 0, r_2 \neq 0$$

  You divide two complex numbers in polar form by dividing the moduli and subtracting the arguments.

- **De Moivre's theorem**

  Given a complex number in polar or Euler's form $z = r\operatorname{cis}\theta = r\mathrm{e}^{\mathrm{i}\theta}, z \neq 0$, $r \in \mathbb{R}^+, \theta \in \mathbb{R}$ then $z^n = r^n \operatorname{cis} n\theta = r^n \mathrm{e}^{\mathrm{i}n\theta}, n \in \mathbb{Z}$.

- **Roots of complex numbers**

  Given the complex number $z = r\operatorname{cis}\theta$, r, $\theta \in \mathbb{R}$, $r \geq 0$ then

  $$\sqrt[n]{z} = \sqrt[n]{r}\operatorname{cis}\frac{\theta + 2k\pi}{n}, k = 0, 1, 2, \ldots, n-1$$

Developing inquiry skills

Return to the chapter opening problem. You should now be able to answer these questions.

What knowledge have you gained that can help you with these questions? What skills have you learned that can help you?

Chapter review

Click here for a mixed review exercise

1 Find the real and the imaginary parts of the complex number $z = 6\mathrm{e}^{-\frac{3\pi}{4}\mathrm{i}}$.

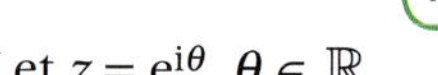

2 Given that $z_1 = r \operatorname{cis}\frac{\pi}{6}$ and $z_2 = 5 - 12\mathrm{i}$, find the value of r such that $|z_1^2 z_2| = 52$.

3 Let $z = \mathrm{e}^{\mathrm{i}\theta}$, $\theta \in \mathbb{R}$. Show that $\frac{1}{1+z} = \frac{1}{2}\left(1 - \mathrm{i}\tan\frac{\theta}{2}\right)$.

4 The equation $z^5 - 1 = 0$ is given.

a Find all the solutions in polar form.

b Hence, show that

$$\cos\frac{2\pi}{5} + \cos\frac{4\pi}{5} + \cos\frac{6\pi}{5} + \cos\frac{8\pi}{5} = -1$$

5 Let $z = \mathrm{e}^{\mathrm{i}\theta}$, $\theta \in \mathbb{R}$.

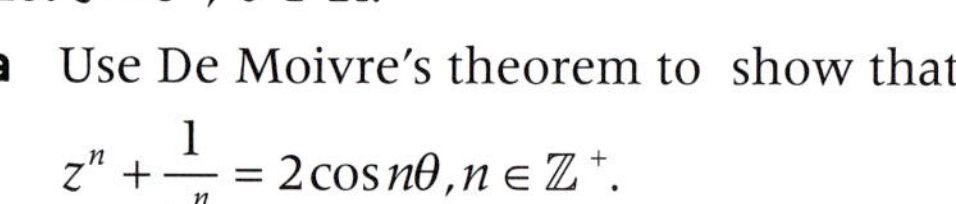

a Use De Moivre's theorem to show that $z^n + \frac{1}{z^n} = 2\cos n\theta, n \in \mathbb{Z}^+$.

b Expand $\left(z + \frac{1}{z}\right)^6$.

c Given that $\cos^6\theta = a\cos 6\theta + b\cos 4\theta + c\cos 2\theta + d$, $a, b, c, d \in \mathbb{Q}^+$, determine the values of a, b, c and d.

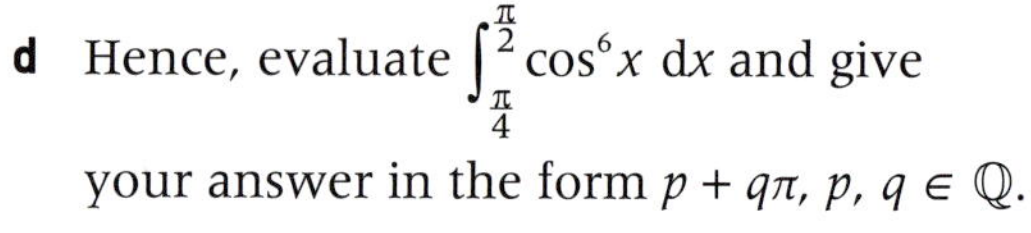

d Hence, evaluate $\int_{\frac{\pi}{4}}^{\frac{\pi}{2}} \cos^6 x \, \mathrm{d}x$ and give your answer in the form $p + q\pi$, $p, q \in \mathbb{Q}$.

6 Find the argument of the complex number $z = 23 - 41\text{i}$, giving your answer in radians correct to 4 significant figures.

7 Given the numbers

$$z_1 = 3 + \text{i},\ z_2 = 5\ \text{cis}\frac{17\pi}{83} \text{ and } z_3 = 2\sqrt{2}\,\text{e}^{-\frac{\pi}{4}\text{i}},$$

calculate $3z_1 - 2z_2 + \frac{1}{2}z_3$, giving your answer correct to 3 significant figures.

8 Find the following limit $\lim\limits_{x\to 0}\frac{\sin 4x}{\sin x}$.

9 Find the area of the regular pentagon whose vertices are solutions of the equation $z^5 = -32\text{i}$, giving your answer correct to 2 decimal places.

10 Let $z = (a + 2\text{i})^2$. Find the value of a such that $\arg z = 1$.

Exam-style questions

11 P1: Given $z_1 = 4\,\text{cis}\left(-\frac{\pi}{3}\right)$ and $z_2 = 3\,\text{cis}\frac{5\pi}{6}$, find expressions for the following.

In each case, give your answers in the form $a + b\text{i}$, $a, b \in \mathbb{R}$.

a $z_1 z_2$ (3 marks)

b $\left(\frac{z_1}{z_2}\right)^3$ (5 marks)

c $\left(z_1^{\,2}\right)^*$ (3 marks)

12 P1: Use de Moivre's theorem to find the value of $(1 + \text{i})^{10}$. (5 marks)

13 P1: a Express $z = 1 - \text{i}\sqrt{3}$ in complex polar form. (3 marks)

b Hence find the smallest value for n such that $z^n \in \mathbb{R}^+$. (3 marks)

c Find the value of $\left(1 - \text{i}\sqrt{3}\right)^{15}$. (3 marks)

14 P1: a Use the binomial theorem to find the expansion of $(\cos\theta + \text{i}\sin\theta)^5$. (3 marks)

b By using de Moivre's theorem and your expansion from part **a**, prove that $\cos 5\theta \equiv 16\cos^5\theta - 20\cos^3\theta + 5\cos\theta$. (4 marks)

15 P1: a Find, in complex polar form, the cube roots of -27i. (8 marks)

b Given that the three roots from part **a** form the vertices of a triangle in an Argand diagram, determine the exact area of the triangle. (3 marks)

16 P1: Use de Moivre's theorem to show that $z^n + \frac{1}{z^n} = 2\cos n\theta$, where z is a complex number of the form $\cos\theta + \text{i}\sin\theta$. (4 marks)

b By considering the binomial expansion of $\left(z + \frac{1}{z}\right)^4$, show that $\cos^4\theta \equiv \frac{1}{16}(2\cos 4\theta + 8\cos 2\theta + 6)$. (6 marks)

c Hence find the exact value of $\int_0^{\frac{\pi}{6}} \cos^4\theta\ \text{d}\theta$. (5 marks)

17 P1: a Express $\frac{\sqrt{3} + \text{i}}{\sqrt{3} - \text{i}}$ in the form $r\text{e}^{\text{i}\theta}$ for $0 \le \theta \le 2\pi$. (5 marks)

b i Show that, for $n \in \mathbb{Z}^+$, $\left(\sqrt{3} + \text{i}\right)^n + \left(\sqrt{3} - \text{i}\right)^n = 2^{n+1}\cos\left(\frac{n\pi}{6}\right)$.

ii Hence find the value of $\left(\sqrt{3} + \text{i}\right)^8 + \left(\sqrt{3} - \text{i}\right)^8$. (8 marks)

18 P1: The complex number ω, where $\omega \ne 1$, is a solution of the equation $z^3 - 1 = 0$.

Simplify the following expressions.

a $(1 + \omega + \omega^*)^2$ (3 marks)

b $(1 + \omega + 3\omega^2)^2$ (3 marks)

c $(1 + 2\omega + 3\omega^2)(1 + 3\omega + 2\omega^2)$ (6 marks)

19 P1: Find all solutions of the equation $(z - 2\text{i})^3 = \text{i}$, $z \in \mathbb{C}$. (10 marks)

Modelling and investigation activity

The cubic formula

Approaches to learning: Critical thinking, Transfer skills

Exploration criteria: Mathematical communication (B), Use of mathematics (E)

IB topic: Polynomials, Complex numbers, roots, Algebraic manipulation

Solving a quadratic equation

The quadratic has had a solution (by formula) since the time of the time of the Babylonians.

Solve this equation using the quadratic formula:

$x^2 - 26x - 27 = 0$

Solving a cubic equation

The method for solving a cubic of the form $ax^3 + bx^2 + cx + d = 0$ was more elusive.

The general solution of the cubic (and the quartic) were not discovered until the sixteenth century after work from a large number of mathematicians including Luca Pacioli, Scipione del Ferro, Antonio Fior, Niccolo Fontana (aka Tartaglia), Gerolamo Cardano and Ludovico Ferrari.

Now, of course, calculators can be used, but let's consider the method devised by these great mathematicians to solve the general cubic $ax^3 + bx^2 + cx + d = 0$.

STEP 1

First eliminate the quadratic term by making a substitution $x = y - \frac{b}{3a}$ to obtain a **depressed cubic** (a cubic with no square powers).

This will be of the form $y^3 + \mathrm{e}y + f = 0$.

ALGEBRA CHALLENGE 1:

Use algebraic manipulation to produce an equation of the form $y^3 + ey + f = 0$.

where $\mathrm{e} = \frac{1}{a}\left(c - \frac{b^2}{3a}\right)$ and $f = \frac{1}{a}\left(d + \frac{2b^2}{27a^2} - \frac{bc}{3a}\right)$.

Worked example:

Consider the equation $x^3 - 9x^2 - 36x - 80 = 0$.

What are the values of a, b, c and d?

Find the values of e and f.

Hence show that the depressed cubic form of the equation $x^3 - 9x^2 - 36x - 80 = 0$ is $y^3 + 9y - 26 = 0$.

STEP 2

You can further reduce this cubic using the substitution $y = z - \frac{\mathrm{e}}{3z}$.

ALGEBRA CHALLENGE 2:

Use algebraic manipulation to reduce $y^3 + \mathrm{e}y + f = 0$ to $z^6 + fz^3 - \frac{\mathrm{e}^3}{27} = 0$.

STEP 3

Make one further substitution of $w = z^3$ to give a the reduced quadratic form $w^2 + fw - \frac{e^3}{27} = 0$.

Worked example:

From the example $x^3 - 9x^2 + 36x - 80 = 0$, show that the the reduced quadratic form of the equation in terms of w can be written as $w^2 - 26w - 27 = 0$.

This quadratic can now be solved using the quadratic formula.

You will get two roots for w, w_1 and w_2.

Solve $w^2 - 26w - 27 = 0$ to find w_1 and w_2 (this is actually the quadratic equation you solved at the beginning).

Undoing the substitutions

Choose one of the two values of w and find three values of z by solving $w = z^2$ from STEP 3.

You will find that two of the roots are complex in this case.

Having obtained these 3 values for z, you can now find three values of y from $y = z - \frac{e}{3z}$ (introduced in STEP 2).

For each of these three values of y, you can find three values of x from $x = y - \frac{b}{3a}$ (introduced in STEP 1).

This gives the three roots of the original equation as x_1, x_2 and x_3.

You have solved the cubic!

Extension

- Investigate Del Ferros general formula

  $$x = \left(\frac{q}{2} + \sqrt{\frac{p^2}{4} + \frac{p^3}{27}}\right)^{\frac{1}{3}} + \left(\frac{q}{2} - \sqrt{\frac{p^2}{4} + \frac{p^3}{27}}\right)^{\frac{1}{3}}$$

  for giving one solution of a depressed cubic.

  How could you obtain the other two solutions once you know this solution?
- Investigate the discriminant of the cubic formula.

  Under what condition will this have:

  three real solutions at least two of which are equal?

  one real root and two conjugate imaginary roots?

  three distinct real roots?

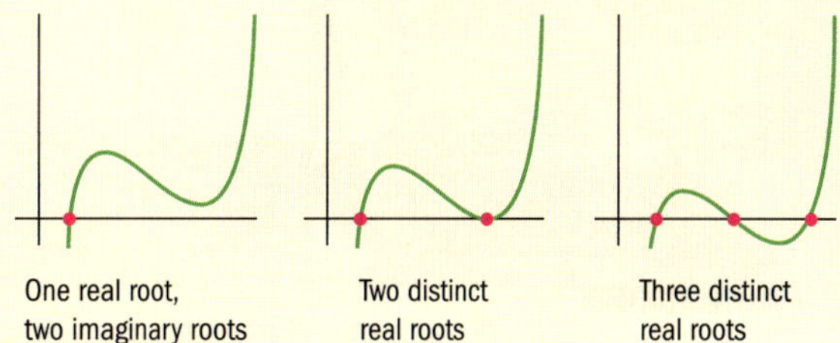

11 Valid comparisons and informed decisions: probability distributions

Concepts

- Quantity
- Representation

Microconcepts

- Axiomatic probability systems
- Bayes' theorem
- Discrete random variables and their probability distributions
- Probability density functions for continuous random variables
- Mode and median
- Mean, $\mathrm{E}(x)$, $\mathrm{E}(x^2)$
- Variance, $\mathrm{Var}(x) = \mathrm{E}(x^2) - [\mathrm{E}(x)]^2$
- Standard deviation $= \sqrt{\mathrm{Var}(x)}$
- The effect of linear transformations of x
- Binomial distribution
- Mean and variance of the binomial distribution
- Normal distributions and curves
- Understanding the natural occurrence of the normal distribution
- Properties of the normal distribution
- Diagrammatic representation
- Expected values
- Normal probability calculations
- Inverse normal calculations

Between 2008 and 2010, Paul the Octopus correctly predicted the outcome of 12 out of 14 football matches by approaching one of two boxes marked with the flag of the national teams playing. This is a success rate of almost 86%. Is this convincing? What is the probability that this would happen purely by chance?

Paul is not the only famous predicting animal. Punxsutawney Phil is a groundhog, immortalized in the film *Groundhog Day*. The legend is that if he sees his shadow on 2 February, six more weeks of winter weather lie ahead; no shadow indicates an early spring. Phil has been forecasting the weather on Groundhog Day for more than 130 years (although we suspect it is not the same groundhog!). Phil has a 39% success rate. Is this low? What success rate would you expect?

In this section you will learn about the axiomatic probability system and how probability can be used in real-life situations. The chapter starts by looking at basic probability and leads into probability distributions with both discrete and continuous variables.

What would persuade you that an octopus could predict the outcomes of soccer matches? Is it just guessing? How about a groundhog predicting the weather?

Consider the following games.

You are initially given 30 tokens. The aim of the game is to collect as many tokens as possible. Which game should you play?

10 tokens to roll 1 dice
If you get a 6 win 30 tokens
If you get a 5 or 4 get your tokens back
If you get a 1, 2 or 3 you get nothing!

20 tokens to toss 4 coins
If you get 4 Heads you win 100 tokens
If you get 3 Heads you get your tokens back
If you get 2 Heads you win 10 tokens
If you get 0 Heads or 1 Head you get nothing!

30 tokens to drop a marble into this machine
If it goes into box 1 or 8 you win 500 tokens
If it goes into 2 or 7 you win 50 tokens
If it goes into 3 or 6 you win 10 tokens
If it goes into 4 or 5 you get nothing!

Developing inquiry skills

How can an understanding of games such as those above help us in real-world situations?

Research some common "cognitive biases" that can be overcome through an understanding of statistics and probability.

Before you start

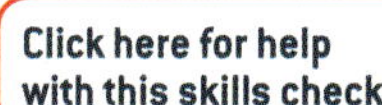
Click here for help with this skills check

You should know how to:

1 Use counting techniques.

eg You won a competition to take 2 friends to a concert. You have 5 friends who want to join you. How many different groups of friends could you take to the concert?

$$^{5}C_{2} = \frac{5!}{2!3!} = \frac{5\times 4}{2\times 1} = 10$$

2 Construct theoretical possibility spaces for single and combined experiments.

eg Two dice are thrown, and the results are noted. Draw a representation of the possible outcomes. Use this representation to determine the probability that the two dice show the same number.

	1	2	3	4	5	6
1	(1, 1)	(1, 2)	(1, 3)	(1, 4)	(1, 5)	(1, 6)
2	(2, 1)	(2, 2)	(2, 3)	(2, 4)	(2, 5)	(2, 6)
3	(3, 1)	(3, 2)	(3, 3)	(3, 4)	(3, 5)	(3, 6)
4	(4, 1)	(4, 2)	(4, 3)	(4, 4)	(4, 5)	(4, 6)
5	(5, 1)	(5, 2)	(5, 3)	(5, 4)	(5, 5)	(5, 6)
6	(6, 1)	(6, 2)	(6, 3)	(6, 4)	(6, 5)	(6, 6)

$$\text{P(2 dice show the same number)} = \frac{6}{36} = \frac{1}{6}$$

Skills check

1 Five friends go to the cinema. Given that Fred wants to sit on the end of the group, how many different seating arrangements can there be?

2 Two dice are numbered 1 1 3 4 6 8. Draw a representation of the possible outcomes. Use this representation to determine the probably that the two dice add to an even number.

Continued on next page

3 Understand set notation and basic probability from the different representations.

3 In a group of 38 students 14 speak Arabic, 18 speak Bulgarian and 22 speak Cantonese. It is also known that 3 speak Arabic and Bulgarian, 8 speak Bulgarian and Cantonese, 6 speak Arabic and Cantonese, and 14 speak exactly two languages.

a Draw a Venn diagram to represent this situation.

b Determine the probability of a student speaking all three languages.

4 Calculate and interpret conditional probabilities through representations using expected frequencies.
eg A group of 30 students were surveyed and the results were represented in a two-way table

	Played tennis	Did not play tennis	
Liked pizza	12	4	16
Did not like pizza	8	6	14
	20	10	30

a Calculate the probability that a randomly selected student plays tennis.

b Calculate the probability that a randomly selected student likes pizza.

c Calculate the probability that a randomly selected student likes pizza given that they played tennis.

a $\text{P(tennis)} = \frac{\text{n(tennis)}}{\text{n}(U)} \frac{20}{30} = \frac{2}{3}$

b $\text{P(pizza)} = \frac{16}{30} = \frac{8}{15}$

c The sample space has now changed from 30 to 20 (number of tennis players). Therefore

$$\text{P(tennis given pizza)} = \frac{12}{20} = \frac{3}{5}$$

4 A group of 29 students were surveyed and the results were represented in a two-way table.

	Takes Physics	Does not take Physics	
Takes History	5	12	17
Does not take History	9	3	12
	14	15	29

a Calculate the probability that a randomly selected student takes History.

b Calculate the probability that a randomly selected student does not take Physics.

c Calculate the probability that a randomly selected student does not take Physics given that they do not take History.

11.1 Axiomatic probability systems

Probability theory is a branch of mathematics which determines the likelihood of an event happening. In order to have an axiomatic probability system, it is important to recall some basic set theory and rules for probability.

Set rules

Recall the definition of the binary operations of union and intersection for two sets A and B:

$A \cup B = \{x \mid x \in A \text{ or } x \in B\}$

$A \cap B = \{x \mid x \in A \text{ and } x \in B\}$

The intersection and union can be represented on a Venn diagram:

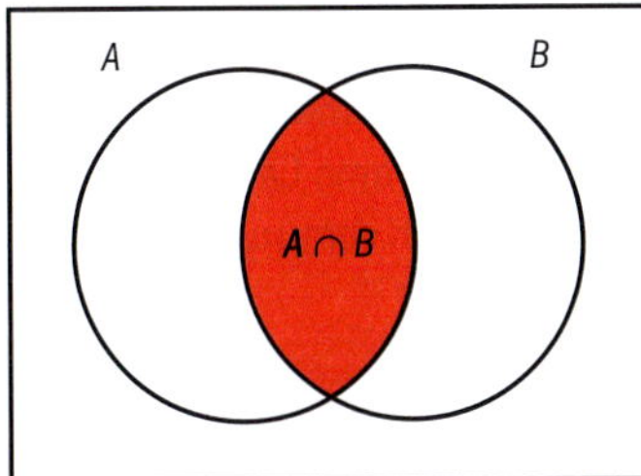

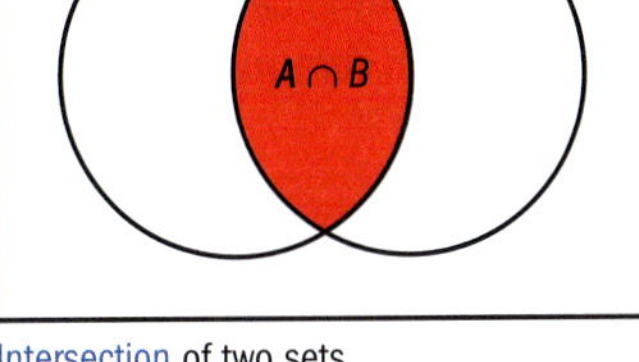

Intersection of two sets
$A \cap B$

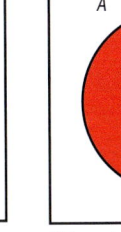

Union of two sets
$A \cup B$

Set theory

1 There exists a universal set (in probability this is called the sample space) U containing all other sets.

For any set $A \subset U$:

$A \cup U = U$ and $A \cap U = A$

2 There exists an empty set $\varnothing$.

For any set $A \subset U$:

$A \cup \varnothing = A$ and $A \cap \varnothing = \varnothing$

3 For any set $A \subset U$ there exists a unique complementary set that is denoted by A'.

$A \cup A' = U$ and $A \cap A' = \varnothing$

These rules are essential for probability theory.

The sample space is a *representation* of the collection of outcomes from a chance experiment and $P(A)$ is a *measure* of the likelihood that event A occurs, calculated from $n(A)$, the number of successful outcomes of A, divided by $n(U)$, the number of outcomes in the sample space.

> The probability of an event A occurring is defined as $P(A) = \frac{n(A)}{n(U)}$ where the sample space U is the set of all possible outcomes.

TOK

"Those who have knowledge, don't predict. Those who predict, don't have knowledge." – Lao Tzu.

Why do you think that people want to believe that an outside influence such as an octopus or a groundhog can predict the future?

HINT

The empty set is a unique set containing no elements and is denoted as $\varnothing$.

International-mindedness

Pierre-Simon Laplace (1749–1827) was a French scholar who published a popular paper entitled " Essai philosophique sur les probabilitiés" in which he set out some principles of probability. The probability of an event happening was the number of favoured events, divided by the sum of the total events, which was later redefined as

Set theory leads to three fundamental axioms of probability.

The probability of any event occurring is a non-negative real number. This means that the smallest value a probability can ever take is zero, and that it cannot be infinite.

The probability of the whole sample space is 1. There are no events outside of the sample space. This implies that something with absolute certainty has a probability of 1. Therefore, the maximum value a probability can take is 1. This, coupled with the previous paragraph, implies the probability of an event occurring must lie between 0 and 1.

These give rise to the formal axioms:

Axiom 1

For any event A, $P(A) \geq 0$

All probabilities take a value which is greater than or equal to zero.

Axiom 2

$P(U) = 1$

The probability of all occurrences is equal to one.

Corollary 1

$0 \leq P(A) \leq 1$

International-mindedness

Probability theory was first studied to increase the chances of winning when gambling. The earliest work on the subject was by Italian mathematician Girolamo Cardano in the 16th century.

Investigation 1

Out of 100 co-workers in a factory, 65 speak English, 27 speak Spanish and 32 speak French.

Some factory workers speak all three of these languages; 5 speak English and French but not Spanish; **1** worker does not speak French, English, or Spanish; 10 speak English and Spanish but not French; no one who does not speak English speaks French and Spanish. Let E represent speaking English, F represent speaking French and S represent speaking Spanish.

1 Represent this information in a Venn diagram.

2 Find the number of factory workers who speak all three languages.

3 Use the numbers given in the Venn diagram to calculate the following probabilities.

a $P(F \cap S \cap E)$ **b** $P(F')$ **c** $P(E \cup F)$ **d** $P(E \cup F')$

4 Use the Venn diagram to show the following results.

a $P(E \cup U) = P(U)$ **b** $P(F \cap U) = P(F)$

c $P(F \cap \varnothing) = P(\varnothing)$ **d** $P(S \cup \varnothing) = P(S)$

e $P(E \cup E') = P(U)$ **f** $P(E \cap E') = P(\varnothing)$

g $P(U) = 1$

5 **Conceptual** How do the results from **3c** relate to the individual probabilities $P(E)$ and $P(F)$.

6 **Conceptual** In words, how do the results from 3c relate the individual probabilities $P(E)$ and $P(F)$? Why do you subtract the intersection when finding the probability of E and F?

From the investigation, it is evident that these definitions hold.

$P(A) = \frac{n(A)}{n(U)}$ also holds for more than one event.

$P(A \cap B) = \frac{n(A \cap B)}{n(U)}$

Question **5** leads into the addition rule theorem for probability:

$$P(A \cup B) = P(A) + P(B) - P(A \cap B)$$

If an event A can occur or an event B can occur, but A and B cannot both occur, then the two events A and B are said to be **mutually exclusive**. In this case, $P(A \cap B) = 0$. A and B are also called disjoint sets.

If A and B are events and there is no intersection between them, then:

$P(A \cup B) = P(A) + P(B)$ $\quad$ $P(A \cap B) = 0$

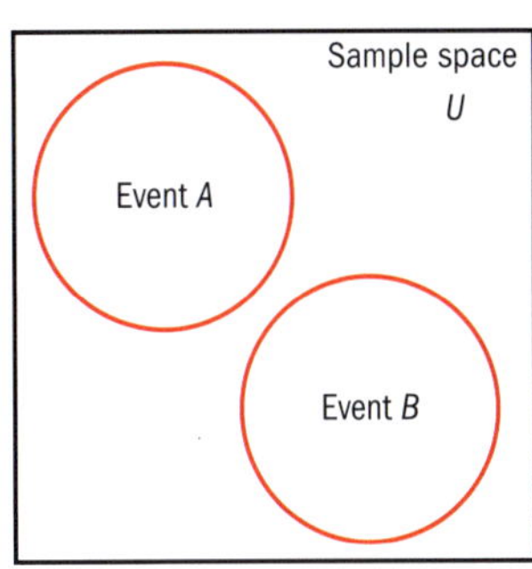

An excellent representation of this is given using the Venn diagram. These events are called mutually exclusive events and there are no overlaps between the two sets. This process also works for a finite number of mutually exclusive events.

Axiom 3

If $A_1, A_2, A_3, \ldots, A_n$ is a set of mutually exclusive events, then

$P(A_1 \cup A_2 \cup A_3 \ldots. \cup A_n) = P(A_1) + P(A_2) + P(A_3) + \ldots + P(A_n)$

These three Axioms are the basis for the axiomatic probability function and prescribe the formal definitions of probability theory. From these and other corollaries, propositions and theorems coupled with the non-empty set U and all partitive subsets of U, a complete axiomatic system of probability space can be built.

Complementary events

If A is an event, then $P(A') = 1 - P(A)$

Proof:

By definition: $P(A \cup A') = P(U)$

By Axiom 2: $P(U) = 1$

$\therefore P(A \cup A') = 1$

A and A' are mutually exclusive

Hence,

$P(A) + P(A') = 1$

and

$P(A') = 1 - P(A)$

Statistics and probability

The probability of the empty set is zero

$P(\emptyset) = 0$

Proof:

- $U' = \emptyset$ definition
- $P(U) = 1$ axiom 2
- U and $\emptyset$ are disjoint and therefore
- $P(U \cup \emptyset) = P(U) + P(\emptyset)$
- $P(U) = P(U) + P(\emptyset)$
- $1 = 1 + P(\emptyset)$
- $P(\emptyset) = 0$

$P(A) = \leq 1$

Proof:

- $P(U) = 1$ Axiom 2
- $P(A) + P(A') = 1$
- $P(A) = 1 - P(A')$
- $P(A') \geq 0$ Axiom 1
- $P(A) \leq 1$

With this information it is possible to answer basic probability questions.

Example 1

Take the letters from the word MATHEMATICS. A letter is drawn at random from the word. Find the probability that the selected letter is:

a A **b** a vowel **c** C **d** the letter A or the letter C

e the letter A or a vowel.

Let A be the event of obtaining the letter A. Let B be the event of obtaining a vowel. Let C be the event of obtaining the letter C. All probabilities are greater than or equal to 0	
The sample set is exhaustive; each letter is included. $n(U) = 11$	$P(A \cup B) = P(A) + P(B) - P(A \cap B)$
a $P(A) = \frac{2}{11}$	There are two letter As in the list of 11 letters.
b $P(B) = \frac{4}{11}$	$n(B) = 4$ $P(B) = \frac{n(B)}{n(U)}$
c $P(C) = \frac{1}{11}$	There is only one letter C in the list of 11 letters.

d $P(A \cup C) = P(A) + P(C) = \frac{3}{11}$	The events A and C are mutually exclusive.
e $P(A \cup B) = P(A) + P(B) - P(A \cap B)$	A is a subset of B
$P(A \cup B) = \frac{2}{11} + \frac{4}{11} - \frac{2}{11} = \frac{4}{11}$	$P(A \cup B) = P(B)$

Combinations and permutations often occur in probability questions.
Recall definitions from Chapter 1:

Definition ${}^nC_r = \frac{n!}{r!(n-r)!}$

Example 2

There are 12 girls and 10 boys in a class. Two students are selected for the school government.

a In how many different ways can two students be selected from the class?

b Find the probability that the selected students are

i both girls **ii** both boys **iii** each of a different gender.

Let G denote the outcome a girl is chosen and B denote the outcome a boy is chosen	
a ${}^{22}C_2 = \frac{22!}{2! \times 20!} = \frac{22 \times 21}{2} = 231$	There are 12 + 10 = 22 students in the class, from whom you choose 2.
b i $P(2G) = \frac{{}^{12}C_2}{231} = \frac{66}{231}$	There are 12 girls and you choose 2 from the girls over the number of all **possible** outcomes.
ii $P(2B) = \frac{{}^{10}C_2}{231} = \frac{24}{231}$	There are 10 boys and you choose 2 over the number of all possible outcomes.
iii $P(GB) = \frac{12 \times 10}{231} = \frac{120}{231}$	You choose 1 girl and 1 boy out of 12 girls and 10 boys over the number of all possible outcomes.

Exercise 11A

1 In a class of 20 students, 10 study French, 8 study German and 5 study neither. One student is selected at random. Find the probability that the selected student studies:

a both French and German language

b exactly one of those two languages.

2 Four letters are picked from the word WONDERFUL. What is the probability that there is at least one vowel among the letters?

3 There are 6 oranges and 3 kiwi fruits in a fruit bowl. Two pieces of fruit are selected at random. Find the probability that the following were selected:

a two kiwi fruits

b two different pieces of fruit.

4 There are 7 green and 5 blue crayons in a pencil case. Three crayons are selected at random. Find the probability that the crayons selected are

a all green

b not all of the same colour.

Statistics and probability

5 The events A and B are such that $P(A) = 0.3$, $P(B) = 0.5$ and $P(A \cap B) = 0.2$. Find the following probabilities:

a $P(A \cup B)$ **b** $P(B \cap A')$ **c** $P(A' \cap B')$

6 The events A and B are such that $P(A) = 0.4$, $P(B) = 0.6$ and $P(A \cup B) = 0.7$. Find the following probabilities:

a $P(A \cap B)$ **b** $P(A \cap B')$ **c** $P(A' \cup B')$

7 The following sets are given.
$U = \{x | x \in \mathbb{Z}^+, x \le 3\}$

$A = \{x \in U: x \text{ is multiple of } 3\}$
$B = \{x \in U: x \text{ is multiple of } 5\}$.

a Use a Venn diagram to represent the information above.

b Find the following probabilities.

i $P(A)$ **ii** $P(B)$

Conditional probability

Tree diagrams allow excellent visual representations of probability calculations. There are two categories and it is important to ascertain which situation you are representing.

If the occurrence of an event A does not influence in any way the probability of an event B, then the event B is said to be independent of event A. Alternatively, the two events A and B are said to be independent.

In tree diagrams for repeated events there are two types: with replacement and without replacement.

1 Tree diagrams for repeated events with replacement

The tree diagram shows the outcome of picking a disc from a bag which contains 7 blue discs and 3 red discs. A disc is chosen, the colour is noted and the disc is returned to the bag before the second choice.

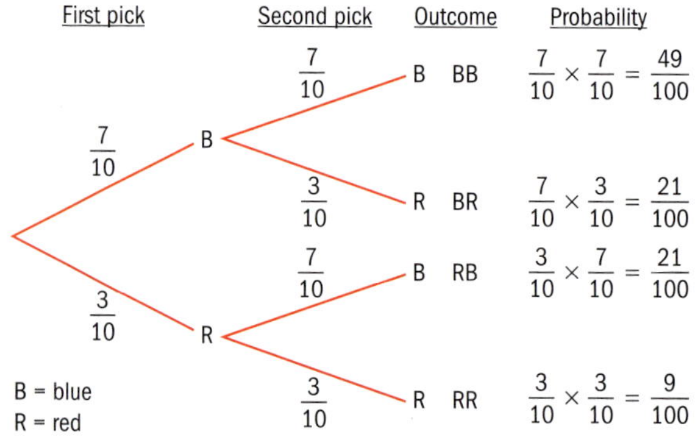

2 Tree diagrams for repeated events without replacement

The tree diagram shows the outcome of picking two discs from a bag which contains 7 blue discs and 3 red discs without returning the first chosen disc to the bag.

In tree diagrams for repeated events with replacement, the outcomes are always independent. In tree diagrams for repeated events without replacement, the outcomes are not independent.

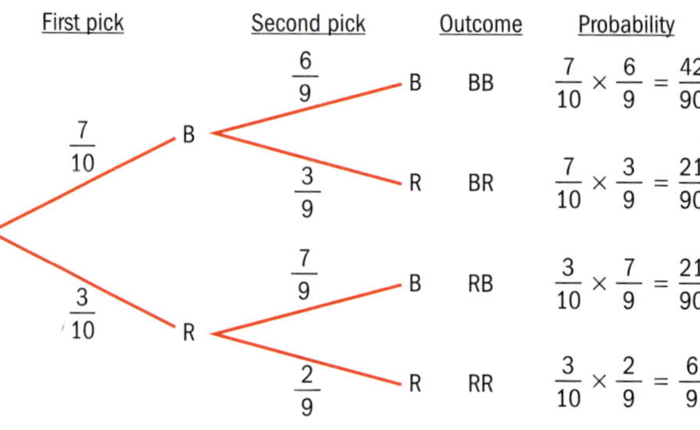

Tree diagrams for successive events

Investigation 2

In tree diagrams for successive events more care is needed.

The diagram illustrates the situation of drawing two cards successively from a deck.

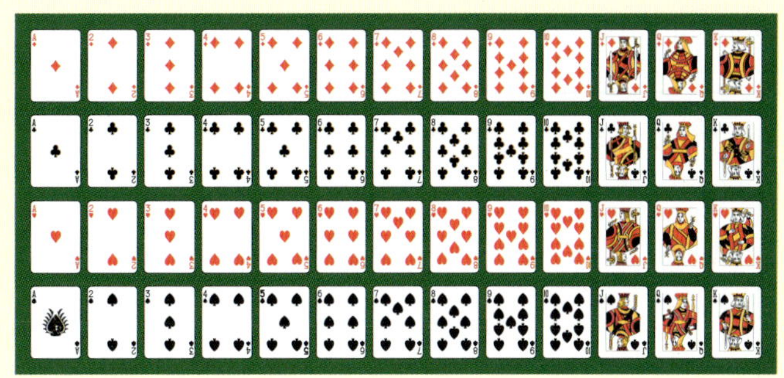

A deck of cards is divided into four suits as shown in the picture. There is an ace in each suit. 26 cards are black and 26 are red.

$P(A)$ is the probability that a black Ace is drawn.

$P(B)$ is the probability that a black card is drawn.

$P(B|A)$ is the probability that a black card is drawn given the first card was an ace.

1 Find $P(A)$ and $P(A')$

2 Copy and complete the tree diagram.

3 What do the four outcomes on the far right represent?

(Explain your answer)

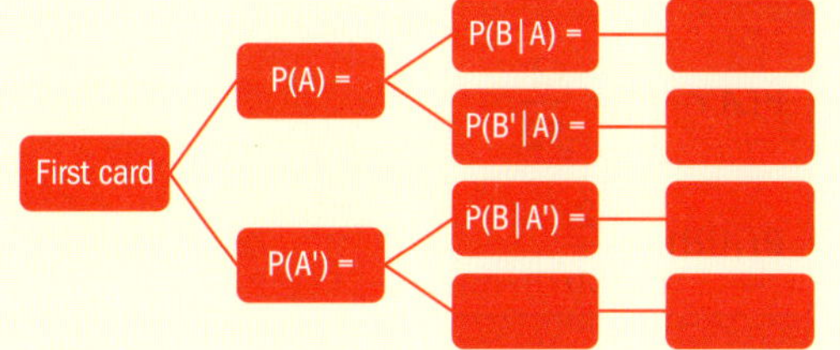

4 How would you calculate the outcomes on the far right?

5 How can you verify the answer you obtain when you add the values for the outcomes on the far right?

In tree diagrams for successive events, more calculations are necessary to ascertain the independence.

The mathematical notation, $P(B|A)$ is read as the probability of event B occurring **given** that A has occurred. This is known as conditional probability.

Therefore, in a successive event tree diagram the outcomes would be as shown in the diagram on the right.

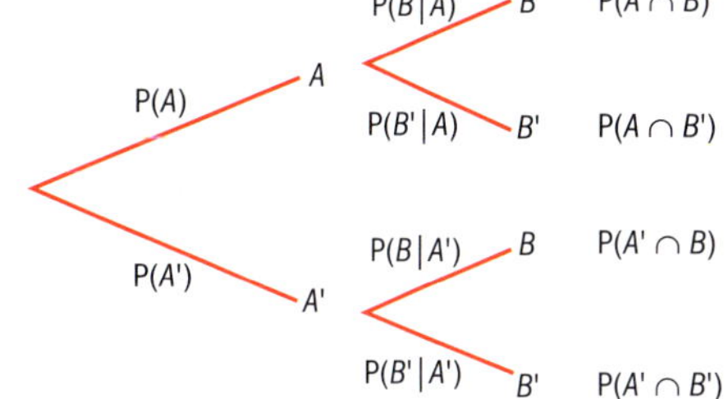

Multiplying along the branches for events A and B gives Axiom 4 and Theorem 3.

Axiom 4

$P(A \cap B) = P(A) \times P(B|A)$

If A and B are independent events $P(B|A) = P(B)$

$\therefore P(A \cap B) = P(A) \times P(B)$

HINT

Due to the symmetry properties of this axiom $P(A|B) = P(A)$

Theorem 3

$P(B|A) + P(B|A') = P(B)$

Statistics and probability

Example 3

There are 52 cards in a regular pack of playing cards. Marie draws one card.

Find the probability that Marie has drawn a black ace.

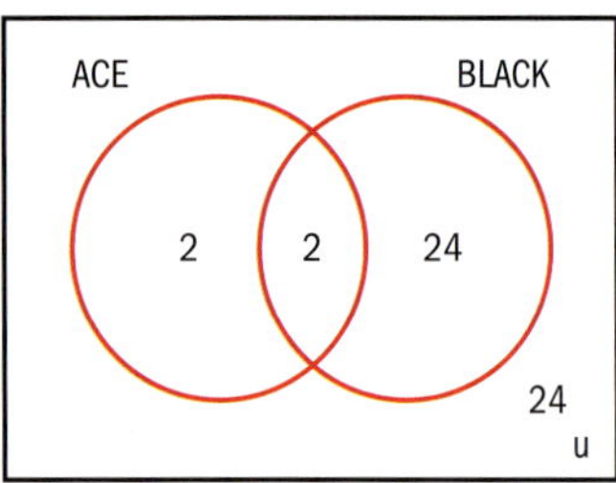	Draw the Venn diagram to show the results.
$P(A) = \frac{4}{52}$	Let A represent an ace and let B represent a black card.
$P(B) = \frac{26}{52}$	Examine the condition being applied.
$P(A \cap B) = \frac{2}{52}$	Sample space has been reduced by condition.
$P(A\vert B) = \frac{2}{26}$	Calculate the conditional probability.

There are many axioms that set up the whole system for probability, however for this course you will only need the following:

Axiom 1

For any event A, $P(A) \geq 0$

All probabilities take a value which is greater than or equal to zero.

Axiom 2

$P(U) = 1$

The probability of all occurrences is equal to one.

Corollary

$0 \leq P(A) \leq 1$

For two events which are mutually exclusive, $P(A \cup B) = P(A) + P(B)$ and $P(A \cap B) = 0$

Axiom 3

If $A_1, A_2, A_3..., A_n$ is a set of mutually exclusive events then

$P(A_1 \cup A_2 \cup A_3...) = P(A_1) + P(A_2) + P(A_3) + ...$

TOK

Do you rely on intuition to help you make decisions?

Axiom 4

$P(A \cap B) = P(A) \times P(B|A)$

$\therefore P(A \cap B) = P(A) \times P(B)$

Theorem 3

$P(B|A) + P(B|A') = P(B)$

Example 4

The probability that a train is late is 0.3. If the train is late, the probability that Florian will make it to work on time is 0.6. If the train is not late, the probability that Florian will be on time for work is 0.9. Let A be the event "the train is late" and let B be the event "arrives at work on time".

a Draw a tree diagram to represent these events.

b Find the probability that the train is late and Florian is on time for work.

c Find the probability that Florian will make it on time.

d Show that the train being late and Florian being on time for work are not independent events.

a Let A be the event "the train is late" let B be the event "Florian is on time for work."

As this is a tree diagram for successive events, the second branch contains conditional probabilities.

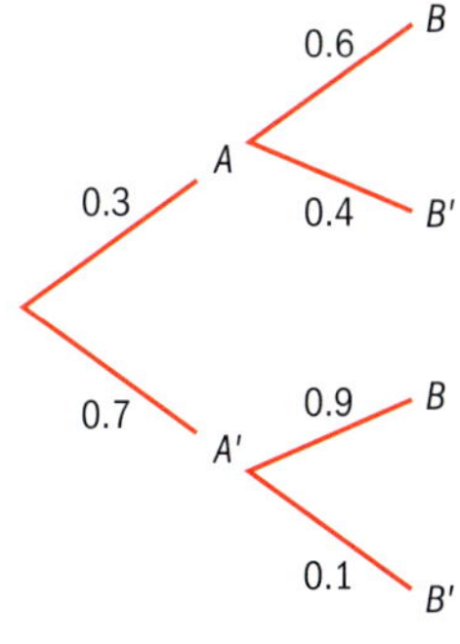

b $P(A \cap B) = P(A)P(B|A) = 0.18$

c $P(B) = P(A \cap B) + P(A' \cap B)$

These are mutually exclusive events, so we can add the probabilities.

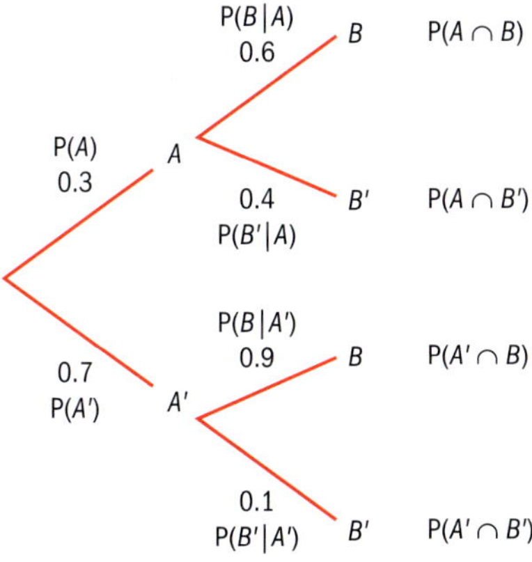

Continued on next page

Statistics and probability

d If A and B are independent events $P(B\|A) = P(B)$ $\therefore P(A \cap B) = P(A) \times P(B)$ In this case $P(B\|A) \neq P(B)$; therefore, they are not independent. $P(B\|A) + P(B\|A') = P(B)$. This is not the case either.	The probability of B differs depending on the event A. So the outcome of event A affects the outcome of B and therefore the events A and B are not independent.

Exercise 11B

1 Draw tree diagrams for the following situations. Determine whether the events are independent or not.

a One coin and two dice are thrown. Let A be the event "getting a tail" and let B be the event "getting two sixes."

b Jar A contains 3 red sweets and 2 blue sweets, Jar B contains 5 red sweets and 4 blue sweets. Norina takes a sweet from Jar A and a sweet from Jar B. Let A be the event "taking a red sweet from Jar A" and B be the event "taking a red sweet from Jar B."

c There are 10 candies in the bag; 5 of them are chocolate and 5 are caramel. Joel takes one candy and eats it, and then he eats a second candy. Let A be the event "the first candy is chocolate" and let B be the event "the second candy is caramel."

d The probability that Pete plays video games given that it is Saturday is 75%. On any other day of the week, the probability is $\frac{3}{4}$. Let B be the event "playing video games" and A be the event "the day is Saturday."

2 On a school field trip the students can choose to do zip lining, orienteering, neither or both.

There are 50 students and their choices are listed below.

The total number of students choosing zip lining is 32.

The total number of students choosing orienteering is 26.

4 choose not to take part.

a Illustrate the information on a Venn diagram.

b Find the probability that:

i at least one student went zip lining

ii a student went orienteering given that they went zip lining

iii a student went orienteering

iv a student went zip lining given that they went orienteering.

3 The Venn diagram shows the homework for 35 students

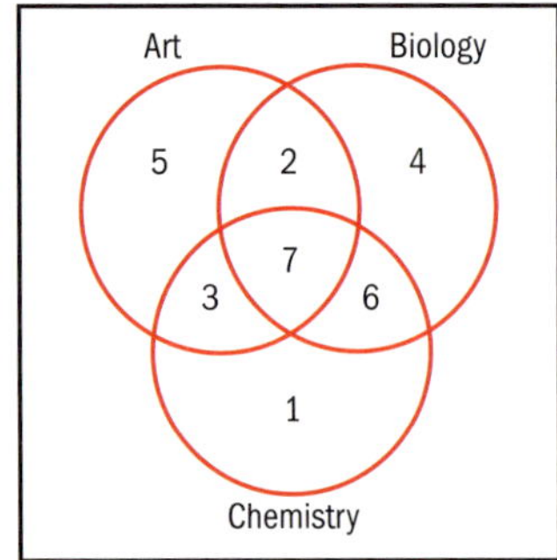

A student is chosen at random. Find the probability that the student:

a had art homework

b had biology homework

c had chemistry homework

d had art homework, given that they had homework in biology

e had biology homework given that they chemistry homework

f had chemistry homework given that they they had not been assigned any homework in art

g had homework in all three subjects, given that they were assigned homework on that day.

4 There are 3 blue and 5 yellow marbles in a jar. Two marbles are drawn simultaneously from the jar at random. Given that the marbles are of the same colour, find the probability that they are blue.

5 There is a shelf with toys, on which there are 8 puzzles and 5 musical instruments. A child takes three toys from the shelf.

a Draw a tree diagram to represent this information.

b Find the probability that:

i all the toys are puzzles

ii at least one is a puzzle.

6 Owen shoots two free throws in a basketball game. The probability that he scores the first shot is 0.85. The probability that he misses the second shot given that he scored the first shot is 0.10. The probability that he is going to score the second shot given that he missed the first shot is 0.75. Find the probability that Owen scores only one shot.

7 Given that

$P(A)=\frac{1}{3}$, $P(B|A)=\frac{3}{5}$ and $P(B|A')=\frac{1}{2}$, find

a $P(B')$

b $P(A' \cup B')$.

Bayes' theorem

Investigation 3

Three brothers Kaita, David and Tom have equal areas in a vegetable garden.

Kaita has 70 strawberry plants and 20 blueberry plants.

David has 40 strawberry plants and 30 blueberry plants.

Tom has 10 strawberry plants and 50 blueberry plants.

Their sister selects a plant at random to harvest.

1 If all the boundaries were ignored, find the probability their sister chooses a strawberry plant.

2 Draw a tree diagram to show the possibilities if she picks one of the plots at random first.

3 Find the probability that she picks a strawberry plant given that she has picked a plot at random.

4 Why are the answers for parts **1** and **3** not equal?

5 Given that she chose a strawberry plant, find the probability that it came from Kaita's area.

You already know that: $P(A \cap B) = P(A) \times P(B|A)$

This can be rearranged to

$$P(B|A) = \frac{P(A \cap B)}{P(A)}$$

In addition:

$P(A \cap B) = P(B) \times P(A|B)$

or

$$P(A|B) = \frac{P(A \cap B)}{P(B)}$$

International-mindedness

A well-known French gambler, Chevalier de Méré consulted Blaise Pascal in Paris in 1650 with questions about some games of chance. Pascal began to correspond with his friend Pierre Fermat about these problems which began their study of probability.

If $A_1, A_2, A_3 \ldots, A_n$ is a finite set of mutually exclusive events, then

$P(A_1 \cup A_2 \cup A_3 \cup \ldots \cup A_n) = P(A_1) + P(A_2) + P(A_3) + \ldots + P(A_n)$

and $A_1 \cup A_2 \cup A_3 \ldots \cup A_n = U$ are exhaustive events.

Axiom 1 $P(U) = 1$

and B is an arbitrary event of S. Then for $i = 1, 2, 3, \ldots, n$ it follows that

$$P(A_i|B) = \frac{P(A_i) \times P(B|A_i)}{P(B|A_1)P(A_1) + P(B|A_2)P(A_2) + \ldots + P(B|A_n)P(A_n)}$$

If 2 events A and B are such that $A \cup B = U$, where U is the total probability space, then $P(A \cup B) = 1$ and the sets A and B are said to be exhaustive.

Although this looks complicated, it is made clearer by the fact that the denominator is just the total probability of B, and often $n = 2$ where A_1 and A_2 are exhaustive events $(A_2 = A_1')$

This is commonly referred to as **Bayes' theorem**. In the examination, you will only be required to use Bayes' theorem for a maximum of three events.

Bayes' theorem can be used in various real life situations, as illustrated in the following examples.

Example 5

In a school there are 60% male teachers. Of the male teachers, 30% teach DP Maths, while 80% of the female teachers teach DP Maths. Given that a randomly selected student is studying DP Maths, find the probability that the teacher is female.

$P(M) = 0.6$ $P(M') = 0.4$ $P(DP\|M) = 0.3$ $P(DP'\|M) = 0.7$ $P(DP\|M') = 0.8$ $P(DP'\|M') = 0.2$	Write down what you know. 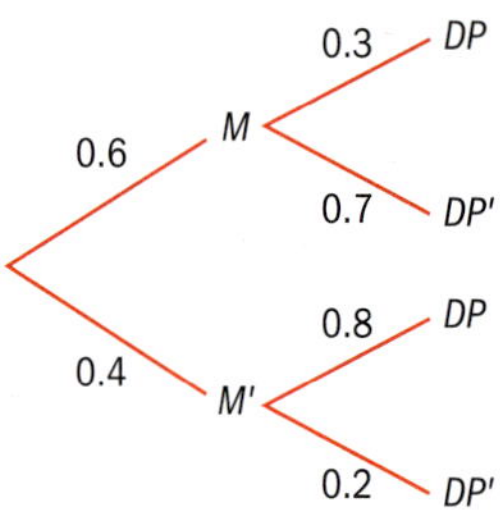
You require: $P(M'\|DP) = \frac{P(M') \times P(DP\|M')}{P(DP)}$	In this case, $n = 2$ and $A_2 = A_1'$
$P(DP) = P(DP\|M)P(M) + P(DP\|M')P(M')$ $P(DP) = (0.3)(0.6) + (0.8)(0.4)$ $P(M'\|DP) = \frac{0.4 \times 0.8}{0.5} = 0.64$	The denominator is the total probability.

Example 6

Three students, Lauren, Madi and Georgi, work in a coffee shop. In any give hour, Lauren will make 55% of the coffees, Madi will make 35% and Georgi will make 10%. The probability that Lauren, Madi and Georgi spill some coffee is 0.6, 0.2 and 0.1, respectively.

Given that a coffee is spilled, find the probability that it was served by Madi.

$P(A_1) = 0.55$ $P(A_2) = 0.35$ $P(A_3) = 0.1$ $P(S\|A_1) = 0.6$ $P(S\|A_2) = 0.2$ $P(S\|A_3) = 0.1$	Let A_1 be the even Lauren makes coffee. Let A_2 be the event Madi makes coffee. Let A_3 be the event Georgi makes coffee.
You require $P(A_2\|S)$. $P(A_2\|S) = \frac{P(A_2) \times P(S\|A_2)}{P(S\|A_1)P(A_1) + P(S\|A_2)P(A_2) + P(S\|A_3)P(A_3)}$ $P(A_2\|S) = \frac{0.35 \times 0.2}{(0.6 \times 0.55) + (0.2 \times 0.35) + (0.1 \times 0.1)}$ $= 0.171$ (3 d.p.)	Apply Bayes' theorem

Exercise 11C

1 There are two jars with yellow and green marbles. In jar A there are 5 yellow and 8 green while in jar B there are 2 yellow and 6 green marbles. Two marbles are transferred from jar A to jar B and then one marble is taken from jar B.

Find the probability that the marble drawn from jar B is yellow.

Given that the marble drawn is yellow, find the probability that it was originally from jar A.

2 In a town, citizens are classified in the following sections: high income, middle income, low income. Of the citizens, 10% are in the high income section and 25% are in the low income section. The percentages of high, middle and low income citizens that are female are 50%, 30% and 20%, respectively.

a Find the probability that a randomly selected citizen from the town is male.

b Given that a randomly selected citizen is male, find the probability that he is classified as high income.

c Given that a randomly selected citizen is high income, find the probability that they are female.

3 Three machines produce transistors. The first machine produces 50% of all the transistors, the second machine produces 45% of all the transistors and the third machine produces 5% of all the transistors. 4% of the transistors the first machine produces are defective, as are 3% of the transistors the second machine produces and 8% of those produced by the third machine. Given that a randomly selected transistor is not defective, find the probability that the transistor was produced by the second machine.

4 There are 20 identical laptops on a trolley, out of which 12 have a hard disk with a capacity of 160 GB and 8 with a capacity of 320 GB. A teacher randomly takes two laptops from the trolley. A student then takes a laptop from the trolley to complete a project. Given that the student took a laptop with 160 GB, find the probability that the teacher took both laptops with 320 GB.

5 In a basketball squad there are 12 players. In the squad, 4 players have a percentage shot of 0.9% 4 players have a percentage shot of 0.6% and the remaining players have a percentage shot of 0.2%. A randomly selected player shoots a basket. Find the probability that the player scores.

6 There are several blood group systems but the most important are ABO (O, A, B and AB) and Rh (+ or –) blood group systems. A population was tested and the following results were obtained. In the population, 45% has O type, out of which 16% are Rh–, 37% has A type, out of which 16% are Rh–, 14% has B type, out of which 17% are Rh–, and finally 4% has AB type, out of which 25% are Rh–.

A person is selected at random from the population.

a Find the probability that the person has Rh+.

b Given that the person has Rh–, find the probability that the person is AB.

11.2 Probability distributions

Discrete random variables

A discrete variable is one which can take values which are distinct and discrete. They are usually integer values. For example, in a class of N students, the variable might be the number of students who are left-handed. The variable could take values 0, or N or any whole number in between, but not 1.5, 21.28 or a negative number.

The actual number of left-handed students in a class of 25 will vary from class to class, hence this is a variable.

Investigation 4

Suppose the art teacher wanted to buy left-handed scissors for all the classes which he taught. To help make his decision of how many pairs of scissors to buy he needs to know the frequency distribution of the number of left-handed students in a class of 25 and then construct a model to predict the situation.

He has researched on the internet that the probability of being left-handed is 0.1. Therefore, he can use a random number generator simulation to gather data where obtaining the digit 0 forms one-tenth of the population and represents the left-handed people, and the digits 1–9 represent the right-handed people.

1 Using your GDC set up a random number generator to generate 35 numbers.

2 Record the occurrences of the digits 1–9 and the digit 0.

3 How does this situation mimic the real-life situation?

4 In your class, take 500 samples of 25 numbers in this way.

Number of occurrences of 0	0	1	2	3	4	5	6	7	>7	
Frequency										500

5 Record the results in a spreadsheet or table.

By taking 500 samples in this way you have generated the frequency distribution of the number of 0s in random samples of 25. The variable is "the number of 0s" in a sample of 25 random numbers.

6 What are the values the variable can take?

7 What relation does it have to the variable the art teacher is really interested in?

8 Use your table to establish the relative frequency for the distribution.

9 From the data determine how many pairs of left-handed scissors he should buy.

In this investigation two assumptions have been made:

1) The proportion of left-handed students is constant from class to class, and
2) the number of left-handed students in each class is random.

This process is time-consuming, and an alternative approach is to construct a theoretical model.

The random variable has been specified and the probability distribution can be derived using probability theory.

You will then be able to make predictions.

As n increases, the experimental probability gets closer to theoretical probability.

The variable X being considered in the investigation is the changing value "number of left-handed students in a class of 25 students." It can only take exact values ($x = 0, 1, 2, 3 \ldots$) and therefore X is a **discrete variable**.

For a discrete random variable the sum of the probabilities must equal 1. Often sigma notation is used to express this:

$$X \text{ is a random variable } \Leftrightarrow \sum_{\text{all } x} \mathrm{P}(X = x) = 1$$

HINT

A discrete random variable can only have exact values. For example, the score on a dice can only be an integer between 1 and 6.

X is the variable

x_i is a value that X can take

p_i is the associated probability for that value $p_i = \mathrm{P}(X = x_i)$

For discrete random variables, the probability distribution function can be represented by a table or a formula. If the data collected from an experiment is large enough, the two approach each other.

Example 7

The probability distribution function of a discrete random variable X is given by
$\mathrm{P}(X = x_i) = \frac{1}{k}x^2$, $x = 0, 1, 2, 3, 4$

a Complete a probability distribution table. Find the constant k.

b Find the probability that X is less than or equal to 2.

c Find the mode of the distribution.

d Represent the distribution with a vertical line graph.

a

x	0	1	2	3	4
p	0	$\frac{1}{k}$	$\frac{4}{k}$	$\frac{9}{k}$	$\frac{16}{k}$

$\sum_{\text{all } x} \mathrm{P}(X = x) = 1$

You are told the discrete variable is random, therefore:

$\mathrm{P}(U) = 1$

$$0 + \frac{1}{k} + \frac{4}{k} + \frac{9}{k} + \frac{16}{k} = 1$$

$$\frac{30}{k} = 1$$

$$k = 30$$

b $\mathrm{P}(X = x_i) = \frac{1}{30}x^2$

The probability distribution function is $\frac{1}{x}x^2$.

and the associated probability distribution table is:

x	0	1	2	3	4
p	0	$\frac{1}{30}$	$\frac{4}{30}$	$\frac{9}{30}$	$\frac{16}{30}$

$$\mathrm{P}(X \le 2) = \mathrm{P}(X = 0) + \mathrm{P}(X = 1) + \mathrm{P}(X = 2) = \frac{5}{30}$$

c The mode is 4.

d

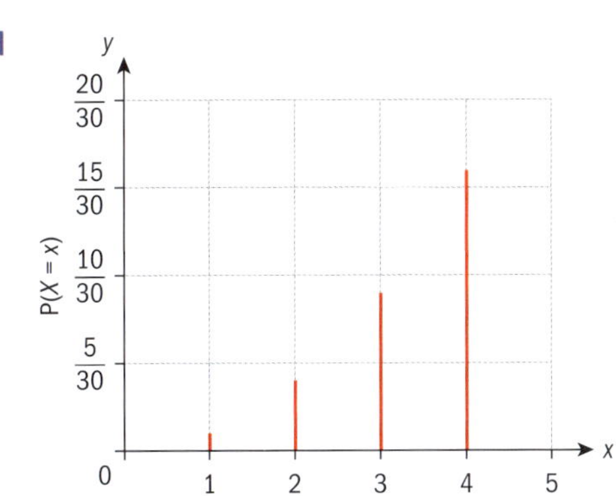

The mode is the value which takes the highest probability.

Exercise 11D

1 Determine whether the following probability distribution functions are valid.

a

x	0	1	2	3	4
$P(X = x)$	0	0.1	0.2	0.3	0.4

b

x	0	1	2	3	4
$P(X = x)$	0.3	0.3	0.3	0.3	0.3

2 By drawing the probability distribution table, determine whether the following functions describe a discrete random variable. If a discrete random variable is described find

a the mode **b** $P(X > 3)$

i $P(X = x_i) = \frac{1}{2}x^3,\ x = 0, 1, 2, 3, 4$

ii $P(X = x_i) = \frac{x!}{5x + 2},\ x = 0, 1, 2, 3, 4$

3 The discrete random variable has PDF given by: $P(X = kx),\ x = 10, 11, 12$

a Find the value of k.

b Find the mode.

c Find the probability that X takes an even number.

Investigation 5

Two tetrahedral dice have sides labelled with 1, 2, 3, 4, and are both rolled. The sum of the two outcomes is recorded. On a tetrahedral die, the outcome recorded is the face which lands on the ground.

1 Draw a sample space for the outcomes.

2 Construct a probability distribution table to represent the associated probabilities.

3 Find the probability distribution function. Note: it may be useful to find two piecewise functions: one for the ascending probabilities and one for the descending probabilities.

4 Verify that X is a random variable.

5 **Factual** Is there more than one probability distribution that X can take?

6 **Conceptual** Why is representing outcome on a sample space useful?

TOK

Is it possible to reduce all human behaviour to a set of statistical data?

Statistics and probability

Expectation

It is important to establish the difference between theoretical probability and experimental probability, which is the ratio between the recorded number of successes to the total number of trials performed. It is assumed that as the number of trials increases, the experimental probability will become closer to the theoretical probability.

This premise works for probability distributions.

If we have a statistical experiment, we record the results in a table and use these values to make assumptions.

For example, an unbiased dice is thrown 60 times and the results are recorded in a frequency distribution table:

x	1	2	3	4	5	6
Freq f	9	9	12	8	11	11

From this you can work out the experimental probability:

$P(X = 1) = \frac{9}{60}$ $P(X = 2) = \frac{9}{60}$ $P(X = 3) = \frac{12}{60}$ and so on.

In addition, you can establish the mean score obtained:

$$\bar{x} = \frac{\sum fx}{\sum f} = \frac{9+18+36+32+55+66}{60} = 3.6$$

In theoretical probability distributions, this value is called expectation. Expectation defines the value you would expect as the number of trials increases, and is written as $E(X)$.

x	1	2	3	4	5	6
$P(X = x)$	$\frac{1}{6}$	$\frac{1}{6}$	$\frac{1}{6}$	$\frac{1}{6}$	$\frac{1}{6}$	$\frac{1}{6}$

$$E(X) = \Sigma\, x \times P(X = x)$$
$$= (1\times\frac{1}{6}) + (2\times\frac{1}{6}) + (3\times\frac{1}{6}) + (4\times\frac{1}{6}) + (5\times\frac{1}{6}) + (6\times\frac{1}{6})$$
$$= 3.5$$

$E(X)$ can also be written as μ.

International-mindedness

In 1933 Russian mathematician Andrey Kolmogorov built up probability theory from fundamental axioms in a way comparable with Euclid's treatment of geometry that forms the basis for the modern theory of probability. Kolmogorov's work is available in an English translation titled "The Foundations of Probability Theory".

Exercise 11E

1 The probability distribution for the random variable X is shown in the table.

x	0	1	2	3	4
$\mathrm{P}(X=x)$	$\frac{1}{6}$	$\frac{1}{6}$	$\frac{1}{3}$	$\frac{1}{4}$	$\frac{1}{12}$

Find $\mathrm{E}(X)$.

2 The probability distribution for the random variable X is shown in the distribution table.

x	4	5	6	7	8
$\mathrm{P}(X=x)$	0.1	0.2	k	$2k$	$k-0.1$

Find k and $\mathrm{E}(X)$.

3 A discrete random variable X can take the values 0, 1, 2, 3 only.

Given
$\mathrm{P}(X \le 2) = 0.85$
$\mathrm{P}(X \le 1) = 0.5$
$\mathrm{E}(X) = 1.45$

a construct a probability distribution table

b find the mode.

4 A student council delegation of three is to be chosen from four females and six males. Find the expected number of females on the delegation if the students are selected at random.

5 The probability distribution table is given.

x	1	2	3	4	5	6
$\mathrm{P}(X=x)$	0	$2c$	c	$3c^2$	$3c^2+c$	c

a Complete the table with the associated probabilities.

b Find $\mathrm{E}(X)$.

Expectation of a linear function of a discrete random variable

Investigation 6

Consider the random variable Y, such that Y is obtained by increasing the values of the random variable X by a constant 3.

1 Write down an expression for Y in terms of X and 3.

2 Will the associated probabilities change?

3 Copy and complete a probability distribution table of X and Y.

x	0	1	2	3	4
y					
P	$\frac{1}{6}$	$\frac{1}{4}$	$\frac{5}{24}$	$\frac{1}{8}$	$\frac{1}{4}$

4 Find $\mathrm{E}(X)$.

5 Find $\mathrm{E}(Y)$.

Continued on next page

6 Generalize a result for $E(R + c)$ where R is a random variable and c is a constant.

7 **Conceptual** What are the effects of linear functions on random variables?

8 **Conceptual** Why is it useful to look at the effects of linear functions on discrete random variables?

International-mindedness

The Dutch scientist Christian Huygens, a teacher of Leibniz, published the first book on probability in 1657.

Developers of games use expectation in order to determine risk and fairness.

Example 8

A game is developed such that a die is thrown and the banker pays out three times the value in dollars for the score on the die. How much would you pay to play this game?

In order to look at this question, it is important to construct a probability distribution table.

X	1	2	3	4	5	6
$Y = 3X$	3	6	9	12	15	18
$P(X = x)$	$\frac{1}{6}$	$\frac{1}{6}$	$\frac{1}{6}$	$\frac{1}{6}$	$\frac{1}{6}$	$\frac{1}{6}$

$$E(Y) = (3\times\frac{1}{6}) + (6\times\frac{1}{6}) + (9\times\frac{1}{6}) + (12\times\frac{1}{6}) + (15\times\frac{1}{6}) + (18\times\frac{1}{6})$$

$$= 10.5$$

You can see from this that multiplying the variable by 3 also multiplies the expected value by 3.

The banker would expect to pay out \$10.50 on average each time.

You could win anything between \$3 and \$18 - you decide.

To play, \$10.50 would be a fair amount.

From the example you can see that $E(2X) = 2E(X)$. This leads to the first probability distribution theorem.

If a and b are constants, a cannot equal 0 and $aX + b$ is a new random variable derived from the discrete random variable X, then

$P(X = x) = p$ and $P(X = aX + b) = p$

This implies

$E(aX + b) = \sum (ax_i + b)p_i$

and

$E(aX + b) = aE(X) + b$

Variance

The variance σ^2 of a set of observations with mean μ is defined as

$$\sigma^2 = \frac{\sum_{i=1}^{k} f_i(x_i-\mu)^2}{n} = \frac{\sum_{i=1}^{k} f_i x_i^2}{n} - \mu^2$$

You can use this to define the variance of a discrete random variable as

$\text{Var}(X) = \sum p_i(x_i - \mu)^2$
where $\mu = \text{E}(X)$.

Alternatively,

$$\begin{aligned}\text{Var}(X) &= \text{E}(X-\mu)^2\\ &= \text{E}(X^2 - 2X\mu + \mu^2)\\ &= \text{E}(X^2) - 2\mu\text{E}(X) + \text{E}(\mu^2) \quad \text{where } \mu = \text{E}(X) \text{ and } \text{E}(\mu^2) = \mu^2\\ &= \text{E}(X^2) - 2\mu^2 + \mu^2\\ &= \text{E}(X^2) - \mu^2\\ &= \text{E}(X^2) - \text{E}^2(X)\end{aligned}$$

The standard deviation is the square root of the variance.

HINT

$(\text{E}(X))^2$ is written $\text{E}^2(X)$ in a similar way that $(\sin(X))^2$ is written $\sin^2(X)$.

TOK

What does it mean to say that mathematics can be regarded as a formal game lacking in essential meaning?

Example 9

Two discs are drawn without replacement from a box containing four blue discs and three red discs. If X is the discrete random variable "the number of blue discs drawn"

a construct a probability distribution table

b find $\text{E}(X)$

c find $\text{E}(X^2)$

d find $\text{Var}(X)$

e find the standard deviation of X.

a $P(X=2) = \frac{12}{42}$

$P(X=1) = \frac{24}{42}$

$P(X=0) = \frac{6}{42}$

X	0	1	2
$P(X=x)$	$\frac{6}{42}$	$\frac{24}{42}$	$\frac{12}{42}$

Draw a tree diagram to show results

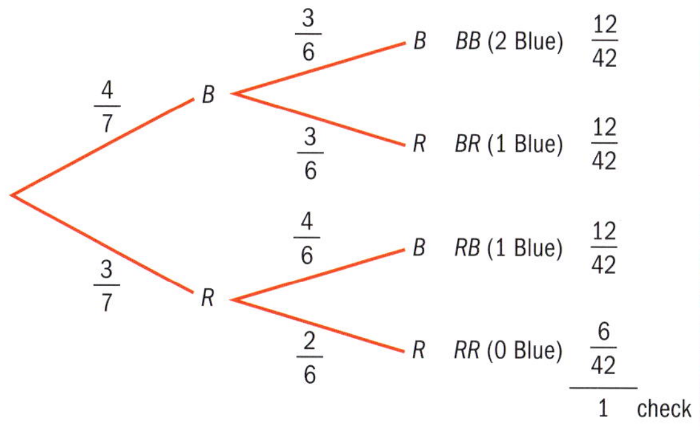

Continued on next page

Statistics and probability

b $E(X)=0\times\frac{6}{42}+1\times\frac{24}{42}+2\times\frac{12}{42}$ $=\frac{8}{7}=1.14(3\text{ s.f})$	Calculate E(X) add formula
c $E(X^2)=0\times\frac{6}{42}+1^2\times\frac{24}{42}+2^2\times\frac{12}{42}$ $=\frac{12}{7}=1.71(3\text{ s.f})$	Calculate E(X^2)
d $\text{Var}(X)=\frac{12}{7}-\left(\frac{8}{7}\right)^2$ $=\frac{20}{49}=0.408(3\text{ s.f})$	$\text{Var}(X)=E(X^2)-(E(X)^2)$
e Standard deviation $=\sqrt{\frac{20}{49}}=0.639(3\text{ s.f})$	Standard deviation $=\sqrt{\text{variance}}$

Proving standard results for linear expectation

Some standard results are very useful.

$\text{Var}(a) = 0$

Proof:

$\text{Var}(a) = E(a^2) - E^2(a)$

$= a^2 - a^2$

$= 0$

$\text{Var}(aX) = a^2(\text{Var}(X))$

Proof:

$\text{Var}(aX) = E((aX)^2) - E^2(aX)$

$= a^2E(X^2) - a^2E^2(X)$

$= a^2(E(X^2) - E^2(X))$

$= a^2(\text{Var}(X))$

Therefore,

$\text{Var}(aX + b) = (E(aX)^2) - E^2(aX)) + (E(b^2) - E^2(b))$

$\text{Var}(aX + b) = a^2(\text{Var}(X))$

TOK

Do ethics play a role in the use of mathematics?

Exercise 11F

1 A discrete random variable X has a PDF given by the table:

x	1	2	3
$P(X = x)$	$\frac{1}{8}$	$\frac{3}{8}$	$\frac{1}{2}$

Find the value of:

a $E(X)$ b $E(5X)$

c $E(X^2)$ d $Var(X)$

e the standard deviation of X.

2 A discrete random variable X can take only the values 0, 1, 2, 3, 4 and 5. The probability distribution of X is given by the following table:

x	0	1	2	3	4
$P(X = x)$	a	a	a	$2a$	b

Given that $P(X \geq 1) = 2P(X \leq 2)$, find

a the values of a and b

b $E(X)$ and $E(X^2)$

c the value of $Var(X)$.

3 A selection of 10 playing cards numbered from 1 to 10 is shuffled.

Two cards are taken, one from the bottom and one from the top of the pack.

Let S be the random variable "sum of the numbers on the top and bottom card."

a Find $P(S = 4)$, $P(S = 8)$ and $P(S = 11)$.

b Construct a table for the probability distribution of S.

c Find $E(S)$ and $Var(S)$.

4 A random variable T that takes only integer values has a probability distribution function defined by

$$f(t) = P(T = t) = \begin{cases} k(8-t)^2 & \text{if } t = 4, 5, 6, 7 \\ kt^2 & \text{if } t = 1, 2, 3 \end{cases},$$

where k is a constant.

a Find the value of k.

b Find $P(T = 4)$, $P(T \leq 4)$ and $P(T = 4 | T \leq 4)$.

c Find $E(T)$ and $Var(T)$.

d Determine the mode of T.

Developing inquiry skills

Consider the first game in the opening problem for the chapter. If you played this game a few times, do you think you would end up with more tokens than you started with? How could we tell?

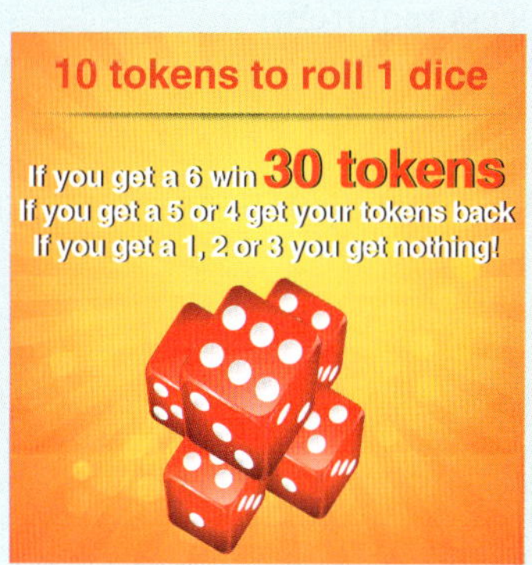

10 tokens to roll 1 dice
If you get a 6 win 30 tokens
If you get a 5 or 4 get your tokens back
If you get a 1, 2 or 3 you get nothing!

Define a random variable for the winnings from playing the game once.

Draw a probability distribution for your random variable.

Calculate the expected winnings from the game.

How much would you expect to win or lose if you played the game 10 times? 100 times?

Is the game "fair"?

How could you define a fair game?

What adjustment could you make to the prizes that would ensure that this is a fair game?

11.3 Continuous random variables

In Chapter 5 we defined continuous values as:

Continuous variables (data) can be measured (for example, weight and height) and can take any numerical value within a given range.

A continuous random variable is a variable which can take continuous values, ie not discrete.

Similar to a discrete random variable forming a probability density function, a continuous random variable forms a probability density function - the difference being that there is a continuous range of values for which x is valid and the function will be represented by a curve. The function is such that the probability of the variable taking any value within a particular interval $[a, b]$ equals the area under the curve.

The behaviour of a continuous random variable X is described by a function on the interval of real numbers called "probability density function of X" or simply "PDF of X."

For a function f to be a probability density function of a continuous random variable X (PDF of X), it must verify the following axioms:

1 $f(x) \geq 0$ for all values $x \in \mathbb{R}$.

2 $\int_{-\infty}^{+\infty} f(x)\, dx = 1$

HINT

This definition of a probability density function is a general definition. In most situations the density function is only positive for values in an interval $[a, b]$ and takes the value zero elsewhere. For this reason, it is common to restrict the domain of PDF functions to the interval $[a, b]$. This is particularly convenient when you need to test condition 2. By considering the restriction, you just need to verify if $\int_a^b f(x)\, dx = 1$.

The PDF is given by a function (curve) defined over an interval and the curve must be positive over the interval. The probabilities are given by the area under the curve. Each probability will still be positive and the sum of the probabilities will still equal 1, and this is equivalent to the area under the whole curve.

In order to establish the associated probabilities, it is necessary to find the area under the curve between the required values and therefore definite integration is the process required.

$$P(x_1 \leq X \leq x_2) = \int_{x_1}^{x_2} f(x)dx$$

the probability is determined by the integration of the curve between the limits.

Example 10

A continuous random variable X has a PDF $f(x)$ such that $f(x) = \begin{cases} k & 0 \leq x < 2 \\ k(2x-3) & 2 \leq x \leq 3,\ k > 0 \\ 0 & \text{otherwise} \end{cases}$

a Find the value of the constant k.

b Sketch a graph of the function.

c Find the following, both geometrically and using integration:

$P(x \geq 2)$

$P(1 \leq X \leq 3)$

$P(x \geq 4)$.

a $\int_0^2 k\mathrm{d}x + \int_2^3 k(2x-3)\mathrm{d}x = 1$	$\int_{-\infty}^{\infty} f(x)\mathrm{d}x = 1$
$k[x]_0^2 + k\left[\frac{2x^2}{2} - 3x\right]_2^3 = 1$ $2k + 2k = 1$ $k = \frac{1}{4}$	Solve for k.
b 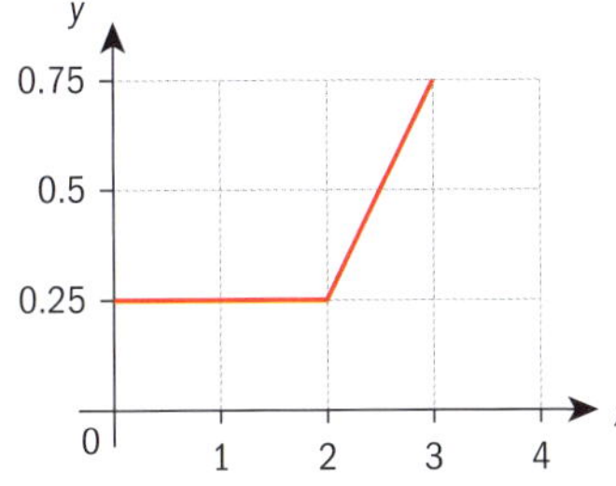	Sketch the graph.
c $\mathrm{P}(x \geq 2)$ 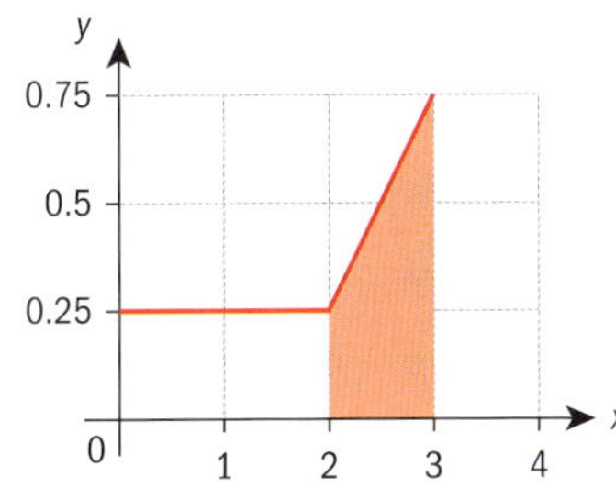	
The area of the trapezium is $\frac{0.25 + 0.75}{2} \times 1 = 0.5$	As $\mathrm{P}(x \geq 2)$, required area lies strictly below the second part of the graph.
$\int_2^3 \frac{1}{4}(2x-3)\mathrm{d}x = 0.5$	Confirm the result of the geometrical figure using integration.
$\mathrm{P}(1 \leq X \leq 3)$ Area of rectangle + area of trapezium $= 1 \times 0.25 + 0.5 = 0.75$	
$\int_1^2 \frac{1}{4}\mathrm{d}x + \int_2^3 \frac{1}{4}(2x-3)\mathrm{d}x = 0.75$	As this is not in the domain, the answer is 0.
$\mathrm{P}(x \geq 4) = 0$	

Statistics and probability

The function shown above is known as a piecewise function.

Example 11

A company produces charcoal pieces which are baked in a furnace. The mass of each piece of charcoal, in kg, is determined by the probability density function:

$$f(x) = \begin{cases} \dfrac{1}{36}x(6-x) & 0 \le x \le 6 \\ 0 & \text{otherwise} \end{cases}$$

Find the probability that a piece of charcoal weighs more than 4 kg.

Let X be the continuous random variable "the mass of the charcoal, in kg."

$$\int_4^6 \frac{1}{36}x(6-x)\,dx$$

$$\int_4^6 \frac{1}{36}(6x - x^2)\,dx$$

$$\frac{1}{36}\left[3x^2 - \frac{x}{3}\right]_4^6$$

0.259 (3 d.p.)

The probability that the mass is more than 4 kg is 0.259.

Define the continuous random variable.

Sketch the graph.

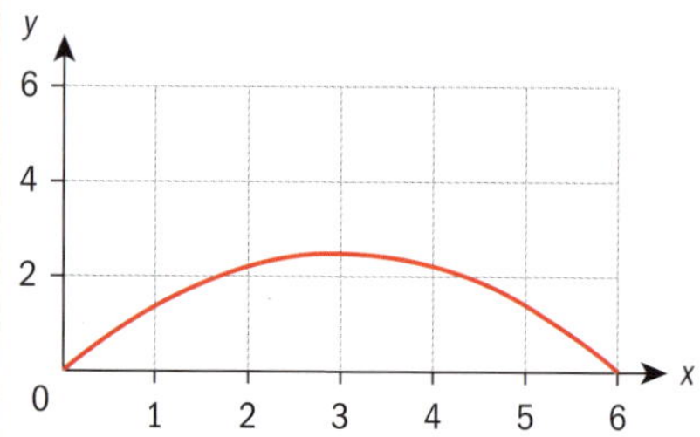

Shade in the required part.

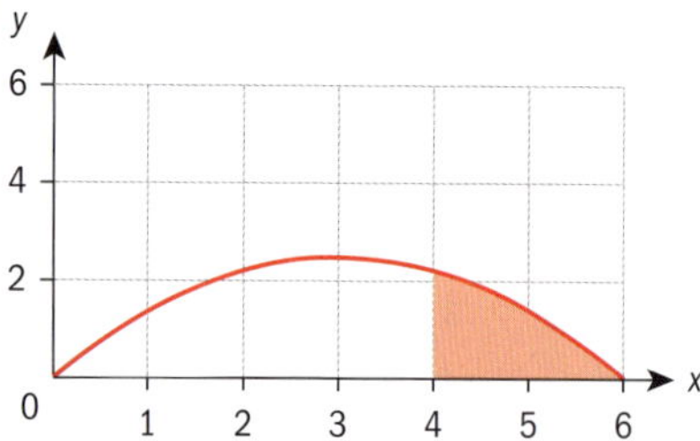

In order to find this area, we need to use integration.

Exercise 11G

1 A continuous random variable X has the PDF $f(x)$ where

$$f(x) = \begin{cases} k(x+2)^2 & -2 \le x < 0 \\ 4k & 0 \le x \le 1\frac{1}{3} \\ 0 & \text{otherwise} \end{cases}$$

a Find the value of the constant k.

b Sketch the PDF.

c Find:

i $P(X \le -1)$

ii $P(0 \le X \le 1)$

iii $P(-1 \le X \le 1)$

2 The continuous random variable X has a PDF. where $f(x) = k(3-x)$ 0 and $0 \le x < 3$, $k > 0$.

a Sketch the PDF.

b Find the value of the constant k.

c Find $P(1.2 \le X \le 2.3)$.

3 The continuous random variable X is given by $f(x) = \begin{cases} 2x^c & 0 \le x < 1 \\ 0 & \text{otherwise} \end{cases}$

a Find the value of the constant.

b Find $P(X < 0.5)$.

Mode, median and mean of a continuous random variable

The mode is the value of x which has the greatest probability. Therefore, with a probability density function $f(x)$, the mode of the continuous random variable X is the value of x for which $f(x)$ has a local maximum. By sketching the probability density function $f(x)$ and hence by differentiating $f(x)$ with respect to x and solving $f'(x) = 0$, you can obtain the mode. You can also find the mode directly from your GDC.

The median is often referred to as the midpoint of a frequency distribution, implying that there is an equal probability of a value being greater than or less than the median. With a probability density function, this requires the value of M that is the solution of the equations

The median is the value M for which
$\int_{-\infty}^{M} f(x)\mathrm{d}x = 0.5$ and $\int_{M}^{+\infty} f(x)\mathrm{d}x = 0.5$

From the definition of the mean for grouped data of a continuous variable it follows that for a continuous random variable X, you can split the range of the variable into strips with a width of δx. The centre of each strip is x_i and the associated probability is $f(x)\ \delta x$.

If you replace the frequency with the probability, it follows that the mean $= \sum_{x \in \mathbb{Z}} xf(x)\delta x$ as δx tends to zero.

The mean of a continuous random variable is given by $E(X) = \int_{-\infty}^{\infty} xf(x)\mathrm{d}x$ which is also expressed as $E(X)$ or μ.

TOK

Do you think that people from very different backgrounds are able to follow mathematical arguments, as they possess deductive ability?

Statistics and probability

Example 12

A continuous random variable has the probability density function:

$$f(x) = \begin{cases} cx(4-x)^2 & 0 \le x \le 4 \\ 0 & \text{otherwise} \end{cases}$$

Find

a the mean **b** the mode **c** the median.

Continued on next page

a $\int_0^4 cx(4-x)^2\,\mathrm{d}x = 1$

In order for the variable to be a continuous random variable, the sum of the probabilities must equal 1.

$\Rightarrow c\int_0^4 x(4-x)^2\,\mathrm{d}x = 1$

c is a constant

$\Rightarrow c\left[8x^2 - \frac{8}{3}x^3 + \frac{x^4}{4}\right]_0^4 = 1$

$\Rightarrow c = \frac{3}{64}$

$\mu = \mathrm{E}(X) = \frac{3}{64}\int_0^4 x \times x(4-x)^2\,\mathrm{d}x$

Use your calculator to find the mean.

$= 1.6$

b The mode occurs at the local maximum.

Sketch the graph.

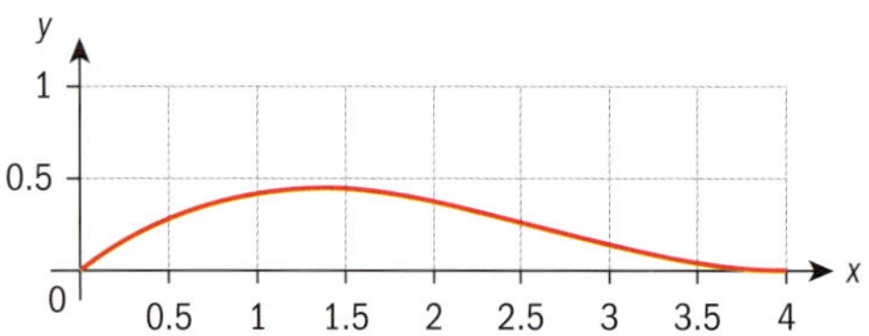

The graph is part of a cubic curve which increases to a maximum as x increases and then decreases until $x = 4$.

$f(x) = \frac{3}{64}x(4-x)^2$

$f(x) = \frac{3}{64}(16x - 8x^2 + x^3)$

$f'(x) = \frac{3}{64}(16 - 16x + 3x^2) = 0$

$(4-x)(4-3x) = 0$

So, the turning points occur at 4 and $\frac{4}{3}$.

The mode occurs at the maximum, which is $x = \frac{4}{3}$.

You can take the second derivative to see which one is a maximum.

At the maximum $f'(x) = 0$ and $f''(x) < 0$

The maximum and therefore the mode occurs when $x = \frac{4}{3}$.

c

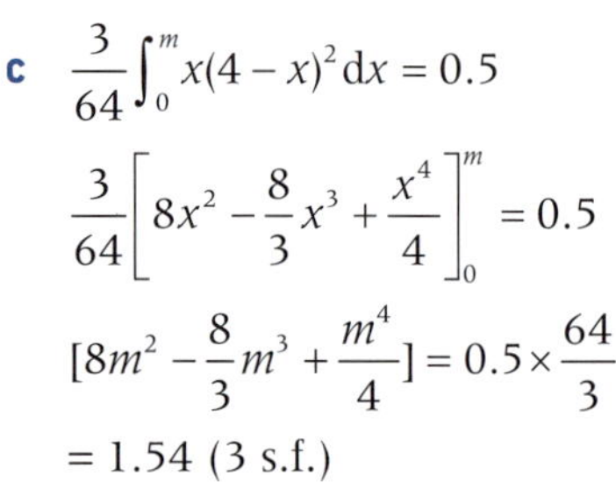

$\frac{3}{64}\int_0^m x(4-x)^2\,\mathrm{d}x = 0.5$

$\frac{3}{64}\left[8x^2 - \frac{8}{3}x^3 + \frac{x^4}{4}\right]_0^m = 0.5$

$[8m^2 - \frac{8}{3}m^3 + \frac{m^4}{4}] = 0.5 \times \frac{64}{3}$

$= 1.54$ (3 s.f.)

Since the density function is zero for $x < 0$, the median is value of m for which $\int_0^m xf(x)\mathrm{d}x = 0.5$.

Use your calculator to solve for m.

In a similar way the variance of a continuous random variable can be discovered.

$$\text{Var}(X) = \int_{all\ x} (x-\mu)^2 f(x)\,dx \text{ where } \mu = \text{E}(X)$$

can also be written

$$\text{Var}(X) = \text{E}(X^2) - (\text{E}(X))^2 \text{ where}$$

$$\text{E}(X^2) = \int_a^b x^2 f(x)\,dx$$

Example 13

A continuous random variable X has PDF $f(x) = \frac{a}{x}$ where $1 \le x \le 4$ and a is a constant.

a Find the value of a.

b Hence, find $\text{E}(X)$ and $\text{Var}(X)$.

a $\int_1^4 \frac{a}{x}\,dx = 1$ $a[\ln x]_1^4 = 1$ $a[(\ln 4) - (\ln 1)] = 1$ $a = \frac{1}{\ln 4}$	$f(x)$ is a PDF, so $\int_{x\in\mathbb{R}} f(x)\,dx = 1$
b $\text{E}(X) = \frac{1}{\ln 4}\int_1^4 \frac{x}{x}\,dx = 2.16$	$\mu = \text{E}(X)$
$\text{Var}(X) = \frac{1}{\ln 4}\int_1^4 \frac{x^2}{x}\,dx - (\frac{1}{\ln 4}\int_1^4 \frac{x}{x}\,dx)^2$	$\sigma^2 = \text{E}(X^2) - (\text{E}(X))^2$
$= 0.727$	Use your calculator to solve.

Exercise 11H

1 Sketch the graph of the function defined by

$$f(x) = \begin{cases} \frac{1}{8}x \text{ if } 0 \le x \le 4 \\ 0 \quad \text{elsewhere} \end{cases}$$

a Explain why functions of this type are probability density functions.

b Find $\text{P}(1 < X < 20)$.

c Find the mean, mode, median and standard deviation.

2 A continuous random variable X has PDF given by

$f(x) = ax(2-x)$ where $0 \le x \le 2$.

Find the value of a and hence find:

a $\text{E}(X)$ **b** $\text{Var}(X)$

c median of X **d** mode of X.

Statistics and probability

3 A continuous random variable X has PDF given by

$f(x) = 2\cos(2x)$ where $0 \le x \le \frac{\pi}{4}$.

a Find $P\left(\frac{\pi}{12} < X < \frac{\pi}{6}\right)$.

b Find the values of the median, the mean and the mode of X.

4 Consider the function defined by

$$f(x) = \begin{cases} k \text{ for } 0 < x < 2 \\ 2k \text{ for } 2 \le x \le 3 \\ 0 \text{ otherwise} \end{cases}$$

a Determine the value of k given that f is a PDF of a random variable X.

b Sketch the graph of $f(x)$. Hence find the median of X.

c Find the values of $E(X)$and $Var(X)$.

5 A continuous random variable X has PDF defined by

$f(x) = ax^2 + b$ where $0 \le x \le 3$ where a and b are constants.

a Find, in terms of a, the value of b.

b Given that the median of X is 1, determine the values of a and b.

c Find the values of $E(X)$and $Var(X)$.

11.4 Binomial distribution

Jacob Bernoulli, a Swiss mathematician, named a specific discrete probability distribution where there are precisely two outcomes to an event. It is often noted with a Boolean-valued outcome eg either 0 or 1.

let X be a discrete random variable, where $X = 0$ for success and $X = 1$ for failure

then $P(X = 0) = p$

and $P(X = 1) = 1 - p = q$

An extension of this is the binomial distribution. In this case the event is repeated a fixed number of times, and the random variable X states the number of successes in the fixed number of trials.

TOK

A model might not be a perfect fit for a real-life situation, and the results of any calculations will not necessarily give a completely accurate depiction. Does this make it any less useful?

There are certain conditions which need to be met in order for a distribution to be classified as binomial.

1. The discrete random variable X is the "number of successes in n trials" where n is a fixed number of trials.
2. Each trial has exactly two possible outcomes: **success** or **failure**.
3. The trials are independent, so the outcome of one trial does not affect the outcome of the others.
4. For each trial, the probability of success p is constant.

International-mindedness

The Galton board, also known as a quincunx or bean machine, is a device for statistical experiments named after English scientist Sir Francis Galton. It consists of an upright board with evenly spaced nails or pegs driven into its upper half, where the nails are arranged in staggered order, and a lower half divided into a number of evenly-spaced rectangular slots. In the middle of the upper edge, there is a funnel into which balls can be poured. Each time a ball hits one of the nails, it can bounce right or left with the same probability. This process gives rise to a binomial distribution in the heights of heaps of balls in the lower slots and the shape of a normal or bell curve.

Investigation 7

Consider the situation of rolling a die, where obtaining a 6 is a success and any other number is a failure. The die is to be rolled five times.

1. State the number of trials n.
2. State the outcomes.
3. Calculate the probability of success in each trial.
4. Define the random variable and write all of its values.
5. **Factual** How many variables can a binomial random variable have?
6. One possible outcome is SFSSF. Where S indicates success and F failure. Calculate the probability of this outcome.
7. **Factual** How many outcomes are there where there are three successes and two failures?
8. Calculate the probability of three successes and two failures.

Recall combinatorics:

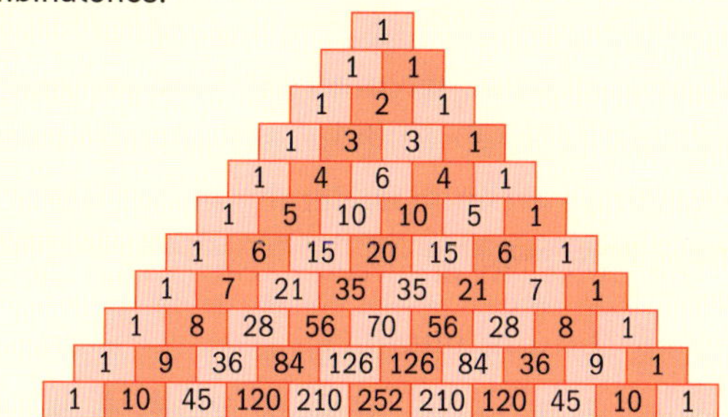

$^{n}C_{r} = \dfrac{n!}{r!(n-r)!}$ or Pascal's triangle

Copy and complete the following table:

x	0	1	2	3	4	5
Number of ways of obtaining x						

9. If p is the probability of success and $(1-p)$ is the probability of failure, find the probability of 0, 1, 2, 3, 4, 5 successes in five independent trials.
10. Hence, generalize a formula for $\mathrm{P}(X=r)$ in a binomial distribution with n fixed trials.
11. **Conceptual** How are Pascal's triangle and binomial theorem related?

Once these binomial probabilities have been established, you can define a binomial distribution.

> If a discrete random variable X follows a bionomial distribution, then the probability distribution function will be given as
>
> $P(X = k) = {}^nC_k\ p^k(1-p)^{n-k},\ \ k = 0, 1, 2, 3, \ldots, n$
>
> Where $0 \le p \le 1$.

If X follows this distribution, then you write $X \sim B(n, p)$. This determines the distribution as binomial and defines the two parameters n and p which give the shape and size of the distribution where n is the number of trials and p is the probability of a successful outcome.

$X \sim B(n, p)$is read as "X is a discrete random variable and X follows a binomial distribution with the parameters n and p."

Example 14

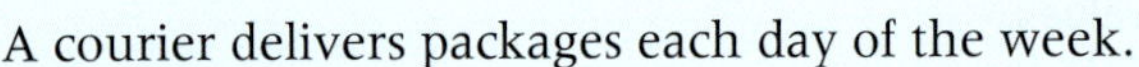

A courier delivers packages each day of the week.

The probability that the courier delivers the packages by 9.00 am is 0.7.

Construct a probability distribution table for the week and write an expression for the probability that on seven days they deliver the packages before 9.00 am on five occasions.

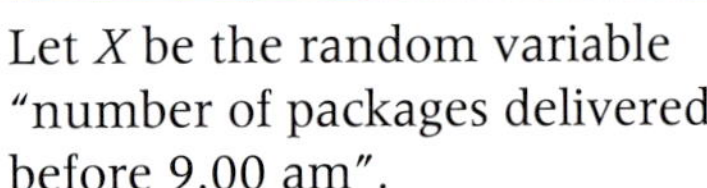

Let X be the random variable "number of packages delivered before 9.00 am". $X \sim B(n, p)$ $X \sim B(7, 0.7)$	Check conditions for binomial distribution. The discrete random variable X is the "number of success in n trials." There are seven fixed trials, $n = 7$. There are only two outcomes: the package is delivered before 9.00 a.m. or not. The trials are independent. The probability of success in each trial is constant, $p = 0.7$. State the distribution. $P(X = x) = {}^nC_x\ p^x(1-p)^{n-x}\ \ x = 0, 1, 2, 3, \ldots, n$
$P(X = 5) = {}^7C_5\ 0.7^5 0.3^2 = 0.318$ Therefore, the probability that the packages arrive before 9.00 am exactly five times during the week is 0.318.	State the PDF. In this case you require the probability that there are five successes. Use your calculator to obtain the solution.

Although it is often useful to calculate exact probabilities, it is more common to answer questions that require cumulative answers.

Example 15

Using the example above, find the probability that the courier delivers the package before 9.00 am at least five times during the week.

$P(X=5)+P(X=6)+P(X=7)$ $= {}^7C_5\,0.7^5 0.3^2 + {}^7C_6\,0.7^6 0.3^1 + {}^7C_7\,0.7^7 0.3^0$ $=0.647$ Therefore, the probability that the packages arrive before 9.00 am at least five times during the week is 0.647.	0, 1, 2, 3, 4, 5, 6, 7 ~~0, 1, 2, 3, 4~~, 5, 6, 7 Write down all possibilities and eliminate what is not required. You require at least 5.

The probability of "at least x" $P(X \geq x) = 1 - P(X \leq (x-1))$

The probability of "more than x" $P(X > x) = 1 - P(X \leq x)$

Example 16

A game is played where a ball is removed from a bag, the colour noted and the ball is replaced in the bag. In the bag there are red and white balls in a ratio of 3:5.

Find how many balls must be chosen so that the probability that there is at least one red ball in the selection is greater than 0.9.

Let X be the random variable "number of red balls."	Check conditions for binomial distribution. Define the variable. $X \sim B(n, p)$
$X \sim B(n, 0.375)$ The probability of "at least 1" $= P(X \geq 1) = 1 - P(X \leq 0)$ $= 1 - P(X=0)$	State the distribution; n is unknown.
$1 - P(X=0) > 0.9$ $1 - {}^nC_0\,0.375^0 0.625^n > 0.9$ $0.1 > 0.625^n$ $\log 0.1 > n \log 0.625$ $n > 4.899$ Therefore, at least five balls must be removed from the bag in order that the probability of at least one of them being red is greater than 90%.	We require $P(X \geq 1) > 0.9$

HINT

On the distribution menu of your calculator you will find various options:

binompdf and binomcdf

where "pdf" stands for probability density function and "cdf" stands for cumulative distribution function and determines $P(X \leq x)$, $x = 0, 1, 2, 3 \ldots n$.

You could obtain the answer to Example 14 directly from your GDC using the 'CDF' function.

Finding the mean and variance of the binomial distribution

Investigation 8

Let X be a discrete random variable such that

$X \sim \mathrm{B}(n, p)$

The probability density function is defined as

$\mathrm{P}(X = x) = {}^nC_x p^x (1-p)^{n-x} \quad x = 0, 1, 2, 3, \ldots, n$

1 Copy and complete the probability distribution table:

x	0	1	2	3	…	n
$\mathbf{P(X = x)}$	${}^nC_0 p^0 (1-p)^n$	${}^nC_1 p^1 (1-p)^{n-1}$				
$\mathbf{P(X = x)}$	$(1-p)^n$	$np(1-p)^{n-1}$				p^n

As X is a discrete random variable, it follows that

$$\mathrm{E}(X) = \sum_{\text{all } x} x\mathrm{P}(X = x)$$

2 Write a sum for $\mathrm{E}(X)$,

3 **Factual** Which factors are present in each term?

4 Use the definition for polynomials to simplify the expression.

5 **Conceptual** How can we calculate the expected value of a binomial distribution?

6 Hence, write an expression for $\mathrm{E}(X)$ of a binomial distribution.

A similar method can be used to show that

$\mathrm{Var}(X) = np(1-p)$

Mean and variance of a binomial distribution.

If $X \sim B(n, p)$ then $\mathrm{E}(X) = np$
$\mathrm{Var}(X) = np(1-p)$

Example 17

On a particular Internet channel, a travel vlog occurs at the same time every day. The probability that Vikki is home from school to watch the travel vlog live is 0.4. Find the probability that in any given school week (5 days) Vikki arrives home to watch the vlog:

a on three days **b** at least three days

c on three days which are consecutive.

d Find the expected number of days and the standard deviation.

Let X be the random variable "number of times Vikki watches the vlog in a week."	Check conditions for binomial distribution.
	Define the variable.
$X \sim \mathrm{B}(5, 0.4)$	State the distribution.
	$X \sim \mathrm{B}(n, p)$
a $\mathrm{P}(X = 3) = {}^5C_3 0.4^3 0.6^2 = 0.2304$	
b $\mathrm{P}(X \geq 3) = 1 - \mathrm{P}(X \leq (2)) = 0.31744$	There are 3 possibilities for consecutive days:
c $0.3 \times 0.2304 = 0.06912$	sssff, fsssf, or ffsss out of the 10 possibilities of 3 successes out of 5 days, therefore we require $\frac{3}{10}$ X $\mathrm{P}(X = 3)$.
d $\mathrm{E}(X) = 2$ standard deviation = square root of variance = 1.10 (3 s.f.)	

Exercise 11I

1 Given $X \sim \mathrm{B}(8, 0.25)$ find:

a $\mathrm{P}(X = 3)$

b $\mathrm{P}(X < 10)$

c $\mathrm{P}(X = 0 \text{ or } 1)$.

2 Given that $X \sim \mathrm{B}(6, 0.4)$ find:

a $\mathrm{E}(X)$

b $\mathrm{Var}(X)$

c The most likely value for X.

3 Given that $X \sim \mathrm{B}(5, 0.5)$:

a show that $\mathrm{P}(X = 5) = \frac{1}{32}$

b construct a probability distribution table

c find $\mathrm{E}(X)$

d find $\mathrm{Var}(X)$

e find the mode.

4 The probability that is will snow in Les Diablerets, Switzerland on any given day in January is 0.85. In any given week in January, find the probability that it will snow on:

a exactly one day

b more than two days

c at least three days

d no more than 4 days.

5 In a French course, students complete 12 tests in a semester. The probability that a student passes the quiz is $\frac{2}{3}$.

a Find the expected number of quizzes a student will fail.

b Find the probability that a student will pass more than half the quizzes.

c Find the most likely number of quizzes that a student will pass.

Developing inquiry skills

Consider the second option from the opening problem for the chapter. If you played this game a few times, do you think you would end up with more money than you started with? How could we tell?

Define a random variable for the winnings from playing the game once.

Does this experiment fit a binomial distribution? What are the parameters?

20 tokens to toss 4 coins
If you get 4 Heads
you win 100 tokens
If you get 3 Heads
you get your tokens back
If you get 2 Heads
you win 10 tokens
If you get 0 Heads
or 1 Head
you get nothing!

Statistics and probability

11.5 The normal distribution

The normal distribution is the most important continuous probability distribution. Many naturally-occurring events in the real world that generate data follow a normal distribution – for example, heights, weights, lengths of leaves. Other distributions can sometimes be approximated to a normal distribution if they satisfy certain conditions.

HINT
The French mathematician Abraham de Moivre is often credited with proposing the normal distribution, but Carl Friedrich Gauss popularized it and in engineering it is known as the Gaussian distribution. It was Karl Pearson who named it the normal distribution.

The PDF of the normal distribution is

$$f(x) = \frac{1}{\sigma\sqrt{2\pi}} e^{\frac{-(x-\mu)^2}{2\sigma^2}}, \text{ where } x \in \mathbb{R}$$

The variable X from the normal distribution is a continuous random variable and the probability density function relies on two parameters: μ the mean of the distribution and σ^2 the variance of the distribution.

If a continuous random variable follows a normal distribution, then you write $X \sim \mathrm{N}(\mu, \sigma^2)$.

Your GDC has an inbuilt normal distribution generator and can draw graphs of normal distributions with the appropriate parameters given.

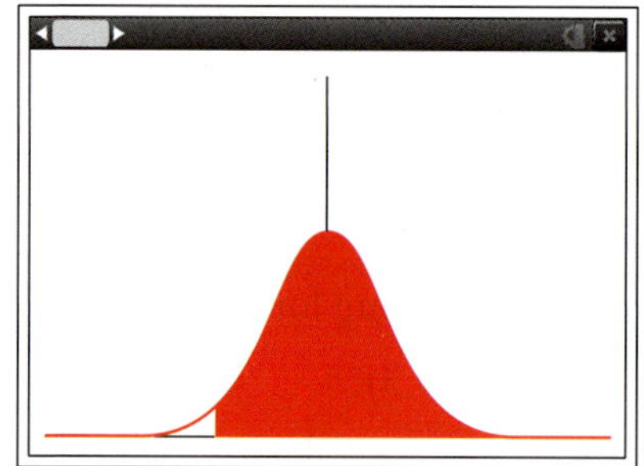

Investigation 9

1 Use your GDC to graph the normal curves for each of the following normal variables:

A $X \sim \mathrm{N}(1, 1)$
B $X \sim \mathrm{N}(2, 1)$
C $X \sim \mathrm{N}(3, 1)$
D $X \sim \mathrm{N}(4, 1)$

Describe differences and similarities between these normal curves. How does the value of the parameter μ affect the normal curve?

2 Graph the normal curves associated with these normal variables:

E $X \sim \mathrm{N}(0, 1)$
F $X \sim \mathrm{N}(0, 2)$
G $X \sim \mathrm{N}(0, 3)$
H $X \sim \mathrm{N}(0, 4)$

Describe differences and similarities between these normal curves. How does the value of the parameter σ^2 affect the normal curve?

3 **Factual** Explore some other normal curves and write down your conclusions about the effects of the parameters on the graphs of the normal variables.

4 **Factual** The normal curves are defined by just two parameters: the mean and the variance of the distribution. Based on your knowledge of the shape of a normal curve, explain why it is not necessary to include the values of the median and the mode of the distribution.

5 **Conceptual** How is representing the normal distribution beneficial?

Properties of a normal curve

A normal curve has the following characteristics:

1 The graph is a bell-shaped curve (often referred to as a Gaussian curve).

2 It is symmetrical about the vertical line $x = \mu$.

3 The maximum value of $f(x)$ occurs when $x = \mu$. Substituting $x = \mu$ into $f(x)$ implies that the mode is given by $\mu = \frac{1}{\sigma\sqrt{2\pi}}$

4 There are two points of inflection on the curve at $x = \mu \pm \sigma$.

5 The actual size of the curve depends on the values for μ and σ^2.

6 The probabilities $\mathrm{P}(a \le X \le b)$ are given by the area under the curve between a and b.

7 The total area under the curve equals 1.

8 The domain of x is from $-\infty$ to ∞ and the x-axis is an asymptote to the curve at both extremes.

9 The mean, median and mode all occur when $x = \mu$.

The data of the normal distribution approximately lies as shown:

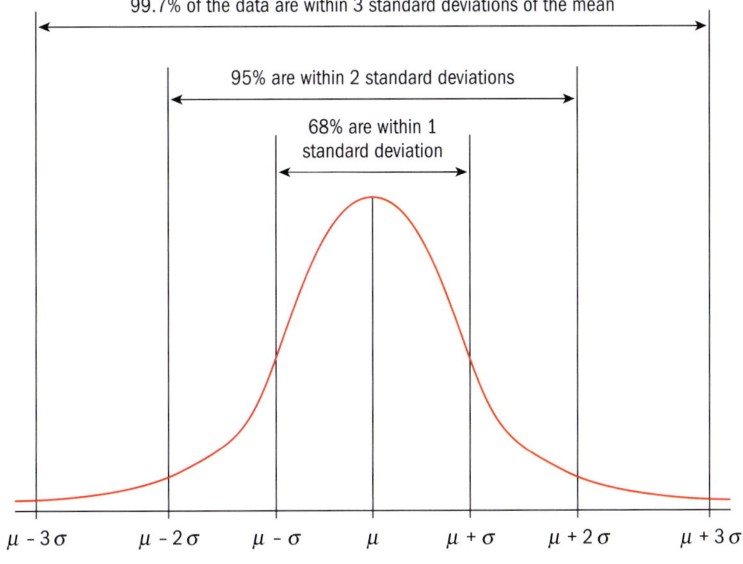

International-mindedness

French mathematician Pierre Laplace used the normal curve to describe the distribution of errors in 1783 and in 1810 to prove the Central Limit Theorem

Statistics and probability

Standardized normal variable

Because each distribution is unique, with a specific μ and σ^2, it is often necessary to "standardize the variable" by examining the situation which could bring the specific distribution to a standard normal distribution with $\mu = 0$ and $\sigma^2 = 1$.

Let $X \sim \mathrm{N}(\mu, \sigma^2)$, you can construct a new random variable Z such that

$$Z = \frac{X - \mu}{\sigma}$$

This allows you to make the following assumptions.

Let Z be the standardized continuous random variable with mean = 0 and variance = 1.

Then

$\mathrm{P}(Z < 0) = 0.5$

$\mathrm{P}(-1 < Z < 1) = 0.68$

$\mathrm{P}(Z < 1) = 0.84$

$\mathrm{P}(-2 < Z < 2) = 0.95$

$\mathrm{P}(Z < 2) = 0.975$

HINT

See the graph above. 68% of the population lies within one standard deviation of the mean.

Using this standardized normal distribution, you can find $\mathrm{E}(X)$ and $\mathrm{Var}(X)$.

EXAM HINT

Proofs of $\mathrm{E}(X)$ and $\mathrm{Var}(X)$ are not examined, but they will help your understanding.

Let Z be a standard variable

$Z = \frac{X - \mu}{\sigma}$ then

$$f(z) = \frac{1}{\sqrt{2\pi}} \mathrm{e}^{\frac{-z^2}{2}}$$

You know the sum of the probabilities equals 1.

Therefore,

$$\int_{-\infty}^{\infty} \frac{1}{\sqrt{2\pi}} \mathrm{e}^{\frac{-z^2}{2}} \mathrm{d}z = 1$$

$$\frac{1}{\sqrt{2\pi}} \int_{-\infty}^{\infty} \mathrm{e}^{\frac{-z^2}{2}} \mathrm{d}z = 1$$

Therefore, $\int_{-\infty}^{\infty} \mathrm{e}^{\frac{-z^2}{2}} \mathrm{d}z = \sqrt{2\pi}$

Recall for continuous random variables $\mathrm{E}(X) = \int_{-\infty}^{\infty} x f(x) \mathrm{d}x$

Therefore, $\mathrm{E}(X) = \int_{-\infty}^{+\infty} x \frac{1}{\sigma\sqrt{2\pi}} \mathrm{e}^{\frac{-(x-\mu)^2}{2\sigma^2}} \mathrm{d}x$

as z is the standard variable $z = \frac{x - \mu}{\sigma}$

$x = z\sigma + \mu$

Example 18

The time taken for a student to complete a language test follows a normal distribution with a mean of 25 minutes and a standard deviation of 3 minutes. They take a test each week (35 weeks in the school year). Estimate the number of tests during the year which are

a longer than 30 minutes **b** less than 23 minutes **c** between 18 and 25 minutes.

a 

X → (follows) $N(25, 3^2)$

Let X be the random variable "time taken to complete the test".

Use a GDC.

You require the probability of X being greater than 30.

Set the lower bound to 30 and the upper bound to ∞.

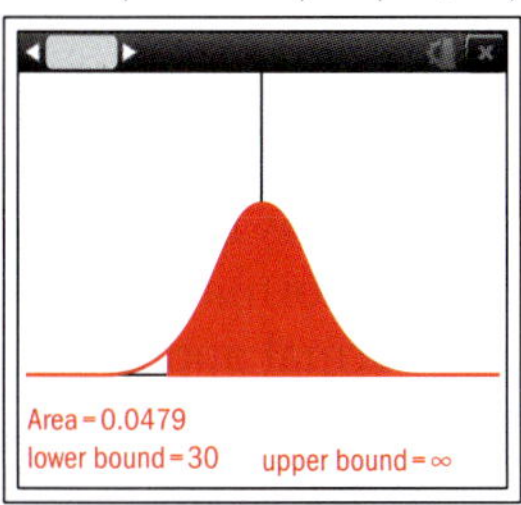

$P(X > 30) = 0.0478$ (3 s.f)

$35 \times 0.0478 = 1.673$

A student could expect 2 tests longer than 35 minutes throughout the year

The probability is 0.0478, you require the estimated number of tests over the whole year (35 weeks).

b

You require the probability that a test is less than 23 minutes.

The lower bound can be set to 0 as there are no values less than 0.

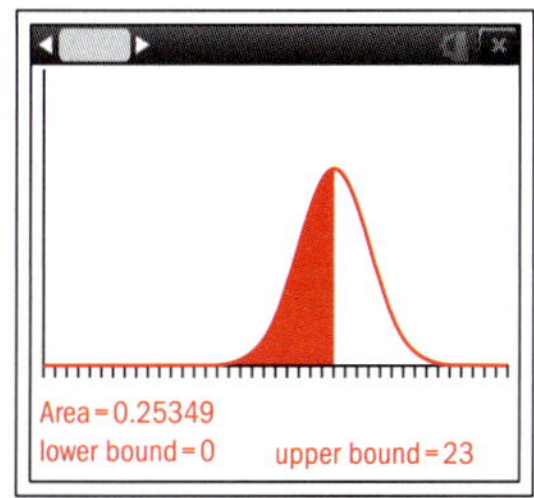

$P(X < 23) = 0.252$ (3 s.f)

$35 \times 0.25349 = 8.87$

An estimate of the number of tests being shorter than 23 minutes is 9 tests.

The probability is 0.25349 for a single test, therefore for the expected value, multiply by the number of tests.

c

You require the probability to be between 18 and 23 minutes.

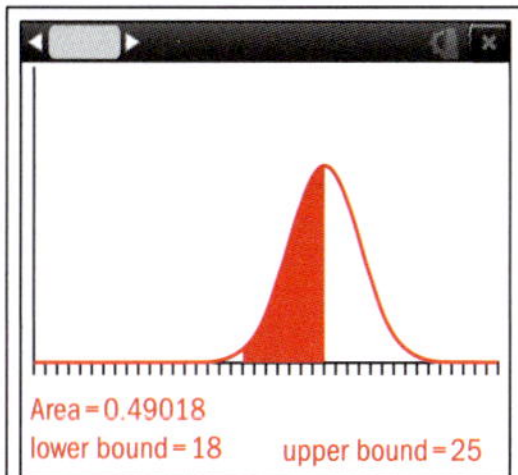

$P(18 < X < 23) = 0.49018$

The number of tests between 18 and 23 minutes would be expected to be

$35 \times 0.49018 = 17$ tests

Statistics and probability

Example 19

The weights of apples follow a normal distribution with a mean of 45 g and a standard deviation of 5 g. Apples are rejected from sales if they do not fall in the central 80% of the distribution.

Find the value of the limits within which the central 80% of the distribution lies.

$$\mathrm{P}\left(\left|\frac{X-45}{5}\right| < a\right) = 0.8$$

$X \sim \mathrm{N}(45, 25)$

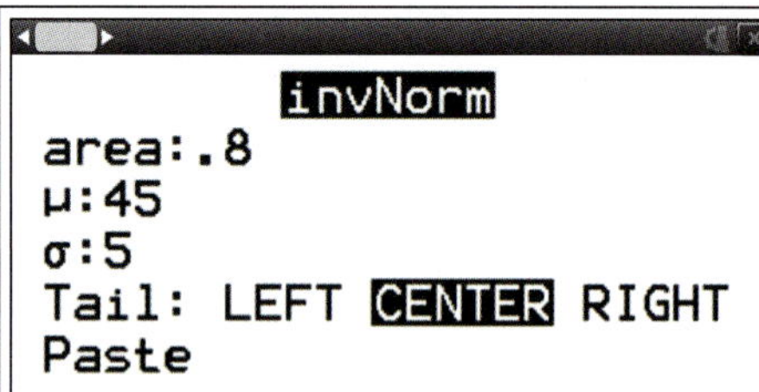

In this case it is necessary to enter the information in the inverse normal application of the calculator.

P("as before")

The limits are therefore

{38.56224217 51.40775783}

The calculator can give the limits

The values are therefore 38.6 and 51.4

Therefore, in this case you are given the area and need to use inv norm in order to find the limits.

Exercise 11J

1 Given $X \sim \mathrm{N}(2, 3^2)$, sketch a diagram, shade the required area and find

a $\mathrm{P}(0.5 < X < 1.5)$

b $\mathrm{P}(X < 0.5)$

c $\mathrm{P}(X \geq 2)$.

2 If $X \sim \mathrm{N}(30, 8^2)$, find the values of:

a $\mathrm{P}(X < 25)$

b $\mathrm{P}(17 \leq X < 35)$

c $\mathrm{P}(X \geq 12)$

3 Given that $X \sim \mathrm{N}(5, 9)$,

a state the value of the mean and standard deviation of X

b find $\mathrm{P}(X < 4)$

c find the values of a and b such that $\mathrm{P}(X < a) = 0.011$ and $\mathrm{P}(X \geq b) = 0.871$.

4 A sample of 100 pears is taken from a fruit farm.

The pears have the following distribution of weights:

Weight to the nearest 20 g	220	240	260	280	300
Frequency	11	21	38	17	13

a Determine the mean and standard deviation of the weights of the pears.

b Given part **a** and assuming that the distribution is approximately normal, find the range of weights of pears to be packed if 5% are to be rejected for being too light and 5% are to be rejected for being too heavy.

5 A packing machine produces potato bags which have a mass normally distributed with mean 150 kg and standard deviation 0.5 kg. If a bag produced by this machine is selected at random, find the probability that its mass is

a less than 149 kilograms

b more than 151.5 kilograms

c between 149 and 151 kilograms.

6 A continuous random variable T follows a normal distribution with mean 13.2 and standard deviation 1.5.

a Find the value of $P(T < 12.1)$ and $P(T > 14.9)$.

b Find the value of t if $P(T < t) = 0.444$.

TOK

How well do models and predictions fit real-life situations?

Often problems arise which require finding the values of μ, σ or both. In this case we have to standardize the random variable, as shown in the next example.

Example 20

Let X be a random variable such that

$X \sim N(\mu, \sigma^2)$

$P(X > 85) = 0.05$ and $P(X > 25) = 0.10$. Find the values of the mean and the standard deviation.

$P(X > 85) = 0.05$ $P\left(Z > \frac{85-\mu}{\sigma}\right) = 0.05$ $\left(\frac{85-\mu}{\sigma}\right) = (\text{InvNorm})(0.95)$ $\left(\frac{85-\mu}{\sigma}\right) = 1.645$ Similarly $P(X < 25) = 0.10$ $P\left(Z < \frac{25-\mu}{\sigma}\right) = 0.1$ $\left(\frac{25-\mu}{\sigma}\right) = \text{InvNorm}(0.1)$ $\left(\frac{25-\mu}{\sigma}\right) = -1.282$	Set up equations for both conditions.
$85 - \mu = 1.645\sigma$ $\mu - 25 = 1.282\sigma$ $\mu = 51.3$ $\sigma = 20.5$ The mean of the mass is 51.3 g and the standard deviation is 20.5g.	You get a pair of simultaneous equations.

Statistics and probability

Exercise 11K

1 If $X \sim N(8, \sigma^2)$, find the value of σ^2 given that $P(X \le 4) = 0.321$.

2 Consider $X \sim N(\mu, \sigma^2)$. Find the values of μ and σ given that $P(X \le 1) = 0.345$ and $P(X \le 3) = 0.943$.

3 A factory has a machine designed to produce 1 kg bags of sugar. It is found that the average weight of sugar in the bags is 1.02 kg. Assuming that the weights of the bags are normally distributed, find the standard deviation if 1.3% of the bags weigh below 1 kg.

Give your answer correct to the nearest 0.1 gram.

4 A random variable X is normally distributed with mean σ and standard deviation σ, such that $P(X > 46.8) = 0.203$ and $P(X < 42.6) = 0.315$.

a Find the values of μ and σ.

b Hence find $P\left(|X - \mu| < \frac{\sigma}{2}\right)$

5 The masses of the broccoli stems sold at Geensgreens online market are normally distributed with mean 320 g and standard deviation 20 g.

a If a broccoli stem is chosen at random, find the probability that its mass lies between 200 and 350 grams.

b Find the mass exceeded by 10% of the broccoli stems.

c In one day, 500 broccoli stems are sold. Estimate the number of broccoli stems that weigh more than 350 grams.

d At the nearby Kelbricks supermarket, 15% of the broccoli stems sold weighed at least 400 grams and not more than 10% weighed less than 370 grams. Assuming that the mass M of the broccoli stems follows a normal distribution, find the expected value and variance of M.

6 Monsterchip produces 1 kg bags of chips. Assuming that the masses M of the bags follows a normal distribution with a mean of 1.03 kg, find the maximum value of the variance if less than 1% of the bags are underweight.

Developing your toolkit

Now do the Modelling and investigation activity on page 730.

Chapter summary

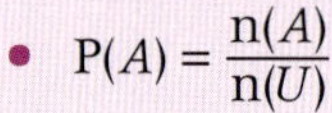

- $P(A) = \frac{n(A)}{n(U)}$

Axiom 1

- For any event A, $P(A) \ge 0$
- All probabilities take a value which is greater than or equal to zero.

Axiom 2

- $P(U) = 1$
- The probability of all occurrences is equal to one.

Corollary

- $0 \le P(A) \le 1$
- For two events which are mutually exclusive, $P(A \cup B) = P(A) + P(B)$ and $P(A \cap B) = 0$

Axiom 3

- If $A_1, A_2, A_3, \ldots, A_n$ is a set of mutually exclusive events, then
  $P(A_1 \cup A_2 \cup A_3 \ldots \cup A_n) = P(A_1) + P(A_2) + P(A_3) + \ldots + P(A_n)$

Axiom 4

- $P(A \cap B) = P(A) \times P(B|A)$
- $\therefore P(A \cap B) = P(A) \times P(B)$ for two independent events.
- $P(B|A) = P(B|A') = P(B)$

Bayes' theorem

- $P(A_i|B) = \dfrac{P(A_i) \times P(B|A_i)}{P(B|A_1)P(A_1) + P(B|A_2)P(A_2) + \ldots + P(B|A_n)P(A_n)}$
- In an exam you will only be required to use Bayes' theorem for a maximum of 3 events.

Discrete random variables and distributions

- The **probability distribution function** (PDF) of a discrete random variable X verifies the properties: $0 \le f(x) \le 1$ and $\sum_{x \in A} f(x) = 1$.
- The **mean** or expected value of X is given by $\mu = \sum x\, P(X = x)$
- The **variance of X** is given by $\sigma^2 = \sum (x - \mu)^2 P(X = x)$.
- $\sigma = \sqrt{\mathrm{Var}(X)}$ is called the standard deviation of X.
- $\sigma^2 = \mathrm{Var}(X) = E(X^2) - (E(X))^2$ where $E(X^2) = \sum x^2 P(X = x)$
- Where the PDF of X has a maximum value, this value is called the **mode** of X.

Binomial distribution

- If $X \sim B(n, p)$, $P(X = k) = {}^nC_k\, p^k (1-p)^{n-k}$, $k = 0, 1, 2, 3, \ldots, n$.
- $E(X) = np$ and $\mathrm{Var}(X) = npq$ where $q = 1 - p$.

Continuous random variables and distributions

- The **probability density function** of a random variable X (PDF of X) has the properties: $f(x) \ge 0$ for all values $x \in \mathbb{R}$ and $\int_{-\infty}^{+\infty} f(x)\, dx = 1$.
- $\mu = E(X) = \int_a^b x f(x)\, dx$ and $\sigma^2 = \mathrm{Var}(X) = \int_a^b (x - \mu)^2 f(x)\, dx$

Continued on next page

- The **median** m is the solution of the equation $\int_a^m f(x)\,dx = \frac{1}{2}$ and the **mode** is the value(s) of x for which f has a maximum.

Normal distribution

- If $X \sim N(\mu, \sigma^2)$ and $Z = \frac{X-\mu}{\sigma}$ then $Z \sim N(0, 1)$.

Developing inquiry skills

Consider the third option from the opening problem for the chapter.

30 tokens to drop a marble into this machine
If it goes into box 1 or 8 you win 500 tokens
If it goes into 2 or 7 you win 50 tokens
If it goes into 3 or 6 you win 10 tokens
If it goes into 4 or 5 you get nothing!

If you played this game a few times, do you think you would end up with more money than you started with?

How could we tell?

When a ball is dropped and it hits one of the pegs it has n equal chance of falling either side. Let "L" and "R" denote whether a ball goes left or right at any given peg.

What path does the ball need to take in order to go in to box number 1?

What is the probability of this happening?

What possible paths can the ball take in order to go into box number 2?

The "game" is called in quincunx or Galton board or bean machine. Sir Edward Galton, a British scientist, originally devised it for probability experiments.

What is the relationship between the quincunx and the binomial distribution?

What is the probability that you will win 500 tokens? 50? 10? 0?

Draw a probability distribution table for your random variable.

Calculate the expected winnings from the game.

How much would you expect to win or lose if you played the game 10 times?

100 times?

Is the game "fair"?

Interestingly, if a large number of balls are dropped in the quincunx then the shape of the resulting balls approximates a normal distribution. This can be seen in the simulation available here:

https://www.mathsisfun.com/data/quincunx.html

You can alter the number of boxes and the speed that the balls drop. It is also possible to change the probability that the ball falls to one side of the other.

What effect does this have on the distribution of the balls?

Chapter review

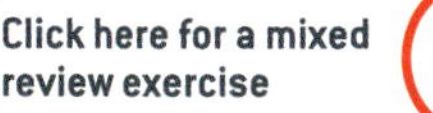
Click here for a mixed review exercise

1 In the following trials, state the universal set and the elements in sets A, B and C. Hence find the probabilities of these events happening.

Two students have their birthdays in the same week:

a both birthdays are on the Monday

b both birthdays are on the same day

c the birthdays are on consecutive days.

2 A digit is chosen at random. Find the probability that it is divisible by 3 or 7.

3 Find the number of times an unbiased dice should be thrown if the probability that obtaining a six at least once is to be greater than $\frac{9}{10}$.

4 Ten counters are in a bag: seven are red and three are white. Two counters are taken out one at a time without replacement. Find, with the aid of a tree diagram, the probability that:

a the first taken is red

b the second is red

c exactly one of the first two is red

d at least one of the first two is red.

5 Of the students in a school, 20% have been vaccinated against a certain virus. In an epidemic, the chance of infection amongst the student who have been vaccinated is 0.1, whereas the risk of infection for the students who have not been vaccinated is 0.75.

a In the case of an epidemic, estimate the proportion of the students in the school that would be infected.

b One student is selected at random, and found to be infected. Find the probability they were vaccinated.

c Find the probability they had not been vaccinated.

6 A game consists of taking eight counters which are blue on one side and green on the other. The counters are thrown into the air and land randomly on the ground.

If the number of counters which show green is an odd number, the thrower wins. If all the counters are the same colour, it counts as two wins. All other scenarios count as a loss. Find the expected number of wins and losses.

7 In a consignment of 20 articles, 4 are defective. If a sample of 5 is taken at random, find the probability they will contain:

a no defective articles

b at least three defective articles.

8 An energy-saving light bulb has a life of H hours, where H is normally distributed with a mean of 1300 hours and a standard deviation of 125 hours.

a Find the probability that a randomly selected lightbulb will last for more than 1500 hours.

b The manufacture offers to replace for free any lightbulb which burns for less than 1050 hours. Find the percentage of light bulbs which will need to be replaced.

c If two lightbulbs are installed at the same time, find the probability that both will burn for less than 1400 hours but more than 1200 hours.

9 Discs produced by a machine are required to have a thickness of between 1.99 mm and 2.01 mm. Of all the discs produced, 7% are rejected for being too narrow and 4% are rejected for being too wide. Assuming the thicknesses are normally distributed, find the mean and the standard deviation. Give your answers correct to 3 significant figures.

10 In a factory, chocolate bars are produced on one of four conveyor belts, A, B, C or D.

A small percentage of bars are not produced correctly for sale.

Conveyer belt	Percentage of faulty chocolate bars	Percentage of total outcome
A	1%	35
B	3%	20
C	2.5%	24
D	2%	21

a Find the probability that a bar chosen at random from the whole output is faulty.

b Find the probability that a faulty bar comes from conveyor belt D.

11 The probability density function $f(x)$ of a variable x is given by

$$f(x) = \begin{cases} kx\sin\pi x & 0 \le x \le 1 \\ 0 & \text{otherwise} \end{cases}$$

Show that $k = \pi$ and find the mean and variance for the distribution.

Exam-style questions

12 P1: Two events A and B are independent. It is given that $P(A) = 0.3$ and $P(B) = 0.8$

a State, with a reason, whether events A and B are mutually exclusive. (2 marks)

b Find the probability of each of the following events.

i $A \cap B$ **ii** $A \cup B$

iii $A|B'$

iv $A' \cap B$ (8 marks)

13 P2: The discrete random variable X has the following probability distribution.

x	0	0.5	1	1.5	2
$P(X=x)$	$\frac{k}{2}$	k	k^2	$2k^2$	$\frac{k}{2}$

a Find the value of k. (4 marks)

b Show that $E(X) = \frac{17}{18}$. (3 marks)

c Find $P(X \ge 1.25)$. (3 marks)

d Find $\text{Var}(X)$. Give your answer correct to 3 significant figures. (3 marks)

14 P2: Approximately 4% of eggs produced and sold by a local farm are cracked.

Jerry buys 24 eggs from the farm.

a Find the probability that exactly two of Jerry's eggs are cracked. (2 marks)

b Find the probability that Jerry buys no more than four cracked eggs. (2 marks)

c Find the probability that Jerry buys at least two cracked eggs. (2 marks)

d Find the variance of the number of cracked eggs. (2 marks)

15 P2: Alison walks to school every day. The time she takes to walk to school is modelled as a normal distribution, with mean 36 minutes and standard deviation 3.12 minutes.

a Find the probability that on any randomly selected day, Alison's journey to school takes longer than 40 minutes. (2 marks)

b Find the probability that on any randomly selected day, Alison's journey takes between 34 and 38 minutes. (4 marks)

c If the probability that Alison walks for longer than M minutes is 0.015, find the value of M. (3 marks)

d Given that Alison walks to school on 195 days of the year, find the number of days on which she can expect to reach school in under 30 minutes. (4 marks)

16 P2: The masses of cans of baked beans are normally distributed, with a mean mass of 415g and standard deviation 12g.

a The probability that a randomly selected can of beans has a mass greater than mg is 0.65. Find the value of m. (3 marks)

b A can of beans is randomly selected. It is known that the mass of the can is more than 420g. What is the probability that it weighs more than 422.5g? (4 marks)

c Ashok buys 144 cans of beans from the supermarket. What is the probability that at least 75 of them have a mass of less than 413.5g? (4 marks)

17 P2: A continuous random variable X has probability density function

$$f(x) = \begin{cases} kx^2(10 - x), & 0 \le x \le 1, \\ 0, & \text{otherwise} \end{cases}$$

(18 marks)

a Show that $k = \frac{12}{37}$. (4 marks)

b Find $E(X)$. (4 marks)

c Find $Var(X)$. (5 marks)

d Find the median value of X. (5 marks)

18 P2: X is a random variable such that $X \sim N(\mu, \sigma^2)$. It is given that $P(X < 110) = 0.10$ and $P(X > 130) = 0.45$.

a Find the value of μ and the value of σ. (7 marks)

b Find the range of values of X such that $P(|X - \mu|) < 0.22$. (4 marks)

19 P1: A continuous random variable X has probability density function

$$f(x) = \begin{cases} k\cos x, & 0 \le x \le \frac{\pi}{2}, \\ 0, & \text{otherwise} \end{cases}$$

a Show that $k = 1$. (4 marks)

b Hence show that $Var(X) = \pi - 3$. (12 marks)

Click here for further exam practice

Random walking

Approaches to learning: Critical thinking
Exploration criteria: Mathematical communication (B), Personal engagement (C), Use of mathematics (E)
IB topic: Probability, Discrete distributions

Modelling and investigation activity

The problem

A man walks down a long, straight road. With each step he either moves left or right with equal probability. He starts in the middle of the road. If he moves 3 steps to the left or 3 steps to the right, he will fall into a ditch on either side of the road. The aim is to find probabilities related to the man falling into the ditch, and in particular to **find the average number of steps he takes before inevitably falling into the ditch**.

Explore the problem

Use a counter to represent the man and a 'board' to represent the scenario:

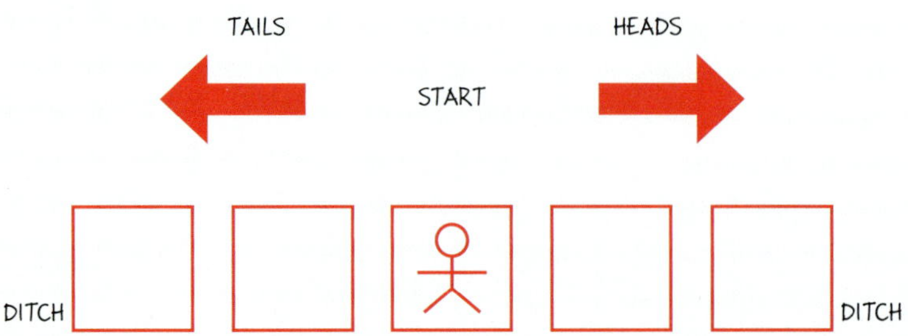

Toss a coin.

Let a tail (T) represent a left step and a head (H) represent a right step.

Write down the number of tosses/steps it takes for the man to fall into the ditch.

Do this a total of 10 times.

Calculate the average number of steps taken.

Construct a spreadsheet with the results from the whole class.

Calculate the average number of steps taken from these results.

How has this changed the result?

Do you know the actual average number of steps required?

How could you be certain what the average is?

Calculate probabilities

Construct a tree diagram that illustrates the probabilities of falling into the ditch within 5 steps.

Use your tree diagram to answer these questions:

What is the probability associated with each sequence in which the man falls into the ditch after a total of exactly 5 steps?

What is the probability that the man falls into the ditch after a total of exactly 5 steps?

What is the minimum number of steps to fall into the ditch?

What is the maximum number?

What is the probability that the man falls into the ditch after a total of exactly 3 steps?

Explain why all the paths have an odd number of steps.

Let x be the number of the steps taken to fall into the ditch.

Copy and complete this table of probabilities:

x	1	2	3	4	5	6	7	8	9	10	11	12	....
$\mathbf{P}(X=x)$													

Look at the numbers in your table.

Can you see a pattern?

Could you predict the next few entries?

Simulation

You could use the table together with the formula $E(X) = \sum_{1}^{\infty} x\mathrm{P}(X = x)$ to try and find $E(X)$, the exact theoretical answer to the problem posed.

However, since there is an infinite number of values of x, calculating the expected number of steps to fall into the ditch would be very complicated.

An alternative approach is to run a computer simulation to generate more results, and to calculate an average from these results.

You can write a code in any computer language available that will run this simulation as many times as needed.

This will allow you to improve on the average calculated individually and as a class.

Although this would not be a proof, it is convincing if enough simulations are recorded.

Extension

Once you have a code written you could easily vary the problem.

What variations of the problem can you think of?

You may also be able to devise your own probability question which you could answer using simulation.

12 Exploration

All IB Diploma subjects have an Internal Assessment (IA). The IA in Mathematics is called an exploration. The exploration will be assessed internally by your teacher and externally moderated by the IB and counts for 20% of your final grade.

This chapter gives you advice on planning your exploration, as well as hints and tips to help you to achieve a good grade by making sure that your exploration satisfies the assessment criteria. There are also suggestions on choosing a topic and how to get started on your exploration.

About the exploration

The exploration is an opportunity for you to show that you can apply mathematics to an area that interests you. It is a piece of written work investigating an area of mathematics.

There are 30 hours in the syllabus for developing your mathematical toolkit and your exploration. The "toolkit" is the inquiry, investigative, problem solving and modelling skills you need to write a good exploration. You can build these skills throughout this book—in particular, in the Investigations, Developing Inquiry Skills and Modelling activities in each chapter.

You should expect to spend around 10–15 hours of class time on your exploration and up to 10 hours of your own time.

During **class time** you will:

- go through the assessment criteria with your teacher
- brainstorm to come up with suitable topics/titles
- look at previous explorations and the grading
- meet with your teacher to discuss your choice of topic and your progress.

During **your own time** you will:

- research the topic you have chosen, to make sure that it is appropriate for an exploration (if not, you will have to conduct further research to help you select a suitable topic)
- collect and organize your information/data and decide which mathematical processes to apply
- write your exploration
- submit a draft exploration to your teacher (your teacher will set a deadline for this)
- present your draft exploration to some of your peers, for their feedback
- submit the final exploration (your teacher will set a deadline for this). If you do not submit an exploration then you receive a grade of "N" and will not receive your IB Diploma.

How the exploration is marked

After you have submitted the final version of your exploration your teacher will mark it. This is "internal assessment" (in school). Your teacher submits these marks to the IB, from which a random sample of explorations is selected automatically. Your teacher uploads these sample explorations to be marked by an external moderator. This external moderation of IA ensures that all teachers in all schools are marking students' work to the same standards.

To begin with, the external moderator will mark three of your school's explorations. If the moderator's mark is within 2 marks of your teacher's mark, then all your teacher's marks stay the same.

If the moderator's mark is more than 2 marks higher or lower than your teacher's mark, the external moderator will mark the remaining explorations in the sample. This may increase the mark if the teacher marked too harshly or decrease the mark if the teacher marked too leniently. The moderator sends a report to the school to explain the reason for any change in the marks.

IA tip

When coloured graphs are uploaded in black and white, they are very difficult to follow, as information such as colour-coded keys is lost. Make sure your exploration is uploaded in colour if it contains colour diagrams.

Internal assessment criteria

Your exploration will be assessed by your teacher, against the criteria given below. The IB external moderator will use the same assessment criteria.

The final mark for each exploration is the sum of the scores for each criterion. The maximum possible final mark is 20. This is 20% of your final mark for Mathematics: analysis and approaches Higher level.

The criteria cover five areas, A to E.

Criterion A	Presentation
Criterion B	Mathematical communication
Criterion C	Personal engagement
Criterion D	Reflection
Criterion E	Use of mathematics

IA tip

Make sure that **all** the pages are uploaded. It is almost impossible to mark an exploration with pages missing.

IA tip

Make sure you understand these criteria. Check your exploration against the criteria frequently as you write it.

Criterion A: Presentation

This criterion assesses the organization, coherence and conciseness of your exploration.

Achievement level	Descriptor
0	The exploration does not reach the standard described by the descriptors below.
1	The exploration has some coherence or some organization.
2	The exploration has some coherence and shows some organization.
3	The exploration is coherent and well organized.
4	The exploration is coherent, well organized and concise.

To get a good mark for Criterion A: Presentation

- A **well organized** exploration has:
    - a **rationale** which includes an explanation of why you chose this topic
    - an **introduction** in which you discuss the context of the exploration
    - a statement of the **aim** of the exploration, which should be clearly identifiable
    - a **conclusion**.
- A **coherent** exploration:
    - is logically developed and easy to follow
    - should "read well" and express ideas clearly
    - includes any graphs, tables and diagrams where they are needed—not attached as appendices to the document.
- A **concise** exploration:
    - focuses on the aim and avoids irrelevancies
    - achieves the aim you stated at the beginning
    - explains all stages in the exploration clearly and concisely.
- References must be cited where appropriate. Failure to do so could be considered academic malpractice.

IA tip

For more on citing references, academic honesty and malpractice, see pages 737–738.

Criterion B: Mathematical communication

This criterion assesses how you:

- use appropriate mathematical language (notation, symbols, terminology)
- define key terms, where required
- use multiple forms of mathematical representation, such as formulae, diagrams, tables, charts, graphs and models, where appropriate.

Achievement level	Descriptor
0	The exploration does not reach the standard described by the descriptors below.
1	There is some relevant mathematical communication which is partially appropriate.
2	The exploration contains some relevant, appropriate mathematical communication.
3	The mathematical communication is relevant, appropriate and is mostly consistent.
4	The mathematical communication is relevant, appropriate and consistent throughout.

IA tip

Only include forms of representation that are relevant to the topic, for example, don't draw a bar chart and pie chart for the same data. If you include a mathematical process or diagram without using or commenting on it, then it is irrelevant.

To get a good mark for Criterion B: Mathematical communication

- Use appropriate mathematical language and representation when communicating mathematical ideas, reasoning and findings.
- Choose and use appropriate mathematical and ICT tools such as graphic display calculators, screenshots, mathematical software, spreadsheets, databases, drawing and word-processing software, as appropriate, to enhance mathematical communication.

IA tip

Use technology to enhance the development of the exploration—for example, by reducing laborious and repetitive calculations.

- Define key terms that you use.
- Express results to an appropriate degree of accuracy.
- Label scales and axes clearly in graphs.
- Set out proofs clearly and logically.
- Define variables.
- Do not use calculator or computer notation.

Criterion C: Personal engagement

This criterion assesses how you engage with the exploration and make it your own.

Achievement level	Descriptor
0	The exploration does not reach the standard described by the descriptors below.
1	There is evidence of some personal engagement.
2	There is evidence of significant personal engagement.
3	There is evidence of outstanding personal engagement.

To get a good mark for Criterion C: Personal engagement

- Choose a topic for your exploration that you are interested in, as this makes it easier to display personal engagement.
- Find a topic that interests you and ask yourself "What if...?"
- Demonstrate personal engagement by using some of these skills and practices from the mathematician's toolkit.
  - Creating mathematical models for real-life situations.
  - Designing and implementing surveys.
  - Running experiments to collect data.
  - Running simulations.
  - Thinking and working independently.
  - Thinking creatively.
  - Addressing your personal interests.
  - Presenting mathematical ideas in your own way.
  - Asking questions, making conjectures and investigating mathematical ideas.
  - Considering historical and global perspectives.
  - Exploring unfamiliar mathematics.

Criterion D: Reflection

This criterion assesses how you review, analyse and evaluate your exploration.

Achievement level	Descriptor
0	The exploration does not reach the standard described by the descriptors below.
1	There is evidence of limited reflection.
2	There is evidence of meaningful reflection.
3	There is substantial evidence of critical reflection.

IA tip

Students often copy their GDC display, which makes it unlikely they will reach the higher levels in this criterion. You need to express results in proper mathematical notation, for example, use 2^x and not $2\text{^}x$

use × not *

use 0.028 and not 2.8E-2.

IA tip

Just showing personal interest in a topic is not enough to gain the top marks in this criterion. You need to write in your own voice and demonstrate your own experience with the mathematics in the topic.

To get a good mark for Criterion D: Reflection

- Include reflection in the conclusion to the exploration, but also throughout the exploration. Ask yourself "What next?"
- Show reflection in your exploration by:
  - discussing the implications of your results
  - considering the significance of your findings and results
  - stating possible limitations and/or extensions to your results
  - making links to different fields and/or areas of mathematics
  - considering the limitations of the methods you have used
  - explaining why you chose this method rather than another.

IA tip

Discussing your results without analysing them is not meaningful or critical reflection. You need to do more than just describe your results. Do they lead to further exploration?

Criterion E: Use of mathematics

This criterion assesses how you use mathematics in your exploration.

Achievement level	Descriptor
0	The exploration does not reach the standard described by the descriptors below.
1	Some relevant mathematics is used.
2	Some relevant mathematics is used. The mathematics explored is partially correct. Some knowledge and understanding are demonstrated.
3	Relevant mathematics commensurate with the level of the course is used. The mathematics explored is correct. Good knowledge and understanding are demonstrated.
4	Relevant mathematics commensurate with the level of the course is used. The mathematics explored is correct. Good knowledge and understanding are demonstrated.
5	Relevant mathematics commensurate with the level of the course is used. The mathematics explored is correct and demonstrates sophistication or rigour. Thorough knowledge and understanding are demonstrated.
6	Relevant mathematics commensurate with the level of the course is used. The mathematics explored is precise and demonstrates sophistication and rigour. Thorough knowledge and understanding are demonstrated.

To get a good mark for Criterion E: Use of mathematics

- Produce work that is commensurate with the level of the course you are studying. The mathematics you explore should either be part of the syllabus, at a similar level or beyond.
- If the level of mathematics is not commensurate with the level of the course you can only get a maximum of two marks for this criterion.
- Only use mathematics relevant to the topic of your exploration. Do not just do mathematics for the sake of it.
- Demonstrate that you fully understand the mathematics used in your exploration.

IA tip

Make sure the mathematics in your exploration is not only based on the prior learning for the syllabus. When you are deciding on a topic, consider what mathematics will be involved and whether it is commensurate with the level of the course.

- Justify **why** you are using a particular mathematical technique (do not just use it).
- Generalize and justify conclusions.
- Apply mathematics in different contexts where appropriate.
- Apply problem-solving techniques where appropriate.
- Recognize and explain patterns where appropriate.

- **Precise** mathematics is error-free and uses an appropriate level of accuracy at all times.
- Demonstrate **sophistication** of mathematics in your exploration.
    - Show that you understand and can use challenging mathematical concepts.
    - Show that you can extend the applications of mathematics beyond those you learned in the classroom.
    - Look at a problem from different mathematical perspectives.
    - Identify underlying structures to link different areas of mathematics.
- Demonstrate **rigour** by using clear logic and language in your mathematical arguments and calculations.

IA tip

Each step in the mathematical development of the exploration needs to be clearly explained so that a peer can follow your working without stopping to think.

Academic honesty

This is very important in all your work. Your school will have an Academic Honesty Policy which you should be given to discuss in class, to make sure that you understand what malpractice is and the consequences of committing malpractice.

According to the IB Learner Profile for Integrity:

"We act with integrity and honesty, with a strong sense of fairness and justice, and with respect for the dignity and rights of people everywhere. We take responsibility for our actions and their consequences."

Academic Honesty means:

- that your work is authentic
- that your work is your own intellectual property
- that you conduct yourself properly during examinations
- that any work taken from another source is properly cited.

Authentic work:

- is work based on your own original ideas
- can draw on the work of others, but this must be fully acknowledged in footnotes and bibliography
- must use your own language and expression
- must acknowledge all sources fully and appropriately in a bibliography.

IA tip

Reference any photographs you use in your exploration, including to decorate the front page.

Malpractice

The IB Organization defines malpractice as "behaviour that results in, or may result in, the candidate or any other candidate gaining an unfair advantage in one or more assessment components."

Malpractice includes:

- plagiarism—copying from others' work, published or otherwise, whether intentional or not, without the proper acknowledgement
- collusion—working together with at least one other person in order to gain an undue advantage (this includes having someone else write your exploration)
- duplication of work—presenting the same work for different assessment components
- any other behaviour that gains an unfair advantage such as taking unauthorized materials into an examination room, stealing examination materials, disruptive behaviour during examinations, falsifying CAS records or impersonation.

IA tip

Plagiarism detection software identifies text copied from online sources. The probability that a 16-word phrase match is "just a coincidence" is $\frac{1}{10^{12}}$.[1]

Collaboration and collusion

It is important to understand the distinction between collaboration (which is allowed) and collusion (which is not).

Collaboration

In several subjects, including mathematics, you will be expected to participate in group work. It is important in everyday life that you are able to work well in a group situation. Working in a group entails talking to others and sharing ideas. Every member of the group is expected to participate equally and it is expected that all members of the group will benefit from this collaboration. However, the end result must be your own work, even if it is based on the same data as the rest of your group.

Collusion

This is when two or more people work together to intentionally deceive others. Collusion is a type of plagiarism. This could be working with someone else and presenting the work as your own or allowing a friend to copy your work.

IA tip

Discussing individual exploration proposals with your peers or in class before submission is collaboration.

Individually collecting data and then pooling it to create a large data set is collaboration. If you use this data for your own calculations and write your own exploration, that is collaboration. If you write the exploration as a group, that is collusion.

References and acknowledging sources

The IB does not tell you which style of referencing you should use—this is left to your school.

IA tip

Be consistent and use the same style of referencing throughout your exploration.

[1]Words & Ideas. The Turnitin Blog. Top 15 misconceptions about Turnitin. Misconception 11: matched text is likely to be completely coincidental or common knowledge (posted by Katie P., March 09, 2010).

The main reasons for citing references are:

- to acknowledge the work of others
- to allow your teacher and moderator to check your sources.

To refer to someone else's work:

- include a brief reference to the source in the main body of your exploration—either as part of the exploration or as a footnote on the appropriate page
- include a full reference in your bibliography.

The bibliography should include a list with full details of **all** the sources that you have used.

IA tip

Cite references to others' work even if you paraphrase or rewrite the original text.

You do not need to cite references to formulae taken from mathematics textbooks.

Choosing a topic

You need to choose a topic that interests you, as then you will enjoy working on the exploration and you will be able to demonstrate personal engagement by using your own voice and demonstrating your own experience.

Discuss the topic you choose with your teacher and your peers before you put too much time and effort into developing the exploration. Remember that the work does not need to go beyond the level of the course which you are taking, but you can choose a topic that is outside the syllabus and is at a commensurate level. You should avoid choosing topics that are too ambitious, or below the level of your course.

IA tip

You must include a brief reference in the exploration as well as in the bibliography. It is not sufficient just to include a reference in the bibliography.

These questions may help you to find a topic for your exploration:

- What areas of the syllabus are you enjoying most?
- What areas of the syllabus are you performing best in?
- Would you prefer to work on purely analytical work or on modelling problems?
- Have you discovered, through reading or talking to peers on other mathematics courses, areas of mathematics that might be interesting to look into?
- What mathematics is important for the career that you eventually hope to follow?
- What are your special interests or hobbies? Where can mathematics be applied in this area?

IA tip

Your exploration should contain a substantial amount of mathematics at the level of your course, and should not just be descriptive. Although the history of mathematics can be very interesting it is not a good exploration topic.

One way of choosing a topic is to start with a general area of interest and create a mind map. This can lead to some interesting ideas on applications of mathematics to explore. The mind map on the following pages shows how the broad topic "Transport" can lead to suggestions for explorations into such diverse topics as baby carriage design, depletion of fossil fuels and queuing theory.

On page 740 there is an incomplete mind map for you to continue, either on your own or by working with other mathematics students.

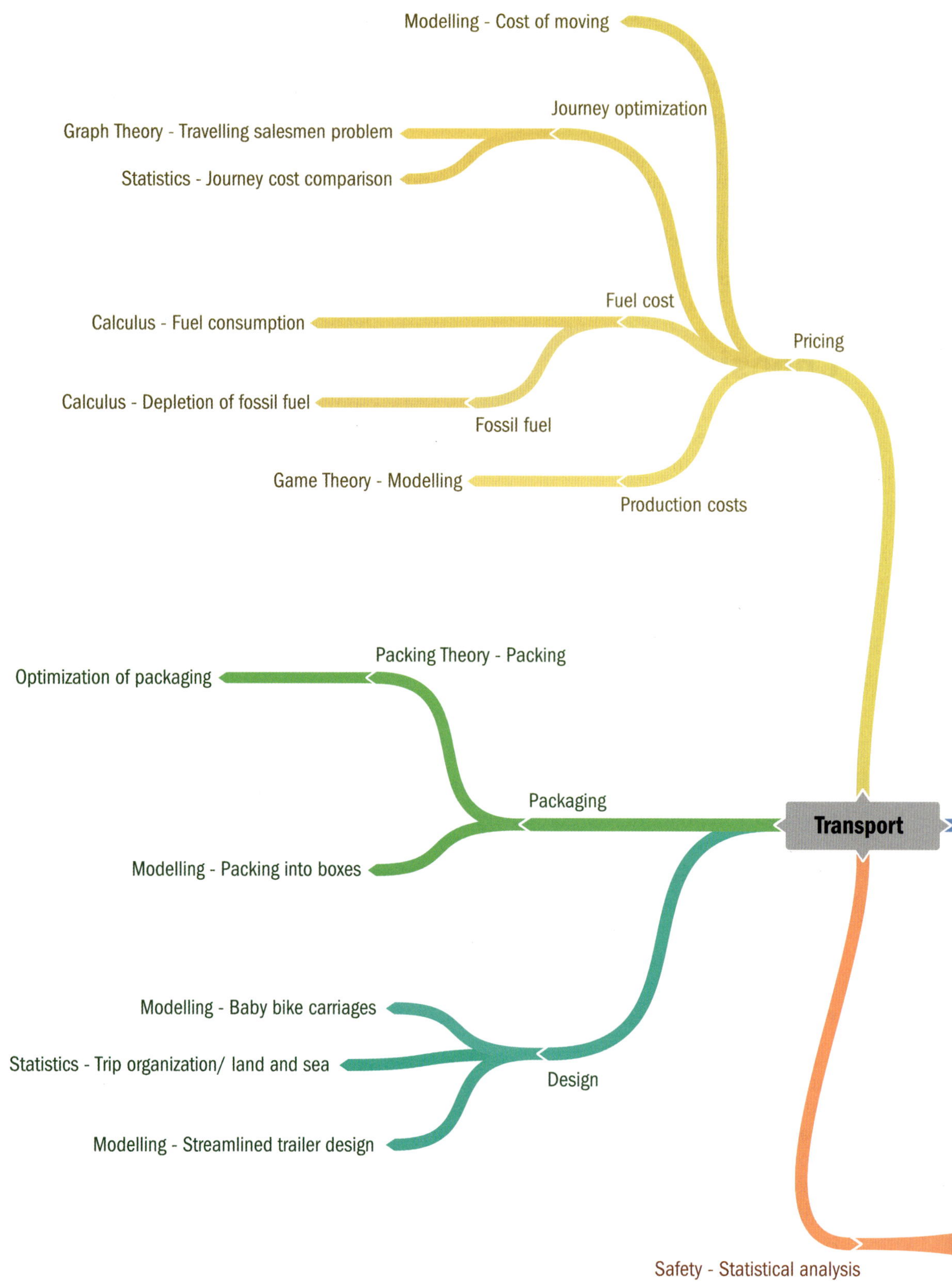
Modelling - Cost of moving
Journey optimization
Graph Theory - Travelling salesmen problem
Statistics - Journey cost comparison
Fuel cost
Calculus - Fuel consumption
Pricing
Calculus - Depletion of fossil fuel
Fossil fuel
Game Theory - Modelling
Production costs
Packing Theory - Packing
Optimization of packaging
Packaging
Transport
Modelling - Packing into boxes
Modelling - Baby bike carriages
Statistics - Trip organization/ land and sea
Design
Modelling - Streamlined trailer design
Safety - Statistical analysis

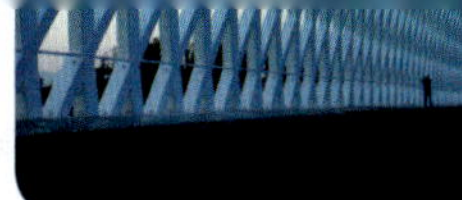

Arcs and trajectories - eg shortest flight path
Calculus - Space travel
Air traffic
Vectors - Airport traffic control
Graph Theory - Pricing a trip
Motion
Statistics - Carbon footprint
Pollution
Statistics - Electric vs hybrid cars
Modelling bridges
treamlined motion
Queuing theory - Bottleneck traffic
Traffic
Modelling: Toll gates
Kinematics - Displacement/ velocity / acceleration
Aeroplanes in flight
Number of bicycles on the road
Height of lorries

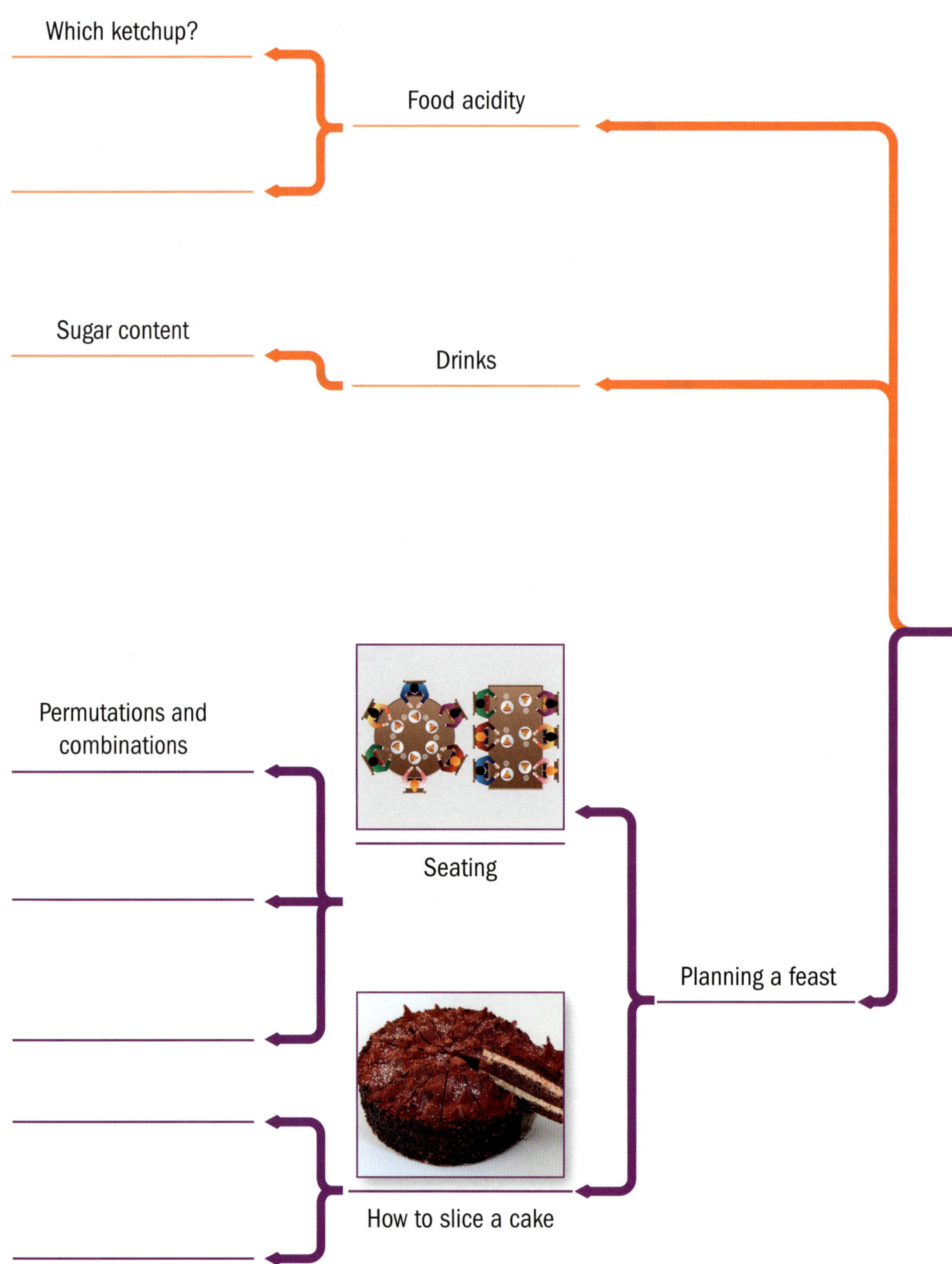
Which ketchup?
Food acidity
Sugar content
Drinks
Permutations and combinations
Seating
Planning a feast
How to slice a cake

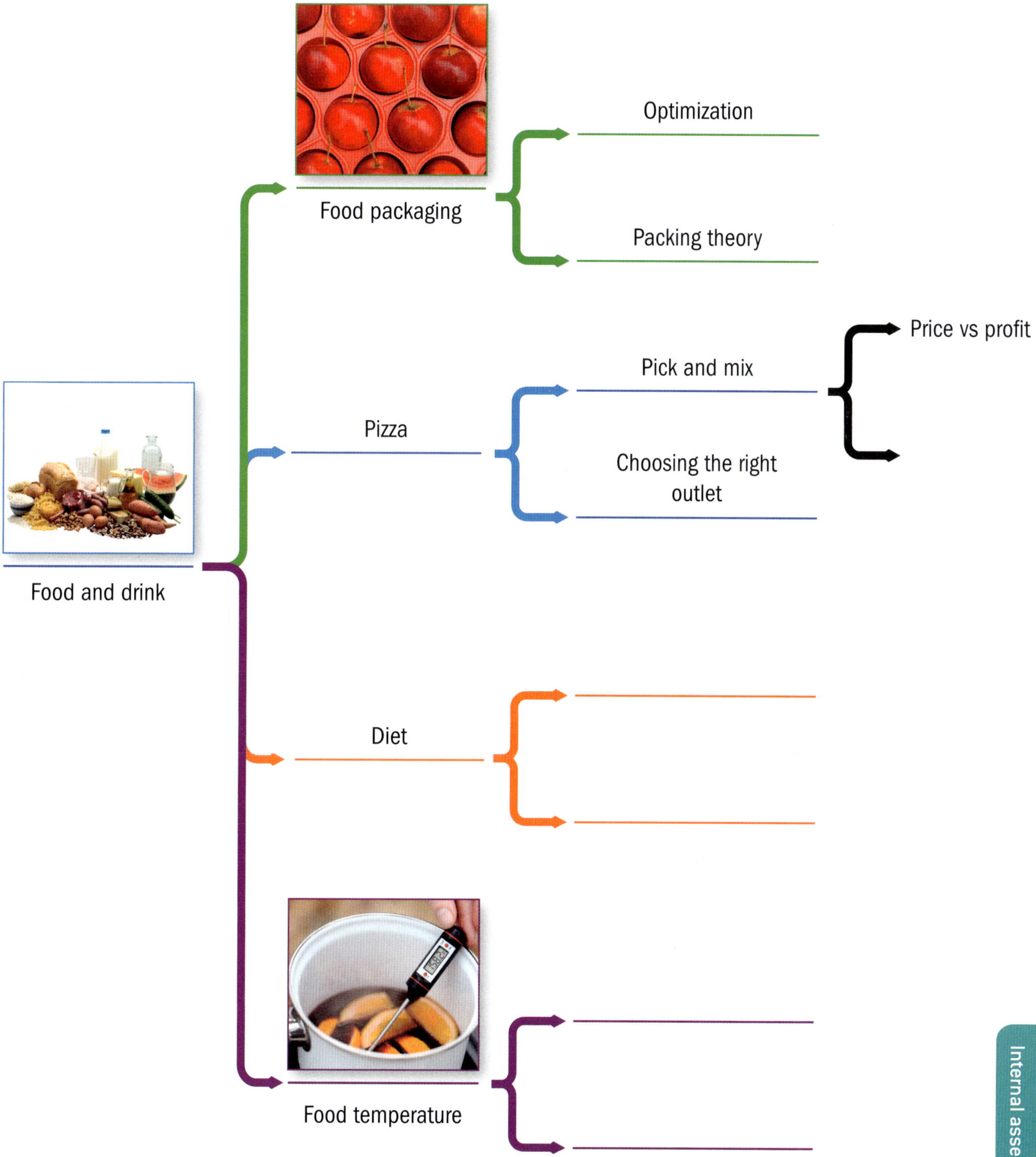

Food and drink

Food packaging

Optimization

Packing theory

Pizza

Pick and mix

Price vs profit

Choosing the right outlet

Diet

Food temperature

Research

Once you have chosen a topic, you will need to do some research. The purpose of this research is to help you determine how suitable your topic is.

- Don't rely on the Internet for all your research—you should also make use of books and academic publications.
- Plan your time wisely—make sure that you are organized.
- Don't put it off—start your research in good time.
- For Internet research: refine your topic so you know exactly what information you are looking for, and use multiple-word searches. It is very easy to spend hours on the Internet without finding any relevant information.
- Make sure that you keep a record of all the websites you use—this saves so much time afterwards. You will need to cite them as sources, and to include them in your bibliography.
- Make sure that the sources are reliable—who wrote the article? Are they qualified? Is the information accurate? Check the information against another source.
- Research in your own language if you find this easier.

These questions will help you to decide whether the topic you have chosen is suitable.

- What areas of mathematics are contained in the topic?
- Which of these areas are contained in the syllabus that you are following?
- Which of these areas are not in the syllabus that you are following but are contained in the other IB mathematics course?
- Which of these areas are in none of the IB mathematics courses? How accessible is this mathematics to you?
- Would you be able to understand the mathematics and write an exploration in such a way that a peer is able to understand it all?
- How can you demonstrate personal engagement in your topic?
- Will you manage to complete an exploration on this topic and meet all the top criterion descriptors within the recommended length of 12 to 20 pages (double spaced and font size 12)?

IA tip

Try to avoid writing a research report in which you merely explain a well-known result that can easily be found online or in textbooks. Such explorations have little scope for meaningful and critical reflection and it may be difficult to demonstrate personal engagement.

Writing an outline

Once you think you have a workable topic, write a brief outline including:

- Why you chose this topic.
- How your topic relates to mathematics.
- The mathematical areas in your topic, for example, algebra, geometry, calculus, etc.
- The key mathematical concepts covered in your topic, for example, modelling data, areas of irregular shapes, analysing data, etc.
- The mathematical skills you will use in the exploration, for example, integration by parts, working with complex numbers, using polar coordinates, etc.
- Any mathematics outside the syllabus that you need to learn.
- Technology you could use to develop your exploration.
- New key terms that you will need to define or explain.
- How you are going to demonstrate personal engagement.
- A list of any resources you have/ will use in the development of your exploration. If this list includes websites you should include the URL and the date when this was accessed.

Share this outline with your teacher and with your peers. They may ask questions that lead you to improve your outline.

IA tip

Learning new mathematics is not enough to reach the top levels in Criterion C: Personal engagement.

IA tip

Popular topics such as the Monty Hall problem, the Birthday paradox, and so on are not likely to score well on all the criteria.

This template may help you write the outline for the exploration when presenting a formal proposal to your teacher.

Mathematics exploration outline

Topic:
Exploration title:
Exploration aim:
Exploration outline:
Resources used:
Personal engagement:

IA tip

As you write your exploration, remember to refer to the criteria below.

Writing your exploration

Now you should be ready to start writing your exploration in detail. You could ask one of your classmates to read the exploration and give you feedback before you submit the draft to your teacher. If your exploration is related to another discipline, for example, economics, it would be better if the peer reading your exploration is someone who does not study economics.

IA tip

Remember that your peers should be able to read and understand your work.

Mathematical exploration checklist

Work through this checklist to confirm that you have done everything that you can to make your exploration successful.

- ☐ Does your exploration have a title?
- ☐ Have you given a rationale for your choice of exploration?
- ☐ Have you ensured your exploration does not include any identifying features—for example, your name, candidate number, school name?
- ☐ Does your exploration start with an introduction?
- ☐ Have you clearly stated your aim?
- ☐ Does your exploration answer the stated aim?
- ☐ Have you used double line spacing and 12-point font?
- ☐ Is your exploration 12–20 pages long?
- ☐ Have you cut out anything that is irrelevant to your exploration?
- ☐ Have you checked that you have not repeated lots of calculations?
- ☐ Have you checked that tables only contain relevant information and are not too long?
- ☐ Is your exploration easy for a peer to read and understand?
- ☐ Is your exploration logically organized?
- ☐ Are all your graphs, tables and diagrams correctly labelled and titled?
- ☐ Are any graphs, tables and diagrams placed appropriately and not all attached at the end?
- ☐ Have you used appropriate mathematical language and representation (not computer notation, eg *, ^, etc.)?
- ☐ Have you used notation consistently through your exploration?
- ☐ Have you defined key terms (mathematical and subject specific) where necessary?
- ☐ Have you used appropriate technology?
- ☐ Have you used an appropriate degree of accuracy for your topic/exploration?
- ☐ Have you shown interest in the topic?
- ☐ Have you used original analysis for your exploration (eg simulation, modelling, surveys, experiments)?
- ☐ Have you expressed the mathematical ideas in your exploration in your own way (not just copy-and-pasted someone else's)?
- ☐ Does your exploration have a conclusion that refers back to the introduction and the aim?
- ☐ Do you discuss the implications and significance of your results?
- ☐ Do you state any possible limitations and/or extensions?
- ☐ Do you critically reflect on the processes you have used?
- ☐ Have you explored mathematics that is commensurate with the level of the course?
- ☐ Have you checked that your results are correct?
- ☐ Have you clearly demonstrated understanding of why you have used the mathematical processes you have used?
- ☐ Have you acknowledged direct quotes appropriately?
- ☐ Have you cited all references in a bibliography?
- ☐ Do you have an appendix if one is needed?

Paper 1

Time allowed: 2 hours

Answer all the questions.

All numerical answers must be given exactly or correct to three significant figures, unless otherwise stated in the question.

Answers should be supported by working and/or explanations. Where an answer is incorrect, some marks may be awarded for a correct method, provided this is shown clearly.

You are not allowed to use a calculator for this paper.

Short questions

1 **a** Find $\int \sin^2 x \, dx$. (3 marks)

b Hence, evaluate $\int_0^{\pi} \sin^2 x \, dx$. (2 marks)

[Total 5 marks]

2 Let $\log x = p$ and $\log y = q$. Find

a $\log(xy)$ (2 marks)

b $\log\left(\frac{x^2}{y}\right)$ (2 marks)

c $\log\sqrt{x}$ (2 marks)

d $\log(100y)$. (2 marks)

[Total 8 marks]

3 A piece of wood is 290 cm long. It is going to be cut into 10 pieces. The length of each piece should be 2 cm longer than the piece cut off before it, and all the wood will be used up.

Find

a the length of the first piece of wood that is cut (3 marks)

b the length of the last piece. (2 marks)

[Total 5 marks]

4 Find $\int_0^4 \frac{1}{2x+1} dx$.

Give your answer in the form $\ln k$, where $k \in \mathbb{N}$. (5 marks)

[Total 5 marks]

5 Let **a, b** and **c** be three, 3-dimensional vectors.

State whether each expression below is a meaningful vector expression, or is meaningless.

Briefly, justify your answer if you claim an expression is meaningless.

- **i** $\mathbf{a}\times(\mathbf{b}\times\mathbf{c})$
- **ii** $\mathbf{a}\cdot(\mathbf{b}\cdot\mathbf{c})$
- **iii** $\mathbf{a}\cdot(\mathbf{b}\times\mathbf{c})$
- **iv** $\mathbf{a}\times(\mathbf{b}\cdot\mathbf{c})$
- **v** $(\mathbf{b}\times\mathbf{c})\mathbf{a}$
- **vi** $(\mathbf{b}\cdot\mathbf{c})\mathbf{a}$ (6 marks)

[Total 6 marks]

6 By repeated use of L'Hôpital's Rule, find $\lim_{x\to 0}\frac{\tan x - x}{x^3}$. At each stage, you should justify that any indeterminant form meets the criteria to use L'Hopital's Rule. (7 marks)

[Total 7 marks]

7 Differentiate the function $f(x)=x^4$ from first principles

(ie using $f'(x)=\lim_{h\to 0}\frac{f(x+h)-f(x)}{h}$). (5 marks)

[Total 5 marks]

8 When the polynomial $p(x)=x^3-x^2+ax+b$ is divided by x the remainder is 9, and when $p(x)$ is divided by $(x+1)$ the remainder is 16.

a Find the values of a and b. (4 marks)

b Factorize $p(x)$ into linear factors. (4 marks)

[Total 8 marks]

9 A polynomial is defined by $p(x)=x^4+10x^3+35x^2+50x+24$.

a For $p(x)=0$, write down (i) the sum of the roots, (ii) the product of the roots. (2 marks)

Another polynomial is defined by $q(x)=p(x-1)$.

b For $q(x)=0$, find (i) the sum of the roots, (ii) the product of the roots. (5 marks)

[Total 7 marks]

Long questions

10 Let $f(x)=x^2+6x+(8+k)$.

a Find the discriminant, D, of this quadratic, in terms of k. (2 marks)

b Given that $f(x)>0$ for all values of x find the set of values that **i** D **ii** k can take. In each case, you must justify your answer. (5 marks)

c Write $f(x)$ in the form $f(x)=(x+p)^2+q$, where $p\in\mathbb{Z}$ and q is an expression involving k. (2 marks)

d Explain how expression you obtained for $f(x)$ in part **c** confirms your findings in part **bii** for the range of k. (2 marks)

e If $k = 4$, write down the minimum point of the quadratic. (2 marks)

[Total 13 marks]

11 a Prove by induction that $\sum_{i=1}^{n} i \times i! = (n+1)! - 1$ for $i, n \in \mathbb{Z}^+$. (9 marks)

b By writing $i \times i!$ as $(i+1-1) \times i!$ construct a direct proof (that is, a proof not using induction) for the result given in part **a**. (4 marks)

[Total 13 marks]

12 a Solve $z^4 - 1 = 0$ for $z \in \mathbb{C}$ by factorizing the left-hand side as far as possible. Give your answers in the form $a + b\text{i}$, where $a, b \in \mathbb{R}$ (4 marks)

b Solve $z^8 = 1$ for $z \in \mathbb{C}$ using De Moivre's theorem. Give your answers in the form $r\text{cis}\theta$ where $r \in \mathbb{R}^+, -\pi < \theta \le \pi$. (6 marks)

c Express the solution to $z^8 = 1$ which lies in the first quadrant of the Argand diagram (with $0 < \theta < \frac{\pi}{2}$) in the form $a + b\text{i}$, where $a, b \in \mathbb{R}$ (2 marks)

d Hence, write down one of the values of $\sqrt{\text{i}}$ in the form $a + bi$, where $a, b \in \mathbb{R}$ (1 mark)

[Total 13 marks]

13 A rumour about Katie is spreading in a college with a large number of students. Let x be the proportion of the college students who have heard the rumour and let t be the time in hours, after 9.00 a.m. This situation is modelled by the differential equation $\frac{\text{d}x}{\text{d}t} = kx(1-x)$, where k is a constant.

a Use partial fractions to solve this differential equation and hence show that $\frac{x}{1-x} = A\text{e}^{kt}$, where A is a constant. (7 marks)

b At 9.00 a.m., one third of the students know about the rumour. Find the value of A. (2 marks)

c At 10.00 a.m., half of the students knew about the rumour. Find the value of e^k. (2 marks)

d Hence, find the proportion of the students who knew about the rumour at 11.00 a.m. (4 marks)

[Total 15 marks]

Paper 2

Time allowed: 2 hours

Answer all the questions.

All numerical answers must be given exactly or correct to three significant figures, unless otherwise stated in the question.

Answers should be supported by working and/or explanations. Where an answer is incorrect, some marks may be awarded for a correct method, provided this is shown clearly.

You need a graphic display calculator for this paper.

Short questions

1 The random variable X is normally distributed, with mean equal to 8. Given that $P(X > 7) = 0.69146$, find the value of the standard deviation of X. (6 marks)

[Total 6 marks]

2 Maria sees a huge monster. The angle of elevation from Maria's feet to the top of the monster's head is 40°. Maria runs 5 m further away from the monster, along horizontal ground. The new angle of elevation from Maria's feet to the top of the monster's head is 35°.

- **a** Sketch a diagram to represent the information given above. (1 mark)
- **b** Find the height of the monster. (6 marks)

[Total 7 marks]

3 An object is initially at the origin. It moves in a straight line with velocity v m/s given by $v = 5\sin\left(t^2\right)$, where t is time in seconds.

- **a** Write down the initial velocity of the object. (1 mark)
- **b** Find the total distance that the object travels in the first 3 seconds. (3 marks)
- **c** Find the displacement of the object from the origin when $t = 3$. (3 marks)

[Total 7 marks]

4 A very old aircraft has four engines, two on each wing. On a particular flight the probability that any engine breaks down is 0.1. All engine breakdowns are independent of each other. The aircraft will be able to complete the flight as long as there is at least one engine operating on each wing. Otherwise it will have to perform an emergency landing. Find the probability that it does not have to perform an emergency landing, giving your answer to 4 decimal places. (4 marks)

[Total 4 marks]

5 A function is given by $f(x) = e^{x^2}$.

a Calculate the area of the region bounded by $y = f(x)$, $y = 0$, $x = -1$ and $x = 1$. (3 marks)

b This region is rotated through 2π about the x-axis. Find the volume of the solid of revolution that is generated. (3 marks)

[Total 6 marks]

6 A triangle ABC has $B\hat{A}C = 40^\circ$, $AB = 5$, $BC = 4$. Find the value of AC given that area $ABC < 5$. (7 marks)

[Total 7 marks]

7 The discrete random variable X satisfies the $B\left(2n, \frac{1}{2}\right)$ distribution.

a By considering symmetry, write down what the mode is. (1 mark)

b Given that $P(X = \text{the mode}) = \frac{35}{128}$, find the value of n. (2 marks)

[Total 3 marks]

8 A forest fire is spreading out in the shape of a circle. The radius is increasing at a rate of 0.1 km per hour.

a Find the rate at which the burned area is increasing when the radius is 5 km. (4 marks)

b Find the radius when the burned area is increasing at a rate of 6 km^2 per hour. (2 marks)

[Total 6 marks]

9 Consider the system of equations

$$x + 2y + 3z = 7$$
$$4x - y - z = 6$$
$$5x + y + kz = 12.$$

a Find the value of k for which there are no solutions. (4 marks)

b For $k = 1$ find the solution. (3 marks)

[Total 7 marks]

Long questions

10 a Find the acute angle, in degrees, between the vectors $\vec{v} = \begin{pmatrix} 1 \\ 2 \\ 3 \end{pmatrix}$ and $\vec{w} = \begin{pmatrix} 2 \\ -1 \\ 2 \end{pmatrix}$. (5 marks)

b Find the acute angle, in degrees, between the planes $\Pi_1 : x + 2y + 3z + 4 = 0$ and $\Pi_2 : 2x - y + 2z - 11 = 0$. (3 marks)

c Find the acute angle, in degrees, between the plane Π_1 and the line $L_1 : \frac{x-5}{2} = \frac{y+6}{-1} = \frac{z-5}{2}$. (3 marks)

d Find the point of intersection between the plane Π_1 and the line L_1. (5 marks)

[Total marks 16]

11 A function is defined by $f(x) = \dfrac{6x+3}{2x-10}, x \in \mathbb{R}, x \neq 5.$

a **i** Write down the equation of the vertical asymptote.

ii Write down the equation of the horizontal asymptote. (2 marks)

b **i** Write down the intercept on the y-axis.

ii Write down the intercept on the x-axis. (2 marks)

c **i** Find $f'(x)$.

ii State what your answer to part **i** tells you about the graph of $f(x)$. (4 marks)

d Sketch the graph of $f(x)$, showing the information found in **a**, **b**, and **c**. (3 marks)

e Write down the equation of the tangent to the curve at $x = 2$. (2 marks)

[Total 13 marks]

12 A very bouncy ball is dropped from a height of 2 m. After each bounce on the floor it rebounds to four-fifths of its previous height.

a Find the height that the ball bounces to after the 5th bounce. (2 marks)

b Find the least number of bounces after which the ball bounces to height less than 0.25 m. (3 marks)

c Calculate the total distance the ball travels before it comes to rest on the floor. (6 marks)

The time t that it takes for the ball to drop a distance of s on each bounce is given by $s = 5t^2$, where t is measured in seconds. For each bounce, the ball takes the same time going up as it does coming down.

d Calculate the total time between the ball being dropped and it coming to rest on the floor. (7 marks)

[Total 18 marks]

13 A continuous random variable has a probability density function defined by

$$f(x) = \begin{cases} 0 & \text{if } x < -\dfrac{1}{2} \\ a\cos(\pi x) & \text{if } -\dfrac{1}{2} \leq x \leq \dfrac{1}{2} \\ 0 & \text{if } x > \dfrac{1}{2}. \end{cases}$$

a Find the exact value of a. (3 marks)

b For this distribution find (i) the mean, (ii) the variance. (4 marks)

c Given that $x > 0$ find the probability that $x < \dfrac{1}{4}$. (3 marks)

[Total 10 marks]

Paper 3

Time allowed: 1 hour

Answer all the questions.

All numerical answers must be given exactly or correct to three significant figures, unless otherwise stated in the question.

Answers should be supported by working and/or explanations. Where an answer is incorrect, some marks may be awarded for a correct method, provided this is shown clearly.

You need a graphic display calculator for this paper.

1 Consider a sequence, with first term of u_1, defined by a recurrence relation of the form $u_{n+1} = au_n + b$, where a and b are constants.

a In the case when $a = 1$

i State what type of sequence this is.

ii Write down an expression for u_n in terms of u_1 and b.

iii Write down an expression for S_n (the sum of the first n terms) in terms of u_1 and b. (3 marks)

b In the case when $b = 0, a \neq 1, a \neq 0$.

i State what type of sequence this is.

ii Write down an expression for u_n in terms of u_1 and a.

iii Write down an expression for S_n (the sum of the first n terms) in terms of u_1 and a. (3 marks)

c The *Towers of Hanoi* is a famous example of a sequence defined by a recurrence relation of the form $u_{n+1} = au_n + b$. In this sequence, $u_1 = 1$ and $u_{n+1} = 2u_n + 1$.

i Find u_2, u_3, u_4, and u_5.

ii Using your answers to part **i**, suggest an explicit expression for u_n in terms of n.

iii Prove your expression from part **ii** using mathematical induction. (9 marks)

d When $a \neq 1$, the recurrence relation $u_{n+1} = au_n + b$ can be rewritten in the form $u_{n+1} + c = a(u_n + c)$ for some constant c.

i Find the constant c in terms of a and b.

A new sequence is defined by $v_n = u_n + c$.

ii Write down the recurrence relation for the sequence v_n, and state what type of sequence v_n is.

iii Hence, write down an explicit formula for v_n in terms of v_1, a and n.

iv Hence, write down an explicit formula for u_n in terms of u_1, a, b, and n. (8 marks)

e Verify that the expression you obtained for u_n in part **cii** for the Towers of Hanoi is consistent with the formula you derived for u_n in part **div**. (2 marks)

[Total 25 marks]

2 In this question you will find the value of the indefinite integral $I = \int \frac{1}{\sqrt{1+x^2}}\,dx$ using different methods.

Suppose that, first, you are asked to find the value of a definite integral. You can do this using technology.

a Write down the value of $\int_0^1 \frac{1}{\sqrt{1+x^2}}\,dx$. (2 marks)

Another integral, which looks similar to I, can be found using a trigonometric substitution.

b Write down $\int \frac{1}{\sqrt{1-x^2}}\,dx$. (1 mark)

c Show how the answer in **b** can be proved by applying the substitution $x = \sin\theta \left(-\frac{\pi}{2} \le \theta \le \frac{\pi}{2}\right)$, to $\int \frac{1}{\sqrt{1-x^2}}\,dx$. (4 marks)

To find the integral I by substitution, we are going to first define two new functions.

$$\sinh x = \frac{e^x - e^{-x}}{2}, \cosh x = \frac{e^x + e^{-x}}{2}.$$

d Show that $(\cosh x)^2 - (\sinh x)^2 = 1$. (2 marks)

e Find the derivative of **i** $\sinh x$ **ii** $\cosh x$. give your answers in terms of $\sinh x$ and $\cosh x$. (4 marks)

The inverse function $\sinh^{-1} x$ is denoted by arsinh x.

f By solving a quadratic in e^y, show and justify that $\operatorname{arsinh} x = \ln\left(x + \sqrt{x^2+1}\right)$. (5 marks)

g Find the indefinite integral I by applying the substitution $x = \sinh u$. (4 marks)

h Hence, find the exact value of $\int_0^1 \frac{1}{\sqrt{1+x^2}}\,dx$ and verify that it agrees with the answer from part **a**. (2 marks)

We will now attempt to see if we can find the integral I just by using the usual trigonometrical functions.

i Apply the substitution $x = \tan\theta$, $\left(-\frac{\pi}{2} < \theta < \frac{\pi}{2}\right)$, to show that I can be changed to $\int \sec\theta\,d\theta$. (2 marks)

j Multiply the $\sec\theta$ by 1, written as $\frac{\sec\theta + \tan\theta}{\sec\theta + \tan\theta}$, to obtain the answer to $\int \sec\theta\,d\theta$ in terms of θ. (2 marks)

k Now express the answer found in part **j** in terms of x and verify this is consistent with the expression you obtained for I in part **g**. (2 marks)

You will now attempt to find the integral I without using a substitution.

Let $f(x) = x + \sqrt{1+x^2}$.

l Find $f'(x)$. (2 marks)

m Multiply $\frac{1}{\sqrt{1+x^2}}$ by 1, written as $\frac{f'(x)}{f'(x)}$ using your expression for $f(x)$ from part **l**. Hence simplify the expression for I into an expression involving only $f(x)$ and $f'(x)$. In this way, find the integral I directly. (3 marks)

[Total 35 marks]

Answers

Chapter 1

Skills check

1 **a** $x=-2$ **b** $x=\frac{1}{5}$

2 **a** $-3-2\sqrt{2}$ **b** $-\sqrt{2}-\sqrt{6}$

3 $\frac{2x^3+3x-1}{(x^2-1)(2x-1)}$

Exercise 1A

1 **a** 9, 10.5, 12, ... $\{u_r\}=\{1.5+1.5r\}$, $r\in\mathbb{Z}^+$

b 5, 2, −1 $\{u_r\}=\{20-3r\}$, $0r\in\mathbb{Z}^+$

c 243, 729, 2187 $\{u_r\}=\{3^r\}$, $r\in\mathbb{Z}^+$

d $\frac{13}{16}$, $\frac{16}{19}$, $\frac{19}{22}$

$\{u_r\}=\left\{\frac{3r-2}{3r+1}\right\}$, $r\in\mathbb{Z}^+$

e $\frac{1}{90}$, $\frac{1}{132}$, $\frac{1}{182}$

$\{u_r\}=\left\{\frac{1}{(2r-1)(2r)}\right\}$, $r\in\mathbb{Z}^+$

2 **a** 1, −1, −3, −5, −7

b $\frac{1}{3}$, $\frac{2}{5}$, $\frac{3}{7}$, $\frac{4}{9}$, $\frac{5}{11}$

c 1, 6, 3, 12, 5

d −2, 2, −2, 2, −2

e 3, $\frac{3}{2}$, $\frac{3}{4}$, $\frac{3}{8}$, $\frac{3}{16}$

3 **a** $\{u_r\}=\{5r\}$, $r\in\mathbb{Z}^+$

b $\{u_r\}=\{8r-2\}$, $r\in\mathbb{Z}^+$

c $\{u_r\}=\left\{\frac{1}{2^r}\right\}$, $r\in\mathbb{Z}^+$

d $\{u_r\}=\left\{\left(-\frac{1}{3}\right)^{r-1}\right\}$, $r\in\mathbb{Z}^+$

e $\{u_r\}=\{r^2-1\}$, $r\in\mathbb{Z}^+$

4 **a** 0 −4 −12 −24

b 0 −1 + 4 −9 +16 −25

c $\frac{1}{2}+\frac{2}{5}+\frac{3}{8}+\frac{4}{11}+\frac{5}{14}$

d 5 + 5 + 5 + 5

f −3 − 2 + 1 + 6

5 **a** $2+\frac{3}{4}+\frac{4}{9}+\frac{5}{16}+\frac{6}{25}$

b $-1+\frac{1}{7}-\frac{1}{17}+\frac{1}{31}-\frac{1}{49}$

c 4 + 8 + 42 + 76 + 120

d −2 −1 +1 +5 +13

e 1 + 4 + 27 + 256 + 3125

6 **a** $\sum_{r=1}^{5}(11-3r)$ **b** $\sum_{r=1}^{5}r(2r+1)$

c $\sum_{r=1}^{6}\frac{r-1}{r+1}$ **d** $\sum_{r=1}^{5}(2r-1)^2$

e $\sum_{r=1}^{5}3kr$

Exercise 1B

1 **a** $5n-2$ **b** $105-4n$

c $4n+a-7$ **d** $15n-35$

2 **a** 89 **b** −60

c $a+32$ **d** $16-4n$

3 **a** 17 **b** 25

c 21 **d** 15

4 **a** $u_1=-2$, $d=5$

b $u_1=14$, $d=3$

c $u_1=-5$, $d=-11$

d $u_1=2a+3$, $d=2$

5 $u_1=2$, $u_n=7n-5$

6 $d=18$, $u_1=-72$

7 −4, 8, 20

8 €46 500 2038

9 **a** −765 **b** 15 070

c $50a+3625$

10 **a** 3425 **b** −39 700

c $420a-20$

11 **a** −465 **b** 390

c −930

12 1275

13 **a** 4 **b** −416

14 3, −3, −9, −15

15 22 500

Exercise 1C

1 **a** $u_5=81$, $u_n=3^{n-1}$

b $u_5=\frac{1}{2}$, $u_n=\frac{1}{2^{n-4}}$

c $u_5=\frac{x^9}{2}$, $u_n=\frac{x^{2n-1}}{2}$

d $u_5=-3$, $u_n=3(-1)^n$

2 **a** $r=\frac{1}{3}$, $u_6=\frac{7}{27}$

b $r=\frac{1}{6}$, $u_7=\frac{1}{192}$

c $r=-\frac{1}{3}$, $u_5=\frac{a}{162}$

3 **a** 10 **b** 14

4 $r=2$, $u_1=\frac{3}{4}$

5 $r=\pm3$, $u_6=\frac{2}{3}(\pm3)^5=\pm162$

(depending on which ratio is used)

6 $u_1=9$

$u_5=u_1r^4=9r^4=16$

$\Rightarrow r^4=\frac{16}{9}\Rightarrow r=\pm\frac{2}{\sqrt{3}}=\pm\frac{2\sqrt{3}}{3}$

$u_7=\frac{64}{3}$

7 $-3, \frac{2}{5}$

8 $r=-\frac{3}{2}$, $u_1=\frac{32}{135}$

9 **a** $S_6 = \frac{182}{81}$ **b** $S_{10} = \frac{1023}{64}$

c $S_{15} = 0.143$ (to 3s.f.)

d $S_{15} = 0.0769$ (3s.f.)

10 **a** 57.2 (to 3s.f.)

b $\frac{5}{9}(10^n - 1)$

11 $u_1 = 3$

$u_7 = u_1 r^6 = 3r^6 = \frac{1}{243}$

$r^6 = \frac{1}{729} \Rightarrow r = \pm\frac{1}{3}$

Therefore there are two possible common ratios, each corresponding to a different sum to infinity.

$r = -\frac{1}{3}$: $S_\infty = \frac{3}{1-\left(-\frac{1}{3}\right)} = \frac{9}{4}$

€

$r = \frac{1}{3}$: $S_\infty = \frac{3}{1-\frac{1}{3}} = \frac{9}{2}$

12 **a** $-\frac{3}{2}, \frac{3}{4}, -\frac{3}{8}$

b The terms are in geometric progression with $r = -\frac{1}{2}$. To see this in general, note

$$u_n = S_n - S_{n-1} = \left(-\frac{1}{2}\right)^n - 1 - \left[\left(-\frac{1}{2}\right)^{n-1} - 1\right] = \left(-\frac{1}{2}\right)^n - \left(-\frac{1}{2}\right)^{n-1}$$

$$= \left(-\frac{1}{2}\right)^{n-1}\left(-\frac{1}{2} - 1\right) = -\frac{3}{2}\left(-\frac{1}{2}\right)^{n-1}$$

i.e. the form of a general term in a geometric progression with first term $-\frac{3}{2}$ and common ratio $-\frac{1}{2}$.

13 $r = \frac{1}{4}$

14 $r = \frac{1}{\sqrt{2}}$ $u_3 = \frac{2}{7}(3 - \sqrt{2})$

15 $\sum_{i=0}^{\infty}(-1)^i\left(\frac{x}{2}+1\right)^i = 1 - \left(\frac{x}{2}+1\right) + \left(\frac{x}{2}+1\right)^2 - \left(\frac{x}{2}+1\right)^3 + \left(\frac{x}{2}+1\right)^4 - \ldots$

Converges when $-4 < x < 0$

Exercise 1D

1 **a** 290 **b** 2040 **c** 2014

2 €42 150 to the nearest euro

3 **a** 30 **b** 19

4 2.9 kg, 9 trials, In general, the geometric model is not reliable, since if Prisana were to carry out a large number of trials then the cake would become excessively sweet (since exponential growth is greater than linear growth).

In fact, the ratio of sugar to flour would eventually become 1 (i.e. the mix is entirely sugar) in the (albeit unrealistic) case that Prisana carries out the trial a large number of times.

5 **a** Second: $\sqrt{2}$, third: 1, fourth: $\frac{1}{\sqrt{2}}$

b $\frac{3(7+3\sqrt{2})}{4}$

c The length converges to a finite value since the common ratio between two consecutive side lengths is less than one.

d 0.996 **e** $S_\infty = 1$

6 **a** \$70.61 **b** \$1695

7 **a** 10 **b** 12.5 m

c 3.75 m

8 **a** Rapid: €300, Quick: €282, Rapid/Quick: €291

b Rapid: \$450, Quick: €473, Rapid/Quick: €461

c 21

9 **a** \$1500 in savings, \$2500 in bonds and \$1000 in shares

b \$6049 **c** \$26

10 **a** $x(1 + 0.375 + 0.375^2 + 0.375^3)$ where x is the amount administered each time.

b 5 mg

c There are 7 mg/ml drug in the bloodstream after the third administration.

Exercise 1E

1 $(a + b)^2 + (a - b)^2 = (a^2 + 2ab + b^2) + (a^2 - 2ab + b^2) = 2a^2 + 2b^2 = 2(a^2 + b^2)$

2 A general odd number can be written in the form $2k + 1$ with $k \in \mathbb{Z}$

$\therefore$ Consider two general odd numbers $2n + 1$ and $2m + 1$

Then $(2n + 1)(2m + 1) = 4nm + 2n + 2m + 1 = 2(2nm + n + m) + 1$ is always an odd number.

3 A four digit number represented by $a_3a_2a_1a_0$ (not to be confused with a product) can be written in the form $N = a_3 \times 10^3 + a_2 \times 10^2 + a_1 \times 10 + a_0$

$\therefore N = (999 + 1)a_3 + (99 + 1)a_2 + (9 + 1)a_1 + a_0$

$= (999a_3 + 99a_2 + 9a_1) + (a_3 + a_2 + a_1 + a_0)$

$= 9(111a_3 + 11a_2 + a_1) + (a_3 + a_2 + a_1 + a_0)$

$\Rightarrow \frac{N}{9} = 111a_3 + 11a_2 + a_1 + \frac{a_3 + a_2 + a_1 + a_0}{9}$

so 9 divides N if $\frac{a_3 + a_2 + a_1 + a_0}{9}$ is a whole number

i.e. 9 divides the sum of the digits.

Hence 3978, 9864 and 5670 are divisible by 9 but 5453 and 7898 are not.

4 $(ad + bc)^2 + (bd - ac)^2$
$= a^2d^2 + 2abcd + b^2c^2 + b^2d^2 - 2abcd + a^2c^2$
$= a^2d^2 + b^2c^2 + b^2d^2 + a^2c^2$
$= a^2(c^2 + d^2) + b^2(c^2 + d^2)$
$= (a^2 + b^2)(c^2 + d^2)$

5 $S = \frac{1}{3} - \frac{2}{9} + \frac{1}{27} - \frac{2}{81} + \frac{1}{243} - \frac{2}{729} + \ldots$

$S = \left(\frac{1}{3} - \frac{2}{9}\right) + \left(\frac{1}{27} - \frac{2}{81}\right) + \ldots$

$= \frac{1}{9} + \frac{1}{81} + \ldots = \frac{1}{9}\frac{1}{1 - \frac{1}{9}} = \frac{1}{8}$

6 Consider an arbitrary integer $n \in \mathbb{Z}$. Then $(n + 1)^2 - n^2 = n^2 + 2n + 1 - n^2 = 2n + 1$ is odd

7 $\frac{1}{n-1} - \frac{1}{n} + \frac{1}{n+1}$

$= \frac{n(n+1) - (n-1)(n+1) + n(n-1)}{n(n-1)(n+1)}$

$= \frac{n^2 + n - (n^2 - 1) + n^2 - n}{n(n^2 - 1)}$

$= \frac{n^2 + 1}{n(n^2 - 1)}$

$\therefore \frac{1}{5} - \frac{1}{6} + \frac{1}{7} = \frac{37}{210}$

8 Area of trapezium:

$\frac{a+b}{2}h = \frac{a+b}{2}(a+b)$

Similarly, the area in terms of the triangles BAE, BEC and EDC are

$\frac{1}{2}ab + \frac{1}{2}c^2 + \frac{1}{2}ab = ab + \frac{1}{2}c^2$

Equating the areas:

$\frac{(a+b)^2}{2} = ab + \frac{1}{2}c^2 \Rightarrow (a+b)^2 = 2ab + c^2$

$\Rightarrow a^2 + 2ab + b^2 = 2ab + c^2$

$\Rightarrow a^2 + b^2 = c^2$

Exercise 1F

1 Suppose for the sake of contradiction that n^2 is odd but n is even

Then $n^2 = 2m + 1$ for some $m \in \mathbb{Z}$ and $n = 2k$ for some $k \in \mathbb{Z}$.

But then $n^2 = (2k)^2 = 4k^2 = 2m + 1$ $4k^2$ is even but $(2m + 1)$ is odd, so this is a contradiction

$\therefore n^2$ is odd $\Rightarrow n$ is also odd.

2 Assume for the sake of contradiction that

$\sqrt{3} = \frac{a}{b}$ where $a, b \in \mathbb{Z}$

are coprime (i.e. they have no common factors).

Then, $3 = \frac{a^2}{b^2} \Rightarrow a^2 = 3b^2$

Therefore, a must be a multiple of $3 \Rightarrow a = 3k$ for some $k \in \mathbb{Z}$. This implies $9k^2 = 3b^2 \Rightarrow b^2 = 3k^2$ so b is also divisible by 3. Therefore 3 is a common factor of a and b. But we assumed that a and b have no common factors, so this is a contradiction.

3 Suppose for the sake of contradiction that $\sqrt[5]{2}$ is rational.

Then $\sqrt[5]{2}$ can be written in the form $\sqrt[5]{2} = \frac{a}{b}$

where $a, b \in \mathbb{Z}^+$ are relatively coprime (i.e. share no common factors)

$\therefore a^5 = 2b^5$ so 2 divides $a \Rightarrow a = 2m$ for some $m \in \mathbb{Z}^+ \Rightarrow b^5 = 2^4m^5$ so 2 divides b

So 2 divides both a and b, but it was assumed that a and b shared no common factors. This is a contradiction.

4 Suppose for the sake of contradiction that there exist p, $q \in \mathbb{Z}$ such that $p^2 - 8q - 11 = 0$
$\Rightarrow p^2 = 8q + 11$ so p is an odd integer
$\therefore p = 2k + 1$ for some $k \in \mathbb{Z}$
$\therefore (2k + 1)^2 = 8q + 11$

$\Rightarrow 4k^2 + 4k + 1 = 8q + 11$

$\Rightarrow 4(k^2 + k - 2q) = 10$

$\Rightarrow 2(k^2 + k - 2q) = 5$

but LHS is even whereas RHS is odd; this is a contradiction.

5 Suppose for the sake of contradiction that for some $a, b \in \mathbb{Z}$, $12a^2 - 6b^2 = 0$

$\therefore 12a^2 = 6b^2 \Rightarrow 2a^2 = b^2$

$\Rightarrow 2 = \frac{a^2}{b^2} = \left(\frac{a}{b}\right)^2 \Rightarrow \sqrt{2} = \frac{a}{b}$,

a contradiction.

6 Suppose for the sake of contradiction that for a, b, $c \in \mathbb{Z}$, the equation $a^2 + b^2 = c^2$ is satisfied and neither a nor b is an even number. c is odd $\Rightarrow c^2$ is odd
$\therefore a^2 + b^2$ is odd
$\therefore$ Either a or b must be odd since an odd number can only be written as the sum of an odd number and an even number. But this implies that either a or b is odd. This

is a contradiction since we assumed that neither a nor b is an even number. Furthermore, this proof shows that precisely one of a or b must be even, with the other number odd.

7 Suppose there exists $n, k \in \mathbb{Z}$ such that $n^2 + 2 = 4k$
Then n must be divisible by 2 and can be written in the form $n = 2m$ with $m \in \mathbb{Z}$
$\therefore 4m^2 + 2 = 4k$
$\Rightarrow m^2 - k = -\frac{1}{2}$
But the left-hand side is an integer whereas the right-hand side is not; this is a contradiction.

8 Suppose p is irrational, q is rational and for the sake of contradiction that $p + q$ is rational. Then
$q = \frac{a}{b}$ and $p + q = \frac{c}{d}$ for some $a, b, c, d \in \mathbb{Z}$
$\Rightarrow p = \frac{c}{d} - q = \frac{c}{d} - \frac{a}{b} = \frac{bc - ad}{bd} \in \mathbb{Z}$
But by assumption, p was irrational. This is a contradiction.

9 Let $m, n \in \mathbb{Z}^+$ and $m > n$ and suppose for the sake of contradiction that $m^2 - n^2 = 1$
Then $m^2 - n^2 = (m + n)(m - n) = 1$
Since $m, n \in \mathbb{Z}^+$, this can only be true if $m + n = m - n = \pm 1$
Neither of these are soluble in the integers. This is a contradiction.

10 **a** Take $m = n = 1$

b Take any prime number: the number is certainly divisible by itself but is still a prime

c Take $n = 4$: $2^4 - 1 = 16 - 1 = 15 = (3)(5)$

d Take the same example as above

e $1 + 2 + 3 = 6$, not divisible by 4

f $1 + 2 + 3 + 4 = 10$, not divisible by 4

Exercise 1G

1 **a** **i** $1 + 3 + 1$
$1 + 3 + 5 + 3 + 1$
$1 + 3 + 5 + 7 + 5 + 3 + 1$

ii $1 + 4 \quad 4 + 9 \quad 9 + 16$

b based on line divisions
$1 + 3 + 5 + 7 + 9 + 7 + 5 + 3 + 1$
$1 + 3 + 5 + 7 + 9 + 11 + 9 + 7 + 5 + 3 + 1$
based on colour
$16 + 25 \qquad 25 + 36$

c Conjecture: P(n):
$2(1 + 3 + 5 \ldots + 2n - 1) + 2n + 1 = n^2 + (n + 1)^2$, $n \in \mathbb{Z}^+, n \geq 2$

d see worked solutions for proof

e see worked solutions for proof

2 **a** $P(n): 1^2 + 2^2 + 3^2 + \ldots + n^2$
$= \frac{1}{3}n(n+1)\left(n + \frac{1}{2}\right)$
The statement $P(1)$ is true:
$1^2 = 1 = \frac{1}{3}1(1+1)(1)\left(1 + \frac{1}{2}\right)$
$= \frac{1}{3}(2)\left(\frac{3}{2}\right) = 1$
Assume the statement is true for $n = N$, where $N \in \mathbb{Z}^+$
Then $1^2 + 2^2 + 3^2 + \ldots + N^3 + (N + 1)^3$
$= \frac{1}{3}N(N+1)\left(N + \frac{1}{2}\right) + (N+1)^2$
$= \frac{1}{3}(N+1)\left[N\left(N + \frac{1}{2}\right) + 3(N+1)\right]$
$= \frac{1}{3}(N+1)\left[N^2 + \frac{1}{2}N + 3N + 3\right]$
$= \frac{1}{3}(N+1)\left[N^2 + \frac{7}{2}N + 3\right]$
$= \frac{1}{3}(N+1)(N+2)\left(N + \frac{3}{2}\right)$
$= \frac{1}{3}(N+1)((N+1)+1)\left((N+1) + \frac{1}{2}\right)$

Since $P(1)$ is true and $P(N) \Rightarrow P(N + 1)$ for $N \in \mathbb{Z}^+$ then by the principle of mathematical induction the statement is true for all positive integers.

b $P(n)$: $1 - 4 + 9 - 16 + \ldots + (-1)^{n+1}n^2 = (-1)^{n+1}\frac{n(n+1)}{2}$
The statement $P(1)$ is true:
$1 = (-1)^{1+1}\frac{1(1+1)}{2}$
Assume the statement P(k) is true for some $k \in \mathbb{Z}^+$ i.e. that $1 - 4 + 9 - 16 + \ldots + (-1)^{k+1}k^2 = (-1)^{k+1}\frac{k(k+1)}{2}$
Then $1 - 4 + 9 - 16 + \ldots + (-1)^{k+1}k^2 + (-1)^{k+2}(k + 1)^2$
$= (-1)^{k+1}\frac{k(k+1)}{2} + (-1)^{k+2}(k+1)^2$
$= (-1)^{k+1}(k+1)\left[\frac{k}{2} - (k+1)\right]$
$= (-1)^{k+1}(k+1)\left(\frac{-k-2}{2}\right)$
$= (-1)^{k+2}\frac{(k+1)((k+1)+1)}{2}$
i.e. $P(k) \Rightarrow P(k + 1)$
Since $P(1)$ is true and $P(k) \Rightarrow P(k + 1)$ for $k \in \mathbb{Z}^+$ then by the principle of mathematical induction, the statement is true for all positive integers.

c $P(n)$: $\sum_{i=0}^{n} 2^i = 2^{n+1} - 1$
The statement $P(0)$ is true:
$\sum_{i=0}^{0} 2^i = 1 = 2^{0+1} - 1$
Assume that $P(k)$ is true for some $k \in \mathbb{Z}^+$
i.e. $\sum_{i=0}^{k} 2^i = 2^{k+1} - 1$

Then $\sum_{i=0}^{k+1} 2^i = \sum_{i=0}^{k} 2^i + 2^{k+1}$

$= 2^{k+1} - 1 + 2^{k+1}$

$= 2 \cdot 2^{k+1} - 1 = 2^{k+1+1} - 1$

i.e. $P(k) \Rightarrow P(k+1)$

Since $P(1)$ is true and $P(k) \Rightarrow P(k+1)$ for $k \in \mathbb{Z}$ then by the principle of mathematical induction, the statement is true for all positive integers.

d $P(n)$: 9^{n-1} is divisible by 8 (for $n \in \mathbb{N}$)

The statement $P(1)$ is true: $9^1 - 1 = 8$ which is clearly divisible by 8 Assume $P(k)$ to be true for some $k \in \mathbb{N}$ i.e. 8 divides $9^k - 1 \Rightarrow 9k - 1 = 8m$ for some $m \in \mathbb{N}$

Then $9^{k+1} - 1 = 9 \cdot 9^k - 1 = 9(8m+1) - 1 = 9(8m) + 9 - 1 = 8(9m) + 8 = 8(9m+1)$ so 8 also divides $9^{k+1} - 1$

i.e. $P(k) \Rightarrow P(k+1)$

Since $P(1)$ is true and $P(k) \Rightarrow P(k+1)$ for $k \in \mathbb{N}$ then by the principle of mathematical induction, the statement is true for all natural numbers.

e $P(n)$: $1^3 + 2^3 + 3^3 + \ldots + n^3 = \frac{n^2(n+1)^2}{4}$

The statement $P(1)$ is true:

$1^3 = \frac{1^2(1+1)^2}{4}$

Assume $P(k)$ is true for some $k \in \mathbb{Z}^+$

i.e. $1^3 + 2^3 + 3^3 + \ldots + k^3 = \frac{k^2(k+1)^2}{4}$

Then $1^3 + 2^3 + \ldots + k^3 + (k+1)^3$

$= \frac{k^2(k+1)^2}{4} + (k+1)^3$

$= \frac{(k+1)^2}{4}\left(k^2 + 4(k+1)\right)$

$= \frac{(k+1)^2(k^2+4k+4)}{4}$

$= \frac{(k+1)^2(k+2)^2}{4}$

$= \frac{(k+1)^2(k+1+1)^2}{4}$

i.e. $P(k) \Rightarrow P(k+1)$

Since $P(1)$ is true and $P(k) \Rightarrow P(k+1)$ for $k \in \mathbb{Z}^+$ then by the principle of mathematical induction, the statement is true for all positive integers.

f $P(n)$: $n^3 - n$ is divisible by 3

The statement $P(1)$ is true: $1^3 - 1 = 0$ and 0 is divisible by 3 Assume $P(k)$ is true for some $k \in \mathbb{Z}$ i.e. $k^3 - k$ is divisible by $3 \Rightarrow k^3 - k = 3m$ for some $m \in \mathbb{Z}$

Then $(k+1)^3 - (k+1)$

$= (k^3 + 3k^2 + 3k + 1) - (k+1)$

$= (k^3 - k) + 3(k^2 + k)$

$= 3m + 3(k^2 + k) = 3(m + k^2 + k)$ which is divisible by 3

i.c. $P(k) \Rightarrow P(k+1)$

Since $P(1)$ is true and $P(k) \Rightarrow P(k+1)$ for $k \in \mathbb{N}$ then by the principle of mathematical induction, the statement is true for all natural numbers.

g $P(n)$: $\frac{1}{1\times 2} + \frac{1}{2\times 3} + \frac{1}{3\times 4} + \ldots + \frac{1}{n(n+1)} = \frac{n}{n+1}$

The statement $P(1)$ is true:

$\frac{1}{1\times 2} = \frac{1}{1+1}$

Assume $P(k)$ is true for some $k \in \mathbb{Z}^+$

i.e. $\frac{1}{1\times 2} + \frac{1}{2\times 3} + \ldots + \frac{1}{k(k+1)} = \frac{k}{k+1}$

Then $\frac{1}{1\times 2} + \frac{1}{2\times 3} + \ldots + \frac{1}{k(k+1)} + \frac{1}{(k+1)(k+2)}$

$= \frac{k}{k+1} + \frac{1}{(k+1)(k+2)}$

$= \frac{1}{k+1}\left(k + \frac{1}{k+2}\right)$

$= \frac{1}{k+1}\left(\frac{k(k+2)+1}{k+2}\right)$

$= \frac{1}{k+1}\left(\frac{k^2+2k+1}{k+2}\right)$

$= \frac{1}{k+1}\left(\frac{(k+1)^2}{k+2}\right)$

$= \frac{k+1}{k+2} = \frac{k+1}{k+1+1}$

i.e. $P(k) \Rightarrow P(k+1)$

Since $P(1)$ is true and $P(k) \Rightarrow P(k+1)$ for $k \in \mathbb{Z}^+$ then by the principle of mathematical induction, the statement is true for all positive integers.

h $P(n)$: $n^3 - n$ is a multiple of 6 for all $n \in \mathbb{Z}^+$

The statement $P(1)$ is true: $1^3 - 1 = 0$ is certainly a multiple of 6

Assume $P(k)$ is true for some $k \in \mathbb{Z}^+$

i.e. $k^3 - k$ is divisible by 6 $\Rightarrow k^3 - k = 6m$ for some $m \in \mathbb{N}$ Then $(k+1)^3 - (k+1) = (k^3 + 3k^2 + 3k + 1) - (k + 1) = (k^3 - k) + 3k^2 + 3k$

$= 6m + 3k(k+1)$ but $k(k+1)$ must be an even number since in any pair of consecutive natural numbers contains an even number,

$\therefore k(k+1) = 2r$ for some $r \in \mathbb{Z}^+$

$\Rightarrow (k+1)^3 - (k+1) = 6(m+r)$ which is divisible by 6

i.e. $P(k) \Rightarrow P(k+1)$

Since $P(1)$ is true and $P(k) \Rightarrow P(k+1)$ for $k \in \mathbb{Z}^+$ then by the principle of mathematical induction, the statement is true for all positive integers.

i $P(n)$: $2^{n+2} + 3^{2n+1}$ is divisible by 7 ($n \in \mathbb{Z}^+$)

The statement $P(1)$ is true: $2^{1+2} + 3^{2+1} = 2^3 + 3^3 = 8 + 27 = 35 = 7(5)$

Assume that $P(k)$ is true for some $k \in \mathbb{Z}^+$

i.e. $2^{k+2} + 3^{2k+1}$ is divisible by 7 $\Rightarrow 2^{k+2} + 3^{2k+1} = 7m$ for some $m \in \mathbb{Z}^+$ Then $2^{k+1+2} + 3^{2(k+1)+1} = 2 \cdot 2^{k+2} + 9 \cdot 3^{2k+1}$

$= 2(7m - 3^{2k+1}) + 9 \cdot 3^{2k+1}$

$= 14m + 7 \cdot 3^{2k+1}$

$= 7(2m + 3^{2k+1})$ which is divisible by 7

so $P(k) \Rightarrow P(k+1)$

Since $P(1)$ is true and $P(k) \Rightarrow P(k+1)$ for $k \in \mathbb{Z}^+$ then by the principle of mathematical induction, the statement is true for all positive integers.

j $P(n): 1^2 + 3^2 + 5^2 + \ldots + (2n-1)^2 = \frac{n(2n-1)(2n+1)}{3}$

The statement $P(1)$ is true:

$1^2 = 1 = \frac{1(2-1)(3)}{3}$

Assume that $P(k)$ is true for some $k \in \mathbb{Z}^+$

i.e. $1^2 + 3^2 + 5^2 + \ldots + (2k-1)^2 = \frac{k(2k-1)(2k+1)}{3}$

Then $1^2 + 3^2 + 5^2 + \ldots + (2k-1)^2 + (2k+1)^2$

$= \frac{k(2k-1)(2k+1)}{3} + (2k+1)^2$

$= \frac{2k+1}{3}\left(k(2k-1) + 3(2k+1)\right)$

$= \frac{2k+1}{3}\left(2k^2 + 5k + 3\right)$

$= \frac{(2k+1)(2k+3)(k+1)}{3}$

$= \frac{(k+1)(2(k+1)-1)(2(k+1)+1)}{3}$

i.e. $P(k) \Rightarrow P(k+1)$ Since $P(1)$ is true and $P(k) \Rightarrow P(k+1)$ for $k \in \mathbb{Z}^+$ then by the principle of mathematical induction, the statement is true for all positive integers.

k $P(n)$:

$\sum_{r=1}^{n} r(r+1) = \frac{n}{3}(n+1)(n+2)$

The statement $P(1)$ is true:

$\sum_{r=1}^{1} r(r+1) = 1(1+1)$

$= \frac{1}{3}(1+1)(1+2)$

Assume $P(k)$ to be true for some $k \in \mathbb{Z}^+$

i.e. $\sum_{r=1}^{k} r(r+1) = \frac{k}{3}(k+1)(k+2)$

Then $\sum_{r=1}^{k+1} r(r+1)$

$= \sum_{r=1}^{k} r(r+1) + (k+1)(k+2)$

$= \frac{k}{3}(k+1)(k+2) + (k+1)(k+2)$

$= \frac{(k+1)(k+2)}{3}(k+3)$

$= \frac{(k+1)(k+1+1)(k+1+2)}{3}$

i.e. $P(k) \Rightarrow P(k+1)$

Since $P(1)$ is true and $P(k) \Rightarrow P(k+1)$ for $k \in \mathbb{Z}^+$ then by the principle of mathematical induction, the statement is true for all positive integers

l $P(n)$: $\sum_{r=1}^{n} \frac{1}{r(r+1)} = \frac{n}{n+1}$

The statement $P(1)$ is true:

$\sum_{r=1}^{1} \frac{1}{r(r+1)} = \frac{1}{1(1+1)}$

$= \frac{1}{2} = \frac{1}{1+1}$

Assume $P(k)$ is true for some $k \in \mathbb{Z}^+$

i.e. $\sum_{r=1}^{k} \frac{1}{r(r+1)} = \frac{k}{k+1}$

Then $\sum_{r=1}^{k+1} \frac{1}{r(r+1)}$

$= \sum_{r=1}^{k} \frac{1}{r(r+1)} + \frac{1}{(k+1)(k+2)}$

$= \frac{k}{k+1} + \frac{1}{(k+1)(k+2)}$

$= \frac{1}{k+1}\left(k + \frac{1}{k+2}\right)$

$= \frac{1}{k+1}\left(\frac{k(k+2)+1}{k+2}\right)$

$= \frac{1}{k+1}\left(\frac{k^2+2k+1}{k+2}\right)$

$= \frac{1}{k+1}\left(\frac{(k+1)^2}{k+2}\right) = \frac{k+1}{k+2}$

3 **a** Best proved by direct argument:

$(4n+3)^2 - (4n-3)^2$

$= (4n + 3 + 4n - 3)$
$(4n + 3 - 4n + 3)$

$= (8n)(6) = 48n = 12(4n)$

so is always divisible by 12 (induction amongst other methods is also valid)

b False: substituting $n = 1$ gives 75 which is not prime

c Best proved by induction:

$P(n)$: $1^3 + 3^3 + \ldots + (2n-1)^3 = n^2(2n^2 - 1)$

The statement $P(1)$ is true:

$1^3 = 1^2(2 \cdot 1^2 - 1)$

Assume the statement $P(k)$ is true for some $k \in \mathbb{Z}^+$

i.e. $1^3 + 3^3 + \ldots + (2k-1)^3 = k^2(2k^2 - 1)$

Then $1^3 + 3^3 + \ldots + (2k-1)^3 + (2k+1)^3$

$= k^2(2k^2 - 1) + (2k+1)^3$

$= 2k^4 - k^2 + 8k^3 + 12k^2 + 6k + 1$

$= 2k^4 + 8k^3 + 11k^2 + 6k + 1$

$= (k+1)(2k^3 + 6k^2 + 5k + 1)$

$= (k+1)(k+1)(2k^2 + 4k + 1)$

$= (k+1)^2(2(k+1)^2 - 1)$

so $P(k) \Rightarrow P(k+1)$

Since $P(1)$ is true and $P(k) \Rightarrow P(k+1)$ for $k \in \mathbb{Z}^+$ then by the principle of mathematical induction, the statement is true for all positive integers.

d Best proved by induction:

$P(n)$: $1\times2+2\times3+3\times4...$

$$+(n-1)\times n=\frac{n(n^2-1)}{3}$$

The statement $P(1)$ is true:

$$0\times1=\frac{1(1^2-1)}{3}$$

Assume the statement $P(k)$ is true for some $k \in \mathbb{Z}^+$

i.e. $1\times2+2\times3+3\times4...$

$$+(k-1)\times k=\frac{k(k^2-1)}{3}$$

Then,

$1\times2+2\times3+3\times4...$
$+(k-1)\times k+k(k+1)$

$$=\frac{k(k^2-1)}{3}+k(k+1)$$

$$=\frac{k(k-1)(k+1)+3k(k+1)}{3}$$

$$=\frac{(k+1)(k(k-1)+3k)}{3}$$

$$=\frac{(k+1)((k+1)^2-1)}{3}$$

so $P(k)\Rightarrow P(k+1)$

Since P(1) is true and P(k) $\Rightarrow$ P(k + 1) for $k \in$ Z$^+$ then by the principle of mathematical induction, the statement is true for all positive integers.

e Best proved by direct argument: $n^3 - n = n(n^2-1) = (n-1)n(n+1)$ i.e. this is the product of three consecutive positive integers (in the case $n = 1$, 0 is divisible by 3 so done) Three consecutive positive integers always include a multiple of 3, so the product is always divisible by 3

Exercise 1H

1

$8!-6!$	$6!(56-1)=39600$
$9!+8!$	$8!(9+1)=403200$
$7!-6!$	$6!(7-1)=4320$
$6!+5!$	$5!(6+1)=840$
$(n+1)!-n!$	$n!(n+1-1)=n(n!)$
$n!-(n-1)!$	$(n-1)!(n-1)=$ $(n-1)((n-1)!)$
$n!+(n-1)!$	$(n-1)!(n+1)$ $=(n+1)(n-1)!$
$(n+1)!+n!$	$n!(n+1+1)$ $=(n+2)n!$

2 **a** 14 **b** $\frac{2}{3}$ **c** $\frac{56}{11}$

3 **a** $-\frac{n+1}{n}$ **b** $n+2$ **c** $1+n!$

4 $\frac{(2n+2)!(n!)^2}{[(n+1)!]^2(2n)!}$ $=\frac{(2n+2)(2n+1)}{(n+1)^2}=\frac{2(2n+1)}{n+1}$

5 $n = 12$

6 $n = 2 (n > 0)$

7 **a** 13! **b** 165 888

8 1404 000

9 **a** 33 649 **b** 32 110

10 **a** 2058 **b** 720 **c** 882 **d** 294

11 ${}^6C_4 = 15$

12 37 800

Exercise 1I

1 **a** $1-\frac{11x}{3}+\frac{55x^2}{9}-\frac{55x^3}{9}+...$

b $1+\frac{7x}{2}+\frac{21x^2}{4}+\frac{35x^3}{8}+...$

c $x^8+16x^6+112x^4+448x^2+...$

2 **a** $-3360a^6b^4$

b $880a^5$

c $-448x^2y^3$

3 ${}^{12}C_4(x)^8\left(-\frac{2}{x^2}\right)^4=7920$

4 $\left(2-\frac{x}{5}\right)^4=16\left(1-\frac{x}{10}\right)^4$

$$=16\left({}^4C_0+{}^4C_1\left(-\frac{x}{10}\right)+{}^4C_2\left(-\frac{x}{10}\right)^2+{}^4C_3\left(-\frac{x}{10}\right)^3+{}^4C_4\left(-\frac{x}{10}\right)^4\right)$$

$$=16\left(1-\frac{2x}{5}+\frac{3x^2}{50}-\frac{x^3}{250}+\frac{x^4}{10000}\right)$$

$$=16-\frac{32x}{5}+\frac{24x^2}{25}-\frac{8x^3}{125}+\frac{x^4}{625}$$

$= 15.68239$

5 $15x^6$

6 **a** $x^5+5x^3y+10xy^2+\frac{10y^3}{x}+\frac{5y^4}{x^3}+\frac{y^5}{x^5}$

b 5

7 **a** $\frac{(n+1)!}{4!(n-3)!}$

b $\frac{4\cdot n!}{3\cdot(n-3)!}$ **c** 31

8 **a** $89\sqrt{3}-109\sqrt{2}$

b $\frac{161}{25}-\frac{44}{25}\sqrt{10}$

c $1664\sqrt{5}$

9 **a** $(-1)^n$ **b** 2^n

Exercise 1J

1 **a** $1-x+x^2-x^3+...$

b $1+4x+12x^2+32x^3+...$

c $2-4x+8x^2-16x^3+....$

d $2+6x+12x^2+20x^3+...$

2 **a** $1+x-\frac{1}{2}x^2+\frac{1}{2}x^3+...$

b $1+\frac{3x}{2}+\frac{3x^2}{8}-\frac{x^3}{16}+...$

c $1+\frac{3x}{2}+\frac{27x^2}{8}+\frac{135}{16}x^3+...$

d $2+\frac{2x}{3}-\frac{2x^2}{9}+\frac{10x^3}{81}+...$

3 $\sqrt{\dfrac{1-x}{1+x}} = (1-x)^{\frac{1}{2}}(1+x)^{-\frac{1}{2}}$

$$= \left(1 - \frac{1}{2}x + \frac{\left(\frac{1}{2}\right)\left(-\frac{1}{2}\right)}{2}x^2 - \frac{\left(\frac{1}{2}\right)\left(-\frac{1}{2}\right)\left(-\frac{3}{2}\right)}{3!}x^3 + ..\right)\left(1 - \frac{1}{2}x + \frac{\left(-\frac{1}{2}\right)\left(-\frac{3}{2}\right)}{2}x^2 + \frac{\left(-\frac{1}{2}\right)\left(-\frac{3}{2}\right)\left(-\frac{5}{2}\right)}{3!}x^3 + ..\right)$$

$$= \left(1 - \frac{x}{2} - \frac{x^2}{8} - \frac{x^3}{16} + ...\right)\left(1 - \frac{x}{2} + \frac{3x^2}{8} - \frac{5x^3}{16} + ...\right)$$

$$= 1 - x + \frac{x^2}{2} - \frac{x^3}{2} + ...$$

---

4 $\dfrac{x}{(1+x)^2} = x(1+x)^{-2}$

$$= x\left[1 + (-2)x + \frac{(-2)(-3)}{2!}x^2 + \frac{(-2)(-3)(-4)}{3!}x^3 + ...\right]$$

$$= x\left[1 - 2x + 3x^2 - 4x^3 + ...\right]$$

$$= x - 2x^2 + 3x^3 - 4x^4 + ...$$

5 $\dfrac{1}{8} + \dfrac{9x}{16} + \dfrac{27x^2}{16} + \dfrac{135x^3}{32} + ...$

6 **a** $1 - 2x - 2x^2 - 4x^3 + ...$

b $\dfrac{2\sqrt{6}}{5}$ **c** 2.44949

7 **a** $1 + x + \dfrac{3x^2}{2} + \dfrac{5x^3}{2} + ...$

b $8 + 44x + 102x^2 + 155x^3 ...$

Chapter review

1 $u_2 = u_1 r = 9 \Rightarrow u_1 = \dfrac{9}{r}$

$S_3 = u_1(1 + r + r^2) = 91$

$\Rightarrow \dfrac{9}{r}(1 + r + r^2) = 91$

$\Rightarrow 9 + 9r + 9r^2 = 91r$

$\Rightarrow 9r^2 - 82r + 9 = 0$

$\Rightarrow (9r - 1)(r - 9) = 0$

$\Rightarrow r = \dfrac{1}{9}$ or $r = 9$

Therefore there are two geometric sequences:

$r = \dfrac{1}{9}$: $u\,4 = \dfrac{1}{9}$

$r = 9$: $u\,4 = 729$

2 1150

3 $a = 48,\ b = 12,\ c = -24,\ d = -36$

4 **a** $\dfrac{1}{1+x} - \dfrac{1}{3\left(1 + \frac{2}{3}x\right)}$

$$= \frac{1}{1+x} - \frac{1}{3+2x}$$

$$= \frac{3 + 2x - (1+x)}{(1+x)(3+2x)}$$

$$= \frac{x+2}{3 + 2x + 3x + 2x^2}$$

$$= \frac{x+2}{2x^2 + 5x + 3}$$

b $\dfrac{2}{3} - \dfrac{7}{9}x + \dfrac{23}{27}x^2 - \dfrac{73}{81}x^3 + ...$

5 **a** ${}^nC_2 + n = \dfrac{n!}{(n-2)!2!} + n$

$$= \frac{1}{2}n(n-1) + n$$

$$= \frac{1}{2}n(n - 1 + 2)$$

$$= \frac{1}{2}n(n+1) = \frac{1}{2}\frac{(n+1)!}{(n-1)!}$$

$$= \frac{(n+1)!}{2!(n-1)!} = {}^{n+1}C_2$$

b ${}^nC_2 \times {}^{n-2}C_{k-2} = \dfrac{n!}{2!(n-2)!}$

$$\times \frac{(n-2)!}{(n-k)!(k-2)!}$$

$$= \frac{n!}{2!(n-k)!(k-2)!}$$

$$= \frac{n!}{(n-k)!} \times \frac{1}{2!(k-2)!}$$

$$= \frac{n!k!}{(n-k)!k!} \times \frac{1}{2!(k-2)!}$$

$$= \frac{n!}{(n-k)!k!} \times \frac{k!}{2!(k-2)!}$$

$$= {}^nC_k \times {}^kC_2$$

6 $(1 + x)^n = {}^nC_0 + {}^nC_1x + {}^nC_2x^2 + ... + {}^nC_rx^r + ... + {}^nC_nx^n$

$\therefore {}^nC_0 + {}^nC_1 \times 3 + {}^nC_2 \times 3^2 + ... + {}^nC_r \times 3^r + ... + {}^nC_n \times 3^n$

$= (1+3)^n = 4^n = (2^2)^n = 2^{2n}$

7 Suppose there exist integers a and b such that $14a + 7b = 1$.

Then, $2a + b = \dfrac{1}{7}$.

But the left-hand side is an integer whereas the right-hand side is not. This is a contradiction. Therefore there are no such integers.

8 Suppose $x = 3$ and $5x - 7 = 13$.

Then, $x = \dfrac{13+7}{5} = 4$. But $x = 3$, so this is a contradiction.

9 **a** Take, for example, $a = 0$ and $b = 1$

b Take, for example, $n = 5$: $3^5 + 2 = 245 = 5(49)$ which is not prime

c Take, for example, $n = 1$: $\sqrt{2(1) - 1} = \sqrt{1} = 1$ which is rational

d Take, for example, $n = 1$: $2^1 - 1 = 1$ and 1 is not prime

10 $P(n)$: $(1 \times 1!) \times (2^2 \times 2!) \times (3^3 \times 3!) \times \ldots \times (n^n \times n!) = (n!)^{n+1}$

The statement $P(1)$ is true: $1 \times 1! = (1!)^{1+1}$

Assume the statement $P(k)$ is true for some $k \in \mathbb{Z}^+$

i.e. $(1 \times 1!) \times (2^2 \times 2!) \times \ldots (k^k \times k!) = (k!)^{k+1}$

Then $(1 \times 1!) \times (2^2 \times 2!) \times \ldots (k^k \times k!) \times ((k+1)^{k+1} \times (k+1)!)$

$= (k!)^{k+1} \cdot (k+1)!(k+1)^{k+1}$

$= ((k+1)k!)^{k+1}(k+1)!$

$= ((k+1)k!)^{k+1}(k+1)!$

$= ((k+1)!)^{k+1}(k+1)!$

$= ((k+1)!)^{k+2}$

so $P(k) \Rightarrow P(k+1)$

Therefore, it has been shown that $P(1)$ is true and that if $P(k)$ is true for some $k \in \mathbb{Z}^+$ then so is $P(k+1)$. Therefore, the statement is true for all positive integers by the principle of mathematical induction.

11 $P(n)$: $n^3 + 2n$ is a multiple of 3

The statement $P(1)$ is true: $1^3 + 2(1) = 3$ which is a multiple of 3.

Assume that $P(k)$ is true for some $k \in \mathbb{Z}^+$

i.e. $k^3 + 2k = 3m$ for some $m \in \mathbb{Z}^+$

Then,

$(k+1)^3 + 2(k+1)$

$= k^3 + 3k^2 + 3k + 1 + 2k + 2$

$= (k^3 + 2k) + 3k^2 + 3k + 3$

$= 3m + 3(k^2 + k + 1)$

$= 3(m + k^2 + k + 1)$

which is a multiple of 3

$\therefore P(k) \Rightarrow P(k+1)$

Therefore, it has been shown that $P(1)$ is true and that if $P(k)$ is true for some $k \in \mathbb{Z}^+$ then so is $P(k+1)$. Therefore, the statement is true for all positive integers by the principle of mathematical induction.

12 a $P(n)$: $\sum_{r=1}^{n} r = \frac{n(n+1)}{2}$

The statement $P(1)$ is true:

$\sum_{r=1}^{1} r = 1 = \frac{1(1+1)}{2}$

Assume that $P(k)$ is true for some $k \in \mathbb{Z}^+$

i.e. $\sum_{r=1}^{k} r = \frac{k(k+1)}{2}$

Then $\sum_{r=1}^{k+1} r = \sum_{r=1}^{k} r + (k+1)$

$= \frac{k(k+1)}{2} + (k+1)$

$= \frac{k+1}{2}(k+2)$

$= \frac{(k+1)(k+2)}{2}$

so $P(k) \Rightarrow P(k+1)$

Therefore, it has been shown that $P(1)$ is true and that if $P(k)$ is true for some $k \in \mathbb{Z}^+$ then so is $P(k+1)$. Therefore, the statement is true for all positive integers by the principle of mathematical induction.

b $P(n)$: $\sum_{r=1}^{n} r^2 = \frac{n(n+1)(2n+1)}{6}$

The statement $P(1)$ is true:

$\sum_{r=1}^{1} r^2 = 1^2 = 1 = \frac{1(1+1)(2+1)}{6}$

Assume that $P(k)$ is true for some $k \in \mathbb{Z}^+$

i.e. $\sum_{r=1}^{k} r^2 = \frac{k(k+1)(2k+1)}{6}$

Then $\sum_{r=1}^{k+1} r^2 = \sum_{r=1}^{k} r^2 + (k+1)^2$

$= \frac{k(k+1)(2k+1)}{6} + (k+1)^2$

$= \frac{k+1}{6}[k(2k+1) + 6(k+1)]$

$= \frac{k+1}{6}(2k^2 + 7k + 6)$

$= \frac{k+1}{6}(2k+3)(k+2)$

$= \frac{(k+1)(k+2)(2(k+1)+1)}{6}$

so $P(k) \Rightarrow P(k+1)$

Therefore, it has been shown that $P(1)$ is true and that if $P(k)$ is true for some $k \in \mathbb{Z}^+$ then so is $P(k+1)$. Therefore, the statement is true for all positive integers by the principle of mathematical induction.

c $P(n): \sum_{r=1}^{n} r^3 = \frac{n^2(n+1)^2}{4}$

The statement $P(1)$ is true:

$\sum_{r=1}^{1} r^3 = 1^3 = \frac{1^2(1+1)^2}{4}$

Assume that $P(k)$ is true for some $k \in \mathbb{Z}^+$

i.e. $\sum_{r=1}^{k} r^3 = \frac{k^2(k+1)^2}{4}$

Then $\sum_{r=1}^{k+1} r^3 = \sum_{r=1}^{k} r^3 + (k+1)^3$

$= \frac{k^2(k+1)^2}{4} + (k+1)^3$

$= \frac{(k+1)^2}{4}[k^2 + 4(k+1)]$

$= \frac{(k+1)^2(k^2+4k+4)}{4}$

$= \frac{(k+1)^2(k+2)^2}{4}$

so $P(k) \Rightarrow P(k+1)$

Therefore, it has been shown that $P(1)$ is true and that if $P(k)$ is true for some $k \in \mathbb{Z}^+$ then so is $P(k+1)$. Therefore, the statement is true for all positive integers by the principle of mathematical induction.

$\therefore \sum_{r=1}^{n} r(r+1)(r+2)$

$= \sum_{r=1}^{n} (r^3 + 3r^2 + 2r)$

$= \sum_{r=1}^{n} r^3 + 3\sum_{r=1}^{n} r^2 + 2\sum_{r=1}^{n} r$

$= \frac{n^2(n+1)^2}{4} + \frac{n(n+1)(2n+1)}{2}$
$+ n(n+1)$

$= \frac{n(n+1)}{4}[n(n+1) + 2(2n+1) + 4]$

$= \frac{n(n+1)}{4}[n^2 + 5n + 6]$

$= \frac{n(n+1)(n+2)(n+3)}{4}$

13 a $9! = 362\,880$

b 233239

c 91

14 a $a^2 - b^2 = (a + b)(a - b) = (2x)(2y) = 4xy$

b $a^3 = x^3 + 3x^2y + 3xy^2 + y^3$
$b^3 = x^3 - 3x^2y + 3xy^2 - y^3$

c $\therefore a^3 - b^3 = (x^3 + 3x^2y + 3xy^2 + y^3) - (x^3 - 3x^2y + 3xy^2 - y^3)$
$= 2(3x^2y + y^3)$
$= (2y)(3x^2 + y^2)$
$= (a - b)(3x^2 + y^2)$
But $a^2 + ab + b^2 = 3x^2 + y^2$
So $a^3 - b^3 = (a - b)(a^2 + ab + b^2)$

d $a^4 = (x + y)^4 = x^4 + 4x^3y + 6x^2y^2 + 4xy^3 + y^4$
$b^4 = (x-y)^4 = x^4 - 4x^3y + 6x^2y^2 - 4xy^3 + y^4$
$a^4 - b^4 = (a-b)(a^3 + a^2b + ab^2 + b^3)$

e Conjecture: $a^n - b^n = (a - b)(a^{n-1} + a^{n-2}b + a^{n-3}b^2 + \ldots + ab^{n-2} + b^{n-1})$

f $P(n): a^n - b^n = (a-b)(a^{n-1} + a^{n-2}b + a^{n-3}b^2 + \ldots + ab^{n-2} + b^{n-1})$
The statement P(2) is true
$a^2 - b^2 = (a - b)(a + b) = a^2 + ab - ab + b^2$
Assume that $P(k)$ is true for some $k \in \mathbb{Z}^+$
i.e. $a^k - b^k = (a - b)(a^{k-1} + a^{k-2}b + a^{k-3}b^2 + \ldots + ab^{k-1} + b^k)$
Then, $a^{k+1} - b^{k+1} = (a - b)(a^k + a^{k-1}b + a^{k-2}b^2 + \ldots + ab^{k-1} + b^{k-1})$
so $P(k) \Rightarrow P(k + 1)$
Therefore, it has been shown that $P(1)$ is true and that if $P(k)$ is true for some $k \in \mathbb{Z}^+$ then so is $P(k + 1)$. Therefore, the statement is true for all positive integers by the principle of mathematical induction.

---

15 The difference between the coefficients must be the same
$\therefore {}^nC_r - {}^nC_{r-1} = {}^nC_{r+1} - {}^nC_r$
$\Rightarrow \frac{n!}{r!(n-r)!} - \frac{n!}{(r-1)!(n-r+1)!} = \frac{n!}{(r+1)!(n-r-1)!} - \frac{n!}{r!(n-r)!}$
Multiplying by $\frac{(r+1)!(n-r+1)!}{n!}$,
$(r + 1)(n - r + 1) - (r + 1)(r) = (n - r + 1)(n - r) - (r + 1)(n - r + 1)$
$\Rightarrow 2(r + 1)(n - r + 1) - r(r + 1) - (n - r + 1)(n - r) = 0$
$\Rightarrow (n - r + 1)(3r + 2 - n) - r^2 - r = 0$
which after expanding and simplifying gives
$n^2 + 4r^2 - 2 - n(4r + 1) = 0$
Solving for $n = 14$, gives $r = 5, 9$, so the coefficients are 1001, 2002, 3003

16 $A = 1, B = -1, C = 2$
$2 + 5x + 5x^2 + 11x^3 + \ldots$

Exam-style questions

17 Require (3 × coefficient of term in x^5) + (1 × coefficient of term in x^4)
$3 \times \binom{8}{5} 4^3(-2x)^5 + 1 \times \binom{8}{4} 4^4(-2x)^4$ (3 marks)
$= 3 \times (-114688) + 1 \times 286720$
$= -57344$ (1 mark)

18 $\binom{n}{2}(1^{n-2})(3x)^2 = 495x^2$ (2 marks)
$\frac{9n(n-1)}{2} = 495$
$n(n - 1) = 110$
$n^2 - n - 110 = 0$ (1 mark)
$(n - 11)(n + 10) = 0$ (2 marks)
So $n = 11$ or $n = -10$ (1 mark)

19 First part is geometric sum, $a = 1, r = 1.6, n = 16$ (1 mark)
Second part is arithmetic sum, $a = 0, d = -12, n = 16$ (1 mark)
Third part is $16 \times 1 = 16$ (1 mark)

Geometric sum:

$$S_{16} = \frac{1.6^{16} - 1}{1.6 - 1} = 3072.791$$

(1 mark)

Arithmetic sum:

$$S_{16} = \frac{16}{2}(2\times 0 + 15\times(-12))$$
$$= -1440$$

(1 mark)

So $\sum_{n=0}^{n=15}(1.6^n - 12n + 1)$

$= 3072.791 - 1440 + 16$

$= 1648.8$ (1 mark)

20 $\binom{n-1}{k} + \binom{n-1}{k-1}$

$$= \frac{(n-1)!}{k!(n-k-1)!} + \frac{(n-1)!}{(k-1)!(n-k)!}$$

(3 marks)

$$= \frac{(n-k)(n-1)! + k(n-1)!}{k!(n-k)!}$$

(1 mark)

$$= \frac{n(n-1)! - k(n-1)! + k(n-1)!}{k!(n-k)!}$$

(1 mark)

$$= \frac{n(n-1)!}{k!(n-k)!}$$ (1 mark)

$$\left(= \frac{n!}{k!(n-k)!}\right)$$

$$= \binom{n}{k}$$

21 Consider multiples of 7:

504 is the first multiple and 1400 is the final multiple

$1400 = 504 + 7(n - 1)$

(1 mark)

$\Rightarrow n = 129$ (1 mark)

So the sum of the multiples of 7 is

$$S_{129} = \frac{129}{2}(2\times 504 + 7\times(129-1))$$
$$= 122\,808$$

(2 marks)

Sum of the integers from 500 to 1400 (inclusive) is

$$S_{901} = \frac{901}{2}(2\times 500 + 1\times(901-1))$$
$$= 855\,950$$

(2 marks)

Therefore require $855\,950 - 122\,808 = 733\,142$ (1 mark)

22 Suppose $n^3 + 3$ is odd.

Assume, for a contradiction, that n is also odd. (1 mark)

Then we can write $n = 2p + 1$ for $p \in \mathbb{Z}^+$ and $n^3 + 3 = 2q + 1$ for $q \in \mathbb{Z}^+$. (1 mark)

So $n^3 + 3 = 2q + 1$

$(2p + 1)^3 + 3 = 2q + 1$ (1 mark)

$8p^3 + 12p^2 + 6p + 1 + 3 = 2q + 1$

(1 mark)

$8p^3 + 12p^2 + 6p + 3 = 2q$

So $q = 4p^3 + 6p^2 + 3p + \frac{3}{2}$

(1 mark)

Since p is an integer, $4p^3 + 3p^2 + 3p$ is also an integer.

Since $\frac{3}{2}$ is a non-integer, then $4p^3 + 3p^2 + 3p + \frac{3}{2}$ is also a non-integer. (1 mark)

This is a contradiction, since q was assumed to be an integer.

(1 mark)

Therefore, the initial assumption is false, and n must be even.

23 Case $n = 1$:

$5^{2(1)-1} + 1 = 5 + 1 = 6$

$= 1 \times 6$ (1 mark)

Therefore true for $n = 1$

Case $n = k$:

Assume the statement is true for some $k \in \mathbb{N}$, $k \geq 0$ (1 mark)

Then $5^{2k-1} + 1 = 6s$ for some positive integer s

Now $5^{2(k+1)-1} + 1$ (1 mark)

$= 5^{2k+2-1} + 1$

$= 5^2 \times 5^{2k-1} + 1$ (1 mark)

$= 5^2 \times (6s - 1) + 1$ (1 mark)

$= 25(6s - 1) + 1$

$= 25 \times 6s - 24$ (1 mark)

$= 6(25s - 4)$

Which is a multiple of 6

(1 mark)

So the statement is true for $n = 1$, and when assumed true for $n = k$, is true for $n = k + 1$.

Therefore the statement is true for all $n \in \mathbb{N}$. (1 mark)

24 a $\sqrt[3]{1-x} = (1-x)^{\frac{1}{3}}$ (1 mark)

$$= 1 - \frac{x}{3} + \frac{\left(\frac{1}{3}\right)\left(-\frac{2}{3}\right)(-x)^2}{2!} + \frac{\left(\frac{1}{3}\right)\left(-\frac{2}{3}\right)\left(-\frac{5}{3}\right)(-x)^3}{3!} + \cdots$$

(2 marks)

$$= 1 - \frac{x}{3} - \frac{x^2}{9} - \frac{5x^3}{81}$$ (1 mark)

b When $x = \frac{1}{64}$, (1 mark)

$$\sqrt[3]{1-x} = \sqrt[3]{1 - \frac{1}{64}}$$
$$= \sqrt[3]{\frac{63}{64}} = \frac{\sqrt[3]{63}}{4}$$

(1 mark)

Therefore, when $x = \frac{1}{64}$, then

$$\sqrt[3]{63} \approx 4\left[1 - \frac{x}{3} - \frac{x^2}{9} - \frac{5x^3}{81}\right]$$

(1 mark)

$$= 4\left[1 - \frac{\left(\frac{1}{64}\right)}{3} - \frac{\left(\frac{1}{64}\right)^2}{9} - \frac{5\left(\frac{1}{64}\right)^3}{81}\right]$$

(1 mark)

$$= 4 - \frac{4}{192} - \frac{4}{36\,864} - \frac{20}{21\,233\,664}$$

$= 3.979057$ (1 mark)

25 a $9! = 362\,880$ (1 mark)

b $2 \times 8! = 80\,640$ (2 marks)

c $9! - 2 \times 8! = 282\,240$

(2 marks)

d We require:

(no. of ways in total) – (no. of ways with one woman separating men) – (no. ways with men together)

(1 mark)

$= 9! - 2 \times 7 \times 7! - 2 \times 8!$

(1 mark)

$= 211\,680$ (1 mark)

Chapter 2

Skills check

1 a

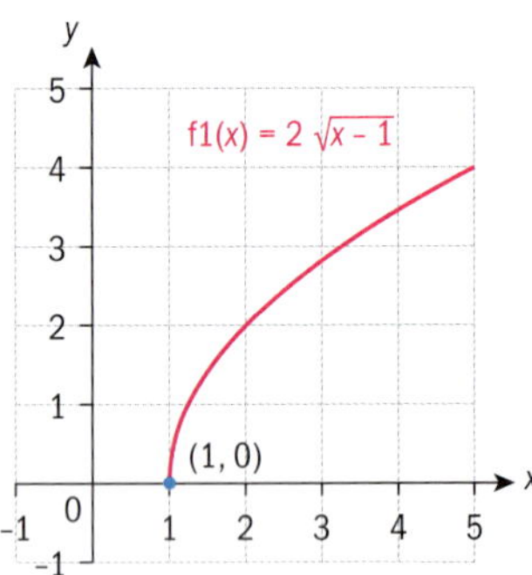

b

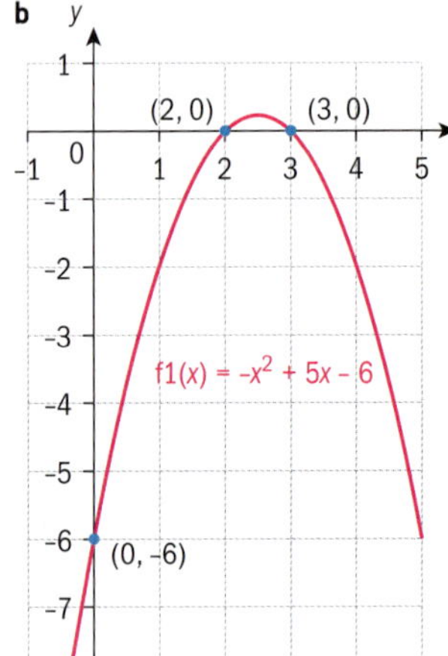

2 a

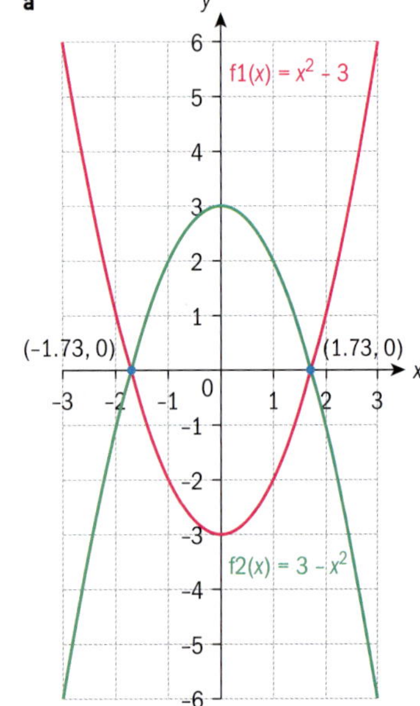

The graphs intersect at (−1.73,0) and (1.73,0), each to 3sf.

b

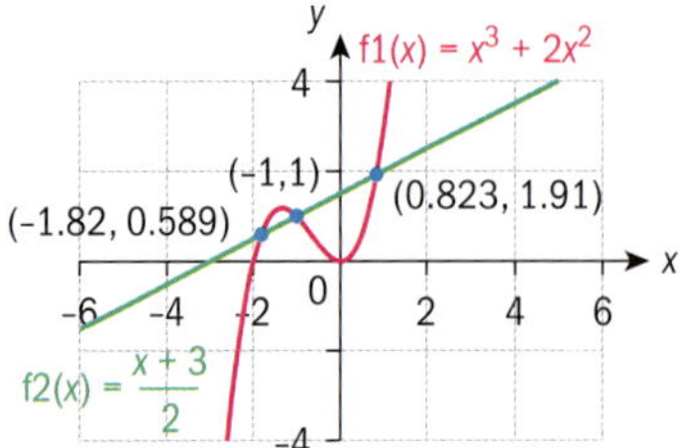

The graphs intersect at (−1,1), (−1.82,0.589) to 3sf, and (0.823,1.91) to 3sf.

3 a $y = (x-1)^2 + 2$

b $y = -(x+3)^2 + 10$

c $y = 3(x+1)^2 - 2$

Exercise 2A

1 a Function: $D_f = \{1, 2, 3, 4\}$, $R_f = \{0, 2, 3, 4\}$

b Function: $D_f = \{-2, -1, 0, 1\}$, $R_f = \{1\}$

c No, this is not a function because it is not well-defined: 2 is mapped to multiple values.

d No, this is not a function because it is not well-defined: π is mapped to both π and π^π.

e Function: $D_f = \{1, 2, 3, 4, 5\}$, $R_f = \{2, 4, 10\}$

f No, this is not a function because it is not well-defined: −5 is mapped to both 0 and 1

g No, this is not a function, since it is does not act on the entire domain 5 has no image.

h No, not a function, because it is not well-defined: 2 is mapped to both 8 and 15.

2 a No, because the graph does not pass the vertical line test.

b Function: $D_f = \mathbb{R}$, $R_f = \{2\}$

c No, because the graph does not pass the vertical line test.

d Function: $D_f = \{x \in \mathbb{R} \mid 1 < x < 6\}$, $R_f = \{y \in \mathbb{R} \mid 1 \le y \le 7\}$

e Function: $D_f = \{-4, -3, -2, -1, 1, 2, 3, 4\}$, $R_f = \{-3, -2, -1, 0, 1, 2, 4\}$

f Function: $D_f = \{x \in \mathbb{R} \mid -4 < x < 3\}$, $R_f = \{y \in \mathbb{R} \mid -2 \le y \le 1\}$

g Function: $D_f = R_f = \mathbb{R}$

Exercise 2B

1 a i $x = -3$ ii $(-3, -1)$

iii Concave up, $D_f = \mathbb{R}$, $R_f = \{y \in \mathbb{R} \mid y \ge -1\}$

b i $x = \frac{3}{2}$ ii $\left(\frac{3}{2}, \frac{49}{4}\right)$

iii Concave down, $D_f = \mathbb{R}$, $R_f = \left\{y \in \mathbb{R} \mid y \le \frac{49}{4}\right\}$

c i $x = 2$ ii $(2, -17)$

iii Concave up, $D_f = \mathbb{R}$, $R_f = \{y \in \mathbb{R} \mid y \ge -17\}$

d i $x = -1$ ii $(-1, 9)$

iii Concave down, $D_f = \mathbb{R}$, $R_f = \{y \in \mathbb{R} \mid y \le 9\}$

2 a $y = (x-2)^2 - 16$

b $y = C(x^2 + 2x - 3)$
$\therefore y = 3 - 2x - x^2$

c $y = C(x^2 - 6x + 5)$
$\therefore y = 4x^2 - 24x + 20$

d $y = 3(x-2)^2 - 6$

e $y = C\left(x^2 + 3x - 10\right)$
$\therefore y = 5 - \frac{3}{2}x - \frac{1}{2}x^2$

f $y = -\frac{3}{5}(x+10)^2 + 60$

Exercise 2C

1 $D_f = \{x \in \mathbb{R} \mid x \ne 2\}$, $R_f = \{y \in \mathbb{R} \mid y \ne 0\}$, asymptotes: $x = 2$ and $y = 0$

2 $D_f = \left\{x \in \mathbb{R} \mid x \neq \frac{1}{2}\right\}$,

$R_f = \{y \in \mathbb{R} \mid y \neq 0\}$,

asymptotes: $x = \frac{1}{2}$ and $y = 0$

3 $D_f = \left\{x \in \mathbb{R} \mid x \neq \frac{1}{2}\right\}$,

$R_f = \left\{y \in \mathbb{R} \mid y \neq -\frac{1}{4}\right\}$,

asymptotes: $x = \frac{1}{2}$ and $y = -\frac{1}{4}$

4 $D_f = \{x \in \mathbb{R} \mid x \neq 1\}$,

$R_f = \{y \in \mathbb{R} \mid y \neq -1\}$,

asymptotes: $x = 1$ and $y = -1$

5 $D_f = \left\{x \in \mathbb{R} \mid x \neq -\frac{1}{2}\right\}$,

$R_f = \{y \in \mathbb{R} \mid y \neq -1\}$,

asymptotes: $x = -\frac{1}{2}$ and $y = -1$

6 $2 - 3x \neq 0 \Rightarrow x \neq \frac{2}{3}$,

$R_f = \left\{y \in \mathbb{R} \mid y \neq \frac{2}{3}\right\}$,

asymptotes: $x = \frac{2}{3}$ and $y = \frac{2}{3}$

Exercise 2D

1 **a** $D_f = \{x \in \mathbb{R} \mid x \geq 2\}$, $R_f = \{y \in \mathbb{R} \mid y \geq 0\}$

b $D_f = \left\{x \in \mathbb{R} \mid x \geq \frac{2}{3}\right\}$,

$R_f = \{y \in \mathbb{R} \mid y \geq 0\}$

c $D_f = \left\{x \in \mathbb{R} \mid x \leq \frac{1}{2}\right\}$,

$R_f = \{y \in \mathbb{R} \mid y \geq 1\}$

d $D_f = \left\{x \in \mathbb{R} \mid x \geq -\frac{1}{2}\right\}$,

$R_f = \{y \in \mathbb{R} \mid y \leq 3\}$

e $D_f = \{x \in \mathbb{R} \mid x \geq 1\}$, $R_f = \{y \in \mathbb{R} \mid y \leq 0\}$

f $D_f = \{x \in \mathbb{R} \mid x \leq 2\}$, $R_f = \{y \in \mathbb{R} \mid y \leq 1\}$

Exercise 2E

1 $D_f = \{x \in \mathbb{R} \mid x \neq 0, x \neq 3\}$, $R_f = \{y \in \mathbb{R} \mid y \neq 0\}$, asymptotes: $x = 0$, $x = 3$, $y = 0$

2 $D_f = \{x \in \mathbb{R} \mid x \neq \pm 3\}$,

$R_f = \left\{y \in \mathbb{R} \mid y > 0 \text{ or } y \leq -\frac{1}{9}\right\}$

asymptotes: $x = 3$, $x = -3$, $y = 0$

3 $D_f = \{x \in \mathbb{R} \mid x \neq 1, x \neq -3\}$,

$R_f = \left\{y \in \mathbb{R} \mid y < 0 \text{ or } y \geq \frac{1}{4}\right\}$,

asymptotes: $x = 1$, $x = 3$, $y = 0$

4 $D_f = \{x \in \mathbb{R} \mid x \neq -2\}$, $R_f = \{y \in \mathbb{R} \mid y > 0\}$, asymptotes: $x = -2$, $y = 0$

5 $D_f = \left\{x \in \mathbb{R} \mid x \neq \frac{3}{2}, x \neq -6\right\}$,

$R_f = \left\{y \in \mathbb{R} \mid y < 0 \text{ or } y \geq \frac{8}{225}\right\}$,

asymptotes: $x = -6$,

$x = \frac{3}{2}$, $y = 0$

6 $D_f = \{x \in \mathbb{R} \mid x > -2\}$, $R_f = \{y \in \mathbb{R} \mid y > 0\}$, asymptotes: $x = -2$, $y = 0$

7 $D_f = \left\{\begin{matrix} x \in \mathbb{R} \mid x > 2 \text{ or} \\ x < \frac{1}{2} \end{matrix}\right\}$,

$R_f = \{y \in \mathbb{R} \mid y > 0\}$,

asymptotes: $x = 2$,

$x = -\frac{1}{2}$, $y = 0$

8 $D_f = \left\{x \in \mathbb{R} \mid x \leq -\frac{5}{2} \text{ or } x \geq \frac{5}{2}\right\}$,

$R_f = \{y \in \mathbb{R} \mid y < 0\}$,

asymptotes: $x = -\frac{5}{2}$,

$x = \frac{5}{2}$, $y < 0$

Exercise 2F

1 $\frac{1}{x+2} - \frac{1}{x+3}$

2 $\frac{1}{x-1} - \frac{2}{x+2}$

3 $\frac{3}{x} + \frac{1}{x-3}$

4 $\frac{1}{2}\left(\frac{1}{x+1} + \frac{1}{x-1}\right)$

5 $\frac{1}{x+3} - \frac{1}{x-2}$

6 $\frac{1}{4x-1} + \frac{2}{2x+1}$

7 $\frac{3}{3x-2} - \frac{1}{2x+3}$

Exercise 2G

1 $D_f = \mathbb{R}$, $R_f = \{y \in \mathbb{R} \mid y \leq 3\}$

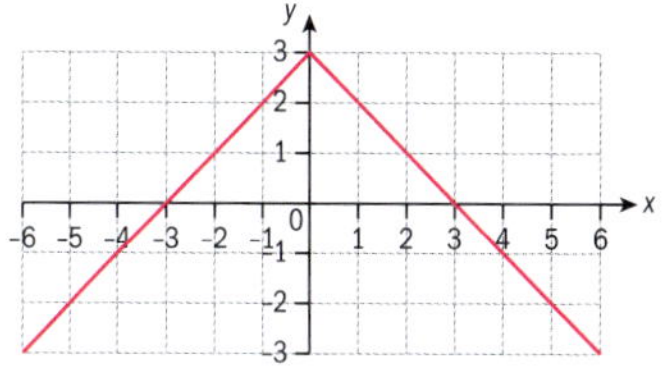

2 $D_f = \mathbb{R}$, $R_f = \{y \in \mathbb{R} \mid y \geq -1\}$

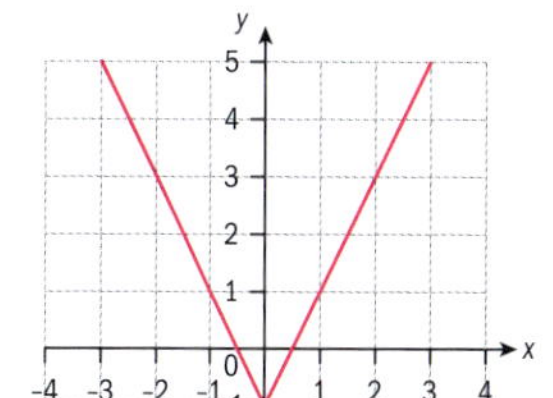

3 $D_f = \mathbb{R}$, $R_f = \{y \in \mathbb{R} \mid y \geq -4\}$

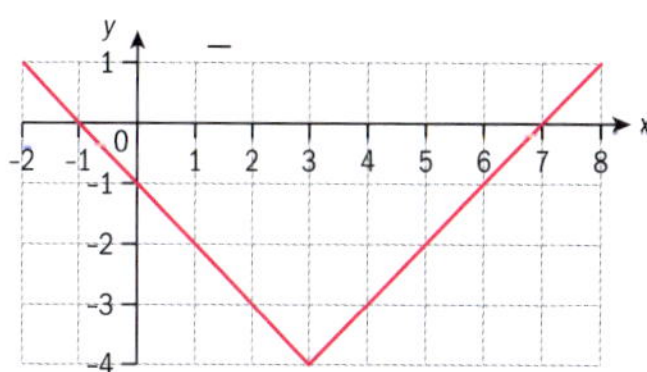

4 $D_f = \mathbb{R}$, $R_f = \{y \in \mathbb{R} \mid y \leq 1\}$

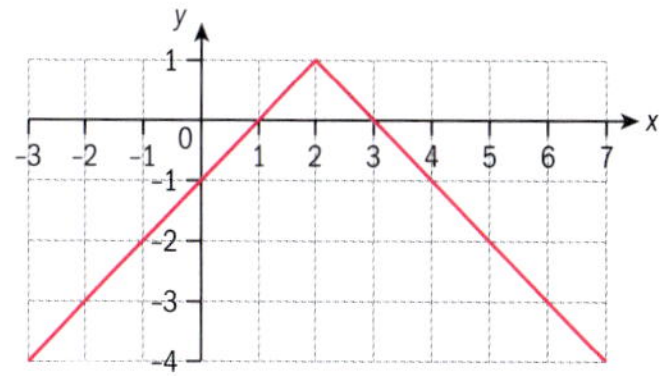

5 $D_f = \mathbb{R}$, $R_f = \{y \in \mathbb{R} \mid y \geq -1\}$

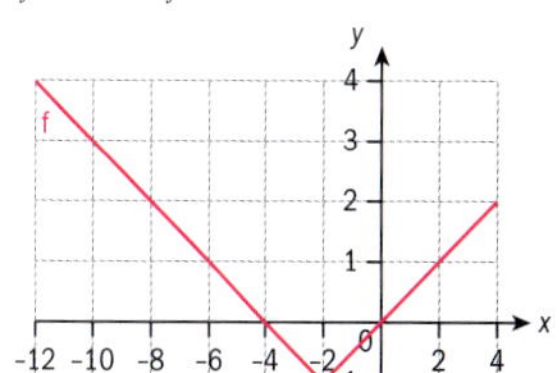

6 $D_f = \mathbb{R}$, $R_f = \{y \in \mathbb{R} \mid y \le 2\}$

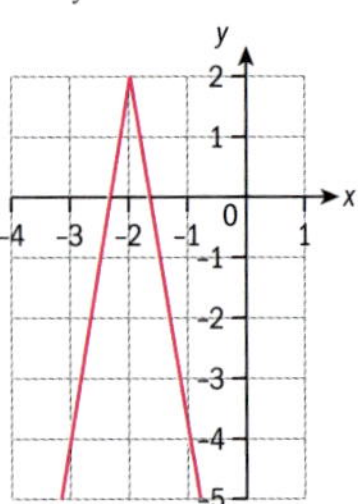

7 $D_f = \mathbb{R}$, $R_f = \{y \in \mathbb{R} \mid y \le 2\}$

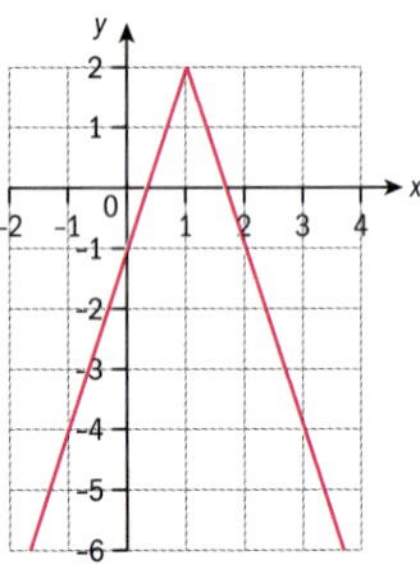

8 $D_f = \mathbb{R}$, $R_f = \{y \in \mathbb{R} \mid y \le -2\}$

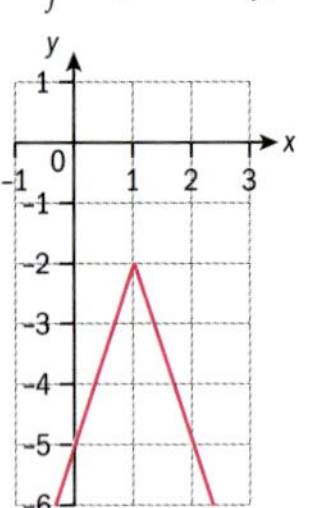

9 $D_f = \mathbb{R}$, $R_f = \{y \in \mathbb{R} \mid y \ge 0\}$

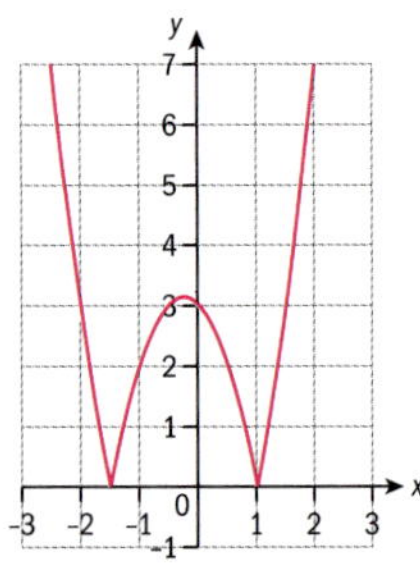

10 $D_f = \mathbb{R}$, $R_f = \{y \in \mathbb{R} \mid y \ge 2\}$

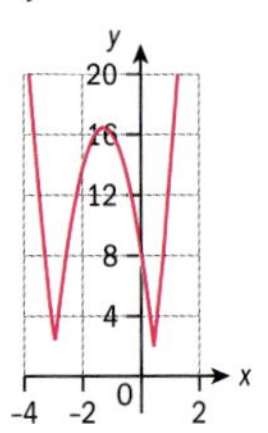

Exercise 2H

1 **a** $x = \frac{1}{3}$ or $-\frac{5}{3}$ (both answers valid)

b $x = -6$ or -8 (both answers valid)

c $x = -3$ or $x = \frac{1}{3}$;

$x = \frac{1}{3}$ only is valid

d $x = 0$ or -2 (both answers valid)

e $x = -5$ or $-\frac{4}{3}$ (both answers valid)

f $x = -\frac{4}{3}$ or $-\frac{2}{7}$ (both answers valid)

g $x = -\frac{22}{5}$ or $-\frac{2}{13}$ (both answers valid)

Exercise 2I

1 **a** 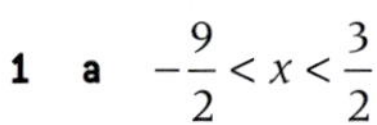 $-\frac{9}{2} < x < \frac{3}{2}$

b $x \le -1$ or $x \ge 4$

c $-1 < x < 4$

d $x \le -\frac{4}{3}$ or $x \ge 2$

e $x < -2$ or $x > 0$ (both valid)

f $-2 < x < 2$

2 **a** $1 < x < 5, x \ne 2$

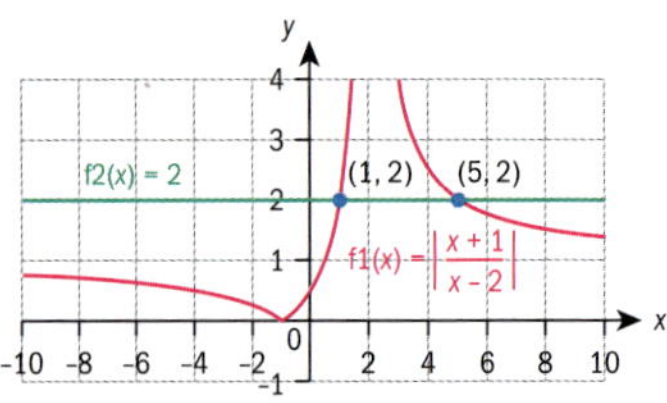

b $0 \le x \le 2, x \ne 1$

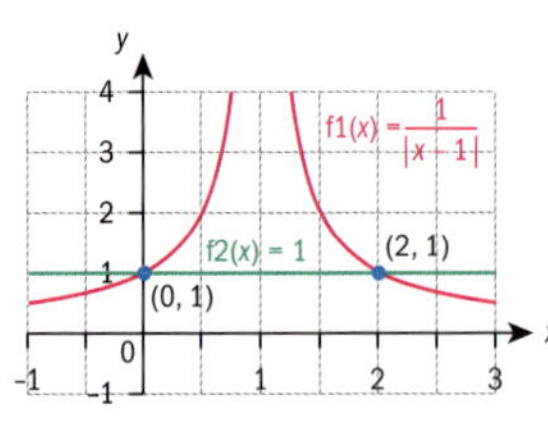

c $-\frac{13}{3} \le x \le 5$

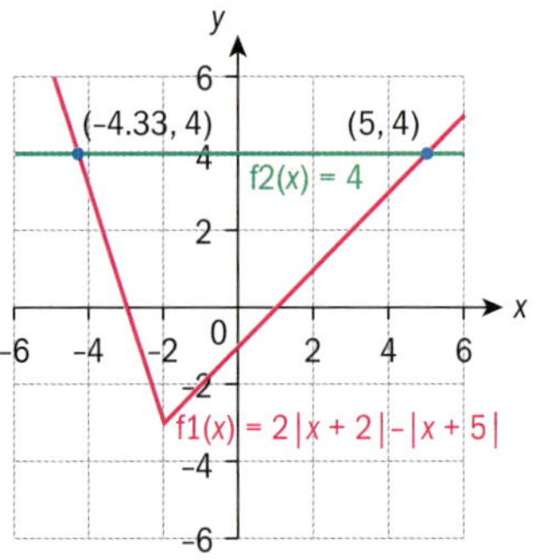

d $-4.54 < x < -3.30$; $0.303 < x < 1.54$

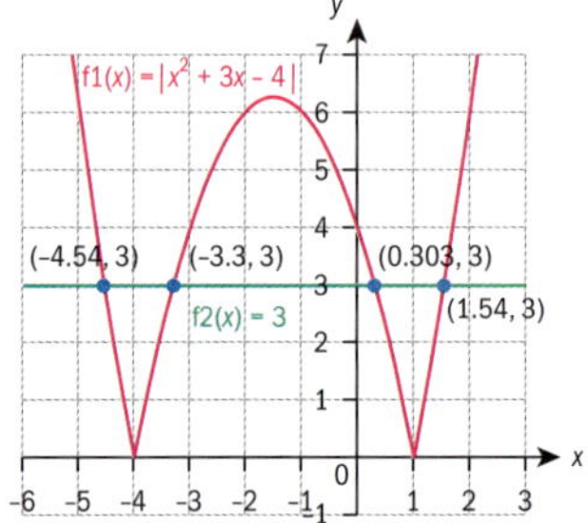

Exercise 2J

1 **a** $f(-9) = 1, f(0) = 1, f(\pi) = -1, f(99) = -1$

b

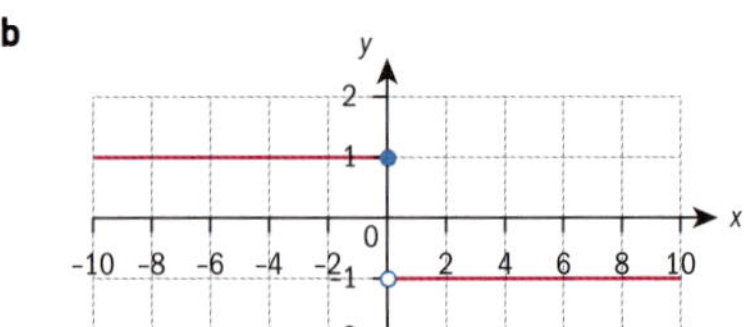

c $D_f = \mathbb{R}$, $R_f = \{-1, 1\}$

2 **a** $f(-4) = 16, f(0) = 0, f(1) = 3$

b

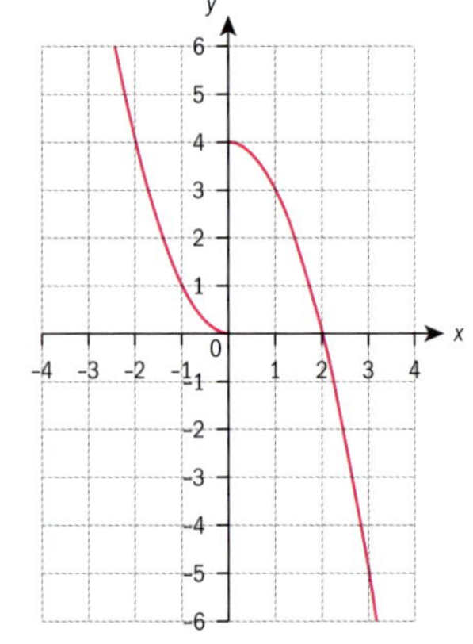

c $D_f = \mathbb{R}$, $R_f = \mathbb{R}$

3 **a** $f(-1) = -4$, $f(0) = 1$, $f(8) = 3$

b

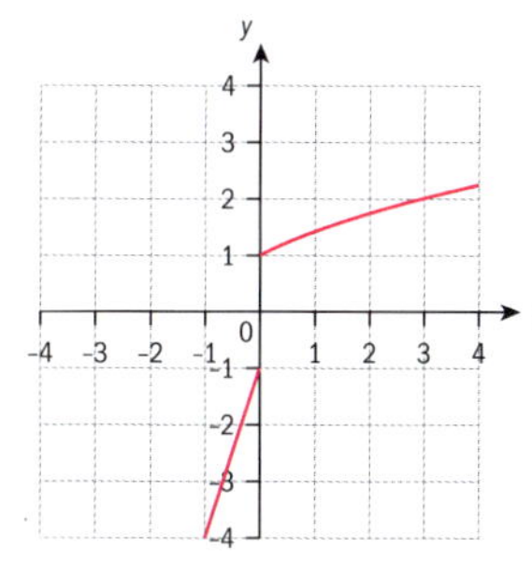

c $D_f = \mathbb{R}$, $R_f = \{y \in \mathbb{R} \mid y > 1$ or $y \le -1\}$

4 **a** $f(-1) = 0$, $f(0) = 1$, $f(-4) = 3$, $f(8) = 3$

b

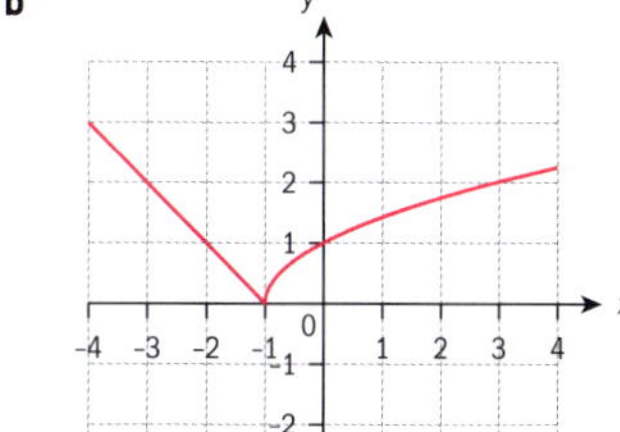

c $D_f = \mathbb{R}$, $R_f = \{y \in \mathbb{R} \mid y \ge 0\}$

5

$$f(x) = \begin{cases} 3x+10, \ x \le -2 \\ 2, -2 < x < 2 \\ -3x+10, \ x \ge 2 \end{cases}$$

6 **a** $f(x) = \begin{cases} 2x+4, \ x \ge -2 \\ -(2x+4), \ x < -2 \end{cases}$

b $f(x) = \begin{cases} 3x-7, \ x \ge 3 \\ 11-3x, \ x < 3 \end{cases}$

Exercise 2K

1 Neither

2 Onto and one-to-one

3 One-to-one, not onto

4 One-to-one, not onto

5 Onto, not one-to-one

6 Onto, not one-to-one

7 Individual response

Exercise 2L

1 **a** Even, one-to-one

b Odd, one-to-one

c Odd, one-to-one

d Neither, many-to-one

e Neither, many-to-one

f Odd, many-to-one

2 $f(x) = 0$ for all x

Exercise 2M

1 **a** **i** $\sqrt{3}$ **ii** $3\sqrt{2}$

iii $3\sqrt{x}$ **iv** $\sqrt{3x}$

b **i** 8 **ii** -19

iii $-3x^2 - 7$

iv $29 - 30x + 9x^2$

c **i** $\sqrt{3}$ **ii** $\sqrt{3}+1$

iii $\sqrt{2x-1}+1$

iv $\sqrt{2x+1}$

2 **a** **i** $D_f = \mathbb{R}$,

$R_f = \left\{y \in \mathbb{R} \mid y \ge -\frac{1}{4}\right\}$,

$D_g = \mathbb{R}$, $R_g = \mathbb{R}$

ii $D_f = \mathbb{R}$, $R_f = \{y \in \mathbb{R} \mid y \ge 0\}$, $D_g = \{x \in \mathbb{R} \mid |x| \ge 2\}$, $R_g = \{y \in \mathbb{R} \mid y \ge 0\}$

b **i** $f \circ g(x) = 3(x-1)(3x-2)$, $D_{f \circ g} = \mathbb{R}$,

$R_{f \circ g} = \left\{y \in \mathbb{R} \mid y \ge -\frac{1}{4}\right\}$

$g \circ f(x) = 2 - 3x - 3x^2$, $D_{g \circ f} = \mathbb{R}$,

$R_{g \circ f} = \left\{y \in \mathbb{R} \mid y \le \frac{11}{4}\right\}$

ii $f \circ g(x) = \left|\sqrt{x^2-4}+1\right|$,

$D_{f \circ g} = \{x \in \mathbb{R} \mid |x| \ge 2\}$,

$R_{f \circ g} = \{y \in \mathbb{R} \mid y \ge 1\}$

$g \circ f(x) = \sqrt{(x+3)(x-1)}$,

$D_{g \circ f} = \{x \in \mathbb{R} \mid x \le -3$ or $x \ge 1\}$, $R_{g \circ f} = \{y \in \mathbb{R} \mid y \ge 0\}$

3 **a** **i** $f(h(x)) = 1 - 2\sqrt{2x+4}$

ii $h(g(x)) = \sqrt{2x^2+2}$

iii $h(h(x)) = \sqrt{2\sqrt{2x+4}+4}$

iv $f(g(h(x))) = -4x - 5$

b **i** $D_{f \circ h} = \{x \in \mathbb{R} \mid x \ge -2\}$, $R_{f \circ h} = \{y \in \mathbb{R} \mid y \le 1\}$

ii $D_{h \circ g} = \mathbb{R}$, $R_{h \circ g} = \{y \in \mathbb{R} \mid y \ge \sqrt{2}\}$

iii $D_{h \circ h} = \{x \in \mathbb{R} \mid x \ge -2\}$, $R_{h \circ h} = \{y \in \mathbb{R} \mid y \ge 2\}$

iv $D_{f \circ g \circ h} = \mathbb{R}$, $R_{f \circ g \circ h} = \mathbb{R}$

4 4

5 Individual response

6 **a** $320h^2 + 420$

This gives the number of bacteria b in food h hours out of the refrigerator.

b 5.47 hours

7 $r(t) = \left(\frac{40+3t+t^2}{500} - 0.1\right)^2 + 0.;$

2 hours

Exercise 2N

1 **a** Inverse does not exist

b $\{(3, 1), (2, -6), (-4, -3), (0, 0), (-5, -5), (-3, -2)\}$

c $\{(-1, -1), (3, -3), (-5, -2), (-4, -4), (1, 1), (3, -5), (-2, 0)\}$

2 **a** $f^{-1}(x) = \frac{x+1}{5}$

b $f^{-1}(x) = 3x + 2$

c $f^{-1}(x) = \sqrt{x+3}$

(must restrict to either positive or negative square root for this to be a function)

d $f^{-1}(x) = \frac{2+3x}{x}$ $(x \ne 0)$

e $f^{-1}(x) = (x-1)^{\frac{1}{3}}$

f $f^{-1}(x) = \frac{x+1}{x-1}$ $(x \ne 1)$

3 **a** $f^{-1}(x) = 2 + \sqrt{x}$,

$D_{f^{-1}} = \{x \in \mathbb{R} \mid x \ge 0\}$,

$R_{f^{-1}} = \{y \in \mathbb{R} \mid y \ge 2\}$

b $f^{-1}(x) = \dfrac{1-x}{x-2}, x \neq 2$

c $f^{-1}(x) = \dfrac{\sqrt{x-1}}{2}, x \geq 1$

4 $(g \circ f) \circ (f^{-1} \circ g^{-1})$
$= g \circ (f \circ f^{-1}) \circ g^{-1}$
$= g \circ (\mathrm{id}) \circ g^{-1} = g \circ (\mathrm{id} \circ g^{-1})$
$= g \circ g^{-1} = \mathrm{id}$
where id is the identity function id $(x) = x \Rightarrow (g \circ f) \circ (f^{-1} \circ g^{-1})(x) = x \Rightarrow (f^{-1} \circ g^{-1})(x) = (g \circ f)^{-1}(x)$ Since this is true in general, it is certainly true for the specified functions.

5 Important: it must be shown that both $f(g(x)) = x$ and $g(f(x)) = x$

a $f(g(x)) = -4\left(1 - \dfrac{x}{4}\right) + 4$

$= -4 + x + 4 = x$ and

$g(f(x)) = 1 - \dfrac{1}{4}(-4x + 4)$

$= 1 + x - 1 = x$

b $f(g(x)) = \dfrac{\dfrac{2}{1-x} + 3 - 5}{\dfrac{2}{1-x} + 3 - 3}$

$= \dfrac{2 - 2(1-x)}{2} = x$

and $g(f(x)) = -\dfrac{2}{\dfrac{x-5}{x-3} - 1} + 3$

$= -\dfrac{2(x-3)}{x-5-(x-3)} + 3$

$= (x-3) + 3 = x$

c $f(g(x)) = \dfrac{\left(2\dfrac{\sqrt[3]{2x}-3}{2} + 3\right)^3}{2}$

$= \dfrac{\left(\sqrt[3]{2x}\right)^3}{2} = \dfrac{2x}{2} = x$ and

$g(f(x)) = \dfrac{\sqrt[3]{2\dfrac{(2x+3)^3}{2}} - 3}{2}$

$= \dfrac{(2x+3)-3}{2} = x$

Exercise 2O

1

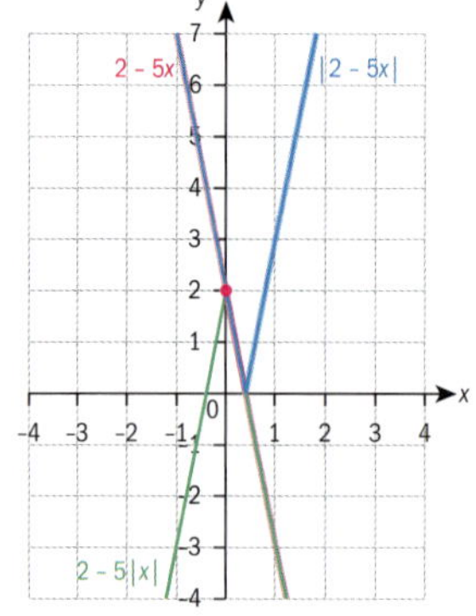

2

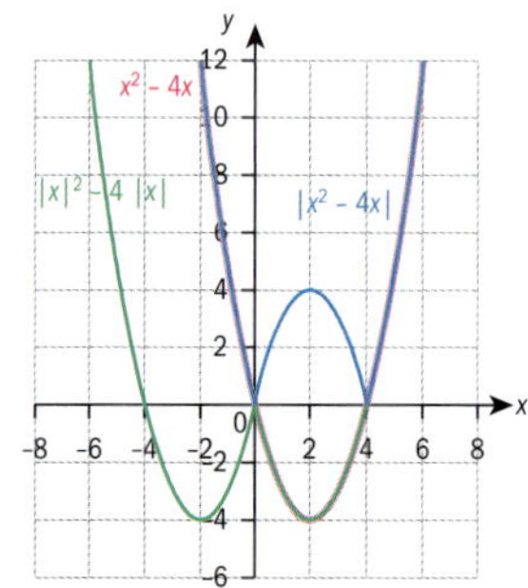

3

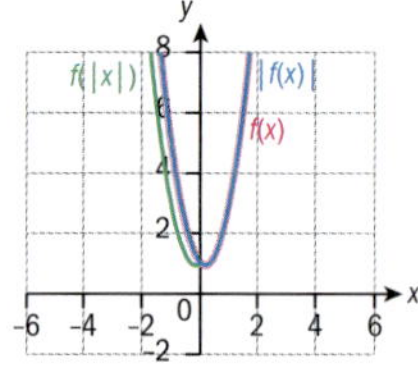

4

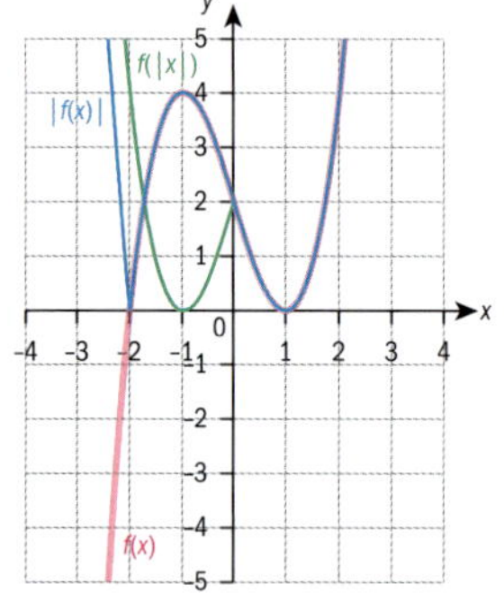

5

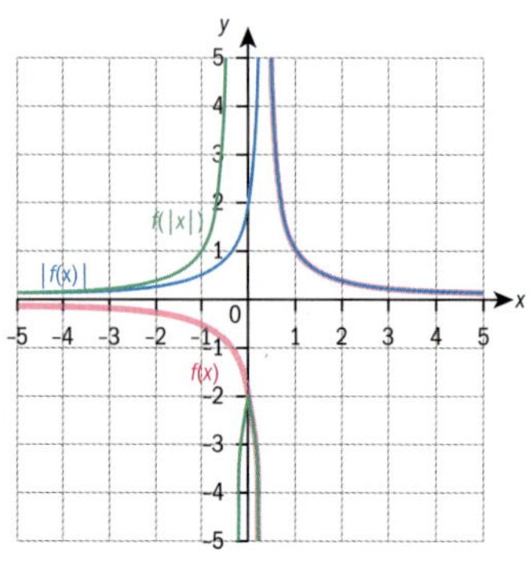

6

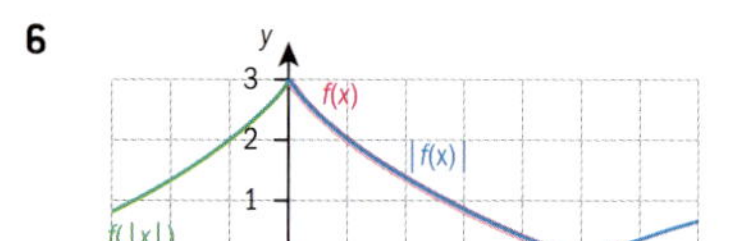

Exercise 2P

1 a

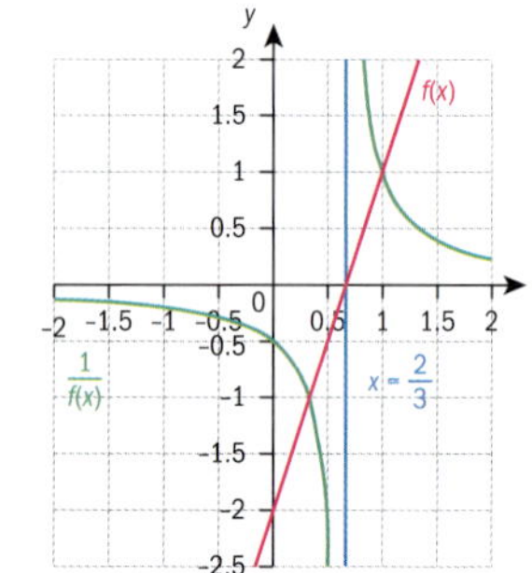

b

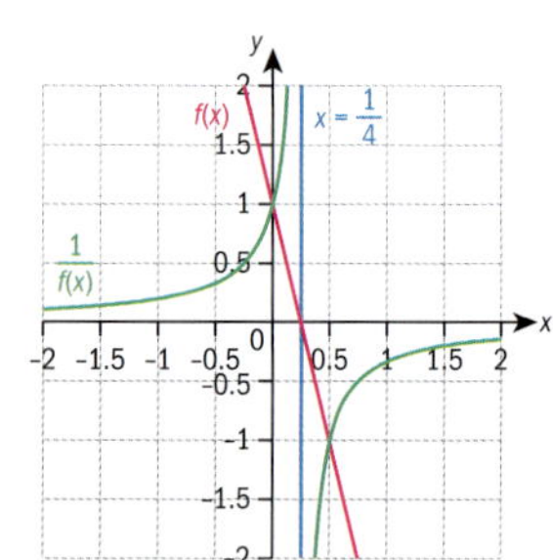

c

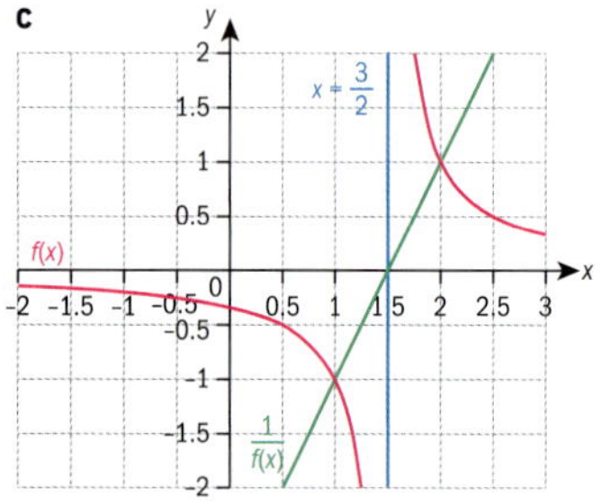

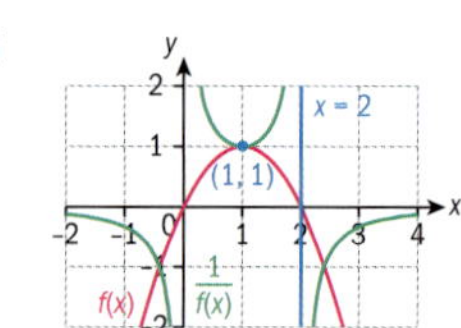

d

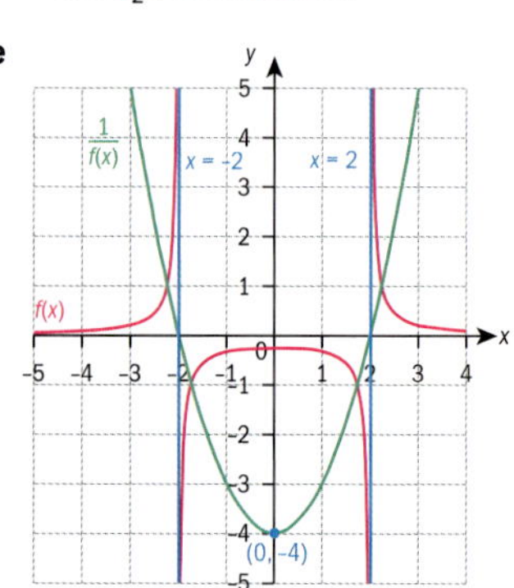

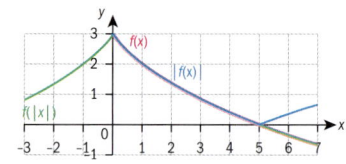

e

2 a

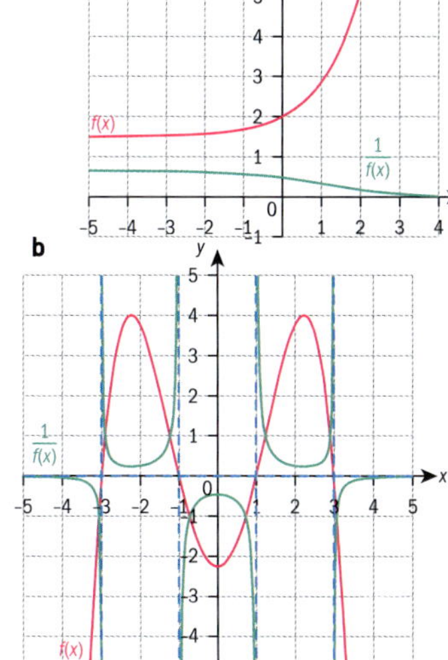

b

Exercise 2Q

1 a

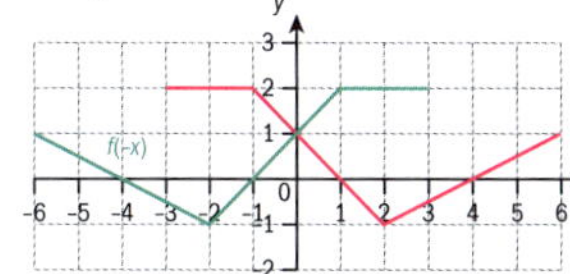

b

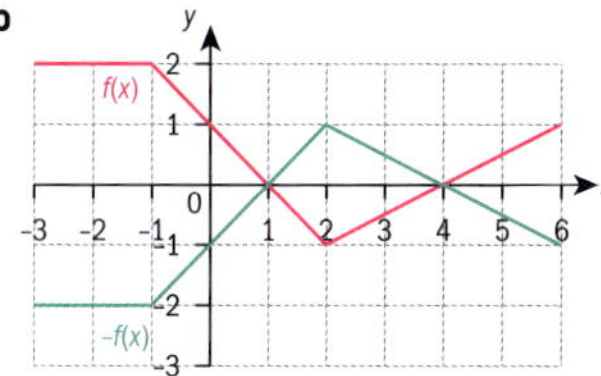

c

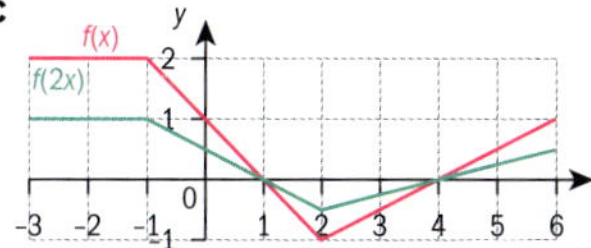

d

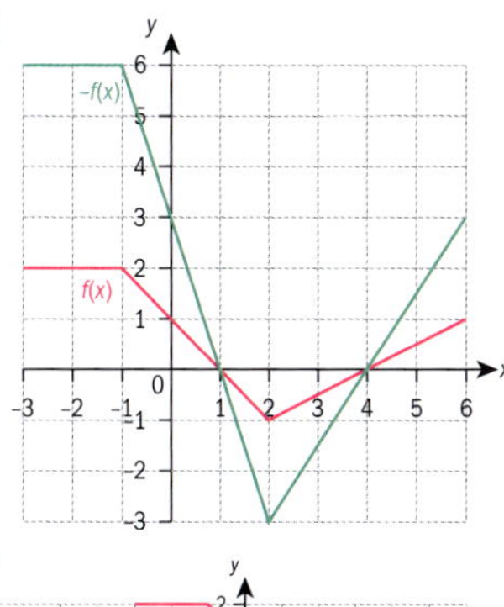

e

f

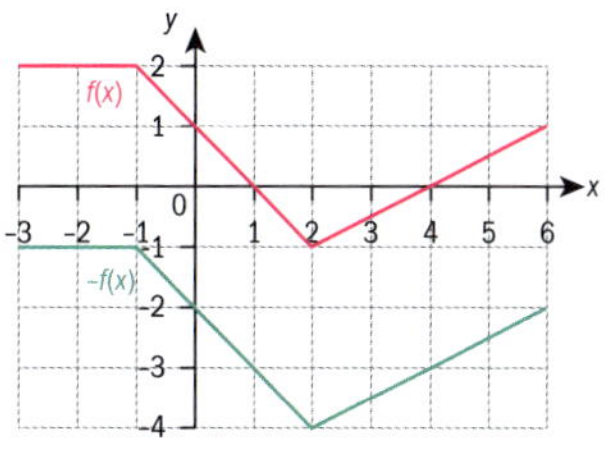

2 a $r(x) = 2f(x)$, $s(x) = -f(x-3)$

b $r(x) = f(-x)$,

$s(x) = f\left(\dfrac{x}{2}\right) - 4$

3 a $R_f = \{y \in \mathbb{R} \mid 0 \le y \le 6\}$

b

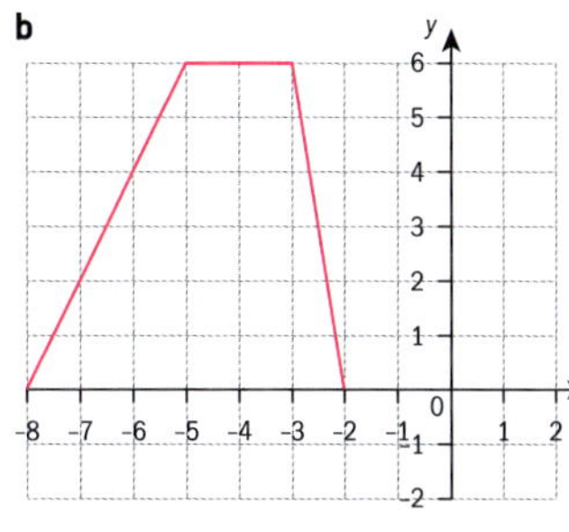

c $D_g = \{x \in \mathbb{R} \mid -8 \le x \le -2\}$

d $h(x) = g(x) - 4$

e $h(x) = f(-x) - 4$

4 a $g(x) = -f(x)$

b $g(x) = f(-x)$

c $g(x) = f(x+3) - 1$

d $g(x) = -f(x) + 1$

e $g(x) = f(-2x)$

f $y = -f(2x)$

Exercise 2R

1 $y = \dfrac{2}{3(x+2)} + 3$, $D_f = \{x \in \mathbb{R} \mid x \ne -2\}$, $R_f = \{y \in \mathbb{R} \mid y \ne 3\}$

2 a Horizontal dilation factor $\frac{1}{3}$, followed by vertical dilation factor 2, then a horizontal translation of 4 units in the positive x-direction, and a vertical translation of 1 unit in the positive y-direction.

b $y = 2f(2(x-4)) + 1$

3 a e.g. translation by $\begin{pmatrix} -5 \\ 0 \end{pmatrix}$, vertical stretch by factor −8, translation by $\begin{pmatrix} 0 \\ 1 \end{pmatrix}$

b e.g. translation by $\begin{pmatrix} -1 \\ 0 \end{pmatrix}$, stretch horizontally by scale factor $\frac{1}{2}$, stretch vertically by scale factor 3, translation by $\begin{pmatrix} 0 \\ 2 \end{pmatrix}$

c e.g. translation by $\begin{pmatrix} -1 \\ 0 \end{pmatrix}$, stretch vertically by scale factor 2, translation by $\begin{pmatrix} 0 \\ 2 \end{pmatrix}$

Exercise 2S

1

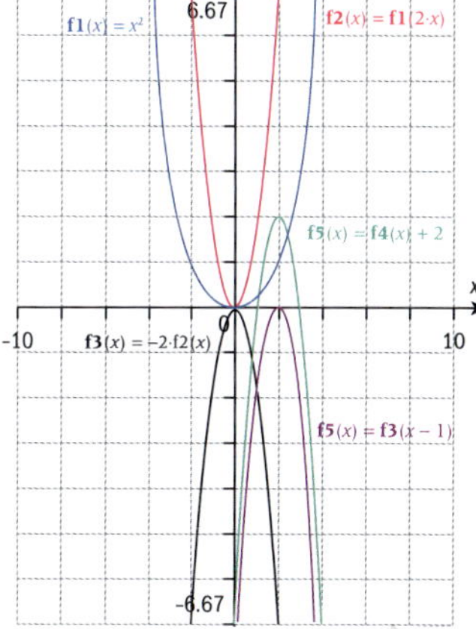

2

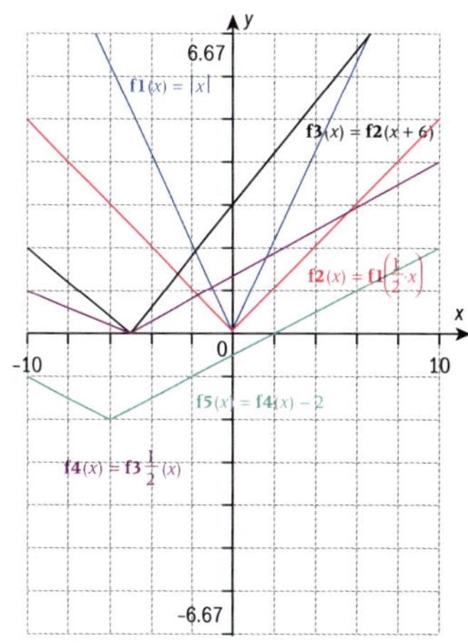

3

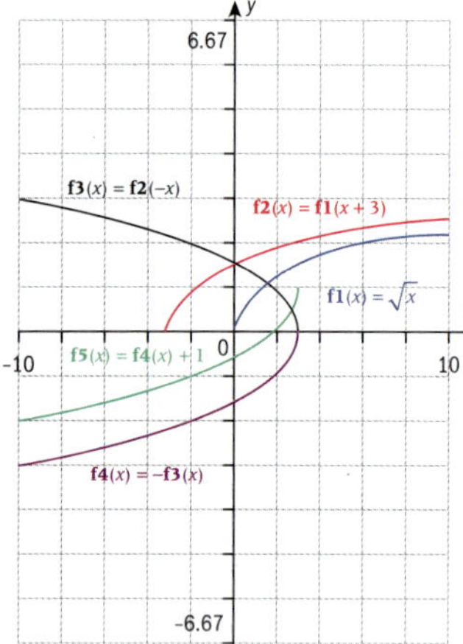

4

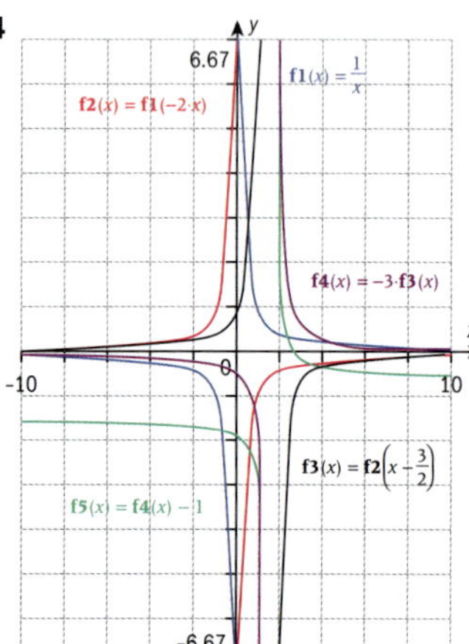

5

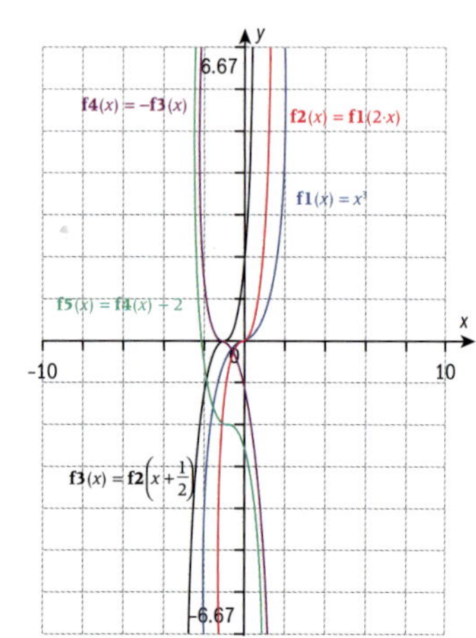

Chapter review

1 a Not a function

b Yes, $D_f = \{-4 \le x \le 5\}$, $R_f = \{-1, 2\}$

c Yes, $D_f = \mathbb{R}$, $R_f = \{y \in \mathbb{R} \mid -1 \le y \le 1\}$

d Not a function

e Not a function

f Yes, $D_f = \{-1, 0, 3, \pi\}$, $R_f = \{\pi\}$

2 a 2 **b** 2

3 a $f^{-1}(x) = 5x + 2$

b $g^{-1}(x) = 1 - x^2$

c $y = \dfrac{2x}{3+x}\ (x \ne -3)$

4 Translate the graph of $y = f(x)$ 3 units in the negative x-direction, reflect in the y-axis, vertical stretch by a factor of 2, vertical translation of 4 units in the positive direction.

5 $-3 \le x \le \dfrac{1}{3}$

6 Take $y = \sqrt{\dfrac{1-x}{x}}$

$$\frac{1}{1+\left(\sqrt{\dfrac{1-x}{x}}\right)^2} = \frac{1}{1+\dfrac{1-x}{x}}$$

$$= \frac{x}{x+1-x} = x$$

and $\sqrt{\dfrac{1-\dfrac{1}{1+x^2}}{\dfrac{1}{1+x^2}}} = \sqrt{\dfrac{1+x^2-1}{1}}$

$= \sqrt{x^2} = |x| = x$ in the domain $[0,1]$

7 a Even

b Odd

c Neither

8 $\dfrac{2}{x+2} - \dfrac{2}{x+3}$

Exam-style questions

9 a No real roots $\Rightarrow \Delta < 0$ (1 mark)

$\Delta = 36 - 4(2k)(k)$

$= 36 - 8k^2$ (1 mark)

$36 - 8k^2 < 0$

$k^2 > \dfrac{36}{8} = \left(\dfrac{9}{2}\right)$

$|k| > \dfrac{3}{\sqrt{2}}$

$k < -\dfrac{3}{\sqrt{2}}$ or $k > \dfrac{3}{\sqrt{2}}$ (2 marks)

b Equation of line of symmetry is

$x = -\dfrac{b}{2a} = -\dfrac{6}{4k} = -\dfrac{3}{2k}$ (2 marks)

Therefore $\dfrac{3}{2k} = 1$

$\Rightarrow k = \dfrac{3}{2}$ (1 mark)

10 a The graph of f is shifted two units in the positive x-direction and one unit in the negative y-direction.

b $y = (2(x-2)^2 + 4(x-2) + 7) - 1$ (3 marks)

$= (2(x^2 - 4x + 4) + 4x - 8 + 7) - 1$

$= 2x^2 - 4x + 6$ (1 mark)

11 a $x = \dfrac{3y-4}{y+2} \Rightarrow yx + 2x = 3y - 4$ (2 marks)

$yx - 3y = -4 - 2x$

$y(x-3) = -(4 + 2x)$

$y = \dfrac{2x+4}{3-x}$

$f^{-1}(x) = \dfrac{2x+4}{3-x}$ (1 mark)

b $x \ne 3$ (1 mark)

12 a $y = \dfrac{k}{x-1} + 1$

$x = \dfrac{k}{y-1} + 1$

$x(y-1) = k + y - 1$ (1 mark)

$xy - x = k + y - 1$ (1 mark)

$xy - y = k + x - 1$

$y(x-1) = k + x - 1$

$y = \dfrac{k+x-1}{x-1}$

$y = \dfrac{k}{x-1} + 1$ (1 mark)

$f^{-1}(x) = \dfrac{k}{x-1} + 1$

So f is self-inverse

b Range is $f(x) > 1$, $f(x) \in \mathbb{R}$ (2 marks)

c

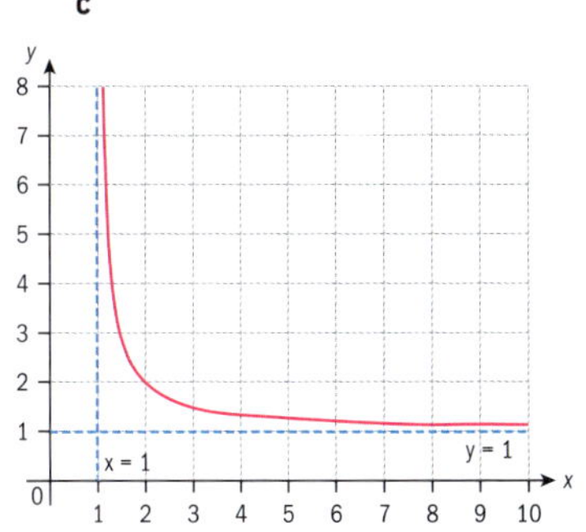

(shape of graph: 1 mark)
(both asymptotes: 1 mark)

13 a $x^2 - 6x + 13 = (x-3)^2 + 4$ (2 marks)

Therefore $k = 3$ (1 mark)

b $y = (x-3)^2 + 4$

$x = (y-3)^2 + 4$

$(y-3)^2 = x - 4$ (2 marks)

$y - 3 = \sqrt{x-4}$

$y = 3 + \sqrt{x-4}$

$f^{-1}(x) = 3 + \sqrt{x-4}$ (1 mark)

c The domain of $f^{-1}(x)$ is $x \geq 4$, $(x \in \mathbb{R})$ (1 mark)

The range of $f^{-1}(x)$ is $f(x) \geq 3$, $(f(x) \in \mathbb{R})$ (1 mark)

14 a $f(x) = \dfrac{17-10x}{2x-1}$

$= \dfrac{12+5-10x}{2x-1}$ (2 marks)

$= \dfrac{12+5(1-2x)}{2x-1}$ (1 mark)

$= \dfrac{12-5(2x-1)}{2x-1}$

$= \dfrac{12}{2x-1} - \dfrac{5(2x-1)}{(2x-1)}$

$= \dfrac{12}{2x-1} - 5$ (1 mark)

b i $x = \dfrac{1}{2}$ (1 mark)

ii $y = -5$ (1 mark)

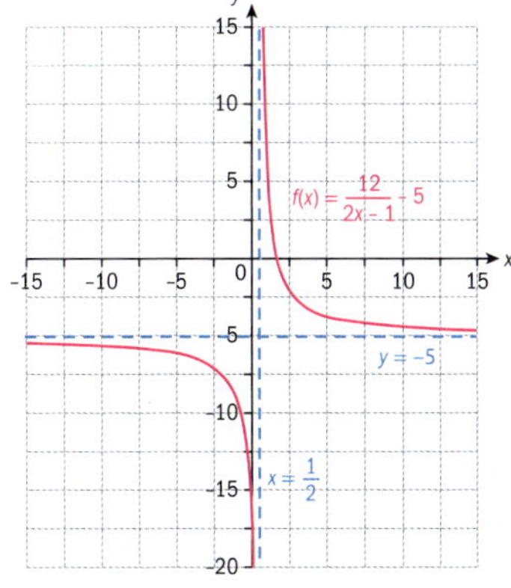

(1 mark for each branch correctly drawn, 1 mark for both asymptotes; 3 marks total)

15 a 6 (1 mark)

b $P = \dfrac{18(1 + 0.82 \times 12)}{3 + (0.034 \times 12)} \approx 57$ (2 marks)

c Solving $100 = \dfrac{18(1+0.82t)}{3+0.034t}$ (1 mark)

$300 + 3.4t = 18(1 + 0.82t)$

$300 + 3.4t = 18 + 14.76t$

$282 = 11.36t$

$t = \dfrac{282}{11.36} = 24.8$ months (1 mark)

OR

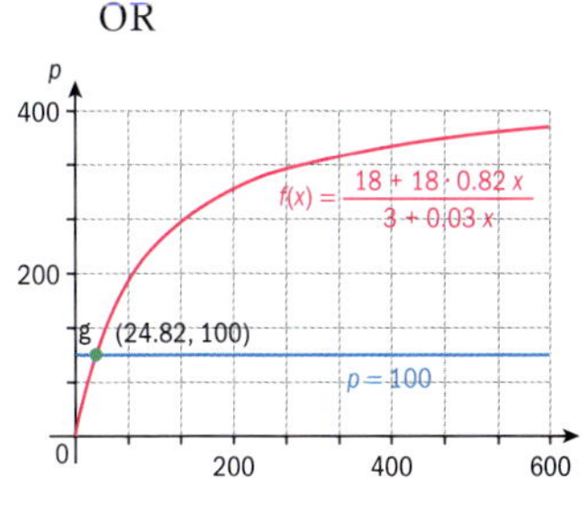

(1 mark)

$t = 24.8$ months (1 mark)

d A horizontal asymptote exists at

$P = \dfrac{18 \times 0.82}{0.034} = 434.12$ (2 marks)

Therefore for $t \geq 0$, $P \leq 434$ (1 mark)

OR

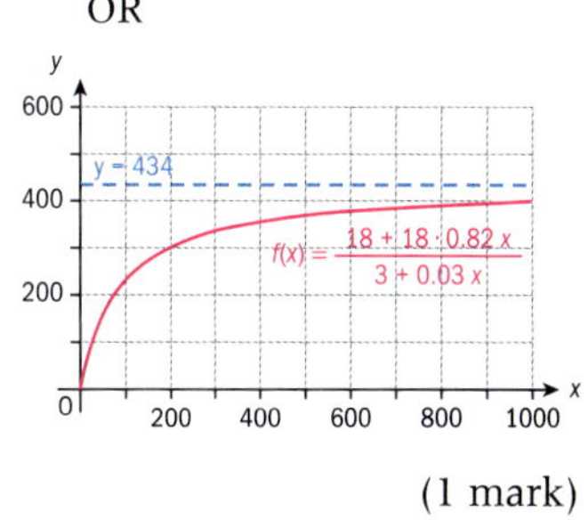

(1 mark)

A horizontal asymptote exists at $P = 434$ (1 mark)

Therefore for $t \geq 0$, $P < 434$ (1 mark)

16 $(x-3)^2 = x^2 - 6x + 9$ (1 mark)

$2(x-3)^2 = 2x^2 - 12x + 18$ (2 marks)

$2(x-3)^2 + 12x = 2x^2 + 18$

Therefore $g(x) = 2x^2 + 12x$ (1 mark)

Chapter 3

Skills check

1 a $x = \pm 13$ **b** $x = \pm 2\sqrt{7}$

c $x = \pm 3\sqrt{11}$

2 a $x = -2, 1$

b $x = -\dfrac{1}{3}, 1$

3 a $x < -3$ **b** $-4 \leq x \leq 2$

c $-\dfrac{1}{4} \leq x \leq 2$

4 a $x = -5, y = -9$

b $y = \dfrac{1}{3} - \dfrac{2x}{3}$

c $x = \dfrac{13}{8}, y = -\dfrac{3}{4}$

d No solutions

Exercise 3A

1 $x = -5$ or $x = -3$

2 $x = -7$ or $x = 2$

3 $x = \frac{1}{3}$ or $x = 2$

4 $x = \frac{5}{2}$

5 $x = -\frac{6}{5}$ or $x = 2$

Exercise 3B

1 **a** $1, -7$ **b** $-3, 10$

c $\frac{1 \pm \sqrt{5}}{2}$ **d** $\frac{1}{3}, 2$

e $-\frac{5}{2}, -\frac{1}{2}$ **f** $1 \pm \frac{\sqrt{15}}{5}$

2 **a** $-2.41, -0.414$

b $0.382, 2.62$

c $-1.00, 1.50$

d $-2.26, -0.736$

Exercise 3C

1 **a** $-6, -3$ **b** $-5, 6$

c $\frac{1 \pm \sqrt{5}}{2}$ **d** $-\frac{1}{2}, 2$

e $\frac{11 \pm \sqrt{121 - 4\sqrt{6}}}{2\sqrt{2}}$

2 **a** $a, -3$ **b** $-b, \frac{1}{2}$

c $k, -2k$ **d** $-\frac{3}{p}, \frac{1}{p}$

Exercise 3D

1 **a** $\Delta = 37 > 0 \Rightarrow$ two distinct real roots

b $\Delta = -7 < 0 \Rightarrow$ no real roots

c $\Delta = 0 \Rightarrow$ one repeated real root

d $\Delta = -37 < 0 \Rightarrow$ no real roots

e $\Delta = \pi^2 - 8 > 0 \Rightarrow$ two distinct real roots

f $\Delta = 0 \Rightarrow$ one repeated real root

2 **a** If

$$\begin{cases} m > -\frac{1}{5} \Rightarrow \text{two distinct real roots} \\ m = -\frac{1}{5} \Rightarrow \text{one repeated real root} \\ m < -\frac{1}{5} \Rightarrow \text{no real roots} \end{cases}$$

b If

$$\begin{cases} t < \frac{73}{16} \Rightarrow \text{two distinct real roots} \\ t = \frac{73}{16} \Rightarrow \text{one repeated real root} \\ t > \frac{73}{16} \Rightarrow \text{no real roots} \end{cases}$$

c Two distinct real roots: $s < 0$ or $s > 4$

One repeated real root: $s = 0$ or $s = 4$

No real roots: $0 < s < 4$

Exercise 3E

1 $x \in \,]-4, -2[$

2 $x \in \mathbb{R} - \{4\}$

3 $x \in \,]-\infty, -2] \cup [15, +\infty[$

4 $x \in \left[\frac{3}{4}, 2\right]$

5 $x \in \left]-\infty, -\frac{4}{5}\right[\cup \,]2, +\infty[$

6 $x \in \left\{\frac{2}{3}\right\}$

Exercise 3F

1 **a** $\text{Re}(z) = 0, \text{Im}(z) = -4$

b $\text{Re}(z) = 5, \text{Im}(z) = 0$

c $\text{Re}(z) = -24, \text{Im}(z) = 7$

d $\text{Re}(z) = \frac{5}{13}, \text{Im}(z) = -\frac{12}{13}$

e $\text{Re}(z) = -\frac{1}{\sqrt{5}}, \text{Im}(z) = \frac{2}{\sqrt{5}}$

2 **a** 4 **b** 5 **c** 25

d 1 **e** 1

3 **a** $-3 - 3\text{i}$ **b** $25 - 13\text{i}$

c $\frac{25}{12} + \frac{13}{6}\text{i}$ **d** $-3 + 10\text{i}$.

Exercise 3G

1 **a** $1 - 5\text{i}$ **b** $-\frac{34}{65} + \frac{1}{65}\text{i}$

c -37i **d** $-1 + \text{i}$

e $\frac{90}{169} + \frac{122}{169}\text{i}$

2 **a** $\text{Re}(z) = 2, \text{Im}(z) = -1$

b $\text{Re}(z) = -1, \text{Im}(z) = 0$

c $\text{Re}(z) = 0, \text{Im}(z) = \frac{8}{5}$

3 **a** $a = 2, b = -1$

b $a = -5, b = -5$

4 **a** $\frac{15}{13} - \frac{10}{13}\text{i}$ **b** $\frac{1}{2} - \frac{1}{2}\text{i}$

c $-1 - \text{i}$ **d** $-\frac{40}{13} + \frac{3}{26}\text{i}$

5 **a** $a = 2b$ **b** $a = -b$

6 **a** $x = \frac{1}{2}, y = 0$ **b** $(x, y) \in \emptyset$

c $x = -2, y = 0$ or $x = 1, y = 0$

Exercise 3H

1 **a** 0 **b** 0 **c** $5 - 5\text{i}$

d $-\frac{6}{25} - \frac{17}{25}\text{i}$ **e** 1 **f** 0

2 **a** $-9 + 46\text{i}$ **b** $28 + 96\text{i}$

c -14 **d** -8i

3 **a** $\pm\left(\frac{\sqrt{2}}{2} + \frac{\sqrt{2}}{2}\text{i}\right)$

b $\pm\left(\frac{\sqrt{2}}{2} - \frac{\sqrt{2}}{2}\text{i}\right)$

c $\pm(2 + 5\text{i})$ **d** $\pm\left(\frac{1}{2} - \frac{1}{3}\text{i}\right)$

4 **a** Use result from Exercise 3G Question 7a (and induction as below).

b $P(n)$: $(z^*)^n = (z^n)^*$, $n \in \mathbb{Z}^+$

$P(1)$ is true

Assume $P(k)$ is true for some $k \in \mathbb{Z}^+$

$(z^*)k + 1 = (z^*)^k z^* = (z^k)^* z^*$

$= (z^k z)^*$ using Exercise 3G Question 8c

$= (z^{k+1})^*$

5 a $n = 4k,\ k \in \mathbb{Z}$

b $n = 4k + 2,\ k \in \mathbb{Z}$

Exercise 3I

1 a $q(x) = 2x^2 - 3x + 1$

b $q(x) = 3x^3 + x^2 + 3$

c $q(x) = x^4 - x^2 - 2$

2 a $q(x) = 3x^2 - 3x - 2,\ r = -3$

b $q(x) = 2x^2 - 5x + 5,\ r(x) = 6x - 15$

c $q(x) = x^2 + x,\ r(x) = -x^2 - x + 1$

Exercise 3J

1 a $q(x) = x^2 + 4x + 5,\ r = 11$

b $q(x) = 2x^2 - 3x - 1,\ r = 1$

c $q(x) = 2x^3 + 2x^2 - 2x + 3,\ r = 1$

d $q(x) = 3x^4 + 2x^3 - 2x^2 - x + 13,\ r = -81$

2 b $(x + 4)(x + 1)(2x - 5)(x - 3)$

Exercise 3K

1 a $q(x) = x^2 - x + 3, r = 1$

b $q(x) = 3x^2 + x + 1, r = 1$

c $q(x) = x^3 + 2x^2 - 2x + 1, r = 4$

d $q(x) = x^4 - 2x^3 + x^2 - x + 3,\ r = -1$

2 $f(x) = 2x^4 - 3x^3 + 5x^2 - 5x - 1$

3 $a = -12$

4 $b = -1$

5 $a = -5$ $b = -6$

6 $a = 4$ $b = 12$

7 $-x + 2$

8 0

Exercise 3L

1 a $x^3 - 9x^2 + 14x$

b $x^4 + x^3 - 11x^2 - 9x + 18$

c $2x^4 - 11x^3 + 23x + 10$

2 a $x^3 - 2x^2 - 2x + 4$

b $2x^5 + x^4 - x^3 - 6x^2 - 3x + 3$

c $x^5 - 2x^4 - 2x^3 - 2x^2 + 4x + 4$

3 a $(x - 1)(x^2 - 2x + 2)$

b $(x + 1)(3x^2 - 4x + 6)$

c $(2x + 1)(x - 1)(x^2 - 2x + 5)$

4 a $f(x) = (x + 2)^2(3x - 5)$

b $f(x) = (3x - 2)^2(x + 4)$

c $f(x) = (x - 1)^2(x + 2)(x - 2)$

d $f(x) = (2x + 1)^3(x + 1)$

e $f(x) = (x - 1)^4(5x + 7)$

5 a $-2\text{i},\ 2$ b $3 + 2\text{i}, \frac{1}{2}$

c $\frac{1}{2} + \frac{\sqrt{3}}{2}\text{i}, -\frac{7}{3}$ d $\text{i},\ 1 \pm 2\text{i}$

e $-2 + \text{i}, \frac{1 \pm \sqrt{7}\,\text{i}}{2}$

f $\frac{-1 - \sqrt{2}\,\text{i}}{3}, \pm\sqrt{6}\,\text{i}$

6 a $a = -2, z = \pm\text{i}$

b $a = 3,\ z = \pm\frac{\sqrt{6}}{2}\text{i}$

c $a = 22,\ b = -10,\ z = -\sqrt{5}$
$z = 1 \pm 4\text{i}$

d $a = -7,\ b = 5,\ z = 1 + 2\text{i},\ z = \frac{1 \pm \sqrt{11}\text{i}}{6}$

Exercise 3M

1 a sum = 3, product = 4

b sum = 3, product = 0

c sum = 0, product = 2

d sum $= \frac{4}{3}$, product $= -\frac{8}{3}$

2 a $\frac{1}{2}$ b $\frac{17}{4}$

c 5 d $\frac{6}{17}$

3 a $-\frac{1}{3}$ b $-\frac{1}{2}$

c -1 d 4

Exercise 3N

1 a $x = -2, -1, 2$

b $x = -3, -2, 3$

c $x = 2$ d $x = 1, 2$

2 a -4

b $-1, 2$

3 a $a = -11,\ b = 12$

b $-\frac{1}{2}$

4 a Let the three roots be $-x_1$, x_1 and x_2

Then, $x_1 - x_1 + x_2$
$= -\frac{a}{1} \Rightarrow x_2 = -a$

and $-x_1^2 + x_1x_2 - x_1x_2$
$= \frac{b}{1} = b \Rightarrow x_1 = -i\sqrt{b}$

So the three zeros are
$-a, \pm i\sqrt{b}$

and the product of the roots is

$(-x_1)(x_1)(x_2) = (i\sqrt{b})(-i\sqrt{b})(-a)$

$= -ab = -\frac{c}{1} = -c$

$\Rightarrow ab = c$

b $-a$

Exercise 3O

1 a $-1 < x < 2$, $x < -2$

b $x > 3, -3 < x < -2$

c $x \le 2$

d $x \ge \frac{1}{2},\quad -\frac{3}{2} \le x \le -1$

e $x > -\frac{1}{3}$

f $x < \frac{-5}{3}, \frac{-1}{2} < x < \frac{7}{2}$

g $x \ge 1,\ -1 \le x \le 1,\ x \le -2$

h $-\frac{1}{2} < x < 1$

2 $x < 1.17227...$

3 a $x > -1$ b $x \le 1$

c $-1 \le x \le 1$

4 a $x > 1.79, -1.8947... < x < 1$

b $0.747 < x < 1.27,\ x < -1.10$

c $-2.06403 < x < -0.88875,\ -0.50531 < x < 0.50533,\ x > 0.86851$

Exercise 3P

1 a 14 b $-2, 3$

2 a 2 b 2

3 a $x = \frac{2s - 1}{3s - 2}, y = \frac{1}{3s - 2}$

b $x = \frac{s^2 - 4}{s^2 + 4s - 5}, y = \frac{2s^2 - s + 8}{s^2 + 4s - 5}$

4 **a** $x = \frac{1}{a+b}, y = 0$

b The system does not have a solution when $a + b = 0$

Exercise 3Q

1 $\begin{pmatrix} x \\ y \end{pmatrix} = \begin{pmatrix} 2+i \\ 2i \end{pmatrix}$

2 $x = \frac{1737 + 2215i}{901}$,

$y = -\frac{109 + 175i}{53}$

Exercise 3R

1 **a** $m \neq -2$

b $m \neq 2$ and $m \neq 3$

2 **a** $k = 6$ **b** $k = 2$

3 $m = 0$ or $m = 1$ or $m = \frac{1}{2}$

4 **a** $a = \frac{9}{4}, b = \frac{21}{4}, c = -\frac{7}{2}$

b $a = \frac{22}{3}, b = -5, c = -\frac{34}{3}$

Chapter review

1 $x(x - a - b) = 1 - ab$

$\Rightarrow x^2 - (a + b)x + ab - 1 = 0$

$\Delta = (a + b)^2 - 4(ab - 1)$

$= a^2 + 2ab + b^2 - 4ab + 4$

$= a^2 - 2ab + b^2 + 4$

$= (a - b)^2 + 4 > 0$

So there are two distinct real solutions for all $a, b \in \mathbb{R}$

2 **a** $(1 - i)^3 + (1 + i)^3$

$= (1 - 3i + 3(-i)^2 - i^3) + (1 + 3i + 3(-i)^2 + i^3)$

$= (1 - 3i - 3 + i) + (1 + 3i - 3 - i)$

$= (-2 - 2i) + (2i - 2)$

$= -4$

b $\omega = \frac{2}{5} - \frac{1}{5}i$

c $f(x) = 5x^3 + 16x^2 - 15x + 4$

3 **a** $a \neq \pm 1, (x, y) = \left(\frac{2}{a-1}, \frac{2}{1-a}\right)$

b **i** $a = 1$

ii $a = -1, y = x + 2$

4 **a** -2 **b** n is even

5 $x \in [-2.45, -1.26] \cup [-0.339, 0.715] \cup [1.34, \infty[$

6 $(x, y, z) = \left(\frac{2}{3}, -\frac{1}{2}, -2\right)$

7 **a** 2 **b** 0.4

8 $z = 1 - 3i, \omega = 3 + 2i$

9 Using sum of a geometric sequence,

$1 - z + z^2 = \frac{(-z)^3 - 1}{-z - 1} = 0 \Rightarrow z^3 = -1$

$\therefore z^{2019} = (z^3)^{673} = (-1)^{673} = -1$

Exam-style questions

10 **a** $x = \frac{-b \pm \sqrt{b^2 - 4ac}}{2a}$ (1 mark)

$x = \frac{6 \pm \sqrt{208}}{2}$ (1 mark)

$x = 3 \pm \sqrt{52}$ (1 mark)

$x = 3 \pm 2\sqrt{13}$ (1 mark)

b Using sketch or table

$3 - 2\sqrt{13} \leq x \leq 3 + 2\sqrt{13}$ (2 marks)

11 **a** $8x^2 + 6x - 5 = 0$

$(4x + 5)(2x - 1) = 0$ (2 marks)

$4x + 5 = 0 \Rightarrow x = -\frac{5}{4}$ (1 mark)

$2x - 1 = 0 \Rightarrow x = \frac{1}{2}$ (1 mark)

b $8x^2 + 6x - 5 - k = 0$

No real solutions $\Rightarrow$ $b^2 - 4ac < 0$ (1 mark)

$36 - 4 \times 8 \times (-5 - k) < 0$ (1 mark)

$36 + 32(5 + k) < 0$

$5 + k < -\frac{36}{32}$

$k < -\frac{36}{32} - 5$

$k < -\frac{9}{8} - \frac{40}{8}$

$k < -\frac{49}{8}$ (1 mark)

12 Two real roots implies $b^2 - 4ac > 0$ (1 mark)

$(k\sqrt{3})^2 - 4(3k)(3) > 0$ (1 mark)

$3k^2 - 36k > 0$

$k^2 - 12k > 0$

$k(k - 12) > 0$ (1 mark)

Critical values are $k = 0$ and $k = 12$ (1 mark)

Solution is $k < 0$ or $k > 12$ (2 marks)

13 **a** Using sketch of $y = (x + 4)(3 - x)$ (1 mark)

Correct sketch or table

Solution is $-4 < x < 3$ (2 marks)

b $2x^2 - 11x + 9 < 0$

$(2x - 9)(x - 1) < 0$ (1 mark)

Using sketch of $y = (2x - 9)(x - 1)$ (1 mark)

Correct sketch or table

Solution is $1 < x < \frac{9}{2}$ (2 marks)

c Comparing answers from **a** and **b** gives $1 < x < 3$ (1 mark)

14 Let $w = a + bi$ where $a, b \in \mathbb{R}$

$(a + bi)^2 = 77 - 36i$ (1 mark)

$a^2 - b^2 + 2abi = 77 - 36i$ (1 mark)

Equating reals: $a^2 - b^2 = 77$ (1) (1 mark)

Equating imaginary

$2ab = -36$ (2) (1 mark)

(2) gives $b = -\frac{18}{a}$

Substitute in (1):

$a^2 - \left(-\frac{18}{a}\right)^2 = 77$ (1 mark)

$a^2 - \frac{324}{a^2} = 77$

$a^4 - 77a^2 - 324 = 0$

Attempting to factorise, or using the quadratic formula:

$(a^2 + 4)(a^2 - 81) = 0$ (1 mark)

Since $a \in \mathbb{R}$, $a^2 = 81$ (1 mark)

$a = \pm 9$

$a = 9 \Rightarrow b = -2$

$a = -9 \Rightarrow b = 2$

So $w = \pm(9 - 2i)$ (2 marks)

15 Let $p(x) = 2x^3 + ax^2 - 10x + b$

$p(1) = 0 \Rightarrow 2 + a - 10 + b = 0$ (2 marks)

$a + b = 8$ **(1)** (1 mark)

$p(-2) = 15 \Rightarrow -16 + 4a + 20 + b = 15$

$4a + b = 11$ **(2)** (2 marks)

Solving equations **(1)** and **(2)** simultaneously: (1 mark)

$a = 1$ (1 mark)

$b = 7$ (1 mark)

16 Since $3 - i$ is a zero, its conjugate is also a zero, i.e. $(3 - i)^* = 3 + i$ is a zero. (1 mark)

By the factor theorem, $[z - (3 - i)]$ is a factor of f, and $[z - (3 + i)]$ is a factor of f. (1 mark)

Therefore $[z - (3 - i)][z - (3 + i)] = z^2 - 6z + 10$ is also a factor of f. (2 marks)

Writing $z^4 - 8z^3 + 48z^2 - 176z + 260 = [z^2 - 6z + 10][z^2 + kz + 26]$

Equating coefficients of z^2 gives $48 = 26 - 6k + 10$ (2 marks)

So $k = -2$ (1 mark)

$z^2 - 2z + 26 = 0$

$$z = \frac{-(-2) \pm \sqrt{(-2)^2 - 4 \times 1 \times 26}}{2}$$

$$= \frac{2 \pm \sqrt{-100}}{2} = \frac{2 \pm 10i}{2} = 1 \pm 5i$$

(2 marks)

The zeros are therefore $(3 \pm i)$ and $(1 \pm 5i)$.

17 a Sum of roots $= -\left(\frac{-4}{5}\right) = \frac{4}{5}$ (2 marks)

Product of roots

$= (-1)^4\left(-\frac{1}{5}\right) = -\frac{1}{5}$ (2 marks)

b Sum of roots $= -\left(\frac{-4}{5}\right) = \frac{4}{5}$ (2 marks)

Product of roots

$= (-1)^5\left(\frac{10}{5}\right) = -2$ (2 marks)

18 Suppose $w = a + bi$ and $z = c + di$ for $a, b, c, d \in \mathbb{R}$ (1 mark)

Then $w^* = a - bi$ and $z^* = c - di$ (1 mark)

So $wz^* - zw^* = (a + bi)(c - di) - (c + di)(a - bi)$

$= [ac + bd + i(bc - ad)] - [ac + bd + i(ad - bc)]$ (1 mark)

$= i(bc - ad) - i(ad - bc)$ (1 mark)

$= i(bc - ad) + i(bc - ad)$

$= 2(bc - ad)i$

which is purely imaginary (1 mark)

20 At $(-1, -5)$: $a - b + c = -5$ (1 mark)

At $(3, -1)$: $9a + 3b + c = -1$ (1 mark)

At $(10, -71)$: $100a + 10b + c = -71$ (1 mark)

Solving simultaneously using GDC: (1 mark)

$a = -1$ (1 mark)

$b = 3$ (1 mark)

$c = -1$ (1 mark)

Chapter 4

Skills check

1 $7x^{\frac{1}{2}}$; $2x^{-3}$; $8x^{\frac{2}{3}}$

2 Vertical asymptote: $x = -3$

Horizontal asymptote: $y = 0$

y-intercept $\left(0, \frac{2}{3}\right)$

3 $S_\infty = 5 \times \frac{1}{1 - \frac{1}{2}} = 5 \times 2 = 10$

Exercise 4A

1 -4 2 6

3 No limit 4 3

5 2 6 0

7 No limit

Exercise 4B

1 Continuous

2 Continuous

3 Not continuous

4 Continuous

5 Continuous at $x = 1$, not continuous at $x = -2$

6 $k = -\frac{1}{3}$ 7 $k = \frac{1}{2}$

8 **a** Not continuous at $x = \pm 3$

b Not continuous at $x = 1$

c continuous

d not continuous at $x = -5, 1$

e not continuous at $x = 1$

f continuous

Exercise 4C

1 **a** 5 **b** 3 **c** 6

d 2 **e** none **f** 0

g $\frac{2}{3}$ **h** $a^2 + b$

2 **a** 3 **b** -2 **c** $-\frac{1}{3}$

d none **e** 0 **f** 0

g none **h** 4

3 **a** $x = \frac{1}{6}$; $y = \frac{1}{2}$

b $x = \pm\sqrt{3}$; $y = 1$

c $x = 1;\ y = -1$

d $x = \pm\sqrt{2};\ y = 0$

e $x = 0$; no horizontal asymptote

f $x = \frac{1}{2};\ x = 1;\ y = \frac{1}{2}$

Exercise 4D

1 **a** Diverges

b Diverges

c Converges

d Converges

2 **a** converges to 1

b converges to $\frac{1}{2}$

c converges to $\frac{1}{2}$

d Diverges

e converges to 0

f converges to 0

3 **a** converges to $\frac{3}{4}$

b converges to 3

c converges to $\frac{20}{9}$

d converges to $\frac{5}{6}$

e converges to $\frac{\pi}{\pi - e}$

f does not converge since $r > 1$.

4 **a** $x < 0$ **b** $x = -3$

5 $-\frac{1}{4} < x < \frac{1}{2}$

6 $\frac{27}{2}$

Exercise 4E

1 **a** -4 **b** -6 **c** -2

d 2 **e** 3 **f** 1

g $-\frac{1}{4}$ **h** 1

2 $2 + h$; 2

3 **a** $6a + 2$ **b** $(-1, 0)$

4 $\left(2, \frac{1}{4}\right)$

Exercise 4F

1 **a** $2x + 1$; 1 **b** $6x - 1$; -7

c $\frac{2}{x^2}$; 2 **d** $\frac{1}{2\sqrt{x+1}}; \frac{1}{4}$

e $-\frac{1}{54}$ **f** $3x^2$; 3

2 **a** $4a + 2h$ **b** $4a$

3 **a** $20t - 3t^2$

b $17\ \text{m s}^{-1}$; $-100\ \text{m s}^{-1}$; After 10 s have elapsed the particle is moving in the opposite direction to the direction it was moving in after 1s.

Exercise 4G

1 **a** $m = 3$ **b** $y = 3x - 1$

c $x + 3y - 7 = 0$

2 $x = 3,\ x = 1;\ y = x - 4;\ y = x$

3 **a** $x = 0$ **b** $x = -1.5$

c $x = 0$ **d** $x = \pm 1$

4 $y = 2;\ x = 1$

Exercise 4H

1 **a** $18x - 6$ **b** $15x^4 - 8x + 2$

c $\frac{1}{2}x + \frac{2}{3}$

d $25x^4 - 12x^2 + 1 + \frac{3}{4x^4} + \frac{1}{x^6}$

e $-\frac{3}{x^2} - 4x + 3x^2$

f $\frac{1}{2\sqrt{x}}$ **g** $-\frac{1}{2\sqrt[2]{x^3}}$ **h** $\frac{2}{5\sqrt[5]{x^3}}$

i $\frac{2}{3\sqrt[3]{x^4}} + \frac{15}{2\sqrt{7}}$

j $\frac{-5\sqrt{x} + 9\sqrt[6]{x} - 2}{6\sqrt[3]{x^2}}$

2 $y = -10x + 2$

3 $x + 3y - 11 = 0$

4 $y = 32x - 22;\ y = -16x - 2$

5 $y = 2x - 13;\ 27y - 54x + 95 = 0$

6 $(3, 1)$

7 $y = 2x;\ y = -54x + 27y + 32 = 0$

Exercise 4I

1 **a** $20(4x - 3)^4$ **b** $\frac{-2}{\sqrt{1 - 4x}}$

c $\frac{-\left(12x^5 + x^2 + 2\right)}{x^2}$

d $-\frac{6x}{\sqrt[2]{1 - 3x^3}}$

e $\frac{3\left(\sqrt{x} - 2\right)\left(\sqrt{x} - 1\right)^2}{2x^4}$

f $\frac{2^{\frac{4}{3}}x}{3\sqrt[3]{\left(x^2 - 2\right)^2}}$

2 $y = -6x + 5$

3 $x + 2y - 5 = 0$

4 $-\frac{1}{4\sqrt{x}\sqrt{2 - \sqrt{x}}}$ **5** $x = \pm 2$

6 **a** $15ax^2 - 4bx + 4c$ **b** $\Delta = 0$

Exercise 4J

1 $8x^3 + 45x^2 + 54x - 27$

2 $-20x^4 + 96x^3 - 27x^2 - 270x + 81$

3 $\frac{2}{(x+1)^2}$ **4** $\frac{4 - 9x}{2\sqrt{2 - 3x}}$

5 $\frac{-3x^2 + 4x - 3}{\left(x^3 - 2x^2 + 3x + 1\right)^2}$

6 $\frac{14x^4 + 36x^3 + 24x^2 - 4x - 6}{\sqrt[3]{3x - 2}}$

7 $\frac{2(x - 11)(2x - 1)^2}{(x - 4)^3}$

8 $-\frac{3x + 4}{\sqrt{\left(3x^2 + 2\right)^3}}$

9 $y = -2x + 2;\ y = \frac{1}{2}x - \frac{1}{2}$

10 **a** $y = \sqrt{x+1}\,(3 - x)^2$

$$y' = \frac{(3-x)^2}{2\sqrt{x+1}} - 2(3-x)\sqrt{x+1}$$

$$= \frac{(3-x)^2 - 4(3-x)(x+1)}{2\sqrt{x+1}}$$

$$= \frac{(3-x)\left((3-x) - 4(x+1)\right)}{2\sqrt{x+1}}$$

$$= \frac{-(3-x)(5x+1)}{2\sqrt{x+1}}$$

b $x = 3;\ -\frac{1}{5}$

Exercise 4K

1 **a** $\frac{16}{(5-x)^2}$ **b** $\frac{x + 2}{2\sqrt{x}(x - 2)^2}$

c $\frac{2 + x}{\left(1 - x^2\right)^{\frac{3}{2}}}$

d $\frac{-3x^2 - 2x + 3}{\left(x^2 + 1\right)^2}$

2 a $\dfrac{-x^4+6x^2-2x}{\left(x^3-1\right)^2}$

b $\dfrac{x^4-2x^3+4x+4}{2\left(x^3+1\right)^{\frac{3}{2}}}$

c $\dfrac{-\left(12x^2-5x+4\right)}{2x^3\left(2x^2-x+1\right)^{\frac{3}{2}}}$

d $-\dfrac{1}{\sqrt{x}\left(\sqrt{x}+1\right)^2}$

e $\dfrac{9x-12\sqrt{x}-1}{4\sqrt{x}(\sqrt{x}-1)^{\frac{3}{2}}}$

f $\dfrac{2}{3\left(x+1\right)^{\frac{1}{3}}\left(x+2\right)^{\frac{5}{3}}}$

3 $y=-\dfrac{1}{4}x$

4 $y=\dfrac{25}{16}x-\dfrac{3}{80}$

5 $(1,4)$

Exercise 4L

1 $-\dfrac{2}{x^3}$ 2 360 3 $a=-3$

4 $x=\pm 2$ 5 $x=2\pm\sqrt{2}$

6 $p=2$; $q=24$; $r=37$

7 a $20\ \text{ms}^{-1}$ b $80\ \text{ms}^{-2}$

c $240\ \text{ms}^{-3}$

8 $\dfrac{-1}{x^2}, \dfrac{2}{x^3}, \dfrac{-6}{x^4}, \dfrac{24}{x^5}, \dfrac{(-1)^n n!}{x^{n+1}}$

9 -384

Exercise 4M

1 a $x=-3, -1, 2, 4$

b $x<-3$; $-1<x<2$; $x>4$

c $-3<x<-1$; $2<x<4$

d $x<-3$; $-1<x<2$; $x>4$

e $x<-2$, $0.5<x<3$

2 a i $-\infty<x<-3$; $x>4$

ii $-3<x<4$

iii $x=-3, 4$

b i $-\infty<x<-1$; $0<x<1$; $2<x<\infty$

ii $-1<x<0$; $1<x<2$

iii $x=-1, 0, 1, 2$

3 a i $(0,-1)$ maximum; $(2,3)$ minimum

ii $x<0$; $x>2$

iii $0<x<2$ $(x\neq 1)$

iv The turning points are local.

b i $(-0.5,-3.69)$ maximum; $(0,-4)$ minimum, $(1,-2)$ maximum

ii $(-\infty,-0.5)$; $(0,1)$

iii $(-0.5,0)$, $(1,\infty)$

iv $(1-2)$ is global extrema, the rest are local.

c i minimum $(-1,0.5)$

ii $(-1,1)$, $(1,\infty)$

iii $(-\infty,-3)$, $(-3,-1)$

iv $(-1,0.5)$ local extrema

d i $(-3,\frac{1}{3})$ maximum, $(1,3)$ minimum

ii $(-\infty,-3)$, $(1,3)$, $(3,\infty)$

iii $(-3,0)$, $(0,1)$

iv both local extema

e i maximum $(0,-1)$; minimum $(2,3)$

ii $x<0$; $x>2$

iii $0<x<2$ $(x\neq 1)$

iv Local

4 $f_{min}=-17$; $f_{max}=7$

5 $y=\pm\infty$

6 $b=21$

7 $a=-3$; $b=-9$; $c=12$

Exercise 4N

1 a $(-1,6)$ is a local maximum; $(1,-6)$ is a local minimum

b $(0,0)$ is a local minimum; $(4,256)$ local maximum

c $(0.5,5)$ is a local minimum; $(-0.5,-3)$ is a local maximum

2 a $b=3$ b $(1,-2)$; $(-1,4)$

c

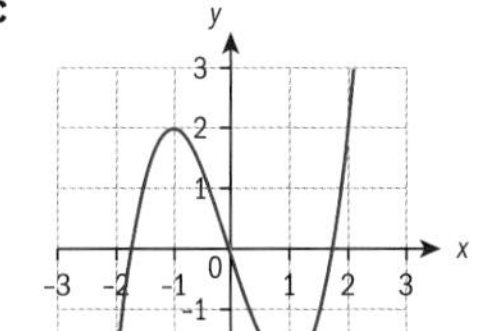

3 a $x=\dfrac{-b}{2a}$

b When $a>0$ the point is a minimum and when $a<0$ the point is a maximum, since $f''(x)=2a$.

4 a $b=-3$; $c=6$ b $(-1,8)$

Exercise 4O

1 a $(0,0)$ b $]0,\infty[$ c $]-\infty,0[$

2 a No points of inflexion

b f is concave up throughout its domain.

c f is never concave down.

3 a $(2,-38)$ b $]2,\infty[$ c $]-\infty,2[$

4 a $\left(-\dfrac{1}{3},-\dfrac{25}{27}\right)$ b $\left]-\dfrac{1}{3},\infty\right[$

c $\left]-\infty,-\dfrac{1}{3}\right[$

5 a $(0,0)$, $(2,16)$ b $0<x<2$

c $x<0$; $x>2$

6 a $(1,0)$ b $x>1$ c $x<1$

7 a $(-0.25,0.992)$, $(0,1)$

b $x<-0.25$; $x>0$;

c $-0.25<x<0$

8 a $(0,-16)$, $(2,0)$

b $x<0$ or $x>2$ c $0<x<2$

9 a $f'(x)=3ax^2+2bx+c$ since there are two distinct stationary points, the discriminant of this quadratic is positive

$\therefore \Delta=(2b)^2-4(3a)(c)>0$

$\Rightarrow 4b^2-12ac>0$

$\Rightarrow b^2>3ac$

b $f''(x)=6ax+2b=0 \Rightarrow$

$x=-\dfrac{b}{3a}$

$(a \neq 0$ by construction since $f(x)$ is a cubic$)$

$f'(x) = 0 \Rightarrow x$

$$= \frac{-2b \pm \sqrt{4b^2 - 4(3a)(c)}}{2(3a)}$$

$$= \frac{-b \pm \sqrt{b^2 - 3ac}}{3a}$$

$\therefore$ Let $x_1 = \dfrac{-b + \sqrt{b^2 - 3ac}}{3a}$ and

$x_2 = \dfrac{-b - \sqrt{b^2 - 3ac}}{3a}$

$$\Rightarrow \frac{x_1 + x_2}{2}$$

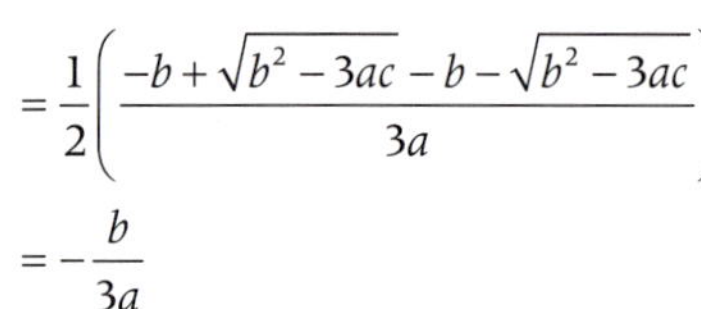

$$= \frac{1}{2}\left(\frac{-b + \sqrt{b^2 - 3ac} - b - \sqrt{b^2 - 3ac}}{3a}\right)$$

$$= -\frac{b}{3a}$$

Exercise 4P

1 **a**

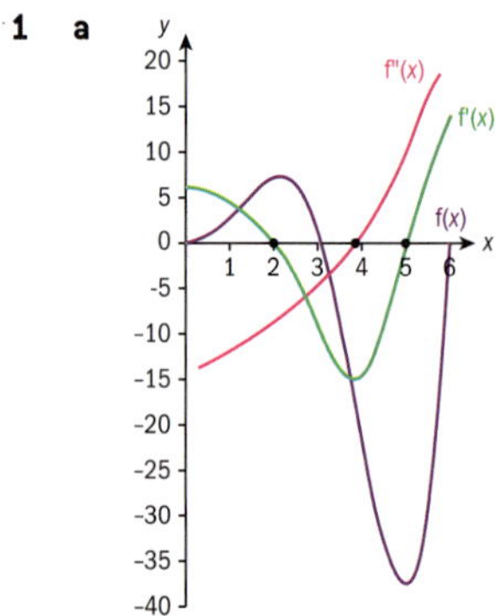

b

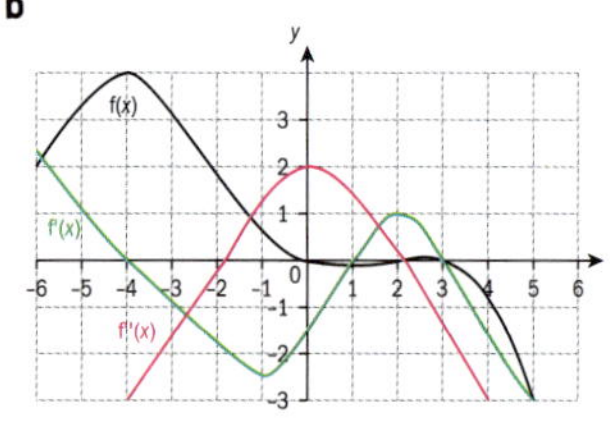

c

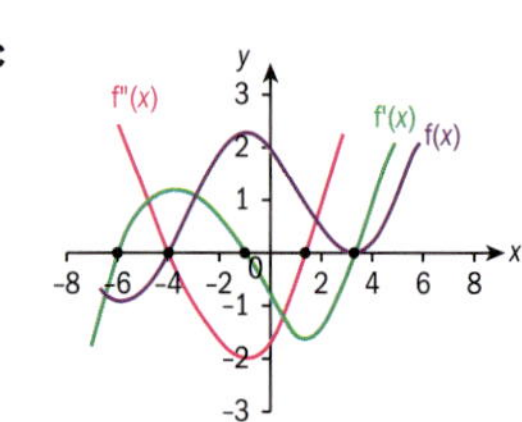

d

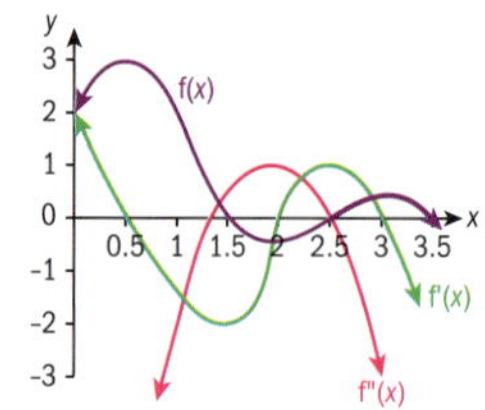

Exercise 4Q

1 **a**

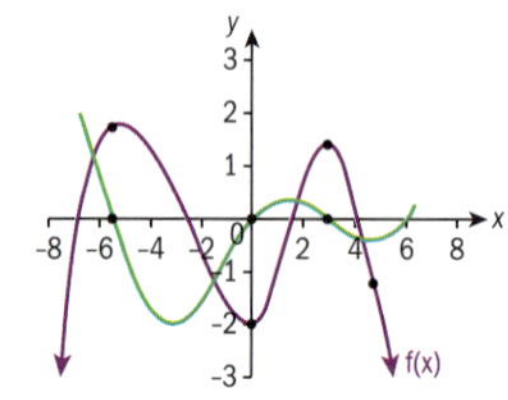

b

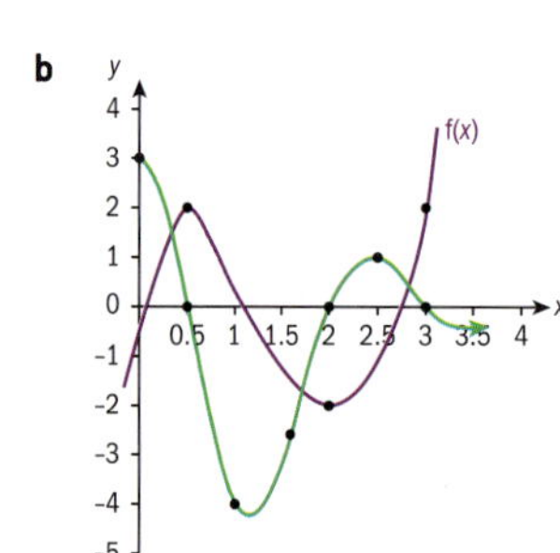

c

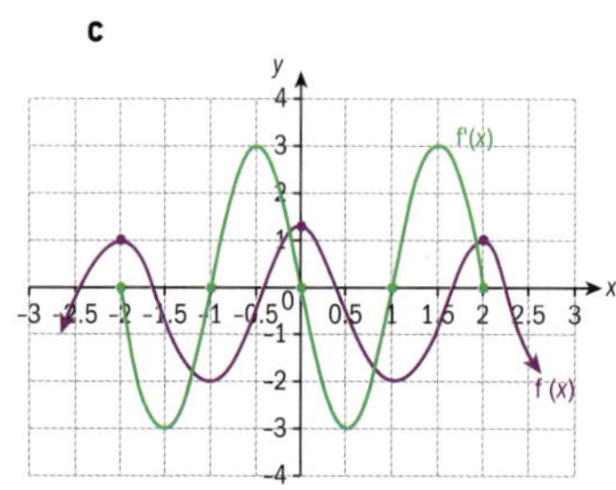

d

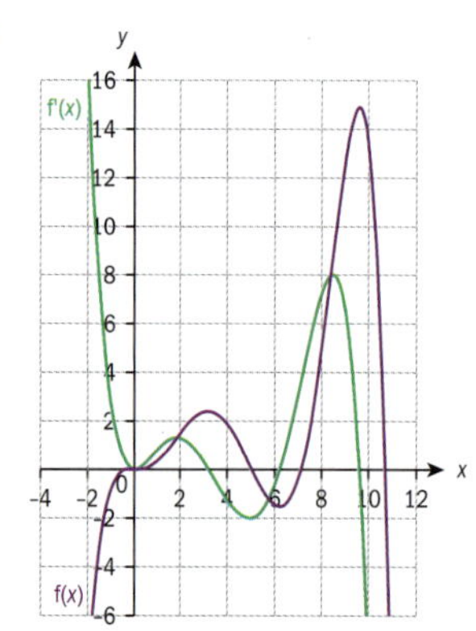

2 **a**

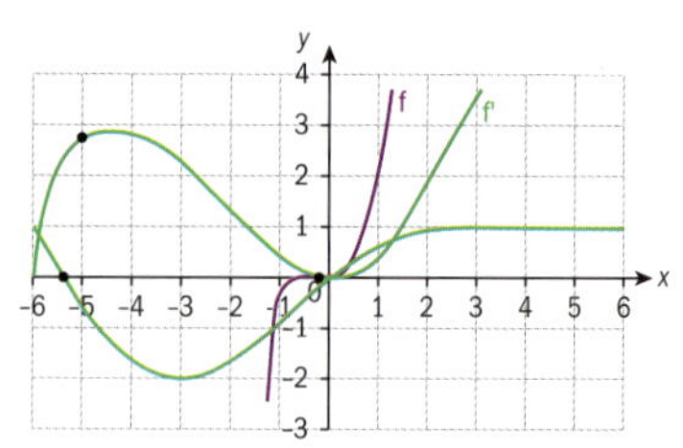

b

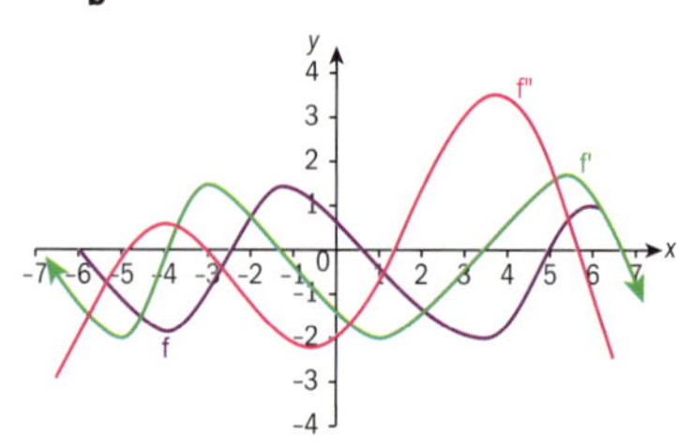

Exercise 4R

1 80,000 m^2

2 $P = 40$ m

3 $r = 1.68$ m; 3.36 m $\times$ 1.68 m

4 $x = 6$

5 $(5)(14)(35) = 2450$

6 $r = h = 1$ m, $V = \pi$ m^3

7 $A = 0.0554$ m^3

8 $V = 1240$ cm^3

9 $\dfrac{\sqrt{5}}{2}$

10 $r = \dfrac{20}{\pi + 2}$

11 16 cm $\times$ 20 cm

12 0.873 km from the point on the shore opposite your boat, or 5.13 km from the town.

Exercise 4S

1 $x = 18$

2 4 cm^3; 4 mmHg

3 **a** $0 \leq x \leq 3500$

b $x = 1000 \pm 500\sqrt{3}$

c $1000 - 500\sqrt{3} < x < 1000 + 500\sqrt{3}$

d 5.5 million euros

4 **a** 272 **b** 367

c Minimizing costs will not necessarily maximize profits, as there are other factors to be taken into account.

5 24.5 m by 40.8 m; 1470 euros

6 300 euros; 220 passengers

Exercise 4T

1 a 96 m s^{-1} b $(3, 256)$

c -128 m s^{-1}

2 a 10 m b 2 s

c The diver hits the water with a velocity of 15 m s^{-1}, and a constant vertical acceleration of -10 m s^{-2}, which is approximately the force of gravity. Since both velocity and acceleration are negative, the diver is speeding up as he/she approaches the water.

3 128 m; 10.2 s

4 a $t = 0, 3, 6, 11$

b i $0 < t < 3$; $6 < t < 11$;

ii $3 < t < 6$

c i $t = 1.5$; ii $t = 4.5$

d $t = 1.5, 4.5$

e i $0 < t < 1.5$; $3 < t < 4.5$; $6 < t < 9$;

ii $1.5 < t < 3$; $4.5 < t < 6$; $9 < t < 11$

5 $v(t) = -3t^2 - 6t + 4$; $a(t) = -6t - 6$; In the interval $0 < t < 0.528$, velocity and acceleration have different signs, hence the particle is slowing down. At $t = 0.528$ it comes to a stop, and for $0.528 < t < 1$, both velocity and acceleration have the same sign, hence the particle is speeding up.

6 a 21 m s^{-1}

b 33 m s^{-1}; 2 m s^{-2}

c Speeding up.

d $t = 0$ s; 6.67 s

7 a $v(t) = 3t^2 - 14t + 11$; $a(t) = 6t - 14$

b $t = 1$; 3.67

c $1 < t < 2.33$; $t > 3.67$

d $t = 1$; 3.67

e 13 m

Exercise 4U

1 a $\frac{dy}{dx} = \frac{3x}{2y}$ b $\frac{dy}{dx} = \frac{4x^3}{3y^2}$

c $\frac{dy}{dx} = \frac{3-4x}{8y-4}$

d $\frac{dy}{dx} = \frac{3y-2x}{-3x+4y}$

e $\frac{dy}{dx} = -\frac{2(x+y)}{2(x+y)+3}$

f $\frac{dy}{dx} = -\frac{2x(x-y)^2+y}{x}$

g $\frac{dy}{dx} = \frac{4x-2\sqrt{2x^2-6y^3}}{18y^2}$

2 $\frac{dy}{dx} = -\frac{y^4}{x^4}$; $y = -x + 2$; $y = x$

3 $\frac{dy}{dx} = -\frac{(y+1)^2}{(x+1)^2}$; $y = -x + 2$; $y = x$

4 $(3, 9)$, $(3, -1)$

5 $\frac{dy}{dx} = \frac{2x-2y-1}{2x-2y+1}$

Exercise 4V

1 a $\frac{dA}{dt} = 2\pi r\frac{dr}{dt}$

b $\frac{dA}{dt} = 2\pi\frac{dr}{dt} + 2\pi\frac{dh}{dt}$

c $\frac{dV}{dt} = \frac{\pi}{3}\left(2rh\frac{dr}{dt} + r^2\frac{dh}{dt}\right)$

d $\frac{dV}{dt} = 4\pi r^2\frac{dr}{dt}$

2 $\frac{l\frac{dl}{dt} + w\frac{dw}{dt} + h\frac{dh}{dt}}{\sqrt{l^2+w^2+h^2}}$

3 a -51 cm s^{-2}

b 0 m s^{-1}

c 2.04 cm s^{-1}

4 0.667 m s^{-1}

5 2 m^2 s^{-1}

6 188.5 cm^2 s^{-1}

7 $-\frac{1}{5}$ cm s^{-1}

8 $\frac{7}{36\pi}$ cm s^{-1}

9 0.424 m per minute

10 $\frac{1}{5\pi}$ cm s^{-1}

Chapter review

1 a No, the left and right hand limits are different.

b 2

c 2.25

d No, a cusp has a vertical tangent.

2 a 4 b $\frac{5}{8}$ c 2

d 2 e 1 f 0

3 No

4 a $a = -1$ b $a = 7$

5 a Converges to $\frac{1}{3}$

b Converges to 0.

c Diverges.

6 Converges to $\frac{3}{2}$

7 For all $n \neq 0$; $1 + n^2$

8 a $x = \pm 3$; $y = 6$

b $x = -3$; $y = 0$

c $x = 0$; $y = x + 1$

9 a $-2(2x-1)^4(3x-2)^5(33x-19)$

b $\frac{3x^2-4x+1}{2(x-1)^{\frac{3}{2}}}$

c $\frac{1}{2}\left(2+\frac{3x^2}{2\sqrt{x^3+1}}\right)\frac{1}{\sqrt{2x+\sqrt{x^3+1}}}$

10 a Vertical: $x = \pm 1$

Horizontal: $y = 0$

b $f(x) = \frac{x}{x^2-1}$

$f(-x) = \frac{-x}{(-x)^2-1} = -\frac{x}{x^2-1} = -f(x)$

c $\frac{dy}{dx} = \frac{(x^2-1)(1)-x(2x)}{(x^2-1)^2}$

$= \frac{-1-x^2}{(x^2-1)^2} = -\frac{1+x^2}{(x^2-1)^2} < 0$

d

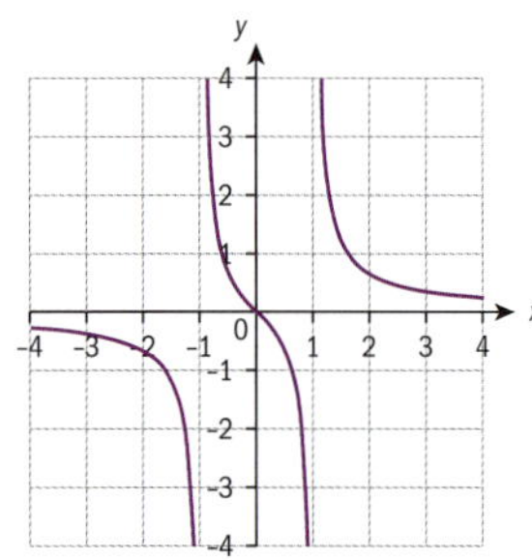

11 a $|OP| = \sqrt{a^2 + (3 - a^2)^2}$

$= \sqrt{a^4 - 5a^2 + 9}$

b $x = \pm\frac{\sqrt{10}}{2}$

12 a $x = -2; x = -1, y = 1$

b $(2,0), (1,0), (0,1)$

c $x = \pm\sqrt{2}$

13 a $x = -1$ **b** $x = 0; y = 0$

c $\left(-2, -\frac{9}{4}\right)$ **d** $]-3, 0[$

e

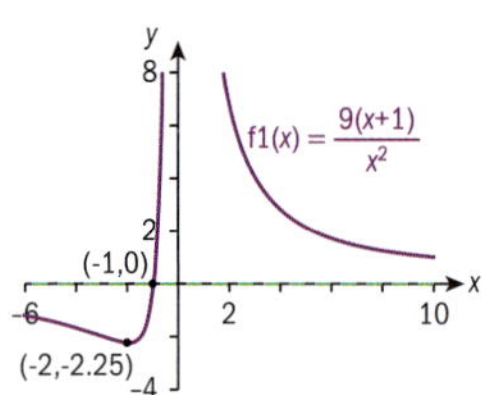

14 a $0, b^2$

b i $x > \frac{b^2}{4}$ **ii** $0 \leq x < \frac{b^2}{4}$

c $y \geq -\frac{b^2}{4}$

d Concave up for all x.

15

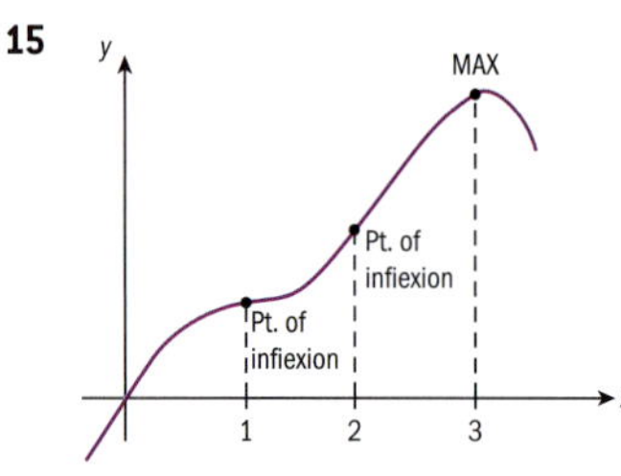

16 -4

17 $-8x + 7y + 1 = 0$

18 $\sqrt{a} + \sqrt{b} = \sqrt{p}$

$$\frac{1}{2\sqrt{x}} + \frac{1}{2\sqrt{y}}\frac{dy}{dx} = 0 \Rightarrow \frac{dy}{dx} = -\sqrt{\frac{y}{x}}$$

$\therefore$ The gradient at (a,b) is $-\sqrt{\frac{b}{a}}$

and the equation of the tangent

is $y - b = -\sqrt{\frac{b}{a}}(x - a)$

$\Rightarrow y = -\sqrt{\frac{b}{a}}x + \sqrt{ab} + b$

The tangent passes through $(0,r)$:

$r = \sqrt{ab} + b$

and $(s,0)$:

$0 = -\sqrt{\frac{b}{a}}s + \sqrt{ab} + b \Rightarrow s = a + \sqrt{ab}$

$\therefore r + s = \sqrt{ab} + b + a + \sqrt{ab}$

$= a + 2\sqrt{ab} + b$

$= \left(\sqrt{a} + \sqrt{b}\right)^2 = \left(\sqrt{p}\right)^2 = p$

as required

19 $2.4\ \text{ms}^{-1}$

20 a $-2\ \text{ms}^{-1}$ **b** $t = 4$

c $0.6\ \text{ms}^{-2}$

d i $t < 4$; **ii** $t > 4$

21 $2.4\pi\ \text{cm s}^{-1}$ or $7.54\ \text{cm s}^{-1}$

22 $6\ \text{cm}^2\ \text{s}^{-1}$

23 $6 - \sqrt{3}$ or 4.27

Exam-style questions

24 a Graphical approach:

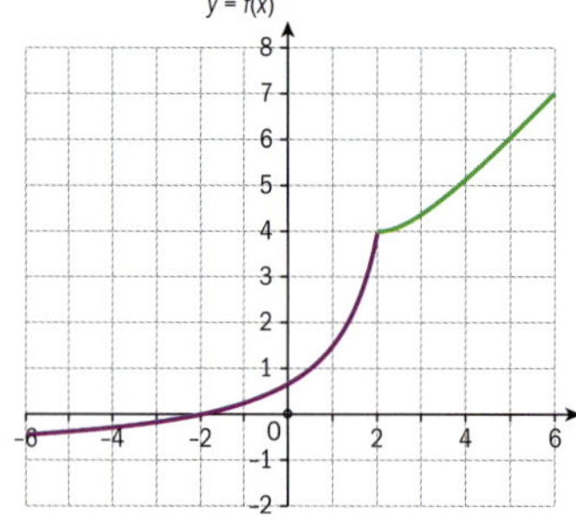

Attempt to draw graph (1 mark)

Each branch correct (2 marks)

Then

$\lim_{x\to 2^-} f(x) = \lim_{x\to 2^+} f(x) = 4$

Hence $\lim_{x\to 2} f(x)$ exists and is equal to 4. (1 mark)

OR: Algebraic approach (note that an algebraic approach will be accepted, but not expected, in examinations)

$\lim_{x\to 2^-} f(x) = \lim_{x\to 2^-} -\frac{x^2 - 4}{x^2 - 5x + 6}$

$= \lim_{x\to 2^-} -\frac{(x-2)(x+2)}{(x-2)(x-3)}$ (1 mark)

$= \lim_{x\to 2^-} \frac{-(x+2)}{x-3} = 4$ (1 mark)

$\lim_{x\to 2^+} f(x) = \lim_{x\to 2^+} e^{2-x}$

$+ x + 1 = 4$ (1 mark)

Then

$\lim_{x\to 2^-} f(x) = \lim_{x\to 2^+} f(x) = 4$

Hence $\lim_{x\to 2} f(x)$ exists and is equal to 4. (1 mark)

b $a^2 + 3a + 6 = 4$ (1 mark)

$(a+1)(a+2) = 0$ (1 mark)

$a = -1, a = -2$ (1 mark)

25 a $x^2 + x - 2 \neq 0 \Rightarrow x \neq 1$, $x \neq -2$ (2 marks)

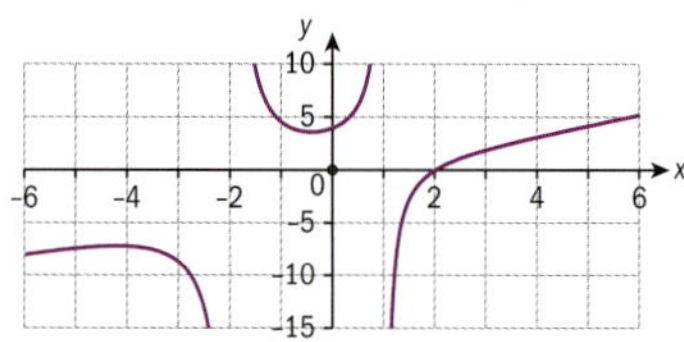

(Shape of each branch correct gains 1 mark) (3 marks)

c i Vertical asymptotes: $x = 1$ and $x = -2$ (2 marks)

ii Using long division, $f(x) = x - 1 + \frac{3x - 10}{x^2 + x - 2}$ (1 mark)

As $x \to \infty$, $f(x) \to x - 1$ which is a slant asymptote. (1 mark)

26 a $g'(x) = \dfrac{\frac{1}{2\sqrt{x}}\left(x^2+1\right)-\sqrt{x}\cdot 2x}{\left(x^2+1\right)^2}$ (2 marks)

$= \dfrac{1-3x^2}{2\sqrt{x}\left(x^2+1\right)^2}$ (1 mark)

b $g(1) = \dfrac{1}{2}$ (1 mark)

$g'(1) = -\dfrac{1}{4}$ (1 mark)

Equation of normal:

$y - \dfrac{1}{2} = 4(x-1)$ or

$y = 4x - 3\dfrac{1}{2}$ (1 mark)

c g' is not defined at $x = 0$ because a derivative is not defined at the end point of a closed interval. (1 mark)

Therefore, there is no tangent to the graph of g at $x = 0$. (1 mark)

27 a

$$2x = \frac{\left(1+\frac{dy}{dx}\right)(x-y)-\left(1-\frac{dy}{dx}\right)(x+y)}{(x-y)^2}$$

(3 marks)

Make $\dfrac{dy}{dx}$ the subject (1 mark)

$2x(x-y)^2 = 2x\dfrac{dy}{dx} - 2y$

$\dfrac{dy}{dx} = (x-y)^2 + \dfrac{y}{x}$ (1 mark)

b $y = 0 \Rightarrow x^2 = 1 \Rightarrow x = \pm 1$ (1 mark)

$\dfrac{dy}{dx} = 1$ (1 mark)

$y = x \mp 1$ (2 marks)

28 a i $\dfrac{101.1-98.5}{2002-2000} = 1.3$ (2 marks)

ii $\dfrac{102.3-101.1}{2004-2002} = 0.6$ (1 mark)

b The average annual profit between 2000 and 2002 was almost double the average annual profit between 2002 and 2004. (2 marks)

---

29 a $f'(x) = \lim\limits_{h\to 0}\dfrac{\left(2(x+h)^2+3(x+h)-4\right)-\left(2x^2+3x-4\right)}{h}$ (2 marks)

$= \lim\limits_{h\to 0}\dfrac{\left(2x^2+4hx+2h^2+3x+3h-4\right)-\left(2x^2+3x-4\right)}{h}$ (1 mark)

$= \lim\limits_{h\to 0}\dfrac{4hx+2h^2+3h}{h}$ (1 mark)

$= \lim\limits_{h\to 0}(4x+2h+3) = 4x+3$ (1 mark)

b $f(1) = 2 + 3 - 4 = 1$ (1 mark)

$f'(1) = 4 + 3 = 7$ (1 mark)

Equation of tangent: $y - 1 = 7(x - 1)$ (or $y = 7x - 6$) (1 mark)

---

30 a $h(4) = 370$ and $h(5) = 438$ (3 s.f.) (2 marks)

b $v(t) = h'(t) = 112 - 9.8t$ (2 marks)

c $v(t) = 0 \Rightarrow 112 - 9.8t = 0$ (1 mark)

$t = 11.4$ (3 *s.f.*) (1 mark)

d double x-coordinate of maximum, or determine zero (1 mark)

22.9 (3 *s.f.*) (1 mark)

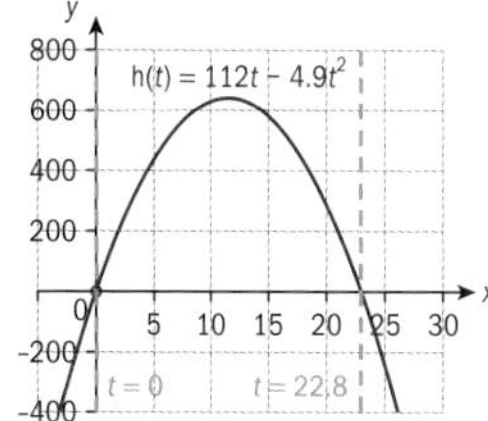

(Shape: 1 mark; Domain: 1 mark; Maximum: 1 mark)

f $v(22.9) = -112\ \text{ms}^{-1}$ (2 marks)

g $a(t) = v'(t) = -9.8$ which is constant (2 marks)

31 a i $\left(\dfrac{f}{g}\right)'(2) = \dfrac{f'(2)g(2)-f(2)g'(2)}{\left(g(2)\right)^2}$ (1 mark)

$= \dfrac{10\times 4 - 9\times\left(-\frac{4}{3}\right)}{4^2}$ (1 mark)

$= \dfrac{52}{16}\left(=\dfrac{13}{4} = 3.25\right)$ (1 mark)

ii $(g\circ f)'(1) = g'(f(1))f'(1)$
$= g'(2)f'(1)$ (1 mark)

$= -\dfrac{4}{3}\times 4 = -\dfrac{16}{3}$ (1 mark)

b i False (1 mark)

as derivative changes sign. (1 mark)

ii False (1 mark)

as the derivatives at these points are not negative reciprocals. (1 mark)

32 **a** Let $d(x)$ be the total length of the pipeline.

$d(x)=\sqrt{x^2+75^2}+100-x$ (2 marks)

b Let $c(x)$ be proportional to the construction costs of the pipeline.

$c(x)=3\sqrt{x^2+75^2}+100-x$ (1 mark)

$\frac{\mathrm{d}c}{\mathrm{d}x}=\frac{3x}{\sqrt{x^2+75^2}}-1$ (2 marks)

$\frac{\mathrm{d}c}{\mathrm{d}x}=0\Rightarrow\frac{3x}{\sqrt{x^2+75^2}}=1$ (1 mark)

Solve equation

$9x^2 = x^2 + 75^2$ (2 marks)

$x=\frac{75}{4}\sqrt{2}$ (1 mark)

c $d(x)=\sqrt{\left(\frac{75}{4}\sqrt{2}\right)^2+75^2}$

$+100-\frac{75}{4}\sqrt{2}$ (2 marks)

$=\sqrt{75^2\left(\frac{2}{16}+1\right)}+100-\frac{75}{4}\sqrt{2}$ (1 mark)

$=75\sqrt{\frac{9}{8}}+100-\frac{75}{4}\sqrt{2}$

$=3\times\frac{75}{4}\sqrt{2}+100-\frac{75}{4}\sqrt{2}$ (1 mark)

$=\frac{75}{2}\sqrt{2}+100$

33 **a** Vertical Asymptote: $x=a$ (1 mark)

Horizontal Asymptote: $y=a$ (1 mark)

b $f'(x)=\frac{a\cdot(x-a)-1\cdot(ax-4)}{(x-a)^2}$

$=\frac{4-a^2}{(x-a)^2}.$

(3 marks)

c $f'(x) = 0$ for turning points (1 mark)

$\frac{4-a^2}{(x-a)^2}=0\Rightarrow$

$a^2=4\Rightarrow a=2$ (1 mark)

For $a = 2$,

$f(x)=\frac{2x-4}{x-2}=\frac{2(x-2)}{x-2}=2$

so the function is constant, and there are no turning points. (1 mark)

For $a \neq 2$, $f'(x) \neq 0$, so the function has no max/min. (1 mark)

d $f'(1)=\frac{4-a^2}{(1-a)^2}$ (1 mark)

gradient of normal is

$m=\frac{(1-a)^2}{a^2-4}$ (1 mark)

$f(1)=\frac{a-4}{1-a}$ (1 mark)

$y-\frac{a-4}{1-a}=\frac{(1-a)^2}{a^2-4}(x-1)$

(or equivalent) (1 mark)

e Asymptotes intersect at (a, a). Substitute (a, a) into normal equation. (1 mark)

$a-\frac{a-4}{1-a}=\frac{(1-a)^2}{a^2-4}(a-1)$

(or equivalent) (1 mark)

Simplify (1 mark)

$(a^2-4)^2=(a-1)^4$

$4a^3-14a^2+4a+15=0$ (1 mark)

f From GDC (1 mark)

$a = 2.5$ or $a = 1.82$ (2 marks)

(For part (f), award 2 marks only if negative root $a = -0.823$ is included)

Chapter 5

Skills check

1 **a** Mean = 728.5, mode = 720, median = 705

b Range = 330, $Q_1 = 690$, $Q_3 = 720$, IQR = 30

2 Mean = 19.21, data is bimodal with modal classes $15 < x \le 20$ and $30 < x \le 35$, the median lies in the $15 \le x \le 20$ interval.

Exercise 5A

1 **a** All celery sticks grown in a certain US state

b Each celery stick

c A list of all celery sticks from the state

d The length of the celery stick

e The positive real numbers

2 **a** All ball bearings manufactured by a company

b Each ball bearing

c A list of all ball bearings enumerated

d The weight of the ball bearing

e The positive real numbers

3 **a** All 1 litre soda bottles from a soft drink factory

b Each 1 litre soda bottle

c All soda bottles enumerated in a list

d The volume of the 1 litre soda bottle

e The natural numbers

4 **a** All crates of 50 oranges

b Each crate of 50 oranges

c An enumerated list of all crates

d The weight of a crate of 50 oranges

e The positive real numbers

Exercise 5B

1 List and enumerate all books, generate a random number x then take books x, $x + 50$, ...

2 a Generate a random number x and then sample x bases from each region.

b Individual response

3 a The number of samples of three from 0, 1, 2, 3, 4 is equal to

$$\binom{5}{3} = \frac{5!}{3!(5-3)!} = \frac{120}{6 \times 2} = 10$$

b (0, 1, 2), (0, 1, 3), (0, 1, 4), (0, 2, 3), (0, 2, 4), (0, 3, 4), (1, 2, 3),(1, 2, 4), (1, 3, 4), (2, 3, 4)

c 1, 1.3333, 1.6667. 1.6667, 2, 2.3333, 2.6667, 3

d Mean of population: $\frac{0+1+2+3+4}{5} = \frac{10}{5} = 2$

Mean of sample means:

$$\frac{1+1.3333+1.6667+1.6667+2+2.3333+2+2.3333+2.6667+3}{10}$$

$$= \frac{20}{10} = 2$$

4 Variable = whether the envelope is sealed correctly, sample = batch of selected envelopes, population = all the envelopes, discrete

5 It is not possible to wait 4000 years to see if they will last that long.

6 Stratified sampling

7 Pick 12.5 students from each grade (13 from two and 11 from another two).

Exercise 5C

1 a $0.5 < x \le 1.5$, $1.5 < x \le 2.5$, $2.5 < x \le 3.5$, $3.5 < x \le 4.5$, $4.5 < x \le 5.5$, $5.5 < x \le 6.5$

b

Number of people	Frequency	Interval on histogram
1	8	$0.5 < x \le 1.5$
2	11	$1.5 < x \le 2.5$
3	6	$2.5 < x \le 3.5$
4	4	$3.5 < x \le 4.5$
5	2	$4.5 < x \le 5.5$
6	2	$5.5 < x \le 6.5$

c

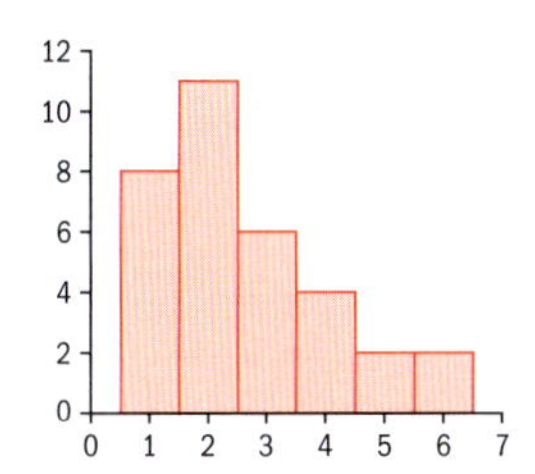

2

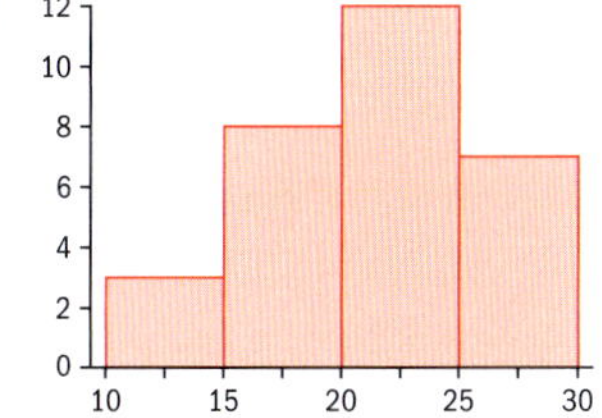

3

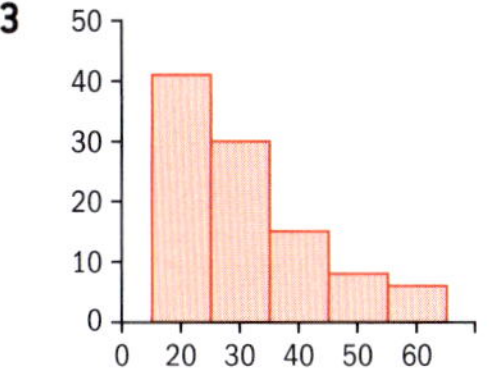

4 a Continuous

b It may show the distribution of lengths of ants in mm

5 a

Hours	Days
4	4
5	5
6	9
7	8
8	4

b 6.1 hours

Exercise 5D

1

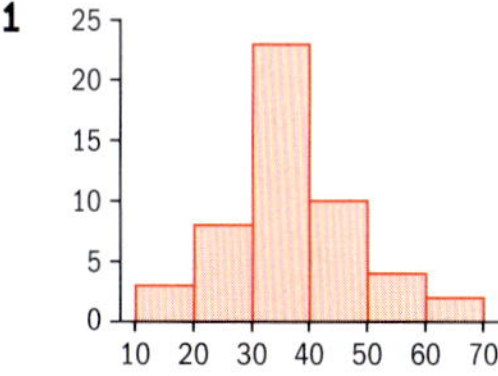

Shape: unimodal, most koalas had a mass of $35 < x \le 40$ kg; centre: midpoint in $35 < x \le 40$ class; spread: mass varies from 17 kg to 61 kg

2 a Each bar represents one day of the week and summarises the data clearly.

b

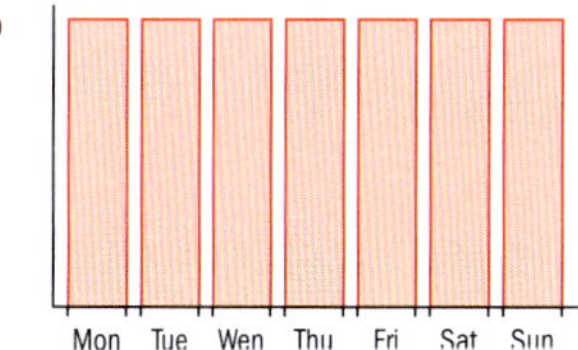

3 a Necessary to compare distributions of two samples from different populations.

b

Time spent per day	Male relative frequency	Female relative frequency
$0 \le x < 15$	0.1667	0.125
$15 \le x < 30$	0.2667	0.1563
$30 \le x < 45$	0.3333	0.2188
$45 \le x < 60$	0.1667	0.4375
$60 \le x < 75$	0.06667	0.0625

c

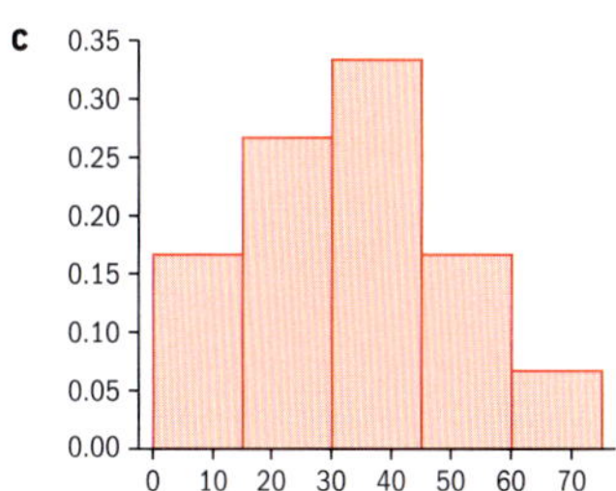

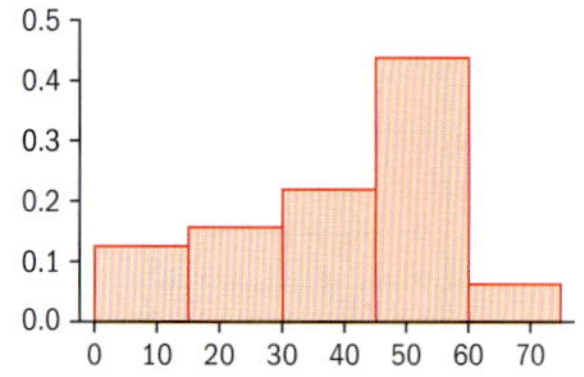

d Male: symmetric unimodal; female: right distorted unimodal

e On average, females spent more time per day on the phone than men.

4 **a** 45 **b** $166.78

c The data is left skewed

5 $0.25 + 0.1875 + 0.125$
$= 0.5625$, 32×0.5625
$= 18$ items

6 **a** Skewed, bimodal, contains an outlier

b Skewed, multimodal, no outliers

c Symmetric, unimodal, no outliers

Exercise 5E

1 **a**

$$\frac{0\times5+1\times10+2\times6+3\times3+4\times1}{25}$$

$$=\frac{35}{25}=1.4$$

b On average, women from Australia have more children.

2 **a** 41.4706

b

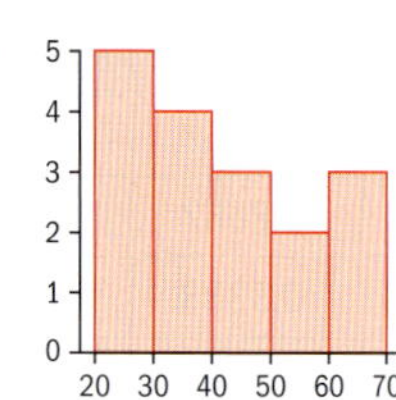

c Left skewed

3 **a** Necessary to compare distributions of two samples of different sizes from different populations

b

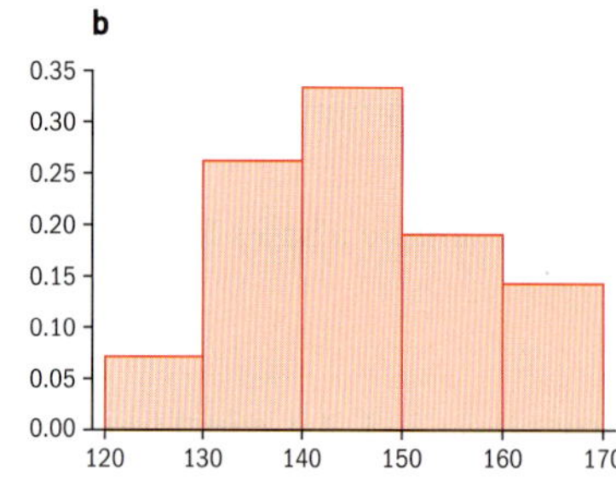

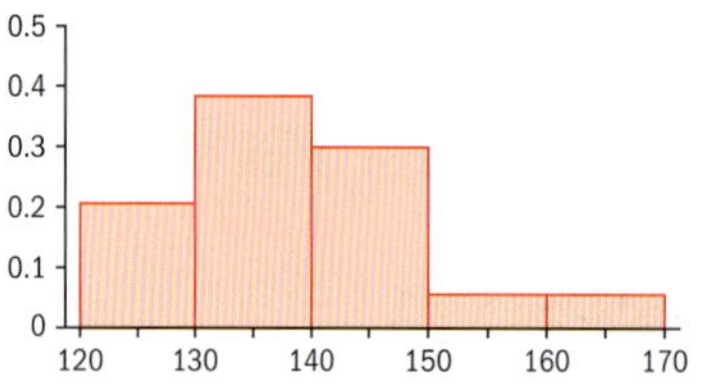

c On average, students from Peru are shorter.

4 **a** A standard histogram

b

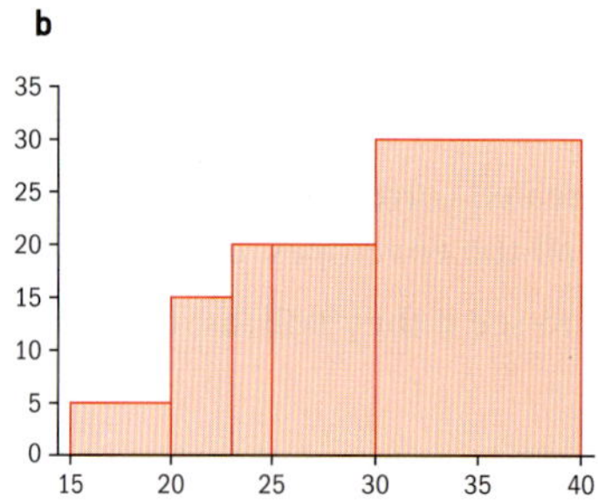

c Mean = 27.6667 years, modal class: $30 < A \le 40$

Exercise 5F

1 **a** Mean = 4.4, standard deviation = 1.3565

b Mean = 24.5 kg, standard deviation = 2.92968 kg

c Mean = 4, standard deviation = 0.75591

d Mean = 13.2941, standard deviation = 6.05632

2 Mean = 28.15, standard deviation = 3.83764

3 **a** Mean = 200.5 g, standard deviation = 2.6551 g

4 **a** Mean = 9.5, standard deviation = 2.48516

b Yes, because mean amount of lead per litre is within 1 standard deviation of the level deemed dangerous.

Exercise 5G

1 $s_{n-1} = 4.35762$

2 Mean = 460.16 kg, $s_{n-1} = 18.72538$ kg

3 Mean = 1550, $s_{n-1} = 39.5285$

4 **a**

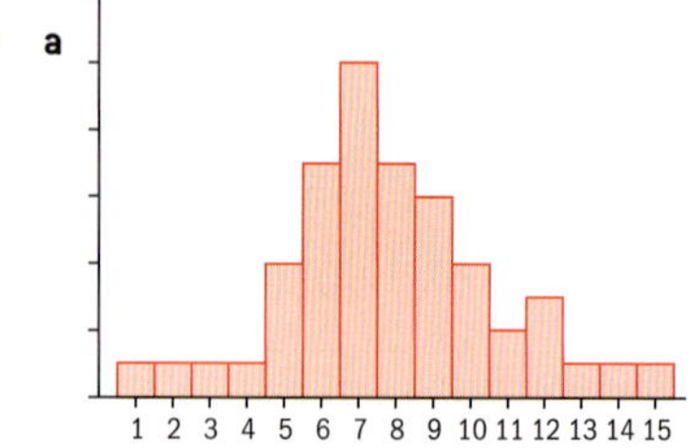

b Symmetric

c Mean = 7.84, $s_{n-1} = 2.8454$

d Individual response

Exercise 5H

1 **a** 80 **b** $a = 55$, $b = 75$

2 **a** $k + 1$ **b** $k - 2$

3 **a** 63 **b** **i** 87 **ii** 73

4 **a** **i** 1.18 m **ii** 0.09 m

b

Class	Frequency
$1.00 \le h < 1.05$	5
$1.05 \le h < 1.10$	8
$1.10 \le h < 1.15$	14
$1.15 \le h < 1.20$	24
$1.20 \le h < 1.25$	18
$1.25 \le h < 1.30$	11

c **i** 0.34 **ii** 0.69

5 **a** **i** Mean =

$$\frac{a+2a+3a+\ldots+na}{n}$$

$$=\frac{a(1+2+3+\ldots+n)}{n}$$

$$=\frac{a}{n}\frac{n(n+1)}{2}$$

$$=\frac{a(n+1)}{2}$$

ii 8

b **i** $M = \dfrac{m(0)+n(1)}{m+n}$

$$=\frac{n}{m+n}$$

$$S=\sqrt{\frac{n}{m+n}-\left(\frac{n}{m+n}\right)^2}$$

$= \frac{\sqrt{mn}}{m+n}$

ii 0.5

Exercise 5I

1 a A b B c C

d E e D

2 a i

b i

c i

ii 65 kg iii 47 s

iv The graphs giving the most accurate predictions are the ones where the data is close to the line of best fit. Graphs B and C are better than graph A.

3 a

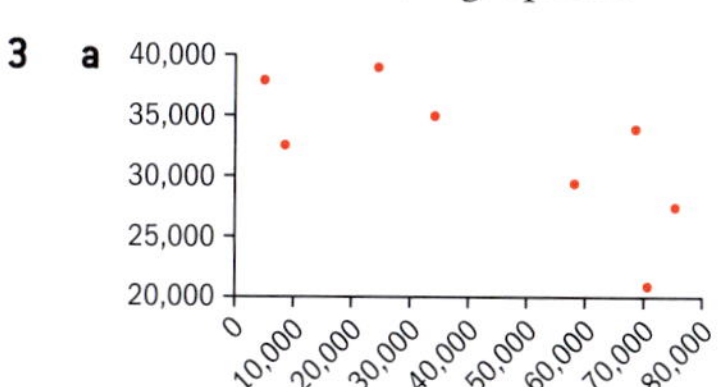

b Moderate negative correlation, $31 031.60

c The make of car or price when new

4 a

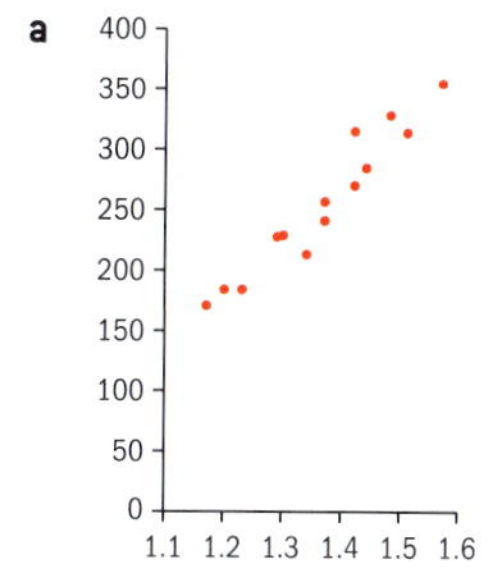

b $y = 473.757x - 390.821$

c Strong positive correlation

d 262.964 kg

Chapter review

1 Pick 15 students at random from each MYP class and 15 students at random from the DP group.

2 a Pick 12 students at random from the whole medical school.

b Pick 1.71 students from each year group at random.

c Pick 2.4 students at random from Year 1 and 1.6 students at random from each of the other year groups.

d Ask for volunteers and pick the first 12.

3 Make sure the questions are clear, make sure no leading questions, ensure possible answers are applicable to everybody and no options are missed

4 a Number of pages

b Height of page

5 a Qualitative, continuous

b Quantitative, continuous

c Quantitative, discrete

d Quantitative, continuous

6

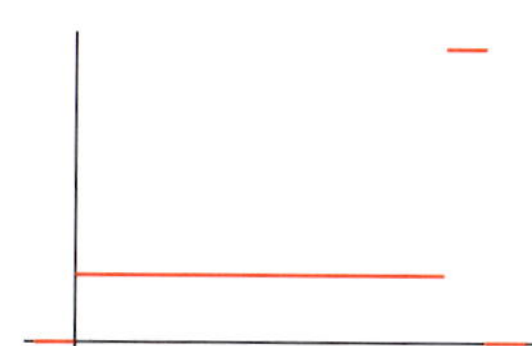

7 a Mean = 35.3 , median interval lies within 31–40, the modal group is 21–30

b 30.2%

Number of words per sentence	Number of sentences	Cumulative frequency
1–10	13	13
11–20	16	29
21–30	146	175
31–40	139	314
41–50	84	398
51–60	32	430
61–70	20	450

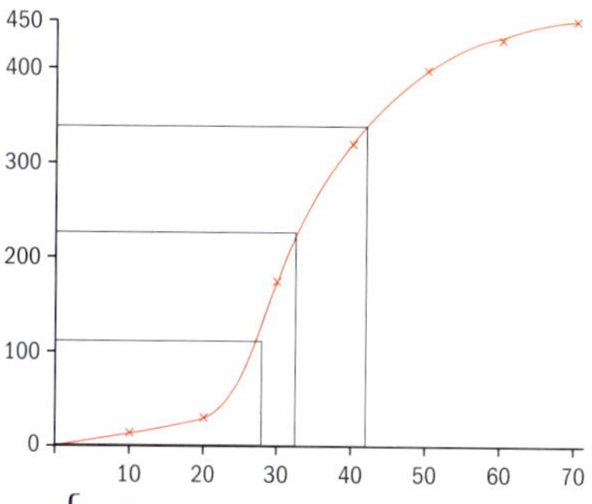

d min = 1
$Q_1 = 28.5$ (approx)
$Q_2 = 33$
$Q_3 = 42$
max = 70

f The number of sentences is uniformly distributed within the interval.

g Individual responses.

8 a

Number of siblings	Frequency
0	14
1	28
2	11
3	5
4	0
5	2

Left skewed

b Mean $= \frac{5}{4}$, standard deviation = 1.105

c

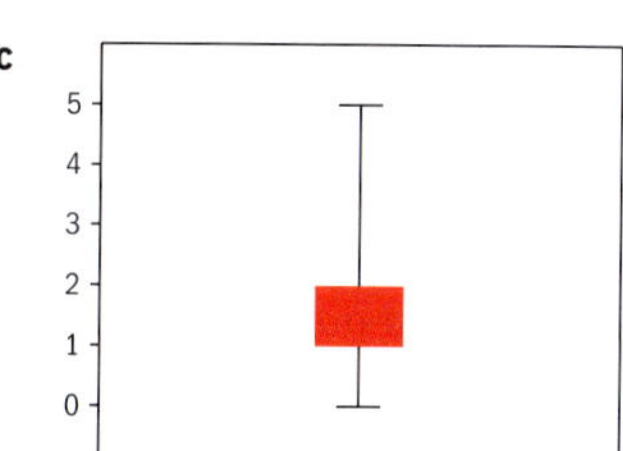

d Notice that the mean of all students is equal to the mean of the original 60 plus the mean of the new 32.

$x = 5$

9 **a**

y^2	0.36	0.2025	0.64	0.7225	1.96	2.7225	5.76	8.1225

b

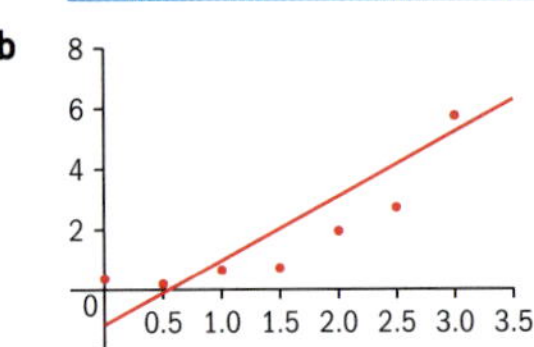

c $y^2 = 2.13357x - 1.1725 \qquad y = \sqrt{2.13357x - 1.1725}$

10 5

Exam-style questions

11 **a** As the mode is 5 there must be at least another 5. (1 mark)

So we have 1, 3, 5, 5, 6 with another number to be placed in order. (1 mark)

The median will be the average of the 3rd and 4th pieces of data. (1 mark)

For this to be 4.5 the missing piece of data must be a 4.

Thus $a = 5$, $b = 4$ (2 marks)

b $\bar{x} = \dfrac{1+3+4+5+5+6}{6}$

$= \dfrac{24}{6} = 4$ (2 marks)

12 **a** $\dfrac{\sum x}{10} = 70 \Rightarrow \sum x = 700$ (1 mark)

Let Steve's mass be s.

$\dfrac{\sum x + s}{11} = 72$ (1 mark)

$700 + s = 792$ (1 mark)

So $s = 92$ kg (1 mark)

b IQR = 10 (1 mark)

$76 + 1.5 \times IQR = 76 + 15$
$= 91$ (1 mark)

So Steve's mass of 92 is greater than $1.5 \times IQR$, so is an outlier. (1 mark)

13 **a** 200 (1 mark)

b 35 (1 mark)

c Using midpoints 5, 15, 25... as estimates for each interval, (1 mark)

i estimate for mean is 22.25 (2 marks)

ii estimate for standard deviation is 11.6 (3 s.f.). (2 marks)

d Median is approximately the 100th piece of data which lies in the interval $20 < h \le 30$. (1 mark)

Will be 15 pieces of data into this interval

Estimate is

$20 + \dfrac{15}{50} \times 10 = 23$ (2 marks)

14 **a** **i** 7.5 (1 mark)

ii 6.125 (2 marks)

b **i** 6 (1 mark)

ii 6.9 (2 marks)

c Sally's had the greater median (1 mark)

d Rob's had the greater mean (1 mark)

15 **a**

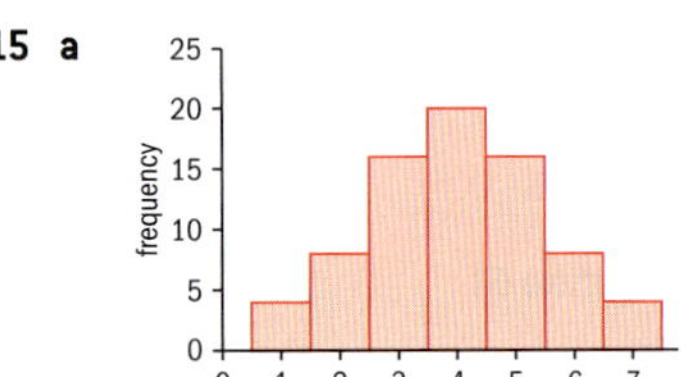

(1 for scale, 1 for correctly drawn graph)

b **i** 4 **ii** 4

iii 4 (3 marks)

c The values of the median and the mean are the same due to the symmetry of the bar chart. (2 marks)

16 **a** $100 = 70m + c$

$140 = 100m + c$

$40 = 30m \quad m = \dfrac{4}{3}$

$c = \dfrac{20}{3}$ (3 marks)

b Positive (1 mark)

c Line goes through $(\bar{x}, \bar{y})$ (1 mark)

$\bar{y} = \dfrac{4}{3} \times 90 + \dfrac{20}{3} = \dfrac{380}{3}$ (2 marks)

d Estimate is $\dfrac{4}{3} \times 60 + \dfrac{20}{3} = \dfrac{260}{3}$ (2 marks)

17 **a**

x	13	14	15	16	16	17	18	18	19	19
y	2	0	3	1	4	1	1	2	1	2

(5 correct: 2 marks; all correct: 3 marks)

b $r = -0.0695\ (3sf)$ (2 marks)

c Very weak (negative) correlation so line of best fit is almost meaningless. (1 mark)

It would be extrapolation to use this data to predict for a 25-year-old. (1 mark)

18 **a** **i** No change; $r = 0.87$ (1 mark)

ii No change; 15 (1 mark)

iii The scatter diagram has just been translated up by 5 and to the left by 4, so the PMCC and the gradient of y on x line of best fit are unchanged. (1 mark)

iv Strong, positive (2 marks)

b i No change; $r = 0.87$ (1 mark)

ii $2 \times 15 = 30$ (1 mark)

iii The scatter diagram has been stretched vertically by scale factor 2, so PMCC remains unchanged, but gradient of y on x line of best fit is doubled. (1 mark)

c i $r = -0.87$ (1 mark)

ii $\frac{15}{-3} = -5$ (1 mark)

iii The scatter diagram has been stretched horizontally by a factor of 3 and then reflected in the y-axis, so gradient becomes −5, but PMCC is unchanged. (2 marks)

iv Strong, negative (2 marks)

19 a i 0.849 (3sf) (2 marks)

ii Strong, positive (2 marks)

iii $y = 0.937x + 0.242$ (2 marks)

b i 0.267 (3sf) (2 marks)

ii Weak, positive (2 marks)

iii The Pearson product moment correlation coefficient is too small to make the line of best fit particularly meaningful when making predictions. (1 mark)

20 a $r = 0.979$ (3sf) (2 marks)

b Strong, positive (2 marks)

c i $y = 1.23x - 21.3$ (2 marks)

ii $x = 0.776y + 20.8$ (2 marks)

d $1.23 \times 105 - 21.3 = 108$ (1 mark)

e $0.776 \times 95 + 20.8 = 95$ (1 mark)

f It is extrapolation (1 mark)

Chapter 6

Skills check

1 250 cm^2

2 All three angles are identical.

Exercise 6A

1 a i (3, 0, 0) ii (3, 4, 0)

iii (3, 0, 2) iv (3, 4, 2)

b (1.5, 2, 1) c 5.4

2 a (0.5, 1.5, 3)

b (−3, 3, 7) c (0.5, −0.5, −6)

d (−1.85, 0.15, 10)

3 a 4.47 b 6.56

c 10.2 d 4.47

4 a

b 46.7° c 56.3°

5 17.1°

6 a 9.80 cm b 54.7°

c 35.3° d 24.1°

7 a 11.3 cm b 5.7 cm

c 9.8 cm d 54.7°

Exercise 6B

1 a $V = 576$ cm^3, $A = 466$ cm^2

b $V = 40$ cm^3, $A = 77.6$ cm^2

c $V = 84.8$ cm^3, $A = 117.8$ cm^2

d $V = 3.14$ cm^3, $A = 13.1$ cm^2

e $V = 2310$ cm^3, $A = 845$ cm^2

f $V = 113.1$ cm^3, $A = 113.1$ cm^2

2 46 cm^3 3 359 cm^3

4 454 cm^3 5 1.44 cm

6 3440 cm^3

Exercise 6C

1 a $\frac{\pi}{4}$ b $\frac{\pi}{3}$ c $\frac{3\pi}{2}$ d 2π

e $\frac{\pi}{10}$ f $\frac{5\pi}{4}$ g $\frac{4\pi}{9}$

h $\frac{10\pi}{9}$ i $\frac{2\pi}{3}$ j $\frac{3\pi}{4}$

2 a 30° b 18° c 150°

d 540° e 63° f 144°

g 315° h 280° i 300°

j 585°

3 a 0.174 b 0.698

c 0.436 d 5.24

e 1.92 f 1.31

g 1.48 h 0.223

i 0.654 j 0.01

4 a 57.3° b 115°

c 36.1° d 80.8°

e 88.8° f 172°

g 20.6° h 73.3°

i 0.573° j 123°

5 a $\frac{(\pi - 1)}{2}$ b $\pi - 2$

Exercise 6D

1 a i 22 cm ii 28.3 m

iii 7.85 m iv 73.3 cm

b i 154 cm^2 ii 170 m^2

iii 11.8 m^2 iv 550 cm^2

2 8.48 cm

3 a $\frac{\pi}{2}$ b 42.8 cm

4 25.1

5 62.8 m

6 $p = 2r + l = 2r + r\theta$,

$r(2 + \theta) = p$, $r = \frac{p}{2 + \theta}$

7 $A_{semi} = 78.54$ cm^2, $p_{semi} = 36.35$ cm

Exercise 6E

1 a Quadrant II, $\theta = 35°$

b Quadrant II, $\theta = \frac{\pi}{4}$

c Quadrant IV, $\theta = \frac{\pi}{2}$

d Quadrant IV, $\theta = \frac{\pi}{6}$

e Quadrant III, $\theta = 24°$

f Quadrant IV, $\theta = 22°$

g Quadrant II, $\theta = \frac{\pi}{3}$

2 **A** $-0.577, \frac{5\pi}{6}$

B $-0.5, -0.577$

C $-0.9204, 2.74$

D $\frac{\sqrt{2}}{2}\ \frac{\pi}{4}$

E $\frac{\sqrt{2}}{2}, \frac{3\pi}{4}$

3 **a** 36.9°, 143.1°

b θ is undefined

c 293°, 113°

4 $\sin\theta = -\frac{4}{5}, \tan\theta = -\frac{4}{3}$

Exercise 6F

1 1.104 and 5.18

2 5.77 and 2.63

3 0.23 and 2.91

4 Undefined

5 $\frac{\pi}{2}, \pi, \frac{3\pi}{2}$

6 $\frac{\pi}{3}, \frac{5\pi}{3}, 0, 2\pi$

7 $\frac{\pi}{6}, \frac{5\pi}{6}, \frac{7\pi}{6}, \frac{11\pi}{6}$

8 3.74, 2.54

9 $2\theta = 5.856, 2.714, 8.997, 12.139$ where the first two angles are the angles for the negative tangent, and the last two are an added rotation to them $(+2\pi)$.

Exercise 6G

1 **a** $p = 4.44, Q = 115.75°, R = 34.25°$

b $y = 6.67, X = 36.7°, Z = 48.3°$

c $A = 30.8°, B = 125.1°$ and $C = 24.1°$

2 $A = 89.2015°$

3 26.38°

4 $x = \frac{8}{3}$, ABC = 18.43°, BCA = 101.57°

5 DAB = BCD and CDA = CBA.
2DAB + 2CDA = 360°,
CDA = 180° − DAB

Use the cosine rule to get relationships

$\cos \text{DAB} = \frac{b^2 + a^2 - q^2}{2ba}$ and

$\cos \text{CDA} = \frac{a^2 + b^2 - p^2}{2ab}$

Note that $\cos \text{CDA} = \cos(180° - DAB) = -\cos \text{DAB}$, so

$\frac{b^2 + a^2 - q^2}{2ba} = \frac{-a^2 - b^2 + p^2}{2ab}$,

which rearranges to
$p^2 + q^2 = 2(b^2 + a^2)$

Exercise 6H

1 **a** $b = 16.4$ cm, $C = 25°$, $c = 8.45$ cm

b $Q = 95°, p = 6.36$ cm, $q = 9.86$ cm

c $A = 30°, B = 110°$, $b = 13.16$ cm

2 38 s

3 300 m

4 MC = 20.5 m, MB = 8.91 m, MA = 8 m

5 ACB = 70°, BAC = 55° or ACB = 110°, BAC = 15°

Exercise 6I

1 197.37 cm^2

2 1063 cm^2

3 19.7 cm^2

4 $CAD = 156.1°, A_{CAD} = 11.8$ m^2

5 **a** $A_{POQ} = \frac{\sqrt{3}}{4}r^2$ and $A_{ROS} = \frac{1}{4}r^2$

b $r^2\left(\frac{\pi}{3} - \frac{\sqrt{3}}{4}\right)$

c $r^2\left(\frac{\pi}{12} - \frac{1}{4}\right)$

d $A_{shaded} = A_{minorPQ} - A_{minorRS}$

$= r^2\left(\frac{\pi}{3} - \frac{\sqrt{3}}{4}\right) - r^2\left(\frac{\pi}{12} - \frac{1}{4}\right)$

$= \frac{r^2}{4}(\pi + 1 - \sqrt{3})$

6 $d = 154$ nautical miles, bearing = 115°

Exercise 6J

1 $\frac{\sqrt{3}}{3}$

2 0

3 **a** $\sin\theta$ **b** $\csc\theta$

c $\sec\theta$ **d** $\tan\theta$

e $\csc\theta$ **f** $\sec\theta$

4 $\cos\theta = \frac{-12}{13}, \cot\theta = -\frac{12}{5}$

5 $\cos\theta = -\frac{4}{5}, \tan\theta = \frac{3}{4}$,

$\sin\theta = -\frac{3}{5}$

Exercise 6K

1 **a** $\cos\theta$ **b** $\tan\theta$

c $\tan\theta$ **d** $\cot\theta$

e $\sec\theta$ **f** $\sin\theta$

2 **a** $a^2\cos^2\theta$ **b** $b^2\frac{\cos\theta}{\sin^2\theta}$

c $\sin\theta$ **d** $\sin\theta$

3 **a** $\frac{\pi}{3}, \frac{5\pi}{3}, 0, 2\pi$ **b** $\frac{\pi}{3}, \frac{5\pi}{3}$

c $\frac{3\pi}{2}$ **d** $\frac{\pi}{4}, \frac{5\pi}{4}$

e $\frac{\pi}{3}, \frac{5\pi}{3}$

f 0.767..., 1.88...

4 **a** $\cot^2\theta$ **b** $\cos\theta + \sin\theta$

c $\sec^2\theta$

Exercise 6L

1 $\sin A \cos B - \sin B \cos A$

2 **a** Using $\cos(A + B)$

$\equiv \frac{OT}{OR} \equiv \frac{OP - TP}{OR}$

$\equiv \frac{OP - SQ}{OR} \equiv \frac{OP}{OR} - \frac{SQ}{OR}$

$\equiv \frac{OP}{OQ}\frac{OQ}{OR} - \frac{SQ}{RQ}\frac{RQ}{OR}$

$\equiv \cos B \cos A - \sin B \sin A$

b $\cos(A + (-B)) = \cos A\cos(-B) - \sin A\sin(-B) = \cos A \cos B + \sin A \sin B$

3 **a** $\tan(A+B)=\dfrac{\sin(A+B)}{\cos(A+B)}$

$=\dfrac{\sin A\cos B+\sin B\cos A}{\cos A\cos B-\sin A\sin B}$

$=\dfrac{\dfrac{\sin A}{\cos A}+\dfrac{\sin B}{\cos B}}{1-\dfrac{\sin A}{\cos A}\dfrac{\sin B}{\cos B}}$

$=\dfrac{\tan A+\tan B}{1-\tan A\tan B}$

b $\tan(A+(-B))$

$=\dfrac{\tan A+\tan(-B)}{1-\tan A\tan(-B)}$

$=\dfrac{\tan A-\tan B}{1+\tan A\tan B}$

4 **a** $\dfrac{\sqrt{2}}{4}(\sqrt{3}+1)$ **b** $-\sqrt{3}+2$

c $\dfrac{1}{2}$ **d** $\dfrac{1}{2}$

5 **a** $-\dfrac{33}{65}$ **b** $-\dfrac{16}{63}$

6 **a** $\cot(A+B)=\dfrac{\cos(A+B)}{\sin(A+B)}$

$=\dfrac{\cos A\cos B-\sin A\sin B}{\sin A\cos B+\sin B\cos A}$

Divide by $\sin A\sin B$ and get

$\dfrac{\dfrac{\cos A\cos B-\sin A\sin B}{\sin A\sin B}}{\dfrac{\sin A\cos B+\sin B\cos A}{\sin A\sin B}}$

$=\dfrac{\cot A\cot B-1}{\cot A+\cot B}$

b $\dfrac{\sin(A+B)}{\cos A\cos B}$

$=\dfrac{\sin A\cos B+\sin B\cos A}{\cos A\cos B}$

$=\tan A+\tan B$

c $\sec A+\tan A$

$=\dfrac{1}{\cos A}+\dfrac{\sin A}{\cos A}=\dfrac{1+\sin A}{\cos A}$

d $\tan A+\cot A$

$=\dfrac{\sin A}{\cos A}+\dfrac{\cos A}{\sin A}$

$=\dfrac{\sin^2 A+\cos^2 A}{\cos A\sin A}$

$=\dfrac{1}{\cos A\sin A}=\sec A\csc A$

e $\sec^2\theta+\csc^2\theta$

$=\dfrac{1}{\cos^2\theta}+\dfrac{1}{\sin^2\theta}$

$=\dfrac{\sin^2\theta+\cos^2\theta}{\sin^2\theta\cos^2\theta}$

$=\dfrac{1}{\sin^2\theta\cos^2\theta}$

$=\sec^2\theta\csc^2\theta$

f $\dfrac{\csc\theta-\cot\theta}{1-\cos\theta}$

$=\dfrac{\dfrac{1}{\sin\theta}-\dfrac{\cos\theta}{\sin\theta}}{1-\cos\theta}=\dfrac{\dfrac{1-\cos\theta}{\sin\theta}}{1-\cos\theta}$

$=\dfrac{1}{\sin\theta}=\csc\theta$

g $\csc x-\sin x$

$=\dfrac{1}{\sin x}-\sin x$

$=\dfrac{1-\sin^2 x}{\sin x}=\dfrac{\cos^2 x}{\sin x}$

$=\cos x\cot x$

h $1+\cos^4 x-\sin^4 x$

$=1+\cos^4 x-1+\cos^4 x$

$=2\cos^4 x$

i $\sec\theta+\tan\theta=\dfrac{1}{\cos\theta}+\dfrac{\sin\theta}{\cos\theta}$

$=\dfrac{1+\sin\theta}{\cos\theta}$

Multiply numerator and denominator by $1-\sin\theta$ and get

$\dfrac{1-\sin^2\theta}{\cos\theta(1-\sin\theta)}=\dfrac{\cos^2\theta}{\cos\theta(1-\sin\theta)}$

$=\dfrac{\cos\theta}{1-\sin\theta}$

j $\dfrac{\sin A\tan A}{1-\cos A}=\dfrac{\sin A\dfrac{\sin A}{\cos A}}{1-\cos A}$

$=\dfrac{\dfrac{\sin^2 A}{\cos A}}{1-\cos A}=\dfrac{\dfrac{1-\cos^2 A}{\cos A}}{1-\cos A}$

$=\dfrac{\dfrac{(1-\cos A)(1+\cos A)}{\cos A}}{1-\cos A}$

$=\dfrac{1+\cos A}{\cos A}=1+\sec A$

7 **a** $\dfrac{\tan A-\tan A\tan B\tan C+\tan B+\tan C}{1-\tan B\tan C-\tan A\tan B-\tan A\tan C}$ **b** $\dfrac{\pi}{4}$

c If A, B and C form the angles of a triangle, then $\tan(A+B+C)$

$=0$, so $\dfrac{\tan A-\tan A\tan B\tan C+\tan B+\tan C}{1-\tan B\tan C-\tan A\tan B-\tan A\tan C}=0$

Exercise 6M

1 **a** $\tan A+\cot A=\dfrac{\sin A}{\cos A}+\dfrac{\cos A}{\sin A}=\dfrac{\sin^2 A+\cos^2 A}{\cos A\sin A}$

$=\dfrac{1}{\cos A\sin A}=\dfrac{2}{2\cos A\sin A}=\dfrac{2}{\sin 2A}=2\csc 2A$

b $\dfrac{\sin 2A+\cos 2A+1}{\sin 2A-\cos 2A+1}=\dfrac{2\sin A\cos A+2\cos^2 A-1+1}{2\sin A\cos A-1+2\sin^2 A+1}$

$=\dfrac{2\cos A(\sin A+\cos A)}{2\sin A(\cos A+\sin A)}=\cot A$

c $\dfrac{\cos 3X-\sin 3X}{1-2\sin 2X}$

$=\dfrac{\cos(2X+X)-\sin(2X+X)}{1-2\sin 2X}$

$=\dfrac{\cos 2X\cos X-\sin 2X\sin X-\sin 2X\cos X-\sin X\cos 2X}{1-4\sin X\cos X}$

$=\dfrac{\cos 2X(\cos X\sin X)-\sin 2X(\sin X+\cos X)}{1-4\sin X\cos X}$

$$= \frac{(\cos^2 X - \sin^2 X)(\cos X - \sin X) - 2\sin X\cos X(\sin X + \cos X)}{1 - 4\sin X\cos X}$$

$$= \frac{(\cos X + \sin X)(\cos X - \sin X)(\cos X - \sin X) - 2\sin X\cos X(\sin X + \cos X)}{1 - 4\sin X\cos X}$$

$$= \frac{(\cos X - \sin X)^2(\sin X + \cos X) - 2\sin X\cos X(\sin X + \cos X)}{1 - 4\sin X\cos X}$$

$$= \frac{(\sin X + \cos X)(\cos^2 X - 2\sin X\cos X + \sin^2 X - 2\sin X\cos X)}{1 - 4\sin X\cos X}$$

$$= \frac{(\sin X + \cos X)(\cos^2 X - 4\sin X\cos X + \sin^2 X)}{1 - 4\sin X\cos X}$$

$$= \frac{(\sin X + \cos X)(1 - 4\sin X\cos X)}{1 - 4\sin X\cos X} = \sin X + \cos X$$

d $\cot x - \csc 2x = \frac{\cos x}{\sin x} - \frac{1}{\sin 2x} = \frac{\cos x}{\sin x} - \frac{1}{2\sin x\cos x}$

$$= \frac{2\cos^2 x - \cos^2 x - \sin^2 x}{2\sin x\cos x} = \frac{\cos^2 x - \sin^2 x}{2\sin x\cos x} = \frac{\cos 2x}{\sin 2x} = \cot 2x$$

2 $\frac{\sin 4A}{\sin A} = \frac{2\sin 2A\cos 2A}{\sin A} = \frac{4\sin A\cos A\cos 2A}{\sin A}$

$$= \frac{4\sin A\cos A\,(2\cos^2 A - 1)}{\sin A} = 8\cos^3 A - 4\cos A$$

---

3 $\sin 2A\sec 2A = \tan 2A = \frac{24}{7}$

4 $\cos 3X = 4\cos^3 X - 3\cos X$, $\sin 3X = -4\sin^3 X + 3\sin X$

5 $8\cos^4 A - 8\cos^2 A + 1$

6 $\frac{4\tan A(1 - \tan^2 A)}{(1 - \tan^2 A)^2 - 4\tan^2 A}$

7 **a** 0.848, 2.29

b $\frac{23\pi}{12}, \frac{11\pi}{12}, \frac{19\pi}{12}, \frac{7\pi}{12}$

c $\pi, \frac{\pi}{2}, \frac{3\pi}{2}$

Exercise 6N

1 At $\theta = \frac{\pi}{2} \pm 2n\pi, n \in \mathbb{Z}$, at $\theta = \frac{3\pi}{2} \pm 2n\pi, n \in \mathbb{Z}$, at $\theta = 0 \pm n\pi$, $n \in \mathbb{Z}$ vertical asymptotes, $\csc\theta \to \infty$ as $\sin\theta \to 0^+$, $\csc\theta \to -\infty$ as $\sin\theta \to 0^-$

2 At $\theta = \frac{\pi}{4} \pm n\pi, n \in \mathbb{Z}$, at $\theta = \frac{3\pi}{4} \pm n\pi, n \in \mathbb{Z}$, at $\theta = 0 \pm n\pi$, $n \in \mathbb{Z}$ vertical asymptotes, $\cot\theta \to \infty$ as $\tan\theta \to 0^+$, $\cot\theta \to -\infty$ as $\tan\theta \to 0^-$

3 **a** Amplitude = 3, period = $\frac{\pi}{2}$

b Same as $\sin(x)$ with vertical shift = $\frac{3\pi}{2}$

c Amplitude = 2, period = 2

d Horizontal shift = $\frac{\pi}{4}$, vertical shift = 1 down

4 **a** Vertical stretch = 2 parallel to y-axis, vertical shift = 2 upwards

b Stretch by a factor of $\frac{1}{3}$ parallel to the x axis, shift vertically downwards by 1, new period is $\frac{2\pi}{3}$

c Reflect along the x-axis, stretch by a factor of 2 parallel to y-axis, stretch by a factor of $\frac{1}{3}$ parallel to the x-axis, new period = $\frac{2\pi}{3}$ and the amplitude is 2.

d Stretch by a factor of 3 parallel to the y-axis, stretch by a factor of $\frac{1}{2}$ parallel to the x-axis, new period = π, shift vertically upwards by 3 and the amplitude is 2.

5 One solution

Exercise 6O

1 **a** Amplitude = 5, horizontal/phase shift = $\frac{\pi}{12}$, vertical shift = 1, period = $\frac{2\pi}{3}$, maximum = 6, minimum = −4

b Amplitude = 2, horizontal/phase shift = $\frac{5\pi}{6}$, vertical shift = −2, period = $\frac{2\pi}{3}$, maximum = 0, minimum = −4

2 **a** 220 V **b** −220 V

c 220 V **d** $\frac{1}{60}$

e

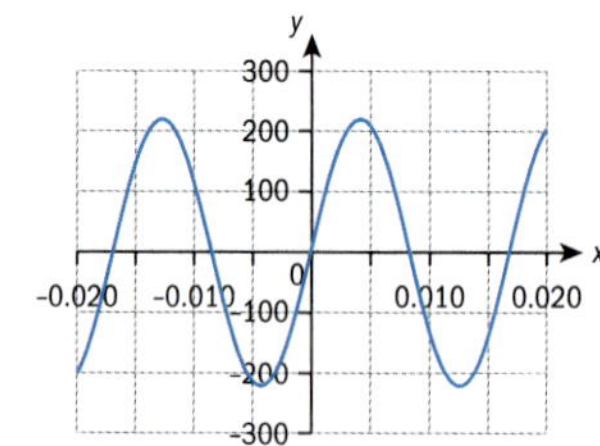

3 **a** $a = 6.6$ m, $d = 7.8$ m, $b = \frac{\pi}{6}$, $c = -5.25$

b First minimum at $t = 2.25$

c $4.41 < t < 12.1$ and $16.4 < t < 24$

4 **a** $f(x) = \frac{1}{2}\cos x - 3$

b $f(x) = 2\cos\left(x - \frac{\pi}{2}\right) + 5$

c $f(x) = 3\cos(2x) - 1$

d $f(x) = -2\sin 3x$

e $f(x) = \tan(x - \frac{\pi}{4})$

f $f(x) = \sec\frac{1}{2}x + 1$

Exercise 6P

1 **a** $\frac{\pi}{6}, \frac{5\pi}{6}$ **b** $\frac{\pi}{4}, \frac{5\pi}{4}$

c π **d** $\frac{\pi}{6}, \frac{7\pi}{6}$

e 0, 2π **f** $\frac{\pi}{4}, \frac{7\pi}{4}$

g $\frac{2\pi}{3}, \frac{4\pi}{3}$

2 $\frac{5\pi}{18}$

3 a $\tan x < 0$ for $\frac{\pi}{2} < x < \pi$ and $\frac{3\pi}{2} < x < 2\pi$

b $\sec 2x < 0$ for $\frac{\pi}{4} < x < \frac{3\pi}{4}$ and $\frac{5\pi}{4} < x < \frac{7\pi}{4}$

c $\sin 4x > 3$ never happens as it is outside the range of sine

4 a $f^{-1}(x)$ is defined for $x \in [0, 1]$, $g^{-1}(x) = 2x$ is well defined for all x

b $g^{-1}\left(\frac{1}{2}\right) = 1$, $f^{-1}g\left(\frac{\pi}{6}\right)$ has no real results

5 a $\frac{2\pi}{5}$ b 6

c The sine function is symmetric about the origin

d Stretch by a factor of 6 parallel to y-axis, stretch by a factor of $\frac{1}{5}$ parallel to the x-axis, new period $= \frac{2\pi}{5}$, there are no vertical shifts

Exercise 6Q

1 a $\theta = 0.464, 3.605$

b $\theta = 1.19, 4.33$ and $\frac{3\pi}{4}, \frac{7\pi}{4}$

c $\theta = \frac{3\pi}{4}, \frac{7\pi}{4}$ and 0.464, 3.61

d $\theta = 1.231, 5.052$

Exercise 6R

1 a $\frac{d}{dx}(\cos x) = \lim_{h\to 0} \frac{\cos(x+h) - \cos(x)}{h} = \lim_{h\to 0} \frac{\cos x \cos h - \sin x \sin h - \cos x}{h} = \lim_{h\to 0}\left(\cos x \frac{(\cos h - 1)}{h} - \sin x \frac{\sin h}{h}\right)$

$= \cos x \lim_{h\to 0} \frac{(\cos h - 1)}{h} - \sin x \lim_{h\to 0} \frac{\sin h}{h} = 0 - \sin x(1) = -\sin x$

b $\frac{d}{dx}(\sin 2x) = \lim_{h\to 0} \frac{\sin\left(2x + \frac{2h}{2}\right) - \sin(2x)}{\frac{h}{2}} = \lim_{h\to 0} \frac{\sin 2x \cos h + \sin 2x \cos 2\text{x} - \sin 2x}{\frac{h}{2}}$

$= \lim_{h\to 0}\left(\sin 2x \frac{\cos h - 1}{\frac{h}{2}} + \cos 2x \frac{\sin h}{\frac{h}{2}}\right) = \sin 2x \lim_{h\to 0} \frac{\cos h - 1}{\frac{h}{2}} + \cos 2x \lim_{h\to 0} \frac{\sin h}{\frac{h}{2}} = \sin 2x \cdot 0 + 2 \times \cos 2x$

$= 2\cos 2x$

c $\frac{d}{dx}\left(\sin \frac{x}{3}\right) = \lim_{h\to 0} \frac{\sin\left(\frac{x}{3} + \frac{3h}{3}\right) - \sin\left(\frac{x}{3}\right)}{3h} = \lim_{h\to 0} \frac{\sin\left(\frac{x}{3}\right)\cos h + \sin h \cos\left(\frac{x}{3}\right) - \sin\left(\frac{x}{3}\right)}{3h}$

$= \lim_{h\to 0}\left(\sin\left(\frac{x}{3}\right)\frac{\cos h - 1}{3h} + \cos\left(\frac{x}{3}\right)\frac{\sin h}{3h}\right) = \sin\left(\frac{x}{3}\right)\lim_{h\to 0}\frac{\cos h - 1}{3h} + \cos\left(\frac{x}{3}\right)\lim_{h\to 0}\frac{\sin h}{3h}$

$= \sin\left(\frac{x}{3}\right) \cdot 0 + \frac{1}{3}\cos\left(\frac{x}{3}\right) \cdot 1 = \frac{1}{3}\cos\left(\frac{x}{3}\right)$

d $\frac{d}{dx}(\sin(2x+3)) = \lim_{h\to 0} \frac{\sin\left(2x + 3 + 2\left(\frac{h}{2} - \frac{3}{2}\right) + 3\right) - \sin(2x+3)}{\frac{h}{2} - \frac{3}{2}}$

$= \lim_{h\to 0} \frac{\sin(2x+3)\cos h + \sin h \cos(2x+3) - \sin(2x+3)}{\frac{h-3}{2}}$

$= \sin(2x+3)\lim_{h\to 0}\frac{\cos h - 1}{\frac{h-3}{2}} + \cos(2x+3)\lim_{h\to 0}\frac{\sin h}{\frac{h-3}{2}}$

$= 0 \cdot \sin(2x+3) + 2\cos(2x+3)\cdot 1 = 2\cos(2x+3)$

2 **a** $\frac{d}{dx}(\tan x) = \lim_{h \to 0} \frac{\tan(x+h) - \tan x}{h} = \lim_{h \to 0} \frac{\frac{\tan x + \tan h}{1 - \tan x \tan h} - \tan x}{h}$

$= \lim_{h \to 0} \frac{\tan h\,(1 + \tan^2 x)}{h} = \sec^2 x \lim_{h \to 0} \frac{\tan h}{h} = \sec^2 x \cdot 1 = \sec^2 x$

b $\frac{d}{dx}(\cot x) = \lim_{h \to 0} \frac{\cot(x+h) - \cot(x)}{h} = \lim_{h \to 0} \frac{\frac{\cos(x+h)}{\sin(x+h)} - \frac{\cos x}{\sin x}}{h} = \lim_{h \to 0} \frac{\frac{\sin x \cos(x+h) - \cos x \sin(x+h)}{\sin x \sin(x+h)}}{h}$

$= \lim_{h \to 0} \frac{\frac{\sin(x - x - h)}{\sin x \sin(x+h)}}{h} = -1 \lim_{h \to 0} \frac{1}{\sin x \sin(x+h)} = \frac{-1}{\sin x} \lim_{h \to 0} \frac{1}{\sin(x+h)} = \frac{-1}{\sin^2 x} = -\csc^2 x$

c $\frac{d}{dx}(\tan 3x) = \lim_{h \to 0} \frac{\tan\left(3x + \frac{h}{3}\right) - \tan 3x}{\frac{h}{3}} = \lim_{h \to 0} \frac{\frac{\tan 3x + \tan\frac{h}{3}}{1 - \tan 3x \tan\frac{h}{3}} - \tan 3x}{\frac{h}{3}} = \lim_{h \to 0} \frac{\tan\frac{h}{3}\,(1 + \tan^2 3x)}{\frac{h}{3}}$

$= \sec^2 3x \lim_{h \to 0} \frac{\tan\frac{h}{3}}{\frac{h}{3}} = \sec^2 3x \cdot 3 = 3\sec^2 3x$

Exercise 6S

1 **a** $-\csc^2 x$ **b** $-\cot x \csc x$

2 **a** $2\cos 2x$ **b** $-2\sin(2x+1)$

c $3\cos(8 - 3x)$

d $\frac{2}{13}\csc^2\left(\frac{7 - 2x}{13}\right)$

3 **a** $-5x^4 \sin(x^5 - 3)$

b $-2x\cot(x^2 + 1)\csc(x^2 + 1)$

c $(12x^2 - 4x + 7)\tan(4x^3 - 2x^2 + 7x + 17)\sec(4x^3 - 2x^2 + 7x + 17)$

d $\frac{1}{2\sqrt{e^x + 1}} e^x \tan\sqrt{e^x + 1} \sec\sqrt{e^x + 1}$

e $-\sec^2 x \sin(\tan x) \cos(\cos(\tan x))$

Exercise 6T

1 **a** $\cos x\,(2x + 1) + 2\sin x$

b $-2(x + x^2)\sin 2x + (1 + 2x)\cos 2x$

c $\frac{-\sin x \times x - \cos x}{x^2} = \frac{-\cos x - x\sin x}{x^2}$

d $\frac{2\sin 2x - (4x + 6)\cos 2x}{\sin^2 2x}$

e $\frac{\sec^2 x}{\sqrt{2 - x}} + \frac{1}{2} \times \frac{\tan x}{(2 - x)^{3/2}}$

2 **a** -3 **b** -2 **c** -2

d $\frac{3\pi}{2}$ **e** $\frac{27\pi^3}{32} - \frac{27\pi^2}{16}$

3 **a** 0 **b** $\tan\beta \sec\beta$

c $\frac{3(-\sin^2 x \cos x + \cos x + 1)}{(\cos x + \cos 2x)^2}$

Exercise 6U

1 **a** $\frac{-2}{\sqrt{1 - 4x^2}}$

b $\frac{3}{\sqrt{4 - 9x^2}}$

c $\frac{1}{2x^2 + 2x + 1}$

2 **a** $2\arccos x - \frac{2x}{\sqrt{1 - x^2}}$

b $\frac{-1}{2x\sqrt{1 - x^2}} - \frac{\arccos x}{2x^2}$

c $2x\arctan 3x + 3 \times \frac{x^2 - 1}{1 + 9x^2}$

3 **a** $\frac{d}{dx}(\arcsin x + \arccos x) = \frac{1}{\sqrt{1 - x^2}} - \frac{1}{\sqrt{1 - x^2}} = 0$ Valid because calculating two angles that add up to π in a right-angled triangle.

b $\frac{d}{dx}(\arctan x + \arctan(-x)) = \frac{1}{1 + x^2} - \frac{1}{1 + x^2} = 0$ Here both inverse tangents correspond to the same angle, in different quadrants (due to the negative sign), so the rate of change between both is zero.

Exercise 6V

1 a Tangent: $3x - \pi$, normal:

$y = -\frac{1}{3}\left(x - \frac{\pi}{3}\right)$

b Tangent: $y = -2.01(x - 0.05) + 1.47$, normal: $y = -\frac{1}{-2.01}(x - 0.05) + 1.47$ or equivalently $y = 0.5(x - 0.05) + 1.47$

c Tangent: $y = -1.382(x + 0.5) + 0.421$, normal: $y = \frac{-1}{1.382}(x+0.5)+0.421$

2 a (0.298, 0.247)

b Normal to $y = x\cos 2x$: $y = -2.03(x - 0.298) + 0.247$

Normal to $y = \tan(3x + \pi) - 1$: $y = -0.131(x - 0.298) + 0.247$

c 56.44°

3 a $\frac{\cos^2 x - x^2 \sin x}{(x + \cos x)^2}$

b Tangent: $y = -\left(x - \frac{\pi}{2}\right)$, normal: $y = x - \frac{\pi}{2}$

4 a $y' = (8x)\arctan x + \frac{4x^2+1}{1+x^2}2$

b $y = 5.14x - 1$

Exercise 6W

1 a 10 m/s

b 12.5 m/s

c Velocity of light beam is increasing and is undetermined at the point where it is exactly parallel

2 a 0.833 rot/s

b 0.491 rot/s

3 a $(0.004y - 0.024)\csc\theta$

b $\frac{0.2y-1}{50}\csc\theta$

4 a 0.002387 cm/min

b 1.333 cm²/s

5 a 0.00829 deg/s

b 0.028 deg/s

6 a 64.9 km/h

b 2.6 degrees per second

7 −0.0212 degrees per second

Chapter review

1 2.48 cm

2 a 754 cm³

b 414.7 cm²

3 a 33.6°

b i 4.516 cm

ii 3.749 cm²

iii 3.348 cm²

4 a $\cos x = \frac{1}{3}$ and $\cos x = -\frac{1}{2}$

b 289.5°, 430.5° and 240°, 480°

5 108° and −108°

6 $\frac{\tan A(3 - \tan^2 A)}{1 - 3\tan^2 A}$

7 $\theta = \frac{2\pi}{7}, \frac{4\pi}{7}, \frac{6\pi}{7}, 0, 2\pi$

If we substitute into the equation $\cos 4\theta = \cos 3\theta$, and apply the identities for $\cos(2\theta + 2\theta)$ and $\cos(2\theta + \theta)$ we get the equation $8\cos^3\theta + 4\cos^2\theta - 4\cos\theta - 1 = 0$ so the roots to the equation are precisely $\cos\frac{2\pi}{7}, \cos\frac{4\pi}{7}, \cos\frac{6\pi}{7}$ where $x = \cos\theta$

8 a $x = 15°, 195°$

b $x = 293.2°, 113.2°$

9 $\sin 3A = \sin(A + 2A)$

$= \sin A\cos 2A + \sin 2A\cos A$

$= \sin A(1 - 2\sin^2 A) + 2\sin A(1 - \sin^2 A)$

$= \sin A - 2\sin^3 A + 2\sin A - 2\sin^3 A = -4\sin^3 A + 3\sin A$

10 $x = \pm 1$

11 $\tan\frac{A}{2}\tan\frac{B-C}{2} = \frac{\sin\frac{A}{2}}{\cos\frac{A}{2}} \times \frac{\sin\left(\frac{B-C}{2}\right)}{\cos\left(\frac{B-C}{2}\right)}$

Note that $A + B + C = \pi$ as ABC are the angles of a triangle. Then $\frac{A}{2} = \frac{\pi - (B+C)}{2}$ so $\sin\frac{A}{2} = \sin\left(\frac{\pi}{2} - \left(\frac{B+C}{2}\right)\right) = \cos\left(\frac{B+C}{2}\right)$ and

$\cos\frac{A}{2} = \cos\left(\frac{\pi}{2} - \left(\frac{B+C}{2}\right)\right) = \sin\left(\frac{B+C}{2}\right)$

Then we substitute back into the first equation to get

$$\frac{\left(\cos\frac{B}{2}\cos\frac{C}{2} - \sin\frac{B}{2}\sin\frac{C}{2}\right)\left(\sin\frac{B}{2}\cos\frac{C}{2} - \sin\frac{C}{2}\cos\frac{B}{2}\right)}{\left(\sin\frac{B}{2}\cos\frac{C}{2} + \sin\frac{C}{2}\cos\frac{B}{2}\right)\left(\cos\frac{B}{2}\cos\frac{C}{2} + \sin\frac{B}{2}\sin\frac{C}{2}\right)}$$

Note that $\sin\frac{B}{2}\cos\frac{B}{2}=\frac{1}{2}\sin B$ and equivalently for C. We multiply the brackets and substitute with the form for $\sin B$ to get

$$\frac{\sin B\left(\cos^2\frac{C}{2}+\sin^2\frac{C}{2}\right)-\sin C\left(\cos^2\frac{B}{2}+\sin^2\frac{B}{2}\right)}{\sin B\left(\cos^2\frac{C}{2}+\sin^2\frac{C}{2}\right)+\sin C\left(\cos^2\frac{B}{2}+\sin^2\frac{B}{2}\right)}=\frac{\sin B-\sin C}{\sin B+\sin C}$$ Finally,

we have that the sine rule holds, so we rewrite in terms of only

$\sin C$ to get $\frac{\frac{b}{c}\times\sin C-\sin C}{\frac{b}{c}\sin C+\sin C}=\frac{\frac{b}{c}-1}{\frac{b}{c}+1}=\frac{b-c}{b+c}$

12 342.5 km, 67°

13 **a** $-2a\cos 2t$ **b** $t=\frac{n\pi}{2}$ **c** $t=\frac{3\pi}{4}$

Exam-style questions

14 **a** $A=\frac{1}{2}\times5\times10\sin 30°=\frac{25}{2}$ (2 marks)

b $BD^2=5^2+10^2-2\times5\times10\cos 30°$ (2 marks)

$BD=\sqrt{125-50\sqrt{3}}$ (1 mark)

$BD=\sqrt{25\left(5-2\sqrt{3}\right)}$ (1 mark)

$BD=5\sqrt{5-2\sqrt{3}}$

c $\frac{\sin C\hat{D}B}{13}=\frac{\sin 45°}{5\sqrt{5-2\sqrt{3}}}$ (2 marks)

$\sin C\hat{D}B=\frac{13\sqrt{2}}{10\sqrt{5-2\sqrt{3}}}$ (1 mark)

d The angle $C\hat{D}B$ can either be acute or obtuse (1 mark) and the two possible values add up to 180°. (1 mark)

15 **a** $l=\sqrt{5^2+3^2}=5.83$ cm (2 marks)

$S=2\times(\pi\times3\times5.83...)$
$=110\text{ cm}^2$ (2 marks)

b $\frac{2\times\frac{1}{3}\times\pi\times3^2\times5}{\pi\times3.05^2\times10.1}\times100=31.9\%$ (2 marks)

16 $2\cos^2x=\sin 2x\Rightarrow 2\cos^2x-2\sin x\cos x=0$ (1 mark)

$2\cos x(\cos x-\sin x)=0$ (1 mark)

$\cos x=0,\ \cos x=\sin x$ (1 mark)

(or $\cos x=0$, $\tan x=1$)

$\cos x=0\Rightarrow x=\frac{\pi}{2},\ x=\frac{3\pi}{2}$ (1 mark)

$\tan x=1\Rightarrow x=\frac{\pi}{4},\ x=\frac{5\pi}{4}$ (1 mark)

17 **a** **i** $-1\le y\le 3$ (1 mark)

ii 2 (1 mark)

b $a=-2$ (1 mark)

$b=\frac{2\pi}{2}=\pi$ (2 marks)

$c=1$ (1 mark)

c $-2\cos\pi x+1=0$

$\Rightarrow\cos\pi x=\frac{1}{2}$ (1 mark)

$\pi x\in\left\{-\frac{\pi}{3},\frac{\pi}{3},\frac{5\pi}{3},\frac{7\pi}{3},\right\}$ (1 mark)

$x\in\left\{-\frac{1}{3},\frac{1}{3},\frac{5}{3},\frac{7}{3},\right\}$ (1 mark)

18 **a** $\frac{\cos x}{1-\sin x}-\tan x$

$=\frac{\cos x}{1-\sin x}-\frac{\sin x}{\cos x}$ (1 mark)

$=\frac{\cos^2x-\sin x(1-\sin x)}{(1-\sin x)(\cos x)}$ (1 mark)

$=\frac{\cos^2x+\sin^2x-\sin x}{(1-\sin x)(\cos x)}$ (1 mark)

$=\frac{1-\sin x}{(1-\sin x)(\cos x)}$ (1 mark)

$=\frac{1}{\cos x}$ (1 mark)

$=\sec x$

b $\frac{\cos 2x}{1-\sin 2x}-\tan 2x=\sec 2x$

So $\sec 2x=\sqrt{2}$ (1 mark)

$\cos 2x=\frac{1}{\sqrt{2}}$ (1 mark)

$2x=\frac{\pi}{4},\frac{7\pi}{4},\frac{9\pi}{4},\frac{15\pi}{4}$ (2 marks)

$x=\frac{\pi}{8},\frac{7\pi}{8},\frac{9\pi}{8},\frac{15\pi}{8}$ (2 marks)

19 **a** $\frac{dy}{dx}=\frac{1}{1+\left(\frac{1}{x}\right)^2}\times\left(-\frac{1}{x^2}\right)$ (2 marks)

$\frac{dy}{dx}=-\frac{1}{1+x^2}$ (1 mark)

b Valid attempt to apply product rule (1 mark)

$\frac{dy}{dx}=2xe^{\arctan x}+\frac{x^2e^{\arctan x}}{1+x^2}$ (3 marks)

$\left(\frac{dy}{dx}=e^{\arctan x}\left(2x+\frac{x^2}{1+x^2}\right)\right)$

20 Valid attempt at implicit differentiation (1 mark)

$(\cos y)\frac{dy}{dx}=-\sin x\left(\sec^2(\cos x)\right)$ (2 marks)

At $\left(\frac{\pi}{2},0\right)$:

$(\cos 0)\frac{dy}{dx} = -\sin\frac{\pi}{2}\left(\sec^2\left(\cos\frac{\pi}{2}\right)\right)$ (1 mark)

$\frac{dy}{dx} = -\sec^2 0$ (1 mark)

$= -1$ (1 mark)

So gradient of normal is

$-\frac{1}{(-1)} = 1$ (1 mark)

So equation is $y - 0 = 1\left(x - \frac{\pi}{2}\right)$,

or $y = x - \frac{\pi}{2}$ (2 marks)

21 a

$S(x) = \underbrace{\sin^2 2x + \cos^2 2x}_{1} + \underbrace{2\sin 2x\cos 2x}_{\sin 4x}$ (3 marks)

$= 1 + \sin 4x$

b

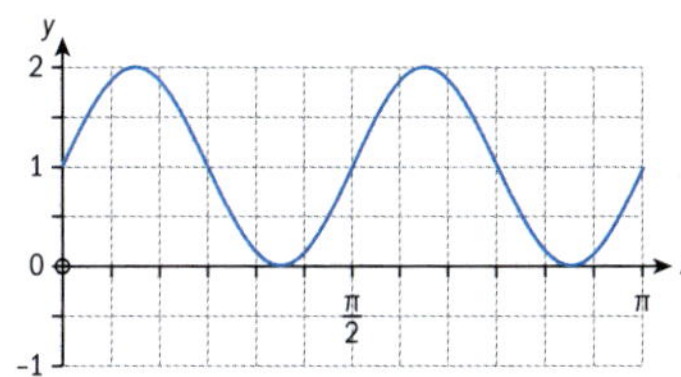

(1) for correct shape, (1) for 2 cycles, (1) for correct max/min

c i $\frac{\pi}{2}$ (1 mark)

ii $0 \le y \le 2$ (1 mark)

d

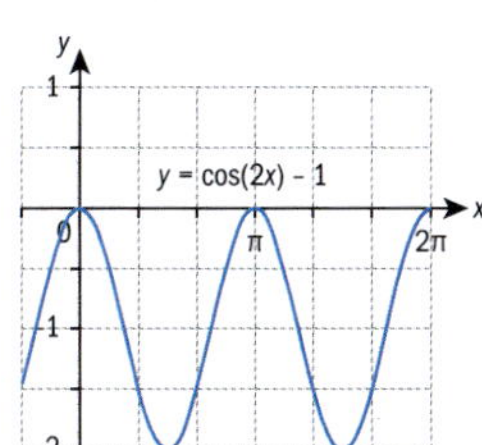

(3 marks)

e i $k = 2$ (1 mark)

ii $p = -\frac{\pi}{4}$, $q = -2$ (2 marks)

22 a $D = \frac{22+12}{2}$ (1 mark)

$= 17$ (1 mark)

$A = \frac{22-12}{2}$ (1 mark)

$= 5$ (1 mark)

The period is $\frac{360}{B} = 24$ (1 mark)

Therefore $B = 15$ (1 mark)

So $T = 5\sin(15(t - C)) + 17$

At (3, 12), $12 = 5\sin(15(3 - C)) + 17$ (1 mark)

$-1 = \sin(15(3 - C))$

$15(3 - C) = -90$ (1 mark)

$C = 9$ (1 mark)

Therefore $T = 5\sin(15(t - 9)) + 17$

b Solving $T = 5\sin(15(t - 9)) + 17$ and $T = 20$ by GDC (1 mark)

Solutions are $T = 18.54$ and $T = 11.46$ (2 marks)

$18.54 - 11.46 = 7.08$ hours (7 hours 5 minutes) (1 mark)

23 $V = \frac{1}{3}\pi r^2 h = \frac{50\pi r^2}{3}$ (1 mark)

$\frac{dV}{dr} = \frac{100\pi r}{3}$ (2 marks)

$\frac{dV}{dt} = \frac{dV}{dr} \times \frac{dr}{dt}$ (1 mark)

$2 = \frac{100\pi r}{3} \times \frac{dr}{dt}$ (1 mark)

$r = 0.4 \Rightarrow 2 = \frac{40\pi}{3} \times \frac{dr}{dt}$

$\Rightarrow \frac{dr}{dt} = \frac{3}{20\pi}$ cm min^{-1} (1 mark)

Chapter 7

Skills check

1

$$\frac{dy}{dx} = \frac{\cos x(3x^2 - 5) + (x^3 - 5x)\sin x}{\cos^2 x} = \frac{(3x^2 - 5) + (x^3 - 5x)\tan x}{\cos x}$$

2 $3xy^2\frac{dy}{dx} + y^3 - 2y\sin x\cos x - \sin^2 x\frac{dy}{dx} = -y\sin x + \cos x\frac{dy}{dx}$

$\Rightarrow (3xy^2 - \sin^2 x - \cos x)\frac{dy}{dx} = y(\sin 2x - \sin x)$

$\Rightarrow \frac{dy}{dx} = \frac{y(\sin 2x - \sin x)}{(3xy^2 - \sin^2 x - \cos x)}$

Exercise 7A

1 a $-\frac{1}{3}x^2 + C$ **b** $\frac{5}{16}x^4 + C$

c $\frac{8}{5}x^{\frac{5}{2}} + C$ **d** $7x^{\frac{1}{2}} + C$

e $-\cos x + C$

2 a $\frac{1}{4}x^4 + x^3 - 2x^2 + 3x + C$

b $\frac{1}{5}x^5 + x^4 + \frac{3}{2}x^{-2} - \frac{1}{3}x^{-3} + C$

c $\frac{1}{5}x^5 - \frac{1}{2}x^4 + \frac{5}{3}x^3 - 2x^2 + 4x + C$

d $8x - 6x^2 + 2x^3 - \frac{1}{4}x^4 + C$

e $\frac{1}{2}x^2 + \sin x - \tan x + C$

3 $y = x^3 - 4x - 1$

4 $f(t) = \frac{1}{2}t^2 - 2t + 2t^{\frac{1}{2}}$

5 $y = 2x^4 - 4x^3 + 3x^2 - x - 8$

6 $\theta - \cos\theta + C$

7 $f(\theta) = 2\theta + 3\cos\theta - 5$

8 a $f(x) = 3x - \sin x + \frac{1}{2}$

b $f(x) = 2\tan x + 3\cos x + 1$

c $f(x) = -2\sin x + \sqrt{2}x + 1$

d $f(x) = x^2 - 3\cos x + 4x + 8$

9 $v(t) = 9t^2 - 2t + 1$, $s(t) = 3t^3 - t^2 + t$

10 $v(t) = 6\sin(t)$, $s(t) = -6\cos t + 5$

11 a $-\frac{1}{21}(1-3x)^7 + C$

b $-2(4-x)^{\frac{3}{2}} + C$

c $\frac{3}{5}\sin(5x+2) - \frac{4}{5}\cos(5x+2) + C$

d $-(2-3x)^{\frac{2}{3}} + \frac{3}{8}(1+2x)^{\frac{4}{3}} + C$

12 $f(\theta) = 1 - \cos\left(2\theta + \frac{\pi}{2}\right)$

Exercise 7B

1 20 **2** 62.5 **3** 9 **4** 20

5 a $\int_{-3}^{1}(x+3)\,dx + \int_{1}^{3}(6-2x)\,dx$

$= \left[\frac{x^2}{2} + 3x\right]_{-3}^{1} + \left[6x - x^2\right]_{1}^{3}$

$= 12$

b The height of the triangle is 4 and the base length is 6 since the lines intersect the x-axis at $x = -3$ and $x = 3$ respectively.

$\therefore$ Area of triangle:

$\frac{1}{2}bh = \frac{1}{2}(6)(4) = 12$

6 $\frac{1}{2}$ **7** $\frac{343}{6}$ **8** $\frac{11}{6}$

9 $4\sin\left(\frac{x}{3} + \frac{\pi}{2}\right) = 0$

$\Rightarrow \frac{x}{3} + \frac{\pi}{2} = n\pi \Rightarrow \frac{x}{3} = n\pi - \frac{\pi}{2}\ (n \in \mathbb{Z})$

a $x = \frac{3\pi}{2}$ for the first and $\frac{9\pi}{2}$ for the second

b $x = -\frac{3\pi}{2}$ for the first and $x = -\frac{9\pi}{2}$ for the second

c 4.13

10 $\frac{40}{3}$

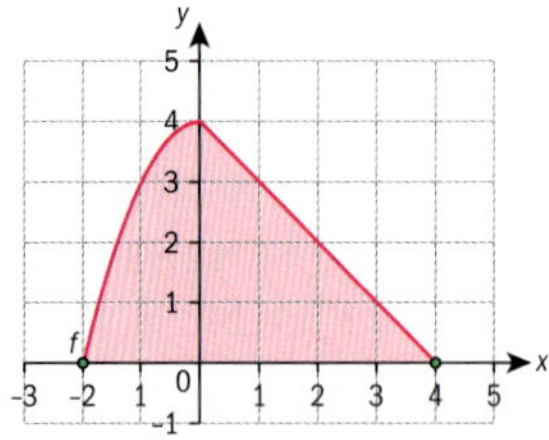

Exercise 7C

1 a 27 **b** $\frac{5}{2}$

c $\frac{4}{9}$

2 a $\left(\frac{a^{12}y^{-3}}{16y}\right)^{\frac{3}{4}} = \left(\frac{a^{12}}{16y^4}\right)^{\frac{3}{4}} = \frac{a^9}{8y^3}$

b $\frac{a^{-2} + 2a^{-1} + 1}{a^{-3}} = \frac{a^3\left(a^{-2} + 2a^{-1} + 1\right)}{a^3\left(a^{-3}\right)}$

$= a + 2a^2 + a^3$

$= a(a^2 + 2a + 1) = a(a+1)^2$

c $\frac{b^4 \times b^{-11}}{b^{-7}} = \frac{b^{-7}}{b^{-7}} = 1$

3 $\frac{3}{2}y^{\frac{5}{6}}$ Therefore, when

$y = 64, \sqrt{9y^3} \div \sqrt[3]{8y^2}$

$= \frac{3}{2}(64)^{\frac{5}{6}}$

$= \frac{3}{2}(32) = 48$

4 $\frac{\left(a^5b^2c^{-3}\right) \times \sqrt{a^{-3}b^3c}}{\sqrt{abc}}$

$= \frac{a^{5-\frac{3}{2}}b^{2+\frac{3}{2}}c^{-3+\frac{1}{2}}}{a^{\frac{1}{2}}b^{\frac{1}{2}}c^{\frac{1}{2}}} = \frac{a^{\frac{7}{2}}b^{\frac{7}{2}}c^{-\frac{5}{2}}}{a^{\frac{1}{2}}b^{\frac{1}{2}}c^{\frac{1}{2}}}$

$= \left(\frac{ab}{c}\right)^{\frac{6}{2}} = \left(\frac{ab}{c}\right)^3$

5 a 0 **b** $3 \times 2^{4n+3}$

6 a $x = \frac{1}{2}$

b $x = -10$

c $y = 9 \Rightarrow x = 2, y = 1 \Rightarrow x = 0$

d $x = -2$ ($y > 0$ so $y \neq -2$)

7 5%

8 a i 5.17% (to 3s.f.)

ii 1.10% (to 3s.f.)

iii The average percentage increase between December 2015 to December 2016 is equal to the average of the percentage increases from December 2015 to June 2016 and June 2016 to December 2016 (3.31% to 3 s.f.).

b Individual response

c 14 cents

9 Paloma: 20571.44 (to 2 d.p.), Concita: 20041.77 (to 2 d.p.)

10 42606.41 Bhat (to 2 d.p.)

Exercise 7D

1 a $\log_3 243$

b $\log_{16} 2 = \frac{1}{4}$

c $\log_q p = 5$

d $-4 = \log_{10}(0.0001)$

e $y = \log_x 11$

2 a $5^4 = 625$

b $64^{\frac{1}{2}} = 8$

c $m^p = n$

d $b^0 = 1$

e $10^{-2} = 0.01$

3 a $x = 2^{\frac{1}{128}}$ **b** $x = 64$

c $x = 16$ **d** $x = 27$

e $x = 7$ $(x > 0)$

4 a $\log_a m - 2\log_a n$

b $\frac{1}{3}\log_a m - 2\log_a n$

5 a $\log 30$ **b** $\log_3 4$

c $\log_a\left(mn^{\frac{3}{2}}\right)$

6 a -5 **b** 4

7 a $a = b^{\frac{3}{4}}$ **b** $a = \frac{b}{2}$

c $a = \left(\frac{10}{b}\right)^{\frac{1}{4}}$

8 a 1 **b** 2 **c** $\frac{3}{2}$

d 2 **e** 1 **f** 1

9 a $x^{\log y} = (10^{\log x})^{\log y}$
$= 10^{\log x \log y}$
$= (10^{\log y})^{\log x} = y^{\log x}$

b $\dfrac{1}{\log_x xy} + \dfrac{1}{\log_y xy}$
$= \log_{xy} x + \log_{xy} y$
$= \log_{xy}(xy) = 1$

10 $\log_x a = \dfrac{1}{p}$ and $\log_y a = \dfrac{1}{q}$

a $\log_{xy} a = \dfrac{1}{\log_a xy}$
$= \dfrac{1}{\log_a x + \log_a y} = \dfrac{1}{p+q}$

b $\log_{\frac{x}{y}} a = \dfrac{1}{\log_a \left(\frac{x}{y}\right)}$
$= \dfrac{1}{\log_a x - \log_a y} = \dfrac{1}{p-q}$

11 a $x = 2.10$ (3s.f.)

b $x = 0.848$ (to 3s.f.)

c $x = -1.30$ (3s.f.)

12 a $x = 125$ or $\dfrac{1}{125}$

b $x = 49$

c $x = 4$

13 $x = \log_5 8$ or 1.29

14 $x = 64$

15 a $x = 625$ or 125

b $x = -0.827$ or 0.712

16 −3.42, 2.71

17 a $x = 25, y = 81$

b $b = 2\ (b > 0), a = 8$

c $n = 3,\ m = \dfrac{3}{4}$

d $x = 3,\ y = \log_4 3$

18 $k_{\min} = 12$

19 a 21.4m (to 3 s.f.)

b $5 + 70\left(1 - \left(\dfrac{7}{8}\right)^{k-1}\right)$

c $k_{max} = 6$

20 20

Exercise 7E

1 Green: $a = \dfrac{1}{4}$, red: $a = 2$

2 a $f(x+1) = ka^{x+1}$
$= aka^x = af(x)$

b $f(x+2) = ka^{x+2}$
$= a^2ka^x = a^2f(x)$

c $f(x-1) = ka^{x-1} = a^{-1}ka^x$
$= a^{-1}f(x)$

d Conjecture: $f(x+n) = a^nf(x)$,
Proof: $f(x+n) = ka^{x+n}$
$= a^nka^x = a^nf(x)$

3

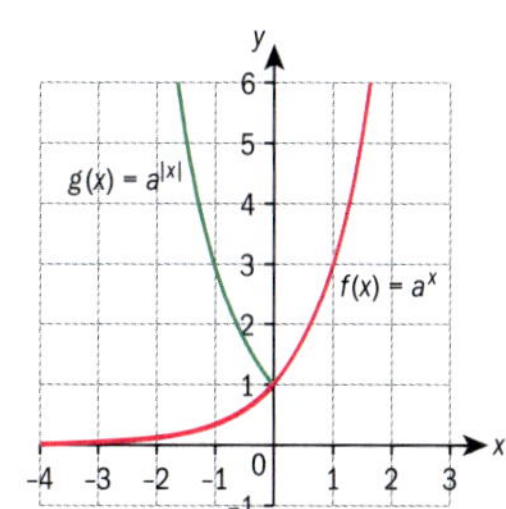

4 $x = 0$

5 a A reflection in the line $x = 0$ (the y-axis) $y = f(-x) = e^{-x}$

b A reflection in the line $y = 0$ (the x-axis) $y = -f(x) = -e^x$

c A reflection in the line $y = 0$ (the x-axis) followed by a reflection in the line $x = 0$ (the y-axis) $y = -f(-x) = -e^{-x}$

6

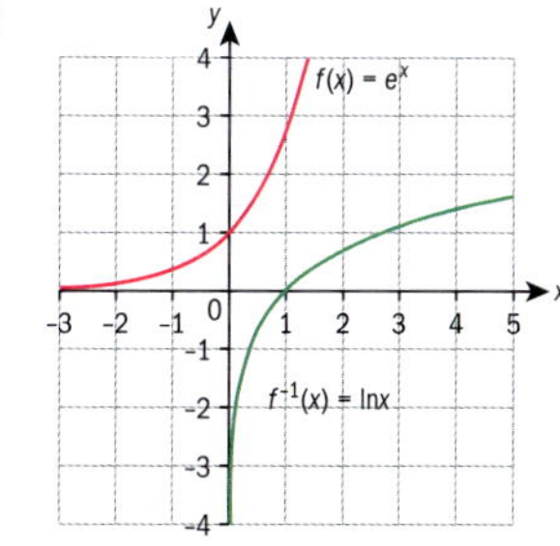

7 $x = a^y \Rightarrow y = \log_a x$ So
$f^{-1}(x) = \log_a x$
$f \circ f^{-1}(x) = a^{\log_a x} = x$ by direct substitution and definition of inverse function

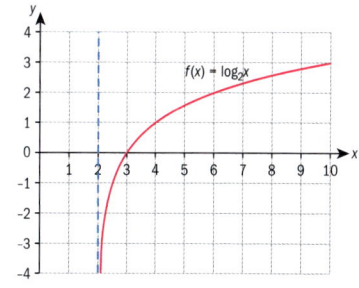

8 a $f(x) = -\ln x$ **b** $g(x) = |\ln x|$

c $h(x) = \ln|x|$

9

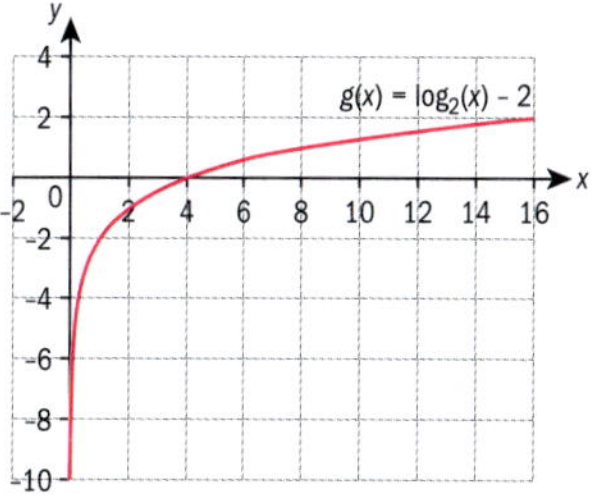

$y = \log_2(x-2)$ is a translation of $y = \log_2 x$ by 2 units in the positive x-direction.

$y = (\log_2 x) - 2$ is a translation of $y = \log_2 x$ by 2 units in the negative y-direction.

10 a Domain: $\{x \in \mathbb{R}: x > -1\}$, asymptote: $x = -1$,

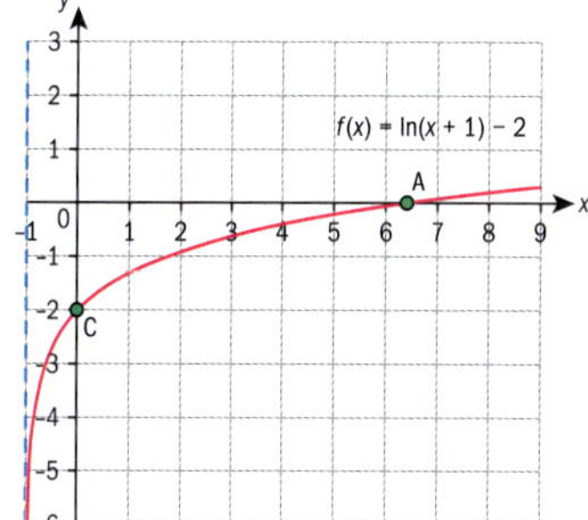

Domain: $\{x \in \mathbb{R}, x > -2\}$, asymptote: $x = -2$

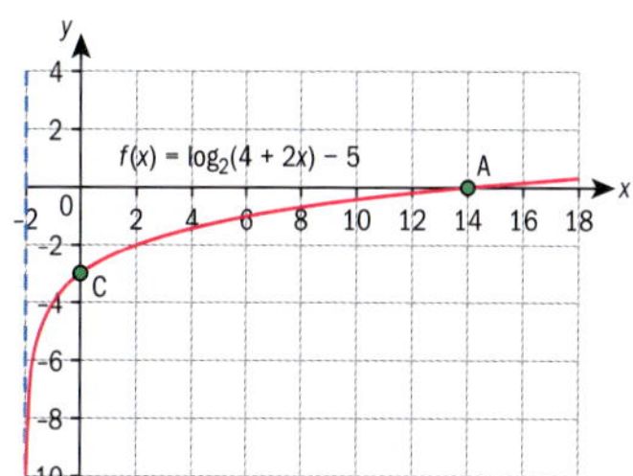

Exercise 7F

1 **a** $15e^{5x+4}$ **b** $6xe^{3x^2}$

c $4(\ln 5)5^{4x}$ **d** $-\sin x\, e^{\cos x}$

e $\dfrac{1}{2x}$ **f** $-\dfrac{5}{x}$

g $\dfrac{5}{5x+4}$ **h** $\dfrac{f'(x)}{f(x)}$

2 **a** $3e^{2x+4}(1+2x)$

b $\dfrac{1}{6}e^{x^{\frac{1}{3}}}\left(3x^{-\frac{1}{2}}+2x^{-\frac{1}{6}}\right)$

c $3x^2\ln(2x+1)+\dfrac{2x^3}{2x+1}$

d $\dfrac{3(\cos x-\sin x)}{e^x}$

e $e^{2x}(2\tan 3x+3\sec^2 3x)$

3 **a** $=\dfrac{1-\sqrt{x}}{2\sqrt{x}e^{\sqrt{x}}}$ **b** $\dfrac{\ln x-2}{2\sqrt{x}(\ln x)^2}$

c $-\dfrac{2e^x}{(1+e^x)^2}$ **d** $\dfrac{1+x(1-\ln x)}{x(1+x)^2}$

e $\dfrac{1+x+2e^x}{e^{-x}}$

4 **a** $\dfrac{e^{\sqrt{x}}}{2\sqrt{x}}-\dfrac{1}{x(\ln x)^2}$

b $(1+\log x)x^x$

5 **a** $\dfrac{\sqrt{2}}{4}$ **b** $\dfrac{4}{3}\ln\dfrac{3}{2}$

6 **a** $f'(x)=\dfrac{1}{x}-1=0\Rightarrow x=1$

Thus there exists only solution to $f'(x)=0$ so thereis only turning point.

b The turning point is located at $(1, f(1))=(1,-1)$

$f''(x)=-\dfrac{1}{x^2}<0$ so the turning point must be a maximum.

c Domain: $x>0$, range: $-\infty<f(x)\le -1$

d $x=0$

e

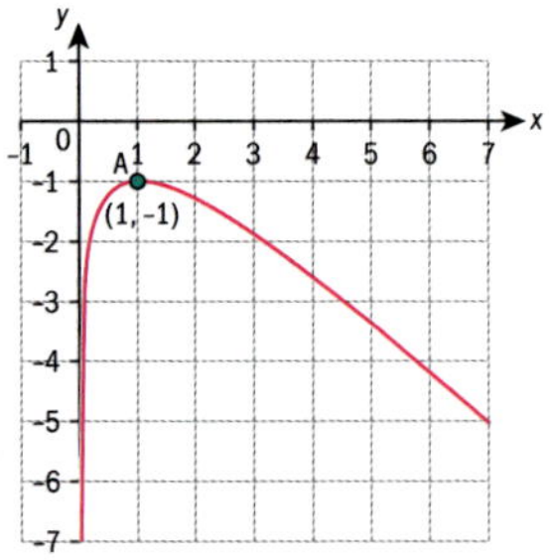

7 **a** $f'(x)=e^x-1=0\Rightarrow x=0$

Thus there exists only solution to $f'(x)=0$ so there is only turning point.

b The coordinates of the turning point are $(0, f(0))=(0,1)$ $f''(x)=e^x>0$ so the turning point must be a minimum.

c Because there is a single turning point and it has just been shown that the point is a minimum.

d Domain: $x\in\mathbb{R}$, Range: $f(x)\ge 1$

e $y=-x$

f

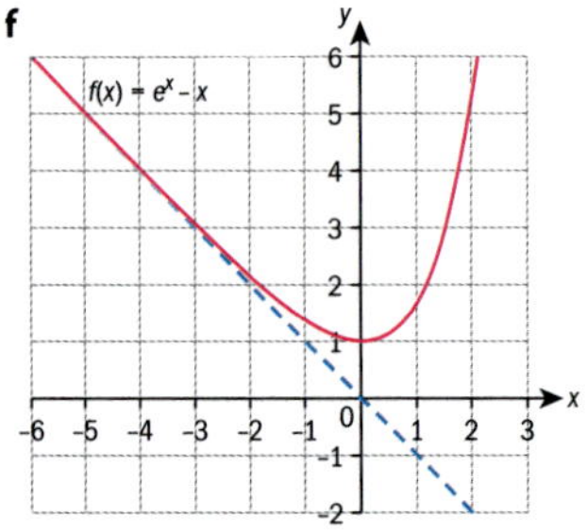

8 **a** Because $\ln|x|$ is not defined at $x=0$

b $f(x)=0\Rightarrow x\ln|x|=0$ or $x=0$ by definition of the function

For $x\ne 0$, $x\ln|x|=0\Rightarrow |x|=1\Rightarrow x=\pm 1$

c For $x>0$, $f'(x)=\ln x+1=0$ $\Rightarrow x=e^{-1}$

For $x<0$, $f(x)=x\ln(-x)\Rightarrow$ $f'(x)=\ln(-x)+1=0$ $\Rightarrow x=-e^{-1}$ So there are two turning points and these are located at $x=\pm e^{-1}$

d Minimum at $(e^{-1},-e^{-1})$ and maximum at $(-e^{-1},e^{-1})$

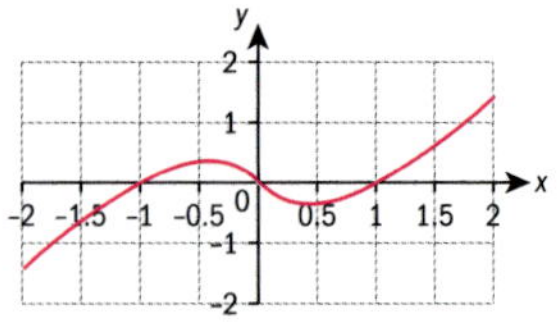

9 **a** $f'(x)=2xe^x+x^2e^x=xe^x(2+x)$

$\therefore f'(x)=0\Rightarrow x=0$ or $x=-2$ so there are two turning points.

$f''(x)=2e^x+2xe^x+2xe^x+x^2e^x=(2+4x+x^2)e^x$

$\therefore f''(x)=0\Rightarrow x^2+4x+2=0$

$\Rightarrow x=-2\pm\sqrt{2}$

So there are two points of inflection.

b $(0, 0)$ is a minimum, $(-2, 4e^{-2})$ is a maximum

c $x=-2\pm\sqrt{2}$ **d** $y=0$

e

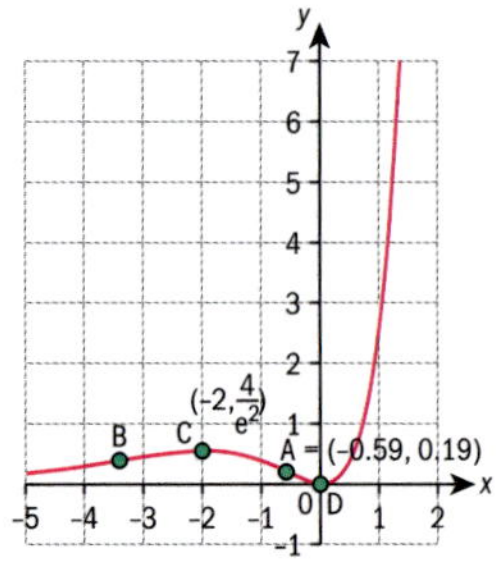

f $y=3ex-2e=e(3x-2)$

g $\dfrac{2e}{3}$

10 **a** $f'(x)=e^x\sin x+e^x\cos x$ $=e^x(\sin x+\cos x)$

$f'(x)=0\Rightarrow e^x(\sin x+\cos x)$ $=0\Rightarrow \tan x=-1$ $(e^x\ne 0)$

so in the range

$-\dfrac{3\pi}{2}\le x\le\dfrac{3\pi}{2}$,

the three roots are located at

$x=-\dfrac{5\pi}{4}$,

$x=-\dfrac{\pi}{4}$ and $x=\dfrac{3\pi}{4}$

$f''(x)=e^x(\sin x+\cos x)$ $+e^x(\cos x-\sin x)=$

$2e^x \cos x \therefore f''(x) = 0$

$\Rightarrow \cos x = 0 \ (e^x \neq 0)$

Therefore there are two points of inflexion, at $x = \pm\frac{\pi}{2}$

b Minimum at $x = -\frac{\pi}{4}$, maximum at $x = \frac{3\pi}{4}$

c $\left(\frac{\pi}{2}, e^{\frac{\pi}{2}}\right)$ and $\left(-\frac{\pi}{2}, e^{-\frac{\pi}{2}}\right)$

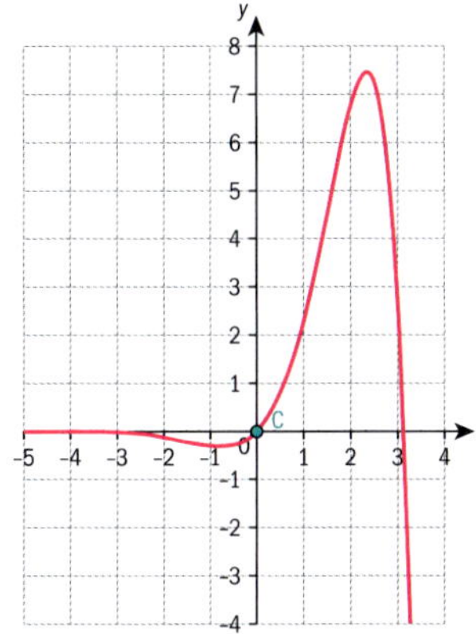

d $f'\left(\frac{\pi}{2}\right) = e^{\frac{\pi}{2}}$ so the normal at this point has gradient $-e^{-\frac{\pi}{2}}$

$f'\left(-\frac{\pi}{2}\right) = -e^{-\frac{\pi}{2}}$ so this is parallel to the normal at $x = \frac{\pi}{2}$

Exercise 7G

1 a $\frac{x^4}{4} - \tan x + C$

b $3e^x + \frac{1}{2}\ln x - \cos x + C$

c $-\frac{1}{2}\cos 2x + C$

d $\frac{1}{3}\tan 3x + C$

e $\frac{3}{2}\ln x + \frac{1}{\ln 3}3^x$

f $-\ln|1 - 3x| + C$

g $x + \ln|x - 1| + \sin x + C$

h $e^{\sin x} - 2\ln|x| + C$

i $e^{\sqrt{x}} + \cos 2x + C$

2 a $x^2 + \frac{1}{4}\cos 4x - \frac{1}{4}$

b $\frac{1}{3}x^3 - e^{-x} + \tan x$

c $\frac{3}{2}\ln(2x-5) - x^3 - 3e^{x-3} + 5$

3 a $-\frac{1}{4}\cos 2x + x + 1$

b $\frac{1}{4}e^{2x-1} - \frac{1}{4}\sin(1-2x) - \frac{x}{2} + \frac{1}{2}$

4 a -2

b $\frac{1}{3}e^{-2}\left(e^{\pi} - 1\right) + \frac{1}{2}$

c $\frac{1}{\ln 16} + \frac{e^2}{2}\left(1 - e^2\right) + \frac{7}{3}$

d $\frac{2}{\ln 3} + \ln 4$ **e** $\frac{1}{2}\left(1 + 3e^{\frac{\pi^2}{4}}\right)$

f $2(2 + \ln 3)$ **g** $-\frac{2}{3}$

h $2 - \sqrt{2}$ **i** $\frac{1}{4}$

j $\frac{1}{\ln 2}\left(\sqrt{2} - 1\right) + \ln 4$

5 $f(x) = -\ln(\cos x)$, range: $f(x) \geq 0$ $(f(x) \in \mathbb{R})$

Tangent: $y = x + \ln\sqrt{2} - \frac{\pi}{4}$,

normal: $y = -x + \ln\sqrt{2} + \frac{\pi}{4}$

Base of triangle (along y-axis):

$\left(\ln\sqrt{2} + \frac{\pi}{4}\right) - \left(\ln\sqrt{2} - \frac{\pi}{4}\right) = \frac{\pi}{2}$

Height of traingle: $\frac{\pi}{4}$ so area is

$\frac{1}{2}\left(\frac{\pi}{2}\right)\left(\frac{\pi}{4}\right) = \frac{\pi^2}{16}$

6 $f(x) = \frac{2}{x-1} - \frac{3}{2x+3}$

$\Rightarrow \int \frac{x+9}{2x^2 + x - 3}dx$

$= 2\ln|x-1| - \frac{3}{2}\ln|2x+3| + C$

7 $\frac{1}{2}\ln\left|\frac{x-1}{x+1}\right| + C$

8 $f(x) = \frac{5x+9}{x^2-9} = \frac{1}{x+3} + \frac{4}{x-3}$

$\Rightarrow \int\left(\frac{5x+9}{x^2-9}\right)dx = \ln(x+3)$
$+ 4\ln(x-3) + C$

9 $\frac{1-2x}{x(x+1)} = \frac{1}{x} - \frac{3}{x+1}$

$\Rightarrow \int_{\frac{1}{2}}^{1}\frac{1-2x}{x+x^2}dx = \int_{\frac{1}{2}}^{1}\left(\frac{1}{x} - \frac{3}{x+1}\right)dx$

$= \left[\ln x - 3\ln(x+1)\right]_{\frac{1}{2}}^{1}$

$= (0 - 3\ln 2) - \left(\ln\frac{1}{2} - 3\ln\frac{3}{2}\right)$

$= \ln\left(\frac{1}{8}\cdot 2\cdot\frac{27}{8}\right) = \ln\frac{27}{32}$

10 $\int_0^{\frac{1}{2}}\left(\frac{2+3x-x^2}{1-x^2}\right)dx$

$= \int_0^{\frac{1}{2}}\left(\frac{1-x^2+1+3x}{1-x^2}\right)dx$

$= \int_0^{\frac{1}{2}}\left(\frac{1+3x}{1-x^2}\right)dx$

$= \frac{1}{2} + \int_0^{\frac{1}{2}}\left(\frac{1+3x}{1-x^2}\right)dx$

$\frac{1+3x}{1-x^2} = \frac{1+3x}{(1-x)(1+x)}$

$= \frac{A}{1-x} + \frac{B}{1+x}$

$\Rightarrow 1 + 3x = A(1 + x) + B(1 - x)$

Set $x = 1$: $4 = 2A \Rightarrow A = 2$

Set $x = -1$: $-2 = 2B \Rightarrow B = -1$

$\therefore \frac{1+3x}{1-x^2} = \frac{2}{1-x} - \frac{1}{1+x}$

So $\int_0^{\frac{1}{2}}\left(\frac{2+3x-x^2}{1-x^2}\right)dx$

$= \frac{1}{2} + \int_0^{\frac{1}{2}}\left(\frac{2}{1-x} - \frac{1}{1+x}\right)$

$= \frac{1}{2} + \left[-2\ln(1-x) - \ln(1+x)\right]_0^{\frac{1}{2}}$

$= \frac{1}{2} + \left(-2\ln\frac{1}{2} - \ln\frac{3}{2}\right) - 0$

$= \frac{1}{2} + \ln\left(4\cdot\frac{2}{3}\right) = \frac{1}{2} + \ln\frac{8}{3}$

Exercise 7H

1 $\frac{2}{3}\left(3x^2+4\right)^{\frac{3}{2}}+C$

2 $\sin(x^3)+C$

3 $e^{2+x-x^2}+C$

4 $-e^{\cos 2x}+C$

5 $-\cos(3^x)+C$

6 $e^{\sqrt{x^3}}+C$

7 $\frac{4}{15}(x-1)^{\frac{3}{2}}(3x+2)+C$

8 $\frac{2}{15}(x+1)^{\frac{3}{2}}(7-3x)+C$

9 $\frac{1}{3}\tan\left(x^3-\frac{3}{2}x^2\right)+C$

10 $-\frac{1}{2\ln 2}2^{\cos 2x}+C$

11 $-\frac{1}{3}\left(1-x^2\right)^{\frac{3}{2}}+C$

12 $-\frac{2}{105}(1-x)^{\frac{3}{2}}\left(8+12x+15x^2\right)+C$

13 $-\frac{2}{15}(1-x)^{\frac{1}{2}}\left(3x^2+4x+8\right)+C$

14 $-\frac{1}{42}(1+x)^6(1-6x)+C$

15 $\ln(1-\cos x)+C$

16 $-\frac{2}{3}e^{\frac{3}{2}(5-x)}+C$

17 $\frac{9}{3-x}+6\ln(3-x)+x-3+C$

Note: it is permissible to incorporate the constant -3 into the arbitrary constant.

18 $-\frac{1}{15}(2x-3)^{\frac{3}{2}}(3x-7)+C$

19 $3\ln|1-x|-\frac{9}{1-x}+C$

20 $\ln|2+\tan x|+C$

Exercise 7I

1 $\frac{20}{3}$ **2** $\ln 16+3$ **3** $\ln 2$

4 $\frac{\pi}{4}$ **5** $\frac{56}{3}+\ln 9$ **6** -1

7 $\frac{1}{3}$ **8** $\frac{1}{4}$ **9** $\frac{1}{3}$ **10** $\frac{52}{3\ln 3}$

11 $\tan^3 x=\tan x\cdot\tan^2 x$
$=\tan x(\sec^2 x-1)$
$=\sec^2 x\tan x-\tan x$

$\int_0^{\frac{\pi}{4}}\tan^3 x\,dx$

$=\int_0^{\frac{\pi}{4}}\sec^2 x\tan x\,dx-\int_0^{\frac{\pi}{4}}\tan x\,dx$

Using question 8, and the fact that $\int_0^{\frac{\pi}{4}}\tan x\,dx=-\left[\ln\left(\frac{1}{\sqrt{2}}\right)\right]$

$\int_0^{\frac{\pi}{4}}\tan^3 x\,dx=\frac{1}{4}+\ln\left(\frac{1}{\sqrt{2}}\right)$

$=\frac{1}{2}\left(1+2\ln\left(\frac{1}{\sqrt{2}}\right)\right)$

$=\frac{1}{2}(1-\ln 2)$ as required

12 $\frac{3}{4}$ or $\frac{9}{4}$

13 $\therefore\int_{-2}^{\frac{2}{\sqrt{3}}-2}\frac{1}{3x^2+12x+16}dx$

$=\frac{1}{4}\int_0^{\frac{\pi}{4}}\frac{1}{1+\frac{3}{4}\left(\frac{2}{\sqrt{3}}\tan\theta\right)^2}\left(\frac{2}{\sqrt{3}}\sec^2\theta\right)d\theta$

$=\frac{1}{2\sqrt{3}}\int_0^{\frac{\pi}{4}}\frac{\sec^2\theta}{1+\tan^2\theta}d\theta=\frac{1}{2\sqrt{3}}$

$\int_0^{\frac{\pi}{4}}d\theta=\frac{\pi}{8\sqrt{3}}$ as required

14 $e^x+e^{-x}=e^{-x}\left(e^{2x}+1\right)=\frac{e^{2x}+1}{e^x}$

$\therefore\int_0^{-\ln\sqrt{3}}\frac{1}{e^x+e^{-x}}dx=\int_0^{-\ln\sqrt{3}}\frac{e^x}{e^{2x}+1}dx$

Let $e^x=\tan\theta\Rightarrow e^x dx=\sec^2\theta\, d\theta$
$=(e^{2x}+1)d\theta$

Limits: $x=-\ln\sqrt{3}\Rightarrow\theta=\frac{\pi}{6}$,

$x=0\Rightarrow\theta=\frac{\pi}{4}$

$\therefore\int_0^{-\ln\sqrt{3}}\frac{1}{e^x+e^{-x}}dx$

$=\int_{\frac{\pi}{4}}^{\frac{\pi}{6}}d\theta=\frac{\pi}{6}-\frac{\pi}{4}=-\frac{\pi}{12}$

15 a There are many ways to do this, the easiest being:

$\tan x=\frac{2\tan\frac{x}{2}}{1-\tan^2\frac{x}{2}}=\frac{2t}{1-t^2}$

Construct RA triangle with opposite side of length $2t$ and adjacent of length $1-t^2$

Then the hypotenuse is of length

$\sqrt{\left(1-t^2\right)+(2t)^2}=\sqrt{1+2t^2+t^4}$

$=\sqrt{\left(1+t^2\right)^2}=1+t^2$

$\sin x=\frac{2t}{1+t^2}$ as required

b Differentiating implicitly,

$\frac{dx}{dt}\cos x=\frac{2\left(1+t^2\right)-(2t)(2t)}{\left(1+t^2\right)^2}$

$=\frac{2-2t^2}{\left(1+t^2\right)^2}=\frac{2\left(1-t^2\right)}{\left(1+t^2\right)^2}$

$\cos x=\frac{1-t^2}{1+t^2}\quad\frac{dx}{dt}$

$=\frac{1+t^2}{1-t^2}\frac{2\left(1-t^2\right)}{\left(1+t^2\right)^2}$

$=\frac{2}{1+t^2}$ as required

c Let $t=\tan\left(\frac{x}{2}\right)\Rightarrow dx=\frac{2dt}{1+t^2}$

Limits: $x=\frac{\pi}{2}\Rightarrow t=1$,

$x=0\Rightarrow t=0$

$\therefore\int_0^{\frac{\pi}{2}}\frac{1}{1+\sin x}dx$

$=\int_0^1\frac{1}{1+\frac{2t}{1+t^2}}\left(\frac{2}{1+t^2}\right)dt$

$= \int_0^1 \frac{2}{1+t^2+2t}dt = \int_0^1 \frac{2}{(1+t)^2}dt$

$= -\left[\frac{2}{1+t}\right]_0^1 = -(1-2)$

$= 1$ as required

Exercise 7J

1 $2e^x(x-1)+C$

2 $3(\sin x - x\cos x)+C$

3 $e^x(3-2x)+C$

4 $(2-x)\cos(2-x)-\sin(2-x)+C$

5 $\frac{2}{3}(1+2x)\tan\frac{x}{2}+\frac{8}{3}\ln\left(\cos\frac{x}{2}\right)+C$

6 $\frac{2^{x+1}}{\ln 2}\left(x-\frac{1}{\ln 2}\right)+C$

7 $-\frac{3^{-x}}{\ln 3}\left(x+\frac{1}{\ln 3}\right)+C$

8 $\frac{x^4}{16}(4\ln x-1)+C$

9 $\left(2x-\frac{5}{2}x^2\right)\ln\frac{x}{3}-2x+\frac{5}{4}x^2+C$

10 $x\arcsin x+\sqrt{1-x^2}+C$ where integral evaluated either by inspection or substitution

11 $\left(x+\frac{3}{2}x^2-\frac{1}{3}x^3\right)\ln 4x-x-\frac{3}{4}x^2$
$+\frac{1}{9}x^3+C$

12 $\frac{x}{\ln a}(\ln x-1)+C$

13 $\frac{x^2}{2}\arccos x+\frac{1}{4}\left(\arcsin x-x\sqrt{1-x^2}\right)+C$

14 $2(x^2+1)\arctan x-2x+C$

15 $\frac{x^3}{3}\arccos x-\frac{\sqrt{1-x^2}}{9}(2+x^2)+C$

16 $\sqrt{3}\arctan\sqrt{3}-\ln 2$

17 $\frac{1}{324}(20-\ln 3)$

18 $\frac{\pi}{4}-\ln\left(\sqrt{2}\right)$

19 $\left(e-e^2+\frac{e^3}{3}-\frac{1}{3}\right)$

$\ln(3)-\frac{e^2}{2}+\frac{2e^3}{9}+\frac{11}{18}$

20 $\frac{\sqrt{2}(4-\pi)}{8}$

Exercise 7K

1 $\frac{e^{2x}}{2}(2x^2-2x+1)+C$

2 $3(x^2-2)\sin x+x(6-x^2)\cos x+C$

3 $(2+x-x^2)\sin x+(1-2x)\cos x$
$+C$

4

$-4(x^2-32)\cos\left(\frac{x}{4}\right)+32x\sin\left(\frac{x}{4}\right)+C$

5 $3e^{\frac{x}{3}}(x^3-9x^2+54x-162)+C$

6 $2(e^2-1)$ 7 $\pi-2$

8 $\frac{\pi^2-4}{4}$ 9 $\frac{5e^3-2}{81}$

10 $\frac{1}{2}\left(1-\frac{5}{e^2}\right)$

Exercise 7L

1 $I=\frac{\tan^2 x}{2}+C$ (redefine I to include constant)

2 $I=\frac{1}{2}\sin^2 x+C$ (redefine I to include constant)

3 $I=\frac{1}{5}(3\sin 2x\sin 3x+2\cos 2x\cos 3x)+C$ (redefine I to include constant)

4 $I=\frac{e^{3x}}{13}(2\sin 2x+3\cos 2x)+C$ (redefine I to include constant)

5 $I=\frac{1}{2}(x-\sin x\cos x)+C$ (redefine I to include constant)

Chapter review

1 a $x=2$ or $x=3$

b $x=1$

2 a $x=10$

b $x=189$

3 a $\frac{1}{4}$ b $x=\frac{1}{49}$

4 $\frac{dy}{dx}=3(2e^{3x}+7e^{-3x})$

$\frac{dy}{dx}=9(2e^{3x}-7e^{-3x})=9y$

5 $\ln x+\ln x^2+\ldots+\ln x^k$
$=\ln x+2\ln x+\ldots+k\ln x$
$=(1+2+\ldots.+k)\ln x$
$=\frac{1}{2}k(k+1)\ln x=2k(k+1)$
$\ln x=4 \quad x=e^4$

6 $\ln x=2\pm\sqrt{5}$

7 $\left(-\ln\sqrt{2},\frac{3\sqrt{2}}{4}\right)$

8 For $x>1$, $y=\ln x\Rightarrow\frac{dy}{dx}$
$=\frac{1}{x}=\frac{\ln x}{x\ln x}=\frac{\ln x}{x|\ln x|}$
For $0<x<1$, $y=|\ln x|=-\ln x$
$\frac{dy}{dx}=-\frac{1}{x}=-\frac{\ln x}{x\ln x}$
$=\frac{\ln x}{x(-\ln x)}=\frac{\ln x}{x|\ln x|}$
So $\frac{dy}{dx}=\frac{\ln x}{x|\ln x|}$

9 a $f'(x)=3\frac{1}{x}(\ln x)^2=\frac{3}{x}(\ln x)^2$

$f''(x)=-\frac{3}{x^2}(\ln x)^2+\frac{3}{x}\frac{2}{x}\ln x$
$=-\frac{3}{x^2}(\ln x)^2+\frac{6}{x^2}\ln x$

$\therefore f''(x)=\frac{3}{x^2}(\ln x)(-\ln x+2)=0$

So $\ln x=0$ or $\ln x=2$

$\ln x=0\Rightarrow x=1$

$\ln x=2\Rightarrow x=e^2>2^2=4$ so this is outside the domain. Therefore the only point of inflection is $(1,0)$

b tan : $y = \frac{3}{e}x - 2$, normal : $y = -\frac{e}{3}x + \frac{e^2}{3} + 1$

c $\frac{e}{2}\left(3 + \frac{e^2}{3}\right)$

10 $2\left(\sqrt{3} - 1\right)$

11 $\frac{\sqrt{3}+1}{4\sqrt{2}}$

12 $y = \frac{1}{2}x$

13 $\frac{x}{2}\sqrt{4-9x^2} - \frac{2}{3}\arccos\left(\frac{3x}{2}\right) + C$

14 $2e^{\frac{x}{2}}\left(x^2 - 4x + 8\right) + C$

15 $\frac{3^x}{1+(\ln 3)^2}\left((\ln 3)\sin x - \cos x\right)$

Exam-style questions

16 a Attempt to factorise (1 mark)

$(3e^x - 1)(e^x + 4) = 0$ (1 mark)

$e^x = \frac{1}{3} \Rightarrow x = \ln\left(\frac{1}{3}\right)$ (1 mark)

$= -\ln 3$ (1 mark)

$e^x = -4$ has no solutions (1 mark)

b Attempt to factorise (1 mark)

$(\ln x - 9)(\ln x + 4) = 0$ (1 mark)

$\ln x = 9 \Rightarrow x = e^9$ (1 mark)

$\ln x = -4 \Rightarrow x = e^{-4}$ (1 mark)

17 a $10^{(5x-1)} = 15$ (1 mark)

$\log_{10}10^{5x-1} = \log_{10}15$ (1 mark)

$5x - 1 = \log_{10}15$ (1 mark)

$x = \frac{1+\log_{10}15}{5}$ (1 mark)

b $\ln\left(3^{2-x}\right) = \ln\left(7^{\frac{x}{2}}\right)$ (1 mark)

$(2-x)\ln 3 = \frac{x}{2}\ln 7$ (1 mark)

$2\ln 3 - x\ln 3 = \frac{x\ln 7}{2}$ (1 mark)

$4\ln 3 - 2x\ln 3 = x\ln 7$

$4\ln 3 = x\ln 7 + 2x\ln 3$

$4\ln 3 = x(\ln 7 + 2\ln 3)$

$x = \frac{4\ln 3}{(\ln 7 + 2\ln 3)}$ (1 mark)

18 a $y = 3\log_{10}(2x + 100)$ (1 mark)

$y = 3\log_{10}[2(x + 50)]$ (1 mark)

The transformations required are therefore:

Translation 50 units to the left (1 mark)

Stretch along the x-axis, scale factor $\frac{1}{2}$ (1 mark)

Stretch along the y-axis, scale factor 3 (1 mark)

b $y = \log_{10}(2x + 100)^3$

$= \frac{\ln(2x+100)^3}{\ln 10}$ (use of change of base formula) (1 mark)

$= \left(\frac{3}{\ln 10}\right)\ln(2x+100)$ (1 mark)

$\frac{dy}{dx} = \left(\frac{3}{\ln 10}\right)\left(\frac{2}{2x+100}\right)$ (2 marks)

$= \frac{6}{(2x+100)\ln 10}$

(or equivalent) (1 mark)

19 a $\log_{16}4 = \frac{\log_4 4}{\log_4 16} = \frac{1}{\log_4 16}$ (1 mark)

$= \frac{1}{\log_4 4^2}$ (1 mark)

$= \frac{1}{2\log_4 4}$

$= \frac{1}{2}$ (1 mark)

b

$\log_{16}(x-4) - \log_{16}(x-12) = \frac{1}{2}$

$\log_{16}(x - 4) - \log_{16}(x - 12) = \log_{16}4$ (1 mark)

$\log_{16}\left(\frac{x-4}{x-12}\right) = \log_{16}4$ (1 mark)

$\frac{x-4}{x-12} = 4$ (1 mark)

$x - 4 = 4x - 48$

$3x = 44$

$x = \frac{44}{3}$ (1 mark)

20 a $\frac{dy}{dx} = -x^3e^{-x} + 3x^2e^{-x}$ (2 marks)

$(= e^{-x}(3x^2 - x^3))$

Substituting $x = 3$ gives $\frac{dy}{dx} = 0$, hence a stationary point. (1 mark)

$\frac{d^2y}{dx^2} = e^{-x}\left(6x - 3x^2\right) - e^{-x}\left(3x^2 - x^3\right)$ (2 marks)

$= e^{-x}(6x - 6x^2 + x^3)$

Substituting $x = 3$ (1 mark)

$\frac{d^2y}{dx^2} = -9e^{-3} < 0$, so a maximum. (1 mark)

b $x = 1 \Rightarrow y = \frac{1}{e}$ (1 mark)

$x = 1 \Rightarrow \frac{dy}{dx} = \frac{2}{e}$ (1 mark)

Equation of tangent is

$y - \frac{1}{e} = \frac{2}{e}(x-1)$ (2 marks)

$ey - 1 = 2x - 2$

$2x - ey - 1 = 0$ (1 mark)

21 $2x^2 + 3x - 2 = (2x - 1)(x + 2)$ (1 mark)

$\frac{5x}{(2x-1)(x+2)} = \frac{A}{2x-1} + \frac{B}{x+2}$

$5x = A(x + 2) + B(2x - 1)$ (1 mark)

$x = \frac{1}{2} \Rightarrow A = 1$ (1 mark)

$x = -2 \Rightarrow B = 2$ (1 mark)

$$\int_1^5 \left(\frac{1}{2x-1}+\frac{2}{x+2}\right) dx$$

$$=\left[\frac{1}{2}\ln|2x-1|+2\ln|x+2|\right]_1^5$$

(2 marks)

$$=\frac{1}{2}\ln 9+2\ln 7-2\ln 3$$ (2 marks)

$= \ln 3 + 2\ln 7 - 2\ln 3$ (1 mark)

$= 2\ln 7 - \ln 3$

$= \ln 49 - \ln 3$

$$= \ln\left(\frac{49}{3}\right)$$ (1 mark)

So $p = \frac{49}{3}$

22 a $$\int_{\frac{1}{6}}^{\frac{1}{3}} \frac{dx}{\sqrt{1-9x^2}} = \frac{1}{3}\int_{\frac{1}{6}}^{\frac{1}{3}} \frac{dx}{\sqrt{\frac{1}{9}-x^2}}$$

(1 mark)

Substitute $x = \frac{1}{3}\sin u$

(1 mark)

$$\frac{dx}{du} = \frac{1}{3}\cos u$$ (1 mark)

$$\frac{1}{3}\int_{\frac{1}{6}}^{\frac{1}{3}} \frac{dx}{\sqrt{\frac{1}{9}-x^2}} = \frac{1}{9}\int_{\frac{\pi}{6}}^{\frac{\pi}{2}} \frac{\cos u\, du}{\sqrt{\frac{1}{9}-\frac{1}{9}\sin^2 u}}$$

(1 mark)

$$=\frac{1}{3}\int_{\frac{\pi}{6}}^{\frac{\pi}{2}} \frac{\cos u\, du}{\sqrt{1-\sin^2 u}}$$

$$=\frac{1}{3}\int_{\frac{\pi}{6}}^{\frac{\pi}{2}} du$$ (1 mark)

$$=\frac{1}{3}\left(\frac{\pi}{2}-\frac{\pi}{6}\right)$$ (2 marks)

$$=\frac{1}{3}\left(\frac{\pi}{3}\right)$$

$$=\frac{\pi}{9}$$ (1 mark)

b 0.349 (1 mark)

23 a $$\int \frac{x}{\sin^2 x}\, dx = \int x\,\mathrm{cosec}^2 x\, dx$$

(1 mark)

Use integration by parts with $u = x$ and $\frac{dv}{dx} = \mathrm{cosec}^2 x$

$u = x \Rightarrow \frac{du}{dx} = 1$ (1 mark)

$\frac{dv}{dx} = \mathrm{cosec}^2 x \Rightarrow v = -\cot x$

(1 mark)

$$\int \frac{x}{\sin^2 x}\, dx = -x\cot x + \int \cot x\, dx$$

(2 marks)

$= -x\cot x + \ln|\sin x| + c$

(1 mark)

b $A = \left[-x\cot x + \ln|\sin x|\right]_{\frac{\pi}{4}}^{\frac{\pi}{2}}$

$\cot x\, dx$ (1 mark)

$$=\left[-\frac{\pi}{2}\cot\frac{\pi}{2}+\ln\left|\sin\frac{\pi}{2}\right|\right] - \left[-\frac{\pi}{4}\cot\frac{\pi}{4}+\ln\left|\sin\frac{\pi}{4}\right|\right]$$

(1 mark)

$$=[-0+0]-\left[-\frac{\pi}{4}+\ln\frac{1}{\sqrt{2}}\right]$$

(2 marks)

$$=\frac{\pi}{4}-\ln\frac{1}{\sqrt{2}}$$

$$=\frac{\pi}{4}-\ln 2^{-\frac{1}{2}}$$ (1 mark)

$$=\frac{\pi}{4}+\frac{1}{2}\ln 2$$ (1 mark)

$$=\frac{\pi}{4}+\frac{1}{4}\ln 4$$ (1 mark)

$$=\frac{1}{4}(\pi+\ln 4)$$

24 $a = \log_2 343$ (1 mark)

$a + 3d = \log_2 1331$ (1 mark)

Solve simultaneously to find d

(1 mark)

$\log_2 343 + 3d = \log_2 1331$

$3d = \log_2 1331 - \log_2 343$

$$3d = \log_2\left(\frac{1331}{343}\right)$$

$$d = \frac{1}{3}\log_2\left(\frac{1331}{343}\right)$$ (1 mark)

$$d = \log_2\left(\frac{1331}{343}\right)^{\frac{1}{3}}$$ (1 mark)

$$d = \log_2\left(\frac{11}{7}\right)$$ (1 mark)

So $\log_2 x = a + d$ (1 mark)

$$= \log_2 343 + \log_2\left(\frac{11}{7}\right)$$

$$= \log_2\left(343\times\frac{11}{7}\right)$$ (1 mark)

$= \log_2(49\times 11)$

$= \log_2 539$ (1 mark)

So $x = 539$

25 Use integration by parts

$u = e^{-x} \Rightarrow \frac{du}{dx} = -e^{-x}$ (1 mark)

$\frac{dv}{dx} = \sin 3x \Rightarrow v = -\frac{1}{3}\cos 3x$ (1 mark)

$$\int e^{-x}\sin 3x\, dx = -\frac{e^{-x}\cos 3x}{3} - \int -\frac{1}{3}\cos 3x\left(-e^{-x}\right)dx$$ (2 marks)

$$= -\frac{e^{-x}\cos 3x}{3} - \frac{1}{3}\int e^{-x}\cos 3x\, dx$$

By using integration by parts a second time,

$$\int e^{-x}\cos 3x\, dx = \frac{e^{-x}\sin 3x}{3} + \frac{1}{3}\int e^{-x}\sin 3x\, dx$$ (2 marks)

$$\text{So} \int e^{-x}\sin 3x\, dx = -\frac{e^{-x}\cos 3x}{3} - \frac{1}{3}\left[\frac{e^{-x}\sin 3x}{3} + \frac{1}{3}\int e^{-x}\sin 3x\, dx\right]$$ (1 mark)

$$\int e^{-x}\sin 3x\, dx = -\frac{e^{-x}\cos 3x}{3} - \frac{e^{-x}\sin 3x}{9} - \frac{1}{9}\int e^{-x}\sin 3x\, dx$$

$$\frac{10}{9}\int e^{-x}\sin 3x\, dx = -\frac{e^{-x}\cos 3x}{3} - \frac{e^{-x}\sin 3x}{9}$$ (1 mark)

$$\int e^{-x}\sin 3x\, dx = \frac{9}{10}\left(-\frac{e^{-x}\cos 3x}{3} - \frac{e^{-x}\sin 3x}{9}\right)$$ (1 mark)

$$= -\frac{e^{-x}}{10}(\sin 3x + 3\cos 3x) + c$$

Chapter 8

Skills check

1 a -8 b 1 c 1

2 a $y=\frac{2}{3}$ b $y=0$

c $y=1$

3 a $\frac{2}{3}\sqrt{(x^2-1)^3}+c$

b $\frac{\sin^2 x}{2}+c$

c $x^2(2\ln x-1)+5x(\ln x-1)+c$

Exercise 8A

1 a $\frac{1}{2}$ b $\frac{1}{5}$ c 16

d 8.59 e 0.878

2 a 10.1 b 7.00 c 1.34

3 0.668

Exercise 8B

1 $\frac{5}{2}$

2 0.163 sq. units

3 11.5 sq. units

4 a $k=1.08$ b $m=2.83$

Exercise 8C

1 a $\frac{15\pi}{2}$ b $\frac{31\pi}{5}$

2 a $\frac{31\pi}{30}$ b $\frac{7\pi}{6}$

c $\frac{\pi}{2}$ d $\frac{\pi(4-\pi)}{4}$

e $\frac{\pi}{2}\left(1-\frac{1}{e^2}\right)$ f $\pi\ln(4)$

3 $\frac{36\pi}{5}$

4 a $\pi(4-\sin(4))$ b $4-\pi$

5 $\frac{\pi^2}{4}$ 6 $\frac{\pi^2}{4}$

7 a 2.35 b 4.18

8 3.58 9 1.31

Exercise 8D

1 a 0.362 m b 0.479 m

2 a i $\frac{\pi}{2};\frac{3\pi}{2};$

ii $\frac{\pi}{2}<t<\frac{3\pi}{2};$

iii $0<t<\frac{\pi}{2};\frac{3\pi}{2}<t<2\pi$

b 0 m c 12 m

3 a i $0;\frac{\pi}{2};$

ii no solution;

iii $0<t<\frac{\pi}{2}$

b 1 m c 1 m

4 a i 2

ii no solution

iii $0<t<2$

b $\frac{4\sqrt{2}}{3}$m c $\frac{4\sqrt{2}}{3}$m

5 a i $0, \pi, 2\pi$

ii $\pi<t<2\pi$ iii $0<t<\pi$

b 0 c 4.70

6 a i 0

ii no solution

iii $0<t\le 3$

b $\frac{\ln(10)}{2}$ or 1.51

c $\frac{\ln(10)}{2}$ or 1.51

7 $s(t)=9\left(1-e^{-\frac{t}{3}}\right)$

8 a 6 cm; b 22 cm

c 6 cm, 22 cm, 15 cm

9 a 5.5 cm to the left of the origin.

b 8.5 cm

Exercise 8E

1 a $y=\sqrt[3]{2x^3+c}$

b $y=-\ln(e^{-x}+c)$

c $y=+-\sqrt{3e^{2x}+c}$

d $y=Ae^{\sin x}$

e $y=\sqrt[5]{\frac{x^2}{2}-\cos x+c}$

f $y=\frac{1}{3}\arctan(3x)+c$

2 a $y=\sqrt[3]{\left(9-x^{\frac{2}{3}}\right)^2}$

b $y=\sqrt{2-e^{x^2}}$

c $y=\frac{1}{2}\ln\left(2e^x-1\right)$

d $y=e^{1-\cos(x^2)}$

e $y=xe^{-x}+e^{-x}+1$

Exercise 8F

1 a $\frac{dT}{dt}=-k(180-T)$

b 20°C

c 53.8°

2 a $\frac{dT}{dt}=k(180-T)$

b 160°C

3 $y=\frac{2}{x}$

4 a $\frac{dP}{dt}=kP$

b $P=500e^{\frac{\ln(20)}{3}t}$

c 737 000

d 7 hours

5 b 10 cm

6 b $A=0.005;\ k=0.007$

c $N=\frac{100(2.01)^t}{(2.01)^t+200};$ 40 students.

7 $P=\frac{2000(1.02)^t}{1.02^t+5.67}$; 315.

8 a $P=\frac{350(1.083)^t}{1.083^t+49};$

b $\lim_{t\to\infty}\frac{350(1.083)^t}{1.083^t+49}=350$

9 a $a=\frac{dv}{dt}=-\frac{1}{2}v$

b $v=20e^{-\frac{1}{2}t}$

10 a $v=\sqrt{2\ln\left|s^2+1\right|+4-2\ln 2}$

b $v=\sqrt{2\ln(13)+4}$

11 $v=40\tan\left(-\frac{4}{5}t+\frac{\pi}{4}\right)$;

$-53.8\ \text{ms}^{-1}$

Exercise 8G

1 a $y=x(\sin x+c)$

b $y=\frac{3x^2+c}{2x}$

c $y=2x\tan(2\ln x+c)$

2 $y=2x\left(2\ln x+\frac{\pi}{4}\right)$

3 $y=\frac{2x}{1-2\ln x}$

4 $y=x(\tan(\ln x)-1)$

Exercise 8H

1 a $y=\frac{1}{2}e^x+ce^{-x}$

b $y=Ae^{-\frac{x^2}{2}-x}$

c $y=\frac{x^3+3x+c}{3x}$

d $y=Ae^{-x}-e^{-x}\cos(e^x)$

e $y=e^{-\frac{x^2}{2}}\left(\frac{1}{3}e^{x^2}+c\right)$

f $y=\frac{\sin x}{x^2}+\frac{c}{x^2}$

g $y=\frac{(x^2-2)\sin x+2x\cos x+c}{x^2}$

2 a $y=x(-x\cos x+\sin x-+)$

b $y=\frac{11}{8}x^2-\frac{1}{x}$

c $y=\sin x+2\cos x$

Exercise 8I

1 0.852

2 158.3

3 1.57; The approximate value is less than the actual value since y' is increasing.

4 2.45

5 a 2.14; Using a smaller value for h would give a more accurate answer.

b $y=x^2+e^{1-x^2}$; 2.19

Exercise 8J

1 a 1 b $\frac{3}{4}$ c 0

d $6e^9$ e -1 f $\frac{1}{\pi}$

g 0 h -2 i 1

j 0 k 0 l $\frac{3}{2}$

2 a $y=0$ b $y=-3$

c No horizontal asymptote

d No horizontal asymptote

Exercise 8K

1 a $x+x^2+\frac{x^3}{2}+\frac{x^4}{6}$

b $1-x+\frac{x^2}{2}-\frac{x^3}{6}+\frac{x^4}{24}$

c $x-\frac{x^3}{6}$

d $1-x^2+\frac{x^4}{3}$

e $1+x+x^2+x^3+x^4$

f $1-x^2+x^4+\dots$

g $x-x^3+\dots$

2 4

Exercise 8L

1 a

$$e^{3x}=1+3x+\frac{9x^2}{2!}+\dots+\frac{3^nx^n}{n!}+\dots$$

b $1-x+x^2-x^3+\dots+(-1)^nx^n+\dots$

c $1-2x+(2x)^2-(3x)^3+\dots+(-1)^n(2x)^n+\dots$

d

$$x^2-\frac{x^6}{3}+\frac{x^{10}}{5}-\dots+(-1)^n\frac{x^{2(2n+1)}}{2n+1}+\dots$$

e

$$\sin^2 x=\frac{2x^2}{2!}-\frac{8x^4}{4!}+\frac{32x^6}{6!}+\dots+\frac{(-1)^{n+1}2^{2n-1}x^{2n}}{(2n)!}+\dots$$

2 $$\sin^2 x=\frac{2x^2}{2!}-\frac{8x^4}{4!}+\frac{32x^6}{6!}+\dots+\frac{(-1)^{n+1}2^{2n-1}x^{2n}}{(2n)!}+\dots$$

3

$$f(x)=\sum_{n=0}^{\infty}3(-1)^n x^n+\sum_{n=0}^{\infty}-2\left(\frac{x}{2}\right)^n$$

$$f(x)=\sum_{n=0}^{\infty}\left(3(-1)^n-\frac{2}{2^n}x^n\right)$$

4 a $-\frac{1}{6}$ b 2 c 1 d $\frac{1}{3}$

5 a

$$\sqrt{1-x}=1-\sum_{n=0}^{\infty}\frac{(2n)!}{\left(2^n n!\right)^2(2n-1)}x^n=1-\frac{1}{2}x-\frac{1}{8}x^2-\frac{1}{16}x^3-\frac{5}{128}x^4-\dots$$

b

$$\frac{1}{(1+x)^3}=\sum_{n=0}^{\infty}(-1)^n\frac{(n+1)(n+2)}{2}x^n=1-3x+6x^2-10x^3+..$$

c $$\frac{1}{\left(1-4x^2\right)^2}=\sum_{n=0}^{\infty}4^n(n+1)x^{2n}=1+8x^2+48x^4+\dots$$

d

$$\frac{1}{\sqrt[4]{1+2x^3}}=1+\sum_{n=0}^{\infty}(-1)^n\frac{1\cdot 5\cdot 9\cdot\ldots\cdot(4n-3)}{2^n n!}x^{3n}=1-\frac{1}{2}x^3+\frac{5}{8}x^6+\dots$$

6 $\sqrt[3]{1+x}=1+\frac{1}{3}x-\frac{1}{9}x^2+\frac{5}{81}x^3+\dots$

$f(0.2)\approx 1.06272$

7 $y = -\dfrac{\pi}{2} + x - \dfrac{\pi}{4}x^2$

8

$y = 1 + x + \dfrac{x^2}{2} + \dfrac{4}{3!}x^3 + \dfrac{14}{4!}x^4 + \dfrac{66}{5!}x^5 + \ldots;$

$y(0.2) = 1.2264$

Chapter review

1 **a** $2\sqrt{2}$ **b** 4

2 **a** $\dfrac{1}{12}$ **b** $\dfrac{1}{5}$ **c** $\dfrac{128}{15}$

d $\dfrac{64}{3}$ **e** $\dfrac{5}{3}$ **f** $2\pi - \dfrac{4}{3}(\approx 4.95)$

3 $2\ln 2 - \dfrac{1}{2}$

4 **a** $y = \dfrac{Ae^x}{x+1}$

b $y = \sqrt{c(x^2+1)}$

c $y = \dfrac{cx^2+1}{1-cx^2}$

d $y = +-\sqrt{2\ln(x+1)+c}$

e

$y = \dfrac{\sin 2x}{2x} + \dfrac{\cos 2x}{4x^2} - \dfrac{\cos 2x}{2} - \dfrac{x^3}{5} + c$

f $y = \tan x + c \sec x$

g $y = \dfrac{1}{2}e^{\frac{1}{2}x}(x+c)$

5 **a** $\dfrac{dy}{dt} = ky$ **b** $y = y_0 e^{-\frac{\ln 2}{5500}t}$

c about 13 000 years

6 **a** $y = 1.84$

b $y = \sqrt{4-x^2}\left(\arcsin\left(\dfrac{x}{2}\right) + \dfrac{1}{2}\right);$

$y = 1.77$

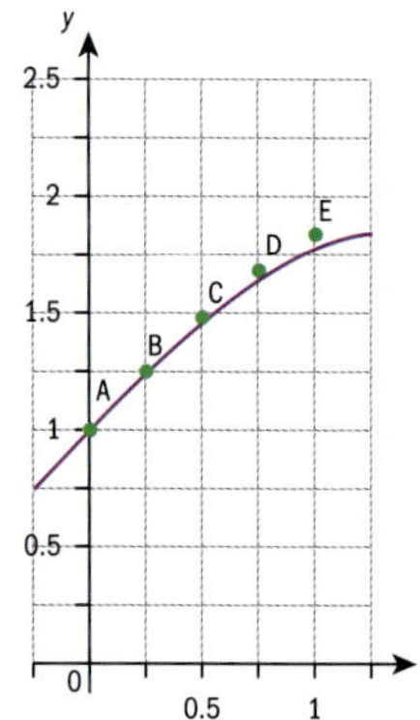

Since y' is decreasing the value of y is greater than the actual value.

7 $x - \dfrac{x^2}{2} + \dfrac{x^3}{6} - \dfrac{x^4}{12} + \dfrac{x^5}{24}$

8 0

9 $\dfrac{1}{2\pi}$

Exam-style questions

10 **a**

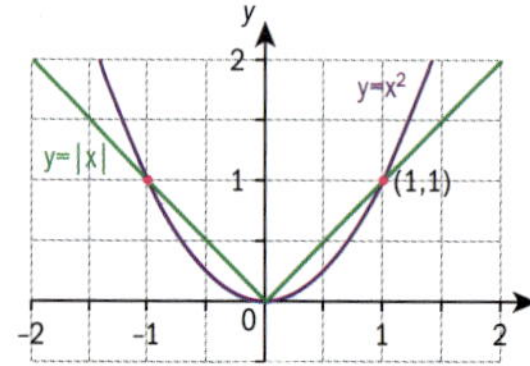

(1) for correct shape, (1) for symmetry about the y-axis, (1) for points of intersection

b $A = 2\int_0^1 (x - x^2)\,dx$ (1 mark)

$= 2\left[\dfrac{x^2}{2} - \dfrac{x^3}{3}\right]_0^1$ (1 mark)

$= 2\left(\dfrac{1}{2} - \dfrac{1}{3}\right)$ (1 mark)

$= \dfrac{1}{3}$ square units (1 mark)

11 **a** $d(t) = |\sin 2t - \sin(t - 0.24)|$ (2 marks)

b Use GDC to find the maximum of $d = 1.88$ (1 mark)

occurs when $t = 2.25$ (1 mark)

c Find intersection of graphs (1 mark)

$t = 1.13$ (1 mark)

12 **a**

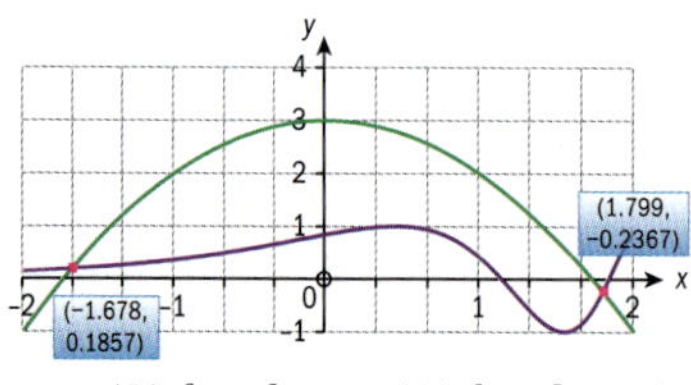

(1) for shape, (1) for domain

b $f(x) = g(x) \Rightarrow x = -1.68,$ $x = 1.80$ (2 marks)

c $-1.68 \le x \le 1.80$ (1 mark)

d $\displaystyle\int_{-1.678\ldots}^{1.799\ldots} (f(x) - g(x))\,dx = 5.68$

(2 marks)

OR can be done using technology

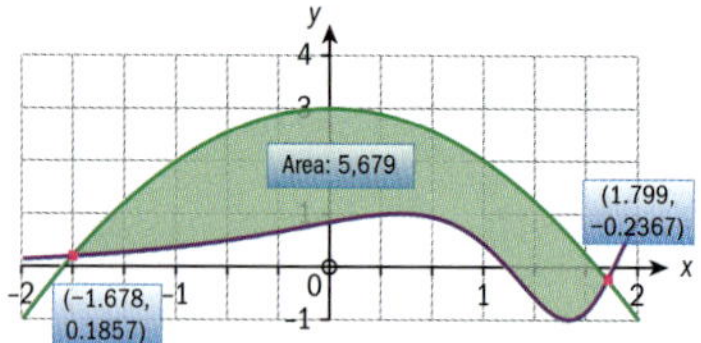

13 Let the number of insects be y.

$\dfrac{dy}{dt} = -ky$ (1 mark)

$\int \dfrac{1}{y}dy = \int -k dt$ (1 mark)

$\ln y = -kt + c$ (1 mark)

$y = e^{-kt+c}$

$y = Ae^{-kt}$

when $t = 0$, $y = 500\,000 \Rightarrow A = 500\,000$ (1 mark)

$y = 500\,000e^{-kt}$

when $t = 5$, $y = 400\,000$

$400\,000 = 500\,000e^{-5k}$ (1 mark)

$\dfrac{4}{5} = e^{-5k}$

$-5k = \ln\dfrac{4}{5}$

$k = -\dfrac{1}{5}\ln\dfrac{4}{5}$ $(= 0.0446)$ (1 mark)

$250\,000 = 500\,000e^{-kt}$ (1 mark)

$\dfrac{1}{2} = e^{-kt}$

$\ln\dfrac{1}{2} = -kt$

$t = \dfrac{5}{\ln\frac{4}{5}}\ln\dfrac{1}{2} = 15.5$ years (1 mark)

14 **a** $\int \sec^2 y\,dy = \int \cos x\,dx$ (2 marks)

$\tan y = \sin x + c$ (1 mark)

$\tan\dfrac{\pi}{4} = \sin \pi + c \Rightarrow c = 1$ (1 mark)

$\tan y = 1 + \sin x$ (1 mark)

$y = \arctan(1 + \sin x)$

b Since the denominator is 0 when $x = \frac{\pi}{2}$, to apply l'Hopital's rule the numerator must also be 0. (1 mark)

Hence $k = \arctan\left(1 + \sin\frac{\pi}{2}\right) = \arctan 2$ (1 mark)

$\lim_{x \to \frac{\pi}{2}} \frac{\arctan(1+\sin x) - \arctan 2}{\left(x - \frac{\pi}{2}\right)^2} = \frac{0}{0}$ so by applying l'Hopital's rule:

$$\lim_{x \to \frac{\pi}{2}} \frac{\arctan(1+\sin x) - \arctan 2}{\left(x - \frac{\pi}{2}\right)^2} = \lim_{x \to \frac{\pi}{2}} \frac{\cos x \cos^2 y}{2\left(x - \frac{\pi}{2}\right)} = \frac{0}{0}$$ (2 marks)

Applying l'Hopital's rule again gives

$$= \lim_{x \to \frac{\pi}{2}} \frac{-\sin\frac{\pi}{2}\cos^2(\arctan(1+\sin\frac{\pi}{2}) - 2\cos(\arctan(1+\sin\frac{\pi}{2})}{2}$$

(3 marks)

15 a $y = \ln(1 + \sin x)$

$y' = \frac{\cos x}{1+\sin x}$ (1 mark)

$y'' = -\frac{1}{1+\sin x}$ (1 mark)

$y^{(3)} = \frac{\cos x}{(1+\sin x)^2}$ (1 mark)

$y^{(4)} = \frac{-\sin x(1+\sin x)^2 x - 2(1+\sin x)\cos^{2x}}{(1+\sin x)^4}$ (2 marks)

$y(0) = 0;\ y'(0) = 1$ (1 mark)

$y''(0) = -1;\ y^{(3)}(0) = 1;\ y^{(4)}(0) = -2$ (2 marks)

$\ln(1 + \sin x) = x - \frac{1}{2}x^2 + \frac{1}{6}x^3 - \frac{1}{12}x^4 + \cdots$

b i $\ln(1 - \sin x) = \ln(1 + \sin(-x))$ (1 mark)

$= -x - \frac{1}{2}x^2 - \frac{1}{6}x^3 - \frac{1}{12}x^4 + \ldots$ (1 mark)

ii $\ln(1 + \sin x) + \ln(1 - \sin x) = \ln(1 - \sin^2 x)$ (1 mark)

$= \ln\cos^2 x$ (1 mark)

$\ln\cos^2 x = -x^2 - \frac{1}{6}x^4 + \ldots$ (1 mark)

$\ln\cos x = \frac{1}{2}\ln\cos^2 x = -\frac{1}{2}x^2 - \frac{1}{12}x^4 + \ldots$ (1 mark)

iii $\frac{d}{dx}(\ln\cos x) = \frac{1}{\cos x} \times (-\sin x)$ (1 mark)

$= -\tan x$ (1 mark)

$\tan x = x + \frac{1}{3}x^3 + \ldots$ (2 marks)

c $\frac{\tan(x^2)}{\ln\cos x} = \frac{x^2 + \frac{x^4}{3} + \ldots}{-\frac{x^2}{2} - \frac{x^4}{12} + \ldots}$

(1 mark)

$= \frac{1 + \frac{x^2}{3} + \ldots}{-\frac{1}{2} - \frac{x^2}{12} + \ldots}$ (1 mark)

$\to -2$ as $x \to 0$ (1 mark)

$\lim_{x \to 0}\left(\frac{\tan(x^2)}{\ln\cos x}\right) = -2$

(1 mark)

Chapter 9

Skills check

1 **a** 29 **b** $2\sqrt{30}$

2 10

3 a and c are parallel, b and d are parallel, and both perpendicular to a and c. $m_e = \frac{2}{3}$ which is neither equal nor opposite reciprocal to the other slopes, hence e is not parallel or perpendicular to any of the other lines.

4 $x = -2,\ y = -3$

Exercise 9A

1 Correct vectors drawn

2 **a** $\mathbf{a} + \mathbf{b}$ **b** $-\mathbf{a} - \mathbf{b}$

c $\mathbf{b} - \mathbf{a}$ **d** $-2\mathbf{a} - \mathbf{b}$

e $-2\boldsymbol{a} + \frac{3}{2}\boldsymbol{b}$ **f** $2\mathbf{b} + \frac{3}{2}\mathbf{a}$

g $-\frac{3\mathbf{a}}{2} + \frac{\mathbf{b}}{2}$ **h** $\frac{1}{2}\mathbf{b} - \frac{1}{2}\mathbf{a}$

3 **a** $AG = \boldsymbol{a} + \boldsymbol{c} + \boldsymbol{b}$

b $CE = \boldsymbol{c} - \boldsymbol{b} - \boldsymbol{a}$

c $DF = \boldsymbol{a} - \boldsymbol{b} + \boldsymbol{c}$

d $MN = \frac{1}{2}\boldsymbol{c} + \boldsymbol{b} - \frac{1}{2}\boldsymbol{a}$

4 **i** $\lambda(\mu a) = \lambda\mu a = (\lambda\mu)a = (\mu\lambda)a = \mu(\lambda a)$

ii We just the distributivity of scalar multiplication $\lambda(\mathbf{a} + \mathbf{b}) = \lambda\mathbf{a} + \lambda\mathbf{b}$

iii We just the distributivity of scalar multiplication $(\lambda + \mu)\mathbf{a} = \lambda\mathbf{a} + \mu\mathbf{a}$

iv 1 is the identity so the operation from the left and the right returns the same vector $1 \cdot \boldsymbol{a} = \boldsymbol{a} \cdot 1 = \boldsymbol{a}$

v multiplying by zero will always return zero vector, as each component of the vector is multiplied by zero. $0 \cdot \boldsymbol{a} = \mathbf{0}$

Exercise 9B

1 We need to show that $BD = \lambda PQ$ so we use the triangle rule for both diagonals to get $BC + CD = BD$ and $PC + CQ = PQ$ where $PC = \frac{1}{2}BC$ and $CQ = \frac{1}{2}CD$
Then $BD = 2(PC + CQ) = 2PQ$ hence they are parallel.

Additionally, $PQ = \frac{1}{2}BD$

2 If PQ is perpendicular to AC, then PQ is parallel to BD, as the diagonals are perpendicular. Then

$PQ = QC + CP$

$AC = AD + DC$

$BD = DC + CB$

$BD = 2QC + 2CP = 2PQ \Rightarrow BD$ and PQ are parallel, hence PQ is orthogonal to AC $AD = 2CP$, $DC = 2QC$

Then $AC = 2CP + 2QC = 2(CP + QC) = 2PQ$ as requested

3 On the side [AB] draw the point E such that **BC** = **ED**. Hence

$\mathbf{NM} = \mathbf{ND} + \mathbf{DC} + \mathbf{CM} = \frac{1}{2}\mathbf{AB}$

$+ \mathbf{DC} + \frac{1}{2}\mathbf{CB} = \frac{1}{2}\mathbf{AB} + \frac{1}{2}\mathbf{DE}$

$+ \mathbf{DC} = \frac{1}{2}\mathbf{AE} + \frac{1}{2}\mathbf{DC} + \frac{1}{2}\mathbf{DC}$

$= \frac{1}{2}\mathbf{AE} + \frac{1}{2}\mathbf{EB} + \frac{1}{2}\mathbf{DC} = \frac{1}{2}$

$\mathbf{AB} + \frac{1}{2}\mathbf{DC}$

We know that [AB] and [DC] are parallel, hence since [NM] is a linear combination of them then all three of them are mutually parallel and the same equation is true for the magnitudes.

$|\mathbf{NM}| = \frac{1}{2}|\mathbf{AB}| + \frac{1}{2}|\mathbf{DC}|$

4 We have that $DC = \frac{5}{3}y$ where y is some length. Then the ratio gives us that $\frac{3}{5}PC = y$

5 **a** $HC = HF + FE + EC$ where $FE = AB$

$HC - HQ = AB$

$HC - \lambda HC = AB$

$HC = (1 - k)AB$

Hence they are parallel. This means we can form a right angled triangle HQA and Pythagoras' theorem gives us $HC = \left(1 + \sqrt{2}\right)AB$

b $MN = ME + ED + DN$

$MN = \frac{1}{2}FE + ED + \frac{1}{2}DC$

$MN = AD + \frac{1}{2}OD + \frac{1}{2}DC$

Hence they are parallel. Again we have a right angled triangle which gives us that

$MN = \left(1 + \frac{\sqrt{2}}{2}\right)AB$

6 $KL = KB + BL$

$NM = ND + DM$

We know that they are the midpoints, so $KL = \frac{AB}{2} + \frac{BC}{2}$

and $NM = \frac{AD}{2} + \frac{DC}{2}$

Then we form a parallelogram.

Exercise 9C

1 **a** $\boldsymbol{a} + \boldsymbol{b} = (2 - 3)\boldsymbol{i} + (-5 + 4)\boldsymbol{j} = -\boldsymbol{i} - \boldsymbol{j}$

b $\boldsymbol{a} - \boldsymbol{b} = (2 + 3)\boldsymbol{i} + (-5 - 4)\boldsymbol{j} = 5\boldsymbol{i} - 9\boldsymbol{j}$

c $5\boldsymbol{a} - 6\boldsymbol{b} = 5(2\boldsymbol{i} - 5\boldsymbol{j}) - 6(-3\boldsymbol{i} + 4\boldsymbol{j}) = 10\boldsymbol{i} - 25\boldsymbol{j} + 18\boldsymbol{i} - 24\boldsymbol{j} = 28\boldsymbol{i} - 49\boldsymbol{j}$

d $7\boldsymbol{b} - 4\boldsymbol{a} = 7(-3\boldsymbol{i} + 4\boldsymbol{j}) - 4(2\boldsymbol{i} - 5\boldsymbol{j}) = -21\boldsymbol{i} + 28\boldsymbol{j} - 8\boldsymbol{i} + 20\boldsymbol{j} = -\boldsymbol{i} + 20\boldsymbol{j}$

e $\frac{3}{5}\boldsymbol{a} + \frac{3}{4}\boldsymbol{b} = \frac{3}{5}(2\boldsymbol{i} - 5\boldsymbol{j}) + \frac{3}{4}(-3\boldsymbol{i} + 4\boldsymbol{j}) = \frac{6}{5}\boldsymbol{i} - \frac{15}{5}\boldsymbol{j} - \frac{9}{4}\boldsymbol{i} + \frac{12}{4}\boldsymbol{j} = -\frac{21}{20}\boldsymbol{i}$

2 **a** Let
$\alpha(3\boldsymbol{i} + 2\boldsymbol{j}) + \beta(\boldsymbol{i} + 5\boldsymbol{j}) = 5\boldsymbol{i} - \boldsymbol{j}$
where α and β are constants. Then
$3\alpha\boldsymbol{i} + 2\alpha\boldsymbol{j} + \beta\boldsymbol{i} + 5\beta\boldsymbol{j} = 5\boldsymbol{i} - \boldsymbol{j}$
This gives two equations
$3\alpha + \beta = 5$ $2\alpha + 5\beta = -1$

Then $\beta = 5 - 3\alpha$
we substitute to get α
$2\alpha + 5(5 - 3\alpha) = -1$
$-13\alpha = -26$
$\alpha = 2$

and we substitute again to get β

$\beta = 5 - 3(2) = -1$

b Let

$\alpha(3\boldsymbol{i} + 2\boldsymbol{j}) + \beta(\boldsymbol{i} + 5\boldsymbol{j}) = 10\boldsymbol{i} + 9\boldsymbol{j}$ where α and β are constants.

Then

$3\alpha\boldsymbol{i} + 2\alpha\boldsymbol{j} + \beta\boldsymbol{i} + 5\beta\boldsymbol{j} = 10\boldsymbol{i} - 9\boldsymbol{j}$

This gives two equations
$3\alpha + \beta = 10$

$2\alpha + 5\beta = 9$

Then
$\beta = 10 - 3\alpha$
we substitute to get
α $2\alpha + 5(10 - 3\alpha) = 9$

$-13\alpha = -41$

$\alpha = \frac{41}{13}$ and we substitute again to get β

$\beta = 10 - 3\left(\frac{41}{13}\right) = \frac{7}{13}$

c Let
$\alpha(3\boldsymbol{i}+2\boldsymbol{j})+\beta(\boldsymbol{i}+5\boldsymbol{j})=-9\boldsymbol{i}+7\boldsymbol{j}$
where α and β are constants. Then
$3\alpha\boldsymbol{i}+2\alpha\boldsymbol{j}+\beta\boldsymbol{i}+5\beta\boldsymbol{j}=-9\boldsymbol{i}+7\boldsymbol{j}$
This gives two equations
$3\alpha+\beta=-9$
$2\alpha+5\beta=7$
Then $\beta=-9-3\alpha$
we substitute to get α
$2\alpha+5(-9-3\alpha)=7$
$-13\alpha=52$
$\alpha=-4$ and
we substitute again to get β
$\beta=-9-3(-4)=3$

d Let
$\alpha(3\boldsymbol{i}+2\boldsymbol{j})+\beta(\boldsymbol{i}+5\boldsymbol{j})=i$
where α and β are constants. Then
$3\alpha\boldsymbol{i}+2\alpha\boldsymbol{j}+\beta\boldsymbol{i}+5\beta\boldsymbol{j}=\boldsymbol{i}$
This gives two equations
$3\alpha+\beta=1$
$2\alpha+5\beta=0$ Then
$\beta=1-3\alpha$
we substitute to get α
$2\alpha+5(1-3\alpha)=0$
$-13\alpha=-5$
$\alpha=\dfrac{5}{13}$ and
we substitute again to get β
$\beta=1-3\left(\dfrac{5}{13}\right)=\dfrac{-2}{13}$

e Let
$\alpha(3\boldsymbol{i}+2\boldsymbol{j})+\beta(\boldsymbol{i}+5\boldsymbol{j})=-\boldsymbol{j}$
where α and β are constants. Then
$3\alpha\boldsymbol{i}+2\alpha\boldsymbol{j}+\beta\boldsymbol{i}+5\beta\boldsymbol{j}=-\boldsymbol{j}$
This gives two equations
$3\alpha+\beta=0$
$2\alpha+5\beta=-1$ Then
$\beta=-3\alpha$
we substitute to get α
$2\alpha+5(-3\alpha)=-1$
$-13\alpha=-1$
$\alpha=\dfrac{1}{13}$ and
we substitute again to get β
$\beta=-3\left(\dfrac{1}{13}\right)=\dfrac{-3}{13}$

f Let
$\alpha(3\boldsymbol{i}+2\boldsymbol{j})+\beta(\boldsymbol{i}+5\boldsymbol{j})=-\dfrac{1}{2}\boldsymbol{i}+\dfrac{2}{3}\boldsymbol{j}$
where α and β are constants. Then
$3\alpha\boldsymbol{i}+2\alpha\boldsymbol{j}+\beta\boldsymbol{i}+5\beta\boldsymbol{j}=-\dfrac{1}{2}\boldsymbol{i}+\dfrac{2}{3}\boldsymbol{j}$
This gives two equations
$3\alpha+\beta=-\dfrac{1}{2}\; 2\alpha+5\beta=\dfrac{2}{3}$
Then $\beta=-\dfrac{1}{2}-3\alpha$ we substitute to get α
$2\alpha+5\left(-\dfrac{1}{2}-3\alpha\right)=\dfrac{2}{3}$
$-13\alpha=\dfrac{19}{6}\; \alpha=\dfrac{-19}{78}$ and
we substitute again to get β
$\beta=-\dfrac{1}{2}-3\left(\dfrac{-19}{78}\right)=\dfrac{3}{13}$

3 Note that $QR\equiv PS$
so we calculate
$QR=-\boldsymbol{i}-3\boldsymbol{j}+4\boldsymbol{i}+\boldsymbol{j}=3\boldsymbol{i}-2\boldsymbol{j}$
and so
$PS=(3-x)\boldsymbol{i}+(3-y)\boldsymbol{j}$
where $P=x\boldsymbol{i}+y\boldsymbol{j}$, corresponding to (x,y) coordinates of P. Then we equate both expressions and get
$3-x=3$
$3-y=-2$
so $x=0$ and $y=5$. Then
$P=(0,5)$

4 In the notation bellow, any vector with a single letter is measured from the origin (e.g. OA = A)

$$A=B+CD=\begin{pmatrix}3\\2\\0\end{pmatrix}\cdot\begin{pmatrix}i\\j\\k\end{pmatrix}+\begin{pmatrix}-1\\-2\\1\end{pmatrix}\cdot\begin{pmatrix}i\\j\\k\end{pmatrix}=\begin{pmatrix}2\\0\\1\end{pmatrix}\begin{pmatrix}i\\j\\k\end{pmatrix}$$

$$E=A+CG=\begin{pmatrix}2\\0\\1\end{pmatrix}\cdot\begin{pmatrix}i\\j\\k\end{pmatrix}+\begin{pmatrix}-2\\0\\-2\end{pmatrix}\cdot\begin{pmatrix}i\\j\\k\end{pmatrix}=\begin{pmatrix}0\\0\\-1\end{pmatrix}\begin{pmatrix}i\\j\\k\end{pmatrix}$$

$$F=B+CG=\begin{pmatrix}3\\2\\0\end{pmatrix}\cdot\begin{pmatrix}i\\j\\k\end{pmatrix}+\begin{pmatrix}-2\\0\\-2\end{pmatrix}\cdot\begin{pmatrix}i\\j\\k\end{pmatrix}=\begin{pmatrix}1\\2\\-2\end{pmatrix}\begin{pmatrix}i\\j\\k\end{pmatrix}$$

$$H-D+CG=\begin{pmatrix}1\\1\\2\end{pmatrix}\cdot\begin{pmatrix}i\\j\\k\end{pmatrix}+\begin{pmatrix}-2\\0\\-2\end{pmatrix}\cdot\begin{pmatrix}i\\j\\k\end{pmatrix}=\begin{pmatrix}-1\\1\\0\end{pmatrix}\begin{pmatrix}i\\j\\k\end{pmatrix}$$

Then $A=(2,0,1)$, $E=(0,0,-1)$, $F=(1,2,-2)$, $H=(-1,1,0)$

5 **i** Commutative: Let $\boldsymbol{a}=x\boldsymbol{i}+y\boldsymbol{j}$, and $\boldsymbol{b}=m\boldsymbol{i}+n\boldsymbol{j}$ for real x, y, m, n. Then for all $\boldsymbol{a}, \boldsymbol{b}$

$\boldsymbol{a}+\boldsymbol{b}=(x+m)\boldsymbol{i}+(y+n)\boldsymbol{j}=(m+x)\boldsymbol{i}+(n+y)\boldsymbol{j}=\boldsymbol{b}+\boldsymbol{a}$ where we have used the commutativity of addition of real numbers.

ii Associative: Let $\boldsymbol{c}=l\boldsymbol{i}+p\boldsymbol{j}$ for real l, p, then for all $\boldsymbol{a}, \boldsymbol{b}, \boldsymbol{c}$
$(\boldsymbol{a}+\boldsymbol{b})+\boldsymbol{c}=(m+x)\boldsymbol{i}+(n+y)\boldsymbol{j}+l\boldsymbol{i}+p\boldsymbol{j}=(m+x+l)\boldsymbol{i}+(n+y+p)\boldsymbol{j}$
$=x\boldsymbol{i}+y\boldsymbol{j}+(m+l)\boldsymbol{i}+(n+p)\boldsymbol{j}=\boldsymbol{a}+(\boldsymbol{b}+\boldsymbol{c})$ where we have used the commutativity and associativity of addition of real numbers.

iii Identity: for $0=0\boldsymbol{i}+0\boldsymbol{j}$ and for all $\boldsymbol{a}$
$0+\boldsymbol{a}=(0+x)\boldsymbol{i}+(0+y)\boldsymbol{j}=(x+0)\boldsymbol{i}+(y+0)\boldsymbol{j}=x\boldsymbol{i}+y\boldsymbol{j}=\boldsymbol{a}$ where we have used the identity and commutativity of addition of real numbers.

iv Let $-\boldsymbol{a} = -x\boldsymbol{i} - y\boldsymbol{j}$. Then for all $\boldsymbol{a}, -\boldsymbol{a}$

$\boldsymbol{a} + (-\boldsymbol{a}) = (x - x)\boldsymbol{i} + (y - y)\boldsymbol{j} = (-x + x)\boldsymbol{i} + (-y + y)\boldsymbol{j}$

$= -\boldsymbol{a} + \boldsymbol{a} = 0\boldsymbol{i} + 0\boldsymbol{j} = 0$

where we have used the identity and commutativity of addition of real numbers.

6 For any real λ, μ and for all $\boldsymbol{a} = x\boldsymbol{i} + y\boldsymbol{j}$ for real x, y we have

i $(\lambda\mu)\boldsymbol{a} = (\lambda\mu)(x\boldsymbol{i} + y\boldsymbol{j}) = \lambda\mu x\boldsymbol{i} + \lambda\mu y\boldsymbol{j} = \lambda(\mu x\boldsymbol{i}) + \lambda(\mu y\boldsymbol{j}) = \lambda(\mu\boldsymbol{a})$

and

$\lambda(\mu x\boldsymbol{i}) + \lambda(\mu y\boldsymbol{j}) = \mu(\lambda x\boldsymbol{i}) + \mu(\lambda y\boldsymbol{j}) = \mu(\lambda\boldsymbol{a})$

where we have used the commutativity of the multiplication of real numbers.

ii Let $\boldsymbol{b} = m\boldsymbol{i} + n\boldsymbol{j}$ for any real n, m

$\lambda(\boldsymbol{a} + \boldsymbol{b}) = \lambda((x + m)\boldsymbol{i} + (y + n)\boldsymbol{j}) = \lambda x\boldsymbol{i} + \lambda m\boldsymbol{i} + \lambda y\boldsymbol{j} + \lambda n\boldsymbol{j}$

$= \lambda x\boldsymbol{i} + \lambda y\boldsymbol{j} + \lambda m\boldsymbol{i} + \lambda n\boldsymbol{j} = \lambda\boldsymbol{a} + \lambda\boldsymbol{b}$

where we have used the commutativity and associativity of the multiplication of real numbers.

iii $(\lambda + \mu)\boldsymbol{a} = (\lambda + \mu)(x\boldsymbol{i} + y\boldsymbol{j}) = \lambda x\boldsymbol{i} + \mu x\boldsymbol{i} + \lambda y\boldsymbol{j} + \mu y\boldsymbol{j}$

$= \lambda x\boldsymbol{i} + \lambda y\boldsymbol{j} + \mu x\boldsymbol{i} + \mu y\boldsymbol{j} = \lambda\boldsymbol{a} + \mu\boldsymbol{b}$

where we have used the commutativity and associativity of the multiplication of real numbers.

iv $1\boldsymbol{a} = 1(x\boldsymbol{i} + y\boldsymbol{j}) = (1 \times x)\boldsymbol{i} + (1 \times y)\boldsymbol{j} = x\boldsymbol{i} + y\boldsymbol{j} = \boldsymbol{a}$

where we have used the identity of multiplication of real numbers

v $0\boldsymbol{a} = 0(x\boldsymbol{i} + y\boldsymbol{j}) = (0 \times x)\boldsymbol{i} + (0 \times y)\boldsymbol{j} = 0\boldsymbol{i} + 0\boldsymbol{j} = 0$ and

$\lambda(0\boldsymbol{i} + 0\boldsymbol{j}) = (\lambda \times 0)\boldsymbol{i} + (\lambda \times 0)\boldsymbol{j} = 0\boldsymbol{i} + 0\boldsymbol{j} = 0$

Exercise 9D

1 a $\hat{a} = \dfrac{7\boldsymbol{i} + 24\boldsymbol{j}}{\sqrt{7^2 + 24^2}} = \dfrac{7\boldsymbol{i} + 24\boldsymbol{j}}{25}$

$= \dfrac{7}{25}\boldsymbol{i} + \dfrac{24}{25}\boldsymbol{j}$

b $\hat{b} = \dfrac{-3\boldsymbol{i} + 2\boldsymbol{j}}{\sqrt{3^2 + 2^2}}$

$= \dfrac{-3\boldsymbol{i} + 2\boldsymbol{j}}{\sqrt{13}}$

$= \dfrac{-3}{\sqrt{13}}\boldsymbol{i} + \dfrac{2}{\sqrt{13}}\boldsymbol{j}$

c $\hat{c} = \dfrac{4\boldsymbol{i} - 5\boldsymbol{j} + 20\boldsymbol{k}}{\sqrt{4^2 + 5^2 + 20^2}}$

$= \dfrac{4\boldsymbol{i} - 5\boldsymbol{j} + 20\boldsymbol{k}}{21}$

$= \dfrac{4}{21}\boldsymbol{i} - \dfrac{5}{21}\boldsymbol{j} + \dfrac{20}{21}\boldsymbol{k}$

d $\hat{d} = \dfrac{-\boldsymbol{i} + 3\boldsymbol{j} + 4\boldsymbol{k}}{\sqrt{1^2 + 3^2 + 4^2}}$

$= \dfrac{-\boldsymbol{i} + 3\boldsymbol{j} + 4\boldsymbol{k}}{\sqrt{26}}$

$= -\dfrac{1}{\sqrt{26}}\boldsymbol{i} + \dfrac{3}{\sqrt{26}}\boldsymbol{j} + \dfrac{4}{\sqrt{26}}\boldsymbol{k}$

2 a $\hat{a} = \dfrac{20\boldsymbol{i} - 21\boldsymbol{j}}{\sqrt{20^2 + 21^2}} = \dfrac{20\boldsymbol{i} - 21\boldsymbol{j}}{29}$

$= \dfrac{20}{29}\boldsymbol{i} - \dfrac{21}{29}\boldsymbol{j}$

All vectors parallel to $\hat{\boldsymbol{a}}$ are of the form $\lambda\hat{\boldsymbol{a}}$ for real λ

b $\hat{b} = \dfrac{\boldsymbol{i} - 3\boldsymbol{j}}{\sqrt{1^2 + 3^2}} = \dfrac{\boldsymbol{i} - 3\boldsymbol{j}}{\sqrt{10}}$

$= \dfrac{1}{\sqrt{10}}\boldsymbol{i} - \dfrac{3}{\sqrt{10}}\boldsymbol{j}$

All vectors parallel to $\hat{\boldsymbol{b}}$ are of the form $\lambda\hat{\boldsymbol{b}}$ for real λ

c $\hat{c} = \dfrac{5\boldsymbol{i} + 6\boldsymbol{j} - 30\boldsymbol{k}}{\sqrt{4^2 + 5^2 + 20^2}}$

$= \dfrac{5\boldsymbol{i} + 6\boldsymbol{j} - 30\boldsymbol{k}}{\sqrt{31}}$

$= \dfrac{5}{\sqrt{31}}\boldsymbol{i} + \dfrac{6}{\sqrt{31}}\boldsymbol{j} - \dfrac{30}{\sqrt{31}}\boldsymbol{k}$

All vectors parallel to $\hat{\boldsymbol{c}}$ are of the form $\lambda\hat{\boldsymbol{c}}$ for real λ

d

$\hat{d} = \dfrac{2\boldsymbol{i} + \boldsymbol{j} - 5\boldsymbol{k}}{\sqrt{2^2 + 1^2 + 5^2}} = \dfrac{2\boldsymbol{i} + \boldsymbol{j} - 5\boldsymbol{k}}{\sqrt{30}}$

$= \dfrac{2}{\sqrt{30}}\boldsymbol{i} + \dfrac{1}{\sqrt{30}}\boldsymbol{j} - \dfrac{5}{\sqrt{30}}\boldsymbol{k}$

All vectors parallel to $\hat{\boldsymbol{d}}$ are of the form $\lambda\hat{\boldsymbol{d}}$ for real λ

3 $\lambda = \dfrac{-5 \pm \sqrt{22}}{3}$

4 a $\hat{a} = \dfrac{5\boldsymbol{i} - \boldsymbol{j}}{\sqrt{5^2 + 1^2}} = \dfrac{5\boldsymbol{i} - \boldsymbol{j}}{\sqrt{26}}$

$= \dfrac{5}{\sqrt{26}}\boldsymbol{i} - \dfrac{1}{\sqrt{26}}\boldsymbol{j}$

Then the required vector is

$m\hat{a} = \dfrac{5 \times 6}{\sqrt{26}}\boldsymbol{i} - \dfrac{6}{\sqrt{26}}\boldsymbol{j}$

$= \dfrac{30}{\sqrt{26}}\boldsymbol{i} - \dfrac{6}{\sqrt{26}}\boldsymbol{j}$

b $\hat{b} = \dfrac{-4\boldsymbol{i} + 5\boldsymbol{j} + 20\boldsymbol{k}}{\sqrt{4^2 + 5^2 + 20^2}}$

$= \dfrac{-4\boldsymbol{i} + 5\boldsymbol{j} + 20\boldsymbol{k}}{21}$

$= \dfrac{-4}{21}\boldsymbol{i} + \dfrac{5}{21}\boldsymbol{j} + \dfrac{20}{21}\boldsymbol{k}$

Then the required vector is

$m\hat{b} = \dfrac{-4 \times 63}{21}\boldsymbol{i} + \dfrac{5 \times 63}{21}\boldsymbol{j} + \dfrac{20 \times 63}{21}\boldsymbol{k}$

$= -12\boldsymbol{i} + 15\boldsymbol{j} + 60\boldsymbol{k}$

5 a This is the same cuboid as in exercise 9C, 4.
A space diagonal could be

$$\mathbf{AG} = \left[\begin{pmatrix}0\\3\\-1\end{pmatrix} - \begin{pmatrix}2\\0\\1\end{pmatrix}\right] \cdot \begin{pmatrix}\boldsymbol{i}\\\boldsymbol{j}\\\boldsymbol{k}\end{pmatrix}$$
$$= -2\boldsymbol{i} + 3\boldsymbol{j} - 2\boldsymbol{k}$$

b Recall that $A = (2, 0, 1)$, $E = (0, 0, -1)$, $F = (1, 2, -2)$, $H = (-1, 1, 0)$

Then $\mathbf{AD} = -\boldsymbol{i} + \boldsymbol{j} + \boldsymbol{k}$

$\mathbf{AE} = -2\boldsymbol{i} + 0\boldsymbol{j} - 2\boldsymbol{k}$

$\mathbf{AB} = \boldsymbol{i} + 2\boldsymbol{j} - \boldsymbol{k}$

$$V = A_{base} \times h = |AD||AE||AB|$$
$$= \sqrt{1^2+1^2+1^2}\sqrt{2^2+2^2}\sqrt{1^2+2^2+1^2}$$
$$= 12$$

Exercise 9E

1 a $\boldsymbol{a}\cdot\boldsymbol{b} = |\boldsymbol{a}||\boldsymbol{b}|\cos\theta$
$= \sqrt{3}\times 4\cos 30° = 6$

b $\boldsymbol{a}\cdot\boldsymbol{b} = |\boldsymbol{a}||\boldsymbol{b}|\cos\theta$
$= 12 \times 8\cos 115°$
$= -40.6$

c $\boldsymbol{a}\cdot\boldsymbol{b} = |\boldsymbol{a}||\boldsymbol{b}|\cos\theta$
$= 3\times 5\cos\frac{\pi}{7} = 13.5$

d $\boldsymbol{a}\cdot\boldsymbol{b} = |\boldsymbol{a}||\boldsymbol{b}|\cos\theta$
$- 5\sqrt{2}\times 17\cos\frac{3\pi}{4} = -85$

2 a $\boldsymbol{a}\cdot\boldsymbol{b} = 3\cdot 6 + (-4)\cdot 5 = -2$

b $|\boldsymbol{a}| = \sqrt{3^2+4^2} = 5$

$|\boldsymbol{b}| = \sqrt{6^2+5^2} = \sqrt{61}$

$\boldsymbol{a}\cdot\boldsymbol{b} = |\boldsymbol{a}||\boldsymbol{b}|\cos\theta$
$= 5\sqrt{61}\cos\theta = -2$

$\cos\theta = \frac{-2}{5\sqrt{61}}$

$\theta = 1.62 \text{ rad} = 93°$

3 a $\boldsymbol{a}\cdot\boldsymbol{b} = 1\cdot(-2) + 4\cdot 3 + (-3)\cdot 1 = 7$

b $|\boldsymbol{a}| = \sqrt{1^2+4^2+3^2} = \sqrt{26}$

$|\boldsymbol{b}| = \sqrt{2^2+3^2+1^2} = \sqrt{14}$

$\boldsymbol{a}\cdot\boldsymbol{b} = |\boldsymbol{a}||\boldsymbol{b}|\cos\theta$
$= \sqrt{26\times 14}\cos\theta = 7$

$\cos\theta = \frac{7}{2\sqrt{91}}$

$\theta = 1.20 \text{ rad} = 68.5°$

4 $\theta = 0.835 = 47.9°$

5 Let $\boldsymbol{a} = x\boldsymbol{i} + y\boldsymbol{j}$ and $\boldsymbol{b} = m\boldsymbol{i} + n\boldsymbol{j}$ for any real x, y, m, n. Then

i $\boldsymbol{a}\cdot\boldsymbol{b} = |\boldsymbol{a}||\boldsymbol{b}|\cos\theta$
$= |\boldsymbol{b}||\boldsymbol{a}|\cos\theta = \boldsymbol{b}\cdot\boldsymbol{a}$

ii $\boldsymbol{a}\cdot\boldsymbol{a} = |\boldsymbol{a}||\boldsymbol{a}|\cos 0 = |\boldsymbol{a}|^2$

We prove it for the two dimensional case.

iii Let $\boldsymbol{a} = x\boldsymbol{i} + y\boldsymbol{j}$, $\boldsymbol{b} = m\boldsymbol{i} + n\boldsymbol{j}$, and $\boldsymbol{c} = s\boldsymbol{i} + t\boldsymbol{j}$ for any real x, y, m, n. Then
$\boldsymbol{a}\cdot(\boldsymbol{b} + \boldsymbol{c}) = x \times (m + s) + y \times (n + t) = xm + xs + yn + yt = xm + yn + xs + yt = \boldsymbol{a}\cdot\boldsymbol{b} + \boldsymbol{a}\cdot\boldsymbol{c}$
We have used multiplicative properties for real numbers, therefore this can be extended to any dimension of vector, as the associativity and distributivity of scalar multiplication holds.

iv Let $\lambda \in \mathbb{R}$, $\lambda(\boldsymbol{a}\cdot\boldsymbol{b})$
$= \lambda(|\boldsymbol{a}||\boldsymbol{b}|\cos\theta) = \lambda|\boldsymbol{a}||\boldsymbol{b}|\cos\theta$
$= (\lambda|\boldsymbol{a}|)|\boldsymbol{b}|\cos\theta = |\boldsymbol{a}|(\lambda|\boldsymbol{b}|)\cos\theta$
Hence
$\lambda(\boldsymbol{a}\cdot\boldsymbol{b}) = (\lambda\boldsymbol{a})\cdot\boldsymbol{b} = \boldsymbol{a}\cdot(\lambda\boldsymbol{b})$

6 i $(\boldsymbol{a} + \boldsymbol{b})\cdot(\boldsymbol{a} + \boldsymbol{b}) = \boldsymbol{a}\cdot\boldsymbol{a} + 2\boldsymbol{a}\cdot\boldsymbol{b} + \boldsymbol{b}\cdot\boldsymbol{b} = |\boldsymbol{a}|^2 + 2|\boldsymbol{a}||\boldsymbol{b}| + |\boldsymbol{b}|^2 = |\boldsymbol{a}|^2 + |\boldsymbol{b}|^2 + 2|\boldsymbol{a}||\boldsymbol{b}|\cos\theta$

ii $(\boldsymbol{a} - \boldsymbol{b})\cdot(\boldsymbol{a} - \boldsymbol{b}) = \boldsymbol{a}\cdot\boldsymbol{a} - 2\boldsymbol{a}\cdot\boldsymbol{b} + \boldsymbol{b}\cdot\boldsymbol{b} = |\boldsymbol{a}|^2 - 2|\boldsymbol{a}||\boldsymbol{b}| + |\boldsymbol{b}|^2$
$= |\boldsymbol{a}|^2 + |\boldsymbol{b}|^2 - 2|\boldsymbol{a}||\boldsymbol{b}|\cos\theta$
Each of these cases correspond to the cosine rule for a triangle with sides $\boldsymbol{a}$, $\boldsymbol{b}$, and $\boldsymbol{a} \pm \boldsymbol{b}$.

7 We use both definitions to write the scalar product between $\boldsymbol{a}$ and $\boldsymbol{b}$, i.e.

$\boldsymbol{a}\cdot\boldsymbol{b} = a_1b_1 + a_2b_2 = |\boldsymbol{a}||\boldsymbol{b}|\cos\theta$
Hence both definitions are equivalent.

8 We form the systems of equations

$(a - b)\cdot(2a + b) = a\cdot 2a - b\cdot 2a + a\cdot b - b\cdot b = 0$ and

$(a - 2b)\cdot(3a + b) = a\cdot 3a + a\cdot b - 2b\cdot 3a - 2b\cdot b = 0$

This simplifies to

$2(\boldsymbol{a}\cdot\boldsymbol{a}) - \boldsymbol{b}\cdot\boldsymbol{b} - \boldsymbol{a}\cdot\boldsymbol{b} = 0$ and

$3(\boldsymbol{a}\cdot\boldsymbol{a}) - 2\boldsymbol{b}\cdot\boldsymbol{b} - 5(\boldsymbol{a}\cdot\boldsymbol{b}) = 0$

We will express $|\boldsymbol{a}|^2$ and $|\boldsymbol{b}|^2$ in terms of the scalar product between $\boldsymbol{a}$ and $\boldsymbol{b}$. This means we solve the system of equations for the norms of $\boldsymbol{a}$ and $\boldsymbol{b}$

$\boldsymbol{b}\cdot\boldsymbol{b} = 2(\boldsymbol{a}\cdot\boldsymbol{a}) - \boldsymbol{a}\cdot\boldsymbol{b}$

$3(\boldsymbol{a}\cdot\boldsymbol{a}) - 4(\boldsymbol{a}\cdot\boldsymbol{a}) + 2(\boldsymbol{a}\cdot\boldsymbol{b}) - 5(\boldsymbol{a}\cdot\boldsymbol{b}) = 0$
$(\boldsymbol{a}\cdot\boldsymbol{a}) = -3(\boldsymbol{a}\cdot\boldsymbol{b})$

Note that the dot product is negative. This will be important as it allows us to take square roots of negative numbers multiplied by the dot product. Then we substitute into form for the norm of $\boldsymbol{b}$

$\boldsymbol{b}\cdot\boldsymbol{b} = -6(\boldsymbol{a}\cdot\boldsymbol{b}) - \boldsymbol{a}\cdot\boldsymbol{b} = -7(\boldsymbol{a}\cdot\boldsymbol{b})$

Then we write

$$\boldsymbol{a}\ \boldsymbol{b} = \sqrt{3(-\boldsymbol{a}\ \boldsymbol{b})}\sqrt{7(-\boldsymbol{a}\ \boldsymbol{b})}\cos\theta$$

or equivalently

$$1 = \sqrt{21}\cos\theta$$

so $\cos\theta = 1/\sqrt{21}$, giving $\theta = 77.4°$

Exercise 9F

1 a $\boldsymbol{d} = (2-0)\boldsymbol{i} + (3-0)\boldsymbol{j} = 2\boldsymbol{i} + 3\boldsymbol{j}$

Then the vector equation of the line is

$$\begin{pmatrix}x\\y\end{pmatrix} = \begin{pmatrix}0\\0\end{pmatrix} + k\begin{pmatrix}2\\3\end{pmatrix}$$

b $d = (-1-2)i + (3-1)j = -3i + 2j$

Then the vector equation of the line is

$$\begin{pmatrix} x \\ y \end{pmatrix} = \begin{pmatrix} -1 \\ 3 \end{pmatrix} + k\begin{pmatrix} -3 \\ 2 \end{pmatrix}$$

c $d = (3+2)i + (-6+5)j = 5i - j$

Then the vector equation of the line is

$$\begin{pmatrix} x \\ y \end{pmatrix} = \begin{pmatrix} 3 \\ -6 \end{pmatrix} + k\begin{pmatrix} 5 \\ -1 \end{pmatrix}$$

d $d = \left(-\frac{1}{2} - \frac{2}{3}\right)i + \left(\frac{3}{4} + 1\right)j$

$= -\frac{7}{6}i + \frac{7}{4}j$

Then the vector equation of the line is

$$\begin{pmatrix} x \\ y \end{pmatrix} = \begin{pmatrix} -\frac{1}{2} \\ \frac{3}{4} \end{pmatrix} + k\begin{pmatrix} -\frac{7}{6} \\ \frac{7}{4} \end{pmatrix}$$

2 a $p = a - \lambda d \Rightarrow \begin{pmatrix} x \\ y \end{pmatrix}$

$= \begin{pmatrix} 2 \\ -7 \end{pmatrix} + \lambda\begin{pmatrix} 1 \\ 1 \end{pmatrix}$

$\Rightarrow \begin{cases} x = 2 + \lambda \\ y = -7 + \lambda \end{cases} \Rightarrow \begin{cases} x - 2 = \lambda \\ y + 7 = \lambda \end{cases}$

$\Rightarrow x - 2 = y + 7 \Rightarrow y = x - 9$

b $n \cdot (p \cdot a) = 0 \Rightarrow n \cdot p = n \cdot a$

$\Rightarrow \begin{pmatrix} 2 \\ -3 \end{pmatrix}\begin{pmatrix} x \\ y \end{pmatrix} = \begin{pmatrix} 2 \\ -3 \end{pmatrix}\begin{pmatrix} 2 \\ -7 \end{pmatrix}$

$\Rightarrow 2x - 3y = 4 + 21$

$2x - 3y = 25$

$y = \frac{2x - 25}{3}$

3 a We obtain the direction vector of L,

$\frac{x-3}{2} = \frac{y+1}{-3}$

Then

$-3x + 9 = 2y + 2 = \lambda$

Then

$-3x + 9 = \lambda \Rightarrow x = 3 - \frac{1}{3}\lambda$ and

$2y + 2 = \lambda \Rightarrow y = -1 + \frac{1}{2}\lambda$

Hence the direction vector

is $d = -\frac{1}{3}i + \frac{1}{2}j$

Then the vector equation parallel to L and passing through T is

$$\begin{pmatrix} x \\ y \end{pmatrix} = \begin{pmatrix} -3 \\ 8 \end{pmatrix} + \lambda\begin{pmatrix} -\frac{1}{3} \\ \frac{1}{2} \end{pmatrix}$$

b The perpendicular line must have a normal vector for its direction vector

$d = -\frac{1}{3}i + \frac{1}{2}j \Rightarrow n = \frac{1}{2}i + \frac{1}{3}j$

Then the vector equation perpendicular to L passing through T is

$$\begin{pmatrix} x \\ y \end{pmatrix} = \begin{pmatrix} -3 \\ 8 \end{pmatrix} + \lambda\begin{pmatrix} \frac{1}{2} \\ \frac{1}{3} \end{pmatrix}$$

c We find the intersection between L and r as

$r = \begin{pmatrix} x \\ y \end{pmatrix} = \begin{pmatrix} 1 \\ -1 \end{pmatrix} + \lambda\begin{pmatrix} 2 \\ 4 \end{pmatrix}$

We write this in Cartesian notation

$x = 1 + 2\lambda$

$y = -1 + 4\lambda$

Then

$\frac{x-1}{2} = \frac{y+1}{4} \Rightarrow 4x - 4$

$= 2y + 2 \Rightarrow y = 2x - 3$

We write L in Cartesian form and get

$\frac{x-3}{2} = \frac{y+1}{3} \Rightarrow 6x - 9$

$= 2y + 2 \Rightarrow y = 3x - \frac{11}{2}$

To find the intersection between the two, we equate both lines and get

$2x - 3 = 3x - \frac{11}{2} \Rightarrow x = \frac{5}{2}$

and $y = 2\left(\frac{5}{2}\right) - 3 = 2$

So we must find a line passing through T and $\left(\frac{5}{2}, 2\right)$, so be obtain the direction vector as

$d = \left(-3 - \frac{5}{2}\right)i + (8-2)j$

$= -\frac{11}{2}i + 6j$

Then the equation of the line passing through T and the intersection between the two lines is

$$\begin{pmatrix} x \\ y \end{pmatrix} = \begin{pmatrix} -3 \\ 8 \end{pmatrix} + \lambda\begin{pmatrix} -\frac{11}{2} \\ 6 \end{pmatrix}$$

4 a For them to be parallel, their direction vectors have to be proportional to each other.

Note that

$r = 3i + 2j + \lambda(-i + aj) \Rightarrow$
$d_1 = -i + aj$

$p = (1 + 2\mu)i + (5\mu - 2)j \Rightarrow$
$d_2 = 2i + 5j$

For them to be parallel, we must have

$d_1 = \gamma d_2$ for real γ. Then the normalised

$-i + aj = \gamma(2i + 5j)$

Note that $2\gamma = -1$ gives

$\gamma = \frac{-1}{2}$, so

$a = 5 \times \frac{-1}{2} = -\frac{5}{2}$

b For them to be perpendicular, their scalar product must be zero, so

$d_1 \cdot d_2 = 0 \Rightarrow -1 \times 2 + a \times 5$

$= 0 \Rightarrow 5a = 2 \Rightarrow a = \frac{2}{5}$

Exercise 9G

1 **a** $d = (4-1)\boldsymbol{i} + (2-3)\boldsymbol{j} + (1+2)k = 3\boldsymbol{i} - \boldsymbol{j} + 3k$

Then we write the vector equation simply as

$$\begin{pmatrix} x \\ y \\ z \end{pmatrix} = \begin{pmatrix} 1 \\ 3 \\ -2 \end{pmatrix} + \lambda \begin{pmatrix} 3 \\ -1 \\ 3 \end{pmatrix}$$

b $d = (5-3)\boldsymbol{i} + (7-0)\boldsymbol{j} + (-2+5)k = 2\boldsymbol{i} + 7\boldsymbol{j} + 3k$

Then we write the vector equation simply as

$$\begin{pmatrix} x \\ y \\ z \end{pmatrix} = \begin{pmatrix} 3 \\ 0 \\ -5 \end{pmatrix} + \lambda \begin{pmatrix} 2 \\ 7 \\ 3 \end{pmatrix}$$

2 **a** We write the form for r as

$$\begin{pmatrix} x \\ y \\ z \end{pmatrix} = \begin{pmatrix} 3 \\ -1 \\ 2 \end{pmatrix} + \lambda \begin{pmatrix} -1 \\ 1 \\ 3 \end{pmatrix}$$

We substitute with P, and get

$0 = 3 - \lambda$

$2 = -1 + \lambda$

$5 = 2 + 3\lambda$

From the first equations, $\lambda = 3$, and then substituting with that value of λ in the last one gives

$2 + 3(3) = 11 \neq 5$

Hence there is a contradiction, and so P does not lie on the line.

b A parallel line has the same direction vector, and now the equation of the line is

$$\begin{pmatrix} x \\ y \\ z \end{pmatrix} = \begin{pmatrix} 0 \\ 2 \\ 5 \end{pmatrix} + \lambda \begin{pmatrix} -1 \\ 1 \\ 3 \end{pmatrix}$$

c We substitute T into the equation of the line to get the system of equations

$-2 = 3 - \lambda$

$4 = -1 + \lambda$

$a = 2 + 3\lambda$

Then $\lambda = 5$, and is consistent in the first two equations. Then

$a = 2 + 3(5) = 17$

3 **a** If the lines are parallel, their direction vectors are proportional to each other. We obtain them by rewriting in the equations for the lines in vector form

$$L_1 : \frac{x-1}{2} = \frac{y-2}{3} = \frac{z+3}{5} = \lambda$$

so

$x = 2\lambda + 1$

$y = 3\lambda + 2$

$z = 5\lambda - 3$

so

$d_1 = 2\boldsymbol{i} + 3\boldsymbol{j} + 5k$

and

$$L_2 : x + 2 = \frac{y-1}{-2} = \frac{z-2}{4} = \mu$$

so

$x = \mu - 2$

$y = -2\mu + 1$

$z = 4\mu + 2$

so

$d_2 = i - 2\boldsymbol{j} + 4k$

The lines are not parallel as there is no real γ for which

$d_1 = \gamma d_2$

b The lines are skew if they are not parallel or perpendicular to each other. We check the scalar product between their direction vectors:

$$d_1 \cdot d_2 = 2 \cdot 1 + 3 \cdot (-2) + 5 \cdot 4 = 16 \neq 0$$

Hence the lines are skew.

4 **a** We rewrite the lines in parametric form

$$L_1 : \begin{pmatrix} x \\ y \\ z \end{pmatrix} = \begin{pmatrix} 3 \\ -2 \\ 1 \end{pmatrix} + \lambda \begin{pmatrix} 5 \\ 4 \\ 3 \end{pmatrix}$$

$$L_2 : \begin{pmatrix} x \\ y \\ z \end{pmatrix} = \begin{pmatrix} 7 \\ 4 \\ -1 \end{pmatrix} + \mu \begin{pmatrix} -1 \\ 2 \\ -3 \end{pmatrix}$$

At the intersection, both lines will take on the same values, so we construct the system of equations

$3 + 5\lambda = 7 - \mu$

$-2 + 4\lambda = 4 + 2\mu$

$1 + 3\lambda = -1 - 3\mu$

From the first equation we get that

$\mu = -3 - 5\lambda + 7 = -5\lambda + 4$

and so

$-2 + 4\lambda = 4 + 2(-5\lambda + 4) \Rightarrow 14\lambda = 14 \Rightarrow \lambda = 1$

so $\mu = -5(1) + 4 = -1$

We check that these values satisfy the third equation

$1 + 3(1) \neq -1 - 3(-1)$

so the lines do not intersect.

b We rewrite the lines in parametric form

$$L_1 : \begin{pmatrix} x \\ y \\ z \end{pmatrix} = \begin{pmatrix} 0 \\ -1 \\ 3 \end{pmatrix} + \lambda \begin{pmatrix} 1 \\ 2 \\ -1 \end{pmatrix}$$

$$L_2 : \begin{pmatrix} x \\ y \\ z \end{pmatrix} = \begin{pmatrix} -7 \\ 0 \\ 7 \end{pmatrix} + \mu \begin{pmatrix} 3 \\ 1 \\ -2 \end{pmatrix}$$

At the intersection, both lines will take on the same values, so we construct the system of equations

$\lambda = -7 + 3\mu$

$-1 + 2\lambda = \mu$

$3 - \lambda = 7 - 2\mu$

We substitute the first equation into the second equation and get

$-1 + 2(-7 + 3\mu) = \mu$

$-1 - 14 + 6\mu = \mu$

$5\mu = 15$

$\mu = 3$ and so

$\lambda = -7 + 3(3) = 2$

which is consistent with the third equation.

c We rewrite the lines in parametric form

$$L_1: \begin{pmatrix} x \\ y \\ z \end{pmatrix} = \begin{pmatrix} 0 \\ 2 \\ 0 \end{pmatrix} + \lambda \begin{pmatrix} 2 \\ 5 \\ 4 \end{pmatrix}$$

$$L_2: \begin{pmatrix} x \\ y \\ z \end{pmatrix} = \begin{pmatrix} 1 \\ -1 \\ -3 \end{pmatrix} + \mu \begin{pmatrix} 3 \\ 2 \\ 1 \end{pmatrix}$$

At the intersection, both lines will take on the same values, so we construct the system of equations

$2\lambda = 1 + 3\mu$

$2 + 5\lambda = -1 + 2\mu$

$4\lambda = -3 + \mu$

The first equation gives us that

$$\lambda = \frac{1+3\mu}{2}$$

We substitute this into the third equation and get

$$4\left(\frac{1+3\mu}{2}\right) + 3 = \mu$$

$$2 + 6\mu + 3 = \mu$$

$$5\mu = -5$$

$$\mu = -1$$

and so

$$\lambda = \frac{1+3(-1)}{2} = -1$$

which is consistent with the second equation.

5 The point of intersection is (1,0.2)

Exercise 9H

1 a We use the provided formula

$$\begin{pmatrix} a_2b_3 & a_3b_2 \\ a_3b_1 & a_1b_3 \\ a_1b_2 & a_2b_1 \end{pmatrix} = \begin{pmatrix} 3 \times 3 - (-5) \times (-2) \\ (-5) \times 1 - 2 \times 3 \\ 2 \times (-2) - 3 \times 1 \end{pmatrix} = \begin{pmatrix} 9 - 10 \\ -5 - 6 \\ -4 - 3 \end{pmatrix} = \begin{pmatrix} -1 \\ -11 \\ -7 \end{pmatrix}$$

b $$\begin{pmatrix} 1\times(-2) - 0 \times 0 \\ 0 \times 3 - 1 \times (-2) \\ 1 \times 0 - 1 \times 3 \end{pmatrix} = \begin{pmatrix} -2-0 \\ 0+2 \\ 0-3 \end{pmatrix} = \begin{pmatrix} -2 \\ 2 \\ -3 \end{pmatrix}$$

c $$\begin{pmatrix} -4 \times 2 - (-1) \times 1 \\ -1 \times 2 - 3 \times 2 \\ 3 \times 1 - (-4) \times 2 \end{pmatrix} = \begin{pmatrix} -8+1 \\ -2-6 \\ 3+8 \end{pmatrix} = \begin{pmatrix} -7 \\ -8 \\ 11 \end{pmatrix}$$

d $$\begin{pmatrix} -\frac{3}{4} \times (-2) - 1 \times \left(-\frac{2}{3}\right) \\ 1 \times 1 - \left(\frac{1}{2}\right) \times (-2) \\ \left(\frac{1}{2}\right) \times \left(-\frac{2}{3}\right) - \left(-\frac{3}{4}\right) \times 1 \end{pmatrix} = \begin{pmatrix} \frac{6}{4} + \frac{2}{3} \\ 1+1 \\ -\frac{2}{6} + \frac{3}{4} \end{pmatrix} = \begin{pmatrix} \frac{13}{6} \\ 2 \\ \frac{5}{12} \end{pmatrix}$$

2 $A = |\boldsymbol{a} \times \boldsymbol{b}|$

$$\boldsymbol{a} \times \boldsymbol{b} = \begin{pmatrix} 3 \times (-4) - (-6) \times (-1) \\ (-6) \times 3 - 2 \times (-4) \\ 2 \times (-1) - (3) \times 3 \end{pmatrix} = \begin{pmatrix} -12-6 \\ -18+8 \\ -2 \; -9 \end{pmatrix} = \begin{pmatrix} -18 \\ -10 \\ -11 \end{pmatrix}$$

Then $|\boldsymbol{a} \times \boldsymbol{b}| = \sqrt{18^2 + 10^2 + 11^2} = 23.3$

3 a $\boldsymbol{AB} = (-2 - 1)\boldsymbol{i} + (0 - 4)\boldsymbol{j} + (3 - 2)\boldsymbol{k} = -3\boldsymbol{i} - 4\boldsymbol{j} + \boldsymbol{k}$

$\boldsymbol{AC} = (-1 - 1)\boldsymbol{i} + (2 - 4)\boldsymbol{j} + (4 - 2)\boldsymbol{k} = -2\boldsymbol{i} - 2\boldsymbol{j} + 2\boldsymbol{k}$

b $A = \frac{1}{2}|\boldsymbol{AB} \times \boldsymbol{AC}|$

$$\boldsymbol{AB} \times \boldsymbol{AC} = \begin{pmatrix} -4 \times 2 - 1 \times -2 \\ 1 \times (-2) - (-3) \times 2 \\ -3 \times (-2) - (-4)(-2) \end{pmatrix} = \begin{pmatrix} -8+2 \\ -2+6 \\ 6-8 \end{pmatrix} = \begin{pmatrix} -6 \\ 4 \\ -2 \end{pmatrix}$$ Then

$$|\boldsymbol{AB} \times \boldsymbol{AC}| = \sqrt{6^2 + 4^2 + 2^2} = \sqrt{36 + 16 + 4} = 7.48$$

Then the area is $A = \frac{1}{2}(7.48) = 3.74$

4 i Let the vectors $\boldsymbol{a}$, $\boldsymbol{b}$, $\boldsymbol{c}$ be well defined. Then

$\boldsymbol{a} \times \boldsymbol{b} = |\boldsymbol{a}||\boldsymbol{b}|\sin\theta = -|\boldsymbol{a}||\boldsymbol{b}|\sin(-\theta) = -|\boldsymbol{b}||\boldsymbol{a}|\sin(-\theta) = -(\boldsymbol{b} \times \boldsymbol{a})$

where we have used the commutativity of real numbers and properties of sines. Note that if the angle from a to b is θ, then the angle from b to a is $-\theta$

ii We calculate

$$\boldsymbol{a} \times (\boldsymbol{b} \times \boldsymbol{c}) = \begin{pmatrix} a_2(b_1c_2 - b_2c_1) \mid a_3(b_1c_3 + a_3c_1) \\ a_3(a_2b_3 - a_3b_2) - a_1(b_1c_2 - b_2c_1) \\ -a_1(b_1c_3 + b_3c_1) - a_2(a_2b_3 - a_3b_2) \end{pmatrix} = (\boldsymbol{a} \cdot \boldsymbol{c})\boldsymbol{b} - (\boldsymbol{a} \cdot \boldsymbol{b})\boldsymbol{c}$$

iii $\lambda\begin{pmatrix} a_2b_3 - a_3b_2 \\ a_3b_1 - a_1b_3 \\ a_1b_2 - a_2b_1 \end{pmatrix} = \begin{pmatrix} \lambda(a_2b_3 - a_3b_2) \\ \lambda(a_3b_1 - a_1b_3) \\ \lambda(a_1b_2 - a_2b_1) \end{pmatrix} = \begin{pmatrix} a_2(\lambda b_3) - a_3(\lambda b_2) \\ a_3(\lambda b_1) - a_1(\lambda b_3) \\ a_1(\lambda b_2) - a_2(\lambda b_1) \end{pmatrix}$

$$= \begin{pmatrix} (\lambda a_2)b_3 - (\lambda a_3)b_2 \\ (\lambda a_3)b_1 - (\lambda a_1)b_3 \\ (\lambda a_1)b_2 - (\lambda a_2)b_1 \end{pmatrix}$$

Hence

$\lambda(a \times b) = (\lambda a) \times b = a \times (\lambda b), \ \lambda \in \mathbb{R}$

iv We can expand out the cross products explicitly as

$$(\boldsymbol{a}+\boldsymbol{b})\times \boldsymbol{c} = \begin{pmatrix} (a_2+b_2)c_3 - (a_3+b_3)c_2 \\ (a_3+b_3)c_1 - (a_1+b_1)c_3 \\ (a_1+b_1)c_2 - (a_2+b_2)c_1 \end{pmatrix} = \begin{pmatrix} a_2c_3 - a_3c_2 \\ a_3c_1 - a_1c_3 \\ a_1c_2 - a_2c_1 \end{pmatrix} + \begin{pmatrix} b_2c_3 - b_3c_2 \\ b_3c_1 - b_1c_3 \\ b_1c_2 - b_2c_1 \end{pmatrix}$$

$$= (\boldsymbol{a}\times\boldsymbol{c}) + (\boldsymbol{b}\times\boldsymbol{c})$$

5 We write out the vectors

$\boldsymbol{AB} = (2-1)\boldsymbol{i} + (-1-1)\boldsymbol{j} + (0-1)\boldsymbol{k} = \boldsymbol{i} - 2\boldsymbol{j} - \boldsymbol{k}$

$\boldsymbol{AC} = (2-1)\boldsymbol{i} + (4-1)\boldsymbol{j} + (2-1)\boldsymbol{k} = \boldsymbol{i} + 3\boldsymbol{j} + \boldsymbol{k}$

$\boldsymbol{AD} = (-2-1)\boldsymbol{i} + (2-1)\boldsymbol{j} + (2-1)\boldsymbol{k} = -3\boldsymbol{i} + \boldsymbol{j} + \boldsymbol{k}$ Then

$$(\boldsymbol{AB}\times\boldsymbol{AC})\cdot\boldsymbol{AD} = \begin{pmatrix} (-2)\times 1 - (-1)\times 3 \\ -1\times 1 - 1\times 1 \\ 1\times 3 - (-2)\times 1 \end{pmatrix}\cdot\begin{pmatrix} -3 \\ 1 \\ 1 \end{pmatrix} = \begin{pmatrix} -2+3 \\ -1-1 \\ 3+2 \end{pmatrix}\cdot\begin{pmatrix} -3 \\ 1 \\ 1 \end{pmatrix} = \begin{pmatrix} 1 \\ -2 \\ 5 \end{pmatrix}\cdot\begin{pmatrix} -3 \\ 1 \\ 1 \end{pmatrix}$$

$= 1 \times (-3) + (-2) \times (1) + 5 \times (1) = 0$

Hence the three points are coplanar.

6 **a** We find D such that AB = DC. Then

$AB = \begin{pmatrix} 2 \\ -1 \\ 3 \end{pmatrix} - \begin{pmatrix} 1 \\ 2 \\ 1 \end{pmatrix} = \begin{pmatrix} 1 \\ -3 \\ 2 \end{pmatrix}$ and $DC = \begin{pmatrix} 4 \\ 5 \\ -1 \end{pmatrix} - \begin{pmatrix} d_1 \\ d_2 \\ d_3 \end{pmatrix} = \begin{pmatrix} 4-d_1 \\ 5-d_2 \\ -1-d_3 \end{pmatrix}$

Then we have the equations

$4 - d_1 = 1$

$5 - d_2 = -3$

$-1 - d_3 = 2$ Then $D = (3, 8, -3)$

b Note that the vectors DC, DA and DH enclose the parallelepiped, so

$$\boldsymbol{DC} = \begin{pmatrix} 1 \\ -3 \\ 2 \end{pmatrix}\cdot\begin{pmatrix} \boldsymbol{i} \\ \boldsymbol{j} \\ \boldsymbol{k} \end{pmatrix}$$

$$\boldsymbol{DA} = \begin{pmatrix} 1-3 \\ 2-8 \\ 1+3 \end{pmatrix}\cdot\begin{pmatrix} \boldsymbol{i} \\ \boldsymbol{j} \\ \boldsymbol{k} \end{pmatrix} = \begin{pmatrix} -2 \\ -6 \\ 4 \end{pmatrix}\cdot\begin{pmatrix} \boldsymbol{i} \\ \boldsymbol{j} \\ \boldsymbol{k} \end{pmatrix}$$

$$\boldsymbol{DH} = \begin{pmatrix} 4-3 \\ 3-8 \\ 6+3 \end{pmatrix}\cdot\begin{pmatrix} \boldsymbol{i} \\ \boldsymbol{j} \\ \boldsymbol{k} \end{pmatrix} = \begin{pmatrix} 1 \\ -5 \\ 9 \end{pmatrix}\cdot\begin{pmatrix} \boldsymbol{i} \\ \boldsymbol{j} \\ \boldsymbol{k} \end{pmatrix}$$

c The volume of the parallelepiped is given by

$V = |(\boldsymbol{DC} \times \boldsymbol{DA})\cdot\boldsymbol{DH}|$

$= |(0)(1) + (-8)(-5) + (-12)(9)|$

$= 68$

7 Assuming D is the apex, we obtain

$\boldsymbol{BC} = \boldsymbol{BD} + \boldsymbol{DC} = (1-2)\boldsymbol{i} + (-2+1)\boldsymbol{j} + (4+5)\boldsymbol{k} = -\boldsymbol{i} - \boldsymbol{j} + 9\boldsymbol{k}$

$V = \frac{1}{3}\,Base\cdot h = \frac{1}{3}\cdot\frac{1}{2}|\boldsymbol{BA}\times\boldsymbol{BD}||\boldsymbol{BC}|$

$$\boldsymbol{BA}\times\boldsymbol{BD} = \begin{pmatrix} -3\times 4 - 2\times(-2) \\ 2\times 1 - (-2)\times 4 \\ -2\times(-2) - (-3)\times 1 \end{pmatrix}$$

$$= \begin{pmatrix} -12+4 \\ 2+8 \\ 4+3 \end{pmatrix} = \begin{pmatrix} -8 \\ 10 \\ 7 \end{pmatrix}$$

Then

$V = \frac{1}{6}\left(\sqrt{8^2+10^2+7^2}\sqrt{1^2+1^2+9^2}\right)$

$= 22.2$

8 $(\boldsymbol{a}\cdot\boldsymbol{b})^2 + |(\boldsymbol{a}\times\boldsymbol{b})\cdot(\boldsymbol{a}\times\boldsymbol{b})|$

$= |\boldsymbol{a}|^2|\boldsymbol{b}|^2\cos^2\theta + |\boldsymbol{a}|^2|\boldsymbol{b}|^2\sin^2\theta$

$= |\boldsymbol{a}|^2|\boldsymbol{b}|^2(\cos^2\theta + \sin^2\theta)$

$= |\boldsymbol{a}|^2|\boldsymbol{b}|^2$

9 We calculate

$\boldsymbol{a}\times(\boldsymbol{b}\times\boldsymbol{c})$

$$= \begin{pmatrix} a_2(b_1c_2 - b_2c_1) + a_3(b_1c_3 + a_3c_1) \\ a_3(a_2b_3 - a_3b_2) - a_1(b_1c_2 - b_2c_1) \\ -a_1(b_1c_3 + b_3c_1) - a_2(a_2b_3 - a_3b_2) \end{pmatrix}$$

$= (\boldsymbol{a}\cdot\boldsymbol{c})\boldsymbol{b} - (\boldsymbol{a}\cdot\boldsymbol{b})$

Exercise 9I

1 **a** $\boldsymbol{p} = \boldsymbol{a} + \lambda\boldsymbol{u} + \mu\boldsymbol{v}$

$$= \begin{pmatrix} 0 \\ 2 \\ -1 \end{pmatrix} + \lambda\begin{pmatrix} 2 \\ 1 \\ 4 \end{pmatrix} + \mu\begin{pmatrix} -1 \\ 2 \\ -1 \end{pmatrix}$$

b $\boldsymbol{p} = \boldsymbol{a} + \lambda\boldsymbol{u} + \mu\boldsymbol{v}$

$$= \begin{pmatrix} 1 \\ -2 \\ 3 \end{pmatrix} + \lambda\begin{pmatrix} 0 \\ -3 \\ 2 \end{pmatrix} + \mu\begin{pmatrix} 1 \\ 4 \\ -2 \end{pmatrix}$$

c $\boldsymbol{p} = \boldsymbol{a} + \lambda\boldsymbol{u} + \mu\boldsymbol{v}$

$$= \begin{pmatrix} -3 \\ 4 \\ 2 \end{pmatrix} + \lambda\begin{pmatrix} -2 \\ 0 \\ 3 \end{pmatrix} + \mu\begin{pmatrix} 2 \\ 1 \\ 5 \end{pmatrix}$$

2 **a** $\boldsymbol{AB} = (-1-0)\boldsymbol{i} + (2-1)\boldsymbol{j} + (0-3)\boldsymbol{k} = -\boldsymbol{i} + \boldsymbol{j} - 3\boldsymbol{k}$

$\boldsymbol{AC} = (3-0)\boldsymbol{i} + (-2-1)\boldsymbol{j} + (4-3)\boldsymbol{k} = 3\boldsymbol{i} - 3\boldsymbol{j} + \boldsymbol{k}$

Then we can write the vector equation of the plane as

$$p = \begin{pmatrix} 0 \\ 1 \\ 3 \end{pmatrix} + \lambda \begin{pmatrix} -1 \\ 1 \\ -3 \end{pmatrix} + \mu \begin{pmatrix} 3 \\ -3 \\ 1 \end{pmatrix}$$

b $x = -\lambda + 3\mu$
$y = 1 + \lambda - 3\mu$
$z = 3 - 3\lambda + \mu$

c We eliminate the parameters in b
$\lambda = 3\mu - x$

which we substitute into the equation for y and z to get
$y = 1 + 3\mu - x - 3\mu \Rightarrow y + x = 1$

$z = 3 - 3(3\mu - x) + \mu$

We cannot express the equation in terms of x, y, and z, so the Cartesian equation is $y + x = 1$

3 The normal vector is

$$\begin{pmatrix} 2 \\ 1 \\ 4 \end{pmatrix} \times \begin{pmatrix} -1 \\ 2 \\ -1 \end{pmatrix} = \begin{pmatrix} -9 \\ -2 \\ 5 \end{pmatrix}$$

$$\begin{pmatrix} -9 \\ -2 \\ 5 \end{pmatrix} \cdot \begin{pmatrix} x \\ y \\ z \end{pmatrix} = \begin{pmatrix} -9 \\ -2 \\ 5 \end{pmatrix} \cdot \begin{pmatrix} 0 \\ 2 \\ -1 \end{pmatrix}$$

$-9x - 2y + 5z = -4 - 5$

$-9x - 2y + 5z = -9$

$$\begin{pmatrix} 0 \\ -3 \\ 2 \end{pmatrix} \times \begin{pmatrix} 1 \\ 4 \\ -2 \end{pmatrix} = \begin{pmatrix} -2 \\ 2 \\ 3 \end{pmatrix}$$

$$\begin{pmatrix} -2 \\ 2 \\ 3 \end{pmatrix} \cdot \begin{pmatrix} x \\ y \\ z \end{pmatrix} = \begin{pmatrix} -2 \\ 2 \\ 3 \end{pmatrix} \cdot \begin{pmatrix} 1 \\ -2 \\ 3 \end{pmatrix}$$

$-2x + 2y + 3z = -2 - 4 + 9$

$-2x + 2y + 3z = 3$

$$\begin{pmatrix} -2 \\ 0 \\ 3 \end{pmatrix} \times \begin{pmatrix} 2 \\ 1 \\ 5 \end{pmatrix} = \begin{pmatrix} -3 \\ 16 \\ -2 \end{pmatrix}$$

$$\begin{pmatrix} -3 \\ 16 \\ -2 \end{pmatrix} \cdot \begin{pmatrix} x \\ y \\ z \end{pmatrix} = \begin{pmatrix} -3 \\ 16 \\ -2 \end{pmatrix} \cdot \begin{pmatrix} -3 \\ 4 \\ 2 \end{pmatrix}$$

$-3x + 16y - 2z = 69$

4 a We substitute the point into the equation of the plane
$3(5) - 4(4) + 2(-2) = -5 \neq 5$
Hence the point is not on the plane

b The normal vector is
$\boldsymbol{d} = 3\boldsymbol{i} - 4\boldsymbol{j} + 2\boldsymbol{k}$
Then we are searching for a plane with the same normal vector but a different point.

$$\begin{pmatrix} 3 \\ -4 \\ 2 \end{pmatrix} \cdot \begin{pmatrix} x \\ y \\ z \end{pmatrix} = \begin{pmatrix} 3 \\ -4 \\ 2 \end{pmatrix} \cdot \begin{pmatrix} 5 \\ 4 \\ -2 \end{pmatrix}$$

$3x - 4y + 2z = 15 - 16 - 4$
$3x - 4y + 2z = -5$

5 We equate solve the equations of a plane as a system of equations
$x + y - z = 1$
$2x - 3y - 9z = 10$
$x + 2y - 3z = -4$
We subtract the third from the first and get
$-y + 2z = 5$
$y = 2z - 5$
We subtract two times the first from the second, and get
$-5y - 7z = 8$
Then substituting our value for y we get that
$-5(2z - 5) - 7z = 8$
$z = 1$
Then
$y = 2(1) - 5 = -3$ and so
$x + (-3) - 1 = 1$
$x = 5$

6 a We express y and z in terms of x
$y = 3 + 2z - x$
$z = 1 + 3y - 2x$
Then
$y = 3 + 2(1 + 3y - 2x) - x$
$y = x - 1$
Which we can then substitute into the equation for z as
$z = 1 + 3(x - 1) - 2x$
this simplifies into
$z = x - 2$

We let $x = \lambda$, and so
$x = \lambda$
$y = \lambda - 1$
$z = \lambda - 2$
We eliminate λ to find the Cartesian equation, as
$x = y + 1 = z + 2$
You take a point from the line in part **a**, e.g. P(0, −1, −2) and find the vector **PA**.

$$\mathbf{PA} = \begin{pmatrix} 2 \\ -4 \\ 1 \end{pmatrix} - \begin{pmatrix} 0 \\ -1 \\ -2 \end{pmatrix} = \begin{pmatrix} 2 \\ -3 \\ 3 \end{pmatrix}$$

b The normal vector of the plane must be perpendicular to **PA** and the direction vector of the line, hence find the normal vector of the plane that is collinear with vector product of those two vectors.

$$\mathbf{PA} \times \mathbf{d} = \begin{pmatrix} 2 \\ -3 \\ 3 \end{pmatrix} \times \begin{pmatrix} 1 \\ 1 \\ 1 \end{pmatrix} = \begin{pmatrix} -6 \\ 1 \\ 5 \end{pmatrix}$$

$$= (-1) \begin{pmatrix} 6 \\ -1 \\ -5 \end{pmatrix}$$

Use one point e.g. P(0, −1, −2) to find the equation of the plane.

$$\begin{pmatrix} 6 \\ -1 \\ -5 \end{pmatrix} \cdot \begin{pmatrix} x \\ y \\ z \end{pmatrix} = \begin{pmatrix} 6 \\ -1 \\ -5 \end{pmatrix} \cdot \begin{pmatrix} 0 \\ -1 \\ -2 \end{pmatrix}$$

$\Rightarrow 6x - y - 5z = 11$

7 If two planes are parallel, their normal vectors are parallel, then
$n_1 \times n_2 = |n_1||n_2|\sin\theta = |n_1||n_2| \times 0 = 0$
If the vector product of the normal vectors is zero, we have
$n_1 \times n_2 = |n_1||n_2|\sin\theta = 0 \Rightarrow \theta = 0$
hence they are parallel

Exercise 9J

1 a $x - 5 = \lambda$
$\frac{y+1}{2} = \lambda$

$\frac{1-z}{3} = \lambda$

or equivalently

$x = \lambda + 5$

$y = 2\lambda - 1$

$z = 1 - 3\lambda$

We substitute in the equation of the plane

$2(\lambda + 5) - 4(2\lambda - 1) + 1 - 3\lambda = -3$

$2\lambda + 10 - 8\lambda + 4 + 1 - 3\lambda = -3$

$-9\lambda = -18$

$\lambda = 2$

There is a unique solution, so the line and the plane intersect at a point. This point is

$x = 2 + 5 = 7$

$y = 2(2) - 1 = 3$

$z = 1 - 3(2) = -5$

So they intersect at $(7, 3, -5)$.

b $1 - 2x = \lambda$

$\frac{y-3}{4} = \lambda$

$\frac{2z+2}{3} = \lambda$

or equivalently

$x = \frac{1-\lambda}{2}$

$y = 4\lambda + 3$

$z = \frac{3\lambda - 2}{2}$

We substitute in the equation of the plane

$$5\left(\frac{1-\lambda}{2}\right) + (4\lambda + 3) - 4\left(\frac{3\lambda - 2}{2}\right) = 3$$

$\lambda = \frac{13}{9}$

There is a unique solution, so the line and the plane intersect at a point. This point is determined by

$x = -\frac{2}{9}$

$y = \frac{79}{9}$

$z = \frac{7}{6}$

c $\frac{x-5}{4} = \lambda$

$\frac{y+2}{-2} = \lambda$

$\frac{z-3}{3} = \lambda$

or equivalently

$x = 4\lambda + 5$

$y = -2\lambda - 2$

$z = 3\lambda + 3$

We substitute in the equation of the plane

$2(4\lambda + 5) + (-2\lambda - 2) - 2(3\lambda + 3) = 3$

This has no solutions, so there is no intersection.

d $\frac{1-x}{2} = \lambda$

$\frac{y+2}{3} = \lambda$

$1 - 3z = \lambda$

or equivalently

$x = 1 - 2\lambda$

$y = 3\lambda - 2$

$z = \frac{1-\lambda}{3}$

We substitute in the equation of the plane

$$2(1 - 2\lambda) + (3\lambda - 2) - 3\left(\frac{1-\lambda}{3}\right) = -1$$

This has infinite solutions, so the line is contained in the plane.

2 The normal of the plane and the direction vector of the line must be orthogonal, so their dot product must be zero. We obtain the parametric equation of the line as

$\frac{x}{m} = \lambda \Rightarrow x = \lambda m$

$\frac{y-1}{2} = \lambda \Rightarrow y = 2\lambda + 1$

$\frac{z+2}{4} = \lambda \Rightarrow z = 4\lambda - 2$

Then

$$\boldsymbol{d} = \begin{pmatrix} m \\ 2 \\ 4 \end{pmatrix}$$

and the normal of the plane is

$$\boldsymbol{n} = \begin{pmatrix} 2 \\ m \\ -3 \end{pmatrix}$$

Then

$$\boldsymbol{d} \cdot \boldsymbol{n} = \begin{pmatrix} m \\ 2 \\ 4 \end{pmatrix} \cdot \begin{pmatrix} 2 \\ m \\ -3 \end{pmatrix} = 2m + 2m - 12 = 0$$

gives

$m = 3$

3 This is precisely what we have calculated above, as

$\boldsymbol{d} \cdot \boldsymbol{n} = |\boldsymbol{d}||\boldsymbol{n}|\cos\theta$

and $\theta = \frac{\pi}{2}$, so $\boldsymbol{d} \cdot \boldsymbol{n} = 0$.

Exercise 9K

1 a Let $x = \lambda$

$3\lambda + y - 2z = -1$

$\lambda - 4y + 2z = 3$

Then

$y = \frac{4\lambda - 2}{3}$ and

$z = \frac{1 + 13\lambda}{6}$

which determine the equation of the line

b $n_1 = (3, 1, -2)$

$n_2 = (1, -4, 2)$

$$\cos\theta = \frac{|n_1 \cdot n_2|}{|n_1||n_2|} = \frac{|3(1) + (1)(-4) + (-2)(2)|}{\sqrt{3^2 + 1^2 + 2^2}\sqrt{1^2 + 4^2 + 2^2}} = \frac{5}{7\sqrt{6}}$$

Then

$\theta = \cos^{-1}\frac{5}{7\sqrt{6}} = 1.275$

2 a (1, 3, 9)

b 1.12 or 64°

3 a We have that

$x = \lambda + 2$

$y = -3\lambda + 1$

$z = 2\lambda + 2$

Substitute in the equation of the plane as

$3(\lambda + 2) + 2(-3\lambda + 1) - (2\lambda + 2) = 1$

$\lambda = 1$

Then the point of intersection is $P = (3, -2, 4)$

b The direction vector of the line is

$d = i - 3j + 2k$

The normal vector of the plane is

$n = 3i + 2j - k$

Then

$$\sin\theta = \frac{|d\cdot n|}{|d||n|}$$

$$= \frac{|(1)(3)+(-3)(2)+(2)(-1)|}{\sqrt{1^2+3^2+2^2}\sqrt{3^2+2^2+1^2}}$$

$$= \frac{5}{14}$$

$\theta = 0.365$

4 We look at the angle between the normal vectors

$n_1 = (a, 0, a)$ and

$n_2 = (b, -b, 0)$

Note that

$n_1 \cdot n_2 = a \cdot b$

$|n_1| = \sqrt{2}|a|$

$|n_2| = \sqrt{2}|b|$

$ab = 2|a||b|\cos\theta$

so

$\cos\theta = \pm\frac{1}{2}$

It is the angle between their normal vectors if it is acute and it is the supplementary angle if it is obtuse, hence for both the positive and the negative case, the angle will be $\frac{\pi}{3}$

Exercise 9L

1 a For A, we have that the direction vector will be

$d_a = (0 + 3000)\boldsymbol{i} + (0 - 5000)\boldsymbol{j}$

and since the speed is 4m/s we have to normalise and multiply by this so that the magnitude holds. Then

$$\boldsymbol{d}_a = \frac{4\times 3000\boldsymbol{i} - 4\times 5000\boldsymbol{j}}{\sqrt{3000^2+5000^2}} = \frac{12}{\sqrt{34}}\boldsymbol{i} - \frac{20}{\sqrt{34}}\boldsymbol{j}$$

and so with the point (−3000,5000), the equation of the position becomes

$$\boldsymbol{a} = (-3000\boldsymbol{i} + 5000\boldsymbol{j}) + t\left(\frac{12}{\sqrt{34}}\boldsymbol{i} - \frac{20}{\sqrt{34}}\boldsymbol{j}\right)$$

similarly for L we have $\boldsymbol{d}_l = (0 - 7000)\boldsymbol{i} + (0 - 9000)\boldsymbol{j}$

and since the speed is 4m/s we have to normalise and multiply by this so that the magnitude holds. Then

$$\boldsymbol{d}_a = \frac{6\times(-7000)\boldsymbol{i} - 6\times(-9000)\boldsymbol{j}}{\sqrt{7000^2+9000^2}} = -\frac{42}{\sqrt{130}}\boldsymbol{i} - \frac{54}{\sqrt{130}}\boldsymbol{j}$$

and so with the point (7000,9000), the equation of the position becomes

$$\boldsymbol{l} = (7000\boldsymbol{i} + 9000\boldsymbol{j}) + t\left(-\frac{42}{\sqrt{130}}\boldsymbol{i} - \frac{54}{\sqrt{130}}\boldsymbol{j}\right)$$

b We check when each boat gets to the point (0, 0). For a

$$x = -3000 + \frac{12}{\sqrt{34}}t$$

$$y = 5000 - \frac{20}{\sqrt{34}}t$$

Then at (0, 0)

$$-3000 + \frac{12}{\sqrt{34}}t = 5000 - \frac{20}{\sqrt{34}}t$$

$$\frac{32}{\sqrt{34}}t = 2000$$

$$t = \frac{125\sqrt{34}}{2} \approx 364.4\,s$$

For L we have

$$x = 7000 - \frac{42}{\sqrt{130}}t$$

$$y = 9000 - \frac{54}{\sqrt{130}}t$$

Then at (0, 0)

$$7000 - \frac{42}{\sqrt{130}}t = 9000 - \frac{54}{\sqrt{130}}t$$

$$\frac{12}{\sqrt{130}}t = 2000$$

$$t = 500\frac{\sqrt{130}}{3} \approx 1900.3\,s$$

Boat A will arrive first and the boat L takes more 1535.9 s to reach the boat in need.

2 a The initial position is given at time $t = 0$ so

$\boldsymbol{p}(0) = 23\boldsymbol{i} + 8\boldsymbol{j} + 43\boldsymbol{k}$

b The speed is given by the magnitude of the direction vector

$$|d| = \sqrt{2^2+1^2+4^2}$$

$$= \sqrt{21} \approx 4.58\frac{\text{m}}{\text{s}}$$

c Intersection between the line given and the plane. The components of p are

$x = 23 + 2t$

$y = 8 - t$

$z = 43 + 4t$

We substitute into the equation of the plane to get

$12(23 + 2t) - 3(8 - t) - 5(43 + 4t) = -2$

$276 + 24t - 24 + 3t - 215 - 20t = -2$

$39 = 7t$

$t = 5.57$ s

d Total distance $= 5.57\sqrt{21} \approx 25.5$ m

3 a Assuming distance is in km and time in hours

Speed of p_1

$$v_1 = \sqrt{8^2 + 9^2 + 0.25^2} \approx 12.04 \text{ km/h}$$

Speed of p_2

$$v_2 = \sqrt{7^2 + 11^2 + 0.2^2} = 13.04 \text{ km/h}$$

b Assume that there is an intersection. We write out the components of p_1 and p_2

$x_1 = 147 - 8t$

$y_1 = -156 + 9t$

$z_1 = 5 + 0.25t$

$x_2 = -118 + 7\mu$

$y_2 = 189 - 11\mu$

$z_2 = 7 + 0.2\mu$

We equate the components to get a value of t

$147 - 8t = -118 + 7\mu$

$-156 + 9t = 189 - 11\mu$

$5 + 0.25t = 7 + 0.2\mu$

This gives $\mu = 15$ and $t = 20$ which is consistent in all three equations. Hence the paths intersect. The point of intersection is given by

$$\begin{pmatrix} x \\ y \\ z \end{pmatrix} = \begin{pmatrix} 147 - 8(20) \\ -156 + 9(20) \\ 5 + 0.25(20) \end{pmatrix} = \begin{pmatrix} -13 \\ 24 \\ 25 \end{pmatrix}$$

c The times at which they reach this point are different, and unique. Hence they will not collide.

Chapter review

1 a $a + b = AB$

Hence the midpoint will have half of that length, so

$$\boldsymbol{m} = \frac{1}{2}(\boldsymbol{a} + \boldsymbol{b})$$

b $AB = \frac{10}{3}DC$, so AD and DC are the parallel sides of the trapezium.

c Midpoints are (6, 1), (4.5, 3.5), (2, −1), (8.5, 5.5), which give two pairs of parallel lines with equal length and thus form a rhombus.

2 a We calculate the Cartesian form

$x = 2 + \lambda - 3\mu$

$y = 2\lambda + \mu$

$z = \mu - 1$

We subtract the second from twice the first

$2x - y = 4 + 2\lambda - 2\lambda - 6\mu - \mu$

Then

$2x - y = 4 - 7\mu$

We add 7 times the third equation as

$2x - y + 7z = 7\mu + 4 - 7 - 7\mu$

$2x - y + 7z = -3$

b We substitute with each of the points, leading to the equations

$2(2) - (0) + 7(a) = -3$

$2(b) - 4 + 7(-1) = -3$

$2(-1) - d + 7(0) = -3$

Then we solve them and get

$$a = \frac{-3 - 4}{7} = -1$$

$$b = \frac{-3 + 4 + 7}{2} = 4$$

$$d = \frac{-3 + 2}{-1} = 1$$

c We write (taking all vectors from the origin)

$C = B - A + D = A = B + D - 2A$

$$C = \begin{pmatrix} 4 \\ 4 \\ -1 \end{pmatrix} + \begin{pmatrix} -1 \\ 1 \\ 0 \end{pmatrix} - 2\begin{pmatrix} 2 \\ 0 \\ -1 \end{pmatrix} = \begin{pmatrix} -1 \\ 5 \\ 1 \end{pmatrix}$$

d We substitute with the point E and get

$2(1) - (-2) + 7(1) = 11 \neq -3$

so the point E does not lie in the plane

e We use the formula for the volume of the pyramid. We calculate

$$AC = \begin{pmatrix} -1 \\ 5 \\ 1 \end{pmatrix} - \begin{pmatrix} 2 \\ 0 \\ -1 \end{pmatrix} = \begin{pmatrix} -3 \\ 5 \\ 2 \end{pmatrix}$$

$$AB = \begin{pmatrix} 4 \\ 4 \\ -1 \end{pmatrix} - \begin{pmatrix} 2 \\ 0 \\ -1 \end{pmatrix} = \begin{pmatrix} 2 \\ 4 \\ 0 \end{pmatrix}$$

$$AE = \begin{pmatrix} 1 \\ -2 \\ 1 \end{pmatrix} - \begin{pmatrix} 2 \\ 0 \\ -1 \end{pmatrix} = \begin{pmatrix} -1 \\ -2 \\ 2 \end{pmatrix}$$

Then

$$V = \frac{1}{6}\left|(AC \times AB) \cdot AE\right|$$

$$V = \frac{1}{6}\left|(-8)(-1) + (4)(-2) + (-22)(2)\right| = 7.33$$

3 a The direction vector of the line will be the normal to the plane

$d = (2, -2, 1)$

Then the equation of the line is

$$p = \begin{pmatrix} 2 \\ -2 \\ 1 \end{pmatrix} + \lambda\begin{pmatrix} 2 \\ -2 \\ 1 \end{pmatrix}$$

b The point of intersection is obtained by substituting

$x = 2 + 2\lambda$

$y = -2 - 2\lambda$

$z = 1 + \lambda$

into the equation of the plane

$2(2 + 2\lambda) - 2(-2 - 2\lambda) + (1 + \lambda) = 0$

$4 + 4\lambda + 4 + 4\lambda + 1 + \lambda = 0$

$\lambda = -1$

Then the point of intersection is

$\begin{pmatrix} 0 \\ 0 \\ 0 \end{pmatrix}$

Then the distance between this point and the plane is

$|OA| = \sqrt{2^2 + 2^2 + 1^2} = 3$

c A point on the plane is $B = (0, 0, 0)$ and we define the vector

$BP = x_0 i y_0 j + z_0 k$

The normal of the plane is n

$n = 2\boldsymbol{i} - 2\boldsymbol{j} + k$

Then the distance we need is

$$d = \frac{|BP \cdot n|}{|n|} = \frac{|2x_0 - 2y_0 + z_0|}{\sqrt{2^2 + 2^2 + 1^2}} = \frac{|2x_0 - 2y_0 + z_0|}{3}$$

4 a Note that

$\boldsymbol{a} \cdot \boldsymbol{b} = pr + 4 + rp$

and

$\boldsymbol{a} \cdot \boldsymbol{b} = (p^2 + 4 + r^2)\cos\theta$

since the components form an arithmetic sequence with common difference d, we have the relationship

$p + d = 2$

$2 + d = r$

We use this to rewrite the formula for the dot product in terms of d and get

$$\frac{(2-d)(2+d) + 4 + (2+d)(2-d)}{(2-d)^2 + 4 + (2+d)^2} = \frac{12 - 2d^2}{12 + 2d^2} = \frac{6 - d^2}{6 + d^2}$$

as required.

b When the angle is 60°, the cosine is $\frac{1}{2}$ so

$\frac{6 - d^2}{6 + d^2} = \frac{1}{2}$

$12 - 2d^2 = 6 + d^2$

$3d^2 = 6$

$d^2 = 2$

Then $d = \pm\sqrt{2}$

5 If these planes are perpendicular, then their normal vectors are always perpendicular, so we check

$n_1 \cdot n_2 = \sin\alpha \cdot \cos\alpha + \cos\alpha \sin\alpha - 1$

These planes are not perpendicular

6 If they are perpendicular, their dot product will be equal to zero. We use the fact that their magnitude is 1 to calculate

$(2\boldsymbol{u} - 3\boldsymbol{v})\cdot(5\boldsymbol{u} + 2\boldsymbol{v}) = 10\boldsymbol{u} \cdot \boldsymbol{u} + 4\boldsymbol{u} \cdot \boldsymbol{v} - 15\boldsymbol{v} \cdot \boldsymbol{u} - 6\boldsymbol{v} \cdot \boldsymbol{v}$

$-11\boldsymbol{u} \cdot \boldsymbol{v} + 4 = -11\cos\theta + 4 = 0$

$\cos\theta = \frac{4}{11}$

$\theta = 69°$

7 a $x = 3\lambda + 4$

$y = 1 - \lambda$

$z = 2\lambda + 5$

$4(3\lambda + 4) - 3(1 - \lambda) + (2\lambda + 5) = 1$

$12\lambda + 16 - 3 + 3\lambda + 2\lambda + 5 = 1$

$\lambda = -1$

Then the point P is at $\lambda = -1$ and so the point is

$x = 3(-1) + 4 = 1$

$y = 1 - (-1) = 2$

$z = 2(-1) + 5 = 3$

b Angle between the line and the plane

$\boldsymbol{d} = (3, -1, 2)$

$\boldsymbol{n} = (4, -3, 1)$

$$\sin\theta = \frac{|\boldsymbol{d} \cdot \boldsymbol{n}|}{|\boldsymbol{d}||\boldsymbol{n}|} = \frac{|(3)(4) + (-1)(-3) + (2)(1)|}{\sqrt{3^2 + 1^2 + 2^2}\sqrt{4^2 + 3^2 + 1^2}} \Rightarrow \theta = 63°$$

8 a $\boldsymbol{a} \cdot \boldsymbol{b} = 2^x(2^x) + (4^x)(0.5^x) + (5)(-4) = 0$

$4^x + 2^x - 20 = 0$

This is true for $x = 2$.

b The equation of the plane is given by

$$\boldsymbol{p} = \begin{pmatrix} 1 \\ 1 \\ -2 \end{pmatrix} + \lambda \begin{pmatrix} 4 \\ 16 \\ 5 \end{pmatrix} + \mu \begin{pmatrix} 4 \\ 0.25 \\ -4 \end{pmatrix}$$

9 a We have the following relations

$\boldsymbol{a} \cdot \hat{\boldsymbol{k}} = |\boldsymbol{a}||\hat{\boldsymbol{k}}|\cos\gamma$

$\boldsymbol{a} \cdot \hat{\boldsymbol{j}} = |\boldsymbol{a}||\hat{\boldsymbol{j}}|\cos\beta$

$\boldsymbol{a} \cdot \hat{\boldsymbol{i}} = |\boldsymbol{a}||\hat{\boldsymbol{i}}|\cos\alpha$

Note that the norm of the unit vectors is one, and

$$|\boldsymbol{a}| = \sqrt{(\boldsymbol{a} \cdot \hat{\boldsymbol{k}})^2 + (\boldsymbol{a} \cdot \hat{\boldsymbol{j}})^2 + (\boldsymbol{a} \cdot \hat{\boldsymbol{i}})^2}$$

We substitute with the relations obtained and get

$|\boldsymbol{a}|^2 = |\boldsymbol{a}|^2\cos^2\gamma + |\boldsymbol{a}|^2\cos^2\beta + |\boldsymbol{a}|^2\cos^2\alpha$

$1 = \cos^2\gamma + \cos^2\beta + \cos^2\alpha$

b The norm of a is

$|\boldsymbol{a}| = \sqrt{3^2 + 6^2 + 2^2} = 7$

We substitute into the relations obtained in a to get

$3 = 7\cos\alpha \Rightarrow \alpha = 64.6°$

$-6 = 7\cos\beta \Rightarrow \beta = 149°$

$2 = 7\cos\gamma \Rightarrow \gamma = 73.4°$

c When the plane passes through zero, the normal vector will correspond precisely to the unit vectors $\boldsymbol{i}, \boldsymbol{j}, \boldsymbol{k}$. As we saw in (a), these can be written as the cosines of the angles. Hence

$\boldsymbol{n} = \cos\alpha\,\boldsymbol{i} + \cos\beta\,\boldsymbol{j} + \cos\gamma\,\boldsymbol{k}$

Then the equation of the plane can be written as

$x\cos\alpha + y\cos\beta + z\cos\gamma = 0$

10 a We calculate the vectors

$$AP = \begin{pmatrix}1\\2\\4\end{pmatrix} - \begin{pmatrix}2\\0\\0\end{pmatrix} = \begin{pmatrix}-1\\2\\4\end{pmatrix}$$

and

$$AQ = \begin{pmatrix}0\\1\\4\end{pmatrix} - \begin{pmatrix}2\\0\\0\end{pmatrix} = \begin{pmatrix}-2\\1\\4\end{pmatrix}$$

These will be the two vectors on the plane equation. Additionally we take a point, choosing for simplicity A = (2, 0, 0). Then the plane equation in vector form is

$$p = \begin{pmatrix}2\\0\\0\end{pmatrix} + \lambda\begin{pmatrix}-1\\2\\4\end{pmatrix} + \mu\begin{pmatrix}-2\\1\\4\end{pmatrix}$$

To write it in Cartesian form, we write out the system of equations

$x = 2 - \lambda - 2\mu$

$y = 2\lambda + \mu$

$z = 4\lambda + 4\mu$

We subtract the third one from twice the second one, to get

$z - 2y = 2\mu$

so $\mu = \frac{z-2y}{2}$ and we add the second one to twice the first one, to get

$y + 2x = 4 + 2\lambda - 2\lambda + \mu - 4\mu$

or equivalently

$y + 2x = 4 - 3\mu$

Then we substitute with our value for μ to get

$$y + 2x = 4 - 3\left(\frac{z-2y}{2}\right)$$

This simplifies to

$4x - 4y + 3z = 8$

b using the equation of the plane written in a. BG gives the direction vector of the line.

$$BG = \begin{pmatrix}0\\0\\4\end{pmatrix} - \begin{pmatrix}2\\2\\0\end{pmatrix} = \begin{pmatrix}-2\\-2\\4\end{pmatrix}$$

Then the equation of the line is written as

$$\boldsymbol{p} = \begin{pmatrix}0\\0\\4\end{pmatrix} - \lambda\begin{pmatrix}-2\\-2\\4\end{pmatrix}$$

c Angle between plane $4x - 4y + 3z = 8$ and line

$$\boldsymbol{p} = \begin{pmatrix}0\\0\\4\end{pmatrix} - \lambda\begin{pmatrix}-2\\-2\\4\end{pmatrix}$$

We have that

$$\sin\theta = \frac{|\boldsymbol{d}\cdot\boldsymbol{n}|}{|\boldsymbol{d}||\boldsymbol{n}|}$$

$$= \frac{|(-2)(4)+(-2)(-4)+(4)(3)|}{\sqrt{2^2+2^2+4^2}\sqrt{4^2+4^2+3^2}}$$

$$= \frac{12}{2\sqrt{246}} = \frac{6}{\sqrt{246}}$$

Then

$\theta = 22.5°$

Exam-style questions

11 a $\overrightarrow{AB} = \begin{pmatrix}2\\0\\-1\end{pmatrix}$ (1 mark)

$\overrightarrow{AC} = \begin{pmatrix}3\\2\\2\end{pmatrix}$ (1 mark)

$\overrightarrow{AB} \times \overrightarrow{AC} = \begin{pmatrix}2\\0\\-1\end{pmatrix} \times \begin{pmatrix}3\\2\\2\end{pmatrix}$ (1 mark)

$= \begin{pmatrix}2\\-7\\4\end{pmatrix}$ (1 mark)

b $\frac{1}{2}|\overrightarrow{AB} \times \overrightarrow{AC}| = \frac{1}{2}\sqrt{2^2+(-7)^2+4^2}$ (2 marks)

$= \frac{\sqrt{69}}{2}$ (1 mark)

c $\mathbf{r}.\begin{pmatrix}2\\-7\\4\end{pmatrix} = \begin{pmatrix}3\\0\\0\end{pmatrix}.\begin{pmatrix}2\\-7\\4\end{pmatrix}$ (2 marks)

$\mathbf{r}.\begin{pmatrix}2\\-7\\4\end{pmatrix} = 6$ (1 mark)

$2x - 7y + 4z = 6$

d $\begin{pmatrix}3\\-5\\1\end{pmatrix} \times \begin{pmatrix}2\\-7\\4\end{pmatrix} = \begin{pmatrix}-13\\-10\\-11\end{pmatrix}$ (2 marks)

$\mathbf{n} = \begin{pmatrix}13\\10\\11\end{pmatrix}$

$y = 0 \Rightarrow x = -\frac{1}{5}, z = \frac{8}{5}$

(or equivalent) (2 marks)

$\mathbf{r} = \begin{pmatrix}-\frac{1}{5}\\0\\\frac{8}{5}\end{pmatrix} + \lambda\begin{pmatrix}13\\10\\11\end{pmatrix}$

(or equivalent) (1 mark)

12

$\overrightarrow{AB} = \overrightarrow{OB} - \overrightarrow{OA} = \begin{pmatrix}-2\\2\\3\end{pmatrix} - \begin{pmatrix}1\\0\\1\end{pmatrix} = \begin{pmatrix}-3\\2\\2\end{pmatrix}$ (1 mark)

$\overrightarrow{AC} = \overrightarrow{OC} - \overrightarrow{OA} = \begin{pmatrix}0\\4\\2\end{pmatrix} - \begin{pmatrix}1\\0\\1\end{pmatrix} = \begin{pmatrix}-1\\4\\1\end{pmatrix}$ (1 mark)

$\overrightarrow{AD} = \overrightarrow{OD} - \overrightarrow{OA} = \begin{pmatrix}3\\1\\3\end{pmatrix} - \begin{pmatrix}1\\0\\1\end{pmatrix} = \begin{pmatrix}2\\1\\2\end{pmatrix}$ (1 mark)

Volume

$= \frac{1}{6}|\overrightarrow{AB}.\overrightarrow{AC} \times \overrightarrow{AD}| = \frac{1}{6}\left|\begin{pmatrix}-3\\2\\2\end{pmatrix}.\begin{pmatrix}7\\4\\-9\end{pmatrix}\right|$ (2 marks)

$= \frac{1}{6}|(-21+8-18)|$

$= \frac{31}{6}$ units². (1 mark)

13 $\overrightarrow{AP} = \mathbf{p} - \mathbf{a}$ (1 mark)

$\overrightarrow{BP} = \mathbf{p} - \mathbf{b}$ (1 mark)

$\overrightarrow{AP}.\overrightarrow{BP} = (\mathbf{p}-\mathbf{a}).(\mathbf{p}-\mathbf{b})$ (1 mark)

$= (\mathbf{p} - \mathbf{a}) . (\mathbf{p} + \mathbf{a})$ (1 mark)

$= \mathbf{p} . \mathbf{p} - \mathbf{a} . \mathbf{p} + \mathbf{a} . \mathbf{p} - \mathbf{a} . \mathbf{a}$ (1 mark)

$= \mathbf{p} . \mathbf{p} - \mathbf{a} . \mathbf{a}$ (1 mark)

$= |\mathbf{p}|^2 - |\mathbf{a}|^2$ (1 mark)

$= 0$ since $|\mathbf{p}| = |\mathbf{a}|$ (1 mark)

Therefore $\overrightarrow{AP}$ is perpendicular to $\overrightarrow{BP}$ and $\angle APB = 90°$

14 a Equation of line perpendicular to Π and passing through P is

$$\mathbf{r} = \begin{pmatrix}1\\0\\2\end{pmatrix} + \lambda\begin{pmatrix}4\\-3\\1\end{pmatrix}$$ (2 marks)

Attempting to solve P and Π simultaneously: (1 mark)

$4(1 + 4\lambda) - 3(-3\lambda) + (2 + \lambda) = 19$

$4 + 16\lambda + 9\lambda + 2 + \lambda = 19$

$26\lambda + 6 = 19$

$\lambda = \frac{1}{2}$ (1 mark)

Therefore

$$\overrightarrow{OQ} = \begin{pmatrix}1\\0\\2\end{pmatrix} + 2\times\frac{1}{2}\times\begin{pmatrix}4\\-3\\1\end{pmatrix}$$ (1 mark)

$$= \begin{pmatrix}5\\-3\\3\end{pmatrix}$$ (1 mark)

b Distance between $P(1, 0, 2)$ and $Q(5, -3, 3)$ is given by

$$\sqrt{(5-1)^2+(-3-0)^2+(3-2)^2}$$ (2 marks)

$$= \sqrt{16+9+1}$$

$$= \sqrt{26}$$ (1 mark)

15 a $4(1 + 6\lambda) + 3(5 - 2\lambda) - (-3 + 2\lambda) = 14$ (1 mark)

$4 + 24\lambda + 15 - 6\lambda + 3 - 2\lambda = 14$

$22 + 16\lambda = 14$

$\lambda = -\frac{1}{2}$ (1 mark)

$$\mathbf{r} = \begin{pmatrix}1\\5\\-3\end{pmatrix} - \frac{1}{2}\begin{pmatrix}6\\-2\\2\end{pmatrix} = \begin{pmatrix}-2\\6\\-4\end{pmatrix}$$ (2 marks)

So $P(-2, 6, -4)$.

b $\begin{pmatrix}-2\\6\\-4\end{pmatrix}$ lies on the plane and

$$\mathbf{n} = \begin{pmatrix}4\\3\\-1\end{pmatrix}$$ (2 marks)

So distance

$$= \frac{\begin{pmatrix}-2\\6\\-4\end{pmatrix}.\begin{pmatrix}4\\3\\-1\end{pmatrix}}{\sqrt{4^2+3^2+(-1)^2}}$$ (1 mark)

$$= \frac{-8+18+4}{\sqrt{26}}$$

$$= \frac{14}{\sqrt{26}}\left(= \frac{14\sqrt{26}}{26} = \frac{7\sqrt{26}}{13}\right)$$ (1 mark)

16 a

$$\overrightarrow{AB} = \overrightarrow{OB} - \overrightarrow{OA} = \begin{pmatrix}2\\0\\6\end{pmatrix} - \begin{pmatrix}8\\2\\0\end{pmatrix} = \begin{pmatrix}-6\\-2\\6\end{pmatrix}$$ (1 mark)

$$\mathbf{r} = \begin{pmatrix}8\\2\\0\end{pmatrix} + \lambda\begin{pmatrix}-6\\-2\\6\end{pmatrix}$$ (2 marks)

b

$$\overrightarrow{CD} = \overrightarrow{OD} - \overrightarrow{OC} = \begin{pmatrix}12\\3\\0\end{pmatrix} - \begin{pmatrix}4\\4\\4\end{pmatrix} = \begin{pmatrix}8\\-1\\-4\end{pmatrix}$$ (1 mark)

$$\mathbf{r} = \begin{pmatrix}4\\4\\4\end{pmatrix} + \mu\begin{pmatrix}8\\-1\\4\end{pmatrix}$$ (2 marks)

c Direction vectors are $\begin{pmatrix}-6\\-2\\6\end{pmatrix}$ and $\begin{pmatrix}8\\-1\\4\end{pmatrix}$

$$\begin{pmatrix}-6\\-2\\6\end{pmatrix} \times \begin{pmatrix}8\\-1\\4\end{pmatrix} = \begin{pmatrix}-2\\72\\22\end{pmatrix}$$ (2 marks)

$(8, 2, 0)$ lies on AB and $(4, 4, 4)$ lies on CD

$$\overrightarrow{AC} = \begin{pmatrix}-4\\2\\4\end{pmatrix}$$ (1 mark)

Projection of $\overrightarrow{AC}$ to the vector $\begin{pmatrix}-2\\72\\22\end{pmatrix}$ is

$$\frac{\left|\begin{pmatrix}-4\\2\\4\end{pmatrix}.\begin{pmatrix}-2\\72\\22\end{pmatrix}\right|}{\sqrt{(-2)^2+72^2+22^2}}$$ (2 marks)

$$= \frac{8+144+88}{\sqrt{(-2)^2+72^2+22^2}}$$ (1 mark)

$$= \frac{240}{\sqrt{5672}}$$ (1 mark)

$$\left(= \frac{240\sqrt{5672}}{5672} = \frac{480\sqrt{1418}}{5672} = \frac{60\sqrt{1418}}{709} (= 3.19)\right)$$

17 Choosing $\lambda = 1$ (say), gives

$$\mathbf{r} = \begin{pmatrix}10\\-4\\4\end{pmatrix} + \begin{pmatrix}1\\2\\1\end{pmatrix} = \begin{pmatrix}11\\-2\\5\end{pmatrix}$$ (1 mark)

Therefore A(5, 8, 0), B(10, −4, 4) and C(11, −2, 5) lie on Π (2 marks)

$$\overrightarrow{AB} = \begin{pmatrix}5\\-12\\4\end{pmatrix} \text{ and } \overrightarrow{AC} = \begin{pmatrix}6\\-10\\5\end{pmatrix}$$

(2 marks)

$$\overrightarrow{AB}\times\overrightarrow{AC} = \begin{pmatrix}5\\-12\\4\end{pmatrix}\times\begin{pmatrix}6\\-10\\5\end{pmatrix} = \begin{pmatrix}-20\\-1\\22\end{pmatrix}$$

(2 marks)

So equation of plane is

$$\mathbf{r}.\begin{pmatrix}-20\\-1\\22\end{pmatrix} = \begin{pmatrix}5\\8\\0\end{pmatrix}.\begin{pmatrix}-20\\-1\\22\end{pmatrix}$$

(2 marks)

$$\mathbf{r}.\begin{pmatrix}-20\\-1\\22\end{pmatrix} = -108$$ (1 mark)

$-20x - y + 22z = -108$

(1 mark)

$$\frac{20}{108}x + \frac{1}{108}y - \frac{22}{108}z = 1$$

(1 mark)

$$\left(\frac{5}{27}x + \frac{1}{108}y - \frac{11}{54}z = 1\right)$$

18 Direction vector of line is $\begin{pmatrix}2\\5\\p\end{pmatrix}$

(1 mark)

Direction normal to plane is $\begin{pmatrix}5\\p\\p\end{pmatrix}$

(1 mark)

If the angle between the line and the plane is θ, then

$$\sin\theta = \frac{\begin{pmatrix}2\\5\\p\end{pmatrix}.\begin{pmatrix}5\\p\\p\end{pmatrix}}{\sqrt{2^2+5^2+p^2}\sqrt{5^2+p^2+p^2}}$$

(3 marks)

$$= \frac{10+5p+p^2}{\sqrt{2^2+5^2+p^2}\sqrt{5^2+p^2+p^2}}$$

(1 mark)

θ is maximum when $\sin\theta$ is maximum. (1 mark)

By GDC, maximum occurs when $p = 6.797$ (1 mark)

So maximum value of $\sin\theta$ is 0.96 (1 mark)

$\Rightarrow \theta_{MAX} = 73.7°$ (1 mark)

19

$$\begin{pmatrix}3\\-1\\1\end{pmatrix} \neq k\begin{pmatrix}1\\-1\\1\end{pmatrix}$$, so L_1 and L_2 are not parallel. (2 marks)

Consider **i** and **j** components:

(1 mark)

$1 + 3\lambda = 2 + \mu$ and $-\lambda = 1 - \mu$

(1 mark)

Solving simultaneously:

(1 mark)

$\lambda = 1$, $\mu = 2$ (1 mark)

Substitute into **k** component:

(1 mark)

$2 + \lambda = 1 + \mu$, $2 + 1 = 1 + 2$ (so equations are consistent).

(1 mark)

Therefore L_1 and L_2 intersect at the point where $\lambda = 1$ and $\mu = 2$, so are not skew.

(1 mark)

Chapter 10

Skills check

1

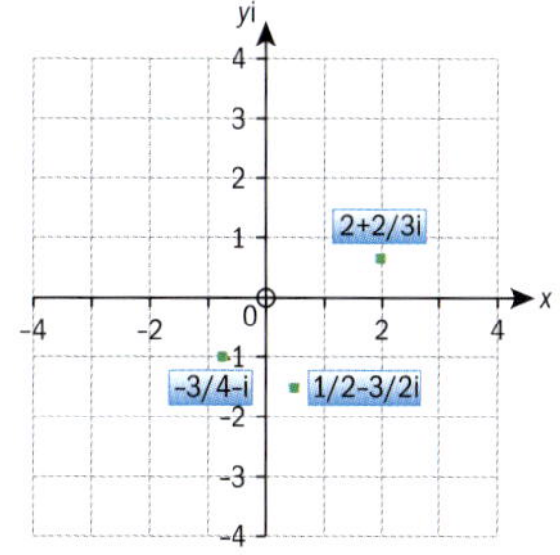

2 $\text{Re}(z_1) = 2, \text{Im}(z_1) = \frac{2}{3}$,

$\text{Re}(z_2) = -\frac{3}{4}, \text{Im}(z_2) = -1$,

$\text{Re}(z_3) = \frac{1}{2}, \text{Im}(z_3) = -\frac{3}{2}$.

3 **a** $1 - 13i$ **b** $-\frac{17}{4} - \frac{7}{4}\text{i}$

4 **a** $z^* = 2+3\text{i}, -z = -2+3\text{i}$,

$\frac{1}{z} = \frac{2}{13} + \frac{3}{13}\text{i}, |z| = \sqrt{13}$

b $z^* = \frac{4}{5} - \frac{3}{5}\text{i}, -z = -\frac{4}{5} - \frac{3}{5}\text{i}$,

$\frac{1}{z} = \frac{4}{5} - \frac{3}{5}\text{i}, |z| = 1$

Exercise 10A

1

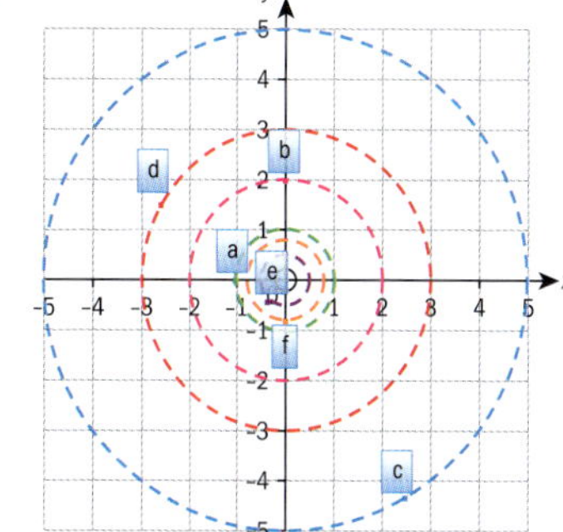

2 **a** $2\text{cis}\left(\frac{\pi}{4}\right)$ **b** $\frac{3}{2}\text{cis}\left(\frac{\pi}{2}\right)$

c $5\text{cis}\left(-\pi + \arctan\left(\frac{3}{4}\right)\right)$

d $29\text{cis}\left(2\pi + \arctan\left(\frac{-20}{21}\right)\right)$

e $2\text{cis}\left(\frac{2\pi}{3}\right)$ **f** $\frac{4}{3}\text{cis}\left(\frac{3\pi}{2}\right)$

g $\frac{5\sqrt{2}}{12}\text{cis}\left(-\arctan\left(\frac{3}{4}\right)\right)$

3 **a** -1 **b** 2i

c $\frac{5}{2} - \frac{5\sqrt{3}}{2}\text{i}$ **d** $-\frac{3\sqrt{3}}{2} + \frac{3}{2}\text{i}$

e $-\frac{\sqrt{2}}{4} - \frac{\sqrt{2}}{4}\text{i}$ **f** $-\frac{4}{5}\text{i}$

4 **a** $\frac{7}{12}\text{cis}\frac{\pi}{9}$ **b** $\frac{7}{12}\text{cis}\left(-\frac{\pi}{9}\right)$

c $-\frac{7}{12}\text{cis}\left(-\frac{\pi}{9}\right)$

Exercise 10B

1 **a** $8e^{\text{i}\frac{7\pi}{12}}$ **b** $30\text{cis}(135°)$

c $\frac{5}{9}e^{\text{i}\left(\frac{17}{14}\right)\pi}$ **d** $\text{cis}(135°)$

2 **a** $-\frac{1}{\sqrt{2}} + \text{i}\frac{1}{\sqrt{2}}$ **b** $e^{\text{i}\frac{2\pi}{3}}$

c $\frac{-\sqrt{6}+\sqrt{2}}{4} - \frac{\sqrt{6}+\sqrt{2}}{4}i$

d See answer to part c.

e $2 + \sqrt{3}$

3 **a** $\theta = \frac{\pi}{6}$ or $\theta = \frac{7\pi}{6}$

$(\text{up to multiples of } 2\pi)$

and less than 3 if $r < \frac{3}{2}$

b Imaginary if $\theta = -\frac{3\pi}{4}$ or

$\theta = \frac{\pi}{4}(\text{up to multiples of } 2\pi)$,

$r > 2\sqrt{2}$

4 -1

Exercise 10C

1 **a** $\frac{3}{4}\text{cis}\left(-\frac{17\pi}{12}\right)$

b $\frac{5}{4}\text{cis}\left(\frac{7\pi}{6}\right)$

c $\frac{3}{20}\text{cis}\left(-\frac{31\pi}{12}\right)$

d $-\frac{5}{12}\text{cis}\left(\frac{3\pi}{4}\right)$

2 **a** $\frac{\sqrt{2}}{2}e^{i\frac{7\pi}{12}}$ **b** $-\sqrt{2}e^{i\frac{\pi}{12}}$

c $\frac{\sqrt{2}}{4}e^{i\frac{\pi}{12}}$ **d** $\frac{1}{2}e^{\frac{2x}{3}}$

3 **a** $\frac{3\sqrt{2}}{4}\text{cis}\left(-\frac{\pi}{4}\right)$

b $\sqrt{2}\text{cis}\left(-\frac{11\pi}{12}\right)$

c $\frac{\sqrt{10}}{2}\text{cis}\left(\frac{3\pi}{2}\right)$

4 **a** $\frac{5}{2} + \frac{5\sqrt{3}}{2}i$ **b** $3\sqrt{2}\text{cis}\frac{\pi}{4}$

c $\frac{5\sqrt{2}}{6}\text{cis}\frac{\pi}{12}$

$= \frac{5}{6}\,\frac{1+\sqrt{3}+i\left(\sqrt{3}-1\right)}{2}$

$= \frac{5\left(1+\sqrt{3}\right)}{12} + i\frac{5\left(\sqrt{3}-1\right)}{12}$

d $\frac{\sqrt{6}+\sqrt{2}}{4}$

e $\frac{\sqrt{6}-\sqrt{2}}{4}$

f $2-\sqrt{3}$

Exercise 10D

1 **a** $8\sqrt{2}e^{i\frac{25\pi}{12}}$ **b** $\frac{\sqrt{2}}{2}e^{i\frac{5\pi}{4}}$

c $128e^{-\frac{5\pi i}{3}}$ **d** $16\sqrt{2}e^{-i\frac{\pi}{4}}$

2 $-\frac{\sqrt{2}}{32} + \frac{\sqrt{6}}{32}i$

3 $-i$

4 **a** $\sqrt{2}e^{i\frac{\pi}{4}}$ **b** $32 + 16i$

c 47⊠14i

Exercise 10E

1 **a** $\omega^3 + 1 = 0$

$\Rightarrow (\omega+1)(\omega^2-\omega+1) = 0$

But $\omega \neq -1$ so it must be the case that $\omega^2 - \omega + 1 = 0$

b Taking the conjugate of the result in part a immediately gives this result.

c -1

2 0

3 **a** $5e^{i\pi\left(\frac{1}{4}+\frac{k}{2}\right)}$ **b** $2^{\frac{1}{5}}e^{i\pi\left(\frac{2}{5}k-\frac{1}{30}\right)}$

c $e^{i\pi\left(\frac{1}{9}+\frac{k}{3}\right)}$

4 Use of calculator

5 **a** $8\left(-\frac{1}{\sqrt{2}} + \frac{i}{\sqrt{2}}\right) = 8\left(e^{i\frac{3\pi}{4}+2ik\pi}\right)$

b $\text{Re}\left(z^{\frac{1}{6}}\right) = \frac{z^{\frac{1}{6}} + (z^*)^{\frac{1}{6}}}{2}$

$= \frac{8^{\frac{1}{6}}\left(e^{i\frac{\pi}{8}} + e^{-i\frac{\pi}{8}}\right)}{2} = \sqrt{2}\cos\frac{\pi}{8}$

$\cos\frac{\pi}{4} = 2\cos^2\frac{\pi}{8} - 1 = \frac{1}{\sqrt{2}}$

$\Rightarrow \cos\frac{\pi}{8} = \sqrt{\frac{1}{2}\left(1+\frac{1}{\sqrt{2}}\right)}$

$= \sqrt{\frac{\sqrt{2}+1}{2\sqrt{2}}} = \sqrt{\frac{2+\sqrt{2}}{4}}$

$\therefore \text{Re}\left(z^{\frac{1}{6}}\right) = \sqrt{2}\sqrt{\frac{2+\sqrt{2}}{4}}$

$= \sqrt{\frac{2\sqrt{2}+4}{4}} = \frac{\sqrt{2\sqrt{2}+4}}{2}$

c Any correct form approximately 0.41...

Exercise 10F

1 $P(n)$: $(\text{cis}\theta)^n = \text{cis}n\theta$

The statement $P(1)$ is true:
$\text{cis}\theta = \text{cis}\theta$

Assume that $P(k)$ is true for some $k \in \mathbb{Z}^+$

i.e. $(\text{cis}\theta)^k = \text{cis}k\theta$

Then $(\text{cis}\theta)^{k+1} = (\text{cis}\theta)^k(\text{cis}\theta) = (\text{cis}k\theta)(\text{cis}\theta)$

$= (\cos k\theta + i\sin k\theta)(\cos\theta + i\sin\theta)$

$= (\cos k\theta\cos\theta - \sin k\theta\sin\theta) + i(\sin k\theta\cos\theta + \sin\theta\cos k\theta)$

$= \cos(k\theta + \theta) + i\sin(k\theta + \theta)$ using the compound angle formula

$= \cos((k+1)\theta) + i\sin((k+1)\theta)$

so $P(k) \Rightarrow P(k+1)$

Therefore it has been shown that $P(1)$ is true and that if $P(k)$ is true for some $k \in \mathbb{Z}^+$ then so is $P(k+1)$. Thus, $P(n)$ is true for all $n \in \mathbb{Z}^+$ by the principle of mathematical induction.

2 **a** $(\cos^4\theta - 6\cos^2\theta\sin^2\theta + \sin^4\theta) + i(4\cos^3\theta\sin\theta - 4\cos\theta\sin^3\theta)$

b **i** $\cos^4\theta - 6\cos^2\theta\sin^2\theta + \sin^4\theta$

ii $4\cos^3\theta\sin\theta - 4\cos\theta\sin^3\theta$

c

$\tan 4\alpha = \frac{\sin 4\alpha}{\cos 4\alpha}$

$= \frac{4\cos^3\alpha\sin\alpha - 4\cos\alpha\sin^3\alpha}{\cos^4\alpha - 6\cos^2\alpha\sin^2\alpha + \sin^4\alpha}$

$$= \frac{4\tan\alpha - 4\tan^3\alpha}{1 - 6\tan^2\alpha + \tan^4\alpha}$$

(to clarify last step, divide top and bottom by $\cos^4\alpha$)

3 **a** $z^4 + 4z^2 + 6 + \frac{4}{z^2} + \frac{1}{z^4}$

b $\frac{1}{8}\cos 4\theta + \frac{1}{2}\cos 2\theta + \frac{3}{8}$

c $\frac{1}{32}\sin 4x + \frac{1}{4}\sin 2x + \frac{3x}{8} + C$

4 **a** $\omega^6 - 1 = (\omega^2 - 1)(\omega^4 + \omega^2 + 1) = 0$ $\omega^2 \neq 1$ so it must be that $1 + \omega^2 + \omega^4 = 0$

b 1

5 $\mathrm{i}\cot\frac{\pi}{16}$

Chapter review

1 $\mathrm{Re}(z) = -3\sqrt{2},\ \mathrm{Im}(z) = -3\sqrt{2}$

2 $r = 2\ (r \geq 0)$

3 $\frac{1}{1+z} = \frac{1+z^*}{(1+z)(1+z^*)}$

$$= \frac{1+z^*}{1+(z+z^*)+|z|^2}$$

$$= \frac{1+z^*}{2+2\mathrm{Re}(z)}$$

$$= \frac{1}{2}\left(\frac{1+\cos\theta - i\sin\theta}{1+\cos\theta}\right)$$

$$= \frac{1}{2}\left(1 - \mathrm{i}\frac{\sin\theta}{1+\cos\theta}\right)$$

$$= \frac{1}{2}\left(1 - \mathrm{i}\frac{2\sin\frac{\theta}{2}\cos\frac{\theta}{2}}{2\cos^2\frac{\theta}{2}}\right)$$

$$= \frac{1}{2}\left(1 - \mathrm{i}\tan\frac{\theta}{2}\right)$$

4

a $z = 1,\ z = e^{\frac{2\pi i}{5}},\ z = e^{\frac{4\pi i}{5}},$ $z = e^{\frac{6\pi i}{5}},\ z = e^{\frac{8\pi i}{5}}$

b The five roots above can be written as

$1, \omega, \omega^2, \omega^3, \omega^4$ i.e. the fifth roots of unity

As a consequence of the fact that the roots of unity sum to zero,

$\mathrm{Re}(1 + \omega + \omega^2 + \omega^3 + \omega^4) = 0$

$$\Rightarrow 1 + \cos\frac{2\pi}{5} + \cos\frac{4\pi}{4} + \cos\frac{6\pi}{5} + \cos\frac{8\pi}{5} = 0$$

$$\Rightarrow \cos\frac{2\pi}{5} + \cos\frac{4\pi}{4} + \cos\frac{6\pi}{5} + \cos\frac{8\pi}{5} = -1$$

5 **a** $z^n + \frac{1}{z^n} = (\cos\theta + \mathrm{i}\sin\theta)^n + (\cos\theta + \mathrm{i}\sin\theta)^{-n}$

$= (\cos n\theta + \mathrm{i}\sin n\theta) + (\cos(-n\theta) + \mathrm{i}\sin(-n\theta))$

$= (\cos n\theta + \mathrm{i}\sin n\theta) + (\cos n\theta - \mathrm{i}\sin n\theta)$

$= 2\cos n\theta$

b

$$z^6 + 6z^4 + 15z^2 + 20 + \frac{15}{z^2} + \frac{6}{z^4} + \frac{1}{z^6}$$

c $a = \frac{1}{32}, b = \frac{3}{16}, c = \frac{15}{32}, d = \frac{5}{16}$

d $\frac{5\pi}{64} - \frac{11}{48}$

6 -1.060 (4s.f.)

7 $2.00 - (4.00)\boldsymbol{i}$ (to 3s.f.)

8 4

9 9.51 to 2 d.p.

10 $2\cot\left(\frac{1}{2}\right)$

Exam-style questions

11 **a** $z_1 z_2 = 4\,\mathrm{cis}\left(-\frac{\pi}{3}\right)3\,\mathrm{cis}\left(\frac{5\pi}{6}\right)$

$= 12\,\mathrm{cis}\left(\frac{5\pi}{6} - \frac{\pi}{3}\right)$ (1 mark)

$= 12\,\mathrm{cis}\left(\frac{\pi}{2}\right)$ (1 mark)

$= 12\boldsymbol{i}$ (1 mark)

b

$\frac{z_1}{z_2} = \frac{4\,\mathrm{cis}\left(-\frac{\pi}{3}\right)}{3\,\mathrm{cis}\left(\frac{5\pi}{6}\right)} = \frac{4}{3}\mathrm{cis}\left(-\frac{\pi}{3} - \frac{5\pi}{6}\right)$ (1 mark)

$= \frac{4}{3}\mathrm{cis}\left(-\frac{7\pi}{6}\right) = \frac{4}{3}\mathrm{cis}\left(\frac{5\pi}{6}\right)$ (1 mark)

So $\left(\frac{z_1}{z_2}\right)^3 = \left(\frac{4}{3}\mathrm{cis}\left(\frac{5\pi}{6}\right)\right)^3$

$= \frac{64}{27}\mathrm{cis}\left(\frac{15\pi}{6}\right)$ (1 mark)

$= \frac{64}{27}\mathrm{cis}\left(\frac{\pi}{2}\right)$ (1 mark)

$= \frac{64}{27}\mathrm{i}$ (1 mark)

c $z_1^{\ 2} = 16\,\mathrm{cis}\left(-\frac{2\pi}{3}\right)$

So $\left(z_1^{\ 2}\right)^* = 16\,\mathrm{cis}\left(\frac{2\pi}{3}\right)$ (1 mark)

$= 16\left(\cos\frac{2\pi}{3} + \mathrm{i}\sin\frac{2\pi}{3}\right)$

$= 16\left(-\frac{1}{2} + \mathrm{i}\frac{\sqrt{3}}{2}\right)$ (1 mark)

$= -8 + 8\sqrt{3}\mathrm{i}$ (1 mark)

12 $|1+\mathrm{i}| = \sqrt{2}$ (1 mark)

$\arg(1+\mathrm{i}) = \frac{\pi}{4}$ (1 mark)

$1+\mathrm{i} = \sqrt{2}\,\mathrm{cis}\frac{\pi}{4}$

$(1+\mathrm{i})^{10} = \left(\sqrt{2}\right)^{10}\mathrm{cis}\frac{10\pi}{4}$ by de Moivre's theorem (1 mark)

$= 2^5\,\mathrm{cis}\frac{5\pi}{2}$ (1 mark)

$= 2^5\,\mathrm{cis}\frac{\pi}{2}$

$= 32\boldsymbol{i}$ (1 mark)

13 **a** $|z| = \sqrt{1+\left(\sqrt{3}\right)^2} = 2$ (1 mark)

$\arg z = -\frac{\pi}{3}$ (1 mark)

$z = 2\,\mathrm{cis}\left(-\frac{\pi}{3}\right)$ (1 mark)

b $z^n = 2^n \operatorname{cis}\left(-\frac{n\pi}{3}\right)$ (1 mark)

$z^n \in \mathbb{R} \Rightarrow -\frac{n\pi}{3} = 2\pi k$

(1 mark)

So $n = 6$ (1 mark)

c $\left(1 - i\sqrt{3}\right)^{15} = 2^{15} \operatorname{cis}\left(-\frac{15\pi}{3}\right)$

(1 mark)

$= 2^{15} cis(-5\pi)$

$= 2^{15} cis(\pi)$ (1 mark)

$= -2^{15} (= -32768)$

(1 mark)

---

14 a $(\cos\theta + i\sin\theta)^5 = \cos^5\theta + 5\cos^4\theta(\boldsymbol{i}\sin\theta) + 10\cos^3\theta(\boldsymbol{i}\sin\theta)^2$

$+ 10\cos^2\theta(\boldsymbol{i}\sin\theta)^3 + 5\cos\theta(\boldsymbol{i}\sin\theta)^4 + (\boldsymbol{i}\sin\theta)^5$ (2 marks)

$= \cos^5\theta + 5\boldsymbol{i}\cos^4\theta\sin\theta - 10\cos^3\theta\sin^2\theta - 10\boldsymbol{i}\cos^2\theta\sin^3\theta +$
$5\cos\theta\sin^4\theta + i\sin^5\theta$ (1 mark)

b By de Moivre's theorem, $(\cos\theta + i\sin\theta)^5 = \cos5\theta + i\sin5\theta$ (1 mark)

Equating real parts of each expression: (1 mark)

$\cos5\theta = \cos^5\theta - 10\cos^3\theta(1 - \cos^2\theta) + 5\cos\theta(1 - \cos^2\theta)^2$ (1 mark)

$= \cos^5\theta - 10\cos^3\theta + 10\cos^5\theta + 5\cos\theta(1 - 2\cos^2\theta + \cos^4\theta)$ (1 mark)

$= 16\cos^5\theta - 20\cos^3\theta + 5\cos\theta$

15 a Let $z^3 = -27\boldsymbol{i}$

$\left|z^3\right| = 27$ (1 mark)

$\arg\left(z^3\right) = -\frac{\pi}{2}$ (1 mark)

$z^3 = 27\left(\cos\left(-\frac{\pi}{2}\right) + i\sin\left(-\frac{\pi}{2}\right)\right)$ (1 mark)

$z^3 = 27\left(\cos\left(-\frac{\pi}{2} + 2\pi k\right) + i\sin\left(-\frac{\pi}{2} + 2\pi k\right)\right)$ (1 mark)

$z^3 = 27\left(\cos\left(\frac{4\pi k - \pi}{2}\right) + i\sin\left(\frac{4\pi k - \pi}{2}\right)\right)$

$z = 3\left(\cos\left(\frac{4\pi k - \pi}{6}\right) + i\sin\left(\frac{4\pi k - \pi}{6}\right)\right)$ (1 mark)

Choosing $k = 1, 2, 3$ (or equivalent)

$z_1 = 3\operatorname{cis}\frac{\pi}{2}$ (1 mark)

$z_2 = 3\operatorname{cis}\frac{7\pi}{6}$ (1 mark)

$z_3 = 3\operatorname{cis}\frac{11\pi}{6}$ (1 mark)

b Area $= 3\times\left(\frac{1}{2}\times3\times3\times\sin\frac{2\pi}{3}\right)$ (2 marks)

$= 3\times\left(\frac{1}{2}\times3\times3\times\frac{\sqrt{3}}{2}\right)$

$= \frac{27\sqrt{3}}{4}$ (1 mark)

16 a $z^n = (\cos\theta + i\sin\theta)n = \cos n\theta$
$+ i\sin n\theta$ (1 mark)

$\frac{1}{z^n} = z^{-n} = \left(\cos\theta + i\sin\theta\right)^{-n}$

$= \cos(-n\theta) + i\sin(-n\theta)$

$= \cos n\theta - i\sin n\theta$

(2 marks)

So

$z^n + \frac{1}{z^n} = \left(\cos n\theta + i\sin n\theta\right)$
$+ \left(\cos n\theta - i\sin n\theta\right)$

(1 mark)

$= 2\cos n\theta$

b $\left(z + \frac{1}{z}\right)^4 = z^4 + 4z^3\left(\frac{1}{z}\right) +$
$6z^2\left(\frac{1}{z}\right)^2 + 4z\left(\frac{1}{z}\right)^3 + \left(\frac{1}{z}\right)^4$

(2 marks)

$= z^4 + 4z^2 + 6 + \frac{4}{z^2} + \frac{1}{z^4}$

(1 mark)

$= z^4 + \frac{1}{z^4} + 4\left(z^2 + \frac{1}{z^2}\right) + 6$

(1 mark)

$= 2\cos4\theta + (2\cos 2\theta) + 6$

(1 mark)

$= 2\cos4\theta + 8\cos2\theta + 6$

Now

$\left(z + \frac{1}{z}\right)^4 = \left(2\cos\theta\right)^4 = 16\cos^4\theta$

(1 mark)

Therefore

$\cos^4\theta \equiv \frac{1}{16}\left(2\cos4\theta + 8\cos2\theta + 6\right)$

c

$\int_0^{\frac{\pi}{6}} \cos^4\theta \, d\theta$

$= \frac{1}{16}\int_0^{\frac{\pi}{6}} \left(2\cos4\theta + 8\cos2\theta + 6\right) d\theta$

(1 mark)

$= \frac{1}{16}\left[\frac{1}{2}\sin4\theta + 4\sin2\theta + 6\theta\right]_0^{\frac{\pi}{6}}$

(1 mark)

$$= \frac{1}{16}\left[\frac{1}{2}\sin\frac{2\pi}{3} + 4\sin\frac{\pi}{3} + \pi\right]_0^{\frac{\pi}{6}}$$ (1 mark)

$$= \frac{1}{16}\left(\frac{1}{2}\left(\frac{\sqrt{3}}{2}\right) + 4\left(\frac{\sqrt{3}}{2}\right) + \pi\right)$$ (1 mark)

$$= \frac{1}{16}\left(\frac{\sqrt{3}}{4} + 2\sqrt{3} + \pi\right)$$

$$= \frac{1}{16}\left(\frac{9\sqrt{3}}{4} + \pi\right)$$ (1 mark)

$$= \frac{\pi}{16} + \frac{9\sqrt{3}}{64}$$

17 a $\left|\frac{\sqrt{3}+i}{\sqrt{3}-i}\right| = \frac{|\sqrt{3}+i|}{|\sqrt{3}-i|} = \frac{2}{2} = 1$ (2 marks)

So $r = 1$

$\arg\left(\frac{\sqrt{3}+i}{\sqrt{3}-i}\right) =$

$\arg\left(\sqrt{3}+i\right) - \arg\left(\sqrt{3}-i\right)$ (1 mark)

$= \frac{\pi}{6} - \left(-\frac{\pi}{6}\right) = \frac{\pi}{3}$ (1 mark)

So $\theta = \frac{\pi}{3}$ (1 mark)

$\frac{\sqrt{3}+i}{\sqrt{3}-i} = e^{i\frac{\pi}{3}}$

b i $\left(\sqrt{3}+i\right)^n + \left(\sqrt{3}-i\right)^n = \left(2\,\text{cis}\frac{\pi}{6}\right)^n + \left(2\,\text{cis}\left(-\frac{\pi}{6}\right)\right)^n$ (1 mark)

$= 2^n\,\text{cis}\frac{n\pi}{6} + 2^n\,\text{cis}\left(-\frac{n\pi}{6}\right)$ (1 mark)

$= 2^n\left(\cos\frac{n\pi}{6} + i\sin\frac{n\pi}{6} + \cos\left(-\frac{n\pi}{6}\right) + i\sin\left(-\frac{n\pi}{6}\right)\right)$

$= 2^n\left(\cos\frac{n\pi}{6} + i\sin\frac{n\pi}{6} + \cos\frac{n\pi}{6} - i\sin\frac{n\pi}{6}\right)$ (1 mark)

$= 2^n\left(\cos\frac{n\pi}{6} + \cos\frac{n\pi}{6}\right)$ (1 mark)

$= 2^n\left(2\cos\frac{n\pi}{6}\right)$ (1 mark)

$= 2^{n+1}\cos\left(\frac{n\pi}{6}\right)$

ii

$\left(\sqrt{3}+i\right)^8 + \left(\sqrt{3}-i\right)^8 = 2^9\cos\left(\frac{8\pi}{6}\right)$ (1 mark)

$= 2^9\cos\left(\frac{4\pi}{3}\right)$

$= 2^9\left(-\frac{1}{2}\right)$ (1 mark)

$= -2^8 = -256$ (1 mark)

18 a $\omega^* = \omega^2$ (1 mark)

$(1 + \omega + \omega^*)^2 = (1 + \omega + \omega^2)^2$

$= \left(\frac{1-\omega^3}{1-\omega}\right)^2$ (1 mark)

$= 0^2 = 0$ (1 mark)

b $(1 + \omega + 3\omega^2)^2 = (1 + \omega^2 + 2\omega^2)^2$ (1 mark)

$= (2\omega^2)^2 = 4\omega^4$ (1 mark)

$= 4\omega$ (1 mark)

c $(1 + 2\omega + 3\omega^2)(1 + 3\omega + 2\omega^2)$

$= (1 + \omega + \omega^2 + \omega + 2\omega^2)$
$(1 + \omega + \omega^2 + 2\omega + \omega^2)$ (1 mark)

$= (\omega + 2\omega^2)(2\omega + \omega^2)$ since $1 + \omega + \omega^2 = 0$ (1 mark)

$= 2\omega^2 + \omega^3 + 4\omega^3 + 2\omega^4$ (1 mark)

$= 2\omega^2 + 5\omega^3 + 2\omega^4$

$= 2\omega^2 + 5 + 2\omega$ (1 mark)

$= 2(1 + \omega + \omega^2) + 3$ (1 mark)

$= 2\times0 + 3$

$= 3$ (1 mark)

19 $i = \cos\frac{\pi}{2} + i\sin\frac{\pi}{2}$ (1 mark)

$= \cos\left(\frac{\pi}{2} + 2\pi k\right) + i\sin\left(\frac{\pi}{2} + 2\pi k\right)$ (1 mark)

$= \cos\left(\frac{4\pi k + \pi}{2}\right) + i\sin\left(\frac{4\pi k + \pi}{2}\right)$

So

$z - 2i = \cos\left(\frac{4\pi k + \pi}{6}\right) + i\sin\left(\frac{4\pi k + \pi}{6}\right)$ (1 mark)

$k = 0 \Rightarrow z - 2i = \cos\frac{\pi}{6} + i\sin\frac{\pi}{6}$ (1 mark)

$k = 1 \Rightarrow z - 2i = \cos\frac{5\pi}{6} + i\sin\frac{5\pi}{6}$ (1 mark)

$k = 2 \Rightarrow z - 2i = \cos\frac{9\pi}{6} + i\sin\frac{9\pi}{6}$ (1 mark)

$z - 2i = \frac{\sqrt{3}}{2} + \frac{1}{2}i$

$z - 2i = -\frac{\sqrt{3}}{2} + \frac{1}{2}i$

$z - 2\mathbf{i} = -\mathbf{i}$ (3 marks)

So roots are $z_1 = \frac{\sqrt{3}}{2} + \frac{5}{2}i$,

$z_2 = -\frac{\sqrt{3}}{2} + \frac{5}{2}i$ and $z_3 = i$ (1 mark)

Chapter 11

Skills Check

1 48

2

3 a

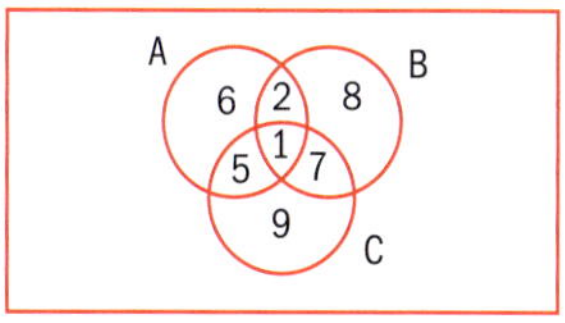

b $\frac{1}{38}$

4 a $\frac{17}{29}$

b $\frac{15}{29}$

c $\frac{1}{4}$

Exercise 11A

1 a $\frac{3}{20}$ **b** $\frac{3}{5}$

2 $\frac{37}{42}$

3 **a** $\frac{1}{12}$ **b** $\frac{1}{2}$

4 **a** $\frac{7}{44}$ **b** $\frac{35}{44}$

5 **a** 0.6

b 0.3

c 0.4

6 **a** 0.3

b 0.1

c 0.7

7 **a**

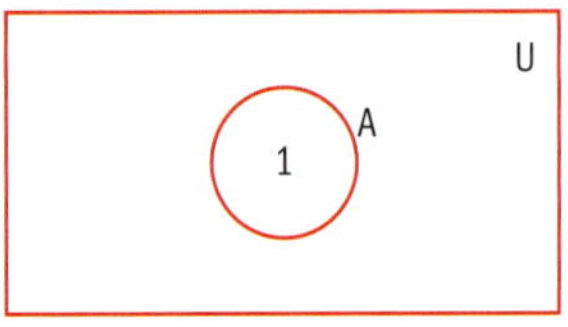

b **i** $\frac{1}{3}$ **ii** 0

Exercise 11B

1 **a** Independent

b Independent

c Not independent

d Independent

2 **a**

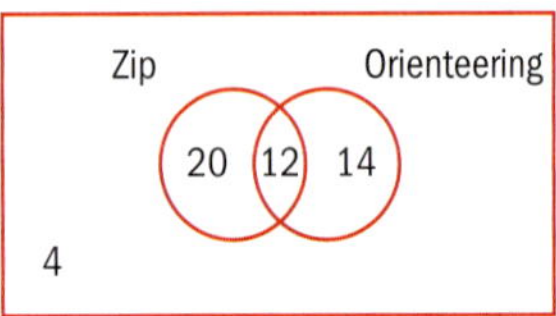

b **i** 1

ii $\frac{3}{8}$ **iii** $\frac{13}{25}$ **iv** $\frac{6}{13}$

3 **a** $\frac{17}{35}$ **b** $\frac{19}{35}$ **c** $\frac{17}{35}$

d $\frac{9}{19}$ **e** $\frac{13}{17}$ **f** $\frac{7}{18}$

g 0

4 $\frac{3}{13}$

5 **a**

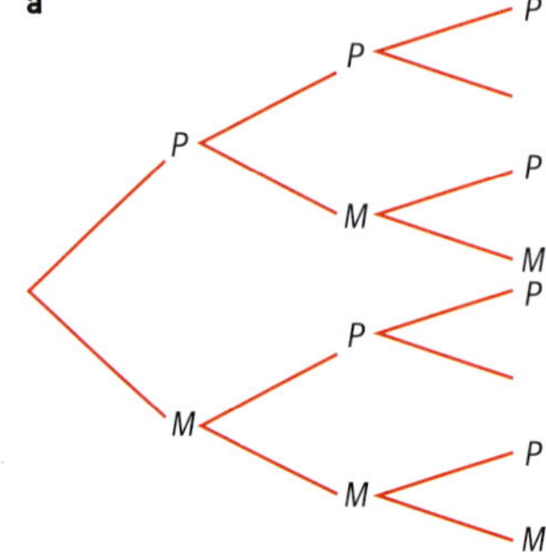

b **i** $\frac{28}{143}$ **ii** $\frac{138}{143}$

6 0.1975

7 **a** $\frac{7}{15}$ **b** $\frac{4}{5}$

Exercise 11C

1 $\text{P(yellow)} = \frac{18}{65}$,

$\text{P(yellow from } A \mid \text{yellow)} = \frac{55}{234}$

2 **a** 0.705 **b** 0.07 **c** 0.5

3 0.4535

4 $\frac{28}{171}$ **5** 0.567%

6 **a** 83.5% **b** 6.06%

Exercise 11D

1 **a** Yes

b No because the sum of the probabilities is not equal to 1

2 **a** Not a probability distribution as some of the probabilities are greater than 1

b Not a probability distribution as one of the probabilities is greater than 1

3 **a** $\frac{1}{33}$ **b** $x = 12$ **c** $\frac{2}{3}$

Exercise 11E

1 $\frac{23}{12}$ **2** $k = \frac{1}{5}$, $\text{E}(X) = 6.2$

3 **a**

x	0	1	2	3
$\text{P}(X=x)$	0.2	0.3	0.35	0.15

b 2

4

x	0	1	2	3
$\text{P}(X=x)$	$\frac{1}{120}$	$\frac{1}{30}$	$\frac{1}{20}$	$\frac{1}{30}$

5 **a** $c = \frac{1}{6}$

b 3.75 (3s.f.)

Exercise 11F

1 **a** 2.375 **b** 11.875 **c** 6.125 **d** 0.484 (3s.f.) **e** 0.696 (3s.f.)

2 **a** $a = b = \frac{1}{6}$ **b** $\text{E}(X) = \frac{13}{6}$, $\text{E}(X^2) = \frac{13}{2}$ **c** $\text{Var}(X) = \frac{65}{36}$

3 **a** $\text{P}(S=4) = \frac{1}{45}, \text{P}(S=8) = \frac{1}{15}, \text{P}(S=11) = \frac{1}{9}$

b

s	3	4	5	6	7	8	9	10	11	12	13	14	15	16	17	18	19
$\text{P}(S=s)$	$\frac{2}{90}$	$\frac{2}{90}$	$\frac{4}{90}$	$\frac{4}{90}$	$\frac{6}{90}$	$\frac{6}{90}$	$\frac{8}{90}$	$\frac{8}{90}$	$\frac{10}{90}$	$\frac{8}{90}$	$\frac{8}{90}$	$\frac{6}{90}$	$\frac{6}{90}$	$\frac{4}{90}$	$\frac{4}{90}$	$\frac{2}{90}$	$\frac{2}{90}$

c $\text{E}(S) = 11$, $\text{Var}(S^2) = \frac{44}{3}$

4 **a** $\frac{1}{44}$ **b** $\text{P}(T=4) = \frac{4}{11}$, $\text{P}(T \le 4) = \frac{15}{22}$, $\text{P}(T=4 \mid T \le 4) = \frac{8}{15}$ **c** $\text{E}(T) = 4$, $\text{Var}(T) = \frac{17}{11}$ **d** $t = 4$

Exercise 11G

1 a $\frac{1}{8}$

b

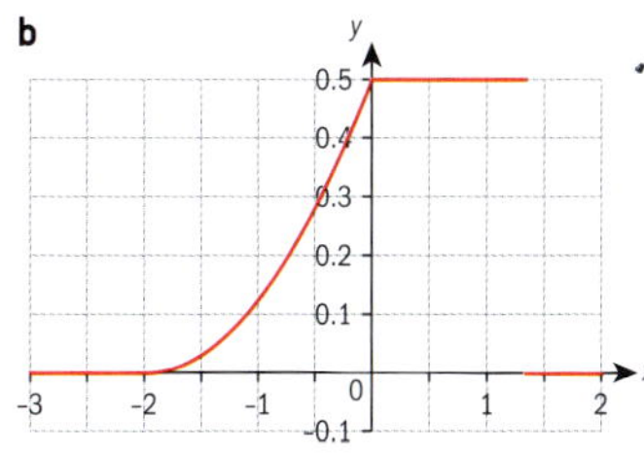

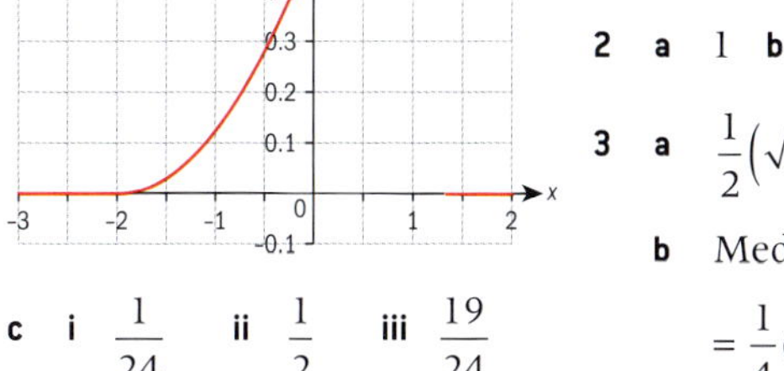

c i $\frac{1}{24}$ **ii** $\frac{1}{2}$ **iii** $\frac{19}{24}$

2 a

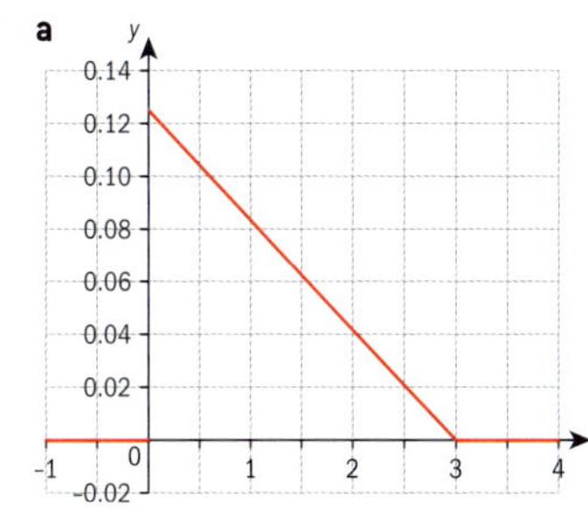

b $\frac{2}{9}$ **c** $\frac{11}{36}$

3 a $c = 1$ **b** $\frac{1}{4}$

Exercise 11H

1

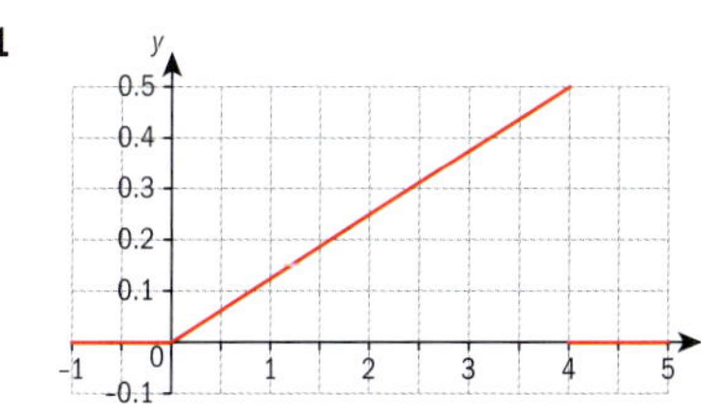

a probability density function because non negative for all possible values and the integral = 1

b $\frac{15}{16}$

c Mean $= \frac{8}{3}$, mode = 4, median $= 2\sqrt{2}$, standard deviation $= \frac{2\sqrt{2}}{3}$

2 a 1 **b** $\frac{1}{5}$ **c** 1 **d** $x = 1$

3 a $\frac{1}{2}\left(\sqrt{3}-1\right)$

b Median $= \frac{\pi}{12}$, mean $= \frac{1}{4}(\pi - 2)$, mode at $x = 0$

4 a $\frac{1}{4}$

b $x = 1$

c $\mathrm{E}(X) = \frac{7}{4}$, $\mathrm{Var}(X) = \frac{37}{48}$

5 a $b = \frac{1}{3} - 3a$

b $a = -\frac{1}{16}$, $b = \frac{25}{48}$

c $\mathrm{E}(X) = \frac{69}{64}$,

$\mathrm{Var}(X) = \frac{9987}{20480}$

Exercise 11I

1 a 0.218 **b** 1

c 0.100

2 a 2.4 **b** 1.44 **c** 2

---

3 a $P(X=5) = \frac{5!}{5!(5-5)!}0.5^5(1-0.5)^{5-5}$

$= \frac{5!}{5!0!}0.5^5 0.5^0 \quad = 0.5^5$

$= \frac{1}{32}$

b

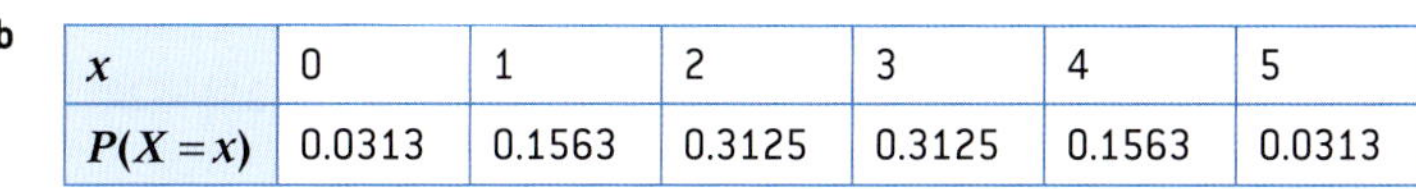

x	0	1	2	3	4	5
$P(X=x)$	0.0313	0.1563	0.3125	0.3125	0.1563	0.0313

c 2.5 **d** 1.25 **e** 2, 3

---

4 a 0.00007

b 0.9988

c 0.9988

d 0.0121

5 a 4

b 0.8223

c 8

Exercise 11J

1 a 0.1253

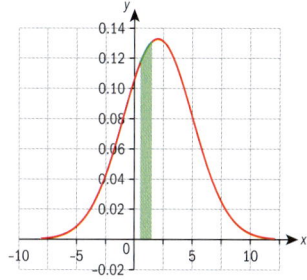

b 0.3085

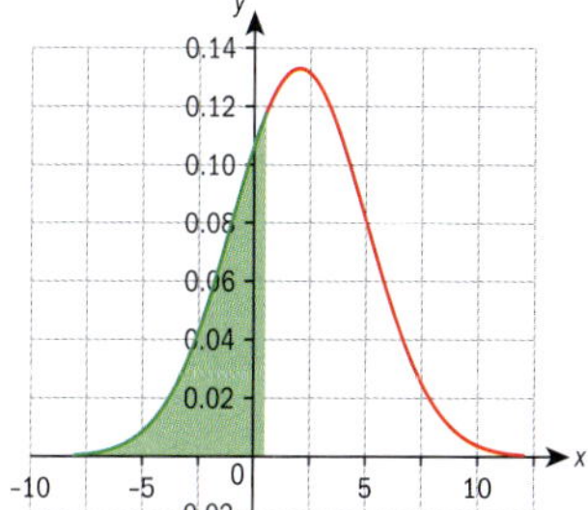

c 0.5

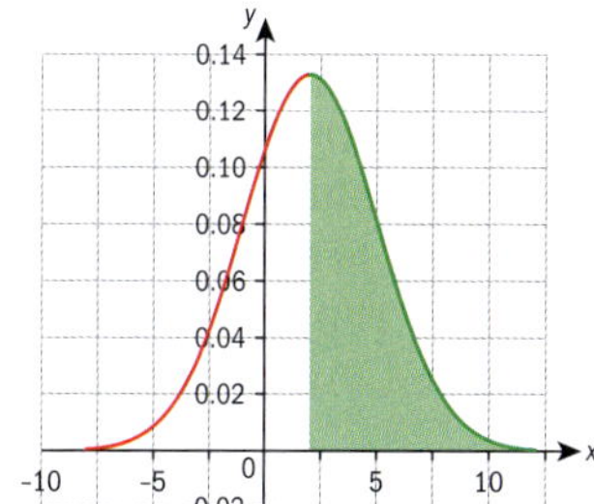

2 a 0.2660 **b** 0.6819

c 0.9878

3 a Mean = 5, standard deviation = 3

b 0.3695

c $\mathrm{P}(X < a) \Rightarrow a = -1.8712$, $\mathrm{P}(X \geq b) \Rightarrow b = 1.607$

4 a Mean = 260, standard deviation = 23.152

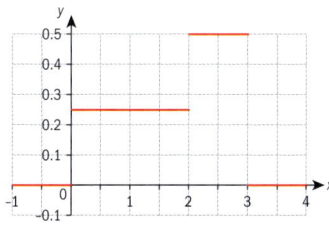

b Between 221.9 g and 291.8 g

5 **a** 0.0228

b 0.00135

c 0.9545

6 **a** $P(T < 12.1) = 0.232$, $P(T > 14.9) = 0.129$

b 12.989

Exercise 11K

1 74.03

2 $\mu = 1.403$, $\sigma = 1.010$

3 $\Sigma = 0.00898$

4 **a** $\mu = 44.141$, $\sigma = 3.200$

b 0.383

5 **a** 0.9332

b 345.63g

c 33.4

d $\mu = 386.59$g, $\sigma = 12.942$g

6 0.0129

Chapter review

1 U = {Mon, Tue, Wed, Thu, Fri, Sat, Sun},
A = {(Mon, Mon)},
B = {(Mon, Mon), (Tue, Tue), (Wed, Wed), (Thu, Thu), (Fri, Fri), (Sat, Sat), (Sun, Sun)},
C = {(Mon, Tue), (Tue, Wed), (Wed, Thu), (Thu, Fri), (Fri, Sat), (Sat, Sun), (Sun, Mon), (Mon, Sun), (Sun, Sat), (Sat, Fri), (Fri, Thu), (Thu, Wed), (Wed, Tue), (Tue, Mon)}

a $\frac{1}{49}$ **b** $\frac{1}{7}$ **c** $\frac{2}{7}$

2 $\frac{2}{5}$

3 13

4 **a** $\frac{7}{10}$ **b** $\frac{7}{10}$

c $\frac{7}{15}$ **d** $\frac{14}{15}$

5 **a** 0.62

b 0.161

c 0.839

6 $\frac{3}{256}$

7 **a** 0.328 **b** 0.0579

8 **a** 0.0548

b 2.28%

c 0.332

9 $\mu = 2.00$, $\sigma = 0.00620$

10 **a** 0.0197 **b** 0.213

---

11 Mean $= \int_0^1 \pi x^2 \sin(\pi x)\,dx = \left[\frac{(2-\pi^2x^2)\cos(\pi x) + 2\pi x\sin(\pi x)}{\pi^2}\right]_0^1 = 1 - \frac{4}{\pi^2}$,

variance

$$= \int_0^1 \pi x^3 sin(\pi x)\,dx - \left(1-\frac{4}{\pi^2}\right)^2 = \left[\frac{\pi x(6-\pi^2x^2)\cos(\pi x) + 3(\pi^2x^2-2)\sin(\pi x)}{\pi^3}\right]_0^1$$

$$-\left(1-\frac{4}{\pi^2}\right)^2 = \left(1-\frac{6}{\pi^2}\right) - \left(1-\frac{4}{\pi^2}\right)^2$$

$$= \frac{2(\pi^2-8)}{\pi^4}$$

Exam-style questions

12 **a** If they were mutually exclusive, then $P(A \cap B) = 0$, (1 mark)

but since they are independent, we have $P(A \cap B) = P(A)P(B) = 0.3 \times 0.8 \neq 0$.

Therefore we have a contradiction, and so A and B are not mutually exclusive. (1 mark)

i $P(A \cap B) = P(A)P(B) = 0.3 \times 0.8 = 0.24$ (2 marks)

ii $P(A \cup B) = P(A) + P(B) - P(A \cap B) = 0.3 + 0.8 - 0.24 = 0.86$ (2 marks)

iii $P(A \mid B') = \frac{P(A \cap B')}{P(B')} = \frac{P(A) - P(A \cap B)}{P(B')} = \frac{0.3-0.24}{0.2} = \frac{0.06}{0.2} = 0.3$ (2 marks)

iv $P(A' \cap B) = P(B) - P(A \cap B) = 0.8 - 0.24 = 0.56$ (2 marks)

---

13 **a** $\frac{k}{2} + k + k^2 + 2k^2 + \frac{k}{2} = 1$ (1 mark)

$3k^2 + 2k - 1 = 0$ (1 mark)

$(3k-1)(k+1) = 0$ (1 mark)

$\Rightarrow k = \frac{1}{3}$ (1 mark)

b

$E(X) = \sum xP(X = x) =$

$0 \times \frac{k}{2} + 0.5 \times k + 1 \times k^2 + 1.5 \times 2k^2 + 2 \times \frac{k}{2}$ (1 mark)

$E(X) = \frac{k}{2} + k^2 + 3k^2 + k$

$= 4k^2 + \frac{3k}{2}$ (1 mark)

$= 4\left(\frac{1}{3}\right)^2 + \frac{3}{2}\left(\frac{1}{3}\right)$

$= \frac{4}{9} + \frac{1}{2}$ (1 mark)

$= \frac{17}{18}$

c $P(X \geq 1.25) = 2k^2 + \frac{k}{2}$ (1 mark)

$= 2\left(\frac{1}{3}\right)^2 + \frac{1}{6}$ (1 mark)

$= \frac{2}{9} + \frac{1}{6}$

$= \frac{7}{18}$ (1 mark)

d $\text{Var}(X) = E(X^2) - [E(X)]^2$ (1 mark)

$= \left[0^2 \times \frac{1}{6} + \left(\frac{1}{2}\right)^2 \times \frac{1}{3} + 1^2 \times \frac{1}{9} + \left(\frac{3}{2}\right)^2 \times \frac{2}{9} + 2^2 \times \frac{1}{6}\right] - \left(\frac{17}{18}\right)^2$ (1 mark)

$= 0.469$ (1 mark)

14 a $X \sim B(24, 0.04)$

$P(X = 2) = \binom{24}{2}(0.04)^2(0.96)^{22}$ (1 mark)

$= 0.180$ (1 mark)

b $P(X \le 4) = 0.998$ (2 marks)

c $P(X \ge 2) = 0.249$ (2 marks)

d $\text{Var}(X) = np(1 - p)$ (1 mark)

$= 24 \times 0.04 \times 0.96$

$= 0.922$ (1 mark)

15 a $X \sim N(36, 3.12^2)$

$P(X > 40) = P\left(Z > \frac{40-36}{3.12}\right)$ (1 mark)

$= 0.1$ (1 mark)

b $P(34 < X < 38)$

$= P\left(\frac{34-36}{3.12} < Z < \frac{38-36}{3.12}\right)$ (1 mark)

$= P(-0.641 < Z < 0.641)$ (1 mark)

$= P(Z < 0.641) - P(Z < -0.641)$ (1 mark)

$= 0.739 - 0.261$

$= 0.478$ (1 mark)

c $P\left(Z > \frac{M-36}{3.12}\right) = 0.015$ (1 mark)

$P\left(Z < \frac{M-36}{3.12}\right) = 0.985$

$\frac{M-36}{3.12} = 2.170$ (1 mark)

$\Rightarrow M = 42.77$

$\Rightarrow M = 42$ minutes, 46 seconds (1 mark)

d $P(X < 30) = P\left(Z < \frac{30-36}{3.12}\right)$ (1 mark)

$= P(Z < -1.923)$

$= 0.027$ (1 mark)

$195 \times 0.027 = 5.3$ (1 mark)

Therefore the expected number of days is 5. (1 mark)

16 a Let X be the discrete random variable 'mass of a can of baked beans'.

Then $X \sim N(415, 12^2)$

$P(X > m) = 0.65 \Rightarrow P(X \le m) = 1 - 0.65 = 0.35$ (1 mark)

Using inverse normal distribution on GDC $\Rightarrow m = 410.4$ (2 marks)

b You require $P(X > 422.5 | X > 420)$. (1 mark)

$P(X > 422.5 | X > 420)$

$= \frac{P(X > 422.5)}{P(X > 420)}$

$= \frac{1 - P(X \le 422.5)}{1 - P(X \le 420)}$ (1 mark)

$= \frac{0.266}{0.338}$ (1 mark)

$= 0.787$ (1 mark)

c Using GDC

$P(X < 413.5) = 0.450$ (2 marks)

Let Y be the random variable 'Number of cans of beans having a mass less than 413.5 g'.

In Ashok's experiment, Y is Binomially distributed across 144 trials with probability of 'success' (i.e. mass less than 413.5 g) being 0.450.

So $Y \sim B(144, 0.450)$ (1 mark)

$P(Y \ge 75) = 0.0524$ (1 mark)

17 a $k\int_0^1 (10x^2 - x^3)\,dx = 1$ (1 mark)

$k\left[\frac{10x^3}{3} - \frac{x^4}{4}\right]_0^1 = 1$ (1 mark)

$k\left(\frac{10}{3} - \frac{1}{4}\right) = 1$ (1 mark)

$\frac{37k}{12} = 1$ (1 mark)

$k = \frac{12}{37}$

b $E(X) = \int_0^1 xf(x)\,dx$

$= \frac{12}{37}\int_0^1 (10x^3 - x^4)\,dx$ (1 mark)

$= \frac{12}{37}\left[\frac{10x^4}{4} - \frac{x^5}{5}\right]_0^1$ (1 mark)

$= \frac{12}{37}\left(\frac{5}{2} - \frac{1}{5}\right)$ (1 mark)

$= \frac{276}{370}\left(= \frac{138}{185}\right)(= 0.746)$ (1 mark)

c $E(X^2) = \int_0^1 x^2 f(x)\,dx$

$= \frac{12}{37}\int_0^1 (10x^4 - x^5)\,dx$ (1 mark)

$= \frac{12}{37}\left[2x^5 - \frac{x^6}{6}\right]_0^1$ (1 mark)

$= \frac{12}{37}\left(2 - \frac{1}{6}\right)$

$= \frac{22}{37}$ (1 mark)

$\text{Var}(X) = E(X^2) - [E(X)]^2$ (1 mark)

$= \frac{22}{37} - \left(\frac{138}{185}\right)^2$

$= 0.0382$ (1 mark)

d $\frac{12}{37}\int_0^m \left(10x^2 - x^3\right) dx = \frac{1}{2}$ (1 mark)

$\frac{12}{37}\left[\frac{10x^3}{3} - \frac{x^4}{4}\right]_0^m = \frac{1}{2}$ (1 mark)

$\frac{12}{37}\left(\frac{10m^3}{3} - \frac{m^4}{4}\right) = \frac{1}{2}$ (1 mark)

$\frac{10m^3}{3} - \frac{m^4}{4} = \frac{37}{24}$

$80m^3 - 6m^4 = 37$

GDC $\Rightarrow m = 0.789$ (2 marks)

18 a $P\left(Z < \frac{110-\mu}{\sigma}\right) = 0.10$

$\Rightarrow \frac{110-\mu}{\sigma} = -1.282$ (2 marks)

$P\left(Z > \frac{130-\mu}{\sigma}\right) = 0.45$

$\Rightarrow \frac{130-\mu}{\sigma} = 0.126$ (2 marks)

Attempt to solve simultaneously: (1 mark)

$\mu = 128$ (1 mark)

$\sigma = 14.2$ (1 mark)

b $P\left(\left|X-\mu\right|\right) < 0.22$

$0.5 - \frac{0.22}{2} = 0.39$ (1 mark)

$P(X < a) = 0.39 \Rightarrow a = 124.2$ (1 mark)

$P(X > b) = 0.39 \Rightarrow b = 132.2$ (1 mark)

So $124.2 < X < 132.2$ (1 mark)

19 a $k\int_0^{\frac{\pi}{2}} \cos x \, dx = 1$ (1 mark)

$k[\sin x]_0^{\frac{\pi}{2}} = 1$ (1 mark)

$k\left(\sin\frac{\pi}{2} - \sin 0\right) = 1$ (1 mark)

$k(1 - 0) = 1$ (1 mark)

$k = 1$

b

$\int_0^{\frac{\pi}{2}} x\cos x \, dx = [x\sin x]_0^{\frac{\pi}{2}} - \int_0^{\frac{\pi}{2}} \sin x \, dx$ (2 marks)

$\int_0^{\frac{\pi}{2}} x\cos x \, dx = [x\sin x]_0^{\frac{\pi}{2}} + [\cos x]_0^{\frac{\pi}{2}}$ (1 mark)

$= \left(\frac{\pi}{2} - 0\right) + (0 - 1)$ (1 mark)

$= \frac{\pi}{2} - 1$ (1 mark)

$\int_0^{\frac{\pi}{2}} x^2\cos x \, dx = \left[x^2 \sin x\right]_0^{\frac{\pi}{2}} - \int_0^{\frac{\pi}{2}} 2x\sin x \, dx$ (2 marks)

$= \left[x^2\sin x\right]_0^{\frac{\pi}{2}} - 2\left[-x\cos x + \sin x\right]_0^{\frac{\pi}{2}}$ (1 mark)

$= \left[x^2\sin x - 2\sin x + 2x\cos x\right]_0^{\frac{\pi}{2}}$

$= \frac{\pi^2}{4} - 2$ $\sum_{i=1}^{n} i \times i! = (n+1)! - 1$ (1 mark)

$\text{Var}(X) = \int_0^{\frac{\pi}{2}} x^2\cos x \, dx - \left(\int_0^{\frac{\pi}{2}} x\cos x \, dx\right)^2$ (1 mark)

$= \frac{\pi^2}{4} - 2 - \left(\frac{\pi}{2} - 1\right)^2$ (1 mark)

$= \frac{\pi^2}{4} - 2 - \left(\frac{\pi^2}{4} - \pi + 1\right)$ (1 mark)

$= \pi - 3$

Paper 1

1 a $\frac{x}{2} - \frac{\sin(2x)}{4} + c$ **b** $\frac{\pi}{2}$

2 a $p + q$ **b** $2p - q$

c $\frac{1}{2}p$ **d** $2 + q$

3 a 20 cm **b** 38 cm

4 ln 3

5 i Meaningful

ii Meaningless; $(\mathbf{b}\cdot\mathbf{c})$ is not a vector, so you cannot take the scalar product of **a** with it.

iii Meaningful

iv Meaningless; $(\mathbf{b}\cdot\mathbf{c})$ is not a vector, so you cannot take vector product of **a** with it.

v Meaningless; both **a** and $(\mathbf{b}\cdot\mathbf{c})$ are vectors, so you cannot perform scalar multiplication.

vi Meaningful

6 Of form $\frac{0}{0}$.

Limit $= \lim_{x\to 0} \frac{\sec^2 x - 1}{3x^2}$

Still of form $\frac{0}{0}$.

Limit $= \lim_{x\to 0} \frac{2\sec^2 x \tan x}{6x}$

Still of form $\frac{0}{0}$.

Limit $=$

$\lim_{x\to 0} \frac{4\sec^2 x\tan^2 x + 2\sec^4 x}{6} = \frac{1}{3}$

7 $f'(x) = \lim_{h\to 0} \frac{(x+h)^4 - x^4}{h}$

$= \lim_{h\to 0}$

$\frac{x^4 + 4x^3h + 6x^2h^2 + 4xh^3 + h^4 - x^4}{h}$

$\lim_{h\to 0} 4x^3 + 6x^2h + 4xh^2 + h^3 = 4x^3$

8 a $a = -9, b = 9$

b $(x - 1)(x - 3)(x + 3)$

9 a i -10 **ii** 24

b i New sum -6 **ii** 0

10 a $D = 36 - 4(8 + k) = 4 - 4k$

b i Concave-up quadratic, always positive, so no roots so $D < 0$

ii $D = 4 - 4k < 0 \Rightarrow k > 1$

c $f(x) = (x + 3)^2 + (k - 1)$

d $f(x) > 0 \Rightarrow (x + 3)^2 + (k - 1) > 0 \Rightarrow k - 1 > 0$ since $(x + 3)^2 \geq 0$ for all x. Hence $k > 1$ as found in **b** part **ii**

e $(-3,3)$

11 a

Let $P(n)$ be the statement

$$\sum_{i=1}^{n} i \times i! = (n+1)! - 1$$

LHS of $P(1)$ is 1. RHS of $P(1)$ is $2! - 1 = 1$. So $P(1)$ is true.

Assume $P(k)$ is true and attempt to prove $P(k+1)$

$$\sum_{i=1}^{k+1} i \times i! = \sum_{i=1}^{k} i \times i! + (k+1) \times (k+1)!$$
$$= (k+1)! - 1 + (k+1) \times (k+1)!$$
$$= (k+2) \times (k+1)! - 1$$
$$= (k+2)! - 1 \text{ as required}$$

Since $P(1)$ is true and $P(k)$ true implies $P(k+1)$ true, by the principle of mathematical induction the statement has been proved for all $n \in \mathbb{Z}^+$.

b

$$\sum_{i=1}^{n}(i+1-1) \times i! = \sum_{i=1}^{n}(i+1) \times i! - i!$$

$$= \sum_{i=1}^{n}(i+1)! - \sum_{i=1}^{n} i! = (n+1)! - 1$$

as almost all the terms cancel.

12 a $z - 1, -1, \text{i}, \text{ or } -\textbf{i}$

b $z = 1\text{cis}0,\ 1\text{cis}\frac{\pi}{4},\ 1\text{cis}\frac{\pi}{2},$

$1\text{cis}\frac{3\pi}{4},\ 1\text{cis}\pi,\ 1\text{cis}-\frac{\pi}{4},$

$1\text{cis}-\frac{\pi}{2},\ 1\text{cis}-\frac{3\pi}{4}$

c $\frac{1}{\sqrt{2}} + \frac{1}{\sqrt{2}}\text{i}$ **d** $\frac{1}{\sqrt{2}} + \frac{1}{\sqrt{2}}\text{i}$

13 a $\int \frac{1}{x(1-x)}\text{d}x = \int k\,\text{d}t$

$\int \frac{1}{x} + \frac{1}{1-x}\text{d}x = kt + c$

$\ln x - \ln(1-x) = kt + c$

$\ln\left(\frac{x}{1-x}\right) = kt + c$

$\frac{x}{1-x} = \text{e}^{kt+c} = \text{e}^c\text{e}^{kt} = A\text{e}^{kt}$

b $\frac{1}{2}$ **c** 2 **d** $\frac{2}{3}$

Paper 2

1 a 2

2 a

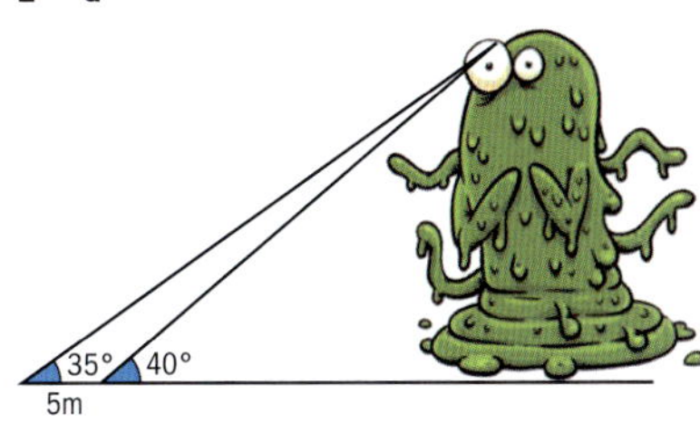

b 21.2 m

3 a 0 **b** 8.51 m

c +3.87 m (3 s.f.)

4 0.9801

5 a 2.93 (3 s.f.) **b** 14.9 (3 s.f.)

6 1.45 (3 s.f.)

7 a $X = n$ **b** 4

8 a 3.14 km^2/h

b 9.55 m (3 s.f.)

9 a 2 **b** $x = 2, y = 1, z = 1$

10 a 57.7° (3s.f.)

b 57.7° (3s.f.)

c 32.3° (3s.f.)

d (1, −4,1)

11 a i $x = 5$ **ii** $y = 3$

b i $\left(0, \frac{-3}{10}\right)$ **ii** $\left(-\frac{1}{2}, 0\right)$

c I $\frac{-66}{(2x-10)^2}$

ii Graph is always decreasing

d

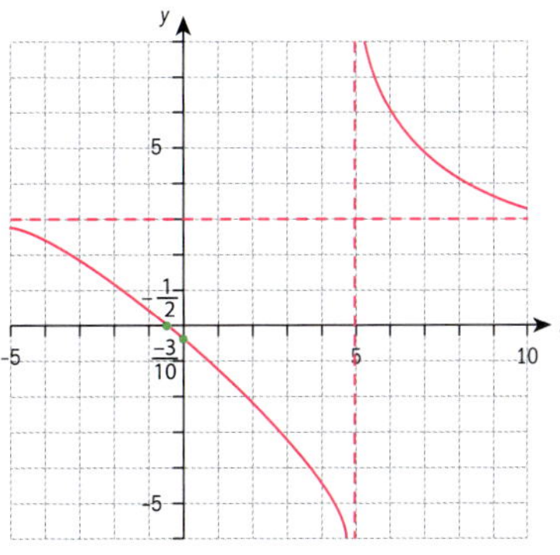

e $y = -1.83x + 1.17$ (3 s.f.)

12 a 0.655 m (3 s.f.) **b** 10

c 18 m **d** 11.3 s

13 a $\frac{\pi}{2}$

b i 0 **ii** 0.0474 (3s.f.)

c 0.707 (3 s.f.)

Paper 3

1 a i Arithmetic progression

ii $u_n = u_1 + (n-1)b$

iii $S_n = \frac{n}{2}\left(2u_1 + (n-1)b\right)$

b i Geometric progression

ii $u_n = u_1 a^{n-1}$

iii $S_n = u_1\frac{(a^n - 1)}{(a-1)}$

c i $u_2 = 3, u_3 = 7, u_4 = 15, u_5 = 31$

ii $u_n = 2^n - 1$

iii Let $P(n)$ be the statement $u_n = 2^n - 1$

$P(1) = 2^1 - 1 = 1$ which is true, since the question gives $u_1 = 1$

Assume the result for $P(k)$ and attempt to prove for $P(k+1)$

By the recurrence relation, $u_{k+1} = 2u_k + 1$
$= 2(2^k - 1) + 1 = 2^{k+1} - 1$
as required

Since $P(1)$ is true and $P(k)$ true implies $P(k+1)$ is also true. Hence, by the principle of mathematical induction, $P(n)$ is true for all $n \in \mathbb{Z}^+$.

d i $c = \frac{b}{a-1}$

ii $v_{n+1} = av_n$, geometric progression

iii $v_n = a^{n-1}v_1$

iv

$u_n + c = a^{n-1}(u_1 + c) \Rightarrow u_n + \frac{b}{a-1}$

$= a^{n-1}u_1 + a^{n-1}\frac{b}{a-1} \Rightarrow u_n$

$= a^{n-1}u_1 + b\frac{(a^{n-1}-1)}{a-1}$

e General formula gives $2^{n-1} \times 1 + 1\frac{(2^{n-1}-1)}{2-1} = 2^{n-1} + 2^{n-1} - 1 = 2^n - 1$ giving agreement

2 a 0.881 (3s.f.)

b $\arcsin x + c$

c Let $x = \sin\theta$, integral becomes $\frac{1}{\sqrt{1-\sin^2\theta}}\cos\theta\,d\theta$

$= \int 1\,d\theta = \theta + c = \arcsin x + c$

d $(\cosh x)^2 - (\sinh x)^2$

$= \frac{e^{2x}+2+e^{-2x}}{4} - \frac{e^{2x}-2+e^{-2x}}{4}$

$= \frac{4}{4} = 1$

e i $\frac{d\sinh x}{dx} = \frac{e^x + e^{-x}}{2} = \cosh x$

ii $\frac{d\cosh x}{dx} = \frac{e^x - e^{-x}}{2} = \sinh x$

f $x \to y = \sinh x = \frac{e^x - e^{-x}}{2}$, inverse function will be given by $x = \frac{e^y - e^{-y}}{2}$

$2x = e^y - e^{-y} \Rightarrow 2xe^y$

$= (e^y)^2 - 1 \Rightarrow (e^y)^2 - 2x(e^y) - 1 = 0$

$e^y = \frac{2x \pm \sqrt{4x^2+4}}{2} = x \pm \sqrt{x^2+1}$

Since $e^y > 0$ we must take the positive square root $y = \operatorname{arsinh} x = \ln\left(x + \sqrt{x^2+1}\right)$

g $\ln\left(x + \sqrt{x^2+1}\right) + c$

h $\ln\left(1 + \sqrt{2}\right)$

i $I = \int \frac{1}{\sqrt{1+\tan^2\theta}}\sec^2\theta\,d\theta$

$= \int \sec\theta\,d\theta$

j $\ln(\tan\theta + \sec\theta) + c$

k $\tan\theta = x,\quad 1 + \tan^2\theta = \sec^2\theta,$

$\sec\theta = \sqrt{1+\tan^2\theta} = \sqrt{1+x^2}$

So $\ln(\tan\theta + \sec\theta) + c$

$= \ln\left(x + \sqrt{1+x^2}\right) + c$

as before.

l $1 + x\left(1+x^2\right)^{-\frac{1}{2}}$

m $\int \frac{f'(x)}{\sqrt{1+x^2}+x}dx,\ \int \frac{f'(x)}{f(x)}dx,$

$\ln\left(x + \sqrt{1+x^2}\right) + c$

Index

2D *see* two dimensions
3D *see* three dimensions

absolute value 93–100
  areas bounded by curves 458, 522
  complex numbers 163–4
  definitions 93–4, 96
  equations 93–100
  functions 93–100, 117, 245
  properties 93
acceleration 285–7, 568–9, 634, 637
acute angles 386, 388, 392, 407, 603
addition
  complex numbers 165, 167, 656
  vectors 576–9, 584, 587
Ain Dubai ferris wheel 367, 378, 381
algebraic definitions 597, 599–601, 608, 617, 619
amplitudes 412–16
angles
  3D space 373–4
  acute 386, 388, 392, 407, 603
  angle equalities 388–89
  compound 403–5, 421, 423
  four quadrants 386–91
  lines and planes 630–3
  measurement 378–83
  obtuse 386–7, 392, 407, 603
  planes and lines 630–3
  reflex 386
  scalar products 598–604, 630–1
  vectors 630–3
antiderivatives 446, 459
antidifferentiation 444–60, 488–90
approximation 546–9, 554–7
arccos 419–20, 427–29, 506
arcs 378–79, 381–2, 422
arcsin 372, 396, 416–20, 427–29, 504
arctan 397, 427–29
areas
  bounded by curves 451–4, 458–9, 521–2
  differential calculus 520–3
  partitioning 522–3
  sectors 381–3
  surface 375–8, 382
  triangles 397–8, 430–1, 456–7, 486–7
Argand diagrams/planes 650–1, 666, 668, 670
arguments
  functions 76, 110, 128, 131
  polar complex numbers 650–3, 656–8, 660–3, 666, 669
arithmetic sequences 10–16, 27–31, 37, 48–9
arithmetic series 10–16, 29–31
arrangements 54–7
associated acute angles 388
associative law 111, 578
ASTC mnemonic 388
asymptotes 87–91, 115, 131, 228–32
  see also horizontal asymptotes; vertical asymptotes
average rate of change 237, 239
average speed 236–7, 242–3
axes of symmetry 82–3, 107
axiomatic probability systems 683–96
axioms 461, 684–6, 689–91, 694
Bayes' theorem 693–6
Bessel's correction 341
best fit line 349–50
bias 314–15
bimodal histograms 324–6
binomial distribution 712–17
binomial series 558–62
binomial theorem 51, 58–65
  combinations 59–60
  complex numbers 172–3
  expansion of 62–5, 151
  generalization 62–5
  polar complex numbers 672, 674
bivariate data/graphs 347, 348–9, 350
box and whisker diagrams 335, 345
break-even point 282–3

Cartesian forms
  complex number 161, 167, 174, 651, 654, 658–65, 670–1
  vector equations 605–6, 610–11, 621–5
causation 347, 354
chain rule
  compound functions 485
  derivatives 271
  differentiation rules 252–4
  functions with form $ax + b$ 449
  implicit differentiation 289–90
  inverse 491, 493
  reverse 449, 491
  trigonometric functions 425
chords 378–79, 381
class intervals/widths 320, 327, 332–3
collinear points 574, 580
collinear vectors 580, 602, 609–10
column vectors 122
combinations 53–8, 59–60, 687
common differences 11–15, 48–9
common ratios 17–21, 23–6, 48–9
commutative law 111, 578
complementary events 685–6
completing the square 149–53
complex numbers 146, 161–75, 648–79
  algebraic approach 161–2
  Cartesian form 161, 167, 174, 651, 654, 658–65, 670–1
  conjugate 168–70, 173, 654–5, 668–9
  division 165, 169–72
  forms of 650–5
  fundamental theorem of algebra 185, 187–9
  geometric approach 161–3
  history of 162
  operations 164–8
  polar form 650–74
  powers 172–5
  roots 172–5
  systems of equations 204–5
complex zeros 185, 187–9
composite functions 109–12, 113, 253–4, 425
compound angle ratios 403–5, 421, 423
compound functions 483–7
compound interest 30–1
compressions 123–5, 135, 413
concavity 82–3, 160, 268–77
concurrent lines/points 609, 613–14
conditional probability 688–92

cones 84, 86, 376–7, 382–3, 524–5
conjugate complex numbers 168–70, 173, 654–5, 668–9
conjugate root/zero theorem 187–9
constant of integration 443, 446
constant of proportionality 535–7
constant terms 85, 151, 443, 446–7
continuity 220, 224–7, 244–6
continuous data 309, 320–1
continuous functions 178, 224–7, 263
continuous random variables 706–12, 718, 720
contradiction, proof by 40–4
contrapositives 40, 42
convenience sampling 311
convergence 220
convergent sequences 233–5
convergent series 22–7, 39, 62
coplanar lines/points 574
coplanar vectors 580, 619
correlation 347–56
cosec (cosecant) 399–3
cosine (cos)
  arccos 419–20, 427–29, 506
  cosine curve 409–12
  cosine ratio 371–4, 386–91
  cosine rule 391–3, 647
  four quadrants 386–91
  see also trigonometric...
costs 282–3
cot (cotangent) 399–3, 497
counterclaims/examples 44
covariance 350, 352
CSOS mnemonic 324
cube roots 173
cubic equations 191–2, 193, 678–9
cubic functions 177
cumulative frequency 343–7
cyclic integration by parts 509

data 309, 315–16, 320–32, 347, 355
decay, exponential 474, 476
deceleration 285
definite integrals 444–60
  area between two curves 522
  integration techniques 490, 492, 501–3
  kinematics 529
  properties 455
degree of function 85, 175–6, 178
degrees 379–80
De Moivre's theorem 664–7, 669, 672–3
dependent variables 76, 348
depreciation 463, 471–2
derivatives 236–48
  exponential functions 476, 483–8
  graphs 262–77
  higher derivatives 259–61
  inverse functions 427–29
  sum or difference of functions 249–52
  tangents, equations 429
  trigonometric functions 422–29
descriptive statistics 317–34
determinants 204
differential calculus 518–69
  applications 278–88
  areas 520–3
  kinematics 284–8, 528–32
  limits 550–3
  ODEs 533–49
  optimization 278–84
  polynomials 554–62
  surface areas 375–8
  volumes 520, 524–8
differential equations 533–49, 560–1
differentiation 249–61
  compound functions 484–7
  differentiability at a point 244–6
  exponential functions 477
  from first principles 243, 249–50, 422–7
  higher derivatives 259–61
  implicit 288–95, 484–5, 542
  power rule 250–1, 290
  product rule 255–6
  quotient rule 257–8, 260
  sum or difference of functions 249–52
  trigonometric functions 422–7
  see also antidifferentiation; chain rule
dilations 123–5, 128–32, 134–6
directed line segments 573–5
direction of vector 576
direction vectors
  angles 630–3
  geometrical representation 578
  lines 605–11, 630–3
  planes 624, 627, 630–3
  vector algebra 590–4
  vector applications 634, 636
direct proof 37, 40, 45, 646
discontinuities 225–6, 228
discrete data 309, 320–32
discrete random variables 696–9, 701–2, 712–14, 716
discriminants 156–9
displacement 285–6
  kinematics 529–31
  straight line motion 568–9
  vectors 575–6, 633
distance 284–5
  distance-time graphs 75
  formula 370–1, 593, 634
  kinematics 529–31
  vector applications 633–6
distributive property 197, 203, 257
divergent sequences 233–4
divergent series 22–7
division
  complex numbers 165, 169–72
  polar complex numbers 662–3
  polynomial functions 178–9
  synthetic 180–3, 185–9
domains
  absolute value functions 94–5
  classifying functions 103–4, 106
  functional relationships 76–9
  function composition 109–10
  function transformations 117–18, 129, 131
  inverse functions 112–15
  periodic functions 412
  piecewise-defined functions 100
  special functions 83–4, 87–91, 94–5, 100
double angle identities 405–7, 421–2, 657, 672

e (Euler's number) 473–4, 477, 479
ellipsoids 378
empty sets 683, 686
"end behaviour" property 178
enlargement 658
equal vectors 587, 590–1
equivalence relations 575
equivalent line segments 573–5
errors in data handling 316
Euler's form of complex number 652–3, 658, 661–5, 671
Euler's method 546, 548
Euler's number (e) 473–4, 477, 479, 504
even functions 106–8, 111
exact differential equations 543–4
exhaustion method 451–2, 550
exhaustive events 694
expectation 700–4
experimental probability 697, 700

explicit formulae 71
explicit relations 288–9
exponential form of complex number 652
exponential functions 473–82
  decay functions 476
  derivatives 476, 483–8
  growth functions 475
  integration by parts 505, 507, 509
  invariance under differentiation 477
  inverses 77
  logarithms 479–81
  polar complex numbers 665
exponential growth/decay 474–6, 537–9
exponents 442, 460–82
  definition 460
  exponential functions 473–82
  financial problems 462–3
  fractional 62–4
  logarithms 465–73, 480
  negative 62–4
  properties 460–1, 466, 468–9
  see also powers
extrapolation 353
extrema *see* maximum points; minimum points

factorials 52–7
factorisation
  complex numbers 174
  fundamental theorem of algebra 184–5, 187–8
  implicit differentiation 291
  polar complex numbers 668
  polynomials 195, 197
  product rule 255–6
  quadratic equations 149, 152, 155
  quadratic inequalities 159
  see also remainder theorem
factor theorem 178–84, 193
financial problems 462–3
finite sequences 5, 7
finite series 8, 13, 21, 23–6
first derivatives 263, 266, 268, 272, 274, 494
first-order differential equations 544–5, 548
floor functions 223
folium of Descartes 289–90
"for all values" notation 413
fractions 62–5, 91–3
frequency histograms 322–32
frequency tables 319–21
functions 72–145
  classifications 102–8
  composition 109–12, 113
  derivatives 236–48
  graphs 75, 77–9, 81–102
  operations 108–16
  relationships 75–80
  special 81–102
  transformations 117–39
  see also individual functions
fundamental theorem of algebra 184–94
fundamental theorem of calculus 459, 490

Gaussian method 207
Gaussian plane 163
general terms
  binomial theorem 60–1
  sequences 5–8, 11, 17, 24
geometric definitions 599, 601–2
geometric sequences 10, 17–27
  curious patterns 17–18
  exponents 463
  integration 489
  limits 234
  logarithms 470
  problem-solving 29–31
  proof by induction 48–9
  sums of 20–2
geometric series 10, 17–27
  binomial theorem 62
  curious patterns 17–18
  limits 234–5
  logarithms 471
  problem-solving 29–31
  proof 39
  sums of 39
geometry 366–441, 573–86
global maxima 263
global minima 263–4
gradients
  derivatives 237–41, 262, 264–6, 268, 270
  gradient functions 241
  graphs 262, 264–6, 268, 270
  normals 247–8, 430
  tangents 247–8
  trigonometric functions 426, 430
graphs
  bivariate 348
  derivatives 262–77
  functions 75, 77–9, 81–102
  polynomial inequalities 198–9
  trigonometric functions 413–18
gravity 637
grouped data 320–2, 325–6, 344–5
grouped frequency tables 319–21
growth 28–31, 474–5, 535–9

harmonic series 5
hemispheres 382
histograms 319–34
homogeneous differential equations 541
horizontal asymptotes 79, 87–8, 228, 231–2, 267
horizontal line test 102–5
horizontal points of inflexion 270–2, 274–5
Horner's algorithm 179–80, 182
hyperbolas 86

identities 384–410
  compound angles 403–5, 421, 422
  double angles 405–7, 421–2, 657, 672
  Pythagorean 401–2, 673
  trigonometric 384–410, 420–1, 499, 500
identity functions 112
identity symbol 92
"if and only if" notation 223, 413
implicit differentiation 288–95, 484–5, 542
indefinite integrals 444, 446–7
independent events 688–9, 691–2
independent variables 76, 348
indeterminate forms 550–3
induction 45–51, 664–5
inequalities 98–9, 175, 196–200
  see also quadratic inequalities
infinite discontinuities 228
infinite sequences 5, 7
infinite series 8, 23–6, 62–3, 559
infinity 18, 144–5, 228–30
integration 442, 444–60
  antidifferentiation 444–60, 488–90
  areas bounded by curves 451–4, 458–9
  by inspection 490–6
  by parts 504–9
  by substitution 496–503
  finding value of constant 448
  functions with form $ax + b$ 448–50
  techniques 488–509
  see also definite integrals
intercept form of quadratic function 82, 85
interest 30–1
interpolation 353
interquartile range 319, 335, 338–9
  see also ranges
inverse chain rule 491–3
inverse functions 112–16, 418–20, 427–29, 477
invertible functions 112
iterative equations 216

kinematics 284–8, 528–32, 568–9
Koch snowflake 3, 9, 27

least squares regression 349–50

Leibniz's formula 259
lemmas 185
L'Hopital's Rule 551–3
limits 220–35
 areas bounded by curves 453–4
 continuity 224–7
 derivatives of functions 239–40
 Euler's number 473
 functions 221–3, 239–40, 423
 indeterminate forms 550–3
 infinity 228–30
 properties 230–3
 quotient rule 257
 sequences 233–5
 trigonometric functions 423
linear combinations 175–6, 580–2, 588–92
linear correlation 349–53
linear equations 200–10
linear functions 101, 701–2
linearization method 546
linear regression 347, 349–50
linear relationships 10
lines 628–33
 angles between lines/planes 631–3
 of best fit 349–50
 line segments 573–5, 582–5
 planes 627, 629–33
 two lines in 2D 612–13
 vectors 605–14, 626–7, 628–33
Little Bézout's theorem 179
local maxima 263–4, 266–7, 276
local minima 263, 266–7, 272, 276
logarithms 442, 460, 465–73
 definition 465
 derivatives 483–8
 exponential functions 479–81
 integration by parts 505
 natural 477, 480–1, 504–5
 properties 467–70, 484, 495
logistic equations 538–9
lower quartiles 318–19, 339

Maclaurin series 554–62, 652
magnitudes
 scalar products 597–8
 vector products 615–16, 617
 vectors 576, 592–5, 634, 636–7
many-to-many mappings 102
many-to-one graphs/mappings 102–3, 106
mathematical induction 45–51, 664–5
mathematical modelling 279, 279–80
maximum points
 absolute value functions 94
 compound functions 486
 continuous random variables 710
 function transformations 119–21
 global maxima 263
 graphs of derivatives 262–6, 269, 273, 276–7
 local maxima 263–4, 266–7, 276
 optimization 281, 283
 periodic functions 412
 quadratic functions 83
mean
 binomial distribution 716–17
 continuous random variables 709–12
 estimator for 341
 histograms 324–6
 normal distribution 721–3
 population 312
 sample 312, 341
 standard deviation 336–41
measures of central tendency 317
measures of dispersion 318–19
median
 continuous random variables 709–12
 cumulative frequency 343–5
 definition 317
 histograms 324–6
 standard deviation 338–9
midpoint formula 370–1
minimum points
 absolute value functions 94
 function transformations 119–21
 global minima 263–4
 graphs of derivatives 262–6, 269, 273, 276–7
 local minima 263, 266–7, 272, 276
 optimization 283
 quadratic functions 83
missing data 316
mixed products 617–20
modal class 327
mode 317, 324–5, 699, 709–12
moduli
 complex numbers 163–4, 168
 of numbers 93, 98–9
 polar complex numbers 650–8, 660–3, 669
modulus-argument form of complex numbers 651, 656
modulus function 245
multimodal histograms 324–5, 327
multiplication
 complex numbers 165–8
 polar complex numbers 656, 657–9
 vectors by scalars 579–80, 583–4, 588–9
multiplicities 186
mutually exclusive events 685, 691, 694

natural logarithms 477, 480–1, 504–5
negative correlation 348
negative exponents/powers 62–5, 172
Newton's 2nd law 533, 542
Newton's Law of Cooling 536–7
Newton's Serpentine curve 258–9
non-linear functions 237
normal distribution 718–24
normals 246–8, 251–2, 256, 429–1, 483
normal vectors
 angles 630–1, 633
 lines 607–8, 610
 planes 622–3, 624–5, 627, 630–1, 633
*n*th terms 11
 see also general terms

oblique asymptotes 229–30
obtuse angles 386–7, 392, 407, 603
odd functions 106–8, 111, 115
ODEs *see* ordinary differential equations
ogives 345–6
one-to-many mappings 102
one-to-one mappings 102–3, 105, 115, 418–19
onto functions 103–6, 115
opposite complex numbers 654
opposite-conjugate complex number 654–5
opposite vectors 576, 583
optimization 278–84
ordered pairs 76
order of equation 533
ordinary differential equations (ODEs) 533–49
orthogonal vectors 602, 607, 609
oscillating sequences 17, 234
outliers 324, 335, 339, 345, 355

parabolas 81–5, 160, 209, 262, 451–2
parallelepipeds 618–19
parallel lines 582–5, 609–10, 612
parallelogram law 573, 576–8
parallel planes 622
parallel vectors 580, 595–6, 602, 616–17, 623, 632

parameters
absolute value functions 94
cubic functions 177
quadratic functions 81–3
real 155, 157, 201–2, 608, 623, 626
parametric equations
lines 605–8, 610, 612–13, 629, 632
planes 621–4, 627, 629, 632
partial fraction decomposition 91–3, 538
partitioning areas 522–3
Pascal's triangle 59–60
paths 284–5
PDFs *see* probability density functions
Pearson product-moment correlation coefficient 351–3, 364–5
percentiles 345
perfect squares 149–52, 195
periodic functions 409–18
periods of functions 412–16
permutations 53–8, 687
perpendicular lines 246–8, 609–10
perpendicular vectors 602, 607
phase shifts 414–15
piecewise functions 100–2
complex numbers 147
folium of Descartes 289
function transformations 123–4
limits 222
probability 699, 708
planes 621–33
angles between lines/planes 631–3
intersecting lines 629–30
three lines in plane 609–10
two/three planes 622–8, 630–1
vector equations 621–8
point-intercept straight line equation 291
points of inflexion 268–72, 274–7, 486
points of intersection
area between two curves 521
lines 608–9, 612–13, 632
planes 626, 632
solving polynomials 198–9
volumes of revolution 525–6
polar complex numbers 650–74
operations 656–64
powers and roots 664–74
trigonometric formulae 672–4
polynomial equations 175, 195–6, 209–10
polynomial functions 85
definition 175
"end behaviour" property 178
factor theorem 178–84
fundamental theorem of algebra 185–94
integration by parts 505, 507
limits 232
rational functions 86–7
remainder theorem 178–84
sum and product of roots 190–4
types of 176–7
polynomial inequalities 175, 196–200
polynomials 554–62
see also polynomial...
populations 309–13, 535–9
position vectors
lines, equations 606, 610–11
planes, equations 622, 624
vector algebra 587–9, 593, 595, 599
positive correlation 348
power functions 176
power rule 250–1, 290
powers 172–5, 664–74
see also exponents
prime notation 542
probability 680–731
"at least"/"more than" 715
axiomatic systems 683–96
axioms/corollaries 684–6, 689–91, 694
continuous random variables 706–12, 718, 720
distributions 680, 696–705
theorems 689, 691, 693–6, 702
probability density functions (PDFs) 706–10, 714–16, 718
probability distributions 680, 696–705
binomial distribution 712–17
functions 698
tables 698–703, 714, 716
product by a constant rule 425
product-moment correlation coefficient 349, 350–4
product of polynomial roots 190–4
product rule 255–6, 271, 426, 485, 504
profit 282–3
proof 2, 33–51, 646–7
by contradiction 40–4
by induction 45–51
compound angles 404
cosine rule 391–3
definition 35
direct 37, 40, 45, 646
layout of 38
sine rule 393–7
trigonometric identities 404, 407–10
types of 35–51
validation 35
proper algebraic fractions 91
proportionality, constant of 536–7
pyramids 376, 619–20
Pythagoras' theorem 294
3D space 370
area of triangle 398
cosine rule 392
proof 35, 42
sine rule 394
surface areas 376
trigonometric identities 400
trigonometric ratios 372–3, 384–5
vectors 594, 635
Pythagorean identities 401–2, 673

quadrants of unit circle 386–91
quadratic equations 149–59
completing the square 149–53
complex numbers 161
discriminants 156–9
exponents 462
factorisation 149, 152, 155
polar complex numbers 668
trigonometric equations 420
quadratic formula 43, 153–6, 679
quadratic functions 81–6
bicycle model 101
forms of 82
graphs of derivatives 262
radical functions 90
systems of equations 209–10
quadratic inequalities 149, 158–60
qualitative data 308
quantitative data 308–9
quartiles 318–19, 343–5
quota sampling 311, 313, 314
quotient of polynomial 179, 180–2
quotient rule 257–8, 260, 267, 425

radians 379–80, 382
radical functions 89–91, 101
random variables 696–9, 706–12
ranges 318
absolute value functions 94–5
classifying functions 103–5
functional relationships 76–9
function composition 109–10
function transformations 117–18, 129, 131
histograms 320, 326
inverse functions 113–15

periodic functions 412
special functions 83, 86–91, 94–5
rank correlation 364–5
rates of change 237, 239, 292–3, 431–4
rates of growth 535–9
rational functions 86–9
limits 229, 232
oblique asymptotes 229
quotient rule 257
transformations 128–9
ratios 384–410
common 17–21, 23–6, 48–9
compound angles 403–5, 421, 423
trigonometric 371–4, 384–91
real-life situations 27–8, 279
real parameters
quadratics 155, 157–8
systems of equations 201–2
vectors 608, 623, 626
real roots 157–8
reciprocal complex numbers 660–1, 665, 673
reciprocal functions
continuity 225
function transformations 119, 128–9
rational functions 86
trigonometric functions 399–3, 430
rectangular hyperbolas 86
recursive equations 11, 17–18
recursive formulas 71
reductio ad absurdum 41
reflections 117–21, 124, 133, 136–8, 478
reflective symmetry 114
reflex angles 386
regression 347, 349–50
related rates 288, 293, 431–3
relative frequency histograms 328–30
reliability of data 315–16, 355
remainder theorem 178–84, 187–9
revenue 282–3
reverse chain rule 449, 491
Riemann sums 454
right-angled triangles 371–4, 382, 384–91
see also Pythagoras' theorem
rigid transformations 123
roots
complex numbers 172–5
conjugate root theorem 187–9
definition 153
integration by substitution 501
polar complex numbers 664, 669–71
polynomial sum/product of 190–4
quadratic equations 153–4, 156–8
radical functions 89
square roots 173–4
trigonometric equations 420–1
rotations 106, 658
row reduction 207

sample means 312, 341
samples 309, 310
sample spaces 683, 685, 690, 699
sampling 308–16, 354
scalar products 597–604
angles 598–604, 630–1
lines, equations 608
magnitudes 597–8
planes, equations 624–5, 627
properties 597, 602
vector applications 635
scalar quantities 633–4
scalars 579–80, 583–4, 588–9, 595–6, 656
scatter diagrams 347–50
secants 237–9, 378–79
second derivatives 266–74, 281, 283, 494, 710
sec (secant) 399–3, 425, 500, 501
sectors 381–3
segments 381–3
self-inverse functions 115
separation of variables 534, 536–8, 542
sequences 2, 4–9
arithmetic 10–16, 27–31, 37, 48–9
definition 4
limits 233–5
see also geometric sequences
series 2, 4–9
arithmetic 10–16, 29–31
definition 5
expanding 8
see also geometric series
set rules/theory 683–4
shifts 134–6, 414–16
see also translations
Sierpinski's triangle 16
sigma notation 4–9, 697
simple random sampling 311, 314
simultaneous equations
complex numbers 168, 174
elimination methods 202–8, 210
exponential functions 478
linear combinations 589
lines 609, 611–13, 628
Newton's Law of Cooling 537
normal distribution 723
planes 623–8
remainder theorem 183
systems of equations 200–10
vector applications 634, 636
see also substitution methods
sine (sin)
arcsin 372, 396, 416–20, 427–29, 504
area of triangle 397–8
four quadrants 386–91
sine curve 409–12, 414
sine ratio 371–4, 386–91, 430
sine rule 393–7, 422
see also trigonometric...
sketching graphs 274–5
skewed histograms 324–5, 327
slant asymptotes 229
slant height of cone 382–3
slope, y-intercept form of equation 605–8
small angle theory 422
solids of revolution 524–8
sound 440–1
Spearman's rank correlation 364–5
speed
average 236–7, 242–3
measuring change 219, 239, 285, 293
vector applications 634, 636–7
spheres 375–6
spheroids 378
spread 319, 324
square roots 173–4
standard deviations 336–43
alternative formula 340
binomial distribution 717
estimator for 341–2
expectation 703–4
histograms 326
normal distribution 719, 721–3
product-moment correlation coefficient 350, 352–3
standard integrals 498–9
standardised normal variables 720, 723
stationary points 268, 270
see also maximum points; minimum points
statistics 306–65
stem-and-leaf diagrams 332
step functions 223
straight lines 247–8, 291, 568–9
stratified sampling 311, 314
stretches 123–34, 414–15
substitution
complex numbers 174
definite integrals 501–3
integration by 496–503
linear equations 201, 205–7
trigonometric functions 501–3
vectors 606, 623, 626–7, 636

subtraction 165, 580, 588–9
successive events 689
sum and difference formulae 446, 673
sums
polynomial roots 190–4
sequences/series 5–6, 12–18, 20–5, 39
surds 152
surface areas 375–8, 382
symmetry
axes of symmetry 82–3, 107
kinematics 531
normal curves 719
reflective 114
rotational 106
trigonometric functions 413–18
synthetic division 180–3, 185–9
systematic sampling 311, 314
systems of equations 200–10
linear equations 200–10
polynomials 209–10
three with three unknowns 205–9
two with two unknowns 200–5
types of solutions 201–2, 208

tangents to curves
derivatives 237–9, 246–8, 252, 483, 486
implicit differentiation 290–1
trigonometric functions 429–1
tangent (tan) 386–91, 397, 427–29
see also trigonometric...
target markets/populations 310, 313
term of sequence, definition 4
theoretical probability 697, 700
theory of the third variable 354
three dimensions (3D) 369–1
angles 374–5
coordinates 590
planes 622
spatial properties 369–1
surface area 375–8, 382
vectors 590–6, 601, 603, 614–22
time 75, 291, 635
transformations
compressions 123–5, 135, 413
dilations 123–5, 128–32, 134–6
enlargement 658
functions 117–39
order of 128, 131, 134
reflections 117–21, 124, 133, 136–8, 478
rotations 658
translations 121–31, 133–8, 481
tree diagrams 688–9, 691, 694, 703, 731
triangle law 577–8, 580–5, 589
triangles
areas 397–8, 430–1, 456–7, 486–7
right-angled 371–4, 382, 384–91
see also Pythagoras' theorem; trigonometry
triangular pyramids 619–20
Trident of Newton curve 258–9
trigonometric equations 390–1, 402, 407, 420–33
trigonometric expressions 431–3
trigonometric formulae 672–4
trigonometric functions 399–29
antiderivatives 446
derivatives 422–29
differentiation 422–7
graphs, properties 413–18
integration 497–503, 506–9
inverse 418–20, 427–29
normals 429–1
periodic 409–18
polar complex numbers 666–7, 673
reciprocal 399–404, 430
tangents to curves 430–1
trigonometric identities 384–410, 420–1, 499, 500
trigonometric ratios 371–4, 384–91
trigonometry 366–441
see also triangles; trigonometric...
turning points 274–6
see also maximum points; minimum points
two dimensions (2D) 590–1, 596, 603, 612–13

uniform sequences 17
unimodal histograms 324–5, 327
unit circle 386–90, 400–1, 410–11
unit vectors 595–7, 616
upper quartiles 318–19, 339

validity of proof 35
variance 336–43
binomial distribution 716–17
continuous random variables 711
expectation 703–4
vector products 614–21
vector quantities 633–4
vectors 570–647
algebra 586–97
algebraic form 586
applications 633–8
column vectors 122, 586–8
geometrical representations 573–86
linear combinations 580–2, 588–92
lines, equations 605–14
operations 576–80
planes, equations 621–8
scalar products 597–604
three lines in plane 609–10
two lines in 2D/3D 612–13
unit vectors 595–7, 616
use in geometry 582–6
vector form of equation 605
vector products 614–21
velocity
kinematics 284–7, 528–31
straight line motion 568–9
vector applications 633–4, 635, 637
Venn diagrams 683, 685, 690
vertex form of quadratic function 82–3, 85
vertical asymptotes 79, 87, 119, 228, 267
vertical line test 78, 113
vertical shifts 415–16
volumes 520, 524–8
3D shapes 376–7
parallelepipeds 618–19
of revolution 524–8
triangular pyramids 619–20

Weierstrass function 246
Witch of Agnesi curve 258–9

x-intercepts 118, 120–1, 480

y-intercepts 118–21, 430, 480

zero polynomial 176
zero product theorem 149, 195
zeros
complex zeros 185, 187–9
definition 153
discriminants 156
function transformations 119
fundamental theorem of algebra 185–94
graphs of derivatives 276
quadratic inequalities 160
remainder theorem 183
solving polynomials 195–9
velocity functions 530–1
zero vector 578, 585, 602, 617